T0238690

Handbook of Formal Languages

Editors: G. Rozenberg A. Salomaa

Springer-Verlag Berlin Heidelberg GmbH

G. Rozenberg   A. Salomaa   (Eds.)

Handbook of Formal Languages

Volume 1
Word, Language, Grammar

With 72 Figures

Springer

Prof. Dr. Grzegorz Rozenberg
Leiden University
Department of Computer Science
P.O. Box 9512
NL-2300 RA Leiden
The Netherlands

Prof. Dr. Arto Salomaa
Turku Centre
for Computer Science
Data City
FIN-20520 Turku
Finland

Library of Congress Cataloging-in-Publication Data
Handbook of formal languages / G. Rozenberg, A. Salomaa, (eds.).
p. cm.
Includes bibliographical references and index.
Contents: v. 1. Word, language, grammar – v.2. Linear modeling:
background and application – v. 3. Beyond words.
ISBN 978-3-642-63863-3   ISBN 978-3-642-59136-5 (eBook)
DOI 10.1007/978-3-642-59136-5
(alk. paper: v. 3)
1. Formal languages.   I. Rozenberg, Grzegorz. II. Salomaa, Arto.
QA267.3.H36 1997
511.3 – DC21   96-47134
                                       CIP

CR Subject Classification (1991): F.4 (esp. F.4.2-3), F.2.2, A.2, G.2, I.3, D.3.1, E.4

ISBN 978-3-642-63863-3

http://www.springer.de

© Springer-Verlag Berlin Heidelberg 1997
Originally published by Springer-Verlag Berlin Heidelberg New York in 1997
Softcover reprint of the hardcover 1st edition 1997
The use of general descriptive names, registered names, trademarks, etc. in this publication does not imply, even in the absence of a specific statement, that such names are exempt from the relevant protective laws and regulations and therefore free for general use.

Cover design: MetaDesign, Berlin
Typesetting: Data conversion by Lewis & Leins, Berlin
SPIN: 11326717   45/3111 - 5 4 3 2 - Printed on acid-free paper

Preface

The need for a comprehensive survey-type exposition on formal languages and related mainstream areas of computer science has been evident for some years. In the early 1970s, when the book *Formal Languages* by the second-mentioned editor appeared, it was still quite feasible to write a comprehensive book with that title and include also topics of current research interest. This would not be possible anymore. A standard-sized book on formal languages would either have to stay on a fairly low level or else be specialized and restricted to some narrow sector of the field.

The setup becomes drastically different in a collection of contributions, where the best authorities in the world join forces, each of them concentrating on their own areas of specialization. The present three-volume Handbook constitutes such a unique collection. In these three volumes we present the current state of the art in formal language theory. We were most satisfied with the enthusiastic response given to our request for contributions by specialists representing various subfields. The need for a Handbook of Formal Languages was in many answers expressed in different ways: as an easily accessible historical reference, a general source of information, an overall course-aid, and a compact collection of material for self-study. We are convinced that the final result will satisfy such various needs.

The theory of formal languages constitutes the stem or backbone of the field of science now generally known as theoretical computer science. In a very true sense its role has been the same as that of philosophy with respect to science in general: it has nourished and often initiated a number of more specialized fields. In this sense formal language theory has been the origin of many other fields. However, the historical development can be viewed also from a different angle. The origins of formal language theory, as we know it today, come from different parts of human knowledge. This also explains the wide and diverse applicability of the theory. Let us have a brief look at some of these origins. The topic is discussed in more detail in the introductory Chapter 1 of Volume 1.

The main source of the theory of formal languages, most clearly visible in Volume 1 of this Handbook, is *mathematics*. Particular areas of mathematics important in this respect are combinatorics and the algebra of semigroups and monoids. An outstanding pioneer in this line of research was

Axel Thue. Already in 1906 he published a paper about avoidable and unavoidable patterns in long and infinite words. Thue and Emil Post were the two originators of the formal notion of a rewriting system or a grammar. That their work remained largely unknown for decades was due to the difficult accessibility of their writings and, perhaps much more importantly, to the fact that the time was not yet ripe for mathematical ideas, where noncommutativity played an essential role in an otherwise very simple setup.

Mathematical origins of formal language theory come also from mathematical logic and, according to the present terminology, computability theory. Here the work of Alan Turing in the mid-1930s is of crucial importance. The general idea is to find models of computing. The power of a specific model can be described by the complexity of the language it generates or accepts. Trends and aspects of mathematical language theory are the subject matter of each chapter in Volume 1 of the Handbook. Such trends and aspects are present also in many chapters in Volumes 2 and 3.

Returning to the origins of formal language theory, we observe next that much of formal language theory has originated from *linguistics*. In particular, this concerns the study of grammars and the grammatical structure of a language, initiated by Noam Chomsky in the 1950s. While the basic hierarchy of grammars is thoroughly covered in Volume 1, many aspects pertinent to linguistics are discussed later, notably in Volume 2.

The *modeling* of certain objects or phenomena has initiated large and significant parts of formal language theory. A model can be expressed by or identified with a language. Specific tasks of modeling have given rise to specific kinds of languages. A very typical example of this are the L systems introduced by Aristid Lindenmayer in the late 1960s, intended as models in developmental biology. This and other types of modeling situations, ranging from molecular genetics and semiotics to artificial intelligence and artificial life, are presented in this Handbook. Words are one-dimensional, therefore linearity is a feature present in most of formal language theory. However, sometimes a linear model is not sufficient. This means that the language used does not consist of words (strings) but rather of trees, graphs, or some other nonlinear objects. In this way the possibilities for modeling will be greatly increased. Such extensions of formal language theory are considered in Volume 3: languages are built from nonlinear objects rather than strings.

We have now already described the contents of the different volumes of this Handbook in brief terms. Volume 1 is devoted to the mathematical aspects of the theory, whereas applications are more directly present in the other two volumes, of which Volume 3 also goes into nonlinearity. The division of topics is also reflected in the titles of the volumes. However, the borderlines between the volumes are by no means strict. From many points of view, for instance, the first chapters of Volumes 2 and 3 could have been included in Volume 1.

We now come to a very important editorial decision we have made. Each of the 33 individual chapters constitutes its own entity, where the subject matter is developed from the beginning. References to other chapters are only occasional and comparable with references to other existing literature. This style of writing was suggested to the authors of the individual chapters by us from the very beginning. Such an editorial policy has both advantages and disadvantages as regards the final result. A person who reads through the whole Handbook has to get used to the fact that notation and terminology are by no means uniform in different chapters; the same term may have different meanings, and several terms may mean the same thing. Moreover, the prerequisites, especially in regard to mathematical maturity, vary from chapter to chapter. On the positive side, for a person interested in studying only a specific area, the material is all presented in a compact form in one place. Moreover, it might be counterproductive to try to change, even for the purposes of a handbook, the terminology and notation already well-established within the research community of a specific subarea. In this connection we also want to emphasize the diversity of many of the subareas of the field. An interested reader will find several chapters in this Handbook having almost totally disjoint reference lists, although each of them contains more than 100 references.

We noticed that guaranteed timeliness of the production of the Handbook gave additional impetus and motivation to the authors. As an illustration of the timeliness, we only mention that detailed accounts about DNA computing appear here in a handbook form, less than two years after the first ideas about DNA computing were published.

Having discussed the reasons behind our most important editorial decision, let us still go back to formal languages in general. Obviously there cannot be any doubt about the mathematical strength of the theory – many chapters in Volume 1 alone suffice to show the strength. The theory still abounds with challenging problems for an interested student or researcher. Mathematical strength is also a necessary condition for applicability, which in the case of formal language theory has proved to be both broad and diverse. Some details of this were already mentioned above. As the whole Handbook abounds with illustrations of various applications, it would serve no purpose to try to classify them here according to their importance or frequency. The reader is invited to study from the Handbook older applications of context-free and contextual grammars to linguistics, of parsing techniques to compiler construction, of combinatorics of words to information theory, or of morphisms to developmental biology. Among the newer application areas the reader may be interested in computer graphics (application of L systems, picture languages, weighted automata), construction and verification of concurrent and distributed systems (traces, omega-languages, grammar systems), molecular biology (splicing systems, theory of deletion), pattern matching, or cryptology, just to mention a few of the topics discussed in the Handbook.

About Volume 1

Some brief guidelines about the contents of the present Volume 1 follow. Chapter 1 is intended as an introduction, where also historical aspects are taken into account. Chapters 2, 3, 5, 6, 8, 9 each give a comprehensive survey of one important subarea of the basic theory of formal languages. The innovative nature of these surveys will become apparent to a knowledgeable reader. Indeed, at least some of these surveys can be classified as the best or first-of-its-kind survey of the area. While the three first-mentioned chapters (2, 3, and 5) are basic for the grammatical or computational aspects of the theory, the remaining three chapters (6, 8, and 9) are basic for the algebraic aspects. Grammatical (resp. algebraic) issues are discussed further in Chapters 4 and 12 (resp. 7, 10, and 11).

Acknowledgements

We would like to express our deep gratitude to all the authors of the Handbook. Contrary to what is usual in case of collective works of this kind, we hoped to be able to keep the time schedule – and succeeded because of the marvellous cooperation of the authors. Still more importantly, thanks are due because the authors were really devoted to their task and were willing to sacrifice much time and energy in order to achieve the remarkable end result, often exceeding our already high expectations.

The Advisory Board consisting of J. Berstel, C. Calude, K. Culik II, J. Engelfriet, H. Jürgensen, J. Karhumäki, W. Kuich, M. Nivat, G. Păun, A. Restivo, W. Thomas, and D. Wood was of great help to us at the initial stages. Not only was their advice invaluable for the planning of the Handbook but we also appreciate their encouragement and support.

We would also like to extend our thanks from the actual authors of the Handbook to all members of the scientific community who have supported, advised, and helped us in many ways during the various stages of work. It would be a hopeless and maybe also unrewarding task to try to single out any list of names, since so many people have been involved in one way or another.

We are grateful also to Springer-Verlag, in particular Dr. Hans Wössner, Ingeborg Mayer, J. Andrew Ross, and Gabriele Fischer, for their cooperation, excellent in every respect. Last but not least, thanks are due to Marloes Boon-van der Nat for her assistance in all editorial stages, and in particular, for keeping up contacts to the authors.

September 1996   Grzegorz Rozenberg, Arto Salomaa

Contents of Volume 1

Chapter 11. **Regularity and Finiteness Conditions**
Aldo de Luca and Stefano Varricchio 747

Chapter 12. **Families Generated by Grammars and L Systems**

Contents of Volume 2

Contents of Volume 3

Authors' Addresses

Jean-Michel Autebert
UFR d'Informatique, Université Denis Diderot-Paris 7
2, place Jussieu, F-75005 Paris, Cedex 05, France
autebert@litp.ibp.fr

Jean Berstel
LITP, IBP, Université Pierre et Marie Curie
Laboratoire LITP, Institut Blaise Pascal
4, place Jussieu, F-75252 Paris, Cedex 05, France
jean.berstel@litp.ibp.fr

Luc Boasson
LITP, IBP, Université Denis Diderot
2, place Jussieu, F-75251 Paris Cedex 05, France
luc.boasson@litp.ibp.fr

Christian Choffrut
LITP, Université de Paris VI
4, place Jussieu, F-75252 Paris Cedex 05, France
cc@litp.ibp.fr

Tero Harju
Department of Mathematics, University of Turku
FIN-20014 Turku, Finland
harju@sara.utu.fi

Helmut Jürgensen
Department of Computer Science, The University of Western Ontario
London, Ontario N6A 5B7, Canada
helmut@uwo.ca
and Institut für Informatik, Universität Potsdam
Am Neuen Palais 10, D-14469 Potsdam, Germany
helmut@mpag-inf.uni-potsdam.de

Juhani Karhumäki
Department of Mathematics, University of Turku
FIN-20014 Turku, Finland
karhumak@utu.fi

Lila Kari
Department of Computer Science, University of Western Ontario
London, Ontario N6A 5B7, Canada
lila@csd.uwo.ca

Stavros Konstantinidis
Department of Mathematics and Computer Science, University of Lethbridge
4401 University Drive, Lethbridge, Alberta T1K 3M4 Canada
stavros@csd.uleth.ca

Werner Kuich
Abteilung für Theoretische Informatik, Institut für Algebra
und Diskrete Mathematik, Technische Universität Wien
Wiedner Hauptstrasse 8-10, A-1040 Wien, Austria
kuich@email.tuwien.ac.at

Aldo de Luca
Dipartimento di Matematica, Università di Roma "La Sapienza"
Piazzale Aldo Moro 2, I-00185 Roma, Italy
deluca@mercurio.mat.uniroma1.it

Alexandru Mateescu
Faculty of Mathematics, University of Bucharest
Academiei 14, RO-70109 Bucharest, Romania
and Turku Centre for Computer Science (TUCS)
Lemninkäisenkatu 14 A, FIN-20520 Turku, Finland
mateescu@sara.utu.fi

Gheorghe Păun
Institute of Mathematics of the Romanian Academy
P.O. Box 1-764, RO-70700 Bucharest, Romania
gpaun@imar.ro

Jean-Eric Pin
LITP, IBP, CNRS, Université Paris VI
4, place Jussieu, F-75252 Paris Cedex 05, France
jep@litp.ibp.fr

Grzegorz Rozenberg
Department of Computer Science, Leiden University
P.O. Box 9512, NL-2300 RA Leiden, The Netherlands
and Department of Computer Science
University of Colorado at Boulder, Campus 430
Boulder, CO 80309, U.S.A.
rozenber@wi.leidenuniv.nl

Arto Salomaa
Academy of Finland and Turku Centre for Computer Science (TUCS)
Lemninkäisenkatu 14 A, FIN-20520 Turku, Finland
asalomaa@sara.cc.utu.fi

Stefano Varricchio
Dipartimento di Matematica, Università di L'Aquila
Via Vetoio Loc. Coppito, I-67100 L'Aquila, Italy
varricch@univaq.it

Sheng Yu
Department of Computer Science, University of Western Ontario
London, Ontario N6A 5B7, Canada
syu@csd.uwo.ca

Formal Languages: an Introduction and a Synopsis

Alexandru Mateescu and Arto Salomaa

1. Languages, formal and natural

What is a language? By consulting a dictionary one finds, among others, the following explanations:

1. The body of words and systems for their use common to people who are of the same community or nation, the same geographical area, or the same cultural tradition.
2. Any set or system of signs or symbols used in a more or less uniform fashion by a number of people who are thus enabled to communicate intelligibly with one other.
3. Any system of formalized symbols, signs, gestures, or the like, used or conceived as a means of communicating thought, emotion, etc.

The definitions 1–3 reflect a notion "language" general and neutral enough for our purposes.

Further explanations are more closely associated with the spoken language and auditory aspects or are otherwise too far from the ideas of this Handbook. When speaking of formal languages, we want to construct formal grammars for defining languages rather than to consider a language as a body of words somehow given to us or common to a group of people. Indeed, we will view a *language* as a set of finite strings of symbols from a finite alphabet. Formal *grammars* will be devices for defining specific languages. Depending on the context, the finite strings constituting a language can also be referred to as *words, sentences, programs*, etc. Such a formal idea of a language is compatible with the definitions 1–3, although it neglects all semantic issues and is restricted to *written* languages.

The idea of a *formal language* being a set of finite strings of symbols from a finite alphabet constitutes the core of this Handbook. Certainly all written languages, be they natural, programming or any other kind, are contained in this idea. On the other hand, formal languages understood in this general fashion have very little form, if any. More structure has to be imposed on them, and at the same time one can go beyond the linear picture of strings. Both approaches will be tried in the present Handbook.

How does one specify a formal language? If we are dealing with a finite set of strings we can, at least in principle, specify the set simply by listing its elements. With infinite languages we have a different situation: we have to invent a finitary device to produce infinite languages. Such finitary devices can be called *grammars, rewriting systems, automata*, etc. Many stories about them will be told in this Handbook. In fact, a major part of formal language theory can be viewed as the study of finitary devices for generating infinite languages.

What is today known as the Theory of Formal Languages has emerged from various origins. One of the sources is *mathematics*, in particular, certain problems in combinatorics and in the algebra of semigroups and monoids. A pioneer in this line of research was Axel Thue at the beginning of the 20th century; in [11, 12] he investigated avoidable and unavoidable patterns in long and infinite words. Together with Emil Post [8], Thue also introduced the formal notion of a rewriting system or a grammar.

Another shade in the mathematical origins of formal language theory comes from logic and, according to the current terminology, the theory of computing. Here the work of Alan Turing [13] is of crucial importance. The general idea is to find models of computing. The power of a specific model can be described by the complexity of the languages it generates or accepts.

Trends and aspects of mathematical language theory will be the subject matter of this Volume I of the Handbook. Same scattered glimpses of it will be presented already in Section 2 of this chapter.

It is quite obvious and natural that much of formal language theory has originated from *linguistics*. Indeed, one can trace this development very far in the past, as will be pointed out later on in this section. Specifically, the study of grammars initiated by Noam Chomsky [1] has opened new vistas in this development.

Many parts of formal language theory have originated from *modeling* certain objects or phenomena. A model can be expressed by or identified with a language. Specific tasks of modeling have given rise to specific types of languages. Very typical examples are L systems of Aristid Lindenmayer [5] intended as models in developmental biology. This and other types of modeling situations, dealing for instance with molecular genetics, semiotics, artificial life and artificial intelligence, will be presented in this Handbook, in particular in Volume 2. Sometimes a linear model is not sufficient. This means that the language used does not consist of strings (words) but rather of trees, graphs or some other *many-dimensional* objects. Such extensions of formal language theory will be considered in Volume 3 of this Handbook: languages are built from many-dimensional objects rather than strings and, thus, the possibilities for modeling will be greatly increased.

The remainder of this Section 1 is devoted to the study of natural languages. We give, very briefly, some glimpses of linguistics from three different points of view: historical studies, genetics and neuroscience. We only hope to give some idea about what is going on. Each of these three aspects of linguistics would require a handbook on its own!

1.1 Historical linguistics

Linguists have tried to investigate, as well as to classify, natural languages existing in the world today, some 6000 in number. The study is becoming increasingly important: it is estimated that about half of the now existing 6000 languages will die out during the next century.

How did a specific language become prevalent in the area in which it is now spoken? It could have been taken to its current territory by farmers, traders, conquerors etc. Often multidisciplinary methods are used to clarify the development. The scientific study of language, *linguistics*, can often penetrate deeper in the past than the oldest written records. Related languages are compared to construct their immediate progenitors. From progenitors one goes to their predecessors and, finally, to their ultimate ancestor or *protolanguage*.

We spoke above of "related" languages. What does this mean? Already for more than 200 years, linguists have recognized that some languages have similarities in vocabulary, grammar, formation of new constructs or the use of sounds strong enough to be grouped in the same family, stemming from a common ancestor. A *language family*, that is a family of natural languages, results by such ancestral alliances that can be traced back in history. This idea is quite different from the notion of a language family in formal language theory met very frequently in this Handbook.

Linguists are far from unanimous about the existing language families. The whole trend of trying to find similarities between different languages has been sometimes condemned as unfruitful. On the contrary, some linguists tend to emphasize the differences that make languages seem unrelated, as well as to use only small independent units in the classification. Such linguists also tend to rule out spurious relationships. The importance of reconstructing protolanguages has been also questioned.

However, certain specific language families have won wide acceptance. Among such families are the *Indo-European* family, which will be discussed in more detail below, the *Hamito-Semitic* (or *Afro-Asiatic*) family, which consists of the Semitic languages and many languages of North Africa, as well as the *Altaic* family whose major representatives are Finnish and Hungarian. We will mention still in Section 1.2 some other language families whose legitimacy has perhaps not won such a wide acceptance.

What does it mean from a practical point of view that two languages, say Finnish and Hungarian, belong to the same family and, moreover, that linguists are unanimous about this state of affairs? Does it imply that a native user of one language is able to use or understand the other language? Certainly not. Some few basic nouns, such as the words for "fish", "water" and "blood", bear easily observable resemblance in Finnish and Hungarian. There is even one word, the word "tuli", having in certain contexts the same meaning in both languages: in "TULIpunainen" and "TULIpiros" the prefix means "flaming", the whole word meaning "flaming red". However, this is merely accidental. In Hungarian "tuli" stems from the Indo-European "tulip", whereas in Finnish it is of old Uralic origin. An often quoted example is the sentence "the train is coming". It is claimed that the sentence is the same in both languages but the Finnish word for "the train" means "is coming" in Hungarian, and vice versa. The claim, however, is quite inaccurate – and so we omit further details. In general one can say that the vocabularies of Finnish and Hungarian lie as far apart as those of English and Persian (which also belong to the same well-recognized family, Indo-European languages). Among easily observable similarities between Finnish and Hungarian are certain grammatical features (no distinction between "he" and "she", no articles for nouns) and a phenomenon called vowel harmony in speech.

Let us go back to the idea of language families. In linguistics the idea dates back for more than 200 years. It is customary to date the field of linguistics to its first significant achievement, the argument propounded in 1786 by Sir William Jones, a British judge at the High Court in Calcutta, who observed relationships between Sanskrit, Greek, Latin, Gothic and Persian. (The year 1786 is such a starting point for linguistics as 1906 is for formal language theory; the first paper of Axel Thue about combinatorics on words appeared in 1906.) Common vocabulary and structural features suggested to Jones that the languages had "sprung from some common source". Nowadays this source language is known as Indo-European. More explicitly, Jones wrote about Sanskrit, Greek and Latin that "no philologer could examine them all three without believing them to have sprung from some common source which, perhaps, no longer exists".

The construction of a protolanguage applies various methods, most of which can be classified as inductive inference. Early linguists in their attempts to reconstruct the Indo-European protolanguage relied heavily on Grimm's law of "sound shift": sets of consonants displace each other in a predictable and regular fashion. This law was suggested in 1822 by Jacob Grimm, who is universally famed for the fairy tales he wrote with his brother Wilhelm. (For instance, "Snow White" and "Cinderella" are universally known as Walt Disney movies.) According to the law, softer "voiced" consonants b, d, g yield to harder "voiceless" consonants p, t, k. A typical development is

$$dhar \qquad draw \qquad tragen$$
$$(Sanskrit) \quad (English) \quad (German)$$

Another development, based on different phenomena, can be depicted as follows, see Figure 1:

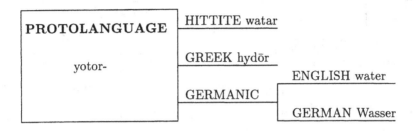

Fig. 1

Linguists have traced back diverging pathways of linguistic transformation, as well as human migration. There are many views about the Indo-European protolanguage and its homeland. Rules such as Grimm's law were used to construct an Indo-European vocabulary. On this basis, conclusions were made about how its speakers lived. The landscape and climate described by the vocabulary was originally placed in Europe between Alps and the Baltic and North seas. More recent evidence [2] places the probable origin of the Indo-European language in western Asia. The language described by the reconstructed Indo-European protolanguage is mountainous – there are many words for high mountains lakes and rapid rivers. According to [2], the vocabulary fits the landscape of eastern Transcaucasia, where the protolanguage flourished some 6, 000 years ago. The protolanguage split into dialects which evolved into distinct languages. The latter split into further dialects, and so forth. The family tree presented in Figure 2 is based on information from [2].

The time depth of the Indo-European protolanguage is only about 6, 000 years. Attempts have been made to go deeper into the origins. If the aim is a single protolanguage, one has to go back probably well beyond 20, 000 years. A rather well-known macrofamily, called *Nostratic*, comprises the Indo-European, Hamito-Semitic and Altaic families and is estimated to date back some 15, 000 years.

The following table, see Figure 3, compares some Indo-European languages, as well as two languages outside the family.

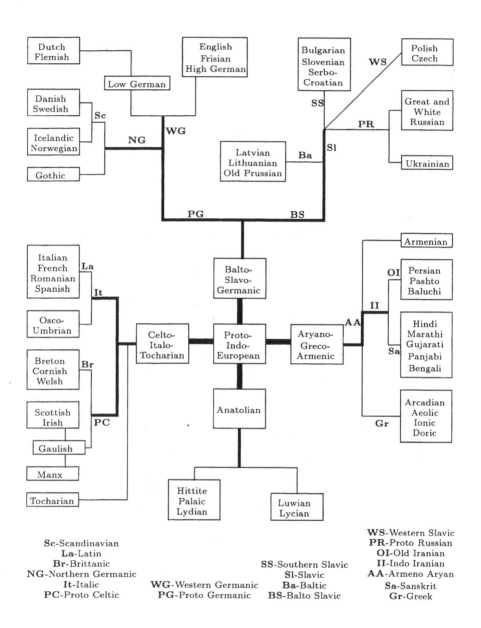

Sc-Scandinavian
La-Latin
Br-Brittanic
NG-Northern Germanic
It-Italic
PC-Proto Celtic

WG-Western Germanic
PG-Proto Germanic

SS-Southern Slavic
Sl-Slavic
Ba-Baltic
BS-Balto Slavic

WS-Western Slavic
PR-Proto Russian
OI-Old Iranian
II-Indo Iranian
AA-Armeno Aryan
Sa-Sanskrit
Gr-Greek

Fig. 2

English	Old German	Latin	Romanian
One	Ains	Unus	Unu
Two	Twai	Duo	Doi
Three	Thrija	Tres	Trei
Four	Fidwor	Quattuor	Patru
Five	Fimf	Quinque	Cinci
Six	Saihs	Sex	Şase
Seven	Sibum	Septem	Şapte
Eight	Ahtau	Octo	Opt
Nine	Niun	Novem	Nouă
Ten	Taihum	Decem	Zece

English	Greek	Sanskrit	Japanese	Finnish
One	Heis	Ekas	Hiitotsu	Yksi
Two	Duo	Dva	Futatsu	Kaksi
Three	Treis	Tryas	Mittsu	Kolme
Four	Tettares	Catvaras	Yottsu	Neljä
Five	Pente	Panca	Itsutsu	Viisi
Six	Heks	Sat	Muttsu	Kuusi
Seven	Hepta	Sapta	Nanatsu	Seitsemän
Eight	Okto	Asta	Yattsu	Kahdeksan
Nine	Ennea	Nava	Kokonotsu	Yhdeksän
Ten	Deka	Dasa	To	Kymmenen

Fig. 3

1.2 Language and evolution

The idea of linking the evolution of languages to biology goes back to Charles Darwin. He wrote in the *Descent of Man* (1871), "The formation of different languages and of distinct species, and the proofs that both have been developed through a gradual process, are curiously parallel." In Chapter 14 of *On the Origin of Species* he also stated that if the tree of genetic evolution were known, it would enable scholars to predict that of linguistic evolution.

If one wants to go deeper in the past, as indicated at the end of Section 1.1, one has to combine genetic evidence with archeological and linguistic evidence. Indeed, molecular genetics can test some elements of proposed theories of the evolution of languages. One compares gene frequencies in various populations and converts the data into a structure, where genetic distance can be represented. One is then able to see to what extent genetic relationships confirm predictions arising from supposed evolution of languages and language families. Genetic and linguistic histories at least roughly correspond because both diverge when populations split apart. The following chart from

[9], see Figure 4, is based on work by Luigi Luca Cavalli-Sforza and Merritt Ruhlen. The genetic closeness of populations is shown by their distance from a common branching point (left). Their linguistic closeness (in terms of language families and superfamilies) is depicted similarly on the right.

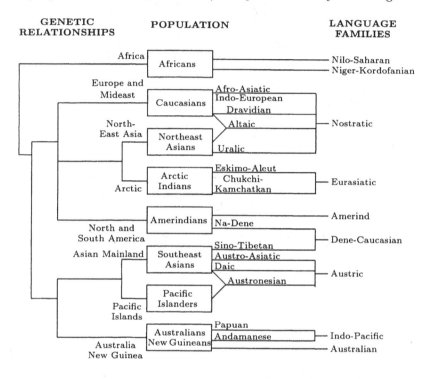

Fig. 4

An inevitable conclusion is that the distribution of genes correlates very well with that of languages. Apparently genes do not control language in a deterministic way, so the correlation is explained by history: the circumstances of birth determine the language to which one is exposed. Human populations are during the evolution fragmented into parts which settle in different locations. Then the fragments evolve both linguistic and genetic patterns ear-marked by the branching points. This explains the correlation.

It is not claimed that evolution transfers simple languages into more complex ones. It is quite widely agreed among linguists that no language, living or dead, is "better" than any other. Language alters, but it neither improves nor degenerates. For scientific discussions, Modern English maybe is better than Old English but the potential resources of both languages are the same. However, as pointed out by [9], earlier generations of linguists had no such reservations about ranking languages. August Schleicher in the 19th century classified languages according to their structures. In Schleicher's view, Chinese is an "isolating" language using simple elements and, consequently, is more "primitive" than Turkish, which builds its words from distinct forms. He put "inflecting" languages like German higher. Sanskrit was ranked highest because its inflections were so elaborate. On these terms, the languages of many hunter-gatherers are elevated above those of the linguists themselves!

1.3 Language and neural structures

How much preprogramming is there in the brain when a child starts learning a language? Any child can learn any language without making as many grammatical and other mistakes as one would expect from a structure without programming, a *tabula rasa*. It can be claimed that language is as innate in the infant as flight is in the eaglet. This means that children do not so much learn language as somehow just develop it in response to a stimulus.

Taking into account the enormous complexity of language phenomena, many researchers have wondered whether the neural machinery involved will ever be understood to any reasonable extent. Indeed, the question about the origin of language in a neuroscientific sense has often been considered hopeless. Linguistic Society of Paris banned its discussion in 1866 because it had generated so much mere talk and so little real knowledge, if any. In spite of the ban, the discussion has continued. Many theories have been presented about the origin of language, starting from onomatopoetic words such as "cuckoo", from emotional interjections, from oral gestures etc. Especially recently the progress in understanding the brain structures responsible for language has accelerated significantly. Tools such as magnetic resonance imaging (MRI) have made it possible to locate brain lesions accurately in patients suffering from aphasia. In such a way specific language deficits have been correlated with damage to particular regions of the brain. Positron emission tomography (PET) makes it possible to study brain activities of healthy persons engaged in linguistic tasks.

Still all important questions remain to be answered about how the brain stores and processes language. However, the motivation to clarify at least some of the issues is great. Language is a superb means of communication, increasing in importance as the concepts become more abstract. Try to communicate without words the background for the events in Eastern Europe in 1989, or the intricacies of Hamlet!

2. Glimpses of mathematical language theory

We now return to the actual topic of our Handbook, the Theory of Formal Languages. It was already indicated in Section 1 above that certain branches of mathematics constitute a major source of origin for formal language theory. The term "mathematical language theory" describes mathematical (algebraic) aspects of formal language theory – the main emphasis is on the mathematical theory rather than any applications. The purpose of this section is to give the reader some scattered glimpses of mathematical language theory. The idea is not to present any coherent theory but rather to give some views about the very basics, about words and languages. We want to give an uninitiated reader some feeling about what is going on. We have tried to select nontrivial problems whose solutions are presentable without too much technical apparatus and yet readable without previous knowledge or consulting other sources. Section 3 will be quite different in nature. It is a telegraphic review of (at least some) basic concepts and results in formal language theory. The purpose has been to collect basic formal language theory in one place for quick reference. Our original idea was to add another section describing briefly the contents of the Handbook. However, such a description already appears in the Prefaces. Due to the marvellous timeliness of the individual authors, it was not possible to write anything more comprehensive without endangering the publication schedule.

2.1 Words and languages

An *alphabet* is a finite nonempty set. The elements of an alphabet Σ are called *letters* or *symbols*. A *word* or *string* over an alphabet Σ is a finite sequence consisting of zero or more letters of Σ, whereby the same letter may occur several times. The sequence of zero letters is called the *empty word*, written λ. Thus, λ, 0, 1, 110, 00100 are words over the "binary" alphabet $\Sigma = \{0, 1\}$. The set of all words (resp. of all nonempty words) over an alphabet Σ is denoted by Σ^* (resp. Σ^+). Observe that Σ^* and Σ^+ are always infinite. If x and y are words over Σ, then so is their *catenation* (or *concatenation*) xy, obtained by juxtaposition, that is, writing x and y after one another. Catenation is an associative operation and the empty word λ acts as an identity: $w\lambda = \lambda w = w$ holds for all words w. Because of the associativity, we may use the notation w^i in the usual way. By definition, $w^0 = \lambda$. In algebraic terms, Σ^* and Σ^+ are the free *monoid* and *semigroup* generated by Σ with respect to the operation of catenation and with the unit element λ.

The notions introduced above will be very basic throughout this Handbook. The notation may vary, for instance, the empty word is often denoted by ε. We will denote alphabets by Σ or V, the latter coming from "vocabulary". The elements are thought as indivisible units; it depends on the viewpoint whether they are called "letters" or, coming from a vocabulary, "words".

We continue with some central terminology concerning words. The *length* of a word w, in symbols $|w|$, is the number of letters in w when each letter is counted as many times as it occurs. Again by definition, $|\lambda| = 0$. A word v is a *subword* of a word w if there are words u_1 and u_2 (possibly empty) such that $w = u_1 v u_2$. If $u_1 = \lambda$ (resp. $u_2 = \lambda$) then v is also called a *prefix* of w (resp. a *suffix* of w). Observe that w itself and λ are subwords, prefixes and suffixes of w. Other subwords, prefixes and suffixes are called *nontrivial*.

Let us write a word w in the form $w = u_1v_1u_2v_2 \ldots u_nv_n$, for some integer n and words u_i, v_i, some of them possible empty. Then the word $v = v_1v_2 \ldots v_n$ is a *scattered subword* of w. (The notion will be the same if we say that the word $u = u_1u_2 \ldots u_n$ is a scattered subword of w. This follows because we may choose some of the u's and v's to be empty.) Thus, a scattered subword is not a coherent segment but may consist of parts picked up from here and there, without changing the order of letters. The French school emphasizes the coherence by using the term "factor" for our "subwords" and saving the term "subword" for our scattered subwords. A nonempty word w is *primitive* if it is not a proper power, that is, the equation $w = u^i$ does not hold for any word u and integer $i \geq 2$. Every word w possesses a unique *primitive root u: u* is the shortest word such that $w = u^i$, for some $i \geq 1$. Obviously, a nonempty word is primitive iff its primitive root equals the word itself.

Words $v = xy$ and $w = yx$ are termed *conjugates*. Thus, a conjugate of a word is obtained if a prefix is transferred to become a suffix. The prefix may be empty; a word is always its own conjugate. Clearly, if u and v are conjugates of w, then also u and v are conjugates among themselves. (Thus, the relation is an equivalence.) If w is primitive, so are its conjugates. Conjugates of a word are often called also *circular variants*.

Thue's work [11, 12] at the beginning of this century dealt with avoidable and unavoidable patterns occurring in long words. Let us consider the pattern $xx = x^2$. A word w is *square-free* if it contains no subword xx, where x is a nonempty word. Thus, a word being square-free means that it avoids the pattern xx. Now the size of the alphabet becomes quite significant. Assume that we are dealing with the binary alphabet $\{0, 1\}$. Can we construct long square-free words? No. The pattern xx is unavoidable as soon as the length of the word is at least 4. The words 010 and 101 of length 3 are indeed square-free – they are the only square-free words of length 3. But we cannot continue them with a further letter without losing square-freeness. Take the word 010. Both continuations 0100 and 0101 contain a square as subword.

Things are different if the alphabet contains at least 3 letters. Thue showed how to construct in this case infinite square-free sequences of letters. He also showed how to construct infinite *cube-free* sequences (that is, avoiding the pattern $xxx = x^3$) over a binary alphabet. The reader is referred to [10] for compact proofs of these nontrivial results.

We now proceed from words to languages. Subsets, finite or infinite, of Σ^* are referred to as (*formal*) *languages* over Σ. Thus,

$$L_1 = \{\lambda, 0, 111, 1001\} \text{ and } L_2 = \{0^p \mid p \text{ prime }\}$$

are languages over the binary alphabet. A finite language can always, at least in principle, be defined as L_1 above: by listing all of its words. Such a procedure is not possible for infinite languages. Some finitary specification other than simple listing is needed to define an infinite language. Much of formal language theory deals with such finitary specifications of infinite languages: grammars, automata etc.

Having read Section 1 of this chapter, the reader might find our terminology somewhat unusual: a language should consist of sentences rather than words, as is the case in our terminology. However, this is irrelevant because we have to choose some terminology. Ours reflects the mathematical origin. But the basic set, its elements and strings of elements could equally well be called "vocabulary", "words" and "sentences".

Various *operations* will be defined for languages: how to get new languages from given ones. Regarding languages as sets, we may immediately define the *Boolean* operations of *union, intersection* and *complementation* in the usual fashion. The operation of *catenation* is extended to concern languages in the natural way:

$$L_1 L_2 = \{w_1 w_2 \mid w_1 \in L_1 \text{ and } w_2 \in L_2\}.$$

The notation L^i is extended to concern languages, now $L^0 = \{\lambda\}$. The *catenation closure* or *Kleene star* (resp. *Kleene plus*) of a language L, in symbols L^* (resp. L^+) is defined to be the union of all nonnegative powers of L (resp. of all positive powers of L). Observe that this definition is in accordance with our earlier notations Σ^* and Σ^+ if we understand Σ as the finite language whose words are the singleton letters.

The operations of union, catenation, and Kleene star are referred as *regular* operations. A language L over Σ is termed *regular* if L can be obtained from the "atomic" languages $\emptyset$ (the empty language) and $\{a\}$, where a is a letter of Σ, by applying regular operations finitely many times. By applying union and catenation, we obviously get all finite languages. The star operation is needed to produce infinite languages. When we speak of the *family* of regular languages, we mean all languages that are regular over some Σ, that is, the alphabet may vary from language to language. This family is very central in formal language theory. It corresponds to strictly finite computing devices, finite automata. Regular languages will be the subject matter of the next chapter of this Handbook. The family has many desirable formal properties. It is *closed* under most of the usual operations for languages, in particular, under all Boolean operations.

A language L over Σ is *star-free* if L can be obtained from the atomic languages $\{a\}$, where $a \in \Sigma$, by applying Boolean operations (complement is taken with respect to Σ^*) and catenation finitely many times. Since regular languages are closed under Boolean operations, it follows that every star-free language is regular. We will return to star-free languages in Section 2.3.

2.2 About commuting

A feature common for all languages, formal or natural, is the noncommutativity of letters and sounds. The English words "no" and "on" are very different. The monoid Σ^* is noncommutative except when Σ consists of one letter a only: the word a^{i+j} results if we catenate a^i and a^j, independently of which one comes first. Because of the noncommutativity, the mathematical problems about formal languages are rather tricky, and results from classical algebra or analysis are seldom applicable.

In this Section 2.2 we will present the basic results about commuting. What can be said about words satisfying certain specific commutativity conditions? The results belong to *combinatorics on words*, an area to which a chapter is devoted in this Handbook. Such results are very useful in situations, where we meet the same word in different positions and want to make conclusions about the word.

The first result is very fundamental: what can be said about a word, where the same word appears both as a prefix and as a suffix?

Theorem 2.1. *Assume that* $xy = yz$, *for some words* x, y, z, *where* $x \neq \lambda$. *Then there are words* u, v *and a nonnegative integer* k *such that*

$$x = uv, \; y = (uv)^k u = u(vu)^k, \; z = vu.$$

Proof. Assume first that $\mid x \mid \geq \mid y \mid$. Then from the equation $xy = yz$ we see, by reading prefixes of length $\mid x \mid$, that $x = yv$, for some word v. (If $\mid x \mid = \mid y \mid$ then $v = \lambda$.) The situation can be depicted as follows:

x		y
y	z	
y	v	y

We may obviously choose $k = 0$ and $u = y$. Then $x = yv = uv$, $y = (uv)^0 u$ and $z = vy = vu$.

If $\mid x \mid < \mid y \mid$, we use induction on the length of y. The basis is clear. If $\mid y \mid = 0$, then $y = \lambda$, and we may choose $u = \lambda$, $v = x = z$, $k = 0$. (Recall that $v^0 = \lambda$.)

Suppose, inductively, that the assertion holds for all $\mid y \mid \leq n$ and consider the case $\mid y \mid = n + 1$. Because $\mid x \mid < \mid y \mid$, we obtain $y = xw$ by reading prefixes of length $\mid y \mid$ from the equation $xy = yz$. By substituting, we infer that $xxw = xwz$, hence $xw = wz$. We may now apply the inductive hypothesis: because $x \neq \lambda$, we have $\mid w \mid < \mid y \mid$ and, hence, $\mid w \mid \leq n$. (Observe that we cannot find u, v as required if $x = z = \lambda$ and $y \neq \lambda$. That's why we have the assumption $x \neq \lambda$ in Theorem 2.1.)

Consequently, for some u, v and k,

$$x = uv, \; y = (uv)^k u = u(vu)^k, \; z = vu.$$

Because $y = xw = uv(uv)^k u = (uv)^{k+1} u$, we have completed the inductive step. □

Our next result answers the basic question about commuting: when is it possible that $xy = yx$ holds between words x and y? The result is that x and y must be powers of the same word, so we are essentially back in the case of a one-letter alphabet. Theorems 2.1 and 2.2 are often referred to as "Lyndon's Theorem", due to [6].

Theorem 2.2. *If $xy = yx$ holds between nonempty words x and y, then there is a word w and nonnegative integers i, j such that $x = w^i$ and $y = w^j$.*

Proof. The proof is by induction on the length of xy. The conclusion is immediate for the case $\mid xy \mid = 2$. Now suppose the Theorem is true for all xy with $\mid xy \mid \leq n$. Let $\mid xy \mid = n + 1$. Then by Theorem 2.1, $x = uv$, $y = (uv)^k u$ for some words u, v and nonnegative integer k. Observe that $uv(uv)^k u = (uv)^k uuv$; hence $uvu = uuv$ and $uv = vu$. Since $\mid uv \mid \leq n$, by the induction hypothesis u and v are powers of a common word w, i.e., $u = w^p$ and $v = w^q$, for some nonnegative integers p and q. It follows that $x = uv = w^{p+q}$ and $y = (uv)^k u = w^{k(p+q)+p}$. Hence, $x = w^i$ and $y = w^j$, where $i = p + q$ and $j = k(p + q) + p$. □

Theorem 2.2 can also be expressed by saying that the equation $xy = yx$ is *periodicity forcing*: the equation admits only "periodic" solutions, where the unknowns are repetitions of the same period. Some other such periodicity forcing equations are known, for instance,

$$xyz = y^i, \; i \geq 2.$$

The equation can hold between x, y, z only if they are powers of the same word.

As an application of our commutativity considerations, we present a result concerning conjugates and primitive words.

Theorem 2.3. *A word w has $\mid w \mid$ different conjugates iff w is primitive.*

Proof. Clearly, w can never have more than $\mid w \mid$ conjugates. The "only if"-part is clear. If $w = u^i$, $i \geq 2$, then w equals the conjugate obtained by transferring one u from the prefix position to the suffix position.

To prove the "if"-part, suppose that w is primitive. Proceed indirectly, assuming that two of w's conjugates are equal: $w_1 = w_2$. Since also w_1 is a conjugate of w_2, we obtain

$$w_1 = xy = yx = w_2, \; x \neq \lambda, \; y \neq \lambda.$$

By Theorem 2.2, x and y are powers of the same word u. Since both x and y are nonempty, $w_1 = w_2 = u^i$, $i \geq 2$, implying that w_1 and w_2 are imprimitive. As a conjugate of w_1, the word w is also not primitive, a contradiction. Hence, all conjugates of w are different. There are $\mid w \mid$ of them, as seen by transferring the letters of w, one after the other, from the beginning to the end. □

2.3 About stars

The operation of Kleene star is a powerful one. It produces infinite languages from finite ones. Together with the other two regular operations, union and catenation, it can lead to very complicated compositions. In fact, for any k, there are regular languages that cannot be represented with fewer than k nested star operations.

We present in this section two properties of the star operation. The first one tells us that, as regards languages over one letter, the star of the language is always very simple, no matter how complicated the original language is. The basic reason for this simplicity is again commutativity. Because words over one letter commute, the intricacies of the star language vanish.

Theorem 2.4. *For every language L over the alphabet $\{a\}$, there is a finite language $L_F \subseteq L$, such that $L^* = L_F^*$.*

Proof. The theorem holds for finite languages L, we just choose $L_F = L$. Assume that L is infinite and a^p is the shortest nonempty word in L, $p \geq 1$. Clearly, $\{a^p\}^* \subseteq L^*$. (This follows by the definition of the star operation. Because a^p is in L, all powers of a^p are in L^*.) If this inclusion is actually an equality (and it must be so if $p = 1$), there is nothing more to prove, we choose $L_F = \{a^p\}$. Otherwise, let a^{q_1} be the shortest word from the difference $L^* - \{a^p\}^*$. (The difference consists of words in L^* but not in $\{a^p\}^*$.) It follows that q_1 can be written in the form

$$q_1 = t_1 p + r_1, \ 0 < r_1 < p, \ t_1 \geq 1.$$

(We cannot have q_1 divisible by p because all exponents of a divisible by p are produced by $\{a^p\}^*$.) Again we see that $\{a^p, a^{q_1}\}^* \subseteq L^*$. If this is not an equality, we let a^{q_2} be the shortest word in the difference $L^* - \{a^p, a^{q_1}\}^*$. It follows that q_2 is of the form

$$q_2 = t_2 p + r_2, \ 0 < r_2 < p, \ r_2 \neq r_1, \ t_2 \geq 1.$$

(All words leaving the remainder r_1 when divided by p are obtained using a^p and a^{q_1}.)

The procedure is continued. At the kth step we obtain the word a^{q_k}, with

$$q_k = t_k p + r_k, \ 0 < r_k < p, \ r_k \neq r_1, r_2, \ \ldots, \ r_{k-1}, \ t_k \geq 1.$$

Since there are at most p possible remainders r_i, including the remainder 0 stemming from p, we actually will have the equality

$$\{a^p, a^{q_1}, \ \ldots, \ a^{q_s}\}^* = L^*, \ \text{for some } s < p. \qquad \square$$

We will now consider languages $\{w\}^*$. Thus, the language consists of all powers of w, we write it simply w^*. The question is: when is w^* star-free? Of course, the star occurs in the definition of w^* but there may be a way of getting around it. The question is not easy. For instance, considering languages

(∗) $(1010)^*, (10101)^*, (101010)^*$

one observes, after considerable reflection, that the middle one is star-free, whereas the others are not. Where does this state of affairs depend on? We will settle this question completely. First we will prove some auxiliary results. We assume that w is over the alphabet Σ and, moreover, Σ is the *minimal* alphabet of w: all letters of Σ actually occur in w.

Lemma 2.1. *The languages $\emptyset$, $\{\lambda\}$, Σ^* and Σ^+ are star-free over Σ.*

Proof. The claim follows by the equations

$$\emptyset = \{a\} \cap \{a^2\}, \text{ where } a \in \Sigma,$$

$$\Sigma^* =\sim \emptyset, \ \Sigma^+ = \Sigma\Sigma^*, \ \{\lambda\} =\sim \Sigma^+. \qquad \square$$

We say that a language L is *noncounting* iff there is an integer n such that, for all words x, y, z,

$$xy^n z \in L \text{ iff } xy^{n+1}z \in L.$$

Lemma 2.2. *Every star-free language is noncounting.*

Proof. Let L be a star-free language, $L \subseteq \Sigma^*$.
   The proof is by induction on the structure of L, i.e. by induction on the number of Boolean operations and catenations used to define L.
   If $L = \emptyset$ or $L = \{a\}$, $a \in \Sigma$, then L is noncounting for $n = 2$.
   Now, assume that $L = L_1 \cup L_2$ and let n_i be the constant for which L_i is noncounting, $i = 1, 2$. Define $n = \max\{n_1, n_2\}$. Assume that $\alpha\beta^n\gamma \in L$. It follows that $\alpha\beta^n\gamma \in L_1$ or $\alpha\beta^n\gamma \in L_2$. If $\alpha\beta^n\gamma \in L_1$, then using the noncounting property $(n - n_1)$ times for L_1, we obtain that $\alpha\beta^{n_1}\gamma \in L_1$. Again, using the same property $(n - n_1 + 1)$ times for L_1 we obtain that $\alpha\beta^{n+1}\gamma \in L_1$. It follows that $\alpha\beta^{n+1}\gamma \in L_1 \cup L_2 = L$. The converse is similar, i.e., if $\alpha\beta^{n+1}\gamma \in L$, we use the same argument to prove that $\alpha\beta^n\gamma \in L$.
   If $L =\sim L_1$ then obviously, L satisfies the noncounting property for $n = n_1$.
   If $L = L_1 \cap L_2$, then observe that $L =\sim (\sim L_1 \cap \sim L_2)$.
   Finally, assume that $L = L_1 L_2$ and define $n = n_1 + n_2 + 1$. Assume that $\alpha\beta^n\gamma \in L = L_1 L_2$. There are $u_i \in L_i$, $i = 1, 2$, such that $\alpha\beta^n\gamma = u_1 u_2$. Observe that at least one of the words u_i, contains a subword β^m, with $m \geq n_i$, $i = 1, 2$. Without loss of generality, assume that $u_1 = \alpha\beta^m u'$ with $m \geq n_1$. Using the noncounting property $(m - n_1)$ times for L_1 we conclude

that $\alpha\beta^{n_1}u' \in L_1$. Again, using the same property $(m - n_1 + 1)$ times for L_1, we deduce that $\alpha\beta^{m+1}u' \in L_1$. Hence, $\alpha\beta^{m+1}u'u_2 = \alpha\beta^{n+1}\gamma \in L_1L_2 = L$. The converse is analogous.

Hence, if L is a star-free language, then L is a noncounting language. □

The converse of Lemma 2.2 holds in the form: every regular noncounting language is star-free. We will need Lemma 2.2 only in the form indicated.

We are now in the position to establish the characterization result. The result also immediately clarifies why only the middle one of the languages $(*)$ is star-free.

Theorem 2.5. *The language w^*, $w \neq \lambda$, is star-free iff w is primitive.*

Proof. Consider first the "only if"-part, supposing that $w = u^i$, $i \geq 2$. We want to show that w^* is not star-free. By Lemma 2.2, it suffices to prove that w^* is not noncounting. Assume the contrary, and let n be the corresponding constant. Choose now, in the definition of "noncounting", $x = u^j$, $y = u$, $z = \lambda$, where j is such that $j + n$ is divisible by i. Then $xy^n z = u^{j+n} \in w^*$, whereas $xy^{n+1}z = u^{j+n+1} \notin w^*$. Consequently, w^* is not noncounting.

For the "if"-part, assume that w is primitive. To prove that w^* is star-free, let us first have a look at what kind of properties are expressible in a star-free form. The set of all words having the word x as a subword, prefix or suffix can be expressed in the form

$$\Sigma^* x \Sigma^*, \quad x\Sigma^*, \quad \Sigma^* x,$$

respectively. By Lemma 2.1, these sets are star-free. Similarly, $\sim (\Sigma^* x \Sigma^*)$ stands for the set of all words *not* having the word x as a subword.

Consider a word in w^*, say w^5. Let us keep most of the word hidden, and move a scanning *window* of length $|w|$ across w^5:

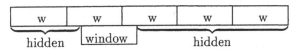

What words do we see through the window? At the beginning and at the end we see w. At all other times, in fact at all times, we see a word in $C(w)$, the set of *conjugates* of w. (Recall that w is its own conjugate.)

Let us now try to express in star-free terms the condition that a word x is in w^*. We assume that $x \neq \lambda$ because we can add $\{\lambda\}$ as a separate term of the union. The word x must *begin* and *end correctly*, that is, it has to belong to the intersection

$$w\Sigma^* \cap \Sigma^* w.$$

Furthermore, all *subwords* of x of length $|w|$ must be conjugates of w, that is, belong to $C(w)$. We express this by saying that no subwords of length $|w|$ and outside $C(w)$ are allowed to appear in x, that is, x belongs to the intersection ("subwords OK")

$$SUBWOK = \bigcap_{y \notin C(w), |y|=|w|} \sim (\Sigma^* y \Sigma^*)$$

(We can compute the number of words y, that is, the number of sets to be intersected. Assume that Σ has k letters. Then there are altogether $k^{|w|}$ words of length $|w|$. By Theorem 2.3, $k^{|w|} - |w|$ are outside $C(w)$.) We are now ready to write a star-free expression for w^*.

$(**)$ $\qquad\qquad w^* = \{\lambda\} \cup (w\Sigma^* \cap \Sigma^* w \cap SUBWOK).$

By Lemma 2.1, the right side of the equation $(**)$ is star-free. We still have to show that our equation is correct.

It is obvious that the left side of the equation is included in the right side: if a nonempty word is in w^*, it begins and ends correctly and has no incorrect subword and, consequently, belongs to all sets in the intersection. To prove the reverse inclusion, we assume the contrary. Let x be the shortest word on the right side of our equation $(**)$ that is not in w^*. Then $x \neq \lambda$ and we may write $x = wx_1$ (because $x \in w\Sigma^*$). We must have $x_1 \neq \lambda$ because $w\lambda = w \in w^*$.

If x_1 has the prefix w, we argue as follows. Since x has w as suffix, so does x_1. The word x_1 has only correct subwords (no subwords of length $|w|$ outside $C(w)$) because x has only correct subwords. Consequently, x_1 belongs to the right side of the equation $(**)$. This implies that $x_1 \in w^*$ because $|x_1| < |x|$ and x was the shortest word belonging to the right side and not to w^*. But now also $x = wx_1 \in w^*$, a contradiction.

Therefore, w is not a prefix of x_1. Assume first that there is an i such that the ith letter a of x_1 differs from the ith letter b of w - let i be the smallest such index. Thus

$$x_1 = w'az, \quad w = w'bw'', \quad a \in \Sigma, \ b \in \Sigma, \ a \neq b,$$

where the words w', w'', z may be empty. Consider now the subword $w''w'a$ of $x = wx_1$:

w'	b	w''	w'	a	z
		window			

This subword is of length $|w|$, so it must belong to $C(w)$. On the other hand, we have clearly $w''w'b \in C(w)$. But both words $w''w'a$ and $w''w'b$ cannot be in $C(w)$ because every letter occurs the same number of times in each word of $C(w)$. Thus our assumption has led to a contradiction and, consequently, no such i exists.

The only remaining possibility is that x_1 is a proper prefix of w, $w = x_1 w_1$, for some $w_1 \neq \lambda$. Since x belongs to the right side of equation $(**)$, the word w is also a suffix of x, leading to the following situation:

$$x = \begin{array}{|c|c|} \hline w & x_1 \\ \hline x_1 & w \\ \hline \end{array}$$

Here w_1 is the overlapping part of the two w's. By Theorem 2.3, w is not primitive, which is a contradiction. (The same conclusion could be reached by Theorem 2.2 as well.)   □

Our proof shows clearly, where and why the primitivity of w is needed. For instance, if $w = u^3$ and $x_1 = u$ then the word wx_1 belongs to the right side of the equation $(**)$ but not to w^*.

2.4 Avoiding scattered subwords

Sometimes you deal with collections of words, none of which is a subword of any other word in the collection. This happens, for instance, when a subword fulfils the purpose you have in mind equally well as the whole word. Then, from whatever language L you originally dealt with, you take only a *basis*: a subset K of L such that (i) no word in K is a subword of another word in K, and (ii) every word in L possesses some word of K as a subword. A desirable situation would be that a basis is finite. Unfortunately, this does not always happen. For the language

$$L = \{ba^i b \mid i \geq 1\},$$

the only basis is L itself because none of the words in L is subword of another word in L.

The situation becomes entirely different if, instead of subwords, scattered subwords are considered. In this case a rather surprising result can be obtained, a result that certainly is not intuitively obvious: If no word in a language K is a scattered subword of another word in K, then K is necessarily finite.

Let us use in this context the notation $v \leq w$ to mean that v is a scattered subword of w. The relation is a *partial order*. This means that the following three conditions hold for all words u, v, w.

(i) $u \leq u$,
(ii) If $u \leq v$ and $v \leq u$ then $u = v$,
(iii) If $u \leq v$ and $v \leq w$ then $u \leq w$.

"Partial" means here that u and v can also be *incomparable*: neither $u \leq v$ nor $v \leq u$. This is the case, for instance, if $u = ab$ and $v = ba$. The notation $u \nleq v$ means that either u and v are incomparable, or else both $v \leq u$ and $v \neq u$. Observe that the similarly defined relation in terms of subwords, rather than scattered subwords, is also a partial order.

We now prove the result already referred to above. The result is due to [3]. An earlier version, based on algebraic considerations, appeared in [4].

Theorem 2.6. *Assume that a language L contains no two distinct words v and w such that $v \leq w$. Then L is finite.*

Proof. We show that a contradiction arises if we assume that there is an infinite language with no two comparable words. This is shown by induction on the size of the alphabet. The statement clearly holds for languages over one-letter alphabet because in this case any two words are comparable. Assume inductively that the statement holds for alphabets up to a certain size but it no longer holds if we add one more letter, to get an alphabet Σ. Thus, there are languages over Σ not satisfying Theorem 2.6, whereas Theorem 2.6 holds true for all languages over smaller alphabets.

For each infinite $L \subseteq \Sigma^*$ consisting of pairwise incomparable words, there is a shortest $x_L \in \Sigma^*$ such that $x_L \not\leq y$, for all $y \in L$. (The existence of x_L can be seen as follows. We first take any $x \in L$ and $a \in \Sigma$. Then $ax \not\leq y$, for all $y \in L$. Otherwise, if there is a word $y' \in L$ with $ax \leq y'$, we infer $x \leq ax \leq y'$, which is a contradiction. We now find out whether there are words shorter than ax with the same property concerning $\leq$ and choose the shortest among them for x_L. The word x_L is not necessarily unique, only its length is unique.)

We now fix the language L in such a way that the corresponding word x_L is shortest possible. Thus, there is no infinite language $K \subseteq \Sigma^*$ consisting of pairwise incomparable words such that $x_K \in \Sigma^*$ is a shortest word with the property $x_K \not\leq y$, for all $y \in K$, and $| x_K |<| x_L |$.

Clearly, $x_L \neq \lambda$ because $\lambda \leq y$, for all y. We express x_L as a catenation of letters, some of them possibly equal:

$$x_L = a_1 a_2 \ldots a_k, \, a_i \in \Sigma.$$

If $k = 1$ and $x_L = a_1$ then the condition $x_L \not\leq y$, for all $y \in L$, implies that L is over the alphabet $\Sigma - \{a_1\}$, which contradicts the inductive hypothesis. Hence, $k \geq 2$ and $a_1 a_2 \ldots a_{k-1} \neq \lambda$.

Clearly, $| a_1 a_2 \ldots a_{k-1} |<| x_L |$.

If $a_1 a_2 \ldots a_{k-1} \not\leq y$ holds for infinitely many words $y \in L$, the subset of L consisting of those infinitely many words would contradict our choice of L and x_L. Thus, by removing a finite subset of L, we get an infinite language $L' = \{y_i \mid i = 1, 2, \ldots\} \subseteq L$ such that

$$a_1 a_2 \ldots a_{k-1} \leq y_i, \text{ for all } y_i \in L' \, .$$

Thus, $a_1 a_2 \ldots a_{k-1}$ appears as a scattered subword in every word of L'. For each $y_i \in L'$, we now find the occurrence of $a_1 a_2 \ldots a_{k-1}$ leftmost in the following sense. First we find the leftmost a_1 in y_i. After this occurrence of a_1, we find the leftmost a_2. After this occurrence of a_2, we find the leftmost a_3, and so forth. Thus, we write each y_i in the form

$$y_i = z_1^i a_1 z_2^i a_2 \ldots z_{k-1}^i a_{k-1} z_k^i,$$

for unique words $z_j^i \in (\Sigma - \{a_j\})^*$, $1 \leq j \leq k - 1$. An important observation now is that, for all i, also $z_k^i \in (\Sigma - \{a_k\})^*$. This follows because, otherwise, $x_L = a_1 a_2 \ldots a_k \leq y_i$, a contradiction. Thus, for each fixed j, all the words z_j^i, $i \geq 1$, are over an alphabet smaller than Σ.

We now construct a decreasing sequence

$$N_1 \supseteq N_2 \supseteq N_3 \supseteq \ldots \supseteq N_k$$

of infinite sets N_j of positive integers such that,

$(*)$ whenever $m, n \in N_j, m < n, 1 \leq j \leq k$, then $z_j^m \leq z_j^n$.

The definition of the sets N_j proceeds inductively. For notational convenience, we let $N_0 = \{i \mid i \geq 1\}$. Assume that the infinite set N_j has already been defined. Here either $j = 0$, or else $1 \leq j \leq k - 1$ and $(*)$ is satisfied. We define the set N_{j+1} as follows. Consider the set

$$Z_{j+1} = \{z_{j+1}^i \mid i \in N_j\}.$$

If the set Z_{j+1} is finite we consider, for each word $w \in Z_{j+1}$, the set

$$M(w) = \{i \mid i \in N_j \text{ and } z_{j+1}^i = w\}.$$

Since, N_j is infinite, at least one of the sets $M(w)$ is infinite. We choose N_{j+1} to be any such set; then $(*)$ is automatically satisfied.

If Z_{j+1} is infinite, we may use the inductive hypothesis because $Z_{j+1} \subseteq (\Sigma - \{a_{j+1}\})^*$. Consequently, there are only finitely many pairwise incomparable elements in Z_{j+1}. From this fact we may conclude that there is an infinite chain of elements of Z_{j+1} satisfying:

$(**)$ $z_{j+1}^{n_1} \leq z_{j+1}^{n_2} \leq z_{j+1}^{n_3} \leq \ldots$, where $n_1 < n_2 < n_3 < \ldots$

Let us now argue how this conclusion can be made.

If we have an infinite chain of elements of Z_{j+1}

$(**)'$ $z_{j+1}^{m_1} \leq z_{j+1}^{m_2} \leq z_{j+1}^{m_3} \leq \ldots,$

we may obviously conclude the existence of $(**)$ by defining $n_1 = m_1$ and

$$n_{r+1} = min\{s \mid m_s > n_r\}, r \geq 1.$$

On the other hand, there must be an infinite chain $(**)'$. For if all maximal chains (that is, chains that cannot be continued further) are finite, then, since the different greatest elements in the maximal chains are pairwise incomparable, infinitely many distinct chains have the same greatest element, say z_{j+1}^{max}. This means that infinitely many words w satisfy $w \leq z_{j+1}^{max}$, which is clearly impossible.

We now define the set N_{j+1} (in the case that Z_{j+1} is infinite) by $N_{j+1} = \{n_r \mid r \geq 1\}$. The definition and (∗∗) guarantee that (∗) is satisfied.

Having defined the set N_k, we choose two distinct numbers $m, n \in N_k$, $m < n$. Consequently, $m, n \in N_j$, for all j, $1 \leq j \leq k$. By (∗),

$$z_j^m \leq z_j^n, \ 1 \leq j \leq k.$$

We infer further that

$$y_m = z_1^m a_1 z_2^m a_2 \ldots z_{k-1}^m a_{k-1} z_k^m \leq z_1^n a_1 z_2^n a_2 \ldots z_{k-1}^n a_{k-1} z_k^n = y_n.$$

Thus, y_m and y_n are not after all incomparable, which is a contradiction. □

Two observations should be made regarding the proof given. We did not only exhibit two comparable words y_m and y_n but rather an infinite chain of words, each of which is a scattered subword of the next one. Indeed, the numbers of N_k in increasing order produce such a chain.

The second observation is that the use of N_0, for which (∗) is not required, makes it possible to combine the basis of induction and the inductive step. It is also instructive to consider the case, where Σ has only two letters. Then each of the sets Z_{j+1} is over one letter (not the same one for all j !), and the ideas of the proof become more straightforward.

Theorem 2.6 has many interesting consequences. By an *infinite word* or *ω-word* over an alphabet Σ we mean an infinite sequence of letters of Σ. For instance, an infinite decimal expansion of a real number is an infinite word over the alphabet $\{0, 1, 2, 3, 4, 5, 6, 7, 8, 9\}$. The following result is an immediate corollary of Theorem 2.6.

Corollary 2.1. *No matter how an infinite word is divided into blocks of finite length, one of the blocks is a scattered subword of another block.* □

Coming back to the discussion at the beginning of the Section 2.4, we say that a subset L_0 of a language L is a *basis* of L if, for every $y \in L$, there is a word $x \in L_0$ such that $x \leq y$, that is, x is a scattered subword of y. Clearly, the set of minimal elements with respect to the relation $\leq$ constitutes a basis. Thus, the following result is immediate.

Corollary 2.2. *Every language possesses a finite basis.* □

Corollary 2.2 is just another formulation of Theorem 2.6. The first documentation of the result appears in [4] in the following form.

Theorem 2.7. *(Higman) If X is any set of words formed from a finite alphabet, it is possible to find a finite subset X_0 of X such that, given a word w in X, it is possible to find w_0 in X_0 such that the letters of w_0 occur in w in their right order, though not necessarily consecutively.* □

An important remark is here in order. We have not in our discussions paid any attention to effectiveness, let alone efficiency. What is for Higman "possible to find" often leads to noncomputable tasks and undecidable problems. Decidability and complexity issues will be frequently discussed later on in this Handbook.

Some further corollaries of Theorem 2.6 could be mentioned. Starting with an arbitrary language L, we define the following two languages:

$$SSW(L, \leq) = \{z \in \Sigma^* \mid y \leq z, \text{ for some } y \in L\},$$

$$SSW(L, \geq) = \{z \in \Sigma^* \mid y \geq z, \text{ for some } y \in L\}.$$

Thus, the latter language consists of all scattered subwords of the words in L, whereas the former consists of all words "generated" by L in regard to the relation $\leq$: every word in $SSW(L, \leq)$ contains some word of L as a scattered subword. The notation SSW comes from "scattered subword".

Corollary 2.3. *For every language L, both of the languages $SSW(L, \leq)$ and $SSW(L, \geq)$ are star-free regular languages.*

Proof. By Corollary 2.2, L possesses a finite basis L_0. For a word $w = a_1 a_2 \ldots a_k$, $a_i \in \Sigma$, $1 \leq i \leq k$, the language

$$SSW(\{w\}, \leq) = \Sigma^* a_1 \Sigma^* a_2 \ldots \Sigma^* a_k \Sigma^*$$

is star-free (see Lemma 2.1 in Section 2.3). We now claim that

$$SSW(L, \leq) = \bigcup_{w \in L_0} SSW(\{w\}, \leq),$$

which shows that the left side is star-free. The claim follows because every word containing some word of L as a scattered subword also contains some word of L_0 as a scattered subword, and the right side of the equation consists of all words having some word of L_0 as a scattered subword.

Denote by K the complement of the language $SSW(L, \geq)$. It suffices to prove that K is star-free. By the first part of the proof we know that $SSW(L, \leq)$ is star-free. But

$$K = SSW(K, \leq).$$

Indeed, the inclusion of the left side in the right side is obvious. To prove the reverse inclusion, assume that there is a word x in the difference $SSW(K, \leq) - K$. This means that

$$x \in SSW(K, \leq) \cap SSW(L, \geq).$$

Thus, $y \leq x$, for some $y \in K$. On the other hand, x is a scattered subword of some word $z \in L$. This implies that also y is a scattered subword of z. But K consists of words that are not scattered subwords of words in L and, hence, $y \notin K$. This contradiction completes the proof.   □

Corollary 2.3 is quite remarkable. Any language generates (in two different ways) a structurally very simple language. However, our previous remark applies also here: in general, we cannot find the languages $SSW(L, \leq)$ and $SSW(L, \geq)$ effectively.

Our final corollary is usually referred to as "König's Lemma". It deals with k-dimensional vectors of nonnegative integers, that is, ordered k-tuples $(m_1, m_2, \ldots, m_k)$, where each m_i is a nonnegative integer. A partial order between such vectors is defined by

$$(m_1, m_2, \ldots, m_k) \leq (n_1, n_2, \ldots, n_k) \text{ iff } m_i \leq n_i, \text{ for all } i, 1 \leq i \leq k.$$

Corollary 2.4. *Every set of pairwise incomparable vectors is finite.*

Proof. Associate with the vector $(m_1, m_2, \ldots, m_k)$ the word $a_1^{m_1} a_2^{m_2} \ldots a_k^{m_k}$ over the alphabet $\{a_1, a_2, \ldots, a_k\}$. Then the relation $\leq$ holds between two vectors iff it holds between the associated words, and Corollary 2.4 now immediately follows by Theorem 2.6.   □

2.5 About scattered residuals

Our last illustration requires some maturity in algebra, although no specific knowledge is needed.

Let A be a set. $\mathcal{P}(A)$ denotes the set of all subsets of the set A. If $a \in \Sigma$, then $|x|_a$ is the number of occurrences of the letter a in x. If $x \in \Sigma^*$, then the *commutative closure* of x is

$$com(x) = \{w \in \Sigma^* | \text{ for all } a \in \Sigma, |w|_a = |x|_a\}.$$

If L is a language, then the *commutative closure* of L is

$$com(L) = \bigcup_{x \in L} com(x).$$

A language L is called *commutative* if and only if $L = com(L)$.

The *shuffle* operation between words, denoted Ш, is defined recursively by

$$(au \text{ Ш } bv) = a(u \text{ Ш } bv) \cup b(au \text{ Ш } v),$$

and

$$(u \text{ Ш } \lambda) = (\lambda \text{ Ш } u) = \{u\},$$

where $u, v \in \Sigma^*$ and $a, b \in \Sigma$.

The shuffle operation is extended in the natural way to languages:

$$L_1 \text{ Ш } L_2 = \bigcup_{u \in L_1, v \in L_2} (u \text{ Ш } v).$$

Let L_1 and L_2 be languages over Σ. The *scattered residual* of L_1 by L_2 is defined as:

$$L_1 \rightarrow_s L_2 = \bigcup_{u \in L_1, v \in L_2} (u \rightarrow_s v)$$

where

$$u \rightarrow_s v = \{u_1 u_2 \ldots u_{k+1} \in \Sigma^* | \ k \geq 1, u = u_1 v_1 u_2 v_2 \ldots u_k v_k u_{k+1},$$
$$v = v_1 v_2 \ldots v_k, \ u_i \in \Sigma^*, 1 \leq i \leq k+1, v_i \in \Sigma^*, 1 \leq i \leq k\}.$$

Thus, the scattered residual $u \rightarrow_s v$ indicates the result of deleting v from u in a scattered way. The following lemma provides an equivalent definition of the scattered residual. The straightforward proof is omitted.

Lemma 2.3. *For any languages L_1 and L_2*

$$L_1 \rightarrow_s L_2 = \{w \in \Sigma^* \mid \exists v \in L_2, (w \amalg v) \cap L_1 \neq \emptyset\}.$$

Moreover, for any words u, $v \in \Sigma^*$,

$$u \rightarrow_s v = \{w \in \Sigma^* \mid u \in w \amalg v\}. \qquad \square$$

A *deterministic finite automaton*, *DFA*, is a construct $\mathcal{A} = (Q, \Sigma, \delta, q_0, F)$, where Q is a finite set of *states*, Σ is an alphabet called the *input alphabet*, $q_0 \in Q$ is the *initial state*, $F \subseteq Q$ is the set of *final states* and

$$\delta : Q \times \Sigma \longrightarrow Q$$

is the *transition function*.

The transition function δ is extended to a function $\overline{\delta}$,

$$\overline{\delta} : Q \times \Sigma^* \longrightarrow Q,$$

by defining:

(i) $\overline{\delta}(q, \lambda) = q$, for all $q \in Q$,
(ii) $\overline{\delta}(q, wa) = \delta(\overline{\delta}(q, w), a)$, for all $w \in \Sigma^*$, $a \in \Sigma$, $q \in Q$.

In the sequel, $\overline{\delta}$ will be denoted simply by δ.

The *language accepted* by the DFA $\mathcal{A}$ is

$$L(\mathcal{A}) = \{w \in \Sigma^* \mid \delta(q_0, w) \in F\}.$$

Theorem 2.8. *Let L be a commutative language. The following conditions are equivalent:*

a) L is accepted by some DFA.
b) L has finitely many scattered residuals.

Proof. Let L be a commutative language over the alphabet Σ. Define the relation $\sim_L$ between languages over Σ as follows:

$$L_1 \sim_L L_2 \text{ if and only if } L \to_s L_1 = L \to_s L_2.$$

It follows immediately that the relation $\sim_L$ is an equivalence relation.

Denote by $\mathcal{P}(\Sigma^*)/_{\sim_L}$ the set of all equivalence classes with respect to the relation $\sim_L$. The equivalence class of a language K will be denoted by $[K]$.

a) $\implies$ *b).* Let $\mathcal{A} = (Q, \Sigma, \delta, q_0, F)$ be a finite deterministic automaton such that $L(\mathcal{A}) = L$. We will assume that all states from Q are *accessible*, i.e. for each $p \in Q$ there is a word $u \in \Sigma^*$ such that $\delta(q_0, u) = p$. Moreover, without loss of generality, we may assume that the automaton $\mathcal{A}$ satisfies the following condition of minimality:

$(*)$ if $\{u \in \Sigma^* | \delta(p, u) \in F\} = \{v \in \Sigma^* | \delta(q, v) \in F\}$, then $p = q$.

Let X be the set $\mathcal{P}(Q)$ and consider the set of all functions defined on X and with values in X, denoted X^X.

Let φ be the function mapping the set $\mathcal{P}(\Sigma^*)/_{\sim_L}$ into the set X^X, defined by

$$\varphi([K])(Y) = \delta(Y, K).$$

Claim 1. The function φ is well defined.

Assume that K_1 and K_2 are languages over Σ such that $K_1 \sim_L K_2$. We prove that in this situation $\delta(Y, K_1) = \delta(Y, K_2)$ for all $Y \subseteq Q$. Let p be a state in $\delta(q, K_1)$ for some $q \in Y$. It follows that there is a word $u_1 \in K_1$, such that $\delta(q, u_1) = p$. Let α, β be words such that $\delta(q_0, \alpha) = q$ and $\delta(p, \beta) \in F$. Therefore, $\alpha u_1 \beta \in L$ and consequently, $\alpha\beta \in L \to_s K_1$. But, $K_1 \sim_L K_2$ and hence, $\alpha\beta \in L \to_s K_2$. Therefore, there exists $u_2 \in K_2$ such that $(\alpha\beta \text{ Ш } u_2) \cap L \neq \emptyset$. It follows that $\alpha u_2 \beta \in com(L) = L$. Moreover, note that for all $\alpha, \beta \in \Sigma^*$,

$$\alpha u_1 \beta \in L \text{ if and only if } \alpha u_2 \beta \in L.$$

Thus, from condition $(*)$, it follows that $\delta(q, u_2) = p \in \delta(Y, K_2)$. The converse inclusion is established analogously. Therefore Claim 1 is true.

Claim 2. φ is an injective function.

We will prove that if for all $Y \subseteq Q$, $\delta(Y, K_1) = \delta(Y, K_2)$, then $K_1 \sim_L K_2$.

Let u be in $L \to_s K_1$. There exists $v_1 \in K_1$ such that $(u \text{ Ш } v_1) \cap L \neq \emptyset$. Hence, $uv_1 \in com(L) = L$. Assume that $\delta(q_0, u) = p$ and observe that $\delta(p, v_1) \in F$ and $v_1 \in K_1$. From the equality $\delta(Y, K_1) = \delta(Y, K_2)$, for all $Y \subseteq Q$, it follows that there exists $v_2 \in K_2$ such that $\delta(p, v_2) \in F$. Hence, $uv_2 \in L$ and thus u is in $L \to_s K_2$. Therefore, $L \to_s K_1 \subseteq L \to_s K_2$ and note that the converse inclusion is analogous. Therefore φ is an injective function.

Now, observe that the set X^X is a finite set and hence the set $P(\Sigma^*)/_{\sim_L}$ is finite, too.

$b) \implies a)$. Define the finite deterministic automaton $\mathcal{A} = (Q, \Sigma, \delta, q_0, F)$, where
$Q = \mathcal{P}(\Sigma^*)/_{\sim_L}$. Note that Q is a finite set. Define $q_0 = [\{\lambda\}]$ and $F = \{[K] | K \cap L \neq \emptyset\}$. Observe that from $[K] = [K']$ and $K \cap L \neq \emptyset$ it follows that $K' \cap L \neq \emptyset$. The transition function, $\delta : Q \times \Sigma \to Q$ is defined by

$$\delta([B], \sigma) = [B \text{ III } \sigma].$$

Observe that δ is well defined, i.e. if $B \sim_L B'$, then $B \text{ III } \sigma \sim_L B' \text{ III } \sigma$. In order to prove that $L(\mathcal{A}) = L$ assume first that $w \in L(\mathcal{A})$. Therefore, $\delta([\lambda], w) \in F$. If $w = w_1 w_2 \ldots w_n$, where $w_i \in \Sigma, i = 1, \ldots, n$, then it is easy to see that $\delta([\lambda], w) = [w_1 \text{ III } w_2 \text{ III } \ldots \text{ III } w_n]$. Hence, there is a language K, such that $K \cap L \neq \emptyset$ and $K \sim_L (w_1 \text{ III } \ldots \text{ III } w_n)$. Consequently, $(w_1 \text{ III } w_2 \text{ III } \ldots \text{ III } w_n) \cap L \neq \emptyset$. Therefore, $w = w_1 w_2 \ldots w_n \in com(L) = L$. Thus $w \in L$. Conversely, assume that $w \in L$. Now, if $w = w_1 w_2 \ldots w_n, w_i \in \Sigma, i = 1, \ldots, n$, then $\delta([\lambda], w) = [w_1 \text{ III } w_2 \text{ III } \ldots \text{ III } w_n]$ and thus $(w_1 \text{ III } w_2 \text{ III } \ldots \text{ III } w_n) \cap L \neq \emptyset$. Hence, $[w_1 \text{ III } w_2 \text{ III } \ldots \text{ III } w_n] \in F$ and therefore $w \in L(\mathcal{A})$. Consequently, $L = L(\mathcal{A})$. $\square$

It will be seen in many connections in this Handbook that L is regular iff L is acceptable by a DFA. Thus, Theorem 2.7 yields,

Theorem 2.9. *Let L be a commutative language. The following conditions are equivalent:*

a) L is a regular language.
b) L has finitely many scattered residuals. $\square$

3. Formal languages: a telegraphic survey

This section will be very different from the preceding ones. Following [7], we will present a "telegraphic survey" of some of the basics of formal language theory. This means that we just quickly run through very many concepts and results, without going into any details. It is recommended that the survey is consulted only when need arises.

3.1 Language and grammar. Chomsky hierarchy

Basic terminology and notation

Alphabet: a finite nonempty set of abstract *symbols* or *letters*.
V^*: the free monoid generated by the alphabet V under the operation of *catenation* or *concatenation* (the catenation of x and y is denoted by xy; the catenation of n copies of $x \in V^*$ is also denoted by x^n).

Word/string: an element of V^*.

λ: the empty string (the identity of the monoid V^*; $V^* - \{\lambda\}$ is denoted by V^+).

$|x|$: the *length* of the string $x \in V^*$.

$|x|_a$: the number of occurrences of $a \in V$ in $x \in V^*$.

Language: a subset of V^*.

λ-free language: a subset of V^+.

Parikh mapping: for $V = \{a_1, a_2, \ldots, a_k\}$, the mapping $\Psi_V : V^* \longrightarrow N^k$ defined by $\Psi_V(x) = (|x|_{a_1}, |x|_{a_2}, \ldots, |x|_{a_k})$.

Linear set: a set $M \subseteq N^k$ such that $M = \{v_0 + \sum_{i=1}^{m} v_i x_i \mid x_i \in N\}$, for some $v_0, v_1, \ldots, v_m$ in N^k.

Semilinear set: a finite union of linear sets.

Semilinear language: a language $L \subseteq V^*$ with semilinear $\Psi_V(L)$.

Letter-equivalent languages: two languages $L_1, L_2 \subseteq V^*$ such that $\Psi_V(L_1) = \Psi_V(L_2)$.

Length set (of a language L): $length(L) = \{|x| \mid x \in L\}$.

Operations with languages

Union, intersection, complement: usual set operations.

Concatenation: $L_1 L_2 = \{xy \mid x \in L_1, y \in L_2\}$.

Kleene star (∗): $L^* = \cup_{i \geq 0} L^i$, where $L_0 = \{\lambda\}$, $L^{i+1} = L^i L, i \geq 0$.

Kleene +: $L^+ = \cup_{i \geq 1} L^i$.

Substitution: a mapping $s : V \longrightarrow 2^{U^*}$, V, U alphabets, extended to V^* by $s(\lambda) = \{\lambda\}$, $s(ax) = s(a)s(x)$, $a \in V, x \in V^*$. For $L \subseteq V^*, s(L) = \cup_{x \in L} s(x)$.

Finite substitution: substitution s with $s(a)$ finite for all symbols a.

Morphism: substitution s with $s(a)$ singleton set for all symbols a (usual monoid morphism).

λ-free substitution/morphism: substitution/morphism such that $\lambda \in s(a)$ for no symbol a.

Inverse morphism: a mapping $h^{-1} : U^* \longrightarrow 2^{V^*}$ defined by $h^{-1}(x) = \{y \in V^* \mid h(y) = x\}, x \in U^*$, for a morphism $h : V^* \longrightarrow U^*$.

Shuffle: $x \text{ Ш } y = \{x_1 y_1 \ldots x_n y_n \mid x = x_1 \ldots x_n, y = y_1 \ldots y_n, x_i, y_i \in V^*, 1 \leq i \leq n, n \geq 1\}$. For $L_1, L_2 \subseteq V^*, L_1 \text{ Ш } L_2 = \{w \in V^* \mid w \in x \text{ Ш } y, \text{ for some } x \in L_1, y \in L_2\}$.

Mirror image: if $x = a_1 a_2 \ldots a_n, a_i \in V, 1 \leq i \leq n$, then $mi(x) = a_n \ldots a_2 a_1$. For $L \subseteq V^*, mi(L) = \{mi(x) \mid x \in L\}$.

Left quotient (of L_1 with respect to L_2, $L_1, L_2 \subseteq V^*$): $L_2 \backslash L_1 = \{w \mid \text{ there is } x \in L_2 \text{ such that } xw \in L_1\}$.

Left derivative (of L with respect to $x, L \subseteq V^*, x \in V^*$): $\partial_x(L) = \{x\} \backslash L$.

Right quotient/derivative: symmetrically.

Given an n-ary operation with languages, α, we say that a family F of languages *is closed* under operation α if $\alpha(L_1, \ldots, L_n) \in F$ for every $L_1, \ldots, L_n \in F$.

A language $L \subseteq V^*$ is called *regular* if it can be obtained from elements of V

and $\emptyset$ using finitely many times the operations of union, concatenation and Kleene star.

A family of languages is *nontrivial* if it contains at least one language different from $\emptyset$ and $\{\lambda\}$. A nontrivial family of languages is called *trio* or a *cone* if it is closed under λ-free morphisms, inverse morphisms, and intersection with regular languages. A trio closed under union is called *semi-AFL* (AFL = abstract family of languages). A semi-AFL closed under concatenation and Kleene $+$ is called *AFL*. A trio/semi-AFL/AFL is said to be *full* if it is closed under arbitrary morphisms (and Kleene $*$ in the case of AFL's). A family of languages closed under none of the six AFL operations is called *anti-AFL*.

1. The family of regular languages is the smallest full trio.
2. Each (full) semi-AFL closed under Kleene $+$ is a (full) AFL.
3. If F is a family of λ-free languages which is closed under concatenation, λ-free morphisms, and inverse morphisms, then F is closed under intersection with regular languages and union, hence F is a semi-AFL. (If F is also closed under Kleene $+$, then it is an AFL.)
4. If F is a family of languages closed under intersection with regular languages, union with regular languages, and substitution with regular languages, then F is closed under inverse morphisms.
5. Every semi-AFL is closed under substitution with λ-free regular languages. Every full semi-AFL is closed under substitution with arbitrary regular languages and under left/right quotient with regular languages.

Chomsky grammars

A *phrase-structure grammar* is a quadruple $G = (N, T, S, P)$, where N, T are disjoint alphabets, $S \in N$, and $P \subseteq V_G^* N V_G^* \times V_G^*$, for $V_G = N \cup T$. The elements of N are called *nonterminal* symbols, those of T are called *terminal* symbols, S is the *start symbol* or the *axiom*, and P the set of *production rules*; $(u, v) \in P$ is written in the form $u \rightarrow v$.

For $x, y \in V_G^*$ we write $x \Longrightarrow_G y$ iff $x = x_1 u x_2, y = x_1 v x_2$, for some $x_1, x_2 \in V_G^*$ and $u \rightarrow v \in P$. If G is understood, we write $x \Longrightarrow y$. One says that x directly derives y (with respect to G). The reflexive and transitive closure of the relation $\Longrightarrow$ is denoted by $\Longrightarrow^*$. The *language generated* by G is $L(G) = \{x \in T^* \mid S \Longrightarrow^* x\}$.

Two grammars G_1, G_2 are called *equivalent* if $L(G_1) = L(G_2)$.

If in $x \Longrightarrow y$ above we have $x_1 \in T^*$, then the derivation is *leftmost* and we write $x \Longrightarrow_{left} y$. The leftmost language generated by G is denoted by $L_{left}(G)$.

A phrase-structure grammar $G = (N, T, S, P)$ is called:

- *monotonous/length-increasing*, if for all $u \rightarrow v \in P$ we have $|u| \leq |v|$.
- *context-sensitive*, if each $u \rightarrow v \in P$ has $u = u_1 A u_2, v = u_1 x u_2$, for $u_1, u_2 \in V_G^*, A \in N, x \in V_G^+$. (In monotonous and context-sensitive grammars a

production $S \to \lambda$ is allowed, providing S does not appear in the right-hand members of rules in P.)

- *context-free*, if each production $u \to v \in P$ has $u \in N$.
- *linear*, if each rule $u \to v \in P$ has $u \in N$ and $v \in T^* \cup T^*NT^*$.
- *right-linear*, if each rule $u \to v \in P$ has $u \in N$ and $v \in T^* \cup T^*N$.
- *left-linear*, if each rule $u \to v \in P$ has $u \in N$ and $v \in T^* \cup NT^*$.
- *regular*, if each rule $u \to v \in P$ has $u \in N$ and $v \in T \cup TN \cup \{\lambda\}$.

The phrase-structure, context-sensitive, context-free, and regular grammars are also called type 0, type 1, type 2, and type 3 grammars, respectively.

1. The family of languages generated by monotonous grammars is equal with the family of languages generated by context-sensitive grammars.

2. The family of languages generated by right- or by left-linear grammars are equal and equal with the family of languages generated by regular grammars, as defined in Section 3.2.

We denote by *RE, CS, CF, LIN, REG* the families of languages generated by arbitrary, context-sensitive, context-free, linear, and regular grammars, respectively (RE stands for *recursively enumerable*).

3. *(Chomsky hierarchy.)* The following strict inclusions hold: $REG \subset LIN \subset CF \subset CS \subset RE$.

4. $L_{left}(G) \in CF$ for each type-0 grammar G; if G is context-free, then $L_{left}(G) = L(G)$.

Closure properties of families in Chomsky hierarchy (Y stands for *yes* and N for *no*).

	RE	*CS*	*CF*	*LIN*	*REG*
Union	Y	Y	Y	Y	Y
Intersection	Y	Y	N	N	Y
Complement	N	Y	N	N	Y
Concatenation	Y	Y	Y	N	Y
Kleene $*$	Y	Y	Y	N	Y
Intersection with regular languages	Y	Y	Y	Y	Y
Substitution	Y	N	Y	N	Y
λ-free substitution	Y	Y	Y	N	Y
Morphisms	Y	N	Y	Y	Y
λ-free morphisms	Y	Y	Y	Y	Y
Inverse morphisms	Y	Y	Y	Y	Y
Left/right quotient	Y	N	N	N	Y
Left/right quotient with regular languages	Y	N	Y	Y	Y
Left/right derivative	Y	Y	Y	Y	Y
Shuffle	Y	Y	N	N	Y
Mirror image	Y	Y	Y	Y	Y

Therefore, *RE, CF, REG* are full AFL's, *CS* is an AFL (not full), and *LIN* is a full semi-AFL.

Decidability problems

Given a class G of grammars, the following basic questions about arbitrary elements G_1, G_2 of G can be formulated:

- *equivalence:* are G_1, G_2 equivalent?
- *inclusion:* is $L(G_1)$ included in $L(G_2)$?
- *membership:* is an arbitrarily given string x an element of $L(G_1)$?
- *emptiness:* is $L(G_1)$ empty?
- *finiteness:* is $L(G_1)$ finite?
- *regularity:* is $L(G_1)$ a regular language?

The languages with decidable membership question are called *recursive* and their family lies strictly intermediate between *CS* and *RE*.

Decidability properties of classes of phrase-structure, context-sensitive, context-free, linear, and regular grammars (on top of columns we have put the corresponding families of languages; N stands for *non-decidable* and D for *decidable*.

	RE	*CS*	*CF*	*LIN*	*REG*
Equivalence	N	N	N	N	D
Inclusion	N	N	N	N	D
Membership	N	D	D	D	D
Emptiness	N	N	D	D	D
Finiteness	N	N	D	D	D
Regularity	N	N	N	N	trivial

Rice theorem: Let P be a nontrivial property of type-0 languages (there are languages having property P and also languages not having property P). Then property P is undecidable for type-0 languages.

3.2 Regular and context-free languages

Characterizations of regular languages

1. For $L \subseteq V^*$, we define the equivalence relation $\sim_L$ over V^* by $x \sim_L y$ iff $(uxv \in L \Leftrightarrow uyv \in L)$ for all $u, v \in V^*$. Then $V^*/\sim_L$ is called the *syntactic monoid* of L.

2. The set *rex* of *regular expressions* (over an alphabet V) is the smallest set of strings containing $\emptyset$ and a, for every $a \in V$, and having the following property: if e_1, e_2 are in *rex*, then $(e_1 e_2), (e_1 \cup e_2)$ and e_1^* are in *rex*, too. To a regular expression e we associate a language $L(e)$ according to the following rules:

(i) $L(\emptyset) = \emptyset, L(a) = \{a\}, a \in V$,
(ii) $L((e_1 e_2)) = L(e_1)L(e_2), L((e_1 \cup e_2)) = L(e_1) \cup L(e_2), L(e_1^*) = (L(e_1))^*$.

The following facts are basic for regular languages.

1. *Kleene theorem:* A language L is in the family REG iff there is a regular expression e such that $L = L(e)$ (hence it is regular in the sense of the definition in the previous section).

2. *Myhill-Nerode theorem:* A language $L \subseteq V^*$ is regular iff $V^* / \sim_L$ is finite.

Pumping lemmas and other necessary conditions

1. *Bar-Hillel (uvwxy, pumping) lemma for context-free languages:* If $L \in CF, L \subseteq V^*$, then there are $p, q \in N$ such that every $z \in L$ with $|z| > p$ can be written in the form $z = uvwxy$, with $u, v, w, x, y \in V^*$, $|vwx| \leq q, vx \neq \lambda$, and $uv^i wx^i y \in L$ for all $i \geq 0$.

2. *Pumping lemma for linear languages:* If $L \in LIN, L \subseteq V^*$, then there are $p, q \in N$ such that every $z \in L$ with $|z| > p$ can be written in the form $z = uvwxy$, with $u, v, w, x, y \in V^*$, $|uvxy| \leq q, vx \neq \lambda$, and $uv^i wx^i y \in L$ for all $i \geq 0$.

3. *Pumping lemma for regular languages:* If $L \in REG, L \subseteq V^*$, then there are $p, q \in N$ such that every $z \in L$ with $|z| > p$ can be written in the form $z = uvw$, with $u, v, w \in V^*$, $|uv| \leq q, v \neq \lambda$, and $uv^i w \in L$ for all $i \geq 0$.

4. *Ogden pumping lemma (pumping with marked positions):* If $L \in CF, L \subseteq V^*$, then there is $p \in N$ such that for every $z \in L$ and for every at least p marked occurrences of symbols in z, we can write $z = uvwxy$, where
    (i)   either each of u, v, w or each of w, x, y contains at least a marked symbol,
    (ii) vwx contains at most p marked symbols,
    (iii) $uv^i wx^i y \in L$ for all $i \geq 0$.

5. *Parikh theorem:* Every context-free language is semilinear.

6. Every context-free language is letter-equivalent with a regular language.

7. *Consequences:* (i) Every context-free language over a one-letter alphabet is regular. (ii) The length set of an infinite context-free language contains an infinite arithmetical progression.

8. All the previous conditions are only necessary, not also sufficient.

Representation theorems

The language generated by the context-free grammar $G = (\{S\}, \{a_1, b_1, \ldots, a_n, b_n\}, S, \{S \rightarrow SS, S \rightarrow \lambda\} \cup \{S \rightarrow a_i S b_i \mid 1 \leq i \leq n\})$ is called *the Dyck language* (over n pairs of symbols that can be viewed as parentheses) and it is denoted by D_n.

1. Every regular language L can be written in the form

$$L = h_4(h_3^{-1}(h_2(h_1^{-1}(a^*b)))),$$

where h_1, h_2, h_3, h_4 are morphisms.

2. *Chomsky-Schützenberger theorem:* Every context-free language L can be written in the form $L = h(D_n \cap R)$, where h is a morphism, D_n is a Dyck language and R is a regular language.

3. Every language $L \in RE$ can be written in the form $L = h(L_1 \cap L_2)$, as well as in the form $L = L_3 \backslash L_4$, where h is a morphism and L_1, L_2, L_3, L_4 are linear languages.

Normal forms

1. *Chomsky normal form:* For every context-free grammar G, a grammar $G' = (N, T, S, P)$ can be effectively constructed, with the rules in P of the forms $A \rightarrow a$ and $A \rightarrow BC$, for $A, B, C \in N, a \in T$, such that $L(G') = L(G) - \{\lambda\}$.

2. *Greibach normal form:* For every context-free grammar G, a grammar $G' = (N, T, S, P)$ can be effectively constructed, with the rules in P of the form $A \rightarrow a\alpha$, for $A \in N, a \in T, \alpha \in (N \cup T)^*$, such that $L(G') = L(G) - \{\lambda\}$.

3. *The super normal form:* For every triple (k, l, m) of nonnegative integers and for every context-free grammar G, an equivalent grammar $G' = (N, T, S, P)$ can be effectively constructed containing rules of the following forms:

(i) $A \rightarrow xByCz$, with $A, B, C \in N, x, y, z \in T^*$, and $|x| = k, |y| = l, |z| = m$,

(ii) $A \rightarrow x$, with $A \in N, x \in T^*, |x| \in length(L(G))$.

Variants of Chomsky and Greibach normal forms can be obtained by particularizing the parameters k, l, m above.

Ambiguity

A context-free grammar G is *ambiguous* if there is $x \in L(G)$ having two different leftmost derivations in G. A context-free language L is *inherently ambiguous* if each context-free grammar of L is ambiguous; otherwise, L is called *unambiguous*.

1. There are inherently ambiguous linear languages; each regular language is unambiguous.

2. The ambiguity problem for context-free grammars is undecidable.

3.3 L Systems

1. An *interactionless Lindenmayer system* (0L system) is a triple $G = (V, P, w)$ where V is an alphabet, P is a finite set of rules $a \rightarrow u, a \in V, u \in V^*$, such that there is at least one rule for each $a \in V$, and $w \in V^*$. For $x, y \in V^*$ we write $x \Longrightarrow_G y$ iff $x = a_1 a_1 \ldots a_k, y = u_1 u_2 \ldots u_k$ and $a_i \rightarrow u_i \in P$, $1 \leq i \leq k$ (*parallel* derivation with each letter developing independently of its neighbours). The generated language is $L(G) = \{z \in V^* \mid w \Longrightarrow_G^* z\}$.

Examples of 0L systems.

System	Axiom	Rules	Language
G_1	a	$a \to a^2$	$\{a^{2^n} \mid n \geq 0\}$
G_2	a	$a \to a, a \to a^2$	a^+
G_3	a	$a \to b, b \to ab$	$\{u_i \mid u_0 = a, u_1 = b, u_{i+2} = u_i u_{i+1}\}$
G_4	aba	$a \to aba, b \to \lambda$	$\{(aba)^{2^n} \mid n \geq 0\}$
G_5	aba	$a \to a, b \to ba^2 b$	$= L(G_4)$

2. *IL Systems* ("interactions") are defined similarly except that the rules are context-dependent: the same letter may be rewritten differently in different contexts.

3. Most capital letters have a standard meaning in connection with L systems. Thus, L refers to parallel rewriting. The character O (or 0) means that information between individual letters ("cells") is zero-sided. The letter P ("propagating") means that λ is never on the right side of a rule, and D ("deterministic") that, for each configuration (a letter in a context), there is only one rule. (The systems G_1, $G_3 - G_5$ are DOL systems and, apart from G_4, also $PDOL$ systems. They are also (degenerate cases of) IL systems. The systems G_4 and G_5 are equivalent.) The letter F ("finite") attached to the name of a system means that, instead of a single axiom, there can be finitely many axioms. The letter T ("tables") means that the rules are divided into subsets. In each derivation step, rules from the same subset have to be used; D in this context means that each subset is deterministic. Finally, E, H, C indicate +1.   language L originally generated is modified in a certain way. E ("⸤   ᵒ"): L is intersected with V_T^*, where V_T ("terminals") is a subset of the original alphabet V. H ("homomorphism"): a morphic image of L under a predefined morphism is taken. C ("coding"): a letter-to-letter morphic image of L is taken. Thus, we may speak of $EDT0L$ and $HPDF0L$ systems. The families of languages generated by such systems are denoted simply by $EDT0L$ and $HPDF0L$.

Convention: two languages are considered equal if they differ only by the empty word.

1. $EIL = ETIL = RE$; $CF \subset E0L \subset EPIL = EPTIL = CS$.

2. $E0L = H0L = C0L = EP0L = HP0L = EF0L = HF0L = CF0L = EPF0L = $ $= HPF0L$; $ET0L = HT0L = CT0L = EPT0L = HPT0L = ETF0L = HTF0L = $ $CTF0L = EPTF0L = HPTF0L \supset CPT0L = CPTF0L$.

3. $HD0L = HPD0L = HDF0L = HPDF0L = CDF0L = CPDF0L \supset CD0L \supset$ $\supset ED0L \supset D0L \supset PD0L$; $EDT0L = HDT0L = CDT0L$.

4. Equivalence is *undecidable* for 0L systems but *decidable* for HD0L systems. Emptiness and infinity are *decidable* for ET0L languages.

5. In general, L families have *week closure properties*, apart from extended families, notably $E0L$ and $ET0L$. For instance, $D0L$ and $0L$ are anti-AFL's but $E0L$ is closed under union, catenation, morphisms, intersection with regular sets, Kleene + (but not under inverse morphisms), whereas $ET0L$ is a full AFL.

Sequences

A D0L system generates its language in a *sequence*, the word lengths in this sequence constitute a *D0L length sequence*. The length sequences of G_1 and G_3 above are the nonnegative powers of 2 and the Fibonacci numbers, respectively. Quite much is known about *D0L* length sequences, their theory being related to the theory of *finite difference equations* with constant coefficients. The length of the nth term can be expressed as an *exponential polynomial*, a finite sum of terms of the form $\alpha n \beta^n$, where α and β are complex numbers. Thus, neither logarithmic nor subexponential (not bounded by a polynomial) growth is possible. (Both types are possible for *DIL* systems. *Decidable characterizations* can be given for *HD0L* length sequences that are not *D0L*, and of strictly increasing *D0L* length sequences that are not *PD0L*. It is *decidable* whether or not two given *HD0L* systems generate the same length sequence, the same question being undecidable for *DIL* systems.

Comment: Parallelism of derivation in L systems reflects the original motivation: L systems were introduced to model the development of (simple filamentous) organisms. Although the L systems discussed above are 1-dimensional (generate only strings), *branching structures* can be described using various conventions in computer graphics.

3.4 More powerful grammars and grammar systems

Regulated rewriting

A *programmed grammar* is a system $G = (N, T, S, P)$, where N, T are disjoint alphabets, $S \in N$, and P is a finite set of quadruples $(b : A \to z, E, F)$, where $A \to z$ is a context-free rule over $N \cup T$, b is a label from a set *Lab*, and E, F are subsets of *Lab*. For $(x, b), (y, b')$ in $V_G^* \times Lab$ we write $(x, b) \Longrightarrow (y, b')$ iff one of the next two cases holds:

(i) $x = x_1 A x_2, y = x_1 z x_2$, for $(b : A \to z, E, F) \in P$ and $b' \in E$,
(ii) $|x|_A = 0, x = y$, and $b' \in F$.

The language generated by G is defined by $L(G) = \{w \in T^* \mid (S, b) \Longrightarrow^* (w, b'), \text{ for some } b, b' \in Lab\}$.

We denote by PR_{ac}^λ the family of languages generated by programmed grammars; the superscript λ is removed when one uses only λ-free grammars; the subscript ac is removed when grammars with only empty sets F in rules $(b : A \to z, E, F)$ are used.

1. $CF \subset PR \subset PR_{ac} \subset CS$, $PR \subset PR^\lambda \subset PR_{ac}^\lambda = RE$.
2. $ET0L \subset PR_{ac}$.
3. PR_{ac} is an AFL (not full).
4. The programmed grammars (of various types) are equivalent with other classes of grammars with regulated rewriting: matrix, controlled, time-varying, random context, state, etc.

Grammar systems

A *cooperating distributed* (CD) *grammar system* of degree $n, n \geq 1$, is an $(n + 3)$-tuple $G = (N, T, S, P_1, \ldots, P_n)$, where N, T are disjoint alphabets, $S \in N$, and $P_1, \ldots, P_n$ are finite sets of context-free rules over $N \cup T$. The derivation relation $x \Longrightarrow_{P_i} y$ is defined in the usual way. Moreover, $x \Longrightarrow_{P_i}^t y$ iff $x \Longrightarrow_{P_i}^* y$ and there is no $z \in (N \cup T)^*$ such that $y \Longrightarrow_{P_i} z$ (maximal derivation). Then we define $L_t(G) = \{w \in T^* \mid S \Longrightarrow_{P_{i_1}}^t w_1 \Longrightarrow_{P_{i_2}}^t \ldots \Longrightarrow_{P_{i_m}}^t w_m = w, m \geq 1, 1 \leq i_j \leq n, 1 \leq j \leq m\}$.

We denote by $CD_n(t)$ the family of languages generated (in the t-mode) by CD grammar systems of degree at most $n, n \geq 1$.

$$CF = CD_1(t) = CD_2(t) \subset CD_3(t) = CD_n(t) = ET0L, n \geq 3.$$

Contextual grammars

An *internal contextual grammar* with F choice (where F is a given family of languages) is a system $G = (V, B, (D_1, C_1), \ldots, (D_n, C_n)), n \geq 1$, where V is an alphabet, B is a finite language over V, D_i are languages in F, and C_i are finite subsets of $V^* \times V^*$, $1 \leq i \leq n$. For $x, y \in V^*$ we write $x \Longrightarrow y$ iff $x = x_1 x_2 x_3, y = x_1 u x_2 v x_3$ for $x_2 \in D_i, (u, v) \in C_i$, for some $1 \leq i \leq n$. Then the language generated by G is $L(G) = \{w \in V^* \mid z \Longrightarrow^* w \text{ for some } z \in B\}$. We denote by $IC(F)$ the family of languages generated by such grammars. We consider here $F \in \{FIN, REG\}$, where FIN is the family of finite languages.

1. $REG \subset IC(FIN) \subset IC(REG) \subset CS$.
2. $IC(FIN)$ and $IC(REG)$ are both incomparable with LIN and CF.
3. Every $L \in RE$ can be written in the form $L = R \backslash L'$, where $L' \in IC(FIN)$ and R is a regular language.

3.5 Books on formal languages

A. V. Aho, J. D. Ullman, *The Theory of Parsing, Translation, and Compiling*, Prentice Hall, Engl. Cliffs, N.J., vol. I 1971, vol. II 1973.

A. V. Aho, J. D. Ullman, *Principles of Compiler Design*, Addison-Wesley, Reading, Mass., 1977.

J. Berstel, *Transductions and Context-Free Languages*, Teubner, Stuttgart, 1979.

W. Brauer, *Automatentheorie*, B. G. Teubner, Stuttgart, 1984.

R. V. Book (ed.), *Formal Language Theory. Perspectives and Open Problems*, Academic Press, New York, 1980.

E. Csuhaj-Varju, J. Dassow, J. Kelemen, Gh. Păun, *Grammar Systems. A Grammatical Approach to Distribution and Cooperation*, Gordon and Breach, London, 1994.

J. Dassow, Gh. Păun, *Regulated Rewriting in Formal Language Theory*, Springer-Verlag, Berlin, 1989.

M. D. Davis, E. J. Weyuker, *Computability, Complexity, and Languages*, Academic Press, New York, 1983.

P. J. Denning, J. B. Dennis, J. E. Qualitz, *Machines, Languages, and Computation*, Prentice-Hall, Engl. Cliffs, N.J., 1978.

S. Eilenberg, *Automata, Languages, and Machines*, Academic Press, New York, vol. A, 1974, vol. B, 1976.

E. Engeler, *Formal Languages*, Markham, Chicago, 1968.

R. W. Floyd, R. Beigel, *The Language of Machines An Introduction to Computability and Formal Languages*, Computer Science Press, 1994.

F. Gécseg, I. Péak, *Algebraic Theory of Automata*, Akadémiai Kiadó, 1972.

S. Ginsburg, *The Mathematical Theory of Context-Free Languages*, McGraw Hill, New York, 1966.

S. Ginsburg, *Algebraic and Automata-Theoretic Properties of Formal Languages*, North-Holland, Amsterdam, 1975.

A. V. Gladkij, *Leçons de linguistique mathématique*, Centre de linguistique quantitative de la faculté des sciences de l'Université de Paris, Dunod, 1970.

M. Gross, A. Lentin, *Notions sur les grammaires formelles*, Gauthier-Villars, Paris, 1967.

M. Harrison, *Introduction to Formal Language Theory*, Addison-Wesley, Reading, Mass., 1978.

G. T. Herman, G. Rozenberg, *Developmental Systems and Languages*, North-Holland, Amsterdam, 1975.

J. E. Hopcroft, J. D. Ullman, *Formal Languages and Their Relations to Automata*, Addison-Wesley, Reading, Mass., 1969.

J. E. Hopcroft, J. D. Ullman, *Introduction to Automata Theory, Languages, and Computation*, Addison-Wesley, Reading, Mass., 1979.

M. Ito (ed.), *Words, Languages, and Combinatorics*, World Scientific, Singapore, 1992.

W. Kuich, A. Salomaa, *Semirings, Automata, Languages*, Springer-Verlag, Berlin, 1986.

H. R. Lewis, C. H. Papadimitriou, *Elements of the theory of computation*, Prentice-Hall, Engl. Cliffs, NJ, 1981.

P. Linz, *An Introduction to Formal Languages and Automata*, D. C. Heath and Co., Lexington, Mass., 1990. M. Lothaire, *Combinatorics on Words*, Addison-Wesley, Reading, Mass., 1983.

S. Marcus, *Grammars and Finite Automata*, The Publ. House of the Romanian Academy, Bucharest, Romania, 1964.

A. Mateescu, D. Vaida, *Discrete Mathematical Structures. Applications*, The Publ. House of the Romanian Academy, Bucharest, Romania, 1989.

H. Maurer, *Theoretische Grundlagen der Programmiersprachen*, Hochschultaschenbücher 404, Bibliographisches Inst., 1969.

Gh. Păun, *Contextual Grammars*, The Publ. House of the Romanian Academy, Bucharest, 1982.

Gh. Păun, *Recent Results and Problems in Formal Language Theory*, The Scientific and Encyclopaedic Publ. House, Bucharest, 1984.

Gh. Păun (ed.) *Mathematical Aspects of Natural and Formal Languages*, World Scientific, Singapore, 1994.

Gh. Păun (ed.) *Artificial Life: Grammatical Models*, Black Sea University Press, Bucharest, Romania, 1995.

Gh. Păun (ed.) *Mathematical linguistics and related topics*, The Publ. House of the Romanian Academy, Bucharest, Romania, 1995.

J. E. Pin, *Varieties of Formal Languages*, Plenum Press, Oxford, 1986.

G. E. Révész, *Introduction to Formal Languages*, McGraw-Hill, New York, 1980.

G. Rozenberg, A. Salomaa, *The Mathematical Theory of L Systems*, Academic Press, New York, 1980.

G. Rozenberg, A. Salomaa, *Cornerstones of Undecidability*, Prentice Hall, New York, 1994.

G. Rozenberg, A. Salomaa (eds.), *Developments in Language Theory*, World Scientific, Singapore, 1994.

A. Salomaa, *Theory of Automata*, Pergamon, Oxford, 1969.

A. Salomaa, *Formal Languages*, Academic Press, New York, London, 1973.

A. Salomaa, *Jewels of Formal Language Theory*, Computer Science Press, Rockville, 1981.

A. Salomaa, *Computation and Automata*, Cambridge Univ. Press, Cambridge, 1985.

A. Salomaa, M. Soittola, *Automata-Theoretic Aspects of Formal Power Series*, Springer-Verlag, Berlin, 1978.

S. Sippu, E. Soisalon-Soininen, *Parsing Theory. Vol. I: Languages and Parsing*, Springer-Verlag, Berlin, 1988.

H. J. Shyr, *Free Monoids and Languages*, Hon Min Book Comp., Taichung, 1991.

D. Wood, *Grammar and L Forms. An Introduction*, Lecture Notes in Computer Science, 91, Springer-Verlag, Berlin, 1980.

D. Wood, *Theory of computation*, Harper& Row, New York, 1987.

References

[1]   N. Chomsky, "Three models for the description of language", *IRE Trans. on Information Theory*, 2: 3 (1956) 113–124.

[2]   T. V. Gamkrelidze, V. V. Ivanov, "The Early History of Indo-European Languages", *Scientific American*, March, (1990), 82–89.

[3]   L. H. Haines, "On free monoids partially ordered by embedding", *J. Combinatorial Theory*, 6 (1969) 94–98.

[4] G. H. Higman, "Ordering by divisibility in abstract algebras", *Proc. London Math. Soc.*, 3, (1952) 326–336.

[5] A. Lindenmayer, "Mathematical models for cellular interactions in development", I, II, *J. Theoret. Biol.*, 18, (1968) 280–315.

[6] R. C. Lyndon, M. P. Schützenberger, "The equation $a^m = b^n c^p$ in a free group", *Michigan Math. J.*, 9, (1962) 289–298.

[7] G. Păun, A. Salomaa, "Formal Languages", in *CRC Handbook of Discrete and Combinatorial Mathematics*, K. H. Rosen (ed.), to appear.

[8] E. Post, "Finite combinatory processes-formulation", I, *J. Symbolic Logic*, 1, (1936) 103–105.

[9] P. E. Ross, "Hard Words", *Scientific American*, April (1991) 70–79.

[10] A. Salomaa, *Jewels of Formal Language Theory*, Computer Science Press, Rockville, 1981.

[11] A. Thue, "Über unendliche Zeichenreihen", *Norske Vid. Selsk. Skr., I Mat. Nat. Kl.*, Kristiania, 7, (1906) 1–22.

[12] A. Thue, "Über die gegenseitige Lage gleicher Teile gewisser Zeichenreihen", *Norske Vid. Selsk. Skr., I Mat. Nat. Kl.*, Kristiania, 1, (1912) 1–67.

[13] A. M. Turing, "On computable numbers with an application to the Entscheidungsproblem", *Proc. London Math. Soc.*, 2, 42, (1936) 230–265. A correction, *ibid.*, 43, 544–546.

Regular Languages

Sheng Yu

Regular languages and finite automata are among the oldest topics in formal language theory. The formal study of regular languages and finite automata can be traced back to the early forties, when finite state machines were used to model neuron nets by McCulloch and Pitts [83]. Since then, regular languages have been extensively studied. Results of early investigations are, for example, Kleene's theorem establishing the equivalence of regular expressions and finite automata [69], the introduction of automata with output by Mealy [86] and Moore [88], the introduction of nondeterministic finite automata by Rabin and Scott [99], and the characterization of regular languages by congruences of finite index by Myhill [90] and Nerode [91].

Regular languages and finite automata have had a wide range of applications. Their most celebrated application has been lexical analysis in programming language compilation and user-interface translations [1, 2]. Other notable applications include circuit design [21], text editing, and pattern matching [70]. Their application in the recent years has been further extended to include parallel processing [3, 37, 50], image generation and compression [9, 28, 29, 33, 116], type theory for object-oriented languages [92], DNA computing [31, 53], etc.

Since the late seventies, many have believed that everything of interest about regular languages is known except for a few very hard problems, which could be exemplified by the six open problems Brzozowski presented at the International Symposium on Formal Language Theory in 1979 [18]. It appeared that not much further work could be done on regular languages. However, contrary to the widespread belief, many new and interesting results on regular languages have kept coming out in the last fifteen years. Besides the fact that three of the six open problems, i.e., the restricted star height problem [52], the regularity of noncounting classes problem [36], and the optimality of prefix codes problem [117], have been solved, there have also been many other interesting new results [65, 82, 102, 111, 120, 124], which include results on measuring or quantifying operations on regular languages. For example, it is shown in [65] that the the "DFA to minimal NFA" problem is PSPACE-complete.

There is a huge amount of established research on regular languages over the span of a half century. One can find a long list of excellent books that include chapters dedicated to regular languages, e.g., [54, 106, 84, 41, 57, 107,

123, 98]. Many results, including many recent results, on regular languages are considered to be highly important and very interesting. However, only a few of them can be included in this chapter. In choosing the material for the chapter, besides the very basic results, we tend to select those relatively recent results that are of general interest and have not been included in the standard texts. We choose, for instance, some basic results on alternating finite automata and complexities of operations on regular languages.

This chapter contains the following five sections: 1. Preliminaries, 2. Finite Automata, 3. Regular Expressions, 4. Properties of Regular Languages, and 5. Complexity Issues.

In the first section, we give basic notations and definitions.

In Section 2, we describe three basic types of finite automata: deterministic finite automata, nondeterministic finite automata, and alternating finite automata. We show that the above three models accept exactly the same family of languages. Alternating finite automata are a natural and succinct representation of regular languages. A particularly nice feature of alternating finite automata is that they are backwards deterministic and, thus, can be used practically [50]. We also describe briefly several models of finite automata with output, which include Moore machines, Mealy machines, and finite transducers. Finite transducers are used later in Section 4 for proving various closure properties of regular languages.

In Section 3, we define regular expressions and describe the transformation between regular expressions and finite automata. We present the well-known star height problem and the extended star height problem. At the end of the section, we give a characterization of regular languages having a polynomial density using regular expressions of a special form.

In Section 4, we describe four pumping lemmas for regular languages. The first two give necessary conditions for regularity; and the other two give both sufficient and necessary conditions for regularity. All the four lemmas are stated in a simple and understandable form. We give example languages that satisfy the first two pumping conditions but are nonregular. We show that there are uncountably many such languages. In this section, we also discuss various closure properties of regular languages. We describe the Myhill-Nerode Theorem and discuss minimization of DFA as well as AFA. We also give a lower bound on the number of states of an NFA accepting a given language.

In the final section, we discuss two kinds of complexity issues. The first kind considers the number of states of a minimal DFA for a language resulting from some operation, as a function of the numbers of states for the operand languages. This function is called the state complexity of the operation. We describe the state complexity for several basic operations on regular languages. The state complexity gives a clear and objective measurement for each operation. It also gives a lower bound on the time required for the operation. The second kind of complexity issue that we consider is the time and space complexity of various problems for finite automata and regular expressions. We list a number of problems, mostly decision problems, together with their time or space complexity to conclude the section as well as the chapter.

1. Preliminaries

An *alphabet* is a finite nonempty set of symbols. A *word* or a *string* over an alphabet Σ is a finite sequence of symbols taken from Σ. The *empty word*, i.e., the word containing zero symbols, is denoted λ. In this chapter, a, b, c, 0, and 1 are used to denote symbols, while u, v, w, x, y, and z are used to denote words.

The *catenation* of two words is the word formed by juxtaposing the two words together, i.e., writing the first word immediately followed by the second word, with no space in between. Let $\Sigma = \{a, b\}$ be an alphabet and $x = aab$ and $y = ab$ be two words over Σ. Then the catenation of x and y, denoted xy, is $aabab$.

Denote by Σ^* the set of all words over the alphabet Σ. Note that Σ^* is a free monoid with catenation being the associative binary operation and λ being the identity element. So, we have

$$\lambda x = x\lambda = x$$

for each $x \in \Sigma^*$. The *length* of a word x, denoted $|x|$, is the number of occurrences of symbols in x.

Let n be a nonnegative integer and x a word over an alphabet Σ. Then x^n is a word over Σ defined by

(i) $x^0 = \lambda$,
(ii) $x^n = xx^{n-1}$, for $n > 0$.

Let $x = a_1 \ldots a_n$, $n \geq 0$, be a word over Σ. The *reversal* of x, denoted x^R, is the word $a_n \ldots a_1$. Formally, it is defined inductively by

(i) $x^R = x$, if $x = \lambda$;
(ii) $x^R = y^R a$, if $x = ay$ for $a \in \Sigma$ and $y \in \Sigma^*$.

Let x and y be two words over Σ. We say that x is a *prefix* of y if there exists $z \in \Sigma^*$ such that $xz = y$. Similarly, x is a *suffix* of y if there exists $z \in \Sigma^*$ such that $zx = y$, and x is a *subword* of y if there exist $u, v \in \Sigma^*$ such that $uxv = y$.

A *language* L over Σ is a set of words over Σ. The *empty language* is denoted $\emptyset$. The *universal language* over Σ, which is the language consisting of all words over Σ, is Σ^*. For a language L, we denote by $|L|$ the cardinality of L.

The *catenation* of two languages $L_1, L_2 \subseteq \Sigma^*$, denoted $L_1 L_2$, is the set

$$L_1 L_2 = \{w_1 w_2 \mid w_1 \in L_1 \text{ and } w_2 \in L_2 \}.$$

For an integer $n \geq 0$ and a language L, the nth power of L, denoted L^n, is defined by

(i) $L^0 = \{\lambda\}$,
(ii) $L^n = L^{n-1}L$, for $n > 0$.

The star (Kleene closure) of a language L, denoted L^*, is the set

$$\bigcup_{i=0}^{\infty} L^i$$

Similarly, we define

$$L^+ = \bigcup_{i=1}^{\infty} L^i$$

Then, the notation Σ^* is consistent with the above definition. The reversal of a language L, denoted L^R, is the set

$$L^R = \{w^R \mid w \in L\}.$$

Note that we often denote a singleton language, i.e., a language containing exactly one word, by the word itself when no confusion will be caused. For example, by xLy, where $x, y \in \Sigma^*$ and $L \subseteq \Sigma^*$, we mean $\{x\}L\{y\}$.

Let Σ and Δ be two finite alphabets. A mapping $h : \Sigma^* \to \Delta^*$ is called a *morphism* if

(1) $h(\lambda) = \lambda$ and
(2) $h(xy) = h(x)h(y)$ for all $x, y \in \Sigma^*$.

Note that condition (1) follows from condition (2), Therefore, condition (1) can be deleted.

For a set S, let 2^S denote the power set of S, i.e., the collection of all subsets of S. A mapping $\varphi : \Sigma^* \to 2^{\Delta^*}$ is called a *substitution* if

(1) $\varphi(\lambda) = \{\lambda\}$ and
(2) $\varphi(xy) = \varphi(x)\varphi(y)$.

Clearly, a morphism is a special kind of substitution where each word is associated with a singleton set. Note that because of the second condition of the definition, morphisms and substitutions are usually defined by specifying only the image of each letter in Σ under the mapping. We extend the definitions of h and φ, respectively, to define

$$h(L) = \{h(w) \mid w \in L\}$$

and

$$\varphi(L) = \bigcup_{w \in L} \varphi(w)$$

for $L \subseteq \Sigma^*$.

Example 1.1. Let $\Sigma = \{a, b, c\}$ and $\Delta = \{0, 1\}$. We define a morphism h : $\Sigma^* \to \Delta^*$ by

$$h(a) = 01, \quad h(b) = 1, \quad h(c) = \lambda.$$

Then, $h(baca) = 10101$. We define a substitution $\varphi : \Sigma^* \to 2^{\Delta^*}$ by

$$\varphi(a) = \{01, 001\}, \quad \varphi(b) = \{1^i \mid i > 0\}, \quad \varphi(c) = \{\lambda\}.$$

Then, $\varphi(baca) = \{1^i 0101, 1^i 01001, 1^i 00101, 1^i 001001 \mid i > 0\}$.   □

A morphism $h : \Sigma^* \to \Delta^*$ is said to be λ-free if $h(a) \neq \lambda$ for all $a \in \Sigma$. A substitution $\varphi : \Sigma^* \to 2^{\Delta^*}$ is said to be λ-free if $\lambda \notin \varphi(a)$ for all $a \in \Sigma$. And φ is called a *finite substitution* if, for each $a \in \Sigma$, $\varphi(a)$ is a finite subset of Δ^*.

Let $h : \Sigma^* \to \Delta^*$ be a morphism. The inverse of the morphism h is a mapping $h^{-1} : \Delta^* \to 2^{\Sigma^*}$ defined by, for each $y \in \Delta^*$,

$$h^{-1}(y) = \{x \in \Sigma^* \mid h(x) = y\}.$$

Similarly, for a substitution $\varphi : \Sigma^* \to 2^{\Delta^*}$, the inverse of the substitution φ is a mapping $\varphi^{-1} : \Delta^* \to 2^{\Sigma^*}$ defined by, for each $y \in \Delta^*$,

$$\varphi^{-1}(y) = \{x \in \Sigma^* \mid y \in \varphi(x)\}.$$

2. Finite automata

In formal language theory in general, there are two major types of mechanisms for defining languages: acceptors and generators. For regular languages in particular, the acceptors are finite automata and the generators are regular expressions and right (left) linear grammars, etc.

In this section, we describe three types of finite automata (FA): deterministic finite automata (DFA), nondeterministic finite automata (NFA), and alternating finite automata (AFA). We show that all the three types of abstract machines accept exactly the same family of languages. We describe the basic operations of union, intersection, catenation, and complementation on the family of languages implemented using these different mechanisms.

2.1 Deterministic finite automata

A finite automaton consists of a finite set of internal states and a set of rules that govern the change of the current state when reading a given input symbol. If the next state is always uniquely determined by the current state and the current input symbol, we say that the automaton is deterministic.

As an informal explanation, we consider the following example[1]. Let A_0 be an automaton that reads strings of 0's and 1's and recognizes those strings

---

[1] A similar example is given in [98].

which, as binary numbers, are congruent to 2 (mod 3). We use $\nu_3(x)$ to denote the value, modulo 3, of the binary string x. For example, $\nu(100) = 1$ and $\nu_3(1011) = 2$. Consider an arbitrary input string $w = a_1 \cdots a_n$ to A_0 where each a_i, $1 \leq i \leq n$, is either 0 or 1. It is clear that for each i, $1 \leq i \leq n$, the string $a_1 \cdots a_i$ falls into one of the three cases: (0) $\nu_3(a_1 \cdots a_i) = 0$, (1) $\nu_3(a_1 \cdots a_i) = 1$ and (2) $\nu_3(a_1 \cdots a_i) = 2$. No other cases are possible. So, A_0 needs only three states which correspond to the above three cases (and the initial state corresponds to the case (0)). We name those three states (0), (1), and (2), respectively. The rules that govern the state changes should be defined accordingly. Note that

$$\nu_3(a_1 \cdots a_{i+1}) \equiv 2 * \nu_3(a_1 \cdots a_i) + a_{i+1} \pmod 3.$$

So, if the current state is (1) and the current input symbol is 1, then the next state is (0) since $2 * 1 + 1 \equiv 0 \pmod 3$. The states and their transition rules are shown in Figure 1.

Clearly, each step of state transition is uniquely determined by the current state and the current input symbol. We distinguish state (2) as the final state and define that A_0 accepts an input w if A_0 is in state (2) after reading the last symbol of w. A_0 is an example of a deterministic finite automaton.

Formally, we define a deterministic finite automaton as follows:

A *deterministic finite automaton* (DFA) A is a quintuple $(Q, \Sigma, \delta, s, F)$, where

> Q is the finite set of states;
> Σ is the input alphabet;
> $\delta : Q \times \Sigma \rightarrow Q$ is the state transition function;
> $s \in Q$ is the starting state; and
> $F \subseteq Q$ is the set of final states.

Note that, in general, we do not require the transition function δ to be total, i.e., to be defined for every pair in $Q \times \Sigma$. If δ is total, then we call A a *complete DFA*.

In the above definition, we also do not require that a DFA is connected if we view a DFA as a directed graph where states are nodes and transitions between states are arcs between nodes. A DFA such that every state is reachable from the starting state and reaches a final state is called a *reduced* DFA. A reduced DFA may not be a complete DFA.

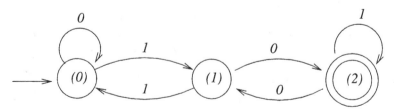

Fig. 1. The states and transition rules of A_0

Example 2.1. A DFA $A_1 = (Q_1, \Sigma_1, \delta_1, s_1, F_1)$ is shown in Figure 2, where $Q_1 = \{0, 1, 2, 3\}$, $\Sigma_1 = \{a, b\}$, $s_1 = 0$, $F_1 = \{3\}$, and δ_1 is defined as follows:

$$\begin{aligned}
\delta_1(0, a) &= 1, & \delta_1(0, b) &= 0, \\
\delta_1(1, a) &= 1, & \delta_1(1, b) &= 2, \\
\delta_1(2, a) &= 1, & \delta_1(2, b) &= 3, \\
\delta_1(3, a) &= 3, & \delta_1(3, b) &= 3.
\end{aligned}$$

The DFA A_1 is reduced and complete. Note that in a state transition diagram, we always represent final states with double circles and non-final states with single circles. □

A *configuration* of $A = (Q, \Sigma, \delta, s, F)$ is a word in $Q\Sigma^*$, i.e., a state $q \in Q$ followed by a word $x \in \Sigma^*$ where q is the current state of A and x is the remaining part of the input. The *starting configuration* of A for an input word $x \in \Sigma^*$ is sx. *Accepting configurations* are defined to be elements of F (followed by the empty word λ).

A computation step of A is a transition from a configuration α to a configuration β, denoted by $\alpha \vdash_A \beta$, where $\vdash_A$ is a binary relation on the set of configurations of A. The relation $\vdash_A$ is defined by: for $px, qy \in Q\Sigma^*$, $px \vdash_A qy$ if $x = ay$ for some $a \in \Sigma$ and $\delta(p, a) = q$. For example, $0abb \vdash_{A_1} 1bb$ for the DFA A_1. We use $\vdash$ instead of $\vdash_A$ if there is no confusion. The kth power of $\vdash$, denoted $\vdash^k$, is defined by $\alpha \vdash^0 \alpha$ for all configurations $\alpha \in Q\Sigma^*$; and $\alpha \vdash^k \beta$, for $k > 0$ and $\alpha, \beta \in Q\Sigma^*$, if there exists $\gamma \in Q\Sigma^*$ such that $\alpha \vdash^{k-1} \gamma$ and $\gamma \vdash \beta$. The transitive closure and the reflexive and transitive closure of $\vdash$ are denoted $\vdash^+$ and $\vdash^*$, respectively.

A *configuration sequence* of A is a sequence of configurations $C_1, \ldots, C_n$, of A, for some $n \geq 1$, such that $C_i \vdash_A C_{i+1}$ for each i, $1 \leq i \leq n - 1$. A configuration sequence is said to be an *accepting configuration sequence* if it starts with a starting configuration and ends with an accepting configuration.

The language accepted by a DFA $A = (Q, \Sigma, \delta, s, F)$, denoted $L(A)$, is defined as follows:

$$L(A) = \{ w \mid sw \vdash^* f \text{ for some } f \in F \}.$$

For convenience, we define the extension of δ, $\delta^* : Q \times \Sigma^* \to Q$, inductively as follows. We set $\delta^*(q, \lambda) = q$ and $\delta^*(q, xa) = \delta(\delta^*(q, x), a)$, for $q \in Q$, $a \in \Sigma$, and $x \in \Sigma^*$. Then, we can also write

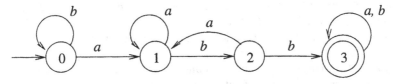

Fig. 2. A deterministic finite automaton A_1

$$L(A) = \{ \ w \mid \delta^*(s, w) = f \text{ for some } f \in F \ \}.$$

The collection of all languages accepted by DFA is denoted $\mathcal{L}_{DFA}$. We call it the *family of DFA languages*. We will show later that the families of languages accepted by deterministic, nondeterministic, and alternating finite automata are the same. This family is again the same as the family of languages denoted by regular expressions. It is called the family of regular languages.

In the remaining of this subsection, we state several basic properties of DFA languages. More properties of DFA languages can be found in Section 4..

Lemma 2.1. *For each $L \in \mathcal{L}_{DFA}$, there is a complete DFA that accepts L.*

Proof. Let $L \in \mathcal{L}_{DFA}$. Then there is a DFA $A = (Q, \Sigma, \delta, s, F)$ such that $L = L(A)$. If A is complete, then we are done. Otherwise, we construct a DFA A' which is the same as A except that there is one more state d and all transitions undefined in A go to d in A'. More precisely, we define $A' = (Q', \Sigma, \delta', s, F)$ such that $Q' = Q \cup \{d\}$, where $d \notin Q$, and $\delta' : Q' \times \Sigma \to Q'$ is defined by

$$\delta'(p, a) = \begin{cases} \delta(p, a), & \text{if } \delta(p, a) \text{ is defined;} \\ d, & \text{otherwise} \end{cases}$$

for $p \in Q'$ and $a \in \Sigma$. It is clear that the new state d and the new state transitions do not change the acceptance of a word. Therefore, $L(A) = L(A')$. $\square$

Theorem 2.1. *The family of DFA languages, $\mathcal{L}_{DFA}$, is closed under union and intersection.*

Proof. Let $L_1, L_2 \subseteq \Sigma^*$ be two arbitrary DFA languages such that $L_1 = L(A_1)$ and $L_2 = L(A_2)$ for some complete DFA $A_1 = (Q_1, \Sigma, \delta_1, s_1, F_1)$ and $A_2 = (Q_2, \Sigma, \delta_2, s_2, F_2)$.

First, we show that there exists a DFA A such that $L(A) = L_1 \cup L_2$. We construct $A = (Q, \Sigma, \delta, s, F)$ as follows:

$Q = Q_1 \times Q_2$,

$s = (s_1, s_2)$,

$F = (F_1 \times Q_2) \cup (Q_1 \times F_2)$, and

$\delta : Q_1 \times Q_2 \to Q_1 \times Q_2$ is defined by $\delta((p_1, p_2), a) = (\delta_1(p_1, a), \delta_2(p_2, a))$. The intuitive idea of the construction is that, for each input word, A runs A_1 and A_2 in parallel, starting from both the starting states. Having finished reading the input word, A accepts the word if either A_1 or A_2 accepts it. Therefore, $L(A) = L(A_1) \cup L(A_2)$.

For intersection, the construction is the same except that $F = F_1 \times F_2$. $\square$

Note that, in the above proof, the condition that A_1 and A_2 are complete is not necessary in the case of intersection. However, if either A_1 or A_2 is incomplete, the resulting automaton is incomplete.

Theorem 2.2. $\mathcal{L}_{DFA}$ *is closed under complementation.*

Proof. Let $L \in \mathcal{L}_{DFA}$. By Lemma 2.1, there is a complete DFA $A = (Q, \Sigma, \delta, s, F)$ such that $L = L(A)$. Then, clearly, the complement of L, denoted $\overline{L}$, is accepted by $\overline{A} = (Q, \Sigma, \delta, s, Q - F)$. □

2.2 Nondeterministic finite automata

Nondeterministic finite automata (NFA) are a generalization of DFA where, for a given state and an input symbol, the number of possible transitions can be greater than one. An NFA is shown in Figure 3, where there are two possible transitions for state 0 and input symbol a: to state 0 or to state 1.

Formally, a *nondeterministic finite automaton A* is a quintuple $(Q, \Sigma, \delta, s, F)$ where Q, Σ, s, and F are defined exactly the same way as for a DFA, and $\delta : Q \times \Sigma \to 2^Q$ is the transition function, where 2^Q denotes the power set of Q.

For example, the transition function for the NFA A_2 of Figure 3 is the following:

$$\delta(0, a) = \{0, 1\}, \qquad \delta(0, b) = \{0\},$$
$$\delta(1, a) = \emptyset, \qquad \delta(1, b) = \{2\},$$
$$\delta(2, a) = \emptyset, \qquad \delta(2, b) = \{3\},$$
$$\delta(3, a) = \{3\}, \qquad \delta(3, b) = \{3\}.$$

A DFA can be considered an NFA, where each value of the transition function is either a singleton or the empty set.

The computation relation $\vdash_A: Q\Sigma^* \times Q\Sigma^*$ of an NFA A is defined by setting $px \vdash_A qy$ if $x = ay$ and $q \in \delta(p, a)$ for $p, q \in Q$, $x, y \in \Sigma^*$, and $a \in \Sigma$. Then the language accepted by A is

$$L(A) = \{ w \mid sw \vdash_A^* f, \text{ for some } f \in F \}.$$

Two automata are said to be *equivalent* if they accept exactly the same language. The NFA A_2 which is shown in Figure 3 accepts exactly the same language as A_1 of Figure 2. Thus, A_2 is equivalent to A_1.

Denote by $\mathcal{L}_{NFA}$ the family of languages accepted by NFA. We show that $\mathcal{L}_{DFA} = \mathcal{L}_{NFA}$.

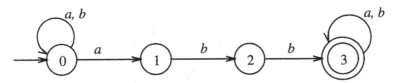

Fig. 3. A nondeterministic finite automaton A_2

Lemma 2.2. *For each NFA A of n states, there exists a complete DFA A'*
of at most 2^n states such that $L(A') = L(A)$.

Proof. Let $A = (Q, \Sigma, \delta, s, F)$ be an NFA such that $|Q| = n$. We construct a
DFA A' such that each state of A' is a subset of Q and the transition function
is defined accordingly. More precisely, we define $A' = (Q', \Sigma, \delta', s', F')$ where
$Q' = 2^Q$; $\delta' : Q' \times \Sigma \to Q'$ is defined by, for $P_1, P_2 \in Q'$ and $a \in \Sigma$,
$\delta'(P_1, a) = P_2$ if

$$P_2 = \{ q \in Q \mid \text{there exists } p \in P_1 \text{ such that } q \in \delta(p, a) \};$$

$s' = \{s\}$; and $F' = \{P \in Q' \mid P \cap F \neq \emptyset \}$. Note that A' has 2^n states.

In order to show that $L(A) = L(A')$, we first prove the following claim.

Claim. For an arbitrary word $x \in \Sigma^*$, $sx \vdash_A^t p$, for some $p \in Q$, if and only
if $s'x \vdash_{A'}^t P$ for some $P \in Q'$ (i.e., $P \subseteq Q$) such that $p \in P$.

We prove the claim by induction on t, the number of transitions. If $t = 0$,
then the statement is trivially true since $s' = \{s\}$. We hypothesize that the
statement is true for $t - 1$, $t > 0$. Now consider the case of t, $t > 0$. Let
$x = x_0 a$, $x_0 \in \Sigma^*$ and $a \in \Sigma$, and $sx_0 a \vdash_A^{t-1} qa \vdash_A p$ for some $p, q \in Q$.
Then, by the induction hypothesis, $s'x_0 \vdash_{A'}^{t-1} P'$ for some $P' \in Q'$ such that
$q \in P'$. Since $p \in \delta(q, a)$, we have $\delta'(P', a) = P$ for some $P \in Q'$ such that
$p \in P$ by the definition of δ'. So, we have $s'x \vdash_{A'}^{t-1} P'a \vdash_{A'} P$ and $p \in P$.
Conversely, let $s'x_0 a \vdash_{A'}^{t-1} P'a \vdash_{A'} P$ and $p \in P$. Then $\delta'(P', a) = P$ and,
therefore, there exists $p' \in P'$ such that $p \in \delta(p', a)$ by the definition of δ'.
By the induction hypothesis, we have $sx_0 \vdash_A^{t-1} p'$. Thus, $sx \vdash_A^{t-1} p'a \vdash_A p$.
This completes the proof of the claim.

Due to the above claim, we have $sw \vdash_A^* f$, for some $f \in F$, if and only if
$s'w \vdash_{A'}^* P$, for some $P \in Q'$, such that $P \cap F \neq \emptyset$, i.e., $P \in F'$. Therefore,
$L(A) = L(A')$. $\square$

The method used above is called the *subset construction*. In the worst case,
all the subsets of Q are necessary. Then the resulting DFA would consist of
2^n states if n is the number of states of the corresponding NFA. Note that if
the resulting DFA is not required to be a complete DFA, the empty subset
of Q is not needed. So, the resulting DFA consists of $2^n - 1$ states in the
worst case. In Section 5., we will show that such cases exist. However, in
most cases, not all the subsets are necessary. Thus, it suffices to construct
only those subsets that are reachable from $\{s\}$. As an example, we construct
a DFA A_6 which is equivalent to NFA A_2 of Figure 3 as follows:

State	a	b
$\{0\}$	$\{0, 1\}$	$\{0\}$
$\{0, 1\}$	$\{0, 1\}$	$\{0, 2\}$
$\{0, 2\}$	$\{0, 1\}$	$\{0, 3\}$
$\{0, 3\}$	$\{0, 1, 3\}$	$\{0, 3\}$
$\{0, 1, 3\}$	$\{0, 1, 3\}$	$\{0, 2, 3\}$
$\{0, 2, 3\}$	$\{0, 1, 3\}$	$\{0, 3\}$

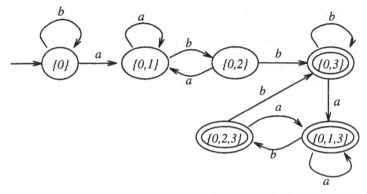

Fig. 4. A DFA A_6 equivalent to NFA A_2

The state transition diagram for A_6 is shown in Figure 4. Only six out of the total sixteen subsets are used in the above example. The other ten subsets of $\{0, 1, 2, 3\}$ are not reachable from $\{0\}$ and, therefore, useless. Note that the resulting DFA can be further minimized into one of only four states. Minimization of DFA is one of the topics in Section 4.

NFA can be further generalized to have state transitions without reading any input symbol. Such transitions are called λ-transitions in the following definition.

A *nondeterministic finite automaton with λ-transitions* (λ-NFA) A is a quintuple $(Q, \Sigma, \delta, s, F)$ where Q, Σ, s, and F are the same as for an NFA; and $\delta : Q \times (\Sigma \cup \{\lambda\}) \rightarrow 2^Q$ is the transition function.

Figure 5 shows the transition diagram of a λ-NFA, where the transition function δ can also be written as follows:

$$\begin{aligned}
\delta(0, a) &= \{0\}, & \delta(0, \lambda) &= \{1\}, \\
\delta(1, b) &= \{1\}, & \delta(1, \lambda) &= \{2\}, \\
\delta(2, c) &= \{2\}.
\end{aligned}$$

and $\delta(q, X) = \emptyset$ in all other cases.

For a λ-NFA $A = (Q, \Sigma, \delta, s, F)$, the binary relation $\vdash_A : Q\Sigma^* \times Q\Sigma^*$ is defined by that $px \vdash_A qy$, for $p, q \in Q$ and $x, y \in \Sigma^*$, if $x = ay$ and $q \in \delta(p, a)$ or if $x = y$ and $q \in \delta(p, \lambda)$. The language accepted by A is again defined as

$$L(A) = \{w \mid sw \vdash_A^* f, \text{ for some } f \in F\}.$$

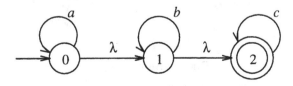

Fig. 5. A λ-NFA A_3

For example, the language accepted by A_3 of Figure 5 is

$$L(A_3) = \{\, a^i b^j c^k \mid i, j, k \geq 0 \,\}.$$

We will show that for each λ-NFA, there exists an NFA that accepts exactly the same language. First, we give the following definition.

Let $A = (Q, \Sigma, \delta, s, F)$ be a λ-NFA. The λ-*closure* of a state $q \in Q$, denoted λ-closure(q), is the set of all states that are reachable from q by zero or more λ-transitions, i.e.,

$$\lambda\text{-closure}(q) = \{\, p \in Q \mid q \vdash_A^* p \,\}.$$

Theorem 2.3. *For each λ-NFA A, there exists an NFA A' such that $L(A) = L(A')$.*

Proof. Let $A = (Q, \Sigma, \delta, s, F)$ be a λ-NFA. We construct an NFA $A' = (Q, \Sigma, \delta', s, F')$ where for each $q \in Q$ and $a \in \Sigma$,

$$\delta'(p, a) = \delta(p, a) \cup \bigcup_{q \in \lambda\text{-closure}(p)} \delta(q, a)\,,$$

and

$$F' = \{\, q \mid \lambda\text{-closure}(q) \cap F \neq \emptyset \,\}.$$

The reader can verify that $L(A) = L(A')$.   □

Consider λ-NFA A_3 which is shown in Figure 5. We have λ-closure$(0) = \{0, 1, 2\}$, λ-closure$(1) = \{1, 2\}$, and λ-closure$(2) = \{2\}$. An equivalent NFA is shown in Figure 6, which is obtained by following the construction specified in the above proof.

Let $M_1 = (Q_1, \Sigma, \delta_1, s_1, F_1)$ and $M_2 = (Q_2, \Sigma, \delta_2, s_2, F_2)$ be two λ-NFA and assume that $Q_1 \cap Q_2 = \emptyset$. Then it is straightforward to construct λ-NFA $M_1 + M_2$, $M_1 M_2$, and M_1^* such that $L(M_1 + M_2) = L(M_1) \cup L(M_2)$, $L(M_1 M_2) = L(M_1)L(M_2)$, and $L(M_1^*) = (L(M_1))^*$, respectively. The constructions are illustrated by the diagrams in Figure 7. Formal definitions of the λ-NFA are listed below:

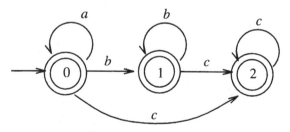

Fig. 6. An NFA A_3'

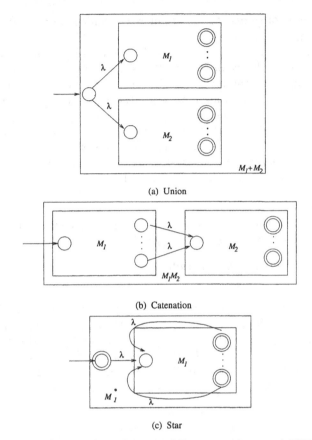

(a) Union

(b) Catenation

(c) Star

Fig. 7. Union, Catenation, and Star operations on λ-NFA

- **Union** $M_1 + M_2 = (Q, \Sigma, \delta, s, F)$ such that $L(M_1 + M_2) = L(M_1) \cup L(M_2)$, where $Q = Q_1 \cup Q_2 \cup \{s\}$, $s \notin Q_1 \cup Q_2$, $F = F_1 \cup F_2$, and
  $$\delta(s, \lambda) = \{s_1, s_2\},$$
  $$\delta(q, a) = \delta_1(q, a) \text{ if } q \in Q_1 \text{ and } a \in \Sigma \cup \{\lambda\},$$
  $$\delta(q, a) = \delta_2(q, a) \text{ if } q \in Q_2 \text{ and } a \in \Sigma \cup \{\lambda\}.$$
- **Catenation** $M_1 M_2 = (Q, \Sigma, \delta, s, F)$ such that $L(M_1 M_2) = L(M_1) L(M_2)$, where $Q = Q_1 \cup Q_2$, $s = s_1$, $F = F_2$, and
  $$\delta(q, a) = \delta_1(q, a) \text{ if } q \in Q_1 \text{ and } a \in \Sigma \text{ or } q \in Q_1 - F_1 \text{ and } a = \lambda,$$
  $$\delta(q, \lambda) = \delta_1(q, \lambda) \cup \{s_2\} \text{ if } q \in F_1,$$
  $$\delta(q, a) = \delta_2(q, a) \text{ if } q \in Q_2 \text{ and } a \in \Sigma \cup \{\lambda\}.$$
- **Star** $M_1^* = (Q, \Sigma, \delta, s, F)$ such that $L(M_1^*) = (L(M_1))^*$, where $Q = Q_1 \cup \{s\}$, $s \notin Q_1$, $F = F_1 \cup \{s\}$, and
  $$\delta(s, \lambda) = \{s_1\},$$
  $$\delta(q, \lambda) = \delta_1(q, \lambda) \cup \{s_1\} \text{ if } q \in F_1,$$
  $$\delta(q, a) = \delta_1(q, a) \text{ if } q \in Q_1 \text{ and } a \in \Sigma \text{ or } q \in Q_1 - F_1 \text{ and } a = \lambda.$$

Intersection and complementation are more convenient to do using the DFA representation.

Another form of generalization of NFA is defined in the following.

A *NFA with nondeterministic starting state* (NNFA) $A = (Q, \Sigma, \delta, S, F)$ is an NFA except that there is a set of starting states S rather than exactly one starting state. Thus, for an input word, the computation of A starts from a nondeterministically chosen starting state.

Clearly, for each NNFA A, we can construct an equivalent λ-NFA A' by adding to A a new state s and a λ-transition from s to each of the starting states in S, and defining s to be the starting state of A'. Thus, NNFA accept exactly the same family of languages as NFA (or DFA or λ-NFA). Each NNFA can also be transformed directly to an equivalent DFA using a subset construction, which is similar to the one for transforming an NFA to a DFA except that the starting state of the resulting DFA is the set of all the starting states of the NNFA. So, we have the following:

Theorem 2.4. *For each NNFA A of n states, we can construct an equivalent DFA A' of at most 2^n states.*   □

Each NNFA has a matrix representation defined as follows [107]: Let $A = (Q, \Sigma, \delta, S, F)$ be an NNFA and assume that $Q = \{q_1, q_2, \ldots, q_n\}$. A mapping h of Σ into the set of $n \times n$ Boolean matrices is defined by setting the (i, j)th entry in the matrix $h(a)$, $a \in \Sigma$, to be 1 if $q_j \in \delta(q_i, a)$, i.e., there is an a-transition from q_i to q_j. We extend the domain of h from Σ to Σ^* by

$$h(w) = \begin{cases} I & \text{if } w = \lambda, \\ h(w_0)h(a) & \text{if } w = w_0 a, \end{cases}$$

where I is the $n \times n$ identity matrix and the juxtaposition of two matrices denotes the multiplication of the two Boolean matrices, where $\wedge$ and $\vee$ are the basic operations. A row vector π of n entries is defined by setting the ith entry to 1 if $q_i \in S$. A column vector ξ of n entries is defined by setting the ith entry to 1 if $q_i \in F$. The following theorem has been proved in [107].

Theorem 2.5. *Let $w \in \Sigma^*$. Then $w \in L(A)$ if and only if $\pi h(w)\xi = 1$.*   □

2.3 Alternating finite automata

The notion of alternation is a natural generalization of nondeterminism. It received its first formal treatment by Chandra, Kozen, and Stockmeyer in 1976 [22, 23, 71]. Various types of alternating Turing machines (ATM) and alternating pushdown machines and their relationship to complexity classes have been studied [24, 37, 61, 62, 79, 72, 94, 103, 38, 55]. Such machines are useful for a better understanding of many questions in complexity theory. For alternating finite automata (AFA – not to be confused with *abstract families of acceptors* defined in [47]), it is proved in [23] that they are precisely as powerful as deterministic finite automata as far as language recognition is concerned. It is also shown in [23] that there exist k-state AFA such that any equivalent complete DFA has at least 2^{2^k} states. A more detailed treatment of alternating finite automata and their operations can be found in [45].

The study of *Boolean automata* was initiated by Brzozowski and Leiss [19] at almost the same time period as AFA were introduced. Boolean automata are essentially AFA except that they allow multiple initial states instead of exactly one initial state in the case of an AFA. In that seminal paper, they also introduced a new type of system of language equations, which can be used to give a clear and comprehensible representation of a Boolean automaton. Boolean automata and the systems of language equations have been further studied in [73, 75, 76, 77].

In the following, we will describe results obtained from both of the above mentioned sources. However, we will use only the term *alternating finite automaton* (AFA). Our basic definitions of AFA follow those in [23]. The equational representation is from [19, 44], and the operations of AFA are from [45].

2.3.1 AFA – the definition

AFA are a natural extension of NFA. In an NFA, if there are two or more possible transitions for the current state and the current input symbol, the outcomes of all the possible computations for the remaining input word are logically **OR**ed. Consider the NFA A_4 shown in Figure 8 with the input $abbb$. When starting at state 0 and reading a, the automaton has two possible moves: to state 1 or to state 2. If we denote by a Boolean variable x_0 whether there is a successful computation for $abbb$ from state 0, and by x_1 and x_2 whether there is a successful computation for the remaining of the input bbb from state 1 and state 2, respectively, then the relation of the computations can be described by the equation

$$x_0 = x_1 \vee x_2.$$

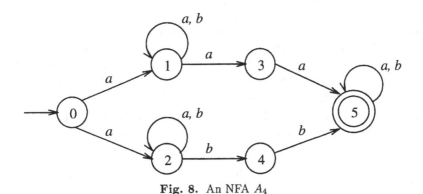

Fig. 8. An NFA A_4

This relation, represented by the equation, captures the essence of nondeterminism. The definition of AFA extends this idea and allows arbitrary Boolean operations in place of the "∨" operation. For example, we may specify that

$$x_0 = (\neg x_1) \wedge x_2.$$

It means that there is a successful computation for $abbb$ from state 0 if and only if there is no successful computation for bbb from state 1 and there is a successful computation for bbb from state 2.

More specifically, an AFA works in the following way: When the automaton reads an input symbol a in a given state q, it will activate all states of the automaton to work on the remaining part of the input in parallel. Once the states have completed their tasks, q will compute its value by applying a Boolean function on those results and pass on the resulting value to the state by which it was activated. A word w is accepted if the starting state computes the value of 1. It is rejected otherwise. We now formalize this idea.

Denote by the symbol B the two-element Boolean algebra $B = (\{0, 1\}, \vee, \wedge, \neg, 0, 1)$. Let Q be a set. Then B^Q is the set of all mappings of Q into B. Note that $u \in B^Q$ can be considered as a vector of $|Q|$ entries, indexed by elements of Q, with each entry being from B. For $u \in B^Q$ and $q \in Q$, we write u_q to denote the image of q under u. If P is a subset of Q then $u|_P$ is the restriction of u to P.

An *alternating finite automaton (AFA)* is a quintuple $A = (Q, \Sigma, s, F, g)$ where

> Q is the finite set of *states;*
> Σ is the *input alphabet;*
> $s \in Q$ is the *starting state;*
> $F \subseteq Q$ is the set of *final states;*
> g is a function of Q into the set of all functions of $\Sigma \times B^Q$ into B.

Note that for each state $q \in Q$, $g(q)$ is a function from $\Sigma \times B^Q$ into B, which we will often denote by g_q in the sequel. For each state $q \in Q$ and $a \in \Sigma$, we define $g_q(a)$ to be the Boolean function $B^Q \to B$ such that

$$g_q(a)(u) = g_q(a, u), \quad u \in B^Q.$$

Thus, for $u \in B^Q$, the value of $g_q(a)(u)$, also $g_q(a, u)$, is either 1 or 0.

We define the function $g_Q : \Sigma \times B^Q \to B^Q$ by putting together the $|Q|$ functions $g_q : \Sigma \times B^Q \to B$, $q \in Q$, as follows. For $a \in \Sigma$ and $u, v \in B^Q$, $g_Q(a, u) = v$ if and only if $g_q(a, u) = v_q$ for each $q \in Q$. For convenience, we will write $g(a, u)$ instead of $g_Q(a, u)$ in the following.

Example 2.2. We define an AFA $A_5 = (Q, \Sigma, s, F, g)$ where $Q = \{q_0, q_1, q_2\}$, $\Sigma = \{a, b\}$, $s = q_0$, $F = \{q_2\}$, and g is given by

State	a	b
q_0	$q_1 \wedge q_2$	0
q_1	q_2	$q_1 \wedge q_2$
q_2	$\overline{q_1} \wedge q_2$	$q_1 \vee \overline{q_2}$

Note that we use $\bar{q}$ instead of $\neg q$ for convenience.   □

We define $f \in B^Q$ by the condition

$$f_q = 1 \iff q \in F,$$

and we call f the *characteristic vector* of F. The characteristic vector for F of A_5 is $f = (f_{q_0}, f_{q_1}, f_{q_2}) = (0, 0, 1)$.

We extend g to a function of Q into the set of all functions $\Sigma^* \times B^Q \to B$ as follows:

$$g_q(w, u) = \begin{cases} u_q, & \text{if } w = \lambda, \\ g_q(a, g(w', u)), & \text{if } w = aw' \text{ with } a \in \Sigma \text{ and } w' \in \Sigma^*, \end{cases}$$

where $w \in \Sigma^*$ and $u \in B^Q$.

Now we define the acceptance of a word and the acceptance of a language by an AFA.

Let $A = (Q, \Sigma, s, F, g)$ be an AFA. A word $w \in \Sigma^*$ is *accepted* by A if and only if $g_s(w, f) = 1$, where f is the characteristic vector of F. The *language accepted* by A is the set

$$L(A) = \{w \in \Sigma^* \mid g_s(w, f) = 1\}.$$

Let $w = aba$. Then w is accepted by A_5 of Example 2.2 as follows:

$$
\begin{aligned}
& g_{q_0}(aba, f) \\
=\ & g_{q_1}(ba, f) \wedge g_{q_2}(ba, f) \\
=\ & (g_{q_1}(a, f) \wedge g_{q_2}(a, f)) \wedge (g_{q_1}(a, f) \vee \overline{g_{q_2}(a, f)}) \\
=\ & (g_{q_2}(\lambda, f) \wedge (g_{q_1}(\lambda, f) \wedge g_{q_2}(\lambda, f))) \wedge (g_{q_2}(\lambda, f) \vee \\
& \overline{g_{q_1}(\lambda, f) \wedge g_{q_2}(\lambda, f)}) \\
=\ & (f_{q_2} \wedge (\overline{f_{q_1}} \wedge f_{q_2})) \wedge (f_{q_2} \vee \overline{\overline{f_{q_1}} \wedge f_{q_2}}) \\
=\ & (1 \wedge (\bar{0} \wedge 1)) \wedge (1 \vee \overline{\bar{0} \wedge 1}) \\
=\ & 1
\end{aligned}
$$

If we denote each $u \in B^Q$ by a vector $(u_{q_0}, u_{q_1}, u_{q_2})$ and write $f = (0, 0, 1)$, then we can rewrite the above:

$$
\begin{aligned}
& g_{q_0}(aba, f) \\
=\ & g_{q_0}(a, g(ba, f)) \\
=\ & g_{q_0}(a, g(b, g(a, f))) \\
=\ & g_{q_0}(a, g(b, g(a, (0, 0, 1)))) \\
=\ & g_{q_0}(a, g(b, (0, 1, 1))) \\
=\ & g_{q_0}(a, (0, 1, 1)) \\
=\ & 1
\end{aligned}
$$

2.3.2 Systems of equations – representations of AFA

Consider again the example of AFA A_5. We may use the following system of equations instead of a table to represent the transitions of A_5:

$$\left\{ \begin{array}{rcl} X_0 & = & a \cdot (X_1 \wedge X_2) + b \cdot 0 \\ X_1 & = & a \cdot X_2 + b \cdot (X_1 \wedge X_2) \\ X_2 & = & a \cdot (\overline{X_1} \wedge X_2) + b \cdot (X_1 \vee \overline{X_2}) + \lambda \end{array} \right.$$

where a variable X_i represents state q_i, $0 \leq i \leq 2$, respectively; and λ appearing in the third equation specifies that q_2 is a final state.

In general, an AFA $A = (Q, \Sigma, s, F, g)$ can be represented by

$$(1) \qquad\qquad X_q = \sum_{a \in \Sigma} a \cdot g_q(a, X) + \varepsilon_q, \quad q \in Q$$

where X is the vector of variables X_q, $q \in Q$, and

$$\varepsilon_q = \left\{ \begin{array}{ll} \lambda & \text{if } q \in F, \\ 0 & \text{otherwise,} \end{array} \right.$$

for each $q \in Q$. Note that all the terms of the form $a \cdot 0$ or 0, $a \in \Sigma$, can be omitted.

For each AFA A, we call such a system of equations the *equational representation* of A. At this moment, we consider the system of equations solely as an alternative form to present the definition of an AFA.

NFA are a special case of AFA. The NFA A_2 of Figure 3 can be represented by

$$\left\{ \begin{array}{rcl} X_0 & = & a \cdot (X_0 \vee X_1) + b \cdot X_0 \\ X_1 & = & b \cdot X_2 \\ X_2 & = & b \cdot X_3 \\ X_3 & = & a \cdot X_3 + b \cdot X_3 + \lambda \end{array} \right.$$

Let Σ be an alphabet. We define the *L-interpretation* as follows:

Notation	Interpretation
0	$\emptyset$
1	Σ^*
$\wedge$	$\cap$
$\vee$	$\cup$
$\neg$	complement
$a, a \in \Sigma$	$\{a\}$
λ	$\{\lambda\}$
$\cdot$	set catenation
$+$	$\cup$
$=$	language equivalence

Under this interpretation, the systems of equations defined above become systems of language equations. Systems of language equations of a different form were studied by Salomaa in [106], where the operations are restricted to catenation, union, and star. The systems of language equations we are considering can be viewed as an extension of the systems of language equations of Salomaa.

Formally, a *system of language equations* over an alphabet Σ is a system of equations of the following form under the L-interpretation:

$$(2) \qquad X_i = \sum_{a \in \Sigma} a \cdot f_i^{(a)}(X) + \varepsilon_i, \quad i = 0, \dots, n$$

for some $n \geq 0$, where $X = (X_0, \dots, X_n)$; for each $a \in \Sigma$ and $i \in \{0, \dots, n\}$, $f_i^{(a)}(X)$ is a Boolean function; and $\varepsilon_i = \lambda$ or 0.

The following result has been proved in [19].

Theorem 2.6. *Any system of language equations of the form (2.3.2) has a unique solution for each X_i, $i = 0, \dots, n$. Furthermore, the solution for each X_i is regular.* □

The following results can be found in [44].

Theorem 2.7. *Let A be an AFA and E the equational representation of A. Assume that the variable X_0 corresponds to the starting state of A. Then the solution for X_0 in E under the L-interpretation is exactly $L(A)$.* □

Theorem 2.8. *For each system of language equations of the form (2.3.2), there is an AFA A such that the solution for X_0 is equal to $L(A)$.* □

It is easy to observe that an AFA is a DFA if and only if each function $g_q(a, X)$, $q \in Q$ and $a \in \Sigma$, in its equational representation (2.3.2) is either a single variable or empty. An AFA is an NFA if and only if each function in its equational representation (2.3.2) is defined using only the $\lor$ operation.

Such systems of language equations and their solutions have been further studied in [74, 76, 77]. Naturally, one may view that each such system of language equations corresponds directly to a set of solutions in the form of extended regular expressions (which will be defined in Section 3.4). However, it remains open how we can solve such a general system of language equations by directly manipulating extended regular expressions without resorting to transformations of the corresponding AFA.

2.3.3 Normal forms

The following results have been proved in [45].

Theorem 2.9. *For any k-state AFA A, $k > 0$, there exists an equivalent k-state AFA A' with at most one final state. More precisely, A' has no final state if $\lambda \notin L(A)$ and A' has one final state otherwise. In the latter case, the starting state is the unique final state.* □

The proof of this theorem relies on the usage of the negation operation in AFA.

Theorem 2.10. *For each AFA $A = (Q, \Sigma, s, F, g)$, one can construct an equivalent AFA $A' = (Q', \Sigma, s', F', g')$ with $|Q'| \leq 2|Q|$ such that $g'_q(a)$ is defined with only the $\wedge$ and $\vee$ operations, for each $q \in Q'$ and $a \in \Sigma$. In other words, A' is an AFA without negations.* $\square$

Theorem 2.11. *Let A be a k-state AFA without negations. One can construct an equivalent $(k + 1)$-state AFA without negations that has one final state if $\lambda \notin L(A)$ and at most two final states otherwise.* $\square$

In the following, we define a special type of AFA, which we call an s-AFA.

An s-AFA $A = (Q, \Sigma, s, F, g)$ is an AFA such that the value of $g_q(a)$, for any $q \in Q$ and $a \in \Sigma$, does not depend on the status of s, that is, in the equational representation of A, the variable X_s does not appear on the righthand side of any equation.

Example 2.3. The following is an equational representation of a 4-state s-AFA which accepts all words over $\{a, b\}$ that do not contain 6 consecutive occurrences of a. We use the convention that the operator $\wedge$ has precedence over $\vee$.

$$
\begin{cases}
X_0 &= a \cdot (\overline{X_1} \vee \overline{X_2}) + b \cdot (\overline{X_1} \vee \overline{X_2} \vee \overline{X_3}) + \lambda, \\
X_1 &= a \cdot (X_1 \vee X_2 \wedge X_3) + b \cdot (X_1 \wedge X_2 \wedge X_3), \\
X_2 &= a \cdot (X_1 \wedge X_2 \vee X_2 \wedge \overline{X_3} \vee \overline{X_2} \wedge X_3) + b \cdot (X_1 \wedge X_2 \wedge X_3), \\
X_3 &= a \cdot (X_1 \wedge X_2 \vee X_1 \wedge \overline{X_3} \vee X_2 \wedge \overline{X_3}) + b \cdot 1 + \lambda.
\end{cases}
$$ $\square$

It is clear that for any AFA, there exists an equivalent s-AFA having at most one additional state.

2.3.4 AFA to NFA – the construction

Let $A = (Q, \Sigma, s, F, g)$ be an AFA and f the characteristic vector of F. We construct an NNFA

$$A_v = (Q_v, \Sigma, \delta_v, S_v, F_v)$$

where

$Q_v = B^Q$,
$S_v = \{u \in B^Q \mid u_s = 1\}$;
$F_v = \{f\}$,
$\delta_v : Q_v \times \Sigma \to 2^{Q_v}$ is defined by $\delta_v(u, a) = \{u' \mid g(a, u') = u\}$, for each $u \in Q_v$ and $a \in \Sigma$.

Claim. $L(A_v) = L(A)$.

Proof. We first prove that for $u \in Q_v (= B^Q)$ and $x \in \Sigma^*$,

$$(3) \qquad\qquad ux \vdash^*_{A_v} f \iff g(x, f) = u$$

by induction on the length of x.

For $x = \lambda$, one has $u = f$ and $g(\lambda, f) = f$. Now assume that the statement holds for all words up to a length l, and let $x = ax_0$ with $a \in \Sigma$ and $x_0 \in \Sigma^l$.

Let $u = g(x, f)$. Then we have $u = g(a, g(x_0, f))$. Let $u' = g(x_0, f)$. By the definition of δ_v, we have $u' \in \delta_v(u, a)$. We also have $u'x_0 \vdash_{A_v}^* f$ by the induction hypothesis. Therefore,

$$u\,x = u\,ax_0 \vdash_{A_v} u'\,x_0 \vdash_{A_v}^* f \ .$$

For the converse, let $u\,x \vdash_{A_v}^* f$. Then

$$u\,x = u\,ax_0 \vdash_{A_v}^* u'\,x_0 \vdash_{A_v}^* f$$

for some $u' \in Q_v$. Thus, $u' = g(x_0, f)$ by the induction hypothesis and $u = g(a, u')$ by the definition of δ_v. Therefore, $u = g(a, u') = g(a, g(x_0, f)) = g(x, f)$. Thus, (3) holds.

By (2.3.4) and the definition of S_v, we have $L(A_v) = L(A)$. □

In the above construction of A_v, the state set is $Q_v = B^Q$, i.e., each state of the NNFA A_v is a Boolean vector indexed by the states of the given AFA A. If the number of states of A is n, then the number of states of A_v is 2^n. Also notice that a computation of an AFA can be viewed as a sequence of calculations of Boolean vectors starting with f, the characteristic vector of F, as the initial vector and proceeding backwards with respect to the input word. At each step of this process, an input symbol is read and a new vector is calculated. Note that at each step, the new vector is uniquely determined. The process terminates when the first input symbol is read. Then the input word is accepted if and only if the resulting vector has a value 1 at the entry that is indexed by the starting state. We have the following results.

Theorem 2.12. *If L is accepted by an n-state AFA, then it is accepted by an NNFA with at most 2^n states.* □

Theorem 2.13. *If L is accepted by an n-state AFA, then L^R is accepted by a DFA with at most 2^n states.* □

2.3.5 NFA or DFA to AFA

NFA and DFA are special cases of AFA. So, the transformations are straightforward.

Let $A = (Q, \Sigma, \delta, s, F)$ be an NFA. We can construct an equivalent AFA $A' = (Q, \Sigma, s, F, g)$, where g is defined as follows: for each $q \in Q$, $a \in \Sigma$, and $u \in B^Q$,

$$g_q(a, u) = 0 \iff u_p = 0 \text{ for all } p \in \delta(q, a) \ .$$

More intuitively, the equational representation of A' is

$$X_q = \sum_{a \in \Sigma} a \cdot \bigvee_{p \in \delta(q,a)} X_p + \varepsilon_q, \quad \text{for } q \in Q,$$

where $\varepsilon_q = \lambda$ if $q \in F$ and $\varepsilon_q = 0$ otherwise. A proof for $L(A) = L(A')$ can be found in [45].

Theorem 2.14. *L is accepted by a complete 2^k-state DFA if and only if L^R is accepted by a $(k+1)$-state s-AFA.*

Proof. The "if"-part is implied by Theorem 2.13. In the following, we describe the construction of an s-AFA for a given DFA but do not give a proof of its correctness. For a detailed proof, the reader can refer to [73, 44]. Let $D = (Q_D, \Sigma, \delta, s_D, F_D)$ be the given 2^k-state complete DFA and $L = L(D)$. We construct a $(k+1)$-state s-AFA $A = (Q_A, \Sigma, s_A, F_A, g)$ as follows. The main idea of the construction is that each of the 2^k states is encoded by a k-bit Boolean vector and each of the k bits is represented by a state of the AFA. In addition to these k states, the s-AFA has one more state, the starting state.

Let $K = \{1, \ldots, k\}$ and $K_0 = K \cup \{0\}$. Then we define K_0 to be the state set of the AFA A, where 0 is the starting state. We define an arbitrary bijection π between Q_D and B^K. The bijection π can be considered as an encoding scheme such that each state in Q_D is encoded by a distinct k-bit vector. For convenience, we simply use $\pi(q)$, i.e., the k-bit vector, to denote q in the following. In particular, we use the vector $(0, \ldots, 0)$ to denote the starting state s_D of D. Note that one can choose any of k-bit vector to encode s_D. We choose $(0, \ldots, 0)$ purely for notational conveniece. Then, we define a $(k+1)$-state s-AFA A as follows: $A = (Q_A, \Sigma, s_A, F_A, g)$ where

$Q_A = K_0,$

$s_A = 0,$ and

$$F_A = \begin{cases} \{0\} & \text{if } s_D \in F_D, \\ \emptyset & \text{otherwise.} \end{cases}$$

The function g is defined by setting, for $a \in \Sigma$ and $u \in B^{Q_A}$,

$$g_0(a, u) = \begin{cases} 1 & \text{if } \delta(u|_K, a) \in F_D, \\ 0 & \text{otherwise} \end{cases}$$

and $v = g(a, u)$, for some $v \in B^{Q_A}$, if and only if $\delta(u|_K, a) = v|_K$. More precisely, we define $g_i(a, u)$, for $i \in K$ and $u \in B^{Q_A}$, in the following. Note that $\theta_z(x)$ denotes either x or $\bar{x}$ depending on the value of z, i.e., $\theta_z(x) = x$ if $z = 1$ and $\theta_z(x) = \bar{x}$ if $z = 0$. Then, for $i \in K$,

$$g_i(a, u) = \bigvee_{v \in B^K} (\delta(v, a)_i \wedge \theta_{v_1}(u_1) \wedge \ldots \wedge \theta_{v_k}(u_k))$$

and

$$g_0(a, u) = \bigvee_{v \in F_D} \theta_{v_1}(g_1(a, u)) \wedge \ldots \wedge \theta_{v_k}(g_k(a, u)). \qquad \square$$

Corollary 2.1. *Let A be an n-state DFA and $L = L(A)$. Then L^R is accepted by an s-AFA with at most $\lceil \log n \rceil + 1$ states.* $\qquad \square$

As an example, we construct a 3-state s-AFA A which is equivalent to the 4-state DFA A_1 of Figure 2 as follows:

$A = (Q_A, \Sigma, s_A, F_A, g)$ where $Q_A = \{0, 1, 2\}$, $s_A = 0$, $F_A = \emptyset$. The encoding of the states of A_1 is shown in the following. Note that we denote a 2-bit Boolean vector as a 2-bit binary number, i.e., we write $X_1 X_2$ instead of (X_1, X_2).

State of A_1	0	1	2	3
Encoding $X_1 X_2$	00	01	10	11

In order to explain intuitively how the function g is defined, we first write $g_1(a, X)$ informally (and in unnecessary detail) as follows:

$$\begin{aligned}
g_1(a, X) &= (\delta(00, a)_1 \wedge \overline{X_1} \wedge \overline{X_2}) \vee (\delta(01, a)_1 \wedge \overline{X_1} \wedge X_2) \\
&\quad \vee (\delta(10, a)_1 \wedge X_1 \wedge \overline{X_2}) \vee (\delta(11, a)_1 \wedge X_1 \wedge X_2) \\
&= ((01)_1 \wedge \overline{X_1} \wedge \overline{X_2}) \vee ((01)_1 \wedge \overline{X_1} \wedge X_2) \vee ((01)_1 \wedge X_1 \wedge \overline{X_2}) \\
&\quad \vee ((11)_1 \wedge X_1 \wedge X_2) \\
&= (0 \wedge \overline{X_1} \wedge \overline{X_2}) \vee (0 \wedge \overline{X_1} \wedge X_2) \vee (0 \wedge X_1 \wedge \overline{X_2}) \vee (1 \wedge X_1 \wedge X_2) \\
&= X_1 \wedge X_2
\end{aligned}$$

Then we have

$$\begin{aligned}
g_1(a, X) &= X_1 \wedge X_2, \\
g_1(b, X) &= (\overline{X_1} \wedge X_2) \vee (X_1 \wedge \overline{X_2}) \vee (X_1 \wedge X_2) = (\overline{X_1} \wedge X_2) \vee X_1 \\
&= X_1 \vee X_2, \\
g_2(a, X) &= (\overline{X_1} \wedge \overline{X_2}) \vee (\overline{X_1} \wedge X_2) \vee (X_1 \wedge \overline{X_2}) \vee (X_1 \wedge X_2) = 1, \\
g_2(b, X) &= ((X_1 \wedge \overline{X_2}) \vee (X_1 \wedge X_2) = X_1, \\
g_0(a, X) &= g_1(a, X) \wedge g_2(a, X) = (X_1 \wedge X_2) \wedge 1 = X_1 \wedge X_2, \\
g_0(b, X) &= g_1(b, X) \wedge g_2(b, X) = (X_1 \vee X_2) \wedge X_1 = X_1.
\end{aligned}$$

So, the equational representation of A is

$$\left\{ \begin{array}{lcl}
X_0 &=& a \cdot (X_1 \wedge X_2) + b \cdot (X_1) \\
X_1 &=& a \cdot (X_1 \wedge X_2) + b \cdot (X_1 \vee X_2) \\
X_2 &=& a \cdot 1 + b \cdot (X_1)
\end{array} \right.$$

and the characteristic vector of F_A is $f = (0, 0, 0)$.

2.3.6 Basic operations

Let

$$A^{(1)} = (Q^{(1)}, \Sigma, s^{(1)}, F^{(1)}, g^{(1)})$$

be an $(m + 1)$-state s-AFA and

$$A^{(2)} = (Q^{(2)}, \Sigma, s^{(2)}, F^{(2)}, g^{(2)})$$

be an $(n + 1)$-state s-AFA. Assume that $Q^{(1)} \cap Q^{(2)} = \emptyset$.

We construct an $(m + n + 1)$-state AFA $A = (Q, \Sigma, s, F, g)$ such that $L(A) = L(A^{(1)}) \cup L(A^{(2)})$ as follows:

$$Q = (Q^{(1)} - \{s^{(1)}\}) \cup (Q^{(2)} - \{s^{(2)}\}) \cup \{s\},$$
$$s \notin Q^{(1)} \cup Q^{(2)},$$

$$F = \begin{cases} F^{(1)} \cup F^{(2)} & \text{if } s^{(1)} \notin F^{(1)} \text{ and } s^{(2)} \notin F^{(2)}, \\ (F^{(1)} \cup F^{(2)} \cup \{s\}) \cap Q & \text{otherwise.} \end{cases}$$

We define g as follows. For $a \in \Sigma$ and $u \in B^Q$,

$$g_s(a, u) = g^{(1)}_{s^{(1)}}(a, u) \vee g^{(2)}_{s^{(2)}}(a, u),$$

and for $q \in Q - \{s\}$,

$$g_q(a, u) = \begin{cases} g^{(1)}_q(a, u) & \text{if } q \in Q^{(1)}, \\ g^{(2)}_q(a, u) & \text{if } q \in Q^{(2)}. \end{cases}$$

An $(m+n+1)$-state AFA $A = (Q, \Sigma, s, F, g)$ such that $L(A) = L(A^{(1)}) \cap L(A^{(2)})$ is constructed as above except the following:

$$g_s(a, u) = g^{(1)}_{s^{(1)}}(a, u) \wedge g^{(2)}_{s^{(2)}}(a, u)$$

and s is in F if and only if both $s^{(1)} \in F^{(1)}$ and $s^{(2)} \in F^{(2)}$.

For complementation, we construct an m-state s-AFA

$$A = (Q^{(1)}, \Sigma^{(1)}, s^{(1)}, F', g)$$

such that $L(A) = \overline{L(A^{(1)})}$, where the function g is the same as $g^{(1)}$ except that $g_{s^{(1)}}(a, u) = \overline{g^{(1)}_{s^{(1)}}(a, u)}$; and $F' = \{s^{(1)}\} \cup F^{(1)}$ if $s^{(1)} \notin F^{(1)}$ and $F' = F^{(1)} - \{s^{(1)}\}$ otherwise.

Let $L_1 = L(A^{(1)})$ and $L_2 = L(A^{(2)})$. We can easily construct an AFA to accept a language which is obtained by an arbitrary combination of Boolean operations on L_1 and L_2, e.g., $L = (L_1 \cup L_2) \cap \overline{(L_1 \cap L_2)}$. The only essential changes are the functions for s and whether s is in the final state set, which are all determined by the respective Boolean operations.

Other AFA operations, e.g., catenation, star, and shuffle, have been described in [45, 44].

2.3.7 Implementation and r-AFA

Although alternation is a generalization of nondeterminism, the reader may notice that AFA are backwards deterministic. We have also shown that a language L is accepted by a 2^n-state DFA if and only if it is accepted by an s-AFA of $n + 1$ states reversely (i.e., words are read from right to left). Due to the above observation, we introduce a variation of s-AFA which we call r-AFA. The definition of an r-AFA is exactly the same as an s-AFA except that the input word is to be read reversely. Therefore, an r-AFA is forward deterministic. Then, for each L that is accepted by a DFA with n states, we can construct an equivalent r-AFA with at most $\lceil \log n \rceil + 1$ states.

An r-AFA $A = (Q, \Sigma, s, F, g)$ can be represented by a *system of right language equations* [19] of the following form:

$$X_q = \sum_{a \in \Sigma} g_q(a, X) \cdot a + \varepsilon_q, \quad q \in Q$$

where X is the vector of variables X_q, $q \in Q$, and

$$\varepsilon_q = \begin{cases} \lambda & \text{if } q \in F, \\ 0 & \text{otherwise,} \end{cases}$$

for each $q \in Q$.

In the following, we present a scheme such that Boolean functions of an r-AFA can be represented by Boolean vectors, and the computation of a Boolean function can be done with bitwise vector operations. Note that for a DFA of n states, its corresponding r-AFA has at most $\lceil \log n \rceil + 1$ states. So, for all practical problems, i.e., those using DFA of up to 2^{31} states, each Boolean vector can be stored in one word. In many cases, this can save space tremendously in comparison to symbolic representations of AFA. Also, each bitwise vector operation can be done with one instruction. So, AFA computations can be done efficiently.

We represent each Boolean function $g_q(a)$, $q \in Q$ and $a \in \Sigma$, in disjunctive normal form. The disjunctive normal form consists of a disjunction of formulas of the type $(Y_1 \wedge \ldots \wedge Y_m)$ where each Y_i is a variable X_i or the negation of a variable, $\overline{X_i}$. We call each such formula of the type $(Y_1 \wedge \ldots \wedge Y_m)$ a term. For example, let $X = (X_1, \ldots, X_8)$. The following Boolean function in disjunctive normal form

$$\mu(X) = (X_2 \wedge \overline{X_4} \wedge X_7) \vee (\overline{X_1} \wedge X_2) \vee (X_3 \wedge X_4 \wedge \overline{X_6})$$

has three terms. We name them $t^{(1)}(X)$, $t^{(2)}(X)$, and $t^{(3)}(X)$, respectively. Each term $t^{(i)}(X)$ can be represented by two 8-bit Boolean vectors $\alpha^{(i)}$ and $\beta^{(i)}$ and the value of $t_i(X)$ can be computed with two bitwise operations. The two Boolean vectors are defined as follows:

$$\alpha_k^{(i)} = 1 \text{ iff } X_k \text{ or } \overline{X_k} \text{ appears in } t^{(i)}$$

and

$$\beta_k^{(i)} = 1 \text{ iff } X_k \text{ appears in } t^{(i)}.$$

For example, the two vectors for $t^{(1)}(X)$ are

$$\alpha^{(1)} = (0, 1, 0, 1, 0, 0, 1, 0),$$
$$\beta^{(1)} = (0, 1, 0, 0, 0, 0, 1, 0).$$

Then, for any instance u of X, $t^{(1)}(u) = 1$ iff $(u \;\&\; \alpha^{(1)}) \uparrow \beta^{(1)} = 0$, where $\&$ and $\uparrow$ are bitwise AND and XOR, respectively, and 0 denotes the all-0 vector.

The above idea is based on the observation that a term $t(X)$ has a value 1 iff all the variables of the form X_i in $t(X)$ have a value 1 and all those of the form $\overline{X_i}$ in $t(X)$ have a value 0. For an instance u of X, $t(u)$ is evaluated with the above defined vectors α and β as follows. First, the vector α changes each u_i such that the variable X_i does not appear in $t(X)$ to 0 and keeps all others unchanged. Then the vector β changes each u_i such that X_i (rather than $\overline{X_i}$) is in $t(X)$ to $\overline{u_i}$, i.e., 1 if $u_i = 0$ and 0 if $u_i = 1$. Finally, $t(u)$ is 1 iff u becomes an all-0 vector.

Note that each term can be evaluated in parallel and each Boolean function of an r-AFA can be evaluated in parallel as well.

2.4 Finite automata with output

In the previous subsections, we have described three basic forms of finite automata: DFA, NFA, and AFA. They are all considered to be language acceptors. In this subsection, we consider several models of finite automata with output, which are not only language acceptors but also language transformers.

A *Moore* machine, informally, is a DFA where each state is associated with an output letter [88, 57]. Formally, a Moore machine A is a 6-tuple $(Q, \Sigma, \Delta, \delta, \sigma, s)$ where Q, Σ, δ, and s are defined as in a DFA; Δ is the output alphabet; and $\sigma : Q \rightarrow \Delta$ is the output function. For an input word $a_1 \cdots a_n$, if the state transition sequence is

$$s = q_0, q_1, \ldots, q_n$$

then the output of A in response to $a_1 \cdots a_n$ is

$$\sigma(q_0)\sigma(q_1) \cdots \sigma(q_n).$$

A *Mealy machine* is a DFA where an output symbol is associated to each transition rather than to each state [86, 57]. Formally, a Mealy machine A is a 5-tuple $(Q, \Sigma, \Delta, \sigma, s)$ where Q, Σ, and s are defined as in a DFA; Δ is the output alphabet; and $\sigma : Q \times \Sigma \rightarrow Q \times \Delta$ is the transition-and-output function. For an input word $x = a_1 \cdots a_n$, $a_1, \ldots, a_n \in \Sigma$, if $\sigma(s, a_1) = (q_1, b_1)$, $\sigma(q_1, a_2) = (q_2, b_2)$, $\ldots$, $\sigma(q_{n-1}, a_n) = (q_n, b_n)$, i.e.,

$$s \xrightarrow{a_1/b_1} q_1 \xrightarrow{a_2/b_2} \ldots\ldots \xrightarrow{a_n/b_n} q_n,$$

then the output of A in response to x is $b_1 \cdots b_n$.

For the above two models, we do not define final states. Final states can be defined such that only those input words that are accepted, i.e., reaching a final state, are associated to an output word. Then the models without final states are only a special case of the corresponding models with final states in the sense that all states are final states.

In the above definitions, we do not require that the transition functions are total. If an input word cannot be completely read, then there is no output word associated to this input word.

Another important model, the *finite transducer* model, is a generalization of the Mealy machines. Many closure properties of regular languages can be easily proved by using various finite transducers. See Section 4.2 for details.

A finite transducer T is a 6-tuple $(Q, \Sigma, \Delta, \sigma, s, F)$ where

Q is the finite set of states;
Σ is the input alphabet;
Δ is the output alphabet;
σ is the transition-and-output function from a finite subset of $Q \times \Sigma^*$ to finite subsets of $Q \times \Delta^*$;

$s \in Q$ is the starting state;

$F \subseteq Q$ is the set of final states.

An example of a finite transducer $T = (\{0, 1, 2\}, \{a, b\}, \{0, 1\}, \sigma, 0, \{2\})$ is shown in Figure 9. The arc from state 0 to state 1 with the label $b/101$ specifies that $(1, 101) \in \sigma(0, b)$.

For a given word $u \in \Sigma^*$, we say that $v \in \Delta^*$ is an output of T for u if there exists a state transition sequence of T, $(q_1, v_1) \in \sigma(s, u_1)$, $(q_2, v_2) \in \sigma(q_1, u_2)$, $\ldots$, $(q_n, v_n) \in \sigma(q_{n-1}, u_n)$, and $q_n \in F$, i.e.,

$$s \xrightarrow{u_1/v_1} q_1 \xrightarrow{u_2/v_2} \ldots\ldots \xrightarrow{u_n/v_n} q_n \in F$$

such that $u = u_1 \cdots u_n$, $u_1, \ldots, u_n \in \Sigma^*$, and $v = v_1 \cdots v_n$, $v_1, \ldots, v_n \in \Delta^*$. We write that $v \in T(u)$, where $T(u)$ denotes the set of all outputs of T for the input word u. Note that $s \in F$ implies that $\lambda \in T(\lambda)$.

T is said to be single-valued if for each input word u, T has at most one distinct output in response to u, i.e., $|T(u)| \leq 1$ for each $u \in \Sigma^*$.

A finite transducer $T = (Q, \Sigma, \Delta, \sigma, s, F)$ is called a *generalized sequential machine* (GSM) if σ is a function from $Q \times \Sigma$ to finite subsets of $Q \times \Delta^*$, i.e., T reads exactly one symbol at each transition. The GSM T is said to be deterministic if its underlying finite automaton (i.e., T without output) is a DFA, i.e., σ is a (partial) function from $Q \times \Sigma$ to $Q \times \Delta^*$. The definition of a GSM is not standardized in the literature. Some authors define GSMs with no final states [51].

Each finite transducer $T = (Q, \Sigma, \Delta, \sigma, s, F)$ defines a *finite transduction* $T : \Sigma^* \to 2^{\Delta^*}$. Note that for an input word $w \in \Sigma^*$, $T(w)$, which is the set of all output words in response to w, may be finite or infinite. $T(w) = \emptyset$ if T cannot reach a final state by reading w. Also note that we use T to denote both the finite transducer and the finite transduction it defines since this clearly will not cause any confusion. For a language $L \subseteq \Sigma^*$, we define

$$T(L) = \bigcup_{w \in L} T(w).$$

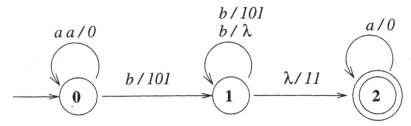

Fig. 9. A finite transducer T

Example 2.4. Let us consider the transducer T of Figure 9. We have

$$T(aabb) = \{010111, 010110111\},$$

$$T(bbba) = \{101110, 101101110, 101101101110\},$$

$$T(\lambda) = \emptyset, \qquad T(aaab) = \emptyset,$$

$$T(\{b, ba\}) = \{10111, 101110\}.$$

Let $L = \{a^i ba^j \mid i, j \geq 0\}$. Then

$$T(L) = \{0^k 101110^l \mid k, l \geq 0\}. \qquad \square$$

A finite transduction T can also be viewed as a relation $R_T \subseteq \Sigma^* \times \Delta^*$ defined by

$$R_T = \{(u, v) \mid v \in T(u)\}.$$

Relations induced by finite transducers are also called *rational relations* in the literature, e.g., [41]. The following is Nivat's Representation Theorem for finite transductions [93].

Theorem 2.15. *Let Σ and Δ be finite alphabets. $R \subseteq \Sigma^* \times \Delta^*$ is a rational relation iff there are a finite alphabet Γ, a regular language $L \subseteq \Gamma^*$ and morphisms $g : \Gamma^* \to \Sigma^*$ and $h : \Gamma^* \to \Delta^*$ such that*

$$R = \{(g(w), h(w)) \mid w \in L\}. \qquad \square$$

Two finite transducers are said to be *equivalent* if they define exactly the same finite transduction. The equivalence problem for finite transducers is undecidable [60]. This holds even for nondeterministic GSMs. However, the equivalence problem for single-valued finite transducers is decidable [114, 32]. This implies that the equivalence problem for deterministic GSMs (DGSMs) is also decidable.

From the above definitions, it is easy to see that morphisms can be characterized by one-state complete DGSMs. By a complete GSM, we mean that its transition-and-output function is a total function. Also, finite substitutions can be characterized by one-state (nondeterministic) GSMs. In both cases, the sole state is both the starting state and the final state.

For a function $T : \Sigma^* \to 2^{\Delta^*}$ (relation $R_T \subseteq \Sigma^* \times \Delta^*$), we define $T^{-1} : \Delta^* \to 2^{\Sigma^*}$ by $T^{-1}(y) = \{x \mid y \in T(x)\}$ ($R_T^{-1} \subseteq \Delta^* \times \Sigma^*$ by $R_T^{-1} = \{(y, x) \mid (x, y) \in R_T\}$). Then, clearly, T (R_T) is a finite transduction (rational relation) iff T^{-1} (R_T^{-1}) is a finite transduction (rational relation). This can be shown by simply interchanging the input and the output of the finite transducer. Then, we have the following:

Theorem 2.16. *Let $T : \Sigma^* \to 2^{\Delta^*}$ be a finite transduction. Then the inverse of T, i.e., $T^{-1} : \Delta^* \to 2^{\Sigma^*}$, is also a finite transduction.* $\qquad \square$

We define the following standard form for finite transducers.

A finite transducer $T = (Q, \Sigma, \Delta, \sigma, s, F)$ is said to be in the *standard form* if σ is a function from $Q \times (\Sigma \cup \{\lambda\})$ to $2^{Q \times (\Delta \cup \{\lambda\})}$. Intuitively, the standard form restricts the input and output of each transition to be only a single letter or λ.

Theorem 2.17. *Each finite transducer can be transformed into an equivalent finite transducer in the standard form.*   □

The transformation of an arbitrary finite transducer to an equivalent one in the standard form consists of two steps: First, each transition that reads more than one letter is transformed into several transitions reading exactly one letter. Second, each transition that has a string of more than one letter as output is transformed into several transitions such that each of them has exactly one letter as output.

More specifically, in the first step, we replace each transition of the form

where $p, q \in Q$, $a_1, \ldots, a_j \in \Sigma$, $j \geq 2$, and $\beta \in \Delta^*$, by the following

where $r_1, \ldots, r_{j-1}$ are new states.

For the second step, each of the transitions of the following form

where $p, q \in Q$, $a \in \Sigma \cup \{\lambda\}$, and $b_1, \ldots, b_k \in \Delta$, $k \geq 2$, is replaced by

where $r_1, \ldots, r_{k-1}$ are new states. It is clear that the two-step transformation results in an equivalent finite transducer in the standard form.

In many cases, the use of the standard form of finite transducers can result in much simpler proofs than the use of the general form. In Section 4.2, we will use the standard form in proving that the family of regular languages is closed under finite transduction.

3. Regular expressions

In the previous section, we have defined languages that are recognized by finite automata. Finite automata in various forms are easy to implement by computer programs. For example, a DFA can be implemented by a case or switch statement; an NFA can be expressed as a matrix and manipulated by corresponding matrix operations; and an AFA can be represented as Boolean vectors and computed by bitwise Boolean operations. However, finite automata in any of the above mentioned forms are not convenient to be specified sequentially by users. For instance, when we specify a string pattern to be matched or define a token for certain identifiers, it is quite cumbersome to write a finite automaton definition for the purpose. In this case, a succinct and comprehensible expression in sequential form would be better suited than a finite automaton definition. For example, the language accepted by the finite automaton A_2 of Figure 3 can be expressed as $(a + b)^*abb(a + b)^*$. Such expressions are called regular expressions and they were originally introduced by Kleene [69]. In practice, regular expressions are often used as user interfaces for specifying regular languages. In contrast, finite automata are better suited as computer internal representations for storing regular languages.

3.1 Regular expressions – the definition

We define, inductively, a *regular expression* e over an alphabet Σ and the language $L(e)$ it denotes as follows:

(1) $e = \emptyset$ is a regular expression denoting the language $L(e) = \emptyset$.
(2) $e = \lambda$ is a regular expression denoting the language $L(e) = \{\lambda\}$.
(3) $e = a$, for $a \in \Sigma$, is a regular expression denoting the languge $L(e) = \{a\}$.
   Let e_1 and e_2 be regular expressions and $L(e_1)$ and $L(e_2)$ the languages they denote, respectively. Then
(4) $e = (e_1 + e_2)$ is a regular expression denoting the language $L(e) = L(e_1) \cup L(e_2)$.
(5) $e = (e_1 \cdot e_2)$ is a regular expression denoting the language $L(e) = L(e_1)L(e_2)$.
(6) $e = e_1^*$ is a regular expression denoting the language $(L(e_1))^*$.

We assume that $*$ has higher precedence than $\cdot$ and $+$, and $\cdot$ has higher precedence than $+$. A pair of parentheses may be omitted whenever the omission would not cause any confusion. Also, we usually omit the symbol $\cdot$ in regular expressions.

Example 3.1. Let $\Sigma = \{a, b, c\}$ and $L \subseteq \Sigma^*$ be the set of all words that contain $abcc$ as a subword. Then L can be denoted by the regular expression $(a + b + c)^* abcc(a + b + c)^*$. □

Example 3.2. Let $L \subseteq \{0, 1\}^*$ be the set of all words that do not contain two consecutive 1's. Then L is denoted by $(10 + 0)^*(1 + \lambda)$. □

Example 3.3. Let $\Sigma = \{a, b\}$ and $L = \{w \in \Sigma^* \mid |w|_b \text{ is odd}\}$. Then L can be denoted by $(a^* ba^* b)^* a^* ba^*$. □

Two regular expressions e_1 and e_2 over Σ are said to be equivalent, denoted $e_1 = e_2$, if $L(e_1) = L(e_2)$. The languages that are denoted by regular expressions are called *regular languages*. The family of regular languages is denoted $\mathcal{L}_{REG}$.

In [69], Kleene has shown that the family of regular languages and the family of DFA languages are exactly the same, i.e., regular expressions are equivalent to finite automata in terms of the languages they define. There are various algorithms for transforming a regular expression to an equivalent finite automaton and vice versa. In the following, we will describe two approaches for the transformation from a regular expression to an equivalent finite automaton and one from a finite automaton to an equivalent regular expression.

3.2 Regular expressions to finite automata

There are three major approaches for transforming regular expressions into finite automata. The first approach, due to Thompson [121], is to transform a regular expression into a λ-NFA. This approach is simple and intuitive, but may generate many λ-transitions. Thus, the resulting λ-NFA can be unnecessarily large and the further transformation of it into a DFA can be rather time and space consuming. The second approach transforms a regular expression into an NFA without λ-transitions. This approach is due to Berry and Sethi [7], whose algorithm is based on Brzozowski's theory of derivatives [16] and McNaughton and Yamada's marked expression algorithm. Berry and Sethi's algorithm has been further improved by Brüggemann-Klein [13] and Chang and Paige [25]. The third approach is to transform a regular expression directly into an equivalent DFA [16, 2]. This approach is very involved and can be replaced by two separate steps: (1) regular expressions to NFA using one of the above approaches and (2) NFA to DFA.

In the following, we give a very brief description of the first approach and give an intuitive idea of the marked expression algorithm [7] that forms the basis of the second approach. Here we will not discuss the above mentioned third approach.

3.2.1 Regular expressions to λ-NFA

The following construction can be found in many introductory books on automata and formal language theory, e.g., [57, 68, 78, 123]. Our approach is different from that of Thompson's [121, 2, 57] in that the number of final states is not restricted to one.

Let e be a regular expression over the alphabet Σ. Then a λ-NFA M_e is constructed recursively as follows:

(i) If $e = \emptyset$, then $M_e = (\{s\}, \Sigma, \delta, s, \emptyset)$ where $\delta(s, a) = \emptyset$ for any $a \in \Sigma \cup \{\lambda\}$.

(ii) If $e = \lambda$, then $M_e = (\{s\}, \Sigma, \delta, s, \{s\})$ where δ is the same as in i).

(iii) If $e = a$, for some $a \in \Sigma$, then $M_e = (\{s, f\}, \Sigma, \delta, s, \{f\})$ where $\delta(s, a) = \{f\}$ is the only defined transition.

(iv) If $e = e_1 + e_2$ where e_1 and e_2 are regular expressions and M_{e_1} and M_{e_2} are λ-NFA constructed for e_1 and e_2, respectively, i.e., $L(M_{e_1}) = L(e_1)$ and $L(M_{e_2}) = L(e_2)$, then $M_e = M_{e_1} + M_{e_2}$, where $M_{e_1} + M_{e_2}$ is defined in Subsection 2.2.

Similarly, if $e = e_1 e_2$, then $M_e = M_{e_1} M_{e_2}$; and if $e = e_1^*$, then $M_e = M_{e_1}^*$, where $M_{e_1} M_{e_2}$ and $M_{e_1}^*$ are defined in Subsection 2.2.

Example 3.4. Following the above approach, the regular expression $a(a + b)a^*b$ would be transformed into the λ-NFA shown in Figure 10. □

3.2.2 Regular expressions to NFA without λ-transitions

The following presentation is a modification of the one given in [14]. An informal description is presented in Figure 11.

Let e be a regular expression over Σ. We define an NFA $\mathcal{M}_e$ inductively as follows:

(0) $\mathcal{M}_\emptyset = (\{s\}, \Sigma, \delta, s, \emptyset)$ where $\delta(s, a) = \emptyset$ for all $a \in \Sigma$.

(1) $\mathcal{M}_\lambda = (\{s\}, \Sigma, \delta, s, \{s\})$ where $\delta(s, a) = \emptyset$ for all $a \in \Sigma$.

(2) For $a \in \Sigma$, $\mathcal{M}_a = (\{s, f\}, \Sigma, \delta, s, \{f\})$ where $\delta(s, a) = \{f\}$ is the only transition.

(3) Assume that $\mathcal{M}_{e_1} = (Q_1, \Sigma, \delta_1, s_1, F_1)$, $\mathcal{M}_{e_2} = (Q_2, \Sigma, \delta_2, s_2, F_2)$, and $Q_1 \cap Q_2 = \emptyset$.

(3.1) $\mathcal{M}_{e_1 + e_2} = (Q, \Sigma, \delta, s_1, F)$ where $Q = Q_1 \cup (Q_2 - \{s_2\})$ (merging s_1 and s_2 into s_1),

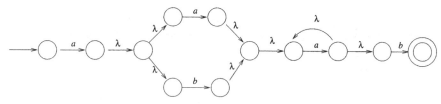

Fig. 10. A λ-NFA constructed for $a(a + b)a^*b$

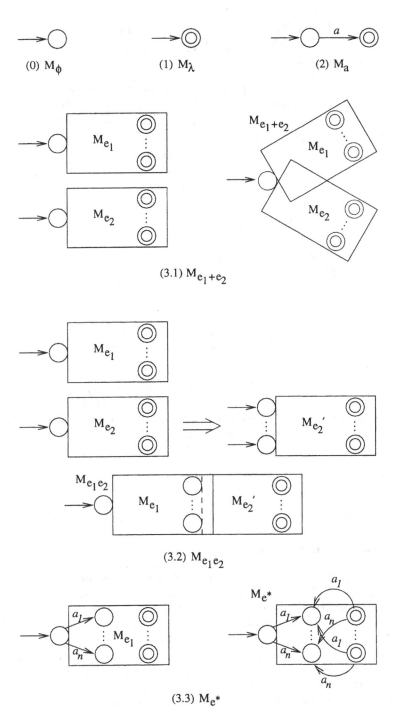

(0) M_ϕ (1) M_λ (2) M_a

(3.1) $M_{e_1+e_2}$

(3.2) $M_{e_1e_2}$

(3.3) M_{e^*}

Fig. 11. Regular expression to an NFA without λ-transitions

$$F = \begin{cases} F_1 \cup F_2 & \text{if } s_2 \notin F_2, \\ F_1 \cup (F_2 - \{s_2\}) \cup \{s_1\} & \text{otherwise;} \end{cases}$$

for $q \in Q$ and $a \in \Sigma$,

$$\delta(q, a) = \begin{cases} \delta_1(s_1, a) \cup \delta_2(s_2, a), & \text{if } q = s_1, \\ \delta_1(q, a) & \text{if } q \in Q_1, \\ \delta_2(q, a) & \text{if } q \in Q_2; \end{cases}$$

(3.2) $\mathcal{M}_{e_1 e_2} = (Q, \Sigma, \delta, s_1, F)$ where
$Q = Q_1 \cup (Q_2 - \{s_2\})$ (merging each state in F_1 with a copy of s_2),

$$F = \begin{cases} F_2 & \text{if } s_2 \notin F_2, \\ F_1 \cup (F_2 - \{s_2\}) & \text{if } s_2 \in F_2; \end{cases} \quad \text{for } q \in Q \text{ and } a \in \Sigma,$$

$$\delta(q, a) = \begin{cases} \delta_1(q, a) & \text{if } q \in Q_1 - F_1, \\ \delta_1(q, a) \cup \delta_2(s_2, a) & \text{if } q \in F_1, \\ \delta_2(q, a) & \text{if } q \in Q_2 - \{s_2\}; \end{cases}$$

(3.3) $\mathcal{M}_{e_1^*} = (Q, \Sigma, \delta, s_1, F)$ where
$Q = Q_1,$
$F = F_1 \cup \{s_1\},$
for $q \in Q$ and $a \in \Sigma$,

$$\delta(q, a) = \begin{cases} \delta_1(q, a) & \text{if } q \in Q_1 - F_1, \\ \delta_1(q, a) \cup \delta_1(s_1, a) & \text{if } q \in F_1. \end{cases}$$

Such NFA are called Glushkov automata in [14] and were first defined by Glushkov in [48]. Note that Glushkov automata have the property that the starting state has no incoming transitions. One may observe that the automaton constructed in step (0), (1), or (2) has no incoming transitions, and each operation in step (3) preserves the property.

A detailed proof of the following result can be found in [7].

Theorem 3.1. *Let e be an arbitrary regular expression over Σ. Then $L(e) = L(\mathcal{M}_e)$.* □

A regular expression e is said to be *deterministic* [14] if $\mathcal{M}_e$ is a DFA.

3.3 Finite automata to regular expressions

Here, we show that for a given finite automaton A, we can construct a regular expression e such that e denotes the language accepted by A. The construction uses *extended finite automata* where a transition between a pair of states is labeled by a regular expression. The technique we will describe in the following is called the *state elimination technique* [123]. For a given finite automaton, the state elimination technique deletes a state at each step and changes the transitions accordingly. This process continues until the FA contains only the starting state, a final state, and the transition between them. The regular expression labeling the transition specifies exactly the language accepted by A.

Let R_Σ denote the set of all regular expressions over the alphabet Σ. An *extended finite automaton* (EFA) is formally defined as follows:

Definition 3.1. *An EFA A is a quintuple $(Q, \Sigma, \delta, s, F)$ where*
Q is the finite set of states;
Σ is the input alphabet;
$\delta : Q \times Q \to R_\Sigma$ is the labeling function of the state transitions;
$s \in Q$ is the starting state;
$F \subseteq Q$ is the set of final states.

Note that we assume $\delta(p, q) = \emptyset$ if the transition from p to q is not explicitly defined.

A word $w \in \Sigma^*$ is said to be accepted by A if $w = w_1 \cdots w_n$, for $w_1, \ldots, w_n \in \Sigma^*$, and there is a state sequence $q_0, q_1, \ldots, q_n$, $q_0 = s$ and $q_n \in F$, such that $w_1 \in L(\delta(q_0, q_1))$, $\ldots$, $w_n \in L(\delta(q_{n-1}, q_n))$. The language accepted by A is defined accordingly.

First we describe the pivotal step of the algorithm, i.e., the elimination of one non-starting and non-final state. Then we give the complete state-elimination algorithm which repeatedly applies the above step and eventually transforms the given EFA to an equivalent regular expression.

Let $A = (Q, \Sigma, \delta, s, F)$ be an EFA. Denote by e_{pq} the regular expression $\delta(p, q)$, i.e., the label of the transition from state p to state q. Let q be a state in Q such that $q \neq s$ and $q \notin F$. Then an equivalent EFA $A' = (Q', \Sigma, \delta', s, F)$ such that $Q' = Q - \{q\}$, i.e., q is eliminated, is defined as follows: For each pair of states p and r in $Q' = Q - \{q\}$,

$$\delta'(p, r) = e_{pr} + e_{pq} e_{qq}^* e_{qr}.$$

We illustrate this step by the diagram in Figure 12, where state 1 is eliminated from the given EFA.

Now, we describe the complete algorithm.

Let $A = (Q, \Sigma, \delta, s, F)$ be an EFA.

(1) (a) If the starting state is a final state or it has an incoming transition, i.e., $s \in F$ or $\delta(q, s) \neq \emptyset$ for some $q \in Q$, then add a new state s' to the state set and define $\delta(s', s) = \lambda$. Also define s' to be the starting state.

(b) If there are more than one final states, i.e., $|F| > 1$, then add a new state f' and new transitions $\delta(q, f') = \lambda$ for each q in F. Then, redefine the final state set to be $\{f'\}$.

Let $A' = (Q', \Sigma, \delta', s', F')$ denote the EFA after the above steps.

(2) If Q' consists only of s' and f', then the resulting regular expression is $e_{s'f'} e_{f'f'}^*$, where $e_{s'f'} = \delta'(s', f')$ and $e_{f'f'} = \delta'(f', f')$, and the algorithm terminates. Otherwise, continue to (3).

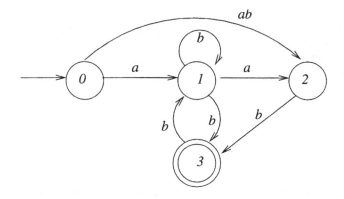

(a) Given EFA

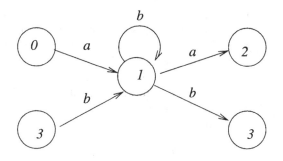

(b) Working sheet for deleting State 1

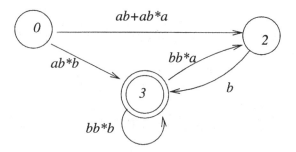

(c) Resulting EFA after deleting State 1

Fig. 12. Deletion of a state from an EFA

(3) Choose $q \in Q'$ such that $q \neq s'$ and $q \neq f'$. Eliminate q from A' following the above description. Then the new state set will be $Q' - \{q\}$ and δ' is changed accordingly. Continue to (2).

Note that every DFA, NFA, or λ-NFA is an EFA. So, the above algorithm applies to all of them.

3.4 Star height and extended regular expressions

Among the three operators of regular expressions, the star operator is perhaps the most essential one. Regular expressions without the star operator define only finite languages. One natural measurement of the complexity of a regular expression is the number of nested stars in the expression, which is called the star height of the expression. Questions concerning star height were considered among the most fascinating problems in formal language theory. Some unsolved problems are still attracting researchers. In the following, we first give the basic definitions and then describe several of the most well known problems and results concerning star height. The interested reader may refer to [98] or [107] for details on the topic.

The star height of a regular expression e over the alphabet Σ, denoted $H(e)$, is a nonnegative integer defined recursively as follows:

(1) $H(e) = 0$, if $e = \emptyset$, λ, or a for $a \in \Sigma$.
(2) $H(e) = \max(H(e_1), H(e_2))$, if $e = (e_1 + e_2)$ or $e = (e_1 e_2)$, where e_1 and e_2 are regular expressions over Σ.
(3) $H(e) = H(e_1) + 1$, if $e = e_1^*$ and e_1 is a regular expression over Σ.

The star height of a regular language R, denoted $H(R)$, is the least integer h such that $H(e) = h$ for some regular expression e denoting R.

Example 3.5. Let $e_1 = (ab(cbc)^*(ca^* + c)^*)^* + b(ca^* + c)^*$. Then $H(e_1) = 3$. Let $e_2 = a(aaa^*)^*$ and $L = L(e_2)$. Then $H(e_2) = 2$ but $H(L) = 1$ because L is denoted also by $a + aaaa^*$ and L is of at least star height one since it is infinite.   □

Concerning the star height of regular languages, one of the central questions is whether there exist languages of arbitrary star height. This question was answered by Eggan in 1963 [39]. He showed that for each integer $h \geq 0$ there exists a regular language R_h such that $H(R_h) = h$. However, in his proof the size of the alphabet for R_h grows with h. Solutions with a two-letter alphabet were given by McNaughton (unpublished notes mentioned in [18]) and later by Dejean and Schützenberger [35] in 1966.

Theorem 3.2. *For each integer $i \geq 0$, there exists a regular language R_i over a two-letter alphabet such that $H(R_i) = i$.*   □

The language R_i, for each $i \geq 0$, is given by a regular expression e_i defined recursively as follows:

$$e_0 = \lambda,$$
$$e_{i+1} = (a^{2^i} e_i b^{2^i} e_i)^*, \ i \geq 0.$$

Thus, for example,

$e_1 = (ab)^*,$
$e_2 = (a^2(ab)^*b^2(ab)^*)^*,$
$e_3 = (a^4(a^2(ab)^*b^2(ab)^*)^*b^4(a^2(ab)^*b^2(ab)^*)^*)^*.$

Clearly, $H(e_i) = i$. This implies that $H(R_i) \leq i$. The proof showing that $H(R_i)$ is at least i is quite involed. Detailed proofs can be found, e.g., in [106, 107].

Since there exist regular languages of arbitrary star height, one may naturally ask the following question: Does there exist an algorithm for determining the star height of a given regular language? This problem, often refered to as "the star height problem", was among the most well-known open problems on regular languages [18]. It had been open for more than two decades until it was solved by Hashiguchi [52] in 1988. The proof of the result is more than 40 pages long. The result by Hashiguchi can be stated as follows.

Theorem 3.3 (The Star Height). *There exists an algorithm which, for any given regular expression e, determines the star height of the language denoted by e.*

Generally speaking, almost all natural important properties are decidable for regular languages. The star height is an example of a property such that, although it is decidable, the proof of decidability is highly nontrivial.

In the following, we discuss the extended regular expressions as well as the extended star height problem.

An *extended regular expression* is one which allows the intersection $\cap$ and the complement $\neg$ operators in addition to the union, catenation, and star operators of a normal regular expression. We specify that the languages denoted by the expressions $(e_1 \cap e_2)$ and $\neg e_1$, respectively, are $L(e_1 \cap e_2) = L(e_1) \cap L(e_2)$ and $L(\neg e_1) = \overline{L(e_1)}$. We assume that $\cap$ has higher precedence than $+$ but lower precedence than $\cdot$ and $*$; and $\neg$ has the lowest precedence. For convenience, we use $\overline{e}$ to denote $\neg e$ in the following. A pair of parentheses may be omitted whenever the omission would not cause any confusion.

For instance, $\overline{\emptyset}$, $\overline{\lambda}$, and $a(a+b)^* \cap \overline{(a+b)^*bb(a+b)^*}$ are all valid extended regular expressions over $\Sigma = \{a, b\}$ denoting, respectively, Σ^*, Σ^+, and the set of all words that start with an a and contain no consecutive b's. Clearly, extended regular expressions denote exactly the family of regular languages.

The definition for the star height of an extended regular expression has the following two additions to the definition for a standard regular expression:

(4) $H(e) = \max(H(e_1), H(e_2))$, if $e = (e_1 \cap e_2)$;

(5) $H(e) = H(e_1)$, if $e = \overline{e_1}$;

where e_1 and e_2 are extended regular expressions over Σ. Similarly, the extended star height of a regular language R, denoted $H(R)$, is the least integer h such that $H(e) = h$ for some extended regular expression e denoting R.

The *star-free languages*, i.e., languages of extended star height zero, form the lowest level of the extended star height language hierarchy. It has been shown that there exist regular languages of extended star height one. However, the following problem which was raised in the sixties and formulated by Brzozowski [18] in 1979 remains open.

Open Problem *Does there exist a regular language of extended star height two or higher?*

Special attention has been paid to the family of star-free languages. The study of star-free languages was initiated by McNaughton [84, 85]. An interesting characterization theorem for star-free languages using noncounting (aperiodic) sets was proved by Schützenberger [113]. A set $S \subseteq \Sigma^*$ is said to be noncounting (aperiodic) if there exists an integer $n > 0$ such that for all $x, y, z \in \Sigma^*$, $xy^n z \in S$ iff $xy^{n+1} z \in S$. We state the characterization theorem below. The reader may refer to [113] or [98] for a detailed proof.

Theorem 3.4. *A regular language is star-free iff it is noncounting (aperiodic).* □

It appears that extended regular expressions correspond to AFA directly. It can be shown that the family of star-free languages can also be characterized by the family of languages accepted by a special subclass of AFA, which we call loop-free AFA. An AFA is said to be *loop-free* if there is a total order $<$ on the states of the AFA such that any state j does not depend on any state i such that $i < j$ or state j itself. The following result can be found in [110].

Theorem 3.5. *A regular language is star-free iff it is accepted by a loop-free AFA.* □

A special sublass of star-free languages which has attracted much attention is the *locally testable languages* [20, 41, 84]. Informally, for a locally testable language L, one can decide whether a word w is in L by looking at all subwords of w of a previously given length k.

For $k \geq 0$ and $x \in \Sigma^*$ such that $|x| \geq k$, denote by $pre_k(x)$ and $suf_k(x)$, respectively, the prefix and the suffix of length k of x, and by $int_k(x)$ (interior words) the set of all subwords of length k of x that occur in x in a position other than the prefix or the suffix. A language $L \subseteq \Sigma^*$ is said to be k-*testable* iff, for any words $x, y \in \Sigma^*$, the conditions $pre_k(x) = pre_k(y)$, $suf_k(x) = suf_k(y)$, and $int_k(x) = int_k(y)$ imply that $x \in L$ iff $y \in L$. A language is said to be *locally testable* if it is k-testable for some integer $k \geq 1$.

Many useful locally testable languages belong to a smaller class of languages, which are called *locally testable languages in the strict sense* [84]. A language $L \subseteq \Sigma^*$ is k-testable in the strict sense if there are finite sets $P, S, I \subset \Sigma^*$ such that, for all $x \in \Sigma^*$ of length at least k, $x \in L$ iff $pre_k(x) \in P$, $suf_k(x) \in S$, and $int_k(x) \subseteq I$. A language is said to be locally testable in the strict sense if it is k-testable in the strict sense for some integer $k > 0$. There are languages that are locally testable but not locally testable in the strict sense. For example, let L be the set of all words over $\{0, 1\}$ that contain either 000 or 111 as an interior word but not both. Then L is locally testable but not locally testable in the strict sense. The class of locally testable languages is closed under Boolean operations. This is not true for the class of languages that are locally testable in the strict sense.

More properties of locally testable languages can be found in [20, 41, 84, 98, 125].

3.5 Regular expressions for regular languages of polynomial density

Given a regular language, it is often useful to know how many words of a certain length are in the language, i.e., the density of the language. The study of densities of regular languages has a long history, see, e.g., [112, 41, 109, 10, 120]. Here, we consider the relationship between the densities of regular languages and the forms of the regular expressions denoting those languages. In particular, we consider the forms of regular expressions that denote regular languages of polynomial density.

For each language $L \subseteq \Sigma^*$, we define the *density function* of L

$$\rho_L(n) = |L \cap \Sigma^n|,$$

where $|S|$ denotes the cardinality of the set S. In other words, $\rho_L(n)$ counts the number of words of length n in L. If $\rho_L(n) = O(1)$, we say that L has a constant density; and if $\rho_L(n) = O(n^k)$ for some integer $k \geq 0$, we say that L has a polynomial density. Languages of constant density are called *slender languages* [34, 115]. Languages that have at most one word for each length are called *thin languages* [34].

The first theorem below characterizes regular languages of polynomial density with regular expressions of a specific form. A detailed proof can be found in [120]. Similar results can be found in [112, 41, 109, 10].

Theorem 3.6. *A regular language R over Σ has a density in $O(n^k)$, $k \geq 0$, iff R can be denoted by a finite union of regular expressions of the following form:*

(4) $$x y_1^* z_1 \ldots y_{k+1}^* z_{k+1}$$

where $x, y_1, z_1, \ldots, y_{k+1}, z_{k+1} \in \Sigma^*$.   □

The following result ([120]) shows that the number of states of a finite automaton A may restrict the order of the density function of $L(A)$.

Theorem 3.7. *Let R be a regular language accepted by a DFA of k states. If R has a polynomial density, then the function $\rho_R(n)$ is $O(n^{k-1})$.* $\qquad\square$

Theorem 3.6 is a powerful tool in proving various properties of regular languages of polynomial density. As an application of Theorem 3.6, we show the following closure properties:

Theorem 3.8. *Let L_1 and L_2 be regular languages over Σ with $\rho_{L_1}(n) = \Theta(n^k)$ and $\rho_{L_2}(n) = \Theta(n^l)$. Then the following statements hold:*

(a) *If $L = prefix(L_1) = \{x \mid xy \in L_1 \text{ for some } y \in \Sigma^*\}$, then $\rho_L(n) = \Theta(n^k)$.*
(b) *If $L = infix(L_1) = \{y \mid xyz \in L_1 \text{ for some } x, z \in \Sigma^*\}$, then $\rho_L(n) = \Theta(n^k)$.*
(c) *If $L = suffix(L_1) = \{z \mid xz \in L_1 \text{ for some } x \in \Sigma^*\}$, then $\rho_L(n) = \Theta(n^k)$.*
(d) *If $L = L_1 \cup L_2$, then $\rho_L(n) = \Theta(n^{\max(k,l)})$.*
(e) *If $L = L_1 \cap L_2$, then $\rho_L(n) = O(n^{\min(k,l)})$.*
(f) *If $L = L_1 L_2$, then $\rho_L(n) = O(n^{k+l})$.*
(g) *If $L = h(L_1)$ where h is an arbitrary morphism[30], then $\rho_L(n) = O(n^k)$.*
(h) *If $L = \frac{1}{m}(L_1) = \{x_1 \mid x_1 \ldots x_m \in L_1, \text{ for } x_1, \ldots, x_m \in \Sigma^*, \text{ and } |x_1| = \ldots = |x_m|\}$, then $\rho_L(n) = \Theta(n^k)$.*

Proof. We only prove (a) as an example. The rest can be similarly proved.

Since $\rho_{L_1}(n) = \Theta(n^k)$, by Theorem 3.6, L_1 can be specified as a finite union of regular expressions of the form:

$$(5) \qquad\qquad xy_1^* z_1 \ldots y_{k+1}^* z_{k+1}$$

where $x, y_1, z_1, \ldots, y_{k+1}, z_{k+1} \in \Sigma^*$. Then clearly, L, where $L = prefix(L_1)$, can be specified as a finite union of regular expressions of the following forms:

x',	x' is a prefix of x,
$xy_1^* z_1 \ldots y_i^* y_i'$,	y_i' is a prefix of y_i, $1 \leq i \leq k+1$,
$xy_1^* z_1 \ldots y_i^* z_i'$,	z_i' is a prefix of z_i, $1 \leq i \leq k+1$.

Then, by Theorem 3.6, the density function $\rho_L(n)$ is in $O(n^k)$. Since L is a superset of L_1, we have $\rho_L(n) \geq \rho_{L_1}(n)$, i.e., $\rho_L(n) = \Omega(n^k)$. Thus, $\rho_L(n) = \Theta(n^k)$. $\qquad\square$

It is clear that all regular languages with polynomial densities are of star height one. But, not all star-height one languages are of polynomial density. For example, the language $(ab+b)^*(a+\epsilon)$ is of exponential density. However, there is a relation between these two subclasses of regular languages, which is stated in the following theorem.

Theorem 3.9. *A regular language is of star height one if and only if it is the image of a regular language of polynomial density under a finite substitution.*

Proof. The *if* part is obvious. For the *only if* part, let E be a regular expression of star height one over an alphabet Σ. Denote by X the set of all regular expressions e (over Σ) such that e^* is a subexpression of E. Choose $\Delta = \Sigma \cup \hat{X}$, where $\hat{X} = \{\hat{e} \mid e \in X\}$. Let $\hat{E}$ be the regular expression over Δ that is obtained from E by replacing each subexpression of the form e^*, $e \in X$, by $\hat{e}^*$. By Theorem 3.6, $L(\hat{E})$ is a regular language of polynomial density. We define a finite substitution $\pi : \Delta^* \to 2^{\Sigma^*}$ as follows. For each $a \in \Sigma$, $\pi(a) \to \{a\}$ and for each $\hat{e} \in \hat{X}$, $\pi(\hat{e}) = L(e)$. Then clearly $\pi(L(\hat{E})) = L(E)$.
□

It is clear that $\rho_\emptyset(n) = 0$ and $\rho_{\Sigma^*}(n) = |\Sigma|^n$. For each $L \subseteq \Sigma^*$, we have $\rho_\emptyset(n) \le \rho_L(n) \le \rho_{\Sigma^*}(n)$ for all $n \ge 0$. It turns out that there exist functions between $\rho_\emptyset(n)$ and $\rho_{\Sigma^*}(n)$ which are not the density function of any regular language. The following two theorems [120] show that, for the densities of regular languages, there is a gap between $\Theta(n^k)$ and $\Theta(n^{k+1})$, for each integer $k \ge 0$; and there is a gap between polynomial functions and exponential functions of the order $2^{\Theta(n)}$. For example, there is no regular language that has a density of the order $\sqrt{n}$, $n \log n$, or $2^{\sqrt{n}}$.

Theorem 3.10. *For any integer $k \ge 0$, there does not exist a regular language R such that $\rho_R(n)$ is neither $O(n^k)$ nor $\Omega(n^{k+1})$.* □

Theorem 3.11. *There does not exist a regular language R such that $\rho_R(n)$ is not $O(n^k)$, for any integer $k \ge 0$, and not of the order $2^{\Omega(n)}$.* □

It is not difficult to show that, for each nonnegative integer k, we can construct a regular language R such that $\rho_R(n)$ is exactly n^k. Therefore, for each polynomial function $f(n)$, there exists a regular language R such that $\rho_R(n)$ is $\Theta(f(n))$; and for each regular language R, either there exists a polynomial function $f(n)$ such that $\rho_R(n) = \Theta(f(n))$, or $\rho_R(n)$ is of the order $2^{\Theta(n)}$.

4. Properties of regular languages

4.1 Four pumping lemmas

There are many ways to show that a language is regular; for example, this can be done by demonstrating that the language is accepted by a finite automaton, specified by a regular expression, or generated by a right-linear grammar. To prove that a language is **not** regular, the most commonly used tools are the *pumping properties* of regular languages, which are usually stated as "pumping lemmas". The term "pumping" intuitively describes the property that any sufficiently long word of the language has a nonempty subword that can be "pumped". This means that if the subword is replaced by an arbitrary number of copies of the same subword, the resulting word is still in the language.

There are many versions of pumping lemmas for regular languages. The "standard" version, which has appeared in many introductory books on the theory of computation, is a necessary but not sufficient condition for regularity, i.e., every regular language satisfies these conditions, but those conditions do not necessarily imply regularity. The first necessary and sufficient pumping lemma for regular languages was introduced by Jaffe [63]. Another necessary and sufficient pumping lemma, which is called "block pumping", was established by Ehrenfeucht, Parikh, and Rozenberg [40]. In contrast, for context-freeness of languages, only some necessary pumping conditions are known, but no conditions are known to be also sufficient.

In the following, we describe four pumping lemmas for regular languages: two necessary pumping lemmas and two necessary and sufficient pumping lemmas. We will give a proof for the first and the third, but omit the proofs for the second and the fourth. Examples will also be given to show how these lemmas can be used to prove the nonregularity of certain languages.

The first pumping lemma below was originally formulated in [5] and has appeared in many introductory books, see, e.g., [57, 108, 123, 27, 58].

Lemma 4.1. *Let R be a regular language over Σ. Then there is a constant k, depending on R, such that for each $w \in R$ with $|w| \geq k$ there exist $x, y, z \in \Sigma^*$ such that $w = xyz$ and*

(1) $|xy| \leq k$,
(2) $|y| \geq 1$,
(3) $xy^t z \in R$ for all $t \geq 0$.

Proof. Let R be accepted by a DFA $A = (Q, \Sigma, \delta, s, F)$ and k be the number of states of A, i.e., $k = |Q|$. For a word $w = a_1 \ldots a_n \in R$, $a_1, \ldots, a_n \in \Sigma$, we denote the computation of A on w by the following sequence of transitions:

$$q_0 \xrightarrow{a_1} q_1 \xrightarrow{a_2} \cdots \xrightarrow{a_n} q_n$$

where $q_0, \ldots, q_n \in Q$, $q_0 = s$, $q_n \in F$, and $\delta(q_i, a_{i+1}) = q_{i+1}$ for all i, $0 \leq i < n$.

If $n \geq k$, the above sequence has states q_i and q_j, $0 \leq i < j \leq k$, such that $q_i = q_j$. Then for each $t \geq 0$, we have the following transition sequence:

$$s = q_0 \xrightarrow{a_1} \cdots \xrightarrow{a_i} \{q_i \xrightarrow{a_{i+1}} \cdots \xrightarrow{a_j}\}^t q_j \xrightarrow{a_{j+1}} \cdots \xrightarrow{a_n} q_n$$

where $\{\alpha\}^t$ denotes that α is being repeated t times. Let $x = a_1 \ldots a_i$, $y = a_{i+1} \ldots a_j$, and $z = a_{j+1} \ldots a_n$. Then $xy^t z \in R$ for all $t \geq 0$, where $|xy| \leq k$ and $|y| \geq 1$. $\square$

The lemma states that every regular language possesses the above pumping property. Therefore, any language that does not possess the property is not a regular language. For example, one can easily show that the language $L = \{a^i b^i \mid i \geq 0\}$ is not regular using the above lemma. The arguments are as follows: Assume that L is regular and let k be the constant for the lemma. Choose $w = a^k b^k$ in L. Clearly, $|w| \geq k$. By the pumping lemma, $w = xyz$ for some $x, y, z \in \Sigma^*$ such that (1) $|xy| \leq k$, (2) $|y| \geq 1$, and (3) $xy^t z \in R$ for all $t \geq 0$. By (1) and (2), we have $y = a^m$, $1 \leq m \leq k$. But $xy^0 z = xz = a^{k-m} b^k$ is not in L. Thus, (3) does not hold. Therefore, L does not satisfy the pumping property of Lemma 4.1.

The pumping lemma has been used to show the nonregularity of many languages, e.g., the set of all binary numbers whose value is a prime [57], the set of all palindromes over a finite alphabet [58], and the set of all words of length i^2 for $i \geq 0$ [123].

However, not only regular languages but also some nonregular languages satisfy the pumping property of Lemma 4.1. Consider the following example.

Example 4.1. Let $L \subseteq \Sigma^*$ be an arbitrary *nonregular* language and

$$L_\# = (\#^+ L) \cup \Sigma^*$$

where $\# \notin \Sigma$. Then $L_\#$ satisfies the conditions of Lemma 4.1 with the constant k being 1. For any word $w \in \#^+ L$, we can choose $x = \lambda$ and $y = \#$, and for any word $w \in \Sigma^*$, we choose $x = \lambda$ and y to be the first letter of w. However, $L_\#$ is not regular, which can be shown as follows. Let h be a morphism defined by $h(a) = a$ for each $a \in \Sigma$ and $h(\#) = \lambda$. Then

$$L = h(L_\# \cap \#^+ \Sigma^*).$$

Clearly, $\#^+ \Sigma^*$ is regular. Assume that $L_\#$ is regular. Then L is regular since regular languages are closed under intersection and morphism (which will be shown in Section 4.2). This contradicts the assumption. Thus, $L_\#$ is not regular. □

Note that $L_\#$ is at the same level of the Chomsky hierarchy as L. So, there are languages at all levels of the Chomsky hierarchy, even non-recursively enumerable languages, that satisfy Lemma 4.1.

Note also that, for each language $L \subseteq \Sigma^*$, we can construct a distinct language $L_\# \subseteq (\Sigma \cup \{\#\})^*$ that satisfies Lemma 4.1. Consequently, there are uncountably many nonregular languages that satisfy the pumping lemma.

Below, we give two more examples of nonregular languages that satisfy the pumping condition of Lemma 4.1. They are quite simple and interesting.

Example 4.2. Let $L \subseteq b^*$ be an arbitrary nonregular language. Then the following languages are nonregular, but satisfy the pumping condition of Lemma 4.1:

(1) $a^+ L \cup b^*$,
(2) $b^* \cup aL \cup aa^+ \{a, b\}^*$.

Note that the first example above is just a simplified version of the language given in Example 4.1, with the alphabet Σ being a singleton. □

Lemma 4.2. *Let R be a regular language over Σ. Then there is a constant k, depending on R, such that for all $u, v, w \in \Sigma^*$, if $|w| \geq k$ then there exist $x, y, z \in \Sigma^*$, $y \neq \lambda$ such that $w = xyz$ and for all $t \geq 0$*

$$[uxy^t zv \in L \text{ iff } uwv \in L. \qquad \qquad \qquad \square$$

Any language that satisfies the pumping condition of Lemma 4.2 satisfies also the pumping condition of Lemma 4.1. This follows by setting $u = \lambda$ and $|w| = k$ in the condition of Lemma 4.2. However, the converse is not true. We can show that there exist languages that satisfy the pumping condition of Lemma 4.1, but do not satisfy that of Lemma 4.2. For example, let $L = \{a^i b^i \mid i \geq 0\}$ and consider the language $L_\# = \#^+ L \cup \{a, b\}^*$ as in Example 4.1. Clearly, $L_\#$ satisfies the pumping condition of Lemma 4.1. However, if we choose $u = \#$, $v = \lambda$, and $w = a^k b^k$ for Lemma 4.2 where k is the constant (corresponding to $L_\#$), it is clear that there do not exist x, y, z as required by the lemma. Therefore, the set of languages that satisfy the pumping condition of Lemma 4.2 is a proper subset of the set of languages that satisfy the condition of Lemma 4.1. In other words, Lemma 4.2 can rule out more nonregular languages. In this sense, we say that Lemma 4.2 is a stronger pumping lemma for regular languages than Lemma 4.1.

Nevertheless, Lemma 4.2 still does not give a sufficient condition for regularity. We show in the following that there exist nonregular languages that satisfy the pumping condition of Lemma 4.2. In fact, the number of such languages is uncountable. A different proof was given in [40].

Example 4.3. Let L be an arbitrary nonregular language over Σ and $\$ \notin \Sigma$. Define

$$L_\$ = \{\$^+ a_1 \$^+ a_2 \$^+ \ldots \$^+ a_m \$^+ \mid a_1 a_2 \ldots a_m \in L, \ a_1, a_2, \ldots, a_m \in \Sigma, m \geq 0\}$$
$$\cup \{\$^+ x_1 \$^+ x_2 \$^+ \ldots \$^+ x_n \$^+ \mid x_1, x_2, \ldots, x_n \in \Sigma^*, \ n \geq 0, |x_i| \neq 1$$
$$\text{for some } i, 1 \leq i \leq n\}.$$

We can easily prove that $L_\$$ is nonregular. Let $\Sigma_\$$ denote $\Sigma \cup \{\$\}$. We now show that $L_\$$ satisfies the pumping condition of Lemma 4.2. Let $k = 3$ be the constant for the pumping lemma. To establish the nontrivial implication of the statement of the lemma, it suffices to show that for any $u, w, v \in \Sigma_\$^*$ with $uwv \in L$ and $|w| \geq 3$, there exist $x, y, z \in \Sigma_\$^*$ with $w = xyz$ and $y \neq \lambda$ such that $uxy^i zv \in L_\$$ for all $i \geq 0$. We can choose $y = \$$ if w contains a $\$$ and $y = a$ for some $a \in \Sigma$ if w does not contain any symbol $\$$. $\qquad \square$

The next pumping lemma, introduced by Jaffe [63], gives a necessary and sufficient condition for regularity. A detailed proof of the following lemma can be found also in [108].

Lemma 4.3. *A language* $L \in \Sigma^*$ *is regular* **iff** *there is a constant* $k > 0$ *such that for all* $w \in \Sigma^*$, *if* $|w| \geq k$ *then there exist* $x, y, z \in \Sigma^*$ *such that* $w = xyz$ *and* $y \neq \lambda$, *and for all* $i \geq 0$ *and all* $v \in \Sigma^*$, $wv \in L$ **iff** $xy^i zv \in L$.

Proof. The *only if* part is relatively straightforward. Let A be a complete DFA which accepts L and k the number of states of A. For any word w of length $l \geq k$, i.e., $w = a_1 a_2 \cdots a_l$, let the state transition sequence of A on w be the following:

$$q_0 \overset{a_1}{\rightarrow} q_1 \overset{a_2}{\rightarrow} \cdots \overset{a_l}{\rightarrow} q_l,$$

where q_0 is the starting state. Since there are at most k distinct states among $q_0, q_1, \ldots, q_l$ and $k < l+1$, it follows that $q_i = q_j$ for some i, j, $0 \leq i < j \leq l$. This implies that the transition from q_i to q_j is a loop back to the same state. Let $x = a_1 \cdots a_i$, $y = a_{i+1} \cdots a_j$, and $z = a_{j+1} \cdots a_l$ ($x = \lambda$ if $i = 1$ and $z = \lambda$ if $j = l$). Then, for all $i \geq 0$,

$$\delta^*(q_0, xy^i z) = q_l,$$

i.e., A is in the same state q_l after reading each word $xy^i z$, $i \geq 0$. Therefore, for all $i \geq 0$ and for all $v \in \Sigma^*$, $xy^i zv \in L$ iff $wv \in L$.

For the *if* part, let L be a language which satisfies the pumping condition of the lemma and k be the constant. We prove that L is regular by constructing a DFA A_L using the pumping property of L and then proving that $L(A_L) = L$.

The DFA $A_L = (Q, \Sigma, \delta, s, F)$ is defined as follows. Each state in Q corresponds to a string w, in Σ^*, of length less than k, i.e.,

$$Q = \{q_w \mid w \in \Sigma^* \text{ and } |w| \leq k - 1\},$$

$s = q_\lambda$ and $F = \{q_w \in Q \mid w \in L\}$. The transition function δ is defined as follows:

(1) If $|w| < k - 1$, then for each $a \in \Sigma$,

$$\delta(q_w, a) = q_{wa}.$$

(2) If $|w| = k - 1$, then by the pumping property of L, for each $a \in \Sigma$, wa can be decomposed into xyz, $y \neq \lambda$, such that for all $v \in \Sigma^*$, $xyzv \in L$ iff $xzv \in L$. There may be a number of such decompositions. We choose the one such that xy is the shortest (and y is the shortest if there is a tie). Then define

$$\delta(q_w, a) = q_{xz}.$$

Now we show that the language accepted by A_L is exactly L. We prove this by induction on the length of a word $w \in \Sigma^*$. It is clear that for all words w such that $|w| < k$, $w \in L(A_L)$ iff $w \in L$ by the definition of A_L. We hypothesize that for all words shorter than n, $n \geq k$, $w \in L(A_L)$ iff $w \in L$. Consider a word w with $|w| = n$. Let $w = w_0 v$ where $|w_0| = k$. By the construction of A_L, we have $\delta^*(s, w_0) = \delta^*(s, xz) = q_{xz}$ for some $x, z \in \Sigma^*$

where $w_0 = xyz$, $y \in \Sigma^+$, and for any $v' \in \Sigma^*$, $w_0v' \in L$ iff $xzv' \in L$. We replace the arbitrary v' by v, then we have that $w \in L$ iff $xzv \in L$. Since xz and w_0 reach the same state in A_L, xzv and $w = w_0v$ will reach the same state, i.e., $w \in L(A_L)$ iff $xzv \in L(A_L)$. Notice that $|xzv| < n$. By the hypothesis, $xzv \in L(A_L)$ iff $xzv \in L$. So, we conclude that $w \in L(A_L)$ iff $w \in L$. □

Example 4.4. Let $L = \{a^i b^i \mid i \geq 0\}$ and $L_\# = (\#^+ L) \cup \{a, b\}^*$. We have shown that $L_\#$ satisfies the pumping condition of Lemma 4.1. Now we demonstrate that $L_\#$ does not satisfy the pumping condition of Lemma 4.3. Assume the contrary. Let $k > 0$ be the constant of Lemma 4.3 for $L_\#$. Consider the word $w = \# a^k b^k$ and any decomposition $w = xyz$ such that $y \neq \lambda$. If y does not contain the symbol $\#$, i.e., $y \in a^+$, $y \in b^+$, or $y \in a^+ b^+$, then let $v = \lambda$ and, clearly, $wv \in L_\#$ but $xy^2 zv \notin L_\#$. If y contains the symbol $\#$, then let $v = a$ and we have $wv = xyzv \notin L_\#$ but $xzv \in L_\#$. So, $L_\#$ does not satisfy the pumping condition of Lemma 4.3. □

Notice that Lemma 4.3 requires a decomposition $w = xyz$ that works for **all** wv, $v \in \Sigma^*$. Another necessary and sufficient pumping lemma for regularity, which does not require this type of global condition, was given by Ehrenfeucht, Parikh, and Rozenberg [40]. The latter is called the block pumping lemma, which is very similar to Lemma 4.2 except that the decomposition of w into xyz has to be along the given division of w into subwords (blocks) $w_1, \ldots, w_k$, i.e., each of x, y, and z has to be a catenation of those subwords.

Lemma 4.4. *(Block pumping)* $L \subseteq \Sigma^*$ *is regular iff there is a constant $k > 0$ such that for all $u, v, w \in \Sigma^*$, if $w = w_1 \cdots w_k$, $w_1, \ldots, w_k \in \Sigma^*$, then there exist m, n, $1 \leq m < n \leq k$, such that $w = xyz$ with $y = w_{m+1} \ldots w_n$, $x, z \in \Sigma^*$, and for all $i \geq 0$,*

$$uwv \in L \text{ iff } uxy^i zv \in L.$$
□

Example 4.5. Let $L = \{a^i b^i \mid i \geq 0\}$ and let $L_\$$ be defined as in Example 4.3. We have shown in Example 4.3 that $L_\$$ satisfies the pumping property of Lemma 4.2. Here we show that $L_\$$ does not satisfy the pumping property of Lemma 4.4. Assume the contrary. Let k be the constant in the lemma and choose $u = \lambda$, $w_1 = \$a$, $w_2 = \$a$, ..., $w_k = \$a$, $v = (\$b)^k \$$, and $w = w_1 \cdots w_k$. Then $uwv \in L_\$$. But, clearly, there do not exist m, n, $1 \leq m < n \leq k$, such that $y = w_{m+1} \ldots w_n$, $w = xyz$, and $uxzv = uw_1 \cdots w_m w_{n+1} \cdots w_k v = (\$a)^{k-n+m} (\$b)^k \$ \in L_\$$. □

In Lemma 4.4, the pumping condition is sufficient for the regularity of L even if we change the statement "for all $i \geq 0$" to "for $i = 0$". Then the pumping property becomes a cancellation property. It has been shown that the pumping and cancellation properties are equivalent [40]. A similar result can also be obtained for Lemma 4.3.

4.2 Closure properties

The following theorem has been established in Section 2. and 3..

Theorem 4.1. *The family of regular languages is closed under the following operations:* (1) *union,* (2) *intersection,* (3) *complementation,* (4) *catenation,* (5) *star, and* (6) *reversal.* □

The next theorem is a remarkably powerful tool for proving other properties of regular languages.

Theorem 4.2. *The family of regular languages is closed under finite transduction.*

Proof. Let L be an arbitrary regular language accepted by a DFA $A = (Q_A, \Sigma, \delta, s, F)$ and $T = (Q_T, \Sigma, \Delta, \sigma_T, s_T, F_T)$ a finite transducer in the standard form. We show that $T(L)$ is regular.

Construct a λ-NFA $R = (Q_R, \Delta, \delta_R, s_R, F_R)$ where
$Q_R = Q_A \times Q_T$;
$s_R = (s_A, s_T)$;
$F_R = F_A \times F_T$;
δ_R is defined by, for $(p, q) \in Q_R$ and $b \in \Delta \cup \{\lambda\}$,

$$\delta_R((p, q), b) = \{(p', q') \mid \text{there exists } a \in \Sigma \text{ such that}$$

$$\delta_A(p, a) = p' \ \& \ (q', b) \in \sigma_T(q, a), \text{or } (q', b) \in \sigma_T(q, \lambda) \ \& \ p = p'\}.$$

Now we show that $L(R) = T(L)$.

Let w be accepted by R. Then there is a state transition sequence of R

$$(s_A, s_T) \xrightarrow{b_1} (p_1, q_1) \xrightarrow{b_2} \ldots\ldots \xrightarrow{b_n} (p_n, q_n)$$

where $w = b_1 \cdots b_n$, $b_1, \ldots, b_n \in \Delta \cup \{\lambda\}$, and $p_n \in F_A$, $q_n \in F_T$. By the definition of R, there exist $a_1, \ldots, a_n \in \Sigma \cup \{\lambda\}$ such that

$$s_T \xrightarrow{a_1/b_1} q_1 \xrightarrow{a_2/b_2} \ldots\ldots \xrightarrow{a_n/b_n} q_n.$$

Let $a_{i_1}, \ldots, a_{i_m}$ be the non-λ subsequence of $a_1, \ldots, a_n$, i.e., $a_{i_1}, \ldots, a_{i_m} \in \Sigma$ and $u = a_{i_1} \cdots a_{i_m} = a_1 \cdots a_n$. Note that if $a_k = \lambda$, then $p_{k-1} = p_k$ (assuming $p_0 = s_A$). Thus, we have

$$s_A \xrightarrow{a_{i_1}} p_{i_1} \xrightarrow{a_{i_2}} \ldots\ldots \xrightarrow{a_{i_m}} p_{i_m} = p_n.$$

So, u is accepted by A and $w \in T(u)$. Therefore, $w \in T(L)$.

Let $u \in L(A)$ and $T(u) = w$. We prove that $w \in L(R)$. Since $T(u) = w$, there is a state transition sequence of T

$$s_T = q_0 \xrightarrow{a_1/b_1} q_1 \xrightarrow{a_2/b_2} \ldots\ldots \xrightarrow{a_n/b_n} q_n$$

for $u = a_1 \cdots a_n$, $a_1, \ldots, a_n \in \Sigma \cup \{\lambda\}$, $w = b_1 \cdots b_n$, $b_1, \ldots, b_n \in \Delta \cup \{\lambda\}$, and $q_n \in F_T$. Let $a_{i_1}, \ldots, a_{i_m}$ be the non-λ subsequence of $a_1, \ldots, a_n$, i.e., $a_{i_1} \cdots a_{i_m} = a_1 \cdots a_n = u$. Since $u \in L(A)$, we have

$$s_A = p_0 \xrightarrow{a_{i_1}} p_1 \xrightarrow{a_{i_2}} \cdots \cdots \xrightarrow{a_{i_m}} p_m,$$

where $p_m \in F_A$. Then, by the construction of R, there exists a state transition sequence of R

$$r_0 \xrightarrow{b_1} r_1 \xrightarrow{b_2} \cdots \cdots \xrightarrow{b_n} r_n$$

where $r_0 = s_R = (p_0, q_0)$ and for each j, $1 \le j \le n$, $r_j = (p_k, q_j)$ if $a_j \ne \lambda$ and $j = i_k$; $r_j = (p_{k-1}, q_j)$ if $a_j = \lambda$ and $i_{k-1} < j < i_k$, $1 \le k \le m$. Thus, $w \in L(R)$.   □

Many operations can be implemented by finite transducers. Thus, the fact that regular languages are closed under those operations follows immediately by the above theorem. We list some of the operations in the next theorem.

Theorem 4.3. *The family of regular languages is closed under the following operations (assuming that $L \subseteq \Sigma^*$):*

(1) prefix$(L) = \{x \mid xy \in L,\ x, y \in \Sigma^*\}$,
(2) suffix$(L) = \{y \mid xy \in L,\ x, y \in \Sigma^*\}$,
(3) infix$(L) = \{y \mid xyz \in L,\ x, y, z \in \Sigma^*\}$,
(4) morphism,
(5) finite substitution,
(6) inverse morphism,
(7) inverse finite substitution.

Proof. (4) and (5) are obvious since they are only special cases of finite transductions: morphisms can be represented as one-state deterministic finite transducers (DGSMs) and finite substitutions can be represented as one-state (nondeterministic) finite transducers without λ-transitions. Note that, in both cases, the sole state is both the starting state and the final state.

(6) and (7) are immediate since, by Theorem 2.16, an inverse finite transduction is again a finite transduction.

Each of the operations (1)–(3) can be realized by a finite transducer given below. We omit the proof showing, in each case, the equality of the transduction and the operation in question. Figure 13 gives the transducers in the case where $\Sigma = \{a, b\}$.

(1) $T_{pre} = (Q_1, \Sigma, \Sigma, \sigma_1, s_1, F_1)$ where $Q_1 = \{1, 2\}$, $s_1 = 1$, $F_1 = Q_1$, and σ_1:
$\sigma_1(1, a) = \{(1, a), (2, \lambda)\}$, for each $a \in \Sigma$;
$\sigma_1(2, a) = \{(2, \lambda)\}$, for each $a \in \Sigma$.
(2) $T_{suf} = (Q_2, \Sigma, \Sigma, \sigma_2, s_2, F_2)$ where $Q_2 = \{0, 1\}$, $s_2 = 0$, $F_2 = Q_2$, and σ_2:
$\sigma_2(0, a) = \{(0, \lambda), (1, a)\}$, for each $a \in \Sigma$;
$\sigma_2(1, a) = \{(1, a)\}$, for each $a \in \Sigma$.

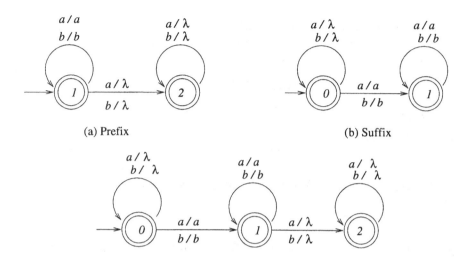

(a) Prefix

(b) Suffix

(c) Subword

Fig. 13. Finite transducers realizing the prefix, suffix, and infix operations

(3) $T_{inf} = (Q_3, \Sigma, \Sigma, \sigma_3, s_3, F_3)$ where $Q_3 = \{0, 1, 2\}$, $s_3 = 0$, $F_3 = Q_3$, and σ_3:

$\sigma_3(0, a) = \{(0, \lambda), (1, a)\}$, for each $a \in \Sigma$;
$\sigma_3(1, a) = \{(1, a), (2, \lambda)\}$, for each $a \in \Sigma$;
$\sigma_3(2, a) = \{(2, \lambda)\}$, for each $a \in \Sigma$.   □

A substitution $\varphi : \Sigma^* \to 2^{\Delta^*}$ is called a regular substitution if, for each $a \in \Sigma$, $\varphi(a)$ is a regular language. The reader can verify that each regular substitution can be specified by a finite transduction. Thus, we have the following:

Theorem 4.4. *The family of regular languages is closed under regular substitution and inverse regular substitution.*   □

Let L be an arbitrary language over Σ. For each $x \in \Sigma^*$, the left-quotient of L by x is the set

$$x \backslash L = \{y \in \Sigma^* \mid xy \in L\},$$

and for a language $L_0 \subseteq \Sigma^*$, the left-quotient of L by L_0 is the set

$$L_0 \backslash L = \bigcup_{x \in L_0} x \backslash L = \{y \mid xy \in L, \ x \in L_0\}.$$

Similarly, the right-quotient of L by a word $x \in \Sigma^*$ is the set

$$L/x = \{y \in \Sigma^* \mid yx \in L\},$$

and the right-quotient of L by a language $L_0 \subseteq \Sigma^*$ is

$$L/L_0 = \bigcup_{x \in L_0} L/x = \{y \mid yx \in L, \ x \in L_0\}.$$

It is clear by the above definition that

$$(L_1 \backslash L) \cup (L_2 \backslash L) = (L_1 \cup L_2) \backslash L$$

for any $L_1, L_2 \subseteq \Sigma^*$. This implies that $\{x, y\} \backslash L = x \backslash L \cup y \backslash L$. Similar equalities hold for right-quotient of languages.

For $L \subseteq \Sigma^*$, we define the following operations:

- $\min(L) = \{w \in L \mid$ there does not exist $x \in L$ such that x is a proper prefix of $w\}$.
- $\max(L) = \{w \in L \mid$ there does not exist $x \in L$ such that w is a propper prefix of $x\}$.

Theorem 4.5. *The family of regular languages is closed under (1) left-quotient by an arbitrary language, (2) right quotient by an arbitrary language, (3) min, and (4) max.*

Proof. Let $L \subseteq \Sigma^*$ be a regular language accepted by a DFA $A = (Q, \Sigma, \delta, s, F)$. We define, for each $q \in Q$, DFA $A_q = (Q, \Sigma, \delta, s, \{q\})$ and $A^{(q)} = (Q, \Sigma, \delta, q, F)$. For each of the four operations, we prove that the resulting language is regular by constructing a finite automaton to accept it. We leave the verifications of the constructions to the reader.

For (1), let $L_0 \subseteq \Sigma^*$ be an arbitrary language. Then $L_0 \backslash L$ is accepted by the NNFA $A_1 = (Q, \Sigma, \delta, S_1, F)$ where Q, δ, and F are the same as in A; and

$$S_1 = \{q \in Q \mid L(A_q) \cap L_0 \neq \emptyset\}$$

is the set of starting states of the NNFA.

For (2), we construct a DFA $A_2 = (Q, \Sigma, \delta, s, F_2)$ where Q, δ, and s are the same as in A; and $F_2 = \{q \in Q \mid L(A^{(q)}) \cap L_0 \neq \emptyset\}$.

For (3), we define $A_3 = (Q, \Sigma, \delta_3, s, F)$ where δ_3 is the same as δ except that all transitions from each final state are deleted.

For (4), we define $A_4 = (Q, \Sigma, \delta, s, F_4)$ where $F_4 = \{f \in F \mid \delta^*(f, x) \notin F$ for all $x \in \Sigma^+\}$. □

Let m and n be two natural numbers such that $m < n$. Then, for a language $L \subseteq \Sigma^*$, $\frac{m}{n}(L)$ is defined to be the language

$$\frac{m}{n}(L) = \{w_1 \cdots w_m \mid w_1 \cdots w_m w_{m+1} \cdots w_n \in$$
$$L, w_1, \ldots, w_n \in \Sigma^*, |w_1| = \cdots = |w_n|\}.$$

Note that the above definition requires that the division of a word into n parts has to be exact. Then, the operations $\frac{m}{n}$ and $\frac{cm}{cn}$ are not equivalent for an integer $c > 1$. For example, let $L = \{\lambda, a, ba, aab, bbab\}$; then $\frac{1}{2}(L) = \{\lambda, b, bb\}$, but $\frac{2}{4}(L) = \{\lambda, bb\}$. We show that the family of regular languages is closed under the $\frac{m}{n}$ operation.

Theorem 4.6. *Let $L \subseteq \Sigma^*$ be a regular language and m, n be two natural numbers such that $m < n$. Then $\frac{m}{n}(L)$ is regular.*

Proof. Let L be accepted by a DFA $A = (Q, \Sigma, \delta, s, F)$. For each $q \in Q$, we construct a variant of an NFA $A(q)$ which reads m symbols at each transition. Such a variant can clearly be transformed into an equivalent standard NFA. More specifically, $A(q) = (Q', \Sigma, \delta', s_q, F_q)$ where $Q' = Q \times Q$; $\delta : Q' \times \Sigma^m \rightarrow 2^{Q'}$ is defined, for $a_1, \ldots, a_m \in \Sigma$,

$$\delta'((p_1, p_2), a_1 \cdots a_m) = \{(p_1', p_2') \mid \delta^*(p_1, a_1 \cdots a_m) = p_1' \text{ and}$$

$$\text{there exists } x \in \Sigma^{n-m} \text{ such that } \delta^*(p_2, x) = p_2'\};$$

$s_q = (s, q)$; and $F_q = \{(q, f) \mid f \in F\}$.

Intuitively, $A(q)$ operates on two tracks, starting with the states s and q, respectively. A word u with $|u| = cm$, for some nonnegative integer c, is accepted by $A(q)$ if $A(q)$, working on the first track, can reach q by reading u and, simultaneously working on the second track, can reach a final state of A from the state q by reading a "phantom" input of length $c(n - m)$.

It is easy to see that $\frac{m}{n}(L) = \cup_{q \in Q} L(A(q))$. We omit the details of the proof. $\qquad\qquad\qquad\qquad\qquad\qquad\qquad\qquad\qquad\qquad\qquad\qquad\qquad\qquad\square$

4.3 Derivatives and the Myhill-Nerode theorem

The notion of derivatives was introduced in [99, 100, 43] (under different names) and was first applied to regular expressions by Brzozowski in [16].

We define derivatives using the left-quotient operation. Let $L \subseteq \Sigma^*$ and $x \in \Sigma^*$. The *derivative* of L with respect to x, denoted $D_x L$, is

$$D_x L = x \backslash L = \{y \mid xy \in L\}.$$

For $L \subseteq \Sigma^*$, we define a relation $\equiv_L \subseteq \Sigma^* \times \Sigma^*$ by

$$x \equiv_L y \text{ iff } D_x L = D_y L$$

for each pair $x, y \in \Sigma^*$. Clearly, $\equiv_L$ is an equivalence relation. It partitions Σ^* into equivalence classes. The number of equivalence classes of $\equiv_L$ is called the *index* of $\equiv_L$. We denote the equivalence class that contains x by $[x]_{\equiv_L}$, i.e.,

$$[x]_{\equiv_L} = \{y \in \Sigma^* \mid y \equiv_L x\}.$$

Clearly, $x \equiv_L y$ iff $[x]_L = [y]_L$. We simply write $[x]$ instead of $[x]_{\equiv_L}$ if there is no confusion.

A relation $R \subseteq \Sigma^* \times \Sigma^*$ is said to be *right-invariant* with respect to catenation if $x \, R \, y$ implies $xz \, R \, yz$, for any $z \in \Sigma^*$. It is clear that the relation $\equiv_L$ is right-invariant.

Lemma 4.5. *Let* $A = (Q, \Sigma, \delta, s, F)$ *be a DFA and* $L = L(A)$. *For each* $q \in Q$, *let* $A_q = (Q, \Sigma, \delta, s, \{q\})$. *Then, for all* $x, y \in \Sigma^*$ *and* $q \in Q$, $x, y \in L(A_q)$ *implies* $x \equiv_L y$.

Proof. Let $x, y \in L(A_q)$. Define $A^{(q)} = (Q, \Sigma, \delta, q, F)$. Then, clearly, $D_x L = L(A^{(q)}) = D_y L$. Thus, $x \equiv_L y$ by the definition of $\equiv_L$. $\qquad \square$

The following is a variant of the theorem which is called the Myhill-Nerode Theorem in [57]. The result was originally given by Myhill [90] and Nerode [91]. A similar result on regular expressions was obtained by Brzozowski [16].

Theorem 4.7. *A language* $L \subseteq \Sigma^*$ *is regular iff* $\equiv_L$ *has a finite index.*

Proof. *Only if:* Let L be accepted by a complete DFA $A = (Q, \Sigma, \delta, s, F)$. As in Lemma 4.5, we define $A_q = (Q, \Sigma, \delta, s, \{q\})$ for each $q \in Q$. Since A is a complete DFA, we have

$$\bigcup_{q \in Q} L(A_q) = \Sigma^*.$$

Thus, $\pi_A = \{L(A_q) \mid q \in Q\}$ is a partition of Σ^*. By Lemma 4.5, for each $q \in Q$, $x, y \in L(A_q)$ implies $x \equiv_L y$, i.e., $L(A_q) \subseteq [x]$ for some $x \in \Sigma^*$. This means that π_A refines the partition induced by $\equiv_L$. Since π_A is a finite partition, the number of the equivalence classes of $\equiv_L$ is finite.

If: We construct a DFA $A' = (Q', \Sigma, \delta', s', F')$ where the elements of Q' are the equivalence classes of $\equiv_L$, i.e., $Q' = \{[x] \mid x \in \Sigma^*\}$; δ' is defined by $\delta'([x], a) = [xa]$, for all $[x] \in Q$ and $a \in \Sigma$; $s' = [\lambda]$; and $F' = \{[x] \mid x \in L\}$. Note that δ' is well-defined because $\equiv_L$ is right-invariant. It is easy to verify that $\delta'(s', x) = [x]$ for each $x \in \Sigma^*$ (by induction on the length of x). Then $x \in L$ iff $\delta'(s', x) = [x] \in F'$. Therefore, $L(A') = L$. $\qquad \square$

Theorem 4.8. *Let* L *be a regular language. The minimal number of states of a complete DFA that accepts* L *is equal to the index of* $\equiv_L$.

Proof. Let the index of $\equiv_L$ be k. In the proof of Theorem 4.7, it is shown that there is a k-state complete DFA that accepts L. We now prove that k is minimum. Suppose that L is accepted by a complete DFA $A = (Q, \Sigma, \delta, s, F)$ of k' states where $k' < k$. Then, for some $q \in Q$, $L(A_q)$ contains words from two distinct equivalence classes of $\equiv_L$, i.e., $x \not\equiv_L y$ for some $x, y \in L(A_q)$. This contradicts Lemma 4.5. $\qquad \square$

Corollary 4.1. *Let* $A = (Q, \Sigma, \delta, s, F)$ *be a complete DFA and* $L = L(A)$. *A is a minimum-state complete DFA accepting* L *iff, for each* $q \in Q$, $L(A_q) = [x]$ *for some* $x \in \Sigma^*$.

Proof. The *if* part follows immediately from Theorem 4.8. For the converse implication, assume that A is a minimum-state complete DFA. By Lemma 4.5, the partition of Σ^* into the languages $L(A_q)$, $q \in Q$, is a refinement of $\equiv_L$. By Theorem 4.8, $|Q|$ equals to the index of $\equiv_L$. Hence, each language $L(A_q)$, $q \in Q$, has to coincide with some class of the relation $\equiv_L$. $\qquad \square$

From the above arguments, one can observe that for minimizing a given DFA A that accepts L, we need just to merge into one state all the states q such that the corresponding languages $L(A_q)$ are in the same equivalence class of $\equiv_L$. Transitions from states to states are also merged accordingly. This can be done because of the right-invariant property of $\equiv_L$.

More formally, for a DFA $A = (Q, \Sigma, \delta, s, F)$, we define an equivalence relation $\approx_A$ on Q as follows:

$$p \approx_A q \text{ iff } L(A^{(p)}) = L(A^{(q)})$$

for $p, q \in Q$. Note that $\approx_A$ is right-invariant in the sense that if $p \approx q$ then $\delta^*(p, x) \approx_A \delta^*(q, x)$, for any given $x \in \Sigma^*$. It is clear that each equivalence class of $\approx_A$ corresponds exactly to an equivalence class of $\equiv_{L(A)}$. Then we can present the following DFA minimization scheme:

(1) Partition Q into equivalence classes of $\approx_A$:

$$\Pi = \{[q] \mid q \in Q\}.$$

(2) Construct $A' = (Q', \Sigma, \delta', s', F')$ where $Q' = \Pi$, $s' = [s]$, $F' = \{[f] \mid f \in F\}$, and $\delta'([p], a) = [q]$ if $\delta(p, a) = q$, for all $p, q \in Q$ and $a \in \Sigma$.

Note that the right-invariant property of $\approx_A$ guarantees that δ' is well defined.

The major part of the scheme is at the step (1), i.e., finding the partition of Q. A straightforward algorithm for step (1) is that we check whether $p \approx_A q$ by simply testing whether $L(A^{(p)}) = L(A^{(q)})$. However, the complexity of this algorithm is too high ($\Omega(n^4)$ where n is the number of states of A).

Many partition algorithms have been developed, see, e.g., [56, 49, 57, 11, 12]. The algorithm by Hopcroft [56], which was redescribed later by Gries [49] in a more understandable way, is so far the most efficient algorithm. A rather complete list of DFA minimization algorithms can be found in [122].

An interesting observation is that, for a given DFA, if we construct an NFA that is the reversal of the given DFA and then transform it to a DFA by the standard subset construction technique (constructing only those states that are reachable from the new starting state), then the resulting DFA is a minimum-state DFA [15, 67, 81, 12]. We state this more formally in the following. First, we define two operations γ and τ on automata. For a DFA $A = (Q, \Sigma, \delta, s, F)$, $\gamma(A)$ is the NNFA $A^R = (Q, \Sigma, \delta^R, F, \{s\})$ where $\delta^R :$ $Q \to 2^Q$ is defined by $\delta^R(p, a) = \{q \mid \delta(q, a) = p\}$; and for an NNFA $M = (Q, \Sigma, \eta, S, F)$, $\tau(M)$ is the DFA $M' = (Q', \Sigma, \eta', s', F')$ where $s' = S$; η' and F' are defined by the standard subset construction technique; and $Q' \subseteq 2^Q$ consists of only those subsets of Q that are reachable from s'.

Theorem 4.9. *Let $A = (Q, \Sigma, \delta, s, F)$ be a DFA with the property that all states in Q are reachable from s. Then $L(\tau(\gamma(A))) = L^R$ and $\tau(\gamma(A))$ is a minimum-state DFA.*

Proof. Let $\gamma(A)$ be the NNFA $A^R = (Q, \Sigma, \delta^R, F, \{s\})$ and $\tau(A^R)$ be the DFA $A' = (Q', \Sigma, \delta', s', F')$ as defined above. Obviously, $L(A') = L^R$. To prove that A' is minimum, it suffices to show that, for any $p', q' \in Q', p' \approx_{A'} q'$ implies $p' = q'$. Notice that p' and q' are both subsets of Q. If $p' \approx_{A'} q'$, then $L(A'^{(p')}) = L(A'^{(q')})$. Let $r \in p'$. Since $\delta^*(s, x) = r$ for some $x \in \Sigma^*$, we have $s \in (\delta^R)^*(r, x)$ and, thus, $x \in L(A^{(p')})$. This implies that $x \in L(A^{(q')})$ and, thus, there exist $t \in q'$ such that $\delta^*(s, x) = t$. Since δ is a deterministic transition function, $r = t$, i.e., $r \in q'$. So, $p' \subseteq q'$. Similarly, we can prove that $q' \subseteq p'$. Therefore, $p' = q'$. $\square$

From the above idea, a conceptually simple algorithm for DFA minimization can be obtained as follows. Given a DFA A, we compute $A' = \tau(\gamma(\tau(\gamma(A))))$. Then A' is a minimum-state DFA which is equivalent to A. The algorithm is descriptively very simple. However, the time and space complexities of the algorithm are very high; they are both of the order of 2^n in the worst case, where n is the number of states of A. This algorithm was originally given by Brzozowski in [15]. Descriptions of the algorithm can also be found in [81, 67, 12, 118, 122]. Watson wrote an interesting paragraph on the origin of the algorithm in [122] (on pages 195–196).

Theorem 4.8 gives a tight lower bound on the number of states of a DFA. Can we get similar results for AFA and NFA? The following result for AFA follows immediately from Theorem 2.14 and Theorem 4.8.

Theorem 4.10. *Let L be a regular language and $k > 1$ be the minimum number of states of a DFA that accepts L^R, i.e., the reversal of L. Then the minimum number of states of an s-AFA accepting L is $\lceil \log k \rceil + 1$.* $\square$

Note that there can be many different minimum-state AFA accepting a given language and they are not necessarily identical or equivalent under a renaming of the states.

NFA are a special case of AFA. Any lower bound on the number of states of an AFA would also be a lower bound on the number of states of an NFA.

Corollary 4.2. *Let L be a regular language and $k > 1$ be the minimum number of states of a DFA that accepts L^R. Then the minimum number of states of an NFA accepting L is greater than or equal to $\lceil \log k \rceil + 1$.*

The above lower bound is reached for some languages, e.g., the languages accepted by the automata shown in Figure 18.

Also by Lemma 2.2, we have the following:

Theorem 4.11. *Let L be a regular language. Let k and k' be the numbers of states of the minimal DFA accepting L and L^R, respectively. Then the number of states of any NFA accepting L is greater than or equal to $\max(\lceil \log k \rceil, \lceil \log k' \rceil)$.* $\square$

Observe that minimum-state NNFA that accept L and L^R, respectively, have exactly the same number of states. A minimum-state NFA requires at most one more state than a minimum-state NNFA equivalent to it. So, this gives another proof for the above lower bound.

5. Complexity issues

In the previous sections, we studied various representations, operations, and properties of regular languages. When we were considering the operations on regular languages, we were generally satisfied with knowing what can be done and what cannot be done, but did not measure the complexity of the operations. In this section, we consider two kinds of measurements: (1) state complexity and (2) time and space complexity. One possibility would have been to discuss these complexity issues together with the various operations in the previous sections. Since this topic has usually not been at all included in earlier surveys, we feel that devoting a separate section for the complexity issues is justified.

5.1 State complexity issues

By the *state complexity* of a regular language, we mean the minimal number of states of a DFA representing the language. By the state complexity of an operation on regular languages we mean a function that associates the sizes of the DFA representing the operands of the operation to the minimal number of states of the DFA representing the resulting language. Note that in this section, by a DFA we always mean a complete DFA.

State complexity is a natural measurement of operations on regular languages. It also gives a lower bound for the time and space complexity of those operations. State complexity is of central importance especially for applications using implementations of finite automata. However, questions of state complexity have rarely been the object of a systematic investigation. Examples of early studies concentrated on this topic are [104, 105] by Salomaa and [89] by Moore. Some recent results can be found in [6, 102, 101, 124, 111]. Most of the results presented in this section are from [111].

By an n-state DFA language, we mean a regular language that is accepted by an n-state DFA. Here, we consider only the worst-case state complexity. For example, for an arbitrary m-state DFA language and an arbitrary n-state DFA language, the state complexity of the catenation of the two languages is $m2^n - 2^{n-1}$. This means that

(1) there exist an m-state DFA language and an n-state DFA language such that any DFA accepting the catenation of the two languages needs at least $m2^n - 2^{n-1}$ states; and

(2) the catenation of an m-state DFA language and an n-state DFA language can always be accepted by a DFA using $m2^n - 2^{n-1}$ states or less.

So, it is a tight lower bound and upper bound. In the following, we first summarize the state complexity of various operations on regular languages. Then we give some details for certain operations. For each operation we consider, we give an exact function rather than the order of the function.

Let Σ be an alphabet, L_1 and L_2 be an m-state DFA language and an n-state DFA language over Σ, respectively. A list of operations on L_1 and L_2 and their state complexity are the following:

- $L_1 L_2 : m2^n - 2^{n-1}$;
- $(L_2)^* : 2^{n-1} + 2^{n-2}$;
- $L_1 \cap L_2 : mn$;
- $L_1 \cup L_2 : mn$;
- $L \backslash L_2 : 2^n - 1$, where L is an arbitrary language;
- $L_2 / L : n$, where L is an arbitrary language;
- $L_2^R : 2^n$.

The state complexity of some of the above operations is much lower if we consider only the case when $|\Sigma| = 1$. For unary alphabets, we have

- $(L_2)^* : (n-1)^2 + 1$;
- $L_1 L_2 : mn$ (if $(m, n) = 1$).

5.1.1 Catenation

We first show that for any $m \geq 1$ and $n > 1$ there exist an m-state DFA A and an n-state DFA B such that any DFA accepting $L(A)L(B)$ needs at least $m2^n - 2^{n-1}$ states. Then we show that for any pair of m-state DFA A and n-state DFA B defined on the same alphabet Σ, there exists a DFA with at most $m2^n - 2^{n-1}$ states that accepts $L(A)L(B)$.

Theorem 5.1. *For any integers $m \geq 1$ and $n \geq 2$, there exist a DFA A of m states and a DFA B of n states such that any DFA accepting $L(A)L(B)$ needs at least $m2^n - 2^{n-1}$ states.*

Proof. We first consider the cases when $m = 1$ and $n \geq 2$. Let $\Sigma = \{a, b\}$. Since $m = 1$, A is a one-state DFA accepting Σ^*. Choose $B = (P, \Sigma, \delta_B, p_0, F_B)$ (Figure 14) where $P = \{p_0, \ldots, p_{n-1}\}$, $F_B = \{p_{n-1}\}$, and $\delta_B(p_0, a) = p_0$, $\delta_B(p_0, b) = p_1$, $\delta_B(p_i, a) = p_{i+1}$, $1 \leq i \leq n - 2$, $\delta_B(p_{n-1}, a) = p_1$, $\delta_B(p_i, b) = p_i$, $1 \leq i \leq n - 1$.

It is easy to see that

$$L(A)L(B) = \{w \in \Sigma^* \mid w = ubv, |v|_a \equiv n - 2 \bmod (n-1)\}.$$

Let $(i_1, \ldots, i_{n-1}) \in \{0, 1\}^{n-1}$ and denote

$$w(i_1, \ldots, i_{n-1}) = b^{i_1} a b^{i_2} \ldots a b^{i_{n-1}}.$$

Then, for every $j \in \{0, \ldots, n-2\}$, $w(i_1, \ldots, i_{n-1})a^j \in L(A)L(B)$ iff $i_{j+1} = 1$. Thus a DFA accepting $L(A)L(B)$ needs at least 2^{n-1} states.

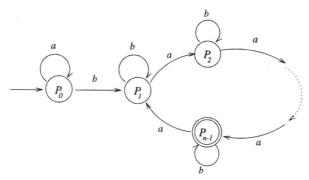

Fig. 14. DFA B

Now we consider the cases when $m \geq 2$ and $n \geq 2$.

Let $\Sigma = \{a, b, c\}$. Define $A = (Q, \Sigma, \delta_A, q_0, F_A)$ where $Q = \{q_0, \ldots, q_{m-1}\}$; $F_A = \{q_{m-1}\}$; for each i, $0 \leq i \leq m - 1$,

$$\delta_A(q_i, X) = \begin{cases} q_j, \ j = (i+1) \bmod m, & \text{if } X = a, \\ q_0, & \text{if } X = b, \\ q_i, & \text{if } X = c. \end{cases}$$

Define $B = (P, \Sigma, \delta_B, p_0, F_B)$ where $P = \{p_0, \ldots, p_{n-1}\}$; $F_B = \{p_{n-1}\}$; and for each i, $0 \leq i \leq n - 1$,

$$\delta_B(p_i, X) = \begin{cases} p_j, \ j = (i+1) \bmod n, & \text{if } X = b, \\ p_i, & \text{if } X = a, \\ p_1, & \text{if } X = c. \end{cases}$$

The DFA A and B are shown in Figure 15 and Figure 16, respectively. The reader can verify that

$$L(A) = \{xy \mid x \in (\Sigma^*\{b\})^*, \ y \in \{a, c\}^* \ \& \ |y|_a = m - 1 \bmod m\},$$

and

$$L(B) \cap \{a, b\}^* = \{x \in \{a, b\}^* \mid |x|_b = n - 1 \bmod n\}.$$

Now we consider the catenation of $L(A)$ and $L(B)$, i.e., $L(A)L(B)$.

Fact 5.1. For $m > 1$, $L(A) \cap \Sigma^*\{b\} = \emptyset$. □

For each $x \in \{a, b\}^*$, we define

$$S(x) = \{i \mid x = uv \text{ such that } u \in L(A), \text{ and } i = |v|_b \bmod n\}.$$

Consider $x, y \in \{a, b\}^*$ such that $S(x) \neq S(y)$. Let $k \in S(x) - S(y)$ (or $S(y) - S(x)$). Then it is clear that $xb^{n-1-k} \in L(A)L(B)$ but $yb^{n-1-k} \notin L(A)L(B)$. So, x and y are in different equivalence classes of $\equiv_{L(A)L(B)}$ where $\equiv_L$ is defined in Section 4.3.

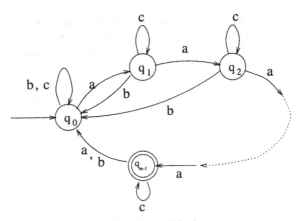

Fig. 15. DFA A

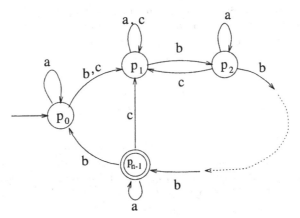

Fig. 16. DFA B

For each $x \in \{a,b\}^*$, define $T(x) = max\{|z| \mid x = yz \ \& \ z \in a^*\}$. Consider $u, v \in \{a,b\}^*$ such that $S(u) = S(v)$ and $T(u) > T(v) \bmod m$. Let $i = T(u) \bmod m$ and $w = ca^{m-1-i}b^{n-1}$. Then clearly $uw \in L(A)L(B)$ but $vw \notin L(A)L(B)$. Notice that there does not exist a word $w \in \Sigma^*$ such that $0 \notin S(w)$ and $T(w) = m-1$, since the fact that $T(w) = m-1$ guarantees that $0 \in S(w)$.

For each subset $s = \{i_1, \ldots, i_t\}$ of $\{0, \ldots, n-1\}$, where $i_1 > \ldots > i_t$, and each integer $j \in \{0, \ldots, m-1\}$ except the case when both $0 \notin s$ and $j = m-1$ are true, there exists a word

$$x = a^{m-1}b^{i_1-i_2}a^{m-1}b^{i_2-i_3}a^{m-1}\ldots a^{m-1}b^{i_t+n}a^j$$

such that $S(x) = s$ and $T(x) = j$. Thus, the relation $\equiv_{L(A)L(B)}$ has at least $m2^n - 2^{n-1}$ distinct equivalence classes.   □

The next theorem gives an upper bound which coincides exactly with the above lower bound result. Therefore, the bound is tight.

Theorem 5.2. *Let A and B be two complete DFA defined on the same alphabet, where A has m states and B has n states, and let A have k final states, $0 < k < m$. Then there exists a $(m2^n - k2^{n-1})$-state DFA which accepts $L(A)L(B)$.*

Proof. Let $A = (Q, \Sigma, \delta_A, q_0, F_A)$ and $B = (P, \Sigma, \delta_B, p_0, F_B)$. Construct $C = (R, \Sigma, \delta_C, r_0, F_C)$ such that

$R = Q \times 2^P - F_A \times 2^{P-\{p_0\}}$ where 2^X denotes the power set of X;
$r_0 = < q_0, \emptyset >$ if $q_0 \notin F_A$, $r_0 = < q_0, \{p_0\} >$ otherwise;
$F_C = \{< q, T > \in R \mid T \cap F_B \neq \emptyset\}$;
$\delta_C(< q, T >, a) = < q', T' >$, for $a \in \Sigma$, where $q' = \delta_A(q, a)$ and $T' = \delta_B(T, a) \cup \{p_0\}$ if $q' \in F_A$, $T' = \delta_B(T, a)$ otherwise.

Intuitively, R is a set of pairs such that the first component of each pair is a state in Q and the second component is a subset of P. R does not contain those pairs whose first component is a final state of A and whose second component does not contain the initial state of B. Clearly, C has $m2^n - k2^{n-1}$ states. The reader can easily verify that $L(C) = L(A)L(B)$. $\quad\square$

We still need to consider the cases when $m \geq 1$ and $n = 1$. We have the following result.

Theorem 5.3. *The number of states that is sufficient and necessary in the worst case for a DFA to accept the catenation of an m-state DFA language and a 1-state DFA language is m.*

Proof. Let Σ be an alphabet and $a \in \Sigma$. Clearly, for any integer $m > 0$, the language $L = \{w \in \Sigma^* \mid |w|_a \equiv m - 1 \bmod m\}$ is accepted by an m-state DFA. Note that Σ^* is accepted by a one-state DFA. It is easy to see that any DFA accepting $L\Sigma^* = \{w \in \Sigma^* \mid \#_a(w) \geq m - 1\}$ needs at least m states. So, we have proved the necessary condition.

Let A and B be an m-state DFA and a 1-state DFA, respectively. Since B is a complete DFA, $L(B)$ is either $\emptyset$ or Σ^*. We need to consider only the case $L(B) = \Sigma^*$. Let $A = (Q, \Sigma, \delta_A, q_0, F_A)$. Define $C = (Q, \Sigma, \delta_C, q_0, F_A)$ where, for any $X \in \Sigma$ and $q \in Q$,

$$\delta_C(q, X) = \begin{cases} \delta_A(q, X), & \text{if } q \notin F_A, \\ q, & \text{if } q \in F_A. \end{cases}$$

The automaton C is exactly as A except that final states are made to be sink-states: when the computation has reached some final state q, it remains there. Now it is clear that $L(C) = L(A)\Sigma^*$. $\quad\square$

5.1.2 Star operation (Kleene closure)

Here we prove that the state complexity of the star operation of an n-state DFA language is $2^{n-1} + 2^{n-2}$.

Theorem 5.4. *For any n-state DFA $A = (Q, \Sigma, \delta, q_0, F)$ such that $|F - \{q_0\}| = k \geq 1$ and $n > 1$, there exists a DFA of at most $2^{n-1} + 2^{n-k-1}$ states that accepts $(L(A))^*$.*

Proof. Let $A = (Q, \Sigma, \delta, q_0, F)$ and $L = L(A)$. Denote $F - \{q_0\}$ by F_0. Then $|F_0| = k \geq 1$. We construct a DFA $A' = (Q', \Sigma, \delta', q'_0, F')$ where

$q'_0 \notin Q$ is a new start state;
$Q' = \{q'_0\} \cup \{P \mid P \subseteq (Q - F_0) \& P \neq \emptyset\} \cup \{R \mid R \subseteq Q \& q_0 \in R \& R \cap F_0 \neq \emptyset\}$;
$\delta'(q'_0, a) = \{\delta(q_0, a)\}$, for any $a \in \Sigma$, and
$$\delta'(R, a) = \begin{cases} \delta(R, a) & \text{if } \delta(R, a) \cap F_0 = \emptyset, \\ \delta(R, a) \cup \{q_0\} & \text{otherwise,} \end{cases}$$
for $R \subseteq Q$ and $a \in \Sigma$;
$F' = \{q'_0\} \cup \{R \mid R \subseteq Q \& R \cap F \neq \emptyset\}$.

The reader can verify that $L(A') = L^*$. Now we consider the number of states in Q'. Notice that in the second term of the union for Q', there are $2^{n-k} - 1$ states. In the third term, there are $(2^k - 1)2^{n-k-1}$ states. So, $|Q'| = 2^{n-1} + 2^{n-k-1}$.   □

Note that if q_0 is the only final state of A, i.e., $k = 0$, then $(L(A))^* = L(A)$. So, the worst-case state complexity of the star operation occurs when $k = 1$.

Corollary 5.1. *For any n-state DFA A, $n > 1$, there exists a DFA A' of at most $2^{n-1} + 2^{n-2}$ states such that $L(A') = (L(A))^*$.*   □

Theorem 5.5. *For any integer $n \geq 2$, there exists a DFA A of n states such that any DFA accepting $(L(A))^*$ needs at least $2^{n-1} + 2^{n-2}$ states.*

Proof. For $n = 2$, it is clear that $L = \{w \in \{a, b\}^* \mid |w|_a \text{ is odd}\}$ is accepted by a two-state DFA, and $L^* = \{\varepsilon\} \cup \{w \in \{a, b\}^* \mid |w|_a \geq 1\}$ cannot be accepted by a DFA with less than 3 states.

For $n > 2$, we give the following construction: $A_n = (Q_n, \Sigma, \delta_n, 0, \{n-1\})$ where $Q_n = \{0, \ldots, n-1\}$; $\Sigma = \{a, b\}$; $\delta(i, a) = (i+1) \bmod n$ for each $0 \leq i < n$, $\delta(i, b) = (i+1) \bmod n$ for each $1 \leq i < n$ and $\delta(0, b) = 0$. A_n is shown in Figure 17.

We construct the DFA $A'_n = (Q'_n, \Sigma, \delta'_n, q'_0, F'_n)$ from A_n exactly as described in the proof of the previous theorem. We need to show that (I) every state is reachable from the start state and (II) each state defines a distinct equivalence class of $\equiv_{L(A_n)^*}$.

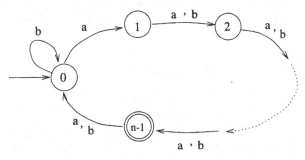

Fig. 17. An n-state DFA A_n: The language $(L(A_n))^*$ requires $2^{n-1} + 2^{n-2}$ states

We prove (I) by induction on the size of the state set. (Note that each state is a subset of Q_n except q'_0.)

Consider all q such that $q \in Q'$ and $|q| = 1$. We have $\{0\} = \delta'_n(q'_0, b)$ and $\{i\} = \delta'_n(i - 1, a)$ for each $0 < i < n - 1$.

Assume that all q such that $|q| < k$ are reachable. Consider q where $|q| = k$. Let $q = \{i_1, i_2, \ldots, i_k\}$ such that $0 \le i_1 < i_2 \ldots < i_k < n - 1$ if $n - 1 \notin q$, $i_1 = n - 1$ and $0 = i_2 < \ldots i_k < n - 1$ otherwise. There are four cases:

(i) $i_1 = n - 1$ and $i_2 = 0$. Then $q = \delta'_n(\{n - 2, i_3 - 1, \ldots, i_k - 1\}, a)$ where the latter state contains $k - 1$ states.

(ii) $i_1 = 0$ and $i_2 = 1$. Then $q = \delta'_n(q', a)$ where $q' = \{n - 1, 0, i_3 - 1, \ldots i_k - 1\}$ which is considered in case (i).

(iii) $i_1 = 0$ and $i_2 = 1 + t$ for $t > 0$. Then $q = \delta'_n(q', b^t)$ where $q' = \{0, 1, i_3 - t, \ldots, i_k - t\}$. The latter state is considered in case (ii).

(iv) $i_1 = t > 0$. Then $q = \delta'_n(q', a^t)$ where $q' = \{0, i_2 - t, \ldots, i_k - t\}$ is considered in either case (ii) or case (iii).

To prove (II), let $i \in p - q$ for some $p, q \in Q'_n$ and $p \ne q$. Then $\delta'_n(p, a^{n-1-i}) \in F'_n$ but $\delta'_n(q, a^{n-1-i}) \notin F'_n$.   □

Note that a DFA accepting the star of a 1-state DFA language may need up to two states. For example, $\emptyset$ is accepted by a 1-state DFA and any complete DFA accepting $\emptyset^* = \{\varepsilon\}$ has at least two states.

It is clear that any DFA accepting the reversal of an n-state DFA language does not need more than 2^n states. But can this upper bound be reached? A result on alternating finite automata ([23], Theorem 5.3) gives a positive answer to the above question if n is of the form 2^k for some integer $k \ge 0$. Leiss has solved this problem in [73] for all $n > 0$. A modification of Leiss's solution is shown in Figure 18. If we reverse all the transitions of this automaton, we will get a good example for showing that, in the worst case, a DFA equivalent to an n-state NFA may need exactly 2^n states.

5.1.3 An open problem

For the state complexity of catenation, we have proved the general result $(m2^n - 2^{n-1})$ using automata with a three-letter input alphabet. We have also given the complexity for the one-letter alphabet case. We do not know whether the result obtained for the three-letter alphabet still holds if the size of the alphabet is two.

5.2 Time and space complexity issues

Almost all problems of interest are decidable for regular languages, i.e., there exist algorithms to solve them. However, for the purpose of implementation, it is necessary to know how hard these problems are and what the time and space complexities of the algorithms are. In the following, we list some basic problems, mostly decision problems, for regular languages together with their

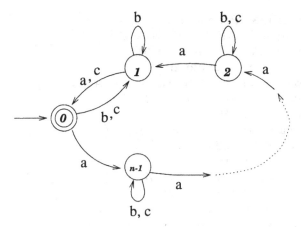

Fig. 18. An n-state DFA such that $L(A)^R$ requires 2^n states

complexity. We give a brief explanation and references for each problem. The reader may refer to [46, 57, 4] for terminology in complexity theory.

One may observe that many of the following problems are NP-complete or even PSPACE-complete, which are categorized as computationally intractable in complexity theory. However, finite automata and regular expressions used in many applications are fairly small in size. In such cases, even exponential algorithms can be practically feasible.

(1) **DFA Membership Problem:**
Given an arbitrary DFA A with the input alphabet Σ and an arbitrary word $x \in \Sigma^*$, is $x \in L(A)$?
Complexity: *DLOGSPACE-complete* [64].

(2) **NFA Membership Problem:**
Given an arbitrary NFA A with the input alphabet Σ and an arbitrary word $x \in \Sigma^*$, is $x \in L(A)$?
Complexity: *NLOGSPACE-complete* [66].

(3) **AFA Membership Problem:**
Given an arbitrary AFA A with the input alphabet Σ and a word $x \in \Sigma^*$, is $x \in L(A)$?
Complexity: *P-complete* [64].

(4) **Regular Expression Membership Problem:**
Given a regular expression e over Σ and a word $x \in \Sigma^*$, is $x \in L(e)$?
Complexity: *NLOGSPACE-complete* [64].

(5) **DFA Emptiness Problem:**
Given an arbitrary DFA A, is $L(A) = \emptyset$?
Complexity: *NLOGSPACE-complete* [66].

(6) **NFA Emptiness Problem:**
Given an arbitrary NFA A, is $L(A) = \emptyset$?
Complexity: *NLOGSPACE-complete* [66].

(7) **AFA Emptiness Problem:**
Given an arbitrary AFA A, is $L(A) = \emptyset$?
Complexity: *PSPACE-complete* [64].

(8) **DFA Equivalence Problem:**
Given two arbitrary DFA A_1 and A_2, is $L(A_1) = L(A_2)$?
Complexity: *NLOGSPACE-complete* [26].

(9) **NFA Equivalence Problem:**
Given two arbitrary NFA A_1 and A_2, is $L(A_1) = L(A_2)$?
Complexity: *PSPACE-complete* [46].

(10) **AFA Equivalence Problem:**
Given two arbitrary AFA A_1 and A_2, is $L(A_1) = L(A_2)$?
Complexity: *PSPACE-complete* [64].

(11) **Regular Expression Equivalence Problem:**
Given two regular expressions e_1 and e_2, is $L(e_1) = L(e_2)$?
Complexity: *PSPACE-complete* [59]. (Note that if one of the regular expressions denotes a language of polynomial density, then the complexity is *NP-complete*.)

The following problems can also be converted into decision problems. However, we prefer to keep them in their natural form:

(i) **DFA Minimization Problem:**
Given a DFA with n states, convert it to an equivalent minimum-state DFA.
Complexity: $O(n \log n)$ [56].

(ii) **NFA Minimization Problem:**
Given an NFA, convert it to an equivalent minimum-state NFA.
Complexity: *PSPACE-complete* [59, 119].

(iii) **DFA to Minimal NFA Problem:**
Given a DFA, convert it to an equivalent minimum-state NFA.
Complexity: *PSPACE-complete* [65].

The following problems remain open:

(a) Is membership for regular expressions over a one-letter alphabet *NLOGSPACE-hard*?

(b) Is membership for extended regular expressions *P-hard*?

Acknowledgements I wish to express my deep gratitute to Kai Salomaa for his significant contributions to this chapter. He has read the chapter many times and made numerous suggestions. I consider him truly the second author of the chapter. However, he has been insisting not to put his name as a co-author. I wish to thank J. Brzozowski for his great help in tracking the status of the six open problems he raised in 1979. Special thanks due K. Culik II, A. Salomaa, J. Shallit, A. Szilard, and R. Webber for their careful reading and valuable suggestions. I wish to express my gratitude to G. Rozenberg and A.

Salomaa for their great idea and efforts in organizing the handbook. Finally, I wish to thank the Natural Sciences and Engineering Research Council of Canada for their support.

References

1. A. V. Aho and J. D. Ullman, *The Theory of Parsing, Translation, and Compiling*, Vol. 1, Prentice-Hall, Englewood Cliffs, N.J., (1972).
2. A. V. Aho, R. Sethi, and J. D. Ullman, *Compilers – Principles, Techniques, and Tools*, Addison-Wesley, Reading, (1986).
3. J. C. M. Baeten and W. P. Weijland, *Process Algebra*, Cambridge University Press, Cambridge, (1990).
4. J. L. Balcázar, J. Díaz, and J. Gabarró, *Structured Complexity* I, II, EATCS Monagraphs on Theoretical Computer Science, vol. 11 and 22, Springer-Verlag, Berlin 1988 and 1990.
5. Y. Bar-Hillel, M. Perles, and E. Shamir, "On formal properties of simple phrase structure grammars", *Z. Phonetik. Sprachwiss. Kommunikationsforsch.* 14 (1961) 143–172.
6. J.-C. Birget, "State-Complexity of Finite-State Devices, State Compressibility and Incompressibility", *Mathematical Systems Theory* 26 (1993) 237–269.
7. G. Berry and R. Sethi, "From Regular Expressions to Deterministic Automata", *Theoretical Computer Science* 48 (1986) 117–126.
8. J. Berstel, *Transductions and Context-Free Languages*, Teubner, Stuttgart, (1979).
9. J. Berstel and M. Morcrette, "Compact representation of patterns by finite automata", *Pixim 89: L'Image Numérique à Paris*, André Gagalowicz, ed., Hermes, Paris, (1989), pp.387–395.
10. J. Berstel and C. Reutenauer, *Rational Series and Their Languages*, EATCS Monographs on Theoretical Computer Science, Springer-Verlag, Berlin (1988).
11. W. Brauer, *Automatentheorie*, Teubner, Stuttgart, (1984).
12. W. Brauer, "On Minimizing Finite Automata", *EATCS Bulletin* 35 (1988) 113–116.
13. A. Brüggemann-Klein, "Regular expressions into finite automata", *Theoretical Computer Science* 120 (1993) 197–213.
14. A. Brüggemann-Klein and D. Wood, "Deterministic Regular Languages", Proceedings of STACS'92, *Lecture Notes in Computer Science* 577, A. Finkel and M. Jantzen (eds.), Springer-Verlag, Berlin (1992) 173–184.
15. J. A. Brzozowski, "Canonical regular expressions and minimal state graphs for definite events", *Mathematical Theory of Automata*, vol. 12 of MRI Symposia Series, Polytechnic Press, NY, (1962), 529–561.
16. J. A. Brzozowski, "Derivatives of Regular Expressions", *Journal of the ACM* 11:4 (1964) 481–494.
17. J. A. Brzozowski, "Developments in the theory of regular languages", *Information Processing 80*, S. H. Lavington edited, Proceedings of IFIP Congress 80, North-Holland, Amsterdam (1980) 29–40.
18. J. A. Brzozowski, "Open problems about regular languages", *Formal Language Theory – Perspectives and open problems*, R. V. Book (ed.), Academic Press, New York, (1980), pp.23–47.

19. J. A. Brzozowski and E. Leiss, "On Equations for Regular Languages, Finite Automata, and Sequential Networks", *Theoretical Computer Science* 10 (1980) 19–35.

20. J. A. Brzozowski and I. Simon, "Characterization of locally testable events", *Discrete Mathematics* 4 (1973) 243–271.

21. J. A. Brzozowski and M. Yoeli, *Digital Networks*, Prentice-Hall, Englewood Cliffs, N. J., (1976).

22. A. K. Chandra and L. J. Stockmeyer, "Alternation", FOCS 17 (1976) 98–108.

23. A. K. Chandra, D. C. Kozen, L. J. Stockmeyer, "Alternation", *Journal of the ACM* 28 (1981) 114–133.

24. J. H. Chang, O. H. Ibarra and B. Ravikumar, "Some observations concerning alternating Turing machines using small space", *Inform. Process. Lett.* 25 (1987) 1–9.

25. C.-H. Chang and R. Paige, "From Regular Expressions to DFA's Using Compressed NFA's", *Proceedings of the Third Symposium on Combinatorial Pattern Matching* (1992) 90–110.

26. S. Cho and D. Huynh, "The parallel complexity of finite state automata problems", *Technical Report* UTDCS-22–88, University of Texas at Dallas, (1988).

27. D. I. A. Cohen, *Introduction to Computer Theory*, Wiley, New York, (1991).

28. K. Culik II and S. Dube, "Rational and Affine Expressions for Image Description", *Discrete Applied Mathematics* 41 (1993) 85–120.

29. K. Culik II and S. Dube, "Affine Automata and Related Techniques for Generation of Complex Images", *Theoretical Computer Science* 116 (1993) 373–398.

30. K. Culik II, F. E. Fich and A. Salomaa, "A Homomorphic Characterization of Regular Languages", *Discrete Applied Mathematics* 4 (1982)149–152.

31. K. Culik II and T. Harju, "Splicing semigroups of dominoes and DNA", *Discrete Applied Mathematics* 31 (1991) 261–277.

32. K. Culik II and J. Karhumäki, "The equivalence problem for single-valued two-way transducers (on NPDTOL languages) is decidable", *SIAM Journal on Computing*, vol. 16, no. 2 (1987) 221–230.

33. K. Culik II and J. Kari, "Image Compression Using Weighted Finite Automata", *Computer and Graphics*, vol. 17, 3, (1993) 305–313.

34. J. Dassow, G. Păun, A. Salomaa, "On Thinness and Slenderness of L Languages", *EATCS Bulletin* 49 (1993) 152–158.

35. F. Dejean and M. P. Schützenberger, "On a question of Eggan", *Information and Control* 9 (1966) 23–25.

36. A. de Luca and S. Varricchio, "On noncounting regular classes", *Theoretical Computer Science* 100 (1992) 67–104.

37. V. Diekert and G. Rozenberg (ed.), *The Book of Traces*, World Scientific, Singapore, (1995).

38. D. Drusinsky and D. Harel, "On the power of bounded concurrency I: Finite automata", *Journal of the ACM* 41 (1994) 517–539.

39. L. C. Eggan, "Transition graphs and the star height of regular events", *Michigan Math. J.* 10 (1963) 385–397.

40. A. Ehrenfeucht, R. Parikh, and G. Rozenberg, "Pumping Lemmas for Regular Sets", *SIAM Journal on Computing* vol. 10, no. 3 (1981) 536–541.

41. S. Eilenberg, *Automata, Languages, and Machines*, Vol. A, Academic Press, New York, (1974).

42. S. Eilenberg, *Automata, Languages, and Machines*, Vol. B, Academic Press, New York, (1974)

43. C. C. Elgot and J. D. Rutledge, "Operations on finite automata", *Proc. AIEE Second Ann. Symp. on Switching Theory and Logical Design*, Detroit, (1961).

44. A. Fellah, *Alternating Finite Automata and Related Problems*, PhD Dissertation, Dept. of Math. and Computer Sci., Kent State University, (1991).

45. A. Fellah, H. Jürgensen, S. Yu, "Constructions for alternating finite automata", *Intern. J. Computer Math.* 35 (1990) 117–132.

46. M. R. Garey and D. S. Johnson, *Computers and Intractability: A Guide to the Theory of NP-Completeness*, Freeman, San Francisco, (1979.)

47. S. Ginsburg, *Algebraic and automata-theoretic properties of formal languages*, North-Holland, Amsterdam, (1975).

48. V. M. Glushkov, "The abstract theory of automata", *Russian Mathematics Surveys* 16 (1961) 1–53.

49. D. Gries, "Describing an Algorithm by Hopcroft", *Acta Informatica* 2 (1973) 97–109.

50. L. Guo, K. Salomaa, and S. Yu, "Synchronization Expressions and Languages", *Proceedings of the Sixth IEEE Symposium on Parallel and Distributed Processing* (1994) 257–264.

51. M. A. Harrison, *Introduction to Formal Language Theory*, Addison-Wesley, Reading, (1978).

52. K. Hashiguchi, "Algorithms for Determining Relative Star Height and Star Height", *Information and Computation* 78 (1988) 124–169.

53. T. Head, "Formal language theory and DNA: An analysis of the generative capacity of specific recombinant behaviors", *Bull. Math. Biol.* 49 (1987) 737–759.

54. F. C. Hennie, *Finite-State Models for Logical Machines*, Wiley, New York, (1968).

55. T. Hirst and D. Harel, "On the power of bounded concurrency II: Pushdown automata", *Journal of the ACM* 41 (1994), 540–554.

56. J. E. Hopcroft, "An $n \log n$ algorithm for minimizing states in a finite automaton", in *Theory of Machines and Computations*, Z. Kohavi (ed.), Academic Press, New York, (1971).

57. J. E. Hopcroft and J. D. Ullman, *Introduction to Automata Theory, Languages, and Computation*, Addison-Wesley, Reading, (1979), 189–196.

58. J. M. Howie, *Automata and Languages*, Oxford University Press, Oxford, (1991).

59. H. B. Hunt, D. J. Rosenkrantz, and T. G. Szymanski, "On the Equivalence, Containment, and Covering Problems for the Regular and Context-Free Languages", *Journal of Computer and System Sciences* 12 (1976) 222–268.

60. O. Ibarra, "The unsolvability of the equivalence problem for epsilon-free NGSM's with unary input (output) alphabet and applications", *SIAM Journal on Computing* 4 (1978) 524–532.

61. K. Inoue, I. Takanami, and H. Tanaguchi, "Two-Dimensional alternating Turing machines", *Proc. 14th Ann. ACM Symp. On Theory of Computing* (1982) 37–46.

62. K. Inoue, I. Takanami, and H. Tanaguchi, " A note on alternating on-line Turing machines", *Information Processing Letters* 15:4 (1982) 164–168.

63. J. Jaffe, "A necessary and sufficient pumping lemma for regular languages", *SIGACT News* (1978) 48–49.

64. T. Jiang and B. Ravikumar, "A note on the space complexity of some decision problems for finite automata", *Information Processing Letters* 40 (1991) 25–31.

65. T. Jiang and B. Ravikumar, "Minimal NFA Problems are Hard", *SIAM Journal on Computing* 22 (1993), 1117–1141. *Proceedings of 18th ICALP*, Lecture Notes in Computer Science 510, Springer-Verlag, Berlin (1991) 629–640.

66. N. Jones, "Space-bounded reducibility among combinatorial problems", *Journal of Computer and System Sciences* 11 (1975) 68–85.

67. T. Kameda and P. Weiner, "On the state minimization of nondeterministic finite automata", *IEEE Trans. on Computers* C-19 (1970) 617–627.
68. D. Kelley, *Automata and Formal Languages – An Introduction*, Prentice-Hall, Englewood Cliffs, N. J., (1995).
69. S. C. Kleene, "Representation of events in nerve nets and finite automata", *Automata Studies*, Princeton Univ. Press, Princeton, N. J., (1996), pp.2–42.
70. D. E. Knuth, J. H. Morris, and V. R. Pratt, "Fast pattern matching in strings", *SIAM Journal on Computing*, vol.6, no.2 (1977) 323–350.
71. D. Kozen, "On parallelism in Turing machines", *Proceedings of 17th FOCS* (1976) 89–97.
72. R. E. Ladner, R. J. Lipton and L. J. Stockmeyer, "Alternating pushdown automata", *Proc. 19th IEEE Symp. on Foundations of Computer Science*, Ann Arbor, MI, (1978) 92–106.
73. E. Leiss, "Succinct representation of regular languages by boolean automata", *Theoretical Computer Science* 13 (1981) 323–330.
74. E. Leiss, "On generalized language equations", *Theoretical Computer Science* 14 (1981) 63–77.
75. E. Leiss, "Succinct representation of regular languages by boolean automata II", *Theoretical Computer Science* 38 (1985) 133–136.
76. E. Leiss, "Language equations over a one-letter alphabet with union, concatenation and star: a complete solution", *Theoretical Computer Science* 131 (1994) 311–330.
77. E. Leiss, "Unrestricted complementation in language equations over a one-letter alphabet", *Theoretical Computer Science* 132 (1994) 71–84.
78. H. R. Lewis and C. H. Papadimitriou, *Elements of the Theory of Computation*, Prentice-Hall, Englewood Cliffs, N. J., (1981).
79. P. A. Lindsay, "Alternation and w-type Turing acceptors", *Theoretical Computer Science* 43 (1986) 107–115.
80. P. Linz, *An Introduction to Formal Languages and Automata*, D. C. Heath and Company, Lexington, (1990).
81. O. B. Lupanow, "Über den Vergleich zweier Typen endlicher Quellen", *Probleme der Kybernetik* 6 (1966) 328–335, and *Problemy Kibernetiki* 6 (1962) (Russian original).
82. A. Mateescu, "Scattered deletion and commutativity", *Theoretical Computer Science* 125 (1994) 361–371.
83. W. S. McCulloch and W. Pitts, "A logical calculus of the ideas immanent in nervous activity", *Bull. Math. Biophysics* 5 (1943) 115–133.
84. R. McNaughton, *Counter-Free Automata*, MIT Press, Cambridge, (1971).
85. R. McNaughton and H. Yamada, "Regular Expressions and State Graphs for Automata", *Trans. IRS* EC-9 (1960) 39–47. Also in *Sequential Machines – Selected Papers*, E. F. Moore ed., Addison-Wesley, Reading, (1964), 157–174.
86. G. H. Mealy, "A method for synthesizing sequential circuits", *Bell System Technical J.* 34: 5 (1955), 1045–1079.
87. A. R. Meyer and M. J. Fischer. "Economy of description by automata, grammars, and formal systems", *FOCS* 12 (1971) 188–191.
88. E. F. Moore, "Gedanken experiments on sequential machines", *Automata Studies*, Princeton Univ. Press, Princeton, N.J., (1966), pp.129–153.
89. F. R. Moore, "On the Bounds for State-Set Size in the Proofs of Equivalence Between Deterministic, Nondeterministic, and Two-Way Finite Automata", *IEEE Trans. Computers* 20 (1971), 1211–1214.
90. J. Myhill, "Finite automata and the representation of events", WADD TR-57-624, Wright Patterson AFB, Ohio, (1957), 112–137.

91. A. Nerode, "Linear automata transformation", *Proceedings of AMS* 9 (1958) 541–544.
92. O. Nierstrasz, "Regular Types for Active Objects", *OOPSLA '93*, 1–15.
93. M. Nivat, "Transductions des langages de Chomsky", *Ann. Inst. Fourier, Grenoble* 18 (1968) 339–456.
94. W. J. Paul, E. J. Prauss and R. Reischuck, "On Alternation", *Acta Inform.* 14 (1980) 243–255.
95. G. Păun and A. Salomaa, "Decision problems concerning the thinness of DOL languages", *EATCS Bulletin* 46 (1992) 171–181.
96. G. Păun and A. Salomaa, "Closure properties of slender languages", *Theoretical Computer Science* 120 (1993) 293–301.
97. G. Păun and A. Salomaa, "Thin and slender languages", *Discrete Applied Mathematics* 61 (1995) 257–270.
98. D. Perrin, (Chapter 1) Finite Automata, *Handbook of Theoretical Computer Science*, Vol. B, J. van Leeuwen (ed.), MIT Press, (1990).
99. M. O. Rabin and D. Scott, "Finite automata and their decision problems", *IBM J. Res.* 3: 2 (1959) 115–125.
100. G. N. Raney, "Sequential functions", *Journal of the ACM* 5 (1958) 177.
101. B. Ravikumar, "Some applications of a technique of Sakoda and Sipser", *SIGACT News*, 21:4 (1990) 73–77.
102. B. Ravikumar and O. H. Ibarra, "Relating the type of ambiguity of finite automata to the succinctness of their representation", *SIAM Journal on Computing* vol. 18, no. 6 (1989), 1263–1282.
103. W. L. Ruzzo, "Tree-size bounded alternation", *Journal of Computer and System Sciences* 21 (1980) 218–235.
104. A. Salomaa, *On the Reducibility of Events Represented in Automata*, Annales Academiae Scientiarum Fennicae, Series A, I. Mathematica 353, (1964).
105. A. Salomaa, *Theorems on the Representation of events in Moore-Automata*, Turun Yliopiston Julkaisuja Annales Universitatis Turkuensis, Series A, 69, (1964).
106. A. Salomaa, *Theory of Automata*, Pergamon Press, Oxford, (1969).
107. A. Salomaa, *Jewels of Formal Language Theory*, Computer Science Press, Rockville, Maryland, (1981).
108. A. Salomaa, *Computation and Automata*, Cambridge University Press, Cambridge, (1985).
109. A. Salomaa and M. Soittola, *Automata-Theoretic Aspects of Formal Power Series*, Springer-Verlag, New York, (1978).
110. K. Salomaa and S. Yu, "Loop-Free Alternating Finite Automata", *Technical Report* 482, Department of Computer Science, Univ. of Western Ontatio, (1996).
111. K. Salomaa, S. Yu, Q. Zhuang, "The state complexities of some basic operations on regular languages", *Theoretical Computer Science* 125 (1994) 315–328.
112. M. P. Schützenberger, "Finite Counting Automata", *Information and Control* 5 (1962) 91–107.
113. M. P. Schützenberger, "On finite monoids having only trivial subgroups", *Information and Control* 8 (1965) 190–194.
114. M. P. Schützenberger, "Sur les relations rationelles", in *Proc. 2nd GI Conference, Automata Theory and Formal languages*, H. Braklage (ed.), *Lecture Notes in Computer Science* 33, Springer-Verlag, Berlin (1975) 209–213.
115. J. Shallit, "Numeration systems, linear recurrences, and regular sets", *Information and Computation* 113 (1994) 331–347.
116. J. Shallit and J. Stolfi, "Two methods for generating fractals", *Computers & Graphics* 13 (1989) 185–191.

117. P. W. Shor, "A Counterexample to the triangle conjecture", *J. Combinatorial Theory*, Series A (1985) 110–112.

118. J. L. A. van de Snepscheut, *What Computing Is All About*, Springer-Verlag, New York, (1993).

119. L. Stockmeyer and A. Meyer, "Word problems requiring exponential time (preliminary report)", *Proceedings of the 5th ACM Symposium on Theory of Computing*, (1973) 1–9.

120. A. Szilard, S. Yu, K. Zhang, and J. Shallit, "Characterizing Regular Languages with Polynomial Densities", *Proceedings of the 17th International Symposium on Mathematical Foundations of Computer Science, Lecture Notes in Computer Science* 629 Springer-Verlag, Berlin (1992) 494–503.

121. K. Thompson, "Regular Expression Search Algorithm", *Communications of the ACM* 11:6 (1968) 410–422.

122. B. W. Watson, *Taxonomies and Toolkits of Regular Language Algorithms*, PhD Dissertation, Department of Mathematics and Computing Science, Eindhoven University of Technology, (1995).

123. D. Wood, *Theory of Computation*, Wiley, New York, (1987).

124. S. Yu and Q. Zhuang, "On the State Complexity of Intersection of Regular Languages", *ACM SIGACT News*, vol. 22, no. 3, (1991) 52–54.

125. Y. Zalcstein, "Locally testable languages", *Journal of Computer and System Sciences* 6 (1972) 151–167.

Context-Free Languages and Pushdown Automata

Jean-Michel Autebert, Jean Berstel, and Luc Boasson

1. Introduction

This chapter is devoted to context-free languages. Context-free languages and grammars were designed initially to formalize grammatical properties of natural languages [9]. They subsequently appeared to be well adapted to the formal description of the syntax of programming languages. This led to a considerable development of the theory.

The presentation focuses on two basic tools: context-free grammars and pushdown automata. These are indeed the standard tools to generate and to recognize context-free languages. A contrario, this means also that we do not consider complexity results at all, neither of recognition by various classes of sequential or parallel Turing machines nor of "succinctness" (see e.g. [52]), that is a measure of the size of the description of a language.

We have chosen to present material which is not available in textbooks [17, 29, 1, 47, 28, 4, 30, 32, 2] (more precisely not available in more than one textbook) because it is on the borderline between classical stuff and advanced topics. However, we feel that a succinct exposition of these results may give some insight in the theory of context-free languages for advanced beginners, and also provide some examples or counter-examples for researchers.

This section ends with notation and examples. In Section 2, we present the relationship between grammars and systems of equations. As an example of the interest of this formalism, we give a short proof of Parikh's theorem.

In the next section, three normal forms of context-free grammars are established. The one with most applications is Greibach's normal form: several variants are given and, in Section 4, we present four such applications. The first three are closely related to each other.

Section 5 is devoted to pushdown automata. We consider carefully the consequences of various restrictions of the general model. The section ends with two results: one concerning the pushdown store language of a pda, the other the output language of a pushdown down transducer.

In the last section, we consider eight important subfamilies of context-free languages. We study in detail linear and quasi-rational languages, and present more briefly the other families.

In the bibliography, we have generally tried to retrieve the references to the original papers, in order to give some flavour of the chronological development of the theory.

1.1 Grammars

As general notation, we use ε to denote the empty word, and $|w|$ for the length of the word w.

A *context-free grammar* $G = (V, P)$ over an alphabet A is composed of a finite alphabet V of *variables* or *nonterminals* disjoint from A, and a finite set $P \subset V \times (V \cup A)^*$ of *productions* or *derivation rules*. Letters in A are called *terminal letters*.

Given words $u, v \in (V \cup A)^*$, we write $u \longrightarrow v$ (sometimes subscripted by G or by P) whenever there exist factorizations $u = xXy$, $v = x\alpha y$, with (X, α) a production. A *derivation* of length $k \geq 0$ from u to v is a sequence $(u_0, u_1, \ldots, u_k)$ of words in $(V \cup A)^*$ such that $u_{i-1} \longrightarrow u_i$ for $i = 1, \ldots, k$, and $u = u_0$, $v = u_k$. If this holds, we write $u \xrightarrow{k} v$. The existence of some derivation from u to v is denoted by $u \xrightarrow{*} v$. If there is a proper derivation (i.e. of length ≥ 1), we use the notation $u \xrightarrow{+} v$. The *language generated* by a variable X in grammar G is the set

$$L_G(X) = \{w \in A^* \mid X \xrightarrow{*} w\}$$

Frequently, grammars are presented with a distinguished nonterminal called the *axiom* and usually denoted by S. The language generated by this variable S in a grammar is then called the language *generated by the grammar*, for short, and is denoted $L(G)$. Any word in $(V \cup A)^*$ that derives from S is a *sentential form*.

A language L is called *context-free* if it is the language generated by some variable in a context-free grammar. Two grammars G and G' are *equivalent* if they generate the same language, i. e. if the distinguished variables S and S' are such that $L_G(S) = L'_{G'}(S')$.

More generally, if $x \in (V \cup A)^*$, we set

$$L_G(x) = \{w \in A^* \mid x \xrightarrow{*} w\} .$$

Context-freeness easily implies that

$$L_G(xy) = L_G(x)L_G(y) .$$

Consider a derivation $u = u_0 \longrightarrow u_1 \longrightarrow \cdots \longrightarrow u_k = v$, with $u, v \in (V \cup A)^*$. Then there exist productions $p_i = X_i \to \alpha_i$ and words x_i, y_i such that

$$u_i = x_i X_i y_i, \quad u_{i+1} = x_i \alpha_i y_i \qquad (i = 0, \ldots, k - 1)$$

The derivation is *leftmost* if $|x_i| \leq |x_{i+1}|$ for $i = 0, \ldots, k - 2$, and *rightmost* if, symmetrically, $|y_i| \leq |y_{i+1}|$ for $i = 0, \ldots, k - 2$. A leftmost (rightmost) derivation is denoted by

$$u \xrightarrow[\ell]{*} v, \qquad u \xrightarrow[r]{*} v$$

It is an interesting fact that any word in a context-free language $L_G(X)$ has the same number of leftmost and of rightmost derivations. A grammar $G = (V, P)$ is *unambiguous* for a variable X if every word in $L_G(X)$ has exactly one leftmost (rightmost) derivation. A language is *unambiguous* if there is an unambiguous grammar to generate it, otherwise it is called *inherently ambiguous*.

A grammar $G = (V, P)$ over A is *trim* in the variable S if the following two conditions are fulfilled :

(i) for every nonterminal X, the language $L_G(X)$ is nonempty;

(ii) for every $X \in V$, there exist $u, v \in A^*$ such that $S \xrightarrow{*} uXv$.

The second condition means that every variable is "accessible", and the first that any variable is "co-accessible". It is not difficult to see that a grammar can always be trimmed effectively. A variation of condition (i) which is sometimes useful is to require that $L_G(X)$ is infinite for every variable X (provided the language generated by the grammar is itself infinite).

A production is *terminal* if its right side contains no variable. A production is called an ε-*rule* if its right side is the empty word. At least one ε-production is necessary if the language generated by the grammar contains the empty word. It is not too difficult to construct, for every context-free grammar G, an equivalent grammar with no ε-production excepted a production $S \longrightarrow \varepsilon$ if $\varepsilon \in L(G)$. The final special kind of grammars we want to mention is the class of proper grammars. A grammar G is *proper* if it has neither ε-productions nor any production of the form $X \longrightarrow Y$, with Y a variable. Again, an equivalent proper grammar can effectively be constructed for any grammar G if $L(G) \not\ni \varepsilon$. These constructions are presented in most textbooks. Normal forms are the topic of the next section.

1.2 Examples

There are several convenient shorthands to describe context-free grammars. Usually, a production (X, α) is written $X \longrightarrow \alpha$, and productions with same left side are grouped together, the corresponding right sides being separated by a '+'. Usually, the variables and terminal letters are clear from the context.

Subsequently, we make use several times of the following notation. Let A be an alphabet. A *copy* of A is an alphabet that is disjoint from A and in bijection with A. A copy is frequently denoted $\bar{A}$ or A'. This implicitly means that the bijection is denoted similarly, namely as the mapping $a \mapsto \bar{a}$ or $a \mapsto a'$. The inverse bijection is denoted the same, that is $\bar{\bar{a}} = a$ (resp. $(a')' = a$), and is extended to a bijection from $(A \cup \bar{A})^*$ into itself (the same for 'bar' replaced by 'prime') by $\overline{xy} = \bar{y}\,\bar{x}$.

The *Dyck languages.* Let A be an alphabet and let $\bar{A}$ be a copy. The *Dyck language* over A is the language D_A^* generated by S in the grammar

$$S \longrightarrow TS + \varepsilon ; \qquad T \longrightarrow aS\bar{a} \quad (a \in A)$$

The notation is justified by the fact that D_A^* is indeed a submonoid of $(A \cup \bar{A})^*$. It is even a free submonoid, generated by the language D_A of *Dyck primes* which is the language generated by the variable T in the grammar above. If A has n letters, then the notation D_n^* is frequently used instead of D_A^*. If $n = 2$, we omit the index.

There is an alternative way to define these languages as follows. Consider the congruence δ over $A \cup \bar{A}$ generated by

$$a\bar{a} \equiv \varepsilon \quad (a \in A)$$

Then

$$D_A^* = \{w \in (A \cup \bar{A})^* \mid w \equiv \varepsilon \bmod \delta\}$$

The *class* of a word w, that is the set of all words x that are congruent to w, is denoted by $[w]_\delta$. Of course, $D_A^* = [\varepsilon]_\delta$. We often omit the subscript δ in this notation.

The *Lukasiewicz language*. Let $A = \{a, b\}$. The Lukasiewicz language is the language generated by the grammar

$$S \longrightarrow aSS + b$$

It is sometimes denoted by L. As we shall see below, $L = D_1^* b$.

2. Systems of equations

This section is devoted to an elementary presentation of systems of equations and their relation to context-free languages. Context-free languages may indeed be defined as the components of the least solution of systems of polynomial equations, whence the term "algebraic" languages introduced by Chomsky and Schützenberger [10]. The same construction was used by Ginsburg and Rice [20]. They preferred to call them ALGOL-like languages because they are "a model for the syntactic classes of the programming language ALGOL". Indeed, one says "an instruction *is...*" rather than "the symbol for instructions *derives...*".

From the methodological point of view, considering equations rather than grammars shifts the induction argument used to prove properties of languages from the number of derivations steps to the length of words. This may frequently simplify exposition, too.

The proofs of the results presented in this section are intentionally from scratch. In fact, most results can be treated differently, in at least two ways: first, they hold in a much more general framework, namely for formal power series over suitable semirings (see the chapter of Kuich[37]); next, there are general results, such as fixed-point theorems in conveniently ordered sets, that imply easily the present results. The present style of exposition was chosen to show what the minimal requirements are to make the arguments work.

The reader should notice that we never assume, in systems of equations, that the right hand sides are finite, and indeed this appears nowhere to be required. Even finiteness of the number of equations is not necessary. Next, the reader should check that all results also hold for partially commutative free monoids (this was observed already by Fliess [15]). Indeed, the argument used in most proofs is just an induction on length, and thus carries over to such monoids.

2.1 Systems

For the definition of equations, we need variables. It will be convenient to number variables. Let $V = \{X_1, \ldots X_n\}$ and A be disjoint alphabets. A *system of equations* over (V, A) is a vector $P = (P_1, \ldots, P_n)$ of subsets of $(V \cup A)^*$, usually written as

$$X_i = P_i \qquad i = 1, \ldots, n \qquad (2.1)$$

Introducing $X = (X_1, \ldots, X_n)$, this can be shortened to

$$X = P$$

We frequently emphasize the dependence of the set V by writing $P_i(X)$ or $P(X)$ instead of P_i and P. An advantage of this is to yield a simple notation for substitution.

Let $L = (L_1, \ldots, L_n)$ be a vector of languages over $V \cup A$. This defines a substitution as follows.

(1) $\varepsilon(L) = \{\varepsilon\}$
(2) $a(L) = a$ $a \in A$
(3) $X_i(L) = L_i$ $i = 1, \ldots, n$
(4) $uv(L) = u(L)v(L)$ $u, v \in (V \cup A)^*$
(5) $Q(L) = \bigcup_{w \in Q} w(L)$ $Q \subset (V \cup A)^*$

Observe that the last equation implies that $Q(L \cup M) = Q(L) \cup Q(M)$, where $L \cup M$ is componentwise union. A vector $L = (L_1, \ldots, L_n)$ of languages over A is a *solution* if

$$L_i = P_i(L) \qquad i = 1, \ldots, n$$

that is if $P_i(L)$ is obtained from $P_i(X)$ by substituting L_j to X_j in any of its occurrences. It is sometimes convenient to write $L = P(L)$ instead $L_i = P_i(L)$ for all i.

Example 2.1. 1) Consider the following system of two equations

$$X = YX + \varepsilon$$
$$Y = aXb$$

Here, the variables are X, Y and the terminal alphabet is $\{a, b\}$. The vector (D_1^*, D_1) is a solution of this system, since indeed

$$D_1^* = D_1 D_1^* + \varepsilon$$
$$D_1 = a D_1^* b$$

2) The system

$$X = (aXb)^*$$
$$Y = aY^*b$$

has right sides that are rational sets. The vector (D_1^*, D_1) is also a solution of the second system, as it follows from elementary properties of the Dyck set. A simple formal proof will be given below.

Solutions are compared componentwise: given two vectors $L = (L_1, \ldots, L_n)$ and $M = (M_1, \ldots, M_n)$, then $L = M$ iff $L_i = M_i$ for all i, and $L \subset M$ iff $L_i \subset M_i$ for all i.

To every context-free grammar over an alphabet A, is canonically associated a polynomial system of equations (i. e. a system where the right sides are finite sets). Assume indeed that the grammar is $G = (V, P)$, with $V = \{X_1, \ldots, X_n\}$. The associated system is

$$X_i = P_i \tag{2.2}$$

with

$$P_i = \{\alpha \in (V \cup A)^* \mid (X_i, \alpha) \in P\}$$

Theorem 2.1. *Let $G = (V, P)$ be a context-free grammar over A with $V = \{X_1, \ldots, X_n\}$. Then the vector*

$$L_G = (L_G(X_1), \ldots, L_G(X_n))$$

is the least solution of the associated system.

We start with a lemma.

Lemma 2.1. *Let $M = (M_1, \ldots, M_n)$ be a solution of (2.2), and let $u, v \in (V \cup A)^*$ be words. If $u \xrightarrow{G} v$ then $v(M) \subset u(M)$.*

Proof. Indeed, if $u \longrightarrow v$, then there exists a production (X_i, α) in G, and two words x, y such that

$$u = xX_iy, \quad v = x\alpha y$$

Thus

$$u(M) = x(M)M_iy(M), \quad v(M) = x(M)\alpha(M)y(M)$$

Since $\alpha \in P_i$ and M is a solution, one has

$$\alpha(M) \subset P_i(M) = M_i$$

and consequently $v(M) \subset u(M)$.   □

Proof of the theorem. Clearly, for each $i = 1, \ldots, n$,

$$L_G(X_i) = \bigcup_{\alpha \in P_i} L_G(\alpha)$$

Now, for any word u in $(V \cup A)^*$,

$$L_G(u) = u(L_G)$$

so that the equation can be written as

$$L_G(X_i) = P_i(L_G)$$

showing that L_G is indeed a solution.

Consider next a solution $M = (M_1, \ldots, M_n)$. To show the inclusion $L_G \subset M$, let $w \in L_G(X_i)$, for some i. Then $X_i \xrightarrow{\;*\;} w$, and by the lemma (extended to derivations)

$$w(M) \subset X_i(M)$$

Since $w \in A^*$, one has $w(M) = \{w\}$, and since M is a solution, $X_i(M) = M_i$. Consequently $w \in M_i$, showing that $L_G(X_i) \subset M_i$.   □

This theorem gives one method for computing the minimal solution of a system of equations, namely by derivations. There is another method, based on iteration. This is the fixed point approach.

Theorem 2.2. *Given a system of equations*

$$X_i = P_i \qquad i = 1, \ldots, n$$

over $V = \{X_1, \ldots, X_n\}$ and A, define a sequence $L^{(h)} = (L_1^{(h)}, \ldots, L_n^{(h)})$ of vectors of subsets of A^* by

$$L^{(0)} = (\emptyset, \ldots, \emptyset)$$
$$L^{(h+1)} = (P_1(L^{(h)}), \ldots, P_n(L^{(h)})) = P(L^{(h)})$$

and set

$$L_i = \bigcup_{h \geq 0} L_i^{(h)} \qquad i = 1, \ldots, n$$

Then the vector

$$L = (L_1, \ldots, L_n)$$

is the least solution of the system.

Proof. First,

$$L_i = \bigcup_{h \geq 0} P_i(L^{(h)}) = P_i(\bigcup_{h \geq 0} L^{(h)}) = P_i(L)$$

showing that L is indeed a solution.

Next, if $M = (M_1, \ldots, M_n)$ is any solution, then $L^{(h)} \subset M$ for all $h \geq 0$. This is clear for $h = 0$, and by induction

$$L_i^{(h+1)} = P_i(L^{(h)}) \subset P_i(M) = M_i \qquad \qquad \square$$

Let us remark that the basic ingredient of the proof is that P_i is "continuous" and "monotone" in the lattice $\mathfrak{P}((V \cup A)^*)^n$, for the order of componentwise inclusion (see also the chapter by Kuich [37]).

A system of equations

$$X_i = P_i \qquad i = 1, \ldots, n$$

over $V = \{X_1, \ldots, X_n\}$ and A is called

- *proper* if, for all i, one has $P_i \cap (\{\varepsilon\} \cup V) = \emptyset$,
- *strict* if, for all i, one has $P_i \subset \{\varepsilon\} \cup (V \cup A)^* A (V \cup A)^*$.

Thus, in a proper system, every word in the right side of an equation is either a terminal letter (in A) or has length at least 2. If a context-free grammar is proper, the associated system of equations is proper. In a strict system, every nonempty word in a right side contains at least one terminal letter.

A solution $L = (L_1, \ldots, L_n)$ is *proper* if $\varepsilon \notin L_i$ for all i.

Theorem 2.3. *A proper system has a unique proper solution. A strict system has a unique solution.*

Before starting the proof, let us give some examples.

Example 2.2. The equation $X = XX$ is proper. Its unique proper solution is the empty set. However, every submonoid is a solution. Thus a proper system may have more than one solution.

Example 2.3. The system

$$X = YX + \varepsilon$$
$$Y = aXb$$

is neither proper nor strict. However, replacing Y by aXb in $YX + \varepsilon$, one sees that the first component of a solution is also a solution of

$$X = aXbX + \varepsilon$$

which is strict. This shows that the system has only one solution.

The system

$$X = (aXb)^*$$
$$Y = aY^*b$$

is strict, so it has a unique solution. It is easily checked that

$$X = (aXb)^* = aXbX + \varepsilon$$

and

$$aXb = a(aXb)^*b$$

Thus the unique solution of this system is equal to the unique solution of the first.

Example 2.4. We claimed earlier that $L = D_1^* b$. Here is the proof. The Lukasiewicz language L is the unique solution of the strict equation $X = aXX + b$, and D_1^* is (the unique) solution of the strict equation $X = aXbX + \varepsilon$. Thus $D_1^* = aD_1^* b D_1^* + \varepsilon$, and multiplying both sides by b, one gets $D_1^* b = aD_1^* b D_1^* b + b$, showing that $D_1^* b$ is a solution of $X = aXX + b$. Since this equation has only one solution, the equality follows.

It is convenient to introduce a notation. For any $k \geq 0$, define an equivalence relation $\sim_k$ for languages $H, H' \subset A^*$ by

$$H \sim_k H' \iff \{w \in H \mid |w| \leq k\} = \{w \in H' \mid |w| \leq k\}$$

Extend these equivalences to vectors componentwise. Then one has the following general lemma:

Lemma 2.2. *Let L and M be two solutions of a system of equations $X = P$. If*

$$L \sim_0 M \tag{2.3}$$

$$L \sim_k M \;\Rightarrow\; P(L) \sim_{k+1} P(M) \tag{2.4}$$

then $L = M$.

Proof. Since $L = P(L)$ and $M = P(M)$, the hypotheses imply that $L \sim_k M$ for all $k \geq 0$, and thus $L = M$. $\square$

Proof of the theorem 2.3. It suffices to show that the conditions of the lemma are fulfilled in both cases.

Consider first the case where L and M are proper solutions of the proper system $X = P$. Then by assumption $L \sim_0 M$. Assume now $L \sim_k M$, and consider any $\alpha \in P_i$ for some i. If $\alpha \in A^+$, then $\alpha(L) = \alpha(M) = \alpha$. Otherwise, there exist non empty words β, γ, such that $\alpha = \beta\gamma$. Clearly $\beta(L) \sim_k \beta(M)$ and $\gamma(L) \sim_k \gamma(M)$, and since the empty word is not in these languages, one has

$$\beta(L)\gamma(L) \sim_{k+1} \beta(M)\gamma(M)$$

Thus $P_i(L) \sim_{k+1} P_i(M)$. This proves (2.4).

Consider now the case where L and M are solutions of the strict system $X_i = P_i$ for $i = 1, \ldots, n$. Since $\varepsilon \in L_i$ for some i iff $\varepsilon \in P_i$, one has $L \sim_0 M$. Next, as before assume $L \sim_k M$, and consider any $\alpha \in P_i$ for some i. If $\alpha \neq \varepsilon$, then $\alpha = \beta a \gamma$ for words β, γ and a letter $a \in A$. Again, $\beta(L) \sim_k \beta(M)$ and $\gamma(L) \sim_k \gamma(M)$, and since a is a terminal letter, this implies that $\alpha(L) \sim_{k+1} \alpha(M)$. This proves (2.4). $\square$

As we have already seen, a system may have a unique solution even if it is neither proper nor strict. Stronger versions of the above theorem exist. For instance, call a system of equations

$$X = P(X)$$

weakly proper (resp. *weakly strict*) if there is an integer k such that the system

$$X = P^k(X)$$

is proper (resp. strict).

Corollary 2.1. *A weakly strict (weakly proper) system has a unique (a unique proper) solution.*

Proof. Let indeed L be a solution of $X = P(X)$. Then $L = P^k(L)$, showing that L is solution of $X = P^k(X)$. Hence the solution of $X = P(X)$ is unique.

□

Observe that, if L is the solution of $X = P^k(X)$, then it is also the solution of the system $X = P(X)$. This may provide an easy way to compute the solution.

Example 2.5. Consider the system $X = P(X)$ given by

$$X = YX + \varepsilon$$
$$Y = aXb$$

Replacing P by P^2, one gets

$$X = aXbYX + aXb + \varepsilon$$
$$Y = aYXb + ab$$

which is not proper but strict. Hence the system is weakly strict.

2.2 Resolution

One popular method for resolution of systems of equations is Gaussian elimination. Consider sets $X = \{X_1, \ldots, X_n\}$ and $Y = \{Y_1, \ldots, Y_m\}$ of variables.

Theorem 2.4. *For any system of equations*

$$X = P(X, Y)$$
$$Y = Q(X, Y) \tag{2.5}$$

over $(X \cup Y, A)$, *let* L'_Y *be a solution of the system of equations*

$$Y = Q(X, Y)$$

over $(Y, A \cup X)$, *and let* L_X *be a solution of the system of equations*

$$X = P(X, L'_Y)$$

over (X, A). *Then* $(L_X, L'_Y(L_X))$ *is a solution of* (2.5).

Proof. Let indeed $L'_Y = L'_Y(X)$ be a solution of the system $Y = Q(X,Y)$. For any vector $L = (L_1, \ldots, L_n)$ of languages over A, one has

$$L'_Y(L) = Q(L, L'_Y(L)) \qquad (2.6)$$

by substitution. Next, let L_X be a solution of

$$X = P(X, L'_Y(X)) \qquad (2.7)$$

and set $L_Y = L'_Y(L_X)$. Then (L_X, L_Y) is a solution of (2.5) since $L_X = P(L_X, L_Y)$ by (2.7) and $L_Y = Q(L_X, L_Y)$ by (2.6).   □

A special case is "lazy" resolution. This means that some variables, or even some occurrences of variables or factors in the right sides are considered as "fixed", the obtained system is solved, and the solution is substituted in the "fixed" part. More precisely,

Proposition 2.1. *The two systems*

$$\begin{array}{ccc} X = P(X,Y) & & X = P(X, Q(X)) \\ Y = Q(X) & and & Y = Q(X) \end{array}$$

have same sets of solutions.   □

As an example, consider the equation $X = aXX + b$, that we write as

$$\begin{array}{l} X = YX + b \\ Y = aX \end{array}$$

The first equation is equivalent to $X = Y^*b$, thus the equations $X = aXX + b$ and $X = (aX)^*b$ have the same solution.

2.3 Linear systems

Left or right linear systems of equations are canonically associated to finite automata. The general methods take here a special form. A system of equations

$$X_i = P_i(X), \qquad i = 1, \ldots, n$$

over (V, A) is *right linear* (resp. *left linear*) if $P_i \subset A^*V \cup A^*$ (resp. $P_i \subset VA^* \cup A^*$). If furthermore it is strict (resp. proper), then $P_i \subset A^+V \cup A^*$ (resp. $P_i \subset A^+V \cup A^+$). A (right) linear system may be written as

$$X_i = \sum_{j=1}^{n} R_{i,j}X_j + S_i \qquad i = 1, \ldots, n \qquad (2.8)$$

where $R_{i,j} \subset A^*$, $S_i \subset A^*$. These sets are the *coefficients* of the system. One may also write

$$X = RX + S$$

by introducing a matrix $R = (R_{i,j})$ and a vector $S = (S_i)$.

Given a finite automaton with state set $Q = \{1, \ldots, n\}$, denote by $R_{i,j}$ the set of labels of edges from state i to state j, and set

$$S_i = \begin{cases} \{\varepsilon\} & \text{if } i \text{ is a final state} \\ \emptyset & \text{otherwise} \end{cases}$$

Then it is easily verified that the least solution of the system (2.8) is the vector $(L_1, \ldots, L_n)$, where L_i is the set of words recognized with initial state i.

Theorem 2.5. *The components of the solution of a strict linear system are in the rational closure of the set of coefficients of the system.*

Proof. There are several proofs of this result. The maybe simplest proof is to consider an alphabet $B = \{r_{i,j} \mid 1 \leq i, j \leq n\} \cup \{s_i \mid 1 \leq i \leq n\}$ and to consider the system obtained in replacing each $R_{i,j}$ by $r_{i,j}$ and similarly for the S_i. Build an automaton with state set $\{0, 1, \ldots, n\}$, having an edge labeled $r_{i,j}$ from state i to state j for $1 \leq i, j \leq n$ and an edge labeled s_i from state i to the unique final state 0. The component L_i of the solution of the system is the set of label of paths from state i to state 0, and therefore is a rational set over B. To get the solution of the original system, it suffices to substitute the sets $R_{i,j}$ and S_i to the corresponding variables.

An equivalent formulation is to say that the vector

$$L = R^* S$$

is the solution, where

$$R^* = \bigcup_{m \geq 0} R^m .$$

One way to solve the original set of equations is to use Gaussian elimination (also called Arden's lemma in the linear case). One rewrites the last equation of the system as

$$X_n = R_{n,n}^* \left(\sum_{j=1}^{n-1} R_{n,j} X_j + S_j \right)$$

and substitutes this expression in the remaining equations.

Another way is to proceed inductively, and to compute the transitive closure R^* from smaller matrices, using the formula

$$\begin{pmatrix} A & B \\ C & D \end{pmatrix}^* = \begin{pmatrix} (A + BD^*C)^* & A^*B(D + CA^*B)^* \\ D^*C(A + BD^*C)^* & (D + CA^*B)^* \end{pmatrix}$$

provided A and D are square matrices.   □

A system (*system*) is called *cycle-free* if none of the diagonal coefficients of the matrix

$$R + R^2 + \cdots + R^n$$

contains the empty word. The terminology is from graph theory : consider the graph over $\{1, \ldots, n\}$ with an edge from i to j iff $\varepsilon \in R_{i,j}$. Then this graph is cycle-free iff the system is. In fact, cycle-free systems are precisely weakly strict right linear systems. Indeed, the graph is cycle-free iff there is no path of length k for $k > n$. This is equivalent to say that in the matrix R^k, none of the coefficients contains the empty word. Thus one has

Proposition 2.2. *A cycle-free (right or left) linear system has a unique solution.* □

2.4 Parikh's theorem

In this section, we prove Parikh's theorem [43]. Our presentation follow [44]. A more general result is given in Kuich's chapter [37]. As already mentioned, all results concerning systems of equations, provided they make sense (e.g. Greibach normal form makes no sense in free commutative monoids) hold also for free partially commutative monoids, since the only argument used is induction on length. Two special cases of partially commutative free monoids are the free monoid and the free commutative monoid. Context-free sets in the latter case are described by Parikh's theorem:

Theorem 2.6. *Any context-free set in the free commutative monoid is rational.*

An equivalent formulation is that the set of Parikh vectors of a context-free language is semi-linear. Indeed, let A be an alphabet, and denote by $A^\oplus$ the free commutative monoid over A. There is a canonical mapping α from A^* onto $A^\oplus$ that associates, to a word w, the element $\Pi_{a \in A} a^{|w|_a}$ in $A^\oplus$, where $|w|_a$ is the number of occurrences of the letter a in A.

Rational sets are defined in $A^\oplus$ as they are in any monoid: they constitute the smallest family of languages containing the empty set, singletons, and closed under union, product and star. Here, product is the product in $A^\oplus$ of course. Because of commutativity, there are special relations, namely

$$(X \cup Y)^* = X^* Y^*, \quad (X^* Y)^* = \{\varepsilon\} \cup X^* Y^* Y$$

Using these, on gets easily the following

Proposition 2.3. *In the free commutative monoid $A^\oplus$, every rational set has star-height at most 1.* □

Proof of the theorem. Consider first the case of a single (strict) equation

$$X = P(X)$$

where $P(X)$ is any rational subset of $(A \cup X)^{\oplus}$. This equation may be rewritten as

$$X = R(X)X + S$$

where $S = P(X) \cap A^{\oplus}$, and $R(X)$ is a rational subset of $(A \cup X)^{\oplus}$. The set $G = R(S)$ is rational, and we show that G^*S is the (rational) solution of the equation.

Consider indeed two subsets K, M of A^*, and set $P = K^*M$. For every $w \in (A \cup X)^{\oplus}$ containing at least one occurrence of X, one has the equality

$$w(P) = w(M)K^*$$

because the set K^* can be "moved" to the end of the expression by commutativity, and $K^*K^* = K^*$. As a consequence, for every word $w \in (A \cup X)^{\oplus}$, one gets $w(P)P = w(M)P$. Thus in particular for $P = G^*S$,

$$S + R(P)P = S + R(S)P = S + GG^*S = G^*S$$

If the system has more than one equation, then it is solved by Gaussian elimination.   □

Example 2.6. Consider the equation

$$X = aXX + b$$

The set $R(X)$ of the proof reduces to aX, and the solution is $(ab)^*b = \{a^n b^{n+1} \mid n \geq 0\}$.

3. Normal forms

In this section, we present three normal forms of context-free grammars. The two first ones are the Chomsky normal form and the Greibach normal form. They are often used to get easier proofs of results about context-free languages. The third normal form is the operator normal form. It is an example of a normal form that has been used in the syntactical analysis.

3.1 Chomsky normal form

A context-free grammar $G = (V, P)$ over the terminal alphabet A is in *weak Chomsky normal form* if each nonterminal rule has a right member in V^* and each terminal rule has a right member in $A \cup \{\varepsilon\}$. It is in *Chomsky normal form* if it is in Chomsky normal form and each right member of a nonterminal rule has length 2.

Theorem 3.1. [28, 9] *Given a context-free grammar, an equivalent context-free grammar in Chomsky normal form can effectively be constructed.*

Proof. The construction is divided into three steps. In the first step, the original grammar is transformed into a new equivalent grammar in weak Chomsky normal form. In the second step, we transform the grammar just obtained so that the length of a right member of a rule is at most 2. In the last step, we get rid of the nonterminal rules with a right member of length 1 (that is to say in V).

Step 1: To each terminal letter $a \in A$, we associate a new variable X_a. In all the right members of the rules of the original grammar, we replace each occurrence of the terminal letters a by the new variable X_a. Finally, we add to the grammar so obtained the set of rules $X_a \longrightarrow a$. Clearly, the resulting grammar so constructed is in weak Chomsky normal form and is equivalent to the original one.

Step 2: We now introduce a new set of variables designed to represent the product of two old variables. More precisely, to each pair $(X, Y) \in V \times V$, we associate a new variable $\langle XY \rangle$. We then construct a new grammar by replacing any product of three or more old variables $XYZ \cdots T$ by $\langle XY \rangle Z \cdots T$. Then we add all the rules $\langle XY \rangle \longrightarrow XY$. This reduces the maximal length of nonterminal rules by 1. This process is repeated until the maximum length of any right member is 2.

Step 3: We finally get rid of nonterminal rules with a right member in V. This is achieved in the same usual way than the one used to get a proper grammar from a general one.   □

3.2 Greibach normal forms

A context-free grammar $G = (V, P)$ over the terminal alphabet A is in *Greibach normal form* iff each rule of the grammar rewrites a variable into a word in $AV^\star$. In particular, the grammar is proper and each terminal rule rewrites a variable in a terminal letter.

It is in *quadratic Greibach normal form* iff it is in Greibach normal form and each right member of a rule of G contains at most 2 variables.

It is in *double Greibach normal form* iff each right member of the rules of G are in $AV^\star A \cup A$. In particular, a terminal rule rewrites a variable in a letter or in a word of length 2.

It is in *cubic double Greibach normal form* (resp. in *quadratic double Greibach normal form* iff it is in double Greibach normal form and each right member of a rule contains at most 3 variables (resp. at most 2 variables).

The fact that any proper context-free grammar G can be transformed in an equivalent grammar G' in Greibach normal form is a classical result [28]. However, the fact that the same result holds with G' in quadratic Greibach normal form is more rarely presented. Nearly never proved is the same result with G' in quadratic double normal form. Hence, we show how such equivalent grammars can effectively be constructed.

Theorem 3.2. *Given a proper context-free grammar G, an equivalent context free grammar in quadratic Greibach normal form can effectively be constructed from G.*

A weaker similar result has originally been proved by Greibach [24]: she showed that, given a proper context-free grammar, an equivalent context-free grammar in Greibach normal form can effectively be constructed. The additional statement stating that this grammar can be in *quadratic* Greibach normal form was proved later by Rosenkrantz [45]. We sketch here the proof he gave; we will see below an alternative proof.

Sketch of the construction:

We may assume that the grammar is proper and in Chomsky normal form, that is that each right-hand side is in $A \cup V^2$. Consider the associated system of equations

$$X_i = P_i \qquad i = 1, \ldots, n$$

This may be written as

$$X = XR + S$$

where

$$S_i = P_i \cap A$$

and

$$R_{j,i} = X_j^{-1} P_i$$

Using lazy evaluation, this system is equivalent to

$$X = SR^*$$

and since

$$R^* = RR^* + I$$

one has

$$X = SY$$
$$Y = RY + I$$

where $Y = (Y_{j,i})$ is a new set of n^2 variables. Observe that each $R_{j,i}$ is a subset of V. Thus, using the system $X = SY$, each $R_{j,i}$ can be replaced by the set

$$\hat{R}_{j,i} = \sum_{X_\ell \in R_{j,i}} (SY)_\ell$$

and the whole system is equivalent to

$$X = SY$$
$$Y = \hat{R}Y + I$$

where $\hat{R} = (\hat{R}_{j,i})$. In order to get the quadratic Greibach normal form, it suffices to eliminate the ε-rules. This is done in the usual way. $\qquad\square$

Theorem 3.3. *Given a proper context-free grammar G, an equivalent context free grammar in quadratic double Greibach normal form can effectively be constructed from G.*

This result has been proved by Hotz [31]. We follow his proof. It should be noted that the same technique allows to give an alternative proof of the previous theorem 3.2.

The proof of theorem 3.3 turns out to be a complement to the proof of

Theorem 3.4. *Given a proper context-free grammar G, an equivalent context free grammar in cubic double Greibach normal form can effectively be constructed from G.*

The construction of the desired grammar is decomposed into four steps. The two first ones will lead to an equivalent grammar in quadratic Greibach normal form. The last two ones complete the construction of an equivalent grammar in quadratic double Greibach normal form.

Let $G = (V, P)$ be a proper context-free grammar in weak Chomsky normal form over the terminal alphabet A.

Step 1 (construction of the set of new variables needed):

To each variable $X \in V$, we associate the sets

$$L(X) = \{am \in AV^\star \mid X \xrightarrow[\ell]{*} \alpha \longrightarrow am \quad \alpha \in V^*\}$$
$$R(X) = \{ma \in V^\star A \mid X \xrightarrow[r]{*} \alpha \longrightarrow ma \quad \alpha \in V^*\}$$

The idea is to construct a new grammar including rules the $X \longrightarrow am$ for each $am \in L(X)$. The difficulty comes from the fact that the sets $L(X)$ are infinite. This difficulty can be overcome using the fact these sets are rational. The sets $R(X)$ will only be used to get the double Greibach normal form. Formally, to each variable $X \in V$ and to each terminal letter $a \in A$, we associate the sets

$$L(a, X) = a^{-1}L(X) = \{m \in V^\star \mid X \xrightarrow[\ell]{*} \alpha \longrightarrow am \quad \alpha \in V^*\}$$
$$R(X, a) = R(X)a^{-1} = \{m \in V^\star \mid X \xrightarrow[r]{*} \alpha \longrightarrow ma \quad \alpha \in V^*\}$$

Clearly, each $L(a, X)$ and each $R(X, a)$ is a rational language over V since $L(X)$ and $R(X)$ are rational: to get a word in $L(X)$, we look at leftmost derivations in the original grammar. Then, such a derivation can be decomposed in a first part where the obtained words all lie in $V^\star$. The second part consists in the last step where the leftmost variable is derived in a terminal letter a. In this process, we then always derive the leftmost variable of the sentential forms. So, this derivation is obtained by using the grammar as if it were left linear. Hence, the set of words so obtained forms a rational set. It then follows immediately that each $L(a, X)$ is rational too.

A similar proof using right linear grammars shows that each $R(X, a)$ is rational.

Next, define two families $\mathcal{L}$ and $\mathcal{R}$ of languages by

$$\mathcal{L} = \{L(a, X) \mid a \in A, X \in V\}, \quad \mathcal{R} = \{R(X, a) \mid a \in A, X \in V\}.$$

We then define $\mathcal{H}$ as the closure of $\mathcal{L} \bigcup \mathcal{R}$ under the right and left quotients by a letter of V. Since each language in $\mathcal{L} \bigcup \mathcal{R}$ is rational, this gives raise to a finite number of new regular languages over V. Thus, the family $\mathcal{H}$ is finite.

The idea is now to use the languages in $\mathcal{H}$ as variables in the grammar to be constructed. The set of new variables will be denoted like this family of languages, that is, an element $L \in \mathcal{H}$ will denote both the language and the new variable.

Example 3.1. Let $G = (V, P)$ be the following grammar in weak Chomsky normal form :

$$S \longrightarrow SXSS + b$$
$$X \longrightarrow a$$

We can now look for the family $\mathcal{L}$. Since

$$L(a, S) = \emptyset \quad L(b, S) = (XSS)^\star = L_0 \quad L(a, X) = \{\varepsilon\} = E \quad L(b, X) = \emptyset$$

$\mathcal{L}$ is formed of the three languages $\{\emptyset, L_0, E\}$.

Similarly, $\mathcal{R} = \{\emptyset, L_1, E\}$ because

$$R(S, a) = \emptyset \quad R(S, b) = (SXS)^\star = L_1 \quad R(X, a) = \{\varepsilon\} = E.$$

Thus, $\mathcal{L} \bigcup \mathcal{R} = \{\emptyset, L_0, L_1, E\}$. From this family, we derive the family $\mathcal{H}$ by closing $\mathcal{L} \bigcup \mathcal{R}$ under left and right quotient by a letter in V. Here are the new languages that appear :

$$
\begin{array}{ll}
X^{-1}L_0 = SS(XSS)^\star = L_2 & L_0 S^{-1} = (XSS)^\star XS = L_3 \\
S^{-1}L_1 = XS(SXS)^\star = L_3 & L_1 S^{-1} = (SXS)^\star SX = L_4 \\
S^{-1}L_2 = S(XSS)^\star = L_5 & L_2 S^{-1} = S(SXS)^\star = L_6 \\
X^{-1}L_3 = (SSX)^\star S = L_6 & L_3 S^{-1} = (XSS)^\star X = L_7 \\
S^{-1}L_4 = (XSS)^\star X = L_7 & L_4 X^{-1} = S(XSS)^\star = L_5 \\
S^{-1}L_5 = (XSS)^\star = L_0 & L_5 S^{-1} = (SXS)^\star = L_1 \\
S^{-1}L_6 = (SXS)^\star = L_1 & L_6 S^{-1} = (SSX)^\star = L_8 \\
X^{-1}L_7 = (SSX)^\star = L_8 & L_7 X^{-1} = (XSS)^\star = L_0 \\
S^{-1}L_8 = SX(SSX)^\star = L_4 & L_8 X^{-1} = (SSX)^\star SS = L_2
\end{array}
$$

Hence, the family $\mathcal{H}$ contains 11 languages: the languages $L_0, \ldots, L_8$, the language $E = \{\varepsilon\}$ and the empty set $\emptyset$. (In the above computations, we have omitted all empty quotients.) Among these languages, E, L_0, L_1 and L_8 contain the empty word.

Step 2 (Construction of an equivalent grammar in quadratic Greibach normal form)

The new grammar has the set of variables $V \cup \mathcal{H}$, and the following rules:

(i)   Each terminal rule of the original grammar is a terminal rule of the new grammar.

(ii) To each variable $X \in V$ of the original grammar is associated the (finite) set of rules $X \longrightarrow aL$ for each $a \in A$, with $L = L(a, X) \in \mathcal{H}$. The rules so created have all their right members in $A\mathcal{H}$.

(iii) Each new variable $L \in \mathcal{H}$ gives raise to the finite set of rules $L \longrightarrow XL'$ for $X \in V$ with $L' = X^{-1}L \in \mathcal{H}$ and to the rule $L \longrightarrow \varepsilon$ if $\varepsilon \in L$. Each such rule has its right member in $V\mathcal{H} \cup \{\varepsilon\}$.

(iv) In each new non ε-rule added just above, the leftmost variable in V is replaced by the right members generated in step (ii); since these right members are in $A\mathcal{H}$, the rules so obtained have all their right members in $A\mathcal{H}\mathcal{H}$.

Hence, the grammar so obtained is almost in quadratic Greibach normal form: each right member is in $A\mathcal{H}\mathcal{H} \cup A \cup \{\varepsilon\}$.

To obtain such a normal form, it suffices to complete a final operation eliminating the ε-rules. We do this in the usual way, that is we replace any occurrence of a variable L giving the empty word by $(L \cup \{\varepsilon\})$ in the right members and erase the ε-rules.

The fact that the new grammar is equivalent to the original one is immediate: it suffices to look at the two grammars as systems of equations. So far, we have proved Theorem 3.2.

Example 3.2. (continued) The rules $S \longrightarrow SXSS + b$ give raise to the new rules $S \longrightarrow bL_0 + bE$. The last rule $X \longrightarrow a$ gives raise to $X \longrightarrow aE$.

The new set of variables gives raise to

$$
\begin{array}{ll}
L_0 \longrightarrow XL_2 + \varepsilon & L_1 \longrightarrow SL_3 + \varepsilon \\
L_2 \longrightarrow SL_5 & L_3 \longrightarrow XL_6 \\
L_4 \longrightarrow SL_7 & L_5 \longrightarrow SL_0 \\
L_6 \longrightarrow SL_1 & L_7 \longrightarrow XL_8 \\
L_8 \longrightarrow SL_4 + \varepsilon & E \longrightarrow \varepsilon
\end{array}
$$

Replacing X by aE and S by $bE + bL_0$, the new grammar becomes

$$
\begin{array}{ll}
S \longrightarrow bL_0 + bE & X \longrightarrow aE \\
L_0 \longrightarrow aEL_2 + \varepsilon & L_1 \longrightarrow bEL_3 + bL_0L_3 + \varepsilon \\
L_2 \longrightarrow bEL_5 + bL_0L_5 & L_3 \longrightarrow aEL_6 \\
L_4 \longrightarrow bEL_7 + bL_0L_7 & L_5 \longrightarrow bEL_0 + bL_0L_0 \\
L_6 \longrightarrow bEL_1 + bL_0L_1 & L_7 \longrightarrow aEL_8 \\
L_8 \longrightarrow bEL_4 + bL_0L_4 + \varepsilon & E \longrightarrow \varepsilon
\end{array}
$$

This is the desired intermediate grammar obtained after step (iv). To obtain the quadratic Greibach normal form, we replace everywhere E, L_0, L_1 and L_8 by themselves plus the empty word in the right members and suppress the ε-rules. Then we get the following grammar (to be compared to the one obtained with Rosenkrantz's method):

$$\begin{aligned}
S &\longrightarrow bL_0 + b & X &\longrightarrow a \\
L_0 &\longrightarrow aL_2 & L_1 &\longrightarrow bL_3 + bL_0L_3 \\
L_2 &\longrightarrow bL_5 + bL_0L_5 & L_3 &\longrightarrow aL_6 \\
L_4 &\longrightarrow bL_7 + bL_0L_7 + bL_7 & L_5 &\longrightarrow bL_0 + bL_0L_0 + b \\
L_6 &\longrightarrow bL_1 + bL_0L_1 + bL_0 + b & L_7 &\longrightarrow aL_8 + a \\
L_8 &\longrightarrow bL_4 + bL_0L_4
\end{aligned}$$

Note that this grammar is not reduced. The only useful variables are S, L_0, L_2 and L_5. The seemingly useless variables and rules will appear to be useful later. Note too that E disappeared because, when ε is removed from the language, E becomes empty.

The next two steps will be devoted to the proof of theorem 3.3.

Step 3 (Construction of an equivalent grammar in cubic double Greibach normal form)

We work on the grammar just obtained above. Each nonterminal rule of this grammar ends with a variable in $\mathcal{H}$. A variable now generates a language that is proper. Thus, the language is not necessarily in $\mathcal{H}$ (considered as a family of languages) because the empty word may be missing. However, the set $\mathcal{H}$ (considered as a set of variables) remains the same. Each variable now generates the associated language of $\mathcal{H}$ up to the empty word.

We first proceed to the same operations as in step 2, using right quotients instead of left ones. This operation is presented below in a slightly different way than we did in step 2. Precisely,

(i) For each language $L \in \mathcal{H}$, the set LX^{-1} is now a language of the family $\mathcal{H}$ up to the empty word. So, each L in $\mathcal{H}$ can be described as the union of $L'X$ for each $X \in V$ with $L' = LX^{-1}$, completed by X as soon as L' contains the empty word. We do this for each L.

Each language generated by a variable $X \in V$ is proper. Hence, in the expression above, X can be replaced by the union of all the Ra for each $a \in A$, with R the variable associated to the language $R(X,a)$. Again, this union has to be completed by a as soon as the language $R(X,a)$ contains the empty word.

This gives a system of equations where each $L \in \mathcal{H}$ is a sum of terms in $\mathcal{H}\mathcal{H}A \cup \mathcal{H}A \cup A$.

(ii) We now go back to the grammar in quadratic normal form resulting from step 2, and replace each rightmost variable of the nonterminal rules in the grammar by the expression in the system obtained in step (i). We thus obtain an equivalent grammar where the nonterminal rules have a terminal letter as rightmost symbol. It should be noted that, in the resulting grammar, the number of variables is increased by at most one in each rule, so that the grammar is cubic. Hence, the so obtained grammar is in cubic double Greibach normal form.

Example 3.3. (continued) The first identities are directly derived from the right quotients computed before. They are

$$L_0 = L_3 S \qquad L_1 = L_4 S$$
$$L_2 = L_6 S \qquad L_3 = L_7 S$$
$$L_4 = L_5 X \qquad L_5 = L_1 S + S$$
$$L_6 = L_8 S + S \qquad L_7 = L_0 X + X$$
$$L_8 = L_2 X + X$$

Replacing now each S by $L_1 b + b$ and each X by a, we obtain

$$L_0 \longrightarrow L_3 L_1 b + L_3 b \qquad L_1 \longrightarrow L_4 L_1 b + L_4 b$$
$$L_2 \longrightarrow L_6 L_1 b + L_6 b \qquad L_3 \longrightarrow L_7 L_1 b + L_7 b$$
$$L_4 \longrightarrow L_5 a \qquad L_5 \longrightarrow L_1 L_1 b + L_1 b$$
$$L_6 \longrightarrow L_8 L_1 b + L_8 b + b \qquad L_7 \longrightarrow L_0 a + a$$
$$L_8 \longrightarrow L_2 a + a$$

Going back to the grammar obtained at the end of step 2, we replace in it each rightmost variable by the finite sum so obtained, giving raise to :

$$S \longrightarrow b L_3 L_1 b + b L_3 b + b$$
$$X \longrightarrow a$$
$$L_0 \longrightarrow a L_6 L_1 b + a L_6 b$$
$$L_1 \longrightarrow b L_7 L_1 b + b L_7 b + b L_0 L_7 L_1 b + b L_0 L_7 b$$
$$L_2 \longrightarrow b L_1 L_1 b + b L_1 b + b b + b L_0 L_1 L_1 b + b L_0 L_1 b + b L_0 b$$
$$L_3 \longrightarrow a L_8 L_1 b + a L_8 b + a L_1 b + a b$$
$$L_4 \longrightarrow b L_0 a + b a + b L_0 L_0 a$$
$$L_5 \longrightarrow b L_3 L_1 b + b L_3 b + b L_0 L_3 L_1 b + b L_0 L_3 b + b$$
$$L_6 \longrightarrow b L_4 L_1 b + b L_4 b + b L_0 L_4 L_1 b + b L_0 L_4 b$$
$$L_7 \longrightarrow a L_2 a + a a + a$$
$$L_8 \longrightarrow b L_5 a + b L_0 L_5 a$$

The steps 1, 2 and 3 allow thus to transform any context-free grammar in weak Chomsky normal form in an equivalent grammar in cubic double Greibach normal form, which proves Theorem 3.4.   □

Step 4 (Construction of an equivalent grammar in quadratic double Greibach normal form)

We use here essentially the same technique of grouping variables that was previously used to derive Chomsky normal form from weak Chomsky normal form. It should be also noted that this technique can be used to transform a grammar in Greibach normal form into quadratic Greibach normal form.

In the grammar obtained in the previous step, no variable of V appears in the right member of a rule. Moreover, any variable of $\mathcal{H}$ represents, up to the empty word, the corresponding language. In particular, a language $L \in \mathcal{H}$ can be described by

– a left quotient description given by the rules of the grammar in quadratic Greibach normal form obtained in step 2.
– a right quotient description obtained in the same way. It is the intermediate description used in step 3 just before transforming the grammar in an equivalent one in double Greibach normal form.

We now enlarge the family $\mathcal{H}$ by adding the family $\mathcal{H}\mathcal{H}$. To this new family of languages is associated a new set of variables W. It should be noted that, each $Y \in W$ represents a product $L \cdot L' \in \mathcal{H}\mathcal{H}$. Hence, replacing L by its left quotient description, and L' by its right quotient description, we get a description of each $Y \in W$ as a finite union of terms in $A\mathcal{H}\mathcal{H}\mathcal{H}\mathcal{H}A \cup A\mathcal{H}\mathcal{H}\mathcal{H}A \cup A\mathcal{H}\mathcal{H}A \cup A\mathcal{H}A \cup AA$.

Each product of four elements of $\mathcal{H}$ can be replaced by a product of two elements of W; similarly, any product of three elements of $\mathcal{H}$ can be replaced by the product of an element in W by an element of $\mathcal{H}$ (or just the contrary as well).

Then using this transformation in the right members of the rules of the grammar in cubic double Greibach normal form and adding the new rules induced by the representation of variables in W just obtained, we get an equivalent grammar which is now in quadratic double Greibach normal form.

Example 3.4. (end) The family W is formed of elements denoted $\langle LL' \rangle$ for $L, L' \in \mathcal{H}$. We first make quadratic the rules of the above obtained grammar by introducing, when necessary, some of our new variables.

$$
\begin{aligned}
S &\longrightarrow bL_3L_1b + bL_3b + b \\
X &\longrightarrow aL_6L_1b + aL_6b + a \\
L_1 &\longrightarrow bL_7L_1b + bL_7b + b\langle L_0L_7\rangle L_1b + bL_0L_7b \\
L_2 &\longrightarrow bL_1L_1b + bL_1b + bb + b\langle L_0L_1\rangle L_1b + bL_0L_1b + bL_0b \\
L_3 &\longrightarrow aL_8L_1b + aL_8b + aL_1b + ab \\
L_4 &\longrightarrow bL_0a + ba + bL_0L_0a \\
L_5 &\longrightarrow bL_3L_1b + bL_3b + b\langle L_0L_3\rangle L_1b + bL_0L_3b + b \\
L_6 &\longrightarrow bL_4L_1b + bL_4b + b\langle L_0L_4\rangle L_1b + bL_0L_4b \\
L_7 &\longrightarrow aL_2a + aa + a \\
L_8 &\longrightarrow bL_5a + bL_0L_5a
\end{aligned}
$$

Doing this, we have introduced the four new variables $\langle L_0L_1\rangle$, $\langle L_0L_3\rangle$, $\langle L_0L_4\rangle$ and $\langle L_0L_7\rangle$. Rather than computing the descriptions of all the elements $\langle L_iL_j\rangle$, we will compute those needed as soon as they appear.

So we begin by computing the description of $\langle L_0L_1\rangle$: for this we use

$$
L_0 = aL_2 \quad L_1 = L_4L_1b + L_4b
$$

which gives raise to the rules

$$
\langle L_0L_1\rangle \longrightarrow a\langle L_2L_4\rangle L_1b + aL_2L_4b
$$

Going on this way, we get the (huge) equivalent grammar in quadratic double Greibach normal form:

$$
\begin{aligned}
S &\longrightarrow bL_3L_1b + bL_3b + b \\
X &\longrightarrow a \\
L_0 &\longrightarrow aL_6L_1b + aL_6b + a \\
L_1 &\longrightarrow bL_7L_1b + bL_7b + b\langle L_0L_7\rangle L_1b + bL_0L_7b
\end{aligned}
$$

$$\begin{aligned}
L_2 &\longrightarrow bL_1L_1b + bL_1b + bb + b\langle L_0L_1\rangle L_1b + bL_0L_1b + bL_0b \\
L_3 &\longrightarrow aL_8L_1b + aL_8b + aL_1b + ab \\
L_4 &\longrightarrow bL_0a + ba + bL_0L_0a \\
L_5 &\longrightarrow bL_3L_1b + bL_3b + b\langle L_0L_3\rangle L_1b + bL_0L_3b + b \\
L_6 &\longrightarrow bL_4L_1b + bL_4b + b\langle L_0L_4\rangle L_1b + bL_0L_4b \\
L_7 &\longrightarrow aL_2a + aa + a \\
\langle L_0L_1\rangle &\longrightarrow a\langle L_2L_4\rangle L_1b + aL_2L_4b \\
\langle L_0L_3\rangle &\longrightarrow a\langle L_2L_7\rangle L_1b + aL_2L_7b \\
\langle L_0L_4\rangle &\longrightarrow aL_2L_5a \\
\langle L_0L_7\rangle &\longrightarrow aL_2L_0a + a \\
\langle L_2L_4\rangle &\longrightarrow b\langle L_0L_5\rangle L_5a + bL_5L_5a \\
\langle L_2L_7\rangle &\longrightarrow b\langle L_0L_5\rangle L_0a + bL_0L_5a + bL_5L_0a + bL_5a \\
\langle L_0L_5\rangle &\longrightarrow a\langle L_2L_1\rangle L_1b + aL_2L_1b + aL_2b \\
\langle L_2L_1\rangle &\longrightarrow b\langle L_0L_5\rangle\langle L_4L_1\rangle b + b\langle L_0L_5\rangle L_4b + bL_5\langle L_4L_1\rangle b + bL_5L_4b \\
\langle L_4L_1\rangle &\longrightarrow ab\langle L_0L_7\rangle\langle L_4L_1\rangle b + b\langle L_0L_7\rangle L_4b + bL_7\langle L_4L_1\rangle b + bL_7L_3b
\end{aligned}$$

Remark 3.1. The only useless variable is now X.

3.3 Operator normal form

We present here another classical normal form, namely the operator normal form. A context-free grammar G over the terminal alphabet A is in *operator normal form* if no right member of a rule contains two consecutive variables. This normal form has been introduced for purposes from syntactical analysis. For these grammars, an operator precedence can be defined which is inspired of the classical precedence relations of usual arithmetic operators. From a general point of view, the following holds :

Theorem 3.5. [28, 16] *Given a context-free language, an equivalent context-free grammar in operator normal form can effectively be constructed.*

Proof. It is very easy. Given a grammar G in Chomsky normal form, to each pair of a terminal letter a and of a variable X is attached a new variable X_a designed to generate the set of words u such that X generates ua, that is to say $X_a = Xa^{-1}$. So, each language $L_G(X)$ is exactly the sum over A of all the languages $L_{X_a}a$, sum completed by $\{\varepsilon\}$ as soon as L_X contains ε. Identifying L_X and X, this can be written:

$$X = (\bigcup_{a\in A} X_aa) \cup (\{\varepsilon\} \cap X) \tag{3.1}$$

In the right members of the original grammar, we now replace each occurrence of the variables X by the right hand side of equation (3.1). This gives raise to a set of rules say P_1. Finally, we add the rules $X_a \longrightarrow \alpha$ for $X \longrightarrow \alpha a \in P_1$. This gives raise to a new grammar which is equivalent to the original one and is in operator normal form. Note that this new grammar may be neither proper nor reduced.   □

Example 3.5. Consider the grammar given by the two rules

$$S \longrightarrow aSS + b.$$

We introduce two new variables S_a and S_b. The set of rules in P_1 is

$$S \longrightarrow aS_a aS_a a + aS_a aS_b b + aS_b bS_a a + aS_b bS_b b + b.$$

We then add the rules

$$S_a \longrightarrow aS_a aS_a + aS_b bS_a \qquad S_b \longrightarrow aS_a aS_b + aS_b bS_b + \varepsilon.$$

and get the desired grammar.

If we reduce the grammar, we note that the variable S_a is useless. So, we get the grammar

$$S \longrightarrow aS_b bS_b b + b \quad S_b \longrightarrow aS_b bS_b + \varepsilon.$$

If we need a proper grammar in operator normal form, we just apply the usual algorithm to make it proper.

Remark 3.2. The theory of *grammar forms* [13] develops a general framework for defining various similar normal forms. These are defined through patterns like $VAV+A$ indicating that ight members have to lie in $VAV \cup A$. From this point of view, the various normal forms presented above appear as particular instances of a very general situation (see [5]).

4. Applications of the Greibach normal form

4.1 Shamir's theorem

We present a first application of Greibach normal form. The presentation given here follows [33]. Recall that, given an alphabet V containing n letters, we denote by D_V^* the Dyck set over the alphabet $(V \cup \overline{V})$. Given a word $m \in (V \cup \overline{V})^\star$, we denote $\widetilde{m}$ the reversal of m. We denote $\mathfrak{P}((V \cup \overline{V})^\star)$ the family of subsets of $(V \cup \overline{V})^\star$. We now state Shamir's theorem [51]:

Theorem 4.1 (Shamir). *For any context-free language L over A, there exists an alphabet V, a letter $X \in V$ and a monoid homomorphism $\Phi : A^\star \to \mathfrak{P}((V \cup \overline{V})^\star)$ such that*

$$u \in L \Longleftrightarrow X\Phi(u) \cap D_V^* \neq \emptyset.$$

Proof. Let $G = (V, P)$ be a context-free grammar in Greibach normal form over A generating L. To each terminal letter $a \in A$, associate the finite set

$$\Phi(a) = \{\overline{X}\widetilde{\alpha} \in (V \cup \overline{V})^* \mid X \longrightarrow a\alpha \in P\}.$$

This defines *Shamir's homomorphism* $\Phi : A^* \longrightarrow \mathfrak{P}((V \cup \overline{V})^*)$. A simple induction allows to prove that, for any terminal word $u \in A^*$ and for any nonterminal word $m \in V^*$

$$X \xrightarrow[\ell]{*} um \iff X\Phi(u) \cap [\widetilde{m}] \neq \emptyset$$

where $[\widetilde{m}]$ represents the class of $\widetilde{m}$ in the Dyck congruence. This shows, in particular, that $X \xrightarrow[\ell]{*} u$ iff $X\Phi(u) \cap D_V^* \neq \emptyset$ which is precisely the theorem. □

For later use, we state another formulation of the theorem.

Given a context-free grammar $G = (V, P)$ over A in Greibach normal form generating L, we associate to each terminal letter $a \in A$ the set $\hat{\Phi}(a) = \Phi(a)a\overline{a}$. As far as Shamir's theorem is concerned, nearly nothing is changed: we use the total alphabet $T = V \cup A$ instead of V, and the same result holds with D_T^* instead of D_V^*, that is

$$X \xrightarrow[\ell]{*} u \iff \exists w \in \hat{\Phi}(u) \; : \; Xw \in D_T^* \; . \tag{4.1}$$

4.2 Chomsky-Schützenberger's theorem

We now show how to use Shamir's theorem 4.1 to prove directly the famous Chomsky-Schützenberger theorem [28, 10], that we recall here :

Theorem 4.2 (Chomsky-Schützenberger). *A language L over the alphabet A is context-free iff there exists an alphabet T, a rational set K over $(T \cup \overline{T})^*$ and a morphism $\psi : (T \cup \overline{T})^* \longrightarrow A^*$, such that*

$$L = \psi(D_T^* \cap K).$$

Proof. We follow again [33]. The "if" part follows from the classical closure properties of the family of context-free languages. Hence, we just sketch the proof of the "only if" part. Let $G = (V, P)$ be a grammar over A and set $T = V \cup A$.

Define a homomorphism ψ from $(T \cup \overline{T})^*$ into A^* by

$$\forall X \in V, \psi(X) = \psi(\overline{X}) = 1$$
$$\forall a \in A, \; \psi(a) = a \text{ and } \psi(\overline{a}) = 1.$$

Using morphism $\hat{\Phi}$ of the reformulation of Shamir's theorem, we note that $w \in \hat{\Phi}(u) \implies \psi(w) = u$. Conversely, if $\psi(w) = u$ and $w \in \hat{\Phi}(A^*)$, then $w \in \hat{\Phi}(u)$. Thus

$$w \in \hat{\Phi}(u) \iff \psi(w) = u \quad \text{for} \quad w \in \hat{\Phi}(A^*).$$

Then, the right hand side of equation (4.1) is equivalent to

$$\exists w \in \hat{\Phi}(A^\star) : \psi(Xw) = u, \quad Xw \in D_T^\star \quad .$$

Thus, setting $K = X\hat{\Phi}(A^\star)$, which is rational, this can be written

$$X \xrightarrow[\ell]{\star} u \Longleftrightarrow \exists w : \psi(Xw) = u, \quad Xw \in D_T^\star \cap K$$

and the Chomsky-Schützenberger theorem is proved.   □

4.3 The hardest context-free language

We now show how to use Shamir's theorem 4.1 to get the hardest context-free language. We begin by some new notions and results.

Given a language L over the alphabet A, we define the *nondeterministic version of L*, denoted $ND(L)$, in the following way: first add to the alphabet A three new letters [,] and +. A word h in $([(A^\star+)^\star A^\star])^\star$ can be naturally decomposed into $h = [h_1][h_2] \cdots [h_n]$, each word h_i being decomposed itself in $h_i = h_{i,1}+h_{i,2}+\cdots+h_{i,k_i}, h_{i,j} \in A^\star$. A *choice* in h is a word $h_{1,j_1} h_{2,j_2} \cdots h_{n,j_n}$ obtained by choosing in each $[h_i]$ a factor h_{i,j_i}. Denote by $\chi(h)$ the set of choices in h. Then, the nondeterministic version of L is defined by:

$$ND(L) = \{h \mid \chi(h) \cap L \neq \emptyset\}.$$

Given an alphabet A, we denote by H_A the nondeterministic version of the Dyck language $D_A^\star$. In the particular case of a two letter alphabet $A = \{a, b\}$, we skip the index, so that H_A is denoted H. By definition, H is the *hardest context-free language*.

The first important observation is given by:

Fact 4.1. *If L is a context-free language, so is its nondeterministic version $ND(L)$.*

This lemma can be easily proved either by using pushdown automata or by showing that $ND(L)$ is obtained by a rational substitution applied to the language L. The terminology nondeterministic version of L comes from the following

Proposition 4.1. *The language $ND(L)$ is deterministic context-free iff L is regular; in this case, $ND(L)$ is regular too.*

For a proof, we refer the reader to [3]. We now end this short preparation by the

Lemma 4.1. *Given an alphabet A, there exists a morphism λ such that $H_A = \lambda^{-1}(H)$.*

Proof. Let $H = H_B$ with $B = \{a, b\}$.

If the alphabet A contains only one letter c, just define the morphism λ by $\lambda([) = [, \lambda(]) =], \lambda(+) = +, \lambda(c) = a$ and $\lambda(\overline{c}) = \overline{a}$.

If A contains $n \geq 3$ letters, λ will be the usual encoding : $\lambda(a_i) = ab^i a$ and $\lambda(\overline{a_i}) = \overline{ab}^i \overline{a}$ for $1 \leq i \leq n$. For the three other letters, we define $\lambda([) = [,$ $\lambda(]) =]$ and $\lambda(+) = +.$ □

We now state the

Theorem 4.3 (Greibach). [23] *A language L over the alphabet A is context-free iff there exists a morphism φ such that $\$L = \varphi^{-1}(H)$, where $\$$ is a new letter.*

Proof. The "if" part follows directly from the closure properties of the family of context-free languages and from the fact that H is context-free. Hence, we turn to the "only if" part, for which we follow once more [33]. Given a context-free grammar $G = (V, P)$ in Greibach normal form, we associate to the morphism Φ used in Shamir's theorem a morphism φ defined by

$$\varphi(a) = [m_1 + m_2 + \cdots m_n] \Longleftrightarrow \Phi(a) = \{m_1, m_2, \ldots, m_n\}$$

Here, $m_1, m_2, \ldots, m_n$ is some arbitrary but fixed enumeration of the words in $\Phi(a)$. Moreover, we define $\varphi(\$) = [X]$ if X is the variable in V generating L. Set now $\theta = \chi\varphi$; hence, $\theta(u)$ will be the set of choices of the word $\varphi(u)$.

It is easy to check that

$$w \in \theta(u) \Longleftrightarrow w \in \Phi(u) \qquad \text{i.e.} \qquad \theta = \Phi.$$

(Just interpret the word $h = \varphi(u)$ as a polynomial representing the set $\Phi(u)$ and develop this polynomial.)

Consequently, Shamir's theorem can be rephrased as

$$X\Phi(u) \cap D_v^* \neq \emptyset \Longleftrightarrow \theta(\$u) \cap D_v^* \neq \emptyset \Longleftrightarrow \$u \in \varphi^{-1}(H_V).$$

Hence, we have $\$L = \varphi^{-1}(H_V)$. The announced result in theorem 4.3 then follows from lemma 4.1. □

Observation. The *membership problem* is the following: given a language L and a word u, does u belong to L? The language H is called the hardest context-free language because, by theorem 4.3, from a complexity point of view, H is the context-free language for which the membership problem is the most difficult. Any algorithm deciding if a given word belongs to H gives raise to a general algorithm for the membership problem for context-free languages; this general algorithm will have the same complexity than the one given for H.

4.4 Wechler's theorem

We end this section by showing another consequence of the Greibach normal form. Given a language L over the alphabet A and a letter $a \in A$, recall that the *left quotient* of L by a is the language $a^{-1}L = \{u \in A^* \mid au \in L\}$. An *algebra* is a family of languages closed under union and product and containing the family Fin of finite languages. An algebra $\mathcal{F}$ is *finitely generated* if there exists a finite family $\mathcal{F}'$ such that any language in $\mathcal{F}$ is obtained from languages in $\mathcal{F}'$ under the algebra operations. It is *stable* if it is closed under left quotient. We may now state the

Theorem 4.4 (Wechler). [54] *A language L is context-free if and only if it belongs to a finitely generated stable algebra.*

Proof. Given a context-free language L, it is generated by a grammar in Greibach normal form. To each variable X is associated the (context-free) language L_X that it generates. Clearly, the left quotient of such a language by a terminal letter a can be described as a finite union of product of languages generated in the grammar. Hence, the algebra generated by all these languages L_X contains L and is stable.

Conversely, if L belongs to a finitely generated stable algebra, the finite set of generators give raise to a finite set of variables and the description of each left quotient as a finite union of product of languages of the generators gives a grammar in Greibach normal form generating L. $\square$

5. Pushdown machines

In this section, we focus on the accepting device for context-free languages, namely pushdown automata with the important subclass induced by determinism, in both classical and less classical presentations. We prove here mainly two beautiful theorems: the first states that the stack language of a pushdown automaton is a rational language; the second says that the output language of a pushdown transducer is context-free when the input is precisely the language recognized by the associated pda.

5.1 Pushdown automata

The classical mechanism of recognition associated to context-free languages is the pushdown automaton. Most of the material presented in this paragraph is already in Ginsburg[17].

A *pushdown machine* over A (a *pdm* for short) is a triple $\mathcal{M} = (Q, Z, T)$ where Q is the set of *states*, Z is the *stack alphabet* and T is a finite subset of $(A \cup \{\varepsilon\}) \times Q \times Z \times Z^* \times Q$, called the set of *transition rules*. A is the *input alphabet*. An element (y, q, z, h, q') of T is a rule, and if $y = \varepsilon$, it is an ε-rule. The first three components are viewed as pre-conditions in the behaviour of a pdm (and therefore the last two components are viewed as post-conditions), T is often seen as a function from $(A \cup \{\varepsilon\}) \times Q \times Z$ into the subsets of $Z^* \times Q$, and we note $(h, q') \in T(y, q, z)$ as an equivalent for $(y, q, z, h, q') \in T$.

A pushdown machine is *realtime* if T is a finite subset of $A \times Q \times Z \times Z^* \times Q$, i.e. if there is no ε-rule. A realtime pdm is *simple* if there is only one state. In this case, the state giving no information, it is omitted, and T is a subset of $A \times Z \times Z^*$.

An *internal configuration* of a pdm $\mathcal{M}$ is a couple $(q, h) \in Q \times Z^*$, where q is the current state, and h is the string over Z^* composed of the symbols in the stack, the first letter of h being the bottom-most symbol of the stack. A *configuration* is a triple $(x, q, h) \in A^* \times Q \times Z^*$, where x is the input word to be read, and (q, h) is an internal configuration.

The *transition relation* is a relation over configurations defined in the following way: let $c = (yg, q, wz)$ and $c' = (g, q', wh)$ be two configurations, where y is in $(A \cup \{\varepsilon\})$, g is in A^*, q and q' are in Q, z is in Z, and w and h are in Z^*. There is a *transition* between c and c', and we note $c \vdash c'$, if $(y, q, z, h, q') \in T$. If $y = \varepsilon$, the transition is called an ε-transition, and if $y \in A$, the transition is said to involve the reading of a letter. A *valid computation* is an element of the reflexive and transitive closure of the transition relation, and we note $c \vdash^* c'$ a valid computation starting from c and leading to c'. A convenient notation is to introduce, for any word $x \in A^*$, the relation on internal configurations, denoted $\models^x$, and defined by:

$$(q, w) \models^x (q', w') \iff (x, q, w) \vdash^* (\varepsilon, q', w').$$

We clearly have: $\models^x \circ \models^y = \models^{xy}$.

An internal configuration (q', w') is *accessible* from an internal configuration (q, w), or equivalently, (q, w) is *co-accessible* from (q', w') if there is some $x \in A^*$ such that $(q, w) \models^x (q', w')$.

A rule $(y, q, z, h, q') \in T$ is an *increasing* rule (respectively a *stationary*, respectively a *decreasing* rule) if $|h| > 1$ (respectively $|h| = 1$, respectively $|h| < 1$). The use of an increasing rule (respectively a stationary, respectively a decreasing rule) in a computation increases (respectively leaves unchanged, respectively decreases) the number of symbols in the stack. A pdm is in *quadratic form* if for all rules $(y, q, z, h, q') \in T$, we have: $|h| \leq 2$.

A pdm is used as a device for recognizing words by specifying starting configurations and accepting configurations. The convention is that there is only one starting internal configuration $i = (q, z)$, where the state q is the initial state, and the letter z is the initial stack symbol. For internal accepting configurations, many kinds make sense, but the set K of internal accepting configurations usually is of the form: $K = \cup_{q \in Q} \{q\} \times K_q$ with $K_q \in Rat(Z^*)$.

A *pushdown automaton* over A (a *pda* for short) is composed of a pushdown machine (Q, Z, T) over A, together with an *initial internal configuration* i, and a set K of *internal accepting configurations*. It is so a 5-tuple $\mathcal{A} = (Q, Z, i, K, T)$, and (Q, Z, T) is called the *pdm associated* to $\mathcal{A}$.

For a pda, an internal configuration is *accessible* if it is accessible from the initial internal configuration, and it is *co-accessible* if it is co-accessible from an internal accepting configuration.

The sets of internal accepting configurations usually considered are:

1. the set $F \times Z^*$ where F is a subset of Q, called the set of *accepting states*.
2. the set $Q \times \{\varepsilon\}$.
3. the set $F \times \{\varepsilon\}$ where F is a subset of Q.
4. the set $Q \times Z^* Z'$ where Z' is a subset of Z.

We call each of these cases a *mode of acceptance*.

A word $x \in A^*$ is *recognized* by a pda $\mathcal{A} = (Q, Z, i, K, T)$ over A with a specified mode of acceptance if there is $k \in K$ such that $i \overset{x}{\vDash} k$. Considering the modes of acceptance defined above, in the first case, the word is said to be recognized by *accepting states* F, in the second case the word is said to be recognized by *empty storage*, in the third case the word is said to be recognized by *empty storage and accepting states* F, and in the last case the word is said to be recognized by *topmost stack symbols* Z'. The *language accepted* by a pda with a given mode of acceptance is the set of all words recognized by this pda with this mode. For any pda $\mathcal{A} = (Q, Z, i, K, T)$, we note $L(\mathcal{A})$ the language recognized by $\mathcal{A}$, and for any set of internal accepting configurations K', we note $L(\mathcal{A}, K')$ the language recognized by the pda $\mathcal{A}' = (Q, Z, i, K', T)$.

Note that, with regards to the words recognized, the names of the states and of the stack symbols are of no importance. Up to a renaming, we can always choose $Q = \{q_1, q_2, \ldots, q_p\}$, and similarly, $Z = \{z_1, z_2, \ldots, z_n\}$. Up to a renaming too, we can always set the initial internal configuration equal to (q_1, z_1).

Example 5.1. Let $\mathcal{A} = (Q, Z, (q_0, t), K, T)$ be the pda over $A = \{a, b\}$, where $Q = \{q_0, q_1, q_2, q_3\}$, $Z = \{z, t\}$ of rules:

$$(a, q_0, t, zt, q_1), \quad (a, q_0, t, zzt, q_2),$$
$$(a, q_1, t, zt, q_1),$$
$$(a, q_2, t, zzt, q_2), \quad (a, q_2, t, zt, q_1),$$
$$(\varepsilon, q_1, t, \varepsilon, q_3), \quad (\varepsilon, q_2, t, \varepsilon, q_3),$$
$$(b, q_3, z, \varepsilon, q_3).$$

In state q_1, each letter a read increases by one the number of symbols z under the top symbol t in the stack. In state q_2, each letter a read increases by two the number of symbols z under the top symbol t in the stack, or increases it by one and changes the state to q_1. The two ε-rules remove the top stack symbol t, changing the state to q_3, in which the only thing possible to do is removing one z in the stack for each b read.

Then we have, for example:

$$L(\mathcal{A}, Q \times \{\varepsilon\}) = \{a^n b^p \mid 0 < n \leq p \leq 2n\},$$
$$L(\mathcal{A}, \{q_3\} \times Z^*) = \{a^n b^p \mid 0 < n \text{ and } 0 \leq p \leq 2n\},$$
$$L(\mathcal{A}, \{q_2\} \times \{\varepsilon\}) = \emptyset,$$
$$L(\mathcal{A}, Q \times Z^* z) = \{a^n b^p \mid 0 < n \text{ and } 0 \leq p < 2n\}.$$

As seen on this example, for a given pda, changing the mode of acceptance changes the languages recognized. Nevertheless, the family of languages that are recognized by pda's, using any of these modes remains the same. This can be proved easily using a useful, though technical, transformation of a pda adding it the bottom symbol testing ability. A pda *admits bottom testing* if there is a partition of the stack alphabet Z in $B \cup B'$ such that for any accessible configuration (q, w), the word w is in BB'^*. In other words, in such an automaton, a symbol at the bottom of the stack always belongs to B and, conversely, a symbol in the stack which belongs to B is the bottom symbol. So, if the topmost symbol of the stack happens to be a symbol in B, it is the only symbol in the stack. Since the only symbol in the stack that may be tested is the topmost symbol, it is then possible to know if it is the bottom symbol of the stack. Under these conditions, a valid computation leads to a configuration with an empty store if and only if the last transition uses a rule of the form: $(y, q, z, \varepsilon, q') \in T$ where z is in B. One construction to transform a pda $\mathcal{A}$ into a pda $\mathcal{A}'$ admitting bottom testing is the following. Let $\mathcal{A} = (Q, Z, i, K, T)$, let $Z' = \{z' \mid z \in Z\}$ be a copy of Z, and define T' by:

$$(y, q, z, \varepsilon, q') \in T \Leftrightarrow (y, q, z, \varepsilon, q') \in T' \wedge (y, q, z', \varepsilon, q') \in T'$$

and

$$(y, q, z, z_1 z_2 \ldots z_r, q') \in T \Leftrightarrow \begin{cases} (y, q, z, z_1 z'_2 \ldots z'_r, q') \in T' \wedge \\ (y, q, z', z'_1 z'_2 \ldots z'_r, q') \in T'. \end{cases}$$

Finally, denoting by $\pi : (Z \cup Z')^* \to Z^*$ the homomorphism that erases the primes, set

$$K' = \{(q, h') \mid (q, \pi(h')) \in K\}$$

and

$$\mathcal{A}' = (Q, Z \cup Z', i, K', T').$$

Proposition 5.1. *The pda $\mathcal{A}'$ admits bottom testing and recognizes the same language as $\mathcal{A}$, for any mode of acceptance.*

Hence there is a common family of languages recognized by pda's using any mode of acceptance which is the family of context-free languages:

Theorem 5.1. *The family of languages recognized by pda's by empty storage and accepting states is exactly the family of context-free languages.*

Proof. Let $\mathcal{A} = (Q, Z, i, K, T)$ be a pda. We denote $[p, w, q]$, for $w \in Z^+$, the language

$$[p, w, q] = \{x \in A^* \mid (p, w) \overset{x}{\vdash\!\!\!=} (q, \varepsilon)\},$$

and set

$$[p, \varepsilon, q] = \begin{cases} \emptyset & \text{if } p \neq q \\ \varepsilon & \text{if } p = q \end{cases}$$

We then have, for $w, w' \in Z^*$:

$$[p, ww', q] = \bigcup_{r \in Q} [p, w', r][r, w, q].$$

We can derive from T that the languages $[q, z, q']$, for $z \in Z$, satisfy the set of equations:

$$[p, z, q] = \bigcup_{(y, p, z, h, q') \in T} y[q', h, q]. \tag{5.1}$$

Hence the languages $[q, z, q']$ are all context-free, and so is the language:

$$\bigcup_{q \in F, \ i = (q_1, z_1)} [q_1, z_1, q]$$

which is exactly the language recognized by $\mathcal{A} = (Q, Z, i, K, T)$ with $K = \{\varepsilon\} \times F$ (i.e. by empty storage and accepting states F).

Conversely, if $G = (V, P)$ is a context-free grammar over A such that $P \subset V \times (A \cup \{\varepsilon\})V^*$, one can construct from P a pdm $\mathcal{M} = (V, T)$ over A without states, where $T \subset (A \cup \{\varepsilon\}) \times V \times V^*$ is defined by: $(y, X, m) \in T \iff X \to y\tilde{m}$. The language $L_G(X)$ is then recognized by the pda associated to $\mathcal{M}$ with initial stack symbol X by empty storage. □

Remark 5.1. If the system of equations (5.1) is replaced by the associated context-free grammar, there is a one to one correspondence between valid computations of the pda and leftmost derivations in the grammar. Hence the number of different valid computations leading to recognize a word x gives the number of different leftmost derivations for x.

For pushdown automata, the mode of acceptance is generally chosen to give the simplest proofs for one's purpose. Other modes of acceptance than the ones quoted above have been investigated. For instance, a result of Sakarovitch [46] shows that if $K = \cup_{q \in Q}\{q\} \times L_q$ with L_q context-free, then the language recognized remains context-free.

The characterization of context-free languages in terms of languages recognized by pda's allows much simpler proofs of certain properties of context-free languages.

Example 5.2. In order to show that the family of context-free languages is closed under intersection with rational languages, consider a context-free language L given by a pda $\mathcal{A}$, and a rational language K given by a finite automaton $\mathcal{B}$. Then a pda recognizing $L \cap K$ can effectively be constructed, using the Cartesian product of the states of $\mathcal{A}$ and of the states of $\mathcal{B}$.

A pushdown automaton is *realtime* (resp. *simple*) if the associated pdm is realtime (resp. simple).

The fact that any proper context-free language can be generated by a context-free grammar in Greibach normal form implies that realtime pda's, (and even simple pda's), recognize exactly proper context-free languages.

The realtime feature is the key to formulate Shamir's and Greibach's theorems (theorems 4.1 and 4.3), and that we rephrase here in an automata-theoretic framework.

In any pdm $\mathcal{M} = (Q, Z, T)$, the set T can be written as a function $\hat{T}$ from $(A \cup \{\varepsilon\})$ into the subsets of $Q \times Z \times Z^* \times Q$. In the case of a realtime pdm, it is a function from A into the subsets of $Q \times Z \times Z^* \times Q$. Let $\overline{Z}$ be a copy of Z and $\overline{Q}$ a copy of Q as well, we can conveniently denote the element (q, z, h, q') in $Q \times Z \times Z^* \times Q$ by the word $\overline{q}.\overline{z}.h.q'$ over the Dyck alphabet $Q \cup Z \cup \overline{Z} \cup \overline{Q}$. Recall that we denote $D^\star_{Q \cup Z}$ the Dyck set over this alphabet. The Shamir function Φ from A into the subsets of $(Q \cup Z \cup \overline{Z} \cup \overline{Q})^*$ is defined by

$$\Phi(a) = \{\overline{q}\overline{z}hq' \mid (q, z, h, q') \in \hat{T}(a)\}.$$

Then extend it in the natural way to a morphism from A^* into the subsets of $(Q \cup Z \cup \overline{Z} \cup \overline{Q})^*$. Thus, Shamir's theorem states that $\Phi(x) \cap z_1 q_1 D^\star_{Q \cup Z} F \neq \emptyset$ iff x is recognized by the realtime pda $\mathcal{A} = (Q, Z, i, F \times \{\varepsilon\}, T)$ by empty storage and accepting states F.

The Shamir function Φ gives raise to a function from A into $(\{[,], +\} \cup Q \cup Z \cup \overline{Z} \cup \overline{Q})^*$, extended to an homomorphism φ, that we call the *Greibach homomorphism*, by setting:

$$\varphi(x) = [m_1 + m_2 + \ldots + m_k] \Longleftrightarrow \Phi(x) = \{m_1, m_2, \ldots, m_k\}.$$

Let $H_{Q \cup Z}$ be the Hardest context-free language over $Q \cup Z$. It follows that $[z_1 q_1] \varphi(x) F \in H_m$ iff x is recognized by the realtime pda $\mathcal{A} = (Q, Z, i, K, T)$ by empty storage and accepting states F. This is theorem 4.3.

The presence of the Dyck set in Shamir's theorem is due to the fact that this language fully describes the behaviour of the stack in a pdm: a letter that is unmarked is pushed on the top of the stack, while a marked letter erases the corresponding letter provided it is the topmost symbol in the stack. Recognition by empty storage means that the stack must be empty at the end of the computation, and D^* is precisely the class of the empty word ε for the Dyck congruence.

5.2 Deterministic pda

We now focus on determinism.

A pdm $\mathcal{M} = (Q, Z, T)$ over A is *deterministic* if the set T of transitions satisfies the following conditions for all $(y, q, z) \in (A \cup \{\varepsilon\}) \times Q \times Z$:

$$\mathrm{Card}(T(y, q, z)) \leq 1$$
$$T(\varepsilon, q, z) \neq \emptyset \implies T(a, q, z) = \emptyset, \ (a \in A).$$

A *deterministic pda* (dpda for short) is a pda with a deterministic associated pdm. The transformation of a pda into a pda admitting bottom testing described above, when applied to a deterministic pda, gives raise to a deterministic pda. Hence, it is possible to prove that the family of languages recognized by dpda's by empty storage is the same as the family of languages recognized by dpda's by empty storage and accepting states, and that this family is included in the family of languages recognized by dpda's by accepting states. On the other hand, it is easy to verify that a language recognized by empty storage by a dpda is prefix, *i.e.* no proper prefix of a word of this language belongs to this language. So, we are left with two families of languages: the family of languages recognized by accepting states, called the family of *deterministic languages*, and the family of languages recognized by empty storage and accepting states, called the family of *deterministic-prefix languages*. It is easy to check the following

Fact 5.1. *The family of deterministic-prefix languages is exactly the family of deterministic languages that are prefix languages.*

The two families are distinct. As an example, the language $L_1 = \{a^n b^p \mid p > n > 0\}$ is deterministic but not prefix. To avoid these problems, a usual trick is to consider languages with an end marker: indeed, $L\#$ is a prefix language which is deterministic if and only if L is deterministic.

One awkward feature about dpda's is that, due to possible ε-transitions that may occur after the input of the last letter of the word, there may be several valid computations for a fixed input word (being the beginning of one each other). This inconvenient can be avoided by a rather technical construction (see e. g. [2]) that transforms a dpda into an other dpda such that an accepting state is reached only if the computation is maximal.

Proposition 5.2. *For any dpda, it is possible to construct a dpda recognizing the same language such that an accepting state cannot be on the left side of an ε-rule.*

Consequently, in such a dpda, for any recognized word, there is only one successful computation. This proves the following

Proposition 5.3. *Deterministic languages are unambiguous.*

To see that the inclusion is strict, consider the language $L_2 = \{a^n b^n \mid n > 0\} \cup \{a^n b^{2n} \mid n > 0\}$. It is unambiguous, and it is not a deterministic language. Indeed, looking carefully at the valid computation used to recognize a word $a^n b^n$, it is not too difficult to prove that it is possible to find a word $a^{n+k} b^{n+k}$ for some $k > 0$ such that the internal configuration reached is the same than for the former word. Now, the valid computation for $a^n b^n$ should be the beginning of the valid computation for the word $a^n b^{2n}$. Hence the automaton must recognize the word $a^{n+k} b^{2n+k}$, which is not in L_2.

By the way, the technical construction invoked in Proposition 5.2 is also the key to prove the following

Theorem 5.2. *The family of deterministic languages is closed under complementation.*

This property is noe for the family of context-free languages: the language $\{a^n b^p c^q \mid n \neq p \text{ or } p \neq q\}$ is a context-free language, and its complement, intersected by the rational language $a^* b^* c^*$, is the language $\{a^n b^p c^q \mid n = p \text{ and } p = q\}$ which is not context-free.

Proposition 5.4. *For any dpda, it is possible to construct a dpda recognizing the same language such that any ε-rule is decreasing.*

This proposition is quoted as an exercise in [28] and [30]. However, no proof has appeared in the standard textbooks. A proof is given below.

The proof is in two steps of independent interest: first, we get rid of nondecreasing ε-rules for dpdas recognizing by topmost stack symbols *and* accepting states. In a second step, we show that such a dpda recognizes a deterministic language. This is achieved by constructing an equivalent ordinary dpda, but without introducing any nondecreasing ε-rule.

Proposition 5.5. *Given a dpda $\mathcal{A}$ recognizing by topmost stack symbols and accepting states, it is possible to construct a dpda $\mathcal{A}'$ recognizing the same language with the same mode of acceptance, and such that any ε-rule is decreasing.*

Proof. Let $\mathcal{A} = (Q, Z, i, K, T)$ be a dpda over A. Observe first that we may always saturate the set K of accepting configurations by adding all configurations $(q, h) \in Q \times Z^+$ such that $(q, z) \overset{\varepsilon}{\Longmapsto} k$ for $k \in K$.

Claim. The number of consecutive nondecreasing ε-transitions in a computation may be assumed to be uniformly bounded.

The proof of the claim is simple, and appears for instance in [17].

Claim. One may assume that there are never two consecutive nondecreasing ε-transitions in a computation.

The idea is to glue together, in a single rule, any maximal (bounded in view of the first claim!) sequence of consecutive nondecreasing ε-transitions appearing in a computation. If such a sequence contains an accepting configuration then, due to the saturation of K, its initial configuration is accepting, too.

Claim. One may assume that, in any computation, there is never a nondecreasing ε-transition followed by a decreasing ε-transition.

Again, the idea is to glue together a nondecreasing ε-rule followed by a decreasing ε-rule into one ε-rule. This decreases the total number of ε-rules. Therefore, the process stops after a finite number of steps. Accepting configurations are handled in the same way than above.

From now on, we may assume, in view of these claims, that any nondecreasing ε-transition either ends the computation or is not followed by an ε-transition.

We now finish the proof. Let $\mathcal{A}' = (Q, Z, i, K, T')$ be the automaton where T' is constructed as follows. T' contains all decreasing ε-rules of T. Next,

- If $T(\varepsilon, q, z) = \emptyset$, then $T'(a, q, z) = T(a, q, z)$ for $a \in A$.
- If $T(\varepsilon, q, z) = (r, m)$, with $m \neq \varepsilon$, then $T'(a, q, z) = (p, h)$, where $(a, r, m) \vdash (\varepsilon, p, h)$.

It is immediate to check that $\mathcal{A}'$ is equivalent to $\mathcal{A}$. By construction, it has only decreasing ε-rules.   □

We now turn to the second step.

Proposition 5.6. *Given a dpda $\mathcal{A}$ recognizing by topmost stack symbols and accepting states, and having only decreasing ε-rules, it is possible to construct a dpda $\mathcal{B}$ recognizing the same language by accepting states, and such that any ε-rule is decreasing.*

Proof. Let $\mathcal{A} = (Q, Z, i, K, T)$ be a dpda over A. We construct $\mathcal{B} = (Q', Z, i, K', T')$ as follows: $Q' = Q \cup P$, where $P = \{q_z \mid (q, z) \in K\}$. Next, $K' = P$. The set of rules T' first contains $T'(\varepsilon, q, z) = (q_z, \varepsilon)$ for $(q, z) \in K$. Furthermore,

- If $T(a, q, z) \neq \emptyset$ for some letter $a \in A$, then $T(a, q, z) = (q', m)$ for some $q' \in Q$ and $m \in Z^*$. In this case,

$$T'(a, q_z, z') = (q', z'm) \text{ for all } z' \in Z.$$

- If $T(\varepsilon, q, z) \neq \emptyset$, then, since the rule is decreasing, $T(\varepsilon, q, z) = (q', \varepsilon)$ for some $q' \in Q$. In this case,

$$T'(y, q_z, z') = T(y, q', z') \text{ for all } y \in A \cup \{\varepsilon\} \text{ and } z' \in Z.$$

By construction, the dpda $\mathcal{B}$ has only decreasing ε-rules. Clearly, $\mathcal{B}$ is equivalent to $\mathcal{A}$.   □

Proof of proposition 5.4. A successive application of propositions 5.5 and 5.6 proves the statement.   □

Remark 5.2. The proposition 5.6, but without reference to ε-rules, is proved in a simpler way in [2]. However, his construction does not apply to the proof of proposition 5.4.

Proposition 5.4 shows that, for deterministic automata, nondecreasing ε-rules are not necessary. On the contrary, decreasing ε-rules cannot be avoided.

Example 5.3. The language

$$L_3 = \{a^n b^p c a^n \mid p, n > 0\} \cup \{a^n b^p d b^p \mid p, n > 0\}$$

is deterministic, and recognized by the dpda with rules:

$$(a, q_1, z_1, z_1 z_2, q_1), \, (a, q_1, z_2, z_2 z_2, q_1), \, (b, q_1, z_2, z_3, q_2),$$
$$(b, q_2, z_3, z_3 z_3, q_2), \, (c, q_2, z_3, \varepsilon, q_3), \quad (d, q_2, z_3, z_3, q_5),$$
$$(\varepsilon, q_3, z_3, \varepsilon, q_3), \quad (a, q_3, z_2, \varepsilon, q_3), \quad (a, q_3, z_1, \varepsilon, q_4)$$
$$(b, q_5, z_3, \varepsilon, q_5), \quad (\varepsilon, q_5, z_2, z_2, q_6).$$

by accepting states, with accepting states q_4 and q_6.

However, L_3 cannot be recognized by any realtime deterministic pda. Indeed, a word starts with $a^n b$, and it is necessary, while reading the factor b^p, to push on the stack an unbounded number of symbols that will be matched when the word ends with db^p, and all these symbols have to be erased when the word ends with ca^n.

This example shows that, contrarily to the general case, the realtime condition induces an effective restriction on the family of recognized languages.

Let $\mathcal{R}$ be the family of languages recognized by deterministic realtime automata by empty storage. There is a Shamir theorem for languages in $\mathcal{R}$, that we state now.

Let Φ be the Shamir function from A^* into the subsets of $(Q \cup Z \cup \overline{Z} \cup \overline{Q})^*$. Since the automaton is deterministic, for all $(q, z) \in Q \times Z$, there is at most one element beginning by $\overline{q}.\overline{z}$ in each image $\Phi(a)$ of $a \in A$. Such an homomorphism is called a "controlled" homomorphism. Shamir's theorem can be rephrased as follows. A language L is in $\mathcal{R}$ if and only if there exists a controlled homomorphism Φ from A^* into the subsets of $(Q \cup Z \cup \overline{Z} \cup \overline{Q})^*$ such that $\Phi(x) \cap z_1 q_1 D_m{}^* F \neq \emptyset \iff x \in L$.

We define a rational set R by: $w \in R$ iff $w = [w_1 + w_2 + \cdots + w_r]$, $w_k \in \overline{Q}\,\overline{Z}\,Z^* Q$ and for all $(q, z) \in Q \times Z$, there is at most one w_k beginning with $\overline{q}\overline{z}$.

Considering the Greibach homomorphism φ from A^* into the monoid $(\{[,],+\} \cup Q \cup Z \cup \overline{Z} \cup \overline{Q})^*$, then $\varphi(a) \in R$, for all a in A. It follows that

$[z_1q_1]\varphi(x) \in H_{Q \cup Z} \cap R^*$ if and only if x is recognized by the realtime dpda $\mathcal{A} = (Q, Z, i, K, T)$ over A by empty storage. (Recall that $H_{Q \cup Z}$ is the Hardest context-free language over $Q \cup Z$.) By the way, we can remark that $H_{Q \cup Z} \cap R^*$ is itself a language in $\mathcal{R}$.

The additional condition of being a simple (realtime) dpda induces also an effective restriction:

Fact 5.2. *The language $L_4 = \{a^n b a^n \mid n > 0\}$ is realtime deterministic, but not simple.*

Proof. L_4 is recognized by empty storage and accepting states by the realtime dpda $\mathcal{A} = (Q, Z, T, i, K, T)$ where $Q = \{q_1, q_2, q_3\}$, $Z = \{B, z\}$, $i = (q_1, B)$, $K = \{(q_3, \varepsilon)\}$ and T is the set of rules:

$$(a, q_1, B, Bz, q_1), \ (a, q_1, z, zz, q_1), \ (b, q_1, z, \varepsilon, q_2),$$
$$(a, q_2, z, \varepsilon, q_2), \quad (a, q_2, B, \varepsilon, q_3).$$

Hence, L_4 is a realtime deterministic language. However, it cannot be recognized by a deterministic simple automata, since it is necessary to know whether an input letter a belongs to the first or to the second factor a^n. $\square$

Given two families of languages $\mathcal{C}$ and $\mathcal{D}$, the *equivalence problem* for $\mathcal{C}$ and $\mathcal{D}$, denoted $Eq(\mathcal{C}, \mathcal{D})$, is the following decision problem:
Instance: A language L in $\mathcal{C}$, and a language M in $\mathcal{D}$.
Question: Is L equal to M?

The *equivalence problem* for $\mathcal{C}$, denoted $Eq(\mathcal{C})$, is the problem $Eq(\mathcal{C}, \mathcal{C})$. It is well known that $Eq(Alg)$ is undecidable for the family Alg of context-free languages, and up to now it is unknown whether $Eq(Det)$ is decidable or not, where Det denotes the family of deterministic languages. So there has been a huge amount of works solving $Eq(\mathcal{C}, \mathcal{D})$ for various subfamilies of Det. We only quote here a few among the results published in the literature, in the positive case. The equivalence problem is decidable for parenthesis languages (see paragraph 6.6 below)[39], for simple languages (see paragraph 6.7 below)[36], for finite-turn languages (see paragraph 6.4 below)[53], realtime languages[42] ,(pre-)NTS languages[49]. A result of Sénizergues[50] shows that if $\mathcal{C}$ is an effective cylinder (i. e. a family of languages effectively closed under inverse homomorphism and intersection with rational sets) containing the family Rat of rational sets for which the equivalence problem is decidable, then so is $Eq(\mathcal{C}, Det)$.

In order to recognize the whole family of deterministic languages by realtime automata, we have to modify the standard model of pdm. We already noticed that, in a dpda, only decreasing ε-rules are necessary. They are necessary because, as seen for the language L_3 defined above, it happens that some unbounded amount of information pushed on the stack has to be erased at the same time. So, if we want to have a realtime device, this leads to use some mechanism that erases an unbounded number of topmost stack symbols in one step. Several such mechanisms have been introduced and studied in the literature (see e.g. Cole[11], Courcelle[12], Greibach[26], Nivat[41], Schützenberger[48]). We present now one such accepting device.

A *jump pdm* over A is a 4-tuple $\mathcal{A} = (Q, Z, J, T)$, where Q and Z have the same meaning as in a pdm, and J is a new alphabet in bijection with Z, the elements of which are called jump stack symbols, or simply *jump symbols*, and T, the set of transitions, is a finite subset of $A \times Q \times Z \times (Z^* \cup J) \times Q$. We denote λ the bijection between J and Z. Observe that, by definition, a jump pdm is a realtime device. A jump pdm is deterministic if for all $(a, q, z) \in A \times Q \times Z$, there is at most one (h, q) such that $(a, q, z, h, q) \in T$.

Configurations of a jump pdm are just the same as configurations of a pdm, but the transition relation is modified: let $c = (ag, q, wz)$ and $c' = (g, q', w')$, where a is in A, g is in A^*, q and q' are in Q, z is in Z, and w and w' are in Z^*. There is a transition between c and c', and we write $c \vdash c'$, if either $(a, q, z, h, q') \in T$ with $h \in Z^*$ and $w' = wh$, just as for pda's, or $(a, q, z, j, q') \in T$ with $j \in J$ if $w = w'zw_2$ and $z = \lambda(j)$ has no occurrence in w_2; in such a transition, an unbounded number of symbols (namely $|zw_2|$) is erased.

A *valid computation* is an element of the reflexive and transitive closure of the transition relation, and we note $c \vdash^* c'$ a valid computation starting from c and leading to c'.

A *jump pda* is to a jump pdm what a pda is to a pdm: it is a 6-tuple $\mathcal{A} = (Q, Z, J, i, K, T)$, where (Q, Z, J, T) is a jump pdm, and i and K have the same significance than in a pda. Observe that jump pda's generalize pda's: a pda is a jump pda with no jump rules. A jump pda is a *deterministic jump pda* (jump dpda for short) if the associated jump pdm is deterministic.

Since, in a jump pda, it is possible to erase an unbounded number of stack symbols in one move, the standard accepting mode is by empty storage. This is the mode considered when we do not specify an other one.

As an example, consider again the deterministic language

$$L_3 = \{a^n b^p c a^n \mid p, n > 0\} \cup \{a^n b^p d b^p \mid p, n > 0\}$$

over $\{a, b, c, d\}$. It is recognized by empty storage and accepting states by the jump pda $\mathcal{A} = (Q, Z, J, i, K, T)$, where $Q = \{q_1, \ldots, q_5\}$, $K = \{(\varepsilon, q_4)\}$, $Z = \{z, A, B\}$, $J = \{j_z, j_A, j_B\}$ and T is the set:

(a, q_1, z, zA, q_1), (a, q_1, A, AA, q_1), (b, q_1, A, AB, q_2), (b, q_2, B, BB, q_2),
$(d, q_2, B, \varepsilon, q_3)$, $(b, q_3, B, \varepsilon, q_3)$, (b, q_3, A, j_z, q_4),
(c, q_2, B, j_A, q_5), $(a, q_5, A, \varepsilon, q_5)$, $(a, q_5, z, \varepsilon, q_4)$.

Indeed a word of L_3 begins with $a^n b^p$. The first four rules just push on the stack $A^n B^p$ over the bottom symbol z. Now, if the word ends with db^p, the three next rules are used to recognize the word: the first two to pop all symbols B while reading db^{p-1}, and the third (with jump symbol j_z) to erase the remaining symbols of the stack, i.e. zA^n. Last, if the word ends with ca^n, the last three rules are used to recognize the word: the first one (with jump symbol j_A) to erase all the top factor AB^p in the stack, the second to pop all symbols A while reading a^{n-1}, and the third to erase the remaining symbol z at the reading of the last a.

It is easy to construct from a (deterministic) jump pdm, a (deterministic) pdm (which will not be in general realtime) that acts in the same way: first, a rule (a, q, z, j, p) is replaced by the rule (a, q, z, z, p_j), where p_j is a new state. This replacement does not change determinism. Then, the following set of rules is added:

$$\{(\varepsilon, p_j, z, \varepsilon, p_j) \mid \lambda(j) \neq z\} \cup \{(\varepsilon, p_j, z, \varepsilon, p \mid \lambda(j) = z\}.$$

Remark that these new rules do not enter in conflict with the older ones, since the states involved are new states, nor with one another. So, determinism is preserved by this construction.

Consequently, in the nondeterministic case, we have the following

Proposition 5.7. *The family of languages recognized by jump pda's is exactly the family of context-free languages.*

A similar statement holds for deterministic languages.

Proposition 5.8. *The family of languages recognized by deterministic jump pda's is exactly the family of deterministic languages.*

In view of the preceding construction, and of the remark concerning the deterministic case, it only remains to prove that a deterministic language can be recognized by a jump dpda. The proof is very technical and lengthy, so we refer the interested reader either to Greibach[25], or to Cole[11].

An other model considered allows to erase rational segments of the stack word. This is clearly a generalization of jump pdm, since in a jump pdm, the erased factors have the form zh with $h \in (Z - \{z\})^*$. Observe that this rational set is recognized by the finite automaton obtained from the rules added in the construction above, (rules of the form: $(\varepsilon, p_j, z, \varepsilon, p_j)$ or $(\varepsilon, p_j, z, \varepsilon, p)$) when skipping first and fourth components (those equal to ε). It is an easy exercise to change the sets of rules added so that the erased factors belong to any rational set. If the rational sets are chosen to be prefix, as it is the case for jump pdm, determinism is still preserved. Hence, this model is equivalent to jump pdm.

Just as the behaviour of the stack in a pdm is described by the Dyck set, the behaviour of the stack in a jump pdm is described by a new set E_Z, which is a generalization of the Dyck set, defined as follows. E_Z is the class of the empty word for the congruence generated by

$$\{z\bar{z} \cong \varepsilon \mid z \in Z\} \bigcup \{zj \cong j \mid z \in Z, \ \lambda(j) \neq z\} \bigcup \{zj \cong \varepsilon \mid z \in Z, \ \lambda(j) = z\}.$$

We name E_m this set if $m = \text{Card}(Z) = \text{Card}(J)$.

It is a result of Greibach[26] that each language E_m cannot be recognized by a deterministic jump pda with $m-1$ jump symbols. Hence, the number of jump symbols induces a hierarchy.

Again, it is possible to state a Shamir-Greibach like theorem for deterministic languages, using jump dpda: let $\mathcal{A} = (Q, Z, J, i, K, T)$ be a deterministic jump pda over A. This time, the Shamir function Φ is a function from A^* into the subsets of $(Q \cup Z \cup T \cup \overline{Z} \cup \overline{Q})^*$, and the Greibach homomorphism φ is a function from A^* into $(\{[,],+\} \cup Q \cup Z \cup T \cup \overline{Z} \cup \overline{Q})^*$. We define a rational set R' by: $w \in R'$ iff $w = [w_1 + w_2 + \cdots + w_r]$, $w_k \in \overline{QZ}(Z^* \cup T)Q$ and for all $(q,z) \in Q \times Z$, there is at most one w_k beginning with $\overline{qz}$.

We have that for all a in A, $\varphi(a)$ is in R'. If $ND(E_{Q \cup Z})$ is the nondeterministic version of $E_{Q \cup Z}$ (see section 3.3), it follows that $[z_1 q_1]\varphi(x) \in ND(E_{Q \cup Z}) \cap R'^*$ if and only if x is recognized by the deterministic jump pda $\mathcal{A} = (Q, Z, J, i, K, T)$.

Again, we can remark that $ND(E_{Q \cup Z}) \cap R'^*$ is itself recognized by a deterministic jump pda.

5.3 Pushdown store languages

In this paragraph, we show that the language composed of the strings that may occur in the pushdown store is rational.

Let $\mathcal{A} = (Q, Z, i, K, T)$ be a pda over A. We call *pushdown store language* of $\mathcal{A}$ the language $P(\mathcal{A})$ over Z of all words u such that there exists some state q for which the internal configuration (q, u) is both accessible and co-accessible. Formally, $P(\mathcal{A})$ is defined by:

$$P(\mathcal{A}) = \{u \in Z^* \mid \exists x, y \in A^*, \ \exists q \in Q, \ \exists k \in K \ : \ i \overset{x}{\vDash} (q, u) \overset{y}{\vDash} k\}.$$

Theorem 5.3. *Given a pda and some mode of acceptance, the pushdown store language of this pda is rational.*

For any state $q \in Q$, we define the two sets:

$$Acc(q) = \{u \in Z^* \mid \exists x \in A^* \ : \ i \overset{x}{\vDash} (q, u)\},$$

$$Co\text{-}Acc(q) = \{u \in Z^* \mid \exists y \in A^*, \exists k \in K \ : \ (q, u) \overset{y}{\vDash} k\}.$$

Clearly:

$$P(\mathcal{A}) = \bigcup_{q \in Q} (Acc(q) \cap Co\text{-}Acc(q)) \tag{5.2}$$

We now show that the languages $Acc(q)$ and $Co\text{-}Acc(q)$ are rational.

Lemma 5.1. *The set $Acc(q)$ is rational.*

Proof. We first consider the particular case of a pda $\mathcal{A} = (Q, Z, i, K, T)$ in quadratic form, i.e. such that for any rule $(a, q, z, h, q') \in T$, $|h| \leq 2$.

Let $u = t_1 \cdots t_{r+1}$, where $t_i \in Z$. A valid computation (x, q_0, y_1) $\overset{*}{\vdash} (\varepsilon, q, u)$ can be decomposed into several steps such that, at the last move of each of these steps, one letter of u is definitively set in the stack. Formally, the whole computation is decomposed into:

$$(x, q_0, y_1) \overset{*}{\vdash} (x_1, q_1, z_1') \vdash (x_1', q_1', t_1 y_2) \overset{*}{\vdash} (x_2, q_2, t_1 z_2') \vdash (x_2', q_2', t_1 t_2 y_3)$$

$$\overset{*}{\vdash} \cdots \overset{*}{\vdash} (x_r, q_r, t_1 t_2 \cdots t_{r-1} z_r') \vdash (x_r', q_r', t_1 t_2 \cdots t_r y_{r+1}) \overset{*}{\vdash} (\varepsilon, q, u),$$

where $y_1, \ldots, y_{r+1}$ and $t_1, \ldots, t_{r+1}$ are in Z. Define now the context-free grammar G_q with terminal alphabet Z, nonterminal alphabet $Q \times Z$, and rules:

$$(p, z) \longrightarrow (p', z') \quad \text{if } \exists x \in A^* \; : \; (p, z) \overset{x}{\models} (p', z')$$
$$(p, z) \longrightarrow t(p', z') \quad \text{if } \exists a \in A \cup \{\varepsilon\} \; : \; (a, p, z, t z', p') \in T$$
$$(p, z) \longrightarrow \varepsilon \qquad \text{if } \exists x \in A^* \; : \; (p, z) \overset{x}{\models} (q, \varepsilon)$$
$$(q, z) \longrightarrow z$$

A straightforward proof by induction on the length of a derivation shows that if there is a derivation $(p, z) \overset{*}{\longrightarrow} u$ in G_q, then there is a valid computation $(p, z) \overset{x}{\models} (q, u)$ in $\mathcal{A}$.

Conversely, if there is a valid computation $(x, q_0, y_1) \overset{*}{\vdash} (\varepsilon, q, u)$ in $\mathcal{A}$, then the decomposition described above of this valid computation gives the rules to be applied to form a derivation $(q_0, y_1) \overset{*}{\longrightarrow} u$ in G_q.

Thus we have:

$$L_{G_q}((q_1, z_1)) = Acc(q),$$

and since G_q is a right linear grammar, $Acc(q)$ is rational.

Note that the grammar G_q can be effectively computed since the condition

$$\exists x \in A^* \; : \; (p, z) \overset{x}{\models} (p', z')$$

is an instance of the emptiness problem for a context-free language.

Considering now the general case, the proof goes along the same lines. However, we have to modify the grammar G_q in order to skip the condition that for any rule $(a, p, z, h, p') \in T$, $|h| \leq 2$.

Indeed, when symbols are definitively set in the stack at a time (there may be more than one), several symbols may be pushed that will have to be erased. The whole computation is now decomposed into:

$$(x, q_0, y_1) \overset{*}{\vdash} (x_1, q_1, z_1') \vdash (x_1', q_1', t_1 y_2) \overset{*}{\vdash} (x_2, q_2, t_1 z_2') \vdash (x_2', q_2', t_1 t_2 y_3)$$

$$\overset{*}{\vdash} \cdots \overset{*}{\vdash} (x_r, q_r, t_1 t_2 \cdots t_{r-1} z_r') \vdash (x_r', q_r', t_1 t_2 \cdots t_r y_{r+1}) \overset{*}{\vdash} (\varepsilon, q, u),$$

where $y_1, \ldots, y_{r+1}$ and $t_1, \ldots, t_{r+1}$ are now nonempty words over Z.

Define now the context-free grammar G_q with terminal alphabet Z, nonterminal alphabet $Q \times Z$, and rules:

$$(p, z) \longrightarrow (p', z') \quad \text{if } \exists x \in A^* \; : \; (p, z) \overset{x}{\models} (p', z')$$
$$(p, z) \longrightarrow t(p', z') \quad \text{if } \exists a \in A \cup \{\varepsilon\}, t, y \in Z^+, x \in A^*, p'' \in Q \; :$$
$$(a, p, z, t y, p'') \in T \text{ and } (p'', y) \overset{x}{\models} (p', z')$$
$$(p, z) \longrightarrow \varepsilon \qquad \text{if } \exists x \in A^* \; : \; (p, z) \overset{x}{\models} (q, \varepsilon)$$
$$(q, z) \longrightarrow z$$

The same proof than before ensures that $L_{G_q}((q_1, z_1)) = Acc(q)$, and since G_q is a right-linear grammar, we get that $Acc(q)$ is rational. $\qquad\square$

We now turn to the proof of

Lemma 5.2. *The set Co-Acc(q) is rational.*

Proof. We first consider the case of the mode of acceptance by empty storage and accepting states. Let F be the set of accepting states.

Consider a valid computation $(x, q, u) \overset{*}{\vdash} (\varepsilon, q', \varepsilon)$ with $q' \in F$ and $u = t_1 \cdots t_{r+1}$, where $t_1, \ldots, t_{r+1}$ are in Z. It can be decomposed into:

$$(x, q, u) \overset{*}{\vdash} (x_r, p_r, t_1 t_2 \cdots t_r) \overset{*}{\vdash} \cdots$$
$$\cdots \overset{*}{\vdash} (x_2, p_2, t_1 t_2) \overset{*}{\vdash} (x_1, p_1, t_1) \overset{*}{\vdash} (\varepsilon, q', \varepsilon).$$

Define now a context-free grammar H over terminal alphabet Z, with non-terminal alphabet Q, and rules:

$$p \longrightarrow p'z \text{ if } \exists x \in A^* \ : \ (p, z) \overset{x}{\vdash} (p', \varepsilon)$$
$$p \longrightarrow \varepsilon \quad \text{if } p \in F.$$

Again, the grammar H can be effectively computed.

A straightforward proof by induction on the length of a derivation shows that if there is a derivation $p \overset{*}{\longrightarrow} u$ in H, then there is a valid computation $(x, p, u) \overset{*}{\vdash} (\varepsilon, q', u)$ with $q' \in F$ in $\mathcal{A}$.

Conversely, if there is a valid computation $(x, q, u) \overset{*}{\vdash} (\varepsilon, q', \varepsilon)$ in $\mathcal{A}$, then the decomposition described above of this valid computation gives the rules to be applied to form a derivation $q \overset{*}{\longrightarrow} u$ in H.

Thus we have:
$$L_H(q) = Co\text{-}Acc(q),$$

and since H is a left linear grammar, $Co\text{-}Acc(q)$ is rational.

It remains to explain how to generalize the result for any mode of acceptance, i.e. how to modify the grammar H in accordance with the mode of acceptance chosen.

Suppose that u is in $Co\text{-}Acc(q)$, i.e. there exists $x \in A^*$, and $(q', u') \in K$ such that $(x, q, u) \overset{*}{\vdash} (\varepsilon, q', u')$. If u' is not empty, there is a longest left factor v of u such that the symbols of v are not involved in this valid computation. This computation can be divided into two subsets: in the first one all but one of the symbols of u above v are deleted, the second one being the rest of the computation. If at the end of the first part, the internal configuration is (p, vz) for some pushdown symbol z, setting $u = vzw$, we then have two valid computations: $(x_1, q, w) \overset{*}{\vdash} (\varepsilon, p, \varepsilon)$ and $(x_2, p, z) \overset{*}{\vdash} (\varepsilon, q', u')$ with $x = x_1 x_2$.

Hence, this leads to the (left linear) grammar H over terminal alphabet Z, with nonterminal alphabet $Q \cup \{\sigma\}$, and rules:

$$q \longrightarrow q'z \qquad \text{if } \exists\, x \in A^* \; : \; (q,z) \xmapsto{x} (q',\varepsilon)$$
$$p \longrightarrow \sigma z \qquad \text{if } \exists\, x \in A^* \; : \; (p,z) \xmapsto{x} (q',u')$$
$$\sigma \longrightarrow \sigma z \; + \; \varepsilon \text{ for all } z \in Z$$
$$q' \longrightarrow \varepsilon \qquad \text{if } (q',\varepsilon) \in K \,.$$

Again we have $L_H(q) = Co\text{-}Acc(q)$, hence we get that $Co\text{-}Acc(q)$ is rational.
□

Note that, also in the general case, the grammars G_q and H can be effectively computed.

From equation (5.2) and lemmas 1 and 2, we get that $P(\mathcal{A})$ is rational, hence the proof of the theorem is complete.
□

5.4 Pushdown transducers

A pda to which is adjoint an output is a pushdown transducer. In this paragraph, we show that the output language of a pushdown transducer is a context-free language when the given input is precisely the language recognized by the associated pda.

A *pushdown machine with output* over A is a 4-tuple $\mathcal{S} = (Q, Z, B, \gamma)$ where B is an alphabet called the output alphabet, γ is a finite subset of $(A \cup \{\varepsilon\}) \times Q \times Z \times Z^* \times Q \times B^*$, and if T is the projection of γ onto $(A \cup \{\varepsilon\}) \times Q \times Z \times Z^* \times Q$, (Q, Z, T) is a pushdown machine, called the pdm associated to $\mathcal{S}$.

We note $(h, q', u) \in \gamma(y, q, z)$ as an equivalent for $(y, q, z, h, q', u) \in \gamma$.

An internal configuration of a pushdown machine with output $\mathcal{S}$ is an internal configuration of the associated pdm. A configuration is a 4-tuple $(x, q, h, g) \in A^* \times Q \times Z^* \times B^*$, where x is the input word to be read, g is the word already output, and (q, h) is an internal configuration.

The transition relation is a relation over configurations defined the following way: there is a transition between $c = (yx, q, wz, g)$ and $c' = (x, q', wh, gu)$, where y is in $(A \cup \{\varepsilon\})$, g is in A^*, q and q' are in Q, z is in Z, w and h are in Z^*, and g and u are in B^*, and we note $c \vdash c'$, if $(y, q, z, h, q', u) \in \gamma$. A valid computation is an element of the reflexive and transitive closure of the transition relation, and we note $c \vdash^* c'$ a valid computation starting from c and leading to c'.

Besides T, we can derive from γ an other function from $A^* \times Q \times Z^*$ into the subsets of B^*, named the *output function* of $\mathcal{S}$, denoted μ, and defined as follows:

$$\mu(x, q, h) = \{g \in B^* \mid \exists\, q' \in Q, h' \in Z^* \; : \; (x, q, h, \varepsilon) \vdash^* (\varepsilon, q', h', g)\} \,.$$

It follows that, for $x \in A^*$, $y \in A \cup \{\varepsilon\}$, $q \in Q$, $z \in Z$ and $w \in Z^*$:

$$\mu(yx, q, wz) = \bigcup_{(y,q,z,h,q',u) \in T} u\mu(x, q', wh) \,.$$

A pushdown transducer is to a pushdown machine with output what a pda is to a pdm, i.e. it is a pushdown machine with output with specified initial and accepting configurations.

A *pushdown transducer* over A (pdt for short in the rest of the text) is a 6-tuple $T = (Q, Z, B, i, K, \gamma)$ where (Q, Z, B, γ) is a pushdown machine with output, i is the internal starting configuration, and $K = F \times \{\varepsilon\}$ where F is a subset of Q, the accepting states.

If T is the projection of γ onto $(A \cup \{\varepsilon\}) \times Q \times Z \times Z^* \times Q$, then $\mathcal{A} = (Q, Z, i, K, T)$ is a pushdown automaton, called the pda associated to T. By convention, the output of T in the initial configuration is the empty word.

The existence of a set K of accepting configurations leads to define a function similar to the function μ, but taking accepting configurations in account:

$$M(x, q, h) = \{g \in B^* \mid \exists\, (q', h') \in K \ : \ (x, q, h, \varepsilon) \overset{*}{\vdash} (\varepsilon, q', h', g)\}.$$

Finally, the *transduction realized* by T is the function Θ from A^* into the subsets of B^* defined by

$$\forall x \in A^*, \ \Theta(x) = M(x, q_1, z_1).$$

Proposition 5.9. *The image through Θ of a rational language is context-free.*

We don't prove this proposition.

Consider now the following example: $T = (Q, Z, B, i, K, \gamma)$ with $A = B = \{a, b, c\}$, $Z = \{z_1, X\}$, $Q = \{q_1, q_2, q_3, q_4\}$, $K = \{(\varepsilon, q_4)\}$ and γ composed of

$$(a, q_1, z_1, z_1 X, q_1, a), \ (a, q_1, X, XX, q_1, a), \ (b, q_1, X, X, q_2, b),$$
$$(b, q_2, X, X, q_2, b), \quad (c, q_2, X, \varepsilon, q_3, c), \quad (c, q_3, X, \varepsilon, q_3, c),$$
$$(c, q_3, z_1, \varepsilon, q_4, c).$$

It is easy to see that, due to the fact that the language recognized by the associated pda is $\{a^i b^j c^{i+1} \mid i, j > 0\}$,

$$\Theta(x) = \begin{cases} a^i b^j c^{i+1} & \text{if } x = a^i b^j c^{i+1} \text{ with } i, j > 0 \\ \emptyset & \text{otherwise.} \end{cases}$$

So, if $L = \{a^i b^i c^j \mid i, j > 0\}$, then $\Theta(L) = \{a^i b^i c^{i+1} \mid i > 0\}$. Hence

Fact 5.3. *The image through Θ of a context-free language is not always context-free.*

Nevertheless,

Theorem 5.4 (Evey). [14] *Given a pushdown transducer T, if L is the context-free language recognized by the associated pda, the image $\Theta(L)$ is a context-free language.*

Proof. Let $T = (Q, Z, B, i, K, \gamma)$ be a pdt over A. Define a new alphabet

$$H = \{\langle y, u \rangle \mid \exists q, z, h, q' \; : \; (y, q, z, h, q', u) \in \gamma\}$$

We can define a set of transitions T in $H \times Q \times Z \times Z^* \times Q$ by:

$$(\langle y, u \rangle, q, z, h, q') \in T \iff (y, q, z, h, q', u) \in a.$$

Setting $\mathcal{A} = (Q, Z, i, K, T)$, we get a (realtime) pda over H recognizing a context-free language N over H. Finally, we consider the two morphisms π and ξ from H^* into A^* and B^* respectively, defined by:

$$\forall \langle y, u \rangle \in H, \; \pi(\langle y, u \rangle) = y \; and \; \xi(\langle y, u \rangle) = u.$$

It is then clear that $\pi(N)$ is the language L recognized by the associated pda, and $\xi(N)$ is equal to $\Theta(L)$. □

6. Subfamilies

We present here some subfamilies among the very numerous ones that have been studied in the literature. We will begin with the probably most classical one, namely

1. the family *Lin* of linear languages.

We then turn to some families derived from it

2. the family *Qrt* of quasi-rational languages
3. the family *Sqrt* of strong quasi-rational languages
4. the family *Fturn* of finite-turn languages.

We then present other subfamilies, namely

5. the families *Ocl* of one-counter languages and *Icl* of iterated counter languages
6. the family of parenthetic languages
7. the family of simple languages
8. the families of *LL* and *LR* languages.

6.1 Linear languages

The simplest way to define the family of linear languages is by grammars: a context-free grammar is *linear* if each right member of the rules contain at most one variable. A context-free language is *linear* if there exists a linear grammar generating it [10, 4].

We denote by *Lin* the family of linear languages. Naturally, the first question that arises is whether *Lin* is a proper subfamily of the family of context-free languages. This is easily seen to be true. Many proofs are possible. Here is an example of a context-free language which is not linear: let Δ be the linear language $\{a^n b^n \mid n \geq 0\}$; the language $\Delta\Delta$ is context-free but not linear. The direct proof naturally leads to a specific iteration theorem:

Theorem 6.1. *Given a linear language L, there exists an integer N_0 such that any word w in L of length at least N_0 admits a factorization $w = xuyvz$ satisfying*

(1) $xu^n yv^n z \in L \quad \forall n \in \mathbb{N}$
(2) $uv \neq 1$.
(3) $|xuvz| \leq N_0$

The proof of this iteration theorem is very similar to the proof of the classical iteration theorem of Bar-Hillel, Perles and Shamir; it uses derivation trees in a grammar generating L. In the usual version, the third condition states that the length of uyv is at most N_0; here, due to the fact the grammar is linear, we may select in the derivation tree the topmost repetition instead of the lowest one. (Note that in a non linear grammar, the notion of topmost repetition does not make sense.) We leave to the reader the proof of the above theorem as well as its use to prove that $\{a^n b^n \mid n \geq 0\}\{a^m b^m \mid m \geq 0\}$ is not linear.

The linear languages can be defined in many various ways. We briefly describe here the most important ones.

6.1.1 Pushdown automata characterization. We begin by some definitions. Given a computation of a pda A, a *turn* in the computation is a move that decreases the height of the pushdown store and is preceded by a move that did not decreased it.

A pda A is said to be *one-turn* if in any computation, there is at most one turn.

Fact 6.1. *A language is linear if and only if it is recognized by a one-turn pda.*

The proof of this fact is easy: the construction of a pda from a grammar presented in the previous section on pda's gives raise to one-turn pda from a linear grammar; similarly, the construction of a grammar from a pda gives raise to a nearly linear grammar from one-turn pda: in the right member of any rule, there is at most one variable generating an infinite language. Such a grammar can easily be transformed in a linear one.

This characterization may help to prove that some languages are linear; it may be easier to describe a one-turn pda than a linear grammar for a given language. This is the case, for example, for the language over $A = \{a, b\}$

$$L = \{a^{n_1} b a^{n_2} b \cdots a^{n_k} \mid k \geq 2, \exists i, j, \ 1 \leq i < j \leq k; n_i \neq n_j\}.$$

The one-turn pda recognizing L can roughly be described as follows: the machine reads an arbitrary number of blocks $a^n b$, then it counts up the number of letters a in a block; it then reads again an arbitrary number of blocks $a^n b$, then it counts down the number of letters a checking it is not equal to the previous number of a's. Clearly, this nondeterministic machine is one-turn and recognized L, hence L is linear.

This characterization also naturally leads to consider the following question: say that a language is in the family $DetLin$ if it is recognized by a deterministic one-turn pda (a one-turn dpda). Clearly, $DetLin \subset Det \cap Lin$. The question raises whether this inclusion is strict or not. The answer is yes. Here is a example : let $A = \{a, b\}$ and consider the language

$$L = \{a^n b^m a^p b^q \mid n = m \text{ or } p = q \quad n, m, p, q \geq 1\}.$$

It is easy to check that L is linear and deterministic:

On one hand, the language L is generated by the linear grammar G given by

$$\begin{array}{lll} S \ \rightarrow T + X & T \rightarrow aT + aT' & T' \rightarrow bT' + bT'' \\ T'' \rightarrow aT''b + ab & X \rightarrow Xb + X'b & X' \rightarrow X'a + X''a \\ X'' \rightarrow aX''b + ab & & \end{array}$$

On the other hand, the language L is recognized by the following dpda: count up the letters in the first block of a's; when entering the first block of b's, check if the number of b's is equal to the number of a's; if these two numbers are equal, read the second block of a's and of b's and accept; if they are not equal, restart the counting of letters a and b in the second block. This shows that $L \in Det \cap Lin$. However, $L \notin DetLin$; there is no deterministic one-turn pda recognizing L. Intuitively, in any deterministic pda recognizing L, the machine has to count in the stack the number of a's in the first block and then to check if the number of b's following these is the same. Hence, after reading the first block in $a^* b^*$, we already got at least one turn. If, at this moment, the number of a's and the number of b's happen to be different, the computation will have to count up the number of a's and to count down the number of b's in the second block, giving raise to a new turn. Hence, any deterministic pda recognizing L will be at least two-turn. It follows that

Proposition 6.1. *The family of deterministic and linear languages strictly contains the family of languages recognized by deterministic one-turn pda (or equivalently, the family of languages simultaneously deterministic and linear).*

The same linear language can be used to prove other results such as:

– L cannot be generated by a linear grammar in Greibach normal form.
– L is unambiguous but cannot be generated by an unambiguous linear grammar (showing that the inclusion $UnAmbLin \subset UnAmb \cap Lin$ is strict).

6.1.2 Algebraic characterization. Given an alphabet A, the *rational subsets* of $A^* \times A^*$ are defined as usual: they are the elements of the least family of subsets of $A^* \times A^*$ containing the finite ones and closed under union, product and star (i.e. generated submonoid). This family is denoted $Rat(A^* \times A^*)$.

To any subset R of $A^\star \times A^\star$, we associate the language L_R over A defined by $L_R = \{u\tilde{v} \mid (u,v) \in R\}$, where $\tilde{v}$ denotes the reversal of v. We may then characterize the family of linear languages by the following

Proposition 6.2. *A language L over A is linear if and only if there exists a rational subset R of $A^\star \times A^\star$ such that $L = L_R$.*

Proof. Given a linear grammar $G = (V, P)$ generating L, we consider the finite alphabet

$$B = \{\langle u, \tilde{v}\rangle \mid \exists X, Y \in V : X \to uYv \in P\} \cup \{\langle u, \varepsilon\rangle \mid \exists X \in V : X \to u \in P\}.$$

We then construct a new grammar over B as follows: to each terminal rule $X \longrightarrow u$ of the original grammar is associated the rule $X \longrightarrow \langle u, \varepsilon\rangle$ in the new grammar; to each nonterminal (linear) rule $X \longrightarrow uYv$ of the original grammar is associated the rule $X \longrightarrow \langle u, \tilde{v}\rangle Y$. This grammar is right linear and generates a rational language K over B. Using the homomorphism transforming each letter $\langle u, \tilde{v}\rangle$ of B in the corresponding element $(u, \tilde{v})$ of $A^\star \times A^\star$, we get an homomorphic image R of K. So, R is rational. Then, it is immediate to prove that $L = L_R$.

Conversely, using the same homomorphism, given a rational subset R of $A^\star \times A^\star$, we can construct a right linear grammar generating R; the rules will be of the form $X \longrightarrow (u, v)Y$ or $X \longrightarrow (u, v)$ for some $u, v \in A^\star$. To such rules we associate $X \longrightarrow uY\tilde{v}$ and $X \longrightarrow u\tilde{v}$ respectively. The new grammar obtained is linear and generates L_R.   □

As the rational subsets of $A^\star \times A^\star$ are exactly the rational transductions from $A^\star$ into $A^\star$, this characterization strongly connects linear languages to the theory of rational transductions and of rational cones [4].

6.1.3 Operator characterization. The above characterization can be reformulated in a slightly different way. Given a rational subset R of $A^\star \times A^\star$ and a language L over A, we define the binary operation *bracket* of R by L, denoted $[R, L]$, by

$$[R, L] = \{um\tilde{v} \mid (u, v) \in R, \ m \in L\}.$$

A family of languages $\mathcal{F}$ is said to be *closed under bracket* if, given a language L in the family $\mathcal{F}$ and any rational set R in $A^\star \times A^\star$, $[R, L]$ is in $\mathcal{F}$. We may then state

Proposition 6.3. *The family Lin of linear languages is the smallest family of languages containing the finite sets and closed under bracket.*

Proof. Denote by Fin the family of finite languages and let $\mathcal{M} = \{[R, F] \mid R \in Rat(A^\star \times A^\star), F \in Fin\}$. Since $[K, [R, F]] = [KR, F]$, $\mathcal{M}$ is closed under bracket and is the smallest family of languages containing the finite sets and closed under bracket. Next, let L be a linear language. By Proposition 6.2, there exists a rational set R of $A^\star \times A^\star$ such that $L = L_R$; this can now be reformulated $L = [R, \{1\}]$ showing that L is in $\mathcal{M}$. Hence, we have $Lin \subset \mathcal{M}$. As we know that the family Lin contains the finite languages and is closed under bracket, we have the reverse inclusion.   □

We shall see later that this characterization leads naturally to define some new subfamilies of the family of context-free languages.

6.2 Quasi-rational languages

One of the oldest families of languages derived from the family Lin is the family Qrt of quasi-rational languages. Again, this family can be defined in various ways, that we present now.

Definition 6.1. *The family Qrt of* quasi-rational languages *is the substitution closure of the family Lin of linear languages.*

This definition can be made more precise: we define, for k in $\mathbb{N}$, the family $Qrt(k)$ by $Qrt(0) = Rat$, and $Qrt(k + 1) = Lin \,\square\, Qrt(k)$, where Rat is the family of rational languages and, for two families of languages $\mathcal{L}$ and $\mathcal{M}$, the family $\mathcal{L} \,\square\, \mathcal{M}$ is the family obtained by substituting languages in $\mathcal{M}$ into languages in $\mathcal{L}$. Clearly,

$$Qrt = \bigcup_{k \in \mathbb{N}} Qrt(k).$$

It follows that $Qrt(1)$ is exactly the family Lin. It should be noted that, due to closure properties of the family Lin, one has $Lin \,\square\, Rat = Lin$. On the contrary, the inclusion $Rat \,\square\, Lin \supset Lin$ is strict.

Example 6.1. Over the alphabet $A = \{a, b\}$, we consider the linear language $L = \{a^n b^n \mid n > 0\}$. We then substitute to the letter a the linear language $L_a = \{x^n y^n \mid n > 0\}$ and to the letter b the finite language $L_b = \{z\}$. This gives raise to a quasi-rational language in $Qrt(2)$, namely $M = \{x^{n_1} y^{n_1} \cdots x^{n_k} y^{n_k} z^k \mid k > 0, n_i > 0, \ i = 1, \ldots, k\}$.

One of the first question solved was: does there exist a context-free language which is not in the family Qrt? The answer is yes, and the first proofs were direct; they proved this and two related results. The first one states that $Qrt(k)$ is strictly included in $Qrt(k + 1)$. The second one states that, similarly to the case of $Lin = Qrt(1)$, we have, for each integer k, $Qrt(k) \,\square\, Rat = Qrt(k)$ and $Rat \,\square\, Qrt(k) \subset Qrt(k + 1)$. We will denote $QRT(k)$ the family $Rat \,\square\, Qrt(k)$. These results can be summarized in the following chain of inclusions

$$Rat = Qrt(0) \subsetneq Lin = Qrt(1) \subsetneq QRT(1) \subsetneq Qrt(2) \subsetneq$$
$$\cdots$$
$$\subsetneq Qrt(k) \subsetneq QRT(k) \subsetneq Qrt(k + 1) \subsetneq$$
$$\cdots$$

Before explaining how these results have been proved, we turn to some characterizations of languages in Qrt used in these proofs and which are of independent interest.

Given a context-free grammar $G = (V, P)$ over A and, for each $a \in A$, a context-free grammar $G_a = (W_a, Q_a)$ over B in which the axiom is a, we construct a new grammar H over B called the *direct sum* of the grammars G and G_a for $a \in A$ as follows. The set of variables of H is the disjoint union of V and of the sets W_a for $a \in A$; the set of rules of H is the union of the sets of rules of G and of all the rules of the grammars G_a.

Using the results of the section considering grammars as equations, it is easy to see that, for each variable $X \in V$, the language generated by X in the grammar H is obtained from $L_G(X)$ by substituting to each letter a the language L_a generated by the grammar G_a. We then may repeat such an operation giving raise to an *iterated sum* of grammars. It then follows immediately that

Proposition 6.4. [40] *A language L is quasi-rational iff there exists a grammar generating L that is an iterated sum of linear grammars.*

Example 6.2. (continued) Consider the language L generated by the linear grammar $S \longrightarrow aSb + ab$ and the languages L_a generated by the linear grammar $a \longrightarrow xay + xy$ and L_b generated by the linear grammar $b \longrightarrow z$. The direct sum of these grammars is:

$$S \longrightarrow aSb + ab \quad a \longrightarrow xay + xy \quad b \longrightarrow z$$

and generates the language M of the previous example.

This characterization leads to the following new approach of quasi-rational languages. A variable S in a context-free grammar G is *expansive* if there exists a derivation $S \overset{*}{\longrightarrow} uSvSw$ for some words u, v, w. A grammar which contains no expansive variable is said to be *nonexpansive*. A language is *nonexpansive* if there exists a nonexpansive grammar generating it. Then,

Proposition 6.5. *A language is quasi-rational iff it is nonexpansive.*

This proposition explains that some authors use the term nonexpansive instead of quasi-rational. Proving that any quasi-rational language of order k is generated by a nonexpansive grammar is straightforward by induction on k: for $k = 1$, we have a linear language, thus generated by a linear grammar; such a grammar is obviously nonexpansive. Given now a language L in $Qrt(k+1)$, by definition, it is obtained by substituting to each letter a a linear language L_a in a language $M \in Qrt(k)$. By induction hypothesis, M is generated by a non-expansive grammar G; each language L_a is generated by a linear grammar. The direct sum of G and of the G_a is clearly nonexpansive.

The converse goes roughly this way: first, given a grammar G, define a preorder relation $\leq$ on the variables by setting $X \leq Y$ if there exists two words u, v and a derivation such that $X \overset{*}{\longrightarrow} uYv$. As usual, this preorder induces an equivalence relation $X \equiv Y$ iff $X \leq Y$ and $Y \leq X$. Verify then that, if G is nonexpansive, in the right member of a rule $X \longrightarrow \alpha$, there is at most one occurrence of a variable Y equivalent to X. Conclude then, using the order relation attached to the preorder $\leq$, that the grammar can be described as an iterated direct sum of linear grammars, so that the generated language is quasi-rational.

Proposition 6.5 is the result that has been used to prove directly that there exists a context-free language which is not quasi-rational (see [40, 55] for example). One of the first languages considered was the Lukasiewicz language generated by the grammar G

$$S \longrightarrow aSS + b.$$

The proofs showed that any grammar generating this language had to be expansive; the proofs were refined to exhibit, for each integer k, a language in $Qrt(k+1)$ not in $Qrt(k)$. They used the following grammars clearly related to G :

$$S_k \longrightarrow aS_kS_{k-1}$$
$$\cdots$$
$$S_i \longrightarrow aS_iS_{i-1}$$
$$\cdots$$
$$S_1 \longrightarrow aS_1S_0$$
$$S_0 \longrightarrow aS_0b + b.$$

These results are now proved as a consequence of a very powerful lemma (the *syntactic lemma*) which will not be presented here (see [4]).

It should be noted that, contrarily to the situation for linear languages, any quasi-rational language can be generated by a nonexpansive grammar in Greibach normal form. This follows from the construction of Rosenkrantz which preserves the nonexpansivity. On the other hand, it is an open problem to know if any unambiguous quasi-rational language can always be generated by an unambiguous nonexpansive grammar (i.e. do we have $NonAmbQrt = NonAmb \cap Qrt$?). A possibly related open problem is the following: given two quasi-rational languages, is it true that their intersection either is quasi-rational or is not context-free?

Proposition 6.5 leads to consider a new notion: the *index of a derivation* is the maximum number of occurrences of variables in the sentential forms composing it. A terminal word u has index k if, among all the derivations generating u, the one of minimum index is of index k. The grammar G is of *finite index* if the index of any generated word is bounded by a fixed integer k. Otherwise, it is of infinite index. It can be proved

Proposition 6.6. *A language is quasi-rational iff it is generated by a grammar of finite index.*

This result can be made even more precise: the family $Qrt(k)$ is exactly the family of languages generated by grammars of index k. We refer the reader to [22, 27, 47, 4] for a proof of this proposition.

6.3 Strong quasi-rational languages

We present now a less usual family of languages. It is derived from the bracket operation defined above. Recall that Lin is the smallest family closed under bracket containing the finite sets. Recall also that $Rat \,\square\, Lin$ denotes the rational closure of Lin, and denote $SQRT(1)$ this family of languages. (This family was denoted $QRT(1)$ just above.) We then define $Sqrt(2)$ as the smallest family of languages containing $SQRT(1)$ and closed under bracket. More generally, for each integer k, we define the families $SQRT(k)$ as the rational closure of $Sqrt(k)$, and $Sqrt(k+1)$ as the smallest family of languages containing $SQRT(k)$ closed under bracket. Hence, we may write

$$SQRT(k) = Rat \,\square\, Sqrt(k) , \quad Sqrt(k+1) = [SQRT(k)] ,$$

where $[\mathcal{L}]$ denotes the bracket closure of the family $\mathcal{L}$. Finally, we denote by $Sqrt$, the infinite union of the families $Sqrt(k)$. This is the family *strong quasi-rational* languages [7]. Clearly, for each k, $Sqrt(k) \subsetneqq Qrt(k)$. The following more precise fact holds

Fact 6.2. *There exists a language in $Qrt(2)$ which is not in $Sqrt$.*

Such a language is the language M of the example above:

$$M = \{x^{n_1} y^{n_1} \cdots x^{n_k} y^{n_k} z^k \mid k > 0, n_i > 0, \quad i = 1, \ldots, k\}.$$

It is in $Qrt(2)$. We want to show that it does not lie in $Sqrt$.

First, we show that if $\mathcal{F}$ is any family of languages such that the language M belongs to $Rat \,\square\, \mathcal{F}$, then M is a finite union of products of languages in $\mathcal{F}$. This follows immediately from the fact that M does not contain any infinite regular set. Hence, if $M \in SQRT(k)$, M is a finite union of product of languages in $Sqrt(k-1)$. Trying to split M into a finite product of languages immediately leads to note that there is exactly one factor in the product very similar to the language M itself. Thus, if $M \in SQRT(k)$, then M belongs to the family $Sqrt(k)$.

Next, we check that if $M = [R, L]$, then R is a finite subset of $\{x, y, z\}^\star \times \{x, y, z\}^\star$. This implies that, if M belongs to the family $Sqrt(k)$, it is a finite union of products of languages lying in $SQRT(k-1)$. Again, there is one factor in this union of products very similar to M leading to the conclusion that M should lie in $SQRT(k-1)$.

Hence, we may conclude that, if M belongs to $Sqrt$, it belongs to $Sqrt(1) = Lin$. As M is not linear, the fact is proved.

Similarly to the situation for quasi-rational languages, we have

Proposition 6.7. *For each k, the family $Sqrt(k)$ is a strict subfamily of $Sqrt(k+1)$.*

6.4 Finite-turn languages

The characterization of linear languages by one-turn pda naturally leads to define finite-turn pda's and languages. A pda is *k-turn* if any computation admits at most k turns. Naturally, a language will be said to belong to the family $Fturn(k)$ if it is recognized by a k-turn pda. Then a *finite-turn* pda is a pda which is k-turn for some integer k. A language is *finite-turn* if it is recognized by a finite-turn pda [21]. It is easy to prove that 0-turn languages are rational.

The family of finite-turn languages can be described using the bracket operation too. This definition is similar to the one of strong quasi-rational languages where the rational closure is replaced by the closure under union and product. More precisely, let $Fturn_1$ be the family Lin and set, for each integer k,

$$FTURN_k = Fin \,\square\, Fturn_k , \quad Fturn_{k+1} = [FTURN_k] ,$$

so that $FTURN_k$ is the closure of $Fturn_k$ under union and product and that $Fturn_{k+1}$ is the closure of $FTURN_k$ under bracket. Finally, we denote by $Fturn$ the infinite union over k in $\mathbb{N}$ the families $Fturn_k$ [21]:

$$Fturn = \bigcup_k Fturn_k = \bigcup_k FTURN_k.$$

It should be noted that the two families $Fturn_k$ and $Fturn(k)$ are not equal. For instance, let $\Delta = \{a^n b^n \mid n \geq 1\}$ and consider the language $L = \Delta^k$. It is easily seen that L is in $FTURN_1$. (So it belongs to $Fturn_2$ also.) Besides, L does not belong to $Fturn(k-1)$. So, $FTURN_1$ is not contained in $Fturn(k)$. However, the infinite union of the families $Fturn_k$ and the infinite union of the families $Fturn(k)$ coincide:

Fact 6.3. *The family $Fturn$ is exactly the family of finite-turn languages.*

Proof. It consists in showing that

(1) if L is a finite-turn language, so is $[R, L]$
(2) the family of finite-turn languages is closed under union and product.

This implies that $Fturn$ is contained in the family of finite-turn languages. Conversely, given a k-turn language, we decompose the computations of the k-turn pda recognizing it to get a description of the language through union, product and the bracket operation of $(k-1)$-turn languages, showing then the reverse inclusion.   □

Remark 6.1. The second part of the above proof shows in fact that, for each k, we have the inclusion $Fturn(k) \subsetneqq Fturn_k$.

The given characterization of finite-turn languages obviously shows that they are all strong quasi-rational languages. Here again, we get a proper subfamily :

Fact 6.4. *There exists a language in Sqrt(1) which is not finite-turn.*

Such a language is, for instance, $\Delta^* = \{a^n b^n \mid n \geq 1\}^*$. As for the above families, we have chains of strict inclusions:

Proposition 6.8. *For each k, the family $Fturn_k$ is a strict subfamily of $Fturn_{k+1}$, and the family $Fturn(k)$ is a strict subfamily of $Fturn(k+1)$.*

6.5 Counter languages

We first present in this section the family of one-counter languages. It is defined through pda's.

Definition 6.2. *A pda is* one-counter *if the stack alphabet contains only one letter. A context-free language is a* one-counter *language if it is recognized by a one-counter pda by empty storage and accepting states.*

We denote by *Ocl* this family of languages. The terminology used here comes from the fact that, as soon as the stack alphabet is reduced to a single letter, the stack can be viewed a counter.

Example 6.3. Over the alphabet $A = \{a, b\}$, we consider the Lukasiewicz language generated by the grammar $S \longrightarrow aSS + b$. It is a one-counter language: each letter a increases the height of the stack by 1, each letter b decreases it by 1. The word is accepted iff it empties the stack.

As in the case of linear languages, the first question is whether *Ocl* is a proper subfamily of the family of context-free languages. The proof that this holds is more technical than in the case of linear languages. The idea is to prove an iteration lemma for one-counter languages and to use it to get the desired strict inclusion [6]. We will give later on such counter-examples, but we will not state this lemma which is too technical and beyond the scope of this presentation.

As in the case of linear languages, the definition of one-counter languages through pda's naturally leads to define the family *DetOcl* as the family of languages recognized by a deterministic one-counter pda. Clearly, $DetOcl \subset Det \cap Ocl$. As in the linear case, the inclusion is strict :

Proposition 6.9. *The family of deterministic and one-counter languages strictly contains the family of languages recognized by a deterministic one-counter pda (or, equivalently, the family of languages simultaneously deterministic and one-counter).*

Proof. Over the alphabet $A = \{a, b, \#\}$, consider the language $L = \{w \# w' \mid w, w' \in \{a, b\}^* \; w' \neq \widetilde{w}\}$. We will show that L is in $Det \cap Ocl$ and is not in *DetOcl*.

It is deterministic: clearly the language $\{w \# \widetilde{w} \mid w \in \{a, b\}^*\}$ is deterministic. So, its complement C is deterministic, too. The language L is exactly the intersection of the language C and of the rational language $\{a,b\}^\star\#\{a,b\}^\star$. It follows that L is deterministic.

It is one-counter: the language L can be described as the (non disjoint) union of the two languages

$$L_1 = \{w\#w' \mid |w| \neq |w'|\}$$
$$L_2 = \{ucv\#u'dv' \mid c,d \in \{a,b\}, c \neq d, |u| = |v'|\}.$$

Clearly, L_1 and L_2 are one-counter languages. Thus, $L = L_1 \cup L_2$ is a one-counter language.

To see that L is not in $DetOcl$, the idea is to observe that, after reading an input w, the length of the stack is polynomially bounded in the length of the input. Since there is only one stack symbol, there exist two distinct words w and x of the same length that lead to the same (unique due to determinism) configuration. Hence, since $w\#\tilde{x}$ is accepted, so is $x\#\tilde{x}$, which is impossible.   □

Remark 6.2. The last argument of the above proof can be also used to show that the linear language $\{w\#\tilde{w} \mid w \in \{a,b\}^\star\}$ is not one-counter; it needs to prove that a one-counter pda may always be supposed to be realtime (see [18]). This shows, in particular, that the family Ocl is a strict subfamily of the family of context-free languages.

As in the case of linear languages, it can be seen that $Ocl \,\square\, Rat = Ocl$ whence the inclusion $Rat \,\square\, Ocl \subset Ocl$ is strict. This new larger family, denoted here OCL, is exactly the family of languages recognized by a one-counter pda which *admits bottom testing*. Again as in the case of linear languages, we may define the family Icl of iterated counter languages as the substitution closure of the family Ocl.

Similarly to what happened for the quasi-rational languages, this definition can be made more precise: we may define, for each k in $\mathbb{N}$, the family $Ocl(k)$ by $Ocl(0) = Rat$, and $Ocl(k+1) = Ocl(k) \,\square\, Ocl$. Then, the family Icl is the infinite union over k in $\mathbb{N}$ of the families $Ocl(k)$. Using such definitions, $Ocl(1) = Ocl$.

The study of the families $Ocl(k)$ leads naturally to prove that $Ocl(k) \,\square\, Rat = Ocl(k)$, whence $Rat \,\square\, Ocl(k) \subsetneq Ocl(k)$. This last family will naturally be denoted $OCL(k)$ and we get the following chain of strict inclusions

$$Rat = Ocl(0) \subsetneq Ocl(1) \subsetneq OCL(1) \subsetneq Ocl(2) \subsetneq$$
$$\cdots$$
$$\subsetneq Ocl(k) \subsetneq OCL(k) \subsetneq Ocl(k+1) \subsetneq$$
$$\cdots$$

(To be compared to the similar chain of inclusions concerning the families $Qrt(k)$ and $QRT(k)$.)

The languages in Icl can be characterized as languages recognized by pda's such that the stack language lies in the bounded set $z_1^\star \cdots z_k^\star$.

Up to now, we may remark that the situation here is very similar to the situation we had when dealing with linear and quasi-rational languages. However, it is worth noticing that, contrarily to the case of linear languages, one-counter languages do not enjoy other characterizations through grammars or operators as linear languages did. This explains that we will not get here subfamilies similar to the strong quasi-rational languages, etc...

If we compare the families Lin and Ocl with respect to inclusion, we see that these two families are incomparable. Even more,

Proposition 6.10. *There is a language in Ocl which is not in Qrt. There is a language in Lin which is not in Icl.*

The Lukasiewicz language given above as an example of one-counter language is precisely the language proved not to be quasi-rational (see previous subsection). The second inclusion can be proved using the linear language $L = \{w\#\widetilde{w} \mid w \in \{a,b\}^\star\}$ (see the previous remark).

Such a result leads to consider the two following problems: is it possible to characterize the languages in $Lin \cap Ocl$ on one hand, and, to describe any context-free language by substituting linear and one-counter languages on the other hand.

The first question is still open (see [8] for results on this theme). The second question has a negative answer: defining the family of *Greibach languages* as the substitution closure of linear and one-counter languages, we get a large strict subfamily of the family of context-free languages: it does not contain the language $D^\star$, the Dyck language over $\{a,b,\overline{a},\overline{b}\}$ (see [4]).

6.6 Parenthetic languages

We now turn to another subfamily of the family of context-free languages. Consider an alphabet A containing two particular letters a and $\overline{a}$. A context-free grammar over the terminal alphabet A is *parenthetic* if each rule of the grammar has the following form $X \longrightarrow a\alpha\overline{a}$ with α containing neither the letter a nor the letter $\overline{a}$. As usual, a language is said to be *parenthetic* if it is generated by some parenthetic grammar.

This family of languages has been introduced by McNaughton [39]. In the particular case where the alphabet A does not contain any other letters than the two special ones a and $\overline{a}$, we speak of *pure parenthetic* grammar or language.

Example 6.4. Over the alphabet $A = \{a, \overline{a}\} \cup \{x\}$, the grammar G given by

$$S \longrightarrow aSS\overline{a} + ax\overline{a}$$

is parenthetic.

Clearly, any pure parenthetic language over $A = \{a, \overline{a}\}$ is included in the Dyck language $D_1^\star$. The following characterization due to Knuth [35] shows, in particular, that D_1^* is not (purely) parenthetic. A word u over the alphabet $A = \{a, \overline{a}\} \cup B$ is *balanced* if it satisfies $|u|_a = |u|_{\overline{a}}$ and, for any prefix v of u, $|v|_a \geq |v|_{\overline{a}}$. It should be noted that a word w is in D_1^* iff it is balanced.

Given a word u over the alphabet $A = \{a, \overline{a}\} \cup B$ and an occurrence of the letter a in u, we factorize u in $u = vaw$. An occurrence of a letter $b \in A$ in w is an *associate* of a iff $u = vaxby$ with xb balanced.

Example 6.5. Let $u = abab\overline{a}ab\overline{a}\overline{a}$. The first a and the first b are associates. The first a and the first $\overline{a}$ are associates too.

A language L over the alphabet $A = \{a, \overline{a}\} \cup B$, is *balanced* iff any word $u \in L$ is balanced. It is said to have *bounded associates* if there exists an integer k such that any occurrence of a letter a in $u \in L$ admits at most k associates. We may then characterize parenthetic languages as follows:

Proposition 6.11. *A context-free language L is parenthetic if and only if it is balanced and has bounded associates.*

The proof of the "only if" part is immediate. The "if" part consists in a careful study of the structure of any grammar generating a language which is balanced and have bounded associates. This study shows that the rules of the grammar may be 'recentered' to fulfill the parenthetic conditions.

Example 6.6. Consider the grammar G over $A = \{a, \overline{a}\} \cup \{b, c, d, e\}$ given by the set of rules

$$S \longrightarrow XY \quad X \longrightarrow aab\overline{a}X\overline{a} + ad \quad Y \longrightarrow aYac\overline{a}\overline{a} + e\overline{a}.$$

Clearly, this grammar is not parenthetic. However, the generated language is balanced and has bounded associates. Hence, it is a parenthetic language.

This characterization can be used to see that the Dyck set D_1^* is not parenthetic: it is balanced but it has unbounded associates. This characterization allows also to prove the following

Proposition 6.12. *If a language L is nonexpansive and parenthetic, there exists a parenthetic nonexpansive grammar generating it.*

This fact contrasts with the Proposition 6.1.

Besides this characterization, parenthetic languages enjoy some other nice properties. In particular, any such language is deterministic. Moreover, the equivalence problem for parenthetic grammars is decidable [35], solving in this particular case the equivalence problem of deterministic languages.

This family of languages can be related to the whole family of context-free languages in the following way. Given a context-free grammar $G = (V, P)$ over B, we associate to it a parenthetic grammar $Par(G)$ as follows : enlarge the terminal alphabet B into the terminal alphabet $A = B \cup \{a, \overline{a}\}$ where a and $\overline{a}$ are two new letters; to each rule $X \longrightarrow \alpha$ of G, associate the rule of $Par(G)$ given by $X \longrightarrow a\alpha\overline{a}$. It is easy to check that to each leftmost derivation in $Par(G)$ generating a word w, corresponds a leftmost derivation in G generating the word u obtained from w by erasing the new letters a and $\bar{a}$. This correspondence is a bijection between derivations. Hence, the *degree of ambiguity* of a word u in the grammar G is the number of words w generated in $Par(G)$ that map to u when the letters a and $\bar{a}$ are erased.

Example 6.7. Consider the grammar H given by $S \longrightarrow SS + x$. The corresponding parenthetic grammar $Par(H)$ is $S \longrightarrow aSS\bar{a} + ax\bar{a}$. The word xxx is the image of the two words $aax\bar{a}aax\bar{a}ax\bar{a}\bar{a}\bar{a}$ and $aaax\bar{a}ax\bar{a}\bar{a}ax\bar{a}\bar{a}$ corresponding to the two (leftmost) derivations in H

$$S \longrightarrow SS \longrightarrow xS \longrightarrow xSS \longrightarrow xxS \longrightarrow xxx$$

and

$$S \longrightarrow SS \longrightarrow SSS \longrightarrow xSS \longrightarrow xxS \longrightarrow xxx.$$

A very similar family of languages has been introduced by Ginsburg [19]. Let $k \geq 1$. Given an alphabet $A = \{a_1, \ldots, a_k\}$, we associate to it the copy $\overline{A} = \{\bar{a}_1, \ldots, \bar{a}_k\}$ and the alphabet $Z = A \cup \overline{A}$. A grammar over $Z \cup B$ with k rules is *completely parenthetic* if the ith rule has the form $X \longrightarrow a_i \alpha \bar{a}_i$ with α containing no letters in Z. As usual, a language is *completely parenthetic* if there exists a completely parenthetic grammar generating it.

Clearly, if we consider the morphism from Z onto $\{a, \bar{a}\}$ erasing the indices, we get from any completely parenthetic language a parenthetic language. Such languages often appear in the study of languages attached to trees.

Example 6.8. Given the completely parenthetic grammar G' given by $S \longrightarrow a_1 SS\bar{a}_1 + a_2 x\bar{a}_2$. The corresponding parenthetic grammar is the grammar G of the above example.

6.7 Simple languages

A context-free grammar $G = (V, P)$ over A is *simple* if it is in Greibach normal form and if, for each pair $(X, a) \in V \times A$, there is at most one rule of the form $X \longrightarrow am$. As usual, a language is *simple* if it can be generated by a simple grammar [28, 36]. It is easy to check that any simple language is deterministic. (It is even $LL(1)$.) It is easy too to check that there does exist deterministic (even $LL(1)$) languages which are not simple. The simple languages are exactly the languages recognized by simple deterministic pda's as defined in the previous section. Moreover, this family of languages enjoys nice properties:

1. Any simple language is prefix (i.e. if the two words u and uv are in L then v is the empty word).

2. The equivalence problem for simple grammars is decidable [36], solving again in a particular case the equivalence problem of deterministic languages.
3. The family of simple languages generates a free monoid.

Similarly to parenthetic and completely parenthetic languages, simple languages give raise to a family, namely the family of *very simple* languages. A grammar is *very simple* if it is simple and such that for any terminal letter a there is at most one rule of the form $S \longrightarrow am$. (Here, a appears as first letter of one rule at most; in the case of simple grammars, it could appear as first letter of various rules, provided they have not the same left member.)

Clearly, any very simple language is simple. The converse is not true: for instance $L = \{a^n cb^n a^m c \mid n \geq 1 \; m \geq 0\}$ is simple but not very simple. It is simple because it is generated by

$$S \longrightarrow aS'XT \quad S' \longrightarrow aS'X + c \quad T \longrightarrow aT + c \quad X \longrightarrow b.$$

To prove that L is not very simple, we show that any grammar in Greibach normal form generating L admits at least two rules whose right member begins with the letter a. Using for instance Ogden's iteration lemma on the word $a^n cb^n a^n c$ where the n first letters a are marked, we get that there is a derivation

$$S \overset{*}{\longrightarrow} a^i X b^j a^n c \quad X \overset{*}{\longrightarrow} a^k X b^k \quad X \overset{*}{\longrightarrow} a^{k'} cb^{k''};$$

from this we derive that there is a rule of the form $X \longrightarrow a\alpha$. Marking now the n last letters a, we get that there is a derivation

$$S \overset{*}{\longrightarrow} a^n cb^n a^{i'} Y a^{j'} c \quad Y \overset{*}{\longrightarrow} a^h Y a^{h'} \quad Y \overset{*}{\longrightarrow} a^{h''};$$

from this we derive that there is a rule of the form $Y \longrightarrow a\beta$.

Clearly the two variables X and Y have to be different: if $X = Y$, we may derive from $X = Y$ the word $a^k a^{h''} b^k$ which is not a factor of L. Thus, we have two different rules with a right member beginning by a, hence, the grammar cannot be very simple.

Any context-free language is an homomorphic image of a very simple language. Indeed, a context-free grammar in Chomsky normal form can be transformed in a very simple grammar by adding new terminal letters. The homomorphism erasing these new letters will reconstruct the original one. Let us mention along the same lines that, to any completely parenthetic grammar is naturally associated a very simple grammar obtained by erasing all the barred letters. Hence, any very simple language is an homomorphic image of a completely parenthetic language.

6.8 LL and LR languages

We end this survey of various classical subfamilies of the family of context-free languages by briefly presenting the two most usual subfamilies appearing in syntactical analysis. Given a word w over the alphabet A, define $First_k(w)$ as the prefix of length k of w; if $|w| < k$, $First_k(w)$ is equal to w. We may now define LL-grammars

Definition 6.3. [1, 38] *A context-free grammar* $G = (V, P)$ *over the terminal alphabet* A *is a* LL(k)-*grammar if*

$$S \xrightarrow[\ell]{*} uXm \longrightarrow u\alpha m \xrightarrow[\ell]{*} uv$$

$$S \xrightarrow[\ell]{*} uXm' \longrightarrow u\alpha'm' \xrightarrow[\ell]{*} uv'$$

(with $u, v, v' \in A^*$ *and* $X \in V$*) and*

$$First_k(v) = First_k(v')$$

imply $\alpha = \alpha'$.

A language is a *LL(k)-language* if it can be generated by a *LL(k)*-grammar. It is a *LL-language* if it is a *LL(k)*-language for some k. The idea is that given a terminal word uv and a leftmost derivation from S into um, the first k letters of v allow to determine what is the next rule to be used in the derivation. We will not develop here this syntactical analysis technique. However, it follows clearly from this remark that any *LL*-language is deterministic. More precisely, the families of *LL(k)*-languages form a hierarchy. Their infinite union is a strict subfamily of the family of deterministic languages.

For instance, the language $L = \{a^n cb^n \mid n \geq 1\} \cup \{a^n db^{2n} \mid n \geq 1\}$ is clearly deterministic. It is not a *LL*-language: an unbounded number of letters a has to be read before it can be decided which rule to apply in an early stage of the leftmost derivation, because it depends on whether the word contains a letter c or a letter d.

Using rightmost derivations instead of leftmost derivations leads to define the LR-grammars:

Definition 6.4. [28, 34] *A context-free grammar* $G = (V, P)$ *over the terminal alphabet* A *is a* LR(k)-*grammar if,*

$$S \xrightarrow[r]{*} mXu \longrightarrow m\alpha u = pv$$

$$S \xrightarrow[r]{*} m'X'u' \longrightarrow m'\alpha'u' = pv'$$

(with $u, u' \in A^*, p \in (V \cup A)^*V$*) and*

$$First_k(v) = First_k(v')$$

imply $X = X'$ *and* $\alpha = \alpha'$.

Again, a language is a *LR(k)-language* if it is generated by a *LR(k)*-grammar. It is a *LR-language* if it is a *LR(k)*-language for some *k*.

The idea is the following: given a sentential form *pv* where *v* is the longest terminal suffix, the first *k* letters of *v* allows to determine the rule that has been applied just before getting the sentential form *pv*. Here again, this remark that we will not develop here, implies that any *LR(k)*-language is deterministic. However, the situation is now very different from the *LL* situation.

Proposition 6.13. *The family of LR(1)-languages is exactly the family of deterministic languages.*

So, from the families of languages point of view, the *LR(k)*-condition does not give raise to an infinite hierarchy. It should be noted that, in terms of grammars, it is indeed an infinite hierarchy. It should be noted also that a grammar which is not *LR* may generate a language which is indeed *LR*. It may even be rational: the grammar $S \longrightarrow aSa$, $S \longrightarrow a$ is not *LR* and it generates the rational language a^+.

References

1. A. V. Aho and J. D. Ullman. *The Theory of Parsing, Translation and Compiling.*, volume 1. Prentice-Hall, 1973.
2. J.-M. Autebert. *Théorie des langages et des automates.* Masson, 1994.
3. J.-M. Autebert, L. Boasson, and I. H. Sudborough. Some observations on hardest context-free languages. Technical Report 81-25, Rapport LITP, April 1981.
4. J. Berstel. *Transductions and Context-Free Languages.* Teubner Verlag, 1979.
5. M. Blattner and S. Ginsburg. Canonical forms of context-free grammars and position restricted grammar forms. In Karpinski, editor, *Fundamentals of Computing Theory*, volume 56 of *Lect. Notes Comp. Sci.*, 1977.
6. L. Boasson. Two iteration theorems for some families of languages. *J. Comput. System Sci.*, 7(6):583–596, December 1973.
7. L. Boasson, J. P. Crestin, and M. Nivat. Familles de langages translatables et fermées par crochet. *Acta Inform.*, 2:383–393, 1973.
8. F. J. Brandenburg. On the intersection of stacks and queues. *Theoret. Comput. Sci.*, 23:69–82, 1983.
9. N. Chomsky. On certain formal properties of grammars. *Inform. and Control*, 2:137–167, 1959.
10. N. Chomsky and M. P. Schützenberger. The algebraic theory of context-free languages. In P. Bradford and D. Hirschberg, editors, *Computer programming and formal systems*, pages 118–161. North-Holland (Amsterdam), 1963.
11. S. V. Cole. Deterministic pushdown store machines and realtime computation. *J. Assoc. Comput. Mach.*, 18:306–328, 1971.
12. B. Courcelle. On jump deterministic pushdown automata. *Math. Systems Theory*, 11:87–109, 1977.
13. A. Cremers and S. Ginsburg. Context-free grammar forms. *J. Comput. System Sci.*, 11:86–117, 1975.
14. R. J. Evey. The theory and application of pushdown store machines. In *Mathematical Linguistics and Automatic Translation*, NSF-IO, pages 217–255. Harvard University, May 1963.

15. M. Fliess. Transductions de séries formelles. *Discrete Math.*, 10:57–74, 1974.

16. R. W. Floyd. Syntactic analysis and operator precedence. *J. Assoc. Comput. Mach.*, 10:313–333, 1963.

17. S. Ginsburg. *The Mathematical Theory of Context-Free Languages*. McGraw-Hill, 1966.

18. S. Ginsburg, J. Goldstine, and S. Greibach. Some uniformely erasable families of languages. *Theoret. Comput. Sci.*, 2:29–44, 1976.

19. S. Ginsburg and M. A. Harrison. Bracketed context-free languages. *J. Comput. System Sci.*, 1:1–23, 1967.

20. S. Ginsburg and H. G. Rice. Two families of languages related to ALGOL. *J. Assoc. Comput. Mach.*, 9:350–371, 1962.

21. S. Ginsburg and E. Spanier. Finite-turn pushdown automata. *SIAM J. Control*, 4:429–453, 1966.

22. S. Ginsburg and E. Spanier. Derivation-bounded languages. *J. Comput. System Sci.*, 2:228–250, 1968.

23. S. Greibach. The hardest context-free language. *SIAM J. Comput.*, 2:304–310, 1973.

24. S. A. Greibach. A new normal form theorem for context-free phrase structure grammars. *J. Assoc. Comput. Mach.*, 12(1):42–52, 1965.

25. S. A. Greibach. Jump pda's, deterministic context-free languages, principal afdl's and polynomial time recognition. In *Proc. 5th Annual ACM Conf. Theory of Computing*, pages 20–28, 1973.

26. S. A. Greibach. Jump pda's and hierarchies of deterministic cf languages. *SIAM J. Comput.*, 3:111–127, 1974.

27. J. Gruska. A few remarks on the index of context-free grammars and languages. *Inform. and Control*, 19:216–223, 1971.

28. M. A. Harrison. *Introduction to Formal Language Theory*. Addison-Wesley, 1978.

29. J. E. Hopcroft and J. D. Ullman. *Formal Languages and Their Relation to Automata*. Addison-Wesley, 1969.

30. J. E. Hopcroft and J. D. Ullman. *Introduction to Automata Theory, Languages and Computation*. Addison-Wesley, 1979.

31. G. Hotz. Normal form transformations of context-free grammars. *Acta Cybernetica*, 4(1):65–84, 1978.

32. G. Hotz and K. Estenfeld. *Formale Sprachen*. B.I.-Wissenschaftsverlag, 1981.

33. G. Hotz and T. Kretschmer. The power of the Greibach normal form. *Elektron. Informationsverarb. Kybernet.*, 25(10):507–512, 1989.

34. D. E. Knuth. On the translation of languages from left to right. *Inform. and Control*, 8:607–639, 1965.

35. D. E. Knuth. A characterization of parenthesis languages. *Inform. and Control*, 11:269–289, 1967.

36. A. J. Korenjack and J. E. Hopcroft. Simple deterministic languages. In *Conference record of seventh annual symposium on switching and automata theory*, pages 36–46, Berkeley, 1966.

37. W. Kuich. *Formal power series*, chapter 9. This volume.

38. P. M. Lewis and R. E. Stearns. Syntax-directed transduction. *J. Assoc. Comput. Mach.*, 15(3):465–488, 1968.

39. R. McNaughton. Parenthesis grammars. *J. Assoc. Comput. Mach.*, 14(3):490–500, 1967.

40. M. Nivat. *Transductions des langages de Chomsky, Ch. VI, miméographié*. PhD thesis, Université de Paris, 1967.

41. M. Nivat. Transductions des langages de Chomsky. *Annales de l'Institut Fourier*, 18:339–456, 1968.

42. M. Oyamaguchi. The equivalence problem for realtime dpda's. *J. Assoc. Comput. Mach.*, 34:731–760, 1987.
43. R. J. Parikh. On context-free languages. *J. Assoc. Comput. Mach.*, 13:570–581, 1966.
44. D. L. Pilling. Commutative regular equations and Parikh's theorem. *J. London Math. Soc.*, 6:663–666, 1973.
45. D. J. Rosenkrantz. Matrix equations and normal forms for context-free grammars. *J. Assoc. Comput. Mach.*, 14:501–507, 1967.
46. Jacques Sakarovitch. Pushdown automata with terminal languages. In *Languages and Automata Symposium*, number 421 in Publication RIMS, Kyoto University, pages 15–29, 1981.
47. A. Salomaa. *Formal Languages*. Academic Press, 1973.
48. M. P. Schützenberger. On context-free languages and pushdown automata. *Inform. and Control*, 6:217–255, 1963.
49. G. Sénizergues. The equivalence and inclusion problems for NTS languages. *J. Comput. System Sci.*, 31:303–331, 1985.
50. G. Sénizergues. Church-Rosser controlled rewriting systems and equivalence problems for deterministic context-free languages. *Inform. Comput.*, 81:265–279, 1989.
51. E. Shamir. A representation theorem for algebraic and context-free power series in noncommuting variables. *Inform. Comput.*, 11:39–254, 1967.
52. S. Sippu and E. Soisalon-Soininen. *Parsing Theory, Vol I*. EATCS Monographs on Theoretical Computer Science. Springer-Verlag, 1988.
53. L. Valiant. The equivalence problem for deterministic finite turn pushdown automata. *Inform. and Control*, 25:123–133, 1974.
54. H. Wechler. Characterization of rational and algebraic power series. *RAIRO Inform. Théor.*, 17:3–11, 1983.
55. M. K. Yntema. Inclusion relations among families of context-free languages. *Inform. and Control*, 10:572–597, 1967.

Aspects of Classical Language Theory

Alexandru Mateescu and Arto Salomaa

1. Phrase-structure grammars

The purpose of this chapter is to give an overview on some types of grammars and families of languages arising in classical language theory and not covered elsewhere in this Handbook. Since we will discuss in this chapter a large number of topics, we cannot penetrate very deeply in any one of them. Topics very related to the ones discussed in this chapter, such as regular languages and context-free languages, have their own chapters in this Handbook, where the presentation is more detailed than in the present chapter. Among the topics covered in this chapter (Section 3 below) will also be the general theory of language families, AFL-theory. In view of the whole language theory, there is a huge number of topics possible for this chapter. It is clear that our choice of topics and the amount of detail in which each of them is presented reflect, at least to some extent, our personal tastes.

Most of the topics presented in this chapter are in some sense modifications of the classical notion of a *rewriting system*, introduced by Axel Thue at the beginning of this century, [130]. A rewriting system is a (finite) set of rules $u \longrightarrow v$, where u and v are words, indicating that an occurrence of u (as a subword) can be replaced by v. As such a rewriting system only transforms words into other words, languages into other languages. After supplementing it with some mechanism for "squeezing out" a language, a rewriting system can be used as a device for defining languages. This is what Chomsky did, with linguistic goals in mind, when he introduced different *types of grammars*, [18, 19, 20], see also [22]. First the classification was not very clear but by mid-60's the four classes of the *Chomsky hierarchy* of grammars and languages had become pretty standard:

(i)   *recursively enumerable* or *type 0*;
(ii) *context-sensitive* or *type 1*;
(iii) *context-free* or *type 2*;
(iv) *regular* or *type 3*.

Thus, in this customary terminology, the type increases when generality decreases. The formal definitions will be given below.

Why are we interested in these four classes, what is the importance of this particular hierarchy? Type 0 grammars and languages are equivalent (in a sense to be made precise) to *computability*: what is in principle computable. Thus, their importance is beyond any question. The same or almost the same can be said about regular grammars and languages. They correspond to strictly finitary computing devices; there is nothing infinite, not even potentially. The remaining two classes lie in-between. From the point of view of rewriting it is natural to investigate what happens if rewriting is context-free: each variable develops on its own, no syntactic class is affected by its neighbouring classes. And similarly: what is the effect of context-dependence?

The motivation given above stems from [115], written a quarter of a century ago. The class of context-sensitive languages has turned out to be of smaller importance than the other classes. The particular type of context-sensitivity combined with linear work-space is perhaps not the essential type, it has been replaced by various complexity hierarchies.

The Chomsky hierarchy still constitutes a testing ground often used: new classes are compared with those in the Chomsky hierarchy. However, it is not any more the only testing ground in language theory. For instance, the basic L classes have gained similar importance. (See the chapter on L systems in this Handbook.) As regards the basic notions and notation in language theory, the reader is referred to Chapter 1 in this Handbook.

1.1 Phrase-structure grammars and Turing machines

Definition 1.1. *A phrase-structure grammar or a type 0 Chomsky grammar is a construct $G = (N, T, S, P)$, N and T are disjoint alphabets, $S \in N$ and P is a finite set of ordered pairs (u, v), where $u, v \in (N \cup T)^*$ and $\mid u \mid_N \neq 0$.* □

Terminology. Elements in N are referred to as *nonterminals*. T is the *terminal* alphabet. S is the *start symbol* and P is the set of *productions* or *rewriting rules*. Productions (u, v) are written $u \longrightarrow v$. The *alphabet* of G is $V = N \cup T$. The *direct derivation* relation induced by G is a binary relation between words over V, denoted $\Longrightarrow_G$, and defined as:

$$\alpha \Longrightarrow_G \beta \text{ iff } \alpha = xuy, \beta = xvy \text{ and } (u \longrightarrow v) \in P,$$

where, $\alpha, \beta, x, y \in V^*$.

The *derivation relation* induced by G, denoted $\Longrightarrow_G^*$, is the reflexive and transitive closure of the relation $\Longrightarrow_G$.

The *language* generated by G, denoted $L(G)$, is:

$$L(G) = \{w \mid w \in T^*, S \Longrightarrow_G^* w\}.$$

Note that $L(G)$ is a language over T, $(L(G) \subseteq T^*)$.

A language L is of type 0 iff there exists a grammar G of type 0 such that $L = L(G)$.

The set of all languages of type 0 over an alphabet T is denoted by $\mathcal{L}_0(T)$ and the family of all languages of type 0, denoted $\mathcal{L}_0$, is

$$\mathcal{L}_0 = \{L \mid \exists G \text{ grammar of type 0 such that } L(G) = L\}.$$

Remark 1.1. Given an alphabet T, obviously $\mathcal{L}_0(T) \subseteq \mathcal{P}(T^*)$. This inclusion is strict and, moreover, one can easily see that $\mathcal{L}_0(T)$ is a denumerable set, whereas $\mathcal{P}(T^*)$ is a nondenumerable set. Thus, in some sense, most of the languages over T cannot be generated by grammars of type 0.

Now we are going to present a fundamental result of Formal Language Theory, that is, the family $\mathcal{L}_0$ is equal with the family RE of all recursively enumerable languages. We introduce the family RE using Turing machines, although there are many other formalisms for defining this family. Turing machines were introduced by Alan Turing in [133].

Definition 1.2. *A Turing machine (TM) is an ordered system* $M = (Q, \Sigma, \Gamma, \delta, q_0, B, F)$ *where* Q *is a finite set of states,* Σ *is the input alphabet,* Γ *is the tape alphabet,* $\Gamma \cap Q = \emptyset$ *and* $\Sigma \subset \Gamma$, $q_0 \in Q$ *is the initial state,* $B \in \Gamma - \Sigma$ *is the blank symbol,* $F \subseteq Q$ *is the set of final states and* δ *is the transition function,*

$$\delta : Q \times \Gamma \longrightarrow \mathcal{P}(Q \times \Gamma \times \{L, R\}).$$ □

Intuitively, a Turing machine has a tape divided into cells that may store symbols from Γ (each cell may store exactly one symbol from Γ). The tape is bounded on the left (there is a leftmost cell) and is unbounded on the right. The machine has a finite control and a tape head that can read/write one cell of the tape during a time instant. The input word is a word over Σ and is stored on the tape starting with the leftmost cell and all the other cells contain the symbol B.

Initially, the tape head is on the leftmost cell and the finite control is in the state q_0. The machine performs moves. A move depends on the current cell scanned by the tape head and on the current state of the finite control. A move consists of: change the state, write a symbol from Γ to the current cell and move the tape head one cell to the left or one cell to the right. An input word is *accepted* iff after a finite number of moves the Turing machine enters a final state.

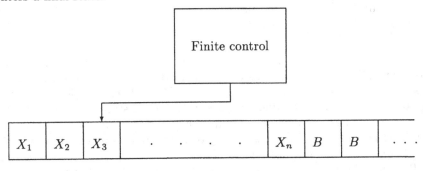

Now we introduce the formal definition of the language accepted by a Turing machine.

An *instantaneous description* (ID) of a Turing machine is a string $\alpha q X \beta$, where $\alpha, \beta \in \Gamma^*, q \in Q, X \in \Gamma$. The string encodes the description of the Turing machine at a time as follows: q is the current state of the finite control, X is the content of the current cell of the tape, α is the content of the tape on the left side of the current cell whereas β is the content of the tape on the right side of the current cell until the rightmost cell that is nonblank, i.e., all the cells to the right of β contain B.

The Turing machine M defines a *direct transition* relation between ID's, denoted $\vdash_M$,

$$\alpha q X \beta \vdash_M \alpha Y p \beta \text{ iff } (p, Y, R) \in \delta(q, X)$$
$$\alpha q \vdash_M \alpha Y p \text{ iff } (p, Y, R) \in \delta(q, B)$$
$$\alpha Z q X \beta \vdash_M \alpha p Z Y \beta \text{ iff } (p, Y, L) \in \delta(q, X)$$
$$\alpha Z q \vdash_M \alpha p Z Y \text{ iff } (p, Y, L) \in \delta(q, B)$$

where, $\alpha, \beta \in \Gamma^*$, $X, Y, Z \in \Gamma$, $p, q \in Q$.

The *transition relation*, denoted $\vdash_M^*$, is the reflexive and transitive closure of $\vdash_M$. The *language accepted* by M, denoted by $L(M)$, is:

$$L(M) = \{w \in \Sigma^* \mid q_0 w \vdash_M^* \alpha p \beta, \text{ for some } p \in F, \alpha, \beta \in \Gamma^*\}.$$

The family of *recursively enumerable* languages is:

$$RE = \{L \mid \exists M, \text{ Turing machine, } L = L(M)\}.$$

A *deterministic Turing machine* is a Turing machine $M = (Q, \Sigma, \Gamma, \delta, q_0, B, F)$ with the property that for each $q \in Q$ and for each $X \in \Gamma$, $card(\delta(q, X)) \leq 1$, i.e., δ is a (partial) function that maps $Q \times \Gamma$ into $Q \times \Gamma \times \{L, R\}$. A Turing machine that is not necessarily deterministic is called a *nondeterministic Turing machine*. It can be shown that nondeterministic Turing machines are equivalent to deterministic Turing machines. That is, a language L is accepted by a nondeterministic Turing machine iff L is accepted by a deterministic Turing machine.

The main result of this section is:

Theorem 1.1. $\mathcal{L}_0 = RE$.

Proof. Firstly, assume that $L \in \mathcal{L}_0$. Let $G = (N, T, S, P)$ be a type 0 grammar such that $L(G) = L$. One can easily define a nondeterministic Turing machine M such that $L(M) = L$. For an input $w \in T^*$, M nondeterministically selects a position i in w and a production $(u \longrightarrow v) \in P$. If starting with position i, v does occur as a subword in w, that is $w = \alpha v \beta$, then v is replaced by u. Note that by replacing v with u, M may do some translation of β to the left, if $|v| > |u|$ or to the right, if $|v| < |u|$.

If after a finite number of such operations, the content of the tape is S, then w is accepted, otherwise not. Clearly, $L(M) = L(G)$.

Conversely, assume that $L \in RE$ and let $M = (Q, \Sigma, \Gamma, \delta, q_0, B, F)$ be a Turing machine such that $L(M) = L$. Define the type 0 grammar $G = (N, \Sigma, S_0, P)$, where

$$N = ((\Sigma \cup \{\lambda\}) \times \Gamma) \cup Q \cup \{S_0, S_1, S_2\}$$

and P is the following set of productions:

1. $S_0 \longrightarrow q_0 S_1$
2. $S_1 \longrightarrow (a, a) S_1$, for all $a \in \Sigma$
3. $S_1 \longrightarrow S_2$
4. $S_2 \longrightarrow (\lambda, B) S_2$
5. $S_2 \longrightarrow \lambda$
6. $q(a, X) \longrightarrow (a, Y)p$ iff $(p, Y, R) \in \delta(q, X)$,
   where $a \in \Sigma \cup \{\lambda\}, p, q \in Q, X, Y \in \Gamma$.
7. $(b, Z)q(a, X) \longrightarrow p(b, Z)(a, Y)$ iff $(p, Y, L) \in \delta(q, X)$,
   where $a, b \in \Sigma \cup \{\lambda\}, p, q \in Q, X, Y, Z \in \Gamma$.
8. $(a, X)q \longrightarrow qaq, q(a, X) \longrightarrow qaq, q \longrightarrow \lambda$, where $q \in F, a \in \Sigma \cup \{\lambda\}$,
   $X \in \Gamma$.

Observe that productions 1–5 lead to a derivation:

$$S_0 \Longrightarrow^* q_0(a_1, a_1) \ldots (a_n, a_n)(\lambda, B)^m$$

where $a_i \in \Sigma, 1 \leq i \leq n$ and $m \geq 0$. The above derivation can be continued with productions of form 6 and/or 7. These productions simulate the transitions of the Turing machine M the tape being encoded in the second component of the symbols (a, b). The derivation continues until a final state $q \in F$ does occur in the string. Then, the rules of form 8 are applicable. The resulting word is $w = a_1 \ldots a_n$.

Clearly, the word $w = a_1 \ldots a_n$ is accepted by M iff w is derivable in G. Therefore, $L(G) = L(M)$. □

A Turing machine defines an effective procedure in the intuitive sense. The converse of this assertion, i.e., that each effective procedure can be defined as a Turing machine is not trivial anymore. The statement to the effect that the two notions are equivalent, the notion of a Turing machine and that of an intuitively effective procedure, goes back to Alonzo Church, [23], and is known as *Church's Thesis*.

Church's Thesis cannot be proved since it identifies a formal notion, the Turing machine, with an informal (intuitive) notion, the effective procedure. However, Church's Thesis is an extremely important assertion. Post, see [108], referred to Church's Thesis not as a definition or axiom but a natural law, a "fundamental discovery" concerning "the mathematicizing power of Homo Sapiens", in need of "continual verification". For a more detailed exposition of this subject see [112].

In the sequel we present some closure properties of the family $\mathcal{L}_0$. See [115] for the proofs of these results.

Theorem 1.2. *The family $\mathcal{L}_0$ is closed under the following operations: union, intersection, catenation, Kleene $*$, Kleene $+$, intersection with regular languages, morphism, inverse morphism, substitution, mirror.* □

Corollary 1.1. *The family $\mathcal{L}_0$ is a full AFL.* □

Theorem 1.3. *The family $\mathcal{L}_0$ is not closed under complementation.* □

1.2 Normal forms for phrase-structure grammars

There are many normal forms for phrase-structure grammars. Here we present only some of them. For a proof of Theorem 1.4, see [115]. For proofs of Theorems 1.5-1.9, see [44].

Theorem 1.4. *Each phrase-structure grammar is equivalent with a phrase-structure grammar having only context-sensitive rules, and a single additional rule $A \longrightarrow \lambda$.* □

Theorem 1.5. *Each phrase-structure grammar is equivalent with a phrase-structure grammar with 5 nonterminal symbols, having only context-free rules of the form $S \longrightarrow v$, where S is the start symbol, and two additional rules $AB \longrightarrow \lambda$, $CD \longrightarrow \lambda$.* □

Theorem 1.6. *Each phrase-structure grammar is equivalent with a phrase-structure grammar with 4 nonterminal symbols, having only context-free rules of the form $S \longrightarrow v$, where S is the start symbol, and two additional rules $AB \longrightarrow \lambda$, $CC \longrightarrow \lambda$.* □

Theorem 1.7. *Each phrase-structure grammar is equivalent with a phrase-structure grammar with 3 nonterminal symbols, having only context-free rules of the form $S \longrightarrow v$, where S is the start symbol, and two additional rules $AA \longrightarrow \lambda$, $BBB \longrightarrow \lambda$.* □

Theorem 1.8. *Each phrase-structure grammar is equivalent with a phrase-structure grammar with 3 nonterminal symbols, having only context-free rules of the form $S \longrightarrow v$, where S is the start symbol, and a single additional rule of the form $ABBBA \longrightarrow \lambda$.* □

Theorem 1.9. *Each phrase-structure grammar is equivalent with a phrase-structure grammar with 4 nonterminal symbols, having only context-free rules of the form $S \longrightarrow v$, where S is the start symbol, and a single additional rule of the form $ABC \longrightarrow \lambda$.* □

1.3 Representations of recursively enumerable languages

Representing the family of recursively enumerable languages through operations on its subfamilies is an important topic in formal language theory. The representation theorems of recursively enumerable languages are useful tools in proving other properties of this family as well as in finding properties of some subfamilies of $\mathcal{L}_0$. In what follows we give a selection of such results. In Theorems 1.10-1.26 existence means that the items in question can be effectively constructed. All of the notions involved are not explained here. In such cases we refer to the literature indicated.

Theorem 1.10. *For every recursively enumerable language L there exist a morphism h and a context-sensitive language L' such that:*

$$L = h(L').$$ □

The following theorem was proved by Savitch, see [124].

Theorem 1.11. *For every recursively enumerable language L there exist a context-free language L' and a morphism h such that*

$$L = h(h^{-1}(D) \cap L'),$$

where D is a Dyck language.   □

Ginsburg, Greibach and Harrison, see [49], obtained:

Theorem 1.12. *For every recursively enumerable language L there exist a morphism h and two deterministic context-free languages L_1 and L_2 such that:*

$$L = h(L_1 \cap L_2).$$ □

A *weak identity* is a morphism $h : \Sigma^* \longrightarrow \Delta^*$ such that for each $a \in \Sigma$, either $h(a)$ is a or $h(a)$ is λ. Let $h_1, h_2 : \Sigma^* \longrightarrow \Delta^*$ be morphisms. The *equality set* of h_1 and h_2 is:

$$E(h_1, h_2) = \{w \in \Sigma^* \mid h_1(w) = h_2(w)\}$$

and the *minimal equality set* of h_1 and h_2 is:

$$e(h_1, h_2) = \{w \in \Sigma^+ \mid w \in E(h_1, h_2) \text{ and}$$

$$\text{if } w = uv, u, v \in \Sigma^+, \text{ then } u \notin E(h_1, h_2)\}.$$

The following three representation theorems were proved by Culik, [28].

Theorem 1.13. *For each recursively enumerable language L there exist a weak identity h_0 and two morphisms h_1 and h_2 such that:*

$$L = h_0(e(h_1, h_2)).$$ □

Theorem 1.14. *For each recursively enumerable language L there exist a deterministic gsm mapping g and two morphisms h_1 and h_2 such that:*

$$L = g(E(h_1, h_2)).$$ □

Theorem 1.15. *For each recursively enumerable language L there exist morphisms h_1, h_2 and regular languages R_1, R_2 and R_3 such that:*

(i) $L = R_1 \backslash e(h_1, h_2) / R_2$,
(ii) $L = (R_1 \backslash E(h_1, h_2) / R_2) \cap R_3$.   □

A sharpening of the above theorem was obtained by Turakainen, see [132].

Theorem 1.16. *For every recursively enumerable language L, there exist a finite alphabet Γ, a symbol B, and two morphisms h, g such that $L = (\Gamma^* B) \backslash e(h, g)$.*   □

A *coding* is a morphism that maps each letter into a letter. A *weak coding* is a morphism that maps each letter into a letter or into λ. Let V and Σ be alphabets such that $\Sigma \subseteq V$. $Pres_{V,\Sigma}$ or simply $Pres_\Sigma$ when V is understood, denotes the weak identity that preserves all letters of Σ and maps all letters from $V - \Sigma$ to λ. If h is a (possibly partial) mapping from Σ^* to Δ^*, then the *fixed point language* of h is

$$Fp(h) = \{w \in \Sigma^* \mid h(w) = w\}.$$

The following representation theorems, Theorem 1.17 – Theorem 1.21, are due to Engelfriet and Rozenberg, see [41].

Theorem 1.17. *For each recursively enumerable language L over an alphabet Σ there exists a deterministic gsm mapping g such that:*

$$L = Pres_\Sigma(Fp(g)).$$ □

Theorem 1.18. *For each recursively enumerable language L there exist a weak identity f, a coding g, a morphism h and a regular language R such that:*

$$L = f(E(g, h) \cap R).$$ □

Theorem 1.19. *For each recursively enumerable language L there exist a weak identity f, a finite substitution g and a regular language R such that:*

$$L = f(Fp(g) \cap R).$$ □

Let Σ be an alphabet. $\bar{\Sigma}$ denotes the alphabet $\bar{\Sigma} = \{\bar{a} \mid a \in \Sigma\}$ and L_Σ is the language obtained by shuffling words $u \in \Sigma^*$ with their barred version $\bar{u} \in \bar{\Sigma}^*$. The language L_Σ is referred to as the *complete twin shuffle over* Σ.

Theorem 1.20. *For each recursively enumerable language L there exist an alphabet Δ, a simple deterministic linear language $K \subseteq \Delta^+ \bar{\Delta}^+$, and a weak identity f such that:*

$$L = f(\{wmi(\bar{w}) \mid w \in \Delta^+\} \cap K).$$ □

Theorem 1.21. *For each recursively enumerable language L:*

(i) *there exist an alphabet Σ, a weak identity f, and a regular language R such that*

$$L = f(L_\Sigma \cap R).$$

(ii) *there exist a weak identity f, a morphism g, and a regular language R such that*

$$L = f(g^{-1}(L_{\{0,1\}}) \cap R).$$

(iii) *there exists a deterministic gsm mapping g such that*

$$L = g(L_{\{0,1\}}). \qquad \square$$

Next representation theorem is due to Geffert, see [43].

Let $h_1, h_2 : \Delta^* \longrightarrow \Gamma^*$ be two morphisms. The *overflow languages of h_1, h_2* are:

$$O(h_1 \backslash h_2) = \{h_1(w) \backslash h_2(w) \mid w \in \Delta^+\},$$

and

$$O(h_1/h_2) = \{h_1(w)/h_2(w) \mid w \in \Delta^+\}.$$

Theorem 1.22. *For each recursively enumerable language $L \subseteq \Sigma^*$, there exist morphisms $h_1, h_2 : \Delta^* \longrightarrow \Gamma^*$, where $\Sigma \subseteq \Gamma$ such that:*

$$L = O(h_1 \backslash h_2) \cap \Sigma^*. \qquad \square$$

Note that the language $O(h_1 \backslash h_2)$ can be replaced in the above theorem by the language $O(h_2/h_1)$.

A simpler proof of Theorem 1.22 can be found in [132].

Theorem 1.23 – Theorem 1.25 were proved by Latteux and Turakainen, [86].

Theorem 1.23. *Every recursively enumerable language $L \subseteq \Sigma^*$ can be represented in the form*

$$L = \{h(w) \backslash g(w) \mid w \in R^+\} \cap \Sigma^*$$

where R is a λ-free regular language, h is a nonerasing morphism, and g is a morphism 3-limited on R^+. $\qquad \square$

Theorem 1.24. *Let a be a fixed letter from an alphabet Σ. For each recursively enumerable language $L \subseteq \Sigma^*$, there exists a minimal linear grammar $G = (\{S\}, \Sigma \cup \{A, \bar{A}, \bar{a}\}, S, P)$ such that $L = \rho(L(G)) \cap \Sigma^*$ and $L = L(G_1)$ where*

$$G_1 = (\{S, A, \bar{A}, \bar{a}\}, \Sigma, P \cup \{a\bar{a} \longrightarrow \lambda, A\bar{A} \longrightarrow \lambda\}, S)$$

and ρ is the Dyck reduction $a\bar{a} \longrightarrow \lambda$, $A\bar{A} \longrightarrow \lambda$. $\qquad \square$

Theorem 1.25. *For every recursively enumerable language $L \subseteq \Sigma^*$:*

(i) there exists a linear grammar G having three nonterminals such that $L = K \backslash L(G)$ where K is the deterministic minimal linear language $K = \{ucmi(u) \mid u \in \Gamma^*\}$ over some alphabet $\Gamma \cup \{c\}$.

(ii) there exist a deterministic minimal linear grammar G_1 and a deterministic linear grammar G_2 having two nonterminals such that $L = Pres_\Sigma(L(G_1) \cap L(G_2))$.   □

Finally, we present a recent result of Păun, [104]. The representation theorem is formulated in terms of the splicing operation. For more details, see the chapter of this Handbook entitled "Language theory and molecular genetics. Generative mechanisms suggested by DNA recombination".

Theorem 1.26. *The family of recursively enumerable languages equals the family of languages generated by extended splicing (EH) systems with a finite set of axioms and a regular set of splicing rules.*   □

1.4 Decidability. Recursive languages

Given a Turing machine $M = (Q, \Sigma, \Gamma, \delta, q_0, B, F)$ and an input word $w \in \Sigma^*$, one of the following situations occurs:

(1) M does halt after a finite number of moves in a state $q \in Q$ and, if $q \in F$, then w is accepted, otherwise, if $q \in Q - F$, then w is not accepted.

(2) M never halts, i.e., M continues the computation forever. In this case w is not accepted.

A Turing machine M *halts on every input* iff for all inputs w only the first situation occurs.

The notion of a Turing machine that halts on every input provides the mathematical (formal) definition of the notion of an *algorithm*. Sometimes this statement is also referred to as *Church's Thesis*, see for instance [32].

A language $L \subseteq \Sigma^*$ is a *recursive language* iff there exists a Turing machine M that halts on every input, such that $L = L(M)$. In other words, using Church's Thesis, a language $L \subseteq \Sigma^*$ is recursive iff there is an algorithm to decide for each $w \in \Sigma^*$, whether $w \in L$ or not.

It is well known, see [115], that the family of all recursive languages is a proper subfamily of the family of all recursively enumerable languages.

Let $\mathcal{L}$ be a family of languages generated by grammars from a family $\mathcal{G}$ of grammars. Each grammar $G \in \mathcal{G}$ is encoded as a word $< G >$ over a fixed alphabet Δ. The encoding method should have good computational properties. That is, there is an algorithm that for a given grammar $G \in \mathcal{G}$ computes $< G >$ and, conversely, there is an algorithm that for a word $v \in \Delta^*$ computes the grammar $G \in \mathcal{G}$ such that $v = < G >$. The word $< G >$ is referred to as an *index* (or a *Gödel number*, if $\Delta = \{0, 1\}$) of the language $L = L(G)$.

A property $\mathcal{P}$ of languages from $\mathcal{L}$ is *trivial* iff $\mathcal{P}$ is true for all languages from $\mathcal{L}$ or if it is false for all languages from $\mathcal{L}$, otherwise the property $\mathcal{P}$ is *nontrivial* for the family $\mathcal{L}$. To a property $\mathcal{P}$ we associate the language

$$L_{\mathcal{P}} = \{< G >| \ G \in \mathcal{G}, \mathcal{P} \text{ is true for } L(G)\}.$$

The property $\mathcal{P}$ is *decidable* for the family $\mathcal{L}$ iff the language $L_{\mathcal{P}}$ is a recursive language. Otherwise, the property $\mathcal{P}$ is *undecidable* for the family $\mathcal{L}$. Informally, this means that $\mathcal{P}$ is decidable iff there is an algorithm such that for a given grammar $G \in \mathcal{G}$ the algorithm says "yes", if $L(G)$ has the property $\mathcal{P}$, and the algorithm says "no", if $L(G)$ does not have the property $\mathcal{P}$. For instance the family $\mathcal{L}$ can be the family of all context-free languages and $\mathcal{G}$ the family of all context-free grammars. As a property $\mathcal{P}$ one can consider the property of a language to be finite, which is decidable, see [115], or the property of a language to be regular, which is undecidable, see [115].

The following theorem was proved by Rice, see [110],

Theorem 1.27. *Any nontrivial property $\mathcal{P}$ of recursively enumerable languages is undecidable.*   □

The above theorem has many negative consequences, as for instance, emptiness, finiteness, regularity, context-freeness, context-sensitivity, recursiviness are undecidable properties for recursively enumerable languages.

Let $\mathcal{P}$ be a property of recursively enumerable languages. Next theorem, also due to Rice, see [111], gives a characterization of those properties $\mathcal{P}$ for which the language $L_{\mathcal{P}}$ is a recursively enumerable language.

Theorem 1.28. *Let $\mathcal{P}$ be a property of recursively enumerable languages. The language $L_{\mathcal{P}}$ is recursively enumerable iff the following three conditions are satisfied:*

(i)   If L has the property $\mathcal{P}$ and $L \subseteq L'$, then L' has also the property $\mathcal{P}$.

(ii) If L is an infinite language such that L has the property $\mathcal{P}$, then there exists a finite language $L_0 \subseteq L$ such that L_0 has also the property $\mathcal{P}$.

(iii) The set of finite languages that have the property $\mathcal{P}$ is enumerable, i.e., there exists a Turing machine that generates the possibly infinite sequence

$$L_1, L_2, \ldots, L_i, \ldots,$$

where L_i is the ith finite language that has the property $\mathcal{P}$.   □

For instance, if $\mathcal{P}$ is one of the following properties of recursively enumerable languages: nonemptiness, the language contains at least k words, where k is a fixed integer, $k \geq 1$, the language contains a word w, where w is a fixed word, then in each case the language associated to $\mathcal{P}$, $L_{\mathcal{P}}$, is a recursively enumerable language.

However, this is not the case if $\mathcal{P}$ is one of the following properties of recursively enumerable languages: emptiness, recursiveness, nonrecursiveness, context-freeness, regularity, the property of a language being a singleton, totality, i.e., the complement of the language being empty.

The following results concern closure properties of the family of recursive languages.

Theorem 1.29. *The family of recursive languages is closed under the following operations: union, intersection, complementation, catenation, Kleene ∗, Kleene +, intersection with regular languages, λ-free morphism, inverse morphism, λ-free substitution, mirror.*   □

Theorem 1.30. *The family of recursive languages is not closed under arbitrary morphism or arbitrary substitution.*   □

In terms of AFL-theory, see section 3 below, the results can be expressed also as follows.

Corollary 1.2. *The family of recursive languages is an AFL but not a full AFL.*   □

Theorem 1.31. *Let $L \subseteq \Sigma^*$ be a language. If both L and its complement $CL = \Sigma^* - L$ are recursively enumerable languages, then L and also CL are recursive languages.*   □

Hence, for a pair of languages consisting of a language L and its complement CL only one of the following situations is possible:

(i) L and CL are not recursively enumerable languages.
(ii) L and CL are recursive languages.
(iii) one of the languages is a recursively enumerable language but is not a recursive language. Then the other language is not a recursively enumerable language.

2. Context-sensitive grammars

2.1 Context-sensitive and monotonous grammars

Definition 2.1. *A context-sensitive (type 1) grammar is a type 0 grammar $G = (N, T, S, P)$ such that each production in P is of the form $\alpha X \beta \longrightarrow \alpha u \beta$, where $X \in N$, $\alpha, \beta, u \in (N \cup T)^*, u \neq \lambda$. In addition P may contain the production $S \longrightarrow \lambda$ and in this case S does not occur on the right side of any production of P.*   □

The language generated by a context-sensitive grammar is defined as for the type 0 grammars. The family of context-sensitive languages is:

$$CS = \{L \mid \exists G \text{ context-sensitive grammar such that } L(G) = L\}.$$

Note that $CS \subseteq \mathcal{L}_0 = RE$.

Definition 2.2. *A length-increasing (monotonous) grammar is a type 0 grammar $G = (N, T, S, P)$ such that for each production $(u \longrightarrow v) \in P$, $\mid u \mid \leq \mid v \mid$. In addition P may contain the production $S \longrightarrow \lambda$ and in this case S does not occur on the right side of any production from P.* □

The next theorem shows that context-sensitive grammars and length-increasing grammars have the same generative power.

Theorem 2.1. *Let L be a language. The following statements are equivalent:*

(i)   there exists a context-sensitive grammar G such that $L = L(G)$.
(ii) there exists a length-increasing grammar G' such that $L = L(G')$.

Proof. $(i) \Longrightarrow (ii)$. Obviously, each context-sensitive grammar is a length-increasing grammar.

$(ii) \Longrightarrow (i)$ We sketch here the main construction. For more details see [115]. Let $G' = (N', T', S', P')$ be a length-increasing grammar. Without loss of generality, we can assume that all productions from P' that contain terminal symbols are of the form $X \longrightarrow a, X \in N', a \in T'$. Now consider a production from P'

$$(*) \qquad\qquad X_1 X_2 \ldots X_m \longrightarrow Y_1 Y_2 \ldots Y_n, \, 2 \leq m \leq n,$$

where $X_i, Y_j \in N', 1 \leq i \leq m, 1 \leq j \leq n$.

The production $(*)$ is replaced by the following productions:

$$X_1 X_2 \ldots X_m \longrightarrow Z_1 X_2 \ldots X_m$$
$$Z_1 X_2 \ldots X_m \longrightarrow Z_1 Z_2 X_3 \ldots X_m$$
$$\vdots$$

$$(**) \qquad\qquad Z_1 Z_2 \ldots Z_{m-1} X_m \longrightarrow Z_1 \ldots Z_m Y_{m+1} \ldots Y_n$$
$$Z_1 \ldots Z_m Y_{m+1} \ldots Y_n \longrightarrow Y_1 Z_2 \ldots Z_m Y_{m+1} \ldots Y_n$$
$$\vdots$$

$$Y_1 \ldots Y_{m-1} Z_m Y_{m+1} \ldots Y_n \longrightarrow Y_1 \ldots Y_m Y_{m+1} \ldots Y_n$$

where $Z_k, 1 \leq k \leq m$ are new nonterminals.

Observe that productions of the form $(**)$ are context-sensitive productions. Productions $(*)$ can be simulated by productions $(**)$ and conversely, the use of a production of type $(**)$ implies the use of the whole sequence with the result exactly as in the use of a production $(*)$.

Hence, the new grammar is equivalent with G'. By repeating the process we obtain a context-sensitive grammar G which is equivalent with G'. □

We present now two examples of grammars generating context-sensitive languages.

Example 2.1. Let $G = (N, T, S, P)$ be the grammar with $N = \{S, B\}$, $T = \{a, b, c\}$ and P :

1. $S \longrightarrow aSBc$
2. $S \longrightarrow abc$
3. $cB \longrightarrow Bc$
4. $bB \longrightarrow bb$

We claim that

$$L(G) = \{a^n b^n c^n \mid n \geq 1\}.$$

Thus, we have to prove that $L(G)$ consists of all words of a specific form. It has happened fairly often in such proofs in the literature that, while the proof is correct in showing that all words of the specific form are generated, it fails to guarantee that no unwanted words are generated. Especially context-sensitive productions easily give rise to "parasitic" derivations, derivations not intended in the design of the grammar. However, the above claim is easy to establish for our grammar G. Productions 3 and 4 can be applied only after 2 has been applied. Then we are dealing with the sentential form

$$(1) \qquad\qquad a^{n+1} bc(Bc)^n,$$

where $n \geq 0$ indicates the number of times the production 1 was applied before applying 2. Only 3 and 4 can be applied after we have introduced the sentential form (1): S has vanished and cannot be created again. The effect of 3 is to move B's to the left, and that of 4 to replace a B, next to b, by b. Thus, the only terminal word resulting is $a^{n+1} b^{n+1} c^{n+1}$. The only nondeterminism lies in the order of applying 3 and 4, but any order gives rise to the same terminal word. Thus, our claim follows.   □

The above example shows that the family CF is strictly contained in the family CS.

Example 2.2. Let T be an alphabet. We define a grammar $G = (N, T, S, P)$ such that

$$L(G) = \{ww \mid w \in T^*\}.$$

The set of nonterminals is:

$$N = \{S\} \cup \{X_i \mid i \in T\} \cup \{L_i \mid i \in T\} \cup \{R_i \mid i \in T\}.$$

and the set of productions is:

1_i. $S \longrightarrow iSX_i$
2_i. $S \longrightarrow L_i R_i$
3_{ij}. $R_i X_j \longrightarrow X_j R_i$
4_i. $R_i \longrightarrow i$

$5_{ij}.\ L_i X_j \longrightarrow L_i R_j$
$6_i.\ L_i \longrightarrow i$

where, $i, j \in T$.

Observe that the use of the first two sets of productions 1_i and 2_i leads to the sentential form:

$$(1) \qquad\qquad j_1 j_2 \ldots j_k L_i R_i X_{j_k} \ldots X_{j_2} X_{j_1}.$$

Using productions of type 3_{ij}, the nonterminal R_i can "emigrate" to the right and R_i should be eliminated only if no X_j has an occurrence on the right of R_i (otherwise, the elimination of X_j is not possible and the derivation is blocked).

Thus, from (1), can be derived

$$j_1 j_2 \ldots j_k L_i X_{j_k} \ldots X_{j_2} X_{j_1} R_i \Longrightarrow_{4_i} j_1 j_2 \ldots j_k L_i X_{j_k} \ldots X_{j_2} X_{j_1} i.$$

Now $L_i X_{j_k}$ should be rewritten using 5_{ij_k} and we obtain

$$j_1 j_2 \ldots j_k L_i R_{j_k} X_{j_{k-1}} \ldots X_{j_1} i.$$

Again, R_{j_k} emigrates to the right of X_{j_1} and is rewritten using 4_{j_k}. Hence, the sentential form is:

$$j_1 j_2 \ldots j_k L_i X_{j_{k-1}} \ldots X_{j_1} j_k i.$$

ting the above derivations, the resulting sentential form is:

$$j_1 j_2 \ldots j_k L_i j_1 j_2 \ldots j_k i.$$

Now using the production 6_i we obtain from the above sentential form the terminal word:

$$j_1 j_2 \ldots j_k i j_1 j_2 \ldots j_k i,$$

that is ww, where $w = j_1 j_2 \ldots j_k i$. Note that the application of a production 6_i before the elimination of all occurrences of X_j leads to a blind alley.

Thus, we conclude that:

$$L(G) = \{ww \mid w \in T^+\}.$$

Observe that G is a length-increasing grammar and therefore the language $\{ww \mid w \in T^+\}$ is a context-sensitive language.   □

2.2 Normal forms for context-sensitive grammars

Here we consider some normal forms for context-sensitive grammars. We assume that the languages considered do not contain the empty word λ.

Definition 2.3. *A grammar G is in the Kuroda normal form if each production of G is of one of the following forms:*

$$A \longrightarrow a,\ A \longrightarrow BC,\ AB \longrightarrow CD. \qquad \square$$

The next theorem was proved by Kuroda, see [83].

Theorem 2.2. *For every context-sensitive grammar there exists (effectively) an equivalent grammar in the Kuroda normal form.*

Proof. The idea of the proof is that each production

$$X_1 X_2 \ldots X_m \longrightarrow Y_1 Y_2 \ldots Y_n,\ 2 \leq m \leq n$$

that is not of the form $AB \longrightarrow CD$, i.e., it is not the case that $m = n = 2$, can be replaced by the following set of productions:

$$X_1 X_2 \longrightarrow Y_1 Z_2$$
$$Z_2 X_3 \longrightarrow Y_2 Z_3$$
$$\vdots$$
$$Z_{m-1} X_m \longrightarrow Y_{m-1} Z_m$$
$$Z_m \longrightarrow Y_m Z_{m+1}$$
$$Z_{m+1} \longrightarrow Y_{m+1} Z_{m+2}$$
$$\vdots$$
$$Z_{n-1} \longrightarrow Y_{n-1} Y_n$$

where $Z_2, Z_3, \ldots, Z_{n-1}$ are new nonterminals. $\qquad \square$

Definition 2.4. *A context-sensitive grammar G is in the one-sided normal form if each production of G is of one of the following forms:*

$$A \longrightarrow a$$
$$A \longrightarrow BC$$
$$AB \longrightarrow AC \qquad \square$$

The next theorem was mentioned by Gladkij, [51], the first proof being due to Penttonen, [106].

Theorem 2.3. *For every context-sensitive grammar there exists effectively an equivalent grammar in the one-sided normal form.* $\qquad \square$

From the above theorem we obtain the following normal form for type 0 grammars:

Theorem 2.4. *Every type 0 grammar is equivalent with a grammar whose productions are of the form:*

$$A \longrightarrow a,\ A \longrightarrow BC,\ AB \longrightarrow AC,\ A \longrightarrow \lambda. \qquad \square$$

2.3 Workspace

Observe that in a derivation of a nonempty sentential form in a context-sensitive grammar the length of the consecutive sentential forms is increasing monotonically. This means that in a derivation of a terminal word w all the sentential forms have the length less than or equal to $| \ w \ |$. Now assume that a language L is generated by a type 0 grammar G such that there is a nonnegative integer k with the property that for each word w, $w \in L(G)$, there exists a derivation of w in G such that the workspace does not exceed $k \ | \ w \ |$, i.e., there is a derivation $S \Longrightarrow_G^* w$ such that all sentential forms from the derivation have the length less than or equal to $k \ | \ w \ |$.

If G has the above property, then $L(G)$ is a context-sensitive language. The formal definition of the notion is:

Definition 2.5. *Let $G = (N, T, S, P)$ be a grammar and consider a derivation D according to G,*

$$D : S = w_0 \Longrightarrow w_1 \Longrightarrow \ldots \Longrightarrow w_n = w.$$

The workspace of w by the derivation D is:

$$WS_G(w, D) = max\{| \ w_i \ | \ | \ 0 \le i \le n\}.$$

The workspace of w is:

$$WS_G(w) = min\{WS_G(w, D) \ | \ D \text{ is a derivation of } w\}.$$

Observe that $WS_G(w) \ge | \ w \ |$ for all G and w.   □

The following theorem, due to Jones, see [73], is a powerful tool in showing languages to be context-sensitive.

Theorem 2.5. *(The workspace theorem) If G is a type 0 grammar and if there is a nonnegative integer k such that*

$$WS_G(w) \le k \ | \ w \ |$$

for all nonempty words $w \in L(G)$, then $L(G)$ is a context-sensitive language.
                                                                                              □

We refer the reader to [115] for the proof of the above theorem. An immediate consequence is the following:

Theorem 2.6. *Let L be a recursively enumerable language that is not a context-sensitive language. Then for every nonnegative integer k and for every grammar G generating L, there is a word $w \in L$ such that $WS_G(w) > k \ | \ w \ |$.*
                                                                                              □

2.4 Linear bounded automata

Linear bounded automata are a special type of Turing machines that accept exactly the context-sensitive languages. Linear bounded automata are closely related to a certain class of Turing space complexity, $NSPACE(n)$.

We start by considering some basic facts concerning Turing space complexity. An *off-line* Turing machine is a Turing machine M that additionally has a read-only input tape with endmarkers # and $, see figure below.

Initially, the input word is stored on the input tape, starting with the endmarker # and ending with the endmarker $. The machine cannot write on the input tape and all the computations are done on the work tape T. Let S be a function from nonnegative integers to nonnegative integers. M is said to be of *space complexity* $S(n)$ if for every input word of length n, M uses at most $S(n)$ cells on the work tape.

The notions of determinism and nondeterminism are extended to concern off-line Turing machines.

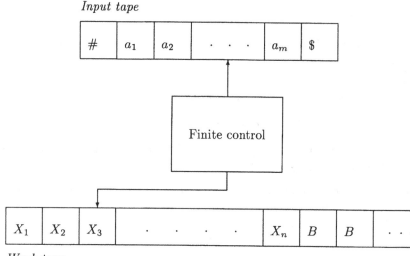

Again, see [112], the nondeterministic off-line Turing machines are equivalent to deterministic off-line Turing machines. That is, a language L is accepted by a nondeterministic off-line Turing machine iff L is accepted by a deterministic off-line Turing machine.

Accordingly, the space complexity classes are defined as follows:

$$NSPACE(S(n)) = \{L \mid \exists M, \text{ off-line Turing machine}$$

$$\text{of space complexity } S(n) \text{ with } L(M) = L\},$$

$DSPACE(S(n)) = \{L \mid \exists M,$ deterministic off-line Turing machine

of space complexity $S(n)$ such that $L(M) = L\}$.

Obviously, $DSPACE(S(n)) \subseteq NSPACE(S(n))$. The following theorem shows that the space complexity can be compressed with any constant factor.

Theorem 2.7. *For any function S and for any constant $c > 0$,*

$$NSPACE(S(n)) = NSPACE(cS(n)).$$

Proof. The idea of the proof is that for a constant c, say $c > 1$, and for a given Turing machine M_1 of space complexity $cS(n)$ one can define a Turing machine M_2 such that M_2 simulates M_1 and, moreover, M_2 encodes r adjacent cells of M_1 into one symbol. M_2 can use its finite control to simulate M_1 inside of such a block of r adjacent cells of M_1. The constant r can be chosen such that M_2 is of space complexity $S(n)$. A similar construction can be done if $0 < c < 1$. □

The above theorem holds true also for deterministic space complexity, i.e., for any function S and for any constant $c > 0$,

$$DSPACE(S(n)) = DSPACE(cS(n)).$$

Definition 2.6. *A linear bounded automaton (LBA) is a Turing machine of space complexity $S(n) = n$. A deterministic linear bounded automaton (DLBA) is a deterministic Turing machine of space complexity $S(n) = n$.* □

Next theorem, see [115] for a detailed proof, reveals the important fact that context-sensitive grammars are equivalent with linear bounded automata.

Theorem 2.8. *Let L be a language. L is generated by a context-sensitive grammar iff L is accepted by a linear bounded automaton. Consequently,*

$$CS = NSPACE(n).$$ □

Open Problem: It is not known whether the inclusion $DSPACE(n) \subseteq CS$ is strict or not.

A *fully space constructible function* is a function $S(n)$ such that there exists a Turing machine M that is $S(n)$ space bounded and for all n, M uses exactly $S(n)$ cells on any input of length n.

Concerning the relation between deterministic and nondeterministic space complexity we have the following:

Theorem 2.9. *(Savitch) If $S(n)$ is a fully space constructible function such that $S(n) \geq \log_2 n$, for all $n \geq 1$, then*

$$NSPACE(S(n)) \subseteq DSPACE(S^2(n)).$$ □

For the proof we refer the reader to [123]. Again it is an open problem whether the above inclusion is strict or not. If a language L is over a one-letter alphabet and if L is in $NSPACE(S(n))$, then L is also in $DSPACE(S^2(n))$ for each function $S(n)$, i.e., no additional assumptions concerning $S(n)$, as in Theorem 2.9, are needed, see [45].

2.5 Closure properties of the family CS

We present here the most important closure properties of the family of context-sensitive languages. More details are provided to the closure of the context-sensitive family under linear erasing and under complementation.

Definition 2.7. *Let k be a nonnegative integer. A morphism h, $h : \Sigma^* \longrightarrow \Delta^*$ is called k-erasing with respect to a language $L \subseteq \Sigma^*$ iff for each $w \in L$*

$$\mid w \mid \leq k \mid h(w) \mid .$$

A family $\mathcal{L}$ of languages is closed under linear erasing iff $h(L) \in \mathcal{L}$ for all $L \in \mathcal{L}$ and for all k-erasing morphisms h, $k \geq 0$.   □

Theorem 2.10. *The family CS is closed under linear erasing.*

Proof. Let L be a context-sensitive language and let h be a k-erasing morphism for some fixed $k \geq 0$. If $k = 0$, then $h(L) = \emptyset$ or $h(L) = \{\lambda\}$ and, obviously $h(L) \in CS$.

Now assume that $k \geq 1$ and let $G = (N, T, S, P)$ be a context-sensitive grammar such that $L(G) = L$. Without loss of generality we can assume that the only productions from P containing terminal letters are of the form $A \longrightarrow a$, $A \in N$, $a \in T$.

Observe that $h(L)$ is generated by the type 0 grammar $G' = (N, T', S, P')$ where P' is obtained from P by replacing the terminal productions $A \longrightarrow a$ with $A \longrightarrow h(a)$.

Note that the workspace of the grammar G', for a nonempty word $w' = h(w)$, satisfies the following inequality:

$$WS_{G'}(w') = max\{WS_G(w), \mid w' \mid\} = max\{\mid w \mid, \mid w' \mid\} \leq k \mid w' \mid .$$

Hence, from the workspace theorem, Theorem 2.5, it follows that $h(L) \in CS$.
□

The closure of the family CS under complementation was a famous open problem for more than 20 years. In 1987, independently, Immerman, [66], and Szelepcsényi, [126], answered positively this problem. Their result is more general and we are going to present their theorem in its full generality.

Theorem 2.11. *If $S(n) \geq \log n$, then the complexity class $NSPACE(S(n))$ is closed under complementation.*

Proof. We sketch the main idea of the proof. Our presentation follows [117]. Let L be in $NSPACE(S(n))$ and let M be a Turing machine of space complexity $S(n)$ such that $L(M) = L$. Consider an input word w of M of length n. The input defines an initial configuration $START$ of M. Note that $START$ has length $n + 1$. We describe a Turing machine M_{test} with the property that $L(M_{test}) = CL$. The Turing machine M_{test} is of space complexity $kS(n)$ for some $k > 0$. Hence, by Theorem 2.7, M_{test} is equivalent with a Turing machine of space complexity $S(n)$ and thus, CL is in $NSPACE(S(n))$. There are two main steps performed by M_{test}.

Firstly, M_{test} computes the total number N of configurations reachable by M from $START$. This can be done as follows. Denote by N_i the number of configurations reachable by M from $START$ in $\leq i$ steps. Then $N = N_i$, where i is the smallest number such that $N_i = N_{i+1}$. Clearly, $N_0 = 1$. Now M_{test} remembers (in one track of its tape) always only the last N_j it has computed. In order to compute N_{j+1}, M_{test} goes through all length $S(n) + 1$ configurations of M, arranged in alphabetical order. M_{test} always remembers only the last configuration C it was working with. After completing the work, M_{test} replaces C by the next configuration $C + 1$. M_{test} can handle a specific configuration, referred in the sequence as the $TARGET$ configuration, by keeping a counter of the number of configurations of M reachable in $\leq j + 1$ steps. The counter (called the first counter) is incremented exactly in case $TARGET$ is a configuration of M reachable in $\leq j + 1$ steps.

Also, M_{test} keeps another counter, namely of configurations reachable in $\leq j$ steps. This second counter is incremented exactly in case a configuration reachable in $\leq j$ steps is found. Given a fixed configuration $SEMITARGET$, M_{test} first guesses whether $SEMITARGET$ is among the N_j configurations. If the guess is "no", M_{test} goes to $SEMITARGET + 1$ and does not increment the second counter. If the guess is "yes", M_{test} guesses M's computation of length $\leq j$ from $START$ to $SEMITARGET$, step by step.

Observe that the length of the computation from $START$ to $SEMITARGET$ is at most exponential in $S(n)$ and thus it can be stored in $S(n)$ space. If M_{test} guessed M's computation correctly, it checks whether $TARGET$ equals $SEMITARGET$ or is reachable from $SEMITARGET$ in one step of M. If neither one of these alternatives holds and N_j can be read from the second counter, M_{test} concludes that $TARGET$ is not reachable in $\leq j + 1$ computation steps by M.

If one of the alternatives holds, $TARGET$ is reachable in $\leq j + 1$ steps and this conclusion is independent of the second counter. M_{test} increments the first counter exactly in the latter case and starts working with $TARGET$ $+1$.

In case M_{test} guesses a wrong computation to a correctly reachable $SEMITARGET$, then M_{test} rejects its input w. Note that we define M_{test} to accept

w if and only if w is not in L. The language of M_{test} is not affected if wrong computations are discontinued early. Observe that M_{test} can choose one correct computation. Clearly, in a correct computation the second counter cannot be less than N_j.

When the last configuration (in alphabetic order) of length $S(n) + 1$ has appeared as $TARGET$, M_{test} knows the number N_{j+1} and consequently, M_{test} knows the number N.

Now M_{test} starts its second step of computation. M_{test} runs through all computations of length $S(n)+1$ keeping at the same time a counter. For each such configuration $FINALTARGET$, M_{test} first guesses whether it is reachable from $START$ by M. If the guess is "no", M_{test} goes to $FINALTARGET$ +1 and does not increment the counter. If the guess is "yes", M_{test} guesses step by step M's computation. When $FINALTARGET$ is reached, M_{test} rejects w if $FINALTARGET$ is an accepting configuration, and goes to $FINALTARGET$ +1 and increments the counter, if $FINALTARGET$ is not accepting, except the case that the counter shows N and $FINALTARGET$ is not accepting. In this latter case, M_{test} accepts w. □

For $S(n) = n$ it follows that

Corollary 2.1. *The family CS (=$NSPACE(n)$) is closed under complementation.* □

Theorem 2.11 was strengthened in [45]. If a language L is over one-letter alphabet and if $L \in NSPACE(S(n))$, then the complement of L is also in $NSPACE(S(n))$, for each function $S(n)$, i.e., it is not necessary for $S(n)$ to be greater than or equal to $\log n$.

The next theorem presents other closure properties of the family of context-sensitive languages. See [115] for a detailed proof of these properties.

Theorem 2.12. *The family CS is closed under: union, intersection, catenation, Kleene $*$, Kleene $+$, intersection with regular languages, λ-free morphism, inverse morphism, λ-free substitution, mirror image, λ-free gsm mapping.*

The family CS is not closed under substitution, morphism, gsm mapping, left (right) quotient with regular languages. □

Corollary 2.2. *The family CS is an AFL which is not a full AFL.* □

2.6 Decidable properties of the family CS

The most important decidable problem for context-sensitive languages is the membership problem.

Theorem 2.13. *Given a context-sensitive grammar $G = (N, T, S, P)$ and a word $w \in T^*$, it is decidable whether $w \in L(G)$ or not.*

Proof. If $w = \lambda$, then $w \in L(G)$ iff $S \longrightarrow \lambda$ is a production in P. Assume now that $w \neq \lambda$ and let n be the length of w. Define the sequence of sets $(\mathcal{V}_i)_{i \geq 0}$ by:

$$\mathcal{V}_0 = \{S\},$$

$$\mathcal{V}_{i+1} = \mathcal{V}_i \cup \{\beta \mid \exists \alpha \in \mathcal{V}_i, \alpha \Longrightarrow_G \beta \text{ and } \mid \beta \mid \leq n\}.$$

Note that for each $i \geq 0$, $\mathcal{V}_i \subseteq \bigcup_{j \leq n}(N \cup T)^j$. The set $\bigcup_{j \leq n}(N \cup T)^j$ is finite, hence there is an index k such that $\mathcal{V}_k = \mathcal{V}_{k+1}$. Clearly, for all $j \geq 0$, $\mathcal{V}_{k+j} = \mathcal{V}_k$. Finally, observe that $w \in L(G)$ if and only if $w \in \mathcal{V}_k$.   □

Corollary 2.3. *The family of context-sensitive languages is a subfamily of the family of recursive languages.*   □

It is known, see [115], that the above inclusion is strict. The membership problem for context-sensitive languages is $PSPACE$-complete, see Karp, [76].

Because context-sensitive languages constitute quite a large family of languages, most of the usual properties are undecidable.

Theorem 2.14. *The following problems are undecidable for the family of context-sensitive languages: emptiness, finiteness, regularity, context-freeness, equivalence, inclusion.*   □

See [115] for a detailed proof of the above theorem.

2.7 On some restrictions on grammars

In this section we consider grammars $G = (N, T, S, P)$ having all productions of the following form:

$$(*) \qquad \alpha_0 A_1 \alpha_1 \ldots \alpha_{n-1} A_n \alpha_n \longrightarrow \alpha_0 w_1 \alpha_1 \ldots \alpha_{n-1} w_n \alpha_n$$

with $n > 0$, $A_i \in N, 1 \leq i \leq n$, $\alpha_i \in T^*, 0 \leq i \leq n$, and $w_i \in (N \cup T)^*, 1 \leq i \leq n$.

Observe that, for instance, each grammar in the Kuroda normal form has all productions of the form $(*)$. Hence, every context-sensitive language can be generated by such a grammar. On the other hand, there are grammars with all productions of the form $(*)$ that are not length-increasing grammars (note that w_i can be λ for some $i, 1 \leq i \leq n$).

Definition 2.8. *A grammar $G = (N, T, S, P)$ with all productions of the form $(*)$ is terminal-bounded if each production is of the form:*

$$\alpha_0 A_1 \alpha_1 \ldots \alpha_{n-1} A_n \alpha_n \longrightarrow \beta_0 B_1 \beta_1 \ldots \beta_{m-1} B_m \beta_m$$

where $\alpha_i, \beta_i \in T^*$, $A_i, B_i \in N$ and either $n = 1$ or for some $j, 0 \leq j \leq m$ and all $k, 1 \leq k \leq n - 1$, $\mid \beta_j \mid > \mid \alpha_k \mid$.   □

The following theorem was proved by Brenda S. Baker, see [9]:

Theorem 2.15. *If G is a terminal-bounded grammar, then $L(G)$ is a context-free language.* □

The above theorem has many consequences. In the sequel, all grammars are supposed to have all productions of the form $(*)$.

Corollary 2.4. [48] *If G is a grammar such that each production is of the form $\alpha \longrightarrow \beta$, $\alpha \in N^+$, $\beta \in (N \cup T)^*$, $\mid \beta \mid_T \neq 0$, then $L(G)$ is a context-free language.* □

Corollary 2.5. [15] *If G is a grammar such that each production is of the form $\alpha \longrightarrow \beta$, $\alpha \in T^*NT^*$, then $L(G)$ is a context-free language.* □

Corollary 2.6. [15] *If G is a grammar such that each production is of the form $\alpha X\beta \longrightarrow \alpha\gamma\beta$, where either $\alpha \in T^*$ and $\mid \alpha \mid \geq \mid \beta \mid$ or $\beta \in T^*$ and $\mid \beta \mid \geq \mid \alpha \mid$, then $L(G)$ is a context-free language.* □

Corollary 2.7. [60] *Let $G = (N, T, S, P)$ be a grammar such that a partial order "$<$" is defined on $N \cup T$. Moreover, assume that each production of G is of the form:*

$$A_1 A_2 \ldots A_n \longrightarrow B_1 B_2 \ldots B_m,$$

where $A_i \in N$ and for some k, $1 \leq k \leq m$, and for all i, $1 \leq i \leq n$, $A_i < B_k$. Then $L(G)$ is a context-free language. □

Definition 2.9. *Let $G = (N, T, S, P)$ be a grammar. If $\alpha \in T^*$, $(\sigma \longrightarrow \gamma) \in P$ and $\beta \in (N \cup T)^*$, then write $\alpha\sigma\beta \Longrightarrow_L \alpha\gamma\beta$ and $\beta\sigma\alpha \Longrightarrow_R \beta\gamma\alpha$. Define*

$$Left(G) = \{w \in T^* \mid S \Longrightarrow_L \theta_1 \Longrightarrow_L \theta_2 \ldots \Longrightarrow_L \theta_n = w\}$$

and

$$Two\text{-}way(G) = \{w \in T^* \mid S \Longrightarrow_G \theta_1 \Longrightarrow_G \theta_2 \ldots \Longrightarrow_G \theta_n = w \text{ and}$$

$$for\ i = 2, \ldots n,\ either\ \theta_{i-1} \Longrightarrow_L \theta_i\ or\ \theta_{i-1} \Longrightarrow_R \theta_i\}.$$ □

Corollary 2.8. [96, 97] *If G is a grammar, then $Left(G)$ and $Two\text{-}way(G)$ are context-free languages.* □

Still we mention some results related to the family of context-sensitive languages. The family of languages consisting of a finite intersection of context-free languages is introduced and investigated in [84] and [52]. This family has useful applications to natural languages, see [85]. In [127] it is shown that both the graph and the range of the Ackermann-Peter's function are context-sensitive languages. The class of elementary functions (EF) is introduced and studied in [67, 68]. This class is the closure under superposition, primitive recursion, and exponentiation of polynomials and exponential functions. It is proved that the graph of a function in EF is always a context-sensitive languages. Using these results, a large subfamily of context-sensitive languages has been found to have the complements also context-sensitive languages. The result is obtained a couple of years before the theorem of Immerman-Szelepcsényi and proved using straightforward methods.

3. AFL-theory

3.1 Language families. Cones and AFL's

In this section we will discuss certain sets of languages, customarily referred to as *language families*. The specific feature we are interested in is *closure under certain operations*. We consider *abstract families of languages* (that's where the abbreviation AFL comes from), without specifying what the languages are. We only require that the families are *closed* under certain operations. Whenever we apply one of the operations to languages in the family (our basic operations are either unary or binary), then the resulting language is also in the family. This is the property defining the family.

The study of language families can be motivated in the following general terms. When we show that a specific family, say the family of all context-free languages, has some particular property, say it is closed under substitution, then we usually need only some other properties, not the strength of the full definition of the family. In this case certain other closure properties of the family of context-free languages will suffice. Thus, the particular property we were interested in is shared by any family that possesses the properties we needed in our proof. This gives rise to the following abstraction: we do not consider any specific language family, such as the family of context-free languages, but rather a collection of language families, each of which possesses certain basic (axiomatic) properties. Using these properties, we then establish other properties which, thus, are shared by all families in the collection. In this way at least parts of the theory will be unified, since it becomes unnecessary to repeat proofs for different families. Also some new insights are gained. A typical example is the fundamental result presented below (Theorem 3.10) concerning the order of closure operations. Over-all unification and insights concerning some general issues – these are the aims of AFL-Theory.

We begin the formal details from the notion of a *family of languages*. It is a set or collection of languages possessing a nonempty language. However, all languages in the collection need not be over the same alphabet Σ. Indeed, such a requirement would seriously restrict many of the customary families, for instance, context-free languages. In a collection of languages, the size of the alphabet of the individual languages may increase beyond all bounds. On the other hand, for every language L in the collection, there must exist a finite alphabet Σ such that L is a language over Σ. (We consider only languages over a finite alphabet, this follows from the definition of a language we have in mind.) It is no loss of generality if we view the individual alphabets Σ as subsets of a fixed infinite set V of symbols. Thus, a *family of languages* is a pair $(V, \mathcal{L})$, where V is a set of letters and $\mathcal{L}$ is a set of languages such that each $L \in \mathcal{L}$ is a language over an alphabet $\Sigma \subseteq V$. Usually we will omit V when speaking of language families.

To say that a family of languages is *closed* under an operation defined for languages means that, whenever the operation is applied to languages in the family, the resulting language is also in the family. To avoid the ambiguity inherent in complementation (see [61]), we make the convention that $\mathcal{L}$ is *closed under complementation* if, for every $L \in \mathcal{L}$ and every Σ such that L is a language over Σ, the difference $\Sigma^* - L$ is in $\mathcal{L}$.

We begin with a construction very typical in AFL-theory. The following result will also be a useful lemma.

Theorem 3.1. *Assume that a language family $\mathcal{L}$ is closed under each of the following operations: union, catenation closure (that is, either Kleene $*$ or Kleene $+$), letter-to-letter morphism, inverse morphism, and intersection with regular languages. Then $\mathcal{L}$ is closed under catenation.*

Proof. We will show that, whenever L_1 and L_2 are in $\mathcal{L}$, so is $L_1 L_2$. Clearly, for some alphabet Σ, both L_1 and L_2 are languages over Σ. Consider two primed versions

$$\Sigma' = \{a' \mid a \in \Sigma\} \text{ and } \Sigma'' = \{a'' \mid a \in \Sigma\},$$

as well as the alphabet $\Sigma_1 = \Sigma \cup \Sigma' \cup \Sigma''$. Let $h : \Sigma_1^* \longrightarrow \Sigma^*$ be the letter-to-letter morphism defined by

$$h(a'') = h(a') = h(a) = a, \, a \in \Sigma.$$

Define

$$L'_1 = h^{-1}(L_1) \cap \{a'x \mid a \in \Sigma, x \in \Sigma^*\},$$

$$L'_2 = h^{-1}(L_2) \cap \{a''x \mid a \in \Sigma, x \in \Sigma^*\}.$$

Then L'_1 (resp. L'_2) results from L_1 (resp. L_2) by replacing in each nonempty word the first letter with its primed (resp. doubly primed) version. Since the second languages in the intersections are regular, we conclude that L'_1 and L'_2 are in $\mathcal{L}$. By the same argument, the language

$$L_3 = (L'_1 \cup L'_2)^+ \cap \{a'xb''y \mid a, b \in \Sigma \text{ and } x, y \in \Sigma^*\}$$

is in $\mathcal{L}$. We may use here Kleene $*$ instead of Kleene $+$ without changing L_3. In both cases we observe that

$$L_3 = \{a'xb''y \mid a, b \in \Sigma, ax \in L_1, by \in L_2\}.$$

The catenation $L_1 L_2$ equals one of the four languages

$$h(L_3),\ h(L_3) \cup L_1,\ h(L_3) \cup L_2,\ h(L_3) \cup L_1 \cup L_2,$$

depending on whether the empty word λ belongs to neither of the languages L_1, L_2, to one of them, or to both of them. Thus, by the assumptions, the catenation $L_1 L_2$ is in $\mathcal{L}$ in any case.   □

Observe that closure under each of the five operations mentioned in the assumption was actually used in the proof. Observe also the following fact valid throughout our exposition concerning AFL-theory. We consider really abstract families $\mathcal{L}$; the languages need not even be recursively enumerable. Consequently, the constructions need not be effective in any sense. For instance, it may be the case that we cannot decide which of the four alternatives holds at the end of the proof.

We are now ready for the fundamental definition.

Definition 3.1. *A family of languages is called a* cone *if it is closed under morphism, inverse morphism and intersection with regular languages. A cone closed under union and Kleene $*$ is a* full AFL. *(The abbreviation comes from "Abstract Family of Languages".) An* AFL *is a family of languages closed under each of the following operations: λ-free morphism, inverse morphism, intersection with regular languages, union and Kleene $+$.*   □

Several remarks, both historical and factual, should be made concerning this definition. Let us first summarize the definitional (axiomatic) closure properties of each of the three types of families. The operation "intersection with regular languages" is abbreviated by $\cap Reg$.

Cone: morphism, inverse morphism, $\cap Reg$.

Full AFL: morphism, inverse morphism, $\cap Reg$, union, catenation, Kleene $*$.

AFL: λ-free morphism, inverse morphism, $\cap Reg$, union, catenation, Kleene $+$.

We have added the operation of catenation to the latter two lists. Indeed, it follows by Theorem 3.1 that full AFL's and AFL's (as defined above) are always closed under catenation. So exactly the same families will result if we include closure under catenation among the definitional closure properties. This has, in fact, been normally done in the past.

At a first glance, the definition above might seem quite arbitrary: we have just picked up a few from the many possible operations, and postulated closure under these operations. The operation $\cap Reg$ might seem particularly artificial in this respect. We could, on seemingly similar grounds, postulate closure under union with context-free languages, for instance. However, there is some strong motivation behind the above choice of operations. As we will see below, the operations of a cone are exactly the ones defining *rational transductions*. The additional AFL-operations, including catenation, are exactly the ones defining regular languages, also called *rational languages*. So cones are associated with rational transductions, and AFL's with rational transductions and rational operations.

The difference between full AFL's and AFL's is that the latter are suitable for λ-*free families:* none of the languages in the family contains the empty word. Also in some other cases, notably in connection with context-sensitive languages, it is important to restrict the closure under morphisms. As we have seen (Theorem 1.10), the closure of context-sensitive languages under unrestricted morphisms equals the whole family of recursively enumerable languages.

Our choice of the terminology is a compromise between the original American (see [50] and [47]) and French (see [102] and [14]) schools. The term "AFL" comes from the Americans, the term "cone" (depicting closure under rational transductions) from the French. The American counterpart of a cone is a "full trio" (reflecting only the fact that three operations are involved), whereas the French use the name "FAL" (Famille Agréable de Langages) for a full AFL. Thus, in addition of using different terms, the two schools position the supplementary adjective ("full" for the Americans) differently. The reader can surely guess what a "trio" is – this concept will not be used in the sequel.

To summarize, the choice of the defining operations has been very well motivated, especially during the time the choice was made, in the late 60's. Rational operations and transductions reflect very well the nature of *sequential* rewriting using nonterminal letters: all families in the Chomsky hierarchy, as well as many of the related families, are AFL's. The situation becomes different if we rewrite *without nonterminals* or *in parallel*, in particular, in connection with L systems. Then the property of being an AFL or a cone is no more characteristic. On the contrary, many of the L families are "anti-AFL's", not closed under any of the six operations.

We now begin the study of some basic properties of cones and AFL's. Also some operations will enter our discussion, other than ones mentioned in the definitions. In many cases it is not necessary to assume that a morphism is λ-free, a weaker type of nonerasing morphisms will suffice. We say that a morphism h is a *linear erasing* with respect to a language L if there is a positive integer k such that whenever $x \in L$, $k \mid h(x) \mid \geq \mid x \mid$. Thus, h may be erasing but it cannot erase too much from words belonging to L. (See Definition 2.7.)

When one tries to show that a specific language family is a full AFL, usually the biggest difficulty consists of showing closure under inverse morphism. The following theorem presents an alternative way of doing this. The proof of the theorem is not difficult – the reader may try to prove it directly or consult [115] for a simple proof. The terminology used should be clear: a substitution is λ-*free* (*regular*) if each language substituted for a letter is λ-free (regular).

Theorem 3.2. *If a family of languages is closed under λ-free regular substitution, linear erasing, union with regular languages and intersection with regular languages, then it is closed under inverse morphism. The same conclusion can be made for λ-free families even without assuming closure under union with regular languages.* □

AFL's can be characterized in many alternative ways. The following two theorems present some criteria where closure under inverse morphism is not needed. Observe that the special case, where the family consists of the language $\{\lambda\}$, has to be excluded. Without inverse morphism one cannot generate new languages from this language using the other operations.

Theorem 3.3. *If a family of languages includes a language containing a nonempty word and is closed under union, Kleene * (or Kleene +), λ-free regular substitution, intersection with regular languages, and (arbitrary) morphism, then the family is a full AFL. If a family of languages contains all regular languages and is closed under substitution, as well as under intersection with regular languages, then the family is a full AFL.* □

Theorem 3.4. *If a family of languages includes a language containing a nonempty word and is closed under union, Kleene +, λ-free regular substitution, intersection with regular languages, and linear erasing, then the family is an AFL. If a λ-free family of languages contains all λ-free regular languages and is closed under λ-free substitution, intersection with regular languages, and linear erasing, then the family is an AFL.* □

Theorem 3.4 is the AFL-version of Theorem 3.3 which concerns full AFL's. Observe the difference in the second sentences. An additional assumption (about closure under linear erasing) is needed in Theorem 3.4. The reason for this difference goes back to Theorem 3.2. Closure under linear erasing is needed to get closure under inverse morphism. In Theorem 3.3 it is obtained directly from closure under substitution, this being not the case in Theorem 3.4, where we have only λ-free substitution.

It is a consequence of Theorem 3.1 that every AFL is closed under catenation. We do not want to state this as a formal theorem for the simple reason that catenation is usually (see [47]) included among the definitional properties. In the definition of an AFL and a full AFL given above, none of the five operations is redundant: all five closure properties are independent in the sense that none of them follows from the others.

We will prove next that a cone (and hence also a full AFL) can never be a very small family: it contains always the lowest level of the Chomsky hierarchy.

Theorem 3.5. *The family of regular languages is contained in every cone.*

Proof. To show that a regular language $L \subseteq \Sigma^*$ is in an arbitrarily given cone $\mathcal{L}$, it suffices to show that Σ^* is in $\mathcal{L}$, since $\mathcal{L}$ is closed under $\cap Reg$

and $L = \Sigma^* \cap L$. Being a language family, $\mathcal{L}$ possesses a nonempty language. From this we conclude that the language $\{\lambda\}$ is in $\mathcal{L}$; we just take the image under the morphism mapping all letters of the language into λ. Consequently, $\Sigma^* = h^{-1}(\{\lambda\})$ is in $\mathcal{L}$, where the morphism h maps all letters of Σ into λ.

□

The final theorem in this subsection gives some further dependencies between closure properties. It turns out that in case of λ-free families four of the operations suffice in defining an AFL.

Theorem 3.6. *If a family of languages is closed under catenation, Kleene $*$, λ-free morphism, inverse morphism, and intersection with regular languages, then it is closed under union. If a λ-free family of languages is closed under catenation, λ-free morphism, and inverse morphism, then it is closed under intersection with regular languages. A λ-free family of languages is an AFL iff it is closed under the following operations: catenation, Kleene $+$, λ-free morphism, and inverse morphism.*   □

3.2 First transductions, then regular operations

Our purpose is two-fold in this subsection. First, we want to show the intimate connection between rational transductions and the operations of morphism, inverse morphism and intersection with regular languages. Second, we want to establish a remarkable result, undoubtedly the most significant one in AFL-theory, concerning the *order* in which closure under the AFL-operations is formed. Given any language family; one can *close* it with respect to any specific operations, that is, form the smallest family that contains the given family and is closed under the operations. If the specific operations are the AFL-operations, then the closure can always be built in such a way that the given family is *first* closed with respect to rational transductions and *after that* the resulting family is closed with respect to regular operations. In other words, after these two steps the final family will always be closed with respect to rational transductions, no matter what the original family was. Such a possibility of ordering the operations in this sense is rather unusual.

We now introduce the notion of a *rational transducer*. It is a finite-state transformation device possessing an input tape and an output tape. At each moment of time, the transducer reads a letter or the empty word from the input tape in some internal state, and goes nondeterministically into another state and writes a letter or the empty word to the output tape. The behaviour is determined by quadruples (a, s, s', b), where s and s' are states, a is a letter of the input alphabet or λ, and b is a letter of the output alphabet or λ. Finitely many such quadruples are specified in the definition of a specific transducer. Among the *states* there is a special *initial* state and a special set of *final* states. At the beginning of each computation, the transducer is in the initial state. It has to be in a final state after having read the whole input word; otherwise the output word is not recorded. In this way the transducer transforms an input word into a set of output words. (Several output words may result because of nondeterminism, even empty and infinite sets are possible.) More generally, an input language is transformed into an output language.

Let us now work out the details formally. We will use here the formalism of quadruples and instantaneous descriptions. Everything can be done also in terms of rewriting systems, see [115]. The latter formalism is often much better for detailed proofs, the former formalism being more intuitive.

A *rational transducer* is a 6-tuple $T = (S, \Sigma_i, \Sigma_o, M, s_0, S_f)$, where

(i) S is a finite nonempty set (called the set of *states*),
(ii) $s_0 \in S, S_f \subseteq S$ (*initial* and *final* states),
(iii) Σ_i and Σ_o are alphabets (*input* and *output* letters),
(iv) $M \subseteq (\Sigma_i \cup \{\lambda\}) \times S \times S \times (\Sigma_o \cup \{\lambda\})$ (the set of *moves*).

Let $\vdash_T$ (or briefly $\vdash$) be the binary relation on $S \times \Sigma_i^* \times \Sigma_o^*$ defined by letting

$$(s, au, v) \vdash (s', u, vb)$$

if $(a, s, s', b) \in M$, $u \in \Sigma_i^*$ and $v \in \Sigma_o^*$.

The triple (s, au, v), referred to as an *instantaneous description*, represents the fact that T is in the state s, au is the still unread portion of the input, and v is the output word produced so far. Since the quadruple (a, s, s', b) is in the set M describing the behaviour of T, it is possible for T to read a, go to the state s' and write b to the output tape. This is exactly what happens and, thus, the next instantaneous description is (s', u, vb). It is possible that some other quadruple in M gives rise to another instantaneous description, following next after (s, au, v).

As usual, we denote by $\vdash^*$ the reflexive and transitive closure of the relation $\vdash$. Thus,

$$(s, u, v) \vdash^* (s', u', v')$$

indicates that it is possible for T to go from the instantaneous description (s, u, v) to the instantaneous description (s', u', v') in a sequence of zero or more moves. The transformation accomplished by a rational transducer T can now be defined formally as follows:

$$T(w) = \{v \mid (s_0, w, \lambda) \vdash^* (s_f, \lambda, v), s_f \in S_f\}.$$

The transformation thus defined maps Σ_i^* into the set of subsets of Σ_o^*. It can be extended to concern languages $L \subseteq \Sigma_i^*$ in the natural additive fashion:

$$T(L) = \bigcup_{w \in L} T(w).$$

Mappings T defined in this way, from the set of languages over Σ_i into the set of languages over Σ_o, are referred to as *rational transductions*.

Some remarks concerning the terminology should be added. Rational transducers are called in [47] "sequential transducers with accepting states" or "a-transducers". Special cases of rational transducers are λ-*free rational transducers* and *generalized sequential machines* (abbreviated *gsm*). The former can never print the output λ, that is, the last component of every element of M is a letter. The latter can never read the input λ, that is, the first component of every element of M is a letter. We have defined a rational transducer in such a way that during one move it never reads or writes more than one letter. One can give a more general definition (see [47]), where both the input read and the output produced can be a word. However, this more general model yields the same rational transductions. Indeed, one can come back to the model presented in our definition by introducing new states in an obvious fashion. Observe, however, that such a reduction is not possible if one deals with the special cases of λ-free rational transducers or generalized sequential machines.

We now establish some auxiliary results. Each of the three operations of morphism, inverse morphism and $\cap Reg$ can be realized by a rational transducer. Consider a morphism h. A rational transducer giving rise to the same mapping as h works as follows. It reads a letter a and outputs λ or the first letter of $h(a)$, depending on whether or not $h(a) = \lambda$. If $\mid h(a) \mid > 1$, the transducer enters a sequence of states specific to a. In each of them it reads the empty word and outputs one letter of $h(a)$. After having produced the last letter of $h(a)$, the transducer is in a final state and also ready to start dealing with the next input letter, if any.

An inverse morphism is handled analogously but now the transducer is nondeterministic. It guesses an image $h(a)$ from the input tape, and reads through it letter by letter, producing finally the output a if the guess was correct. If there are letters a with $h(a) = \lambda$, then λ is always a correct guess and, thus, such letters a may be inserted anywhere in the output. It is clear by the definition that the operation $\cap Reg$ is a rational transduction. For a given regular language L, the following transducer T does the job. T reads the input letter by letter, and produces the same output. At the same time, T simulates by its states the finite automaton for L. The output is recorded only in a final state.

Rational transductions are closed under *composition*. Whenever T_1 and T_2 are rational transducers, there is a rational transducer T_3 such that $T_1(T_2(L)) = T_3(L)$ holds for any language L. (We assume that the output alphabet of T_2 is contained in the input alphabet of T_1. Otherwise, the composition need not be well-defined.) First dummy quadruples (λ, s, s, λ), where s ranges over all states, are added to both T_1 and T_2. Apparently this does not affect the mappings in any way. The states of T_3 are pairs (r, s), where r is a state of T_2 and s is a state of T_1. The quadruple $(a, (r, s), (r', s'), b)$

belongs to T_3 iff, for some c, (a, r, r', c) is in T_2, and (c, s, s', b) is in T_1. We have, thus, established the following result.

Theorem 3.7. *Each of the three operations defining a cone is a rational transduction. Rational transductions are closed under composition.* □

We will now establish the converse of the first sentence of Theorem 3.7: every rational transduction can be expressed in terms of the three cone operations, The following theorem, generally known as *Nivat's Theorem*, [102], can also be viewed as a *normal form* result for rational transductions.

Theorem 3.8. *Given a rational transducer T, two morphisms g and h and a regular language R can be constructed such that*

$$(*) \qquad\qquad T(L) = g(h^{-1}(L) \cap R)$$

holds for all languages L. If T is λ-free, so is g.

Proof. Assume that $T = (S, \Sigma_i, \Sigma_o, M, s_0, S_f)$ is the given rational transducer. We view the quadruples (a, s, s', b) in M as individual inseparable letters. Let Σ be the alphabet consisting of all of these letters. Then h (resp. g) will be the morphism of Σ^* into Σ_i^* (resp. Σ_o^*), defined by

$$h(a, s, s', b) = a \text{ (resp. } g(a, s, s', b) = b).$$

Thus, h and g pick up from the quadruples the input and output parts, respectively. In $(*)$, h^{-1} selects an input word from L, and g produces the corresponding output. The correspondence is guaranteed by the underlying sequence of quadruples (word over Σ), selected nondeterministically by h^{-1}.

The purpose of R is to assure that the underlying sequence of quadruples is indeed a correct computation of T. Let R be the language over Σ consisting of all words w satisfying each of the following conditions (i)-(iii). Recall that the letters of Σ are actually quadruples (a, s, s', b). So we may speak of the first and second state in a letter. (i) The first state in the first letter of w equals s_0. (ii) The second state in the last letter of w is in S_f. (iii) In every two consecutive letters of w, the first state of the second equals the second state of the first. Thus, (i) and (ii) guarantee that the sequence of quadruples begins and ends correctly, whereas a correct sequence of states is obtained due to (iii).

Consequently, $(*)$ is satisfied. It remains to be shown that R is a regular language. It is easy to construct a regular expression or type 3 grammar for R (see [115]). Another possibility is to use the procedure of Theorem 2.5 in the First Chapter of this Handbook. Whether or not a word w is in R can be tested by moving an open window of length 2 across the (elsewhere covered) word w. □

Nivat's Theorem has remarkable consequences. If $\mathcal{L}$ is a family of languages and OP is a set of operations defined for languages, the notation

$Cl_{OP}(\mathcal{L})$) stands for the *closure* of $\mathcal{L}$ under the set OP of operations, that is, the smallest family containing $\mathcal{L}$ and closed under each operation in OP. A special case is that OP consists of a single operation. Then the index for Cl will be just the name of the operation. For instance, $Cl_{h^{-1}}(\mathcal{L})$ stands for the closure of $\mathcal{L}$ under inverse morphism and $Cl_{tr}(\mathcal{L})$ for the closure of $\mathcal{L}$ under rational transductions. Sets OP we will be particularly interested in are the cone, full AFL-, and regular operations:

$$Cone = \{h, h^{-1}, \cap Reg\},$$

$$fAFL = \{\cup, \text{catenation, Kleene } *, h, h^{-1}, \cap Reg\},$$

$$Reg = \{\cup, \text{catenation, Kleene } *\}.$$

Thus, $Cl_{fAFL}(\mathcal{L})$ means the closure of $\mathcal{L}$ under (full) AFL-operations. To say that $\mathcal{L}$ is a full AFL is equivalent to saying that

$$Cl_{fAFL}(\mathcal{L}) = \mathcal{L}.$$

The following theorem is now an immediate consequence of Theorems 3.7 and 3.8.

Theorem 3.9. *Every cone is closed under rational transductions: if $\mathcal{L}$ is a cone then $Cl_{tr}(\mathcal{L}) = \mathcal{L}$. Moreover, for every language family $\mathcal{L}$,*

$$Cl_{Cone}(\mathcal{L}) = Cl_{tr}(\mathcal{L}) =$$

$$= \{T(L) \mid L \in \mathcal{L} \text{ and } T \text{ is a rational transducer } \} =$$

$$= \{g(h^{-1}(L) \cap R) \mid L \in \mathcal{L}, g \text{ and } h \text{ are morphisms and}$$

$$R \text{ is a regular language } \}. \qquad \square$$

Theorem 3.9 can also be interpreted as follows. The cone-closure of any family $\mathcal{L}$ can be obtained by closing $\mathcal{L}$ with respect to the three operations in a specific order: first with respect to inverse morphism, then with respect to $\cap Reg$ and, finally, with respect to morphism. Such a result is very seldom valid for a set of three operations. Concerning this particular set $Cone = \{h, h^{-1}, \cap Reg\}$, the result is not valid if the three closures are not formed in the order mentioned above. The reader might think about this. If the closure is first formed with respect to $\{h, \cap Reg\}$ and, finally, with respect to h^{-1}, then in general a much smaller family results. For instance, if the original family consists of the single language $\{a\}$, then the new order of closures yields all inverse morphic images of singleton languages $\{w\}$, whereas the order of closures stemming from Theorem 3.9 yields in this case the whole family of regular languages. An early application of the closure under $\cap Reg$ keeps "small" families small. It is also an interesting problem (see [29], [56]) how many alternating applications of h and h^{-1} yield the whole family of regular languages from a singleton language $\{w\}$.

We are now in the position to establish the result, referred to at the beginning of this section, concerning the order in which the closure under AFL-operations can always be taken. The result can now be stated explicitly in the following form. For every language family $\mathcal{L}$,

$$(**) \qquad\qquad Cl_{fAFL}(\mathcal{L}) = Cl_{Reg}(Cl_{Cone}(\mathcal{L})).$$

Thus, the closure can first be formed with respect to the cone operations, after which the resulting family is closed with respect to the regular operations. By Theorem 3.9, $(**)$ can also be expressed in the form

$$Cl_{fAFL}(\mathcal{L}) = Cl_{Reg}(Cl_{tr}(\mathcal{L})) = Cl_{Reg}(Cl_h Cl_{\cap Reg} Cl_{h^{-1}}(\mathcal{L})).$$

Although our proof of $(**)$ does not contain major theoretical innovations, it is simpler than the customary proofs (see [47]).

We will now show that $(**)$ is valid. By Theorem 3.9, it suffices to prove that the right side is closed under rational transductions. An arbitrary language on the right side is obtained by regular operations from languages in $Cl_{tr}(\mathcal{L})$. Consequently, an arbitrary language on the right side results from a language R by a substitution σ, defined for each letter a in the alphabet of R as follows:

$$\sigma(a) = T_a(L_a) = L'_a, \ L_a \in \mathcal{L}, \ T_a \text{ is a rational transducer.}$$

Let T be another rational transducer. We have to show that $T(\sigma(R))$ belongs to the right side of $(**)$. To do this, it suffices to prove that there is a regular language R_1 and a substitution σ', where $\sigma'(b) = T'_b(L_b)$ with $L_b \in \mathcal{L}$ and T'_b being a transducer, such that

$$(**)' \qquad\qquad T(\sigma(R)) = \sigma'(R_1),$$

that is, to present the language $T(\sigma(R))$ in the general form of the languages belonging to the right side of $(**)$.

We now define the language R_1. The alphabet of R_1 consists of all triples (s, a, s'), where a is a letter of R and s, s' are states of T. By definition, R_1 consists of all words of the form

$$(s_0, a_1, s_1)(s_1, a_2, s_2) \ldots (s_{n-1}, a_n, s_n), a_1 a_2 \ldots a_n \in R, s_n \in S_f,$$

where s_0 is the initial state and S_f the final state set of T. The language R_1 is regular because it can be obtained from the regular language R by taking first an inverse morphic image (this replaces a letter by a corresponding triple with arbitrary states), and then intersecting the result with a suitable regular language similarly as in the proof of Theorem 3.9 (this guarantees that only legal state sequences from the initial state to a final state pass through).

Let s and s' be states of T, not necessarily distinct. The transducer $T_{s,s'}$ has s as its initial and s' its only final state. The quadruples of $T_{s,s'}$ are obtained from those of T, for instance (a, r, r', b), by replacing the output

b with (r, b, r'). Let g be the morphism mapping the triples to their middle component:

$$g(r, b, r') = b.$$

Finally, the substitution σ' is defined by

$$\sigma'(s, a, s') = gT_{s,s'}(L'_a) = gT_{s,s'}T_a(L_a).$$

Since $L_a \in \mathcal{L}$ and $gT_{s,s'}T_a$ is a rational transduction, the substitution σ' is of the required form. Moreover, $(**)$ holds true. The core element on both sides is a word $a_1 a_2 \ldots a_n \in R$. Considering the left side, T goes from s_0 to s_1 when reading a word in L'_{a_1}. The same output is produced on the right side from (s_0, a_1, s_1) by gT_{s_0,s_1}.

We still state the established equation $(**)$ in the following form.

Theorem 3.10. *The full AFL-closure of any language family $\mathcal{L}$ is obtained by first closing $\mathcal{L}$ under rational transductions, and then closing the resulting family under regular operations.*   □

The reader is referred to [47] for further details and bibliographical remarks concerning AFL-theory. *Principal AFL's* constitute a widely studied topic we have not discussed here. By definition, a full AFL $\mathcal{L}$ is *full principal* if it is generated by a single language L:

$$\mathcal{L} = Cl_{fAFL}(L).$$

Principal AFL's and cones are defined similarly. No nondenumerable family can be principal. Recursive languages constitute a nontrivial example of an AFL that is not principal.

4. Wijngaarden (two-level) grammars

4.1 Definitions, examples. The generative power

Wijngaarden grammars or *two-level grammars* were introduced in [136] in order to define the syntax and the semantics of the programming language ALGOL 68, see [138]. They are a very natural extension of context-free grammars. In a Wijngaarden, or shortly W-grammar, the idea to extend a context-free grammar is very different from the idea to extend a context-free grammar to a context-sensitive grammar. A W-grammar has nonterminals with an inner structure, i.e., the name of the nonterminal may depend on some variables, called *hypervariables*. Consequently, the rules that contain such hypervariables are referred to as *hyperrules*. The hyperrules are not used directly for derivations. In order to obtain the rules for derivations, one has to substitute consistently (uniformly) all occurrences of hypervariables in hyperrules by values from some fixed sets associated to hypervariables. These sets could be context-free languages, too. The resulting set of derivation rules could be infinite. A similar phenomenon occurs in mathematical logic, when the propositional logic is extended to the predicate logic by assigning to propositional variables an inner structure that depends on some variables. Note that the variables that occur in predicates and in formulas are consistently substituted by values from a set (the domain of the model).

A further classical method to define a model is to consider all ground terms as the domain, the well-known Herbrand model. The domain of a Herbrand model has a structure very similar to a context-free language.

The theory of W-grammars was developed slowly. At the first glance, the formalism looks complicated enough. We present here a short and straight way to introduce the notion of a W-grammar. A *context-free scheme* is a context-free grammar without the start symbol, i.e., the start symbol is not defined.

Definition 4.1. *A W-grammar is a triple, $G = (G_1, G_2, z)$, where $G_1 = (M, V, R_M)$ and $G_2 = (H, \Sigma, R_H)$ are context-free schemes such that the elements of H are of the form $< h >$, where $h \in (M \cup V)^*$, $(M \cup V \cup \Sigma) \cap \{<, >\} = \emptyset$ and $z \in H$, $z = < v >$, $v \in V^*$.*   □

Terminology. G_1 is the *metalevel* of G and the elements of the sets M, V, R_M are the *metavariables*, the *metaterminals* and respectively the *metarules*. G_2 is the *hyperlevel* of G and the elements of the sets H, Σ, R_H are the *hypervariables*, the *terminals* and respectively the *hyperrules*. The *start symbol* of G is z.

In order to define the language generated by a W-grammar, some preliminaries are necessary. Denote by L_X the context-free language generated by the context-free grammar $G_X = (M, V, X, R_M)$, where $X \in M$, i.e., G_X is the context-free grammar that is obtained from the context-free scheme G_1 by defining the start symbol $X \in M$. Consider the sets: $A = M \cup V \cup \Sigma \cup \{<, >, \longrightarrow\}$ and $B = A - M$. Any function $t : A \longrightarrow B^*$, with the property that, for every $X \in M$, $t(X) \in L_X$, and for every $v \in A - M$, $t(v) = v$, can be extended to a unique morphism $t' : A^* \longrightarrow B^*$. All these morphisms t' define the set of *admissible substitutions*, denoted by T.

Definition 4.2. *The set of strict rules is $R_S = \{t(r) \mid t \in T, r \in R_H\}$, and the set of strict hypervariables is $H_S = \{t(x) \mid t \in T, x \in H\}$.*

If r is a hyperrule, $r \in R_H$, then the set of strict rules corresponding to r is

$$R_S(r) = \{t(r) \mid t \in T\}.$$ □

Note that the strict rules are context-free rules, where the nonterminals are strict hypervariables. However H_S can be infinite and, consequently, the set of strict rules, R_S, can be infinite, too. Perhaps this is the most important difference between context-free grammars and W-grammars.

Definition 4.3. *The derivation relation is defined as for context-free grammars, but using the strict rules instead of the classical rules of a context-free grammar, i.e., for any x, y from $(H_S \cup \Sigma)^*$, $x \Longrightarrow^* y$ iff there exist $n \geq 0$ and u_i in $(H_S \cup \Sigma)^*$, $i = 0, \ldots, n$, such that $u_0 = x, u_n = y$ and for each i, $i = 0, \ldots, n-1$, there are α, β and a strict rule $\gamma \longrightarrow \delta$ in R_S, with $u_i = \alpha\gamma\beta$ and $u_{i+1} = \alpha\delta\beta$.*

The leftmost (rightmost) derivation is defined as in the case of context-free grammars. □

Definition 4.4. *The language generated by a W-grammar G, is*

$$L(G) = \{x \mid x \in \Sigma^*, z \Longrightarrow^* x\}.$$ □

In sequel we consider some examples. Assume that, for all of the next three examples, the metalevel of the W-grammar is in each case the same, i.e., $M = \{I, J, K\}$, $V = \{q, s, \#, t, t', t''\}$ and for any $X \in M$, $L_X = q^*$. Moreover, assume that $z = < s >$, $\Sigma = \{a, b, c\}$ and therefore, for each of the following W-grammars, one has to define only the set R_H of hyperrules.

Example 4.1. As a first example, assume that R_H is:

h1. $< s > \longrightarrow < \#q >$
h2. $< I\#J > \longrightarrow < tI > | < J\#IJ >$
h3. $< tqI > \longrightarrow a < tI >$
h4. $< t > \longrightarrow \lambda$

It is easy to observe that the strict rules are: h1, h4,

$$\text{h2}_{i,j}. \quad < q^i \#q^j > \longrightarrow < tq^i > | < q^j \#q^{i+j} >,$$

and

$$\text{h3}_k. \quad < tq^{k+1} > \longrightarrow a < tq^k >,$$

where $i, j, k \geq 0$.
Note that a strict hypervariable $< tq^n >$ is producing a^n and nothing else, with $n \geq 0$. A derivation starts with z and if the next rules are h1 and the second case of $\text{h2}_{i,j}$ for a number of applications, then one obtains a derivation such as the following:

$$z \Longrightarrow < q^0 \#q^1 > \Longrightarrow < q^1 \#q^1 > \Longrightarrow < q^1 \#q^2 > \Longrightarrow < q^2 \#q^3 > \Longrightarrow$$

$$\Longrightarrow < q^3 \#q^5 > \Longrightarrow < q^5 \#q^8 > \Longrightarrow < q^8 \#q^{13} > \Longrightarrow \ldots$$

At any step in the above derivation, one can apply the first case of a strict rule $\text{h2}_{i,j}$ that replaces a strict hypervariable $< q^i \#q^j >$ with a strict hypervariable $< tq^i >$ that derives finally a^i. Therefore, the language generated is:

$$L_1 = \{a^f \mid f \text{ in the Fibonacci sequence}\}.$$ □

Example 4.2. Proceed as in the above example, except that now R_H is:

h1. $< s > \longrightarrow < q \# q >$
h2. $< I \# J > \longrightarrow < tI > < t'J > | < Iq \# JqII >$
h3. $< tqI > \longrightarrow a < tI >$
h4. $< t > \longrightarrow \lambda$
h5. $< t'qJ > \longrightarrow b < t'J >$
h6. $< t' > \longrightarrow \lambda$

Note that a derivation is of the form:

$$z \Longrightarrow < q^1 \# q^1 > \Longrightarrow < q^2 \# q^4 > \Longrightarrow < q^3 \# q^9 > \Longrightarrow$$
$$\Longrightarrow < q^4 \# q^{16} > \Longrightarrow < q^5 \# q^{25} > \Longrightarrow \cdots$$

Using the first part of h2 and h3–h6 a strict hypervariable $< q^n \# q^m >$ derives the terminal word $a^n b^m$. Hence, it is easy to note that

$$L_2 = \{a^n b^m \mid n > 0 \text{ and } m = n^2\}. \qquad \square$$

Example 4.3. Let R_H be:

h1. $< s > \longrightarrow < tIq > < t'Jq > < Iq \# Jq >$
h2. $< I \# IJ > \longrightarrow < I \# J >$
h2'. $< IJ \# I > \longrightarrow < I \# J >$
h2''. $< I \# I > \longrightarrow < t''I >$
h3. $< tqI > \longrightarrow a < tI >$
h3'. $< t'qI > \longrightarrow b < t'I >$
h3''. $< t''qI > \longrightarrow c < t''I >$
h4. $< t > \longrightarrow \lambda$
h4'. $< t' > \longrightarrow \lambda$
h4''. $< t'' > \longrightarrow \lambda$

The reader should observe that h1 starts a derivation and h2, h2', h2'' describe the algorithm of Euclid to compute the greatest common divisor of two (nonzero) natural numbers. The remaining rules are used to generate the terminal word. Hence, it is easy to see that the language generated is

$$L_3 = \{a^n b^m c^k \mid n, m > 0 \text{ and } k = gcd(n, m)\}. \qquad \square$$

The generative power of W-grammars was established by Sintzoff, [125]:

Theorem 4.1. *A language L is recursively enumerable iff there exists a W-grammar G, such that $L = L(G)$.* $\qquad \square$

Aad Van Wijngaarden improved the above result in [137]:

Theorem 4.2. *A language L is recursively enumerable iff there exists a W-grammar G such that $L = L(G)$ and G has at most one metavariable.* $\qquad \square$

4.2 Membership problem and parsing

Restrictions on W-grammars such that the language generated is a recursive language were studied by P. Deussen and K. Mehlhorn, see [36] and [37]:

Definition 4.5. *A hyperrule* $< u_0 > \longrightarrow a_0 < u_1 > a_1 \ldots < u_n > a_n$ *is*

- *left bounded (lb) iff for any X in M, if X has at least one occurrence in u_0, then X has at least one occurrence in $< u_1 > \ldots < u_n >$.*
- *right bounded (rb) iff for any X in M, if X has at least one occurrence in $< u_1 > \ldots < u_n >$, then X has at least one occurrence in $< u_0 >$.*
- *strict left bounded (slb) iff for any X in M, the number of occurrences of X in u_0 is less than or equal to the number of occurrences of X in $< u_1 > \ldots < u_n >$ and, moreover, the length of u_0 with respect to V is less than or equal to the length of $u_1 \ldots u_n$ with respect to V.*
- *strict right bounded (srb) iff for any X in M, the number of occurrences of X in u_0 is greater than or equal to the number of occurrences of X in $< u_1 > \ldots < u_n >$ and, moreover, the length of u_0 with respect to V is greater than or equal to the length of $u_1 \ldots u_n$ with respect to V.* □

Definition 4.6. *A W-grammar G is of type L (R), iff all the hyperrules are λ-free and lb (rb) and all the hyperrules of the form $< u > \longrightarrow < v >$ are slb (srb). The class of languages generated by W-grammars of type L (R) is denoted by L_L (L_R).* □

Theorem 4.3.
$$L_L = L_R = EXSPACE,$$

the class of languages acceptable by off-line Turing machines in exponential space. □

L. Wegner used the above restrictions to develop a parsing algorithm for a subclass of W-grammars, see [135]. The algorithm of Wegner can be used for a very large class of languages. In the sequel we present this parsing algorithm. We start by considering some definitions and terminology. Let $G = (G_1, G_2, z)$ be a W-grammar, where $G_1 = (M, V, R_M)$ and $G_2 = (H, \Sigma, R_H)$. A sequence $E_L = r_{i_1}, r_{i_2}, \ldots, r_{i_n}$ of hyperrules is called a *left parse* of $w \in (H_S \cup \Sigma)^*$ if there exists at least one leftmost derivation D of w in G such that $D = \alpha_0, \alpha_1, \ldots, \alpha_n = w$ and for $1 \leq j \leq n - 1$, $\alpha_{j-1} \Longrightarrow \alpha_j$ by use of a strict rule from $R_S(r_{i_j})$. Note that for context-free grammars there is a one to one correspondence between leftmost derivations and left parses. However, for W-grammars this is not the case. Moreover, a stronger result does hold:

Theorem 4.4. *It is undecidable whether or not an arbitrary W-grammar G has the property that for each leftmost derivation there is at most one left parse.*

Proof. Let $G_i = (V_N^i, V_T^i, S_i, P_i)$, $i = 1, 2$ be two arbitrary context-free grammars and define the W-grammar G such that $M = V_N^1 \cup V_N^2$, $V = V_T^1 \cup V_T^2$, $R_M = P_1 \cup P_2$,
$H = \{z, < S_1 >, < S_2 >\}$, $\Sigma = \{a\}$, $R_H = \{z \longrightarrow < S_1 >, < S_1 > \longrightarrow a$,
$< S_2 > \longrightarrow a\}$, where the start symbol of G is z.

Note that there is a leftmost derivation with more than one left parse iff $L(G_1) \cap L(G_2) \neq \emptyset$. But the emptiness of the intersection of context-free languages is undecidable and hence the theorem holds.   □

Definition 4.7. *Let $G = (G_1, G_2, z)$ be a W-grammar, where $G_1 = (M, V, R_M)$ and $G_2 = (H, \Sigma, R_H)$. Let h be in H. The hypernotion system associated to h is the 4-tuple $HS_h = (M, V, h, R_M)$, and h is referred to as the axiom of HS_h. The language defined by HS_h is*

$$L(HS_h) = \{t(h) \mid t \in T\}.$$ □

The hypernotion systems are extended in the natural way for axioms $h \in (H \cup \Sigma)^+$.

Definition 4.8. *Let $HS_h = (M, V, h, R_M)$ be a hypernotion system, where $h = X_1 X_2 \ldots X_n$, $X_i \in (M \cup V)$, $1 \leq i \leq n$. HS_h is uniquely assignable (u.a.) if for all $w \in L(HS_h)$ there is exactly one decomposition $w = w_1 w_2 \ldots w_n$ such that $t(X_i) = w_i$, $1 \leq i \leq n$, $t \in T$.*   □

Definition 4.9. *A W-grammar G is left-hand side uniquely assignable (lhs u.a.) if for every hyperrule $(h_0 \longrightarrow h_1 h_2 \ldots h_n) \in R_H$ the hypernotion system (M, V, h_0, R_M) is uniquely assignable.*

A W-grammar G is right-hand side uniquely assignable (rhs u.a.) if for every hyperrule $(h_0 \longrightarrow h_1 h_2 \ldots h_n) \in R_H$ the hypernotion system $(M, V \cup \Sigma, h_1 h_2 \ldots h_n, R_M)$ is uniquely assignable.   □

Definition 4.10. *A W-grammar G is locally unambiguous if*

(i) for every leftmost derivation D of α in G there is exactly one left parse E of α in G and given D one can effectively find E, and
(ii) for every left parse E of α in G there is exactly one leftmost derivation D of α in G and given E one can effectively find D.   □

Theorem 4.5. *For every W-grammar G one can effectively find a locally unambiguous W-grammar G' such that $L(G) = L(G')$.*   □

The next step is to define a canonical context-free grammar which simulates the derivations in a W-grammar under special assumptions.

Definition 4.11. *Let $G = (G_1, G_2, z)$ be a W-grammar, where $G_1 = (M, V, R_M)$ and $G_2 = (H, \Sigma, R_H)$ and let r be a hyperrule,*

$$r : h_0 \longrightarrow h_1 h_2 \ldots h_m, \text{ where } h_i \in H \cup \Sigma, 1 \leq i \leq m.$$

A cross-reference of r is an m-tuple $(x_1, x_2, \ldots, x_m)$ such that

(i) *if $h_i \in \Sigma$, then $x_i = h_i$; otherwise $x_i = h'_0$ for some hyperrule $h'_0 \longrightarrow \alpha$ in R_H or $x_i = \lambda$, if $\lambda \in L(M, V, h_i, R_M)$, and*
(ii) *$L(M, V, h_1 h_2 \ldots h_m, R_M) \cap L(M, V, x'_1 x'_2 \ldots x'_m, R_M) \neq \emptyset$ where x'_i is obtained from x_i by renaming metanotions in x_i such that they are distinct from those in x_j for $i \neq j$, $1 \leq i, j \leq m$.* $\square$

Definition 4.12. *Let $G = (G_1, G_2, z)$ be a W-grammar, where $G_1 = (M, V, R_M)$ and $G_2 = (H, \Sigma, R_H)$. The skeleton (canonical) grammar associated to G is the context-free grammar $G_{sk} = (V_N, \Sigma, z, P)$ defined as follows:*

$$V_N = \{h_0 \mid (h_0 \longrightarrow \alpha) \in R_H\},$$

$$P = \bigcup_{r \in R_H} R_{sk}(r), \quad \text{where}$$

$$R_{sk}(r) = \{h_0 \longrightarrow x_1 x_2 \ldots x_m \mid r = (h_0 \longrightarrow h_1 h_2 \ldots h_m) \in R_H, \ h_i \in (H \cup \Sigma)$$

for $1 \leq i \leq m$ and $(x_1, x_2, \ldots, x_m)$ a cross-reference of r}. $\square$

Definition 4.13. *A skeleton grammar G_{sk} associated to a W-grammar G is proper if for all $r, r' \in R_H$ with $r \neq r'$, $R_{sk}(r) \cap R_{sk}(r') = \emptyset$.* $\square$

Theorem 4.6. *A skeleton grammar G_{sk} associated to a W-grammar G is proper iff all hyperrules $r = (h_0 \longrightarrow h_1 h_2 \ldots h_m)$ and $r' = (h'_0 \longrightarrow h'_1 h'_2 \ldots h'_n)$ have pairwise distinct cross-references.* $\square$

Definition 4.14. *Let G_{sk} be the skeleton grammar associated to a W-grammar G. Let $E = r_{i_1}, r_{i_2}, \ldots, r_{i_n}$ be a left parse of $w \in L(G)$ and let $E' = r'_{i_1}, r'_{i_2}, \ldots, r'_{i_n}$ be a left parse of $w \in L(G_{sk})$. E' is referred as corresponding to E if for all $1 \leq j \leq n$, $r'_{i_j} \in R_{sk}(r_{i_j})$.* $\square$

Theorem 4.7. *For every W-grammar G and for every left parse E of $w \in L(G)$ in G, there exists exactly one corresponding left parse E' of $w \in L(G_{sk})$ in G_{sk}.* $\square$

Corollary 4.1. *If G is a W-grammar and if G_{sk} is its associated skeleton grammar, then $L(G) \subseteq L(G_{sk})$.* $\square$

Corollary 4.2. *If G is a W-grammar such that the skeleton grammar G_{sk} of G is proper, then for each two left parses E_1, E_2 of $w \in L(G)$*

$$E_1 = E_2 \text{ iff } E'_1 = E'_2,$$

where E'_i denotes the corresponding left parse of E_i, $i = 1, 2$. $\square$

Definition 4.15. *A W-grammar G is ambiguous if for some $w \in L(G)$ there is more than one leftmost derivation of w in G. Otherwise G is unambiguous.*
□

Theorem 4.8. *If G is a locally unambiguous W-grammar and if the skeleton grammar associated to G, G_{sk}, is proper and unambiguous, then G is unambiguous.*
□

The above results lead to the following parsing algorithm for a restricted class of W-grammars, see [135]:

Algorithm:

Input: A W-grammar G locally unambiguous, the proper and unambiguous skeleton grammar G_{sk} associated to G, a word $w \in \Sigma^*$.

Output: "yes" if $w \in L(G)$, "no" otherwise.

Method:

Step 1. Apply any of the known parsing algorithms for context-free grammars, see for instance [2, 3], to G_{sk} and w to obtain the left parse E'. If $w \notin L(G_{sk})$ then the output is "no".

Step 2. Compute the left parse E in G starting from the start symbol z (if G is left-hand side u.a. and rightbound) or from w (if G is right-hand side u.a. and leftbound). Apply hyperrule r to those strict notions which correspond to the terminals and nonterminals used in the derivation step in which skeleton rule $r' \in R_{sk}(r)$ was applied. Because G is locally unambiguous, giving the handle and hyperrule is sufficient to reduce α_i to α_{i-1} or, respectively, to extend α_i to α_{i+1}. The output is "no" and stop if r cannot be applied or, otherwise, if the derivation in G is complete, then the output is "yes" and stop.

Observe that the above algorithm requires as input the unambiguous skeleton-grammar G_{sk}. This condition cannot be removed, since G_{sk} is a context-free grammar and for context-free grammars it is undecidable whether the grammar is unambiguous or not.

There are other interesting results in the area of W-grammars. Some of these results concern normal forms for W-grammars, see for example S. Greibach, [55], J. van Leeuwen, [87], P. Turakainen, [131]. The parsing problem is studied also in J. L. Baker, [11], P. Dembinsky and J. Maluszynski, [33], A. J. Fisher, [42], C. H. A. Koster, [80]. The generative power of W-grammars is studied in J. de Graaf and A. Ollongren, [54]. Many other formalisms can be expressed in terms of W-grammars, such as the first order logic, P. Deussen, [36], W. Hesse, [59], the recursively enumerable functions, I. Kupka, [82]. For a bibliography of van Wijngaarden grammars the reader is referred to [38].

4.3 W-grammars and complexity

W-grammars provide a framework to evaluate the complexity of recursively enumerable languages, see [92, 91, 95]. Our presentation below requires some basic knowledge about recursive functions on the part of the reader.

Assume that S is a fixed alphabet and let ω be the set of natural numbers.

Definition 4.16. *A criterion over S is a recursive sequence, $C = (C_n)_{n\in\omega}$, such that for any $n \in \omega$, $C_n \subseteq S^*$ is a recursive language.*   □

Let $D = (D_n)_{n\in\omega}$ be a given class of generative devices such that, for any language L over S, L is a recursively enumerable language iff there is a d in D such that $L(d) = L$, where $L(d)$ is the language generated by d.

Definition 4.17. *A generative Blum space (GBS) is an ordered system $\mathcal{B} = (S, D, A, C)$ where S, D, C are as above and $A = (A_i)_{i\in\omega}$ is the set of cost functions, i.e., a set of partial recursive functions that satisfy the following two axioms:*

GB1. $A_i(n) < \infty$ iff $L(d_i) \cap C_n \neq \emptyset$.
GB2. the predicate $R(i, x, y) = 1$ iff $A_i(x) = y$ and
      $R(i, x, y) = 0$ otherwise, is a total recursive predicate.   □

Let W be the Gödel enumeration, $W = (w_i)_{i\in\omega}$, of all recursively enumerable languages over S using W-grammars as generative devices. For a W-grammar, G, let K_n be the n-th Kleene's iteration of the W system associated to G (the extension for W-grammars of the algebraic system of equations associated to a context-free grammar, see [81]). Define the cost functions, $TW = (TW_i)_{i\in\omega}$, such that they measure the "time" necessary for the device w_i with respect to the criterion C, to generate the language $L(w_i)$. If there is an $r \in \omega$, such that $K_r^i \cap C_n \neq \emptyset$ then $TW_i(n) = \min \{r \mid K_r^i \cap C_n \neq \emptyset\}$, else $TW_i(n) = \infty$.

Note that the ordered system $\mathcal{B}_T = (S, W, TW, C)$ is a GBS.

Definition 4.18. *Let $\mathcal{B} = (S, D, A, C)$ be a GBS and let f be a recursive function. The generative complexity class bounded by f is*

$$C_\mathcal{B}(f(n)) = \{L \mid \text{ there is an } i \in \omega \text{ such that }$$

$$L(d_i) = L \text{ and } A_i(x) \leq f(x) \text{ almost everywhere }\}.$$ □

If the GBS is $\mathcal{B}_T$ then the generative complexity class bounded by f is denoted by $TIMEW(f(n))$.

Theorem 4.9. *(the linear speed-up theorem) For every GBS, $\mathcal{B}_T = (S, W, TW, C)$, for every recursive function f and for every constant $q > 0$,*

$$TIMEW(f(n)) = TIMEW(qf(n)).$$ □

Definition 4.19. *A W-grammar G is related to the graph of a function f :* $\omega \longrightarrow \omega$ *if* $\{i, \#\} \subseteq V$, $L_X = i^*$ *for every* $X \in M$, $R_H \subset H \times H^*$ *and, for all* $n, m \in \omega$

$$z \Longrightarrow_G^* < i^n \# i^m > \quad \text{iff } f(n) = m. \qquad \square$$

Definition 4.20. *A function* $f : \omega \longrightarrow \omega$ *is TW-self bounded if* f *is a constant function or there is a W-grammar G related to the graph of f and there are constants* $n_0, c_1, c_2 \in \omega$ *such that for every* $n \geq n_0$, $f(n) > 1$ *and, moreover, if the hyperrule* $< i^n \# i^{f(n)} > \longrightarrow a$ *is added to the hyperrules of G, then for all* $p \in \omega$, $K_p = \{a\}$, *where* $p = c_1 f(n) + c_2$.

$$\square$$

Let TWSB be the class of all *TW*-self bounded functions. Intuitively, a function f in TWSB has the property that the graph of f can be generated by a W-grammar that is W-time bounded by f.

Theorem 4.10. *(the TW-separation property) Let f be in TWSB such that $f(n) > \log_2 n$ almost everywhere and let g be a function such that* $\inf_{n \to \infty}(g(n)/f(n)) = 0$. *There exists a language L such that, with respect to the criterion* $C = (a^n b^*)_{n \in \omega}$, *$L$ is in $TIMEW(f(n))$ but L is not in $TIMEW(g(n))$.*

Proof. Consider the language

$$L = \{a^n b^m \mid m = 2^{f(n)}, n \in \omega\}.$$

We prove that $L \in TIMEW(f(n))$. Since $f \in TWSB$, there exists a W-grammar $G = (G_1, G_2, z)$, where $G_1 = (M, V, R_M)$ and $G_2 = (H, \Sigma, R_H)$, such that G is related to the graph of f and let $n_0, c_1, c_2 \in \omega$ be the constants for which the Definition 4.20 is satisfied for f.

Define the W-grammar G', such that $M' = M \cup \{N, N_1\}$, $V' = V \cup \{\alpha, \beta\}$, $\Sigma' = \Sigma \cup \{a, b\}$, $R_{M'} = R_M \cup \{A \longrightarrow B \mid A \in \{N, N_1\}, B \in M\}$, $z' = z$ and $R'_H = R_H \cup \{h_1, \ldots, h_6\}$ where:

h_1. $< N \# N_1 > \longrightarrow < \alpha N >< \beta N_1 >$
h_2. $< \alpha N N > \longrightarrow < \alpha N >< \alpha N >$
h_3. $< \alpha N N i > \longrightarrow a < \alpha N >< \alpha N >$
h_4. $< \alpha > \longrightarrow \lambda$
h_5. $< \beta N i > \longrightarrow < \beta N >< \beta N >$
h_6. $< \beta > \longrightarrow b$

Observe that $L(G') = L$ and note that a terminal derivation in G' is equivalent with a derivation of the following form:

$$z' \Longrightarrow_{G'}^* < i^n \# i^{f(n)} > \Longrightarrow_{h_1} < \alpha i^n >< \beta i^{f(n)} > \Longrightarrow_{h_2 - h_6}^* a^n b^{2^{f(n)}}.$$

One can easily see that $K'_{p_1}(< \alpha i^n >) = \{a^n\}$ and $K'_{p_2}(< \beta i^{f(n)} >) = \{b^{2^{f(n)}}\}$, where $p_1 = \log_2 n + 1$ and $p_2 = f(n) + 1$. Let p_3 be $max\{p_1, p_2\} + 1$.

Since $f(n) \geq \log_2 n$, it follows that $p_3 = f(n) + 2$. Now observe that for all $n \geq n_0$, $K_p' = \{a^n b^{2^n}\}$, where $p = (c_1 + 1)f(n) + c_2 + 2$ and hence $L \in TIMEW(f(n))$.

Now we show that $L \notin TIMEW(g(n))$. Assume by contrary that there exists a W-grammar G'' such that $L(G'') = L$ and $TW_{G''}(n) \leq g(n)$ almost everywhere. Define the constants:

$$j_1 = max\{| \sigma_0 \ldots \sigma_k \| (< u_0 > \longrightarrow \sigma_0 < u_1 > \sigma_1 \ldots < u_k > \sigma_k) \in R_H''\}$$

$$j_2 = max\{| \alpha \| (< u > \longrightarrow \alpha) \in R_H'', \alpha \in \Sigma^*\}$$

$$j_3 = max\{q \mid (< u_0 > \longrightarrow \sigma_0 < u_1 > \sigma_1 \ldots < u_q > \sigma_q) \in R_H''\}.$$

Note that for every $n \geq 1$, for every strict hypernotion $< h >$ and for every $v \in K_n''(< h >)$

$$| v | \leq j_3^{n-1} j_2 + j_1 \sum_{i=0}^{n-2} j_3^i.$$

Let j be $max\{j_1, j_2, j_3\}$. It follows that

$$| v | \leq j^n + j \sum_{i=0}^{n-2} j^i = \sum_{i=1}^{n} j^i \leq nj^n \leq 2^n j^n = (2j)^n.$$

Denote $s = 2j$ and note that for every $v \in K_n''$, $| v | \leq s^n$. Therefore, for each $v \in K_{g(n)}''$ it follows that $| v | \leq s^{g(n)}$. Since $TW_{G''}(n) \leq g(n)$ almost everywhere we deduce that $K_{g(n)}'' \cap C_n \neq \emptyset$ almost everywhere. Hence $a^n b^{2^{f(n)}} \in K_{g(n)}''$ almost everywhere. Thus

$$2^{f(n)} <| a^n b^{2^{f(n)}} | \leq s^{g(n)} \text{ almost everywhere.}$$

Hence $\log_s 2 \leq g(n)/f(n)$ almost everywhere. But this is contrary to the assumption that $inf_{n \to \infty}(g(n)/f(n)) = 0$ and, consequently $L \notin TIMEW(g(n))$. □

Theorem 4.11. *(a dense hierarchy) If* $C = (a^n b^*)_{n \in \omega}$, *then for all real numbers* $0 < \alpha < \beta$, $TIMEW(n^\alpha)$ *is a proper subset of* $TIMEW(n^\beta)$. □

5. Attribute grammars

5.1 Definitions and terminology

Attribute grammars were introduced by D. Knuth, see [78, 79]. Informally, *attribute grammars* consist of:

1) a context-free grammar $G = (V_N, V_T, S, P)$ such that S does not occur on the right side of any production.
2) two disjoint sets of symbols associated with each symbol of G, called *inherited* and *synthesized attributes*, where S must have at least one synthesized attribute and no inherited attributes.
3) a *semantic domain* D_α for each attribute α.
4) a set of *semantic rules* associated with each production. Each semantic rule is a function

$$\varphi_{pi\alpha} : D_{\alpha_1} \times \ldots \times D_{\alpha_k} \longrightarrow D_\alpha$$

which defines the value of an attribute α of an instance i of a symbol appearing in a production p in terms of the values of attributes $\alpha_1, \ldots, \alpha_k$ associated with other instances of symbols in the same production.

Attribute grammars are also referred to as *K-systems*. In the sequel, symbols with indices denote (rename) the symbols without indices. For instance, X_1, X_2 rename the symbol X.

We consider the first example in [78]. The same example is considered also in [17, 26, 27].

Example 5.1. Let $G = (V_N, V_T, S, P)$ be the context-free grammar with: $V_N = \{B, L, N\}$, $V_T = \{0, 1, \cdot\}$, $S = N$. The productions from P are listed in sequence together with the semantic rules associated:

Productions ; Semantic rules
1. $B \longrightarrow 0$; $v(B) = 0$
2. $B \longrightarrow 1$; $v(B) = 2^{s(B)}$
3. $L \longrightarrow B$; $v(L) = v(B)$, $s(B) = s(L)$, $l(L) = 1$
4. $L_1 \longrightarrow L_2 B$; $v(L_1) = v(L_2) + v(B)$, $s(B) = s(L_1)$,
$\qquad\qquad s(L_2) = s(L_1) + 1$, $l(L_1) = l(L_2) + 1$
5. $N \longrightarrow L$; $v(N) = v(L)$, $s(L) = 0$
6. $N \longrightarrow L_1 \cdot L_2$; $v(N) = v(L_1) + v(L_2)$, $s(L_1) = 0$, $s(L_2) = -l(L_2)$ $\quad\square$

It is easy to observe that the language generated by G is the set of all rational numbers in binary notation.

The main purpose of the semantic rules is to define the meaning of each string from $L(G)$.

The synthesized attributes are $v(B)$, $v(L)$, $l(L)$ and $v(N)$, based on the attributes of the descendants of the nonterminal symbol. Inherited attributes are $s(B)$ and $s(L)$, and they are based on the attributes of the ancestors.

Synthesized attributes are evaluated bottom up in the tree structure, while the inherited attributes are evaluated top down in the tree structure.

For instance, consider the string 1101.01 from $L(G)$.

The evaluated structure corresponding to this string is:

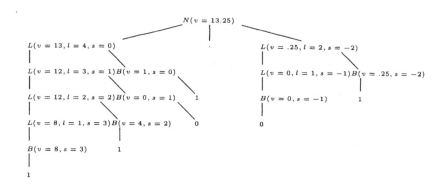

Note that $v(N) = 13.25$ and that the binary representation of 13.25 is the string 1101.01.

Here the semantic domains are: Z (the set of integer numbers) for $s(B)$, $s(L)$, $l(L)$ and Q (the set of rational numbers) for $v(B)$, $v(L)$ and $v(N)$.

5.2 Algorithms for testing the circularity

The semantic rules are *well defined* or *noncircular* iff they are formulated in such a way that all attributes can always be computed at all nodes, in any possible derivation tree.

Note that the definition of semantic rules can lead to circular definitions of attributes and consequently such attributes cannot be evaluated. Moreover, since in general there are infinitely many derivation trees, it is important to have an algorithm for deciding whether or not a given grammar has well-defined semantic rules. An algorithm for testing this property was given in [78, 79].

Let G be an attribute grammar. Without loss of generality, we may assume that the grammar contains no useless productions, i.e., that each production of G appears in the derivation of at least one terminal string. Let T be any derivation tree obtainable in the grammar, allowed to have any symbol of V as the label of the root. Define the directed graph $\mathcal{D}(T)$ corresponding to T by taking the ordered pairs (X, α) as vertices, where X is a node of T and α is an attribute of the symbol which is the label of the node X. The arcs of $\mathcal{D}(T)$ go from (X_1, α_1) to (X_2, α_2) if and only if the semantic rule for the value of the attribute α_2 depends directly on the value of the attribute α_1.

For instance, assume G is the attribute grammar from Example 5.1 and consider the derivation tree T:

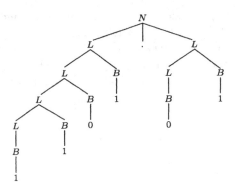

The directed graph $\mathcal{D}(T)$ is:

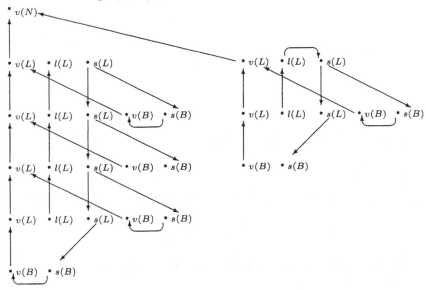

The semantic rules are well defined iff no directed graph $\mathcal{D}(T)$ contains an oriented cycle.

Assume that for some $\mathcal{D}(T_1)$ there is an oriented cycle. Since G contains no useless productions, there is an oriented cycle in some $\mathcal{D}(T)$ for which the root of T has the label S. T is a derivation tree of the language for which it is not possible to evaluate all the attributes.

Let p be the production

$$X_{p_0} \longrightarrow X_{p_1} \ldots X_{p_{n_p}}.$$

Corresponding to p, consider the directed graph $\mathcal{D}_p$ defined as follows:
the vertices are (X_{p_j}, α) where α is an attribute associated to X_{p_j}, and
arcs emanate from $(X_{p_{k_j}}, \alpha_i)$ to (X_{p_j}, α) for $0 \leq j \leq n_p$ and α is a synthesized
attribute, if $j = 0$, and α is an inherited attribute of X_{p_j}, if $j > 0$.

For example, the six productions of G from Example 5.1 have the following
six directed graphs:

$$D_1 : \bullet\, v(B) \quad \bullet\, s(B) \qquad D_2 : \bullet\, v(B) \quad \bullet\, s(B)$$

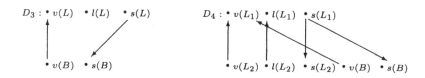

$$D_5 : \bullet\, v(N) \qquad\qquad D_6 : \bullet\, v(N)$$

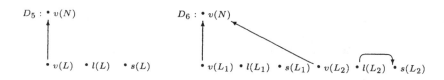

Note that each directed graph $\mathcal{D}(T)$ is obtained as the superposition of
such directed graphs $\mathcal{D}_p$.

Let A_X be the set of attributes of X. Consider a production p and assume
that G_j, $1 \leq j \leq n_p$, is any directed graph whose vertices are a subset of
$A_{X_{p_j}}$.

Denote by $\mathcal{D}[G_1, \ldots, G_p]$ the directed graph obtained from $\mathcal{D}_p$ by adding
an arc from (X_{p_j}, α) to (X_{p_j}, α') whenever there is an arc from α to α' in
G_j.

For instance, assume that

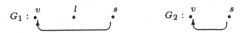

and if $\mathcal{D}_4$ is the directed graph defined above, then $\mathcal{D}_4[G_1, G_2]$ is

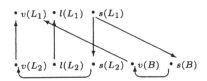

Theorem 5.1. *There exists an algorithm to decide whether or not the semantic rules of an attribute grammar are well defined.*

Proof. Let X be in $V = V_N \cup V_T$ and let $S(X)$ be a set of directed graphs on the vertices A_X. Initially $S(X) = \emptyset$ if $X \in V_N$, and $S(X)$ is the single directed graph with vertices A_X and no arcs if $X \in V_T$.

New directed graphs are added to the sets $S(X)$ until no further directed graphs can be added.

Let m be $card(P)$ and let the k-th production be:

$$X_{k0} \longrightarrow X_{k1} X_{k2} \dots X_{kn_k}.$$

For each integer p, $1 \le p \le m$, and for each j, $1 \le j \le n_p$, choose a directed graph $\mathcal{D}'_j$ in $S(X_{pj})$. Add to the set $S(X_{p0})$ the directed graph whose vertices are $A_{X_{p0}}$ and whose arcs go from α to α' if and only if there is an oriented path from (X_{p0}, α) to (X_{p0}, α') in the directed graph

$$\mathcal{D}_p[\mathcal{D}'_1, \dots, \mathcal{D}'_{n_p}].$$

Note that only finitely many directed graphs are possible and hence this procedure cannot continue forever.

If T is a derivation tree with root X, let $\mathcal{D}'(T)$ be the directed graph with vertices A_X whose arcs go from α to α' iff there is an oriented path from (X, α) to (X, α') in $\mathcal{D}(T)$.

We claim that for all $X \in V$, $S(X)$ is the set of $\mathcal{D}'(T)$, where T is a derivation tree with root X.

To prove this claim, note that the construction adds to $S(X)$ only such directed graphs $\mathcal{D}'(T)$. Moreover, from each such $\mathcal{D}'(T)$ one can obtain an appropriate derivation tree T.

Conversely, if T is a derivation tree, we can show by induction on the number of nodes in T that $\mathcal{D}'(T)$ is in the corresponding $S(X)$.   □

Corollary 5.1. *The semantic rules of an attribute grammar are well defined iff none of the graphs $\mathcal{D}_p[\mathcal{D}'_1, \ldots, \mathcal{D}'_{n_p}]$, for any choice of p and $\mathcal{D}'_j \in S(X_{p_j})$, $1 \leq j \leq n_p$ contains an oriented cycle.* □

Example 5.2. Consider the attribute grammar G from Example 5.1. Using the above algorithm we obtain

$$S(N) = \{ \overset{v}{\bullet} \}, \qquad S(L) = \{ \overset{v}{\bullet} \quad \overset{l}{\bullet} \quad \overset{s}{\bullet}\overset{v}{\bullet} \quad \overset{l}{\bullet} \quad \overset{s}{\bullet} \},$$

$$S(B) = \{ \overset{v}{\bullet} \quad \overset{s}{\bullet}\overset{v}{\bullet} \quad \overset{s}{\bullet} \}, \qquad S(0) = S(1) = \{ \cdot \}.$$

Now one can apply the above corollary in order to verify that the semantic rules of G are well defined. □

The above algorithm, also referred to as the algorithm for testing the noncircularity of definitions in an attribute grammar, requires exponential time, see [69, 74].

5.3 Restricted attribute grammars

A more restrictive property of attribute grammars, called the *strong non-circularity* was studied in [26, 27]. The property of strong noncircularity is decidable in polynomial time. Most attribute grammars arising in applications such as compiler construction have this property, see [34].

Notation. Let $G = (V_N, V_T, S, P)$ be an attribute grammar. The set of synthesized attributes associated to a symbol $X \in V = V_N \cup V_T$ is denoted by A_X^s and the set of inherited attributes associated to X is A_X^h. Moreover, $A_X = A_X^s \cup A_X^h$, $A^s = \bigcup_{X \in V} A_X^s$, $A^h = \bigcup_{X \in V} A_X^h$, $A = A^s \cup A^h$.

Definition 5.1. *Let $p \in P$ be a production*

$$p : X \longrightarrow X_1 X_2 \ldots X_n.$$

Consider the sets

$$W_1(p) = \{a(X_i) \mid 1 \leq i \leq n, a \in A_X\},$$

$$W_0(p) = \{a(X) \mid a \in A_X\},$$

and

$$W(p) = W_1(p) \cup W_0(p).$$

Define on $W(p)$ a binary relation, denoted $\longrightarrow_p$, by
$w \longrightarrow_p w'$ iff w' occurs on the right-hand side of a semantic rule defining w and the semantic rule is associated to p. □

Definition 5.2. *An argument selector is a mapping*

$$\gamma : A^s \times V_N \longrightarrow \mathcal{P}(A^h)$$

such that:
    $\gamma(a, X) \subseteq A_X^h$, *if* $a \in A_X^s$ *and* $\gamma(a, X) = \emptyset$, *otherwise.*   □

Argument selectors are ordered componentwise, i.e., $\gamma \subseteq \gamma'$ iff $\gamma(a, X) \subseteq \gamma'(a, X)$ for all $a \in A^s$ and all $X \in V_N$.

Definition 5.3. *For each argument selector* γ, *we define the binary relation,* $\longrightarrow_\gamma$, *on* $W_1(p)$ *as:*
    $w \longrightarrow_\gamma w'$ *iff* $w = a(X_i)$, $w' = y(X_i)$ *and* $y \in \gamma(a, X_i)$ *for some* $1 \leq i \leq n$,
$a \in A_{X_i}^s$ *and* $y \in A_{X_i}^h$.   □

Denote by $\longrightarrow_{p,\gamma}$ the relation $\longrightarrow_p \cup \longrightarrow_\gamma$ considered only on $W_1(p)$.

Definition 5.4. *An attribute grammar is strongly noncircular if there exists an argument selector* γ *such that*

(1) γ *is closed, i.e., for all* $p \in P$, $p : X \longrightarrow X_1 X_2 \dots X_n$, *for all* $y \in A_X^h$ *and*
    $a \in A_X^s$, *whenever* $a(X) \longrightarrow_p y(X)$ *or* $a(X) \longrightarrow_p w \longrightarrow_{p,\gamma}^* w' \longrightarrow_p y(X)$
    for some $w, w' \in W_1(p)$, *then* $y \in \gamma(a, X)$.
(2) γ *is noncircular, i.e., for all* $p \in P$ *as in* (1), *the relation* $\longrightarrow_{p,\gamma}$ *has no cycles, that is, there exists no* $w \in W_1(p)$ *such that* $w \longrightarrow_{p,\gamma}^+ w$.   □

Definition 5.5. *Let* t *be a derivation tree and let* $W(t)$ *be the set of all attributes associated to vertices of* t. *The binary relation* γ_t *is defined on* $W(t)$ *by* $w \longrightarrow_t w'$ *iff* w' *occurs on the right-hand side of an equation defining* w.   □

We still consider some related subclasses of attribute grammars.
Let a be in A^s, let y be in A^h and let X be in $V_N^a \cap V_N^y$. The attribute a *may call* y at X if $a(X) \longrightarrow_t^* y(X)$ for some derivation tree t.
Let $CALL(a, X)$ denote the set of inherited attributes that a may call at $X \in V$. It is easy to see that $CALL(a, X) \subseteq \gamma(a, X)$ whenever $\gamma(a, X)$ is a closed argument selector.

Definition 5.6. *An attribute grammar is benign if* $CALL$, *as an argument selector, is noncircular.*
    An attribute grammar is ordered if it is noncircular and if there exists a family $(\theta_X)_{X \in V_N}$, *where* θ_X *is a partial order on* A_X *such that for all derivation trees* t, *for all occurrences* u *in* t, *whenever* $a(u) \longrightarrow_t^* b(u)$, *then* $b\theta_Y a$, *where* Y *is the label of the root of* t.   □

The following theorems, see [26, 27], show the relations between these families of attribute grammars.

Theorem 5.2. *Ordered attribute grammars are contained in strongly noncircular ones. Strongly noncircular attribute grammars are contained in benign ones. Benign attribute grammars are contained in noncircular ones. All of the above containments are proper.*   □

Theorem 5.3. *The property of an attribute grammar being ordered (strongly noncircular, benign, noncircular) is decidable and moreover,*

(i) *whether an attribute grammar is ordered or strongly noncircular can be decided in polynomial time,*

(ii) *deciding whether an attribute grammar is benign or noncircular requires exponential time.*   □

Theorem 5.4. *For any attribute grammar there exists a minimal closed argument selector γ_0. The attribute grammar is strongly noncircular iff γ_0 is noncircular.*   □

Using the above theorem, the following polynomial time algorithm for deciding the property of strong noncircularity was introduced in [26]. In the sequel an argument selector γ is considered as a set of triples:

$$R = \{(y, a, X) \mid y \in \gamma(a, X)\} \subseteq A \times A \times V_N.$$

The argument selector corresponding to R is denoted by $\gamma(R)$.

Algorithm:

Step 1. Set $R_0 = \emptyset$.

Step 2. For all $i \geq 1$ compute:

$$R_i = R_{i-1} \cup \{(y, a, X) \mid \exists p \in P, p : X \longrightarrow \alpha, \text{ such that } y \in A_X^h, a \in A_X^s$$
$$\text{and } a(X) \longrightarrow_{p, \gamma(R_{i-1})}^* y(X)\}.$$

Step 3. Stop as soon as $R_i = R_{i-1}$, and the result is $\gamma_0 = \gamma(R_i)$.

After the computation of γ_0, one has to check its noncircularity which can be also done in polynomial time.

Definition 5.7. *An attribute grammar is purely synthesized if $A^h = \emptyset$. An attribute grammar is nonnested if inherited attributes are defined only in terms of inherited attributes.*   □

Clearly, every purely synthesized attribute grammar is a nonnested attribute grammar.

Theorem 5.5. *Every nonnested attribute grammar is strongly noncircular.*

Proof. Assume G is a nonnested attribute grammar and let γ be the argument selector:

$$\gamma(a, X) = A_X^h.$$

Note that γ is closed. Consider now the sequence:

$$w_1 \longrightarrow_{p,\gamma} w_2 \longrightarrow_{p,\gamma} \cdots \longrightarrow_{p,\gamma} w_m.$$

It may be either $a(X) \longrightarrow_p y(X)$ for some $a \in A^s$ and $y \in A^h$, or

$$a(X) \longrightarrow_p b(X_i) \longrightarrow_{p,\gamma} y(X_j) \longrightarrow_p z(X)$$

for some $a, b \in A^s$, $y, z \in A^h$, $X, X_i, X_j \in V_N$, or it is a subsequence of the last one. Consequently, there cannot be cycles. Therefore, γ is noncircular. Hence, G is strongly noncircular. □

Corollary 5.2. *Every purely synthesized attribute grammar is strongly non-circular.* □

5.4 Other results

In [17] it is proved that each attribute grammar can be converted to an equivalent purely synthesized one that possibly uses the μ operator, i.e., the least fixed point of a function. The idea is to consider the semantic rules as a system of equations. After this the inherited attributes are eliminated using the so-called recursion theorem.

Purely synthesized attribute grammars can be related to initial algebra semantics, see [17] and see [53] for initial algebra semantics.

Attribute grammars have been used in parsing theory and practice. *Attribute translation grammars* were introduced in [89] as a device to specify a translation. The translated symbols can have a finite set of attributes. An attributed translation grammar is in some sense a generalization of a context-free grammar. A context-free grammar is first extended to a translation grammar and then the attributes are added. Attributed translation grammars apply also ideas from syntax directed translations.

Attribute grammars were extended to *attributed tree grammars* in [12]. An algorithm for testing noncircularity of attributed tree grammars is described in [12].

LR-attributed grammars are a combination between $LR(k)$ and attribute grammars. They have both features, an efficient parsing algorithm and an algorithm to evaluate the attributes. Informally, an *LR*-attributed grammar is an attribute grammar that has the possibility of attribute evaluation during the *LR* parsing by an *LR parser/evaluator*. An *LR parser/evaluator* is a deterministic algorithm which is an extension of a shift/reduce *LR* parsing algorithm in such a way that it evaluates all synthesized attribute instances of a node in a derivation tree as soon as it has recognized this node and has parsed the subtree rooted in this node. The *LR* parser/evaluator also computes all inherited attributes of this node.

Various families of *LR*-attributed grammars are studied and compared in [4]. See also [134] for an early approach of this subject.

Attribute grammars are used to study natural languages, too. They can be used to encode an algebraic specification of a natural language and attributed translation is used to compute representations of the "meaning" of a sentence at different levels of abstraction, including a treatment of semantic ambiguities. For more details, as well as for an example of a parser that translates a significant subset of English into predicate logic, the reader is referred to [107].

In [88] a system is presented, called *Vinci*, for generating sentences and phrases in natural languages. The system is based on attribute grammars and has been used for experiments in linguistics, and for teaching linguistic theory. Attribute grammars provide the basis of the syntax definition and the use of attributes ensures the "semantic agreement" between verbs and their subjects and objects, nouns and adjectives, etc. For instance, the sentence "The bear ate the banana" is accepted, whereas the sentence "The table ate colourless green ideas" is not accepted.

For a large bibliography on attribute grammars see [34].

6. Patterns

6.1 Erasing and nonerasing patterns

The study of patterns goes back to the beginning of this century. Some of the classical results of Axel Thue, see [128], can be expressed in terms of patterns and pattern languages. Recent research concerning inductive inference, learning theory, combinatorics on words and term rewriting is also related to the study of patterns. Classical devices such as grammars and automata characterize a language completely. Patterns provide an alternative method to define languages, giving more leeway.

Let Σ and V be two disjoint alphabets. Σ is the alphabet of *terminals* and V is the alphabet of *variables*. A *pattern* is a word over $\Sigma \cup V$. The language defined by a pattern α consists of all words obtained from α by leaving the terminals unchanged and substituting a terminal word for each variable X. The substitution has to be *uniform*, i.e., different occurrences of X have to be replaced by the same terminal word. In the case when variables have to be replaced always by *nonempty* words, the pattern will be referred to as a *nonerasing pattern*, or *NE-pattern*. These patterns have been studied systematically first by Dana Angluin, see [5, 6].

However, the situation is essentially different if the empty word is allowed in the substitutions. The study of such patterns, referred to as *erasing patterns*, or *E-patterns*, was initiated in [70].

The formal definitions are the following. Denote by $H_{\Sigma,V}$ the set of all morphisms from $(\Sigma \cup V)^*$ to $(\Sigma \cup V)^*$.

Definition 6.1. *The language generated by an E-pattern $\alpha \in (\Sigma \cup V)^*$ is defined as*

$$L_{E,\Sigma}(\alpha) = \{w \in \Sigma^* \mid w = h(\alpha) \text{ for some } h \in H_{\Sigma,V}$$

$$\text{such that } h(a) = a \text{ for each } a \in \Sigma\}.$$

The language generated by an NE-pattern $\alpha \in (\Sigma \cup V)^*$ is defined as

$$L_{NE,\Sigma}(\alpha) = \{w \in \Sigma^* \mid w = h(\alpha) \text{ for some } \lambda\text{-free } h \in H_{\Sigma,V}$$

$$\text{such that } h(a) = a \text{ for each } a \in \Sigma\}. \qquad \square$$

Whenever Σ is understood we use also the notations $L_E(\alpha)$ and $L_{NE}(\alpha)$. Note that in the sequel the symbol "$\subset$" denotes the proper inclusion.

A *sample* is a finite nonempty language. For many purposes, both theoretical and practical, given a sample F it is useful to find a pattern α such that $F \subset L_{NE}(\alpha)$ or $F \subset L_E(\alpha)$ and, furthermore, α describes F as closely as possible. Consider the following example, see [70].

Example 6.1. Assume that the terminal alphabet is $\Sigma = \{0,1\}$ and let F be the following sample

$$F = \{0000, 000000, 001100, 00000000, 00111100, 0010110100, 0001001000\}.$$

For each E-pattern α from the following list, F is a subset of the language $L_E(\alpha)$.

$$\alpha_1 = XYX, \quad \alpha_2 = XYYX, \quad \alpha_3 = XXYXX,$$

$$\alpha_4 = XXYYXX, \quad \alpha_5 = 00X00, \quad \alpha_6 = 00XX00. \qquad \square$$

The above list can be continued with other examples of patterns as well as with descriptions of the form XX^R, where X^R is the mirror image of X, $00XX^R00$ or for instance XX^P, where X^P is some permutation of X. In the sequel we will not consider the mirror image or the permutation of variables.

Finding a pattern common to all words in a given sample is especially appropriate if the sample set is growing, for instance through a *learning process*, or if one may enquire whether or not some specified words belong to the set. For instance, a positive answer to the query 00011000 will reject the patterns α_4 and α_6, but not the patterns α_1, α_2, α_3, α_5.

Also, trying to infer a pattern common to all words in a given sample is a very typical instance of the process of *inductive inference*, that is, the process of inferring general rules from specific examples. For more details on this line see for instance [7] or [65].

For an interconnection between patterns and *random numbers* see [70].

Let $\mathcal{P}$ be a class of patterns. Given a sample F, a pattern $\alpha \in \mathcal{P}$ is *descriptive* of F within the class $\mathcal{P}$ provided that F is contained in the language of α and there is no other pattern $\beta \in \mathcal{P}$ such that F is contained in the language of β and the language of β is properly contained in the language of α. That is, no other pattern β gives a strictly "closer fit" to the sample F than α does.

Formally, a pattern $\alpha \in \mathcal{P}$ is descriptive of a sample F within $\mathcal{P}$ if the following two conditions are satisfied.

(i) $F \subseteq L(\alpha)$,
(ii) there is no pattern $\beta \in \mathcal{P}$ such that

$$F \subseteq L(\beta) \subset L(\alpha).$$

Moreover, when specifying the class $\mathcal{P}$, it is also stated whether we are dealing with E- or NE-patterns, and also the terminal alphabet Σ is specified.

Example 6.2. Consider again the sample F from Example 6.1 and let $\mathcal{P}$ be the set of all E-patterns. The reader can easily verify that the E-pattern α_6 is descriptive of F within $\mathcal{P}$. $\square$

Note that for a given sample F and for a given class $\mathcal{P}$ of patterns it may be the case that there are zero, one or more descriptive patterns of F within $\mathcal{P}$. On the other hand, if $\mathcal{P}$ is the class of all patterns, then for every sample F there exists at least one descriptive pattern of F within $\mathcal{P}$. For the following theorem we refer the reader to [6].

Theorem 6.1. *There is an algorithm which, given a sample F as input, outputs an NE-pattern α that is descriptive of F within the class of all NE-patterns.* $\square$

The algorithm used in the proof of the above theorem gives a pattern α of the maximum possible length that is descriptive of F.

Theorem 6.1 concerns E-patterns as well, see [70].

Theorem 6.2. *There is an algorithm which, given a sample F as input, outputs an E-pattern α that is descriptive of F within the class of all E-patterns.* $\square$

Both the algorithm described in [5] for NE-patterns, and the algorithm described in [70] for E-patterns are non-efficient.

6.2 The equivalence problem

Given two NE-patterns (E-patterns) α and β, the equivalence problem consists in deciding whether or not $L_{NE}(\alpha) = L_{NE}(\beta)$ ($L_E(\alpha) = L_E(\beta)$).

We start by defining some related notions. A morphism $h \in H_{\Sigma,V}$ is *stable for terminals* if $h(a) = a$ for all $a \in \Sigma$. If h is letter-to-letter, injective and stable for terminals, then h is a *renaming*.

Consider the following relations between patterns:

$\alpha \le \beta$ iff $\alpha = h(\beta)$, for some h stable for terminals.
$\alpha \le_+ \beta$ iff $\alpha = h(\beta)$, for some λ-free h, stable for terminals.
$\alpha \equiv \beta$ iff $\alpha = h(\beta)$, for some renaming h.

Observe that the language defined by an E- , NE-pattern α can be expressed as:

$$L_{E,\Sigma}(\alpha) = \{w \in \Sigma^* \mid w \le \alpha\},$$

and

$$L_{NE,\Sigma}(\alpha) = \{w \in \Sigma^* \mid w \le_+ \alpha\}.$$

The following properties of the above relations are immediate:

1. the relations $\le_+$ and $\le$ are transitive relations,
2. if $\alpha \le_+ \beta$, then $L_{NE}(\alpha) \subseteq L_{NE}(\beta)$ and, if $\alpha \le \beta$, then $L_E(\alpha) \subseteq L_E(\beta)$,
3. $\alpha \equiv \beta$ if and only if $\alpha \le_+ \beta$ and $\beta \le_+ \alpha$.

In the sequel we assume that all patterns have as the set of variables a finite subset V of the infinite set $\{X_i \mid i \ge 1\}$. Let α be a pattern over $\Sigma \cup V$ and let a, b be two fixed letters from Σ. We introduce a particular sample, $F(\alpha)$, that is fundamental for deciding the equivalence of two NE-patterns. Consider the following set of terminally stable morphisms defined for all $i, j \ge 1$:

$$f_a(X_i) = a, \; f_b(X_i) = b,$$

$$g_j(X_i) = a, \text{ if } i = j \text{ and } g_j(X_i) = b \text{ otherwise.}$$

$F(\alpha)$ is the set of words

$$F(\alpha) = \{f_a(\alpha), f_b(\alpha)\} \cup \{g_i(\alpha) \mid i \ge 1\}.$$

Obviously, $F(\alpha)$ is a finite and nonempty language, hence $F(\alpha)$ is a sample.

Example 6.3. Assume that $\Sigma = \{a, b\}$ and let α be the pattern $\alpha = X_1 X_1 aa X_2$. One can easily see that

$$F(\alpha) = \{aaaaa, bbaab, aaaab, bbaaa\}. \qquad \Box$$

Note that if α contains no variables then $F(\alpha) = L(\alpha) = \{\alpha\}$, if α contains only one variable, then $card(F(\alpha)) = 2$ and if α contains $n \ge 2$ variables, then $card(F(\alpha)) = n + 2$. Moreover, note that $F(\alpha) \subseteq L(\alpha)$ and that $F(\alpha)$ can be constructed in polynomial time with respect to the length of α.

The following results were established in [5, 6].

Theorem 6.3. *Let α, β be two patterns such that $\mid \alpha \mid = \mid \beta \mid$. Then,*

i) If $F(\alpha) \subseteq L_{NE}(\beta)$, then $\alpha \leq_+ \beta$,
ii) $L_{NE}(\alpha) \subseteq L_{NE}(\beta)$ if and only if $\alpha \leq_+ \beta$,
iii) α is descriptive of $F(\alpha)$ within the class of NE-patterns. $\square$

Theorem 6.4. *For all NE-patterns α and β,*

$$L_{NE}(\alpha) = L_{NE}(\beta) \ \text{if and only if} \ \alpha \equiv \beta.$$ $\square$

As a consequence, it follows that the equivalence problem for NE-patterns is decidable and, moreover, the equivalence may be tested by a linear-time algorithm.

However, for E-patterns the situation seems to be more complicated. In fact, at the time of this writing, the following problem is still open: Does there exist an algorithm to decide whether two given E-patterns are equivalent or not? Although it is still open, there is much progress on this line, mainly done in [70, 72] and in the recent papers [30], [103]. Two E-patterns α and β over $\Sigma \cup V$ are referred to as *similar*, if $\alpha = \alpha_0 u_1 \alpha_1 \ldots u_n \alpha_n$, $\beta = \beta_0 u_1 \beta_1 \ldots u_n \beta_n$, $n \geq 0$, $\alpha_i, \beta_i \in V^+$, $1 \leq i \leq n-1$, $\alpha_0, \beta_0, \alpha_n, \beta_n \in V^*$ and $u_j \in \Sigma^+$, $1 \leq j \leq n$.

Theorem 6.5. *Let Σ be a terminal alphabet with $card(\Sigma) \geq 3$. If two E-patterns over $\Sigma \cup V$ are equivalent, then they are similar.* $\square$

The result does not hold in the case $card(\Sigma) = 2$. The similarity condition is not a necessary condition in this case. For instance, assume that $\Sigma = \{a, b\}$ and consider the following patterns:

$$\alpha = X_1 a X_2 b X_3 a X_4, \ \beta = X_1 a b X_2 a X_3, \ \gamma = X_1 a X_2 b a X_3.$$

One can easily verify that

$$L_E(\alpha) = L_E(\beta) = L_E(\gamma),$$

but no two distinct E-patterns from α, β, γ are similar. The same holds true with respect to the E-patterns $X_1 a a a X_2 b X_3$ and $X_1 a a a b X_2$.

6.3 The inclusion problem

The inclusion problem for NE-patterns consists of: does there exist an algorithm to decide for two given NE-patterns α and β, whether or not $L_{NE}(\alpha) \subseteq L_{NE}(\beta)$?

This problem was open for a long time and solved in [71, 72].

It turns out that the inclusion problem for NE-patterns is undecidable. Thus, for the family of languages generated by NE-patterns, equivalence is decidable but inclusion undecidable. A similar situation is known for some other language families as well. However, in other known similar cases, the equivalence problem is quite hard, whereas it is almost trivial for NE-pattern languages.

We start our presentation by explaining why the inclusion problem for NE-patterns is hard. The inclusion $L_{NE}(\alpha) \subseteq L_{NE}(\beta)$ can hold for two patterns α and β without α and β having seemingly any connection. The inclusion holds if $\alpha = h(\beta)$, for some nonerasing morphism h that keeps the terminals fixed, but the existence of such a morphism is by no means necessary for the inclusion to hold. For instance, if $\Sigma = \{0, 1\}$ and α is an arbitrary pattern with length at least 6, then $L_{NE}(\alpha) \subseteq L_{NE}(XYYZ)$. Observe that the pattern α is not necessarily a morphic image of $XYYZ$.

But the main reason why the inclusion problem is hard comes from its expressive power. Thue's classical results, [128], can be expressed in terms of non-inclusion among NE-pattern languages as follows:

If Σ consists of two letters then, for all n, $L_{NE}(X_1 X_2 \ldots X_n)$ contains words not belonging to $L_{NE}(XYYYZ)$, i.e., $L_{NE}(X_1 X_2 \ldots X_n)$ is not included in $L_{NE}(XYYYZ)$. If Σ contains at least three letters then, for all n, $L_{NE}(X_1 X_2 \ldots X_n)$ contains words not belonging to $L_{NE}(XYYZ)$.

The above results concerning squares and cubes have been extended to arbitrary patterns in [13]. We present the definitions in our terminology.

A pattern α is termed *unavoidable* (on an alphabet Σ) iff, for some n,

$$L_{NE}(X_1 X_2 \ldots X_n) \subseteq L_{NE}(X \alpha Z),$$

where the X_i's are distinct variables and X and Z are variables not occurring in α. Otherwise, α is *avoidable*.

Thus, YY is unavoidable on two letters but avoidable on three letters. The pattern YYY is avoidable also on two letters.

The paper [13] gives a recursive characterization of unavoidable terminal-free patterns. A rather tricky example is the pattern

$$\alpha = XYXZX'YXY'XYX'ZX'YX',$$

which is unavoidable for a suitable terminal alphabet Σ. Given a pattern α that is unavoidable on Σ, it is of interest to find the *smallest* n such that every word of length at least n possesses a subword of pattern α. The following result, which is obvious from the definitions, shows the interconnection with the inclusion problem.

Any algorithm for solving the inclusion problem for NE-pattern languages can be converted into an algorithm for computing the smallest n such that a given unavoidable pattern cannot be avoided on words of length at least n.

Finally, note that if we allow the use of variables of other types, such as the variables X^P, then also other extensions of Thue's results can be reduced to the inclusion problem. For instance, the important result of [77] can be expressed as follows:

If Σ contains at least four letters then, for all n, $L_{NE}(X_1 X_2 \ldots X_n)$ contains words not belonging to $L_{NE}(XYY^P Z)$.

The above result is not valid for smaller alphabets. Indeed, if Σ consists of three letters and α is a pattern of length at least 10, then

$$L_{NE}(\alpha) \subseteq L_{NE}(XYY^P Z).$$

The following result was obtained in [71], see also [72] for a more detailed proof.

Theorem 6.6. *Given a terminal alphabet Σ, a set of variables V, and two arbitrary E-patterns $\alpha, \beta \in (\Sigma \cup V)^*$, it is undecidable whether*

$$L_{E,\Sigma}(\alpha) \subseteq L_{E,\Sigma}(\beta).$$

Proof. We give here only the idea of the proof. The reader is referred to [72] for the complete proof of Theorem 6.6.

The inclusion problem for E-patterns is proved to be undecidable by reducing to it the emptiness problem for nondeterministic 2-counter automata.

A *nondeterministic 2-counter automaton without input*, see [99], is denoted as $M = (Q, q_0, Q_F, 0, \delta)$, where Q is the finite set of *states*, $q_0 \in Q$ is the *initial state*, $Q_F \subseteq Q$ is the set of *final states*, $0 \notin Q$ is the only symbol for the stack alphabet, and $\delta \subseteq Q \times \{0,1\}^2 \times Q \times \{-1,0,1\}^2$ is the *transition relation*. The set of configurations of M is 0^*Q0^*. If uqv is a configuration, $u, v \in 0^*, q \in Q$, then u, v denote the contents of the counters and q is the current state of the finite control. The relation $\Rightarrow_M$ between configurations is determined by the transition relation δ as follows. If $q_i \in Q$ and $u_i, v_i \in 0^*, i = 1, 2$, then

$$u_1 q_1 u_2 \Rightarrow_M v_1 q_2 v_2$$

iff there exists $(q_1, x_1, x_2, q_2, y_1, y_2) \in \delta$, $x_1, x_2 \in \{0,1\}, y_1, y_2 \in \{-1,0,1\}$, such that for $i = 1, 2$, $x_i = 0$, if $u_i = \lambda$, and $x_i = 1$, otherwise. Moreover, $\mid v_i \mid = \mid u_i \mid + y_i$. It is assumed in the definition of δ that if $x_i = 0$ then $y_i \geq 0, i = 1, 2$.

Let $\#$ be a new symbol not appearing in $Q \cup \{0\}$. An *accepting computation* of M without input is a word W_c:

$$W_c = \#C_1 \# C_2 \# \ldots \# C_m \#,$$

where $m \geq 1$, $C_1 = q_0$, $C_m = u_1 q u_2$, $q \in Q_F$, $u_1, u_2 \in 0^*$, C_i is a configuration of M, and $C_i \Rightarrow_M C_{i+1}$, $1 \leq i \leq m - 1$. It is known that the emptiness problem for nondeterministic 2-counter automata without input is undecidable, see [10, 64, 99].

For a given 2-counter automaton M, one can now construct two patterns α and β such that $L_E(\alpha) \subseteq L_E(\beta)$ exactly in the case M has no accepting computation. Both α and β contain three occurrences of a special terminal letter $, acting as a marker. The subwords between the first two occurrences of $ and the last two occurrences of $ are referred to as the *first* and respectively *second segment*. Apart from $, β contains a considerably large number of variables. Some of these variables occur in both segments. The essential part of the first segment of α is the only occurrence of a variable X. The second segment of α is a terminal word, in fact a catenation of really many words w_i, separated by markers. The pattern α contains also one occurrence of another variable Y. Now, whatever one substitutes for X and Y in α, the same word can be obtained by some substitution of the variables of β. An exception occurs in the case that X is a successful computation of M, and Y is a long enough string of 0's. In this case the word is not in the language $L_E(\beta)$, and the claim concerning the inclusion follows: the inclusion holds if and only if there is no successful computation of M. Why this works is because the words w_i describe all possible deviations from a successful computation. If a deviation occurs in X, some variables get values in the first segment of β that correspond to some w_i in the second segment. There is no way to get such a correspondence when there are no deviations, i.e., when X is a successful computation. Therefore, the words w_i describe all possible types of errors. The key idea is that for an arbitrary 2-counter automaton there are only a finite number of such types of errors and thus they can be encoded in a pattern. The initial and final parts of α and β, those before the first \$ and after the last \$, are still used to guarantee that, for each assignment for X and Y in α, only few variables in β can be active. $\qquad\square$

For the case of NE-patterns a similar result does hold. The inclusion problem between NE-patterns is shown to be undecidable by reducing to it the inclusion problem for E-patterns. The reader is referred to [71, 72] for the details. Thus, we have

Theorem 6.7. *Given a terminal alphabet Σ, a set V of variables, and two NE-patterns α, $\beta \in (\Sigma \cup V)^*$ it is undecidable in general whether or not*

$$L_{NE,\Sigma}(\alpha) \subseteq L_{NE,\Sigma}(\beta).$$
$\qquad\square$

If we restrict our attention to the family of terminal-free patterns, i.e., patterns containing only variables, the situation is essentially different, see [71, 72].

Theorem 6.8. *Given a terminal alphabet Σ, a set of variables V, and two terminal-free E-patterns $\alpha, \beta \in V^+$ it is decidable whether*

$$L_{E,\Sigma}(\alpha) \subseteq L_{E,\Sigma}(\beta)$$

or not.
$\qquad\square$

However, in the case of terminal-free NE-patterns the decidability status of the inclusion problem is open, i.e., it is not known whether the inclusion problem between terminal-free NE-patterns is decidable or not.

6.4 The membership problem

The membership problem, both for NE-patterns and for E-patterns, is decidable and, moreover, in both cases it is known to be NP-complete. This result for NE-patterns can be found in [5] whereas for the case of E-patterns see [70].

Theorem 6.9. *The problem of deciding whether $w \in L_{NE}(\alpha)$ ($w \in L_E(\alpha)$) for an arbitrary terminal word w and NE-pattern α (E-pattern α) is NP-complete.*

Proof. We present here the proof for the case of NE-patterns. The proof for the case of E-patterns is quite similar. Let Σ be an alphabet and let V be a set of variables. Consider an NE-pattern $\alpha \in (\Sigma \cup V)^*$ and let $w \in \Sigma^*$ be a terminal word. Clearly, the problem is in NP. One can easily define a Turing machine that guesses an assignment for the variables in α and checks whether w results. To prove that the problem is complete in NP, we describe a polynomial-time reduction to it of the 3-satisfiability (3-SAT) problem. The 3-SAT problem consists of deciding whether a propositional formula in conjunctive normal form with three literals per clause is satisfiable or not.

Let φ be a given propositional formula in conjunctive normal form with three literals per clause. Assume that the literals of φ are $P_i, 1 \leq i \leq n$ and the clauses are $C_j, 1 \leq j \leq m$, where each clause is a triple of three terms each of which is either a literal or the negation of a literal. Let $0, 1 \in \Sigma$ be two fixed, distinct symbols (note that always $card(\Sigma) \geq 2$). Define the pattern α, containing $2(m+n)$ variables $X_i, Y_i, Z_j, U_j, 1 \leq i \leq n, 1 \leq j \leq m$ as follows.

$$\alpha = 0\alpha_1 0\alpha_2 \ldots 0\alpha_n 0\beta_1 0\beta_2 \ldots 0\beta_m 0\gamma_1 0\gamma_2 \ldots 0\gamma_m 0,$$

such that $\alpha_i = X_i Y_i, 1 \leq i \leq n$, $\gamma_j = Z_j U_j$, and $\beta_j = b_{j,1} b_{j,2} b_{j,3} Z_j, 1 \leq j \leq m$, where $b_{j,k} = X_i$ if the kth literal in C_j is P_i and $b_{j,k} = Y_i$ if the kth literal in C_j is the negation of $P_i, 1 \leq j \leq m, 1 \leq k \leq 3$.

Now we define the terminal word w as being

$$w = 0w_1 0w_2 \ldots 0w_n 0t_1 0t_2 \ldots 0t_m 0v_1 0v_2 \ldots 0v_m 0,$$

where $w_i = 1^3$, $t_j = 1^7$ and $v_j = 1^4$, $1 \leq i \leq n$ and $1 \leq j \leq m$.

One can easily verify that $w \in L_{NE}(\alpha)$ if and only if φ is satisfiable, hence the membership problem for NE-patterns is NP-complete. $\square$

6.5 Ambiguity

Pattern languages reflect some basic language-theoretic issues concerning degrees of ambiguity. The idea of ambiguity in patterns was formulated in [75]. Our presentation here follows [93, 94].

Given a pattern α and a word w in $L_E(\alpha)$ or $L_{NE}(\alpha)$, it may happen that w has several "representations", that is, there are several morphisms h

satisfying $w = h(\alpha)$. For instance, the terminal word $w = a^7ba^7$ possesses 8 representations in terms of the pattern $\alpha = XYX$. (The number is 7 if α is viewed as an NE-pattern.) We express this by saying that the *degree of ambiguity* of w with respect to α equals 8. Whenever important, we indicate whether we are dealing with the E- or NE-case. The *degree of ambiguity of a pattern* α equals the maximal degree of ambiguity of words w in the language of α, or infinity (∞) if no such maximal degree exists. More formally, we associate to a pattern α over $\Sigma \cup V$ and a word $w \in \Sigma^+$ the subset $S(\alpha, w, \Sigma)$ of $H_{\Sigma,V}$, consisting of morphisms h such that $h(\alpha) = w$. The *cardinality* of this subset is denoted by $card(\alpha, w, \Sigma)$. We make here the convention that morphisms differing only on variables not present in α are not counted as different. The *degree of ambiguity* of α equals $k \geq 1$ iff

$$card(\alpha, w, \Sigma) \leq k, \text{ for all } w \in L_E(\alpha),$$

and

$$card(\alpha, w', \Sigma) = k, \text{ for some } w' \in L_E(\alpha).$$

If there is no such k, then the *degree of ambiguity* of α equals ∞. For $k = 1$, α is termed also *unambiguous* and, for $k > 1$, α is termed *ambiguous*.

The terminals actually appearing in α constitute a subset Σ', maybe empty, of Σ. One can easily observe that it suffices to specify only α because the degree of ambiguity of α is independent of the choice of Σ.

The notions are now naturally extended to concern languages. A *pattern language* L is *ambiguous of degree* $k \geq 1$ if $L = L_E(\alpha)$, for some pattern α ambiguous of degree k, but there is no pattern β of degree less than k such that $L = L_E(\beta)$. Here k is a natural number or ∞. Again, if $k = 1$ we say that L is *unambiguous*. Otherwise, L is *(inherently) ambiguous*.

As a consequence of the fact that two NE-patterns are equivalent exactly in case they are identical up to a renaming of variables, see Theorem 6.4, it follows that the degree of ambiguity of an NE-pattern α equals the degree of ambiguity of the pattern language $L_{NE}(\alpha)$. This is not the case for E-patterns: $L_E(X) = L_E(XY)$ but the degrees of ambiguity of the patterns X and XY are 1 and ∞, respectively. Moreover, for every E-pattern (containing at least one variable), there is an equivalent E-pattern whose degree of ambiguity is ∞. Conversely, there are E-patterns such that the degree of ambiguity of every equivalent E-pattern is ∞, for instance $XYYX$. One can easily see that every pattern containing occurrences of only one variable is unambiguous. Also, every pattern containing occurrences of at least two variables but of at most one terminal is ambiguous of degree ∞.

For the proofs of the following theorems the reader is referred to [94].

Theorem 6.10. *Every pattern α satisfying the following two conditions is of ∞ degree of ambiguity:*

i) α contains occurrences of at least two variables.
ii) Some variable occurs in α only once. □

Note that the pattern $\alpha = XaYXbY$ is unambiguous. Using this fact, the second sentence of our next theorem can be inferred.

Theorem 6.11. *Compositions of unambiguous patterns are unambiguous. Unambiguous patterns of arbitrarily many variables can be effectively constructed.* □

Using the general theorem of Makanin, [90], the following results are obtained.

Theorem 6.12. *The following problems are decidable, given a pattern α and a number $k \geq 1$. Is the degree of ambiguity of α equal to k, greater than k or less than k? Consequently, it is decidable whether or not α is unambiguous.* □

By our results so far "most" of all patterns are either unambiguous or of degree ∞. The existence of patterns of a finite degree of ambiguity greater than one constitutes a rather tricky problem. Such patterns must have at least two variables and at least two terminals. Moreover, each of the variables must occur at least twice in the pattern.

The main result concerning finite degrees of ambiguity is the next theorem. See [94] for a long combinatorial argument used in the proof.

Theorem 6.13. *For any $m \geq 0$ and $n \geq 0$, a pattern of degree of ambiguity $2^m 3^n$ can be effectively constructed.* □

It is known, [93, 94], that the pattern $\alpha = XabXbcaYabcY$ has the degree of ambiguity 2, whereas a pattern with the degree of ambiguity 3 constructed in [94] has length 324 and the shortest word that actually has 3 different decompositions with respect to this pattern has length 1018.

An immediate consequence of Theorem 6.13 is that for any k there are patterns with finite degree of ambiguity greater than or equal with k. However, the following problem is open,

Open Problem: Does there exist for every $k \geq 1$ a pattern whose degree of ambiguity is exactly k?

It suffices to carry out the construction for prime numbers k. Another interesting question is:

Open Problem: Does there exist an algorithm to decide whether the degree of ambiguity of a pattern is finite or not?

"Almost all" patterns seem to be of degree ∞, and yet Makanin's Theorem is not directly applicable to this case.

6.6 Multi-pattern languages. Term rewriting with patterns

Another natural generalization of the notion of a pattern language is a *multi-pattern language*, [75]; instead of one basic pattern, there are finitely many of them. The study of multi-patterns has been initiated in [75]. A *multi-pattern* is a finite set of patterns. The *language generated* by a multi-pattern is the union of all languages generated by the patterns from the multi-pattern. An NE-multi-pattern is a finite set of NE-patterns. The notion of an E-multi-pattern is defined similarly. Let $MPLNE$ ($MPLE$) be the family of languages that are generated by NE-multi-patterns (E-multi-patterns). One can easily see that $MPLE = MPLNE$. Hence, in the sequel we refer to these families as the *family of multi-pattern languages*, denoted by MPL.

Theorem 6.14. [75] *The equivalence and the inclusion problems are undecidable for the family MPL. The membership problem is NP-complete for languages in this family.* □

It is also undecidable whether or not an arbitrary context-free language is in MPL, see [75]. The family of erasing multi-patterns of *degree n*, $MPLE(n)$, is the family of languages generated by erasing multi-patterns consisting of exactly n E-patterns. Analogously, one introduces the family $MPLNE(n)$ for the nonerasing case.

Theorem 6.15. [75] *The number of patterns defines an infinite hierarchy:*

$$MPLE(n) \subset MPLE(n+1), \; n \geq 1,$$

$$MPLNE(n) \subset MPLNE(n+1), \; n \geq 1.$$ □

Theorem 6.16. [75] *The family MPL is an anti-AFL, but it is closed under right/left derivatives. MPL is not closed under right/left quotients with regular languages, intersection or complement.* □

There are various natural ways to modify the definition of pattern languages, as presented above. One may start with a set of terminal words ("axioms"), and first replace the variables in a given pattern or a set of patterns by some axioms. This results in a larger set of terminal words. The process is continued with the larger set, and so on. Such generative devices were studied in [31], see also [121]. There are similarities, as well as differences, when this definition of $L(G)$ is compared with the customary definition, see [115], of context-free languages as solutions of algebraic systems of equations. The customary definition has a stronger generative power because different variables are associated with different languages in the generative process. On the other hand, the generative capacity of pattern grammars is stronger because several occurrences of the same word can be enforced to be present simultaneously. Another way to define grammars based on patterns is studied in [105]. An extended model that covers these generative devices was introduced and studied in [100]. Interrelations between pattern languages and parallel communicating grammar systems are studied in [39]. Infinite multi-patterns (regular and context-free) are introduced in [40]. Sixteen new families of languages are defined and their mutual relationships are presented. Descriptional measures for pattern grammars and languages are introduced and studied in [46]. It is proved that these measures induce infinite hierarchies of languages.

Another line of research on patterns is on avoidable and unavoidable patterns, initiated in [13]. Although we did not follow very close this topic here, we refer the reader to [16] for recent results in this area. A complete classification of unavoidable binary patterns is presented there as well as other concepts of avoidability.

7. Pure and indexed grammars. Derivation languages

In this subsection we will describe very briefly some classes of *grammars* and their languages. Our aim is just to give an idea about what is going on, as well as some references. Our presentation will be rather informal.

7.1 Pure grammars

In the Chomsky hierarchy and related types of grammars, it has become customary to divide the alphabet into two parts: *nonterminal* letters and *terminal* letters. Only words consisting entirely of terminal letters are considered to be in the language generated. Basically, this distinction stems from linguistic motivation: nonterminals represent syntactic classes of a language. One can view the use of nonterminals also as an auxiliary definitional mechanism for generative devices. Not everything derived from the axiom or axioms is accepted as belonging to the language. Nonterminals are used to filter out unwanted words.

On the other hand, such a distinction is not made in the original rewriting systems of Thue, [130]. It is also not made in the original work concerning L systems (see the chapter about L systems in this Handbook). Indeed, since L systems were originally used to model the development of filamentous organisms, it would have been unnatural to exclude some stages of development as "improper".

The study of *pure grammars* continues this original line of research concerning rewriting systems. In a pure grammar there is only one undivided alphabet. The starting point, *axiom*, is a word or, more commonly, a finite set of words. Rewriting is defined in the standard way, and a specific pure grammar is obtained by giving a finite set of rewriting rules or productions. All words derivable from some of the axioms belong to the *language* of the pure grammar. A pure grammar is *context-free*, (resp. *length increasing*) if the length of the left side of every production is equal to one (resp. less than or equal to the length of the right side). Abbreviations PCF and PLI will be used for these special classes of pure grammars and their languages. (One can define also pure context-sensitive grammars in the standard way. The resulting class of languages will be strictly smaller than the class of PLI languages. The ordinary simulation of length-increasing productions by context-sensitive ones does not work because one is not able to "hide" the intermediate words needed for the simulation.)

The language $\{a^n c b^n \mid n \geq 1\}$ is generated by the PCF grammar with the axiom acb and the only production $c \longrightarrow acb$. The language $\{a^n b^n \mid n \geq 1\}$ is not PCF but is generated by the PLI grammar with the axiom ab and the only production $ab \longrightarrow a^2 b^2\}$. Neither one of the languages

$$\{a^n b^n c^n \mid n \geq 1\} \text{ and } \{a^n b^n \mid n \geq 1\} \cup \{a^n b^{2n} \mid n \geq 1\}$$

is pure, that is, they are not generated by any pure grammar whatsoever.

Every regular language is PLI and, hence, pure. The family of pure languages over a one-letter alphabet $\{a\}$ coincides with the family of regular languages over $\{a\}$, whereas PCF languages constitute a proper subfamily. There are nonrecursive pure languages. Each of the differences $\mathcal{L}_i - \mathcal{L}_{i+1}$, $0 \leq i \leq 2$, where $\mathcal{L}_i$, $0 \leq i \leq 3$, are the language families in the Chomsky hierarchy, contains both pure and nonpure languages. As far as length sets are concerned, PCF languages coincide with context-free languages.

The reader is referred to [98] for proofs of the results mentioned here, as well as for further references and historical remarks concerning pure grammars. [24] represents more recent work. The set of sentential forms of a context-free grammar constitutes a PCF language. See [114] for related matters.

We still mention some typical decidability results concerning pure languages, [98] and [57]. The inclusion $R \subseteq L$ is undecidable but the equation $R = L$ is decidable, given a regular language R and a PCF language L. The equation $R = L$ is undecidable, given a regular language R and a pure language L. It is undecidable whether or not a given PLI grammar generates a regular language. For a pure grammar G with the alphabet $\{a, b\}$, it is decidable whether or not b occurs in some word in $L(G)$. For PLI grammars G with the alphabet $\{a, b, c\}$ it is undecidable whether or not c occurs in some word in $L(G)$. It is decidable whether or not $L(G) = \Sigma^*$ holds for a given PLI grammar G, but undecidable whether or not $L(G)$ is infinite.

7.2 Indexed grammars

An *indexed grammar* is a quintuple $G = (N, T, S, P, F)$, where the first four items are as in a context-free grammar, and F is a finite set of *indices* or *flags*. Each of the flags, in turn, is a finite set of *index productions* of the form $A \longrightarrow \alpha$, $A \in N$, $\alpha \in (N \cup T)^*$. The productions in P are of the form

$$A \longrightarrow \alpha, \ A \in N, \alpha \in (NF^* \cup T)^*.$$

Thus, nonterminals in a word can be followed by arbitrary lists of flags. If a nonterminal, with flags following, is replaced in a derivation by one or more nonterminals, the flags follow each of the new nonterminals. If a nonterminal is replaced by terminals, the flags disappear. If a nonterminal A is followed by the sequence of flags $f_1 f_2 \ldots f_n$, and f_1 contains an index production $A \longrightarrow \alpha$, we can replace A by α and attach each nonterminal in α by the sequence $f_2 \ldots f_n$. The flags can be "consumed" in this way.

Let us work out formally the definition of the yield relation $\Longrightarrow$. By definition, the relation $x \Longrightarrow y$ holds between words x and y in $(NF^* \cup T)$ iff either

(i) $x = x_1 A \varphi x_2$, $A \in N$, $\varphi \in F^*$; $A \longrightarrow A_1 \psi_1 \ldots A_n \psi_n$, $A_i \in N \cup T$, $\psi_i \in F^*$, is a production in P; and $y = x_1 A_1 \varphi_1 \ldots A_n \varphi_n x_2$, where $\varphi_i = \psi_i \varphi$ if $A_i \in N$, and $\varphi_i = \lambda$ if $A_i \in T$; or else

(ii) $x = x_1 A f \varphi x_2$, where A and φ are as before, $f \in F$, $A \longrightarrow A_1 \ldots A_n$ is an index production in the index f, and $y = x_1 A_1 \varphi_1 \ldots A_n \varphi_n x_2$, where $\varphi_i = \varphi$ if $A_i \in N$, and $\varphi_i = \lambda$ if $A_i \in T$.

As usual, $\Longrightarrow^*$ stands for the reflexive transitive closure of $\Longrightarrow$, and

$$L(G) = \{w \in T^* \mid S \Longrightarrow^* w\}.$$

For instance, if G is defined by letting

$$P = \{S \longrightarrow A f_1, A \longrightarrow A f_2, A \longrightarrow BCB\},$$
$$f_1 = \{B \longrightarrow a, C \longrightarrow b, D \longrightarrow b\},$$
$$f_2 = \{B \longrightarrow aB, C \longrightarrow bCDD, D \longrightarrow bD\},$$

it can be verified that

$$L(G) = \{a^n b^{n^2} a^n \mid n \geq 1\}.$$

Indeed, after the introduction of flags we have a word of the form $B f_2^n f_1 C f_2^n f_1 B f_2^n f_1$. The flag-elimination phase replaces the B-parts with a^{n+1} and the C-part with $b^{(n+1)^2}$. The latter claim follows inductively: we consider the introduction of one more f_2 and use the identity $(n+1)^2 - n^2 = 2n + 1$.

Every indexed grammar can be reduced to an equivalent grammar in a *normal form*, where all index productions in P are of one of the forms

$$A \longrightarrow BC, \ A \longrightarrow Bf, \ A \longrightarrow \alpha$$

with $A, B, C \in N$; $f \in F$, $\alpha \in T \cup \{\lambda\}$. The family of indexed languages is a full AFL. It is closed, further, under substitution but not under intersection or complementation. Emptiness and finiteness are decidable for indexed languages. These results are established in the basic paper for indexed grammars, [1]. The results gained further importance after it was noticed in the early 70's that the most comprehensive among context-free L families, the family of $ET0L$ languages, is contained in the family of indexed languages (See the chapter on L systems in this Handbook. More details about indexed grammars will be given also in Volume II of this Handbook.)

7.3 The derivation (Szilard) language

The last topic in this section is a notion that can be associated to many kinds of grammars. We consider here only grammars in the Chomsky hierarchy. Let us *label* the productions, give each of them a name. What this means is that we consider pairs (G, F), where $G = (N, T, S, P)$ is a grammar and F is an alphabet of the same cardinality as P and, moreover, a one-to-one mapping f of P onto F is given. For $p \in P$, $f(p)$ is referred to as the *label* of p. We now consider "successful" derivations according to G, that is, derivations from S to a word over the terminal alphabet T. To each successful derivation we associate the word over F obtained by writing the labels of the productions applied in the derivation, in the order of their application. In this way a language $Sz(G)$ over F results, referred to as the *derivation language* or *Szilard language* of G. For instance, let G be the context-free grammar with the labeled productions

$$f_1 : S \longrightarrow aSb, \ f_2 : S \longrightarrow ab.$$

Then $Sz(G)$ is the regular language $f_1^* f_2$.

It is easy to see that $Sz(G)$ is regular if G is of type 3. For a context-free grammar G, $Sz(G)$ is not necessarily context-free. For instance, let G be the context-free grammar with the labeled productions

$$f_1 : S \longrightarrow ASB, \ f_2 : S \longrightarrow \lambda, \ f_3 : A \longrightarrow \lambda, \ f_4 : B \longrightarrow \lambda.$$

Then the intersection of $Sz(G)$ with the regular language $f_1^* f_2^* f_3^* f_4^*$ is the non-context-free language $\{f_1^n f_2 f_3^n f_4^n \mid n \geq 1\}$. This implies that $Sz(G)$ cannot be context-free. This example shows also another feature characteristic for the study of Szilard languages: the right sides of all terminating productions can be replaced by λ. This is due to the fact that we are not interested in the terminal word itself but only in its being a terminal word. Observe also that $Sz(G)$ may be empty (this happens iff $L(G)$ is empty) but $Sz(G)$ never contains the empty word λ.

The latter example can be generalized: every nonempty context-free language is generated by a grammar G such that $Sz(G)$ is not context-free. However, the Szilard language of every grammar is context-sensitive. This is seen roughly as follows. Consider an arbitrary type 0 grammar $G = (N, T, S, P)$. Construct a length-increasing grammar G_1 such that $L(G_1)$ contains, for every $w \in L(G)$ and every u which is a control word of a derivation of w, a word $u \# w_1 \#$, where w results from w_1 by erasing occurrences of an auxiliary letter. Moreover, $|w_1|$ is linearly bounded by $|u|$. Thus, the Workspace Theorem can be applied to show that the language consisting of all words u is context-sensitive. It is always even deterministic context-sensitive.

For early references about Szilard languages, see [115], [101]. [109] contains an algebraic treatment of the matter, with further references.

References

1. A. V. Aho, "Indexed grammars - An extension to context-free grammars", *Journal of the ACM*, 15 (1968) 647–671.
2. A. V. Aho, J. D. Ullman, *The Theory of Parsing, Translation, and Compiling*, Prentice Hall, Engl. Cliffs, N. J., vol. I 1971, vol. II 1973.
3. A. V. Aho, J. D. Ullman, *Principles of Compiler Design*, Addison-Wesley, Reading, Mass., 1977.
4. R. op den Akker, B. Melichar, J. Tarhio, "The Hierarchy of LR-attributed Grammars", *in Lecture Notes in Computer Science 461*, Springer-Verlag, Berlin, 1990, 13–28.
5. D. Angluin, "Finding Patterns Common to a Set of Strings", *17th Symposium on Theory of Computation 1979*, 130–141.
6. D. Angluin, "Finding Patterns Common to a Set of Strings", *Journal of Computer and System Sciences*, 21 (1980) 46–62.
7. D. Angluin, "Inductive inference of formal languages from positive data", *Information and Control*, 45 (1980) 117–135.
8. J. W. Backus, "The syntax and semantics of the proposed international algebraic language of the Zurich", *ACM-GAMM*, Paris, 1959.
9. B. S. Baker, "Context-sensitive grammars generating context-free languages", in *Automata, Languages and Programming*, M. Nivat (ed.), North-Holland, Amsterdam, 1972, 501–506.
10. B. S. Baker, R. V. Book, "Reversal-bounded multipushdown machines", *Journal of Computer and System Sciences*, 8 (1974) 315–332.
11. J. L. Baker, "Grammars with structured vocabulary: A model for the ALGOL 68-definition", *Information and Control*, 20 (1972) 351–395.
12. K. Barbar, "Attributed tree grammars", *Theoretical Computer Science*, 119 (1993) 3–22.
13. D. R. Bean, A. Ehrenfeucht, G. F. McNulty, "Avoidable patterns in strings of symbols", *Pacific Journal of Mathematics*, 85, 2 (1979) 261–294.
14. L. Boasson, *Cônes Rationnels et Familles Agréables de Langages-Application au Langage à Compteur, Thesis, Univ. de Paris*, 1971.
15. R. V. Book, "Terminal context in context-sensitive grammars", *SIAM Journal of Computing*, 1 (1972) 20–30.
16. J. Cassaigne, "Unavoidable binary patterns", *Acta Informatica* 30 (1993) 385–395.
17. L. M. Chirica, D. F. Martin, "An Order-Algebraic Definition of Knuthian Semantics", *Mathematical Systems Theory*, 13 (1979) 1–27.
18. N. Chomsky, "Three models for the description of language", *IRE Trans. on Information Theory*, 2, 3 (1956) 113–124.
19. N. Chomsky, "Syntactic Structures", *Mouton, Gravenhage* (1957).
20. N. Chomsky, "On certain formal properties of grammars", *Information and Control* 2 (1959) 137–167.
21. N. Chomsky, "Context-free grammars and pushdown storage", *M.I.T. Res. Lab. Electron. Quart. Prog., Report 65* (1962).
22. N. Chomsky, "Formal properties of grammars", *Handbook of Math. Psych.*, Vol. 2, (1963) 323–418.
23. A. Church, "An unsolvable problem for elementary number theory", *The American Journal of Mathematics*, 58 (1936) 345–363.
24. G. Ciucar, G. Păun, "On the syntactical complexity of pure languages", *Foundations of Control Engineering*, 12, 2 (1987) 69–74.
25. S. A. Cook, "The complexity of theorem proving procedures", *Proc. Third Annual ACM Symposium on the Theory of Computing*, (1971) 151–158.

26. B. Courcelle, P. Franchi-Zannettacci, "Attribute grammars and recursive program schemes I", *Theoretical Computer Science*, 17 (1982) 163–191.
27. B. Courcelle, P. Franchi-Zannettacci, "Attribute grammars and recursive program schemes II", *Theoretical Computer Science*, 17 (1982) 235–257.
28. K. Culik II, "A Purely Homomorphic Characterization of Recursively Enumerable Sets", *Journal of the ACM*, 26, 2 (1979) 345–350.
29. K. Culik II, F. E. Fich, A. Salomaa, "A homomorphic characterization of regular languages", *Discrete Applied Mathematics*, 4 (1982) 149–152.
30. G. Dányi, Z. Fülöp, "A note on the equivalence problem of E-patterns", *to appear in Information Processing Letters*.
31. J. Dassow, G. Păun, A. Salomaa, "Grammars based on patterns", *International Journal of Foundations of Computer Science*, 4 (1993) 1–14.
32. M. Davis, "Unsolvable Problems", in *Handbook of Mathematical Logic*, J. Barwise (ed.), North-Holland, Amsterdam, 1977, 567–594.
33. P. Dembinsky, J. Maluszynski, "Two-level grammars: CF grammars with equation schemes", *Lecture Notes in Computer Science 71*, Springer-Verlag, 1979, 171–187.
34. P. Deransart, M. Jourdan, B. Lorho, Attribute Grammars, *Lecture Notes in Computer Science 323*, Springer-Verlag, 1988.
35. P. Deransart, M. Jourdan, (eds.) Attribute Grammars and their Applications, *Lecture Notes in Computer Science 461*, Springer-Verlag, 1990.
36. P. Deussen, "A decidability criterion for van Wijngaarden grammars", *Acta Informatica* 5 (1975) 353–375.
37. P. Deussen, K. Melhorn, "Van Wijngaarden grammars and space complexity class EXSPACE", *Acta Informatica* 8 (1977) 193–199.
38. P. Deussen, L. Wegner, "Bibliography of van Wijngaarden grammars", *Bulletin of the EATCS* 6 (1978).
39. S. Dumitrescu, G. Păun, A. Salomaa, "Pattern languages versus parallel communicating grammar systems", *to appear*.
40. S. Dumitrescu, G. Păun, A. Salomaa, "Languages associated to finite and infinite sets of patterns", *to appear*.
41. J. Engelfriet, G. Rozenberg, "Fixed Point Languages, and Representation of Recursively Enumerable Languages", *Journal of the ACM*, 27, 3 (1980) 499–518.
42. A. J. Fisher, "A "yo-yo" parsing algorithm for a large class of van Wijngaarden grammars", *Acta Informatica* 29, 6 (1992) 461–482.
43. V. Geffert, "A representation of recursively enumerable languages by two homomorphisms and a quotient", *Theoretical Computer Science*, 62 (1988) 235–249.
44. V. Geffert, "Normal forms for phrase-structure grammars", *RAIRO Informatique théorique et Applications*, 25, 5 (1991) 473–496.
45. V. Geffert, "Tally versions of the Savitch and Immerman-Szelepcsényi theorems for sublogarithmic space", *SIAM Journal of Computing*, 22, 1 (1993) 102–113.
46. G. Georgescu, "Infinite Hierarchies of Pattern Languages", *to appear*.
47. S. Ginsburg, *Algebraic and Automata-Theoretic Properties of Formal Languages*, North-Holland, Amsterdam, 1975.
48. S. Ginsburg, S. Greibach, "Mappings which preserve context-sensitive languages", *Information and Control*, 9 (1966) 563–582.
49. S. Ginsburg, S. Greibach, M. Harrison, "One-way stack automata", *Journal of the ACM*, 14 (1967) 389–418.
50. S. Ginsburg, S. Greibach, J. Hopcroft, "Studies in Abstract Families of Languages", *Memoirs of the American Mathematical Society*, 87 (1969).

51. A. V. Gladkij, *Formal Grammars and Languages*, Izdatelstvo Nauka, Moscow, 1973.
52. I. Gorun, "A Hierarchy of Context-Sensitive Languages", *MFCS-76 Proceedings, Lecture Notes in Computer Science 45*, Springer-Verlag, Berlin, 1976, 299–303.
53. J. A. Gougen, J. W. Thatcher, E. G. Wagner, J. B. Wright, "Initial Algebra Semantics and Continuous Algebras", *Journal of the ACM*, 24, 1 (1977), 6 95.
54. J. De Graaf, A. Ollongren, "On two-level grammars", *International Journal of Computer Mathematics* 15 (1984) 269–290.
55. S. A. Greibach, "Some restrictions on W-grammars", *International Journal of Computer and Information Science* 3 (1974) 289–327.
56. T. Harju, J. Karhumäki, H. C. M. Kleijn, "On morphic generation of regular languages", *Discrete Applied Mathematics*, 15 (1986) 55–60.
57. T. Harju, M. Penttonen, "Some decidability problems of sentential forms", *International Journal of Computer Mathematics*, 7 , 2 (1979) 95–107.
58. M. Harrison, *Introduction to Formal Language Theory*, Addison-Wesley, Reading, Mass., 1978.
59. W. Hesse, "A correspondence between W-grammars and formal systems of logic", *TUM-INFO-7727*, (1977).
60. T. Hibbard, "Scan Limited Automata and Context Limited Grammars", *Doctoral dissertation, University of California at Los Angels* (1966).
61. J. Honkala, A. Salomaa, "How do you define the complement of a language", *Bulletin of EATCS*, 20 (1983) 68–69.
62. J. E. Hopcroft, J. D. Ullman, *Formal Languages and Their Relations to Automata*, Addison-Wesley, Reading, Mass., 1969.
63. J. E. Hopcroft, J. D. Ullman, *Introduction to Automata Theory, Languages, and Computation*, Addison-Wesley, Reading, Mass., 1979.
64. O. Ibarra, "Reversal-bounded multicounter machines and their decision problems", *Journal of the ACM*, 25 (1978) 116–133.
65. O. Ibarra, T. Jiang, "Learning regular languages from counterexamples", *Journal of Computer and System Sciences*, 2, 43 (1991) 299–316.
66. N. Immerman, "Nondeterministic space is closed under complementation", *SIAM Journal of Computing*, 17, 5 (1988) 935–938.
67. S. Istrail, "Elementary Bounded Languages", *Information and Control*, 39 (1978) 177–191.
68. S. Istrail, "On Complements of Some Bounded Context-Sensitive Languages", *Information and Control*, 42 (1979) 283–289.
69. M. Jazayeri, W. F. Ogden, W. C. Rounds, "The Intrinsically Exponential Complexity of the Circularity Problem for Attribute Grammars" *Communications of the ACM*, 18, 12 (1975) 697–706.
70. T. Jiang, E. Kinber, A. Salomaa, K. Salomaa, S. Yu, "Pattern languages with and without erasing", *International Journal of Computer Mathematics*, 50 (1994) 147–163.
71. T. Jiang, A. Salomaa, K. Salomaa, S. Yu, "Inclusion is Undecidable for Patterns", in *ICALP 93 Proceedings, Lecture Notes in Computer Science 700*, Springer-Verlag, Berlin, 1993, 301–312.
72. T. Jiang, A. Salomaa, K. Salomaa, S. Yu, "Decision Problems for Patterns", *Journal of Computer and System Sciences*, 50 (1995) 53–63.
73. N. D. Jones, "A survey of formal language theory" *Technical Report 3*, University of Western Ontario, Computer Science Department (1966).
74. N. D. Jones, "Circularity testing of attribute grammars requires exponential time: a simpler proof" *Report DAIMI-PB-107*, Aarhus University (1980).

75. L. Kari, A. Mateescu, G. Păun, A. Salomaa, "Multi-pattern languages", *Theoretical Computer Science* 141 (1995) 253–268.

76. R. M. Karp, "Reducibility among combinatorial problems", *Complexity of Computer Computations*, Plenum Press, New York, (1972) 85–104.

77. V. Keränen, "Abelian squares can be avoided on four letters" in *ICALP 92 Proceedings, Lecture Notes in Computer Science 623*, Springer-Verlag, Berlin, 1992, 41–52.

78. D. Knuth, "Semantics of context-free languages", *Mathematical Systems Theory*, 2 (1968) 127–145.

79. D. Knuth, "Semantics of context-free languages: correction", *Mathematical Systems Theory*, 5 (1971) 95–96.

80. C. H. A. Koster, "Two-level grammars in compiler constructions, an advanced course" *Lecture Notes in Computer Science 21*, Springer-Verlag, Berlin, 1974, 146–156.

81. W. Kuich, A. Salomaa, *Semirings, Automata, Languages*, Springer-Verlag, Berlin, 1986.

82. I. Kupka, "Van Wijngaarden grammars as a special information processing model", *Lecture Notes in Computer Science 88*, 1980, 367–401.

83. S. Y. Kuroda, "Classes of languages and linear bounded automata", *Information and Control*, 7 (1964) 207–223.

84. L. Y. Liu, P. Weiner, "An infinite hierarchy of intersections of context-free languages", *Mathematical Systems Theory*, 7 (1973) 187–192.

85. M. Latta, R. Wall, "Intersective Context-Free Languages", *9th Congress on Natural and Formal Languages, Proceedings*, C. Martin-Vide (ed), Tarragona, Spain, (1993).

86. M. Latteux, P. Turakainen, "On characterizations of recursively enumerable languages", *Acta Informatica* 28 (1990) 179–186.

87. J. van Leeuwen, "Recursively enumerable languages and van Wijngaarden grammars", *Indagationes Mathematicae*, 39 (1977) 29–39.

88. M. Levison, G. Lessard, "Application of Attribute Grammars to Natural Language Sentence Generation", *in Lecture Notes in Computer Science 461*, Springer-Verlag, Berlin, 1990, 298–312.

89. P. M. Lewis, D. J. Rosenkrantz, R. E. Stearns, "Attributed Translations", *Journal of Computer and System Sciences*, 9 (1974) 279–307.

90. G. S. Makanin, "The problem of solvability of equations in a free semigroup", *Matematiceskij Sbornik*, 103, 145 (1977) 148–236.

91. A. Mateescu, "Van Wijngaarden Grammars and Systems", *Ann. University of Bucharest*, 2 (1988) 75–81.

92. A. Mateescu, "Van Wijngaarden grammars and the generative complexity of recursively enumerable languages", *Ann. University of Bucharest*, 2 (1989) 49–54.

93. A. Mateescu, A. Salomaa, "Nondeterminism in patterns", *Proc. of STACS'94, Lecture Notes in Computer Science 775*, Springer-Verlag, Berlin, 1994, 661–668.

94. A. Mateescu, A. Salomaa, "Finite Degrees of Ambiguity in Pattern Languages", *RAIRO Informatique théorique et Applications*, 28, 3–4 (1994) 233–253.

95. A. Mateescu, D. Vaida, *Discrete Mathematical Structures, Applications*, Publ. House of Romanian Academy, Bucharest, (1989).

96. G. Matthews, "A note on symmetry in phrase structure grammars", *Information and Control*, 7 (1964) 360–365.

97. G. Matthews, "Two-way languages", *Information and Control*, 10 (1967) 111–119.

98. H. Maurer, A. Salomaa, D. Wood, "Pure Grammars", *Information and Control*, 44 (1980) 47–72.

99. M. L. Minsky, "Recursive unsolvability of Post's problem of "Tag" and other topics in theory of Turing machines", *Ann. of Math.*, 74 (1961) 437–455.

100. V. Mitrana, G. Păun, G. Rozenberg, A. Salomaa, "Pattern Systems", *to appear in Theoretical Computer Science*.

101. E. Moriya, "Associate Languages and Derivational Complexity of Formal Grammars and Languages", *Information and Control*, 22 (1973) 139–162.

102. M. Nivat, "Transduction des langages de Chomsky", *Ann. Inst. Fourier Grenoble*, 18 (1968) 339–455.

103. E. Ohlebusch, E. Ukkonen, "On the equivalence problem for E-pattern languages", *Proc. of MFCS'96, Lecture Notes in Computer Science 1113*, Spribger-Verlag, Berlin, 457–468.

104. G. Păun, "Regular Extended H Systems are Computationally Universal", *to appear in Journal Inform. Proc. Cybern., EIK*.

105. G. Păun, G. Rozenberg, A. Salomaa, "Pattern Grammars", *to appear in Journal Inform. Proc. Cybern., EIK*.

106. M. Penttonen, "One-Sided and Two-Sided Context in Formal Grammars", *Information and Control*, 25 (1974) 371–392.

107. J. Pitt, R. J. Cunningham, "Attributed Translation and the Semantics of Natural Language", *in Lecture Notes in Computer Science 461*, Springer-Verlag, Berlin, 1990, 284–297.

108. E. L. Post, "Finite combinatory processes-formulation I", *Journal of Symbolic Logic*, 1 (1936) 103–105.

109. G. E. Révész, *Introduction to Formal Languages*, McGraw-Hill Book Comp., New York, 1980.

110. H. G. Rice, "Classes of recursively enumerable sets and their decision problems", *Transactions of the AMS*, 89 (1953) 25–59.

111. H. G. Rice, "On completely recursively enumerable classes and their key arrays", *Journal of Symbolic Logic*, 21 (1956) 304–341.

112. G. Rozenberg, A. Salomaa, *Cornerstones of Undecidability*, Prentice Hall, New York, 1994.

113. A. Salomaa, *Theory of Automata*, Pergamon Press, 1969.

114. A. Salomaa, "On sentential forms of context-free grammars", *Acta Informatica*, 2 (1973) 40–49.

115. A. Salomaa, *Formal Languages*, Academic Press, New York 1973.

116. A. Salomaa, *Computation and Automata*, Cambridge Univ. Press, Cambridge, 1985.

117. A. Salomaa, "The Formal Languages Column" *Bulletin of EATCS*, 33 (1987) 42–53.

118. A. Salomaa, "Pattern Languages: Problems of Decidability and Generation" *Lecture Notes in Computer Science 710*, Springer-Verlag, Berlin, 1993, 121–132.

119. A. Salomaa, "Patterns" *Bulletin of EATCS*, 54 (1994) 194–206.

120. A. Salomaa, "Patterns and Pattern Languages", in *Salodays in Auckland, Proceedings*, Auckland University, (1994) 8–12.

121. A. Salomaa, "Return to Patterns" *Bulletin of EATCS*, 55 (1995) 144–157.

122. A. Salomaa, M. Soittola, *Automata-Theoretic Aspects of Formal Power Series*, Springer-Verlag, Berlin, 1978.

123. W. J. Savitch, "Relationships between nondeterministic and deterministic tape complexities", *Journal of Computer and System Sciences*, 4, 2 (1970) 177–192.

124. W. J. Savitch, "How to make arbitrary grammars look like context-free grammars", *SIAM Journal of Computing*, 2, 3 (1973) 174–182.

125. M. Sintzoff, "Existence of a van Wijngaarden syntax for every recursively enumerable set", *Annales Soc. Sci., Bruxelles*, 81 (1967) 115–118.
126. R. Szelepcsényi, "The Method of Forced Enumeration for Nondeterministic Automata", *Acta Informatica*, 26 (1988) 279–284.
127. M. Tătărîm, "Ackermann-Peter's function has primitive recursive graph and range", *Found. Control Engineering*, 9, 4 (1984) 177–180.
128. A. Thue, "Über unendliche Zeichenreihen", *Norske Vid. Selsk. Skr., I Mat. Nat. Kl.*, Kristiania, 7, (1906) 1–22.
129. A. Thue, "Über die gegenseitige Lage gleicher Teile gewisser Zeichenreihen", *Norske Vid. Selsk. Skr., I Mat. Nat. Kl.*, Kristiania, 1, (1912) 1–67.
130. A. Thue, "Probleme über Veränderungen von Zeichenreihen nach gegebenen Regeln.", *Skrifter utgit av Videnskapsselskapet i Kristiania I*, 10, (1914) 34 pp.
131. P. Turakainen, "On characterization of recursively enumerable languages in terms of linear languages and VW-grammars", *Indagationes Mathematicae*, 40 (1978) 145–153.
132. P. Turakainen, "A Unified Approach to Characterizations of Recursively Enumerable Languages", *Bulletin of the EATCS*, 45 (1991) 223–228.
133. A. M. Turing, "On computable numbers with an application to the Entscheidungsproblem", *Proc. London Math. Soc.*, 2, 42, (1936) 230–265. A correction, *ibid.*, 43, 544–546.
134. D. A. Watt, "The parsing problem for affix grammars", *Acta Informatica*, 8 (1977) 1–20.
135. L. M. Wegner, "On parsing two-level grammars", *Acta Informatica*, 14 (1980) 175–193.
136. A. van Wijngaarden, "Orthogonal design and description of formal languages", Mathem. Centrum Amsterdam, 1965.
137. A. van Wijngaarden, "The generative power of two level grammars", *Lecture Notes in Computer Science, 14*, Springer-Verlag, Berlin, 1974, 9–14.
138. A. van Wijngaarden(ed.), *Revised report on the algorithmic language ALGOL 68*, Springer-Verlag, Berlin, 1976.

L Systems

Lila Kari, Grzegorz Rozenberg, and Arto Salomaa

1. Introduction

1.1 Parallel rewriting

L systems are *parallel* rewriting systems which were originally introduced
in 1968 to model the development of multicellular organisms [L1]. The basic ideas gave rise to an abundance of language-theoretic problems, both
mathematically challenging and interesting from the point of view of diverse
applications. After an exceptionally vigorous initial research period (roughly
up to 1975; in the book [RSed2], published in 1985, the period up to 1975
is referred to as "when L was young" [RS2]), some of the resulting language
families, notably the families of D0L, 0L, DT0L, E0L and ET0L languages,
had emerged as fundamental ones in the *parallel or L hierarchy*. Indeed, nowadays the fundamental L families constitute a similar testing ground as the
Chomsky hierarchy when new devices (grammars, automata, etc.) and new
phenomena are investigated in language theory.

L systems were introduced by Aristid Lindenmayer in 1968 [L1]. The original purpose was to model the development of simple filamentous organisms.
The development happens in parallel everywhere in the organism. Therefore,
parallelism is a built-in characteristic of L-systems. This means, from the
point of view of rewriting, that everything has to be rewritten at each step of
the rewriting process. This is to be contrasted to the "sequential" rewriting
of phrase structure grammars: only a specific part of the word under scan
is rewritten at each step. Of course, the effect of parallelism can be reached
by several sequential steps in succession. However, the *synchronizing control mechanism* is missing from the customary sequential rewriting devices.
Therefore, parallelism cannot be truly simulated by sequential rewriting.

Assume that your only rule is $a \longrightarrow a^2$ and you start with the word a^3.
What do you get? If rewriting is sequential, you can replace one a at a time
by a^2, obtaining eventually all words words a^i, $i \geq 3$. If rewriting is parallel,
the word a^6 results in one step. It is not possible to obtain a^4 or a^5 from a^3.
Altogether you get only the words $a^{3 \cdot 2^i}$, $i \geq 0$. Clearly, the language

$$\{a^{3 \cdot 2^i} \mid i \geq 0\}$$

is not obtainable by sequential context-free or interactionless rewriting: letters are rewritten independently of their neighbours. Observe also how the

dummy rule $a \longrightarrow a$ behaves dramatically differently in parallel and sequential rewriting. In the latter it has no influence and can be omitted without losing anything. In parallel rewriting the rule $a \longrightarrow a$ makes the simulation of sequential rewriting possible: the "real" rule $a \longrightarrow a^2$ is applied to one occurrence of a, and the "dummy" rule $a \longrightarrow a$ to the remaining occurrences. Consequently, all words a^i, $i \geq 3$, are obtained by parallel rewriting from a^3 if both of the rules $a \longrightarrow a^2$ and $a \longrightarrow a$ are available.

The present survey on L systems can by no means be encyclopedic; we do not even claim that we can exhaust all the main trends. Of the huge bibliography concerning L systems we reference only items needed to illustrate or augment an issue in our presentation. We are fully aware that also some rather influential papers have not been referenced. In the early years of L systems it was customary to emphasize that the exponential function 2^n described the yearly growth in the number of papers in the area. [MRS], published in 1981, was the latest edition of a bibliography on L systems intended to be comprehensive. After that nobody has undertaken the task of compiling a comprehensive bibliography which by now would contain at least 5 000 items.

On the other hand, L systems will be discussed also elsewhere in this Handbook. They form the basic subject matter in the chapter by P. Prusinkiewicz in Volume III but are also covered, for instance, in chapter 9 by W. Kuich and chapter 12 by G. Păun and A. Salomaa in this volume.

When we write "L systems" or "ET0L languages" without a hyphen, it is not intentionally vicious orthography. We only follow the practice developed during the years among the researchers in this field. The main reason for this practice was that in complicated contexts the hyphen was often misleading. We want to emphasize in this connection the notational uniformity followed by the researchers of L systems, quite exceptional in comparison with most areas of mathematical research. In particular, the different letters have a well-defined meaning in the names of the L language families. For instance, it is immediately understood by everyone in the field what HPDF0L languages are. We will return to the terminology in Section 2. below.

1.2 Callithamnion roseum, a primordial alga

We begin with a mathematical model of the development of a red alga, *Callithamnion roseum*. The attribute "primordial" can be associated to it because the model for its development appears in the original paper [L1] by Aristid Lindenmayer. The mathematical model is a PD0L system, according to the terminology developed later. Here 0 indicates that the interaction between individual cells in the development is *zero-sided*, that is, there is no interaction; rewriting is context-free. The letter D indicates *determinism*: there is only one possibility for each cell, that is, there is only one rule for each letter. Finally, the letter P stands for *propagating*: there is no cell death, no rule $a \longrightarrow \lambda$ indicating that a letter should be rewritten as the empty word. We will return to this terminology in Section 2. below.

Following the general plan of this Handbook, our present chapter deals exclusively with words, that is, with one-dimensional L systems. Two-dimensional systems (map L systems, graph L systems, etc.) were introduced quite early, see [RS1]. An interpretation mechanism is needed for one-dimensional systems: how to interpret words as pictures depicting stages of development? In the case of *filamentous organisms* this normally happens using *branching structures*. In the model below the matching brackets [,] indicate branches, drawn alternately on both sides of the stem.

We are now ready to present the mathematical model for Callithamnion roseum. The alphabet is $\Sigma = \{1, 2, 3, 4, 5, 6, 7, 8, 9, [,]\}$ and the rules for the letters: $1 \longrightarrow 23$, $2 \longrightarrow 2$, $3 \longrightarrow 24$, $4 \longrightarrow 25$, $5 \longrightarrow 65$, $6 \longrightarrow 7$, $7 \longrightarrow 8$, $8 \longrightarrow 9[3]$, $9 \longrightarrow 9$. Beginning with the word $w_0 = 1$, we get the following *developmental sequence*:

$$
\begin{aligned}
w_0 &= \quad 1 \\
w_1 &= \quad 23 \\
w_2 &= \quad 224 \\
w_3 &= \quad 2225 \\
w_4 &= \quad 22265 \\
w_5 &= \quad 222765 \\
w_6 &= \quad 2228765 \\
w_7 &= \quad 2229[3]8765 \\
w_8 &= \quad 2229[24]9[3]8765 \\
w_9 &= \quad 2229[225]9[24]9[3]8765 \\
w_{10} &= \quad 2229[2265]9[225]9[24]9[3]8765 \\
w_{11} &= \quad 2229[22765]9[2265]9[225]9[24]9[3]8765 \\
w_{12} &= \quad 2229[228765]9[22765]9[2265]9[225]9[24]9[3]8765 \\
w_{13} &= \quad 2229[229[3]8765]9[228765]9[22765]9[2265]9[225]9[24]9[3]8765
\end{aligned}
$$

Selected developmental stages $(w_0, w_6, w_7, \ldots, w_{15})$ are shown in the following picture [PK].

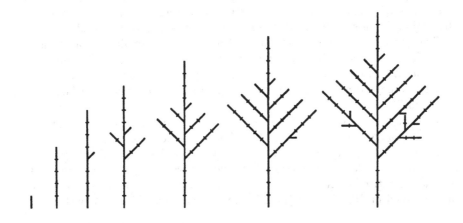

1.3 Life, real and artificial

We conclude this Introduction with some observations that, in our estimation, are rather important in predicting the future developments of L systems. L systems were originally introduced to model the development of multicellular organisms, that is, the development of some form of "real" life. However, there have been by now numerous applications in computer graphics, where L systems have been used to depict *imaginary* life forms, imaginary *gardens of L* and also non-living specimens ranging from toys and birthday cakes to real-estate ads (see, for instance, [PL]). Their utmost simplicity and flexibility to small changes tailor-made according to individual wishes make L systems very suitable to model phenomena of *artificial life*.

Artificial life is customarily understood as the study of man-made constructs that exhibit, in some sense or in some aspects, the behavior of existing living organisms. Artificial life extends the traditional biology that is concerned with the carbon-chain-type of life evolved on Earth. Artificial life tries to synthesize life-like behavior within computers and other man-made devices. It is also more extensive than *robotics*. Robots are constructed to do some specific tasks, whereas the "creatures" of artificial life are only observed. As often explained, artificial life paints the picture of *life-as-it-could-be*, contrasted to the picture of traditional biology about *life-as-we-know-it*.

It is very difficult to draw a strict border between living and nonliving, animate and inanimate. No definition of "life", satisfactory in a mathematical or common sense, has so far been given. Perhaps it is better to view the set of living beings as a fuzzy set rather than to try to define it in crisp mathematical terms. Another possibility is to give lists of properties typical for living beings as contrasted to inanimate objects. However, so far none of such lists seems satisfactory. Also many individual properties, such as growth, give rise to doubts. Although growth is typical for living beings, it can be observed elsewhere.

However, one feature very characteristic for the architecture of all living beings is that life is *fundamentally parallel*. A living system may consist of millions of parts, all having their own characteristic behavior. However, although a living system is highly distributed, it is massively parallel.

Thus, any model for artificial life must be capable of simulating parallelism – no other approach is likely to prove viable. Among all grammatical models, L systems are by their very essence the most suitable for modeling parallelism. L systems may turn out to be even more suitable for modeling artificial life than real life.

Indeed, the utmost simplicity of the basic components and the ease of affecting changes tailor-made for clarifying a specific issue render L systems ideal for modeling artificial life. A good example is the so-called French Flag Problem: does polarized behavior at the global level imply polarization (that is, nonsymmetric behavior) at the local level? The idea behind this problem (as described in [H2] and [HR]) comes from a worm with three parts (like the

French Flag): head, middle, tail. If one of the ends is cut, the worm grows again the missing part, head for head and tail for tail. The behavior of the uncut part is polarized – the remaining organism knows which end it assists to grow. However, such a global behavior can be reached by a fully symmetric local behavior. It can be modeled by an L system, where the rules for the individual cells are fully symmetric – there is no distinction between right and left [HR]. While such a construction does not in any way prove that the real-life phenomenon is locally symmetric – the cells of the worm used in experiments can very well be polarized – it certainly constitutes an important fact of artificial life. We can have living species with polarized global behavior but with fully symmetric behavior at the level of individual cells – and possibly having some other features we are interested in implementing.

Other grammatical models, or grammar-like models such as cellular automata, seem to lack the versatility and flexibility of L systems. It is not easy to affect growth of the interior parts using cellular automata, whereas L-filaments grow naturally everywhere. If you want to make a specific alteration in the species you are interested in, then you very often find a suitable L system to model the situation. On the other hand, it seems that still quite ·much post-editing is needed in graphical real-life modeling. The variations of L systems considered so far seem to be too simple to capture some important features of real-life phenomena. We now proceed to present the most common of these variations, beginning with the basic ones.

2. The world of L, an overview

2.1 Iterated morphisms and finite substitutions: D0L and 0L

We will now present the fundamentals of L systems and their basic properties. Only very few notions of language theory are needed for this purpose. We follow Section 2.1 of Chapter 1 in this Handbook as regards the core terminology about letters, words and languages. The most important language-theoretic notion needed below will be a *finite substitution* over an alphabet Σ, as well as its special cases.

Definition 2.1. *A* finite substitution σ *over an alphabet Σ is a mapping of Σ^* into the set of all finite nonempty languages (possibly over an alphabet Δ different from Σ) defined as follows. For each letter $a \in \Sigma$, $\sigma(a)$ is a finite nonempty language, $\sigma(\lambda) = \lambda$ and, for all words $w_1, w_2 \in \Sigma^*$,*

$$\sigma(w_1 w_2) = \sigma(w_1)\sigma(w_2).$$

If none of the languages $\sigma(a)$; $a \in \Sigma$, contains the empty word, the substitution σ is referred to as λ-free or nonerasing. *If each $\sigma(a)$ consists of a single word, σ is called a* morphism. *We speak also of* nonerasing *and* letter-to-letter morphisms. □

Some clarifying remarks are in order. Morphisms were earlier called *homomorphisms* – this fact is still reflected by the notations h and H used in connection with morphisms. Usually in our considerations the target alphabet Δ equals the basic alphabet Σ – this will be the case in the definition of D0L and 0L systems. Then we speak briefly of a finite substitution or morphism *on the alphabet Σ*. In the theory of L systems letter-to-letter morphisms are customarily called *codings*; *weak codings* are morphisms mapping each letter either to a letter or to the empty word λ. Codings in this sense should not be confused with codes discussed elsewhere in this Handbook, also in Section 6. below.

By the above definition a substitution σ is applied to a word w by *rewriting* every letter a of w as some word from $\sigma(a)$. Different occurrences of a may be rewritten as different words of $\sigma(a)$. However, if σ is a morphism then each $\sigma(a)$ consists of one word only, which makes the rewriting *deterministic*. It is convenient to specify finite substitutions by listing the *rewriting rules* or *productions* for each letter, for instance,

$$a \longrightarrow \lambda, \quad a \longrightarrow a^2, \quad a \longrightarrow a^7,$$

this writing being equivalent to

$$\sigma(a) = \{\lambda, a^2, a^7\}.$$

Now, for instance,

$$\sigma(a^2) = \sigma(a)\sigma(a) = \{\lambda, a^2, a^4, a^7, a^9, a^{14}\}.$$

Similarly, the substitution (in fact, a morphism) defined by

$$\sigma_1(a) = \{b\}, \quad \sigma_1(b) = \{ab\}, \quad \Sigma = \{a, b\},$$

can be specified by listing the rules

$$a \longrightarrow b, \quad b \longrightarrow ab.$$

In this case, for any word w, $\sigma_1(w)$ consists of a single word, for instance,

$$\sigma_1(a^2ba) = \{b^2ab^2\} = b^2ab^2,$$

where the latter equality indicates only that we often identify singleton sets with their elements.

The above results for $\sigma(a^2)$ and $\sigma_1(a^2ba)$ are obtained by applying the rules *in parallel*: every occurrence of every letter must be rewritten. Substitutions σ (and, hence, also morphisms) are in themselves *parallel* operations. An application of σ to a word w means that something happens everywhere in w. No part of w can remain idle except that, in the presence of the rule $a \longrightarrow a$, occurences of a may remain unchanged.

Before our next fundamental definition, we still want to extend applications of substitutions (and, hence, also morphisms) to concern also languages.

This is done in the natural "additive" fashion. By definition, for all languages $L \subseteq \Sigma^*$,

$$\sigma(L) = \{u|\ u \in \sigma(w), \text{ for some } w \in L\}.$$

Definition 2.2. *A 0L system is a triple* $G = (\Sigma, \sigma, w_0)$, *where* Σ *is an alphabet,* σ *is a finite substitution on* Σ, *and* w_0 *(referred to as the* axiom*) is a word over* Σ. *The 0L system is* propagating *or a* P0L *system if* σ *is nonerasing. The 0L system* G *generates the language*

$$L(G) = \{w_0\} \cup \sigma(w_0) \cup \sigma(\sigma(w_0)) \cup \ldots = \bigcup_{i \geq 0} \sigma^i(w_0). \qquad \square$$

Consider, for instance, the 0L system

$$\text{MISS}_3 = (\{a\}, \sigma, a) \text{ with } \sigma(a) = \{\lambda, a^2, a^5\}.$$

(Here and often in the sequel we express in the name of the system some characteristic property, rather than using abruptly an impersonal $G-$ notation.) The system can be defined by simply listing the productions

$$a \longrightarrow \lambda, \quad a \longrightarrow a^2, \quad a \longrightarrow a^5$$

and telling that a is the axiom. Indeed, both the alphabet Σ and the substitution σ can be read from the productions. (We disregard the case, where Σ contains letters not appearing in the productions.) L systems are often in the sequel defined in this way, by listing the productions.

Going back to the system MISS$_3$, we obtain from the axiom a in one "derivation step" each of the words λ, a^2, a^5. Using customary language-theoretic notation, we denote this fact by

$$a \Longrightarrow \lambda, \quad a \Longrightarrow a^2, \quad a \Longrightarrow a^5.$$

A second derivation step gives nothing new from λ but gives the new words a^4, a^7, a^{10} from a^2 and the additional new words $a^6, a^8, a^9, a^{11}, a^{12}, a^{13}, a^{14}, a^{15}, a^{16}, a^{17}, a^{19}, a^{20}, a^{22}, a^{25}$ from a^5. In fact,

$$a^5 \Longrightarrow a^k \text{ iff } k = 2i + 5j, \quad i + j \leq 5, \quad i \geq 0, \quad j \geq 0.$$

(Thus, we use the notation $w \Longrightarrow u$ to mean that $u \in \sigma(w)$.) A third derivation step produces, in fact in many different ways, the missing words $a^{18}, a^{21}, a^{23}, a^{24}$. Indeed, it is straightforward to show by induction that

$$L(\text{MISS}_3) = \{a^i|\ i \neq 3\}.$$

Let us go back to the terminology used in defining L systems. The letter L comes from the name of Aristid Lindenmayer. The story goes that Aristid was so modest himself that he said that it comes from "languages". The number 0 in "0L system" indicates that interaction between individual cells in the development is zero-sided, the development is *without interaction*. In language-theoretic terms this means that rewriting is *context-free*. The system MISS$_3$ is not *propagating*, it is not P0L. However, it is a *unary* 0L system, abbreviated U0L system: the alphabet consists of one letter only. The following notion will be central in our discussions.

Definition 2.3. *A 0L system* (Σ, σ, w_0) *is* deterministic *or a* D0L *system iff* σ *is a morphism.* □

Thus, if we define a D0L system by listing the productions, there is exactly one production for each letter. This means that rewriting is completely deterministic. We use the term *propagating*, or a *PD0L system*, also here: the morphism is nonerasing. The L system used in Section 1.2 for modeling Callithamnion roseum was, in fact, a PD0L system.

D0L systems are the simplest among L systems. Although most simple, D0L systems give a clear insight into the basic ideas and techniques behind L systems and parallel rewriting in general. The first L systems used as models in developmental biology, as well as most of the later ones, were in fact D0L systems. From the point of view of artificial life, creatures modeled by D0L systems have been called "Proletarians of Artificial Life", briefly PALs [S7]. In spite of the utmost simplicity of the basic definition, the theory of D0L systems is by now very rich and diverse. Apart from providing tools for modeling real and artificial life, the theory has given rise to new deep insights into language theory (in general) and into the very basic mathematical notion of an endomorphism on a free monoid (in particular), [S5]. At present still a wealth of problems and mysteries remains concerning D0L systems.

Let $G = (\Sigma, h, w_0)$ be a D0L system – we use the notation h to indicate that we are dealing with a (homo)morphism. The system G generates its language $L(G)$ in a specific order, as a *sequence*:

$$w_0, \quad w_1 = h(w_0), \quad w_2 = h(w_1) = h^2(w_0), \quad w_3, \ldots$$

We denote the sequence by $S(G)$. Thus, in connection with a D0L system G, we speak of its language $L(G)$ and sequence $S(G)$. Indeed, D0L systems were the first widely studied *grammatical devices generating sequences*. We now discuss five examples, paying special attention to sequences. All five D0L systems are propagating, that is, PD0L systems. The first one is also unary, that is, a UPD0L system.

Consider first the D0L system EXP$_2$ with the axiom a and rule $a \longrightarrow a^2$. It is immediate that the sequence $S(\text{EXP}_2)$ consists of the words a^{2^i}, $i \geq 0$, in the increasing length order. Secondly, consider the D0L system LIN with the axiom ab and rules $a \longrightarrow a$, $b \longrightarrow ab$. Now the sequence $S(\text{LIN})$ consists of the words $a^i b$, $i \geq 1$, again in increasing length order. The notation LIN refers to linear growth in word length: the j'th word in the sequence is of length $j + 2$. (Having in mind the notation w_0, w_1, w_2, we consider the axiom ab to be the 0th word etc.)

Our next D0L system FIB has the axiom a and rules $a \longrightarrow b$, $b \longrightarrow ab$. The first few words in the sequence $S(\text{FIB})$ are

$$a, b, ab, bab, abbab, bababbab.$$

From the word ab on, each word results by catenating the two preceding ones. Let us establish inductively this claim,

$$w_n = w_{n-2}w_{n-1}, \quad n \geq 2.$$

Denoting by h the morphism of the system FIB, we obtain using the definition of h:

$$w_n = h^n(a) = h^{n-1}(h(a)) = h^{n-1}(b) = h^{n-2}(h(b)) =$$
$$= h^{n-2}(ab) = h^{n-2}(a)h^{n-2}(b) = h^{n-2}(a)h^{n-2}(h(a)) =$$
$$= h^{n-2}(a)h^{n-1}(a) = w_{n-2}w_{n-1}.$$

The claim established shows that the word lengths satisfy the equation

$$|w_n| = |w_{n-2}| + |w_{n-1}|, \quad n \geq 2.$$

Hence, the length sequence is the well-known *Fibonacci sequence* 1, 1, 2, 3, 5, 8, 13, 21, 34, ...

The D0L system SQUARES has the axiom a and rules

$$a \longrightarrow abc^2, \quad b \longrightarrow bc^2, \quad c \longrightarrow c.$$

The sequence $S(\text{SQUARES})$ begins with the words

$$a, abc^2, abc^2bc^2c^2, abc^2bc^2c^2bc^2c^2c^2.$$

Denoting again by w_i, $i \geq 0$, the words in the sequence, it is easy to verify inductively that

$$|w_{i+1}| = |w_i| + 2i + 3, \text{ for all } i \geq 0.$$

This shows that the word lengths consist of the consecutive sequence: $|w_i| = (i+1)^2$, for all $i \geq 0$. Similar considerations show that the D0L system CUBES with the axiom a and productions

$$a \longrightarrow abd^6, \quad b \longrightarrow bcd^{11}, \quad c \longrightarrow cd^6, \quad d \longrightarrow d$$

satisfies $|w_i| = (i+1)^3$, for all $i \geq 0$.

Rewriting being deterministic gives rise to certain *periodicities* in the D0L sequence. Assume that some word occurs twice in a sequence: $w_i = w_{i+j}$, for some $i \geq 0$ and $j \geq 1$. Because of the determinism, the words following w_{i+j} coincide with those following w_i, in particular,

$$w_i = w_{i+j} = w_{i+2j} = w_{i+3j} = \cdots$$

Thus, after some "initial mess", the words start repeating periodically in the sequence. This means that the language of the system is finite. Conversely, if the language is finite, the sequence must have a repetition. This means that the occurrence of a repetition in $S(G)$ is a necessary and sufficient condition for $L(G)$ being finite. Some other basic periodicity results are given in the following theorem, for proofs see [HR], [Li2], [RS1]. The notation $\text{alph}(w)$ means the minimal alphabet containing all letters occurring in w, $\text{pref}_k(w)$ stands for the prefix of w of length k (or for w itself if $|w| < k$), $\text{suf}_k(w)$ being similarly defined.

Theorem 2.1. *Let w_i, $i \geq 0$, be the sequence of a D0L system $G = (\Sigma, h, w_0)$. Then the sets $\Sigma_i = \text{alph}(w_i)$, $i \geq 0$, form an ultimately periodic sequence, that is, there are numbers $p > 0$ and $q \geq 0$ such that $\Sigma_i = \Sigma_{i+p}$ holds for every $i \geq q$. Every letter occurring in some Σ_i occurs in some Σ_j with $j \leq \text{card}(\Sigma) - 1$. If $L(G)$ is infinite then there is a positive integer t such that, for every $k > 0$, there is an $n > 0$ such that, for all $i \geq n$ and $m \geq 0$,*

$$\text{pref}_{k-1}(w_i) = \text{pref}_{k-1}(w_{i+mt}) \text{ and } \text{suf}(w_i) = \text{suf}_{k-1}(w_{i+mt}). \qquad \square$$

Thus, both prefixes and suffixes of any chosen length form an ultimately periodic sequence. Moreover, the period is independent of the length chosen; only the initial mess depends on it.

Definition 2.4. *An infinite sequence of words w_i, $i \geq 0$, is* locally catenative *iff, for some positive integers k, $i_1, \ldots i_k$ and $q \geq \max(i_1, \ldots, i_k)$,*

$$w_n = w_{n-i_1} \ldots w_{n-i_k} \text{ whenever } n \geq q.$$

A D0L system G is locally catenative *iff the sequence $S(G)$ is locally catenative.* $\qquad \square$

Locally catenative D0L systems are very important both historically and because of their central role in the theory of D0L systems: their study has opened up new branches of the theory. A very typical example of a locally catenative D0L system is the system FIB discussed above.

Also the system EXP_2 is locally catenative: the sequence $S(\text{EXP}_2)$ satisfies

$$w_n = w_{n-1}w_{n-1} \quad \text{for all } n \geq 1.$$

A celebrated problem, still open, is to decide whether or not a given D0L system is locally catenative. No general algorithm is known, although the problem has been settled in some special cases, for instance, when an upper bound is known for the integers $i_1, \ldots i_k$, as well as recently for binary alphabets, [Ch].

An intriguing problem is the avoidability or unavoidability of *cell death*: to what extent are rules $a \longrightarrow \lambda$ necessary in modeling certain phenomena? What is the real difference between D0L and PD0L systems and between 0L and P0L systems? There are some straightforward observations. The word length can never decrease in a PDOL sequence, and a POL language cannot contain the empty word. We will see in the sequel, especially in connection with growth functions, that there are remarkable differences between D0L and PD0L systems. The theory of D0L systems is very rich and still in many respects poorly understood.

As an example of the necessity of cell death, consider the D0L system DEATH$_b$ with the axiom ab^2a and rules

$$a \longrightarrow ab^2a, \quad b \longrightarrow \lambda.$$

The sequence $S(\text{DEATH}_b)$ consists of all words $(ab^2a)^{2^i}$, $i \geq 0$, in the increasing order of word length. We claim that the language L consisting of these words cannot be generated by any PD0L system G. Indeed, ab^2a would have to be the axiom of such a G, and $ab^2a \Longrightarrow ab^2aab^2a$ would have to be the first derivation step. This follows because no length-decrease is possible in $S(G)$. The two occurrences of a in ab^2a must produce the same subword in ab^2aab^2a. This happens only if the rule for a is one of the following: $a \longrightarrow \lambda$, $a \longrightarrow ab^2a$, $a \longrightarrow a$. The first two alternatives lead to non-propagating systems. But also the last alternative is impossible because no rule for b makes the step $b^2 \Longrightarrow b^2aab^2$ possible. We conclude that it is not possible to generate the language $L(\text{DEATH}_b)$ using a PD0L system.

A slight change in DEATH$_b$ makes such a generation possible. Consider the D0L system G_1 with the axiom aba and rules $a \longrightarrow aba, b \longrightarrow \lambda$. Now the PD0L system G_2 with the axiom aba and rules $a \longrightarrow a, b \longrightarrow baab$ generates the same sequence, $S(G_1) = S(G_2)$.

We say that two D0L systems G_1 and G_2 are *sequence equivalent* iff $S(G_1) = S(G_2)$. They are *language equivalent*, briefly *equivalent*, iff $L(G_1) = L(G_2)$. Instead of D0L systems, these notions can be defined analogously for any other class of L systems – sequence equivalence of course only for systems generating sequences. The two systems G_1 and G_2 described in the preceding paragraph are both sequence and language equivalent. Clearly, sequence equivalence implies language equivalence but the reverse implication is not valid. Two D0L systems may be (language) equivalent without being sequence equivalent, they can generate the same language in a different order. A simple example is provided by the two systems

$$(\{a,b\}, \{a \longrightarrow b^2, b \longrightarrow a\}, b) \text{ and } (\{a,b\}, \{a \longrightarrow b, b \longrightarrow a^2\}, a).$$

Among the most intriguing mathematical problems about L systems is the *D0L equivalence problem*: construct an algorithm for deciding whether or not two given D0L systems are equivalent. Equivalence problem is a fundamental decision for any family of generative devices: decide whether or not two given devices in the family generate the same language. For D0L systems one can consider, in addition, the *sequence equivalence problem*: Is $S(G_1) = S(G_2)$, given D0L systems G_1 and G_2? The D0L equivalence problems were celebrated open problems for most of the 70's. They were often referred to as the most simply stated problems with an open decidability status. We will return to them and related material in Section 4.. It was known quite early, [N], that a solution to the D0L sequence equivalence problem yields a solution to the D0L language equivalence problem, and vice versa.

2.2 Auxiliary letters and other auxiliary modifications

A feature very characteristic for D0L systems and 0L systems is that you have to accept everything produced by the machinery. You have the axiom and the rules, and you want to model some phenomenon. You might want to exclude something that comes out of from the axiom by the rules because it is alien to the phenomenon, does not fit it. This is not possible, you have to include everything. There is no way of hiding unsuitable words. Your D0L or 0L models have no filtering mechanism.

It is customary in formal language theory to use various *filtering mechanisms*. Not all words obtained in derivations are taken to the language but the terminal language is somehow "squeezed" from the set of all derived words. The most typical among such filtering mechanisms, quite essential in grammars in the Chomsky hierarchy, is the use of *nonterminal letters*. An occurrence of a nonterminal in a word means that the word is not (maybe not yet) acceptable. The generated language contains only words without nonterminals. Thus, the language directly obtained is *filtered* by intersecting it with Σ_T^*, where Σ_T is the alphabet consisting of terminal letters. This gives the possibility of rejecting (at least some) unsuitable words.

The same mechanism, as well as many other filters, can be used also with L systems. However, a word of *warning* is in order. The original goal was to model the development of a species, be it real or artificial. Each word generated is supposed to represent some stage in the development. It would be rather unnatural to exclude some words and say that they do not represent proper stages of the development! This is exactly what filtering with nonterminals does.

However, this objection can be overruled because the following justification can be given. Some other squeezing mechanisms, notably *codings* (that is, letter-to-letter morphisms), can be justified from the point of view of developmental models: more careful experiments or observations can change the interpretation of individual cells, after which the cells are assigned new names. This amounts to applying a letter-to-letter morphism to the language. A rather amazing result concerning parallel rewriting, discussed below in more detail, is that coding is in important cases, [ER1], equivalent to the use of nonterminals in the sense that the same language is obtained, and the transition from one squeezing mechanism to the other is effective. By this result, also the use of nonterminals is well-motivated.

The letter E ("extended") in the name of an L system means that the use of nonterminals is allowed. Thus, an *E0L system* is a 0L system, where the alphabet is divided into two disjoint parts, *nonterminals* and *terminals*.

EOL and OL systems work in the same way but only words over the terminal alphabet are in the language of an EOL system. Thus, an EOL system G can be also viewed as a OL system, where a subalphabet Σ_T is specified and the language of the OL system is intersected with Σ_T^* to get the language of the EOL system.

The following EOL system SYNCHRO is very instructive. In our notation, *capital* letters are nonterminals and *small* letters terminals. The axiom of SYNCHRO is ABC, and the rules as follows:

$$
\begin{array}{llllll}
A \longrightarrow AA' & A \longrightarrow a & A' \longrightarrow A' & A' \longrightarrow a & a \longrightarrow F \\
B \longrightarrow BB' & B \longrightarrow b & B' \longrightarrow B' & B' \longrightarrow b & b \longrightarrow F \\
C \longrightarrow CC' & C \longrightarrow c & C' \longrightarrow C' & C' \longrightarrow c & c \longrightarrow F \\
F \longrightarrow F
\end{array}
$$

It is easy to verify that

$$L(\text{SYNCHRO}) = \{a^n b^n c^n \mid n \geq 1\}.$$

This follows because our EOL system is *synchronized* in the sense that all terminals must be reached simultaneously. Otherwise, the failure symbol F necessarily comes to the word and can never be eliminated. The language obtained is a classical example in language theory: a context-sensitive language that is not context-free.

Filtering mechanisms provide families of L languages with a feature very desirable both language-theoretically and mathematically: *closure* under various operations. Without filtering, the "pure" families, such as the families of OL and D0L languages, have very weak closure properties: most of the customary language-theoretic operations may produce languages outside the family, starting with languages in the family. For instance, $L = \{a, a^3\}$ is the union of two OL languages. However, L itself is not OL, as seen by a quick exhaustive search over the possible axioms and rules.

We refer the reader to the preceding chapter in this Handbook for a more detailed discussion concerning the following definition. In particular, the six operations listed are not arbitrary but exhaust the "rational" operations in language theory.

Definition 2.5. *A family $\mathcal{L}$ of languages is termed a full AFL ("abstract family of languages") iff $\mathcal{L}$ is closed under each of the following operations: union, catenation, Kleene star, morphism, inverse morphism, intersection with regular languages. The family $\mathcal{L}$ is termed and anti-AFL iff it is closed under none of the operations above.* □

Most of the following theorem can be established by exhibiting suitable examples. [RS1] should be consulted, especially as regards the nonclosure of EOL languages under inverse morphisms.

Theorem 2.2. *The family of 0L languages is an anti-AFL, and so is the family of D0L languages. The family of E0L languages is closed under all AFL operations except inverse morphism.*   □

Theorem 2.2 shows clearly the power of the E-mechanism in transforming a family of little structure (in the sense of weak closure properties) into a structurally strong family. The power varies from L family to another. The difference between D0L and ED0L languages is not so big. The periodicity result of Theorem 2.1 concerning alphabets holds also for ED0L sequences and, thus, the words that are filtered out occur periodically in the sequence, which is a considerable restriction.

We now mention other filtering mechanisms. The *H-mechanism* means taking a morphic image of the original language. Consider the case that the original language is a 0L language. Thus, let $G = (\Sigma, \sigma, w_0)$ be a 0L system and $h : \Sigma^* \longrightarrow \Delta^*$ a morphism. (It is possible that the target alphabet Δ equals Σ.) Then $h(L(G))$ is an *H0L language*. The *HD0L languages* are defined analogously.

The *N-mechanism* refers similarly to *nonerasing* morphisms. Thus, N0L languages are of the form $h(L(G))$ above, with the additional assumption that h is nonerasing. The *C-mechanisms* refers to *codings* and *W-mechanism* to *weak codings*.

A further variation of L systems consists of having a *finite set of axioms* – instead of only one axiom as we have had in our definitions so far. This variation is denoted by including the *letter F* in the name of the system. Thus, every finite language L is an F0L language: we just let L be the set of axioms in an F0L system, where the substitution is the identity.

When speaking of language families, we denote the family of E0L languages simply by E0L, and similarly with other families. Consider the family D0L. We have introduced five filtering mechanisms: E, H, N, C, W. This gives six possibilities – either some filtering or the pure family. For each of the six possibilities, we may still add one or both of the letters P and F, indicating that we are dealing with the propagating or finite-axiom variant. This gives altogether 24 families. The following remarkable theorem gives an exhaustive characterization of the mutual relashionship between these 24 families. That such a complicated hierarchy is completely understood is a rather rare situation in language theory. Many of the proofs are rather involved – we return to a tricky question in Section 3.. Most of the results are originally from [NRSS], see also [RS1]. In comparing the families, we follow the λ-convention: two languages are considered equal if they differ by the empty word only. Otherwise, propagating families would be automatically different from nonpropagating ones.

Theorem 2.3. *The following diagram characterizes mutual relations between deterministic L families. Arrows denote strict inclusion. Families not connected by a path are mutually incomparable.*

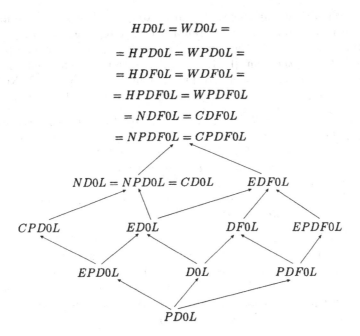

$$HD0L = WD0L =$$
$$= HPD0L = WPD0L =$$
$$= HDF0L = WDF0L =$$
$$= HPDF0L = WPDF0L$$
$$= NDF0L = CDF0L$$
$$= NPDF0L = CPDF0L$$

$$ND0L = NPD0L = CD0L \qquad EDF0L$$

$$CPD0L \qquad ED0L \qquad DF0L \qquad EPDF0L$$

$$EPD0L \qquad D0L \qquad PDF0L$$

$$PD0L$$

In the nondeterministic case there is much more collapse in the hierarchy because the C-mechanism has in most cases the same generative capacity as the E-mechanism. (In the deterministic case the former is much stronger.) A rather surprising fact in the nondeterministic case is that, although E0L = C0L, CP0L is properly contained in EP0L, which is the opposite what one would expect knowing the deterministic case. The key results are given in the next theorem, [NRSS] and [RS1].

Theorem 2.4. *Each of the following families equals E0L:*

$$E0L = \quad C0L = N0L = W0L = H0L = NP0L = EP0L = WP0L =$$
$$HP0L = EF0L = CF0L = NF0L = WF0L = HF0L =$$
$$EPF0L = NPF0L = WPF0L = HPF0L.$$

The family E0L lies strictly between context-free and context-sensitive languages and contains properly the mutually incomparable families CP0L and F0L.   □

Thus, the family E0L contains properly the family of context-free languages. This fact follows because a context-free grammar can be transformed into an E0L system, without affecting the language, by adding the production $x \longrightarrow x$ for each letter x. That the containment is proper can be seen by considering the language generated by the E0L system SYNCHRO. This fact should be contrasted with the fact that most finite languages are outside the family of 0L languages.

It is customary in language theory to try to reduce grammars into *normal forms*, that is, to show that every grammar can be replaced by an equivalent (generating the same language) grammar possessing some desirable properties. The following theorem, [RS1], is an illustration of such a reduction for L systems. Observe that the special property of the E0L system SYNCHRO concerning terminals is, in fact, a general property of E0L languages.

Theorem 2.5. *Every E0L language is generated by an E0L system satisfying each of the following conditions: (i) The only production for each terminal letter a is a $\longrightarrow F$, where F is a nonterminal having $F \longrightarrow F$ as the only production. (ii) The axiom is a single nonterminal not occurring on the right side of any production. (iii) The right side of every production is either a terminal word or consists only of nonterminals. (iv) A terminal word is reachable from every nonterminal apart from F (and the axiom if the language is empty).*   □

Usually it is difficult to show that a given language is *not* in a given family, because, in principle, one has to go through all the devices defining languages in the family. E0L languages possess certain combinatorial properties and, consequently, a language not having those properties cannot be an E0L language, [RS1]. For instance, the language

$$\{a^m b^n a^m \mid 1 \leq m \leq n\}$$

is not an E0L language. It is very instructive to notice that the languages

$$\{a^m b^n a^m \mid 1 \leq n \leq m\} \text{ and } \{a^m b^n a^m \mid m, n \geq 1\}$$

are E0L languages. Finally, the language

$$\{a^{3^n} \mid n \geq 1\} \cup \{b^n c^n d^n \mid n \geq 1\} .$$

is an EP0L language but not a CP0L language.

2.3 Tables, interactions, adults, fragmentation

A feature very characteristic for parallel rewriting is the use of *tables*, [R1], [R2]. A table is simply a set of rewriting rules. A system has several tables, always finitely many. It is essential that, at each step of the rewriting process, always rules from the same table must be used. This reflects the following state of affairs in modeling the development of organisms, real or artificial. There may be different conditions of the environment (day and night, varying heat, varying light, pollution, etc.) or different developmental stages, where it is important to use different rules. Then we consider all sets of rules, tables, obtained in this fashion. Observe that tables do not make sense in sequential rewriting. Because only one rule is used at each derivation step, it suffices to consider the total set of rules. We now define the variations of 0L and D0L systems, resulting by augmenting the system with tables.

Definition 2.6. *A* T0L *system is a triple* $G = (\Sigma, S, w_0)$, *where S is a finite set of finite substitutions such that, for each $\sigma \in S$, the triple (Σ, σ, w_0) is a 0L system. The language of the T0L system, $L(G)$, consists of w_0 and of all words in all languages $\sigma_1 \ldots \sigma_k(w_0)$, where $k \geq 1$ and each σ_i belongs to S – some of the σ_i's may also coincide. If all substitutions in S are, in fact, morphisms then G is* deterministic *or a* DT0L *system.* □

Thus, D indicates that all substitutions (all tables) are deterministic. However, according to the definition above, there is no control in the use of the tables – the tables may be used in an arbitrary order and multitude. Thus, a DT0L language is not generated in a sequence. A definite sequence results only if the order in the use of the tables is specified in a unique way.

The letter F has the same meaning as before: finitely many axioms instead of only one axiom. Also the filtering mechanisms E, H, C, N, W are the same as before – we will use them below without further explanations.

Following our earlier practice, we will define a T0L system by specifying the axiom and each table, a table being a set of productions included in brackets to indicate that they belong together. There are no restrictions, the same production may appear in several tables.

As an example consider the DT0L system PAL with the axiom a and two tables

$$T_d = [a \longrightarrow b, b \longrightarrow b^2, c \longrightarrow a] \text{ and } T_n = [a \longrightarrow c, b \longrightarrow ac, c \longrightarrow c].$$

(The indices d and n come from "day" and "night", the meaning will become clear below.) Instead of a linear sequence, derivations can be represented as a tree:

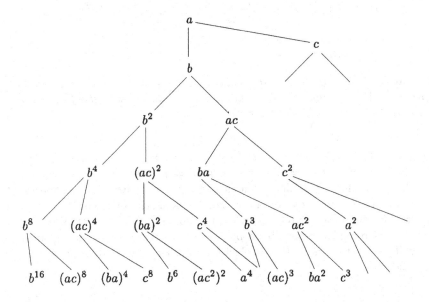

Here the branch indicates which of the tables was used: the left descendant results from an application of T_d, the right descendant from an application of T_n. If a descendant is not marked down (like a and c on the third level), it indicates that it occurrs already at an earlier level. The continuation can, therefore, be ignored if one is only interested in determining the language. The left extremity of the tree contains the powers b^{2^i}, $i \geq 0$. However, all powers b^i, $i \geq 1$, occur somewhere in the tree and so do all powers a^i and c^i, $i \geq 1$. This is seen as follows. Observe first that, for all $i \geq 0$, ba^{i+1} results from ba^i by applying first T_n, then T_d. Now from ba^i the word b^{i+2} results by applying T_d, the word c^{i+2} by applying T_n twice, and the word a^{i+2} from c^{i+2} by T_d.

Although seemingly simple, PAL has a rich structure. An interested reader should have no difficulties in specifying $L(\text{PAL})$ explicitly.) If the order of the application of the tables is given, a unique sequence of words results. One might visualize the two tables as giving rules for the day and night. The alternation is the natural order of application: $T_d T_n T_d T_n T_d \ldots$ (we agree that we begin with T_d). Another possibility is to consider "eternal daylight" (only T_d is used) or "eternal night". Let us still make free the choice of the axiom: instead of the axiom a, we have an arbitrary nonempty word w over $\{a, b, c\}$ as the axiom. (Sometimes the term L *scheme*, instead of an L system, is used to indicate that the axiom is not specified.) Denote by DAY-AND-NIGHT-PAL(w), DAY-PAL(w), NIGHT-PAL(w) the modifications of PAL thus obtained. Each of them generates a specific *sequence* of words. In fact, DAY-PAL(w) and NIGHT-PAL(w) are PD0L systems, whereas DAY-AND-NIGHT-PAL(w) can be viewed as a CPD0L system. For $w = abc$, the sequences look as follows.

$$\begin{aligned}
\text{DAY-NIGHT}: &\quad abc, b^3a, (ac)^3c, (ba)^3a, (ac^2)^3c, (ba^2)^3a, \\
&\quad (ac^3)^3c, (ba^3)^3a, (ac^4)^3c, (ba^4)^3a, \ldots \\
\text{DAY}: &\quad abc, b^3a, b^7, b^{14}, b^{28}, \ldots \\
\text{NIGHT}: &\quad abc, cac^2, c^4, c^4, \ldots
\end{aligned}$$

Definition 2.7. *For an infinite sequence w_i, $i \geq 0$, of words, the function $f(n) = |w_n|$ (mapping the set of nonnegative integers into itself) is termed the* growth function *of the sequence.* □

Thus, for NIGHT-PAL(abc), $f(0) = 3$, $f(n) = 4$ for $n \geq 1$. For DAY-PAL(abc), $f(0) = 3$, $f(1) = 4$, $f(i+2) = 2^{7 \cdot 2^i}$ for $i \geq 0$. Finally, for DAY-NIGHT-PAL(abc), $f(0) = 3$, $f(1) = 4$, $f(2i) = f(2i+1) = 3i+1$ for $i \geq 1$.

Thus, the three growth functions possess, respectively, the property of being bounded from above by a constant, exponential or linear. In fact, the growth function of any NIGHT-PAL(w) (resp. DAY-PAL(w), DAY-NIGHT-PAL(w)) is bounded from above by a constant (resp. exponential, linear). We will return to this matter in Section 5., where growth functions will be discussed.

The following two theorems summarize results concerning mutual relashionships between "table families", [NRSS], [RS1]. The first of the theorems deals with deterministic families, and the second with nondeterministic families.

Theorem 2.6. *The following inclusions and equalities hold:*
DT0L $\subset$ CDT0L= NDT0L= EDT0L= WDT0L= HDT0L,

PDT0L $\subset$ CPDT0L $\subseteq$ NPDT0L $\subseteq$ EPDT0L= WPDT0L= HPDT0L,

DTF0L $\subset$ CDTF0L= NDTF0L= EDTF0L= WDTF0L= HDTF0L,

PDTF0L $\subset$ CPDTF0L $\subseteq$ NPDTF0L $\subseteq$ EPDTF0L= WDPTF0L= HPDTF0L.

The "pure" families (without any filtering) satisfy the following inclusion diagram:

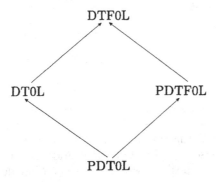

$\square$

Theorem 2.7. *Each of the following families equals ET0L:*

ET0L=	CT0L=	NT0L=	WT0L=	HT0L=	
=	NPT0L=	EPT0L=	WPT0L=	HPT0L=	
=	CTF0L=	NTF0L=	ETF0L=	WTF0L=	HTF0L
=	NPTF0L=	EPTF0L=	WPTF0L=	HPTF0L	

 The families EOL, TF0L and CPT0L= CPTF0L are all stricly included in ET0L. $\square$

The family ET0L is the largest widely studied L family, where rewriting is context-free (no cell interactions are present). It is also very pleasing mathematically and has strong closure properties. (It is, however, not closed under the shuffle operation.) It was observed already in the early 70's that ET0L is contained in the family of indexed languages (see the preceding chapter of this Handbook for a description of indexed languages and see [En] for more general hierarchies) and, consequently, facts concerning indexed languages hold also for ET0L languages. The facts in the following theorem can be established rather easily.

Theorem 2.8. *The family ET0L is a full AFL, whereas the family T0L is an anti-AFL. Every ET0L language is generated by an ET0L system with two tables. Every ET0L language is generated by an ET0L system such that $a \longrightarrow F$ is the only rule in every table for every terminal a, and $F \longrightarrow F$ is the only rule in every table for the nonterminal F.*   □

The two normal form results stated in Theorem 2.8, the two-table condition and synchronization, cannot always be reached simultaneously. A number of deep combinatorial results [RS1] can be established for EDT0L and ET0L languages. Using such results, relatively simple examples can be given of languages *not* in the families. Let Σ contain at least two letters and $k \geq 2$ be a fixed integer. Then neither of the languages

$$\{w \in \Sigma^* | \ |w| = k^n, n \geq 0\} \text{ and } \{w \in \Sigma^* | \ |w| = n^k, n \geq 0\}$$

is in EDT0L. The 0L system with the axiom a and rules

$$a \longrightarrow ab, b \longrightarrow bc, b \longrightarrow bd, c \longrightarrow c, d \longrightarrow d$$

generates a language not in EDT0L. None of the languages

$$\{(ab^n)^m | \ 1 \leq m \leq n\}, \quad \{(ab^n)^m | \ 1 \leq n \leq m\},$$

$$\{(ab^n)^m | \ 1 \leq m = n\}, \quad \{w \in \{a, b\}^* | \ |w|_b = 2^{|w|_a}\}$$

is in ET0L. Here $|w|_x$ denotes the number of occurrences of the letter x in w.

In the remainder of this subsection 2.3 we survey briefly some areas in the theory of L systems that are important both historically and language-theoretically in the sense that they have built bridges between sequential and parallel rewriting. Our exposition will be informal. More details can be found in [RS1], [HWa], [RRS].

In all L systems discussed so far the individual cells develop without any interaction with their neighbours. In language-theoretic terms, rewriting has been context-free. We now discuss *L systems with interactions, IL systems.* First some terminology.

In an $(m, n)L$ *system*, $m, n \geq 0$, the rules look like

$$(\alpha, a, \beta) \longrightarrow w, \quad |\alpha| = m, |\beta| = n.$$

This means that, between the words α and β, the letter a can be rewritten as the word w. Also now, *parallelism* applies: all letters must be rewritten simultaneously. In order to get sufficiently many letters on both sides of any given letter, the whole rewriting takes place between "environment symbols" #, m of them to the left and n to the right. The rules have to be provided also for the case that some prefix of α or suffix of β consists of #'s.

The *IL system* is a collective name for all $(m, n)L$ systems. $(1, 0)L$ and $(0, 1)L$ systems are referred to as *1L systems*: cell interaction is 1-sided, the rewriting of a letter depends always on its left neighbour only (or always on its right neighbour).

The use of letters such as D, E, P is the same as before. In particular, *determinism* means that, for each configuration consisting of a letter and an (m, n)-neighbourhood, there is exactly one rule. The following D(1, 0)L system is known as *Gabor's Sloth*, due to Gabor Herman. It is very important in the theory of growth functions. The alphabet is $\{a, b, c, d\}$ and the axiom ab. The rules are defined by the following table, where the row indicates the left neighbour and the column the letter to be rewritten:

	a	b	c	d
#	c	b	a	d
a	a	b	a	d
b	a	b	a	d
c	b	c	a	ad
d	a	b	a	d

Here again, because of determinism, the language is produced in a *sequence*, the beginning of which is:

$$ad, cd, aad, cad, abd, cbd, acd, caad, abad, cbad,$$

$$acad, cabd, abbd, cbbd, acbd, cacd, abaad, \ldots$$

Such a growth in word length is not possible for D0L sequences. The intervals in which the growth function stays constant grow beyond all limits – they even do so exponentially in terms of the constant mentioned. The entire growth is logarithmic.

It is obvious that the generative capacity increases if the interaction becomes more extensive: $(m + 1, n)$L systems generate more than (m, n)L systems. A similar result concerns also the right context, leading to an infinite hierarchy of language families. It is, however, very interesting that only the *amount of context matters*, not its distribution. All the following families coincide:

$$(4, 1)L = (3, 2)L = (2, 3)L = (1, 4)L.$$

The families $(5, 0)$L and $(5, 0)$L are mutually incomparable, contained strictly in $(4, 1)$L, and incomparable with $(3, 1)$L. Analogous results hold true in general.

Since already E$(1, 0)$L systems (as well as E$(0, 1)$L systems) generate all recursively enumerable languages, further modifications such as tables are studied for some special purposes only. EPIL systems produce, as one would expect, the family of context-sensitive languages. There are many surprises in the deterministic case. For instance, there are nonrecursive languages in D1L (showing that the family is big), whereas ED1L does not contain all regular languages (showing that even the extended family is small).

At some stage also the organisms modeled by L systems are expected to become *adults*. It has become customary in the theory of L systems to define adults as follows. A word belongs to the *adult language* of a system exactly in case it derives no words but itself. For instance, assume that

$$a \longrightarrow ab, \quad b \longrightarrow c, \quad c \longrightarrow \lambda, \quad d \longrightarrow dc$$

are the only rules for a, b, c, d in a 0L system. Then all words of the form $(abc)^i(dc)^j$ belong to the adult language. Adult languages of 0L systems are called A0L *languages*. Of course, A can be used in connection with other types of L systems as well.

We have presented adult languages following their customary definition. It is maybe not a proper way to model adults in artificial life, perhaps it models better some kind of "stagnant stability". It is, however, a very interesting language-theoretic fact that A0L equals the family of context-free languages. Similarly, $A(1,0)L$ equals the family of recursively enumerable languages.

L systems with fragmentation, JL systems, should be quite useful for modeling artificial life. The mechanism of fragmentation provides us with a new formalism for blocking communication, splitting the developing filament and also for cell death. (The letter J comes from the Finnish word JAKAUTUA, to fragment. At the time of its introduction, all letters coming from suitable English words already had some other meaning for L systems!).

The basic idea behind the fragmentation mechanism is the following. The right-hand sides of the rules may contain occurrences of a special symbol q. The symbol induces a *cut* in the word under scan, and the derivation may continue from any of the parts obtained. Thus, if we apply the rules

$$a \longrightarrow aqa, \quad b \longrightarrow ba, \quad c \longrightarrow qb$$

to the word abc, we obtain the words a, aba and b. The derivation may continue from any of them. The J0L system with the axiom aba and rules $a \longrightarrow a$, $b \longrightarrow abaqaba$ generates the language

$$\{aba^n \mid n \geq 1\} \cup \{a^n ba \mid n \geq 1\}.$$

An interesting language-theoretic fact is that E0L languages, in addition to numerous other closure properties, are closed under fragmentation:

$$\text{EJ0L} = \text{JE0L} = \text{E0L}.$$

J0L systems have been used recently, [KRS], for obtaining a compact representation of certain regular trees.

We have already referred to the quite unusual uniformity in the basic terminology about L systems, in particular, the letters used for naming systems. The following summarizing *glossary* is intended to assist the reader. It is not exhaustive – also other letters have been used but without reaching a similar uniformity.

A. adult, adult word, adult language
C. coding, letter-to-letter morphism, image under such morphism
D. deterministic, only one choice, only one choice in each table
E. extended, intersection with a terminal vocabulary is taken

F. finite set of axioms, rather than only one axiom
H. homomorphism, morphism, image under morphism
I. interactions, neighbouring cells affect the development of the cell
J. fragmentation, the mechanism of inducing cuts
L. Lindenmayer, appears in the name of all developmental systems
N. nonerasing morphism, image under such a morphism
O. actually number 0 but often read as the letter, information 0-sided no
    interaction, rewriting context-free
P. propagating, no cell death, empty word never on the right-hand side of a
    rule
T. tables, sets of rules, diversification of developmental instructions
U. unary, alphabet consisting of one letter
W. weak coding, a morphism mapping each letter to a letter or the empty
    word, image under such a morphism

3. Sample L techniques: avoiding cell death if possible

It is clear that, due to space restrictions, we are only able to state results,
not to give proofs or even outline proofs. The purpose of this section is to
give some scattered samples of the techniques used. The study of L systems
has brought many new methods, both for general language theory and for
applications to modeling, notably in computer graphics. Therefore, we feel it
proper to include some examples of the methods also in this Handbook. For
computer graphics, we refer to the contribution of P. Prusinkiewicz in this
Handbook.

A very tricky problem, both from the point of view of the theory and the
phenomenon modeled, is *cell death*. Are rules of the form $a \longrightarrow \lambda$ really necessary? Here you have to be very specific. Consider the D0L system $DEATH_b$
defined in subsection 2.1. We observed that $S(DEATH_b)$ cannot be generated
by any PD0L system. In this sense cell death is necessary. However, the situation is different if we are interested only in the sequence of word lengths. The
very simple PD0L system with the axiom a^4 and the rule $a \longrightarrow a^2$ possesses
the same growth function as $DEATH_b$.

In general, the growth function of every PD0L system is monotonously
increasing. There can be no decrease in word length, because in a PD0L
system every letter produces at least one letter. Assume that you have a
monotonously strictly increasing D0L growth function f, that is, $f(n) <
f(n+1)$ holds for all n. Can such an f always be generated by a PD0L system?
It is quite surprising that the answer is negative [K3]. There are monotonously
strictly increasing D0L growth functions that cannot be generated by any
PD0L system. For some types of D0L growth cell death is necessary, although
it cannot be observed from the length sequence.

If one wants to construct a D0L-type model without cell death, one has to pay some price. The productions $a \longrightarrow \lambda$ certainly add to the generative capacity and cannot be omitted without compensation. The amount of the price depends on what features one wishes to preserve. We are now interested in languages. We want to preserve the language of the system but omit productions $a \longrightarrow \lambda$. We can achieve this by taking finitely many axioms and images under codings, that is, letter-to-letter morphisms. In other words, the C- and F-features are together able to compensate the effect of the productions $a \longrightarrow \lambda$. *If we add C and F to the name of the system, we may also add P.* Specifically, we want to prove that

(*)   D0L $\subseteq$ CPDF0L.

The inclusion

$$\text{CD0L} \subseteq \text{CPDF0L}$$

is an immediate consequence of (*). (In fact, since the composition of two codings is again a coding, we can essentially multiply (*) from the left by C.) Moreover, (*) is an important building block in establishing the hierarchy presented in Theorem 2.3, in particular, in showing the high position of some P-families. The earliest (but very compactly written) proof of (*) appears in [NRSS].

We now begin the proof of (*). Let

$$G = (\Sigma, h, w_0)$$

be a given D0L system. We have to show that $L = L(G)$ is a CPDF0L language, that is, L is a letter-to-letter morphic image of a language generated by a PD0L system with finitely many axioms. This is obvious if L is finite. In this case we take L itself to be the finite axiom set of our PDF0L system. The identity rule $a \longrightarrow a$ will be the only rule for every letter, and the coding morphism is the identity as well. If L contains the empty word λ, then L is necessarily finite. This follows because, whenever λ appears in the sequence, no non-empty word can appear in the sequence afterwards. This observation implies that, for the proof of (*), it is not necessary to make the λ-convention considered in connection with Theorem 2.3. If the empty word appears in the given D0L language, it can be taken as an axiom of the PDF0L system.

Thus, from now on we assume that the given D0L language $L = L(G)$ is *infinite*. We use our customary notation w_i, $i \geq 0$, for the sequence $S(G)$. Our argumentation will make use of *periodicities* in $S(G)$, already mentioned in Theorem 2.1 (However, we will not make use of Theorem 2.1, our argumentation will be self-contained.) We will first illustrate periodicities by a simple case needed only for a minor detail. Consider the minimal alphabets of the words w_i,

$$\Sigma_i = \text{alph}(w_i), \quad i \geq 0.$$

Each Σ_i is a subset of the finite set Σ. Hence, there are only finitely many alphabets Σ_i and, consequently, numbers $q \geq 0$ and $p \geq 1$ such that $\Sigma_q =$

Σ_{q+p}. (We may take q to be the smallest number such that Σ_q occurs later and p the smallest number such that $\Sigma_q = \Sigma_{q+p}$. This choice defines q and p unambiguously.)

We now use the obvious fact that

$$\text{alph}(w) = \text{alph}(w') \text{ implies alph}(h(w)) = \text{alph}(h(w')),$$

for all words w and w'. Consequently, $\Sigma_{q+1} = \Sigma_{q+p+1}$ and, in general,

$$\Sigma_{q+i} = \Sigma_{q+p+i}, \quad \text{for all } i \geq 0.$$

Thus, the alphabets form the ultimately periodic sequence

$$\underbrace{\Sigma_0, \ldots, \Sigma_{q-1}}_{\text{initial mess}}, \underbrace{\Sigma_q, \ldots, \Sigma_{q+p-1}}_{\text{period}}, \Sigma_q, \ldots, \Sigma_{q+p-1}, \ldots$$

From the point of view of a developmental model, we can visualize this as follows. In the early history, there may be some primaeval cell types which vanish and never come back again. Strange phenomena are possible, for instance, one can have in the words of the initial mess any length sequence of positive integers one wants just by using sufficiently many cell types (letters). After the early history, p different phenomena have been merged together. (The merged phenomena should not be too different, for instance, the growth order of the p subsequences must be the same. Section 5. explains this further.)

The following example might illustrate the situation. Consider the PD0L system with the axiom $d_1 d_2^4$ and productions

$$d_1 \longrightarrow d_3 d_4^2, \quad d_2 \longrightarrow d_4, \quad d_3 \longrightarrow ad, \quad d_4 \longrightarrow \lambda,$$
$$a \longrightarrow a_1, \quad b \longrightarrow b_1, \quad c \longrightarrow c_1, \quad d \longrightarrow d,$$
$$a_1 \longrightarrow abc^2, \quad b_1 \longrightarrow bc^2, \quad c_1 \longrightarrow c.$$

The beginning of the word sequence is

$$d_1 d_2^4, d_3 d_4^6, ad, a_1 d, abc^2 d, a_1 b_1 c_1^2 d,$$
$$abc^2 bc^4 d, a_1 b_1 c_1^2 b_1 c_1^4 d, abc^2 bc^4 bc^6 d, \ldots,$$

the corresponding part of the length sequence being

$$5, 7, 2, 2, 5, 5, 10, 10, 17, \ldots.$$

In this case the lengths of the initial mess and period are: $q = 4$, $p = 2$. The reader might have already noticed that our old friend SQUARES is lurking in this D0L system and, consequently, the growth functon satisfies

$$f(0) = 5, f(1) = 7, f(2i) = f(2i+1) = i^2 + 1, \text{ for } i \geq 1.$$

The initial mess is used to generate two exceptional values initially, and the period of length 2 serves the purpose of creating an idle step after each new square has been produced.

We now go back to the proof of (*), considering again the D0L system $G = (\Sigma, h, w_0)$ generating an infinite language $L = L(G)$. For each $a \in \Sigma$, we denote by u_i^a, $i \geq 0$, the word sequence of the D0L system with the axiom a and morphism h. Thus, $u_i^a = h^i(a)$. We divide Σ into two disjoint parts, $\Sigma = \Sigma^{fin} \cup \Sigma^{inf}$ by defining

$$\Sigma^{fin} = \{a \in \Sigma \mid \{u_i^a \mid i \geq 0\} \text{ is finite }\}, \quad \Sigma^{inf} = \Sigma \setminus \Sigma^{fin}.$$

The set Σ^{fin} consists of those letters that generate a finite language. In other words, the sequence u_i^a, $i \geq 0$, is ultimately periodic, with the period p_a and threshold q_a. Of course, the letters a with the rule $a \longrightarrow \lambda$ are in Σ^{fin}. Since the language L is infinite, the set Σ^{inf} must be nonempty.

We now choose a *uniform period* P satisfying

$$u_P^a = u_{iP}^a, \quad \text{for all } i \geq 0 \text{ and } a \in \Sigma^{fin}.$$

Conditions sufficient for such a choice are that P exceeds all thresholds q_a and is divisible by all periods p_a, a rude estimate for P followed in [NRSS] is to take the product of all periods and thresholds. Further, let M be the maximum length of any word derived in G in P steps from a single letter:

$$M = \max \{|u_P^a| \mid a \in \Sigma\}.$$

Our true period will be a multiple of P, PQ. Specifically, $Q \geq 2$ is an integer such that

$$|u_{P(Q+1)}^a| > M, \text{ for all } a \in \Sigma^{inf}.$$

Observe that by definition all sequences $\{u_i^a\}$, $a \in \Sigma^{inf}$, contain arbitrarily long words. However, also length decrease occurs in the sequences. It might look questionable that we can find a number Q as required. Since D0L sequences are full of surprises, let us prove this in detail. We prove a somewhat stronger statement which is of interest also on its own: *For any D0L sequence $\{v_i\}$ containing infinitely many different words and any number t, there is a bound i_0 such that*

$$|v_i| > t \text{ whenever } i \geq i_0.$$

This claim follows by considering the sequence of the minimal alphabets of the words v_i. As seen above, this sequence is ultimately periodic, with period p and threshold q. Each of the D0L sequences

$$\{\bar{v}_i^k = v_{ip+q+k} \mid i \geq 0\}, \quad 0 \leq k < p$$

is *conservative*, that is, the minimal alphabet of all words in the sequence is the same. The morphism defining these sequences is the pth power of the original morphism. This means that all p sequences $\{\bar{v}_i^k\}$ are monotonously length-increasing. Thus, to exceed the length t, we only have to watch that the word length in each of the p sequences, merged together to obtain $\{v_i\}$, exceeds t. Intuitively, the statement just established says that there is some over-all growth in the sequence because all of the subsequences are growing.

There are three key ideas in the following formal proof. *(i)* The effect of the λ-rules can be eliminated in long enough derivation steps. The new productions will simulate steps of length PQ in G. This means that we must have PQ starting points. But this will be no problem because we have the F-feature available. Our first new system G' may still have λ-rules but they can be applied to the axioms only. Thus, a straightforward modification of augmenting the axiom set by words derivable in one step in G gives a second new system G'' without λ-rules. *(ii)* Each letter develops according to G in the same way in a specific number of steps, quite independently of the location of the steps. Our simulation uses the interchange between the steps from 0 to PQ and from P to $P(Q+1)$. *(iii)* In our new system G' an occurrence of a letter, say b, is replaced by information concerning where this particular b comes from P steps earlier in G. The letter b has a definite ancestor on each preceding level. Say c is the ancestor P steps earlier. We cannot recover b if we know only c. However, c generates at most M (occurrences of) letters in P steps. Hence, we should also tell the number i, $1 \le i \le M$, indicating that the particular occurrence of b is the ith letter (counting from left to right) generated by c in P steps. Altogether b is replaced by the pair (c, i); the alphabet of G' includes pairs of this form. The letter b is recovered for the final language by the coding g:

$$g((c, i)) = b.$$

The procedure enables us to get one "free" step (of length P) in the derivations. The situation can be depicted as follows:

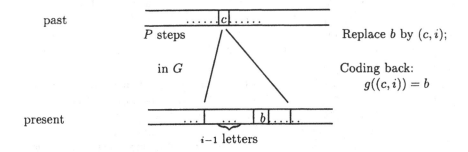

past

P steps

in G

present

$i-1$ letters

Replace b by (c, i);

Coding back:
$$g((c, i)) = b$$

We are now ready for the formal details. Define first a DF0L system G'. The alphabet is

$$\Sigma \cup \Sigma_{place}, \text{ where } \Sigma_{place} = \Sigma \times \{1, 2, \ldots, M\}.$$

Thus, letters of Σ_{place} are pairs (a, i), where $a \in \Sigma$ and $1 \le i \le M$. The axiom set of G' is $\{w_i \mid 0 \le i < PQ\}$. (The words w_i are from the sequence of the original G.) For each $a \in \Sigma$, the right-hand side of its rule in G' is a word over Σ_{place}, obtained from u_{PQ}^a by replacing any occurrence of a letter, say b, with the letter $(c, i) \in \Sigma_{place}$ iff the particular b is the ith letter (from left to right) generated, in P steps, from an occurrence of the letter c in $u_{P(Q-1)}^a$. If $u_{PQ}^a = \lambda$ then $a \longrightarrow \lambda$ is the production for a in G'.

To complete the definition of G', we still have to define the rules for the letters of Σ_{place}. Each right-hand side will be a *nonempty* word over Σ_{place}. Consider first a letter (a, i) for which $i < |u_P^a|$. The rule is $(a, i) \longrightarrow (b, j)$ iff the ith letter of $u_{P(Q+1)}^a$ is derived as the jth letter from an occurrence of the letter b in u_{PQ}^a. Consider next the letter $(a, |u_P^a|)$. The right-hand side of its rule is the following nonempty (by the definition of Q) word over Σ_{place} of length

$$|u_{P(Q+1)}^a| - |u_P^a| + 1.$$

The ith letter in this word is defined as the letter (b, j) for which the $(|u_P^a| - 1 + i)$th letter of $u_{P(Q+1)}^a$ is derived as the jth letter from an occurrence of the letter b in u_{PQ}^a. (The letters (a, i) with $|u_P^a| < i \le M$ are useless in the sense that they never occur in $L(G')$. We may take the identity production $(a, i) \longrightarrow (a, i)$ for such letters. It is also irrelevant how the coding g is defined for such letters.)

We now define the letter-to-letter morphism g as follows:

If $a \in \Sigma$ then $g(a) = a$,
If $(a, i) \in \Sigma_{place}$ and $1 \le i \le |u_P^a|$ then $g((a, i))$ equals the ith letter of u_P^a.

It now follows by our construction that $g(L(G')) = L(G)$. The explanations given before the construction should be helpful in establishing this fact. G' is a DF0L system surely having λ-rules if G has them. However, the λ-rules of G' can be applied only to the axioms. After the first derivation step λ-rules are no longer applicable because all words are over the alphabet Σ_{place} for which all rules are propagating. Let w_i', $0 \le i < PQ$, be the word derived in G' from w_i in one step.

We now transform G' to a PDF0L system G'' by augmenting the axiom set of G' with the set $\{w_i' \mid 0 \le i < PQ\}$ and replacing the rules of G' for letters $a \in \Sigma$ by the identity rules $a \longrightarrow a$. In this way all λ-rules are removed. The effect of the rules of G' for letters of Σ, that is, the introduction of the letters in Σ_{place} is in G'' taken care of by the additional axioms w_i', $0 \le i < PQ$. The original axioms of G' generate themselves in G''. Thus, the PDF0L system G'' satisfies $L(G'') = L(G')$. We may use the coding g defined above to obtain the final result

$$L = L(G) = g(L(G'')).$$

This shows that L is a CPDF0L language, which completes the proof of (*).

$\square$

The construction presented above is quite complicated. This is understandable because the result itself is perhaps the most sophisticated one about the elimination of λ-rules in the theory of L-systems. The analogous construction is much simpler, [SS], if one only wants to preserve length sequences, not the language, as we have preserved.

We mention finally that, to eliminate λ-rules, we need *both* of the features C and F. It is seen from Theorem 2.3 that neither the family CPD0L nor the family PDF0L contains the family D0L.

4. L decisions

4.1 General. Sequences versus languages

We will now present results concerning *decision problems* in the theory of L systems. This is a rich area – our overview will be focused on the celebrated *D0L equivalence problem* and variations. More information is contained especially in [CK2] and [RS3]. We will begin with a general overview.

Customary decision problems investigated in language theory are the *membership, emptiness, finiteness, equivalence* and *inclusion problems*. The problems are stated for a specific *language family*, sometimes in a *comparative* sense between two families, [S4]. When we say that the equivalence problem between 0L and D0L languages is decidable, this means the existence of an algorithm that receives as its input a pair (G, G') consisting of a 0L and a D0L system and tells whether or not $L(G) = L(G')$. "Equivalence" without further specifications always refers to language equivalence, in case of deterministic systems we also speak of *sequence equivalence*. Given two language families $\mathcal{L}$ and $\mathcal{L}'$, we also speak of the $\mathcal{L}'$-ness problem for $\mathcal{L}$, for instance, of the D0L-ness (resp. regularity) problem for 0L languages. The meaning of this should be clear: we are looking for an algorithm that decides of a given 0L system G whether or not $L(G)$ is a D0L language (resp. a regular language).

We now present some basic decidability and undecidability results, dating mostly already from the early 70's, see [Li1], [CS1], [La1], [La2], [RS1], [RS3], [S2], [S5] for further information. It was known already in 1973 that ET0L is included in the family of indexed languages, for which membership, emptiness and finiteness were known to be decidable already in the 60's. For many subfamilies of ET0L (recall that ET0L is the largest L family without interactions) much simpler direct algorithms have been developed.

Theorem 4.1. *Membership, emptiness and finiteness are decidable for ET0L languages, and so are the regularity and context-freeness problem for D0L languages and the D0L-ness problem for context-free languages. Also the equivalence problem between D0L and context-free languages is decidable, and so is the problem of whether or not an E0L language is included in a regular language.* □

Theorem 4.2. *The equivalence problem is undecidable for P0L languages and so are the context-freeness, regularity and 0L-ness problems for E0L languages.*   □

Because even simple subfamilies of IL contain nonrecursive languages (for instance, the family D1L), it is to be expected that most problems are undecidable for the L families with interactions. For instance, language and sequence equivalence are undecidable for PD1L systems.

For systems without interactions, many undecidable problems become decidable in the case of a one-letter alphabet. Decision problems can be reduced to arithmetical ones in the *unary* case. Sample results are given in the following theorem, [La1], [La2], [S2]. TU0L and U0L are written briefly TUL and UL.

Theorem 4.3. *The equivalence problem between TUL and UL languages, as well as the equivalence problem between TUL languages and regular languages are decidable. the regularity and UL-ness problems are decidable for TUL languages, and so is the TUL-ness problem for regular languages.*   □

The cardinality $d_G(n)$ of the language generated by an L system G using *exactly n derivation steps* has given rise to many decision problems. The earliest result is due to [Da1], where the undecidability of the equation $d_{G_1}(n) = d_{G_2}(n)$ for two DT0L systems is shown. The undecidability holds for 0L systems as well. However, it is decidable, [Ni], whether or not a given 0L system G is *derivation-slender*, that is, there is a constant c such that $d_G(n) \leq c$ holds for all n. All derivation-slender 0L languages are *slender* in the sense that there is a constant k such that the language has at most k words of any given length, [NiS]. The converse does not hold true: the language $\{b^i ab^i | \ i \geq 0\} \cup \{b^{2i+1} | \ i \geq 0\}$ is a slender 0L language not generated by any derivation-slender 0L system. Further results concerning slender 0L languages are presented in [NiS]. However, at the time of this writing, the general problem of deciding the slenderness of a given 0L language remains open.

We now go into different variations of *D0L equivalence problems*, this discussion will be continued in the next subsection. This problem area has given rise to many important new notions and techniques, of interest far beyond the theory of L systems. We denote briefly by *LE-D0L* and *SE-D0L* the language and sequence equivalence problems of D0L systems. It was known already in the early 70's, [N], that the two problems are "equivalent" in the sense that any algorithm for solving one of them can be transformed into an algorithm for solving the other. We will now prove this result. Our proof follows [S6] and is another illustration of the techniques used in L proofs.

Theorem 4.4. *SE-D0L and LE-D0L are equivalent.*
Proof from LE-D0L to SE-D0L.

We show how any algorithm for solving LE-D0L can be used to solve SE-D0L. Given two D0L systems

$$G = (\Sigma, g, u_0) \text{ and } H = (\Sigma, h, v_0),$$

we define two new D0L systems

$$G_b = (\Sigma \cup \{b\}, g_b, bu_0), \quad H_b = (\Sigma \cup \{b\}, h_b, bv_0), b \notin \Sigma,$$

where $g_b(b) = h_b(b) = b^2$, and $g_b(a) = g(a)$, $h_b(a) = h(a)$, for all $a \in \Sigma$. Clearly, $S(G) = S(H)$ iff $L(G_b) = L(H_b)$.   □

Proof from SE-D0L to LE-D0L.

We assume without loss of generality that the given languages $L(G)$ and $L(H)$ are infinite. Their finiteness is easily decidable and so is their equality if one of them is finite. The tools we will be using are *decompositions of D0L systems* and *Parikh vectors*. Decompositions refer to *periodicities* already discussed in Subsection 3.1 above. Given a D0L system $G = (\Sigma, g, u_0)$ and integers $p \geq 1$, $q \geq 0$ (period and initial mess), we define the D0L system

$$G(p, q) = (\Sigma, g^p, g^q(u_0)).$$

The sequence $S(G(p, q))$ is obtained from $S(G)$ by taking every pth word after the initial mess.

For $w \in \Sigma^*$, the *Parikh vector* $\Psi(w)$ is a card(Σ)-dimensional vector of nonnegative integers whose components indicate the number of occurrences of each letter in w. For two words w and w', the *notation* $w \leq_p w'$ (resp. $w <_p w'$) means that the Parikh vectors satisfy $\Psi(w) \leq \Psi(w')$ (resp. $\Psi(w) < \Psi(w')$). (The ordering of vectors is taken componentwise, two vectors can be incomparable.)

We use customary notations for the given D0L systems:

$$G = (\Sigma, g, u_0), \quad H = (\Sigma, h, v_0).$$

(We may assume that the alphabets coincide.) The notation w_i refers to words in one of the two D0L sequences u_i and v_i we are considering. We omit the proofs of the following properties (i)–(iv). They are straightforward, apart from (iv) which depends on the theory of growth functions discussed in the following section.

(i)   There are words such that $w_i <_p w_j$.
(ii)   We cannot have both $i < j$ and $w_j \leq_p w_i$.
(iii) Whenever $w_i <_p w_j$, then $w_{i+n} <_p w_{j+n}$ for all n.
(iv)   Assume that x_i and y_i, $i \geq 0$, are D0L sequences over an alphabet with n letters such that $\Psi(x_i) = \Psi(y_i)$, $0 \leq i \leq n$. Then $\Psi(x_i) = \Psi(y_i)$ for all i. (We call x_i and y_i *Parikh equivalent.*)

We now give an algorithm for deciding whether or not $L(G) = L(H)$. The algorithm uses a parameter m (intuitively, the length of the discarded initial mess). Originally we set $m = 0$.

Step 1. Find the smallest integer $q_1 > m$ for which there exists an integer p such that

$$u_{q_1-p} <_p u_{q_1}, \quad 1 \le p \le q_1 - m.$$

Let p_1 be the smallest among such integers p. Determine in the same way integers q_2 and p_2 for the system $H(1,m)$. (Step 1 can be accomplished by property (i).)

Step 2. If the two finite languages

$$\{u_i| \ m \le i < q_1\} \text{ and } \{v_i| \ m \le i < q_2\}$$

are different, stop with the conclusion $L(G) \ne L(H)$. (Otherwise, $q_1 = q_2$.)

Step 3. Apply the algorithm for SE-DOL. Check whether or not $p_1 = p_2$ and there is a permutation Π of the set of indices $\{0, 1, \ldots, p_1 - 1\}$ such that

$$S(G(p_1, q_1 + j)) = S(H(p_1, q_1 + \Pi(j))), \text{ for all } j = 0, \ldots, p_1 - 1.$$

If "yes", stop with the conclusion $L(G) = L(H)$. If "no", take q_1 as the new value of m and go back to Step 1.

Having completed the definition of the algorithm, we establish its correctness and termination.

Correctness. We show first that the conclusion in Step 2 is correct. When entering Step 2, we must have $\{u_i| \ i < m\} = \{v_i| \ i < m\}$. Indeed, this holds vacuously for $m = 0$, from which the claim follows inductively. If some word w belongs to the first and not to the second of the finite languages in Step 2 and still $L(G) = L(H)$, then $w = v_i$ for some $i \ge q_2$. (We cannot have $i < m$ because then w would occur twice in the u-sequence.) By (iii), for some $j \ge m$, $v_j <_p w$. By the choice of q_1 and (ii), $v_j \notin L(G)$, which is a contradiction.

Also the conclusion in Step 3 is correct. When entering Step 3, we know that

$$\{u_i| \ 0 \le i < q_1\} = \{v_i| \ 0 \le i < q_1\}.$$

The test performed in Step 3 shows that also

$$\{u_i| \ i \ge q_1\} = \{v_i| \ i \ge q_1\}.$$

Termination. If $L(G) \ne L(H)$, a word belonging to the difference of the two languages is eventually detected in Step 2 because the parameter m becomes arbitrarily large. Assuming that $L(G) = L(H)$, we show that the equality is detected during some visit to Step 3. It follows by property (iii) that neither p_1 nor p_2 is increased during successive visits to Step 1. Thus, we only have to show that the procedure cannot loop by producing always the same pair (p_1, p_2). If n is the cardinality of the alphabet, there cannot be $n + 2$ such consecutive visits to Step 1.

Assume the contrary: the same pair (p_1, p_2) is defined at $n+2$ consecutive visits. We must have $p_1 = p_2$. Otherwise, the larger of the two numbers has to be decreased at the next visit to Step 1, because of property (iii) and the fact that Step 2 was passed after the preceding visit. The same argument shows that p_1 must assume the maximal value $q_1 - m$. If m is the initial mess at the first of our $n + 2$ visits, the sequences $S(G)$ and $S(H)$ contain $n + 1$ segments of length p_1, beginning with u_m and v_m, such that each segment in $S(H)$ is a permutation of the corresponding segment in $S(G)$. Moreover, it is always the same permutation because otherwise, by property (iii), the value of p_1 will be decreased at the next visit to Step 1. This implies, by property (iv), that for some permutation Π and all $j = 0, \ldots, p_1 - 1$, the sequences

$$S(G(p_1, m + j)) \text{ and } S(H(p_1, m + \Pi(j)))$$

are Parikh equivalent. Since termination did not occur in Step 3, there is a j such that the two sequences are not equivalent. Thus, for some $u \in L(G)$ and $v \in L(H)$, we have $\Psi(u) = \Psi(v)$ but $u \neq v$. Since $L(G) = L(H)$, we have also $v \in L(G)$. Hence, $S(G)$ has two words with the same Parikh vector, which contradicts property (ii). $\qquad\qquad\qquad\qquad\qquad\qquad\qquad\qquad\qquad\qquad\qquad\qquad\qquad$ □

4.2 D0L sequence equivalence problem and variations

The decidability of SE-D0L was first shown in [CF]. A much simpler proof based on *elementary morphisms*, [ER2], was given in [ER3], a detailed exposition of it appears in [RS1]. Our argument below follows [CK2] and uses two results, the correctness of the so-called *Ehrenfeucht Conjecture* and *Makanin's Algorithm*. We refer the reader to [RS1] and [CK2] for further information concerning the history of SE-D0L and related problems.

SE-D0L has given rise to many important new notions that have been used widely in language theory: *morphic equivalence, morphic forcing, test set, elementary morphism, bounded balance, equality set.* All these notions are related to the problem of studying the equation $g(x) = h(x)$ for a word x and morphisms g and h.

We say that the morphisms g and h defined on Σ are *equivalent* on a language $L \subseteq \Sigma^*$, in symbols $g \equiv_L h$, iff $g(x) = h(x)$ holds for all $x \in L$. The *morphic equivalence problem* for a family $\mathcal{L}$ of languages, [CS1], consists of deciding, given a language L in $\mathcal{L}$ and morphisms g and h, whether or not $g \equiv_L h$. Observe that SE-D0L is a special case of the morphic equivalence problem for the family of D0L languages: two D0L systems G and H with morphisms g and h are sequence equivalent iff g and h are equivalent on $L(G)$ (or on $L(H)$). So we have the special case where $L(G)$ is generated by one of the morphisms whose equivalence on $L(G)$ is to be tested.

Let us go back to the morphic equivalence problem. It would be desirable to have a *finite* set $F \subseteq L$ such that, to test $g \equiv_L h$, it suffices to test $g \equiv_F h$. Formally, we say that a language L is *morphically forced* by its subset L_1 iff, whenever two morphisms are equivalent on L_1, they are equivalent on L. A *finite* subset F of a language L is termed a *test set* for L iff L is morphically forced by F. The *Ehrenfeucht Conjecture* claims that every language possesses a test set. It was shown correct in the middle 80's. [AL] is usually quoted as the first proof but there were several independent ones about the same time – see [CK2], [RS3] for details.

The notion of an *equality set* occurs at least implicitly in earlier algorithms for SE-D0L, [CF], [ER3]. The equality set (also called *equality language*) of two morphisms is defined by

$$E(g, h) = \{x \in \Sigma^* \mid g(x) = h(x)\}.$$

SE-D0L amounts to deciding whether or not one of the languages, say $L(G)$, is contained in $E(g, h)$. This again amounts to deciding the emptines of the intersection between $L(G)$ and the complement of $E(g, h)$. If $E(g, h)$ is regular (implying that also its complement is regular), the latter problem becomes decidable because D0L is contained in E0L for which the emptiness is decidable, and E0L is closed under intersection with regular languages. Therefore, one should aim towards the case that $E(g, h)$ is actually regular. This situation was reached in the proof of [ER3] by using *elementary morphisms*. Essentially, a morphism being elementary means that it cannot be "simplified" by presenting it as a composition going via a smaller alphabet, [ER2].

Assume that the morphisms g and h are equivalent on L. We say that the pair (g, h) has a *bounded balance* on L iff there is a constant C such that

$$\|g(x)\| - |h(x)\|| \le C$$

holds for all prefixes x of words in L. (We know that $g(w) = h(w)$ holds for words $w \in L$. Therefore, if x is a prefix of w, then one of $g(x)$ and $h(x)$ is a prefix of the other. Having bounded balance means that the amount by which one of the morphisms "runs faster" is bounded by a constant.) For testing morphic equivalence, the property of having bounded balance gives the possibility of storing all necessary information in a finite buffer. Thus, the situation is similar to the equality set being regular.

Consider two disjoint alphabets Σ and N (N is the alphabet of *nonterminals* or *variables*). An *equation* over Σ with unknowns in N is a pair (u, v), usually written $u = v$, where u and v are words over $\Sigma \cup N$. A set T (possibly infinite) of equations is referred to as a *system of equations*. A *solution* to a system T is a morphism $h : (\Sigma \cup N)^* \longrightarrow \Sigma^*$ such that $h(u) = h(v)$, for all $(u, v) \in T$, and $h(a) = a$ for all $a \in \Sigma$. Thus, solutions can be viewed as card(N)-tuples of words over Σ. For instance, the morphism defined by

$$x \longrightarrow a, \quad y \longrightarrow ba, \quad z \longrightarrow ab$$

is a solution of the equation $xy = zx$ over $\{a, b\}$. This solution can also be represented as the triple (a, ba, ab).

Makanin, [Ma], has presented an algorithm for deciding whether or not a given finite system of equations possesses a solution. It is shown in [CK1] how any such algorithm can be extended to concern finite systems of equations and inequalities $u \ne v$. From this follows easily the decidability of the *equivalence problem for two finite systems of equations*, that is, whether or not the systems have the same sets of solutions. This, in turn, leads directly to the decidability of the problem whether or not a *subset of a finite language F is a test set for F*. We are now ready to establish the following main result.

Theorem 4.5. *The morphic equivalence problem is decidable for DOL languages. Consequently, SE-DOL is decidable.*

Proof. Because the Ehrenfeucht Conjecture is correct, we know that the language $L(G)$ generated by a given D0L system $G = (\Sigma, h, w_0)$ possesses a test set F. We only have to find F effectively. Define the finite languages $L_i, i \geq 0$, by

$$L_0 = \{w_0\}, \quad L_{i+1} = L_i \cup h(L_i).$$

We now determine an integer i_0 such that L_{i_0} is a test set for L_{i_0+1}. Such an integer surely exists because eventually the whole F is contained in some L_i. The integer i_0 can be found because we can decide whether a subset of a finite language is a test set. We claim that L_{i_0} is a test set for $L(G)$, which completes the proof. Indeed, since L_{i_0} is a test set for L_{i_0+1}, also $L_{i_0+1} \setminus \{w_0\}$ is a test set for $L_{i_0+2} \setminus \{w\}$. (Obviously, whenever F' is a test set for L', then $h(F')$ is a test set for $h(L')$.) Consequently, L_{i_0+1} is a test set for L_{i_0+2}. Since "being a test set for" is obviously transitive, we conclude that L_{i_0} is a test set for L_{i_0+2}, from which our claim follows inductively.   □

If the alphabet consists of two letters, it suffices to test the first four words in the given D0L sequences in order to decide sequence equivalence. This result is optimal, as shown by the example

$$w_0 = ab, \quad g(a) = abbaabb, g(b) = a, h(a) = abb, h(b) = aabba.$$

This has given rise to the *2n-conjecture*: in order to decide SE-D0L, it suffices to test the first $2n$ words in the sequences, where n is the cardinality of the alphabet. No counterexamples to this conjecture have been found, although the only known bound of this kind is really huge [ER4].

Theorems 4.4 and 4.5 yield immediately

Theorem 4.6. *LE-D0L is decidable. We are able to establish now also the following very strong result.*

Theorem 4.7. *The HD0L sequence equivalence problem is decidable.*

Proof. Without loss of generality, we assume that the two given HD0L systems are defined by the morphisms f_1 and f_2 and D0L systems $H = (\Sigma, h, w_0)$ and $G = (\Sigma, g, w_0)$, respectively. We consider a "barred copy" $\overline{\Sigma}$ of the alphabet Σ: $\overline{\Sigma} = \{\overline{a} \mid a \in \Sigma\}$. The "barred version" $\overline{w}$ of a word $w \in \Sigma^*$ is obtained by barring each letter. We define three new morphisms, using the descriptive notations $\overline{f_1}, \overline{f_2}$ and $h \cup \overline{g}$, as follows:

$$\overline{f_1}(a) = f_1(a), \overline{f_2}(a) = \lambda, (h \cup \overline{g})(a) = h(a) \text{ for } a \in \Sigma,$$

$$\overline{f_1}(\overline{a}) = \lambda, \overline{f_2}(\overline{a}) = f_2(a), (h \cup \overline{g})(\overline{a}) = \overline{g(a)} \text{ for } a \in \Sigma.$$

Clearly, the original HD0L sequences are equivalent iff the morphisms $\overline{f_1}$ and $\overline{f_2}$ are equivalent on the D0L language defined by the morphism $h \cup \overline{g}$ and axiom $w_0\overline{w_0}$. Thus, Theorem 4.7 follows by Theorem 4.5.   □

The decidability of the HD0L sequence equivalence is a very nontrivial generalization of the decidability of SE-D0L. For instance, no techniques based on bounded balance can be used because two morphisms may be equivalent on a D0L language L without having bounded balance on L. Our simple argument for Theorems 4.5 and 4.7 is no miraculous "deus ex machina". Two very strong tools (Ehrenfeucht and Makanin) have been used. (We want to mention in this connection that, in spite of their efforts, the editors failed in including in this Handbook a reasonably detailed exposition of Makanin's theory.) For a classroom proof, where all tools are developed from the beginning, the proof based on elementary morphisms, presented in [RS1] or [S5], is still to be recommended.

Some related results are collected in our last theorem. See also [CK2], [Ru2], [Ru3], [Ru4].

Theorem 4.8. *The morphic equivalence is decidable for HDT0L languages, and so is the equivalence between D0L and F0L and between D0L and DT0L, as well as the inclusion problem for D0L. The equivalence is undecidable for DT0L, and so is the PD1L sequence equivalence.* □

5. L Growth

5.1 Communication and commutativity

This section will discuss *growth functions* associated to sequences generated by L systems. The theory will be presented in an unconventional way, as a discussion between a mathematician, a language theorist and a wise layman. We have two reasons for this unconventional approach. First, there already exist good detailed expositions of the theory in a book form [SS], [RS1]. In a conventional exposition, we could not do much better here than just repeat, even in a condensed form, what has been already said in [SS] or [RS1]. Secondly, growth functions require considerably more factual knowledge in mathematics than do other parts of the L system theory. Since L systems in general are of interest to a wide audience not otherwise working with mathematics, conventional mathematical formulation might scare away many of the readers of this chapter. While we are convinced that this is less likely to happen with our exposition below, we also hope that our presentation will open new perspectives also for a mathematically initiated reader.

Dramatis personae

   Bolgani, a formal language theorist,
   Emmy, a pure mathematician,
   Tarzan, a lonesome traveller.

Emmy. I often wonder why we mathematicians always write texts incomprehensible practically to everybody. Of course every science has its own terminology and idioms, but our pages laden with ugly-looking formulas are still different. I have the feeling that in many cases the same message could be conveyed much more efficiently using less Chinese-resembling text.

Bolgani. Just think how Fermat got his message across, in a passage that has influenced the development of modern mathematics perhaps more than any passage of comparable length. *Cubum autem in duos cubos, aut quadratoquadratum in duos quadratoquadratos, et generaliter nullam in infinitum ultra quadratum potestatem in duos ejusdem nominis fas est dividere: cujus rei demonstrationem mirabilem sane detexi. Hanc marginis exiguitas non caperet*[1]. Isn't it like beautiful poetry? And think of the philosophical implications. That something is correct is expressed by saying "fas est" – follows the divine law. Not even the gods among themselves can break mathematical laws.

Emmy. It still remains a problem whether some "demonstratio mirabilis" exists for Fermat's Last Theorem. One can hardly imagine that Fermat had in mind anything even remotely resembling the fancy methods of Andrew Wiles. But coming back to poetry: the words sound beautiful but one has to know something in order to appreciate the beauty.

Tarzan. Some basic knowledge is always necessary but in many cases rather little will suffice. Take a passage from Horace, also with references to divine activities. *Pone me pigris ubi nulla campis arbor aestiva recreatur aura, quod latus mundi nebulae malusque Juppiter urget. Pone sub curru nimium propinqui solis, in terra domibus negata: dulce ridentem Lalagen amabo, dulce loquentem*[2]. Certainly it sings much better than Fermat. But still you have to have some background in order to fully appreciate it, a mere understanding of the language is not enough.

Bolgani. I have an idea. Tarzan already read about the basics of L systems. (This refers to Sections 1. and 2. above.) Let us try to explain to him the theory of growth functions, using as few formulas as possible. This might open some interesting perspectives, even if no poetry would come up.

Tarzan. I already know something about the topic. Take, for instance, a D0L system. It generates a sequence of words. One considers the length of each word, as a function of the position of the word in the sequence. This function will be the *growth function* of our D0L system. One starts counting the positions from 0, the axiom having the position 0. This means

---

[1] "It is improper (by divine right) to divide a cube into two cubes, a fourth power into two fourth powers and, generally, any power beyond a square into two powers with the same exponent as the original. I really found a miraculous proof for this fact. However, the margin is too small to contain it."

[2] "Place me where never summer breeze/ Unbinds the glebe, or warms the trees/ Where ever-lowering clouds appear,/ And angry Jove deforms th'inclement year:// Place me beneath the burning ray,/ Where rolls the rapid car of day;/ Love and the nymph shall charm my toils,/ The nymph, who sweetly speaks and sweetly smiles." (Ode XXII, tr. by P. Francis *Horace, Vol.I*, London, printed by A. J. Valpy, M.A., 1831)

that every growth function maps the set of nonnegative integers into the same set. Some things are obvious here. The growth function of a PD0L system is monotonously increasing, not necessarily strictly increasing at every step, whereas decrease is possible in the D0L case. Whenever 0 appears as a value of a D0L growth function, then all values afterwards, that is values for greater arguments, are also equal to 0. Whenever we have the empty word in the sequence, then only the empty word appears afterwards. Once dead, always dead. No resurrection is possible in this model of life, be it real or artificial. However, this is the only restriction I have. If I can take as many letters as I want, I can get the first million, billion or any number of values exactly as I like: 351, 17, 1934, 1, 1964, 15, 5, 1942,... I just introduce new letters at every step, getting the lengths as I like. Since my alphabet is altogether finite, at some stage there will be repetitions of letters and, accordingly, patterns in the growth sequence. The over-all growth cannot be more than exponential. The length of the ith word is bounded from above by c^{i+1}, where the constant c equals the maximum of the lengths of the axiom and any right-hand side of the rules. This bound is valid for DIL systems as well.

Bolgani. We are dealing here with issues concerning *communication and commutativity*. I think we have already agreed that we can communicate with each other in a less formal manner – even about technically complicated topics. *Commutativity is the key to D0L growth functions.* It also opens the possibility of using strong mathematical tools. Why do we have commutativity? Because the order of letters is completely irrelevant from the point of view of growth. We only have to keep track of the number of occurrences of each letter or, as the saying goes, of the *Parikh vector* of the word. Only the number is significant, we can commute the letters as we like. But we have to know the Parikh vector, it is not sufficient to know the total length of the word. Two different letters may grow very differently.

Emmy. All information we need can be stored in the *growth matrix* of the D0L system. It is a square matrix, with nonnegative integer entries and dimension equal to the cardinality of the alphabet, say n. We have a fixed ordering of the letters in mind, and the rows and columns of the matrix are associated to the letters according to this ordering. The third entry in the second row indicates how many times the third letter occurs on the right side of the rule for the second letter. In this way the matrix gives all the information we need for studying growth.

Take our old friend SQUARES as an example. The rules were $a \longrightarrow abc^2$, $b \longrightarrow bc^2$, $c \longrightarrow c$. The growth matrix is, accordingly,

$$\begin{pmatrix} 1 & 1 & 2 \\ 0 & 1 & 2 \\ 0 & 0 & 1 \end{pmatrix}$$

See that? The first entry in the second row is 0, because there are no a's on the right side of the rule for b. To complete the information, we still need the

Parikh vector of the axiom. In this case it is $\varPi = (1,0,0)$ since the axiom is a.

So far we have been able to store the information in matrix form. We now come to the point which makes the play with matrices really worthwhile. *One derivation step amounts to multiplication by the growth matrix* if we are only interested in the length sequence. This is a direct consequence of the rule for matrix multiplication. Let us go back to SQUARES and take the word abc^2bc^4 in the sequence, two steps after the axiom. Its Parikh vector is $(1, 2, 6)$. Compute the matrix product $(1, 2, 6)M$. It is another Parikh vector, obtained as follows. You take a column of M, multiply its entries by the corresponding entry of $(1, 2, 6)$, and sum up the products. If your chosen column is the first one, this gives the result $1 \cdot 1 + 2 \cdot 0 + 6 \cdot 0 = 1$, and $1 \cdot 1 + 2 \cdot 1 + 6 \cdot 0 = 3$ and $1 \cdot 2 + 2 \cdot 2 + 6 \cdot 1 = 12$ in the other two cases. Altogether we get the new Parikh vector $(1, 3, 12)$. What does, for instance, the last entry tell us? The third column

$$\begin{pmatrix} 2 \\ 2 \\ 1 \end{pmatrix}$$

indicates how many c's each of the letters a, b, c produces. The numbers of occurrences of the letters were $1, 2, 6$ to start with. Thus, the last entry $12 = 1 \cdot 2 + 2 \cdot 2 + 6 \cdot 1$ tells how many c's we have after an additional derivation step. Similarly the entries 1 and 3 tell how many a's and b's we have.

There is nothing in our conclusion pertaining to the particular example SQUARES. We have shown that after i derivation steps we have the Parikh vector $\varPi\, M^i$. (This holds true also for $i = 0$.) Let η be the n-dimensional column vector with all components equal to 1. The values $f(i)$ of the growth function are obtained by summing up the components of the Parikh vector $\varPi\, M^i$,

$$f(i) = \varPi\, M^i\, \eta, \quad i \geq 0.$$

We have derived the basic interconnection with the theory of matrices, and have now all the classical results available, [MT], [HW].

Tarzan. I know some of them. For instance, the Cayley-Hamilton Theorem says that every matrix satisfies its own characteristic equation. If we have n letters in the alphabet, we can express M^n in terms of the smaller powers:

$$M^n = c_1 M^{n-1} + \ldots + c_{n-1} M^1 + c_n M^0,$$

in this case with integer constants c_j. By the matrix representation for the growth function, we can use the same formula to express each growth function value in terms of values for smaller arguments:

$$f(i + n) = c_1 f(i + n - 1) + \ldots + c_n f(i), \quad i \geq 0.$$

I see it now quite clearly. I can do my arbitrary tricks at the beginning. But once the number of steps reaches the size of the alphabet, I am stuck with this recursion formula.

Bolgani. We will see that many mysteries remain in spite of the formula. But the formula surely gives rise to certain regularities. For instance, the so-called "Lemma of Long Constant Intervals". Assume that f is a D0L growth function and, for every k, there is an i such that

$$f(i) = f(i+1) = \ldots = f(i+k).$$

Then f is ultimately constant: there is an i_0 such that $f(i) = f(i_0)$ for all $i \geq i_0$.

The matrix representation gives also a very simple method for deciding *growth equivalence*: two D0L systems G and G', with alphabet cardinalities n and n', possess the same growth function iff the lengths of the first $n + n'$ terms are the same in both sequences. The bound is the best possibe. This algorithm is very simple, both as regards the proof of its validity and as regards the resulting procedure. This striking contrast to the difficulty of the D0L sequence equivalence problem is again due to commutativity. Consider an example of language-theoretic interest. Two D0L systems have both the axiom a. The rules of the first system are $a \longrightarrow ab^3$, $b \longrightarrow b^3$, and the rules of the second system:

$$a \longrightarrow acde, \; b \longrightarrow cde, \; c \longrightarrow b^2 d^2, \; d \longrightarrow d^3, \; e \longrightarrow bd.$$

You can verify that the first seven numbers in both length sequences are 1, 4, 13, 40, 121, 364, 1093. Hence, the systems are growth equivalent.

Tarzan. The D0L system DEATH$_b$ had the axiom ab^2a and the rules $a \longrightarrow ab^2a$, $b \longrightarrow \lambda$. We noticed that cell death is necessary to generate its language. However, the very simple system with the axiom a^4 and rule $a \longrightarrow a^2$ is growth equivalent, and is without cell death. What is really the role of cell death, that is rules $a \longrightarrow \lambda$, in D0L growth? Does it consist only of making decreases possible? What is the difference between D0L and PD0L growth? If I have a non-decreasing D0L growth function, can I always get the same function using a PD0L system, perhaps at the cost of taking vastly many new letters?

Emmy. That would be too good to be true. The effects of cell death are deeper, also from the growth point of view. The answer to your question is "no", and we still come to the difference between D0L and PD0L growth. At the moment, I would like to dwell on more immediate matters.

The interconnection with matrices, in particular the recurssion formula you are stuck with, open the possibility of using a lot of classical results from the theory of matrices and difference equations. The values $f(i)$ can be expressed as "exponential polynomials" in i, that is, finite sums of terms of the form

$$(\alpha_0 + \alpha_1 i + \ldots + \alpha_{t-1} i^{t-1}) \rho^i.$$

Indeed, here ρ is a root, with multiplicity t, of the characteristic equation of M. Maybe you like an example. Take the system with the axiom abc and the rules

$$a \longrightarrow a^2, \; b \longrightarrow a^5b, \; c \longrightarrow b^3c,$$

yielding

$$M = \begin{pmatrix} 2 & 0 & 0 \\ 5 & 1 & 0 \\ 0 & 3 & 1 \end{pmatrix}$$

Since the matrix is in lower diagonal form, the roots of the characteristic equation are seen immediately: 2 and 1 (double). This results in the expression

$$f(i) = (\alpha_0 + \alpha_1 i) \cdot 1^i + \alpha_2 \cdot 2^i$$

for the growth function. Finally, the coefficients α are determined by considering the first few values (here first three) in the length sequence. Observe that here the axiom becomes important – so far we did not use it at all! The final result is

$$f(i) = 21 \cdot 2^i - 12i - 18.$$

This solves the growth *analysis problem* for any system: we can write down the exponential polynomial representing the growth function of the system. Although the values $f(i)$ are nonnegative integers, the numbers ρ and α are complex numbers, sometimes not even expressible by radicals. Since the growth function is always an exponential polynomial, some growth types can never occur as D0L growth. D0L growth cannot be logarithmic: whenever the growth function is not bounded by a constant, the growth is at least linear. Similarly, nothing faster than exponential or between polynomially bounded and exponential, something like the $2^{\sqrt{n}}$, is possible.

Bolgani. Such in-between growth orders are possible in the DIL case [K1]. As Tarzan already observed, super-exponential growth is impossible also in the DIL case.

Emmy. Still a few things. The theory says a lot also about the converse problem, *growth synthesis*. We have in mind a function, obtained experimentally or otherwise, and we want to construct a D0L system having the function as its growth function. For instance, we can synthesize any polynomial $p(i)$ with integer coefficients and assuming integer values for all integers $i \geq 0$. ($p(i)$ may assume negative values for non-integral values of i.) We cannot synthesize $i^2 - 4i + 4$ because it represents death at $i = 2$, but $i^2 - 41 + 5$ is OK. It is the growth function of the D0L system with the axiom $a_1^3 a_2 a_3$ and rules

$$a_1 \longrightarrow \lambda, a_2 \longrightarrow a_4, a_3 \longrightarrow a_5, a_4 \longrightarrow a_6, a_5 \longrightarrow \lambda,$$

$$a_6 \longrightarrow ad, a \longrightarrow abc^2, b \longrightarrow bc^2, c \longrightarrow c, d \longrightarrow d.$$

Observe that our old friend SQUARES is lurking in the stomach of this new D0L creature, where Tarzan's technique of disposable letters is also visible.

Making use of periodicities, we can also synthesize several polynomials in a single D0L system. Explicitly, this technique of *merging* means the following. Assume that we have some synthesizable polynomials, say $p_0(i), p_1(i), p_2(i),$

of the same *degree*. Then we can construct a D0L system whose growth function satisfies, for all i,

$$f(3i) = p_0(i), \quad f(3i+1) = p_1(i), \quad f(3i+2) = p_2(i).$$

It is essential that the polynomials are of the *same degree*; in general, the quotient of two mergeable functions should be bounded from above by a constant.

Recall the matrix representation of D0L growth functions, $f(i) = \Pi \, M^i \, \eta$. Functions of this form, where M is a square matrix and Π and η are row and column vectors of the same dimension, all with *integer* entries, are termed *Z-rational*. (This terminology is quite natural, see [SS].) If the entries are *nonnegative* integers, the function is termed *N-rational*. *D0L growth functions* are a special case of N-rational functions: it is required that all entries in ρ are equal to 1. *PD0L growth functions* are a further special case: it is also required that every row of M has a positive entry. The part of the matrix theory that comes into use here is customarily referred to as the Perron-Frobenius theory.

5.2 Merging and stages of death

Bolgani. We now face the challenging task of discussing the differences between the four classes of functions Emmy just introduced. By definition, the classes constitute an increasing hierarchy from PD0L growth functions (smallest class) to Z-rational functions (biggest class). That the inclusions between the classes are strict is also obvious. N-rational functions cannot assume negative values, as Z-rational functions clearly can. 0, 1, 1,... is an N-rational but not a D0L length sequence. Decrease is possible in D0L but not in PD0L length sequences. Such examples can be viewed as trivial, forced by definitions. However, in each case we can show the strictness of the inclusion also by a nontrivial example. Moreover, mathematically very nice characterizations are known for each of the four classes of functions. Emmy should say more about it, let me just give the following rough idea.

The difference between Z-rational and N-rational functions stems from the idea of *merging* or *mergeability*. An N-rational sequence can always be obtained by merging sequences with a "clean dominant term", whereas this is not necessarily possible for Z-rational sequences. By a "clean dominant term" I mean that the expression for an individual member of the sequence has a term $\alpha i^k \rho^i$, where α, ρ are constants and k is a nonnegative integer, completely determining the growth, that is, the member is asymptotically equal to this term. One can say that the *growth order* of the term, as a function of its position i, equals $i^k \rho^i$. As you recall, ρ is a root of the characteristic equation and can in general be a complex number but is real in a clean dominant term.

The difference between D0L and PD0L growth functions can be said to happen at the *primary stage of death*: death affected by the rewriting model itself. We will see that mathematically this difference amounts again to the difference between Z-rational and N-rational functions. Finally, the difference between N-rational and D0L comes from the fact that a *secondary stage of death* is possible for the former. Because some entries of the final vector ρ may equal 0, letters can be erased after the actual D0L sequence has been created. We can also speak of death occurring in the system, contrasted with death caused by outside forces.

Tarzan. I can see the difference between N-rational and D0L very clearly. The secondary stage of death can be viewed also as the *invisibility* of some letters: the letters themselves cannot be seen, although they sometimes produce visible offsprings. Take the very simple D0L system with the axiom ac and rules

$$a \longrightarrow b, \quad b \longrightarrow a, \quad c \longrightarrow d^2, \quad d \longrightarrow c.$$

Take

$$\eta = \begin{pmatrix} 1 \\ 0 \\ 0 \\ 1 \end{pmatrix}$$

so only a and d are kept visible. See? In the underlying D0L sequence a and b alternate, similarly c and d, and the latter also affect exponential growth. But because of death caused by outside forces, the exponential growth is visible only at even stages, the length sequence being 1, 2, 1, 4, 1, 8, ... This is obtained by merging a constant and an exponential sequence. Only sequences with the same growth order can be merged in a D0L system. This follows because clearly $\frac{f(i+1)}{f(i)}$ is bounded from above by a constant, an obvious choice being the maximal length of the right sides of the rules. It is clear by definition that N-rational functions coincide with HD0L growth functions. It is also easy to prove directly that if all entries of η are *positive* then the N-rational function is, in fact, a D0L growth function: a D0L system can be constructed by use of a dummy "filling letter" that takes care of the components of η exceeding 1. This means that the *secondary stage of death*, caused by the entries 0 in η, is really the essential characteristic of N-rational functions, when contrasted with D0L growth functions. The secondary stage of death can be used to eliminate the primary stage, HPD0L growth functions are the same as HD0L ones. This should be easy enough to establish directly; it is of course also a consequence of the characterization result, Theorem 2.3.

Emmy. Essentially everything has already been said. I still want to elaborate the fundamental issues. Take first the role of N-rational functions among Z-rational ones; what Bolgani talked about clean dominant terms. As we have seen, an individual member of a Z-rational sequence can be expressed as an exponential polynomial. Such an exponential polynomial has a part determining its growth order: in the asymptotic behavior only this part is significant. This part is determined by terms $i^t \rho^i$, where ρ has the greatest possible modulus and t is the greatest (for this modulus). Thus, among the roots of the characteristic equation of the growth matrix, we are interested in those with the greatest modulus. There may be many of them since in general the ρ's are complex numbers.

Tarzan. What Bolgani spoke about clean dominant terms means of course that in the decomposition, in the parts to be merged, the dominant terms are real numbers, implying that the original ρ's are real numbers multiplied by roots of unity. This is equivalent to saying that some integral power of ρ is real. I have seen that the term *rational in degrees* is used for such ρ's: the argument of ρ is a rational multiple of π – now the classical π, not your initial vector!

Emmy. You have seen many things during your trips, also from our fields. It was shown in [Be] that the phenomenon described by you happens for ρ's in the N-rational case, the converse result being due to [So2]. Parts of this theory go back to Kronecker. What would be an example? Here it is good that you know some classical analysis. Consider a number α such that

$$\cos 2\pi\alpha = 3/5, \quad \sin 2\pi\alpha = 4/5.$$

Then α must be irrational: the imaginary part of $(3+4i)^t$ is always congruent to 4 modulo 5. Thus, we have here irrationality in degrees. Along these lines one can show that, for instance, the function assuming positive values, defined by

$$f(2m) = 30^m, \quad f(2m+1) = 25^m\cos^2 2\pi m\alpha$$

is Z-rational but not N-rational. It is an interesting detail that Z-rational and N-rational functions coincide in the polynomially bounded case [SS].

The difference between Z-rational and N-rational functions turns out to be decisive also in the characterization of PD0L growth functions among D0L growth functions [So1]. For any D0L growth function $f(m)$, the differences $d(m) = f(m+1) - f(m)$ constitute a Z-rational function. The function $f(m)$ is a PD0L growth function iff the differences $d(m)$ constitute an N-rational function. (We exclude here the empty word as the axiom.) This result gives the possibility of using any Z-rational function $g(m)$ assuming nonnegative values and being not N-rational to obtain a non-decreasing D0L growth function that is not a PD0L growth function. The idea is to merge the sequences R^m and $R^m + g(m)$, where R is a large enough integer. If you do not like the cosine present in the preceding example, I will give you another, still closely related example:

$$f(2m) = 10^m, \quad f(2m+1) = 10^m + m \cdot 5^m + (3+4i)^m + (3-4i)^m.$$

Thus, although the total size of this D0L creature keeps increasing all the time, it must have cells which die immediately! This holds independently of the number of cell types (letters) we are using.

Bolgani. The D0L system of this example has an estimated 200 letters. You are not likely to find such examples just by playing around with D0L systems. The method works here in a backward manner. You begin with some roots ρ, algebraic numbers not rational in degrees. You then construct your characteristic equation, and then the matrix. When you have expressed the resulting Z-rational sequence as the difference of two PD0L sequences, you start to be in business and can work with L systems from that point on. These new aspects have brought with themselves several questions and decision problems not normally asked and studied in classical mathematics. For instance, how to decide of a given polynomial equation with integer coefficients whether or not it has a root outside the unit circle? This information can be used to decide whether or not growth is exponential.

Emmy. I failed to mention that every Z-rational function can be expressed as the difference of two PD0L growth functions. This result, with PD0L replaced by N-rational, was known already before the theory of L systems, and the strengthening is an application of the merging techniques, [SS]. We have clarified just about everything concerning the four classes of functions. We still have to be explicit about the relation between D0L growth functions and N-rational functions. Tarzan already observed that $\frac{f(i+1)}{f(i)}$ is bounded from above by a constant whenever f is a D0L growth function. Also the converse holds: whenever f is an N-rational function assuming positive values and there is a constant c such that $\frac{f(i+1)}{f(i)} \leq c$ holds for all i, then f is a D0L growth function. The proof of this converse, if presented in detail, is the most involved among the proofs of the results concerning growth functions we have been talking about. The most elegant approach is the use of generating functions, originally due to [KOE] and presented in detail in [RS1]. Another consequence of this converse is that a finite number of D0L growth functions with the same growth order can always be merged into one D0L growth function.

Bolgani. Although we have a basic understanding of D0L growth functions, a lot of open problems still remains. We should still talk about them. D0L systems have quite amazing features and capabilities behind their very simple and innocent-looking outer appearance. Let me mention an example.

Recall the Lemma of Long Constant Intervals. If a function stays constant in arbitrarily long intervals then it must be ultimately constant in order to qualify as a D0L growth function. A D0L length sequence can stagnate only for as many consecutive steps as the alphabet size indicates. If we see only the length sequence, we do not know the alphabet size but we know it must have some specific size. It is very natural to expect that a statement analogous to the Lemma of Long Constant Intervals holds true also for intervals, where the length increases strictly at every step. Indeed, when I was first asked this question, I felt sure when I gave a positive answer, thinking that it was only a matter of straightforward matrix calculation to prove the result. But not at all! The opposite result holds true, as shown in [K3]. There is a D0L system SURPRISE whose growth function f satisfies $f(i) < f(i-1)$ for infinitely many values of i. Thus, SURPRISE shrinks every now and then, and this continues forever. Thinking about the periodicities in D0L sequences, you would not expect SURPRISE to have arbitrarily long phases of strict growth. But it does. For each k, there is an i such that

$$f(i) < f(i+1) < f(i+2) < \ldots < f(i+k).$$

No matter how long a phase you need for SURPRISE to become bigger and stronger during the whole phase, a proper timing gives you such a phase!

5.3 Stagnation and malignancy

Tarzan. Cancer is sometimes identified with exponential growth. I would like to call growth in a D0L system *malignant* if, for some rational $t > 1$ and $i_0 \geq 0$,

$$f(i) > t^i \text{ whenever } i \geq i_0.$$

We know that, for D0L growth functions but not in the DIL case, this is equivalent to saying that there is no polynomial p such that $f(i) \leq p(i)$ holds for all i.

Suppose I want to play a doctor for creatures of artificial life, modeled by D0L systems. One of them comes to me and I have to tell whether or not he/she will eventually grow cancer. How can I do it? I know how he/she looked as an infant (the axiom) and the rules for the cell development. It is clear to me that if there is some letter that derives in some number of steps two copies of itself and possibly some other letters as well, then the growth is malignant. Here I take into account only letters actually occurring in the sequence, a condition that can be easily tested. So the occurrence of such an *expanding* letter is sufficient for malignancy but is it also necessary? Could the creature develop cancer without having any expanding cell? Probably not. If the condition is also necessary and if I can test of any letter whether or not it is expanding, then I can diagnose all creatures and tell each of them either good or bad news!

Bolgani. You sure can. The necessity of your condition is fairly easy to establish by purely language-theoretic means. If there are no expanding letters, a growth-equivalent system can be constructed having letters $a_1, \ldots, a_n$ such that the right side of the rule for a_i has at most one a_i and no letters a_j with $j < i$. The growth in this new system is bounded by a polynomial of degree $n - 1$. The system SQUARES is of this type with $n = 3$. In the early years of L systems D0L systems of this type were referred to as systems with "rank". Essentially, the rank r is equivalent with polynomial growth of degree $r - 1$.

How to test whether a given letter a is expanding? The following algorithm works. Studying the rules, form a sequence of sets $S_0, S_1, S_2, \ldots$ Each set S_i will be a subset of a finite set T, consisting of all letters of the given D0L system, as well as of all unordered pairs of letters, pairs (b, b) being included. Explicitly, S_0 consists of the pair (a, a). For all i, S_{i+1} is the union of S_i and S_i', where S_i' is obtained as follows. Take an element y of S_i. If y is a single letter c appearing on the right side of the rule for b, include b to S_i'. Assume next that y is a pair (b, c). Then we include the letter d to S_i' if both b and c (or b twice if $b = c$) occur on the right side of the rule for d. Finally, we include the pair (d, e) to S_i' if the right sides of the rules for d and e, put together, contain both b and c.

The sets S_i form an increasing sequence of subsets of a finite set T. There must be an index j such that $S_{j+1} = S_j$. The letter a is expanding exactly in case it belongs to S_j. Indeed, we begin in S_0 with the situation where two a's occur, and investigate backwards all possibilities where they might have come from. In S_j we have reached the point where there are no more possibilities. Since $S_{j+1} = S_j$, nothing new can come up later. So it is decisive whether we have reached a single a latest by S_j. Whenever we reach a single a, we stop. This is bad news for our creature. If we reach S_j without seeing a single a, it is not yet good news. Every other letter has to be tested in the same way.

Tarzan. In the system DEATH$_b$, a is already in S_1, so it is immediately bad news. If you diagnose b in FIB, you get both (a, a) and (a, b) in S_1 and, hence, you get b in S_2. The letter a is also expanding but you have to go up to S_4 to reach bad news. As you described D0L systems with rank, it seems that for them no diagnosis is needed: it follows by the very definition that the system has no expanding letters. I can immediately tell the good news to SQUARES.

Emmy. The diagnosis can also be based on powers of the growth matrix, [K2]. You compute powers M^i, $i = 1, 2, \ldots$. Whenever you find a diagonal element greater than 1, you diagnose malignancy. If you have reached the value $i = 2^n + n - 1$, where n is the dimension, and have found only 0's and 1's in the main diagonal, you may tell the good news.

Tarzan. Recall our friends DAY-PAL and NIGHT-PAL. The growth matrix of the former has the diagonal entry 2, corresponding to the letter b. The letter b is reachable from any nonempty axiom, in fact all letters eventually become b's. Thus, it is bad news for any DAY-PAL(w).

I would like to talk about another notion. Bolgani used the term "stagnation" in connection with the Lemma of Long Constant Intervals. Let us say that a D0L sequence *stagnates* if it has two words w_t and w_{t+p}, $p > 0$, with the same Parikh vector. Clearly, this implies that

$$f(i) = f(i + p) \text{ for all } i \geq t,$$

that is, the lengths start repeating periodically with the period p after the threshold t. The condition is certainly decidable. For instance, we decide whether the language is finite. Every NIGHT-PAL(w) stagnates. Any axiom becomes a power of c in two steps.

What about a much weaker requirement? A D0L sequence *stagnates momentarily* if $f(t) = f(t + 1)$ holds for some t. I can decide momentary stagnation for PD0L sequences. I just look at the subalphabet Σ' consisting of letters mapped to a letter by the morphism. Momentary stagnation is equivalent to the fact that some subset of Σ' appears in the sequence of minimal alphabets. Since the latter sequence is ultimately periodic, I can decide the fact. But things are different in the general D0L case, and momentary stagnation can take place in many ways. It sure must be decidable but how can I decide it?

Bolgani. You got me. This happens to be a celebrated open problem. Given a D0L growth function f, one has to decide whether or not $f(t) = f(t+1)$ holds for some t. The term "constant level" had also been used, instead of your "momentary stagnation". The following formulation is equivalent in the sense that an algorithm for either one of the problems can be converted into an algorithm for the other problem. Given two PD0L growth functions f and g, decide whether or not $f(t) = g(t)$ holds for some t. An algorithm is known in both cases, [SS], if "for some t" is replaced by "for infinitely many values of t".

Emmy. Also the following is an equivalent formulation for the problem of momentary stagnation. Given a square matrix M with integer entries, decide whether or not the number 0 appears in the upper right-hand corner of some power of M. So far all attempts to determine an upper bound for the exponents to be tested have failed. The problem is decidable for 2×2 matrices. Examples are known of 3×3 matrices where you have to go to the exponent 50 to find the first 0. Again, this problem is decidable if you ask for infinitely many 0s instead of just one 0.

Bolgani. The following problems of *proper thinness* and *continuing growth* are more involved than the problem of momentary stagnation, in the sense that an algorithm for either one of them yields an algorithm for the problem of momentary stagnation but not necessarily vice versa [PaS], [KS]. Decide whether or not a given D0L language has two words of the same length. (A language is called *properly thin* if all its words are of different lengths. This is a special case of slenderness.) Decide whether or not a given D0L growth function f satisfies the inequality $f(i) \leq f(i + 1)$ for all i.

Tarzan. The whole theory of growth functions can apparently be extended to concern DT0L systems as well. Take the system PAL, with some specific axiom. Once we have fixed an order of the tables T_n and T_d, that is a word over the alphabet $\{n, d\}$, we end up with a unique word. From the growth point of view we get a mapping f of the set of words over the alphabet of tables, in the example $\{n, d\}$, into the set of nonnegative integers. In the D0L case the alphabet of tables consists of one letter only. In the DT0L case we can ask basically the same questions as before. Growth equivalence makes sense only if the table alphabets are of the same cardinality. Momentary stagnation takes place if there is a word x and letter a such that $f(xa) = f(x)$. We are now dealing with several growth matrices, one for each letter of the table alphabet; M_n and M_d for PAL. We get the expression $\Pi M_d M_n M_d M_n \rho$ for the length of the DAY-AND-NIGHT-PAL after 48 hours – you understand what I mean. One can also start with a fixed DT0L system, for instance PAL, and investigate what kind of growth phenomena are possible or likely to happen.

Emmy. One can also introduce Z-rational and N-rational functions as before, now the domain of the functions will be some Σ^* instead of N. From the point of view of power series, [SS], [KS], this means several noncommuting variables. Several undecidability results are obtained by a reduction to Hilbert's Tenth Problem, for instance, the undecidability of the problems of momentary stagnation and continuing growth. On the other hand, growth equivalence can be shown decidable by a nice argument using linear spaces, [RS1].

Tarzan. It seems to me that you can be proud of many nice and certainly nontrivial results. Many challenging problems remain open. After some progress in areas such as artificial life you surely will ask entirely new kinds of questions about L growth. I have to go.

6. L codes, number systems, immigration

6.1 Morphisms applied in the way of a fugue

In the remainder of this chapter, we will present some recent work dealing with L systems. Two problem areas will be presented in more detail in Sections 6. and 7., whereas the final Section 8. will run through results in different areas in a rather telegraphic way. Unavoidably, the choice of material reflects at least to some extent our personal tastes. However, we have tried to choose material significant beyond L systems, as well as to present techniques showing the power of language-theoretic methods in completely different areas, a typical example being the results concerning *number systems* presented below.

The purpose of this Section 6. is to present some recent work in L systems, dealing with several developmental processes that have started at different times. One can visualize the different processes as chasing one another like the tunes in a fugue. We will restrict our attention to D0L processes. In fact, apart from some suggestions, very little work concerning other types of L systems has been done in this problem area, although most questions can be readily generalized to concern also other types of L systems. The issues discussed below will concern also generalized number systems, combinatoricss on words, codes and cryptography.

We consider again morphisms $h : \Sigma^* \longrightarrow \Sigma^*$. The morphisms can be used as an *encoding* in the natural way: words w over Σ are encoded as $h(w)$. If h is *injective*, *decoding* will always be unique. This is not the case for the morphism h defined by

$$h(a) = ab \ , h(b) = ba, \ h(c) = a.$$

The word *aba* can be decoded both as *ac* and *cb* because h is not injective. Because of unique decodability, injective morphisms are referred to as *codes*. There is a chapter below in this Handbook dealing with the theory of codes. Usually codes are defined as sets of words rather than morphisms. For finite codes, our definition is equivalent to the customary definition in the following sense. A morphism h is a code iff the set $\{h(a)|\ a \in \Sigma\}$ is a code, provided h is *non-identifying*, that is, $a \neq b$ implies $h(a) \neq h(b)$ for all letters a and b. We still repeat the decodability aspect in terms of cryptography. The morphism h being a code means that every "cryptotext" w' can be "decrypted" in at most one way, that is, there is at most one "plaintext" w that is "encrypted" by h into w': $h(w) = w'$. Most "monoalphabetic" cryptosystems in classical cryptography are codes in this sense.

We now come back to the morphism $h : \Sigma^* \longrightarrow \Sigma^*$ (not necessarily injective) and apply it in the "fugue way". This means that h is applied to the first letter, h^2 to the second, h^3 to the third, and so on, the results being catenated. This gives rise to a mapping $\overline{h} : \Sigma^* \longrightarrow \Sigma^*$, referred to as the L *associate* of h. We now give the formal definition.

Definition 6.1. *Given a morphism* $h : \Sigma^* \longrightarrow \Sigma^*$, *its L associate* $\overline{h}$ *is defined to be the mapping of* Σ^* *into* Σ^* *such that always*

$$\overline{h}(a_1 a_2 \ldots a_n) = h(a_1)h^2(a_2) \ldots h^n(a_n),$$

where the a's are (not necessarily distinct) letters of Σ. *By definition,* $\overline{h}(\lambda) = \lambda$. *The morphism* h *is termed an L code iff its L associate is injective, that is, there are no distinct words* w_1 *and* w_2 *such that* $\overline{h}(w_1) = \overline{h}(w_2)$. $\quad\square$

The L associate $\overline{h}$ is rather seldom a morphism itself. In fact, $\overline{h}$ is a morphism exactly in case h is idempotent, that is, $h^2 = h$. Also the equation $h\overline{h} = \overline{h}h$ is not valid in general, which makes many problems concerning L codes rather involved. Consider the classical "Caesar cipher" h affecting a circular permutation of the English alphabet:

$$h(a) = b, h(b) = c, \ldots, h(y) = z, h(z) = a.$$

For the L associate we obtain, for instance,

$$\overline{h}(aaaa) = bcde, \quad \overline{h}(dcba) = eeee.$$

The L associate is not a morphism but h is both a code and an L code, satisfying the equation $h\overline{h} = \overline{h}h$. The proof of the next result is straightforward.

Theorem 6.1. *Every code is an L code but not vice versa.* $\quad\square$

A significant class of L codes that are not codes results by considering *unary* morphisms. By definition, a morphism h is *unary* iff there is a specific letter a such that $h(b)$ is a power of a for every letter b. (Unary morphisms should not be confused with unary L systems; for the former, the alphabet

Σ may still contain several letters.) Consider the unary morphism h defined by

(*) $$h(a) = a^2, \ h(b) = a.$$

Clearly, h is not a code, for instance, $h(bb) = h(a)$. However, h is an L code. Indeed, this follows by considering *dyadic representations* of integers, with the digits 1 and 2. Compute the exponents of a in $\overline{h}(aaa)$, $\overline{h}(baaba)$, $\overline{h}(bbbb)$:

$$2 \cdot 2^0 + 2 \cdot 2^1 + 2 \cdot 2^2, \ 1 \cdot 2^0 + 2 \cdot 2^1 + 2 \cdot 2^2 + 1 \cdot 2^3 + 2 \cdot 2^4, \ 1 \cdot 2^0 + 1 \cdot 2^1 + 1 \cdot 2^2 + 1 \cdot 2^3.$$

Observe that the original letters b and a behave exactly as the digits 1 and 2, respectively, in the *dyadic representation* of positive integers. For instance, *baaba* represents the number 53, written in dyadic notation $21221 = 2 \cdot 2^4 + 1 \cdot 2^3 + 2 \cdot 2^2 + 2 \cdot 2^1 + 1 \cdot 2^0$.

See [S1] for an exposition concerning *n-adic* and *n-ary* number systems. In both cases n is the base of the representation, but the digits are 1, 2, ...,n in the n-adic and 0, 1,..., $n - 1$ in the n-ary representation; 1, 2 in dyadic, and 0,1 in binary representation. The presence of 0 among the digits renders n-ary systems ambiguous, because an arbitrary number of initial 0's can be added to any word without changing the represented number. n-adic representation is unambiguous. There is a one-to-one correspondence between positive integers and nonempty words over $\{1, 2\}$ when the latter are viewed as dyadic representations. Normally in number systems the leftmost digit is defined to be the most significant one, and we will follow this practice below. The definition of an L associate, which is the natural one because words are normally read from left to right, makes the rightmost letter most significant because there the morphism is iterated most. Thus, actually the mirror image 12212 of the dyadic word 21221 corresponds to the original word *baaba*. This technical inconvenience (of going from words to their mirror images) is irrelevant as regards ambiguity. The morphism h defined by (*) is an L code. If there were two distinct words w and w' such that $\overline{h}(w) = \overline{h}(w')$, then the mirror images of w and w' would be (interpreting b as the digit 1 and a as the digit 2) two distinct dyadic representations for the same number, which is impossible.

Similarly, one can associate a number system to any unary morphism. The morphism is an L code iff the number system is unambiguous. This quite remarkable interconnection between number systems and L systems will now be presented more formally.

Definition 6.2. *A number system is a $(v + 1)$-tuple*

$$N = (n, m_1, \ldots, m_v)$$

of positive integers such that $v \geq 1$, $n \geq 2$, and $1 \leq m_1 < m_2 < \ldots < m_v$. The number n is referred to as the base *and the numbers m_i as* digits. *A nonempty word $m_{i_k} m_{i_{k-1}} \ldots m_{i_1} m_{i_0}$, $1 \leq i_j \leq v$ over the alphabet $\{m_1, \ldots, m_v\}$ is said to* represent *the integer*

$$[m_{i_k} \ldots m_{i_0}] = m_{i_0} + m_{i_1} n + m_{i_2} n^2 + \ldots + m_{i_k} n^k.$$

The set of all represented integers is denoted by $S(N)$. A set of positive integers is said to be representable *by a number system (briefly, RNS), if it equals the set $S(N)$ for some number system N. Two number systems N_1 and N_2 are called* equivalent *if $S(N_1) = S(N_2)$. A number system N is called* complete *if $S(N)$ equals the set of all positive integers, and* almost complete *if there are only finitely many positive integers not belonging to $S(N)$. A number system is termed* ambiguous *if there are two distinct words w_1 and w_2 over the alphabet $\{m_1, \ldots, m_v\}$ such that $[w_1] = [w_2]$. Otherwise, N is termed* unambiguous. *An RNS set is termed* unambiguous *if it equals $S(N)$, for some unambiguous number system N. Otherwise, it is termed* inherently ambiguous.   □

Number systems as defined above are often referred to as "generalized number systems". They are more general than the customary notion because the number and size of the digits is independent of the base. Some integers may also have several representations or none at all. On the other hand, we have excluded the digit 0 (ordinary binary and decimal systems do not qualify) because it immediately induces ambiguity. For each $n \geq 2$, the number system $N = (n, 1, 2, \ldots, n)$ is complete and unambiguous. It is customarily referred to as the *n-adic* number system.

Let $h : \Sigma^* \longrightarrow \Sigma^*$ be a unary morphism. Thus, there is a letter $a \in \Sigma$ such that all h-values are powers of a. Assume that $h(a) = a^n$, $n \geq 2$. The *associated number system* is defined by

$$N(h) = (n, m_1, \ldots, m_v),$$

where the m_i's are the exponents, arranged in increasing order, in the equations $h(b) = a^{m_i}$ where b ranges over the letters of Σ. (We assume that h is non-erasing and non-identifying. The assumptions about h are needed to ensure that $N(h)$ is indeed a number system. On the other hand, if $h(a) = a$ or h is erasing or identifying, it is not an L code.) Observe that the base is always among the digits in a number system $N(h)$. By the above discussion, the following theorem is an immediate consequence of the definitions.

Theorem 6.2. *A unary morphism is an L code iff the associated number system is unambiguous.*   □

L codes were introduced in [MSW2], where also the basic interconnections with number systems were discussed. The ambiguity problem for number systems was shown decidable in [Ho1]. The theory of generalized number systems was initiated in [CS2]. We focus here our attention to basic issues and issues connected with L codes. The reader is referred to [Ho2] – [Ho9] for related problems and some quite sophisticated results (concerning, for instance, regularity, degrees of ambiguity, changes of the base and negative digits). The next theorem summarizes some basic results from [MSW2] and [CS2].

Theorem 6.3. *A number system $N = (n, m_1, \ldots, m_v)$ is ambiguous if $v > n$. N is unambiguous if the digits m_j lie in different residue classes modulo n. No finite set is RNS, whereas every cofinite set is RNS, Every union of some residue classes modulo n, $n \geq 2$, is RNS. Consequently, both even and odd numbers form RNS sets. The former is unambiguous, whereas the latter is inherently ambiguous. There is an RNS set possessing representations with different bases m and n such that it has an unambiguous representation with m, whereas every representation with the base n is ambiguous. There are a 0L systems G and a DT0L system G_1, both with the alphabet $\{a, b\}$ such that*

$$L(G) = L(G_1) = \{b\} \cup \{ba^i \mid i \in S(N)\}. \qquad \square$$

Equivalence is undecidable for 0L systems, whereas it is decidable for U0L systems (see Theorems 4.2 and 4.3). The last sentence of Theorem 6.3 gives a somewhat stronger decidability result. This follows by the decidability results for number systems, discussed still in Section 6.2.

Is an arbitrary given morphism an L code? At the time of this writing, the decidability of this problem is open in its general form, There is every reason to believe the problem to be decidable, especially because the remaining class of morphisms with an open decidability status seems rather small. For unary morphisms, the decidability follows by Theorem 6.2. The decidability was established for *permutation-free* morphisms in [Ho4], that is, for morphisms h such that h permutes no subalphabet Σ_1 of Σ. The next theorem gives the decidability result in its strongest known form, [Ho7].

Given a morphism $h : \Sigma^* \longrightarrow \Sigma^*$, we call a letter a *bounded* (with respect to h) if there is a constant k such that $|h^n(a)| \leq k$ holds for all n. Otherwise, a is said to be *growing*. A letter a is *pumping* if $h^n(a) = uav$ holds for some n and words u and v such that uv is nonempty but contains only bounded letters.

Theorem 6.4. *It is decidable whether or not a morphism for which no letter is pumping is an L code.* $\qquad \square$

6.2 An excursion into number systems

This subsection deals exclusively with number systems. We feel that such a brief excursion is appropriate to show the versatility of L systems. The work dealing with L codes has led to problems dealing with the representation of positive integers in arbitrary number systems. Typical questions concern the equivalence and ambiguity of number systems. In spite of their fundamental number-theoretic nature and also in spite of the fact that the representation of integers is fundamental in the theory of computing, very little was known about the solution of such problems before the interconnection with L codes was discovered. Many of the results below are interesting also withing a general language-theoretic setup: no other than a language-theoretic proof is known for these number-theoretic results. Our exposition will be illustrated by some examples.

Consider first the number system $N_1 = (2, 2, 3, 4)$. We claim that $S(N_1)$ consists of all positive integers that are not of the form $2^k - 3$, for some $k \geq 2$. (Thus, 1, 5, 13, 29, 61 are the first few numbers missed.) To show that no number of this form is in $S(N_1)$ we proceed indirectly. Let $x = 2^k - 3$ be the smallest such number in $S(N_1)$, and consider the representation $[a_1 \ldots a_m] = x$. We must have $m \geq 2$ and $a_m = 3$ because, otherwise, the represented number is even. But now obviously $[a_1 \ldots a_{m-1}] = 2^{k-1} - 3$, contradicting the choice of x. On the other hand, for any $k \geq 1$, an arbitrary integer x satisfying $2^{k+1} - 2 \leq x \leq 2^{k+2} - 4$ is represented by some word of length k, the upper and lower bounds being represented by 2^k and 4^k, respectively. Thus, our claim concerning $S(N_1)$ follows. The system N_1 is ambiguous: $[32] = [24] = 8$. The first sentence of Theorem 6.3 can be used to show that $S(N_1)$ is inherently ambiguous. In the dyadic number system $(2, 1, 2)$, $S(N_1)$ is represented by all words over $\{1, 2\}$ that are not of the form $2^i 1$, for some $i \geq 0$. (Thus, a regular expression can be given for the set of words representing $S(N_1)$ in dyadic notation. A general statement of this fact is contained in the Translation Lemma below.)

The number system $N_2 = (2, 1, 4)$ is unambiguous, by the second sentence of Theorem 6.3. We claim that $S(N_2)$ equals the set of numbers incongruent to 2 modulo 3. Indeed, all numbers in $S(N_2)$ are of this type. This is clearly true of numbers represented by words of length 1 over the digit alphabet. Whenever x is congruent to 0 (resp.1) modulo 3, then both $2x + 1$ and $2x + 4$ are congruent to 1 (resp.0) modulo 3. Hence, by induction, every number in $S(N_2)$ is incongruent to 2 modulo 3. That all such numbers are in $S(N_2)$ is again seen indirectly. If $x = 3k$ or $x = 3k + 1$ is the smallest such number outside $S(N_2)$, we can construct by a simple case analysis a still smaller number outside $S(N_2)$.

Very simple number systems can lead to tricky situations not yet clearly understood. Consider, for $k \geq 3$, the number system $N(k) = (2, 2, k)$. When is $N(k)$ unambiguous? This happens if k is odd or if $k = 2m$ with an even m. The remaining case is not so clear. The first odd values of m yielding an unambiguous $N(K)$ are: 11, 19, 23, 27, 35, 37, 39, 43, 45, 47, 51, 53, 55, 59, 67, 69, 71, 75, 77, 79, 83, 87, 89, 91, 93, 95, 99. [MSW2] contains a more comprehensive theory about this example.

We now introduce a tool basic for decidability results. It consists of viewing the sets $S(N)$ as regular languages.

Translation Lemma. *Given a number system $N = (n, m_1, \ldots, m_v)$, a regular expression $\alpha(N)$ over the alphabet $\{1, \ldots, n\}$ can be effectively constructed such that the set of words in the regular language denoted by $\alpha(N)$, when the words are viewed as n-adic numbers, equals $S(N)$.* □

The reader is referred to [CS2] for details of the proof. The argument is a very typical one about regular languages and finite automata. It is easy to get a rough idea. A finite automaton (sequential machine) translates words $w\#$, where $w \in \{m_1, \ldots, m_v\}^+$, into words over $\{1, \ldots, n\}$. The word w is viewed as a number represented according to N in reverse notation, and the translate will be the representation of the same number in reverse n-adic notation. The "carry" (due to digits exceeding the base) is remembered by the state of the automaton. (A reader not familiar with this terminology is referred to Chapter 2 of this Handbook.) It turns out that $2 \max(n, m_v)$ states will suffice. In one step, when reading the letter j in the state i, the automaton outputs j' and goes to the state i', where i' and j' are unique integers satisfying

$$i + j = j' + i'n, \quad 1 \le j' \le n.$$

When reading the boundary marker $\#$ in the state i, the automaton produces the output i in reverse n-adic notation. This is the only step, where the output may consist of more than one letter.

As an example, consider the number system $N = (2, 13, 22)$. We use the notation $a = 13, b = 22$ to avoid confusion. The computation of the automaton for the input $abaa\#$ looks as follows:

state	0	6	13	12	12
input	a	b	a	a	$\#$
output	1	2	2	1	212
new state	6	13	12	12	

The mirror image $aaba$ of the input $abaa$ represents in N the number 213:

$$[aaba] = 13 \cdot 2^3 + 13 \cdot 2^2 + 22 \cdot 2 + 13 = 213.$$

The mirror image 2121221 of the output is the dyadic representation of 213. We have already pointed out the notational inconvenience caused by the fact that in the number-system notation the leftmost digit is the most significant one.

The claims in the following theorem are either immediate consequences of the Translation Lemma, or can be inferred from it using decidability results concerning regular languages.

Theorem 6.5. *It is decidable whether or not a given number system is ambiguous, complete or almost complete. The equivalence problem is decidable for number systems. It is undecidable whether or not a given recursively enumerable set is RNS. It is also undecidable, given a recursively enumerable set S and an integer $n \ge 2$, whether or not there is a number system N with base n such that $S = S(N)$. It is decidable of a given number system N and an integer $n \ge 2$ whether or not there exists an unambiguous number system N' with base n satisying $S(N) = S(N')$.*   □

For the number system $N = (3, 1, 3, 4, 6, 7)$ and $n = 2$, the decision method of the last sentence produces the number system $N' = N_2 = (2, 1, 4)$

already discussed above. On the other hand, every representation of $S(N)$ with base 3 is ambiguous. Thus, although the property characterizing $S(N)$ is intimately connected with the number 3, one has to choose a different base in order to get an unambiguous representation of $S(N)$!

Using more sophisticated methods, dealing with recognizable sets and formal power series, the decidability of the equivalence can be extended to concern number systems with arbitrary integer digits [Ho5]. Also the final theorem in this subsection [Ho9] results from an application of such methods. A weaker version of the theorem was established in [Ho6].

Theorem 6.6. *Given a number system, it is decidable whether or not there exists an equivalent unambiguous number system.* □

6.3 Bounded delay and immigration

We now return to codes and L codes. For easier reference, we use certain bold letters to denote classes of morphisms: **C** stands for the class of codes and **L** for the class of L codes. The notation **P** refers to *prefix codes*, that is, morphisms h for which there are no distinct letters a and b for which $h(a)$ is a prefix of $h(b)$. (It is clear that morphisms satisfying this condition are codes.) We now present the idea of *bounded delay*.

With cryptographic connotations in mind, let us refer to the argument w as *plaintext* and to the encoded version $h(w)$ or $\overline{h}(w)$ as *cryptotext*. The idea behind the *bounded delay* is the following. We do not have to read the whole cryptotext on order to start writing the plaintext but rather always a certain prefix of the cryptotext determines the beginning of the plaintext.

The notation $\mathrm{pref}_k(w)$ for the prefix of w of length $k \geq 1$ was already introduced in Section 2.. The notation $\mathrm{first}(w)$ stands for the first letter of a nonempty word w. A morphism h is of *bounded delay* k if, for all words u and w, the equation

$$\mathrm{pref}_k(h(u)) = \mathrm{pref}_k(h(w))$$

implies the equation $\mathrm{first}(u) = \mathrm{first}(w)$. The morphism h is of *bounded delay*, in the class **B**, if it is of bounded delay k, for some k. It is again easy to see that the property of bounded delay implies the property of being a code. The morphism h defined by

$$h(a) = aa, \quad h(b) = ba, \quad h(c) = b$$

is a code but not of bounded delay. Indeed, one might have to read the whole cryptotext in order to determine whether the first plaintext letter is b or c. Different notions of bounded delay are compared in [Br] and [BeP]. The same class **B** is obtained no matter which definition is chosen, but different definitions may lead to different minimal values of k.

The idea of bounded delay is the same for codes and L codes: first k letters of the cryptotext determine the first plaintext letter. For codes the situation remains unaltered after handling the first letter a, because the cryptotext is still of the form $h(w)$ when $h(a)$ has been removed from the beginning. However, for L codes, the remainder of the cryptotext equals $h\overline{h}(w)$ rather than $\overline{h}(w)$. This means that we obtain different notions of bounded delay, depending on whether we are interested in finding only the first plaintext letter (*weak* notion, **W**), or the first letter at each stage of decryption (*strong* or *medium strong* notion). The difference between the two latter notions depends on the way of bounding the delay: is the bound kept constant (*strong* notion **S**), or is it allowed to grow according to the stage of decryption (*medium strong* notion, **M**). We now give the formal definition.

Definition 6.3. *A morphism h is of* weakly bounded delay $k \geq 1$ *if, for all words u and w, the equation*

$$pref_k(\overline{h}(u)) = pref_k(\overline{h}(w))$$

implies the equation $first(u) = first(w)$. If for all $i \geq 0$ and all u and w, the equation

$$pref_k(h^i\overline{h}(u)) = pref_k(h^i\overline{h}(w))$$

implies the equation $first(u) = first(w)$, then h is of strongly bounded delay k. *In general, h is of weakly or strongly bounded delay if it is so for some k. The notations* **W** *and* **S** *are used for the corresponding classes of morphisms. Finally, h is of* medium bounded delay *(notation* **M***) if, for some recursive function f and all $i \geq 0$, u and w, the equation*

$$pref_{f(i)}(h^i\overline{h}(u)) = pref_{f(i)}(h^i\overline{h}(w))$$

implies the equation $first(u) = first(w)$. □

Observe that we do not require h to be an L code in these definitions. The situation is analogous to that concerning ordinary codes. However, a morphism being in **B** implies that it is in **C**, whereas **L** and **W** are incomparable. All inclusion relations between the classes introduced are presented in the following theorem [MSW3].

Theorem 6.7. *The mutual inclusion relations between the families introduced are given by the following diagram, where an arrow denotes strict inclusion and two families are incomparable if they are not connected by a path:*

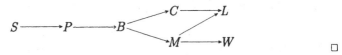

□

Medium bounded delay can be viewed in the theory of L codes as the most natural counterpart of bounded delay codes. It is natural to require that only a bounded amount of lookahead at each stage of the decryption process is needed. If the amount of lookahead remains the same throughout the process, the resulting notion is a very restricted one, as will be seen below. On the other hand, the drawback in the definition of **M** is that, in general, the construction of the sequence of values $f(i) = k_i$, $i = 0, 1, 2, \ldots$, seems to be an infinitary task. The following theorem, [HoS], is pleasing because it shows that it suffices to construct values only up to $\text{card}(\Sigma) - 2$.

We say that a morphism $h : \Sigma^* \longrightarrow \Sigma^*$ is an the set $\mathbf{M'}$ if, for some $k > 0$ and all i with $0 \leq i \leq \text{card}(\Sigma) - 2$, the equation

$$\text{pref}_k(h^i \overline{h}(u)) = \text{pref}_k(h^i \overline{h}(w))$$

always implies the equation $\text{first}(u) = \text{first}(w)$. Thus, we consider the sequence $f(i) = k_i$ only up to $\text{card}(\Sigma) - 2$, and take the maximum of the resulting numbers.

Theorem 6.8. $\mathbf{M'} = \mathbf{M}$.   □

We now present a simple characterization for the family $\mathbf{S}$.

Theorem 6.9. *A morphism h is in $\mathbf{S}$ iff, for any distinct letters a and b, $\text{first}(h(a)) \neq \text{first}(h(b))$.*

Proof. Consider the "if"-part. The assumption means that there is a permutation Π of the alphabet Σ such that, for all a,

$$\text{first}(h(a)) = \Pi(a).$$

Consequently, for all $i \geq 0$ and a,

$$\text{first}(h^i(a)) = \Pi^i(a).$$

Therefore,

$$\text{pref}_1(h^i \overline{h}(a)) = \text{first}(h^{i+1}(a)) = \Pi^{i+1}(a)$$

uniquely determines a, that is, h is of strongly bounded delay 1.

For the "only if" part, we need two auxiliary results that are simple exercises concerning morphisms. Growing letters were defined at the end of Subsection 6.1.

Claim 1. Assume that h is a code and a is a letter. Then either a is growing, or else $|h^i(a)| = 1$, for all i.

Claim 2. If h is a prefix code and a is growing, then $\text{first}(h(a))$ is growing.

Let h be in $\mathbf{S}$. By Theorem 6.7, h is a prefix code. We proceed indirectly and assume that there are two distinct letters a and b such that $\text{first}(h(a)) = \text{first}(h(b))$. Since h is a prefix code, we may write

$$h(a) = cxdy \text{ and } h(b) = cxez,$$

where x, y, z are (possibly empty) words and c, d, e are letters such that $d \neq e$. By Claim 1, a and b are growing. By Claim 2, also c is growing. Hence, for every k, there is an i such that

$$\text{pref}_k(h^i \overline{h}(a)) = \text{pref}_k(h^{i+1}(a)) = \text{pref}_k(h^{i+1}(b)),$$

which contradicts the assumption that h is in $\mathbf{S}$.   □

At the time of this writing, no general decision method is known for testing membership in **M** or **W**.

Some more sophisticated decidability results are obtained, [Ho7], by considering *ambiguity sets* closely connected with the theory of L codes. Essentially, a morphism being an L code means that the ambiguity set is empty. It can be shown that in most cases the ambiguity set is regular, which leads to decidability.

Variations in degrees of iteration constitutes the basic idea behind L codes. One can view the definition also as developmental processes started at different times. The same idea has led to the notion of *L systems with immigration*. One begins with a finite set of words consisting, intuitively, of "beginnings", "origins", "atoms", "founding fathers" that start to develop in an environment. An L system is used to model the development in the environment. The more time has elapsed since immigration, the more steps have been taken in the developmental process.

For any types of L systems, the corresponding system with immigration can be defined. However, so far only D0L systems have been investigated in this respect. We conclude with a few remarks about this area. We begin with a definition.

Definition 6.4. *A D0L system with immigration (shortly, ImD0L system) is a triple $G = (\Sigma, h, B)$ where Σ is an alphabet, $h : \Sigma^* \longrightarrow \Sigma^*$ a morphism and B is a finite subset of Σ^*. Its language is defined by*

$$L(G) = \{b_0 h(b_1) \ldots h^n(b_n) \mid n \geq 0, b_i \in B\}.$$

An ImD0L system is growing *if each word of B contains a growing letter.*   □

As a construct, an ImD0L system is a DF0L system. However, the definition of its language is entirely different. Intuitively, the words of B describe the various possibilities of immigration to the population, and the words of $L(G)$ describe various developmental stages of the immigrants. Mathematically ImD0L systems constitute a very natural generalization of D0L languages. Especially results concerning *subword complexity*, with some best-possible bounds, are of interest [Ho8]. We conclude with some further results from [Ho8].

Theorem 6.10. *Every ImD0L language is an HDT0L language and possesses effectively a test set. Regularity is decidable for languages generated by growing ImD0L systems.*   □

7. Parallel insertions and deletions

L systems are based on the language-theoretic notion of *finite substitution* over an alphabet Σ. So far, a substitution has been defined as an operation on an alphabet. A substitution is never applied to λ (except for the convention that λ is always mapped into λ). The work on parallel insertions [Ka1], [Ka2], [KMPS] can be viewed as an attempt to understand the substitution on the empty word.

Let L_1, L_2 be two languages over an alphabet Σ. The operation of *parallel insertion* of a language L_2 into a language L_1, can be viewed as a nonstandard modification of the notion of substitution. It maps all letters of Σ into themselves and the empty letter into L_2, with the following additional convention. For each word w, between all the letters and also at the extremities, only one λ occurs. The effect of the substitution applied to L_1 will be the insertion of words belonging to L_2 between all letters and also at the extremities of words belonging to L_1.

The exact effect of the classical substitution that maps all letters into themselves and λ into a language L_2 would be the insertion of arbitrary many words of L_2 between letters and at the extremities of words in L_1. According to the definitions mentioned above, this would amount to the parallel insertion of L_2^* into L_1.

While preserving the type of parallelism characteristic to L systems, parallel insertion has a greater degree of nondeterminism: words of L_2 are indiscriminately inserted between all letters of the target words in L_1. One way to regulate the process of insertion is by introducing the notion of *controlled parallel insertion*: each letter determines what can be inserted after it.

Another way to look at operations of controlled insertion is the following. Such an operation can be viewed as a *production* of the form $a \longrightarrow aw$, where the word w comes from the language to be inserted next to a. The mode of controlled insertion determines how the productions are going to be applied. The controlled parallel insertion resembles, thus, the rewriting process of 0L systems. However, it gives rise to something different from 0L systems because the productions are always of the special form and, on the other hand, there are infinitely many productions.

Formally, if L_1, L_2 are languages over the alphabet Σ, the parallel insertion of L_2 into L_1 is defined as:

$$L_1 \Longleftarrow L_2 = \bigcup_{u \in L_1} (u \Longleftarrow L_2),$$

where

$$u \Longleftarrow L_2 = \{v_0 a_1 v_1 a_2 v_2 \ldots a_k v_k | \ k \geq 0, a_j \in \Sigma, 1 \leq j \leq k,$$

$$v_i \in L_2, 0 \leq i \leq k \text{ and } u = a_1 a_2 \ldots a_k\}.$$

The case $k = 0$ corresponds to the situation $u = \lambda$, when only one word $v_0 \in L_2$ is inserted.

As parallel insertion is associative, it induces a monoid structure on 2^{Σ^*}, the set of all subsets of Σ^*, with $\{\lambda\}$ as the neutral element. The monoid is not commutative. For example, $a \Longleftarrow b = \{bab\}$ whereas $b \Longleftarrow a = \{aba\}$. The families of regular, context-free and context-sensitive languages are closed under parallel insertion. Indeed, for L_1, L_2 languages over Σ we have that $L_1 \Longleftarrow L_2 = L_2 s(L_1)$, where s is the λ-free substitution defined by

$$s : \Sigma^* \longrightarrow 2^{\Sigma^*}, \quad s(a) = aL_2, \text{ for every } a \in \Sigma.$$

The assertion above now follows as the families of regular, context-free and context-sensitive languages are closed under catenation and λ–free substitutions.

The parallel insertion amounts thus to the application of a single substitution. If, as in the case of L systems, we consider iterated applications of the substitution, the obtained operation is much more powerful than the parallel insertion: starting with two one-letter words the iterated parallel insertion can produce a non-context-free language. Formally, if L_1, L_2 are languages over Σ, the parallel insertion of order n of L_2 into L_1 is inductively defined by the equations:

$$\begin{aligned} L_1 \Longleftarrow^0 L_2 &= L_1, \\ L_1 \Longleftarrow^{k+1} L_2 &= (L_1 \Longleftarrow^i L_2) \Longleftarrow L_2, i \geq 0. \end{aligned}$$

The *iterated parallel insertion* of L_2 into L_1 is then defined as

$$L_1 \Longleftarrow^* L_2 = \bigcup_{n=0}^{\infty} (L_1 \Longleftarrow^n L_2).$$

The iterated parallel insertion is not commutative: the word bbb belongs to $\lambda \Longleftarrow^* b$ but not to $b \Longleftarrow^* \lambda$. It is not associative either as the word $cbcab$ belongs to $a \Longleftarrow^* (b \Longleftarrow^* c)$ but not to $(a \Longleftarrow^* b) \Longleftarrow^* c$.

The iterated parallel insertion of the letter b into itself is

$$b \Longleftarrow^* b = \{b^{2^k-1} \mid k > 0\},$$

which proves that the families of regular and context-free languages are not closed under iterated parallel insertion. However, the family of context-sensitive languages is still closed under it. Indeed, given two context-sensitive grammars G_1 and G_2 generating the languages L_1 and L_2, a grammar G generating $L_1 \Longleftarrow^* L_2$ can be obtained as follows. G contains the rules of G_2, the rules of G_1, and the rules of G_1 modified in such a way that next to each letter, the axiom of G_2 is attached.

An interesting variation on the theme of parallel insertion is to combine it with the commutative closure. The *commutative closure* of a language L is the smallest commutative language containing L . The *commutative closure* can be viewed as a unary operation associating to every language L its commutative closure com(L). The *permuted parallel insertion* of the word v into u consists thus of the parallel insertion into u of all words which are letter-equivalent to v. (Two words are letter-equivalent if one of them is obtained by permuting the letters of the other.)

More precisely, $L_1 \Longleftarrow_p L_2 = \bigcup_{u \in L_1, v \in L_2} (u \Longleftarrow \mathrm{com}(v))$. Obviously, the permuted parallel insertion can be expressed as $L_1 \Longleftarrow_p L_2 = L_1 \Longleftarrow \mathrm{com}(L_2)$. As expected , the fact that the families of regular and context-free languages are not closed under the commutative closure implies that they are not closed under permuted parallel insertion. On the other hand, being closed under both parallel insertion and commutative closure, the family of context-sensitive languages is closed under permuted parallel insertion.

We have dealt so far with operations where the same language is inserted in parallel between all the letters and at the extremities of a given word. The process resembles more the type of rewriting characteristic to L systems if every letter determines what can be inserted after it. Let L be a language over Σ and $\Delta : \Sigma \longrightarrow 2^{\Sigma^*}$ a so-called *control function* satisfying $\Delta(a) \neq \emptyset$, $\forall a \in \Sigma$. The Δ*-controlled parallel insertion* into L (shortly controlled parallel insertion) is defined as:

$$L \Longleftarrow_c \Delta = \bigcup_{u \in L} (u \Longleftarrow_c \Delta),$$

where

$$u \Longleftarrow_c \Delta = \{a_1 v_1 a_2 v_2 \ldots a_k v_k \mid u = a_1 \ldots a_k, k \geq 1, a_i \in \Sigma,$$

$$\text{and } v_i \in \Delta(a_i), 1 \leq i \leq k\}.$$

Note that in the above definition, the control function cannot have the empty set as its value. This condition has been introduced because of the follwing reason. If there would exist a letter $a \in \Sigma$ such that $\Delta(a) = \emptyset$, then all the words $u \in L$ which contain a would give $u \Longleftarrow_c \Delta = \emptyset$. This means that these words would not contribute to the result of the controlled parallel insertion. Consequently we can introduce, without loss of generality, the condition $\Delta(a) \neq \emptyset$, $\forall a \in \Sigma$.

If we impose the restriction that for a distinguished letter $b \in \Sigma$ we have $\Delta(b) = L_2$, $L_2 \subseteq \Sigma^*$, and $\Delta(a) = \lambda$ for every letter $a \neq b$, we obtain a particular case of controlled parallel insertion: *parallel insertion next to the letter b*. It is a binary language operation, whereas the arity of the Δ-controlled parallel insertion equals $\mathrm{card}(\Sigma) + 1$.

The families of regular, context-free and context-sensitive languages are closed under controlled parallel insertion. This follows as the result of the controlled parallel insertion $L \Longleftarrow_c \Delta$, where $L \subseteq \Sigma^*$ and $\Delta : \Sigma \longrightarrow 2^{\Sigma^*}$, $\Delta(a) \neq \emptyset$, $\forall a \in \Sigma$, can be accomplished by the λ-free substitution

$$\sigma : \Sigma^* \longrightarrow 2^{\Sigma^*}, \quad \sigma(a) = a\Delta(a), \forall a \in \Sigma.$$

For each of the above mentioned variants of insertion, a dual deletion operation can be defined. Take, for example, the *parallel deletion*. Given words

u and v, the parallel deletion of v from u consists of the words obtained by simultaneously erasing from u all the nonoverlapping occurrences of v. The definition is extended to languages in the natural way. Given a word u and a language L_2, the parallel deletion $u \Longrightarrow L_2$ consists of the words obtained by erasing from u all the nonoverlapping occurrences of words in L_2:

$$u \Longrightarrow L_2 = \{u_1 u_2 \ldots u_k u_{k+1} \mid k \geq 1, u_i \in \Sigma^*, 1 \leq i \leq k+1 \text{ and}$$
$$\exists v_i \in L_2, 1 \leq i \leq k \text{ such that } u = u_1 v_1 \ldots u_k v_k u_{k+1},$$
$$\text{where } \{u_i\} \cap [\Sigma^*(L_2 \setminus \{\lambda\})\Sigma^*] = \emptyset, 1 \leq i \leq k+1\}.$$

Note that, besides the fact that the parallel deletion erases from u nonoverlapping occurrences of words from L_2, a supplementary condition has to be fulfilled: between two occurrences of words of L_2 to be erased, no nonempty word from L_2 appears as a subword. This assures that *all* occurrences of words from L_2 have been erased from u, and is taken care of by the last line of the definition. The reason why λ had to be excluded from L_2 is clear. If this wouldn't be the case and λ would belong to L_2, the condition $\{u_i\} \cap \Sigma^* L_2 \Sigma^* = \emptyset$ would imply $\{u_i\} \cap \Sigma^* = \emptyset$ – a contradiction. Note that words from L_2 can still appear as subwords in $u \Longrightarrow L_2$, as the result of catenating the remaining pieces of u. If L_1, L_2 are languages over Σ, we can define now $L_1 \Longrightarrow L_2 = \cup_{u \in L_1} (u \Longrightarrow L_2)$.

If L, R are languages over the alphabet Σ, L a λ-free language and R a regular one, then there exists [Ka1] a rational transducer g, a morphism h and a regular language R' such that $L \Longrightarrow R = h(g(L) \cap R')$. As a corollary, the families of regular and context-free languages are closed under parallel deletion with regular languages. On the other hand, the family of context-free languages is not in general closed under parallel deletion. For example, consider $\Sigma = \{a, b\}$ and the context-free languages:

$$L_1 = \#\{a^i b^{2i} \mid i > 0\}^*,$$

$$L_2 = \#a\{b^i a^i \mid i > 0\}^*.$$

If the language $L_1 \Longrightarrow L_2$ would be context-free, then also the language

$$(L_1 \Longrightarrow L_2) \cap b^+ = \{b^{2^n} \mid n > 0\},$$

would be context-free, which is a contradiction.

A similar nonclosure situation happens in the case of context-sensitive languages: there exists a context-sensitive language L_1 and a word w over an alphabet Σ such that $L_1 \Longrightarrow w$ is not a context-sensitive language. Indeed, let L be a recursively enumerable language (which is not context-sensitive) over an alphabet Σ and let a, b be two letters which do not belong to Σ. Then there exists a context-sensitive language L_1 such that (see [S1]):

(i) L_1 consists of words of the form $a^i b \alpha$, where $i \geq 0$ and $\alpha \in L$;
(ii) For every $\alpha \in L$, there is an $i \geq 0$ such that $a^i b \alpha \in L_1$.

It is easy to see that $aL_1 \Longrightarrow \{a\} = bL$, which is not a context-sensitive language. We have catenated a to the left of L_1 in order to avoid the case $i = 0$, when the corresponding words from L would have been lost.

In a similar way we have defined the iterated parallel insertion, we can define the *iterated parallel deletion*. Despite of the simplicity of the languages involved, it is an open problem whether or not the family of regular languages is closed under iterated parallel deletion. Not only that, but it is still an open problem whether the iterated parallel deletion of a singleton word from a regular language still belongs to the family of regular languages. The answer to both of the previous questions is negative in the case of context-free and context-sensitive languages. For example, if we consider the alphabet $\Sigma = \{a, b\}$, the context-free language $L = \{a^i \# b^{2i} | \ i > 0\}^*$, and the word $w = ba$ then

$$(L \Longrightarrow^* ba) \cap a\#^+ b^+ = \{a\#^n b^{2^n} | \ n > 0\},$$

which implies that $L \Longrightarrow^* ba$ cannot be context-free.

The *permuted parallel deletion* of a language L_2 from L_1 is obtained by erasing from words in L_1 words that are letter-equivalent to words in L_2:

$$L_1 \Longrightarrow_p L_2 = \bigcup_{u \in L_1, v \in L_2} (u \Longrightarrow_p v),$$

where $u \Longrightarrow_p v = u \Longrightarrow \mathrm{com}(v)$. If $\Sigma = \{a, b\}$ and $L_1 = \$a^* b^* \#\# a^* b^* \$$ and $L_2 = \#\$(ab)^*$ then the permuted parallel deletion of L_2 from L_1 is

$$L_1 \Longrightarrow_p L_2 = \begin{cases} \{\$a^n b^m \# | \ m, n \geq 0, m \neq n\} \cup \\ \{\# a^n b^m \$ | \ m, n \geq 0, m \neq n\} \cup \{\lambda\}, \end{cases}$$

which shows that the result of permuted parallel deletion between two regular languages is not necessarily regular. If L_1 is the context-free language

$$L_1 = \{a_1^n b_1^m c_1^l \# c_2^l b_2^m a_2^n \# | \ n, m, l \geq 0\}$$

and L_2 is the regular language $L_2 = \#\#(a_2 b_2 c_2)^*$ then

$$L_1 \Longrightarrow_p L_2 = \{a_1^n b_1^n c_1^n | \ n \geq 0\}.$$

This shows that the family of context-free languages is not closed under permuted parallel deletion with regular languages. By using an argument similar to the one used for parallel deletion, one can show that the family of context-sensitive languages is not closed under permuted parallel deletions with singleton languages.

A *controlled* variant of the *parallel deletion* can be defined as follows. Let $u \in \Sigma^*$ be a word and $\Delta : \Sigma \longrightarrow 2^{\Sigma^*}$ be a control function which does not have $\emptyset$ as its value. The set $u \Longrightarrow_c \Delta$ is obtained by finding all the nonoverlapping occurrences of av_a, $v_a \in \Delta(a)$, in u, and by deleting v_a from them. Between any two ocurrences of words of the type av_a, $v_a \in \Delta(a)$, in u, no other words of this type may remain.

If one imposes the restriction that for a distinguished letter $b \in \Sigma$ we have $\Delta(b) = L_2$, and $\Delta(a) = \lambda$ for any letter $a \neq b$, a special case of controlled parallel deletion is obtained: *parallel deletion next to the letter b*. The parallel deletion next to the letter b is denoted by $\overset{b}{\Longrightarrow}$. Let us examine the set $u \overset{b}{\Longrightarrow} L_2$, where u is a nonempty word and L_2 is a language over an alphabet Σ. If $u = b^k$, $k > 0$, and no word of the form bv, $v \in L_2$ occurs as a subword in u, the set $u \overset{b}{\Longrightarrow} L_2$ equals the empty set. If u contains at least one letter different from b, u is retained in the result as we can erase λ near that letter. The other words in $u \overset{b}{\Longrightarrow} L_2$ are obtained by finding all the nonoverlapping occurrences of words of the type bv_i, $v_i \in L_2$ in u, and deleting v_i from them. There may exist more than one possibility of finding such a decomposition of u into subwords.

One can use similar techniques as for the parallel deletion to show that the family of regular languages is closed under controlled parallel deletion, while the families of context-free and context-sensitive languages are not (except when the languages to be deleted are regular).

Besides studying closure properties, one can get more insight into the power of parallel insertion and deletion operations by investigating classes of languages that contain simple languages and are closed under some of the operations. Such a class should be closed under a parallel insertion operation, a parallel deletion one and an iterated parallel insertion one. Particular controlled operations will be chosen in order to allow an increase as well a decrease of the length of the words in the operands. The iterative operation has been included to provide an infinite growth of the strings. Finally, we have to introduce the mirror image and the union with λ for technical reasons.

Thus, let $\mathcal{P}$ be the smallest class of languages which contains the empty set, the language $\{\lambda\}$, the singleton letters and is closed under mirror image, union with $\{\lambda\}$, controlled parallel insertion, iterated controlled parallel insertion and controlled parallel deletion with singletons. Then [Ka2] $\mathcal{P}$ is contained in the family of context-sensitive languages and properly contains the family of regular languages. If we replace in $\mathcal{P}$ the controlled parallel deletion with singletons by unrestricted controlled parallel deletion, the new family $\mathcal{P}'$ is a Boolean algebra properly containing the family of regular languages.

The study of parallel insertion and deletion operations is closely connected to the study of equations involving them. If $\diamond$ denotes a parallel insertion or deletion operation, the simplest equations to consider are of the form $L \diamond X = R$, $Y \diamond L = R$, where R is a regular language, L is an arbitrary language, and X, Y are the unknown languages.

If $\diamond$ denotes parallel insertion, permuted parallel insertion or parallel insertion next to a letter, the problem whether there exists a singleton solution $X = \{w\}$ to the equation $L \diamond X = R$ is decidable for regular languages L and R. Indeed, if $\diamond$ denotes parallel insertion, let L, R be regular languages and m be the length of the shortest word in R. If there exists a word w such that $L \Longleftarrow w = R$ then it must satisfy the condition $\lg(w) \leq m$. As the family of regular languages is closed under parallel insertion, our problem is decidable. The algorithm for deciding it will consist of checking whether or not $L_1 \Longleftarrow w = R$ for all words w with $\lg(w) \leq m$. The answer is YES if such a word exists and NO otherwise. A similar proof can be used to show that, for $\diamond$ denoting parallel insertion, controlled parallel insertion and parallel insertion next to a letter, the existence of a singleton solution Y to the equation $Y \diamond L = R$ is decidable for regular languages L and R.

In contrast, given context-free languages L and regular languages R, the existence of both a solution and a singleton solution X, to the equation $L \diamond X = R$ is undecidable for $\diamond$ denoting parallel insertion, iterated parallel insertion, permuted parallel insertion or parallel insertion next to a letter. Let $\#$ be a letter which does not belong to Σ. We can actually prove a stronger statement: there exists a fixed regular language $R = \Sigma^*\#$ such that the problem above is undecidable for context-free languages L_1. This follows by noticing that the equation $(L_1\#) \Longleftarrow X = \Sigma^*\#$ holds for languages L_1, X over Σ exactly in case $L_1 = \Sigma^*$ and $X = \lambda$. Hence, if we could decide our problem, we would be deciding the problem "Is $L_1 = \Sigma^*$?" for context-free languages L_1, which is impossible. A similar proof, taking $R = \Sigma^* \cup \{\#\}$ and $L_1 = L \cup \{\#\}$, can be used to show that the existence of both a solution and singleton solution Y to the equation $Y \Longleftarrow L = R$ is undecidable for context-free languages L and regular languages R. Analogously it can be shown that the existence of both a solution and singleton solution to the equation $Y \Longleftarrow \Delta = R$ is undecidable for context-free control functions and regular languages R.

If we consider the same problems for parallel deletion operations, special attention has to be paid to the fact that, while the result of parallel insertion is always a word or set of words, the result of parallel deletion can be the empty set. If $\diamond$ denotes a binary deletion operation and L is a given language, a word y is called *right-useful with respect to L and $\diamond$* if there exists $x \in L$ such that $x \diamond y \neq \emptyset$. A language X is called right-useful with respect to L and $\diamond$ if it consists only of right-useful words with respect to L and $\diamond$. The notion of *left useful* can be similarly defined.

In the following, when we state the undecidability of the existence of a solution to a certain equation, we will mean in fact that the existence of a *useful* solution is undecidable. For example, if $\diamond$ denotes the parallel deletion, permuted parallel deletion or iterated parallel deletion, the existence of a solution to the equation $L \diamond X = R$ is undecidable for context-free languages L and regular languages R. Let us consider the case of parallel deletion. If $\$, \#$ are letters that do not occur in Σ, there exists a regular language $R = \#\Sigma^+\# \cup \$\Sigma^*\$$ such that our problem is undecidable for context-free languages L. Indeed, for a context-free language L, consider the language $L_1 = \#\Sigma^+\# \cup \$L\$$. For all languages $L_2 \subseteq \Sigma^*$, the equation

$$\#\Sigma^+\# \cup \$L\$ \Longrightarrow X = \#\Sigma^+\# \cup \$\Sigma^*\$$$

holds if and only if $X = \{\lambda\}$ and $L = \Sigma^*$. If we could decide our problem, we could decide whether for given context-free languages L, the equation $L = \Sigma^*$ holds, which is impossible.

The problem remains undecidable even if instead of considering a binary operation, like parallel deletion, we consider a $\mathrm{card}(\Sigma)+1$ – ary operation like controllel parallel deletion. It can be namely shown that the existence of a singleton control function with the property $L \Longrightarrow_c \Delta = R$ is undecidable for context-free languages L and regular languages R. The proof is similar to the preceding one and is based on the fact that the equation

$$L_1 \$ \cup \{\#a\$| \ a \in \Sigma\} \Longrightarrow_c \Delta = \Sigma^+ \cup \{\#a| \ a \in \Sigma\}$$

holds for $L_1 \subseteq \Sigma^*$ and singleton control functions Δ if and only if

$$\Delta(\#) = \Delta(\$) = \lambda, \Delta(a) = \$, a \in \Sigma \text{ and } L_1 = \Sigma^+.$$

The operations of parallel insertion and deletion are associated with several notions rather basic in the combinatorics of words. A *parallel deletion set* is a set of the form $w \Longrightarrow L$, $w \in \Sigma^*$, $L \subseteq \Sigma^*$. Parallel deletions sets are universal in the sense that every finite language can be viewed as a parallel deletion set (see [KMPS]). If we fix the nonempty finite set F in the equation

$$w \Longrightarrow L = F,$$

we can ask for an algorithm deciding for a given context-free language L whether or not a solution w to the equation $w \Longrightarrow L = F$ exists. If such an algorithm exists, we say that F is *CF-decidable*, otherwise *CF-undecidable*. F is called *CF-universal* if for any (nonempty) context-free language L, there is a word w such that $w \Longrightarrow L = F$. We have that the set $\{\lambda\}$ is CF-universal and it is the only CF-universal set.

In spite of the fact that parallel deletion sets coincide with finite sets, every finite nonempty set $F \neq \{\lambda\}$ is CF-undecidable ([KMPS]).

The *parallel deletion number* [KMPS] associated to a word w equals the cardinality of the largest parallel deletion set arising from w, that is

$$pd(w) = \max\{\mathrm{card}(w \Longrightarrow L)| \ L \subseteq \Sigma^*\}.$$

For the alphabet with only one element, $pd(w)$ can be computed, but for the general case the question seems not to be simple at all. Indeed, one can show that if $w = a^n$, $n \geq 1$, then $pd(w) = n$. In the case of arbitrary alphabets with at least two symbols the following surprising result [KMPS] is obtained. If $\mathrm{card}(\Sigma) \geq 2$, then there is no polynomial f such that for every $w \in \Sigma^*$ we have $pd(w) \leq f(|w|)$.

Let us prove this result. It suffices to show that, given a polynomial f (in one variable), there are strings w such that $pd(w) > f(|w|)$.

Take a polynomial f of degree $n \geq 1$ and consider the strings

$$w_{n,m} = (a^m b^m)^n.$$

Moreover, take
$$L_m = \{a^i b^j \mid 1 \le i, j \le m - 1\}.$$
and evaluate the cardinality of $w_{n,m} \Longrightarrow L_m$.

As each string in L_m contains at least one occurrence of a and one occurrence of b, we can delete from $w_{n,m}$ exactly n strings of L_m, which implies

$$w_{n,m} \Longrightarrow L_m = \{a^{m-i_1} b^{m-j_1} a^{m-i_2} b^{m-j_2} \ldots a^{m-i_n} b^{m-j_n} \mid$$

$$1 \le i_s, j_s \le m - 1, 1 \le s \le n\}.$$

Consequently,
$$\text{card}(w_{n,m} \Longrightarrow L_m) = (m-1)^{2n}.$$

Clearly, because $2n$ is a constant, for large enough m we have

$$pd(w_{n,m}) \ge (m-1)^{2n} > f(2nm) = f(|w_{n,m}|),$$

which completes the proof.

We observed that, for every word w, $w \Longrightarrow \Sigma^* = \{\lambda\}$. We can express this by saying that every word *collapses* to the empty word when subjected to parallel deletion with respect to Σ^*. We speak also of the *collapse set* of Σ^*. Thus, the collapse set of Σ^* equals Σ^*. In general, we define the *collapse set* of a nonempty language $L \subseteq \Sigma^*$ by

$$cs(L) = \{w \in \Sigma^* \mid w \Longrightarrow L = \{\lambda\}\}.$$

This language is always nonempty because it contains each of the shortest words in L. For example, $cs(\{a^n b^n \mid n \ge 1\}) = (ab)^+$ and $cs(\{a, bb\}) = a^* bb(a^* bb)^* a^* \cup a^+$, hence $cs(L)$ can be infinite for finite L. On the other hand, $cs(\{ab\} \cup \{a^n b^m a^p \mid n, m, p \ge 1\}) = \{ab\}$, hence $cs(L)$ can be finite for infinite L. We have that [KMPS] there is a linear language L such that $cs(L)$ is not context-free. Indeed, take

$$L = \{dda^n b^m c^n \mid n, m \ge 1\} \cup \{da^n b^m c^p \mid n, m, p \ge 1, m \ge p\}.$$

While it is clear that L is linear,

$$cs(L) \cap d^2 a^+ b^+ c^+ = \{d^2 a^n b^m c^n \mid 1 \le m \le n\},$$

which, being non-context-free, shows that $cs(L)$ cannot be a context-free language. In fact, the collapse set of a language can be characterized as follows. If $L \subseteq \Sigma^*$ then
$$cs(L) = L^+ - M,$$
where
$$M = (\Sigma^* L \cup \{\lambda\})(\Sigma^+ \setminus \Sigma^* L \Sigma^*)(L \Sigma^* \cup \{\lambda\}).$$

As a corollary, we deduce that if L is a regular (context-sensitive) language, then $cs(L)$ is also a regular (context-sensitive) language.

Another natural way to define a parallel deletion operation is to remove exactly k strings, for a given k. Namely, for $w \in \Sigma^*$, $L \subseteq \Sigma^*$, $k \geq 1$, write

$$w \Longrightarrow_k L = \{u_1 u_2 \ldots u_{k+1} \mid u_i \in \Sigma^*, 1 \leq i \leq k+1,$$
$$w = u_1 v_1 u_2 v_2 \ldots u_k v_k u_{k+1}, \text{ for } v_i \in L, 1 \leq i \leq k\}.$$

Sets of this form will be referred to as k-*deletion sets*; for given $k \geq 1$ we denote by E_k the family of k-deletion sets. For all $k \geq 1$, $E_k \subset E_{k+1}$, strict inclusion. Moreover, for every finite set F, there is a k such that $F \in E_k$, and given a finite set F and a natural number k, it is decidable whether $F \in E_k$ or not.

The operations of insertion and deletion seem to occur time and again in modeling natural phenomena. We have seen how L systems model the growth of multicellular organisms. The related notions of parallel insertion and deletion have recently become of interest in connection with the topic of DNA computing.

The area of DNA computing was born in 1994 when Adleman [Ad1] succeeded in solving an instance of the Directed Hamiltonian Path solely by manipulating DNA strands. This marked the first instance where a mathematical problem could be solved by biological means and gave rise to a couple of interesting problems: a) can *any* algorithm be simulated by means of DNA manipulation, and b) is it possible, at least in theory, to design a programmable DNA computer?

To answer these questions, various models of DNA computation have been proposed, and it has been proven that these models have the full power of a Turing machine. The models based on the bio-operations proposed by Adleman can be already implemented in laboratory conditions, but the fact that most bio-operations rely on mainly manual handling of tubes prevents the large scale automatization of the process.

There are two ways to overcome this obstacle and to further the research in DNA computing. The first approach is to try to speed-up and automatize the existing operations.

The second one is to try to design a model based entirely on operations that can be carried out by enzymes. Indeed, already in [Ad2] the idea was advanced to "design a molecular computer (perhaps based on entirely different "primitives" than those used here) which would accomplish its task by purely chemical means inside of a single tube".

As DNA strands can be viewed as strings over the four letter alphabet $\Sigma = \{A, C, G, T\}$ (Adenine, Cytosine, Thymine and Guanine), it is natural to pursue this second approach within the frame of formal language theory (see the chapter in this Handbook written by Head, Paun, and Pixton for a formal language model of DNA computing, based on splicing). The site-specific insertions of DNA sequences could then be modeled by controlled parallel insertions and deletions. Further study of the power of these operations could assist in proving the claim in [SmS] that " only a finite collection of rewrite rules which insert or delete single U's in specified contexts of an mRNA sequence will suffice to get universality."

For attaining this goal, and with biological motivations in mind, the notion of control needs strenghtening: a word can be inserted into a string only if certain *contexts* are present. Formally, given a set $C \in \Sigma^* \times \Sigma^*$ called a *context set*, the *parallel contextual insertion* (a generalization of the contextual insertion operation studied in [KT]) of a word v into the word u is

$$u \Longleftarrow_C v = \{u_1 x_1 v y_1 u_2 x_2 v y_2 \ldots u_k x_k v y_k u_{k+1} | \; k \geq 0,$$

$$u = u_1 u_2 \ldots u_k u_{k+1}, (x_i, y_i) \in C, 1 \leq i \leq k\}.$$

The proof that controlled parallel insertions and deletions are enough to simulate the action of a Turing machine, would thus open a possible way for designing a molecular computer with all the operations carried out by enzymes.

8. Scattered views from the L path

We conclude this chapter with some scattered remarks, mostly about recent work. We have already emphasized that, in such a vast field as L systems, one cannot hope to be encyclopedic in a chapter of this size, especially if one tries to survey also the methods as we have done. Some of the areas neglected by us could have deserved even a section of their own.

Inductive inference certainly constitutes one such area. How much knowledge of the language or sequence is needed to infer the L system behind it, or one of the many possible L systems? This is relevant, for instance, in modeling experimental data. Inductive inference is a part of the theory of *learning*: one tries to get a complete picture from scattered information. Here the setup may vary. New data may come independently of the observer, or the latter may want some specific information, for instance, whether or not a specific word is in the language. Inductive inference (or *syntactic inference* as it is sometimes called) has been studied for L systems from the very beginning [HR], [RSed1], [H1]. The reader is referred to [Yo2] and its references for some recent aspects.

There is a special chapter in this Handbook about *complexity*. [Ha], [JS], [vL] are early papers on the complexity of the membership problem for systems with tables, [vL] being especially interesting in that it provides a very nice class-room example of a reduction to an NP-complete problem. [Ke], [SWY], [LaSch] are examples of various areas of current interest. Most of the complexity issues studied in connection with L systems have been on the *metalevel*: complexity of questions *about* the system (membership, equivalence) rather than things happening *within* the system (length of derivations, workspace). This reflects perhaps the lack of really suitable machine models for L systems [RS1].

If an L system is viewed as an *L form*, then it produces a family of languages rather than just a single language. The original L system is used as a propotype giving rise to a collection of systems resembling it. The study of L forms was originated in [MSW1]; they are also discussed in a later chapter in this Handbook.

[Sz], [V], [K2], [PS], [S3] represent different aspects of the early work on *growth functions*. A related more recent approach has been the application of L operations to power series, [Ku], [KS]. This approach will be dealt with in the chapter by W. Kuich in this Handbook. Also the recent very intensively investigated [He] questions about *splicing* and DNA will be discussed elsewhere in this Handbook.

Many recent invetigations concerning length have dealt with *slenderness* [APDS], [DPS1]. *Restricted parallelism* has been the object of many recent studies; the reader is referred to [Fe] and the references given therein.

Since the L families have weak closure properties, we face the following decision problem. Assume, for instance, that a family $\mathcal{L}$ is not closed under union. Given two languages from $\mathcal{L}$, their union may or may not belong to $\mathcal{L}$. Can we decide which alternative holds? The problem is decidable for PD0L languages and for U0L languages but undecidable for P0L languages [DPS2]. For D0L languages the problem is open, due to difficulties discussed above in connection with growth functions.

It is quite common in language theory that an undecidable problem becomes decidable if sharper restrictions are imposed on the phenomena under study. For instance, equivalence is undecidable for context-free grammars, whereas *structural equivalence*, where also the derivation structures of the two grammars have to be the same, is decidable. The situation is exactly the same for E0L systems [SY]. The papers [OW], [Nie], [SWY] present related results.

For *stochastic* variants of 0L systems, the reader is referred to [JM] and the references there. Stochastic variants have been useful in picture generation [PL]. The area of *cooperating grammars and L systems*, [Pa], is discussed elsewhere in this Handbook.

[Yo1] and [Da2] introduce *control mechanisms* for 0L and DT0L systems, respectively. For instance, a table can be used if some other table has not been used. [IT], [Lan], [Har1], [Har2], [Ko] represent various D0L-related studies. In [Ko] conditions are deduced for D0L-simulation of DIL systems. In the *piecewise D0L systems* (or PD0L systems) of [Har2], the set Σ^* is partitioned, and the morphism depends on which part of the partition the word derived so far belongs to. A sequence of words still results. "Dynamical properties" such as finiteness and periodicity are decidable if the sets in the partition are regular languages.

The monographs [HR] and [RS1] and the collections of papers [RSed1] – [RSed4] contain many further references. The collections [RSed1] – [RSed3] also reflect the state of the art after roughly ten-year intervals. Of special value are the survey articles [L3], [L4], [LJ], where Aristid Lindenmayer was the author or coauthor.

References

[Ad1]   L. Adleman, Molecular computation of solutions to combinatorial problems. *Science* 266, (Nov.1994) 1021–1024.

[Ad2]   L. Adleman, On constructing a molecular computer. Manuscript in circulation.

[AL]   M. Albert and J. Lawrence, A proof of Ehrenfeucht's conjecture. *Theoret. Comput. Sci.* 41 (1985) 121–123.

[APDS]   M. Andrasiu, G. Paun, J. Dassow and A. Salomaa, Language-theoretic problems arising from Richelieu cryptosystems. *Theoret. Comput. Sci.* 116 (1993) 339–357.

[Be]   J. Berstel, Sur les pôles et le quotient de Hadamard de séries N-rationelles, *C. R. Acad.Sci., Sér.A* 272 (1971) 1079–1081.

[BeP]   J. Berstel and D. Perrin, *Theory of codes.* Academic Press, New York (1985).

[Br]   V. Bruyere, Codes prefixes. Codes a delai dechiffrage borne. Nouvelle thèse, Université de Mons (1989).

[Ch]   C. Choffrut, Iterated substitutions and locally catenative systems: a decidability result in the binary case. In [RSed3], 49–92.

[CF]   K. Culik II and I. Fris, The decidability of the equivalence problem for D0L systems. *Inform. and Control* 35 (1977) 20–39.

[CK1]   K. Culik II and J. Karhumäki, Systems of equations over a free monoid and Ehrenfeucht's conjecture. *Discrete Math.* 43 (1983) 139–153.

[CK2]   K. Culik II and J. Karhumäki, A new proof for the D0L sequence equivalence problem and its implications. In [RSed2], 63–74.

[CS1]   K. Culik II and A. Salomaa, On the decidability of homomorphism equivalence for languages. *J. Comput. Systems Sci.* 17 (1978) 163–175.

[CS2]   K. Culik II and A. Salomaa, Ambiguity and decision problems concerning number systems. *Inform. and Control* 56 (1983) 139–153.

[Da1]   J. Dassow, Eine Neue Funktion für Lindenmayer-Systeme. *Elektron. Informationsverarb. Kybernet.* 12 (1976) 515–521.

[Da2]   J. Dassow, On compound Lindenmayer systems. In [RSed2], 75–86.

[DPS1]   J. Dassow, G. Paun and A. Salomaa, On thinness and slenderness of L languages. *EATCS Bull.* 49 (1993) 152–158.

[DPS2]   J. Dassow, G. Paun and A. Salomaa, On the union of 0L languages. *Inform. Process. Lett.* 47 (1993) 59–63.

[ER1]   A. Ehrenfeucht and G. Rozenberg, The equality of E0L languages and codings of 0L languages. *Internat. J. Comput. Math.* 4 (1974) 95–104.

[ER2]   A. Ehrenfeucht and G. Rozenberg, Simplifications of homomorphisms, *Inform. and Control* 38 (1978) 298–309.

[ER3]   A. Ehrenfeucht and G. Rozenberg, Elementary homomorphisms and a solution of the D0L sequence equivalence problem. *Theoret. Comput. Sci.* 7 (1978) 169–183.

[ER4]   A. Ehrenfeucht and G. Rozenberg, On a bound for the D0L sequence equivalence problem. *Theoret. Comput. Sci.* 12 (1980) 339–342.

[En]   J. Engelfriet, The ET0L hierarchy is in the OI hierarchy. In [RSed2], 100–110.

[Fe]   H. Fernau, Remarks on adult languages of propagating systems with re-
         stricted parallelism. In [RSed4], 90–101.
[HW]   G. Hardy and E. M. Wright, *An introduction to the Theory of Numbers.*
         Oxford Univ. Press, London (1954).
[Ha]   T. Harju, A polynomial recognition algorithm for the EDT0L languages,
         Elektron. Informationsverarb. Kybernet. 13 (1977) 169–177.
[Har1]   J. Harrison, Morphic congruences and D0L languages. *Theoret. Comput.
         Sci.* 134 (1994) 537–544.
[Har2]   J. Harrison, Dynamical properties of PWD0L systems. *Theoret. Comput.
         Sci.* 14 (1995) 269–284.
[He]   T. Head, Splicing schemes and DNA. In [RSed3], 371 – 384.
[H1]   G. T. Herman, The computing ability of a developmental model for fila-
         mentous organisms. *J. Theoret. Biol.* 25 (1969) 421 – 435.
[H2]   G. T. Herman, Models for cellular interactions in development without
         polarity of individual cells. *Internat. J. System Sci.* 2 (1971) 271 –289; 3
         (1972) 149–175.
[HR]   G. T. Herman and G. Rozenberg, *Developmental Systems and Languages*
         North-Holland, Amsterdam (1975).
[HWa]   G. T. Herman and A. Walker, Context-free languages in biological sys-
         tems. *Internat. J. Comput. Math.* 4 (1975) 369 – 391.
[Ho1]   J. Honkala, Unique representation in number systems and L codes. *Dis-
         crete Appl. Math.* 4 (1982) 229–232.
[Ho2]   J. Honkala, Bases and ambiguity of number systems. *Theoret. Comput.
         Sci.* 31 (1984) 61–71.
[Ho3]   J. Honkala, A decision method for the recognizability of sets defined by
         number systems. *RAIRO* 20 (1986) 395–403.
[Ho4]   J. Honkala, It is decidable whether or not a permutation-free morphism
         is an L code. *Internat. J. Computer Math.* 22 (1987) 1–11.
[Ho5]   J. Honkala, On number systems with negative digits. *Annales Academiae
         Scientiarum Fennicae, Series A. I. Mathematica* 14 (1989) 149–156.
[Ho6]   J. Honkala, On unambiguous number systems with a prime power base.
         Acta Cybern. 10 (1992) 155–163.
[Ho7]   J. Honkala, Regularity properties of L ambiguities of morphisms. In
         [RSed3], 25–47.
[Ho8]   J. Honkala, On D0L systems with immigration. *Theoret. Comput. Sci.* 120
         (1993) 229 –245.
[Ho9]   J. Honkala, A decision method for the unambiguity of sets defined by
         number systems. *J. of Univ. Comput. Sci.* 1 (1995) 648–653.
[HoS]   J. Honkala and A. Salomaa, Characterization results about L codes.
         RAIRO 26 (1992) 287–301.
[IT]   M. Ito and G. Thierrin, D0L schemes and recurrent words. In [RSed2],
         157–166.
[JS]   N. Jones and S. Skyum, Complexity of some problems concerning L sys-
         tems. *Lecture Notes in Computer Science* 52 Springer-Verlab, Berlin
         (1977) 301–308.
[JM]   H. Jürgensen and D. Matthews, Stochastic 0L systems and formal power
         series. In [RSed2], 167–178.
[K1]   J. Karhumäki, An example of a PD2L system with the growth type 2 1/2.
         Inform. Process. Lett. 2 (1974) 131–134.
[K2]   J. Karhumäki, On Length Sets of L Systems, Licentiate thesis, Univ. of
         Turku (1974).
[K3]   J. Karhumäki, Two theorems concerning recognizable N-subsets of σ^*.
         Theoret. Comput. Sci. 1 (1976) 317– 323.

[Ka1]   L. Kari, On insertions and deletions in formal languages. Ph.D. thesis, University of Turku, Finland, 1991.

[Ka2]   L. Kari, Power of controlled insertion and deletion. *Lecture Notes in Computer Science*, 812 Springer-Verlag, Berlin (1994), 197–212.

[KMPS]   L. Kari, A. Mateescu, G. Paun, A. Salomaa. On parallel deletions applied to a word, *RAIRO – Theoretical Informations and Applications*, vol.29, 2(1995), 129–144.

[KRS]   L. Kari, G. Rozenberg and A. Salomaa, Generalized D0L trees. *Acta Cybern.*, 12 (1995) 1–9.

[KT]   L. Kari, G. Thierrin, Contextual insertions/deletions and computability. To appear in *Information and Computation*. Submitted.

[KOE]   T. Katayama, M. Okamoto and H. Enomoto, Characterization of the structure-generating functions of regular sets and the D0L growth functions. *Inform. and Control* 36 (1978) 85–101.

[Ke]   A. Kelemenová, Complexity of 0L systems. In [RSed2], 179–192.

[Ko]   Y. Kobuchi, Interaction strength of DIL systems. In [RSed3], 107–114.

[Ku]   W. Kuich, Lindenmayer systems generalized to formal power series and their growth functions. In [RSed4], 171–178.

[KS]   W. Kuich and A. Salomaa, *Semirings, Automata, Languages.* Springer-Verlag, Berlin (1986).

[Lan]   B. Lando, Periodicity and ultimate periodicity of D0L systems. *Theoret. Comput. Sci.* 82 (1991) 19–33.

[LaSch] K. J. Lange and M. Schudy, The complexity of the emptiness problem for E0L systems. In [RSed3], 167–176.

[La1]   M. Latteux, Sur les T0L systémes unaires. *RAIRO* 9 (1975) 51–62.

[La2]   M. Latteux, Deux problèmes décidables concernant les TUL langages. *Discrete Math.* 17 (1977) 165–172.

[L1]   A. Lindenmayer, Mathematical models for cellular interaction in development I and II. *J. Theoret. Biol.* 18 (1968) 280–315.

[L2]   A. Lindenmayer, Developmental systems without cellular interactions, their languages and grammars. *J. Theoret. Biol.* 30 (1971) 455–484.

[L3]   A. Lindenmayer, Developmental algorithms for multicellular organisms: a survey of L systems. *J. Theoret. Biol.* 54 (1975) 3–22.

[L4]   A. Lindenmayer, Models for multi-cellular development: characterization, inference and complexity of L-systems. *Lecture Notes in Computer Science* 281 Springer-Verlag, Berlin (1987) 138–168.

[LJ]   A. Lindenmayer and H. Jürgensen, Grammars of development: discrete-state models for growth, differentiation and gene expression in modular organisms. In [RSed3], 3–24.

[Li1]   M. Linna, The D0L-ness for context-free languages is decidable. *Inform. Process. Lett.* 5 (1976) 149–151.

[Li2]   M. Linna, The decidability of the D0L prefix problem. *Intern. J. Comput. Math.* 6 (1977) 127–142.

[Ma]   G. Makanin, The problem of solvability of equations in a free semigroup. *Math. USSR Sb.* 32 (1977) 129–138.

[MSW1]   H. Maurer, A. Salomaa and D. Wood, E0L forms. *Acta Inform.* 8 (1977) 75–96.

[MSW2]   H. Maurer, A. Salomaa, D. Wood, L codes and number systems. *Theoret. Comput. Sci.* 22 (1983) 331–346.

[MSW3]   H. Maurer, A. Salomaa and D. Wood, Bounded delay L codes. *Theoret. Comput. Sci.* 84 (1991) 265–279.

[MT]   L. M. Milne-Thompson, *The Calculus of Finite Differences.* Macmillan, New York (1951).

[MRS]   J. Mäenpää, G. Rozenberg and A. Salomaa, Bibliography of L systems. Leiden University Computer Science Technical Report (1981).

[N]   M. Nielsen, On the decidability of some equivalence problems for D0L systems. *Inform. and Control* 25 (1974) 166 – 193.

[NRSS]   M. Nielsen, G. Rozenberg, A. Salomaa and S. Skyum, Nonterminals, homomorphisms and codings in different variations of 0L systems, I and II. *Acta Inform.* 3 (1974) 357-364; 4 (1974) 87–106.

[Nie]   V. Niemi. A normal form for structurally equivalent E0L grammars. In [RSed3], 133–148.

[Ni]   T. Nishida, Quasi-deterministic 0L systems. *Lecture Notes in Computer Science* 623 Springer-Verlag, Berlin (1992) 65–76.

[NiS]   T. Nishida and A. Salomaa, Slender 0L languages. *Theoret. Comput. Sci.* 158 (1996) 161–176.

[OW]   Th. Ottman and D. Wood, Simplifications of E0L grammars. In [RSed3], 149–166.

[Pa]   G. Paun, Parallel communicating grammar systems of L systems. In [RSed3], 405–418.

[PaS]   G. Paun and A. Salomaa, Decision problems concerning the thinness of D0L languages. *EATCS Bull.* 46 (1992) 171–181.

[PS]   A. Paz and A. Salomaa, Integral sequential word functions and growth equivalence of Lindenmayer systems. *Inform. and Control* 23 (1973) 313–343.

[PK]   P. Prusinkiewicz, L. Kari, Subapical bracketed L systems. *Lecture Notes in Computer Science* 1073 Springer-Verlag, Berlin (1994) 550–565.

[PL]   P. Prusinkiewicz and A. Lindenmayer, *The Algorithmic Beauty of Plants.* Springer-Verlag, Berlin (1990).

[R1]   G. Rozenberg, T0L systems and languages. *Inform. and Control* 23 (1973) 262–283.

[R2]   G. Rozenberg, Extension of tabled 0L systems and languages. *Internat. J. Comput. Inform. Sci.* 2 (1973) 311–334.

[RRS]   G. Rozenberg, K. Ruohonen and A. Salomaa, Developmental systems with fragmentation. *Internat. J. Comput. Math.* 5 (1976) 177–191.

[RS1]   G. Rozenberg and A. Salomaa, *The Mathematical Theory of L Systems.* Academic Press, New York (1980).

[RS2]   G. Rozenberg and A. Salomaa, When L was young. In [RSed2], 383–392.

[RS3]   G. Rozenberg and A. Salomaa, *Cornerstones of Undecidability.* Prentice Hall, New York (1994).

[RSed1]   G. Rozenberg and A. Salomaa (eds.), L systems. *Lecture Notes in Computer Science* 15 Springer-Verlag, Berlin (1974).

[RSed2]   G. Rozenberg and A. Salomaa (eds.), *The Book of L.* Springer-Verlag, Berlin (1985).

[RSed3]   G. Rozenberg and A. Salomaa (eds.), *Lindenmayer Systems.* Springer-Verlag, Berlin (1992).

[RSed4]   G. Rozenberg and A. Salomaa (eds.), *Developments in Language Theory.* World Scientific, Singapore (1994).

[Ru1]   K. Ruohonen, Zeros of Z-rational functions and D0L equivalence, *Theoret. Comput. Sci.* 3 (1976) 283–292.

[Ru2]   K. Ruohonen, The decidability of the F0L-D0L equivalence problem. *Inform. Process. Lett.* 8 (1979) 257–261.

[Ru3]   K. Ruohonen, The inclusion problem for D0L langugaes. *Elektron. Informationsverarb. Kybernet.* 15 (1979) 535–548.

[Ru4]   K. Ruohonen, The decidability of the D0L-DT0L equivalence problem. *J. Comput. System Sci.* 22 (1981) 42–52.

[S1]   A. Salomaa, *Formal Languages*. Academic Press, New York (1973).

[S2]   A. Salomaa, Solution of a decision problem concerning unary Lindenmayer systems. *Discrete Math.* 9 (1974) 71–77.

[S3]   A. Salomaa, On exponential growth in Lindenmayer systems. *Indag. Math.* 35 (1973) 23–30.

[S4]   A. Salomaa, Comparative decision problems between sequential and parallel rewriting. *Proc. Symp. Uniformly Structured Automata Logic,* Tokyo (1975) 62–66.

[S5]   A. Salomaa, *Jewels of Formal Language Theory*. Computer Science Press, Rockville (1981).

[S6]   A. Salomaa, Simple reductions between D0L language and sequence equivalence problems. *Discrete Appl. Math.* 41 (1993) 271 – 274.

[S7]   A. Salomaa, Developmental models for artificial life: basics of L systems. In G. Paun (ed.) Artificial life: Grammatical Models. Black Sea University Press (1995) 22–32.

[SS]   A. Salomaa and M. Soittola, *Automata-Theoretic Aspects of Formal Power Series*. Springer-Verlag, Berlin (1978).

[SWY]   K. Salomaa, D. Wood and S. Yu, Complexity of E0L structural equivalence. *Lecture Notes in Computer Science*. 841 Springer-Verlag, Berlin (1994) 587–596.

[SY]   K. Salomaa and S. Yu, Decidability of structural equivalence of E0L grammars. *Theoret. Comput. Sci.* 82 (1991) 131 – 139.

[SmS]   W. Smith, A. Schweitzer, DNA computers in vitro and in vivo. NEC Research Report, Mar.31, 1995.

[So1]   M. Soittola, Remarks on D0L growth sequences. *RAIRO* 10 (1976) 23–34.

[So2]   M. Soittola, Positive rational sequences. *Theoret. Comput. Sci.* 2 (1976) 317–322.

[Sz]   A. Szilard, Growth Functions of Lindenmayer Systems, Tech. Rep., Comput. Sci, Dep., Univ. of Western Ontario (1971).

[vL]   J. van Leeuwen, The membership question for ET0L languages is polynomially complete. *Inform. Process. Lett.* 3 (1975) 138–143.

[V]   P. Vitany, Structure of growth in Lindenmayer Systems. *Indag. Math.* 35 (1973) 247–253.

[Yo1]   T. Yokomori, Graph-controlled systems–an extension of 0L systems. In [RSed2], 461–471.

[Yo2]   T. Yokomori, Inductive inference of 0L languages. In [RSed3], 115–132.

Combinatorics of Words

Christian Choffrut and Juhani Karhumäki

1. Introduction

The basic object of this chapter is a *word*, that is a sequence – finite or infinite – of elements from a finite set. The very definition of a word immediately imposes two characteristic features on mathematical research of words, namely *discreteness* and *noncommutativity*. Therefore the combinatorial theory of words is a part of noncommutative discrete mathematics, which moreover often emphasizes the *algorithmic* nature of problems.

It is worth recalling that in general noncommutative mathematical theories are much less developed than commutative ones. This explains, at least partly, why many simply formulated problems of words are very difficult to attack, or to put this more positively, mathematically challenging.

The theory of words is profoundly connected to numerous different fields of mathematics and its applications. A natural environment of a word is a finitely generated free monoid, therefore connections to algebra are extensive and diversified. Combinatorics, of course, is a fundamental part of the theory of words. Less evident but fruitful connections are those to probability theory or even to topology via dynamical systems. Last but not least we mention the close interrelation of the theory of words and the theory of automata, or more generally theoretical computer science.

This last relation has without any doubt emphasized the algorithmic nature of problems on words, but even more importantly has played a major role in the process of making the theory of words a mature scientific topic of its own. Indeed, while important results on words were until the 1970s only scattered samples in the literature, during the last quarter-century the research on words has been systematic, extensive, and we believe, also successful.

Actually, it was already at the beginning of this century when A. Thue initiated a systematic study on words, cf. [Be6] for a survey of Thue's work. However, his fundamental results, cf. [T1], [T2] and also [Be8], remained quite unnoticed for decades, mainly due to the unknown journals he used. Later many of his results were discovered several times in different connections.

The modern systematic research on words, in particular words as elements of free monoids, was initiated by M. P. Schützenberger in the 1960. Two influencial papers of that time are [LySc] and [LeSc]. This research created also the first monograph on words, namely [Len], which, however, never became widely used.

Year 1983 was important to the theory: the first book *Combinatorics on Words* [Lo] covering major parts on combinatorial problems of words appeared. Even today it is the most comprehensive presentation of the topic.

The goal of this presentation is to consider combinatorial properties of words from the point of view of formal languages. We do not intend to be exhaustive. Indeed, several important topics such as theory of codes, several problems on morphisms of free monoids, as well as unavoidable regularities like Shirshov's Theorem, are not considered in this chapter, but are discussed in other chapters of the Handbook. Neither are the representations of the topics chosen supposed to be encyclopedic.

On the other hand, the criteria we have had in our minds when choosing the material for this chapter can be summarized as follows. In addition to their relevance to formal languages we have paid special attention to select topics which are not yet considered in textbooks, or at least to have a fresh presentation of older topics. We do not prove many of the results mentioned. However, we do prove several results either as examples of proof techniques used, or especially if we can give a proof which has not yet appeared in textbooks. We have made special efforts to fix the terminology.

The contents of our chapter are now summarized.

In Section 2 we fix our terminology. In doing so we already present some basic facts to motivate the notions. Section 3 deals with three selected problems. These problems – mappings between word monoids, binary equality languages, and a separation of words by a finite automaton – are selected to illustrate different typical problems on words.

Section 4 deals with the well-known defect effect: if n words satisfy a nontrivial relation, then they can be expressed as products of at most $n - 1$ words. We discuss different variations of this result some of which emphasizing more combinatorial and some more algebraic aspects. We point out differences of these results, including the computational ones, as well as consider the defect effect caused by several nontrivial relations.

In Section 5 we consider equations over words and their use in defining properties of words, including several basic ones such as the conjugacy. We also show how to encode any Boolean combination of properties, each of which expressable by an equation, into a single equation. Finally, a survey of decidable and undecidable logical theories of equations is presented.

Section 6 is devoted to a fundamental property of periodicity. We present a proof of the Theorem of Fife and Wilf which allows one to analyse its optimality. We also give an elegant proof of the Critical Factorization Theorem from [CP], and finally discuss an interesting recent characterization of ultimately periodic words due to [MRS].

In Section 7 we consider partial orderings of words and finite sets of words. As we note there, normally such orderings are not finitary either in the sense that all antichains or in the sense that all chains would be finite. There are two remarkable exceptions. Higman's Theorem restricted to words states that the subword ordering, i.e., the ordering by the property being a (sparse) subword, allows only finite antichains, and is thus a well-ordering. We also consider several extensions of this ordering defined using special properties of words.

The other finiteness condition is obtained as a consequence of the validity of the Ehrenfeucht Compactness Property for words, which itself states that each system of equations with a finite number of unknowns is equivalent to one of its finite subsystems. As an application of this compactness property we can define a natural partial ordering on finite sets of words, such that it does not allow infinite chains. This, in turn, motivates us to state and solve some problems on subsemigroups of a free semigroup.

Section 8 is related to the now famous work of Thue. We give a survey on results which repetitions or abelian repetitions are avoidable in alphabets of different sizes. We also estimate the number of finite and infinite cube-free and overlap-free words over a binary alphabet, as well as square-free words over a ternary alphabet. We present, as an elegant application of automata theory to combinatorics of words, an automata-theoretic presentation due to [Be7] of Fife's Theorem, cf.[F], characterizing one-way infinite overlap-free (or 2^+-free) words over a binary alphabet. Finally, we recall the complete characterization of binary patterns which can be avoided in infinite binary words.

In Section 9, last of this chapter, we consider the complexity of an infinite word defined as the function associating to n the number of factors of length n in the considered word. Besides examples, we present a complete classification, due to [Pan2], of the complexities of words obtained as fixed points of iterated morphisms.

Finally, as a technical matter of our presentation we note that results are divided into two categories: Theorems and Propositions. The division is based on the fact whether the proofs are presented here or not. Theorems are either proved in details or outlined in the extend that an experienced reader can recover those, while Propositions are stated with only proper references to the literature.

2. Preliminaries

In this section we recall basic notions of words and sets of words, or languages, used in this chapter. The basic reference on combinatorics of words is [Lo], see also [La] or [Shy]. The notions of automata theory are not defined here, but can be found in any textbook of the area, cf. e.g. [Be1], [Harr], [HU] or [Sal1], or in appropriate chapters of this Handbook.

2.1 Words

Let Σ be a finite *alphabet*. Elements of Σ are called *letters*, and sequences of letters are called *words*, in particular, the *empty word*, which is denoted by 1, is the sequence of length zero. The set of all words (all nonempty words, resp.) is denoted by Σ^* (Σ^+, resp.). It is a monoid (semigroup, resp.) under the operation of *concatenation* or *product* of words. Moreover, obviously each word has the unique representation as products of letters, so that Σ^* and Σ^+ are *free*, referred to as the *free monoid* and *semigroup generated by* Σ.

Although we may assume for our purposes that Σ is finite we sometimes consider infinite words as well as finite ones: a *one-way infinite* word, or briefly an infinite word, can be identified with a mapping $\mathbb{N} \to \Sigma$, and is normally represented as $w = a_0 a_1 \ldots$ with $a_i \in \Sigma$. Accordingly, *two-way infinite*, or *bi-infinite*, words over Σ are mappings $\mathbb{Z} \to \Sigma$. We denote the sets of all such words by Σ^ω and $^\omega\Sigma^\omega$, respectively, and set $\Sigma^\infty = \Sigma^* \cup \Sigma^\omega$. The notions $\mathbb{Z}$ and $\mathbb{N}$ are used to denote the sets of integers and nonnegative integers, respectively.

Let u be a word in Σ^*, say $u = a_1 \ldots a_n$ with $a_i \in \Sigma$. We define $u(i)$ to denote the ith letter of u, i.e., $u(i) = a_i$. We say that n is the *length* of u, in symbols $|u|$, and note that it can be computed by the morphism $|\,| : \Sigma^* \to \mathbb{N}$ defined as $|a| = 1 \in \mathbb{N}$, for $a \in \Sigma$. The sets of all words over Σ of length k, or at most k are denoted by Σ^k and $\Sigma^{\leq k}$, respectively. By $|u|_a$, for $a \in \Sigma$, we denote the total number of the letter a in u. The *commutative image* $\pi(u)$ of a word u, often referred to as its *Parikh image*, is given by the formula $\pi(u) = (|u|_{a_1}, \ldots, |u|_{a_{\|\Sigma\|}})$, where $\|\Sigma\|$ denotes the cardinality of Σ and Σ is assumed to be ordered. The *reverse* of u is the word $u^R = a_n \ldots a_1$, and u is called a *palindrome* if it coincides with its reverse. For the empty word 1 we pose $1^R = 1$. By alph(w) we mean the minimal alphabet where w is defined.

Finally a *factorization* of u is any sequence $u_1, \ldots, u_t$ of words such that $u = u_1 \ldots u_t$. If words u_i are taken from a set X, then the above sequence is called an *X-factorization* of u. A related notion of an *X-interpretation* of u is any sequence of words $u_1, \ldots, u_t$ from X satisfying $\alpha u \beta = u_1 \ldots u_t$ for some words α and β, with $|\alpha| < |u_1|$ and $|\beta| < |u_t|$. These notions can be illustrated as in Figure 2.1.

Fig. 2.1. An X-factorization and an X-interpretation of u

For a pair (u, v) of words we define four relations:

u is a *prefix* of v, if there exists a word z such that $v = uz$;

u is a *suffix* of v, if there exists a word z such that $v = zu$;

u is a *factor* of v, if there exist words z and z' such that $v = zuz'$;

u is a *subword* of v, if v as a sequence of letters contains u as a subsequence, i.e., there exist words $z_1, \ldots, z_t$ and $y_0, \ldots, y_t$ such that $u = z_1 \ldots z_t$ and $v = y_0 z_1 y_1 \ldots z_t y_t$.

Sometimes factors are called *subwords*, and then subwords are called *sparse subwords*. We, however, prefer the above terminology. Each of the above relations holds if $u = 1$ or $u = v$. When these trivial cases are excluded the relations are called *proper*. A factor v of a word u can occur in u in different *positions* each of those being uniquely determined by the length of the prefix of u preceding v. For example, ab occurs in $abbaabab$ in positions 0, 4 and 6.

If $v = uz$ we write $u = vz^{-1}$ or $z = u^{-1}v$, and say that u is the *right quotient of v by z*, and that z is the *left quotient of v by u*. Consequently, the operations of right and left quotients define partial mappings $\Sigma^* \times \Sigma^* \to \Sigma^*$. Note that the above terminology is motivated by the fact that the free monoid Σ^* is naturally embedded into the free group generated by Σ. We also write $u \leq v$ ($u < v$, resp.) meaning that u is a prefix (a proper prefix, resp.) of v. Further by $\mathrm{pref}_k(v)$ and $\mathrm{suf}_k(v)$, for $k \in \mathbb{N}$, we denote the prefix and the suffix of v of length k. Finally, we denote by $\mathrm{pref}(x)$, $\mathrm{suf}(x)$, $F(x)$ and $SW(x)$ the sets of all prefixes, suffixes, factors and subwords of x, respectively.

It follows immediately that Σ^* satisfies, for all words $u, v, x, y \in \Sigma^*$ the condition

(1) $uv = xy \Rightarrow \exists t \in \Sigma^* : u = xt$ and $tv = y$, or $x = ut$ and $v = ty$.

Similarly, as we already noted, the length function of Σ^* is a morphism into the additive monoid $\mathbb{N}$:

(2) $$h : \Sigma^* \to \mathbb{N} \quad \text{with} \quad h^{-1}(0) = 1.$$

Conditions (1) and (2) are used to characterize the freeness of a monoid, cf. [Lev]. Consequently, Σ^* is indeed free as a monoid.

For two words u and v neither of these needs to be a prefix of another. However, they always have a unique *maximal common prefix* denoted by $u \wedge v$. Similarly, they always have among their common factors longest ones. Let us denote their lengths by $l(u, v)$. These notions allow us to define a *metric* on the sets Σ^* and Σ^ω. For example, by defining *distance functions* as

$$d(u, v) = |uv| - 2l(u, v) \quad \text{for} \quad u, v \in \Sigma^*,$$

and

$$d_\infty(u, v) = 2^{-|u \wedge v|} \quad \text{for} \quad u, v \in \Sigma^\omega,$$

(Σ^*, d) and $(\Sigma^\omega, d_\infty)$ become metric spaces.

As we shall see later the above four relations on words are *partial orderings*. The most natural *total orderings* of Σ^* are the *lexicographic* and *alphabetic* orderings, in symbols $\prec_l$ and $\prec_a$, defined as follows. Assume that the alphabet Σ is totally ordered by the ordering $\prec$. This is extended to Σ^* in the following ways:

$$u \prec_l v \text{ iff } u^{-1}v \in \Sigma^+ \text{ or } \text{pref}_1((u \wedge v)^{-1}u) \prec \text{pref}_1((u \wedge v)^{-1}v)$$

and

$$u \prec_a v \text{ iff } |u| < |v| \text{ or } |u| = |v| \text{ and } u \prec_l v.$$

Consequently, u is lexicographically smaller than v if, and only if, either u is a proper prefix of v, or the first symbol after the maximal common prefix $u \wedge v$ is smaller in u than in v. It follows that the orderings $\prec_a$ and $\prec_l$ coincide on words of equal length. In some respects they, however, behave quite differently: each word u is preceded only by finitely many words in $\prec_a$, while for $\prec_l$ this holds only for words composed on the smallest letter of Σ.

It follows directly from the definition that the alphabetic ordering $\prec_a$ is *compatible* with the product on two sides: for all words $u, v, z, z' \in \Sigma^*$ we have

$$u \prec_a v \text{ iff } zuz' \prec_a zvz'.$$

For the lexicographic ordering $\prec_l$ the situation is slightly more complicated. As is straightforward to see we have for all $u, v, z, z' \in \Sigma^*$,

$$u \prec_l v \text{ iff } zu \prec_l zv,$$

and

$$u \prec_l v \text{ and } u \notin \text{pref}(v) \text{ implies that } uz \prec_l vz'.$$

2.2 Periods in words

We continue by defining some further notions of words, in particular those connected to periodicity.

We say that words u and v are *conjugates*, if they are obtainable from each other by the cyclic permutation $c : \Sigma^* \to \Sigma^*$ defined as

$$\begin{aligned} c(1) &= 1, \\ c(u) &= \text{pref}_1(u)^{-1}u\,\text{pref}_1(u) \text{ for } u \in \Sigma^+. \end{aligned}$$

Consequently, u and v are conjugates if, and only if, there exists a k such that $v = c^k(u)$. It follows that the conjugacy is an equivalence relation, each class consisting of words of the same length. It also follows that the equivalence class $[u]$ is included in $F(uu)$, or even in $F(\text{pref}_1(u)^{-1}uu)$.

Next we associate periods to each word $u \in \Sigma^+$. Let $u = a_1 \ldots a_n$ with $a_i \in \Sigma$. A *period* of u is an integer p such that

$$(1) \qquad\qquad a_{p+i} = a_i \quad \text{for } i = 1, \ldots, n - p.$$

The smallest p satisfying (1) is called *the period* of u, and it is denoted by $p(u)$. It follows that any $q \geq |u|$ is a period of u, and that

$u \in \mathrm{pref}(\mathrm{pref}_{p(u)}(u))^{\omega}$ and $u \notin F(v^{\omega})$ for any $v \in \Sigma^{\leq p(u)-1}$.

It also follows that the conjugates have the same periods. The words in the conjugacy class $[\mathrm{pref}_{p(u)}(u)]$ are called *cyclic roots* of u. Note that not all cyclic roots of u need to be factors of u, but at least one, namely the prefix of u of length $p(u)$, is so.

We say that a word $u \in \Sigma^{+}$ is *primitive*, if it is not a proper integer power of any of its cyclic roots. We claim that this is equivalent to the following condition (often used as the definition of the primitiveness):

(2) $\qquad \forall z \in \Sigma^{*} : u = z^{n}$ implies $n = 1$ (and hence $u = z$).

Clearly, (2) implies the primitiveness. To see the reverse we assume that u is primitive and $u = z^{n}$ with $n \geq 2$. Then denoting $r = \mathrm{pref}_{p(u)}(u)$ we have the situation depicted as

u :

Since $|r|$ is the period necessarily $|z| \geq |r|$. Moreover, by the primitiveness $z \notin r^{*}$. Consequently, comparing the prefixes of length $|r|$ in the first two occurrences of z we can write

(3) $\qquad\qquad r = ps = sp \quad \text{with} \quad p, s \neq 1.$

The identity (3) is the most basic on combinatorics of words, and implies – after a few line proof, cf. Corollary 4.1 – that p and s are powers of a nonempty word. Therefore u would have a smaller period than $|r|$, a contradiction.

We derive directly from the above argumentation the following representation result of words.

Theorem 2.1. *Each word $u \in \Sigma^{+}$ can be uniquely represented in the form $u = \rho(u)^{n}$, with $n \geq 1$ and $\rho(u)$ primitive.* $\qquad\qquad\square$

The word $\rho(u)$ in Theorem 2.1 is called the *primitive root* of the word u.

There exist two particularly interesting subcases of primitive words: unbordered and Lyndon words. A word $u \in \Sigma^{+}$ is said to be *unbordered*, if none of its proper prefix is one of its suffixes. In terms of the period $p(u)$ this can be stated as

$\qquad u \in \Sigma^{+}$ is unbordered if, and only if, $p(u) = |u|$.

It follows that unbordered words are primitive. Moreover, unbordered words have the following important property: different occurrences of an unbordered factor u in a word w never *overlap*, i.e., they are separate:

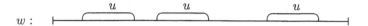

w :

On the other hand, if $u \in \Sigma^+$ is not unbordered, i.e., is *bordered*, then it contains an overlap:

(4)

Consequently, bordered words are sometimes called *overlapping*.

As we noted the situation depicted in (4) is impossible for unbordered words. If u is only primitive, then a variant of (4) is as follows: no primitive word u can be an *inside* factor of uu, i.e., whenever $uu = pus$, then necessarily $p = 1$ or $s = 1$. Being an inside factor can, of course, be illustrated as

This, indeed, is impossible for primitive words by the argument used in (3).

We note that this simple lemma of primitive words is extremely useful in many concrete considerations. As a general example fast algorithms for testing the primitiveness can be based on that. Indeed, use any (linear time) pattern matching algorithm, cf. [CR], to test whether the pattern u is a factor in uu in a nontrivial way, and if "no" the primitiveness of u has been verified.

Now, we go to the second important subcase of the primitive words. A *Lyndon word* $u \in \Sigma^+$ is a word which is primitive and the smallest one in its conjugacy class $[u]$ with respect to the lexicographic ordering.

It is easy to see that a Lyndon word is unbordered. This follows since of the words vuv, vvu and uvv, with $u, v \in \Sigma^+$ and vuv primitive, the first one is never the smallest one. Indeed, by the primitiveness of vuv, we can use the argumentation of (3) to conclude that $vuv \notin \mathrm{pref}(v^\omega)$. Consequently, vuv deviates from v^ω before its end, and so uvv does it earlier and vvu later, if ever, than vuv. Therefore if $vuv \prec_l v^\omega$, then $uvv \prec_l vuv$, and otherwise $vvu \prec_l vuv$.

Let $\mathcal{L}$ denote the set of all Lyndon words. A fundamental property of these words is the following representation result:

Proposition 2.1. *Each word $u \in \Sigma^+$ admits the unique factorization as a product of nonincreasing Lyndon words, i.e., in the form*

$$u = l_1 \ldots l_n, \quad \text{with} \quad l_j \in \mathcal{L} \quad \text{and} \quad l_n \preceq_l l_{n-1} \preceq_l \ldots \preceq_l l_1.$$

The proof of Proposition 2.1 can be found in [Lo], which studies extensively Lyndon words and their applications to factorizations of free monoids. Algorithmic aspects of Lyndon words can be found in [Du2] and [BePo].

2.3 Repetitions in words

One of the most intensively studied topics of combinatorics of words is that of repetitions in words initiated already by Thue in [T1] and [T2]. This differs from the above periodicity considerations in the sense that the focus is on factors of words instead of words themselves. We state the basic definitions here to be used later in Section 8.

A nonprimitive word is a proper power of another, and hence contains a repetition of order at least 2. More generally, a word u is said to contain a *repetition of order k*, with a rational $k > 1$, if it contains a factor of the form

$$z \in \mathrm{pref}(r^\omega), \quad \text{with } \frac{|z|}{|r|} = k.$$

In particular, if $|z| = 2|r|$ and $u = z_1 r r z_2$, with $z_1, z_2 \in \Sigma^*$, u contains a repetition of order 2, i.e., a *square* as a factor.

Special emphasis has been put to study *repetition-free* words. We define three different variants of this notion as follows. Let $k > 1$ be a *real* number. We say that $u \in \Sigma^\infty$ is

k-free, if it does not contain as a factor a repetition of order at least k;
k^+-free, if, for any $k' > k$, it is k'-free;
k^--free, if it is k-free, but not k'-free for any $k' < k$.

It follows that the k^--freeness implies the k-freeness, which, in turn, implies the k^+-freeness. The reverse implications are not true in general, cf. Example 8.1 and Theorem 8.1. There exist commonly used special terms for a few most frequently studied cases: 2-free, 2^+-free and 3-free words are often called *square-free*, *overlap-free* and *cube-free* words, respectively.

In order to illustrate further the above notions we note that in the case $k = 2$, the 2-freeness means that u does not contain as a factor any square, the 2^+-freeness means that it does not contain any factor of the form $vwvwv$, with $v, w \in \Sigma^+$, and the 2^--freeness means that it does not contain any square, but does contain repetitions of order $2 - \varepsilon$, for any $\varepsilon > 0$. As an example, for the word $u = babaabaabb$ the highest order of repetitions is $2\frac{2}{3}$, since it contains the factor $(aba)^{2\frac{2}{3}} = abaabaab$. Note that although u does not contain a factor of the form $v^{2\frac{3}{5}}$ it is not $2\frac{3}{5}$-free, since it contains a repetition of order $2\frac{2}{3} > 2\frac{3}{5}$.

The above notions were generalized in [BEM], and independently in [Z], to arbitrary patterns as follows. Let Ξ be another alphabet, and P a word over Ξ, so-called *pattern*. We say that $u \in \Sigma^\infty$ *avoids* the pattern $P \neq 1$ in Σ, if u does not contain a factor of the form $h(p)$, where h is a morphism

$h : \Xi^* \to \Sigma^*$ with $h(x) \neq 1$ for all x in Ξ. Further a pattern P is called *avoidable* in Σ, if there exists an infinite word $u \in \Sigma^\omega$ avoiding P.

For example, the pattern xx is avoidable in Σ if there exists an infinite square-free word over Σ, and as we already indicated, the pattern $xyxyx$ is avoidable in Σ if there exists an infinite 2^+-free word over Σ. It is worth noting here that the existence of a factor of the form $vwvwv$, with $v, w \in \Sigma^+$, in u is equivalent to the existence of an overlapping factor in u, i.e., of two occurrences of a factor overlapping in u. This explains the term of overlap-free.

Natural commutative variants of the above notions can be defined, when $k \in \mathbb{N}$ and only the k-freeness is considered: we say that $u \in \Sigma^\infty$ is *abelian k-free*, if it does not contain a factor of the form $u_1 \dots u_k$ with $\pi(u_i) = \pi(u_j)$, for $i, j = 1, \dots, k$.

In order to motivate the use of infinite words in connection with avoidable words we note the following simple equivalence: *for each pattern P there exist infinitely many words in Σ^* avoiding P if, and only if, there exists an infinite word in Σ^ω avoiding P*. This follows directly from the finiteness of Σ. Indeed, in one direction the implication is trivial. In the other direction it follows since, by the above reason, from any infinite set L of words, each of which contains a prefix v, we can choose an infinite subset L' and a letter $a \in \Sigma$ such that each element of L' contains va, as a prefix.

We conclude this subsection by listing all the cases when the number of k-free or abelian k-free words, for an integer k, is finite. It is an exhaustive search argument which shows that this is the case for the 2-freeness in the binary alphabet, as well as the abelian 3-freeness in the binary and the abelian 2-freeness in the ternary alphabets. Figure 2.2 describes the corresponding trees $T_{2,2}$, $AT_{2,3}$ and $AT_{3,2}$, respectively. All the words of the required types (up to symmetry) are found from the paths of these trees starting at the roots.

As we shall see in Section 8, in all the other cases there exists an infinite word avoiding corresponding k-repetitions or abelian k-repetitions. By these trees all binary words of length at least 4 contain a square, and all binary words of length at least 10 contain an abelian cube. Finally, all ternary words of length at least 8 contain an abelian square.

2.4 Morphisms

As we shall see, or in fact have already seen, morphisms play an important role in combinatorics of words. *Morphisms* are mappings $h : \Sigma^* \to \Delta^*$ satisfying

$$h(uv) = h(u)h(v) \text{ for all } u, v \in \Sigma^*.$$

In particular, necessarily $h(1) = 1$, and the morphism h is completely determined by the words $h(a)$, with $a \in \Sigma$. Therefore, as a finite set $h(\Sigma)$ of words a morphism is a very natural combinatorial object.

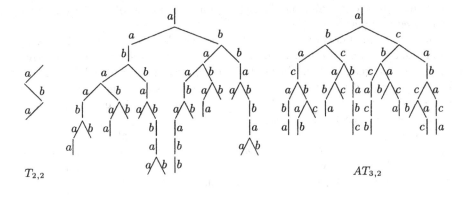

Fig. 2.2. Trees $T_{2,2}$, $AT_{2,3}$ and $AT_{3,2}$

We shall need different kinds of special morphisms in our later considerations. We say that a morphism $h : \Sigma^* \to \Delta^*$ is

binary, if $\|\Sigma\| = 2$;
periodic, if there exists a z such that $h(\Sigma) \subseteq z^*$;
1-free or *nonerasing*, if $h(a) \neq 1$ for each $a \in \Sigma$;
uniform, if $|h(a)| = |h(b)|$ for all $a, b \in \Sigma$;
prolongable, if there exists an $a \in \Sigma$ such that $h(a) \in a\Sigma^+$;
a *prefix*, if none of the words of $h(\Sigma)$ is a prefix of another;
a *suffix*, if none of the words of $h(\Sigma)$ is a suffix of another;
a *code*, if it is injective;
of *bounded delay* p, if for each $a, b \in \Sigma$, $u, v \in \Sigma^*$ we have: $h(au) \leq h(bv)$ with $u \in \Sigma^p \Rightarrow a = b$;
simplifiable, if there exist morphisms $f : \Sigma^* \to \Gamma^*$ and $g : \Gamma^* \to \Delta^*$, with $\|\Gamma\| < \|\Sigma\|$, such that $h = g \circ f$;
elementary, if it is not simplifiable.

In many cases the alphabets Σ and Δ coincide. In the case of equations or patterns we consider morphisms $h : \Xi^* \to \Sigma^*$, i.e., the set of unknowns is denoted by Ξ. For a uniform morphism h we define its *size* as the number $|h(a)|$, with $a \in \Sigma$. Sometimes periodic morphisms are called *cyclic*. Finally, as an example of a morphism with an unbounded delay we give the morphism defined as $h(x) = a$, $h(y) = ab$ and $h(z) = bb$. Then, indeed, the word ab^ω can be factorized as $h(y)h(z)^\omega$ or $h(x)h(z)^\omega$ in $\{h(x), h(y), h(z)\}^+$.

2.5 Finite sets of words

In this last subsection we turn our attention to sets of finite words, i.e., to languages. Indeed, our main interest will be, on one hand, in words including the infinite ones, and on the other hand, on finite, or at most finitely generated, languages.

Many of the operations defined above for words extend, in a natural way, to languages. Consequently, we may talk about, for instance, a commutative image of a language, or quotients of a language by another one. As an example, let us remind that the set of all factors of words in a language X can be expressed as $F(X) = \Sigma^{*^{-1}} X \Sigma^{*^{-1}}$. We define the *size* $s(X)$ of a finite set X by the identity $s(X) = \sum_{x \in X} |x|$.

Let $X \subseteq \Sigma^*$ and $u_1, \ldots, u_t \in X$. We already said that such a sequence $u_1, \ldots, u_t$ is an X-factorization of u if $u = u_1 \ldots u_t$. Exactly as Σ was extended to Σ^* or Σ^+, we can extend the set X to a *monoid* or *semigroup it generates* by considering all X-factorizations:

$$X^* = \{u_1 \ldots u_t \mid t \geq 0, \ u_i \in X\},$$

and

$$X^+ = \{u_1 \ldots u_t \mid t \geq 1, \ u_i \in X\}.$$

Algebraically, such semigroups are subsemigroups of a finitely generated free semigroup Σ^+, and are called *F-semigroups*. Note that $1 \in X^+$ if, and only if, $1 \in X$. For convenience we concentrate to the semigroup case, and normally assume that $1 \notin X$.

Contrary to Σ^+ the semigroup X^+ need not be *free* in the sense that each $u \in X^+$ would have only one X-factorization. However, what is true is that X^+ (as a set) has the unique *minimal generating set*, namely the set Y defined by

$$Y = (X^+ - \{1\}) - (X^+ - \{1\})^2, \ \text{or simply} \ Y = X^+ - X^{+^2}, \ \text{if} \ 1 \notin X.$$

Indeed, any set Z generating X^+, i.e., satisfying $Z^+ = X^+$, must contain Y. On the other hand, any element of X^+, i.e., a product of elements of X, can be expressed as a product of elements of Y, so that Y generates X^+.

If each word of X^+ has exactly one Y-factorization then the semigroup X^+ is *free*, and its minimal generating set Y is a *code*, cf. [BePe].

One of our goals is to measure the complexity of a finite set $X \subseteq \Sigma^+$. A coarse classification is obtained by associating X with a number, referred to as its *combinatorial rank* or *degree*, in symbols $d(X)$, defined as

$$d(X) = \min\{\|F\| \mid X \subseteq F^*\}.$$

Consequently, $d(X)$ tells how many words are needed to build up all words of X. The simplest case corresponds to *periodic* sets, when all words of X are powers of a same word. The above notion will be compared to, but must not be confused with other notions of a rank of a set which will be called in Section 4 *algebraic ranks*, cf. [Lo].

Another way of measuring the complexity of X is to consider all relations satisfied by X. In this approach codes, i.e., those sets which satisfy only trivial relations, are the "simplest" ones. We prefer to consider these as the largest ones, since, indeed, $\|X^n\|$ assumes the maximal value namely $\|X\|^n$, for all $n \geq 1$.

To formalize the above let $X = \{u_1, \ldots, u_t\} \subseteq \Sigma^+$ be an *ordered* set of words and let $\Xi = \{x_1, \ldots, x_t\}$ be an *ordered* set of unknowns. Let $h_X : \Xi^* \to \Sigma^*$ be a morphism defined as $h_X(x_i) = u_i$. Then the set

$$R_X = \ker(h_X) \subseteq \Xi^* \times \Xi^*$$

defines all the relations in X^+. Further the subrelation

$$\min(R_X) = \{(y, z) \in R_X \mid \forall y', z' \in \Xi^+ : y' < y, z' < z \Rightarrow (y', z') \notin R_X\}$$

corresponds to *minimal* relations in X^+. Note that obviously R_X is a submonoid of the product monoid $\Xi^* \times \Xi^*$, and $\min(R_X)$ is the minimal generating set of it, i.e., $\min(R_X)$ generates R_X, and no element of $\min(R_X)$ is a nontrivial product of two elements of R_X.

Now we define a partial ordering $\preceq_r$ on the set of ordered subsets $X \subseteq \Sigma^+$ of a *fixed* cardinality as follows:

$$X \preceq_r Y \text{ if, and only if, } R_Y \subseteq R_X,$$

or equivalently if, and only if, $\min(R_Y) \subseteq \min(R_X)$. We notice that $\preceq_r$ is a partial ordering, where codes are maximal elements, i.e., for any n-element set X and any n-element code C we have $X \preceq_r C$. We also note that the equality under this ordering means the isomorphism of the corresponding F-semigroups. We call this ordering a *relation ordering*, and shall see in Section 7 that is has quite interesting properties.

We conclude this section of preliminaries with an example illustrating the above definitions.

Example 2.1. Consider the following four ordered sets

$$
\begin{aligned}
X_1 &= \{a, abb, bba, baab, babb, baba\}, \\
X_2 &= \{a, abb, bba, bb, babb, baba\}, \\
X_3 &= \{a, abb, bba, bb, bbb, baba\}, \\
X_4 &= \{a, abb, bba, bb, bbb, ba\}.
\end{aligned}
$$

Using finite transducers, cf. [Bel], we can compute all words of X_1^+ having two X_1-factorizations, i.e., all nontrivial relations in X_1^+, as explicitly noticed in [Sp1]. All *minimal* such relations are computed by a transducer τ_{X_1} shown in Figure 2.3. The idea of the construction of τ_{X_1} is obvious: τ_{X_1} searches for minimal double X_1-factorizations systematically remembering at its states which of the factorizations is "ahead" and by "how much". Such a transducer contains always two isomorphic copies, so that in our illustration we can omit half of the transducer (the spotted lines in τ_{X_1}).

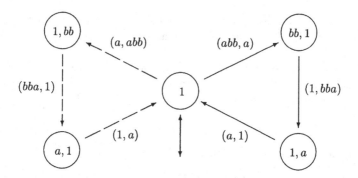

Fig. 2.3. Transducer τ_{X_1}

Let us denote by id_{X_1} the identity relation of X_1^+. Then, τ_{X_1} can be transformed to compute $\min(R_{X_1}) - id_{X_1}$ simply by relabelling the transitions as shown in Figure 2.4. Let us denote this transducer by τ_1. Similarly, we can compute, as shown in Figure 2.5, the transducers τ_i defining the relations $\min(R_{X_i}) - id_{X_i}$, for $i = 2, 3, 4$.

It follows that $X_4 \prec_r X_3 \prec_r X_2 \prec_r X_1 \prec_r C_6$, where C_6 is any six element code. As we shall see in Section 7, the above procedure cannot be continued for ever, i.e., each proper chain is finite.   □

3. Selected examples of problems

In this section we consider three different problems which, we believe, illustrate several important aspects and techniques used in combinatorics of words. The problems deal with different possibilities of mapping Σ^* into Δ^*, a characterization of binary equality languages, and a problem of separating two words by a finite automaton.

3.1 Injective mappings between F-semigroups

In this subsection we consider a problem of mapping a word monoid, or more generally a finitely generated F-semigroup, into another one. Moreover, we require that such a mapping satisfy either some algebraic or automata-theoretic properties. The properties we require are that mappings are

isomorphisms;
embeddings mapping generators into generators;
general embeddings;
bijective sequential mappings.

In particular, all mappings are injective.

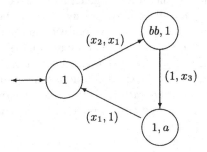

Fig. 2.4. Transducer τ_1 accepting $\min(R_{X_1})$

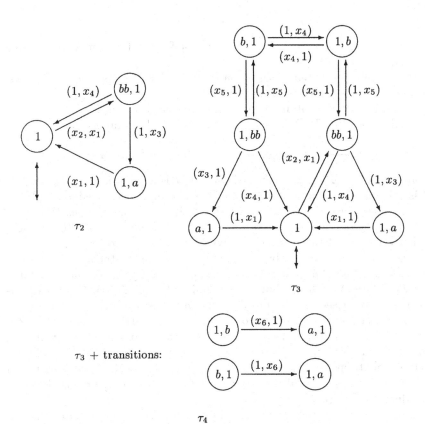

Fig. 2.5. Transducers τ_2, τ_3 and τ_4

Isomorphisms. For finitely generated free semigroups a required isomorphism exists if, and only if, the minimal generating sets of the semigroups are of the same cardinality, and in such a case any bijection between those would work. Also for F-semigroups a necessary condition is that the cardinalities of the minimal generating sets are equal. Therefore, for F-semigroups the problem reduces to that of testing whether a given bijection between generating sets is an isomorphism. How this can be done is shown in Section 7.

Embeddings preserving generators. This problem can be solved by the method of the first problem: guess the bijection, and test whether it is an isomorphism.

Embeddings. Interestingly this is always possible, if only the target semigroup is not *cyclic*, i.e., a subsemigroup of u^*, for some $u \in \Sigma^+$. In order to see this we consider first free semigroups Σ_∞^+ and Σ_2^+ with countably many and two generators, respectively. Let $\Sigma_\infty = \{a_i \mid i \in \mathbb{N}\}$ and $\Sigma_2 = \{a, b\}$. Then the morphism $f : \Sigma_\infty^+ \to \Sigma_2^+$ defined as

$$f(a_i) = a^i b \quad \text{for } i \in \mathbb{N},$$

gives a required embedding. This is due to the fact that f is injective, or even a prefix.

For finitely generated F-semigroups X^+ and Y^+ we proceed as follows. We allow X^+ to be countably generated, say $X = \{u_i \mid i \in \mathbb{N}\} \subseteq \Sigma^+$, and require that Y contains two noncommuting words $\alpha, \beta \in \Delta^+$. Then a required embedding $h : X^+ \to Y^+$ is obtained as the composition

$$u_i \overset{\pi}{\longmapsto} a_{i_1} \ldots a_{i_t} \overset{f}{\longmapsto} a^{i_1} b \ldots a^{i_t} b \overset{c}{\longmapsto} \alpha^{i_1} \beta \ldots \alpha^{i_t} \beta,$$

where $\pi : X^+ \to \Sigma^+$ is a natural projection, f is as above, and $c : \{a, b\}^* \to \Delta^*$ is defined by $c(a) = \alpha$ and $c(b) = \beta$. The mapping h is indeed a morphism, and moreover, injective as a composition of injective morphisms. Note that the injectivity of c follows, since α and β are assumed to be noncommuting, so that they do not satisfy any nontrivial identity, cf. Corollary 5.1.

Next we move from algebraic mappings to automata-theoretic ones.

Bijective sequential mappings. We search for a bijective sequential mapping $T : \Sigma^* \to \Delta^*$, where Σ and Δ are finite alphabets. Recall that *sequential mappings*, or sequential transductions in terms of [Be1] or deterministic generalized sequential mappings of [GR], cf. also [Sal1], are realized by deterministic finite automata over Σ, without final states and equipped with outputs in Δ^*, i.e., for each transition an output word of Δ^* is produced. Such automata are called *sequential transducers* in [Be1]. As an illustration we consider the sequential mapping $T : \{a, b, c\}^* \to \{x, y\}^*$ realized by the transducer of Figure 3.1.

The requirement that τ has to realize a bijection, implies that the underlying automaton with respect to inputs must be a *complete* deterministic automaton. Consequently, the inputs can be ignored (if only there are $\|\Sigma\|$ outgoing transitions from each state), and so we are left with the problem,

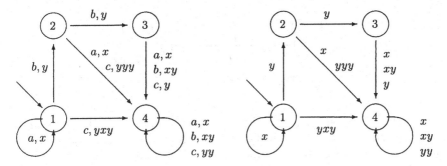

Fig. 3.1. A sequential transducer τ and its underlying output automaton $\mathcal{A}$

whether the underlying output automaton of τ, say $\mathcal{A}$, accepts *unambiguously* Δ^*. Therefore we are led to the theory of finite automata with multiplicities, or in terms of [Ei] to the theory of N-rational subsets. Now, using the Equality Theorem of [Ei] it is easy to check that the above T, constructed by Schützenberger, is actually a required bijection $\{a, b, c\}^* \rightarrow \{x, y\}^*$.

Next we introduce a systematic method from [Ch] for constructing sequential bijections $\Sigma^* \rightarrow \Delta^*$, and illustrate it in the case when $\Sigma = \{a, b, c\}$ and $\Delta = \{x, y\}$. We start from a *maximal suffix code* X over Δ, cf. [BePe]. Such sets are exactly those represented by binary trees, each node of which contains either 0 or 2 descendants. It follows that if S is the subset of all proper suffixes of words in X, then each word $u \in \Delta^*$ has the unique representation in the form $u = sx$, with $s \in S \cup \{1\}$ and $x \in X^*$. In other words, we have the following relation on N-subsets (where we use $+$ instead of $\cup$):

$$(1) \qquad \Delta^* = (1 + S)X^*.$$

Now, let us return to our specific case, and fix X to be the smallest three-element maximal suffix code, i.e., $X = \{x, xy, yy\}$ (or its renaming). Consequently, $S = \{y\}$, and hence using standard properties of N-subsets, cf. [Ei] chapter 3, we transform (1) as follows:

$$
\begin{aligned}
\Delta^* &= (1 + y)X^* = 1 + (X + y)X^* \\
&= 1 + (x + xy + yy + y)X^* \\
&= 1 + x(1 + y)X^* + (y + yy)X^* \\
&= 1 + x\Delta^* + (y + yy)X^*.
\end{aligned}
$$

This relation leads to the two state unambiguous automaton $\mathcal{A}_X$ of Figure 3.2 accepting $\{x, y\}^*$:
Here, $\mathcal{A}_X$ can be obtained, for example, by reversing the method of computing the behaviour of an N-automaton using linear systems of equations, cf. [Ei] chapter 7.

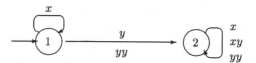

Fig. 3.2. Automaton $\mathcal{A}_X$

As we mentioned a sequential bijection $\{a, b, c\}^* \to \{x, y\}^*$ is obtained from $\mathcal{A}_X$ by labelling, for each state, the outgoing transitions bijectively by $\{a, b, c\}$. We also note that the automaton $\mathcal{A}$ of Figure 3.1 can be derived from the above $\mathcal{A}_X$ by unrolling the loop of state 2 once, and redistributing the loop-free unrolled paths between two states in a suitable way:

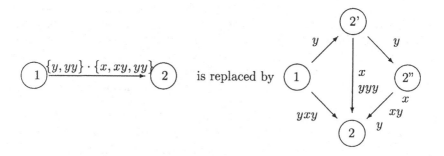

More generally, for details cf. [Ch], if $\|\Sigma\| - 1$ divides $\|\Delta\| - 1$, as in our above considerations, by choosing a maximal suffix code X of the cardinality $(\|\Delta\| - 1)/(\|\Sigma\| - 1)$, one can construct a rational sequential bijection $\Sigma^* \to \Delta^*$ realized by a two state automaton of Figure 3.3, where x is an arbitrary letter of Δ and $S = \mathrm{suf}(X) - (\{1\} \cup X)$.

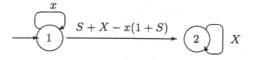

Fig. 3.3. A two state automaton realizing a bijection $\Sigma^* \to \Delta^*$

An elaboration of the previous considerations, cf. [Ch], leads to the following result.

Theorem 3.1. *There exists a bijective sequential mapping $\Sigma^* \to \Delta^*$ if, and only if, $\|\Sigma\| = \|\Delta\|$ or $\|\Sigma\| > \|\Delta\| > 1$. Moreover, if this is the case, then such a mapping is realized by a two state sequential transducer.* □

The trivial parts of Theorem 3.1 are as follows. First, if $\|\Sigma\| = \|\Delta\|$, then the identity mapping (or a renaming) works. Second, if $1 = \|\Delta\| < \|\Sigma\|$ or $\|\Sigma\| < \|\Delta\|$, then simple cardinality arguments show that no required bijection exists.

Finally, we mention that a related problem searching for sequential transductions mapping a given regular set onto another regular one was considered in [MN] and [McN1].

The issues presented in this subsection deserve some comments. Due to the embedding $f : \Sigma_\infty^+ \to \Sigma_2^+$, for many problems in formal language theory, as is well-known, it is irrelevant what the cardinality of the alphabet is, as long as it is at least two. Certainly this is the case when *the property $\mathcal{P}$ to be studied is preserved under the encoding f* in the following sense. The encoded instance of a problem is still an instance of the original problem, and it has the property $\mathcal{P}$ if, and only if, the original instance has the property $\mathcal{P}$.

Let us take an example. Consider a property $\mathcal{P}$ of languages accepted by finite automata. Clearly, rational languages are closed under the encoding f, and moreover many of the natural properties, such as the finiteness, for example, holds for L if, and only if, it holds for $f(L)$. However, if we would consider $\mathcal{P}$ on languages accepted by n-state finite automata, then $\mathcal{P}$ would not be preserved under the encoding f, and hence the cardinality might matter.

There are even more natural cases when the size of the alphabet is decisive. This happens, for instance, when the problem asks something about the domain of morphisms. For example, whether for morphisms $h, g : \Sigma^* \to \Delta^*$ there exists a word $w \in \Sigma^+$ such that $h(w) = g(w)$ – this is the well-known Post Correspondence Problem for lists of length $\|\Sigma\|$, cf. [HK2]. On the other hand, this problem is independent of the target alphabet, as long as it contains at least two letters. Another example is the avoidability of a pattern in infinite words. Of course no embedding from an alphabet of at least three letters into a binary one preserves the square-freeness. Therefore, avoidability problems depend, in general, crucially on the size of the alphabet.

Finally, we note that normally it is enough that an encoding is injective instead of bijective. However, if bijective encodings were needed the solutions of the last problem might be useful, especially because they are defined in terms of automata theory.

3.2 Binary equality sets

As the second example we consider a simple combinatorial problem connected to the above Post Correspondence Problem. For two morphisms $h, g : \Sigma^* \to \Delta^*$ we define their *equality set* as

$$E(h, g) = \{w \in \Sigma^* \mid h(w) = g(w)\}.$$

In the next result we present a partial characterization from [EKR2] of equality sets of binary morphisms.

Theorem 3.2. *The equality set of two nonperiodic binary morphisms is always of one of the following forms*

$$\{\alpha,\beta\}^* \quad \text{or} \quad (\alpha\gamma^*\beta)^* \quad \text{for some} \quad \alpha,\beta,\gamma \in \Sigma^*.$$

In particular, such a set is rational.

Proof. By the considerations of the previous subsection we may assume that h and g are morphisms from $\{a,b\}^*$ into itself. The proof uses essentially the following simple lemma which, we believe, is interesting on its own right.

Lemma 3.1. *Let $X = \{x,y\} \subseteq \Sigma^+$ be a nonperiodic set. Then, for each word $u,v \in X^+$, we have*

$$u \in xX^+, v \in yX^+; |u|,|v| \geq |xy \wedge yx| \Rightarrow u \wedge v = xy \wedge yx.$$

Proof of Lemma. By symmetry, we may assume that $|y| > |x|$. Let $z = xy \wedge yx$, so that, by the nonperiodicity of X, we have $|z| < |xy|$, cf. Corollary 4.1. Now, if $|z| < |x|$ we are done. In the other case we have the situation depicted as

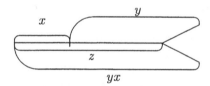

Now, $v \in yyX^+ \cup yxX^+$ and $y \in x\Sigma^*$, so that $|v \wedge yx| > |z|$.

We shall show that also $|u \wedge xy| > |z|$, from which the claim follows, i.e., $|u \wedge v| = |z| = |xy \wedge yx|$. To see this we first note, by the identity $xy \wedge yx = z$, that z has a period $|x|$, i.e., $z \in \text{pref}(x^\omega)$. Second, by the inequality $|y| > |z| - |x|$, we conclude that y has a prefix of length $|z| - |x| + 1$ in $\text{pref}(x^\omega)$. Therefore, the words $u \in xX^+$ and xy have a common prefix of length $|z| + 1$ (in $\text{pref}(x^\omega)$). So our proof is complete. □

Proof of Theorem (continued). We are going to use this lemma to show that in the exhaustive search for elements in the equality set of the pair (h,g) there exists a unique situation when this search does not go deterministically. Before doing this we need some terminology.

Referring to the Post Correspondence Problem, let us call elements of $E(h,g)$ *solutions*, elements of $(E(h,g) - \{1\}) - (E(h,g) - \{1\})^2$ *minimal solutions*, and prefixes of solutions *presolutions*. Further with each presolution w we associate its *overflow* $o(w)$ as an element of the free group generated by $\{a,b\}$:

$$o(w) = h^{-1}(w)g(w).$$

Finally, we say that a presolution w (or the overflow it defines) *admits a c-continuation*, with $c \in \{a, b\}$, if wc is a presolution as well.

Now, let us consider a fixed overflow $o(w)$. Depending on whether it is an element of $\{a, b\}^*$ or not we can illustrate the situation by the following figures:

or

Assuming that we have the first case (the other being symmetric) we now analyse what it means that w admits both a- and b-continuations. Since $E(h, g)$ is closed under the product this can be stated that there exist words w_a and w_b, which can be chosen as long as we want, such that waw_a and wbw_b are solutions. This is illustrated in Figure 3.4.

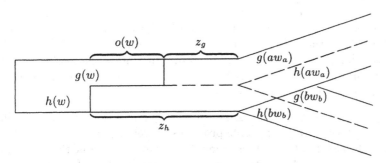

Fig. 3.4. a- and b-continuations

By our choice, $g(waw_a) = h(waw_a)$ and $g(waw_b) = h(wbw_b)$. Now, by the lemma, necessarily

$$g(aw_a) \wedge g(bw_b) = g(ab) \wedge g(ba) = z_g$$

and

$$h(aw_a) \wedge h(bw_b) = h(ab) \wedge h(ba) = z_h.$$

Consequently, both waw_a and wbw_b can be solutions only if

$$o(w) = z_h z_g^{-1},$$

as already depicted in Figure 3.4. This value of $o(w)$ is the unique element of the free group depending only on the pair (h, g). In the case considered it is an element of $\{a, b\}^*$, and in the symmetric case the inverse of an element of $\{a, b\}^*$.

So we have proved that only one particular value of the overflow may allow two ways to extend presolutions into minimal solutions. Let us call such an overflow *critical*. Having this property the completion of the proof is an easy case analysis.

First, if the critical overflow does not exist. Then the presolution 1, if it is such, can be extended to a minimal solution in a unique way. Therefore $E(h, g) = \alpha^*$ for some word $\alpha \in \{a, b\}^*$. If the overflow 1 is the critical one, then the above argumentation shows that $E(h, g) = \{\alpha, \beta\}^*$ for some words $\alpha, \beta \in \{a, b\}^*$.

Finally, if the critical overflow exists, and is different from the empty word, we proceed as follows. Let w be the prefix of a minimal solution such that $o(w)$ is critical. Clearly such a w is unique. We call a letter c *repetitive*, if there exists a word $\overline{w}_c$ such that

$$o(w) = o(wc\overline{w}_c) \text{ and } wcw' \notin E(h, g) \text{ for any } w' \in \mathrm{pref}\overline{w}_c.$$

Now, if neither a nor b is repetitive, then by the definition of the critical overflow, $E(h, g) = \{\alpha, \beta\}^*$ for some words $\alpha, \beta \in \{a, b\}^+$. If exactly one of the letters a and b is repetitive, then $E(h, g) = (\alpha\gamma^*\beta)^*$ for some words $\alpha, \beta, \gamma \in \{a, b\}^*$. Indeed, if a is the repetitive letter, then $\alpha = w$, $\gamma = a\overline{w}_a$, and β equals to the unique word $\hat{w}_b$ such that $wb\hat{w}_b$ is a minimal solution. Again the definition of the critical overflow guarantees the existence of $\hat{w}_b$. A similar argumentation rules out the case that both a and b are repetitive. This completes the proof of Theorem 3.2. □

Theorem 3.2 motivates a number of comments, which, we believe, illustrate nicely how intriguing simple problems of words can be.

First, the cases ruled out in Theorem 3.2, when at least one the morphisms is periodic are easy. If just one is periodic, then, by the defect theorem, cf. Theorem 5.1, the other is injective, and therefore the equality set may contain at most one minimal solution, i.e., is of the form α^* for some $\alpha \in \{a, b\}^*$. If both, in turn, are periodic, then the equality set, if not equal to $\{1\}$, consists of all words containing the letters a and b in a fixed ratio $q \in \mathbb{Q}_+ \cup \{\infty\}$. Such languages are sometimes denoted by L_q.

Second, the idea of the proof of Theorem 3.2 is not extendable into larger alphabets, since the Lemma 3.1, which is the basis of the proof, does not seem to have counterparts in larger alphabets. Note that this lemma allows to construct from a given binary nonperiodic morphism h a so-called *marked* morphism h', i.e., a morphism h' satisfying $\mathrm{pref}_1 h'(a) \neq \mathrm{pref}_1 h'(b)$, simply by applying the cyclic permutation c of Section 2.2 $|h(ab) \wedge h(ba)|$ times simultaneously to $h(a)$ and $h(b)$.

Third, and most interestingly, we compare the result of Theorem 3.2 to the problem of testing whether, for two binary morphisms h and g, there exists a word $w \neq 1$ such that $h(w) = g(w)$, i.e., to the decidability problem of PCP(2). Certainly, our proof of Theorem 3.2 is nonconstructive. As such it is, however, if not very short, at least elementary and drastically shorter than the existing decidability proofs of PCP(2), cf. [EKR1], [Pav] or also [HK2] in this handbook, which are about 20 pages long. As shown in [HKK] our existential proof of Theorem 3.2 can be made constructive, if an algorithm for PCP(2), or in fact for its slight generalization so-called GPCP(2), for definitions cf. [HK2], is known. Moreover, the arguments used in [HKK] to conclude this are short.

As a conclusion from above, we know that the equality set of two binary morphisms h and g is always of one of the three different forms, namely L_q, for some $q \in \mathbb{Q}_+ \cup \{\infty\}$, $\{\alpha, \beta\}^*$ or $(\alpha\gamma^*\beta)^*$ for some words $\alpha, \beta, \gamma \in \{a, b\}^*$. Moreover, we can effectively find it, i.e., find a k or a finite automaton accepting the equality set. Still we do not know whether the third possibility can take place!

We consider this open problem as a very nice example of challenging problems of words. Although we think this problem is difficult it is worth noting that in free groups the sets of the form $(\alpha\gamma^*\beta)^*$, with $\alpha, \beta, \gamma \in \Sigma^*$, are generated by two elements only: $\alpha\gamma^i\beta = (\alpha\gamma\beta(\alpha\beta)^{-1})^i\alpha\beta$ for $i \geq 0$.

3.3 Separating words via automata

Given two distinct words $x, y \in \Sigma^*$ we want to measure by how much they differ when processed by a finite automaton. More precisely, we want to compute the minimal size $s(x, y)$ of an automaton, i.e., the minimal cardinality of the set of states, that accepts one word and rejects the other. That this integer exists results from the fact that the free monoid is residually finite: an automaton of size $|x|$ accepting only x separates the two words.

It is easy to check that $d(x, y) = 2^{-s(x,y)}$ defines an ultrametric distance on the free monoid, once we pose $d(x, x) = 0$. Indeed, if this were not the case then for some x, y, z we would have $d(x, z) > \max\{d(x, y), d(y, z)\}$ or equivalently $s(x, z) < \min\{s(x, y), s(y, z)\}$. Then in a minimum size automaton $\mathcal{A}$ separating x and z, the words x and y are indistinguishable, i.e., they take the initial state to the same state. But this means that y and z are distinguished by $\mathcal{A}$, contradicting the minimality of $s(y, z)$.

For a fixed integer n we denote by S_n the maximum of all $s(x, y)$'s for x, y of lengths bounded by n, and we study S_n as a function of n. Here finite automaton means deterministic finite automaton, but it can be replaced by finite non-deterministic automaton, finite permutation-automaton (all letters induce a permutation of the set of states), finite monoids, finite groups etc. This question was implicitly posed in [Jo].

Surprisingly enough, it is difficult to come up with two words which would require a large automaton to be separated, say an infinite family of pairs of words for which the size of the automaton would increase as n^α for some $\alpha > 0$. Actually, using elementary number theory, it is easy to verify that two words of different lengths bounded by n can be separated by an automaton whose size is of the order of $\mathcal{O}(\log n)$. So in particular, two words of different commutative images can be separated by an automaton of size

$\mathcal{O}(\log \max\{|x|, |y|\})$. This observation can be drawn further. Indeed, assume that a factor z of length k occurs a different number of times in x and y. The above argument shows that counting the occurrences of z modulo m, for some $m = \mathcal{O}(\log n)$, discriminates x and y. As a consequence, if it is true that two different words of length n differ on the number of occurrences of some factor of length $\log n$, then these two words can be separated by a finite automaton of size $\mathcal{O}(\log^2 n)$.

The first non-trivial contribution to this problem is due to [GK], where it was proved that S_n/n tends to 0 as n tends to infinity. Approximately at the same time in [Rob1] it was proved that $S_n = \mathcal{O}(n^{2/5} \log^{3/5} n)$, and then that only a slightly worse upper bound holds when dealing with permutation automata, to with $\mathcal{O}(n^{1/2})$, see [Rob2]. We reproduce from [Rob1] a weaker result.

Theorem 3.3. *Given two words u and v of length n there exists an automaton of size $\mathcal{O}(n \log n)^{1/2}$ that accepts u if, and only if, it rejects v.*

Proof. Let us first present the proof intuitively. Let w be the shortest prefix of u that is not a prefix of v. The discriminating automaton aims at recognizing some suffix z (as an occurrence) of w by counting its position in u modulo some integer. Clearly, z may not be too large since the automaton performs a string-matching based on z. But it may not be too small either, else it might have many occurrences and the modulo counting that identifies unambiguously this occurrence might envolve a large integer. Furthermore, the length of z does not by itself guarantee a small number of occurrences. It's its period that counts, so z has to be chosen with a long period compared to its length. The proof consists in solving this trade-off.

Let $\pi(n)$ be the number of primes that are less than or equal to n. The prime number theorem asserts that there exists a constant c for which $\pi(n) > c\frac{n}{\log n}$ holds, see e.g. [HW], Theorem 6. The first claim is of pure number-theoretic nature.

Claim 1. For sufficiently large n there exists a prime number $p \leq \frac{3}{c}(n \log n)^{1/2}$ such that the following holds. Let $I \subseteq [1, n]$ be a subset of less than $n^{1/2} \log^{-1/2} n$ elements and let $i \in I$ be a fixed element. Then we have

$$i \neq j \bmod p, \text{ for all } i \neq j \text{ and } j \in I.$$

Proof of Claim 1. We first observe that the number of primes greater than $n^{1/2} \log^{-1/2} n$ dividing $j - i$, for some $j \in I$, is less than $2n^{1/2} \log^{-1/2} n$. This follows from the facts that $|j - i|$ is less than n and that I contains at most $n^{1/2} \log^{-1/2} n$ elements. Now the prime number theorem implies

$$\pi(\frac{3}{c} n^{1/2} \log^{1/2} n) > 3\frac{n^{1/2} \log^{1/2} n}{\frac{1}{2}\log n + \log \frac{3}{c} + \frac{1}{2}\log\log n}.$$

Here for sufficiently large n the numerator is smaller than $\log n$, i.e.,

$$\pi(\frac{3}{c}n^{1/2}\log^{1/2}n) > 3n^{1/2}\log^{-1/2}n.$$

Clearly, among these primes there is one that is greater than $n^{1/2}\log^{-1/2}n$ and that divides no $j - i$. □

The second claim concerns the period $p(w)$ of a word w, cf. Section 2.2.

Claim 2. For all $w \in \Sigma^*$, $\max\{p(wa), p(wb)\} > \frac{|w|}{2}$ holds.

Proof of Claim 2. Assume to the contrary that $p(wa), p(wb) \leq \frac{|w|}{2}$. Clearly, $p(wa) \neq p(wb)$. Now wa and wb have a common prefix w of length greater than or equal to $p(wa)+p(wb)$. By the Theorem of Fine and Wilf, cf. Theorem 6.1, this contradicts the minimality of $p(wa)$ and $p(wb)$. □

The last claim gives an estimate on the size of an automaton that carries out a string-matching algorithm, see, Chapter on string-matching in this handbook.

Claim 3. Let $0 \leq i < p$, be two integers and let $w \in \Sigma^*$ be a word of length $k < p$. Then there exists an automaton of size less than $2p$ that recognizes the set of words ending in w, having no other occurrence of w and for which this occurrence starts in position i modulo p.

Proof of Claim 3. We let $w = w_1 \ldots w_k$ and $[p - 1] = \{0, \ldots, p - 1\}$. The set of states of the automaton equals $[p - 1] \cup \{w_1 \ldots w_j | 1 \leq j \leq k\}$, the initial state is 0 and the final state is w. The transition function satisfies

$$w_1 \ldots w_j.c = \begin{cases} w_1 \ldots w_{j+1}, & \text{if} \quad c = w_{j+1}, \\ i + j + 1 \bmod p, & \text{otherwise,} \end{cases}$$

and

$$\alpha.c = \alpha + 1 \bmod p,$$

if $\alpha \in [p - 1] - \{i\}$ and $c \in \Sigma$ or if $\alpha = i$ and $c \neq w_1$. □

Proof of Theorem (continued). Now we contruct an automaton that separates u and v. We denote by w their maximal common prefix: $u = wau_1$ and $v = wbv_1$ for some $u_1, v_1 \in \Sigma^*$ and $a, b \in \Sigma$ with $a \neq b$.

We first rule out an easy case where $|w| < 2(n \log n)^{1/2}$. It suffices to consider the automaton accepting all words having wa as a prefix. It recognizes u if, and only if, it rejects v.

Thus we may assume that $|w| \geq 2(n \log n)^{1/2}$, and consider the suffix z of w of length $2(n \log n)^{1/2} - 1$. We have $u = w_1 z a u_1$, $v = w_1 z b v_1$ and $w = w_1 z$ for some $w_1 \in \Sigma^*$. By Claim 2, we may assume without loss of generality that $p(za) > \frac{|za|}{2}$. In particular this means that two occurrences of za are at least $\frac{|za|}{2}$ apart and therefore that there are less than $\frac{2n}{2(n \log n)^{1/2}} = (n \log n)^{1/2}$ occurrences of za in u.

If v has no occurrence of za then it suffices to construct the automaton that performs the string-matching with za as the string to be matched (see, Chapter on string-matching). We know that this can be achieved with an automaton of size $|za| = 2(n \log n)^{1/2}$.

We are left with the case where za has also an occurrence in v, i.e., $v = w_2 z a v_2$ where $|v_2| < |v_1|$. Let I be the set of positions in u where the occurrences of za end. Let p be as in Claim 1 and let i be the position modulo p of the first occurrence of za in u. Then the automaton $\mathcal{A}$ accepting u and rejecting v consists of two subautomaton $\mathcal{A}_1$ and $\mathcal{A}_2$. Automaton $\mathcal{A}_1$ perfoms as prescribed by Claim 3. When the first occurrence of za is spotted then $\mathcal{A}_1$ switches to $\mathcal{A}_2$, which separates the suffixes u_1 and v_2. We know that $\mathcal{A}_2$ has size bounded by $\log n$. Thus, the automaton $\mathcal{A}$ has size $||\mathcal{A}_1|| + ||\mathcal{A}_2|| < 4(n \log n)^{1/2} + \lambda \log n$, where λ is some constant independent of n. □

4. Defect effect

The defect theorem is one of the important results on words. It is often considered to be a folklore knowledge in mathematics. This may be, at least partially, due to the fact that there does not exist just one result, but, as we shall see, rather many different results which formalize the same *defect effect* of words: *if a set X of n words over a finite alphabet satisfies a nontrivial relation E, then these words can be expressed simultaneously as products of at most $n - 1$ words.* One of the older papers where this is proved is [SkSe]. It was also known in [Len].

The defect effect can be considered from different perspectives. One may concentrate on a set X satisfying one (or several) equation(s), or one may concentrate on an equation E (or a set of equations), and try to associate the notion of a "rank" with the objects studied. Our emphasis is in combinatorial aspects of words, so we concentrate on the first approach.

It follows already from the above informal formulation of a defect theorem, that it can be seen as a dimension property of words: if n words are "dependent" they belong to a "subspace of dimension" at most $n - 1$. However, as we shall see in Section 4.4, words possess only very restricted dimension properties in this sense.

4.1 Basic definitions

Assume that $X \subseteq \Sigma^+$ is a finite set of words having the defect effect. This means that X is of a "smaller" size than $||X||$, but "how much smaller" depends on what properties are required from words used to build up the words of X. This is what leads to different formulations of the defect theorem.

A combinatorial formulation is based on the notion of the *combinatorial rank* or *degree* of $X \subseteq \Sigma^+$, which we already defined in Section 2.5 by the condition

(1)
$$d(X) = \min\{\|F\| \mid X \subseteq F^+\}.$$

It follows immediately that $d(X) \leq \min(\|X\|, \|\Sigma\|)$, so that the finiteness of X is irrelevant. Note also that the degree of a set is not preserved under injective encodings – emphasizing the combinatorial nature of the notion.

In order to give more algebraic formulations we consider the following three conditions. Let $X \subseteq \Sigma^+$ be a finite set and S an F-semigroup. We define three properties of S as follows:

(p) $\forall p, w \in \Sigma^+$: $p, pw \in S \Rightarrow w \in S$;

(f) $\forall p, q, w \in \Sigma^+$: $p, q, wp, qw \in S \Rightarrow w \in S$;

(u) $\forall p, q, w \in \Sigma^+$: $pwq \in X^+, pw, wq \in S \Rightarrow w \in S$.

Conditions (p) and (f) are very basic in the theory of codes, cf. [BePe]. The first one characterizes those F-semigroups having a prefix code as the minimal generating set. Such semigroups are often called *right unitary*. The second condition, which is often referred to as the *stability condition*, characterizes those F-semigroups which are *free*, i.e., have a code as the minimal generating set, cf. [LeSc] or [BePe]. The third condition, which differs from the others in the sense that *it depends also on X*, is introduced here mainly to stress the diversified nature of the defect theorem. As shown in [HK1] it characterizes those F-semigroups, where X^+ factorizes uniquely.

For the sake of completeness we prove the following simple

Lemma 4.1. *An F-semigroup S is right unitary if, and only if, it satisfies* (p).

Proof. Assume first that S is right unitary. This means that the minimal generating set, say P, of S is a prefix code. Let $p = u_1 \ldots u_n$ and $pw = v_1 \ldots v_m$, with $u_i, v_j \in P$, be elements of S. Now, since P is a prefix code we have $u_i = v_i$, for $i = 1, \ldots, n$, and therefore $v_{n+1} \ldots v_m \in P^+ = S$.

Conversely, assume that the F-semigroup S satisfies (p). Let v and q be in the minimal generating set of S. If $v < q$, then we can write $q = vt$ with $t \in \Sigma^+$. Hence, by (p), t is in S, and q is a product of two nonempty words, a contradiction with the fact that q is in the minimal generating set of S. □

For each $i = p, f, u$, F-semigroups satisfying (i) are trivially closed under arbitrary intersections. Therefore the semigroups

$$\hat{X}(i) = \bigcap_{\substack{X \subseteq S \\ S \text{ sat. } (i)}} S$$

are well-defined, and by the definition, the smallest F-semigroups of type (i) containing X. The semigroups $\hat{X}(i)$, for $i = p, f, u$, are referred to as *free hull*, *prefix hull* and *unique factorization hull* of X. Denoting by $X(i)$ the minimal generating set of $\hat{X}(i)$ we now define three different notions of an *algebraic rank* of a finite set $X \subseteq \Sigma^+$:

$$p(X) = \|X(p)\|, \quad r(X) = \|X(f)\| \text{ and } u(X) = \|X(u)\|.$$

These numbers are called *prefix rank* or *p-rank*, *rank* or *f-rank* and *unique factorization rank* or *u-rank* of X, respectively.

The most commonly used notion of a rank of X in the literature is that of our *f*-rank, cf. [BPPR], or [Lo]. From the purely combinatorial point of view *p*-rank is often more natural. The reason we introduced all these variants, which by no means are all the possibilities, cf. [Sp2], is that they can be used to illustrate the subtle nature of the phenomenon called the defect effect.

Our next example modified from [HK1] shows that all the four different notions of a rank may lead to a different quantity.

Example 4.1. Consider the set

$$X = \{aa, aaaaba, aababac, baccd, cddaa, daa, baa\}.$$

The only minimal nontrivial relation satisfied by X is

$$aa.aababac.cddaa = aaaaba.baccd.daa.$$

Now, applying (u) we see that $aaba, bac, cd \in \hat{X}(u)$. Replacing the words $aababac, cddaa, aaaaba$ and $baccd$ of X by the above three words we obtain a set, where X^+ factorizes uniquely, i.e.,

$$X(u) = \{aa, aaba, bac, cd, daa, baa\}.$$

However, $X(u)^+$ is not free, since we have

(2) $$aa.bac.daa = aaba.cd.aa,$$

which actually is the only nontrivial minimal relation in $X(u)^+$. It follows that $\hat{X}(u)$ is a proper subset of $\hat{X}(f)$. Applying now condition (f) to (2) we conclude that $\hat{X}(f)$ contains the words ba, c and d. But now the set

$$X(f) = \{aa, ba, c, d, baa\}$$

is a code, so that $X(f)$ is this set, as already denoted. Finally, $X(f)$ is not a prefix code, so that applying (p) to $X(f)$, or alternatively the procedure described in a moment to the original X, we obtain that

$$X(p) = \{a, ba, c, d\}.$$

Consequently, we have concluded that $p(X) < r(X) < u(X) < \|X\|$. In this example, the degree of X equals to $p(X)$. However, if we replace X by $X' = e(X)$, where $e : \{a, b, c, d\}^* \to \{a, b, c\}^*$ is a morphism defined as $e(d) = bb$ and $e(x) = x$, for $x \in \{a, b, c\}$, then the degree decreases to 3, while all the algebraic ranks remain unchanged, as is easy to conclude. Therefore we have an example of a set X' satisfying

$$3 = d(X') < p(X') < r(X') < u(X') < \|X'\| = 7. \qquad \square$$

Although we called our three notions of the rank algebraic, they do not have all desirable algebraic properties like being invariant under an isomorphism. Indeed, our next example shows that free hulls (or prefix hulls) of two finite sets generating isomorphic F-semigroups need not be isomorphic, i.e., need not have the same number of minimal generators. On the other hand, as a consequence of results in the next subsection, one can conclude that all the algebraic ranks, we defined, are closed under the encodings which are prefix codes.

Example 4.2. Consider the sets

$$X = \{a, baabba, abaab, ba\} \ \text{ and } \ Y = \{a, abb, bbba, ba\}.$$

Then X^+ and Y^+ are isomorphic, since both of these semigroups satisfy only one minimal relation, which, moreover, is the same one under a suitable orderings of sets X and Y:

$$X \ : \ a.baabba = abaab.ba$$
$$Y \ : \ a.bbba = abb.ba.$$

From these relations we conclude, either by definitions or methods of the next subsection, that $X(p) = X(f) = \{a, b\}$, while $Y(p) = Y(f) = \{a, bb, ba\}$.   □

4.2 Defect theorems

In this subsection we show that each of our notions of a rank of a finite set X can be used to formalize the defect effect. In our algebraic cases the words from which the elements of X are built up are, by definitions, unique, while in the case of the degree the minimal F of (1) in Section 4.1 need not be unique.

Let $X \subseteq \Sigma^+$ be finite. We introduce the following procedure using simple transformations to compute the prefix hull of X. Such transformations were used already in [Ni] in connection with free groups.

Procedure P. Given a finite $X \subseteq \Sigma^+$, considered as an unambiguous multiset.

1. Find two words $p, q \in X$ such that $p < q$. If such words do not exist go to 4;
2. Set $X := X \cup \{p^{-1}q\} - \{q\}$ as a multiset;
3. If X is ambiguous identify the equal elements, and go to 1;
4. Output $X(p) := X$.

We obtain the following formulation of the defect theorem.

Theorem 4.1. *For each finite $X \subseteq \Sigma^+$, the minimal generating set $X(p)$ of the prefix hull of X satisfies $\|X(p)\| \leq \|X\|$, and moreover $\|X(p)\| < \|X\|$, if X satisfies a nontrivial relation.*

Proof. First of all, by the definition of the prefix hull and Lemma 4.1, the Procedure P computes $X(p)$ correctly. Hence, $\|X(p)\| \leq \|X\|$ always.

The second sentence of the theorem is seen as follows. Whenever an identification is done in step 3 a required decrease in the size of $\|X\|$ is achieved. Such an identification, in turn, is unavoidable since, if it would not occur, steps 2 and 3 would lead from a set X satisfying a nontrivial relation to a new set of strictly smaller size still satisfying a nontrivial relation. Indeed, the new nontrivial relation is obtained from the old one by substituting $q = pt$, with $t = p^{-1}q$, and by cancelling one p from the left in the old relation. Clearly, such a new relation is still nontrivial.   □

Theorem 4.1 motivates a few comments. By the definition of the prefix hull as an intersection of certain free semigroups, it is not obvious that $\|X(p)\| \leq \|X\|$. Indeed, the intersection of two finitely generated free semigroups, need not be even finitely generated, cf. [Ka2]. On the other hand, the finiteness of $\|X(p)\|$ is obvious, since it must consist of factors of X.

As the second remark we note that although the proof of Theorem 4.1 is very simple, it has a number of interesting corollaries.

Corollary 4.1. *Two words $u, v \in \Sigma^+$ are powers of a word if, and only if, they commute if, and only if, they satisfy a nontrivial relation.*   □

Note that the argumentation of the proof of Theorem 4.1, gives a few line proof for this basic result.

Procedure P can be applied to derive the following representation result for 1-free morphisms $h : \Sigma^* \to \Delta^*$. In order to state it let us call a morphism $e : \Sigma^* \to \Sigma^*$ *basic*, if there are two letters $a, b \in \Sigma$ such that $e(a) = ba$ and $e(c) = c$ for $c \in \Sigma - \{a\}$. Then when applied P to $h(\Sigma)$ in such a way that the identifications are done only at the end we obtain

Corollary 4.2. *Each 1-free morphism $h : \Sigma^* \to \Delta^*$ has a representation*

$$h = p \circ c \circ \pi,$$

where $p : \Delta^* \to \Delta^*$ is a prefix, $c : \Sigma^* \to \Delta^*$ is length preserving and $\pi : \Sigma^* \to \Sigma^*$ is a composition of basic morphisms.   □

Obviously, Corollary 4.2 has also a two-sided variant, where p is a biprefix, and in the definition of the basic morphism the condition $e(a) = ba$ is replaced by $e(a) = ba \vee ab$.

Corollary 4.3. *The prefix hull of a finite set $X \subseteq \Sigma^+$ can be computed in polynomial time in the size $s(X)$ of X.*

Proof. Even by a naive algorithm each step in Procedure P can be done in time $\mathcal{O}(s(X)^3)$. So the result follows since the number of rounds in P is surely $\mathcal{O}(s(X))$.   □

As a final corollary we note a strengthening of Theorem 4.1.

Corollary 4.4. *If a finite set $X \subseteq \Sigma^+$ satisfies a nontrivial 1-way infinite relation, i.e., X does not have a bounded delay (from left to right), then $\|X(p)\| < \|X\|$.*

Proof. Indeed, it is not the property of being a noncode, but the property of not having a bounded delay (from left to right), which forces that at least one identification of words takes place in step 3 of Procedure P.   □

It is interesting to note that Corollary 4.4 does not extend to 2-way infinite words.

Example 4.3. The set $X = \{abc, bca, c\}$ satisfies a nontrivial 2-way infinite relation $^\omega(abc)^\omega = ^\omega(bca)^\omega$, but still even $d(X) = 3 = \|X\|$.   □

Next we turn from a prefix hull of a finite set X to two other types of hulls defined at the beginning of this subsection. Actually, from the algebraic point of view the free hull $X(f)^+$ is the most natural one, and is considered in details in [BPPR] and [Lo]. It yields the following variant of the defect theorem.

Theorem 4.2. *For each finite $X \subseteq \Sigma^+$, which satisfies a nontrivial relation, we have*
$$\|X(f)\| < \|X\|.$$

Without giving a detailed proof of this result we only mention the basic ideas, from which the reader can reconstruct the whole proof, cf. [Sp2]. Actually, the proof given in [BPPR] and [Lo] is even sharper defining precisely the set $X(f)$.

We start from a double X-factorization depicted as

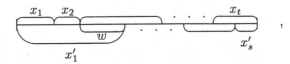

where $x_i, x'_j \in X$ and $w \in \Sigma^+$. Then, by the property (f) and the definition of the free hull, w is in the free hull, i.e., in the construction of $X(f)$ we can replace x'_1 of X by w. Now, the new set obtained may be ambiguous yielding a required defect effect, or it is not a code. However, in any case it is of a smaller size than the old one guaranteeing that the procedure terminates.

Note that we already used these ideas in Examples 4.1 and 4.2.

It follows immediately that Corollary 4.3 extends to free hulls, while Corollary 4.4, of course, does not have a counterpart here. Moreover, the free hull of X satisfies the following important property, cf. [BPPR].

Proposition 4.1. *Let $X \subseteq \Sigma^+$ be finite and $X(f)$ the minimal generating set of its free hull. Then, for each $x \in X(f)$, we have $xX(f)^* \cap X \neq \emptyset$.*

The above result states that each word of $X(f)$ occurs as the first one in some $X(f)$-factorization of a word of X, a property which is, by Procedure P, also true for the prefix hull, i.e., for $X(p)$.

What we said above about free hulls extends to unique factorization hulls. The details can be found in [HK1].

Now, we are in the position to summarize our considerations of this subsection. For a finite $X \subseteq \Sigma^+$ we have

(1) $d(X) \leq p(x) \leq r(x) \leq u(X) \leq \|X\|,$

where, moreover, the last inequality is proper if X is not a code. Here the first inequality is trivial, the second follows, by the definitions, from the fact that $X(f) \subseteq X(p)^+$, and the third similarly from the fact that $X(u) \subseteq X(f)^+$. As we saw in Example 4.1, each of the inequalities in (1) can be proper simultaneously. They, of course, can be equalities as well.

Example 4.4. Let $X = \{a, ab, cc, bccdd, dda\}$. Then the only nontrivial minimal relation is

$$a.bccdd.a = ab.cc.dda$$

from which we conclude that $X(u) = \{a, b, cc, dd\}$. But this is already a prefix code so that $X(p) = X(f) = X(u)$. Finally, the exhaustive search shows that $d(X) = 4$. Therefore we have an example for which $d(X) = p(X) = r(X) = u(X) = \|X\| - 1$.   □

Although we formulated everything in this subsection for sets X not containing the empty word 1, i.e., for free semigroups, the results hold for free monoids, as well. This is because, if $1 \in X$, then trivially any rank of X is strictly smaller that $\|X\| - 1$.

4.3 Defect effect of several relations

In this subsection we consider possibilities of generalizing the above defect theorems to the case of several nontrivial relations. A natural question is: if a set of n words satisfies two "different" nontrivial relations, can these words be expressed as products of $n - 2$ words? Unfortunately, the answer to this question is negative, as we shall see in a moment.

We formalize the term "different" as follows. Let $X \subseteq \Sigma^+$ be a finite set. relations in X^+ are considered as equations with Ξ as the set of unknowns and having X as a solution, cf. Section 2.5. This requires to consider X as an ordered set, and that $\|\Xi\| = \|X\|$. This allows to state the set of all relations of X^+ as a set of equations over Ξ having X as a solution. In Section 2.5 this was referred to as R_X. Here we consider its subset consisting only of so-called *reduced* equations, i.e., equations $(u, v) \in \Xi^+ \times \Xi^+$ satisfying

$\mathrm{pref}_1(u) \neq \mathrm{pref}_1(v)$ and $\mathrm{suf}_1(u) \neq \mathrm{suf}_1(v)$. For simplicity, we prefer to denote the set of all reduced equations of X by $E(X)$.

We say that a system E of equations over the set Ξ of unknowns is *independent* in Σ^+, if no proper subset E' of E is *equivalent* to E, i.e., has exactly the same solutions as E has. Now, identities of X^+ are "different" if their corresponding equations form an independent system of equations.

Example 4.5. The pair

$$xzy = yzx \quad \text{and} \quad xzzy = yzzx$$

of equations is independent, since the former has a solution

$$x = aba, \ y = a \quad \text{and} \quad z = b,$$

which is not a solution of the latter, while the latter has a solution

$$x = abba, \ y = a \quad \text{and} \quad z = b,$$

which is not a solution of the former. However, they have a common solution of degree two, namely $x = y = a$ and $z = b$.   □

Despite of Example 4.5 there are some nontrivial conditions which force sets satisfying these conditions to be of at most certain degree. Particularly useful such results are, if they guarantee that the sets are periodic.

In our subsequent considerations, unlike in those of the previous subsection, it is important that *equations are over free semigroups and not over free monoids*.

Let $\{u_1, \ldots, u_n\} = X \subseteq \Sigma^+$ be finite and $E(X) \subseteq \Xi^+ \times \Xi^+$ the set of all (reduced) equations satisfied by X. This means that $X = h(\Xi)$ for some morphism $h : \Xi^+ \to \Sigma^+$ satisfying $h(\alpha) = h(\beta)$ for all (α, β) in $E(X)$. With each equation in $E(X)$, say

$$e : x\alpha = y\beta \quad \text{with} \ x \neq y, \ x, y \in \Xi, \ \alpha, \beta \in \Xi^*$$

we associate $\pi(e) = \{h(x), h(y)\}$, and with the system $E(X)$ we associate the following graph $G_{E(X)}$:

the set of nodes of $G_{E(X)}$ is X; and
the edges of $G_{E(X)}$ are defined by the condition: (u, v) is an edge in
$G_{E(X)} \Leftrightarrow \exists e \in E(X) : \pi(e) = \{u, v\}$.

It follows that $G_{E(X)}$ defines via its compoments an equivalence relation on X. Now, the degree of X is bounded by the number of *connected components* of $G_{E(X)}$, which we denote by $c(G_{E(X)})$, cf. [HK1]. Note that in above X maybe a multiset, and this indeed is needed in the next proof.

Theorem 4.3. *For each finite $X \subseteq \Sigma^+$, we have*

$$d(X) \leq p(X) \leq c(G_{E(X)}).$$

Proof. We already know that the first inequality holds. To prove the second we proceed as in Procedure P of subsection 4.2.

Let $u - v$ be an edge in $G_{E(X)}$. Then assuming, by symmetry, that $u \leq v$ we have two possibilities:

(i) if $u = v$ we identify the nodes u and v;
(ii) if $v = ut$ with $t \in \Sigma^+$, we replace X by $(X \cup \{t\}) - \{v\}$.

Let $X' \subseteq \Sigma^+$ be a multiset obtained from X by applying either (i) or (ii) once. Note that due to (ii) X' indeed can be a multiset although X would be unambiguous. Our claim is that

(2) $$c(G_{E(X')}) \leq c(G_{E(X)}).$$

If the operation performed is (i) there is nothing to be proved. So we have to analyse what happens to the graph $G_{E(X)}$ when (ii) is performed. In particular, we have to consider what happens to a subgraph of it of the form:

(3)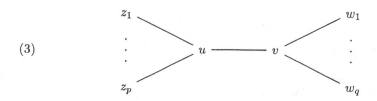

Clearly, the connections $z_i - u$ remain, and connections $v - w_j$ are replaced by $u - w_j$. Moreover, v disappears, and the new node t will be connected in $G_{E(X')}$ to all y_k's in X such that $uy_k\alpha = v\beta$, with $\alpha, \beta \in X^*$, are in $E(X)$. In addition, the introduction of the new t may create some completely new edges to $G_{E(X')}$. But what is important is that, if $G_{E(X)}$ contains the subgraph (3), then $G_{E(X')}$ contains the following subgraphs

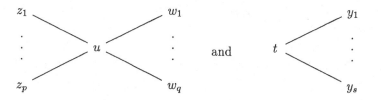

where, moreover, the nodes y_k are nodes of $E_{G(X)}$, i.e., belong to some of the components of $E_{G(X)}$. Therefore, the replacement of v by t does not increase the number of the components, so that we have proved (2).

By the construction $s(X') < s(X)$, and therefore an iterative application of the rules (i) and (ii) leads finally to the discrete graph, the edges of which are labelled by a set $\hat{X}$. It follows from the arguments of the proof of Theorem

4.1, that $\hat{X}$ is contained in the minimal generating set of the prefix hull of X. Therefore, by Theorem 4.1, $\|X(p)\| \leq \|\hat{X}\|$. But, by the discreteness of $G_{E(\hat{X})}$, we have

$$\|\hat{X}\| = c(G_{E(\hat{X})}) \leq c(G_{E(X)}),$$

and hence our proof is complete.   □

Theorem 4.3 has a number of interesting consequences. First, we have a counterpart of Corollary 4.4: if in (1) equations are replaced by ω-equations, i.e., one-way infinite equations, but otherwise the graph – let us denote it now by $G_{E_\omega(X)}$ – is defined as $G_{E(X)}$ we obtain

Corollary 4.5. *For each finite set $X \subseteq \Sigma^+$ we have*

$$d(X) \leq p(X) \leq c(G_{E_\omega(X)}).$$ □

More concrete and useful corollaries are obtained, when the graph $G_{E(X)}$ is connected:

Corollary 4.6. *Let $X \subseteq \Sigma^+$ be finite. If $G_{E(X)}$ is connected, then X is periodic.*   □

Corollary 4.7. *If a three element set $X = \{u, v, w\} \subseteq \Sigma^+$ satisfies the relations $ux = vy$ and $uz = wt$, with $x, y, z, t \in X^\infty$, then u, v and w are powers of a same word.*   □

Corollary 4.7 should be compared to Example 4.5. It also has to be noticed that in our above considerations it is essential that X consists of only nonempty words. Indeed, the graph of the equations

$$x = yx \quad \text{and} \quad z = yz$$

is connected, but it possesses a solution of degree 2, namely $x = a$, $y = 1$ and $z = b$.

As the final application of Theorem 4.3 we give an example from [HK1], which shows that also the inside occurrence of the equations may cause some defect effect.

Example 4.6. Assume that words of X satisfy the following two equations

$$\alpha u \gamma = \beta v \delta \quad \text{and} \quad \alpha w \varepsilon = \beta z \rho,$$

where $u, v, w, z \in X$ and $\alpha, \beta, \gamma, \delta, \varepsilon, \rho \in X^+$. We claim that $d(X) \leq \|X\| - 2$, which cannot be concluded simply by considering the first occurrences of the unknowns in these equations.

There are two cases to be considered. First, if $\alpha = \beta$ (in Σ^+), then u and v, as well as w and z, are in the same component proving the claim. Otherwise assuming, by symmetry, that $\alpha = \beta t$, with $t \in \Sigma^+$, and denoting $X' = X \cup \{t\}$, we see that $G_{E(X')}$ contain the edges $t - z$ and $t - v$, and still one more connecting two words of X, due to the relation $\alpha = \beta t$. Therefore $d(X) \leq d(X') \leq c(G_{E(X')}) \leq \|X'\| - 3 = \|X\| - 2$ as claimed.   □

4.4 Relations without the defect effect

This subsection is in a sense dual to the previous one, where we looked for conditions which would enforce an as large as possible defect effect. Here, motivated by Example 4.5, we try to construct as large as possible independent systems of equations having only a defect effect of a certain size, i.e., still a solution of certain degree d. Two extreme cases, namely those where $d = \|X\| - 1$ or $d = 2$, are of a particular interest. The former asks, what is the maximal number of independent equations forcing only the minimal defect effect, while the latter poses the question how many, if any, would allow only periodic solutions.

The first observation here is that there does not exist any infinite independent system of equations (with a finite number of unknowns). This is due to the validity of the Ehrenfeucht compactness property for free semigroups, considered in Section 7. Whether there can be unboundedly large such systems is an open problem.

Nontrivial bounds for the numbers of independent equations in our above problems are given in the following two examples from [KaPl2].

Example 4.7. Let $\Xi = \{x, y\} \cup \{u_i, v_i, w_i | i = 1, \dots, n\}$ be the set of unknowns and S the following system of equations over Ξ

$$S : \quad x u_j w_k v_j y = y u_j w_k v_j x \quad \text{for } j, k = 1, \dots, n.$$

Then clearly $\|S\| = n^2$ and $\|\Xi\| = 3n + 2$. We claim that

(i) S has a solution of degree $3n + 1$; and
(ii) S is independent.

The condition (i) is easy to fulfill: choose $x = y$, whence all the equations become trivial, so that a required solution can be found in a free semigroup of $3n + 1$ generators.

That the set S is independent is more difficult to see. We have to show that, for each pair (j, k), there exists a solution of the system

$$S(j, k) = S - \{x u_j w_k v_j y = y u_j w_k v_j x\},$$

which is not a solution of S. To find out such a solution is not obvious, however, here is such a solution:

(1)
$$
\begin{cases}
x & = & b^2 ab, \\
y & = & b, \\
u_t & = & \begin{cases} ba & \text{if } t = j, \\ bab & \text{otherwise,} \end{cases} \\
w_{t'} & = & \begin{cases} bab^2 & \text{if } t' = k, \\ b & \text{otherwise,} \end{cases} \\
v_t & = & \begin{cases} ba & \text{if } t = j, \\ a & \text{otherwise.} \end{cases}
\end{cases}
$$

Then if $t = j$ and $t' = k$, we compute

$$xu_jw_kv_jy = b^2ab.ba\ldots \neq b.ba.bab^2\ldots = yu_jw_kv_jx$$

to note that (1) is not a solution of S. The verification that it is a solution of $S - S(j,k)$ is a matter of simple calculations:

$$
\begin{aligned}
t \neq j \wedge t' \neq k &\quad : \quad b^2ab.bab.b.a.b = b.bab.b.a.b^2ab, \\
t \neq j \wedge t' = k &\quad : \quad b^2ab.bab.bab^2.a.b = b.bab.bab^2.a.b^2ab, \\
t = j \wedge t' \neq k &\quad : \quad b^2ab.ba.b.ba.b = b.ba.b.ba.b^2ab.
\end{aligned}
$$

$\square$

Example 4.8. Let $\Xi = \{x_i, y_i, u_i, w_i, v_i | i = 1, \ldots, n\}$ be a set of $5n$ unknowns, and S' the following system of n^3 equations

$$S' \; : \; x_iu_jw_kv_jy_i = y_iu_jw_kv_jx_i \quad \text{for } i, j, k = 1, \ldots, n.$$

Hence S' is obtained from the system S of Example 4.7 by introducing index i for x and y, and by allowing it to range from $1, \ldots, n$. The solution (1) of S is extended by setting

$$
(2) \quad
\begin{cases}
x_{t''} &= \begin{cases} b^2ab & \text{if } t'' = i, \\ a & \text{otherwise,} \end{cases} \\
y_{t''} &= \begin{cases} b & \text{if } t'' = i, \\ a & \text{otherwise.} \end{cases}
\end{cases}
$$

It follows directly from the computations of Example 4.7, that the solution described by (1) and (2) satisfies all the equations of S' except one, namely $x_iu_jw_kv_jy_i = y_iu_jw_kv_jx_i$.

Note that S' has a nonperiodic solution in Σ^+. $\square$

The message of Example 4.7 is that in a free semigroup there can be $\Omega(n^2)$ independent equations in n unknowns without forcing larger than the minimal defect effect, i.e., has still a solution of degree $n - 1$. Similarly, Example 4.8 shows that there can be $\Omega(n^3)$ independent equations in n unknowns having a nonperiodic solution.

Examples 4.7 and 4.8 motivate several comments and open problems. First, one may think that in Example 4.7 the requirement that Σ contains at least $3n+1$ generators makes the whole example artificial. However, if instead of the degree, i.e., the combinatorial rank, for example the prefix rank would be considered, then the example can be encoded into the binary alphabet. Indeed, encoding the alphabet of $3n+1$ letters into the binary one by a prefix encoding, we can find for S a solution over the binary alphabet having the p-rank equal to $3n + 1$.

Second, if the systems of equations are solved in a free monoid, instead of a free semigroup, then the bounds of Examples 4.7 and 4.8 can be improved to $\Omega(n^3)$ and $\Omega(n^4)$, respectively, cf. [KaPl2].

Third, we state two open problems.

Problem 4.1. Improve the bounds $\Omega(n^2)$ and $\Omega(n^3)$ in Examples 4.7 and 4.8. In particular, can they be exponential?

Problem 4.2. Does there exist an independent system of three equations with three unknowns having a nonperiodic solution in Σ^+ ?

Problem 4.2 is connected to Example 4.5, as well as to Corollary 4.7. Our guess is that the answer to this problem is "no". However, the problem does not seem to be easy.

4.5 The defect theorem for equations

In this subsection we turn our focus explicitly from sets to equations, i.e., from solutions of equations to equations itself. The *rank* of an equation $u = v$, with the unknowns Ξ is defined as the maximal rank of its solutions $h : \Xi^+ \to \Sigma^+$ over all free semigroups Σ^+. Consequently, different notions of the rank of a finite set seem to lead to different notions of the rank of an equation. Fortunately, this is not true, at least as long as the rank of a set is defined in one of the four ways we did. To establish this is the goal of this subsection.

We start by comparing the combinatorial rank d and the prefix rank p. This is done in two lemmas, the first one being obvious from the definitions.

Lemma 4.2. *Each solution* $h : \Xi^+ \to \Sigma^+$ *of an equation over* Ξ *satisfies* $d(h(\Xi)) \leq p(h(\Xi))$.

The second lemma is less obvious, and shows that, with each solution h, we can associate so-called *principal* solution of [Len].

Lemma 4.3. *Let* $u = v$ *be an equation over* Ξ. *For each solution* $h : \Xi^+ \to \Sigma^+$ *of the equation* $u = v$, *there exists another solution*, $h' : \Xi^+ \to \Sigma'^+$ *such that* $d(h'(\Xi)) = p(h(\Xi))$.

Proof. Let the minimal generating set of the prefix hull of $h(\Xi)$ be $U = \{u_1, \ldots, u_d\}$. Consequently, for each $x \in \Xi$, $h(x)$ has a U-factorization, say

$$(1) \qquad\qquad h(x) = u_{i_1} \ldots u_{i_t}.$$

Let $\theta : \Sigma' \leftrightarrow U$ be a one-to-one mapping, where Σ' is an new alphabet and denote by $c_i \in \Sigma'$ the element corresponding to $u_i \in U$ in this mapping. Next we define a morphism $h' : \Xi^+ \to \Sigma'^+$ by setting, for each $x \in \Xi$,

$$h'(x) = c_{i_1} \ldots c_{i_t} \Leftrightarrow h(x) = u_{i_1} \ldots u_{i_t} \quad \text{with} \quad u_{i_j} \in U.$$

By construction $\theta(h'(x)) = h(x)$ holds for all $x \in \Xi$ and since θ is injective, we have $h'(x) = h'(v)$ showing that h' is a prefix and by its definition, the minimal generating set of the prefix hull of $h'(\Xi)$ is Σ'^+. Consequently, $d(h'(\Xi)) \leq d = p(h(\Xi))$.

If $d(h'(\Xi)) < d$, there would be at most $d - 1$ words of Σ'^+, such that each word $h'(a)$ could be expressed as a product of these words. Therefore also words in (1) could be expressed as products of at most $d - 1$ words of U^+. This, however, contradicts with the fact that each u_i must be the first factor in at least one of the factorizations (1), cf. Proposition 4.1. Hence, necessarily $d(h'(\Xi)) = p(h(\Xi))$, as required. $\Box$

Both of the Lemmas 4.2 and 4.3 can be extended to the other algebraic ranks. The detailed proofs, using Proposition 4.1 and its counterpart for the u-rank, are left to the reader.

Now we are ready to formulate our main result of this section.

Theorem 4.4. *Let $u = v$ be an equation over Ξ. The rank of the equation $u = v$, defined as the maximal rank of its solutions, is independent of which of our four ranks is used to define the rank of a solution.* $\Box$

Theorem 4.4 allows to denote the *rank* of an equation simply by $r(E)$, as well as restate the defect theorem for equations.

Theorem 4.5. *For each nontrivial equation E over the unknowns Ξ, the rank $r(E)$ of E satisfies $r(E) < \|\Xi\|$.* $\Box$

Note that, as shown by the proof of Theorem 4.4, for all algebraic ranks the rank of an equation can be defined over a fixed free semigroup Σ^+ containing at least two generators, but the combinatorial rank requires it to be defined over all free semigroups Σ^+.

We already noted that the p-rank and the f-rank of a finite set of words can be computed in a polynomial time. The same holds for the u-rank, but as we shall see in the next subsection, is known not to hold for the combinatorial rank. Computing the rank of an equation is essentially more complicated. However, as shown in the next section, this can be achieved by applying Makanin's algorithm.

4.6 Properties of the combinatorial rank

We conclude Section 4 by pointing out some further differences between the combinatorial rank and the algebraic ranks.

First, however, we emphasize the usefulness of the notion of the combinatorial rank, or of the degree. The most important cases are the both extremes, namely when a degree of a finite set $X \subseteq \Sigma^+$ equals 1 or $\|X\|$. The former corresponds to periodic sets, and the usefulness of the notion of the degree in connection with periodic sets was already seen, for instance, in Theorem 4.3. In the other extreme we call a finite $X \subseteq \Sigma^+$ *elementary*, if $d(X) = \|X\|$, and *simplifiable* otherwise. Note that this definition is consistent with that of an elementary morphism defined in Section 2.4.

A striking example of the usefulness of the above notions is an elegant proof of the D0L equivalence problem in [ER1], cf. also [RoSa1]. A crucial step in this proof was the following result.

Theorem 4.6. *An elementary morphism has a bounded delay.*

Proof. Follows directly from Corollary 4.4. Indeed, a morphism $h : \Sigma^+ \to \Delta^+$ having an unbounded delay satisfies $d(h(\Sigma)) < \|\Sigma\|$ so that it is not elementary.   □

Our next example shows that the elementary sets are not closed under composition of sets in the sense of codes, cf. [BePe].

Example 4.9. Let $X = \{b, cab, cabca\}$. Then its composition with itself is

$$X \circ X = \{cab, cabcabcab, cabcabcabcabcab\} \subseteq (cab)^+.$$

Consequently, $d(X \circ X) = 1$, while $d(X) = 2$.   □

As shown in [Ne2] it is not difficult to modify Example 4.9 to show that, for each $n \in \mathbb{N}$, there exists a set $X_n \subseteq \Sigma^+$ such that $d(X_n \circ X_n) - d(X_n) \geq n$. In [Ne2] it is also considered how the degree of a set behaves with respect to certain operations, in particular with respect to rational operations.

Finally, we deal with the problem of computing the degree of a given set. This seems to be computationally very difficult, as a contrast to Corollary 4.3 (or its variants to the other algebraic ranks), which shows that the algebraic ranks are computable in polynomial time. This also explains why we didn't give any procedure to compute a set F in the definition of the degree: no fast method for that is known, or even likely to be discovered, as we now demonstrate.

The complexity results for the degree, due to [Ne1], are as follows:

Theorem 4.7. *(i) The problem of deciding, for a given finite set $X \subseteq \Sigma^+$ and for a given number k, whether $d(X) \leq k$ is NP-complete.*

(ii) The problem of deciding whether a given finite set is simplifiable is NP-complete.

Actually, the problem of (i) remains NP-complete even if k is fixed to be any number larger than 2. The choice $k = 2$ makes the problem computationally easy: as shown in [Ne3] it can be solved in time $\mathcal{O}(n \log^2 m)$, where $n = s(X)$ and $m = \max\{|x| \mid x \in X\}$. Note also that (ii) is equivalent to saying that the elementariness problem is in the class of co-NP-complete problems, cf. [GJ]. In particular, it is not likely that a polynomial time algorithm will be found for it.

We do not present the proof of Theorem 4.7 here, but in order to give an intuition why the result holds, we show, in the next example, that a related problem is NP-complete. Actually, the NP-completeness of this is the first step in the proof of Theorem 4.7 in [Ne1].

Example 4.10. (Strong Factorizability Problem.) The problem asks to decide, for a finite set $X \subseteq \Sigma^+$ and for a number k, whether there exists a set $Y \subseteq \Sigma^+$ such that

$$X \subseteq Y^+, \ \|Y\| \le k \text{ and } X \cap Y = \emptyset.$$

If such a Y exists, we say that X is *strongly k-factorizable*, and we refer this problem to as the *SF-problem*. Note that if we drop from the SF-problem the requirement $X \cap Y = \emptyset$, we obtain the problem (i) of Theorem 4.7.

Obviously the SF-problem is in NP. So to prove its NP-completeness we have to reduce it to some known NP-complete problem, which will be the following variant of the *vertex cover problem*, referred to as the *special* vertex cover problem, or the *SVC-problem* for short. For the NP-completeness of this, which is a straightforward modification of the NP-completeness of the ordinary vertex cover problem, we refer to [Ne2] or [GJ].

The *SVC-problem* asks to decide for a given graph $G = (V, E)$, with $\|V\| = \|E\|$ and having no isolated points, and for a given natural number k, whether there exists a subset V' of V, with $\|V'\| \le k$, such that the set of edges connected to V' equals that of all edges of G. In other words, the SVC-problem asks whether a graph of the required type has a *vertex cover* of size at most k. Now let

$$((V, E), k) \ \text{ with } \ \|V\| = \|E\| = n \ \text{ and } \ 1 \le k \le n - 1$$

be an instance of the SVC-problem. We associate it with an instance

$$(X, k + n)$$

of the SF-problem by defining a subset $X \subseteq VTV$, where T is a renaming of E under the mapping $c : E \to T$, as follows

(1) $$\alpha a \beta \in X \Leftrightarrow \alpha, \beta \in E \ \text{ and } \ a = c(\alpha, \beta).$$

We have to show that

$G = (V, E)$ has a vertex cover V' with $\|V'\| \le k$ if, and only if, X is $(k + n)$-strongly factorizable.

First, assume that G has a vertex cover of size at most k. Let $\alpha a \beta$ be a word in X. It is factorized as $\alpha.a\beta$, if $\alpha \in V'$, and $\alpha a.\beta$, if $\alpha \in V'$. Now let B be the set of all words of length 2 in these factorizations. Then, by (1), $\|B\| = n$, so that $\|V' \cup B\| = \|V'\| + n \le k + n$. Therefore, X is $(k+n)$-strongly factorizable.

Second, assume that X is $(k + n)$-strongly factorizable via Y. We define a partition of X

$$X = X_1 \cup X_2 \ \text{ with } \ X_1 \cap X_2 = \emptyset$$

as follows. The word $\alpha a \beta \in X_1$ if, and only if, it is factorized in Y as $\alpha a.\beta$ or $\alpha.a\beta$, and therefore $\alpha a \beta \in X_2$ if, and only if, it is factorized in Y as $\alpha.a.\beta$.

Let V_i, for $i = 1, 2$, consists of those letters of V which occur in the above factorizations of words of X_i. Similarly, let $T_i \subseteq T \cup TV \cup VT$, for $i = 1, 2$, consists of those words in $Y - V$ which occur in these factorizations of words of X_i. Finally, for each $w \in X_2$, i.e., w being factorized as $\alpha.a.\beta$, we pick up either α or β from V_2, and denote by V_2' the set of all letters picked up when w ranges over X_2. Now, we set

$$K = V_1 \cup V_2'.$$

Then, by the construction, K is a vertex cover. It also follows that the sets T_1, T_2 and $V_1 \cup V_2$ are pairwise disjoint, and moreover, by (1), we have $\|T_i\| = \|X_i\|$, for $i = 1, 2$. Consequently, we obtain the following relation

$$
\begin{aligned}
\|Y\| &= \|V_1 \cup V_2 \cup T_1 \cup T_2\| = \|V_1 \cup V_2\| + \|T_1\| + \|T_2\| \\
&= \|V_1 \cup V_2\| + \|X_1\| + \|X_2\| = \|V_1 \cup V_2\| + \|X\|
\end{aligned}
$$

implying, since $\|Y\| \le k + n = k + \|X\|$, that

$$\|K\| = \|V_1 \cup V_1'\| \le \|V_1 \cup V_2\| \le k.$$

Therefore, the graph (V, E) has a vertex cover of size at most k, completing our proof.   □

5. Equations as properties of words

Two elements x and y of a group are said conjugate if there exists an element z such that equation $x = zyz^{-1}$ holds. In order to extend this definition to monoids, one has to eliminate the inverses which can be easily achieved by multiplying two handsides by the element z to the right yielding equation

$$xz = zy$$

The purpose of this section is to discuss the connection between equations in words and some properties of words. We think that little is known so far and that much remains to be done.

5.1 Makanin's result

We already noted that the p-rank and the f-rank of a finite set of words can be computed in a polynomial time. The same holds for the u-rank, but as we have seen in Section 4.6, it does not hold for the combinatorial rank. Computing the rank of an equation is essentially more complicated since we aim at computing the maximal rank over a (usually) infinite set of solutions. However, this can be achieved by applying Makanin's algorithm which is one of the major advances in combinatorial free monoid theory.

We recall that given an alphabet Ξ of unknowns and an alphabet Σ of constants, Ξ and Σ being disjoint, an *equation with constants* is a pair $(u, v) \in (\Xi \cup \Sigma)^* \times (\Xi \cup \Sigma)^*$, also written $u = v$. A *solution* is a morphism

$h : (\Xi \cup \Sigma)^* \to \Sigma^*$ leaving Σ invariant, i.e., satisfying $h(a) = a$ for all $a \in \Sigma$, for which the following holds

$$h(u) = h(v).$$

For example, the equation $ax = xb$ with $a \neq b \in \Sigma$ and $x \in \Xi$ has no solution since the left handside has one more occurrence of a than the right handside, and the equation $ax = xa$ has the solution $x = a$.

We have the famous result of Makanin, cf. [Mak].

Proposition 5.1. *There exists an algorithm for solving an equation with constants.*

The exact complexity of the problem is unknown but several authors have contributed to lower the complexity of the original algorithm which was an exponential function of height 5. Actually, this complexity depends on the complexity of computing the minimal solutions of diophantine equations. We refer the interested reader to [Ab1], [Do] and [KoPa] for the latest results on this topic. Several sofware packages have been produced which work relatively well up to length, see e.g., [Ab2].

5.2 The rank of an equation

One of the most direct consequences of Makanin's result is the fact that the rank of an equation may be effectively computed, cf. [Pec].

Theorem 5.1. *Given an equation without constants $u = v$, its rank can be effectively computed.*

Proof. The idea of the proof is as follows. Let Ξ be the set of unknowns and denote by ι some mapping of Ξ onto some disjoint subset Σ with $\|\Sigma\| < \|\Xi\|$. Consider the morphism $\theta : \Xi^* \to (\Xi \cup \Sigma)^*$ defined for all $x \in \Xi$ by $\theta(x) = \iota(x)x$. Then the rank of $u = v$ is the maximum cardinality of $\|\iota(\Xi)\|$ for which the equation with unknowns $\theta(u) = \theta(v)$ has a solution. For example, starting with the equation $xyz = zyx$ we would be led to define the 4 equations $axayaz = azayax$, $axaybz = bzayax$, $axbyaz = azbyax$, $axbybz = bzbyax$ and to apply Makanin's result to each of these equation.

More precisely, assume the rank of $u = v$ is r, i.e., there exists a morphism $h : \Xi^* \to \Sigma^*$ such that $h(u) = h(v)$ and $r(X) = r$ where $X = h(\Xi)$. Deleting, if necessary, some unknowns it is always possible to assume that the morphism is nonerasing. Furthermore, without loss of generality, we may assume that the free hull $X(f)^* = \Sigma^*$. Indeed, let $\alpha : \Sigma' \hookrightarrow h(\Xi)$ be an one-to-one mapping, where Σ' is a new alphabet. Then there exists an unique solution $h' : \Xi^+ \to \Sigma'^+$ such that $\alpha(h'(x)) = h(x)$ holds for all $x \in \Xi$. We have $X(f) = \Sigma'$ and $r(h'(\Xi) = r(h(\Xi)$. Let ι be the mapping that associates the initial letter of $h(x)$ to each x. By Proposition 4.1, we know that $\Sigma = \{\iota(x) | x \in \Xi\}$. Consider the morphism $\theta : \Xi^* \to (\Xi \cup \Sigma)^*$ satisfying

$\theta(x) = \iota(x)x$. Then the morphism $g(x) = (\iota(x))^{-1}h(x)$ satisfies the equation with constants $\theta(u) = \theta(v)$.

Example 5.1. With $\Xi = \{x, y, z\}$ and $xyz = zyx$, we have the solution $x = a, y = bab, z = aba$. Then by θ we obtain an equation with unknowns $axbyaz = azbyax$ for which $g(x) = 1, g(y) = ab, g(z) = ba$ is a solution.   □

Proof of Theorem (continued). Conversely, let ι be a mapping of Ξ onto some Σ with $\Xi \cap \Sigma = \emptyset$ and $||\Sigma|| < ||\Xi||$. Consider the morphism $\theta : \Xi^* \to (\Xi \cup \Sigma)^*$ defined for all $x \in \Xi$ by $\theta(x) = \iota(x)x$, and assume that the equation with unknowns $\theta(u) = \theta(v)$ has a solution g. The morphism $h(x) = \iota(x)g(x)$ is clearly a solution of $u = v$. Now we claim that its rank is greater than or equal to $||\Sigma||$. Indeed, let $X \subseteq \Sigma^*$ be the minimal generating set of the free hull of $h(\Xi^*)$: $h(x) \in X^*$ for all $x \in \Xi$. Every element in X appears as the leftmost factor in the decomposition of some $h(x)$. If $||X|| < ||\Sigma||$, then some letter of Σ does not appear in the leftmost position contradicting the definition of h □

Actually, this result carries over to the rank of equations with constants, after a suitable extension of the notion of rank.

5.3 The existential theory of concatenation

Makanin's result can be interpreted either as a statement on systems of equations and inequations, or equivalently as a statement of formulae of the existential theory of concatenation. More precisely, it has been observed that at the cost of introducing new unknowns, negations and disjunctions can be expressed as conjunctions of equations and further that all conjunctions are equivalent to a single equation. In other words, starting from a Boolean combination of equations on the unknowns Ξ, it is possible to define a single equation on the unknowns $\Xi \cup \Xi'$ for some Ξ', whose set of solutions restricted to the unknowns Ξ equals the set of solutions of the Boolean combination.

It is worthwhile considering the power of equations in expressing properties or n-ary relations on words, for some integer n. Following the tradition, we call *diophantine* a relation on words $R(x_1, \ldots, x_n)$ that is equivalent to a formula of the form

(1) $\exists y_1, \ldots, \exists y_m \lambda(x_1, \ldots, x_n, y_1, \ldots, y_m) = \rho(x_1, \ldots, x_n, y_1, \ldots, y_m)$

with $\lambda = \rho$ an equation. For example, "x is imprimitive" can be expressed as

$$\exists y, z : x = 1 \vee (x = yz \wedge yz = zy \wedge y \neq 1 \wedge z \neq 1)$$

and "x and y are conjugate" can be expressed with two extra unknowns as

$$\exists u, v : x = uv \wedge y = vu,$$

or with one extra unknown only as

(2) $$\exists z : xz = zy.$$

These formulae are diophantine. No characterization of diophantine relations seems to exist in the literature. There is no available tool either for showing that a given property is not diophantine, a natural candidate would be, e.g., primitivity. Neither do we know which are the properties that are diophantine and whose negation also is diophantine. Intuitively, this imposes very strong restrictions on the property, one such example being "x is a prefix of y". Yet another area of research is to study the hierarchy of diophantine formulae where the number of existential quantifiers is taken into account, i.e., the integer m of equation (1). In this vein, it was shown in [Sei] that the relation "x is a prefix of y" can not be expressed without an extra variable.

Let us now briefly show how to reduce a Boolean combination of equations to a single equation. Assuming that Σ contains two different constants a and b, the system consisting of the two equations $x = y$ and $u = v$ is equivalent to the single equation $xauxbu = yavybv$ as noticed in [Hm]. To check this, identify the unknowns with their images under the solution h and observe that xau, xbu, yav and ybv have all the same length, to wit half the common length of the left- and right-handsides. Thus $xau = yav$ and $xbu = ybv$ holds. If $x \neq y$, say $|x| < |y|$ without loss of generality, then the first equation says that there is an occurrence of a in position $|x|$ in y, while the second says that this occurrence is equal to b.

Similarly, as noted, e.g., in [CuKa2], introducing new unknowns, the inequation $x \neq y$ is equivalent to a disjunction of equations saying that x and y are prefixes of each other or that their maximum common prefix is a proper prefix of both. Hence three new unknowns are needed here in this reduction. Finally, with the help of more unknowns a disjunction of equations can be expressed as a conjunction of equations as we show in a moment. So, in terms of logics, Makanin's result implies that the existential fragment of the theory of concatenation is decidable. We formulate the above as.

Theorem 5.2. *For any Boolean combination B of equations with Ξ as the set of unknowns we can construct a single equation E with $\Xi \cup \Xi'$ as the set of unknowns such that solutions of B and those of E restricted to Ξ are exactly the same.* □

As we said, in the process of reducing a Boolean combination to a single equation new unknowns are introduced. A more precise computation of how many are needed has been studied though the issue of the exact number is not yet settled. In particular reducing a disjunction to conjunctions has received various solutions. Büchi and Senger used 4 new unknowns in [BS], Senger in his thesis needs 3, while Serge Grigorieff achieves the same result with 2. It is an open question whether or not one unknown suffices though it is suspected it does not.

We reproduce here the unpublished proof of S. Grigorieff.

Theorem 5.3. *The disjunction $x = u \vee y = v$ is equivalent to a formula of the form*

$$\exists z \exists t \lambda(x, y, u, v) = z\rho(x, y, u, v)t$$

where $\lambda(x, y, u, v)$ and $\rho(x, y, u, v)$ are words over the alphabet $\{x, y, u, v, a, b\}$ and z, t are new variables.

Proof. First by observing that $x = u \vee y = v$ is equivalent to $xv = uv \vee uy = uv$, without loss of generality we may start with a disjunction of the form $x = u \vee x = v$. By making the further observation that $x = u \vee x = v$ is equivalent to $xa = ua \vee xa = va$ we may assume that x, u, v are nonempty words.

Now we use the pairing function $< x, y >= xayxby$. We set

$$\rho(x, u, v) =< uu, vv >^2 xx < uu, vv >^3 x < uu, vv >^3$$
$$\lambda(x, u, v) =< uu, vv >^3 uu < uu, vv >^3 u < uu, vv >^3 vv$$
$$< uu, vv >^3 v < uu, vv >^3$$

Making use of the primitivity of $< uu, vv >$, a case study shows that the factor ρ fits in λ in only two places, either

$$\lambda =< uu, vv > \rho vv < uu, vv >^3 v < uu, vv >^3$$

implying that $x = u$, or

$$\lambda =< uu, vv >^3 uu < uu, vv >^3 u < uu, vv > \rho$$

implying that $x = v$.   □

Finally we note that Makanin's result is on the borderline between the decidability and the undecidability. Indeed, [Marc] established the undecidability of the fragment $\forall \exists^4$-positive of the concatenation theory, further improved to $\forall \exists^3$-positive. The previous reduction of disjunctions yields the undecidability of the theory consisting of formulae of the form

$$\forall \exists^5 \lambda(x_1, \ldots, x_6) = \rho(x_1, \ldots, x_6),$$

where $\lambda = \rho$ is an equation.

5.4 Some rules of thumb for solving equations by "hand"

There is unfortunately no method, in the practical sense of the word, for solving equations. We list here just a few simple-minded tricks that are widely used when dealing with real equations. Most of them lead to proving that the equation has only cyclic solutions by reducing the initial equation to the equations that are well-known, such as Levi's Lemma, cf. (1) in Section 2.1, the conjugacy, cf. e.g. (2), or the commutativity, cf. Corollary 4.1.

First of all, conditions on the lengths of the unknowns are expressed as linear equations over the positive integers. When some of these unknowns have length 0 then the number of unknowns reduces. An elaboration of this idea is exemplified by the following well-known fact that appears when solving the general equation $x^n y^m = z^p$ for $n, m, p \geq 2$. Let us verify that $x^2 y^2 = z^2$ implies that x, y and z are powers of the same elements. Indeed, observing that $xy^2 x$ is a conjugate of z^2, there exists a conjugate z' of z such that $xy^2 x = z'^2$. Since xy and yx have same length, they are both equal to z' implying that $x, y \in t^*$ for some word t and thus that $z \in t^*$ also.

Splitting represents another approach. In the easy cases, there is a prefix of the left- and right-handsides that have the same length, i.e., $zxyxzy = yxxzyz$ splits into $zxyx = yxxz$ and $zy = yz$. This ideal situation is rare, however a variant of it is not so seldom. Assume a primitive word x has an occurrence in both handsides of the equation, say $uxv = u'xxv'$ where $u, u', v, v' \in \Xi^*$ are products of unknowns. Assume further $|u'| \leq |u| \leq |u'x|$. Then the equation splits into $u = u'x$ and $v = v'$ or into $u = u'$ and $v = xv'$. Combinatorial problems on words in the theory of finite automata, rational relations, varieties etc..., usually come up as families of equations involving a parameter, e.g., $xy^n z = zy^n t$ with $x, y, z, t \in \Xi$ and $n \in \mathbb{N}$. Then the above condition on the lengths can be enforced by choosing an appropriate value of n.

Another technique proves useful in some very special cases. It was the starting point of the theory developped in [Len] and it consists, for fixed lengths of a solution, to compute the "freest" solution with these lengths. As an illustration let us consider the equation

$$(3) \qquad\qquad xyz = zyx$$

and assume $|x| = 3$, $|y| = 5$, $|z| = 1$, with a total length of 9. Write

$$x = x_1 x_2 x_3, y = y_1 y_2 y_3 y_4 y_5, z = z_1.$$

The idea is to identify the positions which bear the same letter in both handsides, such as 3 and 9 (carrying x_3) and 5 and 3 (carrying y_2).

x_1	x_2	x_3	y_1	y_2	y_3	y_4	y_5	z_1
z_1	y_1	y_2	y_3	y_4	y_5	x_1	x_2	x_3

More precisely, define a graph whose 9 vertices are in one-one correspondence with the 9 occurrences of letters in the solution, and whose non-oriented edges are the pairs $(i, j), 0 \leq i, j \leq 9$, where the letter in position i in the left handside is equal to the letter in position j in the right handside of (3) or vice versa. Then each connected component of the graph is associated with a distinct letter in the target alphabet. In other words, the "richest" alphabet for a solution of (3) has cardinality equal to the number of connected components of the graph.

If we had chosen $|x| = 2$, $|y| = 4$ and $|z| = 1$ for a total length of 7, then we would have found one connected component, actually one Hamiltonian path $1, 5, 4, 3, 2, 6$, i.e., the richest solution would be cyclic.

Fixing the lengths may look like too strong a requirement, however, this very technique allows us to prove in the next section that the Theorem of Fine and Wilf is sharp, i.e., that on a binary alphabet there exist only two words of length p and q, p and q coprimes, whose powers have a common prefix of length exactly equal to $p + q - 2$.

6. Periodicity

Periodicity is one of the fundamental properties of words. Depending on the context, and traditions, the term has had several slightly different meanings. What we mean by it in different contexts is recalled here, cf. also Section 2.2. The other goals of this section is to present three fundamental results on periodicity of words, namely the Periodicity Theorem of Fine and Wilf, the Critical Factorization Theorem, and recent characterizations of ultimately periodic 1-way infinite words and periodic 2-way infinite words.

6.1 Definitions and basic observations

We noted in Section 2.2 that each word $w \in \Sigma^+$ has the unique period $p(w)$ as the length of the minimal u such that

$$(1) \qquad\qquad w \in F(u^\omega).$$

Such a $p(w)$ is called *the period* of w as a distinction of *a period* of w which is the length of any u satisfying (1). When the period refers to a word, and not to the length, then *the periods* of w are all the conjugates of the minimal u in (1), often called *cyclic roots* of w. Similarly *periods* of w are all conjugates of words u satisfying (1). Finally, we call w *periodic*, if $|w| \geq 2p(w)$, i.e., w contains at least two consecutive factors of its same cyclic root. *Local* variants of these notions are defined in Section 6.3.

In connection with infinite words *periodic* 1-way and 2-way infinite words are defined as words of the forms u^ω and $^\omega u^\omega$, with $u \in \Sigma^+$, respectively. By an *ultimately periodic* 1-way infinite word we mean a word of the form uv^ω, with $u \in \Sigma^*$ and $v \in \Sigma^+$. Formally, the word $^\omega u^\omega$, for instance, is defined by the condition

$$^\omega u^\omega(i) = u(i \bmod |u|), \text{ for all } i \in \mathbb{Z}.$$

Finally, a language $L \subseteq \Sigma^*$ is *periodic*, if there exists a $z \in \Sigma^*$ such that $L \subseteq z^*$.

There should be no need to emphasize the importance of periodicity either in combinatorics of words or in formal language theory. Especially in the latter theory periodic objects are drastically simpler than the general ones: the fundamental difficulty of the noncommutativity is thus avoided. Therefore one tries to solve many problems of languages by reducing them to periodic languages, or at least to cases where a "part" of the language is periodic.

Based on the above it is important to search for the *periodicity forcing conditions*, i.e., conditions which forces that the words involved form a periodic language. We have already seen several such conditions, cf. Section 4:

- any nontrivial relation on $\{x, y\} \subseteq \Sigma^*$;
- any pair of nontrivial identities on $X = \{x, y, z\} \subseteq \Sigma^+$ of the form $x\alpha = y\beta$, $y\gamma = z\delta$ with $\alpha, \beta, \gamma, \delta \in X^*$;
- any condition on $X = \{x_1, \ldots, x_n\} \subseteq \Sigma^+$ satisfying: the transitive closure of the relation ρ defined as

$$x\rho y \Leftrightarrow xX^\omega \cap yX^\omega \neq \emptyset$$

equals $X \times X$.

Another classical example of a periodicity forcing condition is the equation, cf. [LySc] or [Lo],
$$x^n y^n = z^k \quad \text{with} \quad n, m, k \geq 2.$$

As we observed in Section 5 many properties of words are expressable in terms of solutions of equations. Thus it is often of interest to know whether such languages, or more generally parts of such languages, either are always periodic or can be periodic. By considerations of Section 5, Makanin's algorithm can be used to test this. Indeed, we only have to add to the system S defining the property suitable predicates of the forms

$$xy = yx \quad \text{or} \quad xy \neq yx,$$

and transform the whole predicate into one equation.

For example, if we want to know, whether there exist words x, y, z, u and v satisfying the equation $\alpha = \beta$ in these unknowns such that x, y and z are powers of a same word, and u and v are not powers of a same word, we consider the system

$$\begin{cases} \alpha & = & \beta \\ xy & = & yx \\ xz & = & zx \\ uv & \neq & vu, \end{cases}$$

and test whether it has a solution.

6.2 The periodicity theorem of Fine and Wilf

Our first result of this section is the classical *periodicity theorem* of Fine and Wilf, cf. [FW]. Intuitively it determines how far two periodic events have to match in order to guarantee a common period. Interestingly, although the result is clearly a result on sequences of symbols, i.e., on words, it was first presented in connection with real functions!

Theorem 6.1. *(Periodicity Theorem). Let $u, v \in \Sigma^+$. Then the words u and v are powers of a same word if, and only if, the words u^ω and v^ω have a common prefix of length $|u| + |v| - \gcd(|u|, |v|)$.*

Proof. We first note that we can restrict to the basic case, where $\gcd(|u|, |v|) = 1$. Indeed, if this is not the case, say $|u| = dp$ and $|v| = dq$, with $\gcd(p, q) = 1$, then considering u and v as elements of $(\Sigma^d)^+$ the problem is reduced to the basic case with only a larger alphabet.

So assume that $|u| = p$, $|v| = q$ and $\gcd(p, q) = 1$. The implication in one direction is trivial. Therefore, we assume that u^ω and v^ω have a common prefix of length $p + q - 1$. Assuming further, by symmetry, that $p > q$ we have the situation depicted in Figure 6.1. Here the vertical dashline denotes how far the words can be compared, the numbers tell the lengths of the words u and v, and the arrows the procedure defined below.

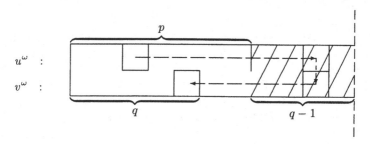

Fig. 6.1. An illustration of the procedure

We denote by i, for $i = 1, \dots, p + q - 1$, the corresponding position in the common prefix of u^ω and v^ω. Next we describe a *procedure* to fix new positions with the same value as a given initial one i_0. Let $i_0 \in [1, q-1]$. Then, by the assumption, the position obtained as follows, cf. arrows in Figure 6.1, gets the same value as i_0

$$(1) \qquad i_0 \xrightarrow{+p} i_0 + p \xrightarrow{\bmod q} i_1 = i_0 + p \pmod{q},$$

where i_1 is reduced to the interval $[1, q]$. Moreover, since $\gcd(p, q) = 1$, i_1 is different from i_0. If i_1 is also different from q we can repeat the procedure, and the new position obtained is different from the previous ones. If the procedure can be continued $q - 1$ steps, then all the positions in the shadowed area will be fixed, so that these together with i_0 make v unary. Hence, so is u, and we are done.

The procedure (1) can indeed be continued $q - 1$ steps if i_0 is chosen as

$$i_0 + (q - 1)p \equiv q \pmod{q}.$$

This is possible since $\gcd(p, q) = 1$. After this choice all the values $i_0 + jp$ (mod q), for $j = 0, \ldots, q - 2$, are different from q, which was the assumption of the procedure (1).   □

In terms of periods of a word and the distance of words, cf. Section 2.1, Theorem 6.1 can be restated in the following forms, the latter of which does not require that the comparison of words has to be started from either ends.

Corollary 6.1. *If a word $w \in \Sigma^+$ has periods p and q, and it is of the length at least $p + q - \gcd(p, q)$, then it also has a period $\gcd(p, q)$.*   □

Corollary 6.2. *For any two words $u, v \in \Sigma^+$, we have*

$$l(u^\omega, v^\omega) \geq |u| + |v| - \gcd(|u|, |v|) \Rightarrow \rho(u) = \rho(v).$$ □

We tried to make the proof of Theorem 6.1 as illustrative as possible. At the same time it shows clearly, why the bound given is optimal, and even more, as we shall see in Example 6.1.

Theorem 6.1 allows, for each pair (p, q) of coprimes, the existence of a word w of length $p + q - 2$ having the periods p and q. Let $W_{p,q}$ be the set of all such words, and define

$$PER = \bigcup_{\gcd(p,q)=1} W_{p,q}.$$

So, we excluded unary words from PER.

Example 6.1. We claim that, for each pair (p, q) of coprimes, $W_{p,q}$ contains exactly one word (up to a renaming), which moreover is binary. These observations follow directly from our proof of Theorem 6.1. The idea of that proof, namely filling positions in the shorter word v, can be illustrated in Figure 6.2. The nodes of this cycle correspond the positions of v, two labelled by ? are those which are missing from the shadowed area of Figure 6.1, and each arrow corresponds one application of the procedure (1). By the construction, starting from any position, and applying (1) the letter in the new position may differ from the previous one, only when to a position labelled by ? is entered. Consequently, during

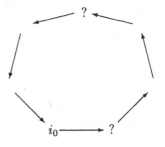

Fig. 6.2. The case PER

the cycle it may change at most twice, but, in fact, the latter change is back to the value of i_0. The fact that all positions are visited is due to the condition $\gcd(p, q) = 1$. Hence, we have proved our claim.

As a concrete example, the word of length 12 in PER starting with a and having the periods 5 and 9 is as depicted below:

a	a	a	b	a	a	a	a	b	a	a	a
2	1	5	4	3	2	1	5	4	3	2	1

Here the word, that is $(aaaba)^2aa = (aaabaaaab)^1aaa$, is described on the upper line, and the order of filling the positions, starting from the second one, on the lower line. Note that the change can take place in steps number 4 and 5, but the latter must assume the same value as the next one encountered in the procedure, which is the value of the second position. □

Example 6.2. Consider a word w of length $dp + dq - d - 1$ with $\gcd(p, q) = 1$, having the periods dp and dq, but not d, for some d. The argumentation of Example 6.1 shows that such a word exists, proving that in all cases the bound given in Theorem 6.1 is optimal. Moreover, for each $i = 1, \ldots, d - 1$, the positions $i + jd$ are filled by the same letter a_i, while in position $d + jd$ the situation is as in Example 6.1: they are uniquely filled by a word in $W_{p,q}$. □

Example 6.1 can be generalized also as follows.

Example 6.3. Let $p, q, k \in \mathbb{N}$, with $p > q$, $\gcd(p, q) = 1$ and $2 \leq k \leq q$. Then there exists a unique word w_k up to a renaming such that

$$|w_k| = p + q - k, \quad \|\text{alph}(w_k)\| = k \text{ and } w_k \text{ has periods } p \text{ and } q.$$

Indeed, the considerations of Example 6.1 extend immediately to this case, when the number of ?'s in Figure 6.2 is k. It follows that all words of length $p + q - k$, with $\gcd(p, q) = 1$, having periods p and q are morphic images of w_k under a length preserving morphism. □

We conclude this section by reminding that the set PER has remarkable combinatorial properties, cf. e.g. [dLM], [dL] and [BdL]. For example, all finite Sturmian words are characterized as factors or words in PER.

Finally we recall a result of [GO], which characterizes the set of all periods of an arbitrary word w.

6.3 The Critical Factorization Theorem

Our second fundamental periodicity result is the *Critical Factorization Theorem* discovered in [CV], and developped into its current form in [Du1], cf. also [Lo]. Our proof is from [CP]. The difference between [CV] and [Du1] was essentially, in terms of Figure 6.3 below, that [CV] considered only the case (i).

Intuitively the theorem says that the period $p(w)$ of a word $w \in \Sigma^+$ is always locally detectable in at least one position of the word. To make this precise we have to define what we mean by a local period of w at some position. We say that p is *a local period of w at the position* $|u|$, if $w = uv$, with $u, v \neq 1$, and there exists a word z, with $|z| = p$, such that one of the following conditions holds for some words u' and v':

$$(1) \quad \begin{cases} (i) & u = u'z \text{ and } v = zv' \text{ ;} \\ (ii) & z = u'u \text{ and } v = zv' \text{ ;} \\ (iii) & u = u'z \text{ and } z = vv' \text{ ;} \\ (iv) & z = u'u = vv' \text{ .} \end{cases}$$

Further *the local period of w at the position* $|u|$, in symbols $p(w, u)$, is defined as the smallest local period of w at the position u. It follows directly from (1), cf. also Figure 6.3, that $p(w, u) \leq p(w)$.

The intuitive meaning of the local period is clear: around that position there exists a factor of w having as its minimal period this local period. The situations of (i), (ii) and (iv) in (1) can be depicted as in Figure 6.3.

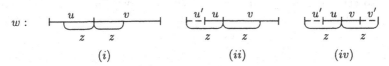

Fig. 6.3. The illustration of a local period

Now, we can say precisely what the above local detectability means. It means that there exists a factorization $w = uv$, with $u, v \neq 1$, such that

$$p(w, u) = p(w),$$

i.e., the local period at position $|u|$ is that of the (global) period of w. The corresponding factorization $w = uv$ is called *critical*. The theorem claims that each word possesses at least one critical factorization (if it possesses any nontrivial factorization at all).

Example 6.4. Consider the words $w_1 = aababaaa$ and $w_2 = a^n ba^n$. The periods of these words are 6 and $n + 1$. The local periods of w_1 in positions $1, 2, \ldots, 7$ are $1, 5, 2, 6, 6, 1, 1$, respectively. For example, at position 4 we have $w = aaba.baaa$ so that $z = baaaba$ contains $baaa$ as a prefix and $aaba$ as a suffix, but no shorter z can be found to satisfy (1). The word w_1 has two critical factorizations. The critical factorizations of w_2 are $a^n b.a^n$ and $a^n.ba^n$, showing that there are none among the first $n - 1$ factorizations.   □

Example 6.5. As an application of Lyndon words we show that, any word $w \in \Sigma^+$ satisfying $|w| \geq 3p(w)$, has a critical factorization. Indeed, we can write

$$w = ullv,$$

where $u, v \in \Sigma^*$ and l is the Lyndon word in the class $[\text{pref}_{p(w)}(w)]$. As we noted in Section 2.2 Lyndon words are unbordered. Consequently, the factorization $w = u.llv$ is critical. Hence, in a critical factorization we can even choose $1 \le |u| \le p(w)$.   □

To extend Example 6.5 for all words is much more difficult.

Theorem 6.2 (Critical Factorization Theorem). *Each word* $w \in \Sigma^+$, *with* $|w| \ge 2$, *possesses at least one factorization* $w = uv$, *with* $u, v \ne 1$, *which is critical, i.e.,* $p(w) = p(w, u)$. *Moreover,* u *can be chosen such that* $|u| < p(w)$.

Proof. Our proof from [CP] not only shows the existence of a critical factorization, but also gives a method to define such a factorization explicitly. We may assume that w is not unary, i.e., $p(w) > 1$.

Let $\preceq_l$ be a lexicographic ordering of Σ^+, and $\preceq_r$ another lexicographic ordering obtained from $\preceq_l$ by reversing the order of letters, i.e., for $a, b \in \Sigma$, $a \preceq_l b$ if, and only if, $b \preceq_r a$. Let v and v' be the maximal suffixes of w with respect to the orderings $\preceq_l$ and $\preceq_r$, respectively. We shall show that one of the factorizations

$$w = uv \quad \text{or} \quad w = u'v'$$

is critical. More precisely, it is the factorization $w = uv$, if $|v| \le |v'|$, and $w = u'v'$ otherwise. In addition, in both the cases

$$(2) \qquad\qquad |u|, \ |u'| < p(w).$$

We need two auxiliary results. The first one holds for any lexicographic ordering $\preceq$.

Claim 1. If v is the lexicographically maximal suffix of w, then no nonempty word t is both a prefix of v and a suffix of $u = wv^{-1}$.

Proof of Claim 1. Assume that $u = xt$ and $v = ty$. Then, by the maximality of v, we have $tv \preceq v$ and $y \preceq v$. Since $v = ty$ these can be rewritten as $tty \preceq ty$ and $y \preceq ty$. Now, from the former inequality we obtain that $ty \preceq y$, which together with the latter one means that $y = ty$. Therefore, t is empty as claimed.   □

The second one, which is obvious from the definitions, claims that the orderings $\preceq_l$ and $\preceq_r$ together define the prefix ordering $\le$.

Claim 2. For any two words $x, y \in \Sigma^+$, we have

$$x \preceq_l y \quad \text{and} \quad x \preceq_r y \Leftrightarrow x \le y, \quad \text{i.e., } x \text{ is a prefix of } y.   □$$

Proof of Theorem (continued). Assume first that $|v| \leq |v'|$. We intend to show that the factorization $w = uv$ is critical. First we show that $u \neq 1$. If this is not the case, and $w = at$, with $a \in \Sigma$, then $w = v = v'$. Therefore, by the definitions of v and v', we have both $t \preceq_l w$ and $t \preceq_r w$. So, by Claim 2, t is a prefix of $w = at$, and hence $t \in a^+$, i.e., $p(w) = 1$. This, however, was ruled out at the beginning. Hence, the word u, indeed, is nonempty.

From now on let us denote $p(w, u) = p$. By Claim 1, we cannot have $p \leq |u|$ and $p \leq |v|$ simultaneously. Hence, if $p \leq |u|$, then necessarily $p > |v|$, implying that v is a suffix of u. This, however, would contradict with the maximality of v, since $v \prec_l vv$. So we have concluded that $p > |u|$. Since p is a local period at the position $|u|$, there exists a word z such that $p = |zu|$, and the words zu and v are comparable in the prefix ordering, i.e., one of the words v and zu is a prefix of another. We consider these cases separately.

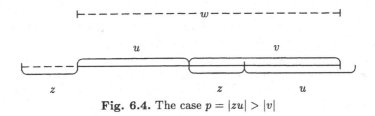

Fig. 6.4. The case $p = |zu| > |v|$

Case I: $p > |v|$. Now, the situation can be depicted as in Figure 6.4. It follows that $|uz|$ is a period of $uv = w$, i.e., $p(w) \leq |uz|$. On the other hand, the period $p(w)$ is always at least as large as any of its local periods, so that $p(w) \geq p(w, u) = p = |uz|$. Therefore, $p(w) = p(w, u)$ showing that the factorization $w = uv$ is critical.

Case II: $p \leq |v|$. Now the illustration is as shown in Figure 6.5, where also the words u' and v' from the factorization $w = u'v'$ are shown.

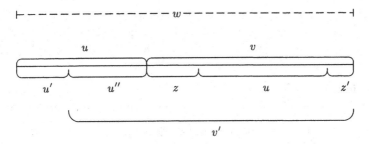

Fig. 6.5. The case $p = |zu| \leq |v|$

Since $p \leq |v|$, and $|v| \leq |v'|$ we indeed have words u', u'' and z' such that $u = u'u''$ and $v = zuz'$. We have to show, as in Case I, that uv has a period $|zu|$.

By the maximality of v', the suffix $u''z'$ of v' satisfies $u''z' \preceq_r v' = u''v$, implying that $z' \preceq_r v$. On the other hand, the maximality of v yields the relation $z' \preceq_l v$. Therefore we conclude from Claim 2 that z' is a prefix of zuz'. It follows that $z' \in \mathrm{pref}(zu)^\omega$, and hence $w = uv = uzuz' \in \mathrm{pref}(uz)^\omega$, showing that w has a period $p = |zu|$. Consequently, also the Case II is completed.

It remains to be proved that (2) holds true, i.e., $|u| < p(w)$ and $|u'| < p(w)$. The former follows from the fact $|u| < p$, which we already proved, and the latter from our assumption $|u'| \leq |u|$.

Finally, to complete the proof of Theorem 6.2 we have to consider the case $|v'| \geq |v|$. But this reduces to the above case by interchanging the orderings $\preceq_l$ and $\preceq_r$.   □

As we already noted we proved more than the existence of a critical factorization. Namely, we proved that such a factorization can be found by computing a lexicographically maximal suffix of a word, or in fact two of those with respect to two different orderings. There exist linear time algorithms for such computations, cf. [CP] or [CR]. For example, one can use the suffix tree construction of [McC].

The Critical Factorization Theorem is certainly a very fundamental result on periodicity of words. It is probably due to its subtle nature, as shown also by the above proof, that it has not been applied as much as it would have deserved.

One application of the theorem, which actually is the source of its discovery, cf. [CV], is as follows. To state it we have to recall the notion of an X-interpretation of a word defined in Section 2.1. An X-interpretation of a word $w \in \Sigma^+$ is a sequence $x, x_1, \ldots, x_n, y$ of words such that

$$xwy = x_1 \ldots x_n,$$

where $x_i \in X$, for $i = 1, \ldots, n$, x is a proper prefix of x_1 and y is a proper suffix of x_n. Two X-interpretations $x, x_1, \ldots, x_n, y$ and $x', x'_1, \ldots, x'_m, y'$ of w are *disjoint*, if for each $i \leq n$ and $j \leq m$, we have $x^{-1}x_1 \ldots x_i \neq x'^{-1}x'_1 \ldots x'_j$. Now an application of Theorem 6.2 yields, cf. [Lo]:

Proposition 6.1. *Let $w \in \Sigma^+$ and $X \subseteq \Sigma^+$ be a finite set satisfying $p(x) < p(w)$ for all $x \in X$. Then w has at most $\|X\|$ disjoint X-interpretations.*

Proposition 6.1 requires two remarks. First the disjointness is essential: if X-interpretations are required to be only different, then taking X to be a noncode the number of different X-interpretations could grow exponentially on $|w|$. In Proposition 6.1 the growth is bounded by a constant.

Second, the bound is close to the optimal one as noted in [Lo]: for each $n \geq 2$, words of the form $w \in (a^{2n-2}b)^+$ have exactly $n - 1$ disjoint X-interpretations for $X = \{a^n, a^iba^i \mid i = 0, \ldots, n - 1\}$.

Another elegant application of Theorem 6.2 was given in [CP], where it was used to describe an efficient pattern matching algorithm.

6.4 A characterization of ultimately periodic words

In this subsection we introduce a recent characterization of ultimately periodic words from [MRS]. The characterization is in terms of local properties of the considered word, or more precisely, in terms of repetitions at the ends of finite prefixes of the considered word. Variants for 2-way infinite words are presented, too.

Clearly, if $w = a_0a_1 \ldots$, with $a_i \in \Sigma$, is ultimately periodic, then the following condition holds for any real number ρ:

(1) $\quad \begin{array}{l} \exists n = n(\rho) \in \mathbb{N} : \forall m \geq n : \operatorname{pref}_m w \text{ contains a repetition of} \\ \text{order at least } \rho \text{ as a suffix.} \end{array}$

Our next simple example shows that infinite words satisfying (1) for $\rho = 2$ need not be ultimately periodic.

Example 6.6. Let $X = \{ab, aba\}$. Note that X is an ω-code, i.e., each word in X^ω has a unique X-factorization, due to the fact that any binary nonperiodic set is such, by Corollary 5.1. We consider infinite words in X^ω satisfying that in their X-factorizations

(i) there are no two consecutive blocks of ab; and
(ii) there are no three consecutive blocks of aba.

Let X_2 be the set of all such words. Obviously, the set X_2 is nondenumerable, and therefore contains words which are not ultimately periodic. Moreover, we claim that words in X_2 satisfy (1) for $\rho = 2$.

To see this we consider all possible sequences of ab- or aba-blocks immediately preceding ab (aba, resp.) in X-factorizations, and note that any position of ab (aba, resp.) is an endpoint of a square in these left extensions of ab (aba, resp.). Luckily there is only a finite number of cases to be considered as illustrated in Figure 6.6.

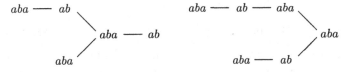

Fig. 6.6. An exhaustive search for left extensions of ab and aba

A concrete example of a word which satisfies (1) for $\rho = 2$, and is not ultimately periodic is obtained by starting from $abaaba$ and extending it on the right nonperiodically by the blocks ab and aba. This particular word satisfies (1) for $\rho = 2$ with $n = 6$.   □

Example 6.6 does not extend to higher integer repetitions, i.e., to the case $\rho = 3$, as we shall see in the next theorem. The proof of it is a modification due to A. Restivo from the proof of a more general result in [MRS].

Theorem 6.3. *A word $w \in \Sigma^\omega$ is ultimately periodic if, and only if, it satisfies (1) for $\rho = 3$, i.e., contains a cube as a suffix of any long enough prefix of w.*

Proof. To prove the nontrivial part we assume that w satisfies (1) with $\rho = 3$. We start with an auxiliary result.

Lemma 6.1. *Let $w = v^2$. If w has a period q satisfying $\frac{2}{3}|v| < q < |v|$, then $w = ux^3$, with $|x| = |v| - q$.*

Proof of Lemma. Denoting $w = zt$, with $|t| = q$, we can illustrate the situation in Figure 6.7.

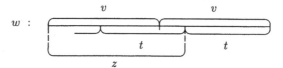

Fig. 6.7. Factorizations of w with $|v| = p$ and $|t| = q$

By the Theorem of Fine and Wilf, v and t has a common period, and therefore $|v| - |t|$ is a period of z. By our assumption $\frac{2}{3}|v| < q = |t|$ implying that

$$3(|v| - |t|) < p.$$

It follows that z contains as a suffix a cube x^3, with $|x| = |v| - |t|$. Now, the lemma follows since any suffix of z is a suffix of w, as well.   □

Proof of Theorem (continued). Let $w = a_0 a_1 \ldots$, with $a_i \in \Sigma$, and set

$$p(n) = \min\{d \mid \exists v \in \Sigma^+ : |v| = d \text{ and } a_0 a_1 \ldots a_n = uv^3\}.$$

Now, let $n(3)$ be the constant of the condition (1), and $m > n(3)$. As a crucial point of the proof we show the following implication:

(2)   if $p(n) < p(m)$, for $n = n(3), \ldots, m - 1$, then $m - n(3) < p(m)$.

To prove (2) we denote $p(m) = p$, and assume that $p(n) < p$ for $n = n(3), \ldots, m - 1$. By the definition of p, we can write $a_0 \ldots a_m = uv^3$, with $|v| = p$. Therefore

$$a_0 \ldots a_{m-p} = uv^2, \quad \text{with} \quad |v| = p.$$

Now, assume contrary to our claim that $m - n(3) \geq p$. Therefore $m - p \geq n(3)$, and so by our assumption, we can write

$$a_0 \ldots a_{m-p} = u'x^3, \quad \text{with} \quad |x| = p(m - p) = q < p.$$

There are two cases to be considered.

First, if $q > \frac{2}{3}p$, then v^2 satisfies the conditions of Lemma 6.1, and so we can write $v^2 = sy^3$, with $|y| = p - q$. This, however, is a contradiction with the choice of q, since $p - q \leq \frac{2}{3}p < q$.

Second, if $q \leq \frac{2}{3}p$, then v^2 has as a suffix a cube of a word of length q. Hence, the same holds for the word $a_0 \ldots a_m$. This, however, is a contradiction since $q < p = p(m)$. This ends the proof of (2).

Next we apply (2) to conclude that

(3) $$\sup\{p(n) \mid n \geq n(3)\} < n(3).$$

Indeed, if (3) does not hold, we choose the smallest m such that $p(m) \geq n(3)$. Then, by (2), we know that $m - n(3) < p(m)$, and therefore $m < p(m) + n(3) \leq 2p(m)$. This, however, contradicts with the fact that $a_0 \ldots a_m = uv^3$ with $|v| = p(m)$. Hence (3) is indeed proved.

Now, we define

$$P = \sup\{p(n) \mid n \geq n(3)\}.$$

By (3), we know that $P \leq n(3)$, and we complete the proof of the theorem by induction on P.

The starting point $P = 1$ is obvious. To prove the induction step there are two possibilities (where actually only the first one relies on induction).

Case I: If there exist only finitely many numbers n such that $p(n) = P$, we can set

$$n(3) := \max\{n \mid p(n) = P\} + 1,$$

and apply induction hypothesis to conclude that w is ultimately periodic.

Case II: If there exist infinitely many integers n such that $p(n) = P$ we proceed as follows. Let the values $n = m_1, m_2, \ldots$ be all such values. We shall prove, again by an induction, that, for $i = 1, 2, \ldots$, the word

(4) $$a_{n(3)} \ldots a_{m_i} \quad \text{has a period} \quad P.$$

The starting point $i = 1$ is clear, since, by Lemma 6.1, $m_1 - n(3) < P$. So assume that the word $a_{n(3)} \ldots a_{m_i}$ has a period P, and consider the word $a_{n(3)} \ldots a_{m_{i+1}}$. Applying again Lemma 6.1, where $n(3)$ is replaced by m_i, we conclude that $m_{i+1} - m_i < P$.

We write

$$a_{n(3)} \ldots a_{m_{i+1}} = uvw,$$

with

$$|w| = m_{i+1} - m_i,$$

and

$$|v| = 2P - 1.$$

Then, by induction hypothesis, uv has a period P. On the other hand, since $|vw| < 3P$ it follows that also vw has a period P. But, since the overlapping factor v is of length at least $P + 1$, it is easy to conclude that also uvw has a period P, which completes the latter induction, as well as the whole proof of Theorem 6.3.   □

Actually, as shown in [MRS], Theorem 6.3 can be sharpened as follows:

Proposition 6.2. *A word $w \in \Sigma^\omega$ is ultimately periodic if, and only if, it satisfies (1) for $\rho = \varphi^2$, where φ is the number of the golden ratio.*

Recall that $\varphi = \frac{1}{2}(\sqrt{5} + 1)$, i.e., the positive root of the equation $\varphi^2 - \varphi - 1 = 0$. It is also shown in [MRS] that Proposition 6.3 is optimal in the sense that the validity of (1) for any smaller ρ than φ^2 does not imply that the word is ultimately periodic. This can be seen from the infinite Fibonacci word w_F considered in Section 8.

Our above considerations deserve two comments. First results extend to 2-way infinite words. Indeed, from the proof of Theorem 6.3 one can directly derive the following characterization.

Theorem 6.4. *A two-way infinite word $w = \ldots a_{i-1}a_i a_{i+1}\ldots$, with $a_i \in \Sigma$ is periodic if, and only if, there exists a constant N such that, for any i, the word $w \ldots a_{i-1}a_i$ contains a cube of length at most N as its suffix.*   □

Note that in Theorem 6.4 the requirement that the cubes must be of a bounded length is necessary, as shown by the next example. In Theorem 6.3 this was not needed, since it dealt with only 1-way infinite words.

Example 6.7. We define a nonperiodic two-way infinite word

$$w = \ldots a_{-1}a_0 a_1 \ldots,$$

which contains a cube as a suffix of any factor $\ldots a_{i-1}a_i$ as follows. We set $w_0 = aaa$ and define

$$w_{2i+1} = \alpha_i w_{2i} a \quad \text{and} \quad w_{2i+2} = \beta_i w_{2i+1}, \quad \text{for} \quad i \geq 0,$$

where $a \in \Sigma$ and the words α_i and β_i are chosen such that both w_{2i+1} and $w_{2i+2}(\text{suf}_{2i}(w_{2i+2}))^{-1}$ are cubes. Clearly, this is possible. It is also obvious that this procedure yields a word of the required form.   □

As the second comment we introduce a modification of the above considerations. Surprisingly the results are quite different.

In above we required that repetitions occurred at any position "immediately to the left from that position". Now, we require that they occur at any position such that this position is the center of the repetition. We obtain the following characterization for periodic 2-way infinite words, in terms of local periods, cf. Section 6.3. Note that the notions of *local periods* extend in a natural way to infinite words, as well.

Theorem 6.5. *A two-way infinite word w is periodic if, and only if, there exists a constant N such that the local period of w at any point is at most N.*

Proof. Clearly, the periodicity of w implies the existence of the required N. The converse follows directly from the Critical Factorization Theorem: periods of all finite factors of w are at most N, and hence by the Theorem of Fine and Wilf w indeed is periodic. □

We note that Theorem 6.5, can be seen as a weak variant of the Critical Factorization Theorem, cf. [Du1]. It is also worth noticing that the boundedness of local periods is crucial, the argumentation being the same as in Example 6.7. Finally, the next example shows the optimality of Theorem 6.5 in a certain sense.

Example 6.8. Theorem 6.5 can be interpreted as follows. If a two-way infinite word w contains at any position a bounded square "centered" at this position, then the word is periodic. The word

$$w = {}^\omega aba^\omega$$

shows that no repetition of smaller order guarantees this. Indeed, for any $\rho < 2$, the word w contains at any position a bounded repetition of order of at least ρ centered at this position. Here, of course, the bound depends on ρ. □

7. Finiteness conditions

In this section we consider partial orderings of finite words and finite languages, and in particular orderings that are finite in either of two natural senses: either each subset contains only finitely many incomparable elements, i.e., each antichain is finite, or each subset contains only finitely many pairwise comparable elements, i.e., each chain is finite. Hence our interest is in two fundamental properties which are dual to each other.

7.1 Orders and quasi-orderings

For the sake of completeness we recall some basic notions on binary relations R over an arbitrary set S.

A binary relation R is a *strict ordering* if it is *transitive*, i.e., $(x, y) \in R$ and $(y, z) \in R$ implies $(x, z) \in R$, and *irreflexive*, i.e., $(x, x) \in R$ holds for no $x \in S$. It is a *quasi-ordering* if it is transitive and *reflexive*, i.e., $(x, x) \in R$ holds for all $x \in S$. It is a *partial ordering* if it reflexive, transitive and *antisymmetric*, i.e., $(x, y) \in R$ and $(y, x) \in R$ implies $x = y$ for all $x, y \in S$.

A *total ordering* is a partial ordering $\preceq$ for which $x \preceq y$ or $y \preceq x$ holds for all $x, y \in S$. An element x of a set S (resp. of a subset $X \subseteq S$) ordered by $\preceq$ is *minimal* if for all $y \in S$ (resp. $y \in X$) the condition $y \preceq x$ implies $x = y$. Of course each subset of a totally ordered set has at most one minimal element.

There is a natural interplay between these three notions. With each quasi-ordering $\preceq$ it is customary to associate the equivalence relation defined as $x \sim y$ if, and only if, $x \preceq y$ and $y \preceq x$ holds. This induces a relation $\leq$ on the quotient $S/\sim$

$$[x] \leq [y] \text{ if, and only if, } x \preceq y,$$

which is a partial ordering on S.

Example 7.1. The relation on Σ^* defined by $x \preceq y$, whenever $|x| \leq |y|$, is a quasi-ordering. The equivalence relation associated with it is: $x \sim y$ if, and only if, $|x| = |y|$.   □

If $\prec$ is a strict ordering then the relation $\leq$ defined by $x \leq y$ if, and only if, $x \prec y$ or $x = y$, is a partial ordering. If $\preceq$ is a quasi-ordering, then the relation $<$ defined by $x < y$ if, and only if, $x \preceq y$ and $y \npreceq x$, is a strict ordering.

Two important notions on partial orderings from the viewpoint of our considerations are those of a chain and an antichain. A subset X of an ordered set S is a *chain* if the restriction of $\preceq$ to X is total. It is an *antichain* if its elements are pairwise incomparable. A partial ordering in which every strictly descending chain is finite is *well-founded* or *Noetherian*. If in addition every set of pairwise incomparable elements is finite it is a *well-ordering*. For example, the set of integers ordered by $n|m$ if, and only if, n divides m is well-founded, but is not a well-ordering.

We concentrate on partial orderings over Σ^* and $\mathrm{Fin}(\Sigma^*)$, the family of finite subsets of Σ^*. We already observed how total orderings like lexicographic or alphabetic orderings are crucial, in considerations envolving words, for example for defining Lyndon words and proving the Critical Factorization Theorem.

Partial quasi-orderings can be defined on Σ^* and $\mathrm{Fin}(\Sigma^*)$ in many ways. Without pretending to be exhaustive, here are a few important examples:

alphabetic quasi-ordering: $x \preceq_a y$ iff $\mathrm{alph}(x) \subseteq \mathrm{alph}(y)$,

length ordering: $x \preceq_l y$ iff $|x| < |y|$ or $x = y$,
commutative image quasi-ordering: $x \preceq_c y$ iff $|x|_a \leq |y|_a$ for all $a \in \Sigma$,
prefix ordering: $x \preceq_p y$ iff there exits z with $xz = y$,
factor ordering: $x \preceq_f y$ iff there exits z, t with $zxt = y$,
subword ordering: $x \preceq_d y$ iff there exist $x_1, \ldots, x_n, u_0, \ldots, u_n \in \Sigma^*$ such
that

$$x = x_1 x_2 \ldots x_n \text{ and } y = u_0 x_1 u_1 x_2 u_2 \ldots x_n u_n.$$

Similarly, for the family $\text{Fin}(\Sigma^*)$ we define the following orderings. Here
the notation R_X denotes the set of relations satisfied by X.

size quasi-ordering: $X \preceq_s Y$ iff $\|X\| \leq \|Y\|$,
inclusion ordering: $X \preceq_i Y$ iff $X \subseteq Y$,
semigroup quasi-ordering: $X \preceq_m Y$ iff $X^+ \subseteq Y^+$ where X and Y are
minimal generating sets,
relation quasi-ordering: $X \preceq_r Y$ iff there exits a bijection $\varphi : X \to Y$ such
that $R_{\varphi(X)} \subseteq R_X$.

We summarize into the following table the facts on how the above partial
orderings behave with respect to our two finiteness conditions, i.e., whether
or not they allow infinite antichains or chains.

Table 7.1. Finiteness conditions of certain quasi-orderings

	$\preceq_a$	$\preceq_l$	$\preceq_p$	$\preceq_f$	$\preceq_d$	$\preceq_s$	$\preceq_i$	$\preceq_m$	$\preceq_r$
no infinite chains	$+$	$-$	$-$	$-$	$-$	$-$	$-$	$-$	$\oplus$
no infinite antichains	$+$	$+$	$-$	$-$	$\oplus$	$+$	$-$	$-$	$-$

There are two particularly interesting entries in this table, namely those
denoted by $\oplus$. These state two fundamental finiteness conditions on words
and finite sets of words we shall be studying in more details later. That the
other entries are correct is, as the reader can verify, easy to conclude. We only
note that the relation ordering $\preceq_r$ is not a well-ordering even in the family
of sets of the same size as shown by the family $\{X_i = \{a, a^i b, b\} | i \geq 1\}$.

7.2 Orderings on words

In this subsection we consider orderings on Σ^*, in particular the subword
ordering and another one related to it.

The subword ordering is called division ordering in [Lo], but this notion
has another use in the literature, where by *division* ordering is meant a partial
ordering satisfying the following two conditions for all $x, y, z, t \in \Sigma^*$

(1) $1 \preceq x$

(2) $x \preceq y$ implies $zxt \preceq zyt$.

Observe that, by (1), we have $1 \preceq z$ and $1 \preceq y$, and hence, by (2), $x \preceq zx, x \preceq xy$ and $zx \preceq zxy$, i.e., $x \preceq zx \preceq zxy$. Thus every word is greater than or equal to each of its factors:

(3)   for all x, y, z, the inequality $x \preceq yxz$ holds.

Actually, the subword ordering is the smallest ordering satisfying the conditions (1) and (2), i.e., for all $x, y \in \Sigma^*$ the relation $x \preceq_d y$ implies $x \preceq y$. Indeed, we have $1 \preceq_d 1$ and $1 \preceq 1$ by condition (1). Now, consider $x = ax' \preceq_d y = by'$ with $a, b \in \Sigma$ and $x', y' \in \Sigma^*$, and let us proceed by induction on $|x| + |y|$. If $a = b$, then $x' \preceq_d y'$, i.e., $x' \preceq y'$ by induction, and by (2), $x = ax' \preceq y = ay'$. On the other hand, if $a \neq b$, then $x \preceq_d y'$, and thus by induction $x \preceq y'$. Condition (1) yields $1 \preceq a$ and condition (2) yields $y' \preceq y = ay'$, so by the transitivity $x \preceq y$.

Total division orderings have been studied in [Mart]. It is proved under a certain assumption, namely the ordering being "tame", that each division ordering is finer than the *strong commutative image* ordering, which is obtained from $\preceq_c$ by replacing inequalities by the strict inequalities in each component. It is also conjectured that the statement holds true even without this condition. However, when the alphabet is binary, each division ordering is tame, and thus the result holds.

Theorem 7.1. *Let $\preceq$ be a total division ordering on the free monoid generated by $\{a, b\}$ and let u, v be two words. Then*

(4) $|u|_a < |v|_a$ *and* $|u|_b < |v|_b$ *implies* $u \prec v$.

Proof. Since $\preceq$ is total, we may assume without loss of generality that $ba \succ ab$ holds. In particular, by commutating the occurrences of a and b in $u \in \Sigma^*$, we have, by equality (2):

(5) $b^n a^m \succeq u \succeq a^m b^n$ with $m = |u|_a$ and $n = |u|_b$.

Now assume that condition (4) is violated: $|u|_a < |v|_a$, $|u|_b < |v|_b$ and $u \succ v$. By setting $|u|_a = m$, $|v|_a = m'$, $|u|_b = n$, $|v|_b = n'$ we have

$$b^n a^m \succeq u \succ v \succeq a^{m'} b^{n'} \succeq a^{m+1} b^{n+1}.$$

Thus we may assume that we have $u = b^n a^m$, $v = a^{m+1} b^{n+1}$ and $u \succ v$. We first observe that $b^n u \succ ub^{n+1}$. Indeed, we have $b^n u = b^n b^n a^m \succ b^n a^{m+1} b^{n+1}$. Now, by (3), we have $a^{m+1} \succ a^m$, i.e., $b^n a^{m+1} b^{n+1} \succ b^n a^m b^{n+1} = ub^{n+1}$.

Assume $b \succ a$ and for all $k > 0$ compute:

$$b^{(k+1)n}b^m \;\succ\; b^{(k+1)n}a^m = b^{kn}u$$
$$\succ\; ub^{k(n+1)}$$
$$\succ\; vb^{k(n+1)} = a^{m+1}b^{n+1}b^{k(n+1)} = a^{m+1}b^{(k+1)n+1}b^k$$
$$\succ\; b^{(k+1)n+1}b^k.$$

This does not hold when $k + 1 > m$. Now, if $a \succ b$, a similar argument leads to the same type of contradiction, proving the theorem. □

The author shows that the inequalities of condition (4) must be strict. Indeed, consider the ordering $\preceq$ on $\{a,b\}^*$, where words are ordered by their number of occurrences first, and then lexicographically with $a \succ b$. Then $u = bababa \succ v = abbabba$, but $|u|_a = |v|_a$ and $|u|_b < |v|_b$.

We turn to consider the subword ordering. We already observed that it is right- and left-invariant, cf. (2). Its second major property, solving one nontrivial entry in Table 7.1, is that it is a well-ordering implying that every subset $X \subseteq \Sigma^*$ has finitely many minimal elements.

Theorem 7.2. *The subword ordering $\preceq_d$ over a finitely generated free monoid is a well-ordering.*

Proof. Clearly subword ordering is well-founded. So we have to prove that any antichain of Σ^* with respect to $\preceq_d$ is finite. Assume to the contrary that $F = \{f_i | i \in \mathbb{N}\}$ is an infinite set of incomparable words. Then, in particular, we have

(6) $$\text{if } i < j, \text{ then } f_i \npreceq_d f_j.$$

Among the sequences satisfying (6) there exist such sequences, where f_1 is the shortest possible. Continuing inductively we conclude that there exists a sequence, say $(g_i)_{i \geq 0}$, which satisfies (6) and none of the sequences $(h_i)_{i \geq 0}$, with $|h_i| < |g_i|$ for some i, satisfies (6).

Now, consider the sequence $(g_i)_{i \geq 0}$. Since Σ is finite there exist a letter a such that, for infinitely many i, we can write $g_i = ag'_i$ with $g'_i \in \Sigma^*$. Say this holds for values $i_1, i_2, \ldots$. Then the sequence

$$g_1, g_2, \ldots, g_{i_1-1}, g'_{i_1}, g'_{i_2}, \ldots$$

satisfies (6) and, moreover, $|g'_{i_1}| < |g_{i_1}|$. This contradicts with the choice of the sequence $(g_i)_{i \geq 0}$. □

This theorem is due to Higman in [Hi], where it is proved in a much more general setting. Subsequently, it has been rediscovered several times, see [Kr] for a complete account. Our proof of Theorem 7.2 is from [Lo]. It is very short and nonconstructive. It is also worth noticing that there is no bound for the size of a maximal antichain in Σ^*, as shown by the antichains $A_n = \{a^i b^{n-i} | i < n\}$ for $n \geq 0$.

We also note that Dickson's Lemma is a consequence of Theorem 7.2. We recall that it asserts that $\mathbb{N}^k$ is well-ordered, where the ordering is the extension of the usual componentwise ordering on $\mathbb{N}$. Indeed, it suffices to interprete the k-tuple $(n_1, \ldots, n_k)$ as the word $a_1^{n_1} \ldots a_k^{n_k}$ over the alphabet $\{a_1, \ldots, a_k\}$.

An interesting formal language theoretic consequence of Theorem 7.2 is the following.

Theorem 7.3. *For each language* $L \subseteq \Sigma^*$ *the languages* $SW(L) = \{w \mid \exists z \in L : w \preceq_d z\}$ *and* $SW_1(L) = \{w \mid \exists z \in L : z \preceq_d w\}$ *are rational.*

Proof. By Theorem 7.2, the set of minimal elements of L with respect to $\preceq_d$ is finite, say F. So, $SW_1(L) = SW_1(F)$, and hence $SW_1(L)$ is rational. A bit more complicated proof for $SW(L)$ is left to the reader.   □

Our above considerations on the subword ordering were purely existential. As an example of algorithmic aspects we state a problem motivated by molecular biology. The problem asks to find, for a given finite set $X = \{x_1, \ldots, x_n\}$ of words, a shortest word z such that $x_i \preceq_p z$ for all $i = 1, \ldots, n$. This problem is usually referred to as the *smallest common supersequence problem*, and it is known to be NP-complete, cf. [GJ].

7.3 Subwords of a given word

In this a bit isolated subsection we consider an interesting problem asking to differentiate two words by a shortest possible subword occurring in one but not in the other.

Example 7.2. The word *bbaa* occurs in *abbaab*, but does not occur in *ababab* as a subword. All words of length 3 occur in both words.   □

We refer the reader to [Lo] for a full exposition of the problem. In particular it is established that a word of length n is determined by the set of its subwords of length $\lceil \frac{n+1}{2} \rceil$, the pair $a^{m-1}ba^m, a^mba^{m-1}$ showing that the bound is sharp. In [Si] it is proved that a shortest subword distinguishing two given different words can be found in linear time. This is not a priori obvious since there may exist exponentially many subwords of a given length in a word. For instance, $(ab)^n$ contains all words of length n as subwords. The linearity of the algorithm is based on several properties among which the fact that, if two words u and v have the same subwords of length m, then they can be merged in a word having also the same subwords of length m.

An elaboration of this question is to consider the subwords with their multiplicities. In the previous example *baab* occurs twice in *abbaab* but once in *ababab*. Milner (personal communication) defines the *k-spectrum* of a word u as the function that associates with each word of length $0 < k \leq |u|$, the number of its occurrences in u. Given an integer k, consider the maximal integer $n = f(k)$ such that two different words of length n have different k-spectrums. The question is to find a reasonable upper bound on n.

Example 7.3. Define 3 sequences of words by

$u_0 = aba, v_0 = 1, w_0 = baa,$

$u_1 = ab, v_1 = 1, w_1 = ba,$

$u_{k+2} = w_k u_{k+1}, v_{k+2} = u_{k+1} u_k, u_{k+2} = u_{k+1} w_k$ if k is even,

$u_{k+2} = u_{k+1} w_k, v_{k+2} = u_k u_{k+1}, u_{k+2} = w_k u_{k+1}$ if k is odd.

Then u_k and v_k are different, and have the same k-spectrum. Their common length $\phi(k)$ grows as a "Fibonacci" type function, starting from the values 2 and 5: $\phi(1) = 2$, $\phi(2) = 5$, $\phi(3) = 7$, $\phi(4) = 12$, $\phi(5) = 19$, $\phi(6) = 31, \dots$.
The exact values for small k are

k	1	2	3	4
$f(k)$	3	5	8	13

,

but for $k = 5$, $\phi(5) = 19$ is far from being optimal due to the following two words of length 16 having the same 5-spectrum:

$$u = abbaaaaabbbaaaab \text{ and } v = baaaabbbaaaaabba. \qquad \square$$

The same questions substituting "factor" for "subword" can be posed. It is shown in [Lo], Exercise 6.2.11, that whenever u is not of the form $(xy)^n x$ with $n \geq 2$, then it is uniquely determined by its factors of length $\lceil \frac{|u|}{2} + 1 \rceil$. If this restriction is relaxed, then the word can not be determined by its proper factors: $(ab)^n a$ and $(ba)^n b$ have the same factors $(ab)^n$ and $(ba)^n$ as occurrences. It is also possible to define the *k-factor* spectrum of a word u which associates with each word of length k the number of its occurrences in u. To our knowledge no nontrivial bounds are known.

7.4 Partial orderings and an unavoidability

In this section we state a generalization of Higman's Theorem. This result is based on the notion of an unavoidable set of words, which *is not* connected to the unavoidability of Section 8. We also consider some other problems connected to this notion of an unavoidability.

We say that a set $X \subseteq \Sigma^*$ is *unavoidable,* if there exists a constant k such that each word $w \in \Sigma^k$ contains a word of X as a factor. For example, the set $X = \{aa, ba, bb\}$ is unavoidable over the free monoid $\{a, b\}^*$, since avoiding a^2 and b^2 obliges a word to be a sequence of a and b alternatively.

This definition was given in [EHR] in connection with an attempt to characterize the rational languages among the context-free ones. In particular, unavoidable subsets are used for extending Theorem 7.2 showing that the subword ordering $\preceq_d$ on words is a *well-ordering.* Actually, saying that a word u is subword of v means that v can be obtained from u by inserting letters. Instead of inserting letters we can insert words from a fixed subset. Given $X \subseteq \Sigma^*$ define $\prec_X$ as the reflexive and transitive closure of the relation

$$\{(u_1 u_2, u_1 x u_2) | x \in X, u_1, u_2 \in \Sigma^*\}.$$

For instance, if $X = \{ab\}$, then we get $1 \prec_X ab \prec_X aabb \prec_X aabbab$.

Then the following is proved in [EHR].

Proposition 7.1. *The ordering $\prec_X$ is a well-ordering if, and only if, X is unavoidable.*

We continue with some elementary properties of unavoidability. It is clear from the definition that from each unavoidable set we can extract a finite unavoidable subset, so the study can be reduced to finite unavoidable sets. It is also easy to verify that a set X is unavoidable if, and only if, it avoids all one-way infinite words if, and only if, it avoids all two-way infinite words. Indeed, let us verify, e.g., that if X is unavoidable, then every two-way infinite word $\ldots a_{-1} a_0 a_1 \ldots$ has a factor in X. By hypothesis, there are infinitely many words avoiding X, so there are infinitely many such words of even length. Now, say a word x is a *central occurrence* of a word y, if $y = y_1 x y_2$ with $|y_1| = |y_2|$. An infinite two-way word avoiding X is constructed as follows. For some $(a_0, b_0) \in \Sigma \times \Sigma$ there are infinitely many words having $x_0 = a_0 b_0$ as a central factor and avoiding X. Now, for some $(a_1, b_1) \in \Sigma \times \Sigma$ there are infinitely many words having $x_1 = a_1 a_0 b_0 b_1$ as a central factor and avoiding X, and so on. The infinite word $\ldots a_2 a_1 a_0 b_0 b_1 b_2 \ldots$ thus defined avoids X.

Testing the unavoidability of X can be done in different ways. We may construct a finite automaton recognizing $\Sigma^* X - \Sigma^* X \Sigma^+$, and then check whether or not there is a loop in the automaton. Another approach is more combinatorial and consists in simplifying X as much as possible. For example, assume that $\{babba, bbb\}$ are elements of a set X. We claim that by substituting $babb$ for $babba$ the set of two-way infinite words that are avoided is unchanged. Indeed, if an infinite word contains $babb$, then this occurrence is either followed by a, and then the word contains $babba$, or it is followed by b, but then it contains the occurrence bbb. The point here is that the occurrence $babbb$ has a suffix in $X - \{babba\}$. This leads to the following definitions.

A set X *immediately left-* (resp. *right-*) *simplifies* to the set Y, if either $Y = X - \{x\}$, where x has a proper factor in X, or $Y = X - \{wa\} \cup \{w\}$ (resp. $Y = X - \{aw\} \cup \{w\}$), where $wa \in X$ (resp. $aw \in X$) with $w \in \Sigma^*$ and $a \in \Sigma$, such that the following holds:

for all $b \in \Sigma, b \neq a$, wb has a suffix (resp. bw has a prefix) in $X - \{wa\}$ (resp. $X - \{aw\}$).

Further a set X *simplifies* (resp. *left-, right-*) *simplifies* to the set Y, if there exists a sequence of $n \geq 0$ sets $X_0 = X, \ldots, X_n = Y$ such that X_i immediately simplifies (resp. left-, right-simplifies) to X_{i+1}, with the convention

$X = Y$, if $n = 0$. Finally, a set X is *simple* (resp. *left-, right-simple*), if there is no $Y \neq X$ such that X simplifies (resp. left-, right-simplifies) to Y.

Above simplifications can be used to test unavoidability, as shown in [Ros2] and also known to J.-P. Duval (private communication).

Proposition 7.2. *A subset X is unavoidable if, and only if, it simplifies (resp. left-simplifies, right-simplifies) to the set consisting of the empty word only.*

Example 7.4. As an illustration, when the above is applied to the set $\{aaa, aba, bb\}$ the following sequence of sets is obtained: $X_0 = \{aaa, \underline{aba}, bb\}$, $X_1 = \{\underline{aaa}, ab, bb\}$, $X_2 = \{aa, \underline{ab}, bb\}$, $X_3 = \{\underline{aa}, a, bb\}$, $X_4 = \{a, \underline{bb}\}$, $X_5 = \{a, \underline{b}\}$, $X_6 = \{\underline{a}, 1\}$, $X_7 = \{1\}$. □

Actually, a more general problem was solved in [Ros2] by showing that for all finite subsets X there exists a unique simple set Y equivalent to X, in the sense that it avoids the same set of infinite words. Furthermore, Y can be obtained by first right-simplifying X as long as possible and then left-simplifying it. More precisely, for each X denote by $\overline{X}$ (resp. $\overline{X^r}$, $\overline{X^l}$) any simple (resp. right-, left-simple) subset which is the last element in a chain of simplification (resp. right-, left-simplification) starting from X. The following asserts a property of confluence saying that the result of a maximal sequence of simplification does not depend on the intermediate choices.

Proposition 7.3. *For all X, there exists a unique simple subset equivalent to it, namely $\overline{X} = \overline{\overline{X^r}}^l = \overline{\overline{X^l}}^r$*

Now we come to the problem that motivated the study of unavoidable sets. Haussler conjectured that every unavoidable set of words X can be extended in the sense that there exists an element $u \in X$ and a letter $a \in \Sigma$ such that substituting ua for u in X yields a new unavoidable set. For instance, in the previous example, the word ba can be replaced by bab (but not by baa, a^2 or b^2 as is easily verified). This conjecture held for some time and was supposed to be true. It was a nontrivial statement since, extending a word need not preserve the avoidability, but all computed examples confirmed that there always existed an extendable word. In [CC] some equivalent statements to the conjecture were given and some particular cases were settled. In fact, the conjecture turned out to be wrong, though it needed some clever efforts to exhibit the following counter-example (with the minimal possible number of elements) from [Ros1]:

$$X = \{aaa, bbbb, abbbab, abbab, abab, bbaabb, baabaab\}.$$

The reader may run the above procedure to check that X is unavoidable, as well as to use an exhaustive case study to show that no word can be extended.

Finally, [Ros2] introduces another interesting notion. Two subsets X and Y are *weakly equivalent*, written $X \sim_w Y$, if the sets of infinite periodic words, i.e., of the form $\ldots uu \ldots$ for some $u \in \Sigma^+$, avoiding them are equal. This notion seems to deserve further research. In particular the proof of the fact that two words $u \neq v \in \{a, b\}^*$, satisfy $u \sim_w v$ if, and only if, $\{u, v\} = \{a^n b, ba^n\}$ or $\{u, v\} = \{b^n a, ab^n\}$ is rather long and should be simplifiable.

7.5 Basics on the relation quasi-ordering $\preceq_r$

We turn to consider orderings on finite sets of words, in particular that of the quasi-ordering $\preceq_r$. By definition it was associated with relations satisfied by words of X, and hence with solutions of equations. This leads us to consider systems of equations with a finite number of unknowns and without constants.

Let Ξ be a finite set of unknowns and

$$S : u_i = v_i \text{ with } u_i, v_i \in \Xi^*, \text{ for } i \in I,$$

be a system of equations over Ξ. We are interested in all solutions of such a system in a given free monoid Σ^*, i.e., all morphisms $h : \Xi^* \to \Sigma^*$ satisfying $h(u) = h(v)$ for all $u = v$ in S. We are going to show that any system S is equivalent to one of its finite subsystems, i.e., any solution of this finite subsystem is also a solution of the whole S. Clearly, this states a fundamental compactness property of systems of equations over free monoids, and hence also of words. This property was conjectured by A. Ehrenfeucht at the beginning of the 1970s in a slightly different form, as we shall see in a moment, cf. also [Ka3].

Let us start with a simple example.

Example 7.5. Consider systems of equations with only two unknowns. Then, by the defect theorem, each solution $h : \{x, y\}^* \to \Sigma^*$ is periodic. Therefore the set of all solutions of a given nontrivial equation consists of morphisms satisfying one of the following conditions:

(i) $h(x) = h(y) = 1$;
(ii) $\exists k \in \mathbb{Q}_+ \cup \{\infty\} : |h(x)|/|h(y)| = k$ and h is periodic;
(iii) $h(x), h(y) \in z^*$ for some $z \in \Sigma^+$, i.e., h is periodic.

Actually, condition (ii) consists of infinitely many different conditions, one for each choice of k. It follows straightforwardly that the set of all solutions of a given system of equations over $\{x, y\}$ is determined by at most two equations. For example, if S contains equations of type (ii) for two different k's, then the only common solution is that of (i), and hence these two equations constitute an equivalent subsystem of two equations.   □

It is interesting to note that no similar analysis is known to work in the case of three unknowns. Indeed, no upper bound for the size of an equivalent finite subsystem is known. This is despite the fact that there exists a finite classification for sets of all equations satisfied by a given morphism $h : \{x, y, z\}^* \to \Sigma^*$, cf. [Sp1].

As we already mentioned the original *Ehrenfeucht's Conjecture* was stated in a slightly different form, more in terms of formal languages. In order to formulate it let us say that two morphisms $h, g : \Sigma^* \to \Delta^*$ *agree* on a word w if $h(w) = g(w)$. Motivated by research on questions when two morphisms agree on all words of a certain language, for more details cf. [Ka3], he conjectured that

$$\forall L \subseteq \Sigma^*, \exists \text{ finite } F \subseteq L : \forall h, g : \Sigma^* \to \Delta^* :$$
$$h(w) = g(w) \text{ for all } w \in L \Leftrightarrow h(w) = g(w) \text{ for all } w \in F.$$

In other words, the conjecture states that, for any language L, there exists a *finite subset* F of L such that to test whether two morphisms agree on words of L it is enough to do that on words of F. Such a finite F is called a *test set* for L. In terms of equations the conjecture states that any system of equations of the form

$$S : u_i = \bar{u}_i, \text{ for } i \in I,$$

where $\bar{u}_i$ is an isomorphic copy of u_i in a disjoint alphabet, is equivalent to one of its finite subsystems. As was first noted in [CuKa1], cf. also [HK2], this restricted formulation of the conjecture is actually equivalent to the general one.

As a result related to Example 7.5 we show next that all languages over a binary alphabet has a very small test set.

Theorem 7.4. *Each binary language possesses a test set of size at most three.*

Proof. The proof is based on Theorem 3.2. Here we present the main ideas of it, but omit a few technical details which can be found in [EKR2].

Let $L \subseteq \{a, b\}^*$ be a binary language. We define the *ratio* of $w \in \{a, b\}^+$ as the quantity $r(w) = |w|_a/|w|_b$. Hence, $r(w) \in \mathbb{Q}_+ \cup \{\infty\}$. A simple length argument shows that no two morphisms h, g, with $h \neq g$, can agree on two words with a different ratio. Consequently, if L contains two words with a different ratio, then they constitute a two-element test set for L.

So we assume that, for some k, $r(w) = k$ for all w in L. Now, each word w in L can be factorized as $w = w_1 \ldots w_n$, where, for each i, $r(w_i) = k$ and, for each prefix w_i' of w_i, we have $r(w_i') \neq k$. Let L_k be the set of all factors in the above factorizations of all words of L. It follows that if L_k has a test set of cardinality at most three, so has L: take a subset of L containing all words of the test set of L_k in the above factorizations.

So it remains to be shown that L_k has a test set of size at most three. If $\|L_k\| \leq 2$, there is nothing to be proved. So, assume that $\|L_k\| \geq 3$. Now, we use the partial characterization of binary equality sets proved in Theorem 3.2. Such a set is always of one of the following forms:

(i) $X_r = \{w|r(w) = r\}$ with $r \in \mathbb{Q}_+ \cup \{\infty\}$,
(ii) $\{\alpha, \beta\}^*$ for some words $\alpha, \beta \in \Sigma^*$,
(iii) $(\alpha\beta^*\gamma)^*$ for some words $\alpha, \beta, \gamma \in \Sigma^*$.

For morphisms having an equality set of form (i) any one-element subset of L_k works as a test set. For morphisms having an equality set of form (ii) any two-element subset of L_k works, since no word in L_k is a product of words having the same ratio. Finally, morphisms having an equality set of the form (iii) (if there are any!) are most complicated to handle. In this case one can show, cf. [EKR2], that if an equality set of form (iii) contains two elements of L_k, then these two elements determine this equality set uniquely. Consequently, for morphisms having equality sets of form (iii) any two-element subset of L_k works for all other pairs of morphisms except for those having as an equality set the one determined by these two words. And for those this two-element set can be extended to a three-element test set by adding a third word from L_k.

Consequently, in all the cases three words are enough. □

Of course, even in Theorem 7.4 a test set cannot be found effectively, in general. However, our above proof indicates that under a rather mild assumptions on L this can be done, cf. [EKR2].

7.6 A compactness property

In this section we prove the compactness property conjectured by Ehrenfeucht, and will later interprete it as a finiteness condition on finite sets of words, as well as consider its consequences.

Theorem 7.5. *Each system of equations with a finite number of unknowns over a free monoid is equivalent to one of its finite subsystems.*

Proof. Let Ξ be a finite set of unknowns in the equations

$$S : u_i = v_i \qquad \text{for } i \in I,$$

and Σ^* a free monoid, where these equations are solved. We exclude the case $\|\Sigma\| = 1$, since this is a trivial exercise in linear algebra. We also note that due the embeddings of Section 3.2 it does not matter what $\|\Sigma\|$ is - it can be even nondenumerable. Let us fix $\Sigma = \{a_0, \ldots, a_{n-1}\}$ with $n \geq 2$.

The basic idea is that we convert equations on words into polynomial equations on numbers. This is possible simply because a word w can be interpreted as the number it presents in n-ary notations.

More precisely, consider an equation

(1) $$u = v \text{ with } u, v \in \Xi^+.$$

Define two copies of Ξ, say Ξ_1 and Ξ_2, and associate with (1) the following pair of polynomial equations

$$(2) \qquad \begin{cases} l(u) - l(v) &=\ 0, \\ n(u) - n(v) &=\ 0, \end{cases}$$

where $l, n : \Xi^* \to (\Xi_1 \cup \Xi_2)^*$ are mappings defined recursively as

$$(3) \qquad \begin{cases} l(a) &= a_1, & \text{for } a \in \Xi, \\ l(wa) &= l(w)a_1, & \text{for } a \in \Xi \text{ and } w \in \Xi^+, \\ n(a) &= a_2, & \text{for } a \in \Xi, \\ n(wa) &= n(w)l(a) + n(a), & \text{for } a \in \Xi \text{ and } w \in \Xi^+. \end{cases}$$

Equations (2) are well-defined, and they are polynomial equations over the set $\Xi_1 \cup \Xi_2$ of *commuting* unknowns. In fact, coefficients of the monomials in (2) are from the set $\{-1, 0, 1\}$. Note also that the function n, as is obvious by induction, satisfies the relation

$$(4) \qquad n(w_1 w_2) = n(w_1)l(w_2) + n(w_2), \text{ for all } w_1, w_2 \in \Xi^+.$$

Now, let $w = a_{i_{k-1}} \dots a_{i_0}$, with $a_{i_j} \in \Sigma$, be a word in Σ^+. We associate with it two numbers

$$\sigma(w) = a_{i_0} + a_{i_1} n + \dots + a_{i_{k-1}} n^{k-1}$$

and

$$\sigma_0(w) = n^k.$$

Hence $\sigma(w)$ is the value of w as the n-ary number and $\sigma_0(w)$ is the value $n^{|w|}$. This guides us to set $\sigma(1) = 0$ and $\sigma_0(1) = n^0 = 1$.

Obviously, the correspondence $w \leftrightarrow (\sigma_0(w), \sigma(w))$ is one-to-one, and we use it to show:

$h : \Xi^* \to \Sigma^*$, i.e., $(h(a_0), \dots, h(a_{n-1}))$, is a solution of (1),

if, and only if,

the $2n$-tuple $(\sigma_0(h(a_0)), \dots, \sigma_0(h(a_{n-1})), \sigma(h(a_0)), \dots, \sigma(h(a_{n-1})))$ is a solution of (2).

To prove this equivalence , let us denote $s = (h(a_0), \dots, h(a_{n-1}))$, $s_1 = \sigma_0(s)$ and $s_2 = \sigma(s)$, where σ_0 and σ are applied to s componentwise. Then, if $h(u) = h(v)$, we conclude that

$$l(u)\Big|_{s_1} = n^{|h(u)|} = n^{|h(v)|} = l(v)\Big|_{s_1},$$

i.e., s_1 is a solution of the equation $l(u) - l(v) = 0$. Similarly, factorizing $u = u_1 u_2$, with $h(u_1), h(u_2) \neq 1$, we conclude from (4) that

$$n(u)\Big|_{s_1, s_2} = n(u_1)\Big|_{s_1, s_2} \cdot l(u_2)\Big|_{s_1, s_2} + n(u_2)\Big|_{s_1, s_2}$$

$$= \sigma(h(u_1))n^{|h(u_2)|} + \sigma(h(u_2)) = \sigma(h(u_1 u_2)) = \sigma(h(u)),$$

where the second equality is due to induction. The above holds also, as the basis of induction, when u does not have the above factorization. Symmetrically, $n(v)\big|_{s_1,s_2} = \sigma(h(v))$, so we have shown that (s_1, s_2) is a solution of (2).

On the other hand, if in above notations $s = (s_1, s_2)$ is a solution of (2) the above calculations show that h is a solution of (1), proving the equivalence.

Now, assume that S is our given system of equations, with $\varXi$ as the set of unknowns, consisting of equations $u_i = v_i$ for $i \in I$. Let

$$p_j(\varXi_1, \varXi_2) = 0 \text{ for } j \in J$$

be a set of polynomial equations, with $\varXi_1 \cup \varXi_2$ as the set of unknowns, consisting of those equations which are obtained in (2) when i ranges over I. For simplicity let $p_j = p_j(\varXi_1, \varXi_2)$ and $\mathcal{P} = \{p_j | j \in J\}$. By Hilbert's Basis Theorem, cf. [Co], $\mathcal{P}$ is finitely based, i.e., there exists a finite subset $\mathcal{P}_0 = \{p_j | j \in J_0\} \subseteq \mathcal{P}$ such that each $p \in \mathcal{P}$ can be expressed as a linear combination of polynomials in $\mathcal{P}_0$:

$$p = \sum_{j \in J_0} g_j p_j \text{ with } g_j \in \mathbb{Z}(\varXi_1 \cup \varXi_2).$$

Consequently, the systems "$\mathcal{P}_j = 0$ for $j \in J$" and "$\mathcal{P}_j = 0$ for $j \in J_0$" have exactly the same solutions. Therefore, by the equivalence we proved, our original system S is equivalent to its finite subsystem containing only those equations of S needed to construct $\mathcal{P}_0$.   □

The proof of Theorem 7.5 deserves a few comments. There are several proofs of this important compactness result, however, all of those rely on Hilbert's Basis Theorem. The two original ones are those by Albert and Lawrence in [AL1] and Guba in [MS]. Our proof is modelled from ideas of Guba presented in [McN2] and [Sal3], cf. also [RoSa2], using n-ary numbers. The other simple possibility of proving this result is to use embeddings of Σ^* into the ring of 2×2-matrices over integers, cf. [Per] or [HK2]. The advantage of the above proof is that it uses only twice as many unknowns as there are in the original system.

It is also worth noticing that we did not need above the full power of Hilbert's Basis Theorem. Indeed, we only needed the fact that the common roots of the polynomials $\mathcal{P}_j$, for $j \in J_0$, are exactly the same as those of the polynomials $\mathcal{P}_j$, for $j \in J$, which is not the Hilbert's Basis Theorem, but only its consequence. Note also that the reduction from word equations to polynomial equations goes only in one direction. Indeed, the existence of a solution of an equation is decidable for word equations, as shown by Makanin, while it is undecidable for polynomial equations, as shown by Matiyasevich, cf. [Mat] and also [Da].

Finally, let us still emphasize one peculiar feature of the above proof. The original problem is, without any doubts, a problem in a very noncommutative algebra, while its solution relies – unavoidably according to the current knowledge – on a result in a commutative algebra.

Of course, a finite equivalent subsystem for a given system of equations cannot be found effectively, in general. However, in several restricted cases this goal can be achieved. The proofs are normally direct combinatorial proofs not relying, for example, on Hilbert's Basis Theorem. We present one such example needed in our later considerations, for other such results we refer to [ACK], [Ka3], [HK2], [KRJ] or [KPR].

We recall that a system of equations in unknowns $\varXi$ is called *rational* if it is a rational relation of $\varXi^* \times \varXi^*$, cf. [Bel].

Theorem 7.6. *For each rational system of equations in a finite number of unknowns one can effectively find an equivalent finite subsystem.*

Proof. Of course, the formulation of Theorem 7.6 silently assumes that the system S is effectively given, for example, defined by a finite transducer τ, cf. [Bel]. Let n be the number of states of τ. Set

$S_0 = \{u = v \in S \mid (u, v)$ has an accepting computation in τ of length at most $2n\}$.

We claim that S_0 is equivalent to S. Assume the contrary that $h : \varXi^* \to \Sigma^*$ is a solution of S, but not of S_0. Choose an equation $u = v$ from S such that $h(u) \neq h(v)$, and moreover, (u, v) is minimal in the sense that there is no such equation in S which would have a shorter computation in τ than what is the shortest one for (u, v).

By the choice of S_0, words u and v factorize as $u = u_1 u_2 u_3 u_4$ and $v = v_1 v_2 v_3 v_4$ such that in τ we have

$$i \xrightarrow{\;u_1, v_1\;} q \xrightarrow{\;u_2, v_2\;} q \xrightarrow{\;u_3, v_3\;} q \xrightarrow{\;u_4, v_4\;} t$$

for some states i, q and t, with i initial and t final. It follows from the minimality of (u, v) that

$$(5) \qquad \begin{cases} h(u_1 u_2) & = & h(v_1 v_2), \\ h(u_1 u_2 u_4) & = & h(v_1 v_2 v_4) \text{ and} \\ h(u_1 u_3 u_4) & = & h(v_1 v_3 v_4). \end{cases}$$

We apply to these identities the following implication on words, the proof of which is straightforward and left to the reader: for any words u, v, w, z, u', v', w', $z' \in \Sigma^*$ we have

$$(6) \qquad \begin{cases} uv & = & u'v' \\ uwv & = & u'w'v' \\ uzv & = & u'z'v' \end{cases} \Rightarrow uwzv = u'w'z'v'.$$

Now, conditions (5) and (6) imply that $h(u) = h(v)$, a contradiction. □

We note that although our above proof does not imply that S_0 can be chosen "small", a more elaborated proof in [KRJ] shows that it can be chosen to be of the size $\mathcal{O}(n)$, where n denotes the number of transitions in τ.

Possibilities of generalizing the fundamental compactness property of Theorem 7.5 are considered in [HKP], cf. also [HK2].

7.7 The size of an equivalent subsystem

Theorem 7.5 leaves it open how large a smallest equivalent subsystem for a given system can be. This is the problem we consider here. Consequently, this section is closely connected to Section 4.4.

Recall that a system S in unknowns Ξ is *independent* if it is not equivalent to any of its proper subsystems. Our problem is to estimate the maximal size of an independent system of equations. Very little seems to be known on this problem. Indeed, we do no know whether the maximal size can be bounded by any function on $\|\Xi\|$.

What we can report here are some nontrivial lower bounds achieved in [KaPl1]. First we note that Example 4.8 introduces an independent system of equations over a *free semigroup* Σ^+ consisting of n^3 equations in $3n$ unknowns. Therefore a lower bound for the maximal size of independent system of equations over a free semigroup is $\Omega(\|\Xi\|^3)$.

Our next example shows that we can do better in a *free monoid*.

Example 7.6. Let $\Xi = \{y_i, x_i, u_i, v_i, \bar{y}_i, \bar{x}_i, \bar{u}_i, \bar{v}_i, \tilde{y}_i, \tilde{x}_i, \tilde{u}_i, \tilde{v}_i \mid i = 1, \ldots, n\}$ and S a system consisting of the following equations

$$S : y_i x_j u_k v_l \bar{x}_j \bar{u}_k \bar{v}_l \tilde{x}_j \tilde{u}_k \tilde{v}_l = x_j u_k v_l \bar{x}_j \bar{u}_k \bar{v}_l \tilde{x}_j \tilde{u}_k \tilde{v}_l y_i \text{ for } i, j, k, l = 1, \ldots, n.$$

Therefore $\|\Xi\| = 12n$ and $\|S\| = n^4$. Let us fix the values i, j, k and l and denote the corresponding equation by $e(i, j, k, l)$. In order to prove that S is independent we have to construct a solution of the system $S - \{e(i, j, k, l)\}$ which is not a solution of $e(i, j, k, l)$. Such a solution is given as follows:

$$\begin{cases} y_i & = ababa, \\ x_j & = u_k = v_l = ab, \\ \bar{x}_j & = \bar{u}_k = \bar{v}_l = a, \\ \tilde{x}_j & = \tilde{u}_k = \tilde{v}_l = ba, \\ z & = 1, \text{ for all other unknowns.} \end{cases}$$

This is not a solution of the equation $e(i, j, k, l)$, since

$$ababa.ab\ldots \neq ab.ab.ab\ldots \quad .$$

However, it is a solution of any other equation since the alternatives are

$y_i \neq ababa$, when the equations become an identity, or
$y_i = ababa$ and $0, 1$ or 2 of the words x_j, u_k and $v_l \neq ab$, when the corresponding relations are:
$ababa = ababa$,
$ababa.ab.a.ba = ab.a.ba.ababa$,
$ababa.ab.ab.a.ba.ba = ab.ab.a.a.ba.ba.ababa$.

Finally, we emphasize that this example uses heavily the empty word 1.   □

We summarize the above considerations to

Theorem 7.7. *(i) A system of equations with n unknowns may contain $\Omega(n^4)$ independent equations over a free monoid.*

(ii) A system of equations with n unknowns may contain $\Omega(n^3)$ independent equations over a free semigroup without the unit element.

A natural problem arises.

Problem 7.1. Does there exist an independent system of equations over a free semigroup or a free monoid consisting of exponentially many equations with respect to the number of unknowns?

We note that if the above question is posed in free groups then the answer is affirmative, although our compactness property is still valid, cf. [HK2]. Even more strongly, in [AL2] it is shown that systems of independent equations in three unknowns over a free group may be unboundedly large.

7.8 A finiteness condition for finite sets of words

In this section we interpret the above compactness result in terms of orderings. We consider relation quasi-ordering $\preceq_r$ defined on finite set of words by the condition

$$X \preceq_r Y \Leftrightarrow \exists \text{ bijection } \varphi : X \to Y \text{ such that } R_{\varphi(X)} \subseteq R_X.$$

Consequently, a finite set X is here considered as a solution of a system of equations, and Y is larger than X if X satisfies all equations Y does.

Now, we obtain as a direct interpretation of Theorem 7.5 our second nontrivial finiteness condition of Table 7.1 in Section 7.1.

Theorem 7.8. *Each chain with respect to relation ordering $\preceq_r$ is finite.* □

Note that Theorem 7.8 states that $\preceq_r$ is well-founded, and moreover, that also the reverse of $\preceq_r$ is well-founded. We also want to emphasize that our two nontrivial finiteness conditions, namely those stated in Theorems 7.2 and 7.8, are different in the sense that in Theorem 7.2 arbitrarily large, although always finite, antichains are known to exist, while it is not known whether there exist arbitrary large chains with respect to $\preceq_r$.

Two natural questions connected to the ordering $\preceq_r$ are to decide, for two given finite sets X and Y of the same cardinality, whether $X \preceq_r Y$ or whether $X = Y$ with respect to $\preceq_r$. These problems have very natural interpretations in terms of questions considered in Section 3.1. The latter asks whether F-semigroups X^+ and Y^+ are *isomorphic*, and the former asks (essentially) whether an F-semigroup can be *strongly embedded* into another one, i.e., whether there exists an injective morphism mapping generators to generators. Recall, as we showed in Section 3.1, that an F-semigroup X^+ can always be embedded into any Y^+ containing two words which do not commute.

As the answer to the above questions we prove, cf. [HK1].

Theorem 7.9. *Given two finite sets $X, Y \subseteq \Sigma^+$ it is decidable whether the F-semigroups X^+ and Y^+ are isomorphic.*

Proof. We may assume that $\|X\| = \|Y\|$, and restrict our considerations to a fixed bijection $\varphi : X \to Y$. We have to decide whether the extension of $\varphi : X^+ \to Y^+$ is an isomorphism, i.e., whether X and $\varphi(X)$ satisfies exactly the same relations. Let the sets of these relations be R_X and $R_{\varphi(X)}$ having a common set Ξ of unknowns, respectively.

It is an easy exercise to conclude that R_X and $R_{\varphi(X)}$ are rational relations, cf. constructions in Example 2.1. Now, deciding whether $R_X = R_{\varphi(X)}$ would solve our problem, but unfortunately the equivalence problem for rational relations is undecidable, cf. [Be1]. So we have to use some other method. Such a method can be found, when noticing that we are asking considerably less than whether R_X and $R_{\varphi(X)}$ are equal, namely we are asking only whether $Y = \varphi(X)$ satisfies R_X, and vice versa. To test this is not trivial, but by Theorem 7.5 it reduces to testing whether Y satisfies a finite subsystem of R_X, and moreover, by Theorem 7.6 such a finite subsystem can be found effectively. Hence, indeed, we have a method to test whether X^+ and Y^+ are isomorphic.   □

Note that the proof of Theorem 7.9 does not need the full power of Theorem 7.5. Only its effective validity for rational systems is needed, and this was easy to prove by direct combinatorial arguments.

Theorem 7.9 and its proof have the following two interesting consequences:

Theorem 7.10. *Given finite sets $X, Y \subseteq \Sigma^+$ it is decidable whether the F-semigroup X^+ is strongly embeddable into the F-semigroup Y^+.*   □

Theorem 7.11. *For finite sets $X, Y \subseteq \Sigma^+$ it is decidable whether (i) $X \preceq_r Y$ or (ii) $X = Y$ with respect to $\preceq_r$.*   □

The proof of Theorem 7.9 is not difficult, however, it contains quite a surprising feature: it does not seem to be extendable to rational subset of Σ^+. This is interesting to note since for many problems finite and rational sets behave in a similar way – due to the fact that rational sets are finite via their syntactic monoids. For instance, in a special case of the above isomorphism problem asking only whether a given F-semigroup X^+ is free, there is no essential difference whether X is finite or rational, cf. [BePe]. In the general isomorphism problem it is not only so that the method of Theorem 7.9 does not seem to work, but we have an open problem:

Problem 7.2. Is it decidable whether two rational subsets of Σ^+ generate isomorphic semigroups?

We conclude this section by considering how equations can be used to describe subsemigroups of Σ^+. These considerations are connected to the validity of Ehrenfeucht's Conjecture.

Let Σ be a fixed finite alphabet and Ξ a denumerable set of unknowns. We say that a system S of equations, with a finite number of unknowns from Ξ, *F-presents* an F-semigroup X^+ if, and only if, the following holds

(i) X is a solution of S; and
(ii) S is equivalent to R_X.

Intuitively this means that X satisfies the equations of S, but nothing else in the sense that any other equation e satisfied by X is dependent on S, i.e., S and $S \cup \{e\}$ are equivalent.

Example 7.7. Consider the following three singleton sets of equations

$$S_1 : xy = zx \; ; \; S_2 : xy = yx \; ; \; S_3 : xyy = yxxx \; .$$

The first one is an F-presentation of X_1^+ with $X_1 = \{a, ba, ab\}$, for example. Indeed, denoting these words by x, y and z in this order, we see that the minimal nontrivial relations of X_1 are $xy^n = z^n x$ for $n \geq 1$. But this set of nontrivial relations is equivalent to the equation $xy = zx$:

$$xy^n = xyy^{n-1} = zxy^{n-1} = zz^{n-1}y = z^n y.$$

On the other hand, S_2 is not an F-presentation. Indeed, assume that $X = \{x, y\}$ satisfies S_2. Then there is a word $z \in \Sigma^+$ and integers p and q such that $x = y^p$ and $y = z^q$. The cases $p = 0$ or $q = 0$ are easy to rule out. In the remaining case R_X is equivalent to the equation $x^q = y^p$, which is not equivalent to S_2. Finally, the above argumentation shows that S_3 is an F-presentation of the semigroup $\{a, aa\}^+$.   □

We did not require in the definition of an F-presentation that the set S is neither finite nor independent. However, such an F-presentation can always been found for any finitely generated F-semigroup.

Theorem 7.12. *For each finite $X \subseteq \Sigma^+$ the F-semigroup X^+ has a finite F-presentation consisting of an independent set of equations. Moreover, such an F-presentation can be found effectively.*

Proof. It is the proof of Theorem 7.6 which allows us to find a finite F-presentation S for X^+. It follows trivially that some of the equivalent subsets of S is independent, and hence a required F-presentation. To find it effectively we proceed as follows. By employing Makanin's algorithm we can test whether two finite systems of equations are equivalent, cf. Section 5. Hence a required F-presentation can be found by an exhaustive search.   □

The problem of characterizing those systems of equations which are F-presentations seems to be so far a neglected research area. Our Example 7.7 shows that not all finite systems are F-presentations. As a related question we state

Problem 7.3. Is it decidable whether a given finite system of equations is an F-presentation?

We note that Problem 7.3 is semidecidable, i.e., if we know that a given finite S is an F-presentation, then an F-semigroup X^+ having S as an F-presentation can be effectively found. This follows by an exhaustive search and arguments presented in the proof of Theorem 7.12.

8. Avoidability

The goal of this section is to give a brief survey on most important results of the theory of avoidable words, or as its special case of repetition-free words. A typical question of this theory asks: does there exist an infinite word over a given finite alphabet which avoids a certain pattern (repetition, resp.), that is does not contain as a factor any word of the form of the pattern (any repetition of that order, resp.). If the pattern is xx all squares must be avoided. It should be clear that, contrary to many other fragments of formal language theory, results of this theory depend on the size of the alphabet.

8.1 Prelude

The theory of avoidable words is among the oldest in formal language theory. A systematic study was carried out by A. Thue at the beginning of this century, see [T1], [T2], [Be6] and [Be8] for a survey of Thue's work. Later these problems have been encountered several times in different connections, and many important results, including most of Thue's original ones, have been discovered or rediscovered, cf. Chapter 3 in [Lo]. The topic has been under a very active research since early 80's, and it is no doubt that this revival is due to a few important papers, such as [BEM], and papers emphasizing a close connection of this theory to the theory of fixed points of iterated morphisms, cf. [Be2] and [CS].

Some basic results of the theory have already been published in details in books like [Lo] and [Sal2]. For survey papers we refer to [Be4] and [Be5]. Finally, applications of the theory especially to algebra, are discussed in [Sap].

To start with our presentation we recall that the basic notions were already defined in Section 2.3. The theory, at its present form, is closely related to an iteration of a morphism $h : \Sigma^* \to \Sigma^*$. For convenience we consider only 1-*free prolongable* morphisms, i.e., 1-free morphisms h satisfying $h(a) = a\alpha$ for some $a \in \Sigma$ and $\alpha \in \Sigma^+$. Then obviously, for each i, $h^{i+1}(a)$ is a proper prefix of $h^i(a)$, so that the unique word

$$w_h = \lim_{i \to \infty} h^i(a)$$

is obtained. Consequently, w_h is a *fixed point* of h, i.e., $h(w_h) = w_h$. Since it is defined by iterating morphism h (at point a) we say that w_h is obtained as a fixed point of *iterated morphism* h. This mechanism, often generalized by a possibility of mapping w_h by another morphism, is by far the most commonly used method to construct avoidable infinite words.

As an illustration let us consider morphisms

$$T : \left\{ \begin{array}{l} a \rightarrow ab \\ b \rightarrow ba \end{array} \right. \quad \text{and} \quad F : \left\{ \begin{array}{l} a \rightarrow ab \\ b \rightarrow a \end{array} \right. .$$

The words they define as iterated morphisms at a are

$$w_T = abbabaabbaababbabaababbaabbabaab \ldots$$

and

$$w_F = abaababaabaababaababaabaababa \ldots$$

The first one played an important role in the considerations of Thue, and later it was made well-known by Morse, cf. [Mor1] and [Mor2]. Therefore it is usually referred to as *Thue–Morse word*, although it was actually considered by Prouhet already in 1851, cf. [Pr]. The latter one is normally referred to as *Fibonacci word*, due to the fact that the lengths of the words $F^i(a)$ form the famous Fibonacci sequence. Accordingly, the morphisms T and F are called *Thue–Morse* and *Fibonacci morphisms*.

It is striking to note that these two words are among the most simple ones obtained by iterated morphisms, and still they have endless number of interesting combinatorial properties. In fact they seem to be the most commonly used counterexamples. For instance, prefixes of w_T of length 2^n show that factors of a word w of length n with multiplicities do not determine w uniquely, cf. Section 7.3. Similarly, w_F can be used to show that Proposition 6.2 is optimal, as well as that prefixes of lengths $p + 2$, $q + 2$, with p and q consecutive Fibonacci numbers, can be used to show the optimality of the Theorem of Fife and Wilf.

As an illustration of another way of defining repetition-free words we note that w_T can be defined recursively by formulas

$$\left\{ \begin{array}{l} u_0 = a, \\ v_0 = b, \end{array} \right. \quad \left\{ \begin{array}{ll} u_{n+1} = u_n v_n & \text{for } n \geq 0, \\ v_{v+1} = v_n u_n & \text{for } n \geq 0, \end{array} \right.$$

since then $T^n(a) = u_n$, as is easy to verify.

8.2 The basic techniques

The following two examples illustrate the basic techniques of proving that a fixed point of an iterated morphism avoids a certain pattern or a certain type of a repetition. In principal, the techniques is very simple, namely that of the infinite descending already used by Fermat, but its implementation might lead to a terrifying case analysis.

Example 8.1. We claim that the fixed point w_h of the iterated morphism

$$h : \begin{array}{l} a \rightarrow aba \\ b \rightarrow abb \end{array}$$

is 3^--free, in other words, does not contain any cube, but does contain repetitions of any order smaller than 3. The latter statement is trivial since any word of the form

$$uuu(\mathrm{suf}_1(u))^{-1}$$

is mapped under h to a word of the same form, and as the starting point w_h contains a factor aab.

To prove the second sentence, assume that w_h contains a cube $v = uuu$, with $|u| = n \geq 2$. Now we consider the four cases depending on the prefix u_2 of u of length 2, and analyse the cutpoints in $\{h(a), h(b)\}$-interpretations of u. It is due to a favourable form of h that, with the exception of the prefix ba, such a cutpoint in u_2 is unique, as depicted in Figure 8.1.

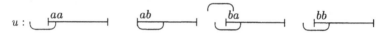

Fig. 8.1. Cutpoints inside u_2

In the cases aa, ab and bb the three prefixes in different occurrences of u have exactly the same cutpoints. Consequently, in the case of ab there exists a word u' such that $h(u') = u$, and in the other two cases there exists a word u' such that $h(u') = \mathrm{suf}_k(u)u\,\mathrm{suf}_k(u)^{-1}$, for $k = 1$ or 2, i.e., $h(u')$ is obtained from u by a shift as illustrated in Figure 8.2 for the prefix aa.

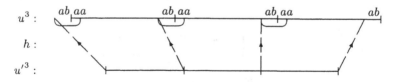

Fig. 8.2. The case $u_2 = aa$

In the case ba is the prefix of u, if the ba prefixes of the first and the second u have the same cutpoint, so have the third one as well, by the length argument. Hence, the above considerations apply. On the other hand, if the first and the second prefix have a different cutpoint, then the third one has, again by the length argument, still a different one. This, however, is not possible.

From above we conclude that, if w_h contains a cube longer than 6, then it contains also a shorter cube, and hence inductively a cube of length at most 6. That this is not the case is trivial to check.   □

Our second example deals with abelian repetitions, and is due to [Dek]. The basic idea of the proof is as above, only the details are more tedious.

Example 8.2. Let w_h be the word defined by the iterated morphism

$$h : \begin{array}{l} a \to abb \\ b \to aaab. \end{array}$$

We intend to show that w_h is abelian 4(-free, i.e., does not contain 4 consecutive commutatively equivalent factors. The idea of the proof is that illustrated in Figure 8.2. Starting from an abelian 4-repetition, we conclude that its small modification by a shift is an image under h of a shorter abelian 4-repetition. Now, the 4 consecutive blocks are only commutatively equivalent, so that it is not clear how to do the shifting. This means that h must possess some strong additional properties. To formalize these we associate with a word $u \in \{a, b\}^*$ a *value* in the group $\mathbb{Z}_5$ (of integers modulo 5) by a morphism $\mu : \{a, b\}^* \to \mathbb{Z}_5$ defined as

$$\mu(a) = 1 \quad \text{and} \quad \mu(b) = 2.$$

It follows that

(i) $\mu(h(w)) = 0 \quad \text{for all} \ w \in \{a, b\}^*.$

Now assume that $B_1 B_2 B_3 B_4$ is an abelian 4-repetition in w_h. We illustrate this, as well as an $\{h(a), h(b)\}$-interpretation of it in Figure 8.3.

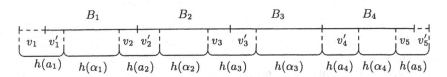

Fig. 8.3. $\{h(a), h(b)\}$-interpretation of $B_1 B_2 B_3 B_4$

Formally, the above means that

$$h(a_1 \alpha_1 \ldots \alpha_4 a_5) = v_1 B_1 B_2 B_3 B_4 v_5' \quad \text{with} \ a_i \in \Sigma, \ \alpha_i \in \Sigma^*,$$

where, for $i = 1, \ldots, 5$ and $j = 1, \ldots, 4$,

$$h(a_i) = v_i v_i' \quad \text{and} \quad B_j = v_j' h(\alpha_j) v_{j+1} \quad \text{with} \ v_i \in \Sigma^*, \ v_i' \in \Sigma^+.$$

Since μ is a morphism we obtain from (i) that, for $j = 1, \ldots, 4$,

$$\begin{aligned}
\mu(v_{j+1}) &= \mu(B_j) - \mu(h(\alpha_j)) - \mu(v_j') \\
&= \mu(B_j) + \mu(v_j) = g + \mu(v_j),
\end{aligned}$$

where g, due to the commutative equivalence of B_j's, denotes a constant element of $\mathbb{Z}_5$. Therefore the sequence

(1) $$\mu(v_1), \mu(v_2), \mu(v_3), \mu(v_4), \mu(v_5)$$

is an arithmetic progression in $\mathbb{Z}_5$. We want to allow only trivial arithmetic progressions, which guides us to require that

(ii) $$S = \{a \in \mathbb{Z}_5 \mid \exists z \in \mathrm{pref}\{h(a), h(b)\} : a = \mu(z)\}$$

is 5-*progression free*, i.e., does not contain any subset $\{a + ng \mid n = 0, \ldots, 4\}$ with $g \neq 0$. That our morphism h satisfies this condition is easy to see: indeed, we have

(2) $$(\mu(a), \mu(ab)) = (1, 3) \quad \text{and} \quad (\mu(a), \mu(aa), \mu(aaa)) = (1, 2, 3),$$

so that $S = \{0, 1, 2, 3\}$, while in $\mathbb{Z}_5$ any arithmetic progression of length 5, with $g \neq 0$, equals the whole $\mathbb{Z}_5$.

Since v_i's in (1) are prefixes of $h(a)$ or $h(b)$ we can write the arithmetic progression (1) in the form

(3) $$\mu(v_1) = \mu(v_2) = \mu(v_3) = \mu(v_4) = \mu(v_5).$$

What we would need, in order to have a shift, is that from (3) we could conclude that either the words v_i or the words v_i' are equal. This is our next condition imposed for h and μ. We say that μ is *h-injective*, if for all factorizations $v_i v_i' \in \{h(a), h(b)\}$, with $i = 1, \ldots, 5$, we have

(iii) $$\mu(v_1) = \cdots = \mu(v_5) \Rightarrow v_1 = \cdots = v_5 \quad \text{or} \quad v_1' = \cdots = v_5'.$$

From our computations in (2) we see that the only case to be checked is the case when $v_1 = ab$ and $v_3 = aaa$. And then indeed $v_1' = b = v_2'$, so that our μ is h-injective.

We are almost finished. We know that the words v_i (or symmetrically the words v_i') coincide. Consequently, the four abelian repetitions can be shifted to match with the morphism h: instead of B_i's we now consider the commutatively equivalent blocks $D_i = v_i B_i v_i^{-1}$ (or $D_i = v_i'^{-1} B_i v_i'$), for $i = 1, \ldots, 4$. Then there are words C_i such that

(4) $$h(C_i) = D_i \quad \text{with} \quad \pi(D_i) = \pi(D_j) \quad \text{for} \; i, j = 1, \ldots, 4 \, ,$$

where π gives the commutative image of a word. If we would know that C_i's were commutatively equivalent, we would be done. Indeed, then by an inductive argumentation w_h would contain either $aaaa$ or $bbbb$ as a factor, and this is clearly not the case.

So to complete the proof we still impose one requirement for h, namely that

(iv) $$M(h) = \begin{pmatrix} |h(a)|_a & |h(a)|_b \\ |h(b)|_a & |h(b)|_b \end{pmatrix} \quad \text{is } invertible.$$

Then, by (4), we would have $\pi(C_i) \cdot M(h) = \pi(D_i)$, or equivalently $\pi(C_i) = \pi(D_i) \cdot M(h)^{-1}$, for $i = 1, \ldots, 4$, so that C_i's would be commutatively equivalent. That $M(h)$ is indeed invertible is clear, since it equals to $\begin{pmatrix} 1 & 2 \\ 3 & 1 \end{pmatrix}$. □

It is worth noticing that conditions (i)–(iv) in the above proof are general ones, which can be used to prove similar results for different values of the size of the alphabet and/or the order of the repetition.

The argumentation of Examples 8.1 and 8.2 was already used by Thue in order to conclude

Theorem 8.1. *The Thue-Morse word w_T is 2^+-free, i.e., does not contain any overlapping factors.*   □

When applied to the Fibonacci word w_F, the above argumentation, with rather difficult considerations, yields the result that it is $(2 + \varphi)^-$-free, where φ is the number of the golden ratio, i.e., $\frac{1}{2}(1 + \sqrt{5})$, cf. [MP].

From Theorem 8.1 we easily obtain

Theorem 8.2. *There exists a 2-free infinite word in the ternary alphabet.*

Proof. Define the morphism $\varphi : \{a, b, c\}^* \to \{a, b\}^*$ by setting $\varphi(a) = abb$, $\varphi(b) = ab$ and $\varphi(c) = a$. Since φ has a bounded delay, the word $\varphi^{-1}(w)$ for $w \in \{a, b\}^\omega$, if defined, is unique, and since it is defined for each w containing no three consecutive b's, it follows that the word

(5) $$w_2 = \varphi^{-1}(w_T) = abcacbabcbacabca\ldots$$

is well-defined. Moreover, it is 2-free since w_T is 2^+-free, and each of the words $\varphi(d)$, with $d \in \{a, b, c\}$, starts with a.   □

The word w_2 can be obtained also as the fixed point of the iterated morphism h defined as $h(a) = abc$, $h(b) = ac$ and $h(c) = b$.

For the sake of completeness we state the result of Example 8.2, and its modification for abelian 3-free words in the ternary alphabet, also due to [Dek], as the following theorem.

Theorem 8.3. *(i) There exists an infinite abelian 4-free word in the binary alphabet.*

(ii) There exists an infinite abelian 3-free word in the ternary alphabet.   □

Table 8.1. Lengths of maximal words avoiding integer repetitions and abelian repetitions

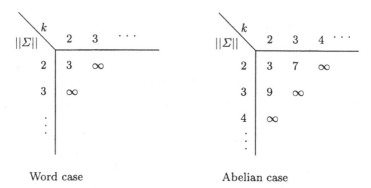

Word case   Abelian case

Now, with the calculations in Section 2.3 we can summarize all avoidable integer repetitions and abelian repetitions to the following table. Here k tells the order of the repetition, and the value of each entry the length of the longest word avoiding this repetition in the considered alphabet.

We note that special cases of (ii) in Theorem 8.3 was solved earlier. The first step was taken in [Ev], where it was shown that the 25th abelian powers were avoidable in the binary case. This was improved to 5 in [Pl] using an iterated *uniform morphism* h *of size* 15, i.e., $|h(a)| = 15$ for each letter a. Later the same result was shown in [Ju] using uniform morphisms of size 5.

Finally, the problem whether abelian squares can be avoided in the 4-letter alphabet, sometimes referred to as Erdös' Problem, was open for a long time, until it was solved affirmatively in [Ke2]. The proof is an interesting combination of a computer checking and of a mathematical reasoning showing that an abelian 4-free word can be obtained as the fixed point of an iterated uniform morphism of size 85. Moreover, it is shown that no smaller uniform morphism works here!

By Table 8.1, all 2-free words in the binary alphabet are finite, while by Theorem 8.1, there exists an infinite 2^+-free binary word. This motivates us to state the following notion explicitly defined in [Bra], cf. also [Dej]. For each $n \geq 2$, the *repetitiveness threshold* in the alphabet of n letters is the number $T(n)$ satisfying:

(i) there exists a $T(n)^+$-free infinite word in the n-letter alphabet; and
(ii) each $T(n)$-free word in the n-letter alphabet is finite.

It follows from the fact that for any irrational number r, the notions of r-free and r^+-free coincide, that the repetitiveness threshold is always rational, if it exists. And it is known to exist for $n \leq 11$: As we noted $T(2) = 2$. The value of $T(3)$ was solved in [Dej], by showing that each ternary $\frac{7}{4}$-free word is finite, and by constructing an infinite $\frac{7}{4}^+$-free ternary word as the fixed point of a uniform morphism of size 19. She also conjectured the values of $T(n)$ correctly up to the current knowledge, which is shown in Table 8.2. For 4 the problem was solved in [Pan1] and for the values from 5 up to 11 in [Mou].

Table 8.2. The repetitiveness thresholds and the lengths $\max(n)$ of longest $T(n)$-free words in the n-letter alphabet

$\|\Sigma\|$	2	3	4	5	6	7	8	9	10	11
$T(n)$	2	7/4	7/5	5/4	6/5	7/6	8/7	9/8	10/9	11/10
$\max(n)$	3	38	122	6	7	8	9	10	11	12

It is interesting to note that, for all $k \geq 2$, only very short words can avoid repetitions of order $\frac{k}{k-1}$. Indeed, any word of length $k+2$ in the k-letter alphabet Σ_k either contains a factor of length k in a $(k-1)$-letter alphabet or an image of the word $1 \ldots k12$ under a permutation of Σ_k. Consequently, such a word contains either a repetition of order at least $\frac{k}{k-1}$ or $\frac{k+2}{k}$, and both of these are at least $\frac{k}{k-1}$, for $k \geq 2$. Consequently, assuming the first line of Table 8.2 the second one follows for k at least 5 by noting that words $1 \ldots k1$ are $\frac{k}{k-1}$-free.

8.3 Repetition-free morphisms

As we have seen, constructions of repetition-free words rely typically on iterated morphisms, which preserve this property when started from a letter a, or in general, from a word having this property. This guides us to state the following definition. A morphism $h : \Sigma^* \to \Delta^*$ is said to be k-*free* if it satisfies:

whenever $w \in \Sigma^+$ is k-free, so is $h(w)$.

Note that the definition of the k-freeness does not require k to be a number – it can also be α^+ or α^- for some number α. For example, the Thue-Morse morphism is 2^+-free. Similarly, a morphism can be abelian k-free, for an integer k, as in Example 8.2.

The problem of deciding, for a given k and a morphism h, whether h is k-free is very difficult. Indeed, even for integer values of k it seems to be still open, cf. [Kel] for partial solutions. On the other hand, computationally feasible sufficient conditions for the k-freeness, with $k \in \mathbb{N}$, are known, an example being the following result from [BEM].

Proposition 8.1. *Let k be an integer ≥ 2. A morphism $h : \Sigma^+ \to \Delta^+$ is k-free if it satisfies the following conditions*

(i)   *h is k-free on k-free words of length at most $k + 1$;*
(ii) *whenever $h(a) \in F(h(b))$, with $a, b \in \Sigma$, then $a = b$; and*
(iii) *whenever $h(b)h(c) = uh(a)v$, with $a, b, c \in \Sigma$, then $u = 1$ and $a = b$, or $v = 1$ and $a = c$.*

The first complete characterization of 2-free morphisms was achieved in [Be3]. Later in [Cr] it was extended to the following sharp form, where $M(h)$ and $m(h)$ denote the maximal and minimal lengths of $h(a)$, when a ranges over the domain alphabet of h.

Proposition 8.2. *(i) A morphism $h : \Sigma^+ \to \Delta^+$ is 2-free if, and only if, it is 2-free on 2-free words of length at most $\max\{3, (M(h) - 3)/m(h)\}$.*
(ii) A morphism $h : \{a, b, c\}^+ \to \{a, b, c\}^+$ is 2-free if, and only if, it is 2-free on 2-free words of length at most 5.

A characterization similar to (ii) – requiring to check words up to length 10 – was shown for 3-free morphism over the binary alphabet in [Ka1]. Note here that not only the decidability of the k-freeness of a morphism, in general, but also the decidability of the 3-freeness in the arbitrary alphabet seems to be open.

We conclude these considerations with two more sharp characterization results. The first one was already known to Thue, cf. also [Harj]. The second one, due to [LeC], considers the problem whether a given morphism $h : \Sigma^+ \to \Delta^+$ is k-free, for all integer values of $k \geq 2$, in other words is *power-free*.

Proposition 8.3. *A binary morphism $h : \{a, b\}^+ \to \{a, b\}^+$ is 2^+-free if, and only if, it is of the form T^k or $T^k \circ \mu$, where T is the Thue-Morse morphism, μ is the permutation of $\{a, b\}$ and k is an integer ≥ 1.*

Proposition 8.4. *A morphism $h : \Sigma^+ \to \Delta^+$ is power-free if, and only if,*

(i) *h is 2-free; and*
(ii) *$h(a^2)$ is 3-free for each $a \in \Sigma$.*

8.4 The number of repetition-free words

In this subsection we study the number of repetition-free words in some special cases. More precisely we consider 2^+- and 3-free words in the binary case and 2-free words in the ternary case. Let us denote by $SF_n(3)$ the set of all 2-free words of length n over the ternary alphabet, where n is allowed to be ∞, as well. Similarly, let $S^+F_n(2)$ and $CF_n(2)$ denote the corresponding sets of 2^+- and 3-free words over the binary alphabet.

We shall show the following result of [Bra], cf. also [Bri].

Theorem 8.4. *$\|SF_n(2)\|$ is exponential, i.e., there exist constants A, B, ρ and σ, with $A, B > 0$ and $\rho, \sigma > 1$, such that*

$$A\rho^n \leq \|SF_n(3)\| \leq B\sigma^n \quad \text{for all } n.$$

Proof. The existence of B and σ are clear. The crucial point in proving the lower bound is to find a 2-free morphism $h : \Sigma^+ \to \{a, b, c\}^+$, with $\|\Sigma\| > 3$. As shown in [Bra] such a morphism exists for each value of $\|\Sigma\|$, and moreover, can be chosen uniform. For small values of $\|\Sigma\|$ it is not difficult to find such a morphism using Proposition 8.2.

Now, let $h : \{a, b, c, \bar{a}, \bar{b}, \bar{c}\}^+ \to \{a, b, c\}^+$ be a uniform 2-free morphism. As shown in [Bra] the smallest size of such a morphism is 22, which means that after having it, the checking of its 2-freeness is computationally easy. Next we define a finite substitution $\tau : \{a, b, c\}^+ \to \{a, b, c, \bar{a}, \bar{b}, \bar{c}\}^+$ by setting

$$\tau(x) = \{x, \bar{x}\} \quad \text{for } x \in \{a, b, c\}.$$

We fix a 2-free word w_k of length k, which by Theorem 8.2 exists, and consider the set $h(\tau(w_k))$ of words. Clearly, words in this set are 2-free, and of length $22k$. Moreover, $\|h(\tau(w_k))\|$ contains 2^k words, since h must be injective, or even a prefix code, by its 2-freeness. So we have concluded that, for each $n \geq 2$, the cardinality of $SF_{22n}(3) \geq 2^n$. This implies that ρ can be chosen to be $2^{\frac{1}{22}} \sim 1{,}032$. $\square$

Theorem 8.4 stimulates for a few comments. First of all, a closer analysis of the problem shows that the constants can be chosen such that

$$6 \cdot 1{,}032^n \leq \|SF_n(3)\| \leq 6 \cdot 1{,}38^n.$$

Moreover, the 20 smallest values of the number of 2-free words of length n over $\{a, b, c\}$ are: 3, 6, 12, 18, 30, 42, 60, 78, 108, 144, 204, 264, 342, 456, 618, 798, 1044, 1392, 1830, 2388.

Second, the above proof immediately extends to infinite words. Starting from a fixed infinite 2-free word over the ternary alphabet Σ_3, say w_2 of Theorem 8.2, τ creates nondenumerably many of those over a six-letter alphabet Σ_6, and h being injective also on Σ_6^ω brings equally many back to Σ_3^ω. So we have

Theorem 8.5. *$SF_\infty(3)$ is nondenumerable.*

Finally, the above ideas can be applied to estimate the number of 3-free words over the binary alphabet Σ_2, if a uniform 3-free morphism $h : \Sigma^+ \to \Sigma_2^+$, with $\|\Sigma\| > 2$, is found. Again, as shown in [Bra], such morphisms exist for each value of $\|\Sigma\| > 2$. Therefore, since the uniformity and the 3-freeness imply a bounded delay, and hence the injectivity on Σ^ω, we obtain

Theorem 8.6. *$CF_n(2)$ is exponential, and $CF_\infty(n)$ is nondenumerable.*

The bounds given for the number of 3-free words of length n in the binary case are

$$2 \cdot 1{,}08^n \leq \|CF_n(2)\| \leq 2 \cdot 1{,}53^n.$$

For 2^+-free words the results are not quite the same as the above for 2- and 3-free words. The result stated as Proposition 8.5 follows from the characterizations of finite and infinite 2^+-free binary words presented in the next subsection.

Proposition 8.5. $S^+F_n(2)$ *is polynomial, while* $S^+F_\infty(2)$ *is nondenumerable.*

Recently, it was shown in [Car2], using the morphism of [Ke2], that the number of abelian 2-free words over the 4-letter alphabet grows exponentially, as well as that of abelian 2-free infinite words is nondenumerable. This seems to be the only estimate for the number of abelian repetition-free words. For repetition-free words over partially commutative alphabets we refer to [CF].

At this point the following remarks are in order. As we saw in all the basic cases the sets of repetition-free infinite words form a nondenumerable set. Consequently, "most" of such words cannot be algorithmically defined. In particular, the by far most commonly used method using iterated morphisms can reach only very rare examples of such words. In the case of 2^+-free words the situation is even more striking: as shown in [See] the Thue-Morse word is the only binary 2^+-free word which is the fixed point of an iterated morphism.

8.5 Characterizations of binary 2^+-free words

In this subsection we present structural characterizations of both finite and infinite binary 2^+-free words. These are obtained by analysing how a given 2^+-free word can be extended preserving the 2^+-freeness. In order to be more precise, let us recall that the recursive definition of the Thue-Morse word was based on two sequences $(u_n)_{n\geq 0}$ and $(v_n)_{n\geq 0}$ of words satisfying

$$u_0 = a \qquad v_0 = b$$
$$u_{n+1} = u_n v_n \quad v_{n+1} = v_n u_n \quad \text{for } n \geq 0.$$

Let us call words u_n and v_n *Morse blocks* of order n, and set $U_n = \{u_n, v_n\}$ and $U = \bigcup_{n=1}^{\infty} U_n$. Clearly, the lengths of Morse blocks are powers of 2, and for instance $v_3 = baababba$.

Now, a crucial lemma in the characterizations is the following implication:

$$(1) \qquad uvwx \in S^+F_{3 \cdot 2^n+1}(2), \ u,v \in U_n, \ |w| = 2^n, \ x \in \Sigma \Rightarrow w \in U_n.$$

This means that, if a product of two Morse blocks of the same order, can be extended to the right, preserving the 2^+-freeness, by a word which is longer than these blocks, then the extension starts with a Morse block of the same order than the original ones.

The proof of (1) is by induction on n. For $n = 0$ there is nothing to be proved. Further the induction step can be concluded from the illustration of Figure 8.4. Indeed, the possible extensions of length 2^n for $u_{n+1}v_{n+1}$ are, by induction hypothesis, words u_n and v_n, and of the two potential extensions of these of length $|v_n|$ one is ruled out in both the cases, since the word must

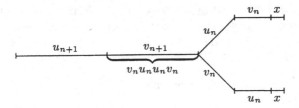

Fig. 8.4. The proof of (1) for $u_{n+1}v_{n+1}$

remain 2^+-free. Consequently, for $u_{n+1}v_{n+1}$, the word w is either u_{n+1} or v_{n+1} as claimed. Similarly, one can prove the other cases of the products.

Based on (1), and a bit more detailed analysis, the following characterization is obtained for 2^+-free finite words in [ReSa]: for each 2^+-free word w there exists a constant k such that w can be written *uniquely* in the form

$$(2) \qquad w = l_0 \ldots l_{k-1} u r_{k-1} \ldots r_0 \quad \text{with } l_i \in L_i, \; r_i \in R_i \text{ and } u \in \bigcup_{i=1}^{12} U_k^i,$$

where $k = \mathcal{O}(|w|)$ and the sets L_i and R_i, for $i = 0, \ldots, k - 1$, are of the cardinality 15.

Denoting $n = |w|$ we obtain from (2) that

$$\|S^+ F_n(2)\| \le \| \bigcup_{i=1}^{12} U_k^i \| \cdot 15^{2k} = \mathcal{O}(n^\alpha),$$

for some $\alpha > 0$. Actually, as computed in [ReSa], α can be chosen to be $\log_2 15 < 4$. Hence, the first sentence of Proposition 8.5 holds.

Note that (2) gives only a necessary condition, and hence only a partial characterization, for finite 2^+-free words. Later a more detailed analysis has improved estimates for the number of binary 2^+-free words of length n, cf. [Kob], [Car1], [Cas1] and [Lep2]. The strictest bounds are given in [Lep2], where, as well as in [Car1], a complete characterization of all finite 2^+-free words is achieved:

$$A \cdot n^{1,22} \le \|S^+ F_n(2)\| \le B \cdot n^{1,37}.$$

On the other hand, in [Cas1] it is shown that the limit

$$\lim_{n \to \infty} \frac{\|S^+ F_n(2)\|}{n^\alpha}$$

does not exist for any α, meaning that the number of 2^+-free binary words of length n behaves irregularly, when n grows.

Now, let us move to a characterization of 1-way infinite binary 2^+-free words. This remarkable result was proved in [F], while our automata-theoretic presentation is from [Be7]. Let us recall that U_n denoted the set of Morse blocks of order n and U the set of all Morse blocks. Further for each binary w let $\bar{w}$ denote its complement, i.e., word obtained from w by interchanging each of its letters to the other. The crucial notion here is the so-called *canonical decomposition* of a word $w \in \Sigma^* U_1$, which is the factorization

$$w = zy\bar{y},$$

where $\bar{y}$ is chosen to be the longest possible $\bar{y}$ in U such that w ends with $y\bar{y}$. Next, three mappings, interpreted as left actions, $\alpha, \beta, \gamma : \Sigma^* U_1 \to \Sigma^* U_1$ are defined based on the canonical decompositions of words:

(3)
$$\begin{cases} w \circ \alpha = zy\bar{y} \circ \alpha = zy\bar{y}yy\bar{y} = wyy\bar{y} \\ w \circ \beta = zy\bar{y} \circ \beta = zy\bar{y}y\bar{y}\bar{y}y = wy\bar{y}\bar{y}y \\ w \circ \gamma = zy\bar{y} \circ \gamma = zy\bar{y}\bar{y}y = w\bar{y}y. \end{cases}$$

We consider

$$A = \{\alpha, \beta, \gamma\}$$

as a ternary alphabet. The mappings α, β and γ extend a word $w = zy\bar{y}$ from the right by words $yy\bar{y}$, $y\bar{y}\bar{y}y$ and $\bar{y}y$, respectively. The use of the canonical decompositions makes these mappings well-defined. It also follows from the fact that w is a proper prefix of $w \circ \delta$, for any $\delta \in A$, that any infinite word $\omega \in A^\omega$ defines a unique word $w \circ \omega \in \Sigma^\omega$. Such an ω is called the *description* of $w \circ \omega$. Of course, the description can be finite, as well.

The mappings α, β and γ are chosen so that, given the canonical description tion $zy\bar{y}$ of w, they add to the end of w two Morse blocks of the same order as $\bar{y}$ in all possible ways the condition (1) allows this to be done preserving the 2^+-freeness. Actually, in the case of α such a block would be yy, but now also one extra $\bar{y}$ is added, since the next block of this length would be $\bar{y}$ in any case, again by (1). Similarly β adds istead of $y\bar{y}$ the word $y\bar{y}\bar{y}y$.

It follows from these considerations that *each 1-way infinite binary 2^+-free word* has a description, which moreover, by (3), *is unique*. Which of the descriptions actually define a 2^+-free infinite word is the contents of the characterization we are looking for. In order to state the characterization we set

$$I = \{\alpha, \beta\}(\gamma^2)^*\{\beta\alpha, \gamma\beta, \alpha\gamma\},$$

and consider the following sets of infinite words over A:

$$F = A^\omega - A^* I A^\omega \quad \text{and} \quad G = \beta^{-1} F.$$

Now, we are ready for the characterization known as Fife's Theorem.

Proposition 8.6. *Let $w \in \Sigma^\omega$.*

(i) A word $w \in ab\Sigma^\omega$ is 2^+-free if, and only if, its description is in F;
(ii) A word $w \in aab\Sigma^\omega$ is 2^+-free if, and only if, its description is in G.

The detailed proof of this result is not very short, cf. e.g. [Be7]. On the other hand, the result provides a very nice example of the usefulness of finite automata in combinatorics. Namely, the set of all descriptions of binary 2^+-free infinite words can be read from the finite automaton of Figure 8.5 accepting any infinite computation it allows.

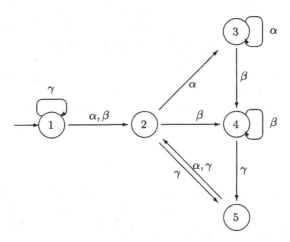

Fig. 8.5. Fife's automaton $\mathcal{A}_F$

Now the second sentence of Proposition 8.5 stating that there exist denumerably many infinite 2^+-free words over the binary Σ is obvious. Indeed, the automaton contains two loops in state 3, for example.

We conclude our discussion on 2^+-free words by recalling a characterization of 2-way infinite binary 2^+-free words. This characterization has interesting interpretations in the theory of symbolic dynamics, cf. [MH].

Proposition 8.7. *A two-way infinite binary word w is 2^+-free if, and only if, there exists a two-way infinite word w' such that $w = T(w')$, where T is the Thue-Morse morphism.*

This characterization was already known to Thue, and it is much easier to obtain than that of Proposition 8.6, by using standard tools presented at the beginning of this section.

We note that no characterization of 2-free words – either finite or infinite – over the three letter alphabet is known. Some results in that direction are obtained in [She], [ShSo1] and [ShSo2]. For example it is shown that the set of such infinite words is perfect in the sense of topology implying immediately Theorem 8.5.

8.6 Avoidable patterns

In this last subsection we consider an interesting problem area introduced in [BEM], and also in [Z], namely that of the avoidability of general patterns. We defined this notion already in Section 2.3, and moreover have used it implicitly several times. Indeed, Theorem 8.2 says that the pattern xx is avoidable in the ternary alphabet, i.e., there exists an infinite ternary word having no square as a factor. It is trivially unavoidable in the binary alphabet, while the pattern $xyxyx$, as shown in Theorem 8.1, is avoidable in this alphabet.

It follows, as expected, that the avoidability of a pattern *depends* on the size of the alphabet considered – contrary to many other problems in formal language theory. More precisely, the pattern $P_2 = xx$ separates the binary and ternary alphabets.

It turned out much more difficult to separate other alphabets of different sizes. A pattern separating 3- and 4-letter alphabets was given in [BMT]. The pattern, containing 7 different letters and being of length 14, is as follows:

$$P_3 = ABwBCxCAyBAzAC.$$

It was shown that any word over $\{a, b, c\}$ of length 131293 (which, however, is not the optimal bound) contains a morphic image of P_3 under a 1-free morphism into $\{a, b, c\}^+$ as a factor. On the other hand, the infinite word obtained – again – as the fixed point of a morphism avoids the pattern P_3. Such a morphism is given by $h(a) = ab$, $h(b) = cb$, $h(c) = ad$ and $h(d) = cd$, i.e., can be chosen uniform of size 2.

We summarize the above as follows.

Proposition 8.8. *For each $i = 1, 2, 3$ there exists a pattern P_i which is unavoidable in the i-letter alphabet, but avoidable in the $(i+1)$-letter alphabet.*

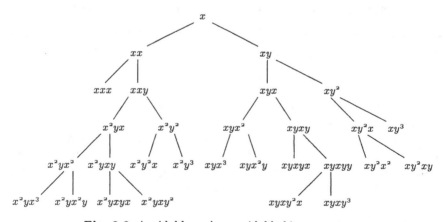

Fig. 8.6. Avoidable and unavoidable binary patterns

It is an open question to settle whether Proposition 8.8 extends to larger alphabets.

As we saw, the problem of settling whether a pattern is avoidable in a given alphabet is not easy at all. However, the case where both the pattern and the alphabet are binary, is completely solved. By a *binary* pattern we, of course, mean a pattern consisting of two letters only, say x and y.

The research leading to this interesting result was initiated in [Sc], continued and almost completed in [Rot], and finally completed in [Cas2].

The result is summarized in Figure 8.6. There the labels of the leaves, and hence also any word obtained as their extensions, are avoidable, while those of inside nodes are unavoidable. Note that the tree covers all the words starting with x, and hence up to the renaming all binary patterns, and yields

Proposition 8.9. *Each binary pattern of length at least 6 is avoidable in the binary alphabet.*

Each of these avoidable patterns was shown to be so by constructing an infinite word avoiding the pattern as the fixed point of an iterated morphism, or as a morphic image of the fixed point of an iterated morphism. For each unavoidable pattern α let $\max(\alpha)$ be the length of the longest finite binary words avoiding α. The values of $\max(\alpha)$, for all unavoidable patterns omitting symmetrical cases, are listed in Table 8.3.

Table 8.3. Unavoidable patterns and maximal lengths of binary words avoiding those

α :	x	xy	x^2	x^2y	xyx	x^2yx	xy^2x	x^2y^2	$xyxy$	x^2yx^2	x^2yxy
$\max(\alpha)$:	0	1	3	4	4	9	10	11	18	18	38

In accordance with Theorem 8.6 and Proposition 8.5 we note the result of [GV] showing that any avoidable binary pattern is avoided by nondenumerably many infinite words.

9. Subword complexity

In this final section we consider a problem area which has attracted quite a lot of attention in recent years, and which provides a plenty of extremely challenging combinatorial problems. A survey of this topic can be found in [Al].

9.1 Examples and basic properties

Let $w \in \Sigma^\omega$ be an infinite word. We define its *subword complexity*, or briefly *complexity*, as the function $g_w : \mathbb{N} \to \mathbb{N}$ by

$$g_w(n) = \|\{u \in \Sigma^n \mid u \in F(w)\}\|.$$

Consequently, $g_w(n)$ tells the number of different factors of length n in w. A very related notion can be defined for languages (consisting of finite words) instead of infinite words in a natural way.

Two problems are now obvious to be asked:

(i) Given a $w \in \Sigma^\omega$, compute its complexity g_w.
(ii) Given a complexity g, find a word having g as its complexity.

In both of these cases one can work either with the *exact* complexity or with the *asymptotic* complexity, i.e., identifying complexities g and g' if they satisfy $g(n) = \theta(g'(n))$. The above problems are natural to call the *analysis problem* and the *synthesis problem* for complexities of infinite words. Mostly only asymptotic versions of these problems are considered here.

We start with two examples.

Example 9.1. Let $w_K \in \{1,2\}^\omega$ be the famous *Kolakoski word*, cf. [Kol], [Lep1] or [CKL],

$$w_K = 2211212212211211212121 \ldots$$

defined by the rule: w_K consists of consecutive blocks of 1's and 2's such that the length of each block is either 1 or 2, and the length of the ith block is equal to the ith letter of w_K. Hence, odd blocks consists of 2's and even ones of 1's. The word is an example of a *selfreading* infinite word, cf. [Sl]. The answer to the analysis problem of w_K is not known, in fact it is not even known whether $g_{w_K}(n) = \mathcal{O}(n^2)$. □

Example 9.2. As an example of the case when the complexity of a word is precisely known we consider the Fibonacci word w_F defined as the fixed point of the Fibonacci morphism: $F(a) = ab, F(b) = a$. We show that its complexity satisfies

(1) $$g_{w_F}(n) = n + 1 \quad \text{for } n \geq 1.$$

This is seen inductively by showing that, for each n, there exists just one word w of length n such that both wa and wb are in $F(w_F)$. Let us call such factors *special*. For $n = 1$ and $n = 2$ the sets of factors of these lengths are $\{a, b\}$ and $\{aa, ab, ba\}$, where a and ba are the special ones. Now consider a factor w of length $n + 1$, with $n \geq 2$. If w ends with b, then, by the form of the morphism F, w admits only the continuation by a, i.e., the a-extension. If $w = xw'a$, with $x \in \{a, b\}$, then by the induction hypothesis of the words $w'a$, with $|w'| = n - 1$, only one is special. Therefore, we are done, when we show that of the words $aw'a$ and $bw'a$, with $|w'| \geq 1$, only one can be a factor of w_F.

Assume to the contrary that both of these words are in $F(w_F)$. Then w' cannot start with b. Therefore both $aaw''a$ and $baw''a$, for some w'', are factors of w_F. By the form of the first of these, w'' is different from 1, as well as cannot start with a. But by the form of the second word, it cannot start with b either, since $bb \notin F(w_F)$. Hence, we have proved that w_F satisfies (1). □

Binary words satisfying (1) are so-called infinite *Sturmian words*. Such words have several equivalent definitions, cf. [MH] and [Ra] emphasizing different aspects of these words, and [Bro] containing a brief survey. Their properties has been studied extensively, cf. [CH], [DG], [Mi] and [Ra], in particular recent works in [BdL], [dL] and [dLM] have revealed their fundamental importance in the theory of combinatorics of words.

Our next simple result, noted already in [CH], shows that the complexity of Sturmian words is the smallest unbounded complexity. In particular, the Fibonacci word is an example of a word achieving this.

Theorem 9.1. *Let* $w \in \Sigma^\omega$ *with* $\|\Sigma\| \geq 2$. *If* g_w *is not bounded, then* $g_w(n) \geq n + 1$ *for all* $n \geq 1$.

Proof. We prove that, if for some $n \geq 1$, $g_w(n + 1) = g_w(n)$, then w is ultimately periodic, and therefore g_w is bounded. Consequently, Theorem 9.1 follows from the fact that the complexity of a word is a nondecreasing function.

Assume now that $g_w(n_0 + 1) = g_w(n_0)$. This implies that each factor u of w of length n_0 admits *one and only one* way to extend it by one symbol on the right such that the result is in $F(w)$. Let the function $E : \Sigma^{n_0} \to \Sigma$ define such extensions. Let now $u_0 \in \Sigma^{n_0}$ be a factor of w, say αu_0 is a prefix of w. We define recursively

$$u_{i+1} = u_i \cdot E(\mathrm{suf}_{n_0}(u_i)) , \quad \text{for } i \geq 0.$$

Then, by the definition of E, αu_i is a prefix of w for all i, implying that $w = \lim_{i \to \infty} \alpha u_i$. But by the pigeon hole principle and the fact that E is a function $\lim_{i \to \infty} \alpha u_i$ is ultimately periodic. □

Theorem 9.1 states that there exists a gap $(\theta(1), \theta(n))$ in the family of complexities of finite words. According to the current knowledge this is the only known gap. We also note that Theorem 9.1 can be reformulated as

Corollary 9.1. *Let* $w \in \Sigma^\omega$ *with* $\|\Sigma\| \geq 2$. *Then* w *is ultimately periodic if, and only if,* g_w *is bounded.* □

Above corollary yields a simple criterium to test whether the complexity of a given word is bounded. Unfortunately, however, it is not trivial to verify this criterium. Indeed, even for fixed points of iterated morphisms the verification is not obvious, although can be done effectively, cf. [HL] and [Pan3].

We continue with another example where the asymptotic complexity can be determined. This is a special case of so-called *Toeplitz words* considered in [CaKa].

Example 9.3. We define an infinite word $w_t \in \{1,2\}^\omega$ as follows. Let $p = 1?2?2$ be a word over the alphabet $\{1, 2, ?\}$, and define recursively

$$w_0 = p^\omega$$
$$w_{i+1} = t(w_i) \quad \text{for } i \geq 0,$$

where $t(w_i)$ is obtained from w_i by substituting w_0 to the positions of w_i filled by the letter ? . Consequently,

$$w_1 = (112?2122?2122121?2221?222)^\omega,$$

and the word $w_t = \lim_{i \to \infty} w_i$ is well-defined over $\{1, 2\}$. The word w can be defined as a selfreading word like the Kolakoski word as follows. For each $i \geq 1$, replace the ith occurrence of ? in w_0 by the ith letter of the word so far defined. Clearly, this yields a unique word w', and moreover $w' = w_t$. These two alternate mechanisms to generate w are referred to as *iterative* and *selfreading*, respectively.

In order to compute g_{w_t} we consider factors of w_t of length $5n$. Each such factor is obtained, by the selfreading definition of w_t, from a conjugate of $u_n = (1?2?1)^n$ by substituting a factor v_n of length $3n$ to the positions filled by ?'s. Therefore,

$$(2) \qquad\qquad g_{w_t}(5n) \leq 5g_{w_t}(3n).$$

It is a straightforward to see that u_n is different from any of its conjugates for $n \geq 2$. Moreover, when different v_n's are substituted to a given conjugate of u_n, different factors of w_t are obtained. Therefore (2) would become the equality, if we could show that each factor v which occurs in w_t occurs in any position modulo 3, i.e., the length of the prefix immediately preceding v can be any number modulo 3. This, indeed, can be concluded from the iterative definition of w_t. First, for each i, the word w_i is periodic over $\{1, 2, ?\}$ with a period 5^i. Second, each factor v of w_t is a factor of w_{i_0} for some i_0. Consequently, since 3 and 5 are coprimes, the above v occurs in all positions modulo 3 in the prefix of length $3 \cdot 5^{i_0}$ of w_t.

So far we concluded the formula

$$g_{w_t}(5n) = 5g_{w_t}(3n) , \quad \text{for } n \geq 2.$$

It is a simple combinatorial exercise to derive from this that $g_{w_t}(n) = \theta(n^r)$ with $r = \log 5/(\log 5 - \log 3)$. $\qquad\square$

Next we say a few words about the synthesis problem.

Example 9.4. We already saw how the smallest unbounded complexity of binary words could be achieved. The largest one is even easier to obtain. Indeed, the complexity of the word $w_{\mathrm{bin}} = \mathrm{bin}(1)\mathrm{bin}(2)\ldots$, where $\mathrm{bin}(i)$ is the binary representation of the number i, equals the exponent function $g(n) = 2^n$. $\qquad\square$

Example 9.5. In [Cas3] the synthesis problem is elegantly solved to all linear function $f(n) = an + b$, with $(a, b) \in \mathbb{N} \times \mathbb{Z}$. Namely, it is shown that such a function is the complexity of an infinite word if, and only if, $a + b \geq 1$ and $2a + b \leq (a+b)^2$, and in the affirmative case a word w having this complexity is constructed as a morphic image of the fixed point of an iterated morphism.

□

9.2 A classification of complexities of fixed points of iterated morphisms

The rest of this section is devoted to a classification of the asymptotic complexities of words obtained as fixed points of iterated morphisms. This research was initiated in [ELR], later continued in [ER2], [ER3], [ER4], and finally completed in [Pan2]. The classification is based on the structure of the morphism, and it allows to decide the asymptotic complexity of such a word, i.e., to solve the analysis problem for iterated morphisms. It also allows an easy way to solve the asymptotic synthesis problem for those complexities which are possible as fixed points of iterated morphisms. As we shall see there exist only five different such possibilities.

Let $h : \Sigma^* \to \Sigma^*$ be a morphism which need not be 1-free, but is assumed to satisfy the condition $a \in \mathrm{pref}(h(a))$, for some a, in order to yield the unique word

$$w_h = \lim_{i \to \infty} h^i(a).$$

Consequently, w_h may be finite or infinite. In the former case the complexity of w_h is $\mathcal{O}(1)$. Of course, we assume here that Σ is minimal, i.e., all of its letters occur in w_h.

The classification of morphisms is based on their growth properties as presented in [SaSo] and [RoSa1]. For a letter $a \in \Sigma$ we consider the function $h_a : \mathbb{N} \to \mathbb{N}$ defined by

$$h_a(n) = |h^n(a)| \quad \text{for} \quad n \geq 0.$$

It follows that there exists a nonnegative integer e_a and an algebraic real number ρ_a such that

$$h_a(n) = \theta(n^{e_a} \rho_a^n),$$

the pair (e_a, ρ_a) being referred to as the *growth index* of a in h.

The set Σ_B of so-called bounded letters plays an important role in the classification. A letter a is called *bounded* if, and only if, the function h_a is so, i.e., its growth index equals either to $(0, 0)$ or $(0, 1)$. We say that h is

nongrowing, if there exists a bounded letter in Σ;
quasi-uniform, if $\rho_a = \rho_b > 1$ and $e_a = e_b = 1$ for each $a, b \in \Sigma$;
polynomially diverging, if $\rho_a = \rho_b > 1$ for each $a, b \in \Sigma$, and $e_a > 1$ for some $a \in \Sigma$;

exponential diverging, if $\rho_a > 1$ for each $a \in \Sigma$, and $\rho_a > \rho_b$ for some $a, b \in \Sigma$.

It is not difficult to conclude, cf. [SaSo] or [RoSa1], that this classification is both exhaustive and unambiguous, i.e., each morphism is in exactly one of these classes. In particular, if we call a morphism *growing*, whenever h_a is unbounded for each $a \in \Sigma$, then the three last notions define a partition on the set of growing morphisms.

The above classification is constructive in the sense that for a given morphism we can decide which of the above types it is. Indeed, the growth index of a letter a can be effectively computed, as well as the questions "$q_a > 1$?" and "$\rho_a > \rho_b$?" can be effectively answered. Details needed to conclude these observations can be found in [SaSo].

Now we are ready for the classification proved in [Pan2]. Unfortunately, it does not depend only on the type of the morphism h, but also on the distribution of the bounded letters in w_h. Even worsely, the complexity can be the smallest possible, namely $\theta(1)$, in each of the four cases, since ultimately periodic words can be fixed points of morphisms of any of the above types. However, as we already mentioned, it is decidable whether an iterated morphism defines an ultimately periodic word.

Proposition 9.1. *Let h be a growing iterated morphism. Then if w_h is not ultimately periodic, its complexity is either $\theta(n)$, $\theta(n \log \log n)$ or $\theta(n \log n)$ depending on whether h is quasi-uniform, polynomially diverging or exponentially diverging, respectively.*

The case of nongrowing morphisms is more complicated, essentially due to the fact that this notion is defined existentially, i.e., a morphism is nongrowing whenever there exists a bounded letter.

Proposition 9.2. *Let h be a nongrowing (not necessarily 1-free) iterated morphism generating a non-ultimately periodic word w_h. Then*

(i) if w_h contains arbitrarily long factors over Σ_B, the complexity of w_h is $\theta(n^2)$;

(ii) if all factors of w_h over Σ_B are shorter than a constant K, the complexity of w_h is that of one of the cases in Proposition 9.1, namely $\theta(n)$, $\theta(n \log \log n)$ or $\theta(n \log n)$, and moreover it is decidable which of these it is.

Propositions 9.1 and 9.2 together with our earlier remarks yield immediately the following important results.

Corollary 9.2. *The asymptotic analysis problem for (not necessarily 1-free) iterated morphisms is decidable.*   □

Corollary 9.3. *The asymptotic synthesis problem for the complexities $\theta(1)$, $\theta(n)$, $\theta(n \log \log n)$, $\theta(n \log n)$ and $\theta(n^2)$ can be solved.*   □

Detailed proofs of Propositions 9.1 and 9.2 can be found in [Pan2]. Here we outline two basic observations of the proofs, as well as give an example of a morphism of each of the above types, and compute the complexities of the corresponding words.

A proof of the fact that the complexity of w_h, for any h, is at most quadratic is not difficult, cf. [ELR]. To see this let us fix n and consider a factor v of w_h of length n. First assume that v is derived in one step from a word v' containing at least one unbounded letter, i.e., the considered (occurrence of) v is a factor in $h(v')$. Let v' be as short as possible and denote $v_0 = v$ and $v_1 = v'$. Obviously, v_1 satisfies automatically our requirement for v, so that we can define inductively $v_0, v_1, v_2, \ldots$ up to v_k with $v_k \in \Sigma$. It follows that $k \leq \|\Sigma\| \cdot n$. Therefore all the factors of length n satisfying our above restriction can be found among the factors of $h(v)$, for $j = k, \ldots, 1$. There are at most $\mathcal{O}(n^2)$ such factors. To cover all the factors of length n, it is enough to note that, for any $v \in \Sigma_B^*$, the language $\{h^i(v) | i \geq 0\}$ contains at most K words for some finite K independent of v. Therefore $\mathcal{O}(n^2)$ is also a valid upper bound for all factors of length n.

Our second remark concerns case (ii) in Proposition 9.2. In [Pan2] this is concluded as follows. Now the factors of w_h in Σ_B^* are shorter than a fixed constant, say K. In particular, each factor v of w_h longer than K contains a growing letter, and therefore for some i independent of v, the word $h^i(v)$ is longer than K. Hence, replacing h by its suitable power, and considering that as a morphism which maps factors of lengths from $K + 1$ to $2K$ into words of factors of these lengths, we can eliminate the bounded letters. Let h' be a new morphism constructed in this way. It follows that h' is growing, and moreover, generates as an iterated morphism a word which consists of certain consecutive factors of w_h. Hence, the original w_h can be recovered from the word $w_{h'}$ by using a 1-free morphism mapping the above factors to the corresponding words of Σ^*. Consequently, the word $w_{h'}$ is nothing but a representation of w_h in a larger alphabet, and therefore the asymptotic complexities of w_h and $w_{h'}$ coincide. This explains how case (ii) in Proposition 9.2 is reduced to Proposition 9.1.

As we already said instead of proving Proposition 9.1 and case (i) in Proposition 9.2, we only analyse one example in each of the complexity classes. First, any ultimately periodic word is a fixed point of an iterated morphism yielding the complexity $\theta(1)$. Second, the Fibonacci word w_F of Example 9.2 has the complexity $\theta(n)$, and indeed the morphism is quasi-uniform with $\rho_a = \rho_b = \frac{1}{2}(1 + \sqrt{5})$. The remaining cases are covered in Examples 9.6–9.8.

Example 9.6. Consider the morphism h defined by $h(a) = aba$ and $h(b) = bb$. Now h is polynomially diverging since

$$|h^i(a)| = (\frac{1}{2}i + 1)2^i \text{ and } |h^i(b)| = 2^i \text{ for } i \geq 0.$$

To prove that $g_{w_h}(n) = \theta(n \log \log n)$ we first note that under the interpretation $a \leftrightarrow 0$ and $b^{2^i} \leftrightarrow i+1$ the word $h^i(a)$ equals to the so-called ith *sesquipower* s_i defined recursively by

$$s_0 = 0,$$
$$s_{i+1} = s_i(i+1)s_i \quad \text{for } i \geq 0.$$

This means that $h^i(a)$ can be described as

$$h^i(a) = \overbrace{s_1 2 s_1 3 s_1 2 s_1}^{s_3} \underbrace{4 s_3}_{} 5 s_4 \ldots s_{i-2} \, i s_{i-1}.$$

with s_4 and s_{i-1} braces below.

We fix integer $n \geq 2$ and choose $i_0 = \lceil \log n - \log \log n \rceil + 2$, where logarithms are at base 2. Then we have

$$|s_{i_0}| \leq i_0 2^{i_0} \leq (\log n + 3) 2^{\log n - \log \log n + 3} \leq (\log n + 3)\frac{8n}{\log n} \leq 32n.$$

Consider now factors of length n occurring in w_h such that they overlap with, or contain as a factor, the first occurrence of i, i.e., b^{2^i}, in w_h. Clearly, any factor of w_h of length n is among these factors for some $i \leq \lfloor \log n \rfloor$. Since, for each i, there are at most $n + 2 \cdot 2^i$ such factors we have

$$g_{w_h}(n) \leq |s_{i_0}| + \sum_{i=i_0+1}^{\lfloor \log n \rfloor} (n + 2 \cdot 2^i) \leq 32n + \sum_{i=i_0+1}^{\lfloor \log n \rfloor} 3n = \mathcal{O}(n \log \log n).$$

On the other hand, of the above factors at least $n - 2^i$, for $i = i_0, \ldots, \lceil \log n \rceil$, are such that they do not occur earlier in w_h. Therefore we also have

$$g_{w_h}(n) \geq \sum_{i=i_0}^{\lceil \log n \rceil} (n - 2^i) \geq \sum_{i=i_0}^{\lfloor \log n \rfloor - 1} \frac{n}{2} = \Omega(n \log \log n).$$

So we have proved that $g_{w_h}(n) = \theta(n \log \log n)$.   □

Example 9.7. Consider the morphism defined by $h(\$) = \$ab$, $h(a) = aa$ and $h(b) = bbb$. Then $|h^i(a)| = 2^i$, $|h^i(b)| = 3^i$ and $\rho_\$ > 1$, so that h is exponentially diverging. Denote

$$\alpha(i) = \$aba^2b^3 \ldots a^{2^i}b^{3^i} \in \text{pref}(w_h).$$

Clearly each factor of w_h of length n occurs in $\alpha(\lfloor \log_3(n) \rfloor)$, so we obtain

$$g_{w_h}(n) \leq |\alpha(\lfloor \log_3 n \rfloor)| = 1 + \sum_{i=0}^{\lfloor \log_3 n \rfloor} (2^i + 3^i) = \mathcal{O}(n \log n).$$

On the other hand, for $i = \lceil \log_3 n \rceil, \ldots, \lfloor \log_2 n \rfloor$, w_h contains at least

$$\sum_{i=\lceil \log_3 n \rceil}^{\lfloor \log_2 n \rfloor} (n - 2^i) \geq \Omega(n \log n)$$

different factors in $b^+ a^+ b^* \cup b^* a^+ b^+$. Therefore we have concluded that $g_{w_h}(n) = \Omega(n \log n)$. □

Example 9.8. Finally consider the word

$$w = abcbccbccc \ldots bc^n \ldots,$$

which is the fixed point of the morphism defined as $h(a) = abc$, $h(b) = bc$ and $h(c) = c$. So h is nongrowing and w contains unboundedly long factors in b^*. Let $\alpha(i) = h^i(a)$. Now, all the factors of w of length n occur in the prefix $\alpha(n + 1)$. On the other hand, all factors of $\alpha(\lceil \frac{n}{2} \rceil 1)$ of length n are different. Therefore the estimate

$$|\alpha(i)| = 1 + \sum_{j=0}^{i}(1 + j) = \theta(i^2)$$

shows that $g_{w_h}(n) = \theta(n^2)$. □

Our above classification can be straightforwardly modified to D0L languages, i.e., to the language of the form $\{h^i(w) \mid i \geq 0\}$, where h is a morphism and w is a finite word, cf. [RoSa1]. Indeed each iterated morphism h, with $a \in \text{pref}(h(a))$, defines a D0L language via the pair (h, a), and each pair (h, w) determines an iterated morphism h' as an extention of h defined by $h'(\$) = \$w$, where $\$$ is a new letter. The classification of complexities of D0L languages leads exactly to the above five classes – although the transformation $(h, w) \rightarrow h'$ might change the class.

Acknowledgement. The authors are grateful to J. Berstel, S. Grigorieff, T. Harju, F. Mignosi, J. Sakarovitch and A. Restivo for useful discussions during the preparation of this work.

References

[Ab1]   H. Abdulrab, Résolution d'équations en mots: étude et implémentation LISP de l'algorithme de Makanin, Ph.D. Thesis, Université de Rouen, 1987.

[Ab2]   H. Abdulrab, Implementation of Makanin's algorithm, Lecture Notes in Computer Science **572**, Springer-Verlag, 1991.

[ACK]   J. Albert, K. Culik II and J. Karhumäki, Test sets for context-free languages and algebraic systems of equations in a free monoid, Inform. Control **52**, 172–186, 1982.

[AL1]   M. H. Albert and J. Lawrence, A proof of Ehrenfeucht's Conjecture, Theoret. Comput. Sci. **41**, 121–123, 1985.

[AL2]   M. H. Albert and J. Lawrence, The descending chain condition on solution sets for systems of equations in groups, Proc. Edinburg Math. Soc. **29**, 69–73, 1985.

[Al]   J.-P. Allouche, Sur la complexité des suites infinies, Bull. Belg. Math. Soc. **1**, 133–143, 1994.

[BMT]   K. A. Baker, G. F. McNulty and W. Taylor, Growth problems for avoidable words, Theoret. Comput. Sci. **69**, 319–345, 1989.

[BEM]   D. R. Bean, A. Ehrenfeucht and G. F. McNulty, Avoidable patterns in strings of symbols, Pacific J. Math. **85**, 261–294, 1979.

[Be1]   J. Berstel, Transductions and Context-Free Languages, Teubner, 1979.

[Be2]   J. Berstel, Mots sans carré et morphismes itérés, Discr. Math. **29**, 235–244, 1979.

[Be3]   J. Berstel, Sur les Mots sans carrés définis par morphisme, Lecture Notes in Computer Science **71**, 16–25, Springer-Verlag, 1979.

[Be4]   J. Berstel, Some recent results on squarefree words, Lecture Notes in Computer Science **166**, 17–25, Springer-Verlag, 1984.

[Be5]   J. Berstel, Properties of infinite words, Lecture Notes in Computer Science **349**, 36–46, Springer-Verlag, 1989.

[Be6]   J. Berstel, Axel Thue's work on repetitions in words, Proc. 4th FPSAC, Montreal, 1992; also LITP Report **70**.

[Be7]   J. Berstel, A rewriting of Fife's Theorem about overlap-free words, Lecture Notes in Computer Science **812**, 19–29, Springer-Verlag, 1994, 1992.

[Be8]   J. Berstel, Axel Thue's papers on repetitions in words: a translation, Publications du Laboratoire de Combinatoire et d'Informatique Mathematique, Université du Québec à Montréal **20**, 1995.

[BdL]   J. Berstel and A. de Luca, Sturmian words, Lyndon words and trees, LITP Report **24**, 1995.

[BePe]   J. Berstel and D. Perrin, Theory of Codes, Academic Press, 1985.

[BPPR]   J. Berstel, D. Perrin, J. F. Perrot and A. Restivo, Sur le Théorème du défaut, J. Algebra **60**, 169–180, 1979.

[BePo]   J. Berstel and M. Pocchiola, Average cost of Duval's algorithm for generating Lyndon words, LITP Report **23**, 1992.

[Bra]   F.-J. Brandenburg, Uniformly growing k-th powerfree homomorphisms, Theoret. Comput. Sci. **23**, 69–82, 1989.

[Bri]   J. Brinkhuis, Non-repetitive sequences on three symbols, Quart. J. Math. Oxford **34**, 145–149, 1983.

[Bro]   T. C. Brown, Descriptions of the characteristic sequence of an irrational, Canad. Math. Bull. **36**, 15–21, 1993.

[BS]   R. Büchi and S. Senger, Coding in the existential theory of concatenation, Arch. Math. Logik **26**, 101–106, 1986/87.

[Car1]   A. Carpi, Overlap-free words and finite automata, Theoret. Comput. Sci. **115**, 243–260, 1993.

[Car2]   A. Carpi, On the number of abelian square-free words on four letter alphabet, manuscript, 1994.

[Cas1]   J. Cassaigne, Counting overlap-free binary words, Lecture Notes in Computer Science **665**, 216–225, Springer-Verlag, 1993.

[Cas2]   J. Cassaigne, Unavoidable binary patterns, Acta Informatica **30**, 385–395, 1993.

[Cas3]   J. Cassaigne, Motifs évitables et régularité dans les mots, Thèse de Doctorat, Université Paris VI, 1994.

[CaKa] J. Cassaigne and J. Karhumäki, Toeplitz words, generalized periodicity and periodically iterated morphisms, manuscript 1995; also Lecture Notes in Computer Science **959**, 244–253, Springer-Verlag, 1995, also Europ. J. Combinatorics (to appear).

[CV] Y. Césari and M. Vincent, Une caractérisation des mots périodiques, C. R. Acad. Sci. Paris **286** A, 1175–1177, 1978.

[Ch] C. Choffrut, Bijective sequential mappings of a free monoid onto another, RAIRO Theor. Inform. Appl. **28**, 265–276, 1994.

[CC] C. Choffrut and K. Culik II, On Extendibility of Unavoidable Sets. Discr. Appl. Math. **9**, 125–137, 1984.

[Co] P. M. Cohn, Algebra, Vol 2, John Wiley and Sons, 1989.

[CF] R. Cori and M. R. Formisano, On the number of partially abelian square-free words on a three-letter alphabet, Theoret. Comput. Sci. **81**, 147–153, 1991.

[CH] E. M. Coven and G. A. Hedlund, Sequences with minimal block growth, Math. Syst. Theory **7**, 138–153, 1973.

[Cr] M. Crochemore, Sharp characterizations of squarefree morphisms, Theoret. Comput. Sci. **18**, 221–226, 1982.

[CP] M. Crochemore and D. Perrin, Two-way string matching, J. ACM **38**, 651–675, 1991.

[CR] M. Crochemore and W. Rytter, Text Algorithms, Oxford University Press, 1994.

[CuKa1] K. Culik II and J. Karhumäki, Systems of equations and Ehrenfeucht's conjecture, Discr. Math. **43**, 139–153, 1983.

[CuKa2] K. Culik II and J. Karhumäki, On the equality sets for homomorphisms on free monoids with two generators, RAIRO Theor. Informatics **14**, 349–369, 1980.

[CKL] K. Culik II, J. Karhumäki and A. Lepistö, Alternating iteration of morphisms and the Kolakoski sequence, in: G. Rozenberg and A. Salomaa (eds.): Lindenmayer Systems, Springer-Verlag, 93–106, 1992.

[CS] K. Culik II and A. Salomaa, On infinite words obtained by iterating morphisms, Theoret. Comput. Sci. **19**, 29–38, 1982.

[Da] M. Davis, Hilbert's tenth problem is undecidable, Amer. Math. Monthly **80**, 233–269, 1973.

[Dej] F. Dejean, Sur un Théorème de Thue, J. Comb. Theor, Ser. A **13**, 90–99, 1972.

[Dek] F. M. Dekking, Strongly non-repetitive sequences and progression-free sets, J. Comb. Theor., Ser. A **27**, 181–185, 1979.

[dL] A. de Luca, Sturmian words: Structure, combinatorics and their arithmetics, Theoret. Comput. Sci., (to appear).

[dLM] A. de Luca and Filippo Mignosi, Some combinatorial properties of Sturmian words, Theoret. Comput. Sci. **136**, 361–385, 1994.

[Do] E. Domenjoud, Solving Systems of Linear Diophantine Equations: An Algebraic Approach, Lecture Notes in Computer Science **520**, Springer-Verlag, 1991.

[DG] S. Dulucq and D. Gougou-Beauchamps, Sur les facteurs des suites de Sturm, Theoret. Comput. Sci. **71**, 381–400, 1990.

[Du1] J. P. Duval, Périodes et répétitions des mots de monoïde libre, Theoret. Comput. Sci. **9**, 17–26, 1979.

[Du2] J. P. Duval, Factorizing words over an ordered alphabet, J. Algorithms **4**, 363–381, 1983.

[EHR] A. Ehrenfeucht, D. Haussler and G. Rozenberg, On Regularity of Context-free Languages. Theoret. Comput. Sci. **27**, 311–322, 1983.

[EKR1] A. Ehrenfeucht, J. Karhumäki and G. Rozenberg, The (generalized) Post Correspondence Problem with lists of consisting of two words is decidable, Theoret. Comput. Sci. **21**, 119–144, 1982.

[EKR2] A. Ehrenfeucht, J. Karhumäki and G. Rozenberg, On binary equality languages and a solution to the test set conjecture in the binary case, J. Algebra **85**, 76–85, 1983.

[ELR]   A. Ehrenfeucht, K. P. Lee and G. Rozenberg, Subword complexities of various classes of deterministic developmental languages without interactions, Theoret. Comput. Sci. **1**, 59–75, 1975.

[ER1]   A. Ehrenfeucht and G. Rozenberg, Elementary homomorphisms and a solution to the D0L sequence equivalence problem, Theoret. Comput. Sci. **7**, 169–183, 1978.

[ER2]   A. Ehrenfeucht and G. Rozenberg, On the subword complexity of square-free D0L-languages, Theoret. Comput. Sci. **16**, 25–32, 1981.

[ER3]   A. Ehrenfeucht and G. Rozenberg, On the subword complexity of D0L-languages with a constant distribution, Inform. Proc. Letters **13**, 108–113, 1981.

[ER4]   A. Ehrenfeucht and G. Rozenberg, On the subword complexity of locally catenative D0L-languages, Inform. Proc. Letters **16**, 121–124, 1983.

[Ei]   S. Eilenberg, Automata, Languages and Machines, vol. A, Academic Press, 1974.

[Ev]   A. A. Evdokimov, Strongly asymmetric sequences generated by a finite number of symbols, Dokl. Akad. Nauk. SSSR **179**, 1268–1271, 1968 (English transl. Soviet Math. Dokl. **9**, 536–539, 1968).

[F]   E. D. Fife, Binary sequences which contain no BBb, Trans. Amer. Math. Soc. **261**, 115–136, 1980.

[FW]   N. J. Fine and H. S. Wilf, Uniqueness theorem for periodic functions, Proc. Am. Math. Soc. **16**, 109–114, 1965.

[GJ]   M. R. Garey and D. S. Johnson, Computers and Intractability: A Guide to the Theory of NP-Completeness, Freeman, 1979.

[GK]   P. Goralčik and V. Koubek, On discerning words by automata, Lecture Notes in Computer Science **226**, 116–122, Springer-Verlag, 1986.

[GV]   P. Goralčik and T. Vaniček, Binary patterns in binary words, Intern. J. Algebra Comput. **1**, 387–391, 1991.

[GR]   S. Ginsburg and G. F. Rosen, A characterization of machine mappings, Can. J. Math. **18**, 381–388, 1966.

[GO]   L. J. Guibas and A. M. Odlyzko, Periods in strings, J. Comb. Theory, Ser. A **30**, 19–42, 1981.

[HW]   G. H. Hardy and E. M. Wright, An Introduction to the Theory of Numbers, 1959.

[Harj] T. Harju, On cyclically overlap-free words in binary alphabets, in: G. Rozenberg and A. Salomaa (eds.), The Book of L, Springer-Verlag, 123–130, 1986.

[HK1]   T. Harju and J. Karhumäki, On the defect theorem and simplifiability, Semigroup Forum **33**, 199–217, 1986.

[HK2]   T. Harju and J. Karhumäki, Morphisms, in this Handbook.

[HKK]   T. Harju, J. Karhumäki and D. Krob, Remarks on generalized Post Correspondence Problem, manuscript 1995; also in Proc. of STACS 95, Lecture Notes in Computer Science **1046**, 39–48, Springer-Verlag, 1996.

[HKP]   T. Harju, J. Karhumäki and W. Plandowski, Compactness of systems of equations in semigroups, Lecture Notes in Computer Science **944**, 444–454, Springer-Verlag, 1995, also Int. J. Algebra Comb. (to appear).

[HL]   T. Harju and M. Linna, On the periodicity of morphisms on free monoids, RAIRO Theor. Inform. Appl. **20**, 47–54, 1986.

[Harr]   M. Harrison, Introduction to Formal Language Theory, Addison-Wesley, 1978.

[Hi]   G. Higman, Ordering with divisibility in abstract algebras, Proc. London Math. Soc. **3**, 326–336, 1952.

[Hm]   Y. I. Hmelevskii, Equations in free semigroups, Proc. Steklov Inst. Math. **107**, 1971; Amer. Math. Soc. Translations, 1976.

[HU]   J. E. Hopcroft and J. D. Ullman, Introduction to Automata Theory, Languages and Computation, Addison-Wesley, 1979.

[Jo]   J. H. Johnson, Rational equivalence relations, Lecture Notes in Computer Science **226**, 167–177, Springer-Verlag, 1986.

[Ju]   J. Justin, Characterization of the repetitive commutative semigroups, J. Algebra **21**, 87–90, 1972.

[Ka1]   J. Karhumäki, On cube-free ω-words generated by binary morphisms, Discr. Appl. Math. **5**, 279–297, 1983.

[Ka2]   J. Karhumäki, A note on intersections of free submonoids of a free monoid, Semigroup Forum **29**, 183–205, 1984.

[Ka3]   J. Karhumäki, The Ehrenfeucht Conjecture: A compactness claim for finitely generated free monoids, Theoret. Comput. Sci. **29**, 285–308, 1984.

[KRJ]   J. Karhumäki, W. Rytter and S. Jarominek, Efficient construction of test sets for regular and context-free languages, Theoret. Comput. Sci. **116**, 305–316, 1993.

[KaPl1]   J. Karhumäki and W. Plandowski, On the size of independent systems of equations in semigroups, Theoret. Comput. Sci., (to appear).

[KaPl2]   J. Karhumäki and W. Plandowski, On defect effect of many identities in free semigroups, in: G. Paun(ed.), Mathematical Aspects of Natural and Formal Languages, World Scientific, 225–232, 1994.

[KPR]   J. Karhumäki, W. Plandowski and W. Rytter, Polynomial size test sets for context-free languages, J. Comput. System Sci. **50**, 11–19, 1995.

[Ke1]   V. Keränen, On the k-freeness of morphisms on free monoids, Ann. Acad. Sci. Fenn. Ser. A I Math. Dissertationes **61**, 1986.

[Ke2]   V. Keränen, Abelian squares are avoidable on 4 letters, Lecture Notes in Computer Science **623**, 41–52, Springer-Verlag, 1992.

[Kob]   Y. Kobayashi, Enumeration of irreducible binary words, Discr. Appl. Math. **20**, 221–232, 1988.

[Kol]   W. Kolakoski, Self generating runs, Problem 5304, Amer. Math. Monthly **72**, 674, 1965; Solution by N. Ücoluk, Amer. Math. Monthly **73**, 681–682, 1966.

[KoPa]   A. Koscielski and L. Pacholski, Complexity of Makanin's algorithm, JACM **43** (1996).

[Kos]   M. Koskas, Complexités de suites de Toeplitz, Discr. Math., (to appear).

[Kr]   J. B. Kruskal, The Theory of Well-Quasi-Ordering: A Frequently Discovered Concept, J. Combin. Theory Ser. A **13**, 297–305, 1972.

[La]   G. Lallement, Semigroups and Combinatorial Applications, Wiley ,1979.

[LeC]   M. Le Conte, A charcterization of power-free morphisms, Theoret. Comput. Sci. **38**, 117–122, 1985.

[Len]   A. Lentin, Equations dans les Monoides Libres, Gauthier-Villars, 1972.

[LeSc]   A. Lentin and M. P. Schützenberger, A combinatorial problem in the theory of free monoids, in: R. C. Bose and T. E. Dowling (eds.), Combinatorial Mathematics, North Carolina Press, 112–144, 1967.

[Lep1]   A. Lepistö, Repetitions in Kolakoski sequence, in: G. Rozenberg and A. Salomaa (eds.), Developments in Language Theory, World Scientific, 130–143, 1994.

[Lep2]   A. Lepistö, Master's thesis, University of Turku, 1995

[Lev]   F. W. Levi, On semigroups, Bull. Calcuta Math. Soc. **36**, 141–146, 1944.

[Lo]   M. Lothaire, Combinatorics on Words, Addison-Wesley, 1983.

[LySc]   R. C. Lyndon and M. P. Schützenberger, The equation $a^m = b^n c^p$ in a free group, Michigan Math J. **9**, 289–298, 1962.

[McN1]   R. MacNaughton, A decision procedure for generalized mappability-onto of regular sets, manuscript.

[McN2]   R. MacNaughton, A proof of the Ehrenfeucht conjecture, Informal memorandum, 1985.

[Mak]   G. S. Makanin, The problem of solvability of equation in a free semigroup, Mat. Sb. **103**, 147–236, 1977 (English transl. in Math USSR Sb. **32**, 129–198).

[Marc]   S. S. Marchenkov, Undecidability of the $\forall\exists$-positive Theory of a free Semigroup, Sibir. Mat. Journal **23**, 196–198, 1982.

[Mart]   U. Martin, A note on division orderings on strings. Inform. Proc. Letters **36**, 237–240, 1990.

[Mat]   Y. Matiyasevich, Enumerable sets are diophantine, Soviet Math. Doklady **11**, 354–357, 1970 (English transl. Dokl. Akad. Nauk. SSSR **191**, 279–282, 1971).

[MN]   H. A. Maurer and M. Nivat, Rational bijections of rational sets, Acta Informatica **13**, 365–378, 1980.

[McC]   E. M. McCreight, A space-economical suffix tree construction algorithm, J. ACM **23**, 262–272, 1976.

[Mi]   F. Mignosi, On the number of factors of Sturmian words, Theoret. Comput. Sci. **82**, 71–84, 1991.

[MP]   F. Mignosi and G. Pirillo, Repetitions in the Fibonacci infinite word, RAIRO Theor. Inform. Appl. **26**, 199–204, 1992.

[MRS]   F. Mignosi, A. Restivo and S. Salemi, A periodicity theorem on words and applications, Lecture Notes in Computer Science **969**, 337–348, Springer-Verlag, 1995.

[Mor1]   M. Morse, Recurrent geodesics on a surface of negative curvature, Trans. Am. Math. Soc. **22**, 84–100, 1921.

[Mor2]   M. Morse, A solution of the problem of infinite play in chess, Bull. Am. Math. Soc. **44**, 632, 1938.

[MH]   M. Morse and G. Hedlund, Symbolic dynamics, Am. J. Math. **60**, 815–866, 1938.

[Mou]   J. Moulin-Ollagnier, Proof of Dejean's conjecture for alphabets with 5, 6, 7, 8, 9, 10 and 11 letters, Theoret. Comput. Sci. **95**, 187–205, 1992.

[MS]   A. A. Muchnik and A. L. Semenov, Jewels of Formal Languages (Russian translation of [Sal2]), Mir, 1986.

[Ne1]   J. Néraud, Elementariness of a finite set of words is co-NP-complete, RAIRO Theor. Inform. Appl. **24**, 459–470, 1990.

[Ne2]   J. Néraud, On the rank of the subset a free monoid, Theoret. Comput. Sci. **99**, 231–241, 1992.

[Ne3]   J. Néraud, Deciding whether a finite set of words has rank at most two, Theoret. Comput. Sci. **112**, 311–337, 1993.

[Ni]   J. Nielsen, Die Isomorphismengruppe der freien Gruppen, Math. Ann. **91**, 169–209.

[Pan1]   J.-J. Pansiot, A propos d'une conjecture de F. Dejean sur les répétitions dans les mots, Discr. Appl. Math. **7**, 297–311, 1984.

[Pan2]   J.-J. Pansiot, Complexité des facteurs des mots infinis engendrés par morphismes itérés, Lecture Notes in Computer Science **172**, 380–389, Springer-Verlag, 1984.

[Pan3]   J.-J. Pansiot, Decidability of periodicity for infinite words, RAIRO Theor. Inform. Appl. **20**, 43–46, 1986.

[Pav]   V. A. Pavlenko, Post Combinatorial Problem with two pairs of words, Dokladi AN Ukr. SSR **33**, 9–11, 1981.

[Pec]   J. P. Pécuchet, Solutions principales et rang d'un système d'équations avec constantes dans le monoïde libre, Discr. Math. **48**, 253–274, 1984.

[Per]   D. Perrin, On the solution of Ehrenfeucht's Conjecture, Bull. EATCS **27**, 68–70, 1985.

[Pl]   P. A. Pleasants, Non repetitive sequences, Mat. Proc. Cambridge Phil. Soc. **68**, 267–274, 1970.

[Pr]   M. E. Prouhet, Mémoire sur quelques relations entre les puissances des numbres, C. R. Acad. Sci. Paris. **33**, Cahier **31**, 225, 1851.

[Ra]   G. Rauzy, Mots infinis en arithmétique, Lecture Notes in Computer Science **95**, 165–171, Springer-Verlag, 1984.

[ReSa]   A. Restivo and S. Salemi, Overlap-free words on two symbols, Lecture Notes in Computer Science **192**, 198–206, Springer-Verlag, 1984.

[Rob1]   J. M. Robson, Separating Strings with Small Automata, Inform. Proc. Letters **30**, 209–214, 1989.

[Rob2]   J. M. Robson, Separating Words with Machines and Groups (to appear), 1995.

[Ros1]   L. Rosaz, Making the inventory of unavoidable sets of words of fixed cardinality, Ph.D. Thesis, Université de Paris 7, 1992.

[Ros2]   L. Rosaz, Unavoidable Languages, Cuts and Innocent Sets of Words, RAIRO Theor. Inform. Appl. **29**, 339–382, 1995.

[Rot]   P. Roth, Every binary pattern of length six is avoidable on the two-letter alphabet, Acta Informatica **29**, 95–107, 1992.

[RoSa1] G. Rozenberg and A. Salomaa, The Mathematical Theory of L Systems, Academic Press, 1980.

[RoSa2] G. Rozenberg and A. Salomaa, Cornerstones of Undecidability, Prentice Hall, 1994.

[Sal1]   A. Salomaa, Formal Languages, Academic Press, 1973.

[Sal2]   A. Salomaa, Jewels of Formal Languages, Computer Science Press, 1981.

[Sal3]   A. Salomaa, The Ehrenfeucht Conjecture: A proof for language theorists, Bull. EATCS **27**, 71–82, 1985.

[SaSo]   A. Salomaa and M. Soittola, Automata-Theoretic Aspects of Formal Power Series, Springer, 1978.

[Sap]   M. Sapir, Combinatorics on words with applications, LITP Report **32**, 1995.

[Sc]   U. Schmidt, Avoidable patterns on two letters, Theoret. Comput. Sci. **63**, 1–17, 1985.

[See]   P. Séébold, Sequences generated by infinitely iterated morphisms, Discr. Appl. Math. **11**, 93–99, 1985.

[Sei]   S. Seibert, Quantifier Hierarchies over Word Relations, in: E. Boerger et al. (eds.), Computer Science Logic, Lecture Notes in Computer Science **626**, 329–338, Springer-Verlag, 1992.

[She]   R. Shelton, Aperiodic words on three symbols I, J. Reine Angew. Math **321**, 195–209, 1981.

[ShSo1] R. Shelton and R. Soni, Aperiodic words on three symbols II, J. Reine Angew. Math **327**, 1–11, 1981.

[ShSo2] R. Shelton and R. Soni, Aperiodic words on three symbols III, J. Reine Angew. Math **330**, 44–52, 1982.

[Shy]   H. J. Shyr, Free Monoids and Languages, Hon Min Book Company, Taiwan, 1991.

[Si]   I. Simon, An Algorithm to Distinguish Words Efficiently by Their Subwords, manuscript, 1983.

[SkSe]   D. Skordev and Bl. Sendov, On equations in words, Z. Math. Logic Grundlagen Math. **7**, 289–297, 1961.

[Sl]   N. J. A. Sloane, A Handbook of Integer Sequences, Academic Press, 1973.

[Sp1]   J.-C. Spehner, Quelques problèmes d'extension, de conjugaison et de presentation des sous-monoïdes d'un monoïde libre, Ph.D. Thesis, Université Paris VII, 1976.

[Sp2]   J.-C. Spehner, Quelques constructions et algorithmes relatifs aux sous-monoïdes d'un monoïde libre, Semigroup Forum **9**, 334–353, 1975.

[T1]   A. Thue, Über unendliche Zeichenreihen, Kra. Vidensk. Selsk. Skrifter. I. Mat.-Nat. Kl., Christiana, Nr. **7**, 1906.

[T2]   A. Thue, Über die gegenseitige Lage gleicher Teile gewisser Zeichenreihen, Kra. Vidensk. Selsk. Skrifter. I. Mat.-Nat. Kl., Christiana, Nr. **12**, 1912.

[Z]   A. I. Zimin, Blocking sets of terms, Math USSR Sb. **47**, 353–364, 1984.

Morphisms

Tero Harju and Juhani Karhumäki

1. Introduction

The notion of a *homomorphism*, or briefly a *morphism* as we prefer to call it, is one of the fundamental concepts of mathematics. Usually a morphism is considered as an *algebraic* notion reflecting similarities between algebraic structures. The most important algebraic structure of this survey is that of a finitely generated *word monoid*, a *free monoid* that is generated by a finite alphabet of letters. Our motivation comes from formal language theory, and therefore the *combinatorial* aspect will be stressed more than the algebraic aspect of morphisms.

Collaborating morphisms will have a dominant role in our considerations. This reflects the *computational* aspect of morphisms, where morphisms of word monoids are used to simulate or even define computational processes. The computational power of two collaborating morphisms was first discovered by E. Post in his seminal paper [P1] of 1946, when he introduced the first simple algorithmically unsolvable combinatorial problem, subsequently referred to as the *Post Correspondence Problem*. The undecidability of this problem is due to the fact that pairs of morphisms can be used to simulate general algorithmic computations.

Later on, and in particular during the past two decades, research on different problems involving morphisms of free monoids has been very active and, we believe, also successful. Indeed, this research has revealed a number of results that are likely to have a lasting value in mathematics. We shall consider some of these results in the present article. Foremost we have, besides several variants of the Post Correspondence Problem, a compactness property of morphisms of free monoids originally conjectured by A. Ehrenfeucht at the beginning of the 1970s, and a morphic characterizations of recursively enumerable languages.

Our goal is to give an overview of the results involving *computational aspects of morphisms* of word monoids. This means that we have overlooked several important research topics on morphisms such as repetition-free words (which are almost exclusively generated by iterating morphisms) and algebraic theory of automata (which essentially uses morphisms as algebraic tools). These topics are considered in other chapters of this Handbook.

In the presentation we have paid special attention on results that have not yet appeared in standard text books. We also have given proofs to some well-known theorems, such as the Post Correspondence Problem, if the proof is different from those in standard text books. Finally, we have made a special effort to list open problems of the present topic. These problems clearly indicate that there remains a plenty of space for interesting research on morphisms. Indeed, typically these problems are easily formulated, but apparently difficult to solve.

Next we shall briefly describe the problems considered here. We start in Section 3 by considering some decidable cases of the Post Correspondence Problem, or PCP. In particular, we outline the proof of the decidability of PCP in the binary case. In Section 4 we deduce the undecidability of PCP from that of the word problem for semigroups and semi-Thue systems. These word problems are discussed in Preliminaries. Here the word problem is chosen not only to obtain less standard proofs, but mainly to obtain sharper undecidable variants of PCP. At the end of the section we give several applications of PCP to different kinds of matrix problems.

In Section 5 we introduce the equality sets, which are defined as the sets of all words that are mapped to a same word by two fixed morphisms. Besides several basic properties we concentrate on those cases where the equality set is regular. For example, we recall a surprising result that for prefix codes the equality set is always regular, but still no finite automaton accepting this language can be found effectively.

Section 6 deals with systems of equations in a word monoid Σ^*. Of course, a solution of a system of equations is just a morphism from the free monoid generated by the variables into Σ^*. We do not consider here methods of solving systems of equations; indeed, that would have been a topic of a whole chapter in this Handbook. Instead we have assumed Makanin's algorithm, which gives a procedure to decide whether an equation possesses a solution and consider its consequences. Furthermore, we concentrate to another fundamental property of systems of equations. Namely, it is shown in details that a free monoid possesses a surprising compactness property stating that each infinite system of equations with a finite number of variables is equivalent to one of its finite subsystems.

In Section 7 we study the effectiveness of the compactness property. We note that Makanin's algorithm together with the (existential) compactness results allows us to prove some decidability results on iterated morphisms, for which no other proof is know at the moment.

In Section 8 representation results are considered for families of languages, as well as for rational transductions. In the spirit of this paper these representation results involve essentially only morphisms. We show, for example, that the family of recursively enumerable languages has several purely morphic characterizations, *i.e.*, each recursively enumerable language can be obtained by applying only few (one or two) simple operations on morphisms. This surely confirms the computational power of morphisms.

In Section 9 we apply results of Section 6 to a problem of deciding whether two mappings (compositions of morphisms and inverse morphisms) are equivalent on a given set of words, *i.e.*, whether each word of the set is mapped to the same word. Certainly, these problems are practically motivated: decide whether two compilers of a programming language are equivalent.

Section 10 is devoted to open problems of the topics of this paper. We believe that these problems are interesting and some of them even fundamental. Some of the problems are also likely to be extremely difficult – but this is what we thought of Ehrenfeucht's Conjecture in 1984.

We conclude this section with a few technical matters on the presentation of this paper. Results mentioned in the text are divided into two categories: Theorems and Propositions. Theorems are those results that are either proved here in detail (possibly using some well-known results like Makanin's algorithm), or at least the proof is outlined in an extent from which an experienced reader can deduce the claim. Propositions, on their turn, are stated only as supplementary material without proofs.

2. Preliminaries

The goal of this section is to fix terminology for the basic notions of words and morphisms, and to recall some results on word problems for semigroups and semi-Thue systems. The word problems are needed as a tool to prove some detailed results on the Post Correspondence Problem in Section 4.

2.1 Words and morphisms

For definitions of language and automata-theoretic notions and basic results not explained here we refer to Berstel [B], Eilenberg [E], Harrison [H] or Salomaa [S6].

For a finite alphabet Σ of letters we denote by Σ^* the word monoid generated by Σ. The identity of this monoid is the empty word, which will be denoted by 1. As is well known Σ^* is a free monoid. The *word semigroup* generated by Σ is $\Sigma^+ = \Sigma^* \setminus \{1\}$. The *length* of a word u is denoted by $|u|$, and for each letter $a \in \Sigma$, $|u|_a$ denotes the number of occurrences of a in the word u.

Let $u, v \in \Sigma^*$ be two words. Then u is a *factor* of v, if $v = w_1 u w_2$ for some words $w_1, w_2 \in \Sigma^*$. Moreover, u is a *prefix* of v, denoted $u \preceq v$, if $v = uw$ for a word $w \in \Sigma^*$. We say that two words u and v are *comparable*, if one of them is a prefix of the other. Further, u is a *suffix* of v, if $v = wu$ for a word $w \in \Sigma^*$. If u is a prefix of the word v, $v = uw$, then we denote $w = u^{-1}v$; and if u is a suffix of v, $v = wu$, then we denote $w = vu^{-1}$. We let also $\mathrm{pref}_1(w)$ denote the first letter of the word w.

A language $L \subseteq \Sigma^*$ is a *code*, if the submonoid L^* is freely generated by L, *i.e.*, if each word $w \in L^*$ has a unique factorization $w = u_1 u_2 \ldots u_k$ in terms of $u_i \in L$. A code L is a *prefix code* (a *suffix code*, resp.), if no code word $u \in L$ is a prefix (a suffix, resp.) of another code word. A code is *biprefix*, if it is both a prefix and a suffix code.

A *morphism* $h: \Sigma^* \to \Delta^*$ between two word monoids Σ^* and Δ^* is a mapping that satisfies the condition: $h(uv) = h(u)h(v)$ for all words $u, v \in \Sigma^*$. In particular, a morphism becomes defined by the images of the letters in its domain, *i.e.*, if $u = a_1 a_2 \ldots a_n$ with $a_i \in \Sigma$, then $h(u) = h(a_1)h(a_2) \ldots h(a_n)$. For the empty word 1, we have always that $h(1) = 1$.

We say that a morphism $h: \Sigma^* \to \Delta^*$ is *periodic*, if there exists a word $w \in \Delta^*$ such that $h(u) \in w^*$ for all words $u \in \Sigma^*$. Therefore h is periodic if and only if there is a word w such that $h(a) \in w^*$ for all letters $a \in \Sigma$. A morphism $h: \Sigma^* \to \Delta^*$ is called *nonerasing*, if $h(a) \neq 1$ for all $a \in \Sigma$; and h is a *projection*, if for all $a \in \Sigma$ either $h(a) = a$ or $h(a) = 1$. An injective morphism can be called a *code* without any confusion. An injective h is a *prefix* (a *suffix, biprefix*, resp.) code, if $h(\Sigma)$ is a prefix (a suffix, biprefix, resp.) code.

Let p be a nonnegative integer. A morphism h is of *bounded delay* p, if for all $u, v \in \Sigma^*$ and $a, b \in \Sigma$

$$h(au) \preceq h(bv) \quad \text{with} \quad |u| \geq p \implies a = b .$$

If h is of bounded delay p for some p, then it is of *bounded delay*. Each morphism of bounded delay is injective, and the morphisms of bounded delay 0 are exactly the prefix morphisms. Note that we have defined bounded delay from "left to right"; similarly it can be defined from "right to left" by considering the suffix relation instead of the prefix relation.

For a morphism $h: \Sigma^* \to \Delta^*$ we denote by $h^{-1}: \Delta^* \to 2^{\Sigma^*}$ the *inverse morphism* of h defined by $h^{-1}(u) = \{v \mid h(v) = w\}$. Thus h^{-1} is a many-valued mapping between word monoids, or a morphism from a word monoid into the monoid of subsets of a word monoid.

For two morphisms $h, g: \Sigma^* \to \Delta^*$ and for a word w, h and g are said to *agree on* w, if $h(w) = g(w)$. Further, h and g agree on a language L, if they agree on all words $w \in L$. If we require only that $|h(w)| = |g(w)|$, then we say that h and g *agree on* w *lengthwise*. These definitions can be extended in an obvious way to other kinds of general mappings.

2.2 Rational transductions

A *finite transducer* $T = (Q, \Sigma, \Delta, \delta, q_T, F)$ consists of a finite set Q of states, an input alphabet Σ, an output alphabet Δ, a transition relation $\delta \subseteq Q \times \Sigma^* \times \Delta^* \times Q$, an initial state q_T, and a final state set F. The finite transducer T is *simple*, if it has a unique final state, which is equal to the initial state, *i.e.*, $F = \{q_T\}$. Further, T is a 1-*free transducer*, if it never writes an empty word in one step, *i.e.*, $\delta \subseteq Q \times \Sigma^* \times \Delta^+ \times Q$. If $\delta \subseteq Q \times \Sigma \times \Delta^* \times Q$, then T is a *nondeterministic sequential transducer (with final states)*. Moreover, if δ is a partial function $Q \times \Sigma \to \Delta^* \times Q$, then T is called a *sequential transducer (with final states)*.

A sequence

$$\alpha = (q_1, u_1, v_1, q_2)(q_2, u_2, v_2, q_3) \ldots (q_k, u_k, v_k, q_{k+1}) \qquad (2.1)$$

of transitions $(q_i, u_i, v_i, q_{i+1}) \in \delta$ is a *computation* of T with input $I(\alpha) = u_1 u_2 \ldots u_k$ and output $O(\alpha) = v_1 v_2 \ldots v_k$. The computation α in (2.1) is *accepting*, if $q_1 = q_T$ and $q_{k+1} \in F$.

Let T be a finite transducer as above. We say that T *realizes* the relation

$$\tau = \{(I(\alpha), O(\alpha)) \mid \alpha \ \text{an accepting computation of } T\} \ .$$

A relation $\tau \subseteq \Sigma^* \times \Delta^*$ is called a *rational transduction*, if it is realized by a finite transducer. We may consider a rational transduction $\tau \subseteq \Sigma^* \times \Delta^*$ also as a function $\tau \colon \Sigma^* \to 2^{\Delta^*}$, where $\tau(u) = \{v \mid (u, v) \in \tau\}$. The *domain* of τ is the set $\mathrm{dom}(\tau) = \{w \mid \tau(w) \neq \emptyset\}$.

A rational transduction $\tau \subseteq \Sigma^* \times \Delta^*$ is *finite-valued*, if there exists a constant k such that for each input word there are at most k different output words, *i.e.*, $|\tau(w)| \leq k$ for all $w \in \Sigma^*$. If here $k = 1$, then τ is a *rational function*.

2.3 Word problem for finitely presented semigroups

Word problems for semigroups and semi-Thue systems will be used as a tool to establish several undecidability results of morphisms in our later considerations.

Let

$$S = \langle a_1, a_2, \ldots, a_n \mid u_1 = v_1, u_2 = v_2, \ldots, u_k = v_k \rangle$$

be a (finite presentation of a) semigroup with the set $\Sigma = \{a_1, a_2, \ldots, a_n\}$ of *generators* and *defining relations* $u_i = v_i$ for $i = 1, 2, \ldots, k$. For two words $u, v \in \Sigma^*$ we say that $u \to v$ is an *elementary operation* of S, if either $u = w u_i w'$ and $v = w v_i w'$ or $u = w v_i w'$ and $v = w u_i w'$ for some i. In particular, the relation $\to$ is symmetric. We say also that the *relation $u = v$ holds in S*, if there exists a finite sequence of words, $u = w_1, w_2, \ldots, w_s = v$ such that $w_i \to w_{i+1}$.

The *word problem* for the presentation S is stated as follows: determine whether for two words $u, v \in \Sigma^*$, $u = v$ holds in S. In the *individual word problem* we are given a fixed word w_0 and we ask for words w whether $w = w_0$ holds in S.

We associate a semigroup S_M to a Turing Machine $M = (Q, \Sigma, \delta, q_M, h)$ with the halting state h as follows. Let $\#$ be the symbol for the empty square. Let the generators of S_M be $\Delta = Q \cup \Sigma \cup \{A, B\}$, where A and B are new letters, and let the defining relations of S_M be

$$qa = pb \qquad \text{if } \delta(q,a) = (p,b) , \qquad qab = apb \qquad \text{if } \delta(q,a) = (p,R) ,$$
$$qaB = ap\#B \qquad \text{if } \delta(q,a) = (p,R) , \qquad aqb = pab \qquad \text{if } \delta(q,b) = (p,L) ,$$
$$Bqa = Bp\#a \qquad \text{if } \delta(q,a) = (p,L)$$

and

$$ha = h , ahB = hB , BhB = A ,$$

where $a,b \in \Sigma$, $q,p \in Q$. In this construction the model for a Turing Machine is a standard one, where the tape head stays in the square if it changes the contents of the square. We write $u \vdash v$ for two configurations u,v of M, if u yields v using one transition of M.

Let $w = BuB$ be a word boarded by the special letter B, where u is a configuration of M. Assume then that $w \to w'$ with $w' \neq A$ is an elementary operation corresponding to one of the first five types of defining relations of S_M. It is easy to see that now $w' = BvB$ for a configuration v of M such that either $u \vdash v$ or $v \vdash u$ in M. Using the fact that M is deterministic and that the state h is a halting state, one can now conclude the following result. We refer to Rotman [Ro] for details of the proof .

Theorem 2.1. *Let M be a Turing Machine with a halting state and let q be its initial state. Then M accepts w if and only if $BqwB = A$ in S_M.*

Since the halting problem for Turing Machines is undecidable, also the word problem for semigroups is undecidable. Moreover, there exists a Turing Machine, for example an universal one, for which the halting problem is undecidable, and thus we have deduced the following result of Markov [Mar] and Post [P2].

Theorem 2.2. *There exists a finitely presented semigroup with an undecidable word problem.*

In fact, Theorem 2.1 yields a stronger result.

Theorem 2.3. *There exists a finitely presented semigroup with an undecidable individual word problem.*

Next we strengthen these results by embedding a finitely presented semigroup into a 2-generator semigroup. Let $S = \langle a_1, a_2, \ldots, a_n \mid u_i = v_i, i = 1, 2, \ldots, k \rangle$ be a finite presentation, and define a mapping $\alpha: \Sigma^* \to \{a,b\}^*$ by $\alpha(a_i) = ab^i a$ for $i = 1, 2, \ldots, n$. Consequently, α encodes the semigroup S into a binary set of generators by using a biprefix morphism. Let

$$S^\alpha = \langle a, b \mid \alpha(u_i) = \alpha(v_i), i = 1, 2, \ldots, k \rangle .$$

We have immediately that $u = v$ in S if and only if $\alpha(u) = \alpha(v)$ in S^α, and

Theorem 2.4. *If S has an undecidable word problem, then so does S^α.*

In particular, the word problem is undecidable for 2-generator semigroups with finitely many relations. Notice that here S and S^α have the same number of relations.

The first concrete example of a finitely presented semigroup with an undecidable word problem was given by Markov [Mar] in 1947. This example has 13 generators and 33 relations. This was later improved by Tzeitin [Tz] (see also [Sco]), who proved that the semigroup $S_7 = \langle a, b, c, d, e \mid R \rangle$ with five generators and seven relations

$$ac = ca \; , \qquad ad = da \; , \qquad bc = cb \; , \qquad bd = db \; ,$$
$$eca = ce \; , \qquad edb = de \; , \qquad cca = ccae$$

has an undecidable word problem. We refer to [L1] for an outline of the proof for Tzeitin's result.

Matiyasevich [Mat] modified the presentation of S_7 to obtain a 2-generator semigroup

$$S_3 = \langle a, b \mid u_1 = u_2, u_1 = u_3, v = w \rangle$$

with only three relations such that S_3 has an undecidable word problem. In the presentation of this semigroup one of the relations has more than 900 occurrences of generators. For the construction of S_3 we refer again to [L1].

Proposition 2.1. *The semigroup S_3 has an undecidable word problem.*

Also, Tzeitin [Tz] used S_7 to construct a rather simple presentation of a semigroup S_I with an undecidable individual word problem. The semigroup S_I has generators a, b, c, d, e and nine relations:

$$ac = ca \; , \qquad ad = da \; , \qquad bc = cb \; ,$$
$$bd = db \; , \qquad eca = ce \; , \qquad edb = de \; ,$$
$$cdca = cdcae \; , \qquad caaa = aaa \; , \qquad daaa = aaa \; .$$

Proposition 2.2. *The individual word problem $w = a^3$ is undecidable in S_I.*

It is still an open problem whether the word problem for 1-relation semigroups $\langle a_1, \ldots, a_n \mid u = v \rangle$ is decidable, see [L2]. For 1-relation groups the word problem was shown to be decidable by Magnus in 1932, see [LS].

2.4 Semi-Thue systems

A *semi-Thue system* $T = (\Sigma, R)$, cf. e.g. [J], consists of an alphabet $\Sigma = \{a_1, a_2, \ldots, a_n\}$ and of a finite set $R \subseteq \Sigma^* \times \Sigma^*$ of *rules*. We write $u \to v$ for words $u, v \in \Sigma^*$, if there are factors u_1 and u_2 such that

$$u = u_1 x u_2 \; , \qquad v = u_1 y u_2 \quad \text{with} \quad (x, y) \in R \; .$$

Let $\to^*$ be the reflexive and transitive closure of $\to$. Hence $u \to^* v$ if and only if either $u = v$ or there exists a finite sequence of words $u = v_1, v_2, \ldots, v_n = v$ such that $v_i \to v_{i+1}$ for each $i = 1, 2, \ldots, n - 1$.

The *word problem* for a semi-Thue system $T = (\Sigma, R)$ is stated as follows: given any two words $w_1, w_2 \in \Sigma^*$ decide whether $w_1 \to^* w_2$ holds in T.

We say that T is a *Thue system*, if its relation R is symmetric, *i.e.*, if $x \to y$ is a rule in R, then so is $y \to x$. In this case the relation $\to^*$ is a congruence of the word monoid Σ^*, and hence each Thue system corresponds to a finite presentation of a semigroup: $S_T = \langle a_1, a_2, \ldots, a_n \mid u = v \text{ for } u \to v \in R \rangle$, where $w_1 = w_2$ in S_T if and only if $w_1 \to^* w_2$ (and $w_2 \to^* w_1$) in T. From this it follows that the word problem for Thue systems, as well as for semi-Thue systems, is undecidable. By the above considerations we have in general

Theorem 2.5. *If S is a semigroup with an undecidable word problem such that S has n generators and m relations, then the corresponding Thue system T has an undecidable word problem and T has $2m$ rules.*

As observed in [Pan] the Matiyasevich semigroup S_3 of Proposition 2.1 can be represented as a semi-Thue systems with only five rules in a 2-letter alphabet: $u_1 \to u_2$, $u_2 \to u_3$, $u_3 \to u_1$, $v \to w$ and $w \to v$. Hence we have

Theorem 2.6. *There is a semi-Thue system $T_5 = (\Sigma, R)$ with $|\Sigma| = 2$ and $|R| = 5$ such that T_5 has an undecidable word problem.*

The status of the word problem of semi-Thue systems with two, three or four rules is still an open problem.

3. Post correspondence problem: decidable cases

One of the most influential papers in formal language theory is the paper of E. Post [P1], where the first algorithmically undecidable combinatorial problem is introduced. Subsequently, this problem has become one of the most suited tool to establish undecidability results in different fields of discrete mathematics.

In this section we shall study the decidable cases of this important problem. Most notably, we shall outline the proof that PCP is decidable when restricted to alphabets of cardinality two.

3.1 Basic decidable cases

In our terms an *instance* (h, g) of the *Post Correspondence Problem*, or of *PCP*, consists of two morphisms $h, g \colon \Sigma^* \to \Delta^*$. The *size* of the instance (h, g) is defined to be the cardinality of the alphabet Σ. We say that a nonempty word $w \in \Sigma^+$ is a (nontrivial) *solution* of the instance (h, g), if $h(w) = g(w)$. The empty word is the *trivial solution* of each instance (h, g). PCP asks for a given instance whether it has a solution or not. We denote by PCP(n) the subproblem of PCP for instances of size at most n.

The original phrasing by Post for PCP is as follows: given elements $(u_i, v_i) \in \Delta^* \times \Delta^*$ for $i = 1, 2, \ldots, n$, decide whether there exists a sequence $i_1, i_2, \ldots, i_k$ of indices such that $u_{i_1} u_{i_2} \ldots u_{i_k} = v_{i_1} v_{i_2} \ldots v_{i_k}$. This formulation of PCP can be restated in terms of monoid morphisms: given a monoid morphism $h \colon \Sigma^* \to \Delta^* \times \Delta^*$, determine whether $h(\Sigma^+) \cap \iota = \emptyset$, where ι is the identity relation on Δ^*.

Given an instance (h, g) we let

$$E(h, g) = \{w \mid w \neq 1,\ h(w) = g(w)\}$$

be the set of all (nontrivial) solutions. The set $E(h, g)$ is called the *equality set* of h and g. Hence PCP can now be restated as: given two morphisms h and g, is $E(h, g) = \emptyset$? We shall not study here the nature of the possible solutions of instances of PCP. For this topic the reader is referred to [MS1], [MS2], [Li1] and [Li2].

By a *minimal solution* of an instance (h, g) we mean a word w, which cannot be factored into smaller solutions. The set of the minimal solutions $e(h, g)$ is the *base* of $E(h, g)$:

$$e(h, g) = E(h, g) \setminus E(h, g)^2 \ .$$

Example 3.1. Define the morphisms h and g as in the following table.

	a	b	c	d
h	a	$bbabb$	ab	a
g	ab	b	bba	aa

Let $\Sigma = \{a, b, c, d\}$. Consider first candidates for a solution $w \in a\Sigma^*$ that begin with the letter a. Since $g(a) = h(a) \cdot b$, the next letter must be b in order for h to cover the missing *overflow* word $b = h(a)^{-1} g(a)$. Now, $h(ab) = g(ab) \cdot abb$, where the overflow is abb in favour of h; therefore the next letter must be a. The result so far is $h(aba) = g(aba) \cdot ba$, and again there is a unique continuation: the following letter must be b, and we have $h(abab) = g(abab) \cdot abbabb$. One can see that this process never terminates, *i.e.*, the result is an infinite periodic word $w = ababab \ldots$, and so no solution of (h, g) begins with a.

On the other hand, in the case $w \in b\Sigma$, we have at first a unique sequence: $h(b) = g(b) \cdot babb$, $h(bb) = g(bb) \cdot abbbbabb$, $h(bba) = g(bba) \cdot bbbabba$, $h(bbab) = g(bbab) \cdot bbabbabbabb$. The situation can now be depicted as follows:

bbabb	bbabbabbab
bbabb	

There is now an alternative how to continue. Either the next letter is b or c, and a systematic search for all solutions becomes rather complex.

This instance (h, g) has at least the following two minimal solutions: $w_1 = bbabcccbaaaccdaaddd$ and $w_2 = dacbbccbaccaadd$. We leave it as an exercise to search for more minimal solutions of this instance.   □

For each alphabet Δ the monoid Δ^* can be embedded into the monoid $\{a, b\}^*$ generated by a binary alphabet. Indeed, if $\Delta = \{a_1, a_2, \ldots, a_n\}$ then the morphism $\alpha \colon \Delta^* \to \{a, b\}^*$ defined by $\alpha(a_i) = ab^i$ for $i = 1, 2, \ldots, n$ is injective. From this we have

Lemma 3.1. *For all instances (h, g) with $h, g \colon \Sigma^* \to \Delta^*$ there exists an instance (h', g') with $h', g' \colon \Sigma^* \to \{a, b\}^*$ such that $E(h', g') = E(h, g)$.*

For our first decidability result, Theorem 3.1, we state the following simple result which follows from the decidability of the emptiness problem for context-free languages, or even for 1-counter languages, see [B].

Lemma 3.2. *Let $\rho \colon \Sigma^* \to \mathbb{Z}$ be a monoid morphism into the additive group $\mathbb{Z}$ of integers, and let $R \subseteq \Sigma^*$ be a regular language. It is decidable whether $\rho^{-1}(0) \cap R \neq \emptyset$.*

As a simple corollary of Lemma 3.2 we obtain

Theorem 3.1. *PCP is decidable for instances (h, g), where h is periodic.*

Proof. Let $h, g \colon \Sigma^* \to \Delta^*$ be two morphisms such that $h(\Sigma^*) \subseteq u^*$ for a word $u \in \Delta^*$. We apply Lemma 3.2 to the morphism defined by $\rho(a) = |h(a)| - |g(a)|$, for $a \in \Sigma$, and to the regular set $R = g^{-1}(u^*) \setminus \{1\}$. Now, $\rho^{-1}(0) = \{v \mid |h(v)| = |g(v)|\}$, and hence $w \in \{v \mid |h(v)| = |g(v)|\} \cap R$ if and only if $w \neq 1$, $g(w) \in u^*$ and $|g(w)| = |h(w)|$. These conditions imply that $g(w) = h(w)$, since $h(w) \in u^*$, and hence there exists a nonempty word w with $h(w) = g(w)$ if and only if $\rho^{-1}(0) \cap R \neq \emptyset$. By Lemma 3.2 the emptiness of $\rho^{-1}(0) \cap R$ can be decided, and hence the claim follows.   □

Note that Theorem 3.1 contains a subcase of PCP, where Δ is unary, or equivalently of the following variant of PCP: for two morphisms decide whether there exists a word on which these morphisms agree lengthwise, see [G2]. This latter problem can be generalized in a number of ways. Instead of asking the lengthwise agreement, we can demand a stronger agreement, but still weaker than the complete agreement of PCP. For instance, we can ask whether there exists a word w such that $h(w)$ and $g(w)$ are *commutatively equivalent*, often referred to as *Parikh equivalent*, or such that $h(w)$ and $g(w)$ contain all factors of length k equally many times. Let us denote these equivalence relations by $\sim_1$ and $\sim_k$, respectively. Note that, by definition, if $|u|, |v| \leq k$, then $u \sim_k v$ just in case $u = v$. With these notations we have

Proposition 3.1. *Let k be a natural number. It is decidable whether for two morphisms h and g there exists a word w such that $h(w) \sim_k g(w)$.*

In case $k = 1$ the above proposition was proved in [IK] by reducing it to the emptiness problem of certain general counter automata, and the general problem was reduced to the case $k = 1$ in [K1].

We conclude this subsection with another decidable variant of PCP. A *restricted PCP* asks whether a given instance (h, g) of PCP has a solution *shorter* than a given number n. The restricted PCP is trivially decidable. However, even this simple problem is computationally hard on average, as shown in [Gu].

3.2 Generalized post correspondence problem

In our proof of the decidability of PCP in the binary case we need the following generalization of it. In the *generalized Post Correspondence Problem*, *GPCP* for short, the instances are of the form

$$(h, g, v_1, v_2, u_1, u_2), \tag{3.1}$$

where $h, g: \Sigma^* \to \Delta^*$ are morphisms and v_1, v_2, u_1, u_2 are words in Δ^*. A *solution* of such an instance is a word $w \in \Sigma^*$ such that $v_1 h(w) v_2 = u_1 g(w) u_2$. Let GPCP($n$) denote the problem: determine whether a given instance of GCPC of size n have a solution, where the size is defined as in PCP.

We shall be using the following shift morphisms in many constructions to prevent the agreement of two morphisms on unwanted words. Let Δ be an alphabet, and $d \notin \Delta$ a letter. Define the *left* and *right shift morphisms* $l, r: \Delta \to (\Delta \cup \{d\})^*$ by

$$l(a) = da \quad \text{and} \quad r(a) = ad \tag{3.2}$$

for all $a \in \Delta$. Hence for all words $w \in \Delta^+$, $l(w) \cdot d = d \cdot r(w)$.

Our next result shows that PCP and GPCP are equivalent from the decidability point of view.

Theorem 3.2. *If GPCP(n) is undecidable, then so is PCP(n + 2).*

Proof. Let $(h, g, v_1, v_2, u_1, u_2)$ be an instance of GPCP with $h, g: \Sigma^* \to \Delta^*$. Further, let c, d and e be new symbols not in $\Sigma \cup \Delta$. The letters c and e are used for marking the beginning and the end of a word. Further, we use d as a shift letter. Let the shift morphisms l and r be defined as in (3.2).

Consider the instance (h', g') of PCP, where the morphisms

$$h', g': (\Sigma \cup \{d, e\})^* \to (\Delta \cup \{c, d, e\})^*$$

are defined by

$$h'(x) = \begin{cases} l(h(x)), & \text{if } x \in \Sigma, \\ c \cdot l(v_1), & \text{if } x = d, \\ l(v_2) \cdot de, & \text{if } x = e, \end{cases} \quad \text{and} \quad g'(x) = \begin{cases} r(g(x)), & \text{if } x \in \Sigma, \\ cd \cdot r(u_1), & \text{if } x = d, \\ r(u_2) \cdot e, & \text{if } x = e. \end{cases}$$

Suppose first that w is a solution of $(h, g, v_1, v_2, u_1, u_2)$. Now,

$$v_1 h(w) v_2 = u_1 g(w) u_2 \implies c \cdot l(v_1 h(w) v_2) \cdot de = c \cdot l(u_1 g(w) u_2) \cdot de$$
$$\implies c \cdot l(v_1) l(h(w)) l(v_2) \cdot de = cd \cdot r(u_1) r(g(w)) r(u_2) \cdot e$$
$$\implies h'(dwe) = g'(dwe) \ ,$$

and hence dwe is a solution of the instance (h', g') of PCP that can be effectively constructed from w.

On the other hand, if w' is a minimal solution of (h', g'), then $w' = dwe$ for a word $w \in \Sigma^*$, because the words $h'(a)$ and $g'(a)$ begin and end with a different letter for each $a \in \Sigma$, and the markers c and e cannot occur in the middle of a minimal solution. We have now

$$h'(dwe) = g'(dwe) \implies c \cdot l(v_1) l(h(w)) l(v_2) \cdot de = cd \cdot r(u_1) r(g(w)) r(u_2) \cdot e$$
$$\implies v_1 h(w) v_2 = u_1 g(w) u_2 \ ,$$

and hence w is a solution of $(h, g, v_1, v_2, u_1, u_2)$ that can be effectively constructed from w'. From these considerations the claim follows.   □

One way to obtain simple, or even decidable, variants of PCP is to restrict the language on which solutions are searched for. We provide an example of such a case here. Actually this result will be used in the solution of PCP(2) that we present below.

Let us call a language L *strictly bounded of order* n, if there are words $r_i, s_i \in \Sigma^*$ such that

$$L = r_0 s_1^* r_1 \ldots s_n^* r_n \ . \tag{3.3}$$

In the proof of Theorem 3.3 we need the following two simple results on combinatorics of words, see [Lo].

Lemma 3.3. *Let h and g be morphisms such that $v_1 h(s)^2 v_2 = u_1 g(s)^2 u_2$, where $|h(s)| = |g(s)|$. If $|v_1 h(s)| > |u_1| \geq |v_1|$, then $v_1 h(s) v_2 = u_1 g(s) u_2$.*

Lemma 3.4. *For words v_1, u, v_2, v there exists an effectively findable constant k depending on the lengths of these words such that if $v_1 u^p$ and $v_2 v^q$ have a common prefix of length at least k, then u and v are powers of conjugate words, i.e., $u = (w_1 w_2)^n$ and $v = (w_2 w_1)^m$ for some w_1 and w_2.*

Theorem 3.3. *Let L be a strictly bounded language of any order n. It is decidable whether a given instance of GPCP has a solution in L.*

Proof. Let $(h, g, v_1, v_2, u_1, u_2)$ be an instance of GPCP, and let L be as in (3.3). We prove the claim by induction on n.

Let $n = 1$. If $|h(s_1)| \neq |g(s_1)|$, then clearly a length argument would give the unique exponent of s_1 for a candidate solution. On the other hand, if $|h(s_1)| = |g(s_1)|$, then Lemma 3.3 gives an upper bound for the minimal power of s_1 needed for a solution.

In the induction step we have the same subcases to consider. Assume $w = r_0 s_1^{k_1} r_1 \ldots s_n^{k_n} r_n$ is a solution. Suppose first that $|h(s_1)| = |g(s_1)|$. By Lemma 3.3, the power of s_1 in a solution can be effectively restricted above, and hence the problem is reduced to a finite number of cases of order $n - 1$. The induction hypothesis can be applied in this case.

On the other hand, suppose that, say $|h(s_1)| > |g(s_1)|$. By Lemma 3.4, if the power of s_1 in a solution is high enough, then $h(s_1)$ and $g(s_1)$ are powers of conjugate words. The same argument can be applied for all images $h(s_i)$ and $g(s_j)$: each of them is a factor of a power of a conjugate of the primitive root w of $h(s_1)$. Of course, this property is easy to check.

Now, if the above property holds, then it is trivial to check whether there exists a solution in L. On the other hand, if it does not hold, then there are only finitely many words that have to be checked. Clearly, this finite set of candidates can be effectively found. This completes the proof.   □

It follows from the previous proof that instead of strictly bounded languages we could consider also *bounded languages*, for which equality in (3.3) is replaced by inclusion, if suitable, very mild properties of L are assumed to be decidable. Notice also that Theorem 3.3 is in accordance with the rather common phenomenon in formal language theory: many problems, which are undecidable in general, become decidable when restricted to bounded languages.

3.3 (G)PCP in the binary case

The decidability status of PCP(2) was a long standing open problem until it was proved to be decidable by Ehrenfeucht, Karhumäki and Rozenberg [EKR1], and Pavlenko [Pav]. Here we shall give a detailed overview of a more general decidability result that was proved in [EKR1].

Theorem 3.4. *GPCP(2) is decidable, and hence also PCP(2) is decidable.*

A detailed proof of Theorem 3.4 would be rather long and some stages even tedious. Consequently, we can present here only the ideas of it. However, we believe that the following pages give a pretty good intuition of the whole proof, as well as, explain why the proof cannot be extended for PCP(3).

Let h and g be morphisms on a binary alphabet $\{a, b\}$. By Theorem 3.1, we may assume that they are both nonperiodic. The first, and very crucial, step is to replace h (and g) by a *marked morphism*, i.e., by a morphism that satisfies $\mathrm{pref}_1(h(a)) \neq \mathrm{pref}_1(h(b))$. This is achieved as follows. For each h define $h^{(1)}$ by

$$h^{(1)}(x) = \mathrm{pref}_1(h(x))^{-1} h(x) \mathrm{pref}_1(h(x)) \quad \text{for } x = a, b .$$

Define recursively $h^{(i+1)} = (h^{(i)})^{(1)}$. It is not difficult to see that $h^{(m)}$ is a marked morphism, where m is the length of the maximum common prefix z_h

of $h(ab)$ and $h(ba)$. Note that since h is nonperiodic, $|z_h| < |h(ab)|$. Similarly we define z_g. Assume, by symmetry, that $|z_h| \geq |z_g|$.

It follows from the above constructions that *the instance (h, g) of PCP has a solution if and only if the instance $(h^{(m)}, g^{(m)}, z_g^{-1}z_h, 1, 1, z_g^{-1}z_h)$ of GPCP has a solution.* Consequently, we have simplified the morphisms to marked ones at the cost of moving from PCP(2) to GPCP(2).

If we start from a generalized instance $(h, g, v_1, v_2, u_1, u_2)$ the above reduction is slightly more complicated. In this case, we search for an x that satisfies the situation illustrated in Fig.3.1.

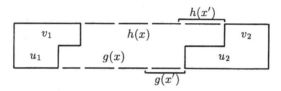

Fig. 3.1. An illustration of a solution of GPCP

First we isolate from the suffixes of $h(x)$ and $g(x)$, for all possible choices of x, words $h(x')$ and $g(x')$ containing the words z_h and z_g as prefixes, respectively. Since z_h and z_g are fixed words there are only a finite number of different cases. Now, the above construction with suitable modifications of the final domino pieces works. It follows from these considerations that the solvability of an instance of GPCP(2) can be reduced to checking whether words of at most certain length are solutions and to solvability of several instances of GPCP(2), where the morphisms are marked. In this way *the decidability of GPCP(2) is reduced to that of GPCP(2) containing only marked morphisms.*

From now on we consider an instance $I = (h, g, v_1, v_2, u_1, u_2)$, where h and g are marked. The advantage of marked morphisms is obvious. Indeed, whenever in an exhaustive search for a solution one of the morphisms is ahead, say h, then the continuation is uniquely determined by the first symbol of the overflow word $(u_1 g(w))^{-1} \cdot v_1 h(w)$.

We apply the idea of exhaustive search to look for a solution. Actually, we do this separately for h and g, thus yielding only *potential solutions*, which have to be checked later on. More precisely, we define sequences $a_1, a_2, \ldots$ and $b_1, b_2, \ldots$ of letters as follows:

1. Set $p := q := 1$;
2. Check whether $u_1 \prec v_1$ ($v_1 \prec u_1$, resp.) and if so define b_q (a_p, resp.) such that $u_1 \mathrm{pref}_1(g(b_q)) \preceq v_1$ ($v_1 \mathrm{pref}_1(h(a_p)) \preceq u_1$, resp.);
3. Set $u_1 := u_1 g(b_q)$, $q := q + 1$ ($v_1 := v_1 h(a_p)$, $p := p + 1$, resp.) and go to **2**.

Clearly, the sequences $(a_p)_{p \geq 1}$ and $(b_q)_{q \geq 1}$ are well-defined. If the sequences are finite, then at some stage $u_1 = v_1$ or u_1 and v_1 are incomparable. If the sequences are infinite, then they are ultimately periodic. Let us refer to these cases as *terminating*, *blocking* and *periodic*, respectively. Of course, it is no problem to decide which of these cases takes place for the instance under consideration.

I (Blocking case) In this case any solution of the instance can be found among the prefixes of an effectively computable *finite* word $a_1 a_2 \ldots a_p$.

II (Periodic case) Now, any solution of the instance can be found among the prefixes of an effectively computable *infinite ultimately periodic* word $a_1 a_2 \ldots$. Hence, by Theorem 3.3, we can decide whether there exists such a solution.

III (Terminating case) We have two finite words $a_1 a_2 \ldots a_p$ and $b_1 b_2 \ldots b_q$ illustrated in Fig. 3.2, where p and q are supposed to be minimal.

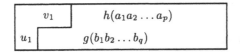

Fig. 3.2. The terminating case

Let μ be the permutation on $\{a, b\}$ such that $\mathrm{pref}_1(h(c)) = \mathrm{pref}_1(g(\mu(c)))$ for $c = a, b$. Now, we carry out the same procedure for the pairs

$$(h(a), g(\mu(a))) \quad \text{and} \quad (h(b), g(\mu(b))) \tag{3.4}$$

as we did for (v_1, u_1). This yields for both of these pairs two finite or infinite words, as well as three cases referred to as blocking, periodic and terminating as above. Hence there are altogether nine different subcases in the main case III.

Eight of these nine cases can be solved by Theorem 3.3. For example, if one of the pairs in (3.4) is terminating and the other is periodic, then any solution is found among the prefixes of words in a language of the form $uv^* rs^*$ for some finite words u, v, r and s. Here u comes from Fig. 3.2, v comes from the terminating pair of (3.4), and r and s from the other pair. More generally, in all the other cases except the one when both pairs in (3.4) are terminating, any solution can be found among the prefixes of words in languages of the forms $uv^* \cup rs^*$ or $uv^* rs^*$ for some (possibly empty) words u, v, r and s. Hence, Theorem 3.3, covers all these cases.

So we are left with the case when both of the pairs in (3.4) are terminating, *i.e.*, there are (minimal) words $c_1, \ldots, c_r$, $d_1, \ldots d_s$, $e_1, \ldots, e_t$ and $f_1, \ldots, f_m$ satisfying the situation in Fig. 3.3. As earlier we can conclude that any solution of the instance is among the prefixes of words from

$(a_1 a_2 \ldots a_p)\{c_1 c_2 \ldots c_r, e_1 e_2 \ldots e_t\}^*$. This means that we (only) know that *the solutions are in a monoid that is freely generated by two words*. Therefore, essentially, we still have the original problem!

$h(c_1)$	$h(c_2)$	$\ldots$	$h(c_r)$
$g(\mu(d_1))g(\mu(d_2))\ldots g(\mu(d_s))$			

$h(e_1)$	$h(e_2)$	$\ldots$	$h(e_t)$
$g(\mu(f_1))g(\mu(f_2))\ldots g(\mu(f_m))$			

Fig. 3.3. Both of the cases in (3.4) are terminating

At this point we need a new approach. We try to define two new instances such that at least one of these has a solution if and only if the original instance has a solution. We refer to these new instances as

$$\overline{I}_i = (\overline{h}, \overline{g}, \overline{v}_1, \overline{v}_2(i), \overline{u}_1, \overline{u}_2(i)) \tag{3.5}$$

for $i = 1, 2$, where

$$\overline{v}_1 = a_1 a_2 \ldots a_p \quad \text{and} \quad \overline{u}_1 = b_1 b_2 \ldots b_q$$

are taken from Fig. 3.2. Accordingly $\overline{h}$ and $\overline{g}$ are defined from Fig. 3.3:

$$\overline{h}(c_1) = c_1 \ldots c_r \;, \qquad \overline{g}(c_1) = \mu(d_1 \ldots d_s) \;,$$
$$\overline{h}(e_1) = e_1 \ldots e_t \;, \qquad \overline{g}(e_1) = \mu(f_1 \ldots f_m) \;.$$

Clearly, $\overline{h}$ and $\overline{g}$ are well-defined.

Now, consider an arbitrary (large enough) solution w of the original instance I, *i.e.*, $v_1 h(w) v_2 = u_1 g(w) u_2$. By our constructions, this can be presented using the blocks of Fig. 3.2 and Fig. 3.3 as follows: we start with the block (3.2), followed by a (unique) sequence of blocks from (3.3). The situation can be written now as in Fig. 3.4, where the blocks represented by the words γ_i are from Fig. 3.3.

Fig. 3.4. Presentation of $v_1 h(w) v_2 = u_1 g(w) u_2$

Further, let us define c_i's such that

$$\gamma_i \in h(c_i)\{h(a), h(b)\}\}^* \;.$$

Since w is a solution of I, the following words are comparable,

$$a_1 a_2 \ldots a_p h^{-1}(\gamma_1) h^{-1}(\gamma_2) \ldots h^{-1}(\gamma_{k-1}) = a_1 a_2 \ldots a_p \overline{h}(c_1) \overline{h}(c_2) \ldots \overline{h}(c_{k-1}),$$
$$b_1 b_2 \ldots b_q g^{-1}(\gamma_1) g^{-1}(\gamma_2) \ldots g^{-1}(\gamma_{k-1}) = b_1 b_2 \ldots b_q \overline{g}(c_1) \overline{g}(c_2) \ldots \overline{g}(c_{k-1}).$$

Therefore $c_1 c_2 \ldots c_{k-1}$ is a solution of a new instance if and only if the final domino piece ($\overline{v}_2(i)$ and $\overline{u}_2(i)$) can be defined properly. This, indeed, can be done. Since w is a solution of (h, g), we have, for example,

$$(u_1 g(w))^{-1}(v_1 h(w)) = u_2 v_2^{-1} .$$

This means that the domino piece matching v_2 and u_2 at their ends has to match with a domino piece constructed from a prefix of a piece corresponding to $\gamma_k \gamma_{k+1} \ldots$. Assuming that this matching piece is minimal, *i.e.*, there are no unnecessary pieces γ_j at the beginning, there are at most two possibilities to obtain this piece, one for each choice of $c_k \in \{a, b\}$. This is, because h and g are marked. Let these choices be

$$u_2 v_2^{-1} = g(k_1(i) \ldots k_{m_i}(i))^{-1} h(l_1(i) \ldots l_{m_i}(i)) \quad \text{for } i = 1, 2 .$$

Then defining

$$\overline{v}_2(i) = l_1(i) \ldots l_{m_i}(i) \quad \text{and} \quad \overline{u}_2(i) = k_1(i) \ldots k_{m_i}(i)$$

for $i = 1, 2$, we have that $c_1 c_2 \ldots c_{k-1}$ is a solution for one of $\overline{I}_i$'s.

It also follows from the construction that, if one of the new instances has a solution, say $c_1 c_2 \ldots c_{k-1}$, so does the original instance, namely

$$w = a_1 a_2 \ldots a_p \overline{h}(c_1) \ldots \overline{h}(c_{k-1}) l_1(i) \ldots l_{m_i}(i) .$$

Note that to make all this work, that is, to be able to define the initial and the final domino pieces, we had to restrict the considerations to solutions that are longer than a computable constant k.

All in all, we have concluded in our most complicated subcase of III the following fact: for a given instance I one can construct two new instances $\overline{I}_i$ for $i = 1, 2$, and a constant k such that I *has a solution longer than k if and only if at least one of $\overline{I}_i$'s has a solution.*

To finish the proof we should be able to show that the pair $(\overline{h}, \overline{g})$ is strictly smaller than (h, g) in some sense. To do this define the *magnitude* $m(h, g)$ of an instance as the sum of different proper suffixes in $\{h(a), h(b)\}$ and in $\{g(a), g(b)\}$. From the construction of the new instance $(\overline{h}, \overline{g})$, see Fig. 3.3, it easily follows that $m(\overline{h}, \overline{g}) \le m(h, g)$. Unfortunately, there may be an equality here as shown by the pair

$h(a) = abb$,	$g(a) = a$,	$\overline{h}(a) = abb$,	$\overline{g}(a) = abb$,
$h(b) = bb$,	$g(b) = bbb$,	$\overline{h}(b) = bbb$,	$\overline{g}(b) = bb$.

In this case, $m(h, g) = 2 + 2 = m(\overline{h}, \overline{g})$.

Fortunately, this happens only in certain very special cases. A detailed and lengthy analysis in [EKR1] shows that $m(\overline{h}, \overline{g}) < m(h, g)$ with the exception of some cases where GPCP(2) can be solved straightforwardly. This completes our presentations of the proof of Theorem 3.4.

As our final remark we emphasize that the above proof is heavily based on the fact that arbitrary binary morphisms can be replaced by marked ones without changing the problem. Our method of doing this does not work in the three letter case, because then no such marked instances exist.

4. Undecidability of PCP with applications

Our goal in this section is to give an overview of the undecidability results of different variants of PCP. We begin this by first reducing PCP to the word problem for semi-Thue systems, and hence to the word problem of finite presentations of semigroups.

4.1 PCP(9) is undecidable

All the existing proofs of the undecidability of PCP use the same basic idea: two collaborative morphisms h and g simulate a computation (or a derivation), that is, a sequence of configurations (or sentential forms) according to an algorithmic system (such as a Turing Machine, a grammar or a Post Normal System). We illustrate this by the following example.

Example 4.1. Consider the morphisms h and g defined by the table

	a	b	c	d	e	f
h	0101	10	c	$d01c$	e	$0c$
g	01	1010	c	d	$c10e$	$c0$

Let us consider solutions that begin with d. At the first step h creates a *marker* c at the end of the word, and g takes care of the symbol d. Thus the overflow word is $g(d)^{-1}h(d) = 01c$. Since g is behind h, it has to parse the overflow word. In doing so g 'computes' something, and the result is given by the corresponding image of h. In this case the computation is simply doubling the factor 01. When g encounters the marker c it has a choice. Either it 'observes' c (using $g(c) = c$), and h creates a new marker c, or g decides (by $g(f) = c0$) that the middle of a solution is at hand after which it starts a different computation: instead of doubling the words it changes the square 1010 into 10. The change of computation after f is obtained by shifting the period of the word between markers. Finally, g reaches h, and the endmarker e tells that a solution if found.

From these observations we can deduce that the words of the form

$$da \cdot c \cdot a^2 \cdot c \cdot a^4 \ldots \cdot c \cdot a^{2^n} \cdot f \cdot b^{2^n} \cdot c \cdot \ldots \cdot c \cdot b^2 \cdot cbe$$

are solutions of (h, g). Moreover, it is easy to check that besides these solutions only the word c is a minimal solution. Consequently, the set of solutions is not a context-free language. Notice that the morphisms h and g are injective; in fact, h is a prefix code and g is a suffix code.

This example illustrates in its modest way, how a pair of morphisms can be used to simulate a computation of a Turing Machine: the initial marker is used by h to create the initial configuration together with a marker; and when g parses this configuration, h creates the next configuration. When g reaches the marker, it either forces h to create a new marker, or g guesses that the configuration is final, in which case g has to catch up h and finally make a complete match as in our illustration.   □

In the following we use the above method to show that PCP is undecidable by reducing it to the word problem of semi-Thue systems. Semi-Thue systems are chosen because Theorem 2.6 gives for these a good upper bound on the number of rules needed to ensure the undecidability of the word problem.

We follow [Cl] in the statement and the proof of Theorem 4.1. The proof is a refinement of the general idea presented in the proof of Theorem 3.2.

Theorem 4.1. *For each semi-Thue system $T = (\Sigma, R)$ and words $u, v \in \Sigma^*$, there effectively exist two morphisms $h_u, g_v \colon \Delta^* \to \Delta^*$ with $|\Delta| = |R|+4$ such that $u \to^* v$ in T if and only if the instance (h_u, g_v) has a solution.*

Proof. For simplicity, as allowed by Theorem 2.4, we restrict ourselves on semi-Thue systems $T = (\{a, b\}, R)$, where the rules $t_i = u_i \to v_i$ in $R = \{t_1, t_2, \ldots, t_m\}$ are such that $u, v, u_i, v_i \in \Phi^*$, where $\Phi = \{aba, ab^2a, \ldots, ab^n a\}$. We consider R also as an alphabet.

Let $d, e \notin \{a, b\} \cup R$ be new letters. To clarify the notations we let $f = aa$ be a special *marker word*. By the form of the (encoded) rules, a derivation of T cannot use any part of the marker f, i.e., if $w_1, \ldots, w_j \in \Phi^*$ such that $w_1 f w_2 f \ldots f w_j \to^* w$ in T, then w has a unique factorization $w = w'_1 f w'_2 f \ldots f w'_j$ such that $w_i \to^* w'_i$ in T for $i = 1, 2, \ldots, j$. In particular, each derivation $u = w_1 \to w_2 \to \ldots \to w_{k+1} = v$ in T can be uniquely encoded as a word $d \cdot w_1 f w_2 f \ldots f w_{k+1} \cdot e$ in $\{a, b, d, e\}^*$.

Again, the shift morphisms $l, r \colon \{a, b\}^* \to \{a, b, d\}^*$ are defined by $l(x) = dx$ and $r(x) = xd$ for $x = a, b$. Next define the morphisms h_u and g_v by

$$
\begin{array}{llll}
h_u(x) = l(x) \;, & g_v(x) = r(x) & \text{for } x = a, b \;, \\
h_u(t_i) = l(v_i) \;, & g_v(t_i) = r(u_i) & \text{for } i = 1, 2, \ldots, m \;, \\
h_u(d) = l(uf) \;, & g_v(d) = d \;, \\
h_u(e) = de \;, & g_v(e) = r(fv)e \;.
\end{array}
$$

We notice first that each minimal solution of (h_u, g_v) is necessarily of the form dze, where z does not contain the letters d and e. Hence we need only to consider words of the form, $w = dw_1 f w_2 f \ldots f w_k e$, where $w_i \in (\{a, b\} \cup R)^*$ does not contain the word f: $w_i = x_{i_0} t_{i_1} x_{i_1} t_{i_2} \ldots t_{i_{p_i}} x_{i_{p_i}}$. Here it is possible that $p_i = 0$ for some i, in which case w_i contains no letters from R. Denote

$$z_i = x_{i_0} v_{i_1} x_{i_1} v_{i_2} \ldots v_{i_{p_i}} x_{i_{p_i}} \quad \text{and} \quad z_i' = x_{i_0} u_{i_1} x_{i_1} u_{i_2} \ldots u_{i_{p_i}} x_{i_{p_i}} \ .$$

Hence $h_u(w_i) = l(z_i)$ and $g_v(w_i) = r(z_i')$, and $z_i \to^* z_i'$ in T. We compute

$$
\begin{aligned}
h_u(w) &= dr(u)r(f)r(z_1)\ldots r(f)r(z_{k-1})r(f)r(z_k)e \,, \\
g_v(w) &= dr(z_1')r(f)r(z_2')\ldots r(f)r(z_k')r(f)r(v)e \ .
\end{aligned}
$$

Therefore, if w is a solution of (h_u, g_v), then $u = z_1'$, $z_1 = z_2'$, $\ldots z_{k-1} = z_k'$ and $z_k = v$, which implies that $u \to^* v$ in T. On the other hand, if $u \to^* v$ holds in T, then a word w as above is easily constructed from this derivation so that $h_u(w) = g_v(w)$. The claim follows from this.   □

When Theorem 4.1 is applied to the 5-rule semi-Thue system T_5 of Theorem 2.6, we obtain

Corollary 4.1. *PCP(9) is undecidable.*

As observed in [HKKr] the constructions of Theorem 4.1 yield a rather sharp undecidability result for the generalized PCP:

Theorem 4.2. *There are two morphisms h, g defined on an alphabet Θ with $|\Theta| = 7$ such that it is undecidable for words $w_1, w_2 \in \Theta^*$ whether or not there exists a word w for which $w_1 h(w) = g(w) w_2$.*

Proof. Consider the semi-Thue system T_5 with five rules of Theorem 2.6 and the morphisms h_u, g_v obtained in the constructions of Theorem 4.1. Let $\Delta = \{a, b, r_1, \ldots, r_5, d, e\}$, and $\Theta = \{a, b, r_1, \ldots, r_5\}$. Hence $|\Theta| = 7$.

We observe that $h_u(x)$ and $g_v(x)$ depend only on T for all $x \in \Theta$, i.e., for any $u_1, u_2 \in \{a, b\}^*$, $h_{u_1}|\Theta^* = h_{u_2}|\Theta^*$, and for any $v_1, v_2 \in \{a, b\}^*$, $g_{v_1}|\Theta^* = g_{v_2}|\Theta^*$. Let $h, g \colon \Theta \to \Delta^*$ be defined by $h = h_u|\Theta^*$ and $g = g_v|\Theta^*$. As noticed in the proof of Theorem 4.1 the minimal solutions of the instances (h_u, g_v) are of the form dwe, where $w \in \Theta^*$, i.e., w does not contain the letters d and e. It follows that it is undecidable for words $u, v \in \{a, b\}^*$ whether or not there exists a word $w \in \Theta^*$ such that $h_u(d)h(w)h_u(e) = g_v(d)g(w)g_v(e)$. Further, $g_v(d)$ is a prefix of $h_u(d)$ and $h_u(e)$ is a suffix of $g_v(e)$, and so $h_u(d)h(w)h_u(e) = g_v(d)g(w)g_v(e)$ just in case when $g_v(d)^{-1}h(d)h(w) = g(w)g_v(e)h_u(e)^{-1}$. The claim follows from this, when we let w_1 vary over the words $g_v(d)^{-1}h(d)$ and w_2 vary over the words $g_v(e)h_u(e)^{-1}$.   □

Corollary 4.2. *GPCP(7) is undecidable.*

Corollary 4.2 together with Theorem 3.4 implies that the decidability status of GPCP(n) is open only in four cases, namely when $n = 3, 4, 5$ and 6. For PCP(n) there are two more open cases, namely $n = 7$ and 8.

The proof of Theorem 4.2 when applied to the semigroup S_I of Proposition 2.2 with an undecidable individual word problem gives the following improvement of Proposition 2.2 at the cost of increasing the cardinality of the domain alphabet of the two morphisms.

Theorem 4.3. (1) *There exists a fixed morphism g such that it is undecidable for a given morphism h whether there exists a word w such that $h(w) = g(w)$.*

(2) *There exist two fixed morphisms h and g such that it is undecidable for a given word z whether $zh(w) = g(w)$ for some word w.*

Proof. Let $S = \langle a, b \mid u_i = v_i$ for $i = 1, 2, \ldots, 9 \rangle$ be the semigroup, which is obtained from S_I of Proposition 2.2 by the construction of Theorem 2.4. Hence S has an undecidable individual word problem with respect to the word $v = abaabaaba$. Now, S can be represented as a semi-Thue system T in the alphabet $\{a, b\}$ with the rules $r_i = u_i \to v_i, s_i = v_i \to u_i$ for $1 \leq i \leq 9$. Let $g = g_v$ be the morphism obtained in the proof of Theorem 4.1 for the fixed word v. Now, by the proof of Theorem 4.1, $u \to^* v$ in T if and only if there exists a word w such that $h_u(dwe) = g(dwe)$, and Claim (1) follows, since the individual word problem for v is undecidable in S.

In addition to above, for any word u, let h be the morphism h_u restricted to $(\{a, b, e\} \cup R)^*$. Clearly, h depends only on S, i.e., it is independent of u. Now, $u \to^* v$ in T if and only if there exists a word w such that $h_u(dwe) = g_u(dwe)$, i.e., if and only if $h_u(d)h(we) = g(dwe)$. Case (2) follows after we observe that $h_u(d)h(we) = g(dwe)$ if and only if $g(d)^{-1}h_u(d) \cdot h(we) = g(we)$. □

One should notice that in these results the cardinality of the domain alphabet of h and g is 22 in Case (1) and 21 in Case (2) of Theorem 4.3.

Another way of obtaining sharper versions of PCP is to restrict the morphisms structurally. We state here two results of this nature. It was shown in [Le] that PCP is undecidable for instances (h, g) of codes, i.e., of injective morphisms. The proof of this result uses reversible Turing Machines, for which each configuration has at most one possible predecessor.

Proposition 4.1. *PCP is undecidable for instances of codes.*

Using the same method of reversible Turing Machines it was proved in [Ru1] that PCP is undecidable already for instances of biprefix codes.

Proposition 4.2. *PCP is undecidable for instances of biprefix codes.*

This is an interesting result also from the viewpoint of equality sets, because as we shall see later on, the equality set $E(h, g)$ of prefix codes h and g is always a regular language. The emptiness problem of a regular language is decidable, and hence, by Proposition 4.2, we cannot determine algorithmically for given biprefix codes h and g which regular language $E(h, g)$ is.

The following variant of PCP is also from [Ru1].

Proposition 4.3. *Given two morphisms h and g it is undecidable whether there exists an infinite word ω on which h and g agree.*

Actually, also in this proposition h and g can be assumed to be biprefix codes.

4.2 A mixed modification of PCP

The Post Correspondence Problem is one of the simplest undecidable problems in mathematics, and for this reason numerous other problems have been shown to be undecidable by reducing these to PCP. As examples we shall consider here some modifications of PCP, as well as some simple problems on multisets. For more classical examples of undecidability results we refer to [B], [S4] and [S6].

In the proof of Theorem 4.1 we used two morphisms h and g, the images of which had been shifted with respect to each other by the shift morphisms l and r. Using the same idea we obtain

Theorem 4.4. *It is undecidable for (injective) morphisms $h, g: \Sigma^* \to \Delta^*$ whether there exists a word $w = a_1 a_2 \ldots a_k$ such that*

$$h_1(a_1)h_2(a_2)\ldots h_k(a_k) = g_1(a_1)g_2(a_2)\ldots g_k(a_k) \ , \tag{4.1}$$

where $h_i, g_i \in \{h, g\}$ and $h_j \neq g_j$ for at least one index j.

Proof. Let $h, g: \Sigma^* \to \Delta^*$ be any two morphisms, and let $c, d, e \notin \Sigma \cup \Delta$ be new letters. Again, let l and r be the morphisms: $l(a) = da$ and $r(a) = ad$ for all $a \in \Sigma$. For each $a \in \Sigma$ define $h_a, g_a: (\Sigma \cup \{d, e\})^* \to (\Delta \cup \{c, d, e\})^*$ by

$$
\begin{aligned}
h_a(x) &= lh(x) \ , & g_a(x) &= rh(x) && \text{for } x \in \Sigma \ , \\
h_a(d) &= c \cdot lh(a) \ , & g_a(d) &= cd \cdot rg(a) \ , \\
h_a(e) &= de \ , & g_a(e) &= e \ .
\end{aligned}
$$

Clearly, the instance (h, g) has a solution $w = au$ if and only if the instance (h_a, g_a) has a solution due. We notice that if h and g are injective, then so are the new morphisms h_a and g_a for all letters $a \in \Sigma$. Therefore we conclude, by Proposition 4.1, that PCP is undecidable for the instances (h_a, g_a), where h_a and g_a are injective morphisms of the above special form. Consequently, we may assume that already $h = h_a$ and $g = g_a$.

Consider now an identity (4.1), where $h_i, g_i \in \{h, g\}$ and $h_j \neq g_j$ for some j. Assume further that k is minimal, *i.e.*, $w = a_1 a_2 \ldots a_k$ is one of the shortest words satisfying (4.1). We show that w is a solution of the instance (h, g), thus proving the claim.

By minimality of k, $h_1 \neq g_1$ and $h_k \neq g_k$ so that necessarily $a_1 = d$ and $a_k = e$, and, moreover, $a_i \notin \{d, e\}$ for $1 < i < k$. Assume, by symmetry, that $h_1 = h$ and $g_1 = g$. We need to show that $h_i = h$ and $g_i = g$ for all $i = 1, 2, \ldots, k$. Assume this does not hold, and let t be the smallest index such that either $g_t = h$ or $h_t = g$. In the first alternative,

$$g(a_1 a_2 \ldots a_{t-1})h(a_t) \in cd \cdot (\Sigma d)^+ (d\Sigma)^+,$$

and so the shortest prefix of the right hand side of (4.1), which is not a prefix in $c(\Sigma d)^\omega$ ends with dd. But no choice of h_i's on the left hand side of (4.1) matches with this prefix: if $h_i \neq g$ for all i, then $h_1(a_1)h_2(a_2)\ldots h_k(a_k) \in$

$c(\Sigma d)^+$, and if $h_i = g$ for some i, then the shortest prefix of the required form in the left hand side of (4.1) is in $c(d\Sigma)^+\Sigma$.

In the second alternative a similar argumentation can be used to derive a contradiction – starting now from the relation $h(a_1a_2\ldots a_{t-1})g(a_t) \in c(d\Sigma)^+(\Sigma d)^+$. □

As an application of Theorem 4.4 we prove a simple undecidability result for multisets of words; for another application, see Theorem 4.8. We remind that a *multiset* is a function $\mu\colon \Sigma^* \to \mathbb{N}$, which gives a nonnegative multiplicity $\mu(w)$ for each word $w \in \Sigma^*$. A multiset μ can also be represented in the set notation as follows $\{\mu(w)w \mid w \in \Sigma^*,\ \mu(w) \neq 0\}$, or equivalently as a formal power series $\sum \mu(w)w$ over Σ, see [BR]. The product of two multisets μ_1 and μ_2 is defined to be the multiset $\mu = \mu_2\mu_1$, for which

$$\mu(w) = \sum_{w=uv} \mu_1(u)\mu_2(v)w\ .$$

The multisets form a semigroup with respect to this product.

We say that μ is a *binary multiset*, if it consists of at most two different words, *i.e.*, if the support $\{w \mid \mu(w) \neq 0\}$ has cardinality at most two.

Theorem 4.5. *It is undecidable whether in a finitely generated semigroup of binary multisets there exists a multiset μ and a word w such that $\mu(w) > 1$.*

Proof. Let $h, g\colon \Sigma^* \to \Delta^*$ be two morphisms, and define for each $a \in \Sigma$ a multiset μ_a by $\mu_a(h(a)) = 1$, $\mu_a(g(a)) = 1$, and $\mu_a(w) = 0$ for all other words. Now, for a multiset $\mu = \mu_{a_k}\mu_{a_{k-1}}\ldots\mu_{a_1}$ and a word $w \in \Sigma^*$, $\mu(w) > 0$ if and only if $w = h_1(a_1)h_2(a_2)\ldots h_k(a_k)$ for some $h_i \in \{h, g\}$. It follows from this that $\mu(w) > 1$ if and only if there are two different sequences $h_1, h_2, \ldots, h_k$ and $g_1, g_2, \ldots, g_k$ with $h_i, g_i \in \{h, g\}$ such that $h_1(a_1)h_2(a_2)\ldots h_k(a_k) = g_1(a_1)g_2(a_2)\ldots g_k(a_k)$. By Theorem 4.4, it is undecidable whether such a sequence exists. □

4.3 Common relations in submonoids

The proof of the next modification of PCP reduce the claim to the injective instances of PCP. It is worth noticing that the claim itself is trivially decidable for injective instances!

Theorem 4.6. *It is undecidable whether for morphisms $h, g\colon \Sigma^* \to \Delta^*$ there are words $u \neq v$ such that $h(u) = h(v)$ and $g(u) = g(v)$.*

Proof. We reduce the claim to Proposition 4.1. For this, let $f\colon \Sigma^* \to \Delta^*$ be any injective morphism, where without restriction we may assume that $\Sigma \cap \Delta = \emptyset$. Further, let $\Gamma = \Sigma \cup \Delta \cup \{c, d, e\}$, and let l and r be again the shift morphisms for the letter $d\colon l(a) = da$ and $r(a) = ad$ for all $a \in \Sigma$. We define for each $a \in \Sigma$ a new morphism $f_a\colon (\Gamma^* \cup \{\bar{a}\})^* \to (\Delta \cup \{c, d, e\})^*$ as follows,

$$f_a(x) = \begin{cases} c \cdot lf(a), & \text{if } x = \overline{a}, \\ lf(x), & \text{if } x \in \Sigma, \\ xd, & \text{if } x = c \text{ or } x \in \Delta, \\ de, & \text{if } x = d, \\ e, & \text{if } x = e. \end{cases}$$

Let (u, v) be a *minimal pair*, that is, assume that u and v are two different words such that $f_a(u) = f_a(v)$ and for no proper prefixes u' and v' of u and v, resp., $f_a(u') = f_a(v')$. In particular, $\mathrm{pref}_1(u) \neq \mathrm{pref}_1(v)$. By symmetry, we may suppose that $\mathrm{pref}_1(u) \neq c$. First we prove that

$$u \in \overline{a}\Sigma^*d \quad \text{and} \quad v \in c\Delta^*e. \tag{4.2}$$

Since $\mathrm{pref}_1(u) \neq \mathrm{pref}_1(v)$, either $\mathrm{pref}_1(u) = \overline{a}$ and $\mathrm{pref}_1(v) = c$, or both $\mathrm{pref}_1(u), \mathrm{pref}_1(v) \in \Sigma$. The latter case leads to a contradiction, since if $w_1, w_2 \in \Sigma^*$ are any two words such that $f_a(w_1)$ is a prefix of $f_a(w_2)$, then $f_a(w_1)^{-1}f_a(w_2) \in (d\Delta)^*$, and this would imply that $u, v \in \Sigma^*$ contradicting the injectivity of f.

So let $\mathrm{pref}_1(u) = \overline{a}$ and $\mathrm{pref}_1(v) = c$. Now, $(f_a(c))^{-1}f_a(\overline{a}) \in (\Delta d)^*\Delta$, and hence v begins with a word v_1, where $v_1 \in c\Delta^*$, and $f_a(\overline{a})^{-1}f_a(v_1) = d$. Assume that we have already shown that u has a prefix $u_i \in \overline{a}\Sigma^*$ and v has a prefix $v_i \in c\Delta^*$ such that $f_a(u_i)^{-1}f_a(v_i) = d$. Now, u begins either with $u_{i+1} = u_i b$ for some $b \in \Sigma$, or with $u_i d$. In the latter case we are done: $u = u_i d$ and $v = v_i e$. In the former case, $f_a(v_i)^{-1}f(u_{i+1}) \in (\Delta d)^*\Delta$, and thus v begins with $v_{i+1} = v_i v'$, where $v' \in \Delta^*$ is such that $f_a(u_{i+1})^{-1}f_a(v_{i+1}) = d$. Thus an induction argument shows (4.2).

We conclude that if $f_a(u) = f_a(v)$ with $u \neq v$, then necessarily $u = \overline{a}wd$ and $v = cw'e$ for some $w \in \Sigma^*$ and $w' \in \Delta^*$. Now, the identities $f_a(u) = c \cdot lf(aw) \cdot de = c \cdot l(w') \cdot de = f_a(v)$ implies that $w' = f(aw)$, and therefore

$$u = \overline{a}wd, \quad v = c \cdot f(aw) \cdot de \quad \text{and} \quad f_a(u) = c \cdot lf(aw) \cdot de. \tag{4.3}$$

We apply the above argumentation to two injective morphisms $h, g \colon \Sigma^* \to \Delta^*$ to prove: the instance (h, g) of PCP has a solution if and only if for some $a \in \Sigma$ there exists words $u \neq v$ such that $h_a(u) = h_a(v)$ and $g_a(u) = g_a(v)$. This, clearly, proves the theorem.

Suppose first that aw is a minimal solution of (h, g) for some letter $a \in \Sigma$ and word $w \in \Sigma^*$. Let h_a and g_a be defined as f_a above. Denote $u = \overline{a}wd$ and $v = c \cdot h(aw) \cdot e$. Now, by the identity $v = c \cdot g(aw) \cdot e$ and the definition of h_a and g_a we have

$$h_a(u) = c \cdot lh(aw) \cdot de = h_a(v) \quad \text{and} \quad g_a(u) = c \cdot lg(aw) \cdot de = g_a(v),$$

and thus there exist u and v as required.

On the other hand, assume that for some $a \in \Sigma$, $h_a(u) = h_a(v)$ and $g_a(u) = g_a(v)$ with $u \neq v$. We first reason that there exists such a pair (u, v) which is minimal with respect to both h_a and g_a. Indeed, if (u, v) is not minimal with respect to, say h_a, then $u = u_1u_2$ and $v = v_1v_2$, where (u_1, v_1) is a minimal pair, and in particular, $h_a(u_1) = h_a(v_1)$ and thus also $h_a(u_2) = h_a(v_2)$. By (4.2), $u_1 \in \bar{a}\Sigma^*d$ and $v_1 \in c\Delta^*e$, and this implies immediately that also $g_a(u_1) = g_a(v_1)$. Thus we may assume that (u, v) is a minimal pair.

Further, again by symmetry, we may assume that $\mathrm{pref}_1(u) = \bar{a}$ and $\mathrm{pref}_1(v) = c$. Therefore, by (4.3), $h_a(u) = c \cdot lh(aw) \cdot de = h_a(v)$ and $g_a(u) = c \cdot lg(aw) \cdot de = g_a(v)$ for some $w \in \Sigma^*$. Now, by (4.2), $v \in c\Delta^*e$, and the definitions of h_a and g_a show that $h_a(v) = g_a(v)$, and hence that $h(aw) = g(aw)$. This completes the proof.   □

Theorem 4.6 has an interesting interpretation. For this suppose that $S = \{w_1, w_2, \ldots w_n\}^*$ is a finitely generated submonoid of Σ^*, and let $X = \{x_1, x_2, \ldots, x_n\}$ be an alphabet. Then $S = h_S(X^*)$, where $h_S: \Sigma^* \to \Delta^*$ is the *natural morphism* defined simply $h_S(x_i) = w_i$ for $i = 1, 2, \ldots, n$. An element of the *kernel* of h_S

$$\ker(h_S) = \{(u, v) \mid h_S(u) = h_S(v)\}$$

is called a *relation* of S. The problem whether S satisfies a nontrivial relation (u, v) with $u \neq v$, is clearly decidable. Indeed, one only has to check whether $w_iS \cap w_jS \neq \emptyset$ for some indices $i \neq j$, and this is decidable since S is a regular subset of Σ^*. This also shows that it is decidable whether S is a free monoid. The next immediate corollary to Theorem 4.6 shows, however, that it is undecidable whether two word monoids S and S' have a common relation.

Corollary 4.3. *It is undecidable whether two finitely generated submonoids of free monoids have a common relation.*

We emphasize that our proof of Corollary 4.3 essentially requires the undecidability of PCP for injective morphisms, and still our problem is not just an "injective variant of something".

4.4 Mortality of matrix monoids

Using coding techniques due to Paterson [Pat] (see also, [Cl], [KBS] and [Kr]) we prove now a beautiful undecidability result for 3×3-integer matrices. In the *mortality problem* we ask whether the zero matrix belongs to a given finitely generated submonoid S of $n \times n$-matrices from $\mathbb{Z}^{n\times n}$, i.e., whether there are generators $M_1, M_2, \ldots, M_k$ of S such that $M_1M_2 \ldots M_k = 0$. We refer to [Sc] for the connections of this problem to other parts of mathematics.

Let $\Gamma = \{a_1, a_2, \ldots, a_n\}$. Define $\sigma: \Gamma^* \to \mathbb{N}$ by $\sigma(1) = 0$ and

$$\sigma(a_{i_1}a_{i_2} \ldots a_{i_k}) = \sum_{j=1}^{k} i_j n^{k-j} .$$

The function σ is injective and it gives an n-adic representation of each word, and moreover,

$$\sigma(uv) = \sigma(v) + n^{|v|}\sigma(u) \ .$$

Define then a monoid morphism $\beta: \Gamma^* \to \mathbb{N}^{2\times 2}$ by

$$\beta(a_i) = \begin{pmatrix} n & 0 \\ i & 1 \end{pmatrix}$$

for all $i = 1, 2, \ldots, n$. Now,

$$\beta(w) = \begin{pmatrix} n^{|w|} & 0 \\ \sigma(w) & 1 \end{pmatrix}$$

for all $w \in \Gamma^*$, as can be seen inductively:

$$\begin{aligned}
\beta(u)\beta(v) &= \begin{pmatrix} n^{|u|} & 0 \\ \sigma(u) & 1 \end{pmatrix}\begin{pmatrix} n^{|v|} & 0 \\ \sigma(v) & 1 \end{pmatrix} = \begin{pmatrix} n^{|u|}n^{|v|} & 0 \\ \sigma(u)n^{|v|} + \sigma(v) & 1 \end{pmatrix} = \\
&= \begin{pmatrix} n^{|uv|} & 0 \\ \sigma(uv) & 1 \end{pmatrix} = \beta(uv) \ .
\end{aligned}$$

The morphism β is injective as already indicated by the $(2,1)$-entry $\sigma(w)$ of the matrix. When two copies of β are applied simultaneously in 3×3-matrices we obtain the following monoid morphism $\gamma_1: \Gamma^* \times \Gamma^* \to \mathbb{N}^{3\times 3}$:

$$\gamma_1(u, v) = \begin{pmatrix} n^{|u|} & 0 & 0 \\ 0 & n^{|v|} & 0 \\ \sigma(u) & \sigma(v) & 1 \end{pmatrix} \ .$$

Here γ_1 is *doubly injective*, i.e., if $\gamma_1(u_1, v_1)_{31} = \gamma_1(u_2, v_2)_{31}$, then $u_1 = u_2$, and if $\gamma_1(u_1, v_1)_{32} = \gamma_1(u_2, v_2)_{32}$, then $v_1 = v_2$.

We present now a somewhat simplified proof due to V. Halava of Paterson's result [Pat].

Theorem 4.7. *The mortality problem is undecidable for the 3×3-matrices with integer entries.*

Proof. First define a special matrix

$$A = \begin{pmatrix} 1 & 0 & 1 \\ -1 & 0 & -1 \\ 0 & 0 & 0 \end{pmatrix} \ .$$

Clearly, A is idempotent, i.e., $A^2 = A$.

As is easily verified, the matrices

$$W(p, q, r, s) = \begin{pmatrix} p & 0 & 0 \\ 0 & r & 0 \\ q & s & 1 \end{pmatrix}, \qquad \text{where } 0 \le q < p, \ 0 \le s < r, \qquad (4.4)$$

with nonnegative integer entries form a monoid. One obtains for such a $W = W(p, q, r, s)$

$$AWA = (p + q - s)A .$$ (4.5)

Let L be a finitely generated monoid of matrices of the type (4.4), and let S be the matrix monoid generated by $\{A\} \cup L$. We show that

$$0 \in S \iff \exists W \in L : AWA = 0.$$ (4.6)

For this, assume that $0 \in S$. Since L consists of invertible matrices, we have

$$AW_1 AW_2 A \ldots AW_t A = 0$$ (4.7)

for some $t \geq 1$ and $W_j \in L$ with $W_j \neq I$. Since A is an idempotent matrix,

$$AW_1 AW_2 A \ldots AW_t A = AW_1 A \cdot AW_2 A \ldots AW_t A = 0 ,$$

which implies by (4.5) that $AW_i A = 0$ for some $i = 1, 2, \ldots, t$. This shows claim (4.6).

Now we use the notations of the beginning of this section. Let (h, g) be an instance of PCP, where $h, g : \Sigma^* \to \Delta^*$ with $\Delta = \{a_2, a_3\}$, and denote $\Gamma = \{a_1, a_2, a_3\}$. Hence $n = 3$. Define

$$W_a = \gamma_1(h(a), g(a)) = W(3^{|h(a)|}, \sigma(h(a)), 3^{|g(a)|}, \sigma(g(a))) ,$$
$$W'_a = \gamma_1(h(a), a_1 g(a)) = W(3^{|h(a)|}, \sigma(h(a)), 3^{|a_1 g(a)|}, \sigma(a_1 g(a))) .$$

for all $a \in \Sigma$. Consider the matrix monoid S generated by A, W_a and W'_a for $a \in \Sigma$. By (4.6), S is mortal if and only if there exists a product W of the matrices W_a and W'_a such that $AWA = 0$. By the definition of the matrices W_a and W'_a the matrix W is of the form

$$W = \begin{pmatrix} 3^{|u|} & 0 & 0 \\ 0 & 3^{|v|} & 0 \\ \sigma(u) & \sigma(v) & 1 \end{pmatrix} ,$$

where $u = h(w)$ for some $w \in \Sigma^*$. Hence by (4.5), $AWA = 0$ is equivalent to

$$3^{|u|} + \sigma(u) = \sigma(v) .$$

This is equivalent to

$$v = a_1 u = a_1 h(w) ,$$

which, by the choice of the matrices W_a and W'_a and by the fact that $a_1 \notin \Delta$, is equivalent to the condition

$$v = a_1 g(w) = a_1 h(w) .$$

Therefore S is mortal if and only if the instance (h, g) has a solution. This proves the theorem.   □

Using the matrix representation γ_1 we obtain also the following result, the second case of which is a slight improvement of a result in [KBS].

Theorem 4.8. (1) *It is undecidable whether two finitely generated subsemigroups of $\mathbb{N}^{3\times3}$ have a common element.*

(2) *It is undecidable whether a finitely generated subsemigroup of triangular matrices from $\mathbb{N}^{3\times3}$ is free.*

Proof. Let $h, g\colon \Sigma^* \to \Delta^*$ be two morphisms, where without restriction we may assume that $\Delta \subseteq \Sigma$. Let S_H and S_G be the semigroups generated by the matrices $H_a = \gamma_1(a, h(a))$ and $G_a = \gamma_1(a, g(a))$ for $a \in \Sigma$, respectively.

For Case (1) we notice that $H_{a_1} \cdot \ldots \cdot H_{a_k} = G_{b_1} \cdot \ldots \cdot G_{b_t}$ if and only if $a_1 \ldots a_k = b_1 \ldots b_t$ and $h(a_1 \ldots a_k) = g(b_1 \ldots b_t)$, since γ_1 is doubly injective. The latter condition, in turn, is equivalent to $h(a_1 a_2 \ldots a_k) = g(a_1 a_2 \ldots a_k)$. Therefore Claim (1) follows now from the undecidability of PCP.

For Claim (2) we take, in the above notations, the matrix semigroup S generated by H_a and G_a for $a \in \Sigma$. The matrices $H(a)$ and $G(a)$ are invertible (in the group of matrices with rational entries), and hence the semigroup S is cancellative. Assume then that $H_{a_1}^{(1)} \cdot \ldots \cdot H_{a_k}^{(t)} = G_{b_1}^{(1)} \cdot \ldots \cdot G_{b_t}^{(s)}$, where $H_{a_i}^{(i)} = \gamma_1(a_i, h_i(a_i))$ and $G_{b_i}^{(i)} = \gamma_1(b_i, g_i(b_i))$ for some $h_i, g_i \in \{h, g\}$. Again it follows that $a_1 a_2 \ldots a_k = b_1 b_2 \ldots b_t$ and $h_1(a_1) \ldots h_t(a_k) = g_1(a_1) \ldots g_t(a_k)$, and conversely. Therefore Claim (2) follows from our Theorem 4.4. □

4.5 Zeros in upper corners

Let $\Gamma = \{a_1, a_2\}$, and define a mapping $\gamma_2\colon \Gamma^* \times \Gamma^* \to \mathbb{Z}^{3\times3}$ by

$$\gamma_2(u, v) = \begin{pmatrix} 1 & \sigma(v) & \sigma(u) - \sigma(v) \\ 0 & 2^{|v|} & 2^{|u|} - 2^{|v|} \\ 0 & 0 & 2^{|u|} \end{pmatrix}.$$

The mapping γ_2 is clearly injective, and it is also a morphism as can be verified by an easy computation.

Using this morphism γ_2 we are able to prove the following result, which is attributed to R.W. Floyd in [Man], see also [Cl].

Theorem 4.9. *It is undecidable whether a finitely generated subsemigroup of $\mathbb{Z}^{3\times3}$ contains a matrix M with $M_{13} = 0$.*

Proof. Let (h, g) be morphisms $\Sigma^* \to \Gamma^*$, and define $M_a = \gamma_2(h(a), g(a))$ for each $a \in \Sigma$. Then $M_{13} = 0$ for a matrix $M = M_{a_1} M_{a_2} \ldots M_{a_m}$ if and only if for $w = a_1 a_2 \ldots a_m$, $\sigma(h(w)) = \sigma(g(w))$, i.e., if and only if $h(w) = g(w)$. Hence the claim follows from the undecidability of PCP. □

Theorem 4.9 has immediate applications in the theory of finite automata with multiplicities, see [E]. Indeed, with a set $M_1, \ldots M_t$ of 3×3-matrices over integers we can associate a 3-state $\mathbb{Z} - \Sigma$-automaton A with $\Sigma = \{a_1, a_2, \ldots, a_t\}$ having an initial state q_1, a final state q_3, and transitions

$$q_n \xrightarrow{\quad a_i \ (M_i)_{nm} \quad} q_m$$

for $n, m = 1, 2, 3$. Hence when A reads a_i in state q_n, it goes to state q_m with multiplicity $(M_i)_{nm}$. From the construction of A, it follows that

$$A(a_{i_1} a_{i_2} \ldots a_{i_r}) = (M_{i_1} M_{i_2} \ldots M_{i_r})_{13} \ ,$$

where the left hand side denotes the multiplicity of the word accepted by A. We obtain

Theorem 4.10. *It is undecidable whether a 3-state finite automaton A with integer multiplicities accepts a word w with multiplicity $A(w) = 0$.*

As Theorem 4.10 was obtained by using the injective morphism γ_2 into $\mathbb{Z}^{3 \times 3}$, we can use β into $\mathbb{N}^{2 \times 2}$ to prove the following result.

Theorem 4.11. *It is undecidable whether two 2-state automata A and B with nonnegative integer multiplicities accepts a word w with the same multiplicity, i.e., $A(w) = B(w)$.*

We conclude this section with some remarks. The problem of Theorem 4.9 is a generalization of *Skolem's Problem*, see Problem 10.3, where one asks for an algorithm to decide whether there exists a power M^k of a given integer $n \times n$-matrix M having a zero in the upper corner, $(M^k)_{1n} = 0$. Surprisingly, a related question whether there exists *infinitely* many such powers is decidable, see [BR]. This result of Berstel and Mignotte is based on Skolem-Mahler-Lech Theorem saying that the set of powers k yielding a zero in $(M^k)_{1n}$ is ultimately periodic, see [Ha] or [BR] for an elementary, but not short, proof of this theorem.

In this context we should also mention the following decidability result of Jacob [Ja] and Mandel and Simon [MaS].

Proposition 4.4. *It is decidable whether or not a finitely generated submonoid of $\mathbb{Z}^{n \times n}$ is finite.*

In contrast to Theorem 4.7, we have the following decidability result for groups of matrices as a corollary to a strong effectiveness theorem of [BCM] for finitely generated rings, see [Mi].

Proposition 4.5. *The word problem is decidable for finitely generated presentations of groups of matrices with entries in a commutative ring.*

5. Equality sets

In this section we derive some basic properties of the equality sets. In particular, we shall study the problem of the regularity of the sets $E(h, g)$ in details.

5.1 Basic properties

Recall that the equality set of two morphisms $h, g\colon \Sigma^* \to \Delta^*$ is the set $E(h, g) = \{w \in \Sigma^+ \mid h(w) = g(w)\}$ of all nontrivial solutions of the instance (h, g) of PCP.

As in PCP we can always restrict ourselves to morphisms into a binary alphabet.

Lemma 5.1. *For each equality set $E \subseteq \Sigma^*$ there are morphisms $h, g\colon \Sigma^* \to \Delta^*$, where $|\Delta| \leq 2$, such that $E = E(h, g)$.*

We start with some simple combinatorial properties of the equality sets.

Lemma 5.2. *For morphisms $h, g\colon \Sigma^* \to \Delta^*$*

(1) if $u, uv \in E(h, g)$, then $v \in E(h, g)$;
(2) if $uv, uwv \in E(h, g)$, then $uw^*v \subseteq E(h, g)$;
(3) if $uv, uw_1v, uw_2v \in E(h, g)$, then also $uw_1w_2v \in E(h, g)$.

Proof. We shall prove only Case (3) of the claim. The assumption $h(uv) = g(uv)$ implies that there exists a word s such that $h(u) = g(u)s$ and $g(v) = sh(v)$ (or symmetrically $g(u) = h(u)s$, $h(v) = sg(v)$). Since $h(uw_iv) = g(uw_iv)$ for $i = 1, 2$, also $g(u)sh(w_i)h(v) = g(u)g(w_i)sh(v)$ and so $sh(w_i) = g(w_i)s$. The following computation proves the claim:

$$
\begin{aligned}
h(uw_1w_2v) &= g(u)sh(w_1)h(w_2)h(v) = g(u)g(w_1)sh(w_2)h(v) \\
&= g(u)g(w_1)g(w_2)sh(v) = g(uw_1w_2v) \ .
\end{aligned}
$$
□

By the first case of Lemma 5.2 we have the following result.

Corollary 5.1. *Let $h, g\colon \Sigma^* \to \Delta^*$ be two morphisms. If $E(h, g) \neq \emptyset$, then $E(h, g)$ is a free semigroup generated by a biprefix code.*

Therefore the base $e(h, g) = E(h, g) \setminus E(h, g)^2$ of $E(h, g)$ is a biprefix code or empty. Of course, the problem whether $e(h, g) = \emptyset$ is undecidable, by PCP.

In Example 4.1 we had already an equality set that is not context-free. Below we give some other examples of equality sets.

Example 5.1. Let $\Sigma = \{a, b, c\}$ and define h, g as below.

	a	b	c
h	$abab$	a	ba
g	ab	a	$baba$

In this case $E(h,g) = \{a^n bc^n \mid n \geq 0\}^+$ is a nonregular context-free language, and the base $e(h,g) = \{a^n bc^n \mid n \geq 0\}$ is an infinite biprefix code. $\square$

Example 5.2. (1) The set $L = \{ab, a^2b^2\}^+$ is not an equality set although its base $L \setminus L^2 = \{ab, a^2b^2\}$ is a biprefix code. In fact, as is easy to see, if $ab, a^2b^2 \in E(h,g)$, then $\{w \mid |w|_a = |w|_b\} \subseteq E(h,g)$. The same argument shows also that $\{a^n b^n \mid n \geq 1\}^+$ is not an equality set.

(2) Let $L = \{a^n b^n c^n \mid n \geq 0\}$. In [EnR1], L^+ is not an equality set. However $L = E(h,g) \cap a^*b^*c^*$ for

	a	b	c
h	a^2	c	c
g	a	a	c^2

Thus $E(h,g)$ is not context-free. Let then $\sigma(x) = d^* x d^*$ for $x = a, b, c$ be a substitution. As shown in [EnR1] $\sigma(E(h,g))$ is an equality set, which cannot be obtained using nonerasing morphisms: if $E(h',g') = \sigma(E(h,g))$, then $h'(d) = 1 = g'(d)$. $\square$

As we have seen, there are equality sets that are not context-free languages. An equality set $E(h,g)$ can, however, be accepted by a deterministic 2-head finite automaton A with a state set

$$Q = \{u \mid u = g(a)^{-1}h(b) \text{ for } a, b \in \Sigma\}$$
$$\cup \; \{\overline{u} \mid u = h(a)^{-1}g(b) \text{ for } a, b \in \Sigma\} \cup \{1\} \; ,$$

where the first head $\downarrow_1$ simulates h and the second head $\downarrow_2$ simulates g so that the automaton is in state u, if $u = g(v_2)^{-1}h(v_1)$ is defined, and in state $\overline{u}$, if $u = h(v_1)^{-1}g(v_2)$ is defined, after $\downarrow_1$ has read v_1 and $\downarrow_2$ has read v_2. Since a Turing Machine can store the positions of the heads of A in an auxiliary tape in space $\log(n)$, we obtain a result from [EnR1]:

Theorem 5.1. *Each equality set $E(h,g)$ can be accepted in $\log(n)$ deterministic space. In particular, the equality sets have deterministic polynomial time complexity.*

The complement $\Sigma^+ \setminus E(h,g)$ is always context-free. In fact, the complement can be accepted by a 1-counter automaton, which on an input w nondeterministically seeks for a position where $h(w)$ and $g(w)$ differ. In particular,

Theorem 5.2. *Each equality set is a complement of a 1-counter language.*

We consider some closure properties of the family of equality sets, *cf.* [EnR1]. Since $E = E(h,g)$ is a (free) semigroup, it is follows that $E = E^+$, and hence the family of equality sets is trivially closed under the operation of iteration. (Of course, the closure under $*$ is impossible, since, by definition $1 \notin E(h,g)$). By the same reason the closure properties of the family of equality sets are rather weak.

Theorem 5.3. *The family of equality sets is closed under inverse morphic images and mirror images, but it is not closed under union, intersection, complement, catenation, or morphic images.*

Proof. Consider three morphisms $f: \Delta^* \to \Sigma^*$ and $h, g: \Sigma^* \to \Delta^*$. Now, $w \in f^{-1}(E(h,g))$ if and only if $f(w) \in E(h,g)$, which means that $f^{-1}(E(h,g)) = E(hf, gf)$. This shows that the equality sets are closed under inverse morphic images. The closure under mirror image is obvious.

The union $a^+ \cup b^+$ is not an equality set, since it is not even a subsemigroup of $\{a,b\}^+$. The same argument applies to catenation. For the intersection let $\Sigma = \{a,b,c\}$, and define $h(a) = a^2$, $h(b) = a = h(c)$. Let g be the morphism $\Sigma^* \to \Sigma^*$ determined by the permutation (a,b,c) of Σ. Now,

$$E(h, hg) \cap E(hg, hg^2) = \{w \mid w \in \Sigma^+, \; |w|_a = |w|_b = |w|_c\}$$

is not an equality set, for details see [EnR1]. For the (nonerasing) morphic images consider the equality set $E = (ab)^+$ and the morphism h, for which $h(a) = a = h(b)$. Clearly, $h(E) = (aa)^+$ is not an equality set. □

5.2 Some restricted cases

If the domain alphabet is restricted to be at most binary, then an equality set has a rather simple structure. It is clear that if $\Sigma = \{a\}$ is unary, then either $E(h,g) = a^+$ or $E(h,g) = \emptyset$. In the binary case the equality sets were partially characterized in [EKR2].

Proposition 5.1. *Let $h, g: \Sigma^* \to \Delta^*$ be two morphisms for $|\Sigma| = 2$. If h and g are nonperiodic, then either $E(h,g) \neq \emptyset$, $E(h,g) = \{u,v\}^+$ or $E(h,g) = (uw^*v)^+$ for some u, w and v. If both h and g are periodic, then there exists a rational number $q \geq 0$ such that $E(h,g) = \{w \in \Sigma^* \mid |w|_a/|w|_b = q\}$ or $E(h,g) = b^+$.*

The (existential) proof of Proposition 5.1 is surprisingly short when compared to the proof of binary PCP.

It is an interesting open problem whether the second possibility is actual in Proposition 5.1, see Problem 10.5. In fact, see [CK1], the only known binary equality sets of the form $\{u,v\}^+$ with $u \neq v$ are the languages $\{a^i b, ba^i\}^+$, while there are quite different types of equality sets of the form u^+. For example, the languages $(a^{mn}b^n)^+$ and $(a^{mn+1}b^n)^+$ for any $m, n \geq 1$ are equality sets. On the other hand, languages $(abaab)^+$ and $(aabab)^+$ are not equality sets.

Instead of restricting the alphabets one can also restrict the morphisms. We shall return to this problem later on. At this point we shall only mention the following result from [HW]. For an endomorphism $h\colon \Sigma^* \to \Sigma^*$ let $\mathrm{Fix}(h) = \{w \in \Sigma^+ \mid h(w) = w\}$ be the set of *fixed points* of h. Clearly, $\mathrm{Fix}(h) = E(h, \iota)$, where ι is the identity mapping on Σ^*, and hence each set of fixed points is an equality set.

Proposition 5.2. *For every endomorphism $h\colon \Sigma^* \to \Sigma^*$ the set of fixed points is finitely generated.*

The proof of this proposition is not too hard. On the other hand, the following related result of Gersten [Ger] for free groups is a deep mathematical theorem.

Proposition 5.3. *Let α be an automorphism of a finitely generated free group. Then the subgroup $\mathrm{Fix}(\alpha)$ of fixed-points is also finitely generated.*

This result has been improved by several authors, for example, the following result on equality sets of free groups was proved in [GT].

Proposition 5.4. *Let F be a finitely generated free group, and let $\alpha\colon F \to F$ be a group morphism and $\beta\colon F \to F$ be a group monomorphism. Then $E(\alpha, \beta)$ is finitely generated.*

5.3 On the regularity of equality sets

Let $h, g\colon \Sigma^* \to \Delta^*$ be two morphisms. We construct now an (infinite state) automaton $A(h, g)$ that accepts $E(h, g) \cup \{1\}$. For this let Δ be an alphabet and let $\Delta^{-1} = \{a^{-1} \mid a \in \Delta\}$ be disjoint from Δ. Set $\Delta^{\pm} = \Delta \cup \Delta^{-1}$. We denote by $\Delta^{(*)}$ the *free group* on Δ. A word $w \in (\Delta^{\pm})^*$ is said to be *reduced*, if it contains no factors aa^{-1} or $a^{-1}a$ for any $a \in \Delta$. As is well known $\Delta^{(*)}$ can be identified with the group of reduced words in $(\Delta^{\pm})^*$. Further, w is a *positive word* (*negative word*), if $w \in \Delta^*$ ($w \in (\Delta^{-1})^*$, resp.).

Let $h, g\colon \Sigma^* \to \Delta^*$ be two morphisms. We define the *overflow function* $\beta\colon \Sigma^* \to \Delta^* \cup (\Delta^{-1})^*$ as follows

$$\beta(w) = g(w)^{-1}h(w) \ .$$

Clearly, β is only a partial function, and we have

$$E(h, g) = \{w \in \Sigma^* \mid \beta(w) = 1\} \ .$$

Next define an (infinite state) automaton $A'(h, g) = (Q, \Sigma, \delta, 1, 1)$ with a state set $Q = \Delta^* \cup (\Delta^{-1})^*$ such that the empty word 1 is the unique initial and final state, and the transition function becomes defined by

$$\delta(u, a) = v \quad \text{if} \quad uh(a) = g(a)v \quad \text{in } \Delta^{(*)} \ .$$

Notice that $A'(h, g)$ is a deterministic automaton.

It is immediate that there is a computation from the initial state 1 to a state v, i.e., $\delta(1, w) = v$, if and only if $h(w) = g(w)v$, where $v = \beta(w)$. Hence, the automaton $A'(h, g)$ accepts $E(h, g) \cup \{1\}$. Consider then the (incomplete) *minimal automaton* $A(h, g)$, which is obtained from $A'(h, g)$ by identifying the equivalent states of $A'(h, g)$, and then removing those states that do not take part in an accepting computation. Here we remove also the 'rubbish state', which is used in the *complete* minimal automaton.

As emphasized by Eilenberg [E], the above minimization process works for infinite state automata, and therefore a language is regular just in case its minimal automaton is finite, and hence we have the following result.

Theorem 5.4. *The automaton $A(h, g)$ accepts $E(h, g) \cup \{1\}$. Further, $E(h, g)$ is regular if and only if $A(h, g)$ is a finite automaton.*

Example 5.3. Consider the morphisms h and g defined as

	a	b	c
h	aba	ba	b
g	a	bab	ab

There are now only finitely many positive and negative overflow words $\beta(w)$. They are the following:

$$ba = \beta(a), \qquad a = \beta(ab), \qquad aba = \beta(aba), \qquad baaba = \beta(abaa),$$
$$ab = \beta(abac), \qquad baba = \beta(abaca), \qquad b = \beta(abacc), \qquad 1 = \beta(abc),$$
$$b^{-1} = \beta(b), \qquad (ab)^{-1} = \beta(bc).$$

These morphisms yield the automaton of Fig. 5.1. After removing the dead ends in the right part of the figure, we obtain a simpler finite automaton illustrated in the framed part of Fig. 5.1. It is immediate that $E(h, g) = (abc \cup bca)^+$. □

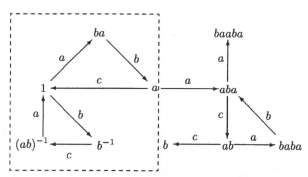

Fig. 5.1. The automata $A'(h, g)$ and $A(h, g)$ for morphisms h and g

If $A(h,g)$ is a finite automaton, then there are only finitely many different overflow words $\beta(w)$ as states of the automaton. This implies that for each regular $E(h,g)$ there exists a constant k such that $E(h,g) = E_k(h,g)$, where

$$E_k(h,g) = \{w \mid h(w) = g(w) \text{ and } |\beta(u)| \leq k \text{ for each } u \preceq w\} \ .$$

Consequently, we have the following characterization of regular equality sets.

Theorem 5.5. $E(h,g)$ *is regular if and only if* $E(h,g) = E_k(h,g)$ *for some* $k \geq 0$.

To continue we call a state u of the minimal automaton $A(h,g)$ *critical,* if $u = 1$ or there are two different letters a, b such that $\delta(u,a)$ and $\delta(u,b)$ are both defined. Hence a noninitial state (or an overflow word) is critical if there is a choice how to continue from this state. Let Q_c be the set of all critical states of $A(h,g)$.

From the minimality of $A(h,g)$ it follows that for each pair (u,v) of critical states, there exists only finitely many words w such that

$$\delta(u,w) = v \text{ in } A(h,g) \text{ and no in-between state is critical.} \tag{5.1}$$

Now, let us define the *critical automaton* $A_c(h,g) = (Q_c, \Sigma, \delta_c, 1, 1)$ as a generalized infinite state automaton, where δ_c becomes defined by (5.1): $\delta_c(u,w) = v$ if and only if w satisfies (5.1). Then $A_c(h,g)$ is well-defined generalized automaton, *i.e.,* for each pair of states (u,v) there exists only finitely many words leading in one step from u to v. Moreover, it is obvious that $A_c(h,g)$ accepts the same language as $A(h,g)$, that is, $E(h,g)$.

We shall now prove, following [ChK1], that the equality sets of bounded delay morphisms are regular.

Theorem 5.6. *If* h,g *are morphisms of bounded delay, then* $E(h,g)$ *is regular.*

Proof. Let $h,g \colon \Sigma^* \to \Delta^*$ be of bounded delay p. If $A_c(h,g)$ is infinite, then there exists a critical state u such that $|u| \geq (p+2)\max\{|h(a)|, |g(a)| \mid a \in \Sigma\}$. We suppose, by symmetry, that u is a positive word. Consider two letters $a \neq b$ from Σ, for which there are transitions $\delta_c(u,a) = v_1$ and $\delta_c(u,b) = v_2$. Let $\delta_c(v_i, w_i) = 1$ be computations for $i = 1, 2$. Hence $uh(aw_1) = g(aw_1)$ and $uh(bw_2) = g(bw_2)$ in $\Delta^{(*)}$. Further, let w'_i be the prefix of w_i of the maximum length such that $g(aw'_1) \preceq u$ and $g(bw'_2) \preceq u$. Hence the words $g(aw'_1)$ and $g(bw'_2)$ are comparable, say $g(aw'_1) \preceq g(bw'_2)$. From the choice of u we have that $|w'_1| \geq p$. However, this contradicts with the assumption that g is of bounded delay p, since $a \neq b$. Therefore Q_c is finite, and the language accepted by $A_c(h,g)$ is regular. $\square$

In [ChK1] the following stronger result was proved.

Proposition 5.5. *For each nonnegative integer p there exists a regular language R_p over an alphabet Γ such that for any pair of morphisms h, g of bounded delay p, $E(h, g) = f(R_p)$ for some morphism $f: \Gamma^* \to \Sigma^*$.*

Theorem 5.6 should be contrasted to Proposition 4.2, which states that PCP is undecidable for biprefix morphisms. Since each biprefix morphism is of bounded delay zero, it follows that one cannot construct effectively the regular sets $E(h, g)$ for bounded delay morphisms.

We have already had several examples of nonregular equality sets. We give now a particularly simple example, where both of the morphisms are injective.

Example 5.4. Let $\Sigma = \{a, b, c, d, e\}$ and consider the following morphisms

	a	b	c	d	e
h	$abbcb$	bb	bcb	cb	e
g	a	bb	c	$bbcb$	$cbbbe$

Both of these morphisms are injective. In fact, h is a prefix code and g is a suffix code which is not of bounded delay from left to right.

One can now show (see, [K3]) that

$$E(h, g) \cap (a\{b, c\}^* d\{b, d\}^* e) = \{abcb^2c \ldots b^n c \cdot b^n \cdot db^{n-1} db^{n-2} \ldots db \cdot dbe \mid n \geq 1\}.$$

Therefore $E(h, g)$ is not even context-free.   □

The above example shows that $E(h, g)$ need not be regular for injective morphisms h and g, which are defined on an alphabet of cardinality 5. The status of regularity in the injective case on an alphabet of cardinality 3 (or 4) is still open, see Problem 10.10.6. By Proposition 5.1 this is not possible in the binary case.

5.4 An effective construction of regular equality sets

As already mentioned PCP is decidable in the binary case. All the known proofs of this are rather complicated – demanding some 20 pages. Moreover, the proofs do not give away any easy construction of the equality set. However, a short proof of Proposition 5.1 in [EKR2] shows that in the binary case the equality sets $E(h, g)$ are of a very simple form, see Proposition 5.1.

Here we conclude, as a consequence of a more general result, that in the binary case the equality set of two morphisms can be effectively found. Our proof is rather short when based on decidability of GPCP(2), and the partial characterization of binary equality sets. Our presentation follows closely that of [HKKr].

We say that an instance (h, g) satisfies the *solvability condition*, if there exists an algorithm which decides for a given $u \in \Delta^*$ whether there exists a nonempty word $w \in \Sigma^*$ such that $uh(w) = g(w)$, i.e., if GPCP is decidable for the instances $(h, g, u, 1, 1, 1)$ with $u \in \Delta^*$. For each $u \in \Delta^*$ let

$$E(h, g; u) = \{w \mid uh(w) = g(w)\}$$

be the *generalized equality set* of the triple $(h, g; u)$. A family $\mathcal{F}$ is said to satisfy the *solvability condition*, if for all $h, g \in \mathcal{F}$, the pair (h, g) satisfies the condition.

Let $E = E(h, g)$ for two morphisms $h, g \colon \Sigma^* \to \Delta^*$, and let $u^{-1}E(h, g) = \{w \mid uw \in E\}$ with $u \in \Sigma^*$. Clearly,

$$u^{-1}E(h, g) = \{w \mid h(u)h(w) = g(u)g(w)\} \ . \tag{5.2}$$

We obtain immediately that

$$u^{-1}E(h, g) = \begin{cases} E(g, h; g(u)^{-1}h(u)), & \text{if } g(u) \text{ is a prefix of } h(u) \ , \\ E(h, g; h(u)^{-1}g(u)), & \text{if } h(u) \text{ is a prefix of } g(u) \ , \end{cases}$$

and therefore

Lemma 5.3. $u^{-1}E(h, g)$ *is a generalized equality set for all h, g and u.*

Two sets in (5.2) are either disjoint or equal:

Lemma 5.4. *For all $u, v \in \Sigma^*$ either $u^{-1}E(h, g) \cap v^{-1}E(h, g) = \emptyset$ or $u^{-1}E(h, g) = v^{-1}E(h, g)$.*

Proof. Denote $E = E(h, g)$. If $w \in u^{-1}E \cap v^{-1}E$, then $h(u)h(w) = g(u)g(w)$ and $h(v)h(w) = g(v)g(w)$. If here $h(u) = g(u)z$ for a $z \in \Delta^*$, then $h(w) = zg(w)$, which implies that also $h(v) = g(v)z$. From this it follows that $u^{-1}E = v^{-1}E$. Symmetrically, the same conclusion follows if $h(u)$ is a prefix of $g(u)$, i.e., $g(u) = h(u)z$. $\square$

We recall the automaton $A(h, g)$ of Section 5.3 accepting $E = E(h, g)$. More precisely, we note that the automaton $B(h, g) = (\mathcal{Q}, \Sigma, \varphi, E, E)$, where $\mathcal{Q}$ is the family of sets in (5.2) and

$$\varphi(u^{-1}E, a) = (ua)^{-1}E \quad \text{for } a \in \Sigma, \ u \in \Sigma^* \ ,$$

is just a renaming of $A(h, g)$. Indeed, $\beta(u) \leftrightarrow u^{-1}E$ gives a correspondence between the states of these two automata. Therefore

Theorem 5.7. $L(B(h, g)) = E(h, g) \cup \{1\}$.

We still need one auxiliary result referred to as *Nerode's theorem*, see [E, Theorem 8.1].

Lemma 5.5. $E(h, g)$ *is regular if and only if the family $\mathcal{Q} = \{u^{-1}E(h, g) \mid u \in \Sigma^*\}$ is finite.*

The following lemma gives a simple condition that allows us to check effectively which one of the cases in Lemma 5.4 holds.

Lemma 5.6. *Suppose $u^{-1}E(h,g) \neq \emptyset$ for some $u \in \Sigma^*$. It is decidable for words $v \in \Sigma^*$ whether or not $u^{-1}E(h,g) = v^{-1}E(h,g)$.*

Proof. By an exhaustive search we can find a word w such that $w \in u^{-1}E(h,g)$, since $u^{-1}E(h,g)$ is assumed to be nonempty. Now, by Lemma 5.4, $u^{-1}E(h,g) = v^{-1}E(h,g)$ if and only if $w \in v^{-1}E(h,g)$, which in turn in equivalent to the condition $vw \in E(h,g)$. Of course, the latter condition is trivial to check.   □

Our main effectiveness result is as follows:

Theorem 5.8. *Let $\mathcal{F}$ be a family of morphisms that satisfies the solvability condition, and let $h, g \in \mathcal{F}$. If $E(h,g)$ is regular then $E(h,g)$ can be effectively found.*

Proof. Let $h, g \colon \Sigma^* \to \Delta^*$ be two morphisms in $\mathcal{F}$, and assume that $E(h,g)$ is regular. Consider the automaton $B(h,g)$ as defined above. By Nerode's theorem, $E(h,g)$ is regular if and only if $B(h,g)$ is a finite automaton. Using the solvability condition for the instance (h,g) and knowing that $B(h,g)$ is a finite automaton, it can be constructed as follows:

(1) Check whether $E(h,g) \neq \emptyset$.
    If $E(h,g) = \emptyset$, then output $B(h,g)$ as the finite automaton having only the initial state and no transitions. Suppose then that $E(h,g) \neq \emptyset$.
(2) For $n \geq 0$ suppose we have already found all nonempty states $u^{-1}E(h,g)$ with $|u| = n$. Let the set of these be $\mathcal{Q}_n$.
    Set $\mathcal{Q}_{n+1} := \mathcal{Q}_n$.
    Check, for each $a \in \Sigma$ and $u \in \mathcal{Q}_n$ whether $(ua)^{-1}E(h,g) \neq \emptyset$. These can be tested by the solvability condition. If the answer is positive, then check by Lemma 5.6 whether $(ua)^{-1}E(h,g) \notin \mathcal{Q}_n$.
(3) When the first n for which $\mathcal{Q}_{n+1} = \mathcal{Q}_n$ is reached then output $B(h,g)$ having the state set $\mathcal{Q}_n$.

Case **(3)** must be eventually reached by Lemma 5.6. By the construction, when this is reached for the first time all the states of $B(h,g)$ have been found, *i.e.*, $B(h,g)$ has been constructed.   □

Corollary 5.2. *Let $\mathcal{F}$ be a family of morphisms for which GPCP is decidable, and let $h, g \in \mathcal{F}$. If $E(h,g)$ is regular, then $E(h,g)$ can be effectively found.*

Since GPCP(2) is decidable, and in this case, by Proposition 5.1, the equality sets $E(h,g)$ are either regular or of a very special form, we have obtained

Corollary 5.3. *In the binary case the equality set of two morphisms can be effectively found.*

Secondly we observe that the proof of Theorem 5.8 extends immediately to the generalized equality sets, when regularity is demanded on the generalized equality set, and the solvability condition is changed to the *strong solvability condition*: given morphisms $h, g \colon \Sigma^* \to \Delta^*$ in a family $\mathcal{F}$, and four words u_1, u_2, v_1, v_2 it is decidable whether the instance $(h, g, u_1, u_2, v_1, v_2)$ of GPCP has a nontrivial solution, *i.e.*, GPCP is decidable for $\mathcal{F}$.

Theorem 5.9. *Let $\mathcal{F}$ be a family of morphisms for which GPCP is decidable. If $E(h, g, u_1, u_2, v_1, v_2) = \{w \in \Sigma^* \mid u_1 h(w) u_2 = v_1 g(w) v_2\}$ is regular for $h, g \in \mathcal{F}$, then it can be effectively found.*

As above, also Theorem 5.9 has an interesting consequence.

Corollary 5.4. *In the binary case the generalized equality set of two morphisms can be effectively found.*

Proof. We need the fact that for two periodic morphisms the corollary holds, as well as, that in other cases the generalized equality set is always regular. Both of these facts can be easily proved, as in the case of ordinary equality sets, see [EKR2].   □

Finally, we notice that, as in the proof of the decidability of PCP(2) in [EKR1], the use of generalized instances is necessary in order to obtain results for nongeneralized instances. Indeed, from the assumptions

$$E(h, g) \text{ is regular, and it is decidable whether } E(h, g) \neq \emptyset$$

we *cannot conclude* Theorem 5.8. To see this concretely let $h, g \colon \Sigma^* \to \Sigma^*$ be any two prefix codes, and define $h', g' \colon (\Sigma \cup \{d\})^* \to (\Sigma \cup \{d\})^*$, with $d \notin \Sigma$, by $h'(d) = d = g'(d)$ and $h'(x) = h(x)$, $g'(x) = g(x)$ for $x \neq d$. Then we have

$$E(h', g') = (E(h, g) \cup \{d\})^+,$$

and hence $E(h', g')$ is a regular set, since $E(h, g)$ is such, see [EKR2]. However, $E(h', g') \neq \emptyset$ by definition, and hence if $E(h, g)$ were effectively computable, we would be able to decide whether $E(h, g) \neq \emptyset$, which would contradict the undecidability of PCP for prefix codes, see Proposition 4.2.

6. Ehrenfeucht's Conjecture and systems of equations

Systems of equations have a natural interest in mathematics, and it is no surprise that this topic has been studied intensively also in combinatorial theory of free semigroups, see *e.g.* [Hm], [K5], [Lo], [M] and [Sch].

In this section we are interested in infinite systems of equations over word monoids, and we shall prove Ehrenfeucht's Conjecture, a variant of which states that every infinite system of equations (over a free monoid) possesses a finite equivalent subsystem.

6.1 Systems of equations

Let Σ be an alphabet, and let X be a finite set of variables (that generates the free monoid X^*). An (*constant-free*) *equation* $u = v$ consists of two words u and v from X^*. A morphism $h\colon X^* \to \Sigma^*$ is a *solution* of the equation $u = v$, if $h(u) = h(v)$, *i.e.*, if (u, v) belongs to the kernel of the morphism h. Therefore h is a solution, if one obtains equal words when each variable x is substituted by the word $h(x)$ in both of the words u and v.

If $\mathcal{E}$ be a *system of equations* $\{u_i = v_i \mid i \in I\}$, then a morphism h is a solution to $\mathcal{E}$, if h is a common solution of each of the equations from $\mathcal{E}$. Also, we say that two systems $\mathcal{E}_1$ and $\mathcal{E}_2$ of equations are *equivalent*, if they have the same set of solutions. In particular, $\mathcal{E}_1$ is an *equivalent subsystem* of $\mathcal{E}_2$, if $\mathcal{E}_1 \subseteq \mathcal{E}_2$ and $\mathcal{E}_1$ is equivalent to $\mathcal{E}_2$.

The most famous result on system of equations is the following theorem due to Makanin [M].

Proposition 6.1. *It is decidable whether a finite system of equations has a nontrivial solution.*

Actually, the original Makanin's result was stated for one equation with constants, see Section 6.3. However, when constants are allowed it does not make any difference whether a finite system of equations is considered instead of only one equation, see Theorem 6.4.

6.2 Ehrenfeucht's Conjecture

We say that a set $T \subset L$ is a *test set* of a language $L \subseteq \Sigma^*$, if for all morphisms $h, g\colon \Sigma^* \to \Delta^*$,

$$L \subseteq E(h, g) \iff T \subseteq E(h, g) \ .$$

A. Ehrenfeucht conjectured at the beginning of 1970s that *every* language has a *finite* test set. This conjecture was solved affirmatively independently by Albert and Lawrence [AL] and Guba, see [MuS], in 1985.

Theorem 6.1 (Ehrenfeucht's Conjecture). *Each language L has a finite test set.*

Before we prove Theorem 6.1 some remarks are in order. First, the terminology is illustrative: if T is a test set of L, then to test whether two morphisms agree on L it is enough to do the testing on a finite subset T. On the other hand, the conjecture is clearly only existential – T cannot be found effectively for arbitrary languages. Finally, the reader (or at least the skeptical one) is encouraged to search experimental or intuitive support for the conjecture by trying to find two morphisms that agree on a given word, and then taking another word, and trying to find two morphisms that agree on both of them, and so forth.

We shall prove Theorem 6.1 on the main lines of Guba's argumentation. The first step towards the proof of Ehrenfeucht's Conjecture was taken in [CK2] by reducing the conjecture to systems of equations over free semigroups.

Theorem 6.2. *Ehrenfeucht's Conjecture is equivalent to the following statement: each system of equations over a finite set of variables has an equivalent finite subsystem.*

Proof. For any alphabet Γ we let $\overline{\Gamma} = \{\overline{a} \mid a \in \Gamma\}$ be a new alphabet with $\Gamma \cap \overline{\Gamma} = \emptyset$. We write for all words $u = a_1 a_2 \ldots a_k \in \Gamma^+$, $\overline{u} = \overline{a}_1 \overline{a}_2 \ldots \overline{a}_k \in \overline{\Gamma}^+$. Hence the function $^-$ is an isomorphism $\Gamma^* \to \overline{\Gamma}^*$.

Let us first assume that Ehrenfeucht's conjecture is true, and let $\mathcal{E}$ be a system of equations in variables $X = \{x_1, x_2, \ldots, x_n\}$. Let $\overline{X}$ be a set of new variables as defined above. Let

$$L = \{u\overline{v} \mid u = v \in \mathcal{E}\} \subseteq X^* \cdot \overline{X}^* \ .$$

By assumption, L has a finite test set $T \subseteq L$: for morphisms $h, g: (X \cup \overline{X})^* \to \Delta^*$, if $h(u\overline{v}) = g(u\overline{v})$ for all $u\overline{v} \in T$, then $h(u\overline{v}) = g(u\overline{v})$ for all $u\overline{v} \in L$.

For each $f: X^* \to \Delta^*$ define two new morphisms $h_1, h_2: (X \cup \overline{X})^* \to \Delta^*$ by

$$h_1(x) = f(x) \ , \qquad h_2(x) = 1 \ ,$$
$$h_1(\overline{x}) = 1 \ , \qquad h_2(\overline{x}) = f(x) \ ,$$

for all $x \in X$. It follows that if f is a solution to $T = \{u = v \mid u\overline{v} \in T\}$, then $f(u) = f(v)$ for all $u = v$ in T, and, consequently, $h_1(u\overline{v}) = h_2(u\overline{v})$ for all $u\overline{v}$ in T. Therefore $h_1(u\overline{v}) = h_2(v\overline{v})$ for all $u\overline{v}$ in L, which implies that $f(u) = f(v)$ for all $u = v$ in $\mathcal{E}$. This proves that $\mathcal{E}$ is equivalent to the finite subsystem T.

Assume then that each system of equations has an equivalent finite subsystem. Let $L \subseteq \Gamma^*$ be any set of words, and let $\overline{\Gamma}$ be a new alphabet as above. We form a system of equations $\mathcal{E}$ from L as follows: $\mathcal{E} = \{u = \overline{u} \mid u \in L\}$. By assumption this system has an equivalent subsystem T. Let $T = \{u \mid u = \overline{u} \in T\}$. Hence T is a finite subset of L. Let $h, g: \Gamma^* \to \Delta^*$ be morphisms such that $h(u) = g(u)$ for all $u \in T$. Let $X = \Gamma \cup \overline{\Gamma}$ be our set of variables, and define a morphism $f: V^* \to \Delta^*$ by

$$f(x) = h(x) \qquad \text{and} \qquad f(\overline{x}) = g(x) \ ,$$

all $x \in \Gamma$. Now, $f(u) = f(\overline{u})$ for all $u \in T$, and hence f is a solution to the finite system T. By assumption, f is a solution to the whole system $\mathcal{E}$, and, consequently, $h(u) = g(u)$ for all $u \in L$. This shows that T is a finite test set of L. □

In the proof of Theorem 6.1 we shall use an immediate corollary of Hilbert's Basis Theorem, see *e.g.* [Co]. Let $\mathbb{Z}[x_1, x_2, \ldots, x_t]$ be the ring of polynomials with integer coefficients on commuting indeterminants $x_1, \ldots, x_t$. A mapping $f: \{x_1, \ldots, x_k\} \rightarrow \mathbb{Z}$ is a *solution* of a polynomial equation $P(x_1, \ldots, x_t) = 0$, if $P(f(x_1), \ldots, f(x_t)) = 0$, or equivalently, if $f(P) = 0$ when f is extended to a ring morphism $\mathbb{Z}[x_1, \ldots, x_t] \rightarrow \mathbb{Z}$.

Proposition 6.2 (Hilbert's Basis Theorem). *Let $P_i \in \mathbb{Z}[x_1, x_2, \ldots, x_t]$ for $i \geq 1$. There exists an integer r such that each P_k can be written as a linear combination $P_k = \sum_{i=1}^{r} Q_i P_i$, where $Q_i \in \mathbb{Z}[x_1, x_2, \ldots, x_t]$. In particular, each system of equations $\{P_i = 0 \mid i \geq 1\}$, has an equivalent finite subsystem $P_1 = 0, \ldots, P_r = 0$.*

In order to make an advantage of Proposition 6.2 we need to transform the word equations $u = v$ into polynomial equations in $\mathbb{Z}[x_1, x_2, \ldots, x_k]$. The difficulty here is that the variables in the word equations do not commute, but the indeterminants in the ring of polynomials are commuting. To overcome this difficulty we use noncommuting matrices of the integer polynomials.

Let $\mathbf{SL}(2, \mathbb{N})$ denote the *special linear monoid*, a multiplicative submonoid of $\mathbb{N}^{2 \times 2}$ consisting of the matrices M with determinants $\det(M) = 1$. Notice that $\mathbf{SL}(2, \mathbb{N})$ is a submonoid of the special linear group $\mathbf{SL}(2, \mathbb{Z})$ consisting of unimodular integer matrices.

Lemma 6.1. $\mathbf{SL}(2, \mathbb{N})$ *is a free monoid generated by the matrices*

$$A = \begin{pmatrix} 1 & 1 \\ 0 & 1 \end{pmatrix} \quad and \quad B = \begin{pmatrix} 1 & 0 \\ 1 & 1 \end{pmatrix} .$$

Proof. The matrices A and B have the inverses

$$A^{-1} = \begin{pmatrix} 1 & -1 \\ 0 & 1 \end{pmatrix} \quad and \quad B^{-1} = \begin{pmatrix} 1 & 0 \\ -1 & 1 \end{pmatrix}$$

in the group $\mathbf{SL}(2, \mathbb{Z})$. Let then

$$M = \begin{pmatrix} m_{11} & m_{12} \\ m_{21} & m_{22} \end{pmatrix}$$

be an arbitrary nonidentity matrix in $\mathbf{SL}(2, \mathbb{N})$. We obtain

$$MA^{-1} = \begin{pmatrix} m_{11} & m_{12} - m_{11} \\ m_{21} & m_{22} - m_{21} \end{pmatrix} \quad and \quad MB^{-1} = \begin{pmatrix} m_{11} - m_{12} & m_{12} \\ m_{21} - m_{22} & m_{22} \end{pmatrix} .$$

If here $m_{11} \leq m_{12}$, then also $m_{21} \leq m_{22}$, since $\det(M) = 1$, and in particular, MA^{-1} is in $\mathbf{SL}(2, \mathbb{N})$, but MB^{-1} is not. Also, in this case MA^{-1} has a strictly smaller sum of entries than M.

Similarly, if $m_{11} > m_{12}$, then $MB^{-1} \in \mathbf{SL}(2, \mathbb{N})$ and $MA^{-1} \notin \mathbf{SL}(2, \mathbb{N})$. In this case, MB^{-1} has a strictly smaller sum of entries than M. Combining these arguments we deduce that there is a unique sequence $A_1, A_2, \ldots, A_k$ of the matrices A and B such that $MA_1^{-1}A_2^{-1}\ldots A_k^{-1} = I$, where I is the identity matrix. It follows that M can be factored uniquely as $M = A_k A_{k-1} \ldots A_1$, which shows also that $\mathbf{SL}(2, \mathbb{N})$ is freely generated by A and B. $\square$

By Lemma 6.1, the morphism $\mu: \{a, b\}^* \rightarrow \mathbf{SL}(2, \mathbb{N})$ defined by

$$\mu(a) = \begin{pmatrix} 1 & 1 \\ 0 & 1 \end{pmatrix} \quad \text{and} \quad \mu(b) = \begin{pmatrix} 1 & 0 \\ 1 & 1 \end{pmatrix} \tag{6.1}$$

is an isomorphism.

Let $X = \{x_1, x_2, \ldots, x_k\}$ be a set of word variables. We introduce for each x_i four new *integer variables* x_{i1}, x_{i2}, x_{i3} and x_{i4}, and denote

$$\bar{X} = \{x_{ij} \mid 1 \leq i \leq k, \ 1 \leq j \leq 4\} \ .$$

Further, for each $i = 1, 2, \ldots, k$ define

$$X_i = \begin{pmatrix} x_{i1} & x_{i2} \\ x_{i3} & x_{i4} \end{pmatrix},$$

and let $\mathbf{M}(\bar{X})$ be the submonoid of the matrix monoid $\mathbb{Z}[\bar{X}]^{2 \times 2}$ generated by the matrices $X_1, X_2, \ldots, X_k$.

Lemma 6.2. *The monoid* $\mathbf{M}(\bar{X})$ *is freely generated by* $X_1, X_2, \ldots, X_k$.

Proof. Consider two elements $M = X_{r_1} X_{r_2} \ldots X_{r_m}$ and $M' = X_{s_1} X_{s_2} \ldots X_{s_t}$ of $\mathbf{M}(\bar{X})$. The upper right corner entry M_{12} of M contains the monomial

$$x_{r_1 1} x_{r_2 1} \ldots x_{r_{j-1} 1} x_{r_j 2} x_{r_{j+1} 4} \ldots x_{r_{m-1} 4} x_{r_m 4} \tag{6.2}$$

for $j = 1, 2, \ldots, m$. It is now easy to see that if these monomials exist in $(M')_{12}$, then $t = m$ and $X_{r_j} = X_{s_j}$ for $j = 1, 2, \ldots, m$. Therefore, $(M)_{12} = (M')_{12}$ implies that $X_{r_1} = X_{s_1}, X_{r_2} = X_{s_2}, \ldots, X_{r_m} = X_{s_m}$. The claim follows from this. $\square$

In particular, the monoid morphism $\varphi: X^* \rightarrow \mathbf{M}(\bar{X})$ defined by

$$\varphi(x_i) = \begin{pmatrix} x_{i1} & x_{i2} \\ x_{i3} & x_{i4} \end{pmatrix} \tag{6.3}$$

is an isomorphism.

The following lemma is now immediate.

Lemma 6.3. *Let* Δ *be a binary alphabet. There exists a bijective correspondence* $h \leftrightarrow \bar{h}$ *between the morphisms* $h: X^* \rightarrow \Delta^*$ *and the monoid morphisms* $\bar{h}: \mathbf{M}(\bar{X}) \rightarrow \mathbf{SL}(2, \mathbb{N})$ *such that the following diagram commutes:*

$$X^* \xrightarrow{\quad h \quad} \Delta^*$$

$$\varphi \downarrow \qquad\qquad \downarrow \mu$$

$$M(\overline{X}) \xrightarrow{\quad \overline{h} \quad} SL(2, \mathbb{N})$$

We shall now prove the main result of this section, from which Theorem 6.1 follows by Theorem 6.2.

Theorem 6.3. *Each system of equations over a finite set of variables has an equivalent finite subsystem.*

Proof. Let X and $\bar{X}$ be as above. Since each monoid Δ^* can be embedded into a free monoid generated by two elements, we can restrict our solutions $h: X^* \to \Delta^*$ to the case where Δ is binary, say $\Delta = \{a, b\}$.

For a word $w \in X^*$ we denote

$$\varphi(w) = \begin{pmatrix} P_{w1} & P_{w2} \\ P_{w3} & P_{w4} \end{pmatrix} ,$$

where $P_{wj} \in \mathbb{Z}[\bar{X}]$. Consider any equation $u = v$, where $u, v \in X^*$. Define a finite system $\mathcal{P}(u, v)$ of polynomial equations as follows:

$$\mathcal{P}(u, v) = \begin{cases} P_{uj} = P_{vj} & \text{for } 1 \le j \le 4 , \\ x_{t1}x_{t4} - x_{t2}x_{t3} = 1 & \text{for } 1 \le t \le k . \end{cases}$$

Let $h: X^* \to \Delta^*$ be a morphism, and let $\overline{h}$ be the corresponding monoid morphism from Lemma 6.3. Now, by Lemmas 6.1, 6.2 and 6.3,

$$h(u) = h(v) \iff \mu h(u) = \mu h(v) \iff \overline{h}\varphi(u) = \overline{h}\varphi(v) . \qquad (6.4)$$

Define then a mapping $h': \bar{X} \to \mathbb{N}$ by

$$\overline{h}(X_i) = \begin{pmatrix} h'(x_{i1}) & h'(x_{i2}) \\ h'(x_{i3}) & h'(x_{i4}) \end{pmatrix} .$$

Now, h' is a solution of the equations on the second line of the definition of $\mathcal{P}(u, v)$. This is because $\overline{h}$ is into $SL(2, \mathbb{N})$. Further, h' extends in a unique way to a ring morphism $h': \mathbb{Z}[\bar{X}] \to \mathbb{Z}$, for which

$$\overline{h}\varphi(w) = \begin{pmatrix} h'(P_{w1}) & h'(P_{w2}) \\ h'(P_{w3}) & h'(P_{w4}) \end{pmatrix}$$

for all $w \in X^*$. Therefore by (6.4), $h(u) = h(v)$ if and only if h' is a solution of the whole $\mathcal{P}(u, v)$.

Let now $\mathcal{E} = \{u_i = v_i \mid i \in I\}$ be a system of equations, and let $\mathcal{P} = \cup_{i \in I} \mathcal{P}(u_i, v_i)$. By Hilbert's Basis Theorem, $\mathcal{P}$ has an equivalent finite subsystem $\mathcal{P}_0 = \cup_{i \in I_0} \mathcal{P}(u_i, v_i)$. Consider the subsystem $\mathcal{E}_0 = \{u_i = v_i \mid i \in I_0\}$ of $\mathcal{E}$. By above,

$$h \text{ is a solution to } \mathcal{E}_0 \iff h' \text{ is a solution to } \mathcal{P}_0$$
$$\iff h' \text{ is a solution to } \mathcal{P} \iff h \text{ is a solution to } \mathcal{E},$$

which proves that $\mathcal{E}_0$ is equivalent to $\mathcal{E}$, and this proves the theorem. □

We want to emphasize that Theorem 6.3, as well as Theorem 6.2, reveals a fundamental compactness property of words. We also note that very little is known about the size of equivalent finite subsystems of given systems of equations. In particular, it is not known whether it can be bounded by any function on the number of unknowns. Some lower bounds are obtained in [KP].

6.3 Equations with constants

Let X be a set of variables and Σ an alphabet of *constants*. An *equation* $u = v$ *with constants* Σ is a pair of words $u, v \in (X \cup \Sigma)^*$. A *solution* of such an equation $u = v$ is a morphism $h: (X \cup \Sigma)^* \to \Delta^*$, where $\Sigma \subseteq \Delta$, $h(u) = h(v)$ and $h(a) = a$ for all $a \in \Sigma$. A morphism h is a solution to a system $\mathcal{E}$ of equations with constants Σ, if it is a common solution to the equations in $\mathcal{E}$.

With each constant $a \in \Sigma$ we associate a new variable x_a and in this way each equation $u = v$ with constants Σ can be presented as a finite system of equations:

$$\begin{cases} u' = v' \\ x_a = a \quad \text{for all } a \in \Sigma \ , \end{cases}$$

where u' and v' are obtained from u and v, resp., by substituting x_a for $a \in \Sigma$. Therefore each system of equations with constants Σ can be presented as a system of equations over the variables $X \cup \{x_a \mid a \in \Sigma\}$ together with a finite number of equations $x_a = a$ containing a unique constant. Therefore Theorem 6.3 can be generalized:

Corollary 6.1. *Each system of equation with constants is equivalent to a finite subsystem.*

Let $u_1 = v_1$ and $u_2 = v_2$ be two equations, and let a, b be two different constants. It is easy to show that h is a solution to the above pair of equations if and only if h is a solution to the single equation $u_1 a v_2 u_1 b v_2 = v_1 a u_2 v_1 b u_2$, see [Hm]. In particular, we have

Theorem 6.4. *Each finite set of equations with constants is equivalent to a single equation with constants.*

Note, however, that the single equation of Theorem 6.4 need not be among the original ones, *i.e.*, Theorem 6.4 is not a compactness result in the sense of Theorem 6.3.

6.4 On generalizations of Ehrenfeucht's Conjecture

Systems of equations and equality sets have been considered in a more general setting of monoids in [dLR], [KP] and [HKP]. Here we shall summarize some of these results.

Let $X = \{x_1, x_2, \ldots, x_n\}$ be a finite set of variables. A *solution* of an equation $u = v$ over X *in a monoid* M is a monoid morphism $\alpha\colon X^* \to M$, for which $\alpha(u) = \alpha(v)$. We say that a monoid M satisfies the *compactness property (for systems of equations)*, or *CP* for short, if for all finite sets of variables X every system $\mathcal{E} \subseteq X^* \times X^*$ is equivalent to one of its *finite* subsystems $\mathcal{T} \subseteq \mathcal{E}$. In particular, Ehrenfeucht's Conjecture states that the finitely generated free monoids satisfy CP.

As in Theorem 6.2 the compactness property can also be restated in terms of test sets.

Theorem 6.5. *For any monoid M the compactness property is equivalent to the condition: for all $L \subseteq \Sigma^*$ there exists a finite subset $T \subseteq L$ such that for any two morphisms $\alpha, \beta\colon \Sigma^* \to M$,*

$$\alpha|T = \beta|T \iff \alpha|L = \beta|L \ .$$

Next we mention four examples, where (the generalization of) Ehrenfeucht's Conjecture, *i.e.*, the compactness property, does not hold.

Example 6.1. Let $Fin(\Sigma^*)$ denote the monoid of all nonempty *finite subsets* of the word monoid Σ^*. It was shown in [La] that the monoid $Fin(\Sigma^*)$ does not satisfy CP even when Σ is a binary alphabet. Indeed, in this case the system $\mathcal{E}$ of equations

$$x_1 x_2^i x_1 = x_1 x_3^i x_1 \quad \text{for } i \geq 1$$

over three variables does not have an equivalent finite subsystem in $Fin(\Sigma^*)$, see Example 6.7. □

Example 6.2. The *bicyclic monoid B* is a 2-generator and a 1-relation semigroup with the presentation $\langle a, b \mid ab = 1 \rangle$. The monoid B is isomorphic to the monoid generated by the functions $\alpha, \beta\colon \mathbb{N} \to \mathbb{N}$:

$$\alpha(n) = \max\{0, n-1\}, \qquad \beta(n) = n+1 \ ,$$

see [L1]. Define $\gamma_i = \beta^i \alpha^i$, for $i \geq 0$. Hence

$$\gamma_i(n) = \begin{cases} i & \text{if } n \leq i \ , \\ n & \text{if } n > i \ . \end{cases}$$

We observe that $\gamma_i \gamma_j = \gamma_{\max\{i,j\}}$. Now, consider the system

$$x_1^i x_2^i x_3 = x_3 \quad \text{for } i \geq 1,$$

of equations over the variables x_1, x_2 and x_3. As is easily seen the morphism δ_j defined by $\delta_j(x_1) = \beta$, $\delta_j(x_2) = \alpha$ and $\delta_j(x_3) = \gamma_j$, is a solution of $x_1^i x_2^i x_3 = x_3$ for all $i \leq j$, but δ_j is not a solution of $x_1^{j+1} x_2^{j+1} x_3 = x_3$. Hence the system $\mathcal{E}$ does not have an equivalent finite subsystem, and therefore the bicyclic monoid B does not satisfy CP, as noted in [HKP]. □

Example 6.3. As shown in [HKP], if a finitely generated monoid M satisfies CP, then M satisfies the *chain condition on idempotents*, i.e., each subset E_1 of idempotents of M contains a maximal and a minimal element with respect to an natural ordering: $e \leq f$, if $fe = e = ef$, see [L1]. From this it follows that the *free inverse monoids* do not satisfy CP. □

Example 6.4. Also by [HKP], if a monoid M satisfies CP, then it is *hopfian*, i.e., M is not isomorphic to any of its proper quotients M/θ. The *Baumslag-Solitar group* [BS], defined by a group presentation $G_{BS} = \langle a, b \mid b^2 a = ab^3 \rangle$, is possibly the simplest non-hopfian group, and consequently, it does not satisfy CP. □

We consider now the compactness property for varieties of monoids. Recall that a class $\mathcal{V}$ of monoids is a *variety*, if it is closed under taking submonoids, morphic images, and arbitrary direct products. We need also another notion: A monoid M is said to satisfy the *maximal condition on congruences*, or *MCC* for short, if each set of congruences of M has a maximal element. Although CP and MCC seem to be quite different notions, they, nevertheless, agree on varieties as shown by the following result from [HKP].

Proposition 6.3. *The monoids in a variety $\mathcal{V}$ satisfy CP if and only if each finitely generated monoid M in $\mathcal{V}$ satisfies the maximal condition on congruences.*

Redei's Theorem [Re] states that the finitely generated commutative monoids satisfy the maximal condition on congruences. For a short proof see [Fr]. Hence we have the following corollary of Proposition 6.3.

Corollary 6.2. *Every commutative monoid satisfies CP.*

It is worth noting that in Proposition 6.3 and its corollary the compactness property holds not only for finitely generated monoids, but for infinitely generated monoids as well. Moreover, if the compactness property holds in a variety, it holds there *uniformly* in the sense that for each system of equations its equivalent finite subsystem can be chosen to be the same for all monoids in the variety. In particular, this holds for commutative monoids. Also the submonoids of free monoids satisfy the compactness property uniformly. This is due to the fact that any free monoid, which is generated by at most countably many elements, can be embedded to a free monoid generated by only two elements, so that the equivalent subsystem in the latter monoid works for all free monoids.

Corollary 6.2 goes beyond varieties in the following result of [dLR], where a *trace monoid* is a monoid having a presentation $\langle A \mid ab = ba \; ((a,b) \in R)\rangle$, where R is an equivalence relation on the set A of generators. The proof of Proposition 6.4 relies again on Hilbert's Basis Theorem.

Proposition 6.4. *The finitely generated trace monoids satisfy CP.*

Finally we note, for more details see [HKP], that it can be shown that the bicyclic monoid B does satisfy the maximal condition on congruences although it does not satisfy CP. On the other hand, the free monoids Σ^* with $|\Sigma| \geq 2$ do not satisfy the maximal condition on congruences, but they do satisfy CP. Consequently the notions of 'compactness property' and 'maximality condition on congruences' are incomparable in general, although they coincide on varieties.

6.5 Ehrenfeucht's Conjecture for more general mappings

We may generalize the problem of Ehrenfeucht's Conjecture on test sets to mappings that are more general than morphisms. Let $\mathcal{F}$ be a family of mappings of words, and let L be a language. We say that a finite subset T of L is a *test set* of L with respect to $\mathcal{F}$, if for all $\varphi_1, \varphi_2 \in \mathcal{F}$,

$$\varphi_1|T = \varphi_2|T \implies \varphi_1|L = \varphi_2|L \; ,$$

where $\varphi|T$ denotes the restriction of φ on the subset T of L.

We consider a few automata-theoretic examples of such generalizations from [K4] and [La]. In one of these cases the conjecture holds, while in two others it fails.

Example 6.5. Let for all $n \geq 1$, $\alpha_n \colon a^* \to a^*$ be the function defined by

$$\alpha_n(a^k) = \begin{cases} a^k, & \text{if } k < n \; , \\ a^{k+1}, & \text{if } k \geq n \; . \end{cases}$$

Clearly, α_n can be realized by a sequential transducer, and hence α_n is a rational function. Denote $a^{<k} = \{1, a, \ldots, a^{k-1}\}$. Clearly, $\alpha_n|a^{<k} = \alpha_m|a^{<k}$ for all $n, m \geq k$, but $\alpha_n|a^* \neq \alpha_m|a^*$. This shows that the language a^* does not have a finite test set with respect to rational functions. $\qquad\square$

Example 6.6. Let $\mathcal{F}_n$ be the set of the set of partial functions $\Sigma^* \to \Delta^*$ defined by sequential transducers *with at most n states*. Then, as shown in [K4], each language $L \subset \Sigma^*$ possesses a finite test set with respect to $\mathcal{F}_n$. The proof of this is not difficult. $\qquad\square$

Example 6.7. We restate now Example 6.4 in terms of finite substitutions. Let $\Sigma = \{a, b\}$. For all $n \geq 1$ define the finite substitutions τ_n, σ_n as follows

$\tau_n(a) = \{a^j \mid 0 \leq j \leq 2n + 2\}$,

$\tau_n(b) = \{a^i b a^j \mid 0 \leq i + j < n$ or $(0 \leq i \leq 2n + 2$ and $n + 1 \leq j \leq 2n + 2)$

   or $(n + 1 \leq i \leq 2n + 2$ and $0 \leq j \leq 2n + 2)\}$,

$\sigma_n(a) = \{a^j \mid 0 \leq j \leq 2n + 2\}$,

$\sigma_n(n) = \{a^i b a^j \mid 0 \leq i, j \leq 2n + 2\}$.

Denote $L_k = \{ab^r a \mid 0 \leq r < k\}$. Then $\tau_n|L_n = \sigma_n|L_n$, but $\tau_n|L_{n+1} \neq \sigma_n|L_{n+1}$, since $(ba^n)^n b \in \sigma_n(ab^{n+1}a)$ and $(ba^n)^n b \notin \tau_n(ab^{n+1}a)$. This shows that language ab^*a does not have a finite test set with respect to finite substitutions.   □

Intuitively the difference between Examples 6.5 and 6.6 is that interpreting the test set problem in terms of equations, the latter requires only finitely many unknowns, while the former requires infinitely many unknowns.

7. Effective subcases

In this section we study possibilities of finding test sets effectively. In particular, as an application of Makanin's algorithm and the validity of Ehrenfeucht's Conjecture we show that some problems on iterated morphisms are decidable. We also point out that there are even such decidable problems, for which no other proof is know.

7.1 Finite systems of equations

As already stated above in Proposition 6.1, it is decidable whether a finite system of equations has a solution. Using this the following result was proved in [CK2].

Theorem 7.1. *Let $\mathcal{E}$ and $\mathcal{E}'$ be finite systems of equations with constants. It is decidable whether there exists a solution of $\mathcal{E}'$ that is not a solution of $\mathcal{E}$.*

Proof. Let $\mathcal{E} = \{u_i = v_i \mid i = 1, 2, \ldots, r\}$ and $\mathcal{E}' = \{u_i' = v_i' \mid i = 1, 2, \ldots, s\}$ be two finite systems of equations over a set X of variables and with constants from Σ. We construct a finite set $\mathcal{E}_1, \mathcal{E}_2, \ldots, \mathcal{E}_n$ of systems of equations such that $\mathcal{E}$ and $\mathcal{E}'$ are equivalent if and only if no one of these systems $\mathcal{E}_i$ has a solution. By Proposition 6.1 the claim then follows.

For each constant $a \in \Sigma$ and integer $k = 1, 2, \ldots, r$ define two systems of equations over variables $X \cup \{x\}$ as follows:

$$\mathcal{E}_{k_1}^a = \mathcal{E}' \cup \{u_k = v_k ax\} \quad \text{and} \quad \mathcal{E}_{k_2}^a = \mathcal{E}' \cup \{u_k ax = v_k\} ,$$

and for each pair (a, b), $a \neq b$, of constants define a system

$$\mathcal{E}_k^{(a,b)} = \mathcal{E}' \cup \{u_k = xay, v_k = xbz\}$$

over variables $X \cup \{x, y, z\}$.

Clearly, if h is a solution of one of these new systems of equations, then h is a solution of $\mathcal{E}'$, but $h(u_k) \neq h(v_k)$. On the other hand, if there exists a solution h of $\mathcal{E}'$, which is not a solution of $\mathcal{E}$, then there is an index k such that $h(u_k) \neq h(v_k)$ and h is a solution of $\mathcal{E}_{k_1}^a$ or $\mathcal{E}_{k_2}^a$ or $\mathcal{E}_k^{(a,b)}$ for some a and b. This shows that we can decide whether there exists a solution of $\mathcal{E}'$ which is not a solution of $\mathcal{E}$.   □

Theorem 7.1 and its proof has a number of interesting corollaries. We state three of them here. The first of these states that a problem which can be called *dual PCP* is decidable, see [CK1].

Corollary 7.1. *It is decidable whether for a word $w \in \Sigma^*$ there exist different nonperiodic morphisms $h, g \colon \Sigma^* \to \Delta^*$ such that $h(w) = g(w)$.*

One should notice that the above statement is trivial without the exclusion of periodic morphisms. Indeed, for all words w there are trivially different periodic morphisms $h, g \colon \Sigma^* \to a^*$ such that $h(w) = g(w)$. Corollary 7.1 remains decidable, when the morphisms h and g are required to be such that just one of them is nonperiodic.

Corollary 7.2. *It is decidable whether a finite set F_1 is a test set of another finite set F_2.*

Corollary 7.3. *The equivalence of two finite systems of equations with constants is decidable.*

7.2 The effectiveness of test sets

In general, finding a finite test set for a given language $L \subseteq \Sigma^*$ is not effective. Indeed, if $\mathcal{L}$ is an effectively given family of languages (say, given by a family of Turing Machines) with an undecidable emptiness problem, then one cannot construct effectively test sets $T_L \subseteq L$ for the languages $L \in \mathcal{L}$, because $T_L \neq \emptyset$ just in case $L \neq \emptyset$. In particular, one cannot find effectively finite test sets in the family of equality sets.

Lemma 7.1. *A language L has an effectively findable finite test set if and only if there is an algorithm that decides whether a finite set T is a test set of L.*

Proof. First of all, if a finite test set $T \subseteq L$ can be effectively found, then we can check whether another finite set T' is a test set of L by Corollary 7.2. On the other hand, if we can decide whether a finite set is a test set of L, then a systematic search will eventually terminate at a test set of L.   □

A closer look of the proof of Theorem 6.2 reveals the following equivalence result, where we say that a system of equations $\mathcal{E}$ is *of type $\mathcal{L}$*, if $\mathcal{E} = \{h(w) = g(w) \mid w \in L\}$ for some morphisms h, g and a language $L \in \mathcal{L}$.

Theorem 7.2. *The following two conditions are equivalent for a family $\mathcal{L}$ of languages:*

(1) *Each (effectively given) language $L \in \mathcal{L}$ possesses an effectively findable test set.*
(2) *Each (effectively given) system $\mathcal{E}$ of equations of type $\mathcal{L}$ has an equivalent finite subsystem that can be effectively found.*

Based on the pumping property of regular languages the following was noticed in [CK2], see also [K2]:

Theorem 7.3. *Each regular language R has a test set consisting of words with lengths $\leq 2n$, where n is the number of states in a finite automaton accepting R.*

Proof. Let $h, g \colon \Sigma^* \to \Delta^*$ be two morphisms. By Lemma 5.2 we have

$$uv, uzv, uwv \in E(h, g) \implies uzwv \in E(h, g)$$

for all words $u, v, z, w \in \Sigma^*$. Consider then an accepting computation of a finite automaton A, which visits a state q at least three times:

$$q_0 \xrightarrow{\ u\ } q \xrightarrow{\ z\ } q \xrightarrow{\ w\ } q \xrightarrow{\ v\ } q_f \ ,$$

where q_1 is the initial state and q_f is a final state. By above, if $h(uzwv) \neq g(uzwv)$, then h and g disagree already on one of the words $uv, uzv, uwv \in L(A)$, and this proves the claim. $\square$

This result was improved in [KRJ] as follows.

Proposition 7.1. *If $L(A)$ is a regular language accepted by a (nondeterministic) finite automaton A with m transitions, then a test set T of $L(A)$ with $|T| \leq m$ can be effectively found.*

We say that a system $\mathcal{E}$ of equations is *rational*, if there exists a finite transducer that realizes $\mathcal{E}$, *i.e.*, if $\mathcal{E}$ is a rational transduction. By Nivat's theorem, see *e.g.* [B], a system $\mathcal{E}$ of equations is rational if and only if it is of type $\mathcal{R}$, where $\mathcal{R}$ denotes the family of the regular languages. Also, a system $\mathcal{E}$ of equations is said to be *algebraic*, if it is of type $\mathcal{CF}$ for the family $\mathcal{CF}$ of the context-free languages.

Theorem 7.1, and consequently Corollary 7.2, was generalized to rational systems of equations in [CK2]. By Theorem 7.3 a regular language R possesses effectively a finite test set $T \subseteq R$, and from this we can deduce the following two corollaries.

Corollary 7.4. *Each rational system of equations possesses an equivalent finite subsystem, which can be effectively found.*

Corollary 7.5. *It is decidable whether two rational systems of equations are equivalent.*

As a consequence of Corollary 7.4 one can prove the following interesting result, *cf.* [ChK2], which should be compared to Theorem 4.6.

Theorem 7.4. *The isomorphism problem is decidable for finitely generated submonoids of a free monoid.*

Theorem 7.3 was improved in [ACK] for context-free languages and, consequently, to algebraic systems of equations.

Proposition 7.2. *Each context-free language has an effectively findable finite test set. In particular, each algebraic system of equations has a finite equivalent subsystem, which can be effectively found.*

An upper bound for the cardinality of a test set for a context-free language L was shown to be of order n^6 in [KPR]. Here n denotes the number of productions in the context-free grammar generating L. On the other hand, a lower bound for the size of such a test set was shown to be of order n^3 also in [KPR].

7.3 Applications to problems of iterated morphisms

One of the most interesting problems on morphisms is the *D0L sequence equivalence problem* (or the *D0L-problem*, for short), which was posed by A. Lindenmayer at the beginning of 1970s. It was a challenging and fruitful problem in the 70s, which created many new results and notions, see [K6]. The D0L-problem was solved by Culik and Fris [CF], and another solution for it was given in [ER]. We give here still another proof from [CK5] based on Ehrenfeucht's Conjecture.

In the *D0L-problem* we are given two endomorphisms $h, g \colon \Sigma^* \to \Sigma^*$ and an initial word $w \in \Sigma^*$, and we ask whether $h^k(w) = g^k(w)$ holds for all $k \geq 0$.

We denote for an endomorphism h

$$h^*(w) = \{h^k(w) \mid k \geq 0\} \ .$$

Clearly, the D0L-problem is equivalent to the problem whether $h(h^k(w)) = g(h^k(w))$ holds for all $k \geq 0$, *i.e.*, whether $h^*(w) \subseteq E(h, g)$. A natural extension of this problem is the following *morphic equivalence problem for D0L languages*: given morphisms $h, g \colon \Sigma^* \to \Delta^*$, $f \colon \Sigma^* \to \Sigma^*$ and a word $w \in \Sigma^*$, does $h(f^k(w)) = g(f^k(w))$ hold for all $k \geq 0$, *i.e.*, does $f^*(w) \subseteq E(h, g)$ hold?

Still a further extension is obtained as follows: given morphisms $h, g \colon \Sigma^* \to \Delta^*$, $f_1, f_2 \colon \Sigma^* \to \Sigma^*$ and a word w, does $h(f_1^k(w)) = g(f_2^k(w))$ hold for all $k \geq 0$? This problem is known as the *HD0L-problem*.

By the validity of Ehrenfeucht's Conjecture and Makanin's algorithm, each of the above three problems can be shown to be decidable.

Theorem 7.5. *For each endomorphism f and word w a finite test set for $f^*(w)$ can be effectively found. Consequently, each of the following problems is decidable: the DOL-problem, the morphic equivalence problem for DOL languages, and the HDOL-problem.*

Proof. Denote $f^{[k]}(w) = \{f^i(w) \mid 0 \le i \le k\}$. By Ehrenfeucht's Conjecture there exists an integer t such that $f^{[t]}(w)$ is a test set of $f^*(w)$, and hence, in particular, $f^{[t]}(w)$ is a test set of $f^{[t+1]}(w)$. The smallest integer k for which $f^{[k]}(w)$ is a test set of $f^{[k+1]}(w)$ can be effectively found, since we know that such a k exists by Ehrenfeucht's Conjecture, and Corollary 7.2 guarantees that the checking can be done effectively. The claim follows when we show that $f^{[k]}(w)$ is a test set of $f^*(w)$.

Indeed, consider the word $f^{k+2}(w)$. Now, since $f^{[k]}(w)$ is a test set of $f^{[k+1]}(w)$, we conclude that $f(f^{[k]}(w)) = f^{[k+1]}(w) \setminus \{w\}$ is a test set of $f(f^{[k+1]}(w)) = f^{[k+2]}(w) \setminus \{w\}$, and hence $f^{[k]}(w)$ is a test set of $f^{[k+2]}(w)$, from which we obtain inductively that $f^{[k]}(w)$ is a test set for $f^*(w)$. This proves the first sentence of the theorem.

To prove the second sentence we proceed as follows. First the decidability of the first two problems are obvious. The third one, in turn, is reduced to the second one by assuming that f_1 and f_2 are defined in disjoint alphabets, say Σ_1 and Σ_2, and extending h and g to $(\Sigma_1 \cup \Sigma_2)^*$ such that $h(\Sigma_1) = \{1\} = g(\Sigma_2)$. $\square$

We remind that the HDOL-problem was shown to be decidable in [Ru2] without using Ehrenfeucht's Conjecture. On the other hand, the following extension of the DOL-problem, called the *DTOL-problem*, is an example of a problem, for which no other proof than the one based on Ehrenfeucht's Conjecture is known. In the *DTOL-problem* we are given a word $w \in \Sigma^*$, two monoids H_1 and H_2 of endomorphisms of Σ^* with equally many generators $\{h_1, h_2, \ldots, h_n\}$ and $\{g_1, g_2, \ldots, g_n\}$, resp., and we ask whether for all sequences $i_1, i_2, \ldots, i_k$ of indices $h_{i_k} \ldots h_{i_1}(w) = g_{i_k} \ldots g_{i_1}(w)$. The proof of the following result, *cf.* [CK5], is a straightforward extension of the proof of Theorem 7.5.

Theorem 7.6. *The DTOL-problem is decidable.*

This result should be contrasted to the undecidability of the *DTOL language equivalence problem*: given a word w and two finitely generated monoids H_1 and H_2 of endomorphisms, it is undecidable whether for all $h \in H_1$ there exists a $g \in H_2$ such that $g(w) = h(w)$. The proof of this result is based on the undecidability of PCP, see [RS1].

The method of the proof of Theorem 7.5 can be applied to prove decidability results on classical automata theory as well. The following was established in [CK4].

Proposition 7.3. *The equivalence problem for finite-valued transducers is decidable.*

We note that while Proposition 7.3 has also a purely automata-theoretic proof, see [W], there are related problems for which only proofs using Ehrenfeucht's Conjecture are known, see [CK4] and [CK6].

8. Morphic representations of languages

In formal language theory representation results for various families of languages have always been a topic of a special interest. Typically such a representation result for a family $\mathcal{L}$ of languages characterizes the languages in $\mathcal{L}$ in terms of generators and operations so that each language $L \in \mathcal{L}$ can be obtained from these generators by using the allowed operations.

The most basic example of such a case is the family of regular languages $\mathcal{R}$, for which the generators are the empty set $\emptyset$ and the singleton sets $\{a\}$ of letters and the operations are union, Kleene iteration, and catenation.

For two morphisms $g: \Delta^* \to \Sigma^*$ and $h: \Delta^* \to \Gamma^*$, we let $hg^{-1}: \Sigma^* \to 2^{\Gamma^*}$ be the *composition* of g^{-1} and h:

$$hg^{-1}(w) = \{h(u) \mid u \in \Delta^*, \ g(u) = w \in \Sigma^*\} \ .$$

We denote by $\mathcal{H}$ the family of morphisms between finitely generated word monoids, and by $\mathcal{H}^{-1}$ the family of inverse morphisms. Each morphism $h: \Sigma^* \to \Delta^*$ induces the language operations $h: 2^{\Sigma^*} \to 2^{\Delta^*}$ and $h^{-1}: 2^{\Delta^*} \to 2^{\Sigma^*}$ in a natural way,

$$h(L) = \{h(u) \in \Delta^* \mid u \in L\} \quad \text{and} \quad h^{-1}(K) = \{w \in \Sigma^* \mid h(w) \in K\} \ .$$

The compositions of such operations are called *morphic compositions*. Obviously, $\mathcal{H}\mathcal{H} = \mathcal{H}$ and $\mathcal{H}^{-1}\mathcal{H}^{-1} = \mathcal{H}^{-1}$, and therefore each morphic composition τ can be written as an alternating composition of morphisms and inverse morphisms, *i.e.*, $\tau = h_n^{\varepsilon_n} h_{n-1}^{\varepsilon_{n-1}} \ldots h_1^{\varepsilon_1}$, where $\varepsilon_j + \varepsilon_{j-1} = 0$ and $\varepsilon_j, \varepsilon_{j-1} \in \{-1, +1\}$ for each $j = 2, \ldots, n$.

8.1 Classical results

Two of the most famous classical representation results in language theory are given for the family $\mathcal{CF}$ of context-free languages. These are due to Chomsky and Schützenberger [CS1] and Greibach [G1]. By the first of these results each context-free language L can be obtained from a Dyck language D, *cf.* [B], using an intersection with a regular set followed by a morphism:

Proposition 8.1. *Each context-free language can be written as $L = h(D \cap R)$, where D is a Dyck language, R a regular language and h a morphism.*

In an operational form this can be written as follows: $CF = \mathcal{H} \circ \cap \mathcal{R}(\mathcal{D})$. where $\mathcal{D}$ consists of the Dyck languages, $\cap \mathcal{R}$ denotes the intersections with regular languages, and $\mathcal{H}$ the family of morphisms. In fact, see *e.g.* [H], the generator set $\mathcal{D}$ can be reduced to the single generator D_2, the Dyck language over one pair of brackets: $CF = \mathcal{H} \circ \cap \mathcal{R} \circ \mathcal{H}^{-1}(D_2)$.

Greibach, on the other hand, showed that each context-free language can be obtained from a single language using only inverse morphisms.

Proposition 8.2. *There exists a context-free language U_2 such that $CF = \mathcal{H}^{-1}(U_2)$.*

Here the context-free language U_2 is known as the *hardest context-free language*, and it is a nondeterministic variant of the Dyck language D_2.

In [CM] a similar result for the family $\mathcal{RE}$ of recursively enumerable languages and for the family CS of context-sensitive languages was established:

Proposition 8.3. *There exist a recursively enumerable language U_0 and a context sensitive language U_1 such that $\mathcal{RE} = \mathcal{H}^{-1}(U_0)$ and $CS = \mathcal{H}^{-1}(U_1)$.*

8.2 Representations of recursively enumerable languages

One of the interesting topics on computational aspects of morphisms is that of morphic characterizations of recursively enumerable languages, initiated in [S3], [C1] and [EnR1]. We begin with the following result from [Ge], which is interesting also from the point of view of equality sets. Indeed, in Theorem 8.1 the recursively enumerable languages are closely related to the overflows of two morphisms. In the proof of Theorem 8.1 we follow the presentation of [T3].

Theorem 8.1. *For each recursively enumerable language $L \subseteq \Sigma^*$ there are two morphisms $h, g: \Delta^* \to \Gamma^*$ with h nonerasing such that*

$$L = \{h^{-1}(w)g(w) \mid w \in \Delta^*\} \cap \Sigma^* .$$

Proof. Let $L = L(G)$, where $G = (V, \Sigma, P, S)$ is a (general) grammar with terminals Σ, nonterminals $N = V \setminus \Sigma$, productions $P \subseteq V^*NV^* \times V^*$ and with the initial nonterminal $S \in N$. We denote by T the *terminal productions* of G, i.e., $T = P \cap (V^* \times \Sigma^*)$.

For any alphabet X we let $\overline{X} = \{\overline{a} \mid a \in X\}$ be an alphabet consisting of new letters. Define

$$\Delta = V \cup \overline{\Sigma} \cup P \cup \overline{T} \cup \{d, e, f\}, \quad \Gamma = V \cup \{c, d, f\} .$$

Let $m, r: \Delta^* \to (\Delta \cup \{c\})^*$ be morphisms defined as follows: for all $x \in \Delta$,

$$m(x) = cxc, \qquad r(x) = xcc .$$

Note that m and r are similar to our shift morphisms of Sections 3 and 4.

The morphisms h and g are defined as follows:

$$
\begin{aligned}
h(d) &= dc \ , & g(d) &= r(df S) \ , \\
h(e) &= m(f)c \ , & g(e) &= 1 \ , \\
h(x) &= m(x) \ , & g(x) &= r(x) & \text{for } x \in V \cup \{f\} \ , \\
h(p) &= m(u) \ , & g(p) &= r(v) & \text{for } p = u \rightarrow v \in P \ , \\
h(\overline{a}) &= h(a) \ , & g(\overline{a}) &= a & \text{for } a \in \Sigma \ , \\
h(\overline{p}) &= h(p) \ , & g(\overline{p}) &= v & \text{for } p = u \rightarrow v \in T \ .
\end{aligned}
$$

Denote $K = \{h^{-1}(w)g(w) \mid w \in \Delta^*\} \cap \Sigma^*$.

Assume first that $S = w_0 \Longrightarrow w_1 \Longrightarrow \ldots \Longrightarrow w_k = w$ is a derivation of a word $w \in L$, where $w_i = r_i u_i s_i \rightarrow r_i v_i s_i = w_{i+1}$ according to a production $p_i = u_i \rightarrow v_i$ for $i = 0, 1, \ldots, k - 1$. Since $w \in \Sigma^*$, the last production p_{k-1} must be terminating, and $r_{k-1}, s_{k-1} \in \Sigma^*$. Now, consider the words

$$
z_1 = df r_0 p_0 s_0 f \ldots f r_{k-2} p_{k-2} s_{k-2} f \quad \text{and} \quad z_2 = \overline{r}_{k-1}\overline{p}_{k-1}\overline{s}_{k-1} e \ ,
$$

where we may assume that $k > 1$. Now,

$$
\begin{aligned}
h(z_1 z_2) &= dc \cdot m(f w_0 f \ldots f w_{k-2} f w_{k-1} f) \cdot c \\
&= dc \cdot cfc \cdot m(w_0) \cdot cfc \ldots cfc \cdot m(w_{k-2}) \cdot cfc \cdot m(w_{k-1}) \cdot cfc \cdot c \\
&= r(df w_0 f \ldots f w_{k-2} f w_{k-1} f) = g(z_1) \ .
\end{aligned}
$$

Since $g(z_2) = w_k$, we have that $h(z_1 z_2)^{-1} g(z_1 z_2) = w$, and hence $w \in K$. We have shown that $L \subseteq K$.

On the other hand, let $w \in K$. Hence there exists a word z such that $g(z) = h(z)w$. Denote $\Omega = V \cup \{d, f\}$ and $\Psi = \Delta \setminus \{d, e\}$.

By the definitions of the morphisms h, g and from the fact that $w \in \Sigma^*$, we conclude that $h(z) \in (\Omega cc)^*$, and that

$$
z = r \cdot ds_1 f s_2 f \ldots f s_k e
$$

for some words $r \in (d\Psi e)^*$ and $s_i \in (\Psi \setminus \{f\})^*$ for $i = 1, 2, \ldots, k$. Now,

$$
\begin{aligned}
h(z) &= h(r) \cdot dc \cdot h(s_1) cf \ldots cfc \cdot h(s_k) cfcc, \\
g(z) &= g(r) \cdot dccfcc Scc \cdot g(s_1) fcc \ldots fcc \cdot g(s_k) \ .
\end{aligned}
$$

From these we have that $g(s_k) = w$, and that $h(s_1) = 1$, i.e., $s_1 = 1$. Moreover,

$$
cSc = h(s_2) \quad \text{and} \quad cg(s_i) = h(s_{i+1})c \ .
$$

Let ε_c be a morphism which erases all occurrences of c from words. By above, $\varepsilon_c(h(s_2)) = S$ and $\varepsilon_c(h(s_i)) \Longrightarrow^* \varepsilon_c(g(s_i))$ for $i = 1, 2, \ldots, k$, and finally $S \Longrightarrow^* \varepsilon(g(s_k)) = w$ showing that $w \in L$. This completes the proof. □

Theorem 8.1 has the following immediate corollary.

Corollary 8.1. *For each recursively enumerable $L \subseteq \Sigma^*$ there exist morphisms h and g such that $L = \{g(w) \in \Sigma^* \mid \exists v \neq 1 : h(vw) = g(v)\}$.*

A purely morphic characterization of computability was proved in [C1] by characterizing recursively enumerable languages as morphic images of the sets of minimal solutions $e(h, g)$ of PCP. The proof of this result refines the basic simulation idea of the proof of the Post Correspondence Problem. For another formulation of the proof, see [T3].

Proposition 8.4. *For each recursively enumerable $L \subseteq \Sigma^*$ there effectively exists morphisms π, h, g such that*

$$L = \pi(e(h, g)),$$

where h is nonerasing and π is a projection.

Finally, we want to mention the following result from [MSSY], which was used to characterize complexity classes in terms of solutions of PCP.

Proposition 8.5. *For each recursively enumerable $L \subseteq \Sigma^*$ there are morphisms $g, h \colon \Gamma^* \to \Delta^*$ so that*

$$L = \pi(E(g, h) \cap \Gamma_1^+ \Sigma^* \Gamma_2^+ \Gamma_3^+) ,$$

where $\Gamma = \Gamma_1 \cup \Gamma_2 \cup \Gamma_3 \cup \Sigma$, $\pi \colon \Gamma^* \to \Sigma$ is the projection onto Σ, and

$$
\begin{aligned}
|g(a)| &> |h(a)| & \text{for } a \in \Gamma_1, \\
|g(a)| &= |h(a)| & \text{for } a \in \Gamma_2 \cup \Sigma, \\
|g(a)| &< |h(a)| & \text{for } a \in \Gamma_3 .
\end{aligned}
$$

In the following result from [EnR1], for an alphabet Δ we let again $\overline{\Delta} = \{\overline{a} \mid a \in \Delta\}$ be a new alphabet disjoint from Δ, and we define the *twin-shuffle* $L(\Delta)$ over Δ by

$$L(\Delta) = \{w \in (\Delta \cup \overline{\Delta})^* \mid \pi_\Delta(w) = \pi_{\overline{\Delta}}(w)\} ,$$

where π_Δ and $\pi_{\overline{\Delta}}$ are projections of $\Delta \cup \overline{\Delta}$ onto Δ and $\overline{\Delta}$, respectively.

Proposition 8.6. *For each recursively enumerable language $L \subseteq \Sigma^*$ there exist an alphabet Δ, a regular language $R \subseteq \Delta^*$, and a morphism $h \colon \Delta^* \to \Sigma^*$ such that $L = h(L(\Delta) \cap R)$.*

Note that $L(\Delta)$ is an equality set, and hence each recursively enumerable language can be obtained from this special equality set by very simple operations.

8.3 Representations of regular languages

As we have seen the families $\mathcal{RE}$, $\mathcal{CS}$ and $\mathcal{CF}$ in the Chomsky hierarchy all have a representation $\mathcal{H}^{-1}(L)$ with a single generator L. The family $\mathcal{R}$ of regular languages is an exception to this rule, *i.e.*, no such simple type of representations exists. Indeed, it was shown in [CM] and [HKK] that the morphic compositions $\mathcal{H} \circ \mathcal{H}^{-1}$ and $\mathcal{H}^{-1} \circ \mathcal{H}$ are not powerful enough to generate $\mathcal{R}$ from a single regular language.

Theorem 8.2. *For all regular L, $\mathcal{H} \circ \mathcal{H}^{-1}(L) \neq \mathcal{R}$ and $\mathcal{H}^{-1} \circ \mathcal{H}(L) \neq \mathcal{R}$.*

Proof. Let L be a regular language accepted by an n-state finite automaton, and let h and g be morphisms. As is easy to verify $h^{-1}(L)$ can be accepted by an n-state finite automaton. From this it follows that $h^{-1}(L)$ is of star height at most n. Further, a morphism g does not increase the star height of $h^{-1}(L)$, and thus the star height of $gh^{-1}(L)$ is at most n. However, as is well-known, there exist regular languages of star height greater than n, see [S1], and this proves the first claim.

We omit the proof of the second claim. We mention only that it was reduced in [HKK] to the following combinatorial result: if $F \subseteq \Sigma^+$ is a finite set and w, w' words with $w' \preceq w$, then either $L = F^* \cap w^* w'$ is infinite or $|L| \leq 2^{|F|}$. □

A positive result by Culik, Fich and Salomaa [CFS] shows, however, that each regular language R can be obtained from the fixed regular language a^*b by using a morphic composition of length four: $\mathcal{R} = \mathcal{H} \circ \mathcal{H}^{-1} \circ \mathcal{H} \circ \mathcal{H}^{-1}(a^*b)$. Hence also the regular languages do have morphic compositional representations with a single generator.

After the existence of such a morphic representation had been proved there followed a sequence of papers improving and generalizing the result of [CFS], see [KL], [LL], [T1] and [T2].

In particular, in [LL] it was proved that every regular language has a compositional representation of length three. The proof of this is a typical application of the techniques of [CFS]; it uses a representation of the accepting computations of a finite automaton through morphisms and inverse morphisms.

Theorem 8.3. $\mathcal{R} = \mathcal{H} \circ \mathcal{H}^{-1} \circ \mathcal{H}(a^*b)$.

Proof. Let $R = L(A) \subset \Sigma^*$ be a regular language accepted by a (nondeterministic) finite automaton A with states $Q = \{q_0, q_1, \ldots, q_m\}$, where q_0 is the initial state, and q_m is the unique final state. Assume further without loss of generality that there are no transitions entering q_0 and leaving q_m. Let

$$\Gamma = \{[q_i, x, q_j] \mid q_i x \to q_j\}$$

be an alphabet that codes the transitions $qx \to p$ of A, and let a, b and d be special symbols. Define $h_1: \{a, b\} \to \{a, b, d\}^*$ by

$$h_1(a) = ad^m \qquad \text{and} \qquad h_2(b) = bd^m,$$

where $m = |Q|$. Hence $h_1(a^n b) = (ad^m)^n \cdot bd^m$. Let $h_2 \colon \Gamma^* \to \{a, b, d\}^*$ be defined by

$$h_2([q_i, x, q_j]) = \begin{cases} d^i a d^{m-j}, & \text{if } j \neq m, \\ d^i b d^m, & \text{if } j = m. \end{cases}$$

So $u \in h_2^{-1} h_1(a^n b)$ if and only if $|u| = n + 1 \; (= |a^n b|)$ and

$$u = [q_0, a_1, q_{r_1}][q_{r_1}, a_2, q_{r_2}] \dots [q_{r_i}, a_{i+1}, q_{r_{i+1}}] \dots [q_{r_n}, a_{n+1}, q_m]$$

for some letters $a_1, a_2, \dots, a_{n+1} \in \Sigma$ and states $q_{r_1}, q_{r_2}, \dots, q_{r_n} \in Q$. In particular, $u \in h_2^{-1} h_1(a^n b)$ if and only if u codes an accepting computation of A for the word $a_1 a_2 \dots a_{n+1}$. Finally, let $h_3 \colon \Gamma^* \to \Sigma^*$ be defined by

$$h_3([q, a, p]) = a$$

for all $[q, a, p] \in \Gamma$. We can now deduce that $L(A) = h_3 h_2^{-1} h_1(a^n b)$. $\qquad \square$

The proof of Theorem 8.3 shows that in the representation of $\mathcal{R}$ as above the morphisms are restricted: the morphism h_1 is *uniform*, i.e., there exists a letter d and an integer m such that $h_1(a) = ad^m$ for all letters a, the morphism h_2 is nonerasing, and h_3 is *length preserving*, i.e., it maps letters to letters.

Using similar techniques it was proven in [LL] that

Proposition 8.7. $\mathcal{R} = \mathcal{H}^{-1} \circ \mathcal{H} \circ \mathcal{H}^{-1}(b)$.

We give another example of this construction method in order to obtain a rather simple representation result for the regular *star languages*, i.e., regular languages satisfying $R = R^*$.

Theorem 8.4. *Let $R \subseteq \Sigma^*$ be any language. Then R^* is regular if and only if there exist a finite set F and a uniform morphism h such that $R^* = h^{-1}(F^*)$.*

Proof. Suppose $R^* = L(A)$ for a deterministic finite automaton A having $Q = \{q_0, q_1, \dots, q_m\}$ as its state set, where q_0 is the initial state, and $T \subseteq Q$ is the set of final states. Let d be a new symbol, and define

$$F = \{d^i a d^{m-j} \mid a \in \Sigma, \; q_i a = q_j\} \cup \{d^i a d^m \mid a \in \Sigma, \; q_i a \in T\} .$$

Define $h \colon \Sigma^* \to (\Sigma \cup \{d\})^*$ by $h(a) = ad^m$ for all $a \in \Sigma$. One can now show, by induction on the length of words, that for all $q_i \in Q$ and $u \in \Sigma^*$,

$$q_i u \in T \iff d^i \cdot h(u) \in F^*,$$

from which it follows that $R^* = h^{-1}(F^*)$.

In the other direction the claim is obvious, since $\mathcal{R}$ is closed under inverse morphic images, and $h^{-1}(F^*)$ is always a star language. $\qquad \square$

8.4 Representations of rational transductions

The ideas of the previous section can be used to prove morphic representations for rational transductions $\tau: \Sigma^* \to 2^{\Delta^*}$. However, such representations require some 'technical' modifications to these constructions.

An *endmarking* is a function $\mu_m: \Sigma^* \to \Sigma^* m$ that adjoins a special letter m at the end of a word: $\mu_m(w) = wm$ for each $w \in \Sigma^*$. We denote by $\mathcal{M}$ the family of all endmarkings.

Clearly, each of the families $\mathcal{H}$, $\mathcal{H}^{-1}$ and $\mathcal{M}$ consist of rational transductions, and since the rational transductions are closed under composition, see [B], also compositions of morphisms, inverse morphisms and endmarkings are again rational transductions. These compositions will be referred to as *rational compositions*.

The operation $\cap \mathcal{R}$ can be obtained as a rational composition as was shown in [T1] and [KL].

Proposition 8.8. $\cap \mathcal{R} \subseteq \mathcal{H} \circ \mathcal{H}^{-1} \circ \mathcal{H} \circ \mathcal{M}$.

As a consequence the Chomsky-Schützenberger theorem can be transformed into a representation result, formulated in [KL], characterizing $\mathcal{CF}$ as the language family generated from D_2 using only rational compositions.

This result on intersection with regular languages can also be applied to the representation theorem for $\mathcal{RE}$ from [EnR1], see also [S4], which is similar to the Chomsky-Schützenberger theorem: each recursively enumerable language can be obtained as the morphic image of a twin-shuffle language intersected with a regular language. This approach yields, in comparison with the Greibach type of representation, a more manageable generator than U_0, or U_2 in the context-free case, and still a fairly simple combination of operations.

By Nivat's theorem [Ni], see [B], a mapping $\tau: \Sigma^* \to 2^{\Delta^*}$ is a rational transduction if and only if there exist a regular set R and two morphisms g and h such that $\tau = \{(g(w), h(w)) \mid w \in R\}$. In other words,

Proposition 8.9. *The family of rational transductions equals $\mathcal{H} \circ \cap \mathcal{R} \circ \mathcal{H}^{-1}$.*

Here $\cap \mathcal{R}$ has a representation from Proposition 8.8, and therefore each rational transduction τ can be representation in the following form:

$$\tau = h h_3 h_2^{-1} h_1 \mu g^{-1} \in \mathcal{H} \circ \mathcal{H}^{-1} \circ \mathcal{H} \circ \mathcal{M} \circ \mathcal{H}^{-1} .$$

Here one can easily move the endmarking to the beginning of the composition in order to obtain

Theorem 8.5. *The family of rational transductions equals*

$$\mathcal{H} \circ \mathcal{H}^{-1} \circ \mathcal{H} \circ \mathcal{H}^{-1} \circ \mathcal{M} .$$

Hence rational compositions of length five suffice for rational transductions. We note here that the compositions in $\mathcal{H}^{-1} \circ \mathcal{H} \circ \mathcal{H}^{-1} \circ \mathcal{H} \circ \mathcal{M}$ can be represented by elements from $\mathcal{H} \circ \mathcal{H}^{-1} \circ \mathcal{H} \circ \mathcal{H}^{-1} \circ \mathcal{M}$. In fact, as a lengthy proof in [LT1] shows these two classes are the same:

Proposition 8.10. *The family of rational transductions is equal to*

$$\mathcal{H} \circ \mathcal{H}^{-1} \circ \mathcal{H} \circ \mathcal{H}^{-1} \circ \mathcal{M} = \mathcal{H}^{-1} \circ \mathcal{H} \circ \mathcal{H}^{-1} \circ \mathcal{H} \circ \mathcal{M} \ .$$

If the endmarking is omitted from above we obtain the class of all morphic compositions. As shown in [T2] the compositions of morphisms and inverse morphisms are exactly those rational transductions that can be realized by simple transducers. Moreover, the following result was proved in a sequence of papers [LL], [T2], and [LT1]:

Proposition 8.11. $\mathcal{H} \circ \mathcal{H}^{-1} \circ \mathcal{H} \circ \mathcal{H}^{-1} = \mathcal{H}^{-1} \circ \mathcal{H} \circ \mathcal{H}^{-1} \circ \mathcal{H}$ *equals the family of rational transductions realized by simple transducers.*

Given a finite transducer realizing τ, we can, in fact, *effectively* construct the morphisms as required in Proposition 8.10 for its representation. Similarly, if τ is realized by a simple transducer, then the representation in Proposition 8.11 is effective. Nevertheless, we have no effective way to decide whether for a rational transduction a representation without endmarkers exists. This was proved in [HKl] using a strong undecidability result from [I].

Proposition 8.12. *It is undecidable whether or not a rational transduction has a representation without endmarker. In other words, it is undecidable whether or not a rational transduction is realized by a simple transducer.*

As proved in [LL] the number of morphisms in Proposition 8.10 cannot be reduced further. In Fig. 8.1 from [LL], we have drawn a diagram of inclusions for the morphic compositions.

If we restrict ourselves, as in [T2], to 1-free transducers, then the representations become even shorter. Below $\mathcal{H}_\varepsilon$ denotes the family of nonerasing morphisms.

Proposition 8.13. *The family of rational transductions realized by 1-free transducers is equal to* $\mathcal{H}^{-1} \circ \mathcal{H}_\varepsilon \circ \mathcal{H}^{-1} \circ \mathcal{M} \ .$

Thus in the general case of Proposition 8.10, we need one morphism to take care of the empty word.

There exists a wealth of interesting subfamilies of rational transductions and many of these have been given a representation by morphisms and inverse morphisms, where the inverse morphisms have been suitably restricted. We list here only a selection of these results from [HKL1] and [HKL2].

$$\mathcal{H} \circ \mathcal{H}^{-1} \circ \mathcal{H} \circ \mathcal{H}^{-1} = \mathcal{H}^{-1} \circ \mathcal{H} \circ \mathcal{H}^{-1} \circ \mathcal{H}$$

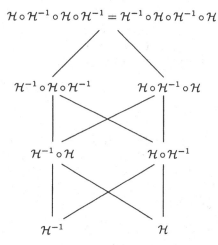

Fig. 8.1. The hierarchy of morphic compositions, where a path upwards means proper inclusion and a horizontal disconnection denotes incomparability.

Proposition 8.14. *The family of rational functions equals* $\mathcal{H} \circ \mathcal{H}_i^{-1} \circ \mathcal{H} \circ \mathcal{M}$, *where* $\mathcal{H}_i$ *denotes the family of injective morphisms.*

Proposition 8.15. *The family of transductions realized by simple sequential transducers is equal to* $\mathcal{H} \circ \mathcal{H}_p^{-1} \circ \mathcal{H}$, *where* $\mathcal{H}_p$ *denotes the family of prefix codes.*

Proposition 8.16. *The family of the subsequential transductions equals*

$$\mathcal{H} \circ \mathcal{H}_p^{-1} \circ \mathcal{H} \circ \mathcal{M} \ .$$

These result together with some of the earlier results were improved in details and depth in [HKLT]. Also, we refer to [DT] for some nice results for rational bijections.

9. Equivalence problem on languages

In this section we consider a problem which is closely connected to our earlier considerations, and which is also well motivated by practical problems. Namely, we want to decide whether two mappings, or translations, $\Sigma^* \to \Delta^*$, are equivalent on Σ^*, or on a subset L of Σ^*. Clearly, such problems are central in the theory of compilers.

9.1 Morphic equivalence on languages

Let again $\mathcal{H}$ denote the family of morphisms from a words monoid into another, and let $\mathcal{L}$ be a family of languages over an alphabet Σ. The *morphic equivalence problem for $\mathcal{L}$*, introduced in [CS1], asks for given two morphisms $h, g \in \mathcal{H}$ and a language $L \in \mathcal{L}$ whether h and g *agree on* L, *i.e.*, whether

$$h(w) = g(w) \quad \text{for all } w \in L ,$$

in symbols $h|L = g|L$.

A natural generalization of this problem is achieved when $\mathcal{H}$ is replaced by a general family $\mathcal{F}$ of mappings, which can, moreover, be many-valued, *i.e.*, from Σ^* into 2^{Δ^*} for some alphabets Σ and Δ. We call this problem the *$\mathcal{F}$-equivalence problem for $\mathcal{L}$*.

Note that we have already considered these problems especially in Section 6.5. We would also like to emphasize the connection to Ehrenfeucht's Conjecture. Indeed, by Ehrenfeucht's Conjecture for each language L there exists a finite $T \subseteq L$ such that any given morphisms h and g agree on L if (and only if) they do so on T.

The effectiveness of this problem on a given language was raised in [CS1], where, among other results, it was shown that the problem $h|L = ?g|L$ is decidable for context-free languages L.

In accordance with Theorem 7.3 and Corollary 7.4 we have

Theorem 9.1. *The morphic equivalence problem is decidable for regular languages.*

Theorem 9.2. *The morphic equivalence problem is decidable for context-free languages.*

Both of these results hold in a stronger form, namely that these problems can be solved in a polynomial time. The polynomial time bound for regular languages was shown in [KRJ], while that for context-free languages is much more difficult to achieve, see [Pl].

It has to be emphasized that the question $h|L = ?g|L$ in the morphic equivalence problem is quite different from the question $h(L) = ?g(L)$ for two given morphisms and a given language L. Indeed, the latter problem is undecidable for the family of context-free languages. This is due to the fact that the equivalence problem is undecidable in this family.

Finally, we summarize the considerations of Section 6.5 in the following result.

Theorem 9.3. *The morphic equivalence problem is decidable for DT0L languages.*

However, contrary to our earlier results, no computationally feasible solution for Theorem 9.3 is known, since the only known proof of this theorem uses Makanin's algorithm.

9.2 More general mappings

We study now the $\mathcal{F}$-equivalence problem for languages in two particular cases. First we recall the basic results on the equivalence problem for transducers, and then we point out an exact borderline when for the families $\mathcal{F}$ of morphic compositions the $\mathcal{F}$-equivalence turns undecidable for regular languages.

Theorem 9.4. *The equivalence problem for sequential transducers is decidable.*

We note that there is a particularly easy way of proving this result using Theorem 9.1. Indeed, let M_1 and M_2 be two sequential transducers, and let A_1 and A_2 be the corresponding underlying finite automata with respect to the input structures. Next let A be the cartesian product of A_1 and A_2, and let $\overline{A}$ be a deterministic automaton recognizing the accepting computations of A. It is now straightforward to define, using the outputs of M_1 and M_2, morphisms h_1 and h_2 such that

$$M_1 \text{ and } M_2 \text{ are equivalent}$$
$$\Longleftrightarrow L(A_1) = L(A_2) \text{ and } h_1|L(\overline{A}) = h_2|L(\overline{A}) \ .$$

As we already noted in Proposition 7.3, Theorem 9.4 extends to finite-valued transducers. However, we have the following undecidability result.

Proposition 9.1. *The equivalence problem for nondeterministic 1-free sequential transducers is undecidable. Moreover, it remains undecidable even when the output alphabet is unary.*

The original proof of the first claim of Proposition 9.1 uses PCP in [Gr], and its remarkable improvement to the unary alphabet is given in [I].

In a moment we introduce an in-between case of Theorem 9.4 and Proposition 9.1, which is still open.

We turn now to consider the equivalence of morphic compositions on regular languages.

Proposition 9.2. *It is decidable whether two mappings from $\mathcal{H}^{-1} \circ \mathcal{H}$ are equivalent on a given regular language.*

The proof of this result in [KK] uses the Cross-Section Theorem of Eilenberg, see [E]. Also the following result, which completes the classification we are looking for, is from [KK].

Theorem 9.5. *It is undecidable whether two mappings from the family $\mathcal{H} \circ \mathcal{H}^{-1}$ are equivalent on a given regular language.*

Proof. We reduce the problem to that of Proposition 9.1. We note first that for any nondeterministic 1-free sequential transducer realizing a transduction τ, we can construct a nondeterministic 1-free simple sequential transducer realizing τ' and a new symbol d such that $\tau(w)d = \tau'(wd)$ for all words w, see [KK].

For two such τ's, say τ_1 and τ_2, let τ_1' and τ_2' be the corresponding transductions as constructed. Further, assume that $\mathrm{dom}(\tau_1) = \mathrm{dom}(\tau_2) = R$. Then $\tau_1|R = \tau_2|R$ if and only if $\tau_1'|(Rd)^* = \tau_2'|(Rd)^*$.

It follows now from Proposition 9.1 that it is undecidable whether two mappings realized by nondeterministic 1-free simple sequential transducer agree on their common domain. Let us denote such transductions simply by τ_1 and τ_2, and their common domain by R.

Next we use a construction from [T2], which allows us to write

$$\tau_1 = h_3^{-1} h_2 h_1^{-1} \in \mathcal{H}^{-1} \circ \mathcal{H} \circ \mathcal{H}^{-1}, \tag{9.1}$$

where, for some $m \geq 1$, $h_3(a) = ad^m$ for each letter a, and moreover,

$$h_1 h_2^{-1} h_3 h_3^{-1} h_2 h_1^{-1}(R) \subseteq R . \tag{9.2}$$

This construction allows us to choose the same constant m for different τ's. Similarly, for τ_2 we obtain

$$\tau_2 = g_3^{-1} g_2 g_1^{-1},$$

where actually $h_3 = g_3$.

Now, by (9.2) and its counterpart for τ_2,

$$h_3^{-1} h_2 h_1^{-1}|R = g_3^{-1} g_2 g_1^{-1}|R$$

if and only if $h_3^{-1} h_2 h_1^{-1}(R) = R_1 = h_3^{-1} g_2 g_1^{-1}(R)$ and $h_1 h_2^{-1} h_3$, $g_1 g_2^{-1} h_3$ agree on R_1. By injectiveness of h_3, this holds if and only if $h_2 h_1^{-1}(R) = g_2 g_1^{-1}(R)$ and $h_1 h_2^{-1}$, $g_1 g_2^{-1}$ agree on R_1. This proves the claim.   □

We conclude this section with a related open problem: Is it decidable whether two finite substitutions agree on a given regular language?

Denoting by $\mathcal{S}$ the family of finite substitutions, and by $\mathcal{C}$ the family of codings, we can write $\mathcal{S} = \mathcal{H} \circ \mathcal{C}^{-1}$. Hence the above open problem asks whether Theorem 9.5 can be sharpened with respect to the first component in the morphic composition. In the light of Example 6.7 this problem is not easy.

Using ideas introduced after Theorem 9.4 the above open problem can be restated as an equivalence problem of rational transductions. Let us call a nondeterministic sequential transducer *semi-deterministic*, if its underlying finite automaton is deterministic with respect to inputs. Then, as stated in [CK5], we have

Theorem 9.6. *The equivalence problem for semi-deterministic transducer is decidable if and only if the S-equivalence problem for regular languages is decidable*

Finally, we refer to [Ma] and [KM] for other simply formulated decidability results on morphic compositions.

10. Problems

We have collected here some of the open problems stated in (or deducible from) our presentation.

In Theorem 3.2 we needed two extra marker symbols to ensure that if $GPCP(n)$ is undecidable then so is $PCP(n + 2)$. Can you do with only one extra symbol?

Problem 10.1. Does undecidability of $GPCP(n)$ imply undecidability of $PCP(n + 1)$ or even of $PCP(n)$?

Problem 10.2. Is $PCP(n)$ decidable for $n = 3, 4, \ldots, 8$? Is $GPCP(n)$ decidable for $n = 3, 4, 5, 6$?

The following problem is known as *Skolem's Problem*:

Problem 10.3. Given an $n \times n$-matrix M with integer entries, is it decidable whether there exists a power k such that $(M^k)_{(1,n)} = 0$?

Problem 10.4. Is the mortality problem decidable for the 2×2-matrices over $\mathbb{Z}$.

In connection with Theorem 5.1 we stated the following

Problem 10.5. For $\Sigma = \{a, b\}$ is it true that if $E(h, g) \neq \emptyset$ and h is non-periodic, then there are two (possibly equal) words $u, v \in \Sigma^+$ such that $E(h, g) = \{u, v\}^+$?

Problem 10.6. Is it true that $E(h, g)$ is regular for all injective morphisms defined on alphabets of cardinality 3 or 4?

It was shown in Example 6.7 that a regular language need not have a finite test set with respect to finite substitutions. This brings us to the following problem, *cf.* also Theorem 9.6.

Problem 10.7. Is it decidable whether two finite substitutions are equivalent on a given regular language?

Consider the basic closure operations on equality sets. For this let L, L_1, L_2 be languages with test sets T, T_1 and T_2, respectively. It is rather easy to prove that T is a test set of L^*, and that $T_1 \cup T_2$ is a test set of $L_1 \cup L_2$. Moreover, if $h \colon \Sigma \to \Delta$ is a morphism, then $h(T)$ is a test set of $h(L)$. With some elaboration we can also prove that if $d \notin \Sigma$ is a new letter and $T_1 d$ and dT_2 are test sets of $L_1 d$ and dL_2, resp., then $T_1 T_2$ is a test set of $L_1 L_2$.

Problem 10.8. Can you determine a finite test set T for the languages $L_1 \cap L_2$ and $L_1 L_2$, when test sets T_1 and T_2 are given for L_1 and L_2.

Proposition 8.14 gives a characterization of the rational functions in terms of rational compositions. If you remove the marking $\mathcal{M}$ of this representation, the resulting family of morphic compositions is *not* equal to the family $\mathcal{F}_*$ of simple rational functions, see [HKL1].

Problem 10.9. Does there exist a natural representation of the family of simple rational functions?

Note added after finishing the chapter: G: Senizergues told us that he together with Yuri Matiyasevich have proved that PCP (7), and hence also GPCP (5), is also undecidable.

References

[ACK]   J. Albert, K. Culik II and J. Karhumäki, Test sets for context-free languages and algebraic systems of equations in a free monoid, *Inform. Control* **52** (1982), 172–186.

[AL]   M. H. Albert and J. Lawrence, A proof of Ehrenfeucht's Conjecture, *Theoret. Comput. Sci.* **41** (1985), 121–123.

[BCM]   G. Baumslag, F. B. Cannonito and C. F. Miller III, Computable algebra and group embeddings, *J. Algebra* **69** (1981), 186–12.

[BS]   G. Baumslag and D. Solitar, Some two-generator one-relator non-hopfian groups, *Bull. Amer. Math. Soc.* **68** (1962), 199–201.

[B]   J. Berstel, Transductions and Context-Free Languages, B. G. Teubner, 1979.

[BR]   J. Berstel and C. Reutenauer, Rational Series and Their Languages, Springer-Verlag, 1988.

[BP]   J. Berstel and D. Perrin, Theory of Codes, Academic Press, 1986.

[Bo]   W. W. Boone, The word problem, *Ann. Math.* **70** (1959), 207–265.

[ChK1]   C. Choffrut and J. Karhumäki, Test sets for morphisms with bounded delay, Discrete Appl. Math. **12** (1985), 93–101.

[ChK2]   C. Choffrut and J. Karhumäki, Combinatorics of words, in this Handbook, 1996.

[CS]   N. Chomsky and M. P. Schützenberger, The algebraic theory of context-free languages, in Computer Programming and Formal Systems (P. Brattfort and D. Hirschberg, eds.), North-Holland, 1963, 118–161.

[Cl]   V. Claus, Some remarks on PCP(k) and related problems, *Bull. EATCS* **12** (1980), 54–61.

[Co]   P. M. Cohn, Algebra, Vol 2, John Wiley & Sons, (2nd ed.) 1989.

[C1]   K. Culik II, A purely homomorphic characterization of recursively enumerable sets, *J. Assoc. Comput. Mach.* **26** (1979), 345–350.

[C2]   K. Culik II, Homomorphisms: decidability, equality and test sets, in Formal Language Theory, Perspectives, and Open Problems (R. Book, ed.), Academic Press, 1980, 167–194.

[CFS]   K. Culik II, F. E. Fich and A. Salomaa, A homomorphic characterization of regular languages, *Discrete Appl. Math.* **4** (1982), 149–152.

[CF]   K. Culik II and I. Fris, The decidability of the equivalence problem for D0L-systems, *Inform. and Control* **35** (1977), 20–39.

[CK1]   K. Culik II and J. Karhumäki, On the equality sets for homomorphisms on free monoids with two generators, *RAIRO Theor. Informatics* **14** (1980), 349–369.

[CK2]   K. Culik II and J. Karhumäki, Systems of equations over free monoids and Ehrenfeucht's Conjecture, *Discrete Math.* **43** (1983), 139–153.

[CK3]   K. Culik II and J. Karhumäki, Decision problems solved with the help of the Ehrenfeucht Conjecture, *Bull. EATCS* **27** (1986), 30–35.

[CK4]   K. Culik II and J. Karhumäki, The equivalence of finite valued transducers (on HDTOL languages) is decidable, *Theoret. Comput. Sci.* **47** (1986), 71–84.

[CK5]   K. Culik II and J. Karhumäki, A new proof of the D0L sequence equivalence problem and its implications, in The Book of L (G. Rozenberg and A. Salomaa, eds.), Springer-Verlag, 1986, 63–74.

[CK6]   K. Culik II and J. Karhumäki, The equivalence problem for single-valued two-way transducers (on HDTOL languages) is decidable, *SIAM J. Comput.* **16** (1987), 221–230.

[CM]   K. Culik II and H. Maurer, On simple representations of language families, *RAIRO Theor. Informatics* **13** (1979), 241–250.

[CS1]   K. Culik II and A. Salomaa, On the decidability of homomorphism equivalence for languages, *J. Comput. System Sci.* **17** (1978), 163–175.

[CS2]   K. Culik II and A. Salomaa, Test sets and checking words for homomorphism equivalence, *J. Comput. System Sci.* **19** (1980), 379–395.

[DT]   D. Derencourt and A. Terlutte, Compositions of codings, in Developments in Language Theory (G. Rozenberg and A. Salomaa, eds.), World Scientific, 1994, 30–43.

[EKR1]   A. Ehrenfeucht, J. Karhumäki and G. Rozenberg, The (generalized) Post Correspondence Problem with lists consisting of two words is decidable, *Theoret. Comput. Sci.* **21** (1982), 119–144.

[EKR2]   A. Ehrenfeucht, J. Karhumäki and G. Rozenberg, On binary equality languages and a solution to the test set conjecture in the binary case, *J. Algebra* **85** (1983), 76–85.

[ER]   A. Ehrenfeucht and G. Rozenberg, Elementary homomorphisms and a solution to D0L sequence equivalence problem, *Theoret. Comput. Sci.* **7** (1978), 169–183.

[E]   S. Eilenberg, Automata, Languages, and Machines, Vol A, Academic Press, New York, 1974.

[EnR1]   J. Engelfriet and G. Rozenberg, Equality languages and fixed point languages, *Inform. Control* **43** (1979), 20–49.

[EnR1]   J. Engelfriet and G. Rozenberg, Fixed point languages, equality languages, and representation of recursively enumerable languages, *J. Assoc. of Comput. Mach.* **27** (1980), 499–518.

[Fr]   P. Freyd, Redei's finiteness theorem for commutative semigroups, *Proc. Amer. Math. Soc.* **19** (1968), 1003.

[Ge]   V. Geffert, A representation of recursively enumerable languages by two homomorphisms and a quotient, *Theoret. Comput. Sci.* **62** (1988), 235–249.

[Ger]   S. M. Gersten, Fixed points of automorphisms of free groups, *Adv. in Math.* **64**, (1987), 51–85.

[GT]   R. Z. Goldstein and E. C. Turner, Fixed subgroups of homomorphisms of free groups, *Bull. London. Math. Soc.* **18** (1986), 468–470.

[G1]   S. Greibach, The hardest CF language, *SIAM J. Comput.* **2** (1973), 304–310.

[G2]   S. Greibach, A remark on code sets and context-free languages, *IEEE Trans. on Computers* **C-24** (1975), 741–742.

[Gr]   T. V. Griffiths, The unsolvability of the equivalence problem for λ-free non-deterministic generalized machines, *J. Assoc. Comput. Mach.* **15** (1968), 409–413.

[Gu]   Y. Gurevich, Average case complexity, *J. Comput. System Sci.* **42** (1991), 346–298.

[Ha]   G. Hansel Une démonstration simple du théorème de Skolem-Mahler-Lech, *Theoret. Comput. Sci.* **43** (1986), 1–10.

[HK]   T. Harju and J. Karhumäki, The equivalence problem of multitape automata, *Theoret. Comput. Sci.* **78** (1991), 347–355.

[HKK]   T. Harju, J. Karhumäki and H. C. M. Kleijn, On morphic generation of regular languages, *Discrete Appl. Math.* **15** (1986), 55–60.

[HKKr] T. Harju, J. Karhumäki and D. Krob, Remarks on generalized Post correspondence problem, *Proceedings of STACS'96*, to appear 1996.

[HKP]   T. Harju, J. Karhumäki and W. Plandowski, Compactness of systems of equations in semigroups, *Lecture Notes in Comput. Sci.* **944**, Springer-Verlag 1995, 444–454.

[HKl]   T. Harju and H. C. M. Kleijn, Decidability problems for unary output sequential transducers, *Discrete Appl. Math.* **32** (1991), 131–140.

[HKL1] T. Harju, H. C. M. Kleijn and M. Latteux, Compositional representation of rational functions, *RAIRO Theor. Informatics* **26** (1992), 243–255.

[HKL2] T. Harju, H. C. M. Kleijn and M. Latteux, Deterministic sequential functions, *Acta Informatica* **29** (1992), 545–554.

[HKLT] T. Harju, H. C. M. Kleijn, M. Latteux and A. Terlutte, Representation of rational functions with prefix and suffix codings, *Theoret. Comput. Sci* **134** (1994), 403–413.

[H]   M. Harrison, Introduction to Formal Language Theory, Addison-Wesley, 1978.

[HW]   G. T. Herman and A. Walker, Context-free languages in biological systems, *Internat. J. Comput. Math.* **4** (1975), 369–391.

[Hm]   Y. I. Hmelevskii, Equations in free semigroups, *Proc. Steklov Inst. Math.* **107** (1971); Amer. Math. Soc. Translations (1976).

[HU]   J. E. Hopcroft and J. D. Ullman, Introduction to Automata Theory, Languages, and Computation, Addison-Wesley, 1979.

[I]   O. H. Ibarra, The unsolvability of the equivalence problem for ε-free NGSM's with unary input (output) alphabet and applications, *SIAM J. Comput.* **7** (1978), 524–532.

[IK]   O. H. Ibarra and C. E. Kim, A useful device for showing the solvability of some decision problems, *Proc. of the Eight Annual ACM Symposium on Theory of Computing* (1976), 135–140.

[Ja]   G. Jacob, La finitude des reprèsentations linéaires de semi-groupes est décidable, *J. Algebra* **52** (1978), 437–459.

[J]   M. Jantzen, Confluent String Rewriting, Springer-Verlag, 1988.

[K1]   J. Karhumäki, Generalized Parikh mapping and homomorphisms, *Information and Control* **47** (1980), 155–163

[K2]   J. Karhumäki, The Ehrenfeucht Conjecture: A compactness claim for finitely generated free monoids, *Theoret. Comput. Sci.* **29** (1984), 285–308.

[K3]   J. Karhumäki, On the regularity of equality languages, *Ann. Univ. Turkuensis, Ser. A I* **186** (1984), 47–58.

[K4]   J. Karhumäki, The Ehrenfeucht Conjecture for transducers, *J. Inform. Process. Cybernet. EIK* **23** (1987), 389–401.

[K5]   J. Karhumäki, Equations over finite sets of words and equivalence problems in automata theory, *Theoret. Comput. Sci.* **108** (1993), 103–118.

[K6]   J. Karhumäki, The impact of the D0L problem, in Current Trends in Theoretical Computer Science. Essays and Tutorials (G. Rozenberg and A. Salomaa, eds.), World Scientific 1993, 586–594.

[KK]   J. Karhumäki and H. C. M. Kleijn, On the equivalence of compositions of morphisms and inverse morphisms on regular languages, *RAIRO Theor.Informatics* **19** (1985), 203–211.

[KM]   J. Karhumäki and Y. Maon, A simple undecidable problem: existential agreement of inverses of two morphisms on a regular language, *J. of Computer and System Sci.* **32** (1986), 315–322.

[KL]   J. Karhumäki and M. Linna, A note on morphic characterization of languages, *Discrete Appl. Math.* **5** (1983), 243–246.

[KP]   J. Karhumäki and W. Plandowski, On the size of independent systems of equations in semigroups, *Lecture Notes in Comput. Sci.* **841**, Springer-Verlag 1994, 443–452; also *Theoret. Comput. Sci.*, to appear.

[KPR]   J. Karhumäki, W. Plandowski and W. Rytter, Polynomial size test sets for context-free languages, *J. Comput. Syst. Sci.* **50** (1995), 11–19.

[KRJ]   J. Karhumäki, W. Rytter and S. Jarominek, Efficient constructions of test sets for regular and context-free languages, *Theoret. Comput. Sci.* **116** (1993), 305–316.

[KBS]   D. A. Klarner, J.-C. Birget and W. Satterfield, On the undecidability of the freeness of integer matrix semigroups, *Int. J. Algebra Comp.* **1** (1991), 223–226.

[Kr]   M. Krom, An unsolvable problem with products of matrices, *Math. System. Theory* **14** (1981), 335–337.

[L1]   G. Lallement, Semigroups and Combinatorial Applications, Wiley, 1979.

[L2]   G. Lallement, Some algorithms for semigroups and monoids presented by a single relation, *Lecture Notes in Math.* **1320** (1988), 176–182.

[LL]   M. Latteux and J. Leguy, On the composition of morphisms and inverse morphisms, *Lecture Notes in Comput. Sci.* **154**, Springer-Verlag 1983, 420–432.

[LT1]   M. Latteux and P. Turakainen, A new normal form for the composition of morphisms and inverse morphisms, *Math. Syst. Theory* **20** (1987), 261–271.

[LT2]   M. Latteux and P. Turakainen, On characterizations of recursively enumerable languages, *Acta Informatica* **28** (1990), 179–186.

[La]   J. Lawrence, The nonexistence of finite test set for set-equivalence of finite substitutions, *Bull. EATCS* **28** (1986), 34–37.

[Le]   M. Y. Lecerf, Récursive insolubilité de l'équation générale de diagonalisation de deux monomorphismes de monoïdes libres $\varphi x = \psi x$, *Comptes Rendus* **257** (1963), 2940–2943.

[Li1]   M. Lipponen, Primitive words and languages associated to PCP, *Bull. EATCS* **53** (1994), 217–226.

[Li2]   M. Lipponen, Post Correspondence Problem: words possible as primitive solutions, *Lecture Notes in Comput. Sci.* **944**, Springer-Verlag 1995, 63–74.

[Lo]   M. Lothaire, Combinatorics on Words, Addison-Wesley, 1983.

[dLR]   A. de Luca and A. Restivo, On a generalization of a conjecture of Ehrenfeucht, *Bull. EATCS* **30** (1986), 84–90.

[LS]   R. C. Lyndon and P. E. Schupp, Combinatorial Group Theory, Springer-Verlag, 1977.

[MaS]   A. Mandel and I. Simon, On finite semigroups of matrices, *Theoret. Comput. Sci.* **5** (1977), 101–111.

[MKS]   W. Magnus, A. Karrass and D. Solitar, Combinatorial Group Theory, Wiley, 1966.

[M]   G. S. Makanin, The problem of solvability of equations in a free semigroup, *Mat. Sb.* **103** (1977), 147–236; *Math. USSR Sb.* **32** (1977), 129–198.

[Man]   Z. Manna, Mathematical Theory of Computations, McGraw-Hill, 1974.

[Ma]   Y. Maon, On the equivalence problem of compositions of morphisms and inverse morphisms on context-free languages, *Theoret. Comput. Sci.* **41** (1985), 105–107.

[Mar]   A. A. Markov, On the impossibility of certain algorithms in the theory of associative systems, *Dokl. Akad. Nauk* **55** (1947), 587–590; **58** (1947), 353–356 (Russian).

[MS1]   A. Mateescu and A. Salomaa, PCP-prime words and primality types, *RAIRO Theor. Informatics* **27** (1993), 57–70.

[MS2]   A. Mateescu and A. Salomaa, On simplest possible solutions for Post Correspondence Problem, *Acta Informatica* **30** (1993), 441–457.

[MSSY]   A. Mateescu, A. Salomaa, K. Salomaa and S. Yu, P, NP and Post Correspondence Problem, *Inform. and Comput.*, **121** (1995), 135–142.

[Mat]   J. Matiyasevich, Simple examples of unsolvable associative calculi, *Dokl. Akad. Nauk* **173** (1967), 1264–1266 (Russian).

[Mi]   C. F. Miller III, Decision problems for groups–Survey and reflections, in Algorithms and Classification in Combinatorial Group Theory (G. Baumslag and C. F. Miller III, eds.), Springer-Verlag, 1992, 1–59.

[MuS]   A. A. Muchnik and A. L. Semenov, *Jewels of Formal Languages*, (Russian translation of [S4]), Mir, 1986.

[Ni]   M. Nivat, Transductions des langages de Chomsky, *Ann. Inst. Fourier* **18** (1968), 339–456.

[No]   P. S. Novikov, On the algorithmic unsolvability of the problem of equality of words in group theory, *Tr. Mat. Inst. Akad. Nauk* **44** (1955), 1–144 (Russian).

[Pan]   J. J. Pansiot, A note on Post's Correspondence Problem, *Inform. Proc. Lett.* **12** (1981), 233.

[Pat]   M. S. Paterson, Unsolvability in 3×3-matrices, *Studies in Appl. Math.* **49** (1970), 105–107.

[Pav]   V. A. Pavlenko, Post combinatorial problem with two pairs of words, *Dokl. Akad. Nauk. Ukr. SSR* (1981), 9–11.

[Pl]   W. Plandowski, Testing equivalence of morphisms on context-free languages, *Lecture Notes in Comput. Sci.* **855**, Springer-Verlag 1994, 460–470.

[P1]   E. Post, A variant of a recursively unsolvable problem, *Bulletin of Amer. Math. Soc.* **52** (1946), 264–268.

[P2]   E. Post, Recursive unsolvability of a problem of Thue, *J. Symb. Logic* **12** (1947), 1–11.

[Re]   L. Redei, The Theory of Finitely Generated Commutative Semi-Groups, Pergamon Press, 1965.

[Ro]   J. J. Rotman, The Theory of Groups, Springer-Verlag, (4th ed.) 1995.

[RS1]   G. Rozenberg and A. Salomaa, The Mathematical Theory of L Systems, Academic Press, 1980.

[RS2]   G. Rozenberg and A. Salomaa, Cornerstones of Undecidability, Prentice Hall, 1994.

[Ru1]   K. Ruohonen, Reversible machines and Post's correspondence problem for biprefix morphisms, *J. Inform. Process. Cybernet. EIK* **21** (1985), 579–595.

[Ru2]   K. Ruohonen, Test sets for iterated morphisms, Report 49, Tampere University of Technology, Tampere, 1986.

[S1]   A. Salomaa, Theory of Automata, Pergamon Press, 1969.

[S2]   A. Salomaa, Formal Languages, Academic Press, 1973.

[S3]   A. Salomaa, Equality sets for homomorphisms of free monoids, *Acta Cybernetica* **4** (1978), 127–139.

[S4]   A. Salomaa, Jewels of Formal Language Theory, Computer Science Press, 1981.

[S5]   A. Salomaa, The Ehrenfeucht Conjecture: A proof for language theorists, *Bull. EATCS* **27** (1985), 71–82.

[S6]   A. Salomaa, Computation and Automata, Cambridge University Press, 1985.

[Sc]   P. Schultz, Mortality of 2 × 2-matrices, *Amer. Math. Monthly* **84** (1977), 463–464.

[Sch]   M. P. Schützenberger, Sur les relations rationelles entre monoides libre, *Theoret. Comput. Sci.* **3** (1976), 243–259.

[Sco]   D. Scott, A short recursively unsolvable problem, *J. Symb. Logic* **21** (1956), 111–112.

[T1]   P. Turakainen, A homomorphic characterization of principal semi-AFLs without using intersection with regular sets, *Inform. Sci.* **27** (1982), 141–149.

[T2]   P. Turakainen, A machine-oriented approach to composition of morphisms and inverse morphisms, *Bull. EATCS* **20** (1983), 162–166.

[T3]   P. Turakainen, A unified approach to characterizations of recursively enumerable languages, *Bull. EATCS* **45** (1991), 223–228.

[Tz]   G. C. Tzeitin, Associative calculus with an unsolvable equivalence problem, *Tr. Mat. Inst. Akad. Nauk* **52** (1958), 172–189 (Russian).

[W]   A. Weber, Decomposing finite-valued transducers and deciding their equivalence, *SIAM J. Comput.* **22** (1993), 175–202.

Codes[1]

Helmut Jürgensen and Stavros Konstantinidis

1. Introduction

Codes are formal languages with special combinatorial and structural properties which are exploited in information processing or information transmission. In this application, codes serve several different purposes. In the following discussion we assume the well-known model of information transmission consisting of a source S sending information to a recipient R via a channel C as illustrated in Fig. 1.1. Before actual transmission, the information is encoded using an encoder γ and, before reception, it is decoded using a decoder δ. During transmission, the encoded information may undergo changes due to environmental conditions or faults in the channel; the potential presence of such changes is modelled by a source N of noise. Moreover, the information may be overheard or even altered during transmission by a hostile participant F.

In this model, S and R may, but need not be distinct physical objects, and C may represent any kind of physical channel, a wire, a compact disk, a computer memory, a nerve connection, a computer program, radio waves – anything accepting an input and producing an output. In this handbook chapter we consider only discrete channels which operate in discrete time steps and which use discrete signals. We represent the signals processed by a channel as symbols over some alphabet. Hence, we model a channel as a device which, for input strings – finite, one-sided infinite, or bi-infinite sequences of symbols –, produces output strings. Input strings are called *messages* and output strings are called *received messages* in the sequel.

The purposes of the encoding γ and the decoding δ include the following:

- translation between the alphabets used by S, C, and R;
- reduction of the effect of noise on C;
- adaptation of the information rates at which S, C, and R operate;

[1] We gratefully acknowledge the support of this work by the Natural Sciences and Engineering Research Council of Canada, Grant OGP0000243. We also thank colleagues and friends, especially F. Gécseg and K. Salomaa, for their comments on an earlier version of this chapter.

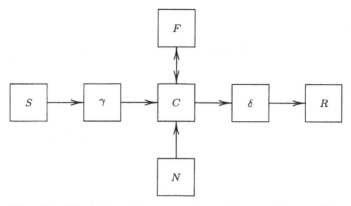

Fig. 1.1. The information processing and transmission model.

- data compression;
- information security.

A very basic requirement is that γ and δ operate correctly, that is,[2] that $\delta(\gamma(w)) = w$ for every message w. This assumes that the channel, without noise and tampering by F, will output w when it gets w as input.[3] This requirement is not sufficient when noise is present or when F might change messages.

For implementation reasons, γ is usually assumed to work by symbols, that is, γ acts as a homomorphism,[4] mapping the output symbols of S onto input words of C. The decoding δ, on the other hand, usually cannot work on a symbol-by-symbol basis. This raises the following issues:

- delay of decoding, that is, how much of an encoded message δ needs to see before it can issue its first output symbol with certainty;
- delay of synchronization, that is, how much of an encoded message δ needs to see after an interruption before it can again start decoding correctly.

These issues are very important from the point of view of applications as they determine the speed at which the system may operate.

In the presence of noise, another set of problems needs to be addressed:

- the presence of errors in a received message needs to be detected;
- some errors in the received message should be corrected.

---

[2] A weaker assumption would be sufficient in general, that is, that $\delta(\gamma(w))$ be what S "intends" it to be. However, mathematically, this does not make a significant difference.

[3] This assumption, while reasonable in most physical situations, can be avoided as shown in Section 5.

[4] Also transducer-based encoding methods are frequently used – for example convolution codes –, in particular for certain error correction or data compression applications; see [99] and [103].

The noise source N is usually modelled by a stochastic source that, somehow, influences the behaviour of the channel. Thus, error detection and error correction can only be expected to be achieved up to a certain probability threshold. Error detection is mainly a matter for δ to accomplish. Error correction may also be achieved by δ, but other system components, like protocols, could be involved as well.

To consider the adaptation of information rates and information compression, one models S and C as probability spaces [22], [36]. Under appropriate assumptions about these spaces, quantities like the average information contents of a transmitted symbol can be defined in a physically meaningful way.

Finally, an encoding γ can be used for encryption to make a message unintelligible for F and to equip it with hidden attributes for δ to detect tampering with the message by F. Beyond codes[5] one also uses transmission protocols to deal with the threats arising from F.

Given this wide range of requirements, the general theory of codes employs concepts and tools from quite a few different areas of mathematics, among them probability and information theory, combinatorics, algebra, and geometry. In this survey – in the spirit of this handbook – we focus on issues which arise when codes are considered as formal languages, that is, mainly on combinatorial issues. Probability theoretic and information theoretic ones are discussed only briefly to set the stage.

The theory of codes has developed into several nearly unrelated fields:

- information theoretic considerations are used to derive the existence and general properties of codes regarding information rate and usability on various channels;
- the theory of error-correcting codes employs a non-probabilistic error model to design and evaluate codes for various types of channels;
- the theory of variable-length codes[6] investigates combinatorial properties of codes with little regard to their error-correcting properties.

We refer to [22] or [36] as sources regarding the information theoretic aspects of coding theory. For the theory of error-correcting codes we refer to [99] or [94]. With very few exceptions, the theory of error-correcting codes concerns codes in which all code words have the same length. In contrast to this, the theory of variable-length codes deals with codes the code words of which may have different lengths, but rarely considers error-correction or error-detection

---

[5] The cryptographic literature has a special meaning for the word *code* which is quite different from its meaning in coding theory (see [67], for example); on the other hand, the notions of encryption in cryptography and encoding in coding and information theory are nearly identical. Hence, we use *encoding* to mean both and, for now, *code* to mean the system employed.

[6] This terminology is quite unfortunate. The lengths of code words in such a code are not variable as the term might suggest, but they vary.

properties of such codes. We refer to [4] and [123] for the theory of variable-length codes.

Even with the focus *just* on language theoretic and combinatorial aspects of variable-length codes, the field is so wide that we cannot cover everything in this survey, not even *all major* aspects. In view of this, we finally decided to present mostly material that is not readily available in any book on codes. *We intend this survey to complement existing book publications and to interpret the material in the context of information transmission systems. We hope to present a unifying view of the field incorporating some of its language theoretic as well as some of its communication theoretic aspects.*

The theory of variable-length codes originated in concrete problems of information transmission. Its language theoretic branch has taken a direction, however, which was often guided more by mathematical considerations than by issues of the application. This is by no means intended to be a criticism of the work or the results; in fact, some very deep insights about the combinatorics on words have been obtained as results of research on codes and many unexpected applications, aside from information transmission, of the theory of codes have been detected. Whenever appropriate, we emphasize the information theoretic considerations that motivate some of the constructions. Also in this respect, this survey will complement the existing literature on codes.

This handbook chapter is structured as follows: After this introductory section, we introduce the notation and some basic notions in Section 2. Some effort is spent to set up a very precise notation for words, languages, and factorizations of words. For example, we distinguish between word schemata, instances of word schemata, and words – the latter being equivalence classes of instances of word schemata. This permits us to give precise meanings to notions like channel, code, factorization, decoding, synchronization even when bi-infinite messages are considered and noise must be assumed to be present on the channel. Of course, for an intuitive understanding of the field, the reader may just substitute their usual language theoretic counterparts, as far as they exist; for a completely unambiguous treatment, however, a formalization like ours is inevitable. In fact, we even omitted a few details that were not absolutely needed for the presentation in this article.

In Section 3, we then introduce the notions of channel and code. We single out the class of P-channels as a class defined by a set of physically convincing, but very general properties. The notions of message, received message, unique decodability are defined. Moreover, we discuss the decidability of unique decodability.

In Section 4, we introduce the general notions of error-correction, synchronization, and decoding. For channels permitting insertions, deletions, and substitutions of symbols it is decidable whether a finite language is error-correcting for finite messages. The notions of synchronization and decoding reveal that the properties of P-channels are not restrictive enough to yield technically realizable results; additional restrictions are needed which, however, may be physically significant.

In Section 5, we explain an alternative approach to the notions of error-correction and decoding. This approach is based on homphonic codes and leads to another formal channel model. We point out the connections and discuss the differences.

In Section 6, we explain the underlying scheme by which, in most cases, *natural* classes of codes can be defined. The definition method has nothing whatsoever to do with formal languages, words, or free monoids. It is based on the purely set theoretic notion of dependence system as proposed in [16]. We review some basic definitions and results of that theory and express the definition of a large number of classes of codes in this framework. For most of the classes, we briefly review some of their prominent properties. We pay particular attention to issues like error-detection and error-correction.

In Section 7, we present an attempt at a systematic understanding of the relations among the classes of codes introduced up to this point. We expand the presentation of dependence theory as required and describe general principles and properties of constructions of hierarchies of classes of codes. In this way, we obtain a uniform treatment of many cases that have been dealt with by special and separate methods so far; moreover, we establish a formal framework for many more potential cases.

In Section 8, we present results about syntactic monoids of codes. This complements various book-form presentations on syntactic monoids of the message sets over codes. The main point in this section is that, under certain formal conditions, the predicate defining a class of dependence systems is *inherited* via syntactic morphisms. In such a case, the properties of the corresponding independent sets, that is, codes, are directly visible in their syntactic monoids. In certain situations this can be exploited to decide properties of codes.

Section 9 focusses on the structure of the decision problem of code properties. We present two non-equivalent general decision procedures, one based on syntactic monoids and one based on transducers. We speculate, why these fail in certain cases and present a survey of known decidability and undecidability results.

In Section 10, we present a summary of many results about maximal codes; we relate some of these to a general theorem pattern about dependence systems. A general meta-theorem on the lines suggested does not exist so far. Using dependence theory, however, we present, a special case of such a meta-theorem which explains and supersedes many special-case proofs.

In Sections 11 and 12, we turn our attention back to codes for noisy channels. In Section 11, we focus on solid codes. Such codes have remarkable synchronization capabilities in the presence of noise. As such they are quite good at error-detection, but not very good at all at error-correction. We discuss results concerning their structure and their cost in terms of communication time. In Section 12, we briefly present codes that can be or are used for certain kinds of noisy channels. We discuss some basic issues about modelling noisy channels and the construction of error-correcting codes for arbitrary noisy channels.

Finally, Section 13 contains a few concluding remarks.

This handbook chapter on codes is not intended to recount the history of the theory and applications of codes. We cite only work which is *directly* relevant to the material presented. We treat the books [4] by Berstel and Perrin and [123] by Shyr as *standard references* for a large part of the earlier results and, in general, do not trace results back to the original authors. In these books one can usually find the references to the originals. Despite this, our list of references is quite long and it is still very far from being complete.

2. Notation and basic notions

This section serves two purposes: We introduce the notation used throughout the handbook chapter, and we review some basic definitions and results.

The symbols $\mathbb{Z}$ and $\mathbb{N}$ denote the sets of integers and of positive integers, respectively; let $\mathbb{N}_0 = \mathbb{N} \cup \{0\}$ and $\mathbb{N}_{\aleph_0} = \mathbb{N} \cup \{\aleph_0\}$. The symbol $\mathbb{R}$ denotes the set of real numbers, and $\mathbb{R}_+$ denotes the set of positive real numbers.

We denote by $|S|$ the cardinality of a set S. For a set $S_1 \times S_2$, the first and second projections are denoted by π_1 and π_2, respectively. If $\varrho \subseteq S_1 \times S_2$ and $s_1 \in S_1$, then $\varrho(s_1) = \{s_2 \mid s_2 \in S_2, (s_1, s_2) \in \varrho\}$. By 2^{S_2} we denote the set of all subsets of S_2, and $2^{S_2}_{\neq \emptyset}$ is the set of all non-empty subsets of S_2. Let σ be a mapping of S_1 into 2^{S_2}. Whenever convenient, we identify σ with the relation $\{(s_1, s_2) \mid s_1 \in S_1, s_2 \in S_2, s_2 \in \sigma(s_1)\}$. If τ is an equivalence relation on a set S, then a *cross section* of τ is a subset of S containing exactly one element of each equivalence class with respect to τ.

For a set S and $n \in \mathbb{N}_{\aleph_0}$, $\mathrm{Sub}_n S$ denotes the set of all subsets of S of cardinality strictly less than n. Thus, $\mathrm{Sub}_1 S = \{\emptyset\}$ and $\mathrm{Sub}_{\aleph_0} S$ is the set of all finite subsets of S.

For $n \in \mathbb{N}_0$, let $\mathfrak{n} = \{0, 1, \ldots, n-1\}$; thus, $\mathfrak{o} = \emptyset$ and $\mathfrak{1} = \{0\}$. Let $\omega = \mathbb{N}_0$, $-\omega = \{-n \mid n \in \mathbb{N}_0\}$, and $\zeta = \omega \cup -\omega = \mathbb{Z}$. We define

$$\mathfrak{o} < \mathfrak{1} < \mathfrak{2} < \cdots < \left\{ {\omega \atop -\omega} \right\} < \zeta.$$

Let $\mathfrak{J} = \{\mathfrak{n} \mid n \in \mathbb{N}_0\} \cup \{-\omega, \omega, \zeta\}$. In this handbook chapter, a set I is said to be an *index set* if $I \in \mathfrak{J}$. For index sets I and J, let

$$I + J = \begin{cases} \{i \mid i \in \mathbb{N}_0, i < n+m\}, & \text{if } I = \mathfrak{n}, J = \mathfrak{m} \text{ with } n, m \in \mathbb{N}_0, \\ \omega, & \text{if } I = \mathfrak{n}, J = \omega \text{ with } n \in \mathbb{N}_0, \\ -\omega, & \text{if } I = -\omega, J = \mathfrak{n} \text{ with } n \in \mathbb{N}_0, \\ \zeta, & \text{if } I = -\omega, J = \omega. \end{cases}$$

The definition of addition could be extended to all pairs of symbols in $\mathfrak{I}$; however, we do not need it for any cases beyond these.[7] For an index set I and an integer n let $n + I = I + n = \{n + i \mid i \in I\}$.

A mapping ψ of an index set I into an index set J is an *index mapping* if it is injective and order-preserving and if the image of I is a convex subset of J, that is, $\psi(i) < j < \psi(i')$ for $i, i' \in I$ and $j \in J$ implies that $j = \psi(i'')$ for some $i'' \in I$.

An *alphabet* is a non-empty set of symbols. In this section, let X be an arbitrary, but fixed alphabet.

A *word schema* w over X is a mapping of an index set I_w into X. A word schema is specified by its index set I_w and the symbols $w(i) \in X$ for $i \in I_w$. A word schema w is *finite* if $|I_w|$ is finite, that is $I_w = \mathbf{n}$ for some $n \in \mathbb{N}_0$ and $n = |I_w|$ is the *length* of w denoted by $|w|$. A word schema w is said to be *right-infinite* if $I_w = \omega$ and *left-infinite* if $I_w = -\omega$. A word schema w is said to be *bi-infinite* if $I_w = \zeta$. An *infinite* word schema is a word schema which is right-infinite or left-infinite or bi-infinite.

Two word schemata w and v over X are said to be equivalent, $w \sim v$, if there is a $j \in \mathbb{Z}$ such that $j + I_w = I_v$ and $w(i) = v(j + i)$ for all $i \in I_w$.

Suppose w and v are equivalent word schemata. If $I_w < \zeta$ then $I_v = I_w$ and $v = w$. If $I_w = \zeta$ then $I_v = \zeta$ and it is possible that $w(i) \neq v(i)$ for some $i \in \zeta$, that is, $w \neq v$.

Let w be a word schema over X. An *instance* of w is a mapping $[\![w, n]\!]$ of $n + I_w$ into X for some $n \in \mathbb{Z}$ such that $[\![w, n]\!](i + n) = w(i)$ for all $i \in I_w$. In particular, $[\![w, 0]\!] = w$. The equivalence $\sim$ of word schemata induces an equivalence of instances of word schemata: Let w and v be word schemata over X and $n, m \in \mathbb{Z}$; the instances $[\![w, n]\!]$ and $[\![v, m]\!]$ are said to be equivalent, $[\![w, n]\!] \approx [\![v, m]\!]$, if $w \sim v$. Some examples of equivalent instances of word schemata are shown in Fig. 2.1. All instances of a given word schema are equivalent. Two word schemata w and v either have no instances in common or the same instances.

A *word* over X is an equivalence class of instances of word schemata. Intuitively, a word schema describes a sequence of symbols fixing some starting point 0. An instance of a word schema is obtained by shifting the starting point. A word is the sequence itself without reference to the starting point. In the context of information transmission, the starting point represents the time at which the symbol at this point is being sent.

- For every word w, let $\mathcal{S}(w)$ be an arbitrary but fixed word schema[8] the instances of which belong to w.

---

[7] The similarity with ordinal numbers is intended.

[8] By the definition of word schemata, there is no arbitrariness in the selection of $\mathcal{S}(w)$ except when w is bi-infinite.

(a) $\quad \begin{cases} I_w = \{0,1,2,3\} \\ w(i) = \begin{cases} 1, & \text{if } i = 1 \\ 0, & \text{if } i \neq 1 \end{cases} \end{cases}$

indices for $[w,5]$	5	6	7	8
symbols	0	1	0	0
indices for $[w,-10]$	-10	-9	-8	-7

(b) $\quad \begin{cases} I_w = \omega \\ w(i) = \begin{cases} 1, & \text{if } i = 1 \\ 0, & \text{if } i \neq 1 \end{cases} \end{cases}$

indices for $[w,-2]$	-2	-1	0	1	...
symbols	0	1	0	0	...
indices for $[w,0]$	0	1	2	3	...

(c) $\quad \begin{cases} I_w = \zeta \\ w(i) = \begin{cases} 1, & \text{if } i = -1 \\ 0, & \text{if } i \neq 1 \end{cases} \end{cases}$

indices for $[w,-2]$	...	-4	-3	-2	-1	...
symbols	...	0	1	0	0	...
indices for $[w,0]$	...	-2	-1	0	1	...

Fig. 2.1. Examples of equivalent instances of word schemata: (a) finite word schema; (b) right-infinite word schema; (c) bi-infinite word schema.

- For a word w let $\mathcal{I}(w)$ be the set of all instances[9] of $\mathcal{S}(w)$; similarly, for a word schema v, let $\mathcal{I}(v)$ be the set of all instances of v; moreover, let $\bar{v}$ be the equivalence class of all instances of v.
- For any instance u of a word schema, let $\mathcal{S}(u)$ be an arbitrary but fixed word schema with $u \in \mathcal{I}(\mathcal{S}(u))$ such that, if u and v are instances of word schemata and $u \approx v$, then $\mathcal{S}(u) = \mathcal{S}(v)$.

This seemingly cumbersome distinction between words, word schemata, and instances of word schemata is needed in some cases in the sequel to avoid ambiguity. *When no ambiguity is possible in the given context, we just use the term word.* In several situations, however, this distinction is essential to make statements precise.

Consider word schemata w and v. Their *concatenation* wv is defined if and only if $I_w + I_v$ is defined and, in this case,

$$
wv(i) = \begin{cases}
w(i), & \text{if } I_w = \mathbf{n} \text{ for some } n \in \mathbb{N}_0 \text{ and } i \in \mathbf{n}, \\
v(i-n), & \text{if } I_w = \mathbf{n} \text{ for some } n \in \mathbb{N}_0 \text{ and } i \geq n, \\
v(i+n-1), & \text{if } I_w = -\omega, I_v = \mathbf{n} \text{ for some } n \in \mathbb{N}_0 \text{ and } i > -n, \\
w(i+n), & \text{if } I_w = -\omega, I_v = \mathbf{n} \text{ for some } n \in \mathbb{N}_0 \text{ and } i \leq -n, \\
w(i+1), & \text{if } I_w = -\omega, I_v = \omega \text{ and } i < 0, \\
v(i), & \text{if } I_w = -\omega, I_v = \omega \text{ and } i \geq 0,
\end{cases}
$$

for $i \in I_{wv} = I_w + I_v$. When using concatenation, we implicitly assume that it is defined.

Consider two instances $[w,n]$ and $[v,m]$ of word schemata such that wv is defined. Then any two instances of the word schema wv are equivalent. The concatenation of word schemata corresponds to the usual definition of concatenation of words.

---

[9] By definition, w itself is the set of all instances of $\mathcal{S}(w)$; we introduce this notation, albeit redundant, to avoid confusion.

Let $\mathfrak{S}$ be the set of symbols

$$\mathfrak{I} \cup \{*, \blacklozenge\} \cup \{\leq I \mid I \in \mathfrak{I}\} \cup \{<I \mid I \in \mathfrak{I}\}.$$

For $I \in \mathfrak{I}$, let X^I be the set of all words w such that $I_{\mathcal{S}(w)} = I$; let $X^{<I} = \bigcup_{J<I} X^J$ and $X^{\leq I} = X^{<I} \cup X^I$; finally, let $X^* = X^{<\omega}$, $X^{\blacklozenge} = X^{\leq \zeta}$, and let $X^+ = X^* \setminus \{\lambda\}$ where λ denotes the *empty word*,[10] that is, the word with $I_{\mathcal{S}(\lambda)} = \emptyset$. With this notation, X^*, X^+ and X^ω get their usual meanings in language theory; the set $X^{\leq \omega}$ is often denoted by X^∞. For $\eta \in \mathfrak{S}$, w is an *η-word schema* if $\overline{w} \in X^\eta$; an *η-language* is a set of η-words. In the rest of this handbook chapter, *word* means $\blacklozenge$-word and *language* means $\blacklozenge$-language.

Many statements about words or languages remain true if one reads the words from right to left instead of from left to right and if one exchanges ω and $-\omega$. We refer to this fact as *duality* and use it frequently.

Let $Y \subseteq \mathcal{S}(X^{\blacklozenge})$ and let φ be a mapping of an index set I_φ into Y. Assume that, for every $j \in I_\varphi$, if $j + 1 \in I_\varphi$ then $\varphi(j)\varphi(j+1)$ is defined. Let $[\varphi]$ be the word schema defined by $[\varphi](i) = \varphi(j)(k)$ where $j \in I_\varphi$, $k \in I_{\varphi(j)}$, and where j and k are determined by

$$i - k = \begin{cases} \sum_{l=0}^{j-1} |\varphi(l)|, & \text{if } i \geq 0 \text{ and } I_\varphi = \zeta \text{ or } I_\varphi \leq \omega, \\ -\sum_{l=j}^{-1} |\varphi(l)|, & \text{if } i < 0 \text{ and } I_\varphi = \zeta, \\ 1 - \sum_{l=j}^{0} |\varphi(l)|, & \text{if } i \leq 0 \text{ and } I_\varphi = -\omega. \end{cases}$$

The index set of $[\varphi]$ is the set of i for which such j and k exist. Intuitively, $[\varphi]$ is the word schema over X which is obtained as the concatenation of the images of φ in the given order such that the occurrence of $\varphi(0)$ starts at position 0. We illustrate the typical situations in Fig. 2.2.

Definition 2.1 *For $y \in \mathcal{S}(X^{\blacklozenge})$ and $Y \subseteq \mathcal{S}(X^{\blacklozenge})$, a factorization[11] of y over Y is a mapping φ of an index set I_φ into Y such that $[\varphi] \sim y$. Two factorizations φ and φ' of y over Y are said to be* equivalent, $\varphi \sim \varphi'$, *if there is an integer n such that $n + I_\varphi = I_{\varphi'}$ and $\varphi(i) = \varphi'(n+i)$ for all $i \in I_\varphi$.*

Remark 2.1 Consider a word schema y and two equivalent factorizations φ and φ' of y over Y. If $I_\varphi < \zeta$ then $\varphi = \varphi'$.

---

[10] In the sequel we use λ also to denote the empty word schema. The precise meaning will be clear from the context.

[11] For simplification we often speak of factorizations over a set K of words rather than word schemata; in such cases, it is implied that the factorization is over $\mathcal{S}(K)$.

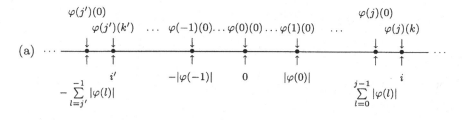

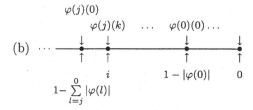

Fig. 2.2. The typical arrangement of symbols in $[\varphi]$: (a) the cases of $I_\varphi = \zeta$ or $I_\varphi \leq \omega$, shown for $i \geq 0$ and $i' < 0$; (b) the case of $I_\varphi = -\omega$. In both cases the upper row shows the symbols; the lower row shows the indices.

Note that a factorization φ of $y \in \mathcal{S}(X^\bullet)$ over $Y \subseteq \mathcal{S}(X^\bullet)$ is itself a word schema over the alphabet Y. With this intuition, the equivalence of factorizations coincides with the equivalence of word schemata. Moreover, for any index set I, Y^I is the set of all word schemata[12] over X having a factorization φ over Y with $I_\varphi = I$.

If $v \in X^+$ then, v^ω denotes the unique ω-word representing $\{\mathcal{S}(v)\}^\omega$. The words v^n for $n \in \mathbb{N}_0$, $v^{-\omega}$, and v^ζ are defined analogously.

A word $w \in X^+$ is said to be *primitive* if $w = u^n$ for $u \in X^+$ and $n \in \mathbb{N}$ implies $n = 1$. The *root* of a word $w \in X^+$ is the unique primitive word u such that $w = u^n$ for some $n \in \mathbb{N}$. By $\sqrt{w}$ we denote the root of w. For a set $L \subseteq X^+$, let $\sqrt{L}$ be the set $\{\sqrt{w} \mid w \in L\}$. For any words $v, w \in X^+$, let $v \sim_{\sqrt{}} w$ if and only if $\sqrt{v} = \sqrt{w}$.

As mentioned before, we use a simplified and more intuitive notation when there is no risk of ambiguity. Thus, with $X = \{0,1\}$, the word $w = 0010$ would be used to describe the word schema w with $I_w = 4$, $w(0) = w(1) = w(3) = 0$, and $w(2) = 1$. Similarly, we write $v = \cdots 010101$ to denote the word schema v with $I_v = -\omega$, $v(0) = v(-2) = v(-4) = 1$, $v(-1) = v(-3) = v(-5) = 0$, and, presumably, $v(-2i) = 1$, $v(-2i - 1) = 0$ for all $i \in \mathbb{N}_0$. The reverse notation is used for word schemata in $\mathcal{S}(X^\omega)$. Even in the case of word schemata in $\mathcal{S}(X^\zeta)$, we sometimes use this simplified notation; for example $u \in \mathcal{S}(X^\zeta)$ with $u(2i) = 1$ and $u(2i - 1) = 0$ may just be written as $u = \cdots 010101010 \cdots$ when the actual values of the indices are not relevant. Moreover, as often done in the literature, we indicate factors by enclosing them in parentheses. For example, factorizations of u over $K = \{01, 10\}$ can

---

[12] When $n \in \mathbb{N}_0$, we usually write Y^n instead of $Y^{\mathbf{n}}$.

be described by

$$u = \cdots (01)(01)(01)(01)(0 \cdots = \cdots 0)(10)(10)(10)(10) \cdots .$$

Strictly speaking, this notation only describes equivalence classes of factorizations.

Example 2.1 Consider the alphabet $X = \{0, 1\}$ and the set $Y = \{010\}$. More precisely, Y is the singleton set consisting of the word schema w with $I_w = 3$, $w(0) = w(2) = 0$, and $w(1) = 1$. Let φ be the mapping of $I_\varphi = \zeta$ into Y with $\varphi(j) = w$ for all $j \in \zeta$ (there is only this mapping). Then

$$[\varphi](i) = \begin{cases} 0, & \text{if } i \text{ or } i-2 \text{ is divisible by 3,} \\ 1, & \text{if } i-1 \text{ is divisible by 3.} \end{cases}$$

Let $v = [\varphi]$ and $Y' = \{w, w'\}$ where $w' = ww$, that is, $I_{w'} = 6$, $w'(0) = w'(2) = w'(3) = w'(5) = 0$, and $w'(1) = w'(4) = 1$. Then v has infinitely many non-equivalent factorizations over Y'. For example, in addition to φ, also φ' with $I_{\varphi'} = \zeta$, $\varphi'(2i) = w'$, and $\varphi'(2i+1) = w$ for $i \in \zeta$ is a factorization of v over Y'. Fig. 2.3 illustrates this situation.

factorization φ	$\cdots$	$\varphi(-1)$			$\varphi(0)$			$\varphi(1)$			$\varphi(2)$			$\cdots$
instances	$\cdots$	$[\![\varphi(-1), -3]\!]$			$[\![\varphi(0), 0]\!]$			$[\![\varphi(1), 3]\!]$			$[\![\varphi(2), 6]\!]$			$\cdots$
word schema $[\varphi]$	$\cdots$	0	1	0	0	1	0	0	1	0	0	1	0	$\cdots$
indices	$\cdots$	-3	-2	-1	0	1	2	3	4	5	6	7	8	$\cdots$
instances	$\cdots$	$[\![\varphi'(-1), -3]\!]$			$[\![\varphi'(0), 0]\!]$			$[\![\varphi'(1), 6]\!]$						$\cdots$
factorization φ'	$\cdots$	$\varphi'(-1)$			$\varphi'(0)$			$\varphi'(1)$						$\cdots$

Fig. 2.3. The word schema $[\varphi]$ for φ, w, φ' and w' as in Example 2.1.

Consider $w \in \mathcal{S}(X^{\leq \omega})$ and $u \in \mathcal{S}(X^*)$. The word schema u is said to be a *prefix* of w if, for ψ the identity mapping on I_u, one has $[w\psi] = u$; it is a *proper prefix* if $0 < I_u < I_w$. Let $\mathrm{Pref}(w)_\lambda$ and $\mathrm{Pref}(w)$ denote the sets of prefixes and of proper prefixes, respectively. By duality, one defines the notion of *suffix* of a word schema $w \in \mathcal{S}(X^{\leq -\omega})$.

Consider $w \in \mathcal{S}(X^{\leq \zeta})$ and $u \in \mathcal{S}(X^*)$. The word schema u is an *infix* of w if there is an index mapping ψ of I_u into I_w such that $[w\psi] = u$; it is a *proper infix* of w if $0 < I_u < I_w$.

The intuition leading to these definitions of prefix, suffix, and infix is illustrated in Fig. 2.4. Note that every prefix and every suffix is an infix.

We assume that the reader is familiar with certain basic notions of formal language theory like *finite automaton, generalized sequential machine, various kinds of acceptors, grammar, regular language, context-free language, linear language, decidability.* If required, definitions can be found in the corresponding chapters of this handbook.

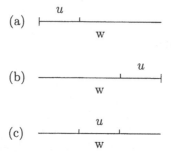

Fig. 2.4. Illustration of prefix, suffix, and infix: (a) u is a prefix of w; (b) u is a suffix of w; (c) u is an infix of w.

For the rest of this handbook chapter, we assume that X is an arbitrary, but fixed *finite* alphabet with $|X| > 1$. Moreover, in examples we usually assume without special mention that the symbols 0 and 1 are distinct elements of X.

3. Channels and codes

In modelling a channel we describe its typical input-output behaviour. In information theory this usually involves probabilities, that is, the conditional probabilities of outputs given certain inputs. In the theory of codes the abstraction usually goes even further: one models the reasonably likely behaviour only, and this behaviour is modelled non-deterministically. We use the latter approach.

Definition 3.1 *A* channel *(over X) is a binary relation on $\mathcal{I}(X^{\blacklozenge})$. A channel γ is said to be* stationary *if, for all $y, y' \in \mathcal{S}(X^{\blacklozenge})$ and for all $n, n' \in \mathbb{Z}$, one has $(\llbracket y', n' \rrbracket, \llbracket y, n \rrbracket) \in \gamma$ if and only if $(\llbracket y', n' - n \rrbracket, \llbracket y, 0 \rrbracket) \in \gamma$.*

If γ is a channel over X then we interpret $(y', y) \in \gamma$ to mean that, upon input y, the channel could output y'. To suggest this interpretation we write $(y' \mid y)$ instead of (y', y) in analogy with the notation used for conditional probabilities.

For $y \in \mathcal{S}(X^{\blacklozenge})$, let

$$\langle y \rangle_\gamma = \left\{ y' \in \mathcal{S}(X^{\blacklozenge}) \mid \exists n, n' \in \mathbb{Z} \left((\llbracket y', n' \rrbracket \mid \llbracket y, n \rrbracket) \in \gamma \right) \right\},$$

that is, $\langle y \rangle_\gamma$ is the set of word schemata that can be obtained as output of γ when y is used as the input. For $Y \subseteq \mathcal{S}(X^{\blacklozenge})$, let

$$\langle Y \rangle_\gamma = \bigcup_{y \in Y} \langle y \rangle_\gamma.$$

Thus $\langle Y \rangle_\gamma$ is the set of all possible outputs of γ when word schemata in Y are used as inputs.

For example, if $y = [\![w, n]\!]$, $y' = [\![w', m]\!]$ with $w, w' \in \mathcal{S}(X^*)$ then $(y' \mid y) \in \gamma$ means the following: if $w(0), w(1), \ldots, w(|w| - 1)$ are the inputs of γ at times $n, n+1, \ldots, n+|w|-1$ then, at times $m, m+1, \ldots, m+|w'|-1$, the output symbols can be $w'(0), w'(1), \ldots, w'(|w'|-1)$.

Stationarity means that the absolute time is not important. Thus, if $[\![\gamma, k]\!]$ is the channel given by

$$[\![\gamma, k]\!] = \Big\{ ([\![w', n']\!] \mid [\![w, n]\!]) \,\Big|\, ([\![w', n'-k]\!] \mid [\![w, n-k]\!]) \in \gamma \Big\}$$

for $k \in \mathbb{Z}$, then γ is stationary if and only if $\gamma = [\![\gamma, k]\!]$ for all $k \in \mathbb{Z}$.

Definition 3.2 *Let γ be a stationary channel over X, let $Y \subseteq \mathcal{S}(X^+)$, $Y' = \langle Y \rangle_\gamma$, $w, w' \in \mathcal{S}(X^\blacklozenge)$, and let φ be a factorization of w over Y. A factorization φ' of $w' \in \langle w \rangle_\gamma$ over Y' is said to be γ-admissible for φ if $I_\varphi = I_{\varphi'}$ and, for all $n, n' \in \mathbb{Z}$ such that $([\![w', n']\!] \mid [\![w, n]\!]) \in \gamma$ and for every non-empty index set I and every index mapping ψ of I into I_φ, one has $([\![[\varphi'\psi], n']\!] \mid [\![[\varphi\psi], n]\!]) \in \gamma$.*

Definition 3.1 seems to be general enough to model most discrete physical channels.[13] In fact, most physical channels seem to satisfy the following additional conditions:

$(\mathcal{P}_0)$ *A channel preserves finiteness and the type of infiniteness,* that is, if $y' \in \langle y \rangle_\gamma$ and $y \in \mathcal{S}(X^\eta)$ for $\eta \in \{*, \omega, -\omega, \zeta\}$ then $y' \in \mathcal{S}(X^\eta)$.

$(\mathcal{P}_1)$ *Input factorizations have corresponding factorizations of the output,* that is, if $y' \in \langle y \rangle_\gamma$ and φ is a factorization of y over a subset Y of $\mathcal{S}(\pi_2(\gamma)) \cap \mathcal{S}(X^+)$ with $I_\varphi \neq \emptyset$ then there is a factorization of y' over $\langle Y \rangle_\gamma$ which is γ-admissible for φ.

$(\mathcal{P}_2)$ *Error-freeness does not depend on the context,* that is,

$$\text{if } ([\![v', n']\!] \mid [\![v, n]\!]) \in \gamma \text{ then } ([\![uv'w, n']\!] \mid [\![uvw, n]\!]) \in \gamma$$

for all $u, w \in \mathcal{S}(\pi_2(\gamma))$ with $[\![uvw, n]\!] \in \pi_2(\gamma)$.

$(\mathcal{P}_3)$ *Empty input can always result in empty output,* that is,

$$([\![\lambda, n']\!] \mid [\![\lambda, n]\!]) \in \gamma$$

for all $n, n' \in \mathbb{Z}$.

The conditions $\mathcal{P}_0$–$\mathcal{P}_3$ correspond, roughly, to the following physical assumptions: Infinite time periods do not exist, that is, for a message of which we don't know the beginning or the end or both, a channel cannot provide that information; on the other hand, if a message has a definite beginning or a

---

[13] Physical channels sometimes have different alphabets for input and output. Modelling this is straightforward, but not necessary in the context of this handbook chapter.

definite end then the channel cannot hide these forever. The noise in a channel is a property of the channel and not of the messages – this is similar to the notion of *additive noise* in information theory. Error-free information transmission is always possible. If there is no input, then it is always possible that there is no corresponding output, regardless of the delay. While most physical channels will have these properties, channels with memory might violate $\mathcal{P}_2$.

On the other hand, conditions $\mathcal{P}_0$–$\mathcal{P}_3$ are really very weak. For instance, in $\mathcal{P}_2$, if $(\llbracket v', n' \rrbracket \mid \llbracket v, n \rrbracket) \in \gamma$ then $(\llbracket uv'w, n' \rrbracket \mid \llbracket uvw, n \rrbracket) \in \gamma$ for all $u, w \in \mathcal{S}(X^{\blacklozenge})$ with $\llbracket uvw, n \rrbracket \in \pi_2(\gamma)$, but it is not guaranteed that the factorization $(u)(v')(w)$ is γ-admissible for the factorization $(u)(v)(w)$. Similarly, if $(u')(v'w')$ and $(v')(w')$ are γ-admissible factorizations of $u'v'w'$ and $v'w'$, respectively, for $(u)(vw)$ and $(v)(w)$, respectively, then one cannot conclude that $(u')(v')(w')$ is a γ-admissible factorization for $(u)(v)(w)$. This situation may seem rather unnatural. It cannot, however, be avoided in general as it permits one to model the case when the noise may depend on the message. Classical error models assume that message and noise are statistically independent; for certain types of modern information transmission media this assumption seems to be not quite adequate.

Definition 3.3 *A P-channel is a stationary channel γ with $\pi_2(\gamma) = \mathcal{I}(X^{\blacklozenge})$ which satisfies $\mathcal{P}_0$, $\mathcal{P}_1$, $\mathcal{P}_2$, and $\mathcal{P}_3$.*

The *noiseless channel* $\{(y \mid y) \mid y \in \mathcal{I}(X^{\blacklozenge})\}$ is a P-channel. A channel, which changes, inserts or deletes up to m symbols in every L consecutive symbols is an example of a noisy P-channel.

In the presence of such errors one cannot, in general, assume that the output resulting from an input sent at time n is received at time n or at least, at time $n + k$ for some fixed $k \in \mathbb{Z}$. For example, if the channel γ could insert or delete 1 symbol in every 3 consecutive symbols, then the input 0010110101101 sent at times $0, 1, \ldots, 12$ could result in the output 0101011011001 received at times $1, 2, \ldots, 13$ as shown in Fig. 3.1. Thus, while it is sometimes convenient to think of the indices in terms of time, this interpretation can be quite misleading when taken literally. Time should not be interpreted as *physical time* in this context. In the case of the channel with insertions and deletions, if $(\llbracket w', n' \rrbracket \mid \llbracket w, n \rrbracket) \in \gamma$, then also $(\llbracket w', n' + k \rrbracket \mid \llbracket w, n \rrbracket) \in \gamma$ for any $k \in \mathbb{Z}$.

Definition 3.2 and the conditions $\mathcal{P}_1$, $\mathcal{P}_2$, and $\mathcal{P}_3$ are formulated in such a way as to take these difficulties into account. For example, in $\mathcal{P}_1$, the input and output factorizations may not have the same factorization points.

Given a channel, a code needs to be found which guarantees[14] that any message can be unequivocally recovered from the corresponding channel output.

---

[14] Strictly speaking, with high enough probability.

input times:	0	1	2	3	4	5	6	7		8	9	10	11		12
input word	0	0	1	0	1	1	0	1		0	1	1	0		1
output word		0	1	0		1	0	1	1	0	1	1	0	0	1
output times:		1	2	3		4	5	6	7	8	9	10	11	12	13

Fig. 3.1. Timing of channel inputs and outputs in the presence of insertions and deletions.

Let $\eta \in \{*, \omega, -\omega, \zeta, \leq\omega, \leq-\omega, \leq\zeta\}$ and $K \subseteq X^+$. An *η-message over K* is a word schema in $\mathcal{S}(K^\eta)$.

Definition 3.4 *A language K is* uniquely *η-decodable if every η-message over K has only a single factorization over $\mathcal{S}(K)$, up to equivalence of factorizations.*

For a stationary channel γ, a *γ-received η-message (over K)* is a word schema w' such that $([\![w', n']\!] \mid [\![w, 0]\!]) \in \gamma$ for some η-message w over K and some $n' \in \mathbb{Z}$. For an η-message w over K, $\langle w \rangle_\gamma$ is the set of γ-received η-messages resulting from input w.

Definition 3.5 *A language K is an η-code if $K \subseteq X^+$ and K is uniquely η-decodable.*

Usually, the term *code* means *-code in the literature on codes, and this is also how we use this term in the rest of this handbook chapter. On the other hand, we often use the terms[15] *message* and *received message* in the generic sense *η-message* and *γ-received η-message* for some implied, but unspecified γ and η. For η as above, let $\mathcal{K}^\eta$ be the class of η-codes.

Example 3.1 We illustrate the concepts with a few examples:

(1) The language $K = \{0, 01, 10\}$ is not a code, not even a *-code as the message 010 has two non-equivalent factorizations over K: $010 = (0)(10) = (01)(0)$.

(2) A language K such that $K \subseteq X^n$ for some $n \in \mathbb{N}$ is called a *uniform code* or a *block code* of *(block) length* n. A uniform code K of length n is *full* if $K = X^n$. Uniform codes are ω-codes and $(-\omega)$-codes, but not necessarily ζ-codes. For example, consider $K = \{01, 10\}$. Then the bi-infinite word $\cdots 010101010 \cdots$ has two non-equivalent factorizations over K, that is,

$$\cdots 010101010 \cdots = \cdots (01)(01)(01)(01)(0 \cdots$$
$$= \cdots 0)(10)(10)(10)(10) \cdots.$$

---

[15] The terminology and notation for codes in the context of infinite messages seems to be still evolving. In [21] two kinds of codes are considered for bi-infinite words, biω-codes and $\mathbb{Z}$-codes. The former are our ζ-codes. The latter require unique factorizations, and not just unique factorizations up to equivalence. As already noted in [21], the notion of $\mathbb{Z}$-code is not really natural in the context of information processing.

(3) The language $K = \{101, 01\}$ is an ω-code. If w is an ω-message over K then the unique factorization of w into word schemata $u_0, u_1, \ldots \in \mathcal{S}(K)$ can be determined as follows: Let $w_0 = w$. For $i \geq 0$, let

$$u_i = \begin{cases} 101, & \text{if } w_i \text{ starts with } 1, \\ 01, & \text{if } w_i \text{ starts with } 0, \end{cases}$$

and let w_{i+1} be the ω-word schema satisfying $w_i = u_i w_{i+1}$. In a similar way one verifies that K is also a $(-\omega)$-code.

This language K is even a ζ-code. To see this, let v be some ζ-message over K. If $v(i) \neq v(i+1)$ for all $i \in \zeta$ then 101 cannot occur in any factorization of v over K as there is no word ending with 0 in K. If $v(i) = v(i+1)$ for some $i \in \zeta$ then $v(i) = 1$. The unique factorization φ of v over K is determined as follows: Let ψ_1 be the index mapping of ω into ζ with $\psi_1(0) = i+1$ and let ψ_2 be the index mapping of $-\omega$ into ζ with $\psi_2(0) = i$. Let φ_1 and φ_2 be the unique factorizations of $[v\psi_1]$ and $[v\psi_2]$, respectively, over K. Let

$$\varphi(j) = \begin{cases} \varphi_1(j), & \text{if } j \geq 0, \\ \varphi_2(j+1), & \text{if } j < 0, \end{cases}$$

for all $j \in \zeta$. Then $[\varphi] = [\varphi_2][\varphi_1]$ and $[\varphi](l) = v(l+i+1)$ for all $l \in \zeta$.

(4) The language $K = \{0, 01, 11\}$ is a $*$-code and even a $(-\omega)$-code, but not an ω-code because of $(0)(11)^\omega = (01)(11)^\omega$.

The η-codes, for $\eta \in \{*, -\omega, \omega, \zeta\}$ form a proper hierarchy (see [21], [128], and [20], for example).

Proposition 3.1 *The following proper inclusions obtain:*

$$\mathcal{K}^\zeta \subsetneq \left\{ \begin{matrix} \mathcal{K}^\omega \\ \mathcal{K}^{-\omega} \end{matrix} \right\} \subsetneq \mathcal{K}^*.$$

Proof: First consider $K \in \mathcal{K}^\zeta$, let w be a word schema with $I_w = \omega$, and let φ_1 and φ_2 be factorizations of w over K. Because of Remark 2.1, we need to show that $\varphi_1 = \varphi_2$. Let $v \in \mathcal{S}(K)$. Define mappings ψ_1 and ψ_2 of ζ into K by

$$\psi_1(i) = \begin{cases} \varphi_1(i), & \text{if } i \geq 0, \\ v, & \text{if } i < 0, \end{cases} \quad \text{and} \quad \psi_2(i) = \begin{cases} \varphi_2(i), & \text{if } i \geq 0, \\ v & \text{if } i < 0, \end{cases}$$

for $i \in \zeta$. Thus, ψ_1 and ψ_2 are factorizations of $[\psi_1] = [\psi_2]$ over K and, therefore, $I_{\psi_1} = I_{\psi_2}$ and $\psi_1(i) = \psi_2(j+i)$ for some $j \in \mathbb{Z}$ and all $i \in \zeta$.

If $j = 0$ then $\psi_1 = \psi_2$, hence $\varphi_1 = \varphi_2$. Therefore, suppose $j \neq 0$. Without loss of generality, we may assume that $j > 0$. The situation is illustrated in Fig. 3.2. Thus, $w = v^j w$ and, by induction, $w = v^{jn} w$ for all $n \in \mathbb{N}$, hence $w = (v^j)^\omega = v^\omega$. Thus, as $v \in \mathcal{S}(K)$ and $K \in \mathcal{K}^\zeta$, $\psi_1(i) = \psi_2(i) = v$ for all $i \in \zeta$, hence $\varphi_1(i) = \varphi_2(i) = v$ for all $i \in \omega$.

This proves the inclusion $\mathcal{K}^\varsigma \subseteq \mathcal{K}^\omega$. The inclusion $\mathcal{K}^\varsigma \subseteq \mathcal{K}^{-\omega}$ follows by duality. The inclusion is proper by Example 3.1(2).

To prove that $\mathcal{K}^\omega \subseteq \mathcal{K}^*$ consider $K \in \mathcal{K}^\omega$ and a word schema $w \in \mathcal{S}(K^*)$. Suppose, w has two non-equivalent factorizations over K. Let $v \in \mathcal{S}(K)$. Then also the word schema wv^ω has two nonequivalent factorizations, a contradiction.

This proves that $\mathcal{K}^\omega \subseteq \mathcal{K}^*$. The inclusion is proper by Example 3.1(4). The inclusion $\mathcal{K}^{-\omega} \subsetneq \mathcal{K}^*$ follows by duality.   □

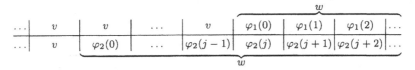

Fig. 3.2. Factorizations in the proof of Proposition 3.1, $\mathcal{K}^\varsigma \subseteq \mathcal{K}^\omega$.

A detailed study of the relation between various classes of η-codes for $\eta \le \omega$ is conducted in [128] and [20].[16] The following result distinguishes between $*$-codes and ω-codes.

Theorem 3.1 [20] *Let $K \subseteq X^+$. The following conditions on K are equivalent:*

(a) *K is a $*$-code.*
(b) *For every $u \in K^+$, u^ω has a unique factorization over K.*
(c) *For every $u \in X^+$ with $u^\omega \in K^\omega$, u^ω has only finitely many different factorizations over K.*
(d) *For every $u \in X^*$ and $v \in X^+$ with $uv^\omega \in K^\omega$, uv^ω has only finitely many different factorizations over K.*

For a given language $K \subseteq X^+$ it can sometimes be quite difficult to determine whether K is an η-code. For the case of K being a finite language and $\eta = *$, an algorithm to solve this problem was first given in [115]. An algorithm to solve this problem for a regular language K can be found in [4], for example. This result was extended to $\eta = \omega$ by [20] and to $\eta = \varsigma$ by [21]. For a complexity theoretic analysis see [88].

Theorem 3.2 *Let $\eta \in \{*, -\omega, \omega, \varsigma\}$. The following question is decidable: Given a deterministic finite automaton accepting a language $K \subseteq X^+$, is K an η-code?*

---

16 In [128], ω-codes are referred to as *ifl-codes*.

Proofs of Theorem 3.2 rely on two main ideas: If a message has two distinct factorizations over K then there are words in K which can be made to *overlap* each other; when K is regular the possible sequences of *overlaps* have certain periodicity properties.

We illustrate the concept of *overlap* by a small example. Consider a language $K \subseteq X^+$ and a $*$-message w having two non-equivalent factorizations φ_1 and φ_2 over K. The typical situation is shown in Fig. 3.3. There the word schemata $u_0, u_1, u_2, u_3, v_0, v_1, v_2$ are assumed to be elements of $\mathcal{S}(K)$. The word schema u_0 is proper prefix of v_0; the rest of v_0 is a proper prefix of u_1; the rest of u_1 is a prefix of v_1; u_2 is an infix of v_1; the rest of v_1 is a prefix of u_3; the rest of u_3 is equal to v_2.

Fig. 3.3. A finite word schema with two distinct factorizations.

Thus, there are two typical situations:

(1) An instance of a word schema $u \in \mathcal{S}(K)$ is an infix of another word schema $v \in \mathcal{S}(K)$. In Section 6 below, we introduce the class of infix codes. A language $K \subseteq X^+$ is an infix code if and only if there are no two distinct word schemata in K such that one is an infix of the other.

(2) The end of an instance of a word schema $u \in \mathcal{S}(K)$ *overlaps* the beginning of a word schema $v \in \mathcal{S}(K)$, where u and v need not be distinct. A class of infix codes for which such overlaps are explicitly excluded is defined in Section 6 and analysed in greater detail in Section 11, the class of solid codes.

These two situations arise in many proofs in coding theory. Several formalizations of the notion of overlap exist in the literature, mainly developed in the context or for the purpose of proving parts of Theorem 3.2 and related results.[17] A definition of overlaps and chains of overlaps that is also applicable in the case of decoding for noisy channels is given in [70].

In Theorem 3.2, the assumption that K be regular is quite important. For context-free and even for linear K it is, in general, even undecidable whether K is a $*$-code.

Theorem 3.3 *There is no algorithm which, given a linear grammar G, always decides whether the language generated by G is a $*$-code.*

We could not find a reference to this result which we believe to have seen proved some twenty years ago. The work in [35] is related to this problem to some extent. The following proof is due to S. Yu [145].

---

[17] See [4], [110] (reprinted in [41]), [40], [12], [127], [74], [78], [76], and the work cited there.

Proof: Let X be an alphabet and let (U, V) be an instance of Post's Correspondence Problem, where

$$U = (u_0, u_1, \ldots, u_{n-1}) \text{ and } V = (v_0, v_1, \ldots, v_{n-1})$$

for some $n \in \mathbb{N}$ and $u_0, u_1, \ldots, u_{n-1}, v_0, v_1, \ldots, v_{n-1} \in X^*$. A solution to (U, V) is a pair (m, I) where $m \in \mathbb{N}$ and I is an m-tuple of integers, $I = (i_0, i_1, \ldots, i_{m-1})$, such that

$$i_j \in \mathfrak{n} \text{ for } j \in \mathfrak{m} \text{ and } u_{i_0} u_{i_1} \cdots u_{i_{m-1}} = v_{i_0} v_{i_1} \cdots v_{i_{m-1}}.$$

Without loss of generality, we may assume that the symbols 0, 1, #, \$, and ¢ are not in X. Let $Y = X \cup \{0, 1, \#, \$, \cent\}$. For any positive integer i, let $\beta(i)$ denote the shortest binary representation of i.

Consider the linear grammar G defined as follows: The terminal alphabet of G is Y. The non-terminal alphabet consists of the symbols S, T_U, and T_V with S the start symbol. The rules are

$$S \to \beta(i)\cent T_U u_i \$, \qquad S \to \beta(i)\cent u_i \$,$$
$$S \to \beta(i)\cent T_V v_i \$\#, \qquad S \to \beta(i)\cent v_i \$\#, \qquad S \to \#,$$
$$T_U \to \beta(i)\cent T_U u_i, \qquad T_U \to \beta(i)\cent u_i, \qquad T_V \to \beta(i)\cent T_V v_i, \qquad T_V \to \beta(i)\cent v_i$$

for $i \in \mathfrak{n}$. Let K be the language generated by G. Then $K = \{\#\} \cup K_U \cup K_V$ where K_U and K_V are languages consisting precisely of all the words

$$\beta(i_{m-1})\cent\beta(i_{m-2})\cent \cdots \beta(i_0)\cent u_{i_0} \cdots u_{i_{m-2}} u_{i_{m-1}} \$$$

and

$$\beta(i_{m-1})\cent\beta(i_{m-2})\cent \cdots \beta(i_0)\cent v_{i_0} \cdots v_{i_{m-2}} v_{i_{m-1}} \$\#,$$

respectively, for all $m \in \mathbb{N}$ and all $i_0, \ldots, i_{m-1} \in \mathfrak{n}$.

If K is not a $*$-code then there is a $*$-message over K that has two different factorizations over K. Let this message be of minimal length. Then the two factorizations start with different word schemata w and w'. Without loss of generality, we may assume that w is a proper prefix of w'. This implies $w \neq \# \neq w'$ and $w \notin K_V$ as no word in K_V is a proper prefix of any word in K. Thus $w \in K_U$ and $w' \in K_V$ as no word in K_U has a proper prefix in K_U. Thus

$$w = \beta(i_{m-1})\cent\beta(i_{m-2})\cent \cdots \beta(i_0)\cent u_{i_0} \cdots u_{i_{m-2}} u_{i_{m-1}} \$$$

and, consequently,

$$w' = \beta(i_{m-1})\cent\beta(i_{m-2})\cent \cdots \beta(i_0)\cent v_{i_0} \cdots v_{i_{m-2}} v_{i_{m-1}} \$\#,$$

that is, the problem (U, V) has (m, I) with $I = (i_0, \ldots, i_{m-1})$ as a solution.

Conversely, if (U, V) has a solution then, with w and w' as above, one has $w\# = w'$, that is, K is not a code.

Thus, K is a $*$-code if and only if (U, V) has no solution. As Post's Correspondence Problem is undecidable in general, also the property of being a $*$-code is undecidable for linear languages. $\qquad\square$

Some of the results discussed in the present section have been extended to *infinitary* or even *bi-infinitary codes*, that is, subsets of $X^{\leq\omega}$ or of $X^\blacklozenge$. To achieve this, the concatenation of words $u, v \in X^\blacklozenge$, so far defined only when $u \in X^{\leq-\omega}$ and $v \in X^{\leq\omega}$, is extended by letting

$$uv = \begin{cases} u, & \text{if } u \in X^\omega \cup X^\varsigma, \\ v, & \text{if } u \in X^{\leq-\omega} \text{ and } v \in X^{-\omega} \cup X^\varsigma. \end{cases}$$

Two notions of *unique decodability* need to be distinguished: the first one is the natural extension of the notion used throughout this handbook chapter; the second one requires unique factorizations – up to equivalence – with finite index sets. In either case, very interesting analogues of classical results in coding theory are obtained [136], [137], [138], [139], [72], [140].

4. Error correction, synchronization, decoding

Unique η-decodability is a necessary property to guarantee that one can read the transmitted message. It is, however, not sufficient:

First, in the presence of noise the channel output may differ from its input; it may not even be a message any more. In this case, the code to be used has to satisfy additional requirements to guarantee correction or at least detection of errors.

For example, consider the uniform code K of Example 3.1(2) and a channel γ in which at most one symbol is erased in every four consecutive symbols. Thus the message 010101 could lead to the channel output 1010 through deletion of the symbols marked by $\times$ in $\aleph1010\chi$. The same channel output could have been obtained from the message 010110 through $\aleph101\chi0$ or from the message 1010 without errors.

Thus, in the presence of noise, unique decodability is quite insufficient. As this example shows, the errors may even be such that the received erroneous message looks like an error-free message.

Second, through noise in the received message, the recipient may not be able to determine where the next received code word starts, that is, in essence the recipient might not have access to the beginning (and end) of the received message.

For example, consider the *-code $K = \{111000, 000111\}$ used on a noisy channel γ that permits up to one deletion among any three consecutive symbols. Then

$$\cdots 0110011001100110 \cdots$$

can be part of a γ-received message over K having the two γ-admissible factorizations

$$\cdots 0)(1100)(1100)(1100)(110 \cdots \quad \text{and} \quad \cdots 011)(0011)(0011)(0011)(0 \cdots.$$

It is impossible to determine which factorization to accept from the available information. Thus, the recipient has lost the *synchronization* with the sender.

Third, the recipient may be unable to afford the cost of decoding as unbounded memory may be required. For example, consider the $*$-code K of Example 3.1(4). For every $n \in \mathbb{N}_0$, $01^n \in K^*$ with the factorization

$$01^n = \begin{cases} (0)(11)(11)\cdots(11), & \text{if } n \text{ is even,} \\ (01)(11)(11)\cdots(11), & \text{if } n \text{ is odd.} \end{cases}$$

To determine even just the first factor, one has to wait until the end of the complete message and, in the meantime, store this message in the decoder.

In this particular example one might argue, that the storage cost grows only as $\log n$; but this is still unbounded and, moreover, just a special property of this example.

One could also argue in this example that the recipient can safely decode everything following 01 in the word above as only 11 is found, possibly shifted by one position. However, the interpretation of 11 may depend on whether the received message starts with 0 or with 01. For example, 0 and 01 might mean *don't use* and *use,* respectively; and the sequence of occurrences 11 might encode what the recipient is supposed not to use or to use. Thus, knowledge of the intermediate code words would be quite useless before the first one is known.

To solve the first problem, one uses codes with prescribed error-correction capabilities. To address the second problem, one introduces the notion of synchronization delay and uses codes which are *uniformly synchronous.* To address the third of these issues, one introduces the notion of decoding delay and uses codes with bounded decoding delay, possibly even with delay 0, the *prefix codes.*

In the formal language theory branch of coding theory, the notions of synchronization delay and decoding delay are only considered for noiseless channels.[18] In this handbook chapter we develop these notions for the general framework of P-channels. They specialize to the usual ones for the case of noiseless channels. Moreover, we show that some fundamental properties of these concepts continue to obtain even in the presence of noise.

Definition 4.1 *Let γ be a P-channel and let $\eta \in \{*, \omega, -\omega, \leq\omega, \leq-\omega, \zeta, \leq\zeta\}$.*

(a) A language K is (γ, η)-detecting if it is an η-code and, for every η-message w over K, $\langle w \rangle_\gamma \cap \mathcal{S}(K^\eta) = \{w\}$. In this case we also say that K is a (γ, η)-detecting code.

(b) A language K is (γ, η)-correcting if it is an η-code and, for all η-messages u and w over K with $u \neq w$, one has $\langle u \rangle_\gamma \cap \langle w \rangle_\gamma = \emptyset$. In this case we also say that K is a (γ, η)-correcting code. Let $\mathcal{K}_\gamma^\eta$ be the class of (γ, η)-correcting codes.

---

[18] [41] is an exception; we briefly discuss that approach in Section 5. In that work a different type of channel model is used.

Intuitively, a code is (γ, η)-detecting if those received messages, different from w, which one is likely to see as output of γ for an input message w are not themselves messages over the code. A code is (γ, η)-correcting if those received messages which one is likely to see as output of γ for an input message w are different from the likely received messages for all other input messages. In this case, in the spirit of *maximum likelihood decoding* (see [99]), the decoder will decode any received message in $\langle w \rangle_\gamma$ as w.

Example 4.1

(1) Consider the uniform code K of Example 3.1(2). Let γ be the channel discussed above, that is, the channel which permits at most one deletion in every four consecutive message symbols. Then $w = 010101$, $v = 010110$, and $u = 1010$ are *-messages over K such that u is contained in all three intersections $\langle w \rangle_\gamma \cap \langle v \rangle_\gamma$, $\langle w \rangle_\gamma \cap \langle u \rangle_\gamma$, and $\langle v \rangle_\gamma \cap \langle u \rangle_\gamma$. Thus, K is not $(\gamma, *)$-correcting.

(2) For the same channel γ, consider the language

$$K = \{0^5 101, 0^7 10011, 0^8 1101, 0^{10} 110011\}.$$

One verifies that $\langle u \rangle_\gamma \cap \langle v \rangle_\gamma = \emptyset$ for any $u, v \in K$ unless $u = v$. Nevertheless, K is not even $(\gamma, *)$-correcting because

$$(\slashed{0} 0^6 \slashed{1} 0011)(0^5 101) = (0^8 110 \slashed{1})(0^4 \slashed{0} 101)$$
$$\in \langle (0^7 10011)(0^5 101) \rangle_\gamma \cap \langle (0^8 1101)(0^5 101) \rangle_\gamma.$$

(3) Consider $K = \{0001, 01011\}$ and a channel γ in which at most one in every five consecutive symbols can be deleted. Then

$$\langle 0001 \rangle_\gamma = \{0001, 001, 000\}$$

and

$$\langle 01011 \rangle_\gamma = \{01011, 1011, 0011, 0111, 0101\}.$$

The set K is an ω-code. Consider an ω-message w over K. Then $v \in \langle w \rangle_\gamma$ can start with any of the word schemata in $\langle 0001 \rangle_\gamma \cup \langle 01011 \rangle_\gamma$.

If v starts with 001 then w could start with 0001 or with 01011. If the next symbol in v is 0 then w starts with 0001; if it is 1 then the initial part of w could be $\slashed{0} 001 \slashed{0} 1011$ or $0 \slashed{1} 0110 \cdots$ with deletions as indicated. The first case is impossible as the two deletions are too close to each other. Therefore, w starts with 01011. Using the fact that γ is a P-channel, we can now remove 01011 from w and 0011 from v and look at the remaining word schemata.

If v starts with 0001 or 000 then w starts with 0001; if it starts with 01011, 1011, 0111, or 0101 then w starts with 01011. As before, we can remove this part from v and the corresponding code word from w.

Extending these considerations one can show that K is even (γ, ζ)-correcting.

(4) The language $K = \{0011, 010111\}$ is $(\gamma, *)$-correcting for the channel γ permitting at most one deletion among any six consecutive symbols. On the other hand, because of

$$(0011)(10111)^\omega = (00111)(01111)^\omega$$
$$\in \langle(0011)(010111)^\omega\rangle_\gamma \cap \langle(010111)^\omega\rangle_\gamma$$

it is not (γ, ω)-correcting. Note that K is an ω-code.

Proposition 4.1 *Let γ be a P-channel and $K \subseteq X^+$. If $|K| > 1$ and $K \in \mathcal{K}_\gamma^\omega \cup \mathcal{K}_\gamma^{-\omega}$ then $K \in \mathcal{K}_\gamma^*$. If $|K| > 2$ and $K \in \mathcal{K}_\gamma^\zeta$ then $K \in \mathcal{K}_\gamma^\omega \cap \mathcal{K}_\gamma^{-\omega}$.*

Proof: First, consider $K \in \mathcal{K}_\gamma^\omega$ with $|K| > 1$. By Proposition 3.1, K is a $*$-code. Let u, v be $*$-messages over K and let $z \in \langle u\rangle_\gamma \cap \langle v\rangle_\gamma$. We show that $u = v$. Let υ and φ be the factorizations of u and v, respectively, over K. If $I_\upsilon = I_\varphi = \text{o}$ then $u = v = \lambda$. Therefore, and without loss of generality, we assume that $I_\varphi \geq I_\upsilon$ and $I_\varphi > \text{o}$. Let $n = \max I_\varphi$. As $|K| > 1$, there is a $w \in S(K)$ such that $w \neq \varphi(n)$. Then $uw^\omega, vw^\omega \in S(\pi_2(\gamma))$ as γ is a P-channel. This implies $zw^\omega \in \langle uw^\omega\rangle_\gamma \cap \langle vw^\omega\rangle_\gamma$ by property $\mathcal{P}_2$, hence $uw^\omega = vw^\omega$ and this word schema has exactly one factorization, χ say, over K as K is an ω-code. Moreover, $\chi(i) = \upsilon(i)$ for $i \in I_\upsilon$ and $\chi(i) = \varphi(i)$ for $i \in I_\varphi$. If $I_\varphi > I_\upsilon$ then $\varphi(n) = w$, contradicting the choice of w. Thus $I_\varphi = I_\upsilon$, hence $u = v$. This proves the first statement for $K \in \mathcal{K}^\omega$. The case of $K \in \mathcal{K}^{-\omega}$ is proved using duality.

Now suppose that $|K| > 2$ and $K \in \mathcal{K}^\zeta$. We show that $K \in \mathcal{K}^\omega$. By duality it then follows that $K \in \mathcal{K}^{-\omega}$. Consider $u, v \in S(K^\omega)$ and $z \in \langle u\rangle_\gamma \cap \langle v\rangle_\gamma$. By $\mathcal{P}_0$, $z \in S(X^\omega)$. Let υ and φ be the factorizations of u and v, respectively, over K. As $|K| > 2$, there is $w \in S(K)$ such that $w \notin \{\varphi(0), \upsilon(0)\}$. Then $w^{-\omega}z \in \langle w^{-\omega}u\rangle_\gamma \cap \langle w^{-\omega}v\rangle_\gamma$, hence $w^{-\omega}u = w^{-\omega}v$. As K is a ζ-code, this word schema has a unique factorization, up to equivalence, over K. Let χ be such a factorization. As before, we can choose χ in such a way that χ extends υ and φ. This implies $\upsilon = \varphi$, hence $u = v$. $\square$

The condition on $|K|$ in Proposition 4.1 is needed. For example, for the channel γ permitting at most one deletion in every six consecutive symbols, the language $\{000000\}$ is (γ, ζ)-correcting and also (γ, ω)-correcting, but not $(\gamma, *)$-correcting.

In the presence of noise, the decidability problem answered for the noiseless channel in Theorem 3.2 takes the following form: For given K, η, and γ decide whether K is a (γ, η)-correcting code. For finite K, $\eta = *$ and a large natural class of channels, one can show that this question is decidable [70]. This result, to be explained in the sequel, extends the Sardinas-Patterson Theorem (Theorem 3.2 for finite languages and $\eta = *$) to noisy channels.

The channels to be considered permit symbol substitutions, symbol insertions, and symbol deletions. We give only an informal description of such channels; a formal syntax and semantics is defined in [59]. For given integers $m \in \mathbb{N}_0$ and $L \in \mathbb{N}$ with $m < L$, the channel will make at most m errors of a given kind in any L consecutive message symbols. The errors could be *substituting a symbol* for another one, *inserting symbols* between message symbols, or *deleting symbols* in the message. The bound of m can be on the total number of errors of all these kinds or on the number of errors for each kind separately. Thus, this model would include a channel permitting up to a total of five insertions and deletions in every twenty consecutive message symbols and, independently, a total of up to four symbol substitutions in every sixteen consecutive message symbols. Any channel of this kind is called an *SID-channel*. The channel considered in Example 4.1(1) is an SID-channel.

The classsical theory of error-correcting codes considers almost exclusively substitution errors. With such an error model, if the starting point of the message is known, synchronization between coding and decoding is not an issue of theoretical interest. For such channels, uniform codes are quite suitable as the uniform code word length affords the synchronization cheaply. For certain classical error models also *convolution codes* have been used successfully; the encoding for these codes is, in essence, computed by a finite automaton, a *gsm;* one of their main purposes is to take message patterns into account for encoding.

When the starting point of the message cannot be assumed to be known, then synchronization is a difficult problem even for the substitution-only error model and even when only block codes are considered (see, for example, [83] and [85]).

For fast modern information transmission systems – like satellite communication, communication via optical fibres, or optical data storage – the substitution-only error model is not adequate. Errors which look like insertions or deletions of symbols – regardless of their true physical nature – are much more common; and *bursts* of errors, that is, errors affecting many consecutive positions are quite likely. In this situation, the synchronization provided for free by uniform codes may no longer help – as shown in Example 4.1(1).

SID-channels correspond more closely to the physical error situations in modern communication systems than the classical error models which usually only involve symbol substitutions.[19] Uniform codes for the correction of insertions and deletions have been investigated by Levenshtein in [80], [79], [81], Sellers in [122], by Varshamov and Tenengol'ts in [141], and by Tenengol'ts in [129]. Recent work on codes for physical SID channels includes [5], [45], and [112], for instance. SID channels can be used to model certain aspects of speech recognition; see [2] and [3] for a probabilistic and information

---

[19] SID channels as defined in [59] do not handle bursts well. A modification of the model is discussed in [58].

theoretic analysis of such channels. Some further related work is discussed in Section 5 of this handbook chapter.

Theorem 4.1 [59] *Every SID-channel is a P-channel.*

Theorem 4.2 [70] *For any finite $K \subseteq X^+$ and any SID-channel γ, it is decidable whether K is a $(\gamma, *)$-correcting code.*

The proof of this extension of Theorem 3.2 for finite K and $\eta = *$ to noisy channels relies on a generalization of the notion of *overlap:* In addition to overlaps resulting from overlapping codewords, one also needs to consider overlaps in the received messages resulting from channel noise. Details are provided in [70].

We now turn to the notions of synchronization, synchronization delay, decodability, and decoding delay. They are intended to capture the following intuition: Suppose, K is an η-code to be used with a P-channel γ and that w' is a γ-received η-message over K. After having received a finite or, rather, a bounded part of w' one wants to be able to start the decoding and error correction process. This process will work on w' in the left-to-right or right-to-left directions or the combination of both; the latter is necessary, for instance, when $\eta = \zeta$. The process may have a definite starting point – the left end of w' if $\eta \leq \omega$ or the right end of w' if $\eta \leq -\omega$. There is no such obvious starting point for $\eta = \zeta$. This latter case models, for example, the situation when parts at the beginning and end of a received finite message have been corrupted beyond repair and the intermediate part is not in $\langle \mathcal{S}(K^*) \rangle_\gamma$ as symbols at the beginning and end may be missing. In this case, as a first step, one will attempt to find a position in w' corresponding to a point in the original message, where two code words were concatenated; this task is known as synchronization. After that, one can decompose w' at this position and decode the parts separately.

In the case of noiseless channels, a clear distinction can be made between synchronization and decoding; for noisy channels involving synchronization errors, this distinction is less clear.

Definition 4.2 Let γ be a P-channel and let $\eta \in \{*, \omega, -\omega, \leq\omega, \leq-\omega, \zeta, \leq\zeta\}$. A (γ, η)-correcting code $K \subseteq X^+$ is said to be uniformly (γ, η)-synchronous if there is an integer $n \in \mathbb{N}_0$ such that, for all $w \in \mathcal{S}(K^n)$, all $w' \in \langle w \rangle_\gamma$, and all $u, v \in \mathcal{S}(X^{\leq\eta})$ the following property obtains: If $uw'v \in \langle \mathcal{S}(K^*) \rangle_\gamma$ then there exist $z_1, z_2 \in \mathcal{S}(X^{\leq\eta})$ such that $z_1w, wz_2 \in \mathcal{S}(K^{\leq\eta})$, $z_1wz_2 \in \mathcal{S}(K^\eta)$ and $(u)(w')(v)$ is γ-admissible for $(z_1)(w)(z_2)$.

The smallest n for which the implication in Definition 4.2 holds true is the (γ, η)-*synchronization delay* of K. If n is the (γ, η)-synchronization delay of K, it is not true in general that the implication in Definition 4.2 holds true for any n' with $n' \geq n$, unless γ satisfies some additional appropriate conditions[20]

---

[20] For a noiseless channel the implication holds true trivially; see [4], for example.

beyond those for a P-channel. With u, w, w', v as in Definition 4.2, if $uw'v$ is a γ-received η-message over K then $(u)(w'v)$ and $(uw')(v)$ are candidates for decompositions of $uw'v$. For a noiseless channel and $\eta = *$, the definition coincides with the usual one of a uniformly synchronous code (see [4]); in this case, $z_1 = u$, $z_2 = v$, $w = w'$. Let $\mathcal{L}_{\text{unif-synch}}$ be the class of *uniformly* $(\gamma, *)$-*synchronous codes* when γ is a noiseless channel.

Definition 4.3 *Let* γ *be a P-channel, let* $\eta \in \{*, \omega, -\omega, \leq\omega, \leq-\omega, \zeta, \leq\zeta\}$, *and let* $K \subseteq X^+$. *A pair* (x, y) *of word schemata* $x, y \in \langle S(K^*) \rangle_\gamma$ *is* (γ, η)-synchronizing *if it has the following two properties:*

(1) *There are word schemata* $w_1, w_2 \in S(K^*)$ *such that* $(x)(y)$ *is* γ-*admissible for* $(w_1)(w_2)$.

(2) *For all such* $w_1, w_2 \in S(K^*)$ *and for all* $u, v \in S(X^{\leq\eta})$ *with* $uxyv \in \langle S(K^\eta) \rangle_\gamma$, *there exist word schemata* $z_1, z_2 \in S(X^{\leq\eta})$ *such that*

$$z_1 w_1, w_2 z_2 \in S(K^{\leq\eta}), \quad z_1 w_1 w_2 z_2 \in S(K^\eta),$$

and $(u)(xy)(v)$ *is* γ-*admissible for* $(z_1)(w_1 w_2)(z_2)$.

For a noiseless channel and $\eta = *$, Definition 4.3 is equivalent to the usual definition [4]. With u, x, y, v as above and K an η-code, the γ-received η-message $uxyv$ can be decomposed into ux and yv and then these parts can be decoded separately.

The assumption in Definition 4.3(1) expresses a subtle, but common problem arising in the modelling of noisy channels: If w_1 and w_2 are $*$-messages over K and if $x \in \langle w_1 \rangle_\gamma$ and $y \in \langle w_2 \rangle_\gamma$, it does not follow in general that $xy \in \langle w_1 w_2 \rangle_\gamma$. For example, if γ is the channel permitting up to one deletion in every two consecutive symbols and $K = \{101, 01\}$ is the ζ-code of Example 3.1(3), then $10 \in \langle 101 \rangle_\gamma$ and $1 \in \langle 01 \rangle_\gamma$, but $101 \notin \langle 10101 \rangle_\gamma$; to obtain 101 from 10101, two deletions in two consecutive positions are required, contrary to the definition of γ. Rather than being exotic, this property is inevitable due to the statistical properties of typical physical channels with independent source of noise. In Section 5, we summarize results which are obtained using a channel model which always permits the concatenation of channel outputs and we indicate how that model can be expressed in our terminology and where the fundamental differences regarding the physical assumptions are to be found.

The following proposition connects the notions of *uniformly* (γ, η)-*synchronous* and (γ, η)-*synchronizing pair*. Its special case for noiseless channels and $\eta = *$ is given in [4], Proposition 2.4. The proof below is based on the ideas used in [4], but requires some additional careful analysis of admissibility properties.

Theorem 4.3 *Let* γ *be a P-channel, let* $\eta \in \{*, \leq\omega, \leq-\omega, \leq\zeta\}$, *and let* $K \subseteq X^+$ *be a* (γ, η)-*correcting code. The following two statements are equivalent:*

(a) *K is uniformly* (γ, η)-*synchronous.*

(b) *There exists an integer* $n \in \mathbb{N}_0$ *such that every pair* (x, y) *with* $x, y \in \langle \mathcal{S}(K^n) \rangle_\gamma$ *is* (γ, η)*-synchronizing provided there exist* $w_1, w_2 \in \mathcal{S}(K^n)$ *with* $(x)(y)$ γ*-admissible for* $(w_1)(w_2)$.

Proof: Because of the choice of η one has $L^\eta = L^{\leq \eta}$ for any subset L of X^*. We can, therefore, use η instead of $\leq \eta$ in this proof.

Assume that statement (a) is true. Let n be the (γ, η)-synchronization delay of K. Let $x, y \in \langle \mathcal{S}(K^n) \rangle_\gamma$ be such that, for some $w_1, w_2 \in \mathcal{S}(K^n)$, $(x)(y)$ is γ-admissible for $(w_1)(w_2)$. Note that, as K is (γ, η)-correcting, w_1 and w_2 are uniquely determined if they exist. Let $u, v \in \mathcal{S}(X^\eta)$ such that $uxyv \in \langle \mathcal{S}(K^\eta) \rangle_\gamma$.

The assumption that K is uniformly (γ, η)-synchronous implies:

(1) There exist $z_1, z_2 \in \mathcal{S}(X^\eta)$ such that $z_1 w_1, w_1 z_2, z_1 w_1 z_2 \in \mathcal{S}(K^\eta)$, and $(u)(x)(yv)$ is γ-admissible for $(z_1)(w_1)(z_2)$.
(2) There exist $s_1, s_2 \in \mathcal{S}(X^\eta)$ such that $s_1 w_2, w_2 s_2, s_1 w_2 s_2 \in \mathcal{S}(K^\eta)$, and $(ux)(y)(v)$ is γ-admissible for $(s_1)(w_2)(s_2)$.

It follows that $z_1 w_1, w_2 s_2 \in \mathcal{S}(K^\eta)$ and, therefore, $z_1 w_1 w_2 s_2 \in \mathcal{S}(K^\eta)$. We need to prove that $(u)(xy)(v)$ is γ-admissible for $(z_1)(w_1 w_2)(s_2)$.

By (1), $u \in \langle z_1 \rangle_\gamma$; by assumption, $xy \in \langle w_1 w_2 \rangle_\gamma$; by (2), $v \in \langle s_2 \rangle_\gamma$.

By (2), $uxy \in \langle s_1 w_2 \rangle_\gamma$ with $s_1 w_2 \in \mathcal{S}(K^\eta)$. As $x \in \langle w_1 \rangle_\gamma$ and $w_1 \in \mathcal{S}(K^n)$, there exist $t_1, t_2 \in \mathcal{S}(X^\eta)$ such that $t_1 w_1, w_1 t_2, t_1 w_1 t_2 \in \mathcal{S}(K^\eta)$ and $(u)(x)(y)$ is γ-admissible for $(t_1)(w_1)(t_2)$. Thus, $uxy \in \langle t_1 w_1 t_2 \rangle_\gamma$, $ux \in \langle t_1 w_1 \rangle_\gamma$, and $xy \in \langle w_1 t_2 \rangle_\gamma$. Moreover, $ux \in \langle z_1 w_1 \rangle_\gamma$ and $xy \in \langle w_1 w_2 \rangle_\gamma$. As K is (γ, η)-correcting, one has $t_1 w_1 t_2 = s_1 w_2$, $t_1 w_1 = z_1 w_1$, and $w_1 t_2 = w_1 w_2$, hence, $t_1 = z_1$, $t_2 = w_2$, and $z_1 w_1 = s_1$. Similarly, using $xyv \in \langle w_1 z_2 \rangle_\gamma$ and $y \in \langle w_2 \rangle_\gamma$, one shows that $w_2 s_2 = z_2$. Hence, $z_1 w_1 w_2 = s_1 w_2$, $w_1 w_2 s_2 = w_1 z_2$, and $z_1 w_1 w_2 s_2 = s_1 w_2 s_2$. Therefore, $uxy \in \langle z_1 w_1 w_2 \rangle_\gamma$, $xyv \in \langle w_1 w_2 s_2 \rangle_\gamma$, and $uxyv \in \langle z_1 w_1 w_2 s_2 \rangle_\gamma$ as needed.

For the converse, assume that statement (b) is true. Let $w \in \mathcal{S}(K^{n+n})$, $w' \in \langle w \rangle_\gamma$, and $u, v \in \mathcal{S}(X^\eta)$ with $uw'v \in \langle \mathcal{S}(K^\eta) \rangle_\gamma$. There are $w_1, w_2 \in \mathcal{S}(K^n)$ with $w_1 w_2 = w$. By property $\mathcal{P}_1$, there are $x, y \in \mathcal{S}(X^\eta)$ such that $w' = xy$ and $(x)(y)$ is γ-admissible for $(w_1)(w_2)$. By assumption (x, y) is a (γ, η)-synchronizing pair. Hence, there exist $z_1, z_2 \in \mathcal{S}(X^\eta)$ such that $z_1 w_1, w_2 z_2, z_1 w_1 w_2 z_2 \in \mathcal{S}(K^\eta)$, and $(u)(xy)(v)$ is γ-admissible for the factorization $(z_1)(w_1 w_2)(z_2)$. Thus, K is uniformly (γ, η)-synchronous. $\square$

Example 4.2 Consider the code $K = \{0001, 01011\}$ over the alphabet $X = \{0, 1\}$. For the noiseless channel γ, K is uniformly $(\gamma, *)$-synchronous with delay 1. On the other hand, if γ is the channel permitting at most one deletion in every five consecutive symbols then K is not uniformly $(\gamma, *)$-synchronous.

First note, that K is $(\gamma, *)$-correcting. A γ-received $*$-message over K will start with 000, 0010, 0011, 010, 011, or 101; in the first two cases, the first code word used is 0001 and in the latter four it is 01011 – due to the characteristics of γ. Moreover, the start of the factorization of the received message is unique as no code word is subject to more than one error.[21] By induction this proves that K is indeed $(\gamma, *)$-correcting.

Now suppose that K is uniformly $(\gamma, *)$-synchronous with delay n and let $x = (1011)^n \in \langle(01011)^n\rangle_\gamma$. Consider $u = 001$ and $v = 1$. Then $uxv = 001(1011)^n 1 = (0011)(0111)^n \in \langle(01011)^{n+1}\rangle_\gamma$, but $xv = (1011)^n 1 \notin \langle\mathcal{S}(K^*)\rangle_\gamma$.

Definition 4.4 *Let $\eta \in \{*, \leq\omega\}$. A (γ, η)-correcting code $K \subseteq X^+$ is said to be right (γ, η)-decodable if there is a constant $d \in \mathbb{N}$ with the following property: For all $w \in \mathcal{S}(K^\eta)$, all $w', x, y \in \mathcal{S}(X^\eta)$, all $u, u' \in \mathcal{S}(K)$, and all $v \in \mathcal{S}(K^0)$, if $xw' \in \langle u'w\rangle_\gamma$, $xy \in \mathrm{Pref}(xw')$, and if $(x)(y)$ is γ-admissible for $(u)(v)$, then $u = u'$ and $w' \in \langle w\rangle_\gamma$.*

The smallest d such that K is right (γ, η)-decodable[22] with delay d is called the *right (γ, η)-decoding delay of K.* Let d be the right (γ, η)-decoding delay of K. As in the case of the synchronization delay, it is not true in general, that K is also right (γ, η)-decodable with delay d' for any $d' \geq d$. This can be guaranteed only when γ satisfies certain additional conditions beyong being a P-channel. It is trivially true for the noiseless channel. Let $\mathcal{L}_{\mathrm{rdecodable}}$ and $\mathcal{L}_{\mathrm{ldecodable}}$ be the classes of *right $(\gamma, *)$-decodable codes* and *left $(\gamma, *)$-decodable codes,* respectively, when γ is a noiseless channel.[23]

Remark 4.1 With K, η, γ, x, u, w, and w' as in Definition 4.4, if K is right (γ, η)-decodable then $(x)(w')$ is γ-admissible for $(u)(w)$.

Assume that K is right (γ, η)-decodable with delay d. By Remark 4.1, once the decoder has seen an initial part of the received message that could have been the output resulting from an input of $d+1$ consecutive code words and if, within this initial part, a factorization into the output for the first code word followed by the output for the next d code words is possible, then output x corresponding to the first code word – and, hence, that code word u – is uniquely identified. Moreover, the remainder w' of the received message is possible as an output for w and the concatenation uw as input could have yielded xw', given the properties of γ. Thus, Definition 4.4 correctly captures the idea of left-to-right decoding with a bounded delay.

---

[21] Note: This argument is only true for this particular type of situation; it cannot be generalized.

[22] Instead of *decodable* and *decoding delay* often also the terms *decipherable* and *deciphering delay* are used; see [4], for example.

[23] The codes in $\mathcal{L}_{\mathrm{rdecodable}}$ are usually referred to as *codes with finite decoding (or deciphering) delay;* see [4]. In [128] they are called *codes with bounded decoding delay;* the *codes with finite decoding delay* of that work form a different and larger class of codes. In [44] and [14] algorithms are presented for determining the decoding and synchronization delays of finite $*$-codes, and the complexity of these algorithms is determined.

Example 4.3 Let $X = \{0,1\}$ and let γ be the channel permitting at most one deletion in every four consecutive symbols. Let $K = \{u,v\}$ with $u = 0011$ and $v = 1100$. K is (γ, ω)-correcting. Consider the input word

$$(uv)^\omega = 0011110000111100\cdots$$

over K with the unique factorization

$$(0011)(1100)(0011)(1100)\cdots.$$

The word $w' = 00111000111000111000\cdots$ is possible as an output of γ for w as input. Note that w' has *two* γ-admissible factorizations, that is, $w' = (001)(110)(001)(110)\cdots$ and $w' = (0011)(100)(011)(100)(011)(100)\cdots$. This illustrates the point that, even if K is (γ, ω)-correcting, a γ-received ω-message need not have a unique factorization over $\langle S(K)\rangle_\gamma$. Nevertheless, the decoding is unique and K is right (γ, ω)-decodable with delay 0.

As in the case of synchronization, the notion of decodability and of decoding delay is equivalent to the usual one when $\eta = *$ and γ is noiseless (see [4], II.8, ans [128]). A more restrictive notion, counting the delay in terms of symbols rather than code words, is used in [57]. That definition implies the conditions of Definition 4.4 when K is finite, albeit with different values for the delays. By duality, one defines the notions of *left (γ, η)-decodability* and *left (γ, η)-decoding delay* for $\eta \in \{*, \leq-\omega\}$.

As in the case of noiseless channels, also for P-channels the synchronization delay is an upper bound of the right decoding delay.

Theorem 4.4 *Let $\eta \in \{*, \leq\omega\}$, let γ be a P-channel and let $K \subseteq X^+$ be (γ, η)-correcting. If K is uniformly (γ, η)-synchronous with delay n then K is right (γ, η)-decodable with delay n.*

Proof: Consider $x, y, w' \in S(X^\eta)$, $w \in S(K^\eta)$, $u, u' \in S(K)$, and $v \in S(K^n)$ such that $xw' \in \langle u'w\rangle_\gamma$, $xy \in \text{Pref}(xw')$, and $(x)(y)$ is γ-admissible for $(u)(v)$. Hence, $x \in \langle u\rangle_\gamma$, $y \in \langle v\rangle_\gamma$, $xy \in \langle uv\rangle_\gamma$, and there is $t \in S(X^\eta)$ with $xw' = xyt$.

As K is uniformly (γ, η)-synchronous with delay n, there are $z_1, z_2 \in S(X^\eta)$ such that $z_1v, vz_2, z_1vz_2 \in S(K^\eta)$ and $(x)(y)(t)$ is γ-admissible for $(z_1)(v)(z_2)$. As K is (γ, η)-correcting, one has $u'w = z_1vz_2$ and $uv = z_1v$, hence $u = z_1$ and $u'w = u(vz_2)$. This implies $u = u'$, $w = vz_2$, and $yt = w' \in \langle w\rangle_\gamma$. $\square$

For noiseless channels γ, the finite right $(\gamma, *)$-decodable codes are precisely the finite ω-codes.

Theorem 4.5 [74], [78] *For a finite set $K \subseteq X^+$, one has $K \in \mathcal{L}_{\text{rdecodable}}$ if and only if $K \in \mathcal{K}^\omega$.*

Theorem 4.6 $\mathcal{L}_{\text{unif-synch}} \subsetneqq \mathcal{K}^{\varsigma}$.

Proof: Let $K \in \mathcal{L}_{\text{unif-synch}}$ with delay d. Then, $K \in \mathcal{L}_{\text{rdecodable}} \cap \mathcal{L}_{\text{ldecodable}}$ [4] and $K \in \mathcal{K}^{-\omega} \cap \mathcal{K}^{\omega}$ [128]. Consider $w \in \mathcal{S}(X^{\varsigma})$ and two factorizations φ and ψ of w over $\mathcal{S}(K)$ with $I_{\varphi} = I_{\psi} = \varsigma$. We show that $\varphi \sim \psi$.

Assume there are indices $k, l \in \varsigma$ such that $\varphi(k)$ and $\psi(l)$ start at the same index in w. Then

$$\cdots \varphi(k-2)\varphi(k-1) = \cdots \psi(l-2)\psi(l-1)$$

and

$$\varphi(k)\varphi(k+1)\cdots = \psi(l)\psi(l+1)\cdots .$$

This implies $\varphi \sim \psi$ as $K \in \mathcal{K}^{-\omega} \cap \mathcal{K}^{\omega}$.

Now assume that there are no such indices k and l. Then there are indices $n, m, i \in \varsigma$ and a proper prefix p of $\varphi(n)$ such that p is a proper suffix of $\psi(m)$, the last symbol of p and $\psi(m)$ are at position i in w, that is, $w(i-|p|+1)\cdots w(i)$ is an instance of p as a prefix of an instance of $\varphi(n)$ and an instance of $\psi(m)$ ends at $w(i)$. The situation is shown more completely in Fig. 4.1. Let s be such that $\varphi(n) = ps$. Then $s \neq \lambda$ and s is proper prefix of $\psi(m+1)$ or $\psi(m+1)$ is a proper prefix of s.

Let $t, r \in \mathbb{N}$ with $t \geq d$ be such that one has the situation shown in Fig. 4.1. Such t and r exist. Let p' be the proper prefix of $\varphi(n+t+1)$ which overlaps or contains $\psi(m+r)$ and let s' be such that $\varphi(n+t+1) = p's'$. Thus,

$$\psi(m+1)\cdots \psi(m+r) = s\varphi(n+1)\cdots \varphi(n+t)p'.$$

Because of $t \geq d$ one has

$$s\varphi(n+1)\cdots \varphi(n+t) \in \mathcal{S}(K^*)$$

and

$$\varphi(n+1)\cdots \varphi(n+t)p' \in \mathcal{S}(K^*).$$

As $p' \neq \lambda$, there is an $h \in \mathbb{N}$ such that

$$\varphi(n+1)\cdots \varphi(n+t)p' = u_0 \cdots u_{h-1}$$

for some $u_0, \ldots, u_{h-1} \in \mathcal{S}(K)$. Thus

$$\varphi(n+1)\cdots \varphi(n+t)\varphi(n+t+1)\cdots = u_0 \cdots u_{h-1}\psi(m+r+1)\cdots$$

and, as $K \in \mathcal{K}^{\omega}$, $u_i = \varphi(n+i+1)$ for all $i \in \mathfrak{h}$, hence $\varphi(i) = \psi(i-n-h+m+r)$ for all i with $n+h+1 \leq i$. On the other hand,

$$\cdots \varphi(n)\cdots \varphi(n+h) = \cdots \psi(m)\cdots \psi(m+r)$$

and $K \in \mathcal{K}^{-\omega}$ imply that $\varphi(i) = \psi(i-n-h+m+r)$ for all i with $i \leq n+h$. Thus $\varphi \sim \psi$ and $K \in \mathcal{K}^{\varsigma}$.

Finally, consider

$$K = \{010, 20\} \cup 2(001)^*.$$

One has $K \notin \mathcal{L}_{\text{rdecodable}}$ by [128], hence $K \notin \mathcal{L}_{\text{unif-synch}}$; however, $K \in \mathcal{K}^{\varsigma}$, hence $\mathcal{L}_{\text{unif-synch}} \subsetneqq \mathcal{K}^{\varsigma}$.   □

Fig. 4.1. The factorization situation in the proof of Theorem 4.6.

5. Homophonic codes

In cryptography, a *homophonic cipher* is a system which, to every source symbol, may assign several encryptions (see [67], for example). The purpose of a homophonic cipher is to obscure statistical properties of the message. For example, for a message in plain English, a homophonic cipher would provide many different encodings of the most frequent letters – like the letter e – so that applying a statistical analysis to the cryptogram would not expose its encryption. Cryptographic aspects of homophonic ciphers are discussed in [37] and [53]; see also [60] for an analysis of the limitations of the method.

A *homophonic encoding* f of an alphabet X into an alphabet Y is a mapping of X into the set of non-empty subsets of Y^+, the idea being that, for encoding a word over X, one independently and non-deterministically encodes each symbol in the word according to f and then concatenates the results. Such encodings have been investigated in [117], [12], [15], and [13] as *multi-valued codes*. Beyond their cryptographic interpretation, homophonic codes can also be considered in the context of error correction. To do so, we construct a channel from a given homophonic code in such a way that the error behaviour of the channel reflects the variations of possible encodings. Our construction is slightly more general than required for homophonic codes.

The foundations of the theory of channels and codes – of error correction, synchronization, and decoding – in the framework to which the homophonic codes belong was first given in a series of articles leading to and later combined in the book [41] of 1974. To our knowledge, this was the first *systematic* attempt to lay the foundations of a comprehensive and uniform theory of variable-length codes for noisy channels. We explain below how that work relates to the model used in this handbook chapter. For detailed results of that research we refer the book [41] itself.

Definition 5.1 *Let X and Y be alphabets and $f \subseteq X \times Y^*$. Let*

$$K = f(X) = \{w \mid \exists x \in X \,((x, w) \in f)\}.$$

A word schema $y \in \mathcal{S}(Y^\blacklozenge)$ is an f-encoding of a word schema $x \in \mathcal{S}(X^\blacklozenge)$ if there is a factorization φ of y over K such that $I_\varphi = I_x$ and $(x(i), \varphi(i)) \in f$ for all $i \in I_x$.

Let $F_f \subseteq X^\blacklozenge \times Y^\blacklozenge$ be the relation

$$F_f = \{(x, y) \mid x \in X^\blacklozenge, y \in Y^\blacklozenge, \mathcal{S}(y) \text{ is an } f\text{-encoding of } \mathcal{S}(x)\}.$$

The relation F_f is the encoding *determined by f.*

In the special case, when f is a mapping and x and y are $*$-words, y is an f-encoding[24] of x if and only if y is the image of x under the morphism of X^* into Y^* induced by f. Definition 5.1 extends this notion to the cases when f is a many-to-many relation[25] and when x and y are infinite. We now exhibit a general construction of channels representing relations in $X \times Y^+$ including homophonic codes.

Definition 5.2 *Let X and Y be alphabets, let K be a $*$-language over Y with $|K| = |X|$, and let g be a bijection of X onto K. Let $f \subseteq X \times Y^+$. The channel $\gamma_{g,f}$ defined as*

$$\gamma_{g,f} = \left\{ (\llbracket y',n' \rrbracket \mid \llbracket y,n \rrbracket) \; \middle| \; \begin{array}{l} y,y' \in \mathcal{S}(Y^\blacklozenge), \; n,n' \in \mathbb{Z}, \\ \exists x \in X^\blacklozenge \left((x,\overline{y}) \in F_g \wedge (x,\overline{y'}) \in F_f \right) \end{array} \right\}$$

is called the channel generated by g and f.

Definition 5.2 exhibits some arbitrariness as to how physical errors are modelled by a relation f – a homophonic code in our special case – due to the required encoding g.

Example 5.1 Let $X = \{0,1\} = Y$ and

$$f = \{(0,111),(0,110),(0,101),(0,011),(1,000),(1,001),(1,010),(1,100)\}.$$

If g maps 0 onto 111 and 1 onto 000 then the resulting channel permits at most one substitution in every code word. Thus, for example, $(110 \mid 111)$ would be a possible channel behaviour. On the other hand, if g maps 0 onto 000 and 1 onto 111 then the channel would be quite different; for example $(111 \mid 000)$ would be a possible channel behaviour.

Not every channel of the form $\gamma_{g,f}$ is an SID channel. This is true even under severe restrictions on g and f. As shown in Example 5.1, the possible errors are determined by a given factorization of the input message whereas, for an SID channel, the possible error patterns – like one insertion in every five consecutive symbols – do not and cannot take a given factorization into account.[26] Conversely, not all SID channels can be modelled in the form $\gamma_{f,g}$. In a strict sense, the SID channel model may be closer to the physical error situation; practically, the differences may be not so important.

In general, channels of the form $\gamma_{g,f}$ may not even be P-channels. However, under some natural assumptions they have most of the essential properties of P-channels.

---

[24] The notion of *encoding* is not intended to imply unique decodability.

[25] A homophonic code can be considered as a one-to-many relation.

[26] The same problem arises, by the way, also with the classical substitution-error model for block codes as, for instance, expressed by the Hamming or Lee metrics; also there, a factorization is assumed to be given.

Proposition 5.1 *Let X and Y be alphabets, let K be a ζ-code (in the usual sense) over Y with $|K| = |X|$, and let g be a bijection of X onto K. Let $f \subseteq X \times Y^+$ such that, $g \subseteq f$. The channel $\gamma_{g,f}$ is stationary and satisfies conditions $\mathcal{P}_0$, $\mathcal{P}_1$, $\mathcal{P}_2$, and $\mathcal{P}_3$.*

Proof: If $(\llbracket w', n' \rrbracket \mid \llbracket w, n \rrbracket) \in \gamma_{g,f}$ then $(\llbracket w', k' \rrbracket \mid \llbracket w, k \rrbracket) \in \gamma_{g,f}$ for every $k, k' \in \mathbb{Z}$ and, therefore, $\gamma_{g,f}$ is stationary.

Condition $\mathcal{P}_0$ is satisfied as, for all $x \in X$, $g(x) \in Y^+$ and $\pi_2(f)$ is a non-empty subset of Y^+.

Condition $\mathcal{P}_2$ is satisfied as $g \subseteq f$ and $\gamma_{g,f}$ is stationary.

Condition $\mathcal{P}_3$ is satisfied as $(\lambda, \lambda) \in F_f \cap F_g$.

We turn to proving that condition $\mathcal{P}_1$ is satisfied. Consider $(y' \mid y) \in \gamma_{g,f}$. Let $y = \llbracket w, n \rrbracket$ and $y' = \llbracket w', n' \rrbracket$ where $w, w' \in \mathcal{S}(Y^\vartheta)$ for some $\vartheta \leq \zeta$. Let φ be a factorization of w over some set $Z \subseteq \mathcal{S}(\pi_2(\gamma_{g,f})) \cap \mathcal{S}(X^+)$ with $I_\varphi \neq \emptyset$. Note that $\mathcal{S}(\pi_2(\gamma_{g,f})) = \mathcal{S}(\pi_2(F_g))$. Thus $[\varphi] \sim w$ and $w \neq \lambda \neq [\varphi]$. As $w \in \mathcal{S}(\pi_2(F_g))$, there exists $x \in \mathcal{S}(X^\vartheta)$ and a factorization κ of w over K such that $I_x = I_\kappa$ and $g(x(i)) = \kappa(i)$ for all $i \in I_x$. As $[\kappa] \sim w$, $I_\kappa \neq \emptyset$.

As K is a ζ-code, there are, for all $i \in I_\varphi$, unique $r_i \in I_\kappa = I_x$ and $l_i \in \mathbb{N}_0$ such that $\varphi(i) = \kappa(r_i) \cdots \kappa(r_i + l_i)$ and $r_{i+1} = r_i + l_i + 1$.

Similarly, there is a factorization κ' of w' over $\mathcal{S}(\pi_2(f))$ such that $I_{\kappa'} = I_x = I_\kappa$ and, for all $i \in I_x$, $(x(i), \kappa'(i)) \in f$.

We define a factorization φ' as follows: Let $I_{\varphi'} = I_\varphi$ and, for $i \in I_\varphi$, let $\varphi'(i) = \kappa'(r_i) \cdots \kappa'(r_i + l_i)$. Then $[\varphi'] \sim [\kappa'] \sim w'$.

Consider an index mapping $\psi : I \to I_\varphi$ with $I \neq \emptyset$. We need to show that, for all $n, n' \in \mathbb{Z}$, $(\llbracket [\varphi' \psi], n' \rrbracket \mid \llbracket [\varphi \psi], n \rrbracket) \in \gamma_{g,f}$. For this, it suffices to show that there is a word $z \in X^{\leq \zeta}$ such that $(z, \overline{[\varphi \psi]}) \in F_g$ and $(z, \overline{[\varphi' \psi]}) \in F_f$.

Let $\chi : I_\varphi \to X^+$ be given by $\chi(i) = x(r_i) \cdots x(r_i + l_i)$ for $i \in I_\varphi$. Then $z = \overline{[\chi \psi]}$ has the required properties. $\qquad \square$

Proposition 5.1 helps to understand and clarify the importance of condition $\mathcal{P}_1$. The condition establishes that any factorization of the input, whichsoever, has a corresponding factorization of the output. A channel $\gamma_{g,f}$ according to Proposition 5.1 is *nearly* a P-channel: The only difference is that, usually, $\pi_2(\gamma_{g,f}) \neq \mathcal{I}(Y^\bullet)$. Thus, a channel of the form $\gamma_{g,f}$ models the input-output behaviour of a P-channel under the assumption that only messages over the chosen and fixed ζ-code K are sent. On the other hand, a P-channel in general models the behaviour of physical channels under arbitrary inputs.

We now turn to the decoding of messages encoded using homophonic codes or – in general – arbitrary relations.

Definition 5.3 *Let X and Y be alphabets and let $f \subseteq X \times Y^+$ such that, for every $x \in X$, $f(x) \neq \emptyset$. For $\eta \in \{*, \omega, -\omega, \zeta, \leq\omega, \leq-\omega, \leq\zeta\}$, f is said to be uniquely η-decodable if, for every $y \in \pi_2(F_f) \cap Y^\eta$ there is one and only one $x \in X^\eta$ such that $(x, y) \in F_f$.*

The connection between the two notions of unique η-decodability introduced in this handbook chapter, that of Definition 3.4 for languages and that of Definition 5.3 for relations, is expressed in the following statement for the case when f is a mapping (see [4] or [123]).

Theorem 5.1 *Let X and Y be alphabets and let f be a mapping of X into Y^+. Then f is uniquely η-decodable if and only if f is injective and $f(X)$ is an η-code.*

Note that, if f is uniquely η-decodable, then f is also uniquely η'-decodable for $\eta' < \eta$. The following theorem explains a connection between decoding for the encoding determined by a relation f and error correction for the channel $\gamma_{g,f}$ for some g.

Theorem 5.2 *Let X and Y be alphabets and let $f \subseteq X \times Y^+$ such that, for every $x \in X$, $f(x) \neq \emptyset$. Let $\eta \in \{*, \omega, -\omega, \zeta, \leq\omega, \leq-\omega, \leq\zeta\}$. The relation f is uniquely η-decodable if and only if there is an η-code K and a bijection g of X onto K such that K is $(\gamma_{g,f}, \eta)$-correcting.*

Moreover, if f is uniquely η-decodable then g can be chosen to satisfy $g \subseteq f$ and any such choice g results in $g(X)$ being $(\gamma_{g,f}, \eta)$-correcting.

Proof: First, assume that f is uniquely η-decodable, hence, also uniquely $*$-decodable. For every $x \in X$, select a $c_x \in Y^+$ such that $(x, c_x) \in f$. Let $K = \{c_x \mid x \in X\}$ and let $g : X \to K$ be the mapping with $g(x) = c_x$. The fact that f is uniquely $*$-decodable implies that $f(x) \cap f(x') = \emptyset$ if $x \neq x'$, $x, x' \in X$. Therefore, g is bijective.

Consider $w \in K^\eta$ and a factorization φ of $\mathcal{S}(w)$ over K. Let ψ be the mapping of I_φ into X defined by $\psi(i) = g^{-1}(\varphi(i))$ for $i \in I_\varphi$. Thus $(\psi(i), \varphi(i)) \in g$ and, by $g \subseteq f$, also $(\psi(i), \varphi(i)) \in f$, hence, $(v, w) \in F_f$ where $v \in X^\eta$ is the word given by $[\psi]$. By assumption, v is unique.

Suppose, φ' is also a factorization of $\mathcal{S}(w)$ over K. Let ψ' be the corresponding mapping of $I_{\varphi'}$ into X defined as above. The uniqueness of v implies $[\psi] \sim \mathcal{S}(v) \sim [\psi']$, hence $I_\varphi = I_\psi = I_{\psi'} = I_{\varphi'}$. Moreover, if $\psi(i) = \psi'(j) = x$ for some $i, j \in I_\psi$, then $\varphi(i) = g(x) = \varphi'(j)$. Therefore, φ and φ' are equivalent. This proves that K is an η-code.

Now consider η-messages u and w over K such that there is a $\gamma_{g,f}$-received η-message $z \in \langle u \rangle_{\gamma_{g,f}} \cap \langle w \rangle_{\gamma_{g,f}}$. There are words $x, y \in X^{\blacklozenge}$ such that $(x, \overline{u}), (y, \overline{w}) \in F_g$ and $(x, \overline{z}), (y, \overline{z}) \in F_f$. As f is uniquely η-decodable, one has $x = y$ and, therefore, $\overline{u} = \overline{w}$. This shows that K is $(\gamma_{g,f}, \eta)$-correcting.

For the proof of the converse implication, assume that there is an η-code and a bijection g of X onto K such that K is $(\gamma_{g,f}, \eta)$-correcting. Consider $y \in \mathcal{S}(Y^\eta)$ and $x, z \in \mathcal{S}(X^\eta)$ such that $(\overline{x}, \overline{y}), (\overline{z}, \overline{y}) \in F_f$. Let $\varphi : I_x \to K$ and $\psi : I_z \to K$ be given by $\varphi(i) = g(x(i))$ and $\psi(i) = g(z(i))$, respectively. Hence, $(\overline{x}, \overline{[\varphi]})$ and $(\overline{z}, \overline{[\psi]})$ are in F_g. Therefore, $y \in \langle [\varphi] \rangle_{\gamma_{g,f}} \cap \langle [\psi] \rangle_{\gamma_{g,f}}$. As K is $(\gamma_{g,f}, \eta)$-correcting, one has $[\varphi] \sim [\psi]$. This implies $x \sim z$ because g is injective. Thus, f is uniquely η-decodable. $\square$

In [13] also the notion of a *fault-tolerant* homophonic code is investigated. This notion can be expressed in our general framework for channels. However, a natural error-model has yet to be developed to capture the type of faults against which these codes are tolerant. These faults include certain insertion and deletion faults – or more generally, synchronization faults – and fault-tolerance is shown to be equivalent with the property that the decoding can be achieved using a finite-state transducer in which every state can serve as the initial state. Such decoders are also considered in [77] and [111] and are presented in some detail in the Section 11.

Theorem 5.3 [117] *Let $f : X \to 2_{\neq \emptyset}^{Y^+}$ be a homophonic encoding such that, for every $x \in X$, the set $f(x)$ is finite. It is decidable, whether f is uniquely $*$-decodable.*

As before, the notions of decoding delay and synchronizability can be introduced in a natural way for homophonic codes. In [12], a characterization of homophonic codes having these properties is provided. The construction of transducers achieving the decoding is shown in [15] (see also [116] and [117]).

6. Methods for defining codes

A language $K \subseteq X^+$ is a $*$-code if every $*$-message over K has a unique factorization over K. Thus, $*$-codes are characterized by the following property:

$$(C_{\text{code}}) \begin{cases} \text{For all } n, m \in \mathbb{N} \text{ and all } u_0, u_1, \ldots, u_{n-1}, v_0, v_1, \ldots, v_{m-1} \in \\ K, \text{ if } u_0 u_1 \cdots u_{n-1} = v_0 v_1 \cdots v_{m-1} \text{ then } n = m \text{ and } u_i = v_i \\ \text{for } i \in \mathbf{n}. \end{cases}$$

In algebraic terms this says that a language K is a code if and only the semigroup generated by K, with concatenation of words as multiplication, is free with K as a free set of generators.[27] Condition C_{code} can be considered as expressing a kind of independence among the elements of K.

Many different classes of codes have been introduced to satisfy various constraints regarding decodability, synchronizability, or fault-tolerance. The classes of infix codes and solid codes mentioned before are examples of such classes – constructed to avoid certain overlap situations which could sometimes render decoding difficult. More about the latter of these two classes of codes is said in Section 11.

In this section we analyse the idea underlying most constructions of natural classes of codes. The key notion is that of *independence* in universal algebra. Independence in the sense of being a free set of generators is the special example defining the class of all codes.

Dependence theory in universal algebra extends ideas from linear algebra to other types of algebras and deals with notions of the following kind: *independent set; an element depends on a set; span of a set; minimal spanning*

---

[27] This property can be simplified by *assuming* already that $n = m$ [123].

set; dimension. In essence, this theory has two branches which have very little in common: one based on a purely set theoretic model and one based on the notion of freeness in algebras. An introduction to the former can be found in [16];[28] for the latter, an introduction is given in [34]. The relation between the two approaches is analysed in [31]. We follow the former, as it is suitable for the problem.[29]

We now outline the basic concepts of dependence theory. We give a few more details than absolutely required for our application to help the reader put the concepts into a broader context.

Definition 6.1 *Let S be a set. A dependence system on S is a subset D of 2^S which has the following property: $L \in D$ if and only if there is a finite, non-empty subset L' of L with $L' \in D$. The subsets of S which are in D are called D-dependent. All other subsets of S are D-independent.*

By definition, every subset of an independent set is independent; in particular, the empty set is independent. Every superset of a dependent set is dependent.

Example 6.1 Let $S = X^+$ for an alphabet X. Let D_{code} be the set of all non-empty subsets L of X^+ which are not codes, that is, which are not freely generating L^+. Thus, a set $K \subseteq X^+$ is D_{code}-independent if and only if there is no finite subset L of K which does not freely generate L^+, that is, if and only if K is a code.

In Definition 6.1, the set L' can be of any size. We also need a notion of dependence for the case when the size of L' is bounded.

Definition 6.2 *Let $n \in \mathbb{N}_{\aleph_0}$. A dependence system D on the set S is said to be an n-dependence system if and only if it satisfies the following condition: $L \in D$ if and only if there is a non-empty subset L' of L with $|L'| < n$ and $L' \in D$.*

Of course, an $\aleph_0$-dependence system is just a dependence system without any restriction. The dependence system of Example 6.1 is an $\aleph_0$-dependence system; as a consequence of results in [51] one finds that it is not an n-dependence for any $n \in \mathbb{N}$. In the context of universal algebra in general, that is, not just that of semigroups, the dependence described in Example 6.1 is sometimes called *standard dependence.*

For $n \in \mathbb{N}_{\aleph_0}$, let $\mathbb{D}^{(n)}(S)$ be the class of n-dependence systems on S. We usually omit n when $n = \aleph_0$. For $D \in \mathbb{D}^{(n)}(S)$, let $\mathcal{L}_D$ be the family of

---

[28] The relevant chapter of that book contains several errors; see [32] for corrections and further explanations.

[29] Note that dependence theory in the sense to be used is part of the general framework of the theory of *matroids* [143]. In our specific case, that of classes of codes, however, the more restrictive properties of matroids *never* obtain.

D-independent subsets of S. Let $\mathbb{L}(S)$ be the class $\{\mathcal{L}_D \mid D \in \mathbb{D}(S)\}$ and let $\mathbb{L}^{(n)}(S) = \{\mathcal{L}_D \mid D \in \mathbb{D}^{(n)}(S)\}$.

Remark 6.1 Let D and D' be an n-dependence system and an n'-dependence system on the set S, respectively. Then $D \cup D'$ is an m-dependence system on S with $m = \max(n, n')$.

In contrast to Remark 6.1, the intersection of two dependence systems need not be a dependence system.

Definition 6.3 *Let D be a dependence system on S, and let $a \in S$ and $X \subseteq S$. The element a is said to* depend *on X if $a \in X$ or if there is an independent subset $X' \subseteq X$ such that $X' \cup \{a\}$ is dependent. The* span *of X, $\langle X \rangle$, is the set of all elements of S which depend on X. If X is independent and $\langle X \rangle = S$ then X is called a* basis *of S. A set X is said to be* closed *if $\langle X \rangle = X$.*

Of course, the standard example of dependence is linear dependence in vector spaces. We are, however, concerned with far more general dependences. We review some of the basic properties of dependence systems needed in our present context.

Recall that a mapping $C : 2^S \to 2^S$ is a *closure operator* if, for every $X \subseteq S$, one has $X \subseteq C(X)$ and $C(C(X)) = C(X)$ and, moreover, $X \subseteq Y$ implies $C(X) \subseteq C(Y)$ for all $X, Y \subseteq S$. A *closure system* on S is a set $\mathfrak{C} \subseteq 2^S$ which is closed under arbitrary intersections. With a closure system $\mathfrak{C}$ one associates the closure operator $C_{\mathfrak{C}}$ by

$$C_{\mathfrak{C}}(X) = \bigcap_{X \subseteq Y \in \mathfrak{C}} Y.$$

Conversely, with a closure operator C one associates a closure system $\mathfrak{C}_C$ by

$$\mathfrak{C}_C = \{X \mid X \subseteq S, C(X) = X\}.$$

This correspondence between closure operators and closure systems is bijective and $C_{\mathfrak{C}_C} = C$, $\mathfrak{C}_{C_{\mathfrak{C}}} = \mathfrak{C}$. A closure operator C is said to be *algebraic* if, for every $X \subseteq S$ and every $a \in S$, $a \in C(X)$ implies that there is a finite subset X' of X such that $a \in C(X')$. A closure system $\mathfrak{C}$ is said to be *algebraic* if the corresponding closure operator $C_{\mathfrak{C}}$ is (see [16]).

Lemma 6.1 *Let D be a dependence system on S.*

(a) *If $a \in S$ depends on $X \subseteq S$ then there is a finite independent subset X' of X such that a depends on X' [16].*

(b) *For every $X \subseteq S$ one has $X \subseteq \langle X \rangle$. For every $X, Y \subseteq S$, $X \subseteq Y$ implies $\langle X \rangle \subseteq \langle Y \rangle$ [32].*

(c) *The set of closed subsets of S forms an algebraic closure system [16].*

In general, the span does not satisfy the equation $\langle X \rangle = \langle \langle X \rangle \rangle$. For example, let $S = \{a, b, c\}$ and $D = \{\{a, b\}, \{b, c\}, \{a, b, c\}\}$. Then $\emptyset$, $\{a\}$, $\{b\}$, $\{c\}$, and $\{a, c\}$ are the independent sets. Clearly, $\langle a \rangle = \{a, b\}$ and $\langle \langle a \rangle \rangle = \{a, b, c\}$.

Definition 6.4 *Let D be a dependence system on S. The dependence is said to be* transitive *if $\langle \langle X \rangle \rangle = \langle X \rangle$ for every $X \subseteq S$.*

Transitive dependence systems have been studied under several different names (see [143], p. 7). The collection of independent sets of a transitive dependence on a finite set is known as a *matroid* [143]. For a discussion of the problems concerning the definition of infinite matroids see Chapter 20 of [143].

Transitivity is a very natural requirement for a dependence relation as shown in the following characterization result:

Theorem 6.1 [16] *Let S be a set.*

(1) If D is a transitive dependence, then the mapping $X \to \langle X \rangle$ is an algebraic closure operator with the exchange property

$$y \notin \langle X \rangle \wedge y \in \langle X \cup \{z\} \rangle \Rightarrow z \in \langle X \cup \{y\} \rangle.$$

(2) If $\alpha : 2^S \to 2^S$ is an algebraic closure operator with the exchange property

$$y \notin \alpha(X) \wedge y \in \alpha(X \cup \{z\}) \Rightarrow z \in \alpha(X \cup \{y\}),$$

then there is a transitive dependence D such that $\alpha(X) = \langle X \rangle$ for all $X \subseteq S$.

For transitive dependence systems, one can define the notions of basis and dimension in a meaningful way. For vector spaces, for example, linear dependence defines dependence systems; the basis and dimension obtained in this way coincide with their linear counterparts.

In the context of the theory of codes, the set S is the free semigroup X^+, that is, the set of all non-empty finite words over X with concatenation as the multiplication operation. The dependence considered in Example 6.1 is an example of a *standard dependence* [16], [31]. In this case, the dependence is not transitive as shown by the following example.

Example 6.2 Let $X = \{0, 1\}$ and consider the words $v = 01$, $w = 0110$, $u = 100110$, and $z = 010110$. Then the sets $\{u, v\}$, $\{v, w\}$, and $\{u, v, z\}$ are codes, while the sets $\{u, v, w\}$ and $\{v, w, z\}$ are not codes. Thus, $w \in \langle u, v \rangle$, $z \notin \langle u, v \rangle$, but $z \in \langle \langle u, v \rangle \rangle$.

We now use the framework of dependence systems to define several classes of codes. For each class, we provide a brief explanation of its properties. Each class is identified by an abbreviation of its name. For example, the dependence system D_u defines the class $\mathcal{L}_u$ of uniform codes and we refer to the condition defining this dependence system as C_u. The connection between these classes of codes is presented in Section 7 where also some additional properties of dependence systems are presented as required. The references given for the classes of codes point to additional information about these classes; usually these references do not employ the notion of dependence system for their definitions and analyses.

- The classes $\mathcal{L}_{n\text{-code}}$ of *n-codes for* $n \in \mathbb{N}$:

$$(C_{n\text{-code}}) \begin{cases} A \text{ set } L \subseteq X^+ \text{ is } D_{n\text{-code}}\text{-dependent if and only if there is} \\ a \text{ non-empty subset } L' \text{ of } L \text{ with } |L'| \leq n \text{ such that } L' \text{ is} \\ D_{\text{code}}\text{-dependent [48], [51].} \end{cases}$$

Thus, a language K is an n-code if and only if every subset of K with at most n elements is a code. The n-codes form a proper hierarchy

$$\mathcal{K}^* = \mathcal{L}_{\text{code}} \subsetneqq \cdots \subsetneqq \mathcal{L}_{(n+1)\text{-code}} \subsetneqq \mathcal{L}_{n\text{-code}} \subsetneqq \cdots \mathcal{L}_{2\text{-code}} \subsetneqq \mathcal{L}_{1\text{-code}} = 2^{X^+}$$

with $\mathcal{K}^*$ and 2^{X^+} as lower and upper bounds, respectively.

If K is a code over X and $w_1, \ldots, w_n$ are distinct elements of K then the set $K \cup \{w_1 \cdots w_n\}$ is an n-code, but not an $(n+1)$-code [48]. This also shows that, for every n, there is an n-code which is not a code.

- The 2-codes are of particular interest due to the fact that a two-element set $\{u, v\}$ is a code if and only if $uv \neq vu$ (see [123]). Thus $D_{2\text{-code}}$ can also be defined by the following condition expressing *anti-commutativity* for the independent sets.

$$(C_{\text{ac}}) \begin{cases} A \text{ set } L \subseteq X^+ \text{ is } D_{\text{ac}}\text{-dependent if and only if } uv = vu \text{ for} \\ some \ u, v \in L \text{ with } u \neq v. \end{cases}$$

Thus $\mathcal{L}_{2\text{-code}} = \mathcal{L}_{\text{ac}}$. No similar characterization is known for any of the classes of n-codes with $n > 2$. Indeed no *simple* characterization of the n-element sets which are codes is known for $n > 2$. In [69] the 3-element sets which are codes are characterized (see also [68]); due to its complexity, this characterization seems to be very difficult to apply in the analysis of 3-codes.

A language K is a 2-code if and only if, for every primitive word w, there is at most one $n \in \mathbb{N}$ such that $w^n \in K$. This is a consequence of the characterization of $\mathcal{L}_{2\text{-code}}$ by $\mathcal{L}_{\text{ac}}$. In contrast, for $n \geq 3$, if K is an n-code then there are infinitely many primitive words w such that $w^n \notin K$ for all $n \in \mathbb{N}$ [48].

The class of n-codes is defined by an $(n+1)$-dependence system. There is no n'-dependence system with $n' < n+1$ by which this class could also be defined. This follows from general properties of dependence systems explained in Section 7 below.

- The class $\mathcal{L}_\mathrm{p}$ of *prefix codes:*

$$(\mathrm{C_p}) \begin{cases} A \ language \ L \subseteq X^+ \ is \ D_\mathrm{p}\text{-}dependent \ if \ there \ are \ u, v \in L \\ such \ that \ u \ is \ a \ proper \ prefix \ of \ v. \end{cases}$$

Prefix codes have been studied extensively – see [4] for an account of the work until about 1985. Every prefix code is a code. However, for example, the set $\{0, 01\}$ is a code, but not a prefix code.

Prefix codes have (right) decoding delay 0 for a noiseless channel. As a consequence, every prefix code is an ω-code. The language $\{0, 10, 11\}$ is a prefix code, but not a $(-\omega)$-code.

For most major results of information theory, the restriction from codes to prefix codes does not make a difference of any significance. Prefix codes have a natural interpretation as labelled trees – each code word describes a path from the root to a leaf – and, hence, a natural description by automata. Their tree description applies to algorithmic problems, like searching or sorting. Decisions in a search algorithm correspond to the choice of the next symbol in a code word and, hence, the number of decisions in a search is bounded from below by the length of the corresponding code word. An excellent account of the application of prefix codes to the analysis of algorithms can be found in [1].

- By duality, one defines the class $\mathcal{L}_\mathrm{s}$ of *suffix codes:*

$$(\mathrm{C_s}) \begin{cases} A \ language \ L \subseteq X^+ \ is \ D_\mathrm{s}\text{-}dependent \ if \ there \ are \ u, v \in L \ such \\ that \ u \ is \ a \ proper \ suffix \ of \ v. \end{cases}$$

Properties of suffix codes can be derived trivially from properties of prefix codes using duality. For example, every suffix code is a $(-\omega)$-code.

In general, prefix codes and suffix codes need not be ζ-codes: Suppose K is a prefix code which is not a $(-\omega)$-code. Let $u \in K^{-\omega}$ be such that u has two non-equivalent factorizations over K and let $v \in K^\omega$; then uv has at least two non-equivalent factorizations over K, that is, K is not a ζ-code.

- The construction leading to the classes of n-codes has no meaningful counterpart for prefix codes or suffix codes alone: If K is a language such that every two-element subset is a prefix code then, clearly, K is a prefix code. However, the following construction starting from both prefix codes and suffix codes yields some objects – and a phenomenon – of mathematical interest, the classes of *n-ps-codes* for $n \in \mathbb{N}$, $n \geq 2$:

$$(\mathrm{C}_{n\text{-ps}}) \begin{cases} A \ language \ L \subseteq X^+ \ is \ D_{n\text{-ps}}\text{-}dependent \ if \ and \ only \ if \ there \ is \\ a \ non\text{-}empty \ subset \ L' \ of \ L \ with \ |L'| \leq n \ such \ that \ L' \ is \ both \\ D_\mathrm{p}\text{-}dependent \ and \ D_\mathrm{s}\text{-}dependent \ [49]. \end{cases}$$

For all n, $\mathcal{L}_{(n+1)\text{-ps}} \subseteq \mathcal{L}_{n\text{-ps}}$ by definition. In contrast to the hierarchy of n-codes, which is infinite, the hierarchy of n-ps-codes collapses at $n = 4$ with

$\mathcal{L}_{4\text{-ps}} = \mathcal{L}_p \cup \mathcal{L}_s$. Thus $\mathcal{L}_{4\text{-ps}} \subsetneq \mathcal{K}^*$. As there are three-element codes which are neither prefix nor suffix codes, also $\mathcal{L}_{3\text{-ps}} \subsetneq \mathcal{L}_{3\text{-code}}$. The class $\mathcal{L}_{2\text{-ps}}$ has an alternative characterization by the following condition:

(C_d) $\left\{ \begin{array}{l} \text{A language } L \subseteq X^+ \text{ is } D_d\text{-dependent if } v = ux = yu \text{ for some} \\ u, v \in L \text{ and } x, y \in X^+. \end{array} \right.$

Every 2-ps-code is a 2-code. However, the language $\{0, 010\}$ is a 2-code, but not a 2-ps-code. Therefore, $\mathcal{L}_{2\text{-ps}} \subsetneq \mathcal{L}_{2\text{-code}}$.

• Between the classes of 3-ps-codes and 2-ps-codes there is another class of languages, interesting for its combinatorial properties, the class of g-3-ps-*codes:*

$(C_{\text{g-3-ps}})$ $\left\{ \begin{array}{l} \text{A language is } D_{\text{g-3-ps}}\text{-dependent if it is } D_{2\text{-ps}}\text{-dependent or if} \\ \text{there are three distinct words } u, v, w \in L \text{ such that } \{u, v\} \text{ is} \\ D_p\text{-dependent and } \{v, w\} \text{ is } D_s\text{-dependent [49].} \end{array} \right.$

Thus, a 2-ps-code is a language $K \subseteq X^+$ such that every two-element subset of K is a prefix code or a suffix code. A g-3-ps-code is 2-ps-code K such that, in every three-element subset of K, at least two elements form a prefix code or a suffix code.

Every 3-ps-code is the disjoint union of a prefix code and a suffix code; on the other hand, the union of the prefix code $\{10, 010\}$ and the suffix code $\{0, 1\}$ is not even a 2-ps-code, let alone a 3-ps-code. Moreover, the language 0^+1^+ is a 2-ps-code which cannot be decomposed into a union of a prefix code and a suffix code. The classes $\mathcal{L}_{3\text{-ps}}$ and $\mathcal{K}^*$ are incomparable. Hence, in general, 3-ps-codes, g-3-ps-codes, or 2-ps-codes need not be codes. Moreover, $\mathcal{L}_{3\text{-ps}} \subsetneq \mathcal{L}_{\text{g-3-ps}} \subsetneq \mathcal{L}_{2\text{-ps}}$ [49].[30]

The general principles of the constructions of the n-codes and the n-ps-codes and the reasons as to why the hierarchy does not collapse in the former case while it does so in the latter are explained in Section 7.

• The class $\mathcal{L}_b$ of *bifix codes:*

(C_b) $\left\{ \begin{array}{l} \text{A language is } D_b\text{-dependent if it is } D_p\text{-dependent or } D_s\text{-de-} \\ \text{pendent, that is, } D_b = D_p \cup D_s. \end{array} \right.$

Bifix codes are also called *biprefix codes* in the literature. Their main merit is that they are both, ω-codes and $(-\omega)$-codes, and that they have some very interesting mathematical properties (see [4]).

The language $K = \{01, 10\}$ is a bifix code, but not a ζ-code. For example, the ζ-message $\cdots 010101010 \cdots$ has two non-equivalent factorizations over K.

In the general context of information transmission via a noisy channel γ, the bifix property has the following interpretation – assuming appropriate

---

[30] That the inclusions are proper follows from results discussed in Section 10 below.

synchronization and error correction capabilities: After synchronization, decoding can take place, from the point identified by the synchronization, both to the right and the left without decoding delay – for as long as there are no errors.

The classes of codes considered so far, have very little – if any – synchronization, error detection, or error correction capabilities *in general*. As we move towards codes with additional structure, some of these issues get addressed.

- There are four basic classes of *shuffle codes of index n* (sh$_n$-codes) for $n \in \mathbb{N}$: the class $\mathcal{L}_{i_n}$ of *infix-shuffle codes of index n* (i$_n$-codes); the class $\mathcal{L}_{o_n}$ of *outfix-shuffle codes of index n* (o$_n$-codes); the class $\mathcal{L}_{p_n}$ of *prefix-shuffle codes of index n* (p$_n$-codes); and the class $\mathcal{L}_{s_n}$ of *suffix-shuffle codes of index n* (s$_n$-codes) [132], [89], [90], [92], [93].[31] The conditions defining these classes are as follows:

(C_{i_n})
$$\begin{cases} A \text{ language } L \subseteq X^+ \text{ is } D_{i_n}\text{-dependent if there are words} \\ u_1, \ldots, u_n, v_0, v_1, \ldots, v_n \in X^* \text{ such that } u = u_1 \cdots u_n \in L, \\ v = v_0 u_1 v_1 u_2 \cdots u_n v_n \in L \text{ and } u \neq v. \end{cases}$$

(C_{o_n})
$$\begin{cases} A \text{ language } L \subseteq X^+ \text{ is } D_{o_n}\text{-dependent if there are words} \\ u_0, u_1, \ldots, u_n, v_1, \ldots, v_n \in X^* \text{ such that } u = u_0 u_1 \cdots u_n \in L, \\ v = u_0 v_1 u_1 \cdots v_n u_n \in L \text{ and } u \neq v. \end{cases}$$

(C_{p_n})
$$\begin{cases} A \text{ language } L \subseteq X^+ \text{ is } D_{p_n}\text{-dependent if there are words} \\ u_1, \ldots, u_n, v_1, \ldots, v_n \in X^* \text{ such that } u = u_1 \cdots u_n \in L, \\ v = u_1 v_1 \cdots u_n v_n \in L \text{ and } u \neq v. \end{cases}$$

(C_{s_n})
$$\begin{cases} A \text{ language } L \subseteq X^+ \text{ is } D_{s_n}\text{-dependent if there are words} \\ u_1, \ldots, u_n, v_1, \ldots, v_n \in X^* \text{ such that } u = u_1 \cdots u_n \in L, \\ v = v_1 u_1 \cdots v_n u_n \in L \text{ and } u \neq v. \end{cases}$$

One has $\mathcal{L}_{p_1} = \mathcal{L}_p$ and $\mathcal{L}_{s_1} = \mathcal{L}_s$. The classes $\mathcal{L}_{i_1}$ and $\mathcal{L}_{o_1}$ are the classes $\mathcal{L}_i$ and $\mathcal{L}_o$ of infix codes and outfix codes, respectively [63], [52], [38], [50]. The corresponding dependences are $D_i = D_{i_1}$ and $D_o = D_{o_1}$.

The relation between shuffle codes is shown in Fig. 6.1. Lines indicate proper inclusions. The language $\{(0^k 1^k)^n 0^k \mid k \in \mathbb{N}\}$ is in $\mathcal{L}_{i_n} \cap \mathcal{L}_{o_n}$, but not in $\mathcal{L}_{p_{n+1}} \cup \mathcal{L}_{s_{n+1}}$, and the language $\{(0^k 1^k)^n \mid k \in \mathbb{N}\}$ is in $\mathcal{L}_{p_n} \cap \mathcal{L}_{s_n}$, but not in $\mathcal{L}_{i_n} \cup \mathcal{L}_{o_n}$ [90]. The language $\{(01)^n 0, 1^n\}$ is in $\mathcal{L}_{o_n}$, but not in $\mathcal{L}_{i_n}$; the language $\{(01)^n 0, 0^{n+1}\}$ is in $\mathcal{L}_{i_n}$, but not in $\mathcal{L}_{o_n}$; $\{1^n, (01)^n\}$ is in $\mathcal{L}_{p_n}$, but not in $\mathcal{L}_{s_n}$; $\{1^n, (10)^n\}$ is in $\mathcal{L}_{s_n}$, but not in $\mathcal{L}_{p_n}$. Some further properties of shuffle codes are mentioned in Section 8 below. Note that $\mathcal{L}_{4\text{-ps}}$ can be considered as the largest member of the families of shuffle codes.

The shuffle codes of index n address the following error situation to some extent: Insertions at up to n positions or deletions at up to n positions cannot

---

[31] The letters i, o, p, and s stand for *infix*, *outfix*, *prefix*, and *suffix*, respectively. The classes of shuffle codes have been introduced also under various other names.

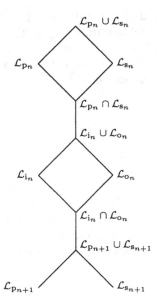

Fig. 6.1. Relations between the shuffle codes. All inclusions indicated are proper.

turn a code word into another code word. In general, the shuffle codes do not, however, have useful synchronization capabilities. If synchronization is not a problem[32] then these codes detect up to n bursts of insertions or up to n bursts of deletions in a code word.

- The class $\mathcal{L}_h$ of *hypercodes* [124], [42], [130]:

$$(\mathrm{C_h}) \begin{cases} \text{A language } L \subseteq X^+ \text{ is } D_h \text{ dependent if it is } D_{i_n}\text{-dependent} \\ \text{for some } n \in \mathbb{N}. \end{cases}$$

As $D_{i_n} \subseteq D_{i_{n+1}}$ and as $\bigcup_{n=1}^{\infty} D_{i_n}$ is a 3-dependence system, one has $D_h = \bigcup_{n=1}^{\infty} D_{i_n}$. A mathematically very interesting generalization of hypercodes is defined and analysed in [104].

Theorem 6.2 *Let* $K \subseteq X^+$. *Each of the following conditions implies that* K *is finite.*

(a) K *is regular and* $K \in \mathcal{L}_o$.
(b) K *is regular and* $K \in \mathcal{L}_{x_n}$ *for* $n \geq 2$ *and* $x \in \{i, o, p, s\}$.
(c) K *is context-free and* $K \in \mathcal{L}_{o_2}$.
(d) K *is context-free and* $K \in \mathcal{L}_{x_n}$ *for* $n \geq 3$ *and* $x \in \{i, o, p, s\}$.
(e) K *is a hypercode.*

---

[32] In some applications, synchronization is provided through a separate signal.

Statement (a) is proved in [50] using the Pumping Lemma for regular languages. Statement (b) follows from $\mathcal{L}_{x_n} \subseteq \mathcal{L}_o$. Statement (c) is a consequence of the Pumping Lemma for context-free languages. Statement (d) follows from $\mathcal{L}_{x_n} \subseteq \mathcal{L}_{o_2}$. For (e), see [123]. This reference contains also further information about hypercodes.[33]

- The class $\mathcal{L}_u$ of *uniform codes* (or *block codes*):

$$(C_u) \left\{ \begin{array}{l} \text{A language } L \subseteq X^+ \text{ is } D_u\text{-dependent if it contains words of} \\ \text{different lengths.} \end{array} \right.$$

Nearly all codes in practical use for error correction or error detection are uniform codes. Moreover, to enable the construction of such codes with predictable properties and simple algorithms for coding and decoding, additional structural properties are assumed. Normally, X is assumed to be a finite field and the code K is assumed to be a finite-dimensional vector space over X. See [94] or [99] for further information about uniform codes.

Uniform codes are good at detecting or correcting substitution errors when synchronization between coder and decoder is not a problem – for example, in the presence of a common error-free clock signal.

- The classes $\mathcal{L}_{\text{inter}_n}$ of *intercodes of index n* for $n \in \mathbb{N}$ [125], [144]:

$$(C_{\text{inter}_n}) \left\{ \begin{array}{l} \text{A language } L \subseteq X^+ \text{ is } D_{\text{inter}_n}\text{-dependent if there are words} \\ u_1, \ldots, u_{n+1}, v_1, \ldots, v_n \in L \text{ and } w_1, w_2 \in X^+ \text{ such that} \\ u_1 \cdots u_{n+1} = w_1 v_1 \cdots v_n w_2. \end{array} \right.$$

Note that D_{inter_n} is a $(2n + 2)$-dependence. From $D_{\text{inter}_{n+1}} \subsetneqq D_{\text{inter}_n}$ it follows that $\mathcal{L}_{\text{inter}_n} \subsetneqq \mathcal{L}_{\text{inter}_{n+1}}$. The class $\mathcal{L}_{\text{inter}_1}$ of intercodes of index 1 is also known as the class $\mathcal{L}_{\text{comma-free}}$ of *comma-free codes*. Comma-free codes are uniformly synchronous bifix codes with synchronization delay 1 for the noiseless channel.[34]

In general, if K is an intercode of index n and if, in an error-free message w over K, n consecutive code words are found then every factorization of w over K will contain those instances of code words; hence these code words can be decoded without regard for the rest of the message.

A language $K \subseteq X^+$ is said to be an *intercode* if it is an intercode of index n for some $n \in \mathbb{N}$; hence, $\mathcal{L}_{\text{inter}} = \bigcup_{n=1}^{\infty} \mathcal{L}_{\text{inter}_n}$ is the class of all intercodes.

It is not known whether a dependence system D exists such that $\mathcal{L}_{\text{inter}}$ is the class of all D-independent languages in X^+. If so, D would probably have to be constructed from the intersection $\bigcap_{n=1}^{\infty} D_{\mathcal{L}_{\text{inter}_n}}$. The intersection of dependence systems is, however, not necessarily a dependence system.

---

[33] Parts of Theorem 6.2 have also been proved independently in [89] and [132].

[34] The term *comma-free* refers to the fact that no special synchronization symbol – a comma – is needed for synchronization without delay. For further information about intercodes and comma-free codes see [123] and [4].

Let $\mathcal{Q}$ be the class of languages containing only primitive words. Then $\mathcal{L}_{\text{inter}}$ is a proper subclass of $\mathcal{L}_b \cap \mathcal{L}_{\text{unif-synch}} \cap \mathcal{Q}$. The language $\{0, 1101011\}$ is an intercode (of index 2) which is not an infix code. On the other hand, one has $\mathcal{L}_{\text{inter}_1} \subsetneqq \mathcal{L}_i$.

- Note that $\mathcal{Q}$ is the set of indepedent sets with respect to the following dependence system:

$(\text{C}_{\text{primitive}})$ $\begin{cases} A \ language \ L \subseteq X^+ \ is \ D_{\text{primitive}}\text{-}dependent \ if \ there \ is \ a \ word \\ in \ L \ which \ is \ not \ primitive. \end{cases}$

- As in the case of $\mathcal{K}^* = \mathcal{L}_{\text{code}}$, one can also consider the classes $\mathcal{L}_{m\text{-inter}_n}$ of m-intercodes of index n and the classes $\mathcal{L}_{m\text{-inter}}$ of m-intercodes for $n, m \in \mathbb{N}$ [123]:

$(\text{C}_{m\text{-inter}_n})$ $\begin{cases} A \ language \ L \subseteq X^+ \ is \ D_{m\text{-inter}_n}\text{-}dependent \ if \ there \ is \\ a \ subset \ L' \ of \ L \ with \ at \ most \ m \ elements \ such \ that \ L' \ is \\ D_{\text{inter}_n}\text{-}dependent. \end{cases}$

$(\text{C}_{m\text{-inter}})$ $\begin{cases} A \ language \ L \subseteq X^+ \ is \ D_{m\text{-inter}}\text{-}dependent \ if \ there \ is \ a \\ subset \ L' \ of \ L \ with \ at \ most \ m \ elements \ such \ that \ L' \ is \ not \ an \\ intercode. \end{cases}$

Details about the relations between these classes of codes are shown in Fig. 6.2. Note that $\mathcal{L}_{1\text{-inter}} = \mathcal{Q}$, that $\mathcal{L}_{1\text{-inter}}$ is a proper subclass of $\mathcal{L}_{2\text{-code}}$, and that $\mathcal{L}_{2\text{-inter}}$ is a proper subclass of $\mathcal{L}_b \cap \mathcal{Q}$. The sets $D_{m\text{-inter}_n}$ and $D_{m\text{-inter}}$ are $(m+1)$-dependence systems.

- The class $\mathcal{L}_{\text{ol-free}}$ of *overlap-free languages*:

$(\text{C}_{\text{ol-free}})$ $\begin{cases} A \ language \ L \subseteq X^+ \ is \ D_{\text{ol-free}}\text{-}dependent \ if \ there \ are \ words \\ u_1, u_2, u_3 \in X^+ \ such \ that \ u_1 u_2, u_2 u_3 \in L. \end{cases}$

Not every $D_{\text{ol-free}}$-independent set is a code. For example, the language $L = \{0, 01, 001\}$ is not a code as $(0)(01) = (001)$, but is overlap-free. As mentioned in the context of Theorem 3.2, the notion of *overlap* is crucial for decoding and synchronization. An instance of a word $u \in X^+$ found in a message over an overlap-free language K can never be both, a proper prefix and a proper suffix of a word in K.

In the literature, a word $w \in X^+$ is called *unbordered* if the set $\{w\}$ is an overlap-free language.

- The class $\mathcal{L}_{\text{solid}}$ of *solid codes*:

$(\text{C}_{\text{solid}})$ $\begin{cases} A \ language \ L \subseteq X^+ \ is \ D_{\text{solid}}\text{-}dependent \ if \ it \ is \ D_i\text{-}dependent \\ or \ D_{\text{ol-free}}\text{-}dependent, \ that \ is, \ D_{\text{solid}} = D_i \cup D_{\text{ol-free}} \ [126], \\ [64], [123]. \end{cases}$

The solid codes are remarkably strong in the presence of errors that result in loss of synchronization. A detailed discussion of properties of solid codes is given in Section 11 below.

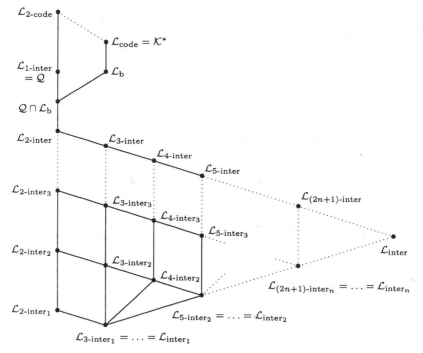

Fig. 6.2. The intercode hierarchy [65].

- The class $\mathcal{L}_{pi}$ of *p-infix codes* [50]:

(C_{pi}) $\left\{ \begin{array}{l} \textit{A language } L \subseteq X^+ \textit{ is } D_{pi}\textit{-dependent if there are words} \\ u_1 \in X^*, \ v \in L, \textit{ and } u_2 \in X^+ \textit{ such that } u_1vu_2 \in L. \end{array} \right.$

The class $\mathcal{L}_{si}$ of *s-infix codes* is defined dually. One has $\mathcal{L}_i = \mathcal{L}_{pi} \cap \mathcal{L}_{si}$. The class $\mathcal{L}_b$ is incomparable with $\mathcal{L}_{pi}$ and $\mathcal{L}_{si}$ [50]. A characterization of the classes of p-infix codes and s-infix codes is given below in the context of semaphore codes.

- The class $\mathcal{L}_{circular}$ of *circular codes*:

$(C_{circular})$ $\left\{ \begin{array}{l} \textit{A language } L \subseteq X^+ \textit{ is } D_{circular}\textit{-dependent if there exist } n, m \in \\ \mathbb{N}, \textit{ and words } u_0, u_1, \ldots, u_{n-1}, v_0, v_1, \ldots, v_{m-1} \in L, \ w \in X^+ \\ \textit{such that } |w| < |v_0| \textit{ and } wu_0u_1 \cdots u_{n-1} = v_0v_1 \cdots v_{m-1}w. \end{array} \right.$

Circular codes were introduced in [78] to investigate synchronization properties of finite codes. Synchronization is expressed there in terms of *finite self-correcting decoding automata*. Such an automaton reads an encoded message, received as output of an SID-channel, and computes, as its output, the decoding. When an error is encountered, the automaton may go through an incorrect sequence of states; it will, however, reach a correct state – a state it should reach in the absence of noise – after a bounded number of steps if no further error occurs.

Theorem 6.3 [78] *Let $K \subseteq X^+$ be finite. There is a finite self-correcting decoding automaton for K if and only if K is a circular code.*

This automaton-based approach is used in [78] and also in [97] and [98] to construct codes that can limit the effect of noise;[35] moreover, for a given memoryless message source, the average code word length of these codes is nearly minimal.

Corollary 6.1 [107] *A finite subset of X^+ is a circular code if and only if it is a uniformly synchronous code.*

A proof of Corollary 6.1 and several further results concerning circular codes can be found in [4].

• The class $\mathcal{L}_{\text{precirc}}$ of *precircular codes:*

$$(\mathrm{C}_{\text{precirc}}) \begin{cases} \textit{A language } L \subseteq X^+ \textit{ is } D_{\text{precirc}}\textit{-independent if, for all} \\ n, m \in \mathbb{N}, \textit{ all } u_0, \ldots, u_{n-1}, v_0, \ldots, v_{m-1} \in L, \textit{ and all } s, t \in X^* \\ \textit{with } st = v_0, \textit{ the equality } u_0 \cdots u_{n-1} = s v_1 \cdots v_{m-1} t \textit{ implies} \\ n = m \textit{ and } u_i = v_{i+h \bmod n} \textit{ for some } h \in \mathbb{N}_0 \textit{ and all } i \in \mathfrak{n}. \end{cases}$$

Precircular codes are introduced in [19] to study ζ-codes for noiseless channels. The class $\mathcal{L}_{\text{precirc}}$ is incomparable with $\mathcal{K}^{-\omega}$ and $\mathcal{K}^\omega$, it is properly contained in $\mathcal{K}^*$ and properly contains $\mathcal{K}^\zeta$ and $\mathcal{L}_{\text{circular}}$. The classes $\mathcal{K}^\zeta$ and $\mathcal{L}_{\text{circular}}$ are incomparable. For further information about precircular codes, see also [21].

• The classes $\mathcal{L}_{\text{rsema}}$ and $\mathcal{L}_{\text{lsema}}$ of *right semaphore codes* and *left semaphore codes,* respectively *cannot* be defined using dependence systems. A language $K \subseteq X^+$ is a right semaphore code if K is a prefix code such that $X^*K \subseteq KX^*$. By duality, K is a left semaphore code if K is a suffix code such that $KX^* \subseteq X^*K$ (see [4], [119], and [123]). One has $\mathcal{L}_{\text{rsema}} \subsetneq \mathcal{L}_{\text{pi}}$ and $\mathcal{L}_{\text{lsema}} \subsetneq \mathcal{L}_{\text{si}}$ [50].

Any right semaphore code K is a maximal prefix code, that is, not properly contained in any prefix code. A proper subset K' of K is a prefix code, but not maximal, hence not a right semaphore code. As all subsets of an independent set are independent, there is no dependence system D such that $\mathcal{L}_{\text{rsema}}$ would be the class of D-independent languages.

Right semaphore codes are interesting for their decoding properties in the absence of noise. A right semaphore code K has the form $X^*S \setminus X^*SX^+$ for some non-empty set $S \subseteq X^+$ of *right semaphores.* In a message over K, the presence of a right semaphore signals the end of a code word. The error correction capabilities of right semaphore codes are not known.

Theorem 6.4 [50] *A language $L \subseteq X^+$ is a p-infix code if and only if it is a subset of a right semaphore code; it is an s-infix code if and only if it is a subset of a left semaphore code.*

---

[35] See also [96].

Thus, the class $\mathcal{L}_{\text{pi}}$ of p-infix codes is the smallest class of languages which can be defined by a dependence system and contains the class $\mathcal{L}_{\text{rsema}}$ of right semaphore codes. By duality, $\mathcal{L}_{\text{si}}$ is the smallest class of languages which can be defined by a dependence system and contains $\mathcal{L}_{\text{lsema}}$.

Theorem 6.5 [39] *A language $L \subseteq X^+$ is an infix code if and only if it is the intersection of a right semaphore code with a left semaphore code.*

- The class $\mathcal{L}_{\text{refl}}$ of *reflective codes* is defined as follows: A language $K \subseteq X^+$ is *reflective* if, for every $u, v \in X^*$ with $uv \in K$, one has $vu \in K$. A *reflective code* is a language which is reflective and a code [106]. The class $\mathcal{L}_{\text{refl}}$ is a proper subset of the class of $\mathcal{L}_{\text{i}}$ of infix codes. As a subset of a reflective code is not necessarily reflective, the class of reflective codes cannot be defined by a dependence system.

Many more classes of codes are considered in the literature – with motivation ranging from purely combinatorial questions to special concerns in information transmission. The classes introduced in this section are meant to serve as paradigms for typical constructions and typical questions.

7. A hierarchy of classes of codes

One of the questions to be raised when one introduces a new class of codes or applies some construction – like the n-code construction – to existing classes of codes is where the new classes are located with respect to the known classes. The theory of dependence systems provides an elegant general framework for dealing with such questions. The required tools are explained in this section.

Before elaborating on the details, we show the relation among the classes of codes or languages introduced in Section 6. In Fig. 7.2, we show the global situation. Details of the hierarchies of shuffle codes and intercodes are shown in Fig. 6.1 and Fig. 6.2, respectively. Some further inclusions between classes of codes are summarized in the next theorem.

Theorem 7.1 *The inclusions among the classes $\mathcal{L}_{\text{unif-synch}}$, $\mathcal{K}^{\varsigma}$, $\mathcal{L}_{\text{ldecodable}}$, $\mathcal{L}_{\text{rdecodable}}$, $\mathcal{L}_{\text{circular}}$, $\mathcal{K}^{-\omega}$, $\mathcal{K}^{\omega}$, $\mathcal{L}_{\text{precirc}}$, and $\mathcal{K}^*$ as shown in Fig. 7.1 are valid and are the only valid ones among these classes.*

Proof: The inclusions $\mathcal{K}^{\varsigma} \subsetneq \mathcal{K}^{\omega} \subsetneq \mathcal{K}^*$ and $\mathcal{K}^{\varsigma} \subsetneq \mathcal{K}^{-\omega} \subsetneq \mathcal{K}^*$ are proved as Proposition 3.1 (see [21], [128], and [20]). For the inclusions

$$\mathcal{L}_{\text{unif-synch}} \subsetneq \mathcal{L}_{\text{rdecodable}}, \quad \mathcal{L}_{\text{unif-synch}} \subsetneq \mathcal{L}_{\text{ldecodable}}, \quad \text{and } \mathcal{L}_{\text{unif-synch}} \subsetneq \mathcal{L}_{\text{circular}}$$

see [4]. By [128],

$$\mathcal{L}_{\text{rdecodable}} \subsetneq \mathcal{K}^{\omega} \quad \text{and} \quad \mathcal{L}_{\text{ldecodable}} \subsetneq \mathcal{K}^{-\omega}.$$

By [21],

$$\mathcal{L}_{\text{circular}} \subsetneq \mathcal{L}_{\text{precirc}} \subsetneq \mathcal{K}^* \quad \text{and} \quad \mathcal{K}^{\varsigma} \subsetneq \mathcal{L}_{\text{precirc}}.$$

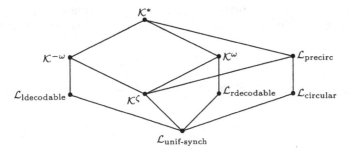

Fig. 7.1. Inclusions for Theorem 7.1. Lines indicate proper inclusion. The diagram does not indicate intersections or unions.

By Theorem 4.6, $\mathcal{L}_{\text{unif-synch}} \subsetneq \mathcal{K}^{\varsigma}$. It remains to prove that no other inclusions exist.

Consider the following seven codes:

$$K_0 = \{00\}, \qquad\qquad K_1 = \{01, 10\},$$
$$K_2 = \{0, 01, 11\}, \qquad\qquad K_3 = \{0, 10, 11\},$$
$$K_4 = \{01\} \cup \{01^n 01^{n+1} \mid n \in \mathbb{N}\}, \quad K_5 = \{10\} \cup \{1^{n+1} 01^n 0 \mid n \in \mathbb{N}\}.$$

By [21],

$$K_1 \in (\mathcal{K}^{-\omega} \cap \mathcal{K}^{\omega}) \setminus \mathcal{L}_{\text{precirc}} \text{ and } K_2 \in \mathcal{L}_{\text{precirc}} \setminus \mathcal{K}^{\omega},$$

hence $K_3 \in \mathcal{L}_{\text{precirc}} \setminus \mathcal{K}^{-\omega}$. Therefore, $\mathcal{L}_{\text{precirc}}$ is incomparable with $\mathcal{K}^{\omega}$ and $\mathcal{K}^{-\omega}$ [21]. By Example 3.1(4), $\mathcal{K}^{\omega}$ and $\mathcal{K}^{-\omega}$ are incomparable. By the same example, also $\mathcal{L}_{\text{ldecodable}}$ and $\mathcal{L}_{\text{rdecodable}}$ are incomparable. Moreover,

$$K_2 \in \mathcal{L}_{\text{precirc}} \setminus \mathcal{L}_{\text{rdecodable}}, \; K_3 \in \mathcal{L}_{\text{precirc}} \setminus \mathcal{L}_{\text{ldecodable}},$$

and

$$K_1 \in (\mathcal{L}_{\text{rdecodable}} \cap \mathcal{L}_{\text{ldecodable}}) \setminus \mathcal{L}_{\text{precirc}}.$$

Thus, $\mathcal{L}_{\text{precirc}}$ is incomparable with $\mathcal{L}_{\text{ldecodable}}$ and $\mathcal{L}_{\text{rdecodable}}$. One has

$$K_0 \in (\mathcal{K}^{\omega} \cap \mathcal{K}^{-\omega}) \setminus \mathcal{L}_{\text{circular}}, \; K_4 \in \mathcal{L}_{\text{circular}} \setminus \mathcal{K}^{\omega},$$

and

$$K_5 \in \mathcal{L}_{\text{circular}} \setminus \mathcal{K}^{-\omega}.$$

Hence, $\mathcal{L}_{\text{circular}}$ is incomparable with $\mathcal{K}^{\omega}$ and $\mathcal{K}^{-\omega}$. Moreover,

$$K_0 \in \mathcal{K}^{\varsigma} \setminus \mathcal{L}_{\text{circular}} \text{ and } K_4 \in \mathcal{L}_{\text{circular}} \setminus \mathcal{K}^{\varsigma}.$$

Thus, $\mathcal{K}^{\varsigma}$ and $\mathcal{L}_{\text{circular}}$ are incomparable. The incomparability of $\mathcal{K}^{\omega}$ and $\mathcal{K}^{-\omega}$ implies that $\mathcal{K}^{\omega} \not\subseteq \mathcal{L}_{\text{ldecodable}}$ $\mathcal{K}^{-\omega} \not\subseteq \mathcal{L}_{\text{rdecodable}}$; by Example 3.1(4),

$\mathcal{L}_{\text{ldecodable}} \not\subseteq \mathcal{K}^\omega$ and, by duality, $\mathcal{L}_{\text{rdecodable}} \not\subseteq \mathcal{K}^{-\omega}$. Hence, $\mathcal{L}_{\text{ldecodable}}$ and $\mathcal{K}^\omega$ are incomparable and, dually, $\mathcal{L}_{\text{rdecodable}}$ and $\mathcal{K}^{-\omega}$ are incomparable. As

$$K_0 \in (\mathcal{L}_{\text{ldecodable}} \cap \mathcal{L}_{\text{rdecodable}}) \setminus \mathcal{L}_{\text{circular}}$$

and $\mathcal{L}_{\text{circular}}$ is incomparable with $\mathcal{K}^\omega$ and $\mathcal{K}^{-\omega}$, $\mathcal{L}_{\text{circular}}$ is also incomparable with $\mathcal{L}_{\text{ldecodable}}$ and $\mathcal{L}_{\text{rdecodable}}$. Finally,

$$K_1 \in \mathcal{L}_{\text{rdecodable}} \setminus \mathcal{K}^\varsigma$$

while the code

$$K = \{010, 20\} \cup 2(001)^*$$

used in the proof of Theorem 4.6 is in $\mathcal{L}_{\text{rdecodable}}$, but not in $\mathcal{K}^\varsigma$; hence, $\mathcal{L}_{\text{rdecodable}}$ and $\mathcal{K}^\varsigma$ are incomparable. By duality also $\mathcal{L}_{\text{ldecodable}}$ and $\mathcal{K}^\varsigma$ are incomparable. $\qquad\square$

Many of the classes of codes introduced in Section 6 can also be defined using partial order relations. For example, for $u, v \in X^+$, let $u \leq_{\text{p}} v$ if u is a prefix of v. A language $K \subseteq X^+$ is a prefix code if and only if, for any two words $u, v \in K$, $u \leq_{\text{p}} v$ implies $u = v$, that is, if and only if K is an anti-chain with respect to the order $\leq_{\text{p}}$. Similar characterizations by partial orders exist for many other natural classes of codes; for some classes of codes, for instance the outfix codes, this type of characterization requires a binary relation, which is not a partial order, however. Moreover, in some cases no such characterizations seem to exist while, on the other hand, there are also partial orders on X^+ – even very natural ones – that lead to classes of languages which are not codes. A summary of most of the research on this approach to the characterization of classes of codes is given in [123]. In an attempt to clarify the underlying principles and to unify the presentation, this approach is extended to a certain very general class of finitary relations in [65]. From that work, one can draw the following three observations: First, many of the properties of code constructions – like building of hierarchies or their collapse – are nearly unrelated to the structure of X^+ and are simply consequences of properties of the abstract construction method itself; second, the specific property of relations – being sets of tuples – are not used at all;[36] third, the key notion in all constructions is that of independence as their principal abstract properties can be derived already at this level of generality.

Dependence theory serves as an appropriate general tool: While, of course, it cannot solve all problems, it can, however, help to clarify their structure and to expose essential properties. In the sequel we show a few applications of this approach.

---

[36] Thus, while it may be quite interesting, in applications of prefix codes and in proofs concerning prefix codes, to know that $\leq_{\text{p}}$ is a partial order, this fact itself is not essential for the construction.

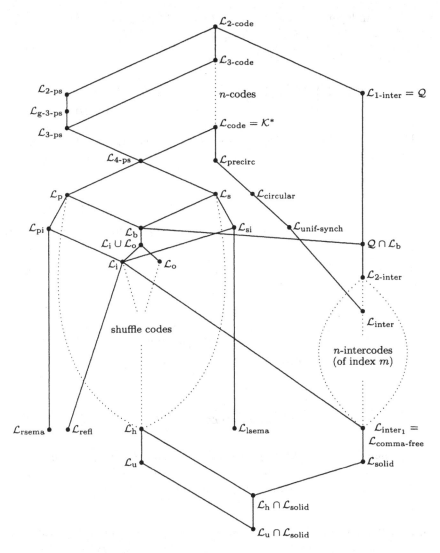

Fig. 7.2. The classes of languages introduced in Section 6. Lines indicate (known) proper inclusions; dotted lines indicate hierarchies. The diagram does not, in general, indicate intersections or unions.

As a preparation for the exploration of the *hierarchies of codes* we review a few simple properties of dependence and n-dependence systems. For this purpose, let S be an arbitrary, but fixed non-empty set. The following statement is an immediate consequence of the definitions.

Remark 7.1 If D_1 and D_2 are dependence systems on a set S and $D_1 \subseteq D_2$ then $\mathcal{L}_{D_2} \subseteq \mathcal{L}_{D_1}$, that is, every D_2-independent set is D_1-independent. A set is both D_1-independent and D_2-independent if and only it is $(D_1 \cup D_2)$-independent, that is, $\mathcal{L}_{D_1} \cap \mathcal{L}_{D_2} = \mathcal{L}_{D_1 \cup D_2}$.

For a family $\mathcal{D}$ of dependence systems, let

$$\bigcap \mathcal{D} = \bigcap_{D \in \mathcal{D}} D \quad \text{and} \quad \bigwedge \mathcal{D} = \bigwedge_{D \in \mathcal{D}} D = \bigcup_{D \in \mathbb{D}(S), D \subseteq \bigcap \mathcal{D}} D.$$

By Remark 6.1, $\bigwedge \mathcal{D}$ is a dependence system; it is the largest dependence system contained in $\bigcap \mathcal{D}$. Similarly, if $\mathcal{D}$ is a family of n-dependence systems then

$$\bigwedge_n \mathcal{D} = \bigcup_{D \in \mathbb{D}^{(n)}(S), D \subseteq \bigcap \mathcal{D}} D$$

is an n-dependence system, hence the largest n-dependence system contained in $\bigcap \mathcal{D}$.

Lemma 7.1 [66] *For $n \in \mathbb{N}_{\aleph_0}$, the algebra $(\mathbb{D}^{(n)}, \bigwedge, \cup)$ is a complete lattice with $\emptyset$ as the minimum and $2^S_{\neq \emptyset}$ as the maximum.*

For every $\mathcal{L} \subseteq 2^S$ and $n \in \mathbb{N}_{\aleph_0}$, let

$$\Psi_n(\mathcal{L}) = \{D \mid D \in \mathbb{D}^{(n)}(S), \mathcal{L} \subseteq \mathcal{L}_D\}$$
$$\psi_n(\mathcal{L}) = \{D \mid D \in \mathbb{D}^{(n)}(S), \mathcal{L}_D \subseteq \mathcal{L}\}$$

and

$$D_{\mathcal{L}}^{(n)} = \bigcup_{D \in \mathbb{D}^{(n)}(S), \mathcal{L} \subseteq \mathcal{L}_D} D.$$

Again, when $n = \aleph_0$, we usually omit n. By Lemma 7.1, $D_{\mathcal{L}}^{(n)}$ is well-defined as the maximal n-dependence D system such that all $L \in \mathcal{L}$ are D-independent. If $\mathcal{L} = \mathcal{L}_D$ for some n-dependence system D then $D = D_{\mathcal{L}}^{(n)}$ and $\mathcal{L}_{D_{\mathcal{L}}^{(n)}} = \mathcal{L}$.

Lemma 7.2 [66] *Let S be a non-empty set, $\mathcal{L} \subseteq 2^S$, and $n, m \in \mathbb{N}_{\aleph_0}$ with $n < m$. Then $D_{\mathcal{L}}^{(n)} \subseteq D_{\mathcal{L}}^{(m)}$. Moreover, if $\mathcal{L} = \mathcal{L}_{D_{\mathcal{L}}^{(n)}}$ then $D_{\mathcal{L}}^{(m)} = D_{\mathcal{L}}^{(n)}$ and, therefore, $\mathcal{L} = \mathcal{L}_{D_{\mathcal{L}}^{(m)}}$.*

A set $\mathcal{L} \subseteq 2^S$ is said to be Sub$_n$-*determined*, with $n \in \mathbb{N}_{\aleph_0}$ if it satisfies the following condition: $L \in \mathcal{L}$ if and only if Sub$_n L \subseteq \mathcal{L}$. If $m \in \mathbb{N}_{\aleph_0}$, $n \leq m$, and if $\mathcal{L}$ is Sub$_n$-determined then $\mathcal{L}$ is also Sub$_m$-determined.

The class $\mathcal{K}^*$, for example, is Sub$_{\aleph_0}$-determined: A language K is a code if and only if every finite subset of K is a code. Most classes of codes introduced so far are Sub$_n$-determined for some $n \in \mathbb{N}_{\aleph_0}$. The classes $\mathcal{L}_{\text{rsema}}$ and $\mathcal{L}_{\text{lsema}}$ are not Sub$_n$-determined for any n as they are not closed under taking subsets.

For an arbitrary set $D \subseteq 2_{\neq \emptyset}^S$, let compl($D$) be the set of all subsets L of S such that $L' \subseteq L$ for some $L' \in D$. We say that compl(D) is the *completion* of D. For $n \in \mathbb{N}_{\aleph_0}$ and $\mathcal{L} \subseteq 2^S$, one has

$$D_{\mathcal{L}}^{(n)} \subseteq \text{compl}(\{L \mid L \notin \mathcal{L}, 0 < |L| < n\})$$

with equality if $\mathcal{L} \in \mathbb{L}^{(n)}(S)$.

For $n \in \mathbb{N}_{\aleph_0}$, the *n-support* of a set $D \subseteq 2_{\neq \emptyset}^S$ is the set $\text{support}_n(D) = D \cap \text{Sub}_n S$. In general, compl($\text{support}_n(D)) \subseteq D$ with equality when D is an n-dependence system.

For example, $\text{support}_3(D_p)$ is the set of all two-element sets $\{x, y\}$ of words $x, y \in X^+$ such that x is a proper prefix of y. Moreover, $D_p = \text{compl}(\text{support}_3(D_p))$.

In general, if D is a set of non-empty subsets of S of cardinality strictly less than n, then compl(D) is an n-dependence system.

Theorem 7.2 [66] *Let S be a non-empty set, let $\mathcal{L} \subseteq 2^S$, and let $n \in \mathbb{N}_{\aleph_0}$. The following properties obtain:*

(a) $(\Psi_n(\mathcal{L}), \wedge, \cup)$ *is a complete lattice. Hence it has a unique maximum and minimum.*

(b) $\psi_n(\mathcal{L})$ *is a $\cup$-complete $\cup$-semilattice. Hence, it has a unique maximum.*

Moreover, if $\mathcal{L}$ is $\text{Sub}_{\aleph_0}$-determined then one has:

(c) $D_{\mathcal{L}}^{(n)} = \text{compl}(\{L \mid L \notin \mathcal{L}, 0 < |L| < n\})$, *and this is the maximum of $\Psi_n(\mathcal{L})$.*

(d) *For every $D \in \psi_n(\mathcal{L})$, there is a minimal $D_\infty \in \psi_n(\mathcal{L})$ with $D_\infty \subseteq D$. This statement relies on Zorn's Lemma.*

(e) *There is an n-dependence system D such that $\mathcal{L} = \mathcal{L}_D$ if and only if $\psi_n(\mathcal{L})$ has a minimum; moreover, in this case, $D = \min \psi_n(\mathcal{L}) = \max \Psi_n(\mathcal{L})$.*

As a consequence of Theorem 7.2 one derives the following *Gap Theorem*; it states a condition under which a class of languages will not be characterizable by any n-dependence system.

Theorem 7.3 [66] *Let S be a non-empty set, let $n \in \mathbb{N}_{\aleph_0}$, and let $\mathcal{L}, \mathcal{L}_1 \subseteq 2^S$. If $\mathcal{L} \subseteq \mathcal{L}_1 \subsetneq \mathcal{L}_{D_{\mathcal{L}}^{(n)}}$ then there is no n-dependence system D with $\mathcal{L}_1 = \mathcal{L}_D$.*

Note that, in Theorem 7.3, it is not excluded that $\mathcal{L} = \mathcal{L}_1$. We show an application of Theorem 7.3 to the hierarchy of n-codes. To unify the notation, we refer to the class $\mathcal{K}^* = \mathcal{L}_{\text{code}}$ of codes as the class $\mathcal{L}_{n\text{-code}}$ of n-codes for $n = \aleph_0$.

Theorem 7.4 *Let $n, m \in \mathbb{N}_{\aleph_0}$ with $n < m$. There is no $(n + 1)$-dependence system D with $\mathcal{L}_{m\text{-code}} = \mathcal{L}_D$.*

Proof: As the class $\mathcal{L}_{\text{code}}$ is $\text{Sub}_{\aleph_0}$-determined, one has

$$D_{\mathcal{L}_{\text{code}}}^{(n+1)} = \text{compl}(\{L \mid L \notin \mathcal{L}_{\text{code}}, 0 < |L| < n+1\})$$

by Theorem 7.2(c). Hence, $\mathcal{L}_{D_{\mathcal{L}_{\text{code}}}^{(n+1)}} = \mathcal{L}_{n\text{-code}}$. By [51],

$$\mathcal{L}_{\text{code}} \subseteq \mathcal{L}_{m\text{-code}} \subsetneq \mathcal{L}_{n\text{-code}}.$$

By Theorem 7.3, there is no $(n+1)$-dependence system characterizing $\mathcal{L}_{m\text{-code}}$. □

For the special case of $m = \aleph_0$, Theorem 7.4 states that the class of codes cannot be characterized by any n-dependence system with n finite. Theorem 7.4 is based on a similar result of [65] which, however, is stated there in terms of finitary relations. It strengthens various non-characterizability results for the classes of codes by binary relations (see [123]). Moreover, its proof clearly separates language theoretic from structural issues.

The constructions used for the classes of languages shown in Fig. 7.2 follow three different patterns:

(1) An infinite or finite strictly increasing sequence of n-dependence systems $D_1 \subsetneq D_2 \subsetneq \cdots$, for a fixed $n \in \mathbb{N}_{\aleph_0}$, is used to define a strictly decreasing sequence of classes of codes $\mathcal{L}_{D_1} \supsetneq \mathcal{L}_{D_2} \supsetneq \cdots$. The limit $\bigcap \mathcal{L}_{D_i}$ exists and is the class of independent sets with respect to $\bigcup D_i$.

The hierarchy of shuffle codes is obtained by a construction of this kind for $n = 3$. It starts with the 4-ps-codes and ends with the hypercodes.

(2) Starting from a dependence system D and the corresponding class $\mathcal{L}_D$ of D-independent sets, one considers the n-dependence systems $D_n = \text{compl}(\text{support}_n(D))$ derived from D and the resulting classes $\mathcal{L}_{D_n}$ of D_n-independent sets, where $n \in \mathbb{N}_{\aleph_0}$.

This is the construction used in the case of the n-codes and of the m-intercodes of index n – for fixed n and varying m in the latter case.

A variant of this construction starts with several dependence systems as in the case of the n-ps-codes of [49] or the n-infix-outfix codes of [91].

(3) An infinite or finite strictly decreasing sequence of dependence systems $D_1 \supsetneq D_2 \supsetneq \cdots$ is used to define a strictly increasing sequence of classes of codes $\mathcal{L}_{D_1} \subsetneq \mathcal{L}_{D_2} \subsetneq \cdots$.

This is the construction used in the case of the intercodes of index n and, with m varying, for m-intercodes of index n.

Construction method (1) can be purely language theoretic and, thus, outside the realm of the structural analysis tools provided by dependence theory. Similarly, construction method (3) may have very few properties permitting a structural analysis. In either case, this depends very much on the specific sequences of dependence systems. On the other hand, method (2) turns out to depend very little on language theoretic specifics. This issue is explored further in the rest of this section. Again, the main results can be proved without using any properties of X^+.

Consider a family $\mathbb{L}$ of families of subsets of S, that is, $\mathbb{L} \subset 2^{2^S}$. For $m \in \mathbb{N}_{\aleph_0}$, let $\mathcal{L}_{m,\mathbb{L}}$ be the family of subsets of S defined by

$$L \in \mathcal{L}_{m,\mathbb{L}}, \text{ if and only if, for all } L' \subseteq L, |L'| < m \text{ implies } L' \in \bigcup_{\mathcal{L} \in \mathbb{L}} \mathcal{L}.$$

When $\mathbb{D}$ is a class of dependence systems, to simplify the notation, we write $\mathcal{L}_{m,\mathbb{D}}$ instead of $\mathcal{L}_{m,\{\mathcal{L}_D | D \in \mathbb{D}\}}$.

For $m = 1$, one has

$$\mathcal{L}_{1,\mathbb{L}} = \begin{cases} \emptyset, & \text{if } \emptyset \notin \bigcup_{\mathcal{L} \in \mathbb{L}} \mathcal{L}, \\ 2^S, & \text{otherwise.} \end{cases}$$

In the sequel, we assume without special mention that $m > 1$.

Example 7.1 Let $\mathbb{L} = \{\mathcal{L}_\mathrm{p}, \mathcal{L}_\mathrm{s}\}$. Then $\mathcal{L}_{n+1,\mathbb{L}}$ is the class $\mathcal{L}_{n\text{-ps}}$ of n-ps-codes introduced in [49]. Note that, in this case,

$$\mathcal{L}_{5+i,\mathbb{L}} = \mathcal{L}_\mathrm{p} \cup \mathcal{L}_\mathrm{s} = \mathcal{L}_{4\text{-ps}} \subsetneqq \mathcal{L}_{4,\mathbb{L}} = \mathcal{L}_{3\text{-ps}} \subsetneqq \mathcal{L}_{3,\mathbb{L}} = \mathcal{L}_{2\text{-ps}}$$

for all $i \geq 0$. In a similar fashion, one defines the class of n-infix-outfix codes [91]. The classes of n-codes are also obtained in this way using $\mathbb{L} = \{\mathcal{L}_{\mathrm{code}}\}$.

We now determine how the construction of $\mathcal{L}_{m,\mathbb{L}}$ depends on the parameters m and $\mathbb{L}$.

Proposition 7.1 [66] *Let* $m, n \in \mathbb{N}_{\aleph_0}$ *and let* $\mathbb{L} \subseteq 2^{2^S}$. *If* $m \leq n$ *then* $\mathcal{L}_{n,\mathbb{L}} \subseteq \mathcal{L}_{m,\mathbb{L}}$.

By Proposition 7.1, a change in the parameter m in the construction $\mathcal{L}_{m,\mathbb{L}}$ may lead to a hierarchy. Further below we establish criteria as to when this hierarchy collapses.

Proposition 7.2 [66] *Let* I *be a non-empty set, let* $m \in \mathbb{N}_{\aleph_0}$, *and, for* $i \in I$, *let* $m_i \in \mathbb{N}_{\aleph_0}$ *and* $\mathbb{L}_i \subseteq 2^{2^S}$. *If* $m \leq m_i$ *for all* $i \in I$ *then* $\bigcup_{i \in I} \mathcal{L}_{m_i,\mathbb{L}_i} \subseteq \mathcal{L}_{m,\bigcup_{i \in I} \mathbb{L}_i}$.

To illustrate Proposition 7.2, consider $I = \{1,2\}$ with $\mathbb{L}_1 = \{\mathcal{L}_\mathrm{p}\}$ and $\mathbb{L}_2 = \{\mathcal{L}_\mathrm{s}\}$, and let $m = m_1 = m_2 = 3$. Then $\bigcup_{i \in I} \mathcal{L}_{3,\mathbb{L}_i} = \mathcal{L}_\mathrm{p} \cup \mathcal{L}_\mathrm{s} = \mathcal{L}_{4\text{-ps}}$ while $\mathcal{L}_{3,\bigcup_{i \in I} \mathbb{L}_i} = \mathcal{L}_{2\text{-ps}}$ is a proper superset.

Theorem 7.5 [66] *Let $m \in \mathbb{N}_{\aleph_0}$, let I be an arbitrary non-empty set and, for $i \in I$, let D_i be an n_i-dependence system where $n_i \in \mathbb{N}_{\aleph_0}$. Let $\mathbb{D} = \{D_i \mid i \in I\}$. The following statements hold true:*

(a) $\mathcal{L}_{m,\mathbb{D}} \subseteq \mathcal{L}_{m,\{\bigwedge_{i \in I} D_i\}}$.
(b) If I and each n_i is finite and $m > \sum_{i \in I} n_i - |I|$ then

$$\mathcal{L}_{m,\mathbb{D}} = \bigcup_{i \in I} \mathcal{L}_{n_i,\{D_i\}}.$$

By Theorem 7.5(b), the sequence

$$\mathcal{L}_{2,\mathbb{D}} \supseteq \mathcal{L}_{3,\mathbb{D}} \supseteq \cdots \supseteq \mathcal{L}_{m,\mathbb{D}} \supseteq \mathcal{L}_{m+1,\mathbb{D}} \supseteq \cdots$$

stabilizes no later than at $\mathcal{L}_{m,\mathbb{D}}$ if $\mathbb{D} = \{D_i \mid i \in I\}$ is a finite family of n_i-dependence systems D_i with each n_i finite, where $m = \sum_{i \in I} n_i - |I| + 1$. Moreover, in this case $\mathcal{L}_{m,\mathbb{D}} = \bigcup_{i \in I} \mathcal{L}_{D_i}$. The sequence may, of course, stabilize earlier than that.

Example 7.2 As in Example 7.1, let $\mathbb{L} = \{\mathcal{L}_p, \mathcal{L}_s\}$. Both $\mathcal{L}_p$ and $\mathcal{L}_s$ can be characterized by a 3-dependence system. Hence, $\mathcal{L}_{5+j,\mathbb{L}} = \mathcal{L}_p \cup \mathcal{L}_s$ for all $j \geq 0$. For the same reason, with $\mathbb{L} = \{\mathcal{L}_i, \mathcal{L}_o\}$, $\mathcal{L}_{5+j,\mathbb{L}} = \mathcal{L}_i \cup \mathcal{L}_o$ for all $j \geq 0$.

The point of Theorem 7.5 and Example 7.2 is that the collapse of the hierarchy construction $\mathcal{L}_{m,\mathbb{D}}$ is not a language theoretic phenomenon at all; it is solely a matter of dependence theory.

Theorem 7.6 [66] *The following statements hold true:*

(a) Let $\mathbb{D}$ be a family of dependence systems on S such that $\bigcup_{D \in \mathbb{D}} \mathcal{L}_D$ is $\mathrm{Sub}_{\aleph_0}$-determined. Then $\lim_{m \to \infty} \mathcal{L}_{m,\mathbb{D}} = \bigcup_{D \in \mathbb{D}} \mathcal{L}_D$.
(b) Let D be a dependence system on S such that $\mathcal{L}_D \neq \mathcal{L}_{D'}$ for every n-dependence system D' with $n < \aleph_0$. Then there are infinitely many proper inclusions in the sequence $\mathcal{L}_{2,\{D\}} \supseteq \mathcal{L}_{3,\{D\}} \supseteq \cdots$.

Generalizing Theorem 7.5(b), $\bigcup_{D \in \mathbb{D}} \mathcal{L}_D$ always is an upper bound for $\lim_{m \to \infty} \mathcal{L}_{m,\mathbb{D}}$ which is achieved if the former is $\mathrm{Sub}_{\aleph_0}$-determined. Theorem 7.6(b) complements Theorem 7.3 in the following sense: Theorem 7.3 permits one to show that a class of languages cannot be characterized by any n-dependence system, but requires the knowledge about the proper inclusion of the language classes involved; Theorem 7.6(b) permits one to show that inclusions are proper, given that non-characterizability properties have been established.

To conclude this section, we consider the classes of m-intercodes of index n. The intercodes of index n can be defined using $(2n + 2)$-dependence systems. One has $\mathcal{L}_{m\text{-inter}_n} = \mathcal{L}_{m+1,\{\mathcal{L}_{\text{inter}_n}\}}$ and, therefore, by Theorem 7.5, that the hierarchy $\mathcal{L}_{m\text{-inter}_n}$, for fixed n and varying m, stabilizes at $m =$

$2n+1$ as shown in Fig. 6.2. Previous proofs of this hierarchy collapse use fairly involved language theoretic arguments; Theorem 7.5 provides a tool which is independent of language theory. On the other hand, language theoretic arguments are definitely needed to prove that the inclusions prior to the predicted collapse are proper.

8. The syntactic monoid of a code

The syntactic monoid of a language often reveals certain combinatorial properties of the language in algebraic terms. In the theory of codes, two types of syntactic monoids are usually considered, the syntactic monoid of the code itself and the syntactic monoid of the set of all *-words over the code. Some important results concerning the latter type are presented in [4], [123], and [71]. In this section we focus on less known work concerning the former type.

Recall that a *semigroup* is a non-empty set, M say, equipped with an associative multiplication; M is a *monoid* if it is a semigroup with an *identity element*, that is, an element 1_M such that $1_M x = x = x 1_M$ for all $x \in M$. An element $y \in M$ is said to be a *zero element* of M if $xy = y = yx$ for all $x \in M$; in this case we write 0_M instead of y. A non-empty subset N of M is an *ideal* if $NM \cup MN \subseteq N$.

Now assume that M is a monoid with zero element and $|M| \geq 2$, hence $1_M \neq 0_M$. Clearly $\{0_M\}$ is an ideal of M and is equal to the intersection of all ideals of M. The intersection of all ideals of M which are different from $\{0_M\}$ – if this intersection is different from $\{0_M\}$ – is called the *core* of M, denoted by $\mathrm{core}(M)$. An element $c \in M$ is called an *annihilator* if $cx = xc = 0_M$ for all $x \in M \setminus \{1_M\}$. Let $\mathrm{annihil}(M)$ be the set of all annihilators of M.

Let M be a monoid and $L \subseteq M$. Let $P_L \subseteq M \times M$ be the relation on M consisting of all the pairs $(u, v) \in M \times M$ such that, for all $x, y \in M$, $xuy \in L$ if and only if $xvy \in L$. The relation P_L is called the *principal congruence* of L – it is indeed a congruence. Instead of $(u, v) \in P_L$ we write $u \equiv v(P_L)$ in the sequel. The set L is said to be *disjunctive* in M if P_L is the equality relation. The set $W_L = \{u \mid u \in M \wedge MuM \cap L = \emptyset\}$ is called the *residue* of L. If $W_L \neq \emptyset$ then W_L is an ideal of M. If L is a singleton set, $L = \{c\}$, we often write c instead of $\{c\}$; thus c being disjunctive means $\{c\}$ is disjunctive, $P_c = P_{\{c\}}$, and $W_c = W_{\{c\}}$.

For $x \in M$, let x/P_L denote the P_L-class of x; this notation is extended to subsets of M in the natural fashion. Note that L is a union of P_L classes. The multiplication on M induces a multiplication on M/P_L via $(x/P_L)(y/P_L) = (xy)/P_L$, and M/P_L is a monoid with $1_M/P_L$ as identity element. The mapping σ_L of M onto M/P_L given by $\sigma_L(x) = x/P_L$ is a morphism and the set L/P_L is disjunctive in M/P_L.

When $M = X^*$ for an alphabet X then P_L is also referred to as the *syntactic congruence* of L and the factor monoid $\mathrm{syn}\, L = X^*/P_L$ is the *syntactic monoid* of L. A language L is regular if and only if $\mathrm{syn}\, L$ is finite.

The morphism σ_L of X^* onto syn L is called the *syntactic morphism* of L. For general information about syntactic monoids of languages see [71] and [24].

Example 8.1 Let $X = \{0, 1, 2\}$ and $L = \{0^n 1^n 2^n \mid n \in \mathbb{N}\}$. The set L is an infix code which is not regular, hence syn L is infinite. The residue of L is non-empty; for instance, $021 \in W_L$. In general, when the residue of a set is non-empty then it is a congruence class with respect to the principal congruence and it is the zero element of the factor monoid. Thus, syn L has a zero element 0 and $0 = W_L/P_L$.

The set $\{\lambda\}$ is also a P_L-class; therefore, the identity element 1 of syn L is $\lambda/P_L = \{\lambda\}$ and syn $L \setminus \{1\}$ is a subsemigroup of syn L.

If $u \in L$ then $xuy \in L$ if and only if $x = y = \lambda$. Therefore, also L is a P_L-class; let $c = L/P_L$. Then $1 \neq c \neq 0$ and c is a disjunctive element of syn L. Moreover, $c = xcy$ implies $x = y = 1$. If $u \in L$ and $v \in X^+$ then $uv, vu \in W_L$. Thus, $c \in \text{annihil(syn } L)$.

If N is any ideal of syn L and $N \neq \{0\}$ then $\{b, 0\} \subseteq N$ for some $b \neq 0$. Let $v \in X^*$ be such that $v/P_L = b$. Then $v \notin W_L$ and there are $x, y \in X^*$ such that $xvy \in L$. Thus, $(x/P_L)b(y/P_L) = c$ and, as N is an ideal, also $c \in N$. Thus syn L has a core and c is an element of it.

It is shown in [101] that the properties derived in Example 8.1 are characteristic of syntactic monoids of infix codes in general. Among these, the property that $c = xcy$ implies $x = y = 1$, where $c = L/P_L$, can be viewed as a way of describing D_i-independence in terms of the elements of syn L. This observation leads to a general treatment of syntactic monoids of codes which is based on abstract formal properties of the predicates defining dependence systems [54]. We outline the main ideas and results of this approach. As a general framework we use the category of pointed monoids.

A *pointed monoid*[37] is a pair (M, L) where M is a monoid and L is a subset of M. Let (M, L) and (M', L') be pointed monoids. A *pointed-monoid morphism* of (M, L) into (M', L') is a semigroup morphism φ of M into M' such that $\varphi^{-1}(L') = L$. Such a pointed-monoid morphism φ is *surjective (or onto), injective, bijective* if it is so as a semigroup morphism of M into M'; it is *non-erasing* if $\varphi^{-1}(1_{M'}) = \{1_M\}$. Let $\mathbb{P}$ denote the category of pointed monoids. A predicate P on $\mathbb{P}$ is said to be *invariant* if, for any pointed monoid (M, L), P satisfies the following conditions:

- For any surjective pointed-monoid morphism φ of (M, L), P is true on $(\varphi(M), \varphi(L))$ if P is true on (M, L).
- For any surjective non-erasing pointed-monoid morphism φ of (M, L), P is true on (M, L) if it is true on $(\varphi(M), \varphi(L))$.

Proposition 8.1 *Let P be an invariant predicate on $\mathbb{P}$ and let $\mathcal{L}_P$ be the class of languages L over X for which P is true on (X^*, L). The following statements are true:*

---

[37] Called *p-monoid* in [113].

(1) If σ_L is non-erasing then $L \in \mathcal{L}_P$ if and only if P is true on the pointed monoid $(\operatorname{syn} L, \sigma_L(L))$.

(2) If P is decidable on finite pointed monoids, L is (constructively) regular, and σ_L is non-erasing, then it is decidable whether $L \in \mathcal{L}_P$.

Proof: The first claim is just a restatement of the definition of invariance. For the second claim, if L is constructively given as a regular language then one can compute the syntactic monoid $\operatorname{syn} L$ and the set $\sigma_L(L)$. Note that σ_L is a pointed-monoid morphism. The fact that L is regular implies that $\operatorname{syn} L$ is finite. Therefore, P is decidable on $(\operatorname{syn} L, \sigma_L(L))$. The invariance of P implies that it is decidable whether $L \in \mathcal{L}_P$. □

To apply Proposition 8.1 to a given predicate P, one has to establish that P is invariant and decidable on finite pointed monoids. In the rest of this section we focus on predicates on $\mathbb{P}$ which express independence conditions in the form of implications as follows. Let V be a set of variable symbols (for monoid elements), and let Λ be a set variable symbol (for subsets of monoids). We consider quantifier formulæ of the form

$$\langle \texttt{quantifier prefix} \rangle (\langle \texttt{formula} \rangle \longrightarrow \langle \texttt{formula} \rangle)$$

in which the quantifier prefix and the formulæ satisfy the following conditions:

- The quantifier prefix involves only universal quantifiers, one for each variable in V actually used.[38]
- The formulæ are disjunctions of conjunctions of equations and inclusions. An equation has the form $u = v$ and an inclusion has the form $u \in \Lambda$, where $u, v \in V^* \cup \{1\}$. We refer to the first formula as the *premiss* and to the second one as the *conclusion*.

For lack of a better term, such implications are called *basic implicational conditions*.

Definition 8.1 *An* implicational condition *is a conjunction of basic implicational conditions. If I is an implicational condition and (M, L) is a pointed monoid, then (M, L) is said to* satisfy I *if I is true on M for $\Lambda = L$.*

In Table 8.1 we show implicational conditions for some of the language classes defined in Section 6; in the sequel, we refer to these conditions by the names assigned to them in that table.

There is a similarity between implicational conditions as introduced and *implications* used to define *implicationally defined classes* or *quasivarieties* of algebras (see [100]); the difference is that disjunctions and inclusions are permitted in our implicational conditions. The full meaning of this similarity is not known.

---

[38] In a more general setting, quantification over the number of variables and restrictions on the domains of variables are also useful [54].

Table 8.1. Implicational conditions for some of the language classes introduced in Section 6.

Family	Name	Implicational condition
$\mathcal{L}_{\mathrm{ac}} = \mathcal{L}_{2\text{-code}}$	I_{ac}	$\forall u, v\big((u \in \Lambda \wedge v \in \Lambda \wedge uv = vu)$ $\rightarrow u = v\big)$
$\mathcal{L}_{\mathrm{d}} = \mathcal{L}_{2\text{-ps}}$	I_{d}	$\forall u, x, y\,\big((u \in \Lambda \wedge ux \in \Lambda \wedge yu \in \Lambda$ $\wedge\, ux = yu) \rightarrow x = y = 1\big)$
$\mathcal{L}_{\mathrm{p}_n}$	I_{p_n}	$\forall x_1, \ldots, x_n, y_1, \ldots, y_n$ $\big((x_1 \cdots x_n \in \Lambda$ $\wedge\, x_1 y_1 x_2 y_2 \cdots x_n y_n \in \Lambda)$ $\rightarrow y_1 = \cdots = y_n = 1\big)$
$\mathcal{L}_{\mathrm{i}_n}, \mathcal{L}_{\mathrm{o}_n}, \mathcal{L}_{\mathrm{s}_n}$	$I_{\mathrm{i}_n}, I_{\mathrm{o}_n}, I_{\mathrm{s}_n}$	analogous to $\mathcal{L}_{\mathrm{p}_n}$
$\mathcal{L}_{\mathrm{p}_n} \cap \mathcal{L}_{\mathrm{s}_n}$	$I_{\mathrm{p}_n, \mathrm{s}_n}$	$I_{\mathrm{p}_n} \wedge I_{\mathrm{s}_n}$
$\mathcal{L}_{\mathrm{b}}$	I_{b}	see $\mathcal{L}_{\mathrm{p}_n} \cap \mathcal{L}_{\mathrm{s}_n}$ for $n = 1$
$\mathcal{L}_{\mathrm{i}_n} \cap \mathcal{L}_{\mathrm{o}_n}$	$I_{\mathrm{i}_n, \mathrm{o}_n}$	$I_{\mathrm{i}_n} \wedge I_{\mathrm{o}_n}$
$\mathcal{L}_{\mathrm{pi}}$	I_{pi}	$\forall u, x, y\big((u \in \Lambda \wedge xuy \in \Lambda) \rightarrow y = 1\big)$
$\mathcal{L}_{\mathrm{si}}$	I_{si}	analogous to $\mathcal{L}_{\mathrm{pi}}$
$\mathcal{L}_{\mathrm{inter}_n}$	I_{inter_n}	$\forall u_1, \ldots, u_{n+1}, v_1, \ldots, v_n, x, y$ $\big((u_1 \in \Lambda \wedge \cdots \wedge u_{n+1} \in \Lambda$ $\wedge\, v_1 \in \Lambda \wedge \cdots \wedge v_n \in \Lambda$ $\wedge\, u_1 \cdots u_{n+1} = xv_1 \cdots v_n y)$ $\rightarrow ((x = 1 \wedge y = u_{n+1})$ $\vee\, (x = u_1 \wedge y = 1)))$
$\mathcal{L}_{\mathrm{ol\text{-}free}}$	$I_{\mathrm{ol\text{-}free}}$	$\forall x, y, z\,\big((xy \in \Lambda \wedge yz \in \Lambda)$ $\rightarrow (x = 1 \vee z = 1 \vee y = 1)\big)$
$\mathcal{L}_{\mathrm{solid}}$	I_{solid}	$I_{\mathrm{i}} \wedge I_{\mathrm{ol\text{-}free}}$

The families $\mathcal{L}_{\mathrm{code}}$, $\mathcal{L}_{n\text{-code}}$ for $n > 2$, $\mathcal{L}_{m\text{-inter}}$, $\mathcal{L}_{\mathrm{inter}}$, $\mathcal{Q}$, $\mathcal{L}_{\mathrm{rsema}}$, $\mathcal{L}_{\mathrm{lsema}}$, and $\mathcal{L}_{\mathrm{h}}$, $\mathcal{L}_{\mathrm{u}}$ are not listed in Table 8.1; it seems that some of these cannot be characterized with implicational conditions as they seem to require existential quantifiers or quantification over the number of variables. For example, $\mathcal{L}_{\mathrm{h}}$ would be defined by

$$I_{\mathrm{h}} = \forall n \forall x_0, \ldots, x_n, y_1, \ldots, y_n$$
$$\big((x_0 \cdots x_n \in \Lambda \wedge x_0 y_1 x_1 y_2 \cdots y_n x_n \in \Lambda) \rightarrow y_1 = \cdots = y_n = 1\big).$$

The classes $\mathcal{L}_{4\text{-ps}}$, $\mathcal{L}_{3\text{-ps}}$, $\mathcal{L}_{g\text{-}3\text{-ps}}$, and $\mathcal{L}_{m\text{-inter}_n}$ are not listed either; they do, however, have characterizations by implicational conditions, albeit rather complicated ones.

Proposition 8.2 [54] *The predicates* I_{p_n}, I_{i_n}, I_{o_n}, I_{s_n}, I_{p_n,s_n}, I_{i_n,o_n}, I_{pi}, I_{si}, I_h, *and* I_{solid} *are invariant.*

Proof idea: Each of the predicates satisfies the following simple conditions:

(1) For any pointed monoid (M, L) and any surjective pointed-monoid morphism φ of (M, L), if a premiss is false for some assignment of values in M to the variables then that premiss is also false for the corresponding assignment in $\varphi(M)$.

(2) For any pointed monoid (M, L) and any surjective non-erasing pointed-monoid morphism φ of (M, L), if a conclusion is false for some assignment of values in M to the variables then that conclusion is also false for the corresponding assignment in $\varphi(M)$.

These conditions are sufficient for an implicational condition (even involving quantification over the number of variables) to be invariant. All the predicates listed satisfy (1) and (2). This is true because the premisses involve only inclusions and the equations involved in the conclusions are of the form $x = 1$.   □

Note that I_{inter_1} does not satisfy condition (1) of the proof of Proposition 8.2, nor does the natural implicational condition I_{code} for $\mathcal{L}_{code}$ – even permitting quantification over the number of variables.

We apply Proposition 8.2 to the classes of codes contained in $\mathcal{L}_i$ to obtain a uniform characterization of their syntactic monoids. This characterization is expressed in terms of the following properties of a monoid M.

(M_0) M is finitely generated.
(M_1) $M \setminus \{1\}$ is a subsemigroup of M.
(M_2) M has a zero.
(M_3) M has a disjunctive element c distinct from 1 and 0 such that $c = xcy$ implies $x = y = 1$.
(M_4) M has a disjunctive zero.
(M_5) There is an element $c \in \text{annihil}(M)$ distinct from 0 such that $\text{core}(M) = \{c, 0\}$.
(M_6) There is an element c distinct from 0 such that

$$c \in \text{core}(M) \cap \text{annihil}(M).$$

Theorem 8.1 [54] *Let P be an invariant predicate on $\mathbb{P}$ such that $\mathcal{L}_P \subseteq \mathcal{L}_i$ where $\mathcal{L}_P$ is the family of languages L over X such that P is true on (X^*, L). The following conditions on a monoid are equivalent:*

(1) M is isomorphic with the syntactic monoid of a (non-empty) code in $\mathcal{L}_P$.
(2) M has the properties M_0, M_1, M_2, and M_3 and P is true on (M, c).
(3) M has the properties M_0, M_1, M_4, and M_5 and P is true on (M, c).
(4) M has the properties M_0, M_1, M_4, and M_6 and P is true on (M, c).

Proof: Suppose M is isomorphic with the syntactic monoid of a code $L \in \mathcal{L}_P$. As the properties M_0 through M_6 and P are preserved under isomorphisms and pointed-monoid isomorphisms, we may assume that $M = \operatorname{syn} L$ for some $L \in \mathcal{L}_P$. Then L is an infix code. For $u \in L$ and $x, y \in X^*$, one has $xuy \in L$ if and only if $x = y = \lambda$. Thus, L is a single P_L-class. Let $\sigma_L(L) = c$. As P is invariant, P is true on (M, c).

As X is finite, M is finitely generated. Thus M_0 holds true.

The empty word forms a P_L-class of its own. For, if $\lambda \equiv v(P_L)$ and $u \in L$ then $u \equiv uv(P_L)$, hence $uv \in L$ and, thus, $v = \lambda$. This implies M_1.

The residue W_L is non-empty as $ux \in W_L$ for any $u \in L$ and $x \in X$. It also forms an equivalence class, the image of which is the zero of $\operatorname{syn} L$. This implies M_2.

The element c is distinct from 1 and 0 as $\lambda \notin L$ and $W_L \cap L = \emptyset$. Suppose that $c = xcy$ for some $x, y \in M$. Then there are words $u, v \in L$ and $x', y' \in X^*$ such that $u = x'vy'$, $\sigma_L(x') = x$, and $\sigma_L(y') = y$, hence $x' = y' = \lambda$, that is, $x = y = 1$. This proves M_3; hence (1) implies (2).

Now suppose that (2) obtains. By M_3, c is disjunctive and, therefore, $W_c = \{0\}$. Let N be an ideal of M with $N \neq \{0\}$ and $a \in N \setminus \{0\}$. Then there are $x, y \in M$ such that $xay = c$. Therefore, $c \in N$ and, consequently, $c \in \operatorname{core}(M)$. Moreover, $\{c, 0\}$ is an ideal, thus $\operatorname{core}(M) = \{c, 0\}$. For any $x \in M \setminus \{1\}$, one has $xc = 0 = cx$ as $W_c = \{0\}$. This proves M_5.

Now consider $a, b \in M$ such that $a \equiv b(P_0)$. We show that $a = b$. Then $xay = 0$ if and only if $xby = 0$. If $a = 0$ then $xay = 0$ for all x, y; hence, in particular for $x = y = 1$, $b = 0$. Therefore, assume that $a \neq 0$, $b \neq 0$, and $a \neq b$. Then there are x, y such that $xay = c$ and $xby \neq c$ as c is disjunctive; moreover, $xay \equiv xby(P_0)$ as P_0 is a congruence. Hence, $xby \neq 0$ as $c \neq 0$. Therefore, there are x' and y' such that $x'xbyy' = c$ and $x' \neq 1$ or $y' \neq 1$. As P_0 is a congruence, one has $x'xayy' \equiv x'xbyy'(P_0)$; moreover, $x'xayy' = x'cy' = 0$ as $c \in \operatorname{annihil}(M)$. Thus, $0 \equiv x'xbyy'(P_0) = c \neq 0$, a contradiction. Therefore, $a = b$. This proves M_4; hence (2) implies (3). Obviously, (3) implies (4).

Now assume that (4) holds true. We show that this implies (1). As a first step we prove that $\operatorname{core}(M) = \{c, 0\}$. Let $a \in \operatorname{annihil}(M) \cap \operatorname{core}(M)$, $a \neq 0$. One has $xay = 0$ if and only if $x \neq 1$ or $y \neq 1$ as $a \in \operatorname{annihil}(M)$. Thus, $a \equiv c(P_0)$ and, as 0 is disjunctive, $a = c$. Moreover, $\{c, 0\}$ is an ideal and, therefore, $\operatorname{core}(M) = \{c, 0\}$. The fact that $c \in \operatorname{annihil}(M)$ implies that $xcy = c$ only if $x = 1 = y$.

Next, we show that c is disjunctive. Consider $a, b \in M$ with $a \equiv b(P_c)$, $a \neq b$. Thus, for all x and y, $xay = c$ if and only if $xby = c$. As $a \neq b$ and 0 is disjunctive, there are $x, y \in M$ such that, without loss of generality, $xay = 0$ and $xby \neq 0$. As $\operatorname{core}(M) = \{0, c\}$, c is contained in the ideal $MxbyM$ and, therefore, there are $x', y' \in M$ such that $x'xbyy' = c$; on the other hand, $x'xayy' = 0$, a contradiction.

Now consider a finite minimal set X of generators of M and let φ be the monoid morphism of X^* onto M which is induced by the inclusion $X \subseteq M$. Note that φ is non-erasing[39] because of M_1. Let $L = \varphi^{-1}(c)$. Then (X^*, L) is a pointed monoid and M is isomorphic with the syntactic monoid of L via the isomorphism ψ of syn L onto M given by $\psi(\sigma_L(x)) = \varphi(x)$ for $x \in X^*$. ψ is well-defined and $\psi(\sigma_L(L)) = c$. Thus, φ is a pointed-monoid morphism of (X^*, L) onto (M, c) and σ_L is also non-erasing.

Consider $u, v \in L$ such that $u = xvy$ for some $x, y \in X^*$. Then

$$\psi(\sigma_L(u)) = c = \psi(\sigma_L(x))c\psi(\sigma_L(y)),$$

hence $\psi(\sigma_L(x)) = 1 = \psi(\sigma_L(y))$. As ψ is an isomorphism, one has $\sigma_L(x) = \lambda/P_L = \sigma_L(y)$ and $x = y = \lambda$ as σ_L is non-erasing. Thus, L is an infix code.

As P is invariant and true on (M, c) it is also true on (X^*, L). Therefore, $L \in \mathcal{L}_P$.   □

For the special case of $\mathcal{L}_P = \mathcal{L}_i$, the clause "and P is true on (M, c)" can be omitted in all statements of Theorem 8.1. In the proof, it is crucial that X is chosen in such a way that σ_L is non-erasing as shown by the following example.

Example 8.2 Consider the monoid $M = \{1, c, 0\}$ with $c^2 = 0$. By Theorem 8.1, M is isomorphic with the syntactic monoid of an infix code. If we choose $X = \{a\}$ and define $\varphi : X^* \to M$ by

$$\varphi(x) = \begin{cases} 1, & \text{if } x = \lambda, \\ c, & \text{if } x = a, \\ 0, & \text{otherwise,} \end{cases}$$

then $L = \varphi^{-1}(c) = \{a\}$ is an infix code. On the other hand, if $X = \{a, b\}$ with $\varphi : X^* \to M$ given by

$$\varphi(x) = \begin{cases} 1, & \text{if } x \in b^*, \\ c, & \text{if } x \in b^*ab^*, \\ 0, & \text{otherwise,} \end{cases}$$

then $L = \varphi^{-1}(c) = b^*ab^*$, and this is not an infix code. This shows that it is important that the morphism φ in the proof of Theorem 8.1 be chosen in such a way that it is non-erasing, and this is guaranteed when X is a minimal set of generators.

---

[39] X is a minimal set of generators of M as a *monoid!* The identity element of M is obtained from the empty word over X.

Theorem 8.1 is applicable, in particular, to $\mathcal{L}_{i_n}$ and $\mathcal{L}_{i_n} \cap \mathcal{L}_{o_n}$ for $n \geq 1$, to $\mathcal{L}_{p_n}$, $\mathcal{L}_{s_n}$, $\mathcal{L}_{p_n} \cap \mathcal{L}_{s_n}$, $\mathcal{L}_{o_n}$ for $n \geq 2$, and to $\mathcal{L}_h$ and $\mathcal{L}_{solid}$. For the classes $\mathcal{L}_i$ and $\mathcal{L}_h$, the statement of Theorem 8.1 is proved in [101]; for a characterization of the syntactic monoids of hypercodes see also [131] and [135]; for $\mathcal{L}_{p_n}$, $\mathcal{L}_{s_n}$, $\mathcal{L}_{i_n}$, and $\mathcal{L}_{o_n}$ with $n \geq 2$ similar results are proved in [90], [89], [93], and [132].

By Theorem 6.2 every regular outfix code and every hypercode is finite. Thus, if L is a regular language in $\mathcal{L}_i \cap \mathcal{L}_o$ then there is an $n \in \mathbb{N}$ such that $|w| < n$ for all $w \in L$. As a consequence, if $w \in X^n X^*$ then $w \in W_L$ and, therefore, $(\operatorname{syn} L \setminus \{1\})^n = \{0\}$, that is, $\operatorname{syn} L \setminus \{1\}$ is *nilpotent*. Recall that a semigroup with at least two elements is *subdirectly irreducible* if it has a unique congruence which differs from the equality and is contained in every other congruence. By a result of [121], if M is a monoid with zero such that $M \setminus \{1\}$ is a nilpotent subsemigroup of M with a least two elements, then the zero is disjunctive if and only if M is subdirectly irreducible. This leads to the following generalization of results due to [131], [135], and [101].

Corollary 8.1 *Let P be an invariant predicate on $\mathbb{P}$ such that $\mathcal{L}_P \subseteq \mathcal{L}_i \cap \mathcal{L}_o$. A finite monoid M is isomorphic with the syntactic monoid of a non-empty regular code in $\mathcal{L}_P$ if and only if M is subdirectly irreducible, $M \setminus \{1\}$ is a nilpotent subsemigroup of M, and there is an element c, distinct from $0 \in M$, in $\operatorname{core}(M) \cap \operatorname{annihil}(M)$ such that P is true on (M, c).*

Theorem 8.1 and Corollary 8.1 provide characterizations of families of syntactic monoids for various types of infix codes. Conversely, for a given class $\mathcal{L}$ of infix codes, one can attempt to characterize the family $\hat{\mathcal{L}}$ of all those languages the syntactic monoid of which is isomorphic with the syntactic monoid of a language in $\mathcal{L}$. For the special case of $\mathcal{L}_h$, a concrete characterization of $\hat{\mathcal{L}}_h$ is given in [135]. This establishes a correspondence between families of languages and families of monoids similar to the one of [24] between *varieties of languages* and *pseudo-varieties of monoids*. In our context, however, the families of monoids under consideration are usually not pseudo-varieties, and completing them to pseudo-varieties will introduce many monoids which correspond to languages of little resemblance to those under investigation. In essence, the construct of *variety* is not fine enough to exhibit the details of the combinatorial structure of codes – let alone of codes low in the hierarchy.

Theorem 8.2 [47] *Let $L \subseteq X^+$, $m \in \operatorname{syn} L$ with $1 \neq m \neq 0$, and let $L' = \sigma_L^{(-1)}(m)$. The following statements hold true for all $n \in \mathbb{N}$:*

(1) If $L \in \mathcal{L}_{o_n}$ then $L' \in \mathcal{L}_{o_n}$.
(2) If $L \in \mathcal{L}_{p_{n+1}} \cup \mathcal{L}_{s_{n+1}}$ then $L' \in \mathcal{L}_{o_n}$.
(3) If $L \in \mathcal{L}_h$ then $L' \in \mathcal{L}_h$.

For $n = 1$, Theorem 8.2(1) is proved in [50]. For the case of n-infix codes a weaker version of (2) is proved in [132]. Statement (3) is proved in [101].

9. Deciding properties of codes

In this section we discuss the following natural question: *Given a class $\mathcal{L}$ of codes over X and given a language $L \subseteq X^+$, is it decidable whether $L \in \mathcal{L}$? And if not, under which additional assumptions is this decidable?* In view of the classes of languages introduced in Section 6, we extend this question to include classes $\mathcal{L}$ of languages which are not codes, but are related to codes in some natural way.

For example, given a language $L \subseteq X^+$ in some constructive way,[40] one could ask whether L is 2-code or whether L is a hypercode.

Clearly, in this generality, the question is undecidable. For example, as shown in Theorem 3.3, if $\mathcal{L}$ is the class $\mathcal{K}^*$ of *-codes and L can be any linear language in X^+ then it is undecidable in general whether $L \in \mathcal{L}$. Thus, restrictions are needed on the class $\mathcal{L}$ and on the types of languages L to be considered; moreover, these restrictions will depend on each other. For the classes $\mathcal{L}$, we essentially restrict the consideration to those classes introduced in Section 6; for the languages L, the restrictions will follow the Chomsky hierarchy.[41] In Table 9.1 we show known decidability results for the classes of languages in Fig. 7.2 and a few others.[42]

The proofs of the decidability results shown in Table 9.1 are quite *ad hoc,* that is, they rely heavily on the specific properties of the class $\mathcal{L}$ at hand. Given that dependence theory provides a uniform framework for the definition of most of the natural classes of codes (and related languages), one can rephrase the decidability problem as follows:

Let $n \in \mathbb{N}_{\aleph_0}$, let D be an n-dependence system on X^+, and let $\mathcal{L} \subseteq 2^{X^+}$. For $L \in \mathcal{L}$, decide whether L is D-independent.

The challenge of this problem is to find a uniform proof method for decidability results or to find general criteria as to when a given abstract proof method is applicable. In view of Theorem 3.3, one cannot expect $\mathcal{L}$ to be larger than the class of regular languages in general; on the other hand, also D will have to satisfy some restrictive conditions, a natural one being that the set $\text{support}_n(D)$ is decidable. Thus it seems appropriate to ask the following question:

---

[40] The language could be given by a grammar or an acceptor or some other computing device; which type, does not matter as long as the device can be effectively simulated by one of these.

[41] See the general exposition on formal languages in this handbook. Other language classes, for example those defined by L systems of various kinds, could be used. As is quite common in formal language theory, very little is known about the boundary separating decidable from undecidable cases.

[42] This table should be used with some care. Empty entries indicate that either there is nothing interesting to report or we do not know of an existing or readily available answer; this does not imply that the answer is not known or is difficult to find.

Table 9.1. Decidability of code properties.

family	decidable for:	undecidable for:	remarks
$\mathcal{K}^{\eta}$	L regular, $\eta \leq \zeta$	L linear, $\eta = *$	see Theorem 3.2 and Theorem 3.3
$\mathcal{K}^{*}_{\gamma}$	L finite, γ any SID channel		see Theorem 4.2
$\mathcal{L}_{2\text{-code}}$	L regular, [48]		
$\mathcal{L}_{n\text{-code}}$	L finite		for L regular and $n > 2$ the problem is open
$\mathcal{L}_{2\text{-ps}}$, $\mathcal{L}_{3\text{-ps}}$, $\mathcal{L}_{g\text{-3-ps}}$	L regular, [49]		
$\mathcal{L}_{p}$	L regular	L linear, Theorem 9.5	emptiness of $L \cap LX^{+}$ is decidable
$\mathcal{L}_{s}$, $\mathcal{L}_{b}$	L regular	L linear	analogous to $\mathcal{L}_{p}$
$\mathcal{L}_{4\text{-ps}}$	L regular	L linear	see Theorem 9.5
$\mathcal{L}_{\text{rsema}}$, $\mathcal{L}_{\text{pi}}$, $\mathcal{L}_{\text{lsema}}$, $\mathcal{L}_{\text{si}}$	L regular, [50]		
$\mathcal{L}_{o}$	L regular, [50]		regular implies finite
$\mathcal{L}_{i}$	L regular, [50]	L linear	see Theorem 9.5
$\mathcal{L}_{p_n}$, $\mathcal{L}_{i_n}$, $\mathcal{L}_{o_n}$, $\mathcal{L}_{s_n}$ for $n > 1$	L regular		regular implies finite
$\mathcal{L}_{h}$, $\mathcal{L}_{u}$	always		always finite
$\mathcal{L}_{\text{inter}}$	L regular, [61]	L linear, [61]	
$\mathcal{L}_{m\text{-inter}}$, $\mathcal{L}_{m\text{-inter}_n}$, $\mathcal{L}_{1\text{-inter}} = \mathcal{Q}$			the problem seems to be open
$\mathcal{L}_{\text{inter}_n}$	L regular, [125]	L linear, [61]	
$\mathcal{L}_{\text{solid}}$	L regular, [64]	L linear	see Theorem 9.5
$\mathcal{L}_{\text{unif-synch}}$, $\mathcal{L}_{\text{circular}}$	L regular, [73]		[78] for finite
$\mathcal{L}_{\text{rdecodable}}$, $\mathcal{L}_{\text{ldecodable}}$	L regular, [20]		[74], [78] for finite
$\mathcal{L}_{\text{precirc}}$	L regular, [19]		

Let D be an n-dependence system on X^+ such that, for every set $L \in$ $\mathrm{Sub}_n X^+$, it is decidable whether $L \in D$. Given an arbitrary regular language L over X^+, is it decidable whether $L \in \mathcal{L}_D$?

Those classes of languages shown in Table 9.1, which can be defined using dependence systems, and for which a decidability result is known, may seem to indicate an affirmative answer to this question. However, the general answer is – not unexpectedly – negative.

Theorem 9.1 [114] *There is an $n \in \mathbb{N}$ and an n-dependence system D on X^+ with the following properties:*

(a) For every $L \in \mathrm{Sub}_n X^+$, it is decidable whether $L \in D$.

(b) It is undecidable whether a given regular language is D-independent.

Proof: Consider an instance (U, V) of Post's Correspondence Problem over the alphabet $\{0, 1\}$, where

$$U = (u_0, u_1, \ldots, u_{n-1}) \text{ and } V = (v_0, v_1, \ldots, v_{n-1})$$

for some $n \in \mathbb{N}$ and $u_0, u_1, \ldots, u_{n-1}, v_0, v_1, \ldots, v_{n-1} \in \{0, 1\}^*$. Let $X = \{0, 1, \cent, \$, @\}$ and let $D_{U,V}$ be the smallest 4-dependence system on X^* containing all sets of the form

$$\{\bar{u}_{i_0} \$ \bar{u}_{i_1} \$ \cdots \$ \bar{u}_{i_{k-1}}, @\bar{v}_{j_0} \$ \bar{v}_{j_1} \$ \cdots \$ \bar{v}_{j_{k-1}}, u_0 \cent v_0 \$ u_1 \cent v_1 \$ \cdots \$ u_{n-1} \cent v_{n-1}\}$$

with $\bar{u}_{i_0} \cdots \bar{u}_{i_{k-1}} = \bar{v}_{j_0} \cdots \bar{v}_{j_{k-1}}$, where $k \in \mathbb{N}$ and for all $l \in \mathfrak{k}$ there is $h \in \mathfrak{n}$ such that $\bar{u}_{i_l} = u_h$ and $\bar{v}_{j_l} = v_h$. For every set $L \subseteq X^+$ with $|L| < 4$, it is decidable whether $L \in D_{U,V}$.

Consider

$$L_{U,V} = (\{0, 1\}^* \$)^* \{0, 1\}^* \cup @(\{0, 1\}^* \$)^* \{0, 1\}^*$$
$$\cup \{u_0 \cent v_0 \$ u_1 \cent v_1 \$ \cdots \$ u_{n-1} \cent v_{n-1}\}.$$

The language $L_{U,V}$ is regular. Moreover, $L_{U,V}$ is $D_{U,V}$-dependent if and only if (U, V) has a solution. □

As a consequence of Theorem 9.1, further restrictions on the n-dependence systems under consideration are required. We discuss two such approaches, the former based on results presented in Section 8 above, the latter based on automaton theoretic considerations.

Theorem 9.2 [54] *Let D be a dependence system satisfying the following conditions:*

(1) There is (effectively) an invariant predicate P_D on $\mathbb{P}$ such that $L \in \mathcal{L}_D$ if and only if P_D is true on (X^*, L).

(2) For every $L \in \mathcal{L}_D$, σ_L is non-erasing.

Then, for any (constructively given) regular language $L \subseteq X^+$, it is decidable whether L is D-independent.

This result is an immediate consequence of Proposition 8.1. By Proposition 8.2, the cases to which Theorem 9.2 is applicable include the following: $\mathcal{L}_{p_n}$, $\mathcal{L}_{s_n}$, $\mathcal{L}_b$, $\mathcal{L}_{i_n}$, $\mathcal{L}_{o_n}$ for $n \geq 1$, $\mathcal{L}_h$, and $\mathcal{L}_{\text{solid}}$. On the other hand, it is not applicable to $\mathcal{L}_{\text{code}}$, and $\mathcal{L}_{\text{inter}_1}$, for example.

For $\mathcal{L}_{\text{inter}}$ we do not know whether Theorem 9.2 applies. Indeed, we do not even know whether the class is definable by a dependence system.

A totally different approach is proposed in [62]. In contrast to the previous one, it works only for n-dependence systems with $n < \aleph_0$. The key idea is, to express $\text{support}_n(D)$ as the set of $(n-1)$-tuples accepted by a multi-input automaton with decidable emptiness problem. We sketch the definitions only and refer the reader to the references provided for details.

An automaton with abstract storage type $\mathfrak{T}$, a $\mathfrak{T}$-*automaton* as defined in [25] and [27], consists of a finite state control, a set C of storage configurations, an initial configuration $c_0 \in C$, and sets T and I of tests and instructions, respectively. The tests are mappings of C into the set 2, and the instructions are partial mappings of C into C.

For instance, in the case of pushdown automata the tests of T are used to check what the topmost stack symbol is and the instructions in I either pop the stack or push a new symbol onto the stack.

The operation of the automaton consists of reading the input, changing the internal state, and applying the tests and instructions to the current storage configuration. The current input symbol, the internal state and the result of the test together non-deterministically determine whether the automaton reads the next input symbol, changes the internal state, and performs a test or instruction on the storage configuration. We assume that the automata accept by final state, that the set of configurations is finitely specified, and that the sets of tests and instructions are effectively given. All automata that we consider are non-deterministic and have one-way read-only inputs. However, we allow them to have more than one input tape.

Let $\mathfrak{T}$ be an abstract storage type, and let $n \in \mathbb{N}$. A $\mathfrak{T}$-$[n, X]$-*automaton* is a $\mathfrak{T}$-automaton with n one-way input tapes and input alphabet X. Let A be a $\mathfrak{T}$-$[n, X]$-automaton. An n-tuple $(w_1, \ldots, w_n) \in X^+ \times \cdots \times X^+$ is accepted by A if there is a finite sequence of computation steps of A which, starting at c_0, leads to some final state for $w_1, \ldots, w_n$ on the input tapes numbered $1, \ldots, n$. Let $L(A)$ be the set of n-tuples accepted by A.

Definition 9.1 *Let $\mathfrak{T}$ be an abstract storage type. The* emptiness problem *for $\mathfrak{T}$ is to determine, for a given $\mathfrak{T}$-$[1, X]$-automaton A, whether $L(A) = \emptyset$. The* membership problem *for $\mathfrak{T}$ is to determine, for a given $\mathfrak{T}$-$[1, X]$-automaton A and a word $w \in X^*$, whether $w \in L(A)$.*

For an abstract storage type, the membership problem is decidable if and only if the emptiness problem is decidable [62]. This equivalence holds true for the definition of abstract storage types as sketched above which is due to [27] and [25], but does not hold true for a more general definition of abstract storage types proposed in [26] and [28].

The abstract storage types to be considered are built from *pushdowns* and *stacks*. The main difference between a pushdown and a stack is that, on a pushdown, only the symbol at the top is accessible while, on a stack the symbols below the top are also accessible, but in read-only mode. Pushdown $[1, X]$-automata accept the context-free languages, and the emptiness problem is decidable for context-free languages, hence, for the abstract storage type of pushdown. A *k-iterated pushdown*, for $k \in \mathbb{N}$, is an abstract storage type in which the storage configurations consist of pushdown configurations in the case of $k = 1$ and of a pushdown of $(k - 1)$-iterated pushdown configurations in the case of $k > 1$. In a similar fashion one defines *k-iterated stacks*. A characterization of the languages accepted by k-iterated pushdown $[1, X]$-automata is given in [17] where it is also shown that the emptiness problem for these languages is decidable. In [25] it is shown that a k-iterated stack can be simulated by a $2k$-iterated pushdown. Therefore, the emptiness problem is also decidable for k-iterated stacks.

A *counter* is a pushdown the tape alphabet of which is a singleton set. A counter is said to have a *reversal bound* of $n \in \mathbb{N}_0$ if, in any computation, the counter may switch between pushing and popping at most n times. A *reversal-bounded multi-counter* consists of k counters, for some $k \in \mathbb{N}$, which have a reversal bound of n for some $n \in \mathbb{N}_0$. Emptiness is decidable for the languages accepted by reversal-bounded multi-counter $[1, X]$-automata; this remains true even when the input tape is two-way reversal-bounded or when the automaton has an additional pushdown, but not both [46].

Theorem 9.3 [62] *Let $\mathfrak{T}$ be an abstract storage type with decidable emptiness problem. Then, for any $\mathfrak{T}$-$[n, X]$-automaton A and any regular languages $L_1, \ldots, L_n$ over X, it is decidable whether $L(A) \cap L_1 \times \cdots \times L_n$ is empty.*

Now we establish the connection to language classes defined by $(n + 1)$-dependence systems with $n < \aleph_0$. For such a dependence system D, consider a set ω_D of n-tuples satisfying the following two conditions:[43]

- If $(w_0, \ldots, w_{n-1}) \in \omega_D$ then $\bigcup_{i \in \mathfrak{n}} \{w_i\} \in D$.
- If $w_0, \ldots, w_{k-1}$ are k distinct words such that $\{w_0, \ldots, w_{k-1}\} \in D$ and $k \leq n$ then $(w_{\pi(0)}, \ldots, w_{\pi(n-1)}) \in \omega_D$ for some mapping π of $\mathfrak{n}$ onto $\mathfrak{k}$.

Thus, a set $L \subseteq X^+$ is D-independent if and only if $\omega_D \cap L \times \cdots \times L = \emptyset$. Applying Theorem 9.3, one obtains the following general decidability result.

Theorem 9.4 [62] *Let D be an $(n+1)$-dependence system on X^+ with $n < \aleph_0$ and let $\mathfrak{T}$ be an abstract storage type with decidable emptiness problem such that there is a $\mathfrak{T}$-$[n, X]$-automaton A with $L(A) = \omega_D$. Then, for any regular language L over X, it is decidable whether $L \in \mathcal{L}_D$.*

---

[43] The details concerning the connection between n-ary relations and $(n + 1)$-dependence systems are explained in [66], but are not essential in the present context.

Table 9.2. Language classes and abstract storage types proving decidability.

families	abstract storage type
$\mathcal{L}_{3\text{-ps}}$, $\mathcal{L}_{4\text{-ps}}$, $\mathcal{L}_{p_n}$, $\mathcal{L}_{i_n}$, $\mathcal{L}_{o_n}$, $\mathcal{L}_{s_n}$, $\mathcal{L}_b$, $\mathcal{L}_h$, $\mathcal{L}_u$, $\mathcal{L}_{\text{solid}}$, $\mathcal{L}_{\text{inter}_n}$	finite state
$\mathcal{L}_{2\text{-ps}}$, $\mathcal{L}_{g\text{-}3\text{-ps}}$, $\mathcal{L}_{2\text{-code}}$	stack

To apply Theorem 9.4 to a given $(n+1)$-dependence system D one has to exhibit an appropriate n-ary relation ω_D and an abstract storage type $\mathfrak{T}$ with decidable emptiness problem such that ω_D is accepted by a $\mathfrak{T}$-$[n, X]$-automaton. We list language classes and abstract storage types establishing the decidability in Table 9.2. Note that non-determinism is used heavily in several of these cases.

The decision method of Theorem 9.4 extends far beyond the hierarchy of codes. The following example is discussed briefly in [62].

Example 9.1 With every word $w \in X^*$ one associates a mapping π_w of X into $\mathbb{N}_0$ such that $\pi_w(x)$, for $x \in X$, is the number of occurrences of x in w; with a slightly different notation, π_w is usually called the *Parikh vector* of w. Consider the dependence system D_{Parikh} such that $L \in D_{\text{Parikh}}$ if there are words $w, v \in L$ with $w \neq v$ and $\pi_w = \pi_v$. To decide D_{Parikh}-dependence, one uses a $[2, X]$-automaton with $|X| + 1$ counters with a reversal bound of 1 for each counter.

The two general decision methods presented differ in two respects: By its very nature, the automaton-based approach[44] cannot handle n-dependence when the only known bound on n is $\aleph_0$, as this approach uses automata with $n - 1$ input tapes; thus, it is not applicable to the class $\mathcal{L}_{\text{code}}$ of codes – more precisely, to the relation or dependence system defining this class. On the other hand, the monoid-based approach seems not to be suitable when syntactic morphisms may be erasing or for classes the definition of which requires the quantification over the number of variables. Neither method seems to be able to deal with the case of intercodes. This is not surprising because, as mentioned before, it is not very likely that the class of intercodes can be defined by a dependence system.

The decidability of $\mathcal{L}_{\text{inter}}$ is proved in [61] and [105] using completely different methods. Neither of these approaches seems to be related to the two proof methods discussed in this handbook chapter.

The monoid-based and automaton-based proof methods are general and can be applied to many cases; they are, however, not equivalent. Whether a universal proof method – or at least a more comprehensive one – exists, continues to be an intriguing open problem.

---

[44] A different transducer-based technique is used in [43] to decide other code-related problems.

We conclude this section with a theorem asserting the undecidability of several properties for linear languages as shown in Table 9.1.

Theorem 9.5 *There is no algorithm which, given a linear grammar G, always decides whether the language generated by G is a prefix code, a suffix code, a bifix code, a 4-ps-code, an infix code, or a solid code.*

Sketch of the proof: The proof is analogous to that of Theorem 3.3. Hence we only sketch the major steps and refer to the proof of Theorem 3.3 for details and unexplained notation. Let (U, V) be an instance of Post's Correspondence Problem, where

$$U = (u_0, u_1, \ldots, u_{n-1}) \text{ and } V = (v_0, v_1, \ldots, v_{n-1}).$$

For $\mathcal{L}_p$, consider the linear language L consisting of precisely the words of the form

$$\beta(i_{m-1})\phi\beta(i_{m-2})\phi\cdots\beta(i_0)\$u_{i_0}\cdots u_{i_{m-2}}u_{i_{m-1}}\#$$

and

$$\beta(i_{m-1})\phi\beta(i_{m-2})\phi\cdots\beta(i_0)\$v_{i_0}\cdots v_{i_{m-2}}v_{i_{m-1}}\#\#.$$

Then (U, V) has a solution if and only if $L \cap L\# \neq \emptyset$. Moreover, $L \cap L\#$ is empty if and only if L is a prefix code. Thus, for linear languages it is undecidable in general whether they are prefix codes. By duality, also the property of being a suffix code is undecidable for linear languages.

For $\mathcal{L}_{4\text{-ps}} = \mathcal{L}_p \cup \mathcal{L}_s$ and $\mathcal{L}_b$, consider the linear language L consisting of precisely the words of the form

$$\#\beta(i_{m-1})\phi\beta(i_{m-2})\phi\cdots\beta(i_0)\$u_{i_0}\cdots u_{i_{m-2}}u_{i_{m-1}}\#,$$
$$\#\beta(i_{m-1})\phi\beta(i_{m-2})\phi\cdots\beta(i_0)\$v_{i_0}\cdots v_{i_{m-2}}v_{i_{m-1}}\#\#,$$

and

$$\#\#\beta(i_{m-1})\phi\beta(i_{m-2})\phi\cdots\beta(i_0)\$v_{i_0}\cdots v_{i_{m-2}}v_{i_{m-1}}\#.$$

If (U, V) has a solution then $L \cap L\# \neq \emptyset$ and $L \cap \#L \neq \emptyset$. Moreover, $L \cap L\#$ is empty if and only if L is a prefix code; $L \cap \#L$ is empty if and only if L is a suffix code. Thus, if (U, V) has a solution then $L \notin \mathcal{L}_p \cup \mathcal{L}_s = \mathcal{L}_{4\text{-ps}}$, hence $L \notin \mathcal{L}_b$. On the other hand, if (U, V) has no solution then $L \in \mathcal{L}_p \cap \mathcal{L}_s = \mathcal{L}_b$, hence $L \in \mathcal{L}_{4\text{-ps}}$. This proves the undecidability for $\mathcal{L}_{4\text{-ps}}$ and $\mathcal{L}_b$.

For $\mathcal{L}_i$, consider the linear language L consisting of precisely the words of the form

$$\#\beta(i_{m-1})\phi\beta(i_{m-2})\phi\cdots\beta(i_0)\$u_{i_0}\cdots u_{i_{m-2}}u_{i_{m-1}}\#$$

and

$$\#\#\beta(i_{m-1})\phi\beta(i_{m-2})\phi\cdots\beta(i_0)\$v_{i_0}\cdots v_{i_{m-2}}v_{i_{m-1}}\#\#.$$

Then (U, V) has a solution if and only if L is not an infix code.

For $\mathcal{L}_{\text{solid}}$, we add a new symbol % to the alphabet. Now consider the linear language L consisting of precisely the words of the form

$$\#\beta(i_{m-1})\phi\beta(i_{m-2})\phi\cdots\beta(i_0)\$u_{i_0}\cdots u_{i_{m-2}}u_{i_{m-1}}\%$$

and

$$\#\#\beta(i_{m-1})\phi\beta(i_{m-2})\phi\cdots\beta(i_0)\$v_{i_0}\cdots v_{i_{m-2}}v_{i_{m-1}}\%\%.$$

Then (U, V) has a solution if and only if L is not a solid code.   □

10. Maximal codes

Let $\mathcal{L}$ be a family of languages over the alphabet X. A language L over X is said to be $\mathcal{L}$-*maximal* if $L \in \mathcal{L}$ and, for $L' \in \mathcal{L}$, $L \subseteq L'$ implies $L = L'$. A language L over X is said to be *dense* if $X^*wX^* \cap L \neq \emptyset$ for every $w \in X^*$. A language which is not dense is called *thin*. A language L is *complete* if and only if L^* is dense. A language L over X is *right dense* if $wX^* \cap L \neq \emptyset$ for every $w \in X^*$.

Maximality and completeness concern the economy of a code. If L is a complete code then every word occurs as part of a message, hence no part of X^* is "wasted." If $\mathcal{L}$ is a class of codes, and L is $\mathcal{L}$-maximal, then L provides for the encoding of $|L|$ distinct symbols and L cannot be "improved" given the constraints imposed by the properties of $\mathcal{L}$. Other important criteria concerning the economy of a code include the *average word length* (see [4]).

We turn our attention to the following two issues: For a given family $\mathcal{L}$ in the hierarchy of codes, one seeks a characterization of the $\mathcal{L}$-maximal, the finite $\mathcal{L}$-maximal, and the regular $\mathcal{L}$-maximal languages. For a given $L \in \mathcal{L}$, can L be embedded in a $\mathcal{L}$-maximal, a finite $\mathcal{L}$-maximal, or a regular $\mathcal{L}$-maximal language?

For most of natural classes $\mathcal{L}$ of codes, the characterization of the $\mathcal{L}$-maximal codes is an open problem. Significant results are known for the classes $\mathcal{L}_{\text{code}}$ of codes, $\mathcal{L}_{\text{p}}$ of prefix codes, $\mathcal{L}_{\text{s}}$ of suffix codes, $\mathcal{L}_{\text{b}}$ of bifix codes (see [4]), and for certain classes which are quite low in the hierarchy. We quote a few classical results from [4]. For a recent survey see [8].

Theorem 10.1 [4] *Let X be an alphabet with $|X| > 1$, and let $L \subseteq X^+$. The following statements hold true:*

(a) *If L is $\mathcal{L}_{\text{code}}$-maximal then L is complete.*
(b) *If L is thin and complete then L is $\mathcal{L}_{\text{code}}$-maximal.*
(c) *If L is thin then the following statements are equivalent:*
  (c1) *L is $\mathcal{L}_{\text{p}}$-maximal;*
  (c2) *$L \in \mathcal{L}_{\text{p}}$ and L is $\mathcal{L}_{\text{code}}$-maximal.*
(d) *If L is thin then the following statements are equivalent:*
  (d1) *L is $\mathcal{L}_{\text{b}}$-maximal;*
  (d2) *$L \in \mathcal{L}_{\text{b}}$ and L is $\mathcal{L}_{\text{code}}$-maximal;*

(d3) L is $\mathcal{L}_p$-maximal and $\mathcal{L}_s$-maximal.

(e) If L is thin then the following statements are equivalent:

    (e1) L is $\mathcal{L}_{circular}$-maximal;

    (e2) $L \in \mathcal{L}_{circular}$ and L is $\mathcal{L}_{code}$-maximal.

(f) If L is thin then the following statements are equivalent:

    (f1) L is $\mathcal{L}_{rdecodable}$-maximal;

    (f2) $L \in \mathcal{L}_{rdecodable}$ and L is $\mathcal{L}_{code}$-maximal.

The class of thin codes is quite large as it contains the class of all regular codes. The proof of this statement given in [4] implies the following stronger result.

Theorem 10.2 *Let X be an alphabet with $|X| > 1$ and let $\mathcal{L}$ be a family of languages over X such that every $L \in \mathcal{L}$ is the subset of a cross section of $\sim_{\sqrt{}}$. Then every regular language in $\mathcal{L}$ is thin.*

By [18] and [48], the 2-codes are precisely the subsets of cross sections of the equivalence relation $\sim_{\sqrt{}}$.

Corollary 10.1 *Every regular language in $\mathcal{L}_{2\text{-code}}$ is thin.*

The statements (c)–(f) of Theorem 10.1 show an interesting pattern, that has been observed in several additional cases. Consider two classes $\mathcal{L}$ and $\mathcal{L}'$ of languages such that $\mathcal{L} \subseteq \mathcal{L}'$. The pattern is as follows:

    Suppose L satisfies a certain condition C; then L is $\mathcal{L}$-maximal if and only if $L \in \mathcal{L}$ and L is $\mathcal{L}'$-maximal.

In Theorem 10.1 the condition C is that L be thin. One part of this type of statement follows, regardless of C, from the following general observation.

Lemma 10.1 *Let $\mathcal{L}$ and $\mathcal{L}'$ be families of sets such that $\mathcal{L} \subseteq \mathcal{L}'$ and let $L \in \mathcal{L}$. If L is $\mathcal{L}'$-maximal then L is $\mathcal{L}$-maximal*

The converse Lemma 10.1 usually requires the special condition C. We list a few similar results.

Theorem 10.3 *Let X be an alphabet with $|X| > 1$ and let $L \subseteq X^+$. The following statements hold true:*

(a) Suppose L is a left (right) dense prefix (suffix) code. Then L is $\mathcal{L}_{3\text{-ps}}$-maximal if and only if L is $\mathcal{L}_p$-maximal ($\mathcal{L}_s$-maximal) [49].

(b) Suppose L is a $\mathcal{L}_p$-maximal ($\mathcal{L}_s$-maximal) suffix (prefix) code. Then L is $\mathcal{L}_s$-maximal ($\mathcal{L}_p$-maximal) if and only if L is $\mathcal{L}_{3\text{-ps}}$-maximal [49].

(c) No 3-ps-code is a maximal 2-ps-code [49].

(d) No 2-ps-code is a maximal 2-code; every maximal 2-ps-code is dense [49].

(e) If L is a maximal prefix code and a maximal suffix code then L is a maximal 3-ps-code [49].

(f) No g-3-ps-code is a maximal 2-ps-code; there are finite maximal g-3-ps-codes and also finite maximal 3-ps-codes which are not maximal g-3-ps-codes [49].

(g) A language is $\mathcal{L}_{\mathrm{pi}}$-maximal ($\mathcal{L}_{\mathrm{si}}$-maximal) if and only if it is a p-infix (s-infix) code which is $\mathcal{L}_{\mathrm{p}}$-maximal ($\mathcal{L}_{\mathrm{s}}$-maximal) [50].

(h) Every non-empty right (left) semaphore code is $\mathcal{L}_{\mathrm{p}}$-maximal ($\mathcal{L}_{\mathrm{s}}$-maximal) [4], hence $\mathcal{L}_{\mathrm{rsema}}$-maximal ($\mathcal{L}_{\mathrm{lsema}}$-maximal).

(i) A language is $\mathcal{L}_{\mathrm{pi}}$-maximal ($\mathcal{L}_{\mathrm{si}}$-maximal) if and only if it is a non-empty right (left) semaphore code [50].

(j) An $\mathcal{L}_{\mathrm{b}}$-maximal language is $\mathcal{L}_{\mathrm{p}}$-maximal or $\mathcal{L}_{\mathrm{s}}$-maximal.

(k) An $\mathcal{L}_{\mathrm{b}}$-maximal infix code is in $\mathcal{L}_{\mathrm{rsema}} \cap \mathcal{L}_{\mathrm{si}}$ or $\mathcal{L}_{\mathrm{lsema}} \cap \mathcal{L}_{\mathrm{pi}}$ [50].

(l) An infix code which is $\mathcal{L}_{\mathrm{p}}$-maximal or $\mathcal{L}_{\mathrm{s}}$-maximal is a uniform code which is $\mathcal{L}_{\mathrm{u}}$-maximal, hence is of the form X^{n} for some $n \in \mathbb{N}$ [50].

(m) An outfix code which is $\mathcal{L}_{\mathrm{p}}$-maximal or $\mathcal{L}_{\mathrm{s}}$-maximal is $\mathcal{L}_{\mathrm{u}}$-maximal [50].

(n) An infix code or outfix code is $\mathcal{L}_{\mathrm{code}}$-maximal if and only if it is $\mathcal{L}_{\mathrm{p}}$-maximal and this holds if and only if it is $\mathcal{L}_{\mathrm{u}}$-maximal [90], [89].

(o) A p-infix (s-infix) code is $\mathcal{L}_{\mathrm{s}}$-maximal ($\mathcal{L}_{\mathrm{p}}$-maximal) if and only if it is $\mathcal{L}_{\mathrm{u}}$-maximal [92].

Note that Theorem 10.3(c)–(d) are used in [49] to show that $\mathcal{L}_{\text{3-ps}} \subsetneq \mathcal{L}_{\text{2-ps}} \subsetneq \mathcal{L}_{\text{2-code}}$.

Numerous results characterizing $\mathcal{L}$-maximal sets of a given family $\mathcal{L}$ exist in the literature (usually satisfying certain additional conditions). Several of these can be found in [6], [146], [120], [118] for $\mathcal{L}_{\mathrm{p}}$ and $\mathcal{L}_{\mathrm{s}}$, [142], [30], [29], [23], [109] for $\mathcal{L}_{\mathrm{code}}$, [7] for $\mathcal{L}_{\mathrm{rdecodable}}$, [124] for $\mathcal{L}_{\mathrm{h}}$, and [51] for $\mathcal{L}_{\text{2-code}}$.

The basic question raised, but not solved by these results is as follows:

Given two dependence systems D and D' on X^+ such that $\mathcal{L}_D \subseteq \mathcal{L}_{D'}$, under which condition C on L and depending on D and D' is it true that L being $\mathcal{L}_D$-maximal implies that L is also $\mathcal{L}_{D'}$-maximal?

Taking into account that $\mathcal{L}_D \subseteq \mathcal{L}_{D'}$ if and only if $D' \subseteq D$, one can consider the following variant of this question:

Given a condition C, for any dependence system D what are the minimal dependence systems D' with $D' \subseteq D$ such that L being $\mathcal{L}_D$-maximal implies that L is $\mathcal{L}_{D'}$-maximal?

Similarly, one could start with D' and ask for a characterization of maximal dependence systems D with the corresponding properties. From the application point of view, any condition C that excludes no regular languages – or at least no finite languages – would still be of interest.

We now turn to the second issue: Let $\mathcal{L}$ be a family of languages and $L \in \mathcal{L}$; can L be embedded in an $\mathcal{L}$-maximal, a finite $\mathcal{L}$-maximal, or a regular $\mathcal{L}$-maximal language? The following rather general result implies that the first problem, that of embedding L in an $\mathcal{L}$-maximal language can always be achieved when $\mathcal{L}$ is the family of D-independent sets for some dependence system D.

Theorem 10.4 [56] *Let D be a dependence system on a set S. Every D-independent set can be embedded in a maximal D-independent set.*

Proof: We use Zorn's lemma. Consider an ascending chain $\{L_i \mid i \in I\}$ of D-independent subsets, where I is an arbitrary, totally ordered index set and $L_i \subseteq L_j$ for $i \leq j$, $i, j \in I$. Let $L = \bigcup_{i \in I} L_i$. Suppose that L is D-dependent. Then there is a finite, non-empty subset L' of L such that $L' \in D$. Hence, $L' \subseteq L_i$ for some $i \in I$, and L_i is D-dependent, a contradiction.   □

From Theorem 10.4 it follows that every code over X is contained in a maximal code over X (see [4], p. 41), every prefix code is contained in a maximal prefix code, etc. Indeed, the proof of Theorem 10.4 captures the abstract pattern of all proofs of such results in the theory of codes.

We now turn to the second and third parts of our question: Suppose $L \in \mathcal{L}$ is finite or regular; can L be embedded in a finite or regular $\mathcal{L}$-maximal language? The answer is only known in few cases.

For the class $\mathcal{L}_{\mathrm{code}}$, an example of [108] shows that there are finite codes which cannot be embedded in finite maximal codes. The smallest known example has 4 elements. On the other hand, every 2-element code can be embedded in a finite maximal code [109]. By [23], every regular code can be embedded in a regular maximal code. Every finite prefix code can be embedded in a finite maximal prefix code [109]. Every regular or thin code with decoding delay[45] d can be embedded in a maximal code with decoding delay d which is also regular or thin, respectively, [7], [9]. In [55], we hope to have established that every finite solid code can be embedded in a finite maximal solid code.

The abundance of results concerning maximality, completeness, denseness, and related properties available about many special classes of codes made it impossible to include or even just mention every single one. Our aim, in this section, was to exhibit the *structure* of typical problems in this field and to show typical examples of results. It would be a very interesting and probably very ambitious project to attempt a unified approach to this whole field along the lines of thought suggested.

11. Solid codes

In this section we examine the class of $\mathcal{L}_{\mathrm{solid}}$ in some detail. Solid codes have some remarkable synchronization properties in the presence of SID-channel errors.

Let $\eta \leq \zeta$. For any word schema $w \in \mathcal{S}(X^\eta)$ and any language $K \subseteq X^+$, an *error decomposition of w over K* is a factorization

$$w = \cdots (u_1)(w_1)(u_2)(w_2) \cdots$$

with the following properties:

(1) $w_i \in \mathcal{S}(K)$ for all i.

---

[45] For the noiseless channel, $*$-words, and right decoding.

(2) $u_i \in \mathcal{S}(X^{\leq \eta})$ for all i.

(3) For all i, no word schema $v \in \mathcal{S}(K)$ is an infix of u_i.

Thus, an error decomposition of a word schema w over K consists of factors belonging to K and factors which do not contain code words as infixes. For example, if K is the uniform code $\{010, 001\}$, then the word schema 0010010 has the two error decompositions $(0)(010)(010)$ and $(001)(001)(0)$ over K.

Theorem 11.1 [64] $K \subseteq X^+$ *is a solid code if and only if every word schema* $w \in \mathcal{S}(X^*)$ *has a unique error decomposition*[46] *over* K.

Suppose a solid code K is used for information transmission using an SID-channel γ and that w' is a γ-received η-message over K. Then w' has a unique error decomposition over K; thus any code word of the original message that was transmitted correctly will also be identified correctly through the error decomposition of w' over K. If γ and K are such that it is highly unlikely that a code word will result from the errors in γ acting on messages over K then, with high probability, all code words detected in w' were in the original message, and in the same order. Any received correctly transmitted code word will restore the synchronization completely. K may not be (γ, η)-correcting, but, if errors have a reasonably low probability and messages contain sufficient redundancy, they can be sufficient to achieve nearly error-free information transmission. In general, the actual quality of the solid code will, of course, depend on how well it matches the channel.

Theorem 11.2 *Let* $\eta \leq \zeta$ *and let* K *be a solid code. Every* η-*word schema over* X *has a unique error decomposition over* K *up to equivalence.*

Proof: Let K be a solid code and $K' = X^* \setminus X^* K X^*$. Let w be an η-word schema over X, and let φ and ψ be factorizations of w over $\mathcal{S}(K \cup K')$ such that

$$\varphi(i), \psi(i) \in \begin{cases} \mathcal{S}(K'), & \text{if } i \text{ is even,} \\ \mathcal{S}(K), & \text{if } i \text{ is odd,} \end{cases}$$

for $i \in \eta$; for the purposes of this proof, a factorization having this last property is said to be *alternating*. We have to show that $\varphi \sim \psi$.

First, we consider the case of $\eta = \omega$. In this case, $\varphi \sim \psi$ if and only if $\varphi = \psi$. Suppose, $\varphi \neq \psi$, and let $i \in \omega$ be minimal with $\varphi(i) \neq \psi(i)$. Then, without loss of generality, $|\psi(i)| < |\varphi(i)|$ and $\psi(i)$ is a prefix of $\varphi(i)$. Let x be the non-empty word schema such that $\varphi(i) = \psi(i)x$. As K is a solid code, i cannot be odd. Thus, i is even and $\psi(i), \varphi(i) \in \mathcal{S}(K')$. Hence, $\varphi(i+1), \psi(i+1) \in \mathcal{S}(K)$. If $|\psi(i+1)| \leq |x|$ then $\psi(i+1)$ is an infix of

---

[46] Using the property of unique error decompositions, solid codes seem to have been introduced first in [126]. The definition used in this handbook chapter was derived as a characterization in [64]. With our present definition, solid codes were introduced already much earlier in [77] as *strongly regular codes* and re-named in [84] into *codes without overlaps*.

$\varphi(i) \in S(K')$, contradicting the definition of K'. Therefore, $|\psi(i+1)| > |x|$. But then $\varphi(i+1)$ is an infix of $\psi(i+1)$ or a proper prefix of $\varphi(i+1)$ is equal to a proper suffix of $\psi(i+1)$, contradicting the assumption that K is a solid code. This proves $\varphi = \psi$. The case of $\eta = -\omega$ is settled by duality.

Now, let $\eta = \zeta$. We distinguish two cases.

Suppose there exists $l \in \zeta$ such that, with respect to both φ and ψ, a factor starts at position l in w. Let these be the factors with numbers i and j respectively. If $\varphi(i) = \lambda$ then i is even and $\varphi(i+1) \in S(K)$, hence $\varphi(i+1) \neq \lambda$; in this case we use $i+1$ instead of i, and similarly for ψ and j. Hence, without loss of generality, we assume that $\varphi(i)$ and $\psi(j)$ are both non-empty. For $n \in \omega$, let

$$\varphi_1(n) = \begin{cases} \lambda, & \text{if } n = 0 \text{ and } \varphi(i) \in S(K), \\ \varphi(i), & \text{if } n = 0 \text{ and } \varphi(i) \in S(K'), \\ \varphi(i+n-1), & \text{if } n > 0 \text{ and } \varphi(i) \in S(K), \\ \varphi(i+n), & \text{if } n > 0 \text{ and } \varphi(i) \in S(K'), \end{cases}$$

$$\psi_1(n) = \begin{cases} \lambda, & \text{if } n = 0 \text{ and } \psi(j) \in S(K), \\ \psi(j), & \text{if } n = 0 \text{ and } \psi(j) \in S(K'), \\ \psi(j+n-1), & \text{if } n > 0 \text{ and } \psi(j) \in S(K), \\ \psi(j+n), & \text{if } n > 0 \text{ and } \psi(j) \in S(K'), \end{cases}$$

and

$$w_1(n) = w(n+l).$$

Then φ_1 and ψ_1 are alternating factorizations of w_1 over $S(K \cup K')$ and, therefore, $\varphi_1 = \psi_1$. Similarly, one obtains alternating factorizations φ_2 and ψ_2 of w_2, given by

$$w_2(n) = w(l+n-1)$$

for $n \in -\omega$, over $S(K \cup K')$ and, again, $\varphi_2 = \psi_2$. For $n \in \zeta$, let

$$\varphi'(n) = \begin{cases} \varphi_2(0)\varphi_1(0), \\ \varphi_1(n), \\ \varphi_2(n), \end{cases} \quad \text{and} \quad \psi'(n) = \begin{cases} \psi_2(0)\psi_1(0), & \text{if } n = 0, \\ \psi_1(n), & \text{if } n > 0, \\ \psi_2(n), & \text{if } n < 0, \end{cases}$$

and let $w' = w_2 w_1$. Then φ' and ψ' are alternating factorizations of w' over $S(K \cup K')$. Moreover, $\varphi' = \psi'$ and $\varphi \sim \psi$.

Finally, suppose there is no $l \in \zeta$ such that, with respect to both φ and ψ, a factor starts at position l in w. Observe that $\varphi(1) \in S(K)$. Then there is a $j \in \zeta$ such that $\psi(j)$ is an infix of a proper suffix of $\varphi(1)$ or a proper prefix of $\psi(j)$ is a proper suffix of $\varphi(1)$, and $\psi(j-1)$ is an infix of a proper prefix of $\varphi(1)$ or a proper suffix of $\psi(j-1)$ is a proper prefix of $\varphi(1)$. One of the two word schemata, $\psi(j)$ and $\psi(j-1)$, is in $S(K)$. As K is a solid code, we obtain a contradiction. □

Corollary 11.1 *Every solid code is a ζ-code.*

Proof: Let w be a ζ-message over K. By Theorem 11.2, w has a unique error decomposition over K. As w is a ζ-message over K, the error decomposition is a factorization over K. Thus K is uniquely ζ-decodable. $\qquad\square$

For finite right decodable codes, the decoding process can be performed by a finite transducer. Let X and Y be alphabets, with $|X| > 1$ and $|Y| > 1$. Let K be a right decodable η-code over Y with $|K| = |X|$. Let f be an arbitrary bijection of X onto K. Let $\vartheta \le \eta$ and $\vartheta < \zeta$. As in Section 5, let F_f be the encoding determined by f. Given these assumptions, F_f is a mapping of $X^\blacklozenge$ into $Y^\blacklozenge$ and its restriction to X^ϑ is an injective mapping of X^ϑ into Y^ϑ. In the absence of noise, the restriction G_f^ϑ of F_f^{-1} to K^ϑ is a ϑ-*decoding* for f. We may assume that $\eta \le \omega$, hence $\vartheta \le \omega$. Consider a ϑ-message w over K. Let $v = \mathcal{S}(G_f^\vartheta(\overline{w}))$. Then $\overline{v}$ is the unique ϑ-word over X such that $(\overline{v}, \overline{w}) \in F_f$. Let φ be the factorization of w over K. Then $v(i) = f^{-1}(\overline{\varphi(i)})$ for all $i \in \vartheta$. As K is right decodable, with delay d say, and finite, the decoding G_f^ϑ of F_f can be computed by a finite automaton which needs to be able to store no more than $d + 1$ code words at any given moment. Such an automaton will output λ during a state transition if the information seen and stored so far is insufficient to determine the next symbol of the decoding; it will output the next symbol when the information suffices. An automaton computing G_f^ϑ is a special kind of deterministic transducer, a ϑ-*decoder* for f. Such ϑ-decoders, called also *decoding automata*, have been studied in many contexts. For the purposes of the present section on solid codes, we refer to [75], [76], [78], [77], [33], and [111]. Some classes of finite right decodable codes can be characterized by the type of transducers required for their decoding. We present such a characterization for finite solid codes [111].

Recall that a *(finite) deterministic ϑ-transducer* is a construct

$$A = (Y, X, Q, \delta, \mu)$$

with the following properties and interpretation:

(1) Y is the *input alphabet* and X is the *output alphabet* of A.
(2) Q is a finite non-empty set, the *set of states (state symbols)* of A.
(3) δ is a mapping of $Q \times Y$ into Q, the *transition function* of A.
(4) μ is a mapping of $Q \times Y$ into X^*, the *output function* of A.

One extends the transition and output functions to input ϑ-words over Y by

$$\delta(q, w) = \begin{cases} q, & \text{if } w = \lambda, \\ \delta(\delta(q, w_0), w_1 w_2 \cdots), & \text{otherwise,} \end{cases}$$

and

$$\mu(q, w) = \begin{cases} \lambda, & \text{if } w = \lambda, \\ \mu(q, w_0)\mu(\delta(q, w_0), w_1 w_2 \cdots), & \text{otherwise,} \end{cases}$$

for $q \in Q$ and $w = w_0 w_1 w_2 \cdots \in Y^{\le \vartheta}$ and $w_0, w_1, w_2, \ldots \in Y$.

Definition 11.1 *Let $\vartheta \leq \omega$ and let X and Y be alphabets with $|X| > 1$ and $|Y| > 1$. Let $K \subseteq Y^+$ such that $|K| = |X|$ and let f be a bijection of X onto K. A state-invariant decoder for f without look-ahead is a finite deterministic ϑ-transducer $A = (Y, X, Q, \delta, \mu)$ with the following properties:*

(1) For all $v \in K$ and all $q \in Q$, $\mu(q, v) = f^{-1}(v)$.
(2) For all $v \in K$, all proper prefixes u of v, and for all $q \in Q$, $\mu(q, u) = \lambda$.

Theorem 11.3 [111] *Let $\vartheta \leq \omega$, let X and Y be alphabets with $|X| > 1$ and $|Y| > 1$, let $K \subseteq Y^+$ such that $|K| = |X|$ and let f be a bijection of X onto K. Then K is a solid code if and only if there is a state-invariant decoder for f without look-ahead.*

Proof: Let K and f be given and let $A = (Y, X, Q, \delta, \mu)$ be a state-invariant decoder for f without look-ahead. Suppose that K is not a solid code. We distinguish two cases.

Case 1: There are $u, v \in K$ such that u is a proper infix of v. In this case, $v = v_1 u v_2$ for some words v_1 and v_2 with at least one of v_1 and v_2 non-empty. For any $q \in Q$, one has $\mu(q, v_1) = \lambda$. Let $q' = \delta(q, v_1)$. If $v_2 \neq \lambda$ then $v_1 u$ is a proper prefix of v and, by this, $\mu(q, v_1 u) = \lambda$; on the other hand,

$$\mu(q, v_1 u) = \lambda \mu(q', u) = f^{-1}(u) \neq \lambda,$$

a contradiction. Finally, if $v_2 = \lambda$, then

$$\mu(q, v) = f^{-1}(v) \neq f^{-1}(u) = \mu(q, v_1 u),$$

again a contradiction. This proves that K is an infix code.

Case 2: There are $u, v \in K$ such that a proper prefix of u is a proper suffix of v. In this case, $u = w u_1$ and $v = v_1 w$ for some non-empty words u_1, v_1, and w. For any state $q \in Q$, one has

$$\mu(q, v) = \mu(q, v_1 w) = \lambda \mu(q', w) = f^{-1}(v)$$

where $q' = \delta(q, v_1)$, hence $\mu(q', w) = f^{-1}(v)$. On the other hand, $\mu(q', w) = \lambda$, as w is a proper prefix of u. This contradiction implies that K is overlap-free.

To prove the converse, we assume that K is a solid code with $|K| = |X|$ and that f is a bijection of X onto K. We define a transducer $A = (Y, X, Q, \delta, \mu)$ as follows. Let $Q = \mathrm{Pref}(K) \cup \{\lambda\}$. For any $w \in X^*$, let $\sigma(w)$ be the longest suffix of w which is in $Q \cup K$. Then

$$\delta(q, y) = \begin{cases} \sigma(qy), & \text{if } \sigma(qy) \notin K, \\ \lambda, & \text{otherwise,} \end{cases}$$

and

$$\mu(q, y) = \begin{cases} \lambda, & \text{if } \sigma(qy) \notin K, \\ f^{-1}(qy), & \text{if } \sigma(qy) \in K, \end{cases}$$

for all $q \in Q$ and $y \in Y$.

One verifies that, for $q \in Q$ and $w \in Y^*$,

$$\delta(q, w) = \begin{cases} \sigma(qw), & \text{if } \sigma(qw) \notin K, \\ \lambda, & \text{if } \sigma(qw) \in K. \end{cases}$$

Moreover, as K is a solid code, if a suffix of qw is in K then no other suffix of qw is in K. Therefore μ is well-defined and $\mu(q, w) = f^{-1}(w)$ for $w \in K$. If v is a proper prefix of a word $w \in K$ then $\sigma(qv) \notin K$ as K is a solid code. Hence $\mu(q, v) = \lambda$. Thus, A has the required properties. □

The two properties defining solid codes, to be an infix code and to be overlap-free, impose severe restrictions on the selection of code words. One has to expect that only very few words of any given length can be used in a solid code and that, as a consequence, the average code word length and the *redundancy* of a solid code will be very large. The next two theorems examine these issues.

Recall that, for functions $f, g : \mathbb{N} \to \mathbb{R}_+$, one writes $f \lesssim g$ (or $f(n) \lesssim g(n)$) if $\limsup_{n \to \infty} \big(f(n)/g(n)\big) \leq 1$; one writes $f \gtrsim g$ (or $f(n) \gtrsim g(n)$) if $\liminf_{n \to \infty} \big(f(n)/g(n)\big) \geq 1$; one writes $f \sim g$ (or $f(n) \sim g(n)$) if $f \lesssim g$ and $g \lesssim f$. Note that $f \sim g$ if and only if $\lim_{n \to \infty} \big(f(n)/g(n)\big)$ exists and is equal to 1.

Theorem 11.4 [77], [84] *For $n \in \mathbb{N}$ and $|X| = r > 1$, let $\mu_{\text{solid}}(n)$ be the maximal number of code words of a solid code in X^n. Then*

$$\left(1 - \frac{1}{n}\right)^{n-1} \frac{r^n}{n} \geq \mu_{\text{solid}}(n) \gtrsim \frac{r^n \ln r}{n r^{r/(r-1)}}.$$

Consider an injective mapping κ of $\mathbb{N}_0$ into X^* such that $|\kappa(i)| \leq |\kappa(i+1)|$ for all $i \in \mathbb{N}_0$, and let $K_\kappa = \kappa(\mathbb{N}_0)$. We refer to such a mapping κ as an *injective, length-monotonic mapping*. Let ϱ_κ be the mapping of $\mathbb{N}$ into $\mathbb{N}$ given by

$$\varrho_\kappa(i) = |\kappa(i)| - \lfloor \log_r i \rfloor - 1$$

for $i \in \mathbb{N}$. Intuitively, if K_κ is considered as a code to encode the non-negative integers, that is, encode i by $\kappa(i)$, then ϱ_κ measures how many more digits are needed by this encoding than by the usual positional number representation at base r (which is optimal). For this reason, ϱ_κ is sometimes referred to as the *redundancy* of κ or of K_κ.

Theorem 11.5 *Let $|X| = r > 1$. The following statements hold true:*

(1) *There exists an injective length-monotonic mapping $\kappa : \mathbb{N}_0 \to X^*$ such that K_κ is a prefix code and $\varrho_\kappa(n) \sim \log_r \log_r n$ as $n \to \infty$ [82].*

(2) *For $r = 2$, there exists an injective length-monotonic mapping $\kappa : \mathbb{N}_0 \to X^*$ such that $K_\kappa \in \mathcal{L}_{\text{p}} \cap \mathcal{L}_{\text{ol-free}}$ and $\varrho_\kappa(n) \lesssim \frac{3}{2} \log_2 \log_2 n$ as $n \to \infty$ [95], [84].*

(3) For any injective length-monotonic mapping $\kappa : \mathbb{N}_0 \to X^*$ such that $K_\kappa \in \mathcal{L}_{\text{solid}}$ one has $\varrho_\kappa(n) \geq c_1 \log_r n + c_2$ for some constants c_1 and c_2 with $c_1 > 0$ [84].

The construction of [95] used in [84] for Theorem 11.5(2) is of considerable interest in its own right.[47] Let $X = \{0, 1\}$. For $i \in \mathbb{N}_0$ and $n \in \mathbb{N}$, we define the language

$$K_{i,n} = \begin{cases} \{0\}, & \text{if } i = 0, \\ \displaystyle\bigcup_{\substack{j_1, j_2, \ldots, j_n \in \mathbb{N}_0 \\ j_1 + j_2 + \cdots + j_n = i - 1}} 1K(j_1, n) \cdots K(j_n, n), & \text{if } i > 0. \end{cases}$$

Let

$$K(n) = \bigcup_{i \in \mathbb{N}_0} K(i, n).$$

Then $K(n)$ is a prefix code; every word in $K(i, n)$ has length $ni + 1$ and contains i occurrences of the symbol 1; every word is a prefix of some word in $K(n)$; moreover,

$$|K(i, n)| = \frac{1}{ni + 1} \binom{ni + 1}{i}.$$

Now consider an injective length-monotonic mapping $\kappa : \mathbb{N}_0 \to X^*$ such that $K_\kappa = K(2)$. One computes that

$$\lim_{i \to \infty} \frac{\varrho_\kappa(i)}{\frac{3}{2} \log_2 \log_2 i} \leq 1$$

as required.

Interpreting ϱ_κ as redundancy, Theorem 11.5 shows that solid codes have a greater redundancy than prefix codes, $\log n$ compared to $\log \log n$, and that this difference is, essentially, *not* due to the overlap-freeness, but to the fact that they are infix codes. Thus, the penalty in redundancy is not really paid for the synchronization capabilities – even in the presence of noise –, but for the ability of infix codes to detect certain kinds of insertion or deletion errors.

For the remainder of this section, we focus on combinatorial properties of solid codes and their relation to other types of codes.

Theorem 11.6 [64] *The class of solid codes is closed under inverse non-erasing morphisms. Moreover, if X and Y are alphabets and φ is a non-erasing morphism of X^* into Y^*, then φ maps solid codes onto solid codes if and only if the restriction of φ to X is injective and, for any $a, b \in X$, the set $\{h(a), h(b)\}$ is a solid code.*

---

[47] It is used in [95] to exhibit the relation among four different "completeness" conditions for prefix codes which are known to be equivalent for finite prefix codes, but turn out not to be equivalent for infinite prefix codes.

Theorem 11.7 [126] $\mathcal{L}_{\text{solid}}$ *is a proper subset of* $\mathcal{L}_{\text{inter}_1}$, *hence a proper subset of* $\mathcal{L}_i$ *and* $\mathcal{Q}$.

Let $N = (N_0, N_1, \ldots)$ be a sequence such that, for $k \in \mathbb{N}_0$, N_k is a set of $2k$-tuples of positive integers. Let $f = (f_0, f_1, \ldots)$ and $g = (g_0, g_1, \ldots)$ be sequences of mappings, such that, for $k \in \mathbb{N}_0$, f_k and g_k map N_k into $\mathbb{N}$. We say that f and g *satisfy the solidity condition* if, for all $k \in \mathbb{N}_0$, all $x, y \in N_k$ where $x = (x_1, \ldots, x_{2k})$ and $y = (y_1, \ldots, y_{2k})$, and t with $1 \le t \le k$, one has

$$0^* 0^{f_k(x)} 1^{x_1} 0^{x_2} \cdots 1^{x_{2t-1}} \cap 0^{y_{2(k-t+1)}} 1^{y_{2(k-t+1)+1}} \cdots 1^{y_{2k-1}} 0^{y_{2k}} 1^{g_k(y)} 1^* = \emptyset.$$

We denote the fact that f and g satisfy the solidity condition for N by $\mathrm{Sol}(N, f, g)$.

Lemma 11.1 [55] *Let* $X = \{0, 1\}$, *let* $C \subseteq 0^+ X^* 1^+$ *be a solid code, and for* $k \in \mathbb{N}_0$ *let*

$$N_k = \{(j_1, i_1, j_2, i_2, \ldots, j_k, i_k) \mid C \cap 0^+ 1^{j_1} 0^{i_1} 1^{j_2} 0^{i_2} \cdots 1^{j_k} 0^{i_k} 1^+ \neq \emptyset\}.$$

There are sequences $f = (f_0, f_1, \ldots)$ *and* $g = (g_0, g_1, \ldots)$ *with the following properties:*

(1) For all $k \in \mathbb{N}_0$, f_k *and* g_k *are mappings of* N_k *into* $\mathbb{N}$.
(2) $\mathrm{Sol}(N, f, g)$ *obtains.*
(3) For all $k \in \mathbb{N}_0$ *and all* $x \in N_k$, *one has*

$$C \cap 0^+ 1^{x_1} 0^{x_2} \cdots 1^{x_{2k-1}} 0^{x_{2k}} 1^+ = \{0^{f_k(x)} 1^{x_1} 0^{x_2} \cdots 1^{x_{2k-1}} 0^{x_{2k}} 1^{g_k(x)}\}.$$

Conversely, let $N = (N_0, N_1, \ldots)$ *be a sequence such that, for* $k \in \mathbb{N}_0$, N_k *is a set of* $2k$-*tuples of positive integers. Let* f *and* g *be sequences of mappings satisfying* (1) *and* (2). *Then, for any* $k \in \mathbb{N}$ *with* $N_k \neq \emptyset$, *the set* $\{0^{f_k(x)} 1^{x_1} 0^{x_2} \cdots 1^{x_{2k-1}} 0^{x_{2k}} 1^{g_k(x)} \mid x \in N_k\}$ *is a solid code.*

For the special case of $k = 0$, N_0 is either empty or the set consisting of the empty tuple. Hence, $|f_0(N_0)|$ and $|g_0(N_0)|$ are both 0 or both 1. Thus $|C \cap 0^+ 1^+| \le 1$. This fact is also proved in [64].

By Lemma 11.1, a solid code in $(0^+ 1^+)^+$ is uniquely given by a triple (N, f, g). Note that, even when $\mathrm{Sol}(N, f, g)$ is true, the language corresponding to such a triple need not be a solid code. This is so, because the solidity condition does not concern the interaction between different components of N. For our purpose this is not causing a serious problem as, in all subsequent applications of Lemma 11.1, N_0 and N_1 are the only potentially non-empty components of N.

Theorem 11.8 [55] *Let* $X = \{0,1\}$ *and let* $C \subseteq 0^+1^+ \cup 0^+1^+0^+1^+$. *For* $k \in \mathbb{N}_0$ *let*

$$N_k = \{(j_1, i_1, j_2, i_2, \ldots, j_k, i_k) \mid C \cap 0^+1^{j_1}0^{i_1}1^{j_2}0^{i_2} \cdots 1^{j_k}0^{i_k}1^+ \neq \emptyset\}.$$

Then C *is a solid code which is not a proper subset of a solid code* $C' \subseteq$ $0^+1^+ \cup 0^+1^+0^+1^+$ *if and only if there are sequences* f *and* g *as in Lemma 11.1 and* C *is of one of the following forms:*

> 0: $N_0 = \{()\}$, $N_1 = N_2 = \ldots = \emptyset$, $f_0() = 1$ *or* $g_0() = 1$.
> FF: $N_0 = \{()\}$, $f_0() > 1$, $g_0() > 1$, $N_1 = \{1, \ldots, g_0() - 1\} \times \{1, \ldots, f_0() - 1\}$, $N_2 = N_3 = \ldots = \emptyset$, *and* $\mathrm{Sol}(N, f, g)$.
> II: $N_0 = \emptyset$, $N_1 = \mathbb{N} \times \mathbb{N}$, $N_2 = N_3 = \ldots = \emptyset$, *and* $\mathrm{Sol}(N, f, g)$.
> IF: $N_0 = \emptyset$, $N_1 = \{1, \ldots, m - 1\} \times \mathbb{N}$ *for some* $m > 1$, $N_2 = N_3 = \ldots = \emptyset$, *and* $\mathrm{Sol}(N, f, g)$ *such that there are infinitely many* i *with* $g_1(j, i) = m$ *for some* j.
> FI: $N_0 = \emptyset$, $N_1 = \mathbb{N} \times \{1, \ldots, n - 1\}$ *for some* $n > 1$, $N_2 = N_3 = \ldots = \emptyset$, *and* $\mathrm{Sol}(N, f, g)$ *such that there are infinitely many* j *with* $f_1(j, i) = n$ *for some* i.

Using Theorem 11.8, one can characterize the maximal solid codes C in $0^+1^+ \cup 0^+1^+0^+1^+$. This characterization again distinguishes the 5 types 0, FF, II, IF, and FI.

Theorem 11.9 [55] *Let* $X = \{0,1\}$ *and let* $C \subseteq 0^+1^+ \cup 0^+1^+0^+1^+$, $C \neq \emptyset$. *For* $k \in \mathbb{N}_0$ *let*

$$N_k = \{(j_1, i_1, j_2, i_2, \ldots, j_k, i_k) \mid C \cap 0^+1^{j_1}0^{i_1}1^{j_2}0^{i_2} \cdots 1^{j_k}0^{i_k}1^+ \neq \emptyset\}.$$

Then C *is a maximal solid code over* X *if and only if there are sequences* f *and* g *as in Lemma 11.1 and* C *satisfies one of the following conditions:*

(1) C is of type 0.

(2) C is of one of the types FF, II, IF, or FI, $f_1(j, i)$ is independent of i and increasing with j, $g_1(j, i)$ is independent of j and increasing with i, and f_1 and g_1 determine each other uniquely. Moreover, for $(j, i) \in N_1$, let

$$G_i = \{l \mid (l, i) \in N_1, \ f_1(l, i) > i\}$$

and

$$F_j = \{l \mid (j, l) \in N_1, \ g_1(j, l) > j\}.$$

The mappings f_1 and g_1 have the following properties depending on the type of C:

(a) Type FF with $n = f_0()$ and $m = g_0()$: For all $(j, i) \in N_1$,

$$g_1(j, i) = \begin{cases} \min G_i, & \text{if } G_i \neq \emptyset, \\ m, & \text{otherwise,} \end{cases}$$

$f_1(j, i) \leq n$, and $g_1(j, i) \leq m$.

(b) *Type II: For all* $(j, i) \in N_1$, $g_1(j, i) = \min G_i$, *and*

$$\lim_{l \to \infty} f_1(l, i) = \lim_{l \to \infty} g_1(j, l) = \infty.$$

(c) *Type IF with* $N_1 = \{1, \ldots, m-1\} \times \mathbb{N}$: *For all* $(j, i) \in N_1$, $f_1(j, i) = \min F_j$, $g_1(j, i) \leq m$, $\lim_{l \to \infty} g_1(j, l) = m$.

(d) *Type FI with* $N_1 = \mathbb{N} \times \{1, \ldots, n-1\}$: *For all* $(j, i) \in N_1$, $g_1(j, i) = \min G_i$, $f_1(j, i) \leq n$, $\lim_{l \to \infty} f_1(l, i) = n$.

Moreover, suppose $N = (N_0, N_1, \emptyset, \ldots)$, f_0, *and* g_0 *have the properties required by one of the types FF, II, IF, or FI. If* f_1 *and* g_1 *have the corresponding properties in condition (2), then* $\mathrm{Sol}(N, f, g)$ *where* $f = (f_0, f_1, \ldots)$ *and* $g = (g_0, g_1, \ldots)$.

Note that in Theorem 11.9 one also has

$$f_1(j, i) = \begin{cases} \min F_j, & \text{if } F_j \neq \emptyset, \\ n, & \text{otherwise,} \end{cases}$$

in the case (2a) and $f_1(j, i) = \min F_j$ in the case (2b). This follows by duality.

By Theorem 11.9, there is a maximal solid code $C \subseteq 0^+1^+ \cup 0^+1^+0^+1^+$ with $k = |C|$ for every $k \in \mathbb{N}$. A maximal solid code of type 0 has exactly 1 element. For $k > 1$, a maximal solid code of type FF with $n = 2$ and $m = k$ has exactly k elements [55].

The following theorem describes the relation of the class of maximal solid codes in the set $0^+1^+ \cup 0^+1^+0^+1^+$ to the classical Chomsky hierarchy of formal languages.

Theorem 11.10 [55] *Let* $X = \{0, 1\}$ *and let* $C \subseteq 0^+1^+ \cup 0^+1^+0^+1^+$ *be a maximal solid code. The following statements hold true:*

(1) C *is finite if and only if* C *is of type 0 or FF.*

(2) C *is regular if and only if it is of one of the types 0, FF, IF, or FI. Moreover, there exist infinitely many infinite regular maximal solid codes in* $0^+1^+0^+1^+$.

(3) *If* C *is context-free, but not regular, then* C *is of type II. Moreover, there exist infinitely many context-free, non-regular maximal solid codes in* $0^+1^+0^+1^+$.

For the remaining statements assume that C *is of type II and*

$$C = \{0^{\varphi(j)} 1^j 0^i 1^{\gamma(i)} \mid j, i \in \mathbb{N}\}.$$

(4) *The following conditions on* C *are equivalent:*

(a) φ *is recursive.*

(b) γ *is recursive.*

(c) C *is recursively enumerable.*

(d) C is recursive.

(5) There are infinitely many maximal solid codes which are context-sensitive, but not context-free.

(6) There is a maximal solid code in $0^+1^+0^+1^+$ which is not recursively enumerable.

Our final theorem in this section relates maximal solid codes to prefix codes and suffix codes.

Theorem 11.11 [55] *Let $X = \{0, 1\}$ and let $C \subseteq 0^+1^+ \cup 0^+1^+0^+1^+$ be an infinite maximal solid code. Then $C = PS$ where P and S are contained in 0^+1^+ and satisfy the following conditions.*

(1) If C is of type II then P is an infinite suffix code and S is an infinite prefix code.

(2) If C is of type IF then P is a finite suffix code and S is an infinite prefix code.

(3) If C is of type FI then P is an infinite suffix code and S is a finite prefix code.

12. Codes for noisy channels

In this section we discuss some aspects of coding for noisy channels. The presentation is relatively informal as it is meant to point out problems rather than present solutions.

A message received via a noisy channel may be distorted in many ways depending on the physical properties of the channel and sometimes also on the structure of the message.[48] Thus, a model of the physical channel needs to be known before an appropriate code can be chosen.

For example, if the channel is known to substitute signals with some probability greater than 0, but to keep synchronization intact with probability 1, one may decide to use a uniform code K of word length n, say. If $w \in K$ is sent and w' is received then w' is assumed to have the correct length. As synchronization errors occur with probability 0, one does not worry about the boundaries between received words in a received message.[49]

Similar considerations apply to other kinds of channels. In essence, if w' is a received message and, for a message v, $\mathrm{Prob}(v \mid w')$ is the probability

---

[48] For example, a channel may become *saturated* by a long sequence of identical input signals and continue reproducing them for some time even after the input has changed.

[49] To be more precise, it is not really the probability of a synchronization error, but the *cost* of a synchronization error that needs to be considered. Even for a uniform code, if synchronization errors are disregarded, but can occur, albeit with a very small probability, the cost can be devastating.

of v having been sent when w' has been received, then *maximum-likelihood decoding* will decode w' as a message w such that

$$\mathrm{Prob}(w \mid w') > \mathrm{Prob}(v \mid w')$$

for all messages v distinct from w. In choosing a code, this is a key condition. Channel models abstract from this probabilistic situation by eliminating all low-probability cases.

Thus, if γ is a channel according to Section 3 then γ models only that part of the physical channel that occurs with a high probability or results in a high cost if it occurs; all these events are treated as equally probable; all other events are treated as impossible, that is, not just as having probability 0. Using this idealization, one can try to replace the maximum-likelihood decoding argument by a *minimum-distance decoding* argument (see [22]), that is, replace probabilistic reasoning by combinatorial reasoning. The classical *Hamming metric* and *Lee metric* are examples of this kind of transition for substitution-only channels. The *Levenshtein distance* serves a similar purpose for certain SID-channels. The Hamming and Lee metrics are defined for words of the same length only: the Hamming distance between two words is the number of positions in which they differ; the Lee metric also takes into account *by how much* corresponding positions differ.[50] The Levenshtein distance between two words, u and v say, is the minimal total number of substitutions, insertions, and deletions that can transform u into v.

If w' is the message received when w was sent, then any distance d to be used for minimum-distance decoding will have to satisfy

$$d(v, w') > d(w, w') \text{ if and only if } \mathrm{Prob}(v \mid w') < \mathrm{Prob}(w \mid w')$$

for all messages v different from w (see [22] for further explanations). Thus, the choice of a metric implies assumptions about the physical channel; these are usually not made explicit in the literature. The extent to which the assumptions are valid determines very much how well theoretical results predict physically observable behaviour. This is, of course, not unique to coding theory; however, much of coding theory is now an area of beautiful mathematical results with very little connection to the true problems in the field. To clarify this statement: first, in the context of mathematics, there are many wonderful and deep results derived in coding theory; second, in the context of coding theory, however, they may be of less value as their assumptions may be physically unrealistic. For example, very-high speed information transmission via optical fibres seems to require an error model that is significantly different from that assumed by the Hamming metric. Thus, the quality of classical codes, as predicted in the literature, should not be taken for granted. Much more fundamental research is needed to understand the abstract connection between error models and codes to correct errors in the context of

---

[50] See [99], for example, for precise definitions.

these models. Most of coding theory, so far, has focussed on noiseless channels (the language theory branch) and channels modelled by the Hamming metric (the error-correcting code branch).

The Hamming and Lee metrics are applicable to communication systems in which the only likely errors are substitutions and where errors are independent. The Levenshtein metric is applicable to physical SID-channels where errors are independent and equally likely. Not all SID-channels – physical or just within our formal model – have these properties.

For the abstraction from probabilities to distances, there have been two essentially different techniques. The first and most common one is based on an errors-per-code-word limit; the channel is assumed not to produce more than k, say, errors of a certain kind per code word. This is the standard approach in the classical theory of error-correcting codes; it is also the approach taken in [41] and [13]. The second technique, used in our presentation, is based on an errors-per-length limit; the channel is assumed not to produce more than k, say, errors of a certain kind in any l consecutive input symbols. The former is considerably easier to use;[51] the latter represents many physical situations more realistically. For example, assume a block code of length $l = 5$ is used and the channel permits one deletion; suppose $(01001)(10100)$ is sent. In the errors-per-code-word model, one could get $0100\text{Ⅰ Ⅰ}0100 = 01000100$ which is impossible in the errors-per-length model; on the other hand, one can get 01001010 in both models, as $0100\text{Ⅰ}101\text{Ⅰ}0$ in the former model and as $0100\text{Ⅰ}1010\text{Ⅰ}$ in both models. We believe that, for codes having only very long code words, these two models may coincide in probability. For codes with realistically short code words, they seem to be vastly different.

As mentioned before, solid codes seem to be extremely good at synchronizing in the presence of errors – any code word transmitted correctly will be decoded correctly. They are not good enough in general, however, to correct incorrectly transmitted parts of a message; in those parts, they detect that errors are present, but do not guide the correction.

For example, suppose the only errors that are reasonably likely to occur are deletions of symbols; these are independent; and it is, by any standards, completely unlikely that there will be more than 1 deletion in any l consecutive message symbols. The corresponding (abstract) channel is the SID-channel $\delta(1, l)$ which permits at most 1 deletion in any l consecutive symbols [59]. Let $\mathbb{M} = (\mathbb{N} \setminus \{1\}) \times (\mathbb{N} \setminus \{1\})$ and let $\mathbb{F}$ be the set

$$\mathbb{F} = \left\{ f_{x,y} \left| \begin{array}{l} x, y \in \mathbb{N} \setminus \{1\}, x \neq y, f_{x,y} : \mathbb{M} \to \mathbb{N}, \\ f_{x,y}(r_1, r_2) = xr_1 + yr_2 \text{ for } (r_1, r_2) \in \mathbb{M} \end{array} \right. \right\}.$$

For $f \in \mathbb{F}$ and $n \in \mathbb{N} \setminus \{1\}$, let

$$G_f(n) = \{r_1 \mid (r_1, r_2) \in \mathbb{M}, f(r_1, r_2) \leq n\}.$$

---

[51] See the admissibility problems in our definitions of synchronization and decodability, Definition 4.2 and Definition 4.4, which are a result of the fact that error situations do not know about the ends of code words.

Using properties explained in Section 11, one can construct codes for the channel $\delta(1,l)$ as follows.

Theorem 12.1 [57] *Let* $f \in \mathbb{F}$ *and* $g : \mathbb{M} \to \mathbb{N} \setminus \{1\}$ *such that, for all* $r = (r_1, r_2) \in \mathbb{M}$, $g(r) > \max G_f(r_2)$. *Let*

$$K = \{0^{f(r_1,r_2)} 1^{r_2} 0^{r_1} 1^{g(r_1,r_2)} \mid (r_1,r_2) \in \mathbb{M}\}.$$

Then K *is a solid code. If* K' *is a finite subset of* K *and* l *is the maximal length of a word in* K' *then* K' *is right* $\delta(1,l)$-*decodable.*

Example 12.1 The language

$$\{0^{2r_1+3r_2} 1^{r_2} 0^{r_1} 1^{\lfloor r_1/2 \rfloor + 1} \mid (r_1,r_2) \in \mathbb{M}\}$$

satisfies the conditions of Theorem 12.1.

In Example 4.1(3), a two-word code is given which is (γ, ζ)-correcting for $\gamma = \delta(1,5)$; That code is not a solid code.

The code words in error-correcting solid codes for SID-channels are quite long; this is undesirable, but could be inevitable. The lower bound of Theorem 11.5(3) does not take error-correction into account. The redundancy – as introduced in Section 11 – of the language in Example 12.1, assuming the words are ordered by increasing length, is asymptotically equal[52] to $\frac{1}{2}\sqrt{112n} - \log n$.

Most work on error-correcting codes for SID-channels has concentrated on block codes, often restricted to the cases of insertion and deletion errors – sometimes referred to as synchronization errors.[53] In [133], block codes of length m, with $m < n + \log_2 n + 4$, for encoding words of length n are constructed which are $(\gamma, *)$-correcting for the channel $\gamma = (\iota \odot \delta)(1, m + 2)$. Moreover, it is shown in [134] that, asymptotically, $n + \log_2 n + 1$ is a lower bound on the achievable length. The codes of [133] are not solid, in general. Non-block codes have been studied even less in the context of error-correction.[54]

Beyond the classical problems of coding theory, discussed in part in earlier sections of this chapter, the presence of noise raises several additional fundamental issues:

---

[52] To prove this, first note that the code words have lengths $3r_1 + 4r_2 + \lfloor \frac{r_1}{2} \rfloor + 1$ with $r_1, r_2 \in \mathbb{N} \setminus \{1\}$. Let $\nu(l)$ be the number of code words of length l, where $l \geq 16$. One shows, with a case distinction according to $l \bmod 4$, that $\frac{l}{14} - 1 \leq \nu(l) \leq \frac{l}{14} + 1$. The integer n is mapped onto a code word of length l if and only if $\sum_{i=16}^{l-1} \nu(i) \leq n < \sum_{i=16}^{l} \nu(i)$. Combining this, one obtains $\frac{-29 + \sqrt{112n+3481}}{2} < l < \frac{29 + \sqrt{112n+9}}{2}$. Thus, $\frac{1}{2}\sqrt{112n+3481} - \log n - \frac{31}{2} < \varrho(n) \leq \frac{1}{2}\sqrt{112n+9} - \log n + \frac{27}{2}$, hence $\varrho(n) \sim \frac{1}{2}\sqrt{112n} - \log n$.

[53] [122], [133], [141], [80], [79], [81], [83], [86], [112], [87], [10],[41], [5].

[54] [102], [45], [59], [57], [58], [70], [13], [41].

- For restricted classes of channels, like SID-channels, determine constructions of classes of codes which are error-correcting, decodable, uniformly synchronous, and efficient.[55]
- For SID-channels and in terms of the channel parameters, describe good upper and lower bounds on redundancy.[56]
- For SID-channels, determine which type of constraint – expressed as independence or in some other systematic formal manner – addresses which error-correction, decodability, or synchronizability problems.

This type of questions is unsolved even for substitution-only channels.[57] Moreover, it seems that these considerations will have to focus on classes of codes in the intersection $\mathcal{L}_i \cap \mathcal{L}_o \cap \mathcal{L}$, where $\mathcal{L}$ is a family of languages which is slightly larger than, but very similar to $\mathcal{L}_{ol\text{-}free}$.

SID-channels, while more realistic than substitution-only channels, are not likely to be adequate for modelling even just the noisy channels arising in present communication technology. We believe that the framework of P-channels is general enough to model many physically realizable channels. However, the following fundamental problems need answers:

- Good and simple channel models for existing communications technology need to be developed, and a programme has to be carried out for these, similar to the one described for SID-channels.[58]
- The model of P-channels, the channel model used in the classical theory of error-correcting codes, and the models assumed in [41] and [13] need to be unified.
- A method for abstracting relevant mathematical channel models from physical channels needs to be developed.[59]

This list of problems is by no means complete. Given the experience of coding theory so far, pursuing any one of them will be very difficult.

---

[55] Efficiency may be achieved through small redundancy or small complexity of encoding and decoding, for example.

[56] For block codes and substitution-only channels, the *Hamming bound* and the *Gilbert-Varshamov bound* are examples of such bounds; see [94].

[57] In the absence of synchronization errors, block codes – of course – are easier to implement; and, asymptotically, they are as efficient as non-block codes.

[58] As a nearly randomly selected example, we mention the problems arising in optical data transmission when *overlapping pulse-position modulation* is used for the physical representation of signals; see [11] for details.

[59] Of course, this question is ill-posed. However, even slight progress along these lines could be extremely important for our understanding of communication systems.

13. Concluding remarks

As mentioned before, the theory of codes has, essentially, three nearly unrelated branches. In writing this handbook chapter, while focussing on aspects of the theory of formal languages, we attempted to provide an overview which includes aspects of all three branches. As much of the language theoretic work on codes is well represented in book form, except, of course, some quite recent results, we decided to focus on general structural issues and on issues concerning codes for noisy channels. We attempted a unified presentation of the theory, hoping that we might succeed in exposing many unanswered questions, albeit sometimes implicitly.

Natural languages are quite fault-tolerant; they have enormous error-correction capabilities. In this spirit, we believe the notion of error-correction deserves a focal spot in the theory of formal languages.

References

[1] R. Ahlswede, I. Wegener: *Suchprobleme*. B. G. Teubner, Stuttgart, 1979.

[2] L. R. Bahl, F. Jelinek: Decoding for channels with insertions, deletions, and substitutions with applications to speech recognition. *IEEE Trans. Inform. Theory* **IT-21** (1975), 404–411.

[3] L. R. Bahl, F. Jelinek, R. L. Mercer: A maximum likelihood approach to continuous speech recognition. *IEEE Trans. Pattern Analysis and Machine Intell.* **5** (1983), 179–190.

[4] J. Berstel, D. Perrin: *Theory of Codes*. Academic Press, Orlando, 1985.

[5] P. A. H. Bours: Construction of fixed-length insertion/deletion correcting runlength-limited codes. *IEEE Trans. Inform. Theory* **IT-40** (1994), 1841–1856.

[6] V. Bruyère: Maximal prefix products. *Semigroup Forum* **36** (1987), 147–157.

[7] V. Bruyère: Maximal codes with bounded deciphering delay. *Theoret. Comput. Sci.* **84** (1991), 53–76.

[8] V. Bruyère, M. Latteux: Variable-length maximal codes. In F. Meyer auf der Heide, B. Monien (editors): *Automata, Languages and Programming, 23rd International Colloquium, ICALP '96, Paderborn, Germany, July 1996, Proceedings. Lecture Notes in Computer Science* **1099**, 24–47, Springer-Verlag, Berlin, 1996.

[9] V. Bruyère, L. Wang, L. Zhang: On completion of codes with finite deciphering delay. *European J. Combin.* **11** (1990), 513–521.

[10] L. Calabi, W. E. Hartnett: A family of codes for the correction of substitution and synchronization errors. *IEEE Trans. Inform. Theory* **IT-15** (1969), 102–106.

[11] A. R. Calderbank, C. N. Georghiades: Synchronizable codes for the optical OPPM channel. *IEEE Trans. Inform. Theory* **IT-40** (1994), 1097–1107.

[12] R. M. Capocelli: A decision procedure for finite decipherability and synchronizability of multivalued encodings. *IEEE Trans. Inform. Theory* **IT-28** (1982), 307–318.

[13] R. M. Capocelli, L. Gargano, U. Vaccaro: Decoders with initial state invariance for multivalued encodings. *Theoret. Comput. Sci.* **86** (1991), 365–375.

[14] R. M. Capocelli, C. M. Hoffmann: Algorithms for factorizing and testing subsemigroups. In A. Apostolico, Z. Galil (editors): *Combinatorial Algorithms on Words*. *NATO ASI Series* **F12**, 59–81, Springer-Verlag, Berlin, 1985.

[15] R. M. Capocelli, U. Vaccaro: Structure of decoders for multivalued encodings. *Discrete Appl. Math.* **23** (1989), 55–71.

[16] P. M. Cohn: *Universal Algebra*. D. Reidel Publishing Co., Dordrecht, revised ed., 1981.

[17] W. Damm: The IO- and OI-hierarchies. *Theoret. Comput. Sci.* **20** (1982), 95–206.

[18] P. H. Day, H. J. Shyr: Languages defined by some partial orders. *Soochow J. Math.* **9** (1983), 53–62.

[19] J. Devolder: Precircular codes and periodic biinfinite words. *Inform. and Comput.* **107** (1993), 185–201.

[20] J. Devolder, M. Latteux, I. Litovsky, L. Staiger: Codes and infinite words. *Acta Cybernet.* **11** (1994), 241–256.

[21] J. Devolder, E. Timmerman: Finitary codes for bi-infinite words. *RAIRO Inform. Théor. Appl.* **26** (1992), 363–386.

[22] J. Duske, H. Jürgensen: *Codierungstheorie*. BI Wissenschaftsverlag, Mannheim, 1977.

[23] A. Ehrenfeucht, G. Rozenberg: Each regular code is included in a maximal regular code. *RAIRO Inform. Théor. Appl.* **20** (1985), 89–96.

[24] Eilenberg: *Automata Languages and Machines, Volume A*. Academic Press, New York, 1974.

[25] J. Engelfriet: Iterated stack automata and complexity classes. *Inform. and Comput.* **95** (1991), 21–75.

[26] J. Engelfriet, H. Vogler: Pushdown machines for the macro tree transducer. *Theoret. Comput. Sci.* **42** (1986), 251–368.

[27] J. Engelfriet, H. Vogler: Look-ahead on pushdowns. *Inform. and Comput.* **73** (1987), 245–279.

[28] J. Engelfriet, H. Vogler: High level tree transducers and iterated pushdown tree transducers. *Acta Inform.* **26** (1988), 131–192.

[29] C. de Felice: Construction of a family of finite maximal codes. *Theoret. Comput. Sci.* **63** (1989), 157–184.

[30] C. de Felice, A. Restivo: Some results on finite maximal codes. *RAIRO Inform. Théor. Appl.* **19** (1985), 383–403.

[31] F. Gécseg, H. Jürgensen: Dependence in algebras. *Fund. Inform.* To appear.

[32] F. Gécseg, H. Jürgensen: Algebras with dimension. *Algebra Universalis* **30** (1993), 422–446.

[33] Y. V. Glebskii: Coding by means of finite automata. *Dokl. Akad. Nauk. SSSR* **141** (1961), 1054–1057, in Russsian. English translation: *Soviet Physics Dokl.* **6** (1962), 1037–1039.

[34] G. Grätzer: *Universal Algebra*. Van Nostrand, Princeton, NJ, 1968.

[35] S. A. Greibach: A remark on code sets and context-free languages. *IEEE Trans. Comput.* **C-24** (1975), 741–742.

[36] S. Guiaşu: *Information Theory with Applications*. McGraw-Hill, London, 1977.

[37] C. G. Günther: A universal algorithm for homophonic coding. In C. G. Günther (editor): *Advances in Cryptology—Proceedings of Eurocrypt'88, Workshop on the Theory and Application of Cryptographic Techniques,*

Davos, 1988. *Lecture Notes in Computer Science* **330**, 405–414, Springer-Verlag, Berlin, 1988.

[38] Y. Q. Guo, H. J. Shyr, G. Thierrin: e-convex infix codes. *Order* **3** (1986), 55–59.

[39] Y. Q. Guo, G. Thierrin, S. H. Zhang: Semaphore codes and ideals. *J. Inform. Optim. Sci.* **9**(1) (1988), 73–83.

[40] W. E. Hartnett: Generalization of tests for certain properties of variable-length codes. *Inform. and Control* **13** (1968), 20–45.

[41] W. E. Hartnett (editor): *Foundations of Coding Theory*. Boston, 1974. D. Reidel Publishing Co.

[42] T. Head, G. Thierrin: Hypercodes in deterministic and slender OL languages. *Inform. and Control* **45**(3) (1980), 251–262.

[43] T. Head, A. Weber: Deciding code related properties by means of finite transducers. In R. Capocelli, A. de Santis, U. Vaccaro (editors): *Sequences II, Methods in Communication, Security, and Computer Science*. 260–272, Springer-Verlag, Berlin, 1993.

[44] C. M. Hoffmann: A note on unique decipherability. In M. P. Chytil, V. Koubek (editors): *Mathematical Foundations of Computer Science 1984; Proceedings, 11th Symposium; Praha, Czechoslovakia; September 3–7, 1984. Lecture Notes in Computer Science* **176**, 50–63, Springer-Verlag, Berlin, 1984.

[45] H. D. L. Hollman: A relation between Levenshtein-type distances and insertion-and-deletion correcting capabilities of codes. *IEEE Trans. Inform. Theory* **IT-39** (1993), 1424–1427.

[46] O. H. Ibarra: Reversal-bounded multicounter machines and their decision problems. *J. Assoc. Comput. Mach.* **25** (1978), 116–133.

[47] M. Ito, H. Jürgensen: Shuffle relations. Manuscript, 1996. In preparation.

[48] M. Ito, H. Jürgensen, H. J. Shyr, G. Thierrin: Anti-commutative languages and n-codes. *Discrete Appl. Math.* **24** (1989), 187–196.

[49] M. Ito, H. Jürgensen, H. J. Shyr, G. Thierrin: n-Prefix-suffix languages. *Internat. J. Comput. Math.* **30** (1989), 37–56.

[50] M. Ito, H. Jürgensen, H. J. Shyr, G. Thierrin: Outfix and infix codes and related classes of languages. *J. Comput. System Sci.* **43** (1991), 484–508.

[51] M. Ito, H. Jürgensen, H. J. Shyr, G. Thierrin: Languages whose n-element subsets are codes. *Theoret. Comput. Sci.* **96** (1992), 325–344.

[52] M. Ito, G. Thierrin: Congruences, infix and cohesive prefix codes. *Theoret. Comput. Sci.* **136** (1994), 471–485.

[53] H. N. Jendal, Y. J. B. Kuhn, J. L. Massey: An information-theoretic treatment of homophonic substitution. In J.-J. Quisquater, J. Vandewalle (editors): *Advances in Cryptology—Proceedings of Eurocrypt'89, Workshop on the Theory and Application of Cryptographic Techniques, Houthalen, 1989. Lecture Notes in Computer Science* **434**, 382–394, Springer-Verlag, Berlin, 1989.

[54] H. Jürgensen: Syntactic monoids of codes. Report 327, Department of Computer Science, The University of Western Ontario, 1992.

[55] H. Jürgensen, M. Katsura, S. Konstantinidis: Maximal solid codes. Manuscript, 1996. In Preparation.

[56] H. Jürgensen, S. Konstantinidis: The hierarchy of codes. In Z. Ésik (editor): *Fundamentals of Computation Theory, 9th International Conference, FCT'93. Lecture Notes in Computer Science* **710**, 50–68, Springer-Verlag, Berlin, 1993.

[57] H. Jürgensen, S. Konstantinidis: Variable-length codes for error correction. In Z. Fülöp, F. Gécseg (editors): *Automata, Languages and Programming, 22nd International Colloquium, ICALP95, Proceedings*. *Lecture Notes in Computer Science* **944**, 581–592, Springer-Verlag, Berlin, 1995.

[58] H. Jürgensen, S. Konstantinidis: Burst error correction for channels with substitutions, insertions and deletions. Manuscript, 1996. In preparation.

[59] H. Jürgensen, S. Konstantinidis: Error correction for channels with substitutions, insertions, and deletions. In J.-Y. Chouinard, P. Fortier, T. A. Gulliver (editors): *Information Theory and Applications 2, Fourth Canadian Workshop on Information Theory. Lecture Notes in Computer Science* **1133**, 149–163, Springer-Verlag, Berlin, 1996.

[60] H. Jürgensen, L. Robbins: Towards foundations of cryptography: Investigation of perfect secrecy. *J. UCS* **2** (1996), 347–379. Special issue: C. Calude (ed.), *The Finite, the Unbounded and the Infinite, Proceedings of the Summer School "Chaitin Complexity and Applications,"* Mangalia, Romania, 27 June – 6 July, 1995.

[61] H. Jürgensen, K. Salomaa, S. Yu: Decidability of the intercode property. *J. Inform. Process. Cybernet., EIK* **29** (1993), 375–380.

[62] H. Jürgensen, K. Salomaa, S. Yu: Transducers and the decidability of independence in free monoids. *Theoret. Comput. Sci.* **134** (1994), 107–117.

[63] H. Jürgensen, G. Thierrin: Infix codes. In M. Arató, I. Kátai, L. Varga (editors): *Topics in the Theoretical Bases and Applications of Computer Science, Proceedings of the 4th Hungarian Computer Science Conference, Györ, 1985.* 25–29, Akadémiai Kiadó, Budapest, 1986.

[64] H. Jürgensen, S. S. Yu: Solid codes. *J. Inform. Process. Cybernet., EIK* **26** (1990), 563–574.

[65] H. Jürgensen, S. S. Yu: Relations on free monoids, their independent sets, and codes. *Internat. J. Comput. Math.* **40** (1991), 17–46.

[66] H. Jürgensen, S. S. Yu: Dependence systems and hierarchies of families of languages. Manuscript, 1996. In preparation.

[67] D. Kahn: *The Codebreakers*. Macmillan Publishing Co., New York, 1967.

[68] J. Karhumäki: On three-element codes. In J. Paredaens (editor): *Automata, Languages and Programming, 11th International Colloquium, ICALP 1984, Proceedings. Lecture Notes in Computer Science* **172**, 292–302, Springer-Verlag, Berlin, 1984.

[69] J. Karhumäki: On three-element codes. *Theoret. Comput. Sci.* **40** (1985), 3–11.

[70] S. Konstantinidis: *Error Correction and Decodability*. Ph. D. thesis, The University of Western Ontario, London, Canada, 1996.

[71] G. Lallement: *Semigroups and Combinatorial Applications*. John Wiley & Sons, Inc., New York, 1979.

[72] N. H. Lâm, D. L. Van: On a class of infinitary codes. *RAIRO Inform. Théor. Appl.* **24** (1990), 441–458.

[73] J. L. Lassez: Circular codes and synchronization. *Internat. J. Comput. Inform. Sci.* **5** (1976), 201–208.

[74] V. I. Levenshtein: Certain properties of code systems. *Dokl. Akad. Nauk. SSSR* **140** (1961), 1274–1277, in Russian. English translation: *Soviet Physics Dokl.* **6** (1962), 858–860.

[75] V. I. Levenshtein: Self-adaptive automata for decoding messages. *Dokl. Akad. Nauk. SSSR* **141** (1961), 1320–1323, in Russian. English translation: *Soviet Physics Dokl.* **6** (1962), 1042–1045.

[76] V. I. Levenshtein: The inversion of finite automata. *Dokl. Akad. Nauk. SSSR* **147** (1962), 1300–1303, in Russian. English translation: *Soviet Physics Dokl.* **7** (1963), 1081–1084.

[77] V. I. Levenshtein: Decoding automata, invariant with respect to the initial state. *Problemy Kibernet.* **12** (1964), 125–136, in Russian.

[78] V. I. Levenshtein: Some properties of coding and self-adjusting automata for decoding messages. *Problemy Kibernet.* **11** (1964), 63–121, in Russian. German translation: Über einige Eigenschaften von Codierungen und von selbstkorrigierenden Automaten zur Decodierung von Nachrichten, *Probleme der Kybernetik* **7** (1966), 96–163. An English translation is available from the Clearinghouse for Federal Scientific and Technical Information, U. S. Department of Commerce, under the title *Problems of Cybernetics, Part II*, document AD 667 849; it was prepared as document FTD-MT-24-126-67 by the Foreign Technology Division, U. S. Air Force.

[79] V. I. Levenshtein: Binary codes capable of correcting deletions, insertions, and reversals. *Dokl. Akad. Nauk. SSSR* **163** (1965), 845–848, in Russian. English translation: *Soviet Physics Dokl.* **10** (1966), 707–710.

[80] V. I. Levenshtein: Binary codes capable of correcting spurious insertions and deletions of ones. *Problemy Peredachi Informatsii* **1**(1) (1965), 12–25, in Russian. English translation: *Problems Inform. Transmission* **1**(1) (1966), 8–17.

[81] V. I. Levenshtein: Asymptotically optimum binary code with correction for losses of one or two adjacent bits. *Problemy Kibernet.* **19** (1967), 293–298, in Russian. English translation: *Systems Theory Research* **19** (1970), 298–304.

[82] V. I. Levenshtein: On the redundancy and delay of decodable coding of natural numbers. *Problemy Kibernet.* **20** (1968), 173–179, in Russian. English translation: *Systems Theory Research* **20** (1971), 149–155.

[83] V. I. Levenshtein: Bounds for codes ensuring error correction and synchronization. *Problemy Peredachi Informatsii* **5**(2) (1969), 3–13, in Russian. English translation: *Problems Inform. Transmission* **5**(2) (1969), 1–10.

[84] V. I. Levenshtein: Maximum number of words in codes without overlaps. *Problemy Peredachi Informatsii* **6**(4) (1970), 88–90, in Russian. English translation: *Problems Inform. Transmission* **6**(4) (1973), 355–357.

[85] V. I. Levenshtein: One method of constructing quasilinear codes providing synchronization in the presence of errors. *Problemy Peredachi Informatsii* **7**(3) (1971), 30–40, in Russian. English translation: *Problems Inform. Transmission* **7**(3) (1973), 215–222.

[86] V. I. Levenshtein: On perfect codes in deletion and insertion metric. *Diskret. Mat.* **3** (1991), 3–20, in Russian. English translation: *Discrete Math. Appl.* **2** (1992), 241–258.

[87] J. E. Levy: Self-synchronizing codes derived from binary cyclic codes. *IEEE Trans. Inform. Theory* **IT-12** (1966), 286–290.

[88] B. E. Litow: Parallel complexity of the regular code problem. *Inform. and Comput.* **86** (1990), 107–114.

[89] D. Y. Long: *k*-Outfix codes. *Chinese Ann. Math. Ser. A* **10** (1989), 94–99, in Chinese.

[90] D. Y. Long: k-Prefix codes and k-infix codes. *Acta Math. Sinica* **33** (1990), 414–421, in Chinese.

[91] D. Y. Long: n-Infix-outfix codes. In *Abstracts, Second International Colloquium on Words, Languages, and Combinatorics, Kyoto, 25–28 August, 1992.* 50–51, Kyoto, 1992.

[92] D. Y. Long: On the structure of some group codes. *Semigroup Forum* **45** (1992), 38–44.

[93] D. Y. Long: k-Bifix codes. *Riv. Mat. Pura Appl.* **15** (1994), 33–55.

[94] F. J. MacWilliams, N. J. A. Sloane: *The Theory of Error-Correcting Codes.* North-Holland, Amsterdam, 1977, 2 vols.

[95] A. A. Markov: Some properties of infinite prefix codes. *Problemy Peredachi Informatsii* **6**(1) (1970), 97–98, in Russian. English translation: *Problems Inform. Transmission* **6**(1) (1973), 85–87.

[96] P. G. Neumann: Codes auf der Grundlage von Schaltfunktionen und ihre Anwendung in der Praxis der Verschlüsselung. *Nachrichtentechn. Z.* **14** (1961), 254–261, 307–312.

[97] P. G. Neumann: Efficient error-limiting variable-length codes. *IEEE Trans. Inform. Theory* **IT-8** (1962), 292–304.

[98] P. G. Neumann: Error-limiting coding using information-lossless sequential machines. *IEEE Trans. Inform. Theory* **IT-10** (1964), 108–115.

[99] W. W. Peterson, E. J. Weldon, Jr.: *Error-Correcting Codes.* MIT Press, Cambridge, MA, second ed., 1972.

[100] M. Petrich: *Lectures in Semigroups.* Akademie-Verlag, Berlin, 1977.

[101] M. Petrich, G. Thierrin: The syntactic monoid of an infix code. *Proc. Amer. Math. Soc.* **109** (1990), 865–873.

[102] P. Piret: Comma free error correcting codes of variable length, generated by finite-state encoders. *IEEE Trans. Inform. Theory* **IT-28** (1982), 764–775.

[103] T. Pratt, W. C. Bostian: *Satellite Communications.* John Wiley & Sons, New York, 1986.

[104] H. Prodinger, G. Thierrin: Towards a general concept of hypercodes. *J. Inform. Optim. Sci.* **4** (1983), 255–268.

[105] C. M. Reis: Intercodes and the semigroups they generate. *Internat. J. Comput. Math.* **51** (1994), 7–13.

[106] C. M. Reis, G. Thierrin: Reflective star languages and codes. *Inform. and Control* **42** (1979), 1–9.

[107] A. Restivo: A combinatorial property of codes having finite synchronization delay. *Theoret. Comput. Sci.* **1** (1975), 95–101.

[108] A. Restivo: On codes having no finite completions. *Discrete Math.* **17** (1977), 309–316.

[109] A. Restivo, S. Salemi, T. Sportelli: Completing codes. *RAIRO Inform. Théor. Appl.* **23** (1989), 135–147.

[110] A. Riley: The Sardinas-Patterson and Levenshtein theorems. *Inform. and Control* **10** (1967), 120–136.

[111] O. T. Romanov: Invariant decoding automata without look-ahead. *Problemy Kibernet.* **17** (1966), 233–236, in Russian.

[112] R. M. Roth, P. H. Siegel: Lee-metric BCH codes and their application to constrained and partial-response channels. *IEEE Trans. Inform. Theory* **IT-40** (1994), 1083–1096.

[113] J. Sakarovitch: Un cadre algébrique pour l'étude des monoïdes syntactiques. In *Séminaire P. Dubreil (Algèbre), 28e année*. 14. Paris, 1974/75.

[114] K. Salomaa. Personal communication, 1996.

[115] A. A. Sardinas, C. W. Patterson: A necessary and sufficient condition for the unique decomposition of coded messages. *IRE Intern. Conven. Rec.* **8** (1953), 104–108.

[116] K. Sato: Decipherability of GSM encoding. *Denshi Tsushin Gakkai Ronbunshi* **57-D** (1974), 181–188, in Japanese. English translation: *Systems-Computers-Controls* **5** (1974), 53–61.

[117] K. Sato: A decision procedure for the unique decipherability of multivalued encodings. *IEEE Trans. Inform. Theory* **IT-25** (1979), 356–360.

[118] M. Satyanarayana: Uniformly synchronous codes. *Semigroup Forum* **46** (1993), 246–252.

[119] M. Satyanarayana, S. Mohanty: Limited semaphore codes. *Semigroup Forum* **45** (1992), 367–371.

[120] M. Satyanarayana, S. Mohanty: Uniformly synchronous limited codes. *Semigroup Forum* **46** (1993), 21–26.

[121] B. M. Schein: Homomorphisms and subdirect decompositions of semigroups. *Pacific J. Math.* **17** (1966), 529–547.

[122] F. F. Sellers, Jr.: Bit loss and gain correction code. *IRE Trans. Inform. Theory* **IT-8** (1962), 35–38.

[123] H. J. Shyr: *Free Monoids and Languages*. Hon Min Book Company, Taichung, second ed., 1991.

[124] H. J. Shyr, G. Thierrin: Hypercodes. *Inform. and Control* **24** (1974), 45–54.

[125] H. J. Shyr, S. S. Yu: Intercodes and some related properties. *Soochow J. Math.* **16** (1990), 95–107.

[126] H. J. Shyr, S. S. Yu: Solid codes and disjunctive domains. *Semigroup Forum* **41** (1990), 23–37.

[127] J. C. Spehner: Quelques constructions et algorithmes relatifs aux sous-monoïdes d'un monoïde libre. *Semigroup Forum* **9** (1975), 334–353.

[128] L. Staiger: On infinitary finite length codes. *RAIRO Inform. Théor. Appl.* **20** (1986), 483–494.

[129] G. Tenengol'ts: Nonbinary codes, correcting single deletion or insertion. *IEEE Trans. Inform. Theory* **IT-30** (1984), 766–769.

[130] G. Thierrin: Hypercodes, right convex languages and their syntactic monoids. *Proc. Amer. Math. Soc.* **83**(2) (1981), 255–258.

[131] G. Thierrin: The syntactic monoid of a hypercode. *Semigroup Forum* **6** (1973), 227–231.

[132] G. Thierrin, S. S. Yu: Shuffle relations and codes. *J. Inform. and Optim. Sci.* **12** (1991), 441–449.

[133] J. D. Ullman: Near-optimal, single-synchronization-error-correcting code. *IEEE Trans. Inform. Theory* **IT-12** (1966), 418–424.

[134] J. D. Ullman: On the capabilities of codes to correct synchronization errors. *IEEE Trans. Inform. Theory* **IT-13** (1967), 95–105.

[135] E. Valkema: Syntaktische Monoide und Hypercodes. *Semigroup Forum* **13** (1976/77), 119–126.

[136] D. L. Van: Codes avec des mots infinis. *RAIRO Inform. Théor. Appl.* **16** (1982), 371–386.

[137] D. L. Van: Sous-monoïdes et codes avec des mots infinis. *Semigroup Forum* **26** (1983), 75–87.

[138] D. L. Van: Ensembles code-compatibles et une généralisation du théorème de Sardinas–Patterson. *Theoret. Comput. Sci.* **38** (1985), 123–132.

[139] D. L. Van: Langages écrits par un code infinitaire. Théorème du défaut. *Acta Cybernet.* **7** (1986), 247–257.

[140] D. L. Van, D. G. Thomas, K. G. Subramanian: Bi-infinitary codes. *RAIRO Inform. Théor. Appl.* **24** (1990), 67–87.

[141] R. R. Varshamov, G. M. Tenengol'ts: Codes capable of correcting single asymmetric errors. *Avtomat. i Telemekh.* **26** (1965), 288–292, in Russian.

[142] M. Vincent: Construction de codes indecomposables. *RAIRO Inform. Théor. Appl.* **19** (1985), 165–178.

[143] D. J. A. Welsh: *Matroid Theory*. Academic Press, London, 1976.

[144] S. S. Yu: A characterization of intercodes. *Internat. J. Comput. Math.* **36** (1990), 39–45.

[145] S. Yu. Personal communication, 1995.

[146] S. Zhang: An equivalence relation on suffix codes defined by generalized regular languages. *Internat. J. Comput. Math.* **35** (1990), 15–24.

Semirings and Formal Power Series: Their Relevance to Formal Languages and Automata

Werner Kuich

1. Introduction

The purpose of Chapter 9 is to develop some classical results on formal languages and automata by an algebraic treatment using semirings, formal power series and matrices. The use of semirings, formal power series and matrices makes the proofs computational in nature and, consequently, more satisfactory from the mathematical point of view than the customary proofs.

Many proofs of language and automata theory ask for the construction of devices with some specified properties. Often, the construction is easily understood by intuitive reasoning while the proof that the construction has the specified properties is inadequate from a mathematical point of view. We try to separate the intuitive idea of the construction from the proof that the construction works. Most of the constructions used in this survey are analogous to the usual constructions given in language and automata theory. But most of our proofs are different from the usual proofs and use tools from semiring theory. In spite of the generality of our results, many proofs are much shorter than the corresponding customary proofs.

This survey should definitely not be considered as a first introduction to language and automata theory. We assume some "mathematical maturity" and some basic knowledge in algebra, formal languages and automata on the part of the reader.

Most of our theorems are proved in full length. Missing proofs can be found in a reference as indicated. Often we will give a reference to a text in language or automata theory to encourage the knowledgeable reader to compare our descriptions of the constructions and their proofs with the customary ones.

The results presented in Chapter 9 cover the basics of semirings and formal power series, and include generalized versions of the Theorem of Kleene, the equivalence of context-free grammars and pushdown automata, the basic theory of abstract families of languages, the Theorem of Parikh, and Lindenmayer systems.

The theory of formal power series is capable of establishing specific language-theoretic results which are difficult if not impossible to establish by other means. In the last section we present some of these results, indicate the mathematical tools needed for their proofs and give references. Rigorous mathematical proofs of these results lie beyond the scope of this survey. Additionally, we refer to papers applying semirings or formal power series to various problems in mathematics and theoretical computer science.

2. Semirings, formal power series and matrices

In our theory, ω-continuous semirings, formal power series and matrices play a central role. So we state the respective definitions and prove some basic theorems.

A *monoid* consists of a set M, an associative binary operation $\circ$ on M and of a neutral element 1 such that $1 \circ a = a \circ 1 = a$ for every a. A monoid is called *commutative* iff $a \circ b = b \circ a$ for every a and b. The binary operation is usually denoted by juxtaposition and often called *product*.

If the operation and the neutral element of M are understood then we denote the monoid simply by M. Otherwise we use the triple notation $\langle M, \circ, 1 \rangle$.

The most important type of a monoid in our considerations is the *free monoid* Σ^* generated by a nonempty set Σ. It has all the finite *strings*, also referred to as *words*,

$$x_1 \ldots x_n, \quad x_i \in \Sigma,$$

as its elements and the product $w_1 \cdot w_2$ is formed by writing the string w_2 immediately after the string w_1. The neutral element of Σ^*, also referred to as the *empty word*, is denoted by ε.

The members of Σ are called *letters* or *symbols*. The set Σ itself is called an *alphabet*. In case of a finite Σ subsets of Σ^* are called *(formal) languages over* Σ.

By a *semiring* we mean a set A together with two binary operations $+$ and $\cdot$ and two constant elements 0 and 1 such that

(i) $\langle A, +, 0 \rangle$ is a commutative monoid,
(ii) $\langle A, \cdot, 1 \rangle$ is a monoid,
(iii) the distribution laws $a \cdot (b + c) = a \cdot b + a \cdot c$ and $(a + b) \cdot c = a \cdot c + b \cdot c$ hold for every a, b, c,
(iv) $0 \cdot a = a \cdot 0 = 0$ for every a.

A semiring is called *commutative* iff $a \cdot b = b \cdot a$ for every a and b.

If the operations and the constant elements of A are understood then we denote the semiring simply by A. Otherwise, we use the notation $\langle A, +, \cdot, 0, 1 \rangle$.

Intuitively, a semiring is a ring (with unity) without subtraction. A typical example is the semiring of nonnegative integers $\mathbb{N}$. A very important semiring in connection with language theory is the *Boolean* semiring $\mathbb{B} = \{0, 1\}$ where $1 + 1 = 1 \cdot 1 = 1$. Clearly, all rings (with unity), as well as all fields, are semirings, e. g., integers $\mathbb{Z}$, rationals $\mathbb{Q}$, reals $\mathbb{R}$, complex numbers $\mathbb{C}$ etc.

Let $\mathbb{N}^\infty = \mathbb{N} \cup \{\infty\}$. Then $\langle \mathbb{N}^\infty, +, \cdot, 0, 1 \rangle$ and $\langle \mathbb{N}^\infty, \min, +, \infty, 0 \rangle$, where $+$, $\cdot$ and min are defined in the obvious fashion (observe that $0 \cdot \infty = \infty \cdot 0 = 0$), are semirings. The first one is usually denoted by $\mathbb{N}^\infty$, the second one is called the *tropical* semiring.

Let $\mathbb{R}_+ = \{a \in \mathbb{R} \mid a \geq 0\}$ and $\mathbb{R}_+^\infty = \mathbb{R}_+ \cup \{\infty\}$. Then $\langle \mathbb{R}_+, +, \cdot, 0, 1 \rangle$ and $\langle \mathbb{R}_+^\infty, +, \cdot, 0, 1 \rangle$ are semirings.

Let Σ be an alphabet and define, for $L_1, L_2 \subseteq \Sigma^*$, the *product* of L_1 and L_2 by

$$L_1 \cdot L_2 = \{w_1 w_2 \mid w_1 \in L_1,\ w_2 \in L_2\}.$$

Then $\langle \mathfrak{P}(\Sigma^*), \cup, \cdot, \emptyset, \{\varepsilon\} \rangle$ is a semiring. In case Σ is finite it is called the *semiring of formal languages over* Σ. Here $\mathfrak{P}(S)$ denotes the power set of a set S and $\emptyset$ denotes the empty set.

If S is a set, $\mathfrak{P}(S \times S)$ is the set of binary relations over S. Define, for two relations R_1 and R_2, the product $R_1 \cdot R_2 \subseteq S \times S$ by

$$R_1 \cdot R_2 = \{(s_1, s_2) \mid \text{there exists an } s \in S \text{ such that}$$
$$(s_1, s) \in R_1 \text{ and } (s, s_2) \in R_2\}$$

and, furthermore, define

$$\Delta = \{(s, s) \mid s \in S\}.$$

Then $\langle \mathfrak{P}(S \times S), \cup, \cdot, \emptyset, \Delta \rangle$ is a semiring, called the *semiring of binary relations over* S.

We now give some basic definitions and results on semirings. A semiring $\langle A, +, \cdot, 0, 1 \rangle$ is *naturally ordered* iff the set A is partially ordered by the relation $\sqsubseteq$: $a \sqsubseteq b$ iff there exists a c such that $a + c = b$. This partial order on A is called *natural order*.

Theorem 2.1. *Let A be a naturally ordered semiring. Then the following conditions are satisfied for all $a, b, c \in A$:*

(i) $0 \sqsubseteq a$,

(ii) *if $a \sqsubseteq b$ then $a + c \sqsubseteq b + c$, $ac \sqsubseteq bc$, $ca \sqsubseteq cb$.*

This means that 0 is the least element of A, and addition and multiplication in A are monotone mappings.

A semiring A is called *complete* iff it is possible to define sums for all families $(a_i \mid i \in I)$ of elements of A, where I is an arbitrary index set, such that the following conditions are satisfied (see Eilenberg [25], Mahr [104], Goldstern [43], Krob [75], Weinert [142], Hebisch [52], Karner [72]):

(i) $\displaystyle\sum_{i \in \emptyset} a_i = 0, \quad \sum_{i \in \{j\}} a_i = a_j, \quad \sum_{i \in \{j,k\}} a_i = a_j + a_k$ for $j \neq k$,

(ii) $\displaystyle\sum_{j \in J} \Big(\sum_{i \in I_j} a_i \Big) = \sum_{i \in I} a_i$, if $\bigcup_{j \in J} I_j = I$ and $I_j \cap I_{j'} = \emptyset$ for $j \neq j'$,

(iii) $\displaystyle\sum_{i\in I}(c\cdot a_i) = c\cdot\left(\sum_{i\in I}a_i\right), \quad \sum_{i\in I}(a_i\cdot c) = \left(\sum_{i\in I}a_i\right)\cdot c.$

This means that a semiring A is complete if it is possible to define "infinite sums" (i) that are an extension of the finite sums, (ii) that are associative and commutative and (iii) that satisfy the distribution laws.

In complete semirings we can define, for each element a, the *star* a^* of a and the *quasi-inverse* a^+ of a by

$$a^* = \sum_{j\geq 0}a^j \quad \text{and} \quad a^+ = \sum_{j\geq 1}a^j.$$

Theorem 2.2. *Let A be a complete semiring. Then, for all $a \in A$,*

(i) $a^+ = aa^* = a^*a,$

(ii) $a^* = 1 + a^+,$

(iii) $a^* = \sum_{0\leq j\leq n}a^j + a^{n+1}a^* = \sum_{0\leq j\leq n}a^j + a^*a^{n+1}.$

Example 2.1. The *length* of a word w over an alphabet Σ, in symbols $|w|$, is defined to be the number of letters of Σ occurring in w, whereby each letter is counted as many times as it occurs in w. By definition, the length of the empty word equals 0. Due to these definitions, Σ^n is the set of all words over Σ of length n. Again, $\Sigma^* = \bigcup_{n\geq 0}\Sigma^n$ is the set of all words over Σ. Moreover, $\Sigma^+ = \bigcup_{n\geq 1}\Sigma^n$ is the set of all nonempty words over Σ.

Given a language $L \subseteq \Sigma^*$, we obtain

$$L^* = \bigcup_{n\geq 0}L^n \quad \text{and} \quad L^+ = \bigcup_{n\geq 1}L^n.$$

If we consider the semiring $\mathfrak{P}(S\times S)$ of binary relations over a set S, and a binary relation $R \subseteq S\times S$, then $R^* = \bigcup_{n\geq 0}R^n$ is the *reflexive and transitive closure of R*, i. e., the smallest reflexive and transitive binary relation over S containing R. Similarly, $R^+ = \bigcup_{n\geq 1}R^n$ is the *transitive closure of R*.   □

A semiring A is called ω-*continuous* iff the following conditions are satisfied:

(i) A is complete,

(ii) A is naturally ordered,

(iii) if $\sum_{0\leq i\leq n}a_i \sqsubseteq c$ for all $n \in \mathbb{N}$ then $\sum_{i\in\mathbb{N}}a_i \sqsubseteq c$ for all sequences $(a_i \mid i \in \mathbb{N})$ in A and all $c \in A$.

In the next theorem, "sup" denotes the least upper bound with respect to the natural order.

Theorem 2.3. (Karner [69]) *Let A be a complete, naturally ordered semiring. Then the following statements are equivalent for all $a_i, b_i, c \in A$ and all countable index sets I:*

$$r = \sum_{w \in M} (r, w)w.$$

The values (r, w) are also referred to as the *coefficients* of the series. The collection of all power series r as defined above is denoted by $S\langle\langle M \rangle\rangle$.

This terminology reflects the intuitive ideas connected with power series. We call the power series "formal" to indicate that we are not interested in summing up the series but rather, for instance, in various operations defined for series. The power series notation makes it very convenient to discuss such operations in case S has enough structure, for instance if S is an ω-continuous semiring A.

Given $r \in A\langle\langle M \rangle\rangle$, the subset of M defined by

$$\{w \mid (r, w) \neq 0\}$$

is termed the *support* of r and denoted by supp(r). The subset of $A\langle\langle M \rangle\rangle$ consisting of all series with a finite support is denoted by $A\langle M \rangle$. Series of $A\langle M \rangle$ are referred to as *polynomials*.

Examples of polynomials belonging to $A\langle M \rangle$ for every A are 0, w, aw, $a \in A$, $w \in M$, defined by:

$(0, w) = 0$ for all w,
$(w, w) = 1$ and $(w, w') = 0$ for $w \neq w'$,
$(aw, w) = a$ and $(aw, w') = 0$ for $w \neq w'$.

Note that w equals $1w$.

We now introduce two operations inducing a semiring structure to power series. For $r_1, r_2 \in A\langle\langle M \rangle\rangle$, we define the *sum* $r_1 + r_2 \in A\langle\langle M \rangle\rangle$ by $(r_1 + r_2, w) = (r_1, w) + (r_2, w)$ for all $w \in M$. For $r_1, r_2 \in A\langle\langle M \rangle\rangle$, we define the *(Cauchy) product* $r_1 r_2 \in A\langle\langle M \rangle\rangle$ by $(r_1 r_2, w) = \sum_{w_1 \circ w_2 = w} (r_1, w_1)(r_2, w_2)$ for all $w \in M$. Clearly, $\langle A\langle\langle M \rangle\rangle, +, \cdot, 0, e \rangle$ and $\langle A\langle M \rangle, +, \cdot, 0, e \rangle$ are semirings.

Moreover, $A\langle\langle M \rangle\rangle$ is an ω-continuous semiring. Sums in $A\langle\langle M \rangle\rangle$ are defined by $(\sum_{j \in J} r_j, w) = \sum_{j \in J} (r_j, w)$ for all $w \in M$ and all index sets J.

For $a \in A$, $r \in A\langle\langle M \rangle\rangle$, we define the *scalar products* $ar, ra \in A\langle\langle M \rangle\rangle$ by $(ar, w) = a(r, w)$ and $(ra, w) = (r, w)a$ for all $w \in M$. Observe that $ar = (ae)r$ and $ra = r(ae)$. If A is commutative then $ar = ra$.

A series $r \in A\langle\langle M \rangle\rangle$, where every coefficient equals 0 or 1, is termed the *characteristic series* of its support L, in symbols, $r = \text{char}(L)$.

The *Hadamard product* of two power series r_1 and r_2 belonging to $A\langle\langle M \rangle\rangle$ is defined by

$$r_1 \odot r_2 = \sum_{w \in M} (r_1, w)(r_2, w)w.$$

In Sections 3–7, we will only consider power series in $A\langle\langle \Sigma^* \rangle\rangle$, where A is an ω-continuous semiring and Σ is an alphabet. It will be convenient to use the notations $A\langle \Sigma \cup \varepsilon \rangle$, $A\langle \Sigma \rangle$ and $A\langle \varepsilon \rangle$ for the collection of polynomials having their supports in $\Sigma \cup \{\varepsilon\}$, Σ and $\{\varepsilon\}$, respectively. Power series in

$A\langle\langle M\rangle\rangle$, where M is a monoid distinct from Σ^*, will only be considered in Sections 8, 9 and 10.

The following notational conventions are valid in the sequel: Σ, possibly provided with indices, will denote a finite alphabet and the letter I (resp. Q), possibly provided with indices, will denote a nonempty (resp. nonempty finite) index set.

Clearly, $\mathfrak{P}(\Sigma^*)$ is a semiring isomorphic to $\mathbb{B}\langle\langle\Sigma^*\rangle\rangle$. Essentially, a transition from $\mathfrak{P}(\Sigma^*)$ to $\mathbb{B}\langle\langle\Sigma^*\rangle\rangle$ and vice versa means a transition from L to $\mathrm{char}(L)$ and from r to $\mathrm{supp}(r)$, respectively. The operation corresponding to the Hadamard product is the intersection of languages. If r_1 and r_2 are the characteristic series of the languages L_1 and L_2 then $r_1 \odot r_2$ is the characteristic series of $L_1 \cap L_2$.

The next example gives an application of Theorem 2.4 (iii).

Example 2.3. Let $r = a\varepsilon + bx \in \mathbb{R}_+^\infty\langle\{x\}^*\rangle$, $a < 1$. Then $r^* = (a^*bx)^*a^* = \sum_{n\geq 0} b^n x^n/(1-a)^{n+1}$. □

We now introduce matrices. Consider two nonempty index sets I and I' and a set S. Mappings M of $I \times I'$ into S are called *matrices*. The values of M are denoted by $M_{i,i'}$, where $i \in I$ and $i' \in I'$. The values $M_{i,i'}$ are also referred to as the *entries* of the matrix M. In particular, $M_{i,i'}$ is called the (i,i')-*entry* of M. The collection of all matrices as defined above is denoted by $S^{I \times I'}$.

If I or I' is a singleton, M is called a *row* or *column vector*, respectively. If both I and I' are finite, then M is called a *finite matrix*.

We introduce some operations and special matrices inducing a monoid or semiring structure to matrices. For $M_1, M_2 \in A^{I \times I'}$ we define the *sum* $M_1 + M_2 \in A^{I \times I'}$ by $(M_1 + M_2)_{i,i'} = (M_1)_{i,i'} + (M_2)_{i,i'}$ for all $i \in I$, $i' \in I'$. Furthermore, we introduce the *zero matrix* $0 \in A^{I \times I'}$. All entries of the zero matrix 0 are 0. By these definitions, $\langle A^{I \times I'}, +, 0\rangle$ is a commutative monoid.

For $M_1 \in A^{I_1 \times I_2}$ and $M_2 \in A^{I_2 \times I_3}$ we define the *product* $M_1 M_2 \in A^{I_1 \times I_3}$ by

$$(M_1 M_2)_{i_1,i_3} = \sum_{i_2 \in I_2} (M_1)_{i_1,i_2}(M_2)_{i_2,i_3} \quad \text{for all } i_1 \in I_1,\, i_3 \in I_3.$$

Furthermore, we introduce the *matrix of unity* $E \in A^{I \times I}$. The diagonal entries $E_{i,i}$ of E are equal to 1, the off-diagonal entries E_{i_1,i_2}, $i_1 \neq i_2$, of E are equal to 0, $i, i_1, i_2 \in I$.

It is easily shown that matrix multiplication is associative, the distribution laws are valid for matrix addition and multiplication, E is a multiplicative unit and 0 is a multiplicative zero. So we infer that $\langle A^{I \times I}, +, \cdot, 0, E\rangle$ is a semiring.

Infinite sums can be extended to matrices. Consider $A^{I \times I'}$ and define, for $M_j \in A^{I \times I'}$, $j \in J$, where J is an index set, $\sum_{j \in J} M_j$ by its entries:

$$\left(\sum_{j\in J} M_j\right)_{i,i'} = \sum_{j\in J} (M_j)_{i,i'}, \quad i \in I, \, i' \in I'.$$

By this definition, $A^{I \times I}$ is an ω-continuous semiring.

Moreover, the natural order on A is extended entrywise to matrices M_1 and M_2 in $A^{I \times I'}$:

$$M_1 \sqsubseteq M_2 \quad \text{iff} \quad (M_1)_{i,i'} \sqsubseteq (M_2)_{i,i'} \text{ for all } i \in I, \, i' \in I'.$$

We now introduce blocks of matrices. Consider a matrix M in $A^{I \times I}$. Assume the existence of a nonempty index set J and of nonempty index sets I_j for $j \in J$ such that $I = \bigcup_{j \in J} I_j$ and $I_{j_1} \cap I_{j_2} = \emptyset$ for $j_1 \neq j_2$. The mapping M, restricted to the domain $I_{j_1} \times I_{j_2}$, i. e., $M : I_{j_1} \times I_{j_2} \to A$ is, of course, a matrix in $A^{I_{j_1} \times I_{j_2}}$. We denote it by $M(I_{j_1}, I_{j_2})$ and call it the (I_{j_1}, I_{j_2})-*block* of M.

We can compute the blocks of the sum and the product of matrices M_1 and M_2 from the blocks of M_1 and M_2 in the usual way:

$$
\begin{aligned}
(M_1 + M_2)(I_{j_1}, I_{j_2}) &= M_1(I_{j_1}, I_{j_2}) + M_2(I_{j_1}, I_{j_2}), \\
(M_1 M_2)(I_{j_1}, I_{j_2}) &= \sum_{j \in J} M_1(I_{j_1}, I_j) M_2(I_j, I_{j_2}).
\end{aligned}
$$

In a similar manner the matrices of $A^{I \times I'}$ can be partitioned into blocks. This yields the computational rule

$$(M_1 + M_2)(I_j, I'_{j'}) = M_1(I_j, I'_{j'}) + M_2(I_j, I'_{j'}).$$

If we consider matrices $M_1 \in A^{I \times I'}$ and $M_2 \in A^{I' \times I''}$ partitioned into compatible blocks, i. e., I' is partitioned into the same index sets for both matrices, then we obtain the computational rule

$$(M_1 M_2)(I_j, I''_{j''}) = \sum_{j' \in J'} M_1(I_j, I'_{j'}) M_2(I'_{j'}, I''_{j''}).$$

In the sequel we will need the following isomorphisms:

(i) The semirings

$$\left(A^{Q \times Q}\right)^{I \times I}, \quad A^{(I \times Q) \times (I \times Q)}, \quad A^{(Q \times I) \times (Q \times I)}, \quad \left(A^{I \times I}\right)^{Q \times Q}$$

are isomorphic by the correspondences between

$$\left(M_{i_1,i_2}\right)_{q_1,q_2}, \quad M_{(i_1,q_1),(i_2,q_2)}, \quad M_{(q_1,i_1),(q_2,i_2)}, \quad \left(M_{q_1,q_2}\right)_{i_1,i_2}$$

for all $i_1, i_2 \in I$, $q_1, q_2 \in Q$.

(ii) The semirings $A^{I \times I}\langle\!\langle \Sigma^* \rangle\!\rangle$ and $\left(A\langle\!\langle \Sigma^* \rangle\!\rangle\right)^{I \times I}$ are isomorphic by the correspondence between $(M, w)_{i_1,i_2}$ and (M_{i_1,i_2}, w) for all $i_1, i_2 \in I$, $w \in \Sigma^*$.

Observe that these correspondences are isomorphisms of complete semirings, i. e., they respect infinite sums. We will use these isomorphisms without further mention. Moreover, we will use the notation M_{i_1,i_2}, $i_1 \in I_1$, $i_2 \in I_2$, where $M \in A^{I_1 \times I_2}\langle\!\langle \Sigma^* \rangle\!\rangle$: M_{i_1,i_2} is the power series in $A\langle\!\langle \Sigma^* \rangle\!\rangle$ such that the coefficient (M_{i_1,i_2}, w) of $w \in \Sigma^*$ is equal to $(M, w)_{i_1,i_2}$. Similarly, we will use the notation (M, w), $w \in \Sigma^*$, where $M \in (A\langle\!\langle \Sigma^* \rangle\!\rangle)^{I_1 \times I_2}$: (M, w) is the matrix in $A^{I_1 \times I_2}$ whose (i_1, i_2)-entry $(M, w)_{i_1,i_2}$, $i_1 \in I_1$, $i_2 \in I_2$, is equal to (M_{i_1,i_2}, w).

The next theorem is central for automata theory (see Conway [21], Lehmann [97], Kuich, Salomaa [92], Kuich [83], Kozen [78], Bloom, Ésik [12]). It allows to compute the blocks of the star of a matrix M by sum, product and star of the blocks of M. For notational convenience, we will denote $M(I_i, I_j)$ by $M_{i,j}$, $1 \leq i, j \leq 3$.

Theorem 2.5. Let $M \in A^{I \times I}$ and $I = I_1 \cup I_2$, $I_1 \cap I_2 = \emptyset$. Then

$$M^*(I_1, I_1) = (M_{1,1} + M_{1,2}M_{2,2}^*M_{2,1})^*,$$
$$M^*(I_1, I_2) = (M_{1,1} + M_{1,2}M_{2,2}^*M_{2,1})^*M_{1,2}M_{2,2}^*,$$
$$M^*(I_2, I_1) = (M_{2,2} + M_{2,1}M_{1,1}^*M_{1,2})^*M_{2,1}M_{1,1}^*,$$
$$M^*(I_2, I_2) = (M_{2,2} + M_{2,1}M_{1,1}^*M_{1,2})^*.$$

Proof. Consider the matrices

$$M_1 = \begin{pmatrix} M_{1,1} & 0 \\ 0 & M_{2,2} \end{pmatrix} \quad \text{and} \quad M_2 = \begin{pmatrix} 0 & M_{1,2} \\ M_{2,1} & 0 \end{pmatrix}.$$

The computation of $(M_1 + M_2M_1^*M_2)^*(E + M_2M_1^*)$ and application of Theorem 2.4 (iv) prove our theorem. $\square$

Corollary 2.6. If $M_{2,1} = 0$ then

$$M^* = \begin{pmatrix} M_{1,1}^* & M_{1,1}^*M_{1,2}M_{2,2}^* \\ 0 & M_{2,2}^* \end{pmatrix}.$$

Corollary 2.7. If $M_{2,1} = 0$, $M_{3,1} = 0$ and $M_{3,2} = 0$ then

$$M^* = \begin{pmatrix} M_{1,1}^* & M_{1,1}^*M_{1,2}M_{2,2}^* & M_{1,1}^*M_{1,2}M_{2,2}^*M_{2,3}M_{3,3}^* + M_{1,1}^*M_{1,3}M_{3,3}^* \\ 0 & M_{2,2}^* & M_{2,2}^*M_{2,3}M_{3,3}^* \\ 0 & 0 & M_{3,3}^* \end{pmatrix}$$

In the next theorem, I is partitioned into I_j, $j \in J$, and j_0 is a distinguished element in J.

Theorem 2.8. *Assume that the only non-null blocks of the matrix $M \in A^{I \times I}$ are $M(I_j, I_{j_0})$, $M(I_{j_0}, I_j)$ and $M(I_j, I_j)$, for all $j \in J$ and a fixed $j_0 \in J$. Then*

$$M^*(I_{j_0}, I_{j_0}) = \left(M(I_{j_0}, I_{j_0}) + \sum_{j \in J, \, j \neq j_0} M(I_{j_0}, I_j)M(I_j, I_j)^*M(I_j, I_{j_0}) \right)^*.$$

Proof. We partition I into I_{j_0} and $I' = I - I_{j_0}$. Then $M(I', I')$ is a block-diagonal matrix and $(M(I', I')^*)(I_j, I_j) = M(I_j, I_j)^*$ for all $j \in J - \{j_0\}$. By Theorem 2.5 we obtain

$$M^*(I_{j_0}, I_{j_0}) = \big(M(I_{j_0}, I_{j_0}) + M(I_{j_0}, I')M(I', I')^* M(I', I_{j_0})\big)^*.$$

The computation of the right side of this equality proves our theorem.   □

3. Algebraic systems

In this section we introduce semiring-polynomials and consider algebraic systems as a generalization of the context-free grammars. Our development of the theory concerning algebraic systems parallels that of Eilenberg [26]. In order to have a sound theoretical basis for solving algebraic systems we give an introduction into the theory of ω-complete partially ordered sets and their Fixpoint Theorem.

In the sequel, $Y = \{y_1, \ldots, y_n\}$ is a finite set of *variables*. We denote by $A(Y)$ the *polynomial semiring over the semiring A in the set of variables Y* (see Lausch, Nöbauer [96], Chapter 1.4). To distinguish the polynomials in $A(Y)$ from the polynomials in $A\langle\langle \Sigma^* \rangle\rangle$, we call them *semiring-polynomials*.

Each semiring-polynomial has a representation as follows. A *product term* t has the form

$$t(y_1, \ldots, y_n) = a_0 y_{i_1} a_1 \ldots a_{k-1} y_{i_k} a_k, \quad k \geq 0,$$

where $a_j \in A$ and $y_{i_j} \in Y$. The elements a_j are referred to as *coefficients* of the product term. Observe that for $k = 0$ we have $t(y_1, \ldots, y_n) = a_0$. If $k \geq 1$, we do not write down coefficients that are equal to 1; e. g., $y_1 y_2$ stands for $1 \cdot y_1 \cdot 1 \cdot y_2 \cdot 1$. Each semiring-polynomial p has a *representation* as a finite sum of product terms t_j, i. e.,

$$p(y_1, \ldots, y_n) = \sum_{1 \leq j \leq m} t_j(y_1, \ldots, y_n).$$

The coefficients of all the product terms t_j, $1 \leq j \leq m$, are referred to as the *coefficients* of the semiring-polynomial p. For a nonempty subset A' of A we denote the collection of all semiring-polynomials with coefficients in A' by $A'(Y)$.

If the basic semiring is given by $A\langle\langle \Sigma^* \rangle\rangle$ then $A\langle (\Sigma \cup Y)^* \rangle$, the set of polynomials over $\Sigma \cup Y$, can be regarded as a subset of the set $A'(Y)$ of semiring-polynomials, where $A' = \{aw \mid a \in A, w \in \Sigma^*\}$.

We are not interested in the algebraic properties of $A(Y)$, but only in the mappings induced by semiring-polynomials. These mappings are called *polynomial functions on A* (see Lausch, Nöbauer [96], Chapter 1.6).

Each product term t (resp. semiring-polynomial p) with variables $y_1, \ldots$ $\ldots, y_n$ induces a mapping $\bar{t}$ (resp. $\bar{p}$) from A^n into A. For a product term t represented as above, the mapping $\bar{t}$ is defined by

$$\bar{t}(\sigma_1, \ldots, \sigma_n) = a_0 \sigma_{i_1} a_1 \ldots a_{k-1} \sigma_{i_k} a_k ;$$

for a semiring polynomial p, represented by a finite sum of product terms t_j as above, the mapping $\bar{p}$ is defined by

$$\bar{p}(\sigma_1, \ldots, \sigma_n) = \sum_{1 \leq j \leq m} \bar{t}_j(\sigma_1, \ldots, \sigma_n)$$

for all $(\sigma_1, \ldots, \sigma_n) \in A^n$.

In the sequel we denote the mapping $\bar{p}$ induced by p also by p. This should not lead to any confusion.

We are now ready to define the basic notions concerning algebraic systems. Let A' be a nonempty subset of A. An A'-*algebraic system* (with variables in Y) is a system of equations

$$y_i = p_i, \quad 1 \leq i \leq n,$$

where each p_i is a semiring-polynomial in $A'(Y)$. A *solution* to the A'-algebraic system $y_i = p_i$, $1 \leq i \leq n$, is given by $(\sigma_1, \ldots, \sigma_n) \in A^n$ such that

$$\sigma_i = p_i(\sigma_1, \ldots, \sigma_n), \quad 1 \leq i \leq n.$$

A solution $(\sigma_1, \ldots, \sigma_n)$ of the A'-algebraic system $y_i = p_i$, $1 \leq i \leq n$, is termed a *least solution* iff

$$\sigma_i \sqsubseteq \tau_i, \quad 1 \leq i \leq n,$$

for all solutions $(\tau_1, \ldots, \tau_n)$ of $y_i = p_i$, $1 \leq i \leq n$.

Often it is convenient to write the A'-algebraic system $y_i = p_i$, $1 \leq i \leq n$, in matrix notation. Defining the two column vectors

$$y = \begin{pmatrix} y_1 \\ \vdots \\ y_n \end{pmatrix} \quad \text{and} \quad p = \begin{pmatrix} p_1 \\ \vdots \\ p_n \end{pmatrix},$$

we can write our A'-algebraic system in the matrix notation

$$y = p(y) \quad \text{or} \quad y = p.$$

A *solution* to $y = p(y)$ is now given by $\sigma \in A^n$ such that $\sigma = p(\sigma)$. A solution σ of $y = p$ is termed a *least solution* iff $\sigma \sqsubseteq \tau$ for all solutions τ of $y = p$.

One of our main results in this section will be that an A'-algebraic system has a unique least solution. This will be shown by the theory of ω-complete partially ordered sets and their Fixpoint Theorem (see Wechler [141], Section 1.5).

Let S be a set partially ordered by $\leq$. A sequence $(s_i \mid i \in \mathbb{N})$ of elements of S is called ω-*chain* iff $s_i \leq s_{i+1}$ for all $i \in \mathbb{N}$. The partially ordered set S is called ω-*complete* iff

(i) S has a least element $\bot$,
(ii) every ω-chain has a least upper bound.

We denote the least upper bound by "sup".

Let S_1 and S_2 be ω-complete partially ordered sets. A mapping $f : S_1 \to S_2$ is called ω-*continuous* iff, for every ω-chain $(s_i \mid i \in \mathbb{N})$ of elements of S_1, the least upper bound of $(f(s_i) \mid i \in \mathbb{N})$ exists and

$$f(\sup\,(s_i \mid i \in \mathbb{N})) = \sup\,(f(s_i) \mid i \in \mathbb{N}).$$

If the mapping f is ω-continuous then it is also monotone. Hence, $(f(s_i) \mid i \in \mathbb{N})$ is an ω-chain again. Observe that the functional composition of ω-continuous mappings is again ω-continuous.

Let S be partially ordered and $f : S \to S$. Then $s \in S$ is a *fixpoint of* f iff $f(s) = s$. A fixpoint $s \in S$ is called *least fixpoint of* f iff $s \leq s'$ for all fixpoints s' of f. If the least fixpoint of f exists, it is unique. We denote it by $\text{fix}(f)$.

Theorem 3.1 (Fixpoint Theorem). *Let S be an ω-complete partially ordered set and $f : S \to S$ be an ω-continuous mapping. Then $(f^i(\bot) \mid i \in \mathbb{N})$ is an ω-chain, $\sup\,(f^i(\bot) \mid i \in \mathbb{N})$ exists and*

$$\text{fix}(f) = \sup\,(f^i(\bot) \mid i \in \mathbb{N}).$$

(Here $\bot$ is the least element of S and f^i, $i \in \mathbb{N}$, is the i-th iterate of f.)

We now apply the theory of ω-complete partially ordered sets to ω-continuous semirings.

Theorem 3.2. *Let $\langle A, +, \cdot, 0, 1 \rangle$ be an ω-continuous semiring. Then A is an ω-complete partially ordered set and each ω-chain $(a_i \mid i \in \mathbb{N})$ is of the form*

$$\Big(\sum_{0 \leq j \leq i} b_j \mid i \in \mathbb{N} \Big).$$

Moreover,

$$\sup(a_i \mid i \in \mathbb{N}) = \sum_{i \in \mathbb{N}} b_i.$$

Proof. Let $(a_i \mid i \in \mathbb{N})$ be an ω-chain in A. Since $a_i \sqsubseteq a_{i+1}$ for all $i \in \mathbb{N}$, there exist $b_i \in A$ such that

$$a_0 = b_0, \quad a_{i+1} = a_i + b_{i+1}, \quad i \in \mathbb{N}.$$

This implies $a_i = \sum_{0 \leq j \leq i} b_j$ for all $i \in \mathbb{N}$. Hence, by Theorem 2.3 (iv),

$$\sup\,(a_i \mid i \in \mathbb{N}) = \sup\,\Big(\sum_{0 \leq j \leq i} b_j \mid i \in \mathbb{N} \Big) = \sum_{i \in \mathbb{N}} b_i.$$

By Theorem 2.3 (vii), the choice of b_i, $i \in \mathbb{N}$, is irrelevant. $\square$

An easy application of Theorem 3.2 yields the next result.

Theorem 3.3. *Let A be an ω-continuous semiring. Then addition and multiplication are ω-continuous mappings from $A \times A$ into A.*

Corollary 3.4. *Let A be an ω-continuous semiring and let p be a semiring-polynomial in $A(Y)$. Then the mapping $p : A^n \to A$ is ω-continuous.*

Let now $p \in A(Y)^{n \times 1}$, i. e., p is a column vector of semiring-polynomials. Then p induces a mapping $p : A^n \to A^n$ by $(p(a_1, \ldots, a_n))_i = p_i(a_1, \ldots, a_n)$, $1 \leq i \leq n$, i. e., the i-th component of the value of p at $(a_1, \ldots, a_n) \in A^n$ is given by the value of the i-th component p_i of p at $(a_1, \ldots, a_n)$.

The next corollary follows by the observation that the least upper bound of a sequence of vectors can be taken componentwise.

Corollary 3.5. *Let A be an ω-continuous semiring and let $p \in A(Y)^{n \times 1}$. Then the mapping $p : A^n \to A^n$ is ω-continuous.*

Consider now an A'-algebraic system $y = p$. The least fixpoint of the mapping p is nothing else than the least solution of $y = p$.

Theorem 3.6. *Let A be an ω-continuous semiring and A' be a nonempty subset of A. Then the least solution of an A'-algebraic system $y = p$ exists in A^n and equals*

$$\mathrm{fix}(p) = \sup(p^i(0) \mid i \in \mathbb{N}).$$

Theorem 3.6 indicates how we can compute an approximation to the least solution of an A'-algebraic system $y = p$. The *approximation sequence* $\sigma^0, \sigma^1, \sigma^2, \ldots, \sigma^j, \ldots$, where each $\sigma^j \in A^{n \times 1}$, *associated to* an A'-algebraic system $y = p(y)$ is defined as follows:

$$\sigma^0 = 0, \quad \sigma^{j+1} = p(\sigma^j), \ j \in \mathbb{N}.$$

Clearly, $(\sigma^j \mid j \in \mathbb{N})$ is an ω-chain and $\mathrm{fix}(p) = \sup(\sigma^j \mid j \in \mathbb{N})$, i. e., we obtain the least solution of $y = p$ by computing the least upper bound of the approximation sequence associated to it.

The collection of the components of the least solutions of all A'-algebraic systems, where A' is a fixed subset of A, is denoted by $\mathfrak{Alg}(A')$. In the sequel, A' denotes always a subset of A containing 0 and 1. But observe that most of the definitions and some of the results involving a subset A' of A are valid without this restriction as well, i. e., are valid for arbitrary subsets A' of A.

We are now ready to discuss the connection between algebraic systems and context-free grammars.

Consider a context-free grammar $G = (Y, \Sigma, P, y_1)$. Here $Y = \{y_1, \ldots, y_n\}$ is the set of variables or nonterminal symbols, Σ is the set of terminal symbols, P is the set of productions and y_1 is the initial variable. The language generated by G is denoted by $L(G)$. Changing the initial variable yields the context-free grammars $G_i = (Y, \Sigma, P, y_i)$, $1 \leq i \leq n$, and the context-free languages $L(G_i)$ generated by them. Clearly, $L(G) = L(G_1)$. We now assume that the basic semiring is given by $\mathfrak{P}(\Sigma^*)$. We define a $\{\{w\} \mid w \in \Sigma^*\}$-algebraic system $y_i = p_i$, $1 \leq i \leq n$, whose least solution is $(L(G_1), \ldots, L(G_n))$:

$$p_i = \bigcup_{y_i \to \gamma \in P} \{\gamma\}.$$

Whenever we speak of a context-free grammar corresponding to a $\{\{w\} \mid w \in \Sigma^*\}$-algebraic system, or vice versa, then we mean the correspondence in the sense of the above definition. The next theorem is due to Ginsburg, Rice [38]. (See also Salomaa, Soittola [123], Theorem IV.1.2 and Moll, Arbib, Kfoury [107], Chapter 6.)

Theorem 3.7. (Ginsburg, Rice [38], Theorem 2) *Assume that $G = (Y, \Sigma, P, y_1)$ is a context-free grammar and $y_i = p_i$, $1 \leq i \leq n$, is the corresponding $\{\{w\} \mid w \in \Sigma^*\}$-algebraic system with least solution $(\sigma_1, \ldots, \sigma_n)$. Let $G_i = (Y, \Sigma, P, y_i)$, $1 \leq i \leq n$. Then*

$$\sigma_i = L(G_i), \quad 1 \leq i \leq n.$$

Corollary 3.8. *A formal language over Σ is context-free iff it is in $\mathfrak{Alg}(\{\{w\} \mid w \in \Sigma^*\})$.*

We now consider the case where the basic semiring is given by $A\langle\langle \Sigma^* \rangle\rangle$, and A is a commutative ω-continuous semiring. Let $A' = \{aw \mid a \in A, w \in \Sigma^*\}$. Then $\mathfrak{Alg}(A')$ is equal to the collection of the components of the least solutions of A'-algebraic systems $y_i = p_i$, $1 \leq i \leq n$, where p_i is a polynomial in $A\langle\langle (\Sigma \cup Y)^* \rangle\rangle$. This is due to the commutativity of A: any polynomial function on $A\langle\langle \Sigma^* \rangle\rangle$ that is induced by a semiring-polynomial of $A'(Y)$ is also induced by a polynomial of $A\langle\langle (\Sigma \cup Y)^* \rangle\rangle$. In this case, $\mathfrak{Alg}(A')$ is usually denoted by $A^{\mathrm{alg}}\langle\langle \Sigma^* \rangle\rangle$. The power series in $A^{\mathrm{alg}}\langle\langle \Sigma^* \rangle\rangle$ are called *algebraic power series*. Whenever we speak of an *algebraic system* $y_i = p_i$, $p_i \in A\langle\langle (\Sigma \cup Y)^* \rangle\rangle$, $1 \leq i \leq n$, in connection with the basic semiring $A\langle\langle \Sigma^* \rangle\rangle$, then we assume that A is *commutative* and mean an A'-algebraic system as described above.

We generalize the connection between algebraic systems and context-free grammars as discussed above. Define, for a given context-free grammar $G = (Y, \Sigma, P, y_1)$, the algebraic system $y_i = p_i$, $p_i \in A\langle\langle (\Sigma \cup Y)^* \rangle\rangle$, $1 \leq i \leq n$, by

$$(p_i, \gamma) = 1 \text{ if } y_i \to \gamma \in P \quad \text{and} \quad (p_i, \gamma) = 0, \text{ otherwise.}$$

Conversely, given an algebraic system $y_i = p_i$, $p_i \in A\langle\langle (\Sigma \cup Y)^* \rangle\rangle$, $1 \leq i \leq n$, define the context-free grammar $G = (Y, \Sigma, P, y_1)$ by

$$y_i \to \gamma \in P \quad \text{iff} \quad (p_i, \gamma) \neq 0.$$

Whenever we speak of a context-free grammar corresponding to an algebraic system $y_i = p_i$, $p_i \in A\langle\langle (\Sigma \cup Y)^* \rangle\rangle$, $1 \leq i \leq n$, or vice versa, then we mean the correspondence in the sense of the above definition. If attention is restricted to algebraic systems with coefficients 0 and 1 then this correspondence is one-to-one. The correspondence between context-free grammars and algebraic systems $y_i = p_i$, $p_i \in A\langle(\Sigma \cup Y)^*\rangle$, $1 \le i \le n$, is a generalization of the correspondence between context-free grammars and $\{\{w\} \mid w \in \Sigma^*\}$-algebraic systems defined earlier. This is seen by taking in account the isomorphism between the semirings $\mathfrak{P}(\Sigma^*)$ and $\mathbb{B}\langle\langle\Sigma^*\rangle\rangle$. The next theorem is due to Chomsky, Schützenberger [20].

Theorem 3.9. (Salomaa, Soittola [123], Theorem IV.1.5) *Assume that $G = (Y, \Sigma, P, y_1)$ is a context-free grammar and $y_i = p_i$, $p_i \in \mathbb{N}^\infty\langle(\Sigma \cup Y)^*\rangle$, $1 \le i \le n$, is the corresponding algebraic system with least solution $(\sigma_1, \ldots, \sigma_n)$. Denote by $d_i(w)$ the number (possibly ∞) of distinct leftmost derivations of w from the variable y_i, $1 \le i \le n$, $w \in \Sigma^*$. Then*

$$\sigma_i = \sum_{w \in \Sigma^*} d_i(w)w, \quad 1 \le i \le n.$$

Corollary 3.10. *Under the assumptions of Theorem 3.9, G is unambiguous iff, for all $w \in \Sigma^*$,*

$$(\sigma_1, w) \le 1$$

Example 3.1. (See Chomsky, Schützenberger [20], Kuich [80].) Consider the context-free grammar $G = (\{y\}, \{x\}, \{y \to y^2, y \to x\}, y)$. If the basic semiring is $\mathfrak{P}(\{x\}^*)$, the corresponding algebraic system is given by $y = y^2 \cup \{x\}$. The j-th element of the approximation sequence is $\{x^{2^{j-1}}, x^{2^{j-1}-1}, \ldots, x\}$, $j \ge 1$. Hence, $\{x\}^+$ is the least solution of $y = y^2 \cup \{x\}$. Observe that $\{x\}^*$ is also a solution.

If the basic semiring is $\mathbb{N}^\infty\langle\langle\{x\}^*\rangle\rangle$, the corresponding algebraic system is given by $y = y^2 + x$. The first elements of the approximation sequence are $\sigma^0 = 0$, $\sigma^1 = x$, $\sigma^2 = x^2 + x$, $\sigma^3 = x^4 + 2x^3 + x^2 + x$. It can be shown that

$$\sum_{n \ge 0} C_n x^{n+1}, \quad \text{where } C_n = \frac{(2n)!}{n!(n+1)!}, \ n \ge 0,$$

is the least solution of $y = y^2 + x$. This means that x^{n+1} has C_n distinct leftmost derivations with respect to G.   □

In an ω-continuous semiring A, the three operations $+$, $\cdot$, * are called the *rational operations*. A subsemiring of A is called *fully rationally closed* iff it is closed under the rational operations.

Theorem 3.11. $\langle\mathfrak{Alg}(A'), +, \cdot, 0, 1\rangle$ *is a fully rationally closed semiring.*

Proof. Let a and a' be in $\mathfrak{Alg}(A')$. Then there exist A'-algebraic systems $y_i = p_i$, $1 \le i \le n$, and $y'_j = p'_j$, $1 \le j \le m$, such that a and a', respectively, are the first components of their least solutions. Assume that the sets of variables are disjoint and consider the A'-algebraic systems

(i) $\begin{aligned} y_0 &= y_1 + y_1' \\ y_i &= p_i \\ y_j' &= p_j' \end{aligned}$ (ii) $\begin{aligned} y_0 &= y_1 y_1' \\ y_i &= p_i \\ y_j' &= p_j' \end{aligned}$ (iii) $\begin{aligned} y_0 &= y_1 y_0 + 1 \\ y_i &= p_i \end{aligned}$

where $1 \leq i \leq n$, $1 \leq j \leq m$. Let σ and σ' be the least solutions of $y_i = p_i$, $1 \leq i \leq n$, and $y_j' = p_j'$, $1 \leq j \leq m$, respectively. Then we claim that $(a+a', \sigma, \sigma')$, (aa', σ, σ') and (a^*, σ) are the least solutions of the A'-algebraic systems (i), (ii) and (iii), respectively.

We prove only the third case. We have

$$(y_1 y_0 + 1)(a^*, a, \sigma_2, \ldots, \sigma_n) = aa^* + 1 = a^*,$$
$$p_i(a^*, a, \sigma_2, \ldots, \sigma_n) = p_i(\sigma_1, \ldots, \sigma_n) = \sigma_i, \quad 1 \leq i \leq n.$$

Hence, (a^*, σ) is solution of (iii). Assume that (b, τ) is the least solution. This means $b \sqsubseteq a^*$ and $\tau \sqsubseteq \sigma$. Since $p_i(b, \tau_1, \ldots, \tau_n) = p_i(\tau_1, \ldots, \tau_n) = \tau_i$, $1 \leq i \leq n$, τ is solution of $y_i = p_i$, $1 \leq i \leq n$. Since σ is the least solution of $y_i = p_i$, $1 \leq i \leq n$, this implies, together with $\tau \sqsubseteq \sigma$, the equality $\tau = \sigma$. Since $\tau_1 = \sigma_1 = a$, the substitution of (b, τ) into the first equation of (iii) yields $ab + 1 = b$. Theorem 2.4 (i) implies now $a^* \sqsubseteq b$. Hence $b = a^*$ and $(b, \tau) = (a^*, \sigma)$ is the least solution of (iii).

Summarizing, we have shown that $a + a'$, aa' and a^* are in $\mathfrak{Alg}(A')$.   □

Corollary 3.12. *The family of context-free languages over Σ is closed under the rational operations.*

According to our convention, the ω-continuous semiring A is commutative in the next corollary.

Corollary 3.13. $\langle A^{\mathrm{alg}} \langle\!\langle \Sigma^* \rangle\!\rangle, +, \cdot, 0, \varepsilon \rangle$ *is a fully rationally closed semiring.*

Our last result in this section shows that $\mathfrak{Alg}$ is an idempotent operator (Berstel [4], Wechler [141]).

Theorem 3.14. $\mathfrak{Alg}(\mathfrak{Alg}(A')) = \mathfrak{Alg}(A')$.

Proof. Let $y_i = p_i$, $1 \leq i \leq n$, be an $\mathfrak{Alg}(A')$-algebraic system with least solution σ. Consider the coefficients a of the semiring-polynomials p_i, $1 \leq i \leq n$, where $a \in \mathfrak{Alg}(A')$ and $a \notin A'$. For each of these coefficients a there exists an A'-algebraic system $z_j^a = q_j^a$ with least solution τ^a whose first component is equal to a. Perform now the following procedure on the $\mathfrak{Alg}(A')$-algebraic system $y_i = p_i$, $1 \leq i \leq n$: each coefficient a, $a \in \mathfrak{Alg}(A')$, $a \notin A'$, in p_i is replaced by the variable z_1^a and the equations $z_j^a = q_j^a$ are added to the system for all these a.

The newly constructed system is then an A'-algebraic system whose least solution is given by σ and all the τ^a. Hence, the components of σ are in $\mathfrak{Alg}(A')$.   □

4. Automata and linear systems

In this section we begin the development of automata theory and prove the Theorem of Kleene. The well-known finite automata are generalized in the following two directions:

(i) An infinite set of states will be allowed in the general definition. When dealing with pushdown automata in Section 6 this will enable us to store the contents of the pushdown tape in the states.

(ii) A single state transition generates an element of the basic semiring A instead of reading a symbol. Thus an automaton generates an element of the basic semiring A.

Our model of an automaton will be defined in terms of a (possibly infinite) transition matrix. The semiring element generated by the transition of the automaton from one state i to another state i' in exactly k computation steps equals the (i, i')-entry in the k-th power of the transition matrix. Consider now the star of the transition matrix. Then the semiring element generated by the automaton, also called the behavior of the automaton, can be expressed by the entries (multiplied by the initial and final weights of the states) of the star of the transition matrix.

An A'-automaton

$$\mathfrak{A} = (I, M, S, P)$$

is given by

(i)   a nonempty set I of *states*,
(ii) a matrix $M \in A'^{I \times I}$, called the *transition matrix*,
(iii) $S \in A'^{1 \times I}$, called the *initial state vector*,
(iv) $P \in A'^{I \times 1}$, called the *final state vector*.

The *behavior* $\|\mathfrak{A}\| \in A$ of the A'-automaton $\mathfrak{A}$ is defined by

$$\|\mathfrak{A}\| = \sum_{i_1, i_2 \in I} S_{i_1} (M^*)_{i_1, i_2} P_{i_2} = SM^*P.$$

An A'-automaton is termed *finite* iff its state set is finite.

Usually, an automaton is depicted as a *directed graph*. The *nodes* of the graph correspond to the states of the automaton. A node corresponding to a state i with $S_i \neq 0$ (resp. $P_i \neq 0$) is called *initial* (resp. *final*). The *edges* (i, j) of the graph correspond to the transitions unequal to 0 and are labeled by $M_{i,j}$.

Consider the semiring $\mathbb{B}$. Then, for an arbitrary $\mathbb{B}$-automaton $\mathfrak{A}$, we obtain $\|\mathfrak{A}\| = 1$ iff there is a path in the graph from some initial node to some final node.

Let now $\mathbb{N}^\infty$ be the basic semiring and let $\mathfrak{A}$ be a $\{0, 1\}$-automaton. Then $\|\mathfrak{A}\|$ is equal to the number (including ∞) of distinct paths in the graph from the initial nodes to the final nodes.

Assume that the basic semiring is the tropical semiring and consider an $\{\infty, 1, 0\}$-automaton $\mathfrak{A} = (I, M, S, P)$ such that the entries of M are in $\{\infty, 1\}$, and the entries of S and P are in $\{\infty, 0\}$. (Observe that a node i is initial or final if $S_i = 0$ or $P_i = 0$, respectively.) Then $\|\mathfrak{A}\|$ is equal to the length of the shortest path in the graph from some initial node to some final node. There is no such path iff $\|\mathfrak{A}\| = \infty$. (See Carré [17].)

Consider now the semiring $\mathbb{R}_+^{\infty}$ and let $[0, 1] = \{a \in \mathbb{R}_+ \mid 0 \le a \le 1\}$. A $[0, 1]$-automaton, whose transition matrix is stochastic, can be considered as a Markov chain (see Paz [110], Seneta [130]).

Example 4.1. Let

$$\mathfrak{A} = \left(\{q_0, q_1\}, \begin{pmatrix} a_1 & a_2 \\ a_3 & a_4 \end{pmatrix}, (\; 1 \quad 0 \;), \begin{pmatrix} 0 \\ 1 \end{pmatrix}\right)$$

Then $\|\mathfrak{A}\| = (M^*)_{q_0, q_1} = (a_1 + a_2 a_4^* a_3)^* a_2 a_4^*$ by Theorem 2.5.

Specializing the semiring A, assume the basic semiring to be $\mathfrak{P}(\Sigma^*)$, where $\Sigma = \{x_1, x_2, x_3, x_4\}$. Then, for $a_i = \{x_i\}$, $1 \le i \le 4$,

$$\|\mathfrak{A}\| = (\{x_1\} \cup \{x_2\}\{x_4\}^*\{x_3\})^*\{x_2\}\{x_4\}^* \qquad \Box$$

An A'-*linear system* is of the form

$$y = My + P,$$

where y is a variable, M is a matrix in $A'^{I \times I}$ and P is a column vector in $A'^{I \times 1}$. A column vector $T \in A'^{I \times 1}$ is called *solution* to $y = My + P$ iff $T = MT + P$. It is called *least solution* iff $T \sqsubseteq T'$ for all solutions T'.

Theorem 4.1. *Let $y = My + P$ be an A'-linear system. Then M^*P is its least solution.*

Proof. By Theorem 2.2 (i), (ii), M^*P is a solution. A proof analogous to the proof of Theorem 2.4 (i) shows that $MT + P \sqsubseteq T$ implies $M^*P \sqsubseteq T$. Hence, M^*P is the least solution. $\qquad \Box$

Corollary 4.2. *Let $\mathfrak{A} = (I, M, S, P)$ be an A'-automaton and let T be the least solution of the A'-linear system $y = My + P$. Then $\|\mathfrak{A}\| = ST$.*

A matrix $M \in (A\langle\langle \Sigma^* \rangle\rangle)^{I \times I}$ is called *cycle-free* iff there exists an $n \ge 1$ such that $(M, \varepsilon)^n = 0$. An $A\langle\langle \Sigma^* \rangle\rangle$-linear system $y = My + P$ is called *cycle-free* iff M is cycle-free.

Theorem 4.3. (Kuich, Urbanek [94], Corollary 3) *The cycle-free $A\langle\langle \Sigma^* \rangle\rangle$-linear system $y = My + P$ has the unique solution M^*P.*

An $A\langle\langle \Sigma^* \rangle\rangle$-automaton $\mathfrak{A} = (I, M, S, P)$ is called *cycle-free* iff M is cycle-free.

Corollary 4.4. *Let $\mathfrak{A} = (I, M, S, P)$ be a cycle-free $A\langle\langle \Sigma^* \rangle\rangle$-automaton and let T be the unique solution of the cycle-free $A\langle\langle \Sigma^* \rangle\rangle$-linear system $y = My + P$. Then $\|\mathfrak{A}\| = ST$.*

Example 4.1 indicated one method to compute the behavior of an A'-automaton: Apply Theorem 2.5. Corollary 4.4 yields another method to compute the behavior of a cycle-free $A\langle\langle \Sigma^* \rangle\rangle$-automaton $\mathfrak{A} = (I, M, S, P)$: Prove that $T \in (A\langle\langle \Sigma^* \rangle\rangle)^{I \times 1}$ is a solution of $y = My + P$. Then ST is the behavior of $\mathfrak{A}$.

Example 4.2. (Kuich, Salomaa [92], Example 7.2) Let $\Sigma = \{x_1, x_2, x_3\}$, $Q = \{q_1, q_2\}$,

$$C = \begin{pmatrix} x_1 & 0 \\ 0 & 0 \end{pmatrix}, \quad D = \begin{pmatrix} 0 & 0 \\ 0 & x_3 \end{pmatrix} \quad \text{and} \quad B_n = \begin{pmatrix} 0 & x_2^n \\ 0 & 0 \end{pmatrix}, \quad n \geq 0.$$

Define $M \in ((A\langle \Sigma^* \rangle)^{Q \times Q})^{N \times N}$, $S \in ((A\langle \varepsilon \rangle)^{1 \times Q})^{1 \times N}$ and $P \in ((A\langle \varepsilon \rangle)^{Q \times 1})^{N \times 1}$ by their non-null blocks:

$$M_{n,n+1} = C, \quad M_{n+1,n} = D, \quad M_{n,n} = B_n, \quad n \geq 0,$$
$$S_0 = (\ \varepsilon \quad 0\ ), \quad P_0 = \begin{pmatrix} 0 \\ \varepsilon \end{pmatrix}.$$

Consider the $A\langle \Sigma^* \rangle$-automaton $\mathfrak{A} = (I, M, S, P)$, where $I = \mathbb{N} \times Q$. (Strictly speaking, we should take the copies of M, S and P in $(A\langle \Sigma^* \rangle)^{(N \times Q) \times (N \times Q)}$, $(A\langle \varepsilon \rangle)^{1 \times (N \times Q)}$ and $(A\langle \varepsilon \rangle)^{(N \times Q) \times 1}$, respectively.) Let

$$T_n = \begin{pmatrix} \sum_{j \geq n} x_1^{j-n} x_2^j x_3^j \\ x_3^n \end{pmatrix}, \quad n \geq 0,$$

be the n-th block of the column vector $T \in ((A\langle\langle \Sigma^* \rangle\rangle)^{Q \times 1})^{N \times 1}$. We claim that T is a solution (and hence, the unique solution) of the cycle-free $A\langle \Sigma^* \rangle$-linear system $y = My + P$ $((M, \varepsilon)^2 = 0)$. This claim is proved by showing the equalities $(MT + P)_n = T_n$, $n \geq 0$:

$$\begin{aligned} (MT + P)_0 &= B_0 T_0 + C T_1 + P_0 = T_0, \\ (MT + P)_n &= D T_{n-1} + B_n T_n + C T_{n+1} = T_n, \quad n \geq 1. \end{aligned}$$

This yields $\|\mathfrak{A}\| = ST = S_0 T_0 = \sum_{j \geq 0} x_1^j x_2^j x_3^j$. $\square$

By definition, an A'-automaton $\mathfrak{A} = (I, M, S, P)$ is *normalized* iff the following conditions (i), (ii) and (iii) are satisfied.

(i) There exists an $i_0 \in I$ such that $S_{i_0} = 1$ and $S_i = 0$ for $i \neq i_0$.
(ii) There exists an $i_f \in I$, $i_f \neq i_0$, such that $P_{i_f} = 1$ and $P_i = 0$ for $i \neq i_f$.
(iii) $M_{i,i_0} = M_{i_f,i} = 0$ for all $i \in I$.

Theorem 4.5. *If a is the behavior of an A'-automaton then a is also the behavior of a normalized A'-automaton.*

Proof. Assume $a = \|\mathfrak{A}\|$, where $\mathfrak{A} = (I, M, S, P)$ is an A'-automaton. Let i_0 and i_f be new states and define the A'-automaton $\mathfrak{A}' = (I \cup \{i_0, i_f\}, M', S', P')$, where

$$S' = (\ 1\ \ 0\ \ 0\ ), \quad M' = \begin{pmatrix} 0 & S & 0 \\ 0 & M & P \\ 0 & 0 & 0 \end{pmatrix}, \quad P' = \begin{pmatrix} 0 \\ 0 \\ 1 \end{pmatrix}.$$

(Here the blocks are indexed by i_0, I and i_f.) By Corollary 2.7, we obtain $(M'^*)_{i_0, i_f} = SM^*P$. Hence, $\|\mathfrak{A}'\| = S'M'^*P' = (M'^*)_{i_0, i_f} = SM^*P = \|\mathfrak{A}\|$. $\square$

By definition, $\mathfrak{Rat}(A')$ is the smallest fully rationally closed subsemiring of A containing A'. Clearly, $\mathfrak{Rat}(A') \subseteq \mathfrak{Alg}(A')$ by Theorem 3.11. We now will prove a generalization of the Theorem of Kleene: $a \in \mathfrak{Rat}(A')$ iff a is the behavior of a finite A'-automaton.

Theorem 4.6. *The set of behaviors of finite A'-automata forms a fully rationally closed semiring that contains A'.*

Proof. Let $\mathfrak{A} = (Q, M, S, P)$ and $\mathfrak{A}' = (Q', M', S', P')$ be finite A'-automata, where Q and Q' are disjoint. Then we construct finite automata $\mathfrak{A}_1$, $\mathfrak{A}_2$ and $\mathfrak{A}_3$ with behaviors $\|\mathfrak{A}\| + \|\mathfrak{A}'\|$, $\|\mathfrak{A}\| \cdot \|\mathfrak{A}'\|$ and $\|\mathfrak{A}\|^*$, respectively.
(i) Define $\mathfrak{A}_1 = (Q \cup Q', M_1, S_1, P_1)$ by

$$M_1 = \begin{pmatrix} M & 0 \\ 0 & M' \end{pmatrix}, \quad S_1 = (\ S\ \ S'\ ), \quad P_1 = \begin{pmatrix} P \\ P' \end{pmatrix}.$$

(ii) Define $\mathfrak{A}_2 = (Q \cup Q', M_2, S_2, P_2)$ by

$$M_2 = \begin{pmatrix} M & PS' \\ 0 & M' \end{pmatrix}, \quad S_2 = (\ S\ \ 0\ ), \quad P_2 = \begin{pmatrix} 0 \\ P' \end{pmatrix}.$$

(iii) Define $\mathfrak{A}_3 = (Q \cup \{q_0\}, M_3, S_3, P_3)$, where q_0 is a new state, by

$$M_3 = \begin{pmatrix} 0 & S \\ P & M \end{pmatrix}, \quad S_3 = (\ 1\ \ 0\ ), \quad P_3 = \begin{pmatrix} 1 \\ 0 \end{pmatrix}.$$

Obviously, we have $\|\mathfrak{A}_1\| = \|\mathfrak{A}\| + \|\mathfrak{A}'\|$. Corollary 2.6 yields at once $\|\mathfrak{A}_2\| = \|\mathfrak{A}\| \cdot \|\mathfrak{A}'\|$. Theorem 2.5 yields $\|\mathfrak{A}_3\| = (SM^*P)^* = \|\mathfrak{A}\|^*$.

If the entries of PS' are in A' then $\mathfrak{A}_2$ is again an A'-automaton. If not, normalize $\mathfrak{A}$ or $\mathfrak{A}'$ by the construction of Theorem 4.5 and perform then construction (ii). The result is an A'-automaton $\mathfrak{A}_2$. Clearly, $\mathfrak{A}_1$ and $\mathfrak{A}_3$ are A'-automata.

For each $a \in A'$, a trivial construction yields a finite A'-automaton whose behavior is a. Hence, constructions (i) and (ii) show that the set of behaviors of finite A'-automata forms a semiring containing A'. Construction (iii) shows that this semiring is fully rationally closed. $\square$

Compare the constructions of Theorem 4.6 with the constructions of Harrison [46], Theorem 2.3.3, Hopcroft, Ullman [64], Theorem 2.3 and Perrin [111], Theorem 4.1.

We now show the converse to Theorem 4.6.

Theorem 4.7. *Let A' be a fully rationally closed subsemiring of A and let Q be finite. Then*

$$M \in A'^{Q \times Q} \quad implies \quad M^* \in A'^{Q \times Q}.$$

Proof. The proof is by induction on the number of elements in Q. For $|Q| = 1$, M^* is in A' since A' is fully rationally closed. For $|Q| > 1$, partition Q into $Q = Q_1 \cup Q_2$, $Q_1 \cap Q_2 = \emptyset$, $1 \le |Q_1|, |Q_2| < |Q|$. Then application of Theorem 2.5 proves our theorem. □

Corollary 4.8. *Let Q be finite and $M \in A'^{Q \times Q}$. Then $M^* \in \mathfrak{Rat}(A')^{Q \times Q}$.*

Corollary 4.9. *Let $\mathfrak{A}$ be a finite $\mathfrak{Rat}(A')$-automaton. Then $\|\mathfrak{A}\| \in \mathfrak{Rat}(A')$.*

Since $A' \subseteq \mathfrak{Rat}(A')$, we have proved the announced generalization of the Theorem of Kleene.

Theorem 4.10. *Let A' be a subset of an ω-continuous semiring A that contains 0 and 1. Then $\mathfrak{Rat}(A')$ coincides with the fully rationally closed semiring of the behaviors of finite A'-automata.*

Theorem 4.7 and some easy computations yield also the following corollary. It states explicitly that rational operations on finite matrices are performed by rational operations on their entries.

Corollary 4.11. *Let Q be finite. Then $\mathfrak{Rat}(A'^{Q \times Q}) = (\mathfrak{Rat}(A'))^{Q \times Q}$.*

The definition of $\mathfrak{Rat}(A')$ yields a result analogous to Theorem 3.14.

Theorem 4.12. $\mathfrak{Rat}(\mathfrak{Rat}(A')) = \mathfrak{Rat}(A')$.

Before specializing Theorem 4.10 to semirings $A\langle\langle \Sigma^* \rangle\rangle$ and $\mathfrak{P}(\Sigma^*)$, we show a result that allows us to delete ε-transitions without changing the behavior of an $A\langle \Sigma \cup \varepsilon \rangle$-automaton. (See also Hopcroft, Ullman [64], Theorem 2.2, and compare the proofs.)

Theorem 4.13. *Consider an $A\langle \Sigma \cup \varepsilon \rangle$-automaton $\mathfrak{A} = (I, M, S, P)$. Then there exists an $A\langle \Sigma \cup \varepsilon \rangle$-automaton $\mathfrak{A}' = (I, M', S, P')$, where $M' \in (A\langle \Sigma \rangle)^{I \times I}$ and $P' \in (A\langle \varepsilon \rangle)^{I \times 1}$, such that $\|\mathfrak{A}'\| = \|\mathfrak{A}\|$.*

Proof. By Theorem 4.5, we assume without loss of generality that $P \in (A\langle \varepsilon \rangle)^{I \times 1}$. Define $M_0 = (M, \varepsilon)\varepsilon$, $M_1 = \sum_{x \in \Sigma}(M, x)x$ and let $M' = M_0^* M_1$, $P' = M_0^* P$. Then, by Theorem 2.4 (iii),

$$\|\mathfrak{A}'\| = S(M_0^* M_1)^* M_0^* P = S(M_0 + M_1)^* P = SM^*P = \|\mathfrak{A}\|. \qquad \square$$

In the next corollary the basic semiring is again $A\langle\langle \Sigma^* \rangle\rangle$. In this case, $\mathfrak{Rat}(\{ax \mid a \in A, x \in \Sigma \cup \{\varepsilon\}\})$ coincides with $A^{\mathrm{rat}}\langle\langle \Sigma^* \rangle\rangle$ (see Kuich [83]). The power series in $A^{\mathrm{rat}}\langle\langle \Sigma^* \rangle\rangle$ are called *rational power series*.

Corollary 4.14. *A power series r is in $A^{\mathrm{rat}}\langle\langle \Sigma^* \rangle\rangle$ iff there exists a finite $A\langle \Sigma \cup \varepsilon \rangle$-automaton $\mathfrak{A} = (Q, M, S, P)$ such that $\|\mathfrak{A}\| = r$ and $M \in (A\langle \Sigma \rangle)^{Q \times Q}$, $P \in (A\langle \varepsilon \rangle)^{Q \times 1}$, and $S_{q_0} = \varepsilon$, $S_q = 0$, $q_0 \neq q$, for some $q_0 \in Q$.*

Corollary 4.14 is a variant of the Theorem of Kleene-Schützenberger (Schützenberger [128]; see also Sakarovitch[124]).

An $\{\{x\} \mid x \in \Sigma \cup \{\varepsilon\}\}$-algebraic system corresponding to a regular grammar and written in matrix notation is nothing else than an $\{\{x\} \mid x \in \Sigma\} \cup \{\{\varepsilon\}, \emptyset\}$-linear system. This yields the last result of this section.

Theorem 4.15. *The following statements on a formal language L over Σ are equivalent:*

(i) L *is a regular language over Σ.*
(ii) L *is the behavior of a finite $\mathfrak{P}(\Sigma \cup \{\varepsilon\})$-automaton $\mathfrak{A} = (Q, M, S, P)$ such that $M \in \mathfrak{P}(\Sigma)^{Q \times Q}$, $P \in \{\{\varepsilon\}, \emptyset\}^{Q \times 1}$, and $S_{q_0} = \{\varepsilon\}$, $S_q = \emptyset$, $q \neq q_0$, for some $q_0 \in Q$.*
(iii) $L \in \mathfrak{Rat}(\{\{x\} \mid x \in \Sigma\})$.

In language theory, a formula telling how a given regular language is obtained from the languages $\{x\}$, $x \in \Sigma$, and $\emptyset$ by rational operations is referred to as a *regular expression*. Hence, item (iii) of Theorem 4.15 means that L is denoted by a regular expression (see Salomaa [120] and [121], Theorem 5.1). The equivalence of items (ii) and (iii) of Theorem 4.15 is called the Theorem of Kleene (Kleene [77]).

5. Normal forms for algebraic systems

In this section, we will show that elements of $\mathfrak{Alg}(A')$ can be defined by A'-algebraic systems that are "simple" in a well-defined sense. In other words, we will exhibit a number of *normal forms* that correspond to well-known normal forms in language theory, e. g. the Chomsky normal form, the operator normal form and the Greibach normal form. Apart from the beginning of this section, we will only consider power series in $A^{\mathrm{alg}}\langle\langle \Sigma^* \rangle\rangle$.

An A'-algebraic system is in the *canonical two form* (see Harrison [46]) iff its equations have the form

$$y_i = \sum_{1 \leq k,m \leq n} a^i_{km} y_k y_m + \sum_{1 \leq k \leq n} a^i_k y_k + a_i, \quad 1 \leq i \leq n,$$

where $a^i_{km}, a^i_k \in \{0, 1\}$ and $a_i \in A'$.

Consider an A'-algebraic system $y_i = p_i$, $1 \leq i \leq n$, whose least solution is given by σ. Perform the following procedure on the product terms $a_0 y_{i_1} a_1 \ldots a_{k-1} y_{i_k} a_k$, $k \geq 1$, of the semiring-polynomials p_i, $1 \leq i \leq n$: replace each coefficient $a_j \neq 1$ by a new variable z and add an additional equation $z = a_j$; shorten now each product term $z_1 z_2 \ldots z_k$, $k > 2$, to $z_1 u_1$ and add additional equations $u_1 = z_2 u_2, \ldots, u_{k-2} = z_{k-1} z_k$, where $u_1, \ldots, u_{k-2}$ are new variables. Then the components σ_i, $1 \leq i \leq n$, of σ are components of the least solution of the newly constructed A'-algebraic system in the canonical two form.

Theorem 5.1. *Each $a \in \mathfrak{Alg}(A')$ is a component of the least solution of an A'-algebraic system in the canonical two form.*

We will now consider a very useful transformation of an A'-algebraic system. In the next theorem, we write an A'-algebraic system in the form $y = My + P$, where $M \in A'^{n \times n}$ and $P \in A'(Y)^{n \times 1}$. Here the entries of My contain product terms of the form $a y_i$, $a \in A'$, $y_i \in Y$.

Theorem 5.2. *The least solutions of the A'-algebraic system $y = My + P$ and of the $\mathfrak{Rat}(A')$-algebraic system $y = M^* P$, where $M \in A'^{n \times n}$ and $P \in A'(Y)^{n \times 1}$, coincide.*

Proof. Let σ and τ be the least solutions of $y = My + P(y)$ and $y = M^* P(y)$, respectively.

(i) By substituting $\tau = M^* P(\tau)$ into $My + P(y)$, we obtain $M\tau + P(\tau) = MM^* P(\tau) + P(\tau) = M^+ P(\tau) + P(\tau) = M^* P(\tau) = \tau$. Hence, τ is a solution of $y = My + P(y)$ and $\sigma \sqsubseteq \tau$.

(ii) Clearly, σ is a solution of the A-linear system $y = My + P(\sigma)$. The least solution of $y = My + P(\sigma)$ is given by $M^* P(\sigma)$. Hence $M^* P(\sigma) \sqsubseteq \sigma$.

(iii) Let $(\sigma^j \mid j \in \mathbb{N})$ be the approximation sequence of $y = My + P(y)$. It is easily shown by induction on j that $\sigma^j \sqsubseteq M^* P(\sigma^j)$, $j \geq 0$. This implies $\sigma \sqsubseteq M^* P(\sigma)$.

By (ii) and (iii) we infer that $\sigma = M^* P(\sigma)$, i. e., σ is a solution of $y = M^* P(y)$. This implies $\tau \sqsubseteq \sigma$. Hence, by (i), $\tau = \sigma$ and the least solutions coincide. $\qquad \Box$

For the remainder of this section, our basic semiring will be $A\langle\langle \Sigma^* \rangle\rangle$, where A is *commutative*, and we will consider algebraic systems $y_i = p_i$, $1 \leq i \leq n$, where $p_i \in A\langle (\Sigma \cup Y)^* \rangle$.

Corollary 5.3. *The least solutions of the algebraic systems $y = My + P$ and $y = M^* P$, where $M \in (A\langle\varepsilon\rangle)^{n \times n}$ and $\operatorname{supp}(P_i) \subseteq (\Sigma \cup Y)^* - Y$, $1 \leq i \leq n$, coincide.*

Observe that the context-free grammar corresponding to the algebraic system $y = M^* P$ has no *chain rules*, i. e., has no productions of the type $y_i \to y_j$. (Compare with Salomaa [121], Theorem 6.3; Harrison [46], Theorem 4.3.2; Hopcroft, Ullman [64], Theorem 4.4.)

We now consider another useful transformation of an algebraic system. It corresponds to the transformation of a context-free grammar for deleting ε-rules (i. e., productions of the type $y_i \to \varepsilon$).

Theorem 5.4. *Let $y_i = p_i$, $1 \le i \le n$, $p_i \in A\langle\langle(\Sigma \cup Y)^*\rangle\rangle$, be an algebraic system with least solution σ. Let $\sigma = (\sigma, \varepsilon)\varepsilon + \tau$, where $(\tau, \varepsilon) = 0$. Then there exists an algebraic system $y_i = q_i$, $1 \le i \le n$, $q_i \in A\langle\langle(\Sigma \cup Y)^*\rangle\rangle$, $(q_i, \varepsilon) = 0$, whose least solution is τ.*

Proof. Substitute $(\sigma_j, \varepsilon)\varepsilon + y_j$ for y_j, $1 \le j \le n$, into p_i and define

$$q_i(y) = \sum_{w \in \Sigma^+} (p_i((\sigma, \varepsilon)\varepsilon + y), w)w, \quad 1 \le i \le n.$$

The equalities

$$p_i((\sigma, \varepsilon)\varepsilon + \tau) = p_i(\sigma) = \sigma = (\sigma, \varepsilon)\varepsilon + \tau, \quad 1 \le i \le n,$$

imply, by comparing coefficients,

$$q_i(\tau) = \tau, \quad 1 \le i \le n.$$

Hence, τ is a solution of the algebraic system $y_i = q_i$, $1 \le i \le n$. Consider now an arbitrary solution τ' of $y_i = q_i$, $1 \le i \le n$. Then $\sigma' = (\sigma, \varepsilon)\varepsilon + \tau'$ is a solution of $y_i = p_i$, $1 \le i \le n$. Since σ is the least solution of $y_i = p_i$, $1 \le i \le n$, we infer that $\sigma \sqsubseteq \sigma'$. But this implies $\tau \sqsubseteq \tau'$. Hence, τ is the least solution of $y_i = q_i$, $1 \le i \le n$. $\qquad\square$

Observe that the context-free grammar corresponding to the algebraic system $y_i = q_i$, $1 \le i \le n$, has no ε-rules. (Compare with Salomaa [121], Theorem 6.2; Harrison [46], Theorem 4.3.1; Hopcroft, Ullman [64], Theorem 4.3.)

An algebraic system $y_i = p_i$, $1 \le i \le n$, $p_i \in A\langle\langle(\Sigma \cup Y)^*\rangle\rangle$, is termed *proper* iff $\operatorname{supp}(p_i) \subseteq (\Sigma \cup Y)^+ - Y$ for all $1 \le i \le n$. Proper algebraic systems correspond to context-free grammars without ε-rules and chain rules.

Corollary 5.5. *Let $r \in A^{\mathrm{alg}}\langle\langle\Sigma^*\rangle\rangle$. Then there exists a proper algebraic system such that $\sum_{w \in \Sigma^+}(r, w)w$ is a component of its least solution.*

Proof. Apply the constructions of Theorem 5.4 and Corollary 5.3, in this order. $\qquad\square$

Corollary 5.6. *For every context-free language L there exists a context-free grammar G without ε-rules and chain rules such that $L(G) = L - \{\varepsilon\}$.*

An algebraic system $y_i = p_i$, $1 \le i \le n$, $p_i \in A\langle\langle(\Sigma \cup Y)^*\rangle\rangle$, is termed *strict* iff $\operatorname{supp}(p_i) \subseteq \{\varepsilon\} \cup (\Sigma \cup Y)^* \Sigma (\Sigma \cup Y)^*$ for all $1 \le i \le n$. For a proof of the next result see Salomaa, Soittola [123], Theorem IV.1.1 and Kuich, Salomaa [92], Theorem 14.11.

Theorem 5.7. *Let* $y_i = p_i$, $1 \leq i \leq n$, $p_i \in A\langle\langle (\Sigma \cup Y)^* \rangle\rangle$, *be an algebraic system with least solution* σ.

If $y_i = p_i$, $1 \leq i \leq n$, *is a proper algebraic system then* $(\sigma, \varepsilon) = 0$ *and* σ *is the only solution with this property.*

If $y_i = p_i$, $1 \leq i \leq n$, *is a strict algebraic system then* σ *is its unique solution.*

Example 5.1. Consider an $A^{\mathrm{rat}}\langle\langle \Sigma^* \rangle\rangle$-algebraic system that can be written in matrix notation in the form $Z = M_1 Z M_2 + M$, where Z is an $n \times n$-matrix of variables and $M_1, M_2, M \in (A^{\mathrm{rat}}\langle\langle \Sigma^* \rangle\rangle)^{n \times n}$. Then, by computing the approximation sequence, it is easily seen that $S = \sum_{i \geq 0} M_1{}^i M M_2{}^i \in (A\langle\langle \Sigma^* \rangle\rangle)^{n \times n}$ is the least solution of $Z = M_1 Z M_2 + M$. (See also Berstel [4], Section V.6 and Kuich [82].)

Consider the linear context-free grammar

$$G = (\{y_1, y_2\}, \{a, b\}, \{y_1 \to a y_2, y_2 \to a y_2, y_2 \to a y_2 b, y_2 \to b\}, y_1)$$

and the corresponding strict algebraic system $y_1 = a y_2$, $y_2 = a y_2 + a y_2 b + b$. Denote its unique solution by $\sigma = (\sigma_1, \sigma_2)$. We infer by Theorem 5.2 that σ is the unique solution of the $A^{\mathrm{rat}}\langle\langle \Sigma^* \rangle\rangle$-algebraic system $y_1 = a y_2$, $y_2 = a^+ y_2 b + a^* b$. Hence,

$$\sigma_2 = \sum_{n \geq 0} (a^+)^n a^* b^{n+1} = \sum_{n \geq 0} a^* (a^+)^n b^{n+1} \quad \text{and} \quad \sigma_1 = \sum_{n \geq 0} (a^+)^{n+1} b^{n+1}.$$

This implies

$$L(G) = \{a\}^* \bigcup_{n \geq 1} \{a\}^n \{b\}^n = \{a^m b^n \mid m \geq n \geq 1\}.$$

If $A = \mathbb{N}^\infty$, $\sigma_1 = \sum_{m \geq n \geq 1} \binom{m-1}{n-1} a^m b^n$. Hence, by Theorem 3.9, the word $a^m b^n \in L(G)$ has $\binom{m-1}{n-1}$ distinct leftmost derivations according to G.   □

We are now ready to proceed to the various normal forms. Our next result deals with the transition to the Chomsky normal form. By definition, an algebraic system $y_i = p_i$, $1 \leq i \leq n$, is in the *Chomsky normal form* (Chomsky [18]) iff $\mathrm{supp}(p_i) \subseteq \Sigma \cup Y^2$, $1 \leq i \leq n$.

Theorem 5.8. *Let* $r \in A^{\mathrm{alg}}\langle\langle \Sigma^* \rangle\rangle$. *Then there exists an algebraic system in the Chomsky normal form such that* $\sum_{w \in \Sigma^+} (r, w) w$ *is a component of its least solution.*

Proof. We assume, by Theorem 5.1, that r is a component of the least solution of an algebraic system in the canonical two form. Apply now the constructions of Theorem 5.4 and Corollary 5.3, in this order. The resulting algebraic system is in the Chomsky normal form.   □

We now introduce operators w^{-1}, for $w \in \Sigma^*$, mapping $A\langle\langle\Sigma^*\rangle\rangle$ into $A\langle\langle\Sigma^*\rangle\rangle$. For $u \in \Sigma^*$, we define $uw^{-1} = v$ if $u = vw$, $uw^{-1} = 0$ otherwise. As usual we extend these mappings to power series $r \in A\langle\langle\Sigma^*\rangle\rangle$ by

$$rw^{-1} = \sum_{u \in \Sigma^*} (r, u)uw^{-1} = \sum_{v \in \Sigma^*} (r, vw)v.$$

Observe that, if $(r, \varepsilon) = 0$ then $r = \sum_{x \in \Sigma} (rx^{-1})x$. In language theory the mappings corresponding to w^{-1} are usually referred to as *right derivatives* with respect to the word w.

Our next result deals with the transition from the Chomsky normal form to the operator normal form. By definiton, an algebraic system $y_i = p_i$, $1 \le i \le n$, is in the *operator normal form* iff $\text{supp}(p_i) \subseteq \{\varepsilon\} \cup Y\Sigma \cup Y\Sigma Y$. Operator normal forms are customarily defined in language theory to be more general: there are no two consecutive nonterminals on the right sides of the productions. (See Floyd [36], Harrison [46].)

Theorem 5.9. *Let $r \in A^{\text{alg}}\langle\langle\Sigma^*\rangle\rangle$. Then there exists an algebraic system in the operator normal form such that r is a component of its unique solution.*

Proof. By Theorem 5.8 we may assume that $\sum_{w \in \Sigma^+} (r, w)w$ is the first component of the least solution σ of an algebraic system $y_i = p_i$, $1 \le i \le n$, in the Chomsky normal form. We write this system as follows:

$$y_i = \sum_{x \in \Sigma} (p_i, x)x + \sum_{1 \le k, m \le n} (p_i, y_k y_m)y_k y_m, \quad 1 \le i \le n.$$

We now define a new algebraic system. The alphabet of new variables will be $Y' = \{z_0\} \cup \{z_i^x \mid x \in \Sigma, 1 \le i \le n\}$. The equations of the new algebraic system are

$$z_0 = (r, \varepsilon)\varepsilon + \sum_{x \in \Sigma} z_1^x x,$$

$$z_i^x = (p_i, x)\varepsilon + \sum_{x' \in \Sigma} \sum_{1 \le k, m \le n} (p_i, y_k y_m)z_k^{x'} x' z_m^x, \quad x \in \Sigma, 1 \le i \le n.$$

We claim that the components of the unique solution of this new algebraic system are given by r (z_0-component) and $\sigma_i x^{-1}$ (z_i^x-component). The claim is proven by substituting the components of the unique solution into the equations:

$(r, \varepsilon)\varepsilon + \sum_{x \in \Sigma} (\sigma_1 x^{-1})x = (r, \varepsilon)\varepsilon + \sigma_1 = (r, \varepsilon)\varepsilon + \sum_{w \in \Sigma^+} (r, w)w = r$,

$(p_i, x)\varepsilon + \sum_{1 \le k, m \le n} (p_i, y_k y_m)(\sum_{x' \in \Sigma} (\sigma_k x'^{-1})x')(\sigma_m x^{-1}) =$
$(p_i, x)xx^{-1} + \sum_{1 \le k, m \le n} (p_i, y_k y_m)(\sigma_k \sigma_m)x^{-1} = \sigma_i x^{-1}, \quad x \in \Sigma, 1 \le i \le n.$

Observe that the equalities are valid for σ because $(\sigma, \varepsilon) = 0$. They are not valid for solutions τ of $y_i = p_i$, $1 \le i \le n$, with $(\tau, \varepsilon) \ne 0$.   □

By definition, an algebraic system $y_i = p_i$, $1 \leq i \leq n$, is in the *Greibach normal form* iff $\mathrm{supp}(p_i) \subseteq \{\varepsilon\} \cup \Sigma \cup \Sigma Y \cup \Sigma YY$. (See Greibach [44], Rosenkrantz [119], Jacob [67], Urbanek [136].)

Theorem 5.10. *Let $r \in A^{\mathrm{alg}} \langle\langle \Sigma^* \rangle\rangle$. Then there exists an algebraic system in the Greibach normal form such that r is a component of its unique solution.*

Proof. By Theorem 5.9 we may assume that r is the first component of the unique solution of an algebraic system $y_i = p_i$, $1 \leq i \leq n$, in the operator normal form. We write this system as follows:

$$y^{\mathrm{T}} = y^{\mathrm{T}} M(y) + P,$$

where $y^{\mathrm{T}} = (y_1, \ldots, y_n)$ is the transpose of y, $M \in (A\langle\langle (Y \cup \Sigma)^* \rangle\rangle)^{n \times n}$,

$$M_{j,i} = \sum_{x \in \Sigma} (p_i, y_j x) x + \sum_{1 \leq m \leq n} \sum_{x \in \Sigma} (p_i, y_j x y_m) x y_m, \quad 1 \leq i, j \leq n,$$

and $P = ((p_1, \varepsilon)\varepsilon, \ldots, (p_n, \varepsilon)\varepsilon)$. Let Z be an $n \times n$-matrix whose (i,j)-entry is a new variable z_{ij}, $1 \leq i, j \leq n$. We now consider the algebraic system in the Greibach normal form

$$
\begin{aligned}
y^{\mathrm{T}} &= PM(y)Z + PM(y) + P \\
Z &= M(y)Z + M(y).
\end{aligned}
$$

We show that $(\sigma, M(\sigma)^+)$ is its unique solution:

$$PM(\sigma)M(\sigma)^+ + PM(\sigma) + P = PM(\sigma)^* = \sigma^{\mathrm{T}},$$

by a row vector variant of Theorem 5.2, and

$$M(\sigma)M(\sigma)^+ + M(\sigma) = M(\sigma)^+.$$

Hence, $r = \sigma_1$ is a component of the unique solution of the new algebraic system in the Greibach normal form. $\qquad\square$

In language theory, the two most important normal forms are the Chomsky normal form and the Greibach normal form. By definition, a context-free grammar $G = (Y, \Sigma, P, y_1)$ is in the *Chomsky normal form* iff all productions are of the two forms $y_i \to y_k y_m$ and $y_i \to x$, $x \in \Sigma$, $y_i, y_k, y_m \in Y$. It is in the Greibach normal form iff all productions are of the three forms $y_i \to x y_k y_m$, $y_i \to x y_k$ and $y_i \to x$, $x \in \Sigma$, $y_i, y_k, y_m \in Y$. (Usually, productions $y_i \to \varepsilon$ are not allowed in the Greibach normal form.)

Corollary 5.11. *For every context-free language L there exist a context-free grammar G_1 in the Chomsky normal form and a context-free grammar G_2 in the Greibach normal form such that $L(G_1) = L(G_2) = L - \{\varepsilon\}$.*

Proof. By Theorem 5.8, and by Theorem 5.10 together with Theorem 5.4. $\quad\square$

If our basic semiring is $N^\infty \langle\langle \Sigma^* \rangle\rangle$, we can draw some even stronger conclusions by Theorem 3.9 and by Lemma IV.2.5 of Salomaa, Soittola [123].

Corollary 5.12. *Let* $d : \Sigma^* \to N$. *Then the following three statements are equivalent:*

(i) *There exists a context-free grammar* G *with terminal alphabet* Σ *such that the number of distinct leftmost derivations of* w, $w \in \Sigma^*$, *from the start variable is given by* $d(w)$.

(ii) *There exists a context-free grammar* G_1 *in the Chomsky normal form with terminal alphabet* Σ *such that the number of distinct leftmost derivations of* w, $w \in \Sigma^+$, *from the start variable is given by* $d(w)$.

(iii) *There exists a context-free grammar* G_2 *in the Greibach normal form with terminal alphabet* Σ *such that the number of distinct leftmost derivations of* w, $w \in \Sigma^+$, *from the start variable is given by* $d(w)$.

Corollary 5.13. *For every unambiguous context-free grammar* G *there exist an unambiguous context-free grammar* G_1 *in the Chomsky normal form and an unambiguous context-free grammar* G_2 *in the Greibach normal form such that* $L(G_1) = L(G_2) = L(G) - \{\varepsilon\}$.

6. Pushdown automata and algebraic systems

We now define A'-pushdown automata and consider their relation to A'-algebraic systems. It turns out that, for $a \in A$, $a \in \mathfrak{Alg}(A')$ iff it is the behavior of an A'-pushdown automaton. This generalizes the language theoretic result due to Chomsky [19] that a formal language is context-free iff it is accepted by a pushdown automaton.

A'-pushdown automata are finite automata (with state set Q) augmented by a pushdown tape. The contents of the pushdown tape is a word over the pushdown alphabet Γ. We consider an A'-pushdown automaton to be an A'-automaton in the sense of Section 4: the state set is given by $\Gamma^* \times Q$ and its transition matrix is in $A'^{(\Gamma^* \times Q) \times (\Gamma^* \times Q)}$. This allows us to store the contents of the pushdown tape and the states of the finite automaton in the states of the A'-pushdown automaton. Because of technical reasons, we do not work in the semiring $A^{(\Gamma^* \times Q) \times (\Gamma^* \times Q)}$ but in the isomorphic semiring $(A^{Q \times Q})^{\Gamma^* \times \Gamma^*}$.

A matrix $M \in (A'^{Q \times Q})^{\Gamma^* \times \Gamma^*}$ is termed an A'-*pushdown transition matrix* iff

(i) for each $p \in \Gamma$ there exist only finitely many blocks $M_{p,\pi}$, $\pi \in \Gamma^*$, that are unequal to 0;

(ii) for all $\pi_1, \pi_2 \in \Gamma^*$,

$$M_{\pi_1, \pi_2} = \begin{cases} M_{p,\pi} & \text{if there exist } p \in \Gamma, \ \pi' \in \Gamma^* \text{ with} \\ & \pi_1 = p\pi' \text{ and } \pi_2 = \pi\pi', \\ 0 & \text{otherwise.} \end{cases}$$

The above definition implies that an A'-pushdown transition matrix has a finitary specification: it is completely specified by its non-null blocks of the form $M_{p,\pi}$, $p \in \Gamma$, $\pi \in \Gamma^*$. Item (ii) of the above definition shows that only the following transitions are possible: if the contents of the pushdown tape is given by $p\pi'$, the contents of the pushdown tape after a transition has to be of the form $\pi\pi'$; moreover, the transition does only depend on the leftmost (topmost) pushdown sympol p and not on π'. In this sense the A'-pushdown transition matrix represents a proper formalization of the principle "last in—first out".

An A'-pushdown automaton

$$\mathfrak{P} = (Q, \Gamma, M, S, p_0, P)$$

is given by

(i)   a finite set Q of *states*,

(ii) a finite alphabet Γ of *pushdown symbols*,

(iii) an A'-*pushdown transition matrix* $M \in (A'^{Q \times Q})^{\Gamma^* \times \Gamma^*}$,

(iv) $S \in A'^{1 \times Q}$, called the *initial state vector*,

(v) $p_0 \in \Gamma$, called the *initial pushdown symbol*,

(vi) $P \in A'^{Q \times 1}$, called the *final state vector*.

The behavior $\|\mathfrak{P}\|$ of the A'-pushdown automaton $\mathfrak{P}$ is defined by

$$\|\mathfrak{P}\| = S(M^*)_{p_0,\varepsilon} P.$$

We now describe the computations of an A'-pushdown automaton. Initially, the pushdown tape contains the special symbol p_0. The A'-pushdown automaton now performs transitions governed by the A'-pushdown transition matrix until the pushdown tape is emptied. The result of these computations is given by $(M^*)_{p_0,\varepsilon}$. Multiplications by the initial state vector and by the final state vector yield the behavior of the A'-pushdown automaton.

Let now $\mathfrak{P}(\Sigma^*)$ be our basic semiring. We connect our definition of an $\mathfrak{P}(\Sigma \cup \{\varepsilon\})$-pushdown automaton $\mathfrak{P} = (Q, \Gamma, M, S, p_0, P)$ to the usual definition of a *pushdown automaton* $\mathfrak{P}' = (Q, \Sigma, \Gamma, \delta, q_0, p_0, F)$ (see e. g., Harrison [46]), where Σ is the *input alphabet*, $\delta : Q \times (\Sigma \cup \{\varepsilon\}) \times \Gamma \to$ finite subsets of $Q \times \Gamma^*$ is the *transition function*, $q_0 \in Q$ is the *initial state* and $F \subseteq Q$ is the set of *final states*.

Assume that a pushdown automaton $\mathfrak{P}'$ is given as above. The transition function δ defines the pushdown transition matrix M of $\mathfrak{P}$ by

$$x \in (M_{p,\pi})_{q_1,q_2} \quad \text{iff} \quad (q_2, \pi) \in \delta(q_1, x, p)$$

for all $q_1, q_2 \in Q$, $p \in \Gamma$, $\pi \in \Gamma^*$, $x \in \Sigma \cup \{\varepsilon\}$. Let now $\vdash$ be the move relation over the *instantaneous descriptions* of $\mathfrak{P}'$ in $Q \times \Sigma^* \times \Gamma^*$. Then $(q_1, w, \pi_1) \vdash^k (q_2, \varepsilon, \pi_2)$ iff $w \in ((M^k)_{\pi_1,\pi_2})_{q_1,q_2}$ and $(q_1, w, \pi_1) \vdash^* (q_2, \varepsilon, \pi_2)$ iff $w \in ((M^*)_{\pi_1,\pi_2})_{q_1,q_2}$ for all $k \geq 0$, $q_1, q_2 \in Q$, $\pi_1, \pi_2 \in \Gamma^*$, $w \in \Sigma^*$. Hence,

$(q_0, w, p_0) \vdash^* (q, \varepsilon, \varepsilon)$ iff $w \in ((M^*)_{p_0,\varepsilon})_{q_0,q}$. Define the initial state vector S and the final state vector P by $S_{q_0} = \{\varepsilon\}$, $S_q = \emptyset$ if $q \neq q_0$, $P_q = \{\varepsilon\}$ if $q \in F$, $P_q = \emptyset$ if $q \notin F$. Then a word w is accepted by the pushdown automaton $\mathfrak{P}'$ by both final state and empty store iff $w \in S(M^*)_{p_0,\varepsilon}P = \|\mathfrak{P}\|$.

In our first theorem we show that an A'-pushdown automaton can be regarded as an A'-automaton.

Theorem 6.1. *For each A'-pushdown automaton $\mathfrak{P}$ there exists an A'-automaton $\mathfrak{A}$ such that $\|\mathfrak{A}\| = \|\mathfrak{P}\|$.*

Proof. Let $\mathfrak{P} = (Q, \Gamma, M, S, p_0, P)$. We define the A'-automaton $\mathfrak{A} = (\Gamma^* \times Q, M', S', P')$ by $M'_{(\pi_1,q_1),(\pi_2,q_2)} = (M_{\pi_1,\pi_2})_{q_1,q_2}$, $S'_{(p_0,q)} = S_q$, $S'_{(\pi,q)} = 0$, if $\pi \neq p_0$, $P'_{(\varepsilon,q)} = P_q$, $P'_{(\pi,q)} = 0$, if $\pi \neq \varepsilon$. Then

$$
\begin{aligned}
\|\mathfrak{A}\| &= S'M'^*P' \\
&= \sum_{(\pi_1,q_1),(\pi_2,q_2)\in\Gamma^*\times Q} S'_{(\pi_1,q_1)}(M'^*)_{(\pi_1,q_1),(\pi_2,q_2)}P'_{(\pi_2,q_2)} \\
&= \sum_{q_1,q_2\in Q} S'_{(p_0,q_1)}(M'^*)_{(p_0,q_1),(\varepsilon,q_2)}P'_{(\varepsilon,q_2)} = \\
&= \sum_{q_1,q_2\in Q} S_{q_1}((M^*)_{p_0,\varepsilon})_{q_1,q_2}P_{q_2} = S(M^*)_{p_0,\varepsilon}P = \|\mathfrak{P}\|. \qquad \square
\end{aligned}
$$

Consider an A'-pushdown automaton with A'-pushdown transition matrix M and let $\pi = \pi_1\pi_2$ be a word over the pushdown alphabet Γ. Then our next theorem states that emptying the pushdown tape with contents π has the same effect (i. e., $(M^*)_{\pi,\varepsilon}$) as emptying first the pushdown tape with contents π_1 (i. e., $(M^*)_{\pi_1,\varepsilon}$) and afterwards (i. e., multiplying) the pushdown tape with contents π_2 (i. e., $(M^*)_{\pi_2,\varepsilon}$). (For a proof see Kuich, Salomaa [92], Theorem 10.5; Kuich [86], Satz 11.4.)

Theorem 6.2. *Let $M \in (A'^{Q\times Q})^{\Gamma^*\times\Gamma^*}$ be an A'-pushdown transition matrix. Then*

$$(M^*)_{\pi_1\pi_2,\varepsilon} = (M^*)_{\pi_1,\varepsilon}(M^*)_{\pi_2,\varepsilon}$$

holds for all $\pi_1, \pi_2 \in \Gamma^*$.

Let $M \in (A'^{Q\times Q})^{\Gamma^*\times\Gamma^*}$ be an A'-pushdown transition matrix and let $\{y_p \mid p \in \Gamma\}$ be an alphabet of variables. We define $y_\varepsilon = \varepsilon$ and $y_{p\pi} = y_p y_\pi$ for $p \in \Gamma$, $\pi \in \Gamma^*$, and consider the $A'^{Q\times Q}$-algebraic system

$$y_p = \sum_{\pi\in\Gamma^*} M_{p,\pi}y_\pi, \quad p \in \Gamma.$$

Moreover, we define $F \in (A'^{Q\times Q})^{\Gamma^*\times 1}$ by $F_\varepsilon = E$ and $F_\pi = 0$ if $\pi \in \Gamma^+$, and consider the $A'^{Q\times Q}$-linear system $y = My + F$.

Given matrices $T_p \in A^{Q \times Q}$ for all $p \in \Gamma$, we define matrices $T_\pi \in A^{Q \times Q}$ for all $\pi \in \Gamma^*$ as follows: $T_\varepsilon = E$, $T_{p\pi} = T_p T_\pi$, $p \in \Gamma$, $\pi \in \Gamma^*$. By these matrices we define a matrix $\tilde{T} \in (A^{Q \times Q})^{\Gamma^* \times 1}$: the π-block of $\tilde{T}$ is given by T_π, $\pi \in \Gamma^*$, i. e., $\tilde{T}_\pi = T_\pi$.

Theorem 6.3. *If* $(T_p)_{p \in \Gamma}$, $T_p \in A^{Q \times Q}$, *is a solution of* $y_p = \sum_{\pi \in \Gamma^*} M_{p,\pi} y_\pi$, $p \in \Gamma$, *then* $\tilde{T} \in (A^{Q \times Q})^{\Gamma^* \times 1}$ *is a solution of* $y = My + F$.

Proof. Since M is an A'-pushdown transition matrix, we obtain, for all $p \in \Gamma$ and $\pi \in \Gamma^*$,

$$
\begin{aligned}
(M\tilde{T})_{p\pi} &= \sum_{\pi_1 \in \Gamma^*} M_{p\pi,\pi_1} \tilde{T}_{\pi_1} = \sum_{\pi_2 \in \Gamma^*} M_{p\pi,\pi_2\pi} \tilde{T}_{\pi_2\pi} = \\
&= \sum_{\pi_2 \in \Gamma^*} M_{p,\pi_2} \tilde{T}_{\pi_2} T_\pi = (M\tilde{T})_p \tilde{T}_\pi.
\end{aligned}
$$

Since $(T_p)_{p \in \Gamma}$ is a solution of $y_p = \sum_{\pi \in \Gamma^*} M_{p,\pi} y_\pi$, $p \in \Gamma$, we infer that $\tilde{T}_p = T_p = \sum_{\pi \in \Gamma^*} M_{p,\pi} T_\pi = \sum_{\pi \in \Gamma^*} M_{p,\pi} \tilde{T}_\pi = (M\tilde{T})_p$. Hence, $(M\tilde{T} + F)_{p\pi} = (M\tilde{T})_{p\pi} = \tilde{T}_p \tilde{T}_\pi = \tilde{T}_{p\pi}$, $p \in \Gamma$, $\pi \in \Gamma^*$. Additionally, we have $\tilde{T}_\varepsilon = E$ and $(M\tilde{T} + F)_\varepsilon = F_\varepsilon = E$. This implies that $\tilde{T}$ is a solution of $y = My + F$. $\square$

Theorem 6.4. *The* $A'^{Q \times Q}$-*algebraic system* $y_p = \sum_{\pi \in \Gamma^*} M_{p,\pi} y_\pi$ *has the least solution* $((M^*)_{p,\varepsilon})_{p \in \Gamma}$.

Proof. We first show that $((M^*)_{p,\varepsilon})_{p \in \Gamma}$ is a solution of the $A'^{Q \times Q}$-algebraic system by substituting $(M^*)_{\pi,\varepsilon}$ for y_π:

$$
\sum_{\pi \in \Gamma^*} M_{p,\pi} (M^*)_{\pi,\varepsilon} = (M^+)_{p,\varepsilon} = (M^*)_{p,\varepsilon}, \quad p \in \Gamma.
$$

Assume now that $(T_p)_{p \in \Gamma}$ is a solution of $y_p = \sum_{\pi \in \Gamma^*} M_{p,\pi} y_\pi$, where $T_p \sqsubseteq (M^*)_{p,\varepsilon}$, $p \in \Gamma$. Then, by Theorem 6.3, $\tilde{T}$ is a solution of $y = My + F$. Since M^*F is the least solution of this $A'^{Q \times Q}$-linear system, we infer that $M^*F \sqsubseteq \tilde{T}$. This implies $(M^*F)_\pi = (M^*)_{\pi,\varepsilon} \sqsubseteq \tilde{T}_\pi = T_\pi$ for all $\pi \in \Gamma^*$. Hence, $T_p = (M^*)_{p,\varepsilon}$ for all $p \in \Gamma$, and $(M^*)_{p,\varepsilon}$ is the least solution of $y_p = \sum_{\pi \in \Gamma^*} M_{p,\pi} y_\pi$, $p \in \Gamma$. $\square$

Let $\mathfrak{P} = (Q, \Gamma, M, S, p_0, P)$ be an A'-pushdown automaton and consider the A'-algebraic system

$$
\begin{aligned}
y_0 &= S y_{p_0} P, \\
y_p &= \sum_{\pi \in \Gamma^*} M_{p,\pi} y_\pi, \quad p \in \Gamma,
\end{aligned}
$$

written in matrix notation: y_p is a $Q \times Q$-matrix whose (q_1, q_2)-entry is the variable $[q_1, p, q_2]$, $p \in \Gamma$, $q_1, q_2 \in Q$; if $\pi = p_1 \ldots p_r$, $r \geq 1$, then the (q_1, q_2)-entry of y_π is given by the (q_1, q_2)-entry of $y_{p_1} \ldots y_{p_r}$, $p_1, \ldots, p_r \in \Gamma$; y_0 is a variable. Hence, the variables of the above A'-algebraic system are y_0, $[q_1, p, q_2]$, $p \in \Gamma$, $q_1, q_2 \in Q$.

Corollary 6.5. *Let* $\mathfrak{P} = (Q, \Gamma, M, S, p_0, P)$ *be an* A'-*pushdown automaton. Then* $\|\mathfrak{P}\|$, $(M^*)_{p,\varepsilon}$, $p \in \Gamma$, *is the least solution of the* A'-*algebraic system*

$$y_0 = S y_{p_0} P,$$

$$y_p = \sum_{\pi \in \Gamma^*} M_{p,\pi} y_\pi, \quad p \in \Gamma.$$

Corollary 6.6. *The behavior of an* A'-*pushdown automaton is an element of* $\mathfrak{Alg}(A')$.

Theorem 3.14 admits another corollary.

Corollary 6.7. *Let* $\mathfrak{P}$ *be an* $\mathfrak{Alg}(A')$-*pushdown automaton. Then* $\|\mathfrak{P}\| \in \mathfrak{Alg}(A')$.

We now want to show the converse of Corollary 6.6.

Theorem 6.8. *Let* $a \in \mathfrak{Alg}(A')$. *Then there exists an* A'-*pushdown automaton* $\mathfrak{P}$ *such that* $\|\mathfrak{P}\| = a$.

Proof. We assume, by Theorem 5.1, that a is the first component of the least solution σ of an A'-algebraic system in the canonical two form

$$y_i = \sum_{1 \leq k,m \leq n} a^i_{km} y_k y_m + \sum_{1 \leq k \leq n} a^i_k y_k + a_i, \quad 1 \leq i \leq n.$$

We define the A'-pushdown transition matrix (with $|Q| = 1$) $M \in A'^{Y^* \times Y^*}$ by

$$M_{y_i, y_k y_m} = a^i_{km}, \quad M_{y_i, y_k} = a^i_k, \quad M_{y_i, \varepsilon} = a_i, \quad 1 \leq i, k, m \leq n,$$

and write the above A'-algebraic system in the form

$$y_i = \sum_{1 \leq k,m \leq n} M_{y_i, y_k y_m} y_k y_m + \sum_{1 \leq k \leq n} M_{y_i, y_k} y_k + M_{y_i, \varepsilon}, \quad 1 \leq i \leq n.$$

By Theorem 6.4, the least solution of this A'-algebraic system is given by $((M^*)_{y_1, \varepsilon}, \ldots, (M^*)_{y_n, \varepsilon})$. Hence, $\sigma_i = (M^*)_{y_i, \varepsilon}$, $1 \leq i \leq n$. Consider now the A'-pushdown automata $\mathfrak{P}_i = (\{q\}, Y, M, 1, y_i, 1)$, $1 \leq i \leq n$. Then we obtain $\|\mathfrak{P}_i\| = (M^*)_{y_i, \varepsilon} = \sigma_i$, $1 \leq i \leq n$. Hence, $\|\mathfrak{P}_1\| = a$ and our theorem is proven. $\qquad\square$

This completes the proof of the main result of Section 6:

Corollary 6.9. *Let $a \in A$. Then $a \in \mathfrak{Alg}(A')$ iff there exists an A'-pushdown automaton $\mathfrak{P}$ such that $\|\mathfrak{P}\| = a$.*

If our basic semiring is $\mathfrak{P}(\Sigma^*)$ then Corollary 6.9 is nothing else than the well-known characterization of the context-free languages by pushdown automata.

Corollary 6.10. *A formal language is context-free iff it is accepted by a pushdown automaton.*

Observe that the construction proving Corollary 6.5 is nothing else than the well-known triple construction. (See Hopcroft, Ullman [64], Theorem 5.4; Harrison [46], Theorem 5.4.3; Bucher, Maurer [16], Sätze 2.3.10, 2.3.30.) By the proof it is clear that $[q_1, p, q_2] \Rightarrow^* w$ in G iff $w \in ((M^*)_{p,\varepsilon})_{q_1, q_2}$ in $\mathfrak{P}$ iff $(q_1, w, p) \vdash^* (q_2, \varepsilon, \varepsilon)$ in $\mathfrak{P}'$, $q_1, q_2 \in Q$, $p \in \Gamma$, $w \in \Sigma^*$. This means that there exists a derivation of w from the variable $[q_1, p, q_2]$ iff w empties the pushdown tape with contents p by a computation from state q_1 to state q_2.

If our basic semiring is $\mathbb{N}^\infty \langle\langle \Sigma^* \rangle\rangle$, we can draw some even stronger conclusions by Theorem 3.9. In Corollary 6.11 we consider, for a given pushdown automaton $\mathfrak{P}' = (Q, \Sigma, \Gamma, \delta, q_0, p_0, F)$, the number of distinct computations from the *initial instantaneous description* (q_0, w, p_0) *for w* to an *accepting instantaneous description* $(q, \varepsilon, \varepsilon)$, $q \in F$.

Corollary 6.11. *Let L be a formal language over Σ and let $d : \Sigma^* \to \mathbb{N}^\infty$. Then the following two statements are equivalent:*

(i) *There exists a context-free grammar with terminal alphabet Σ such that the number (possibly ∞) of distinct leftmost derivations of w, $w \in \Sigma^*$, from the start variable is given by $d(w)$.*

(ii) *There exists a pushdown automaton with input alphabet Σ such that the number (possibly ∞) of distinct computations from the initial instantaneous description for w, $w \in \Sigma^*$, to an accepting instantaneous description is given by $d(w)$.*

A pushdown automaton with input alphabet Σ is termed *unambiguous* iff, for each word $w \in \Sigma^*$ that is accepted, there exists a unique computation from the initial instantaneous description for w to some accepting instantaneous description.

Corollary 6.12. *A formal language is generated by an unambiguous context-free grammar iff it is accepted by an unambiguous pushdown automaton.*

Example 6.1. (Hopcroft, Ullmann [64], Example 5.3) Let $\Sigma = \{a, b\}$, $Q = \{q_0, q_1\}$, $\Gamma = \{p_0, p\}$ and $\mathfrak{P}' = (Q, \Sigma, \Gamma, \delta, q_0, p_0, \emptyset)$ be a pushdown automaton, where δ is given by

$$\delta(q_0, a, p_0) = \{(q_0, pp_0)\}, \quad \delta(q_0, a, p) = \{(q_0, pp)\}, \quad \delta(q_1, \varepsilon, p_0) = \{(q_1, \varepsilon)\},$$
$$\delta(q_0, b, p) = \{(q_1, \varepsilon)\}, \quad \delta(q_1, b, p) = \{(q_1, \varepsilon)\}, \quad \delta(q_1, \varepsilon, p) = \{(q_1, \varepsilon)\}.$$

We construct a $\mathfrak{P}(\Sigma \cup \{\varepsilon\})$-pushdown automaton $\mathfrak{P} = (Q, \Gamma, M, S, p_0, P)$ such that $w \in \Sigma^*$ is accepted by $\mathfrak{P}'$ by empty store iff $w \in \|\mathfrak{P}\|$:

$$M_{p_0,pp_0} = M_{p,p^2} = \begin{pmatrix} \{a\} & \emptyset \\ \emptyset & \emptyset \end{pmatrix}, \quad M_{p,\varepsilon} = \begin{pmatrix} \emptyset & \{b\} \\ \emptyset & \{\varepsilon, b\} \end{pmatrix}, \quad M_{p_0,\varepsilon} = \begin{pmatrix} \emptyset & \emptyset \\ \emptyset & \{\varepsilon\} \end{pmatrix},$$

$$S = (\{\varepsilon\}\ \emptyset), \quad P = \begin{pmatrix} \{\varepsilon\} \\ \{\varepsilon\} \end{pmatrix}.$$

By the construction of Corollary 6.5, the following algebraic system in matrix notation corresponds to $\mathfrak{P}$:

$$y_0 = S y_{p_0} P, \quad y_{p_0} = M_{p_0,pp_0} y_p y_{p_0} + M_{p_0,\varepsilon}, \quad y_p = M_{p,p^2} y_p^2 + M_{p,\varepsilon}.$$

Hence, we obtain

$$y_0 = [q_0, p_0, q_0] + [q_0, p_0, q_1]$$

$$\begin{pmatrix} [q_0, p_0, q_0] & [q_0, p_0, q_1] \\ [q_1, p_0, q_0] & [q_1, p_0, q_1] \end{pmatrix} =$$

$$= \begin{pmatrix} \{a\}[q_0, p, q_0] & \{a\}[q_0, p, q_1] \\ \emptyset & \emptyset \end{pmatrix} \begin{pmatrix} [q_0, p_0, q_0] & [q_0, p_0, q_1] \\ [q_1, p_0, q_0] & [q_1, p_0, q_1] \end{pmatrix} + \begin{pmatrix} \emptyset & \emptyset \\ \emptyset & \{\varepsilon\} \end{pmatrix}$$

$$\begin{pmatrix} [q_0, p, q_0] & [q_0, p, q_1] \\ [q_1, p, q_0] & [q_1, p, q_1] \end{pmatrix} =$$

$$= \begin{pmatrix} \{a\}[q_0, p, q_0] & \{a\}[q_0, p, q_1] \\ \emptyset & \emptyset \end{pmatrix} \begin{pmatrix} [q_0, p, q_0] & [q_0, p, q_1] \\ [q_1, p, q_0] & [q_1, p, q_1] \end{pmatrix} + \begin{pmatrix} \emptyset & \{b\} \\ \emptyset & \{\varepsilon, b\} \end{pmatrix}$$

Inspection shows that the components of $[q_0, p_0, q_0]$, $[q_0, p, q_0]$, $[q_1, p_0, q_0]$, $[q_1, p, q_0]$ in the least solution are equal to $\emptyset$. This yields

$$y_0 = [q_0, p_0, q_1],$$
$$[q_0, p_0, q_1] = \{a\}[q_0, p, q_1][q_1, p_0, q_1], \qquad [q_1, p_0, q_1] = \{\varepsilon\},$$
$$[q_0, p, q_1] = \{a\}[q_0, p, q_1][q_1, p, q_1] \cup \{b\}, \quad [q_1, p, q_1] = \{\varepsilon, b\}.$$

Denoting $[q_0, p_0, q_1]$ and $[q_0, p, q_1]$ by y_1 and y_2, respectively, and simplifying yields the algebraic system

$$y_1 = \{a\}y_2, \quad y_2 = \{a\}y_2\{\varepsilon, b\} \cup \{b\}.$$

The context-free grammar of Example 5.1 corresponds to this algebraic system. Hence, $\mathfrak{P}'$ accepts exactly the words $a^m b^n$, $m \geq n \geq 1$, by empty store. Moreover, the number of distinct computations from the initial instantaneous description $(q_0, a^m b^n, p_0)$, $m \geq n \geq 1$, to the accepting instantaneous description $(q_1, \varepsilon, \varepsilon)$ is $\binom{m-1}{n-1}$. (There are no computations leading to $(q_0, \varepsilon, \varepsilon)$.) $\square$

7. Transductions and abstract families of elements

We now generalize the concept of a rational transducer by replacing the rational representations in the usual definition by certain complete rational semiring morphisms. Moreover, we introduce algebraic transducers. It turns out that $\mathfrak{Rat}(A')$ and $\mathfrak{Alg}(A')$ are closed under rational and algebraic transductions, respectively.

These generalized rational transducers lead to the generalization of the concept of a full AFL (abbreviation for "abstract family of languages") to the concept of an AFE (abbreviation for "abstract family of elements"): these are fully rationally closed semirings that are also closed under the application of generalized rational transducers. These AFEs are then characterized by automata of a certain type.

Additionally, the concept of an AFE generalizes the concepts of an AFL, an AFP (abbreviation for "abstract family of power series") and a full AFP.

We start with some basic definitions. Throughout this section, A and $\hat{A}$ will denote ω-continuous semirings. A mapping $h : \hat{A} \to A$ is termed a *semiring morphism* iff $h(0) = 0$, $h(1) = 1$, $h(a_1 + a_2) = h(a_1) + h(a_2)$ and $h(a_1 \cdot a_2) = h(a_1) \cdot h(a_2)$ for all $a_1, a_2 \in \hat{A}$. It is a *complete semiring morphism* iff, for all families $(a_i \mid i \in I)$ in $\hat{A}$, $h(\sum_{i \in I} a_i) = \sum_{i \in I} h(a_i)$. Note that a complete semiring morphism is an ω-continuous mapping.

Given a mapping $h : \hat{A} \to A^{Q_1 \times Q_2}$, we define the mappings $h' : \hat{A}^{I_1 \times I_2} \to A^{(I_1 \times Q_1) \times (I_2 \times Q_2)}$ and $h'' : \hat{A}^{I_1 \times I_2} \to (A^{Q_1 \times Q_2})^{I_1 \times I_2}$ by

$$h'(M)_{(i_1,q_1),(i_2,q_2)} = (h''(M)_{i_1,i_2})_{q_1,q_2} = h(M_{i_1,i_2})_{q_1,q_2}$$

for $M \in \hat{A}^{I_1 \times I_2}$, $i_1 \in I_1$, $i_2 \in I_2$, $q_1 \in Q_1$, $q_2 \in Q_2$. In the sequel, we use the same notation h for the mappings h, h' and h''.

Consider the two mappings $h : A \to A^{Q \times Q}$ and $h_1 : A \to A^{Q_1 \times Q_1}$. The *functional composition* $h \circ h_1 : A \to A^{(Q_1 \times Q) \times (Q_1 \times Q)}$ is defined by $(h \circ h_1)(a) = h(h_1(a))$ for $a \in A$.

Easy computations yield the following two technical results.

Theorem 7.1. *Let $h : \hat{A} \to A^{Q \times Q}$ be a complete semiring morphism. Then $h : \hat{A}^{I \times I} \to A^{(I \times Q) \times (I \times Q)}$ and $h : \hat{A}^{I \times I} \to (A^{Q \times Q})^{I \times I}$ are again complete semiring morphisms.*

Theorem 7.2. *Let $h : \hat{A} \to A$ be a complete semiring morphism. Then, for all $a \in \hat{A}$, $h(a^*) = h(a)^*$.*

In the sequel, A' and $\hat{A}'$ denote subsets of A and $\hat{A}$, respectively, both containing the respective neutral elements 0 and 1.

A semiring morphism $h : \hat{A} \to A^{Q \times Q}$ is called $(\hat{A}', A')$-*rational* or $(\hat{A}', A')$-*algebraic* iff, for all $a \in \hat{A}'$, $h(a) \in \mathfrak{Rat}(A')^{Q \times Q}$ or $h(a) \in \mathfrak{Alg}(A')^{Q \times Q}$, respectively. If $\hat{A} = A$ and $\hat{A}' = A'$, these morphisms are called A'-*rational* or A'-*algebraic*, respectively.

Theorem 7.3. *Let $h : A \to A^{Q \times Q}$ and $h' : A \to A^{Q' \times Q'}$ be complete A'-rational semiring morphisms. Then the functional composition $h \circ h' : A \to A^{(Q' \times Q) \times (Q' \times Q)}$ is again an A'-rational semiring morphism.*

Proof. Clearly, $h'(a)_{q_1, q_2} \in \mathfrak{Rat}(A')$ for $a \in A'$, $q_1, q_2 \in Q'$. Since $\mathfrak{Rat}(A')$ is generated by the rational operations from A', Theorems 7.1 and 7.2 imply that the entries of $h(h'(a))$, $a \in A'$, are again in $\mathfrak{Rat}(A')$. □

Before we show a similar result for A'-algebraic semiring morphisms, some considerations on A'-algebraic systems are necessary.

Let $h : A \to A^{Q \times Q}$ be a semiring morphism and extend it to a mapping $h : A(Y) \to A^{Q \times Q}(Y)$ as follows:

(i) If a semiring-polynomial is represented by a product term $a_0 y_{i_1} a_1 \ldots a_{k-1} y_{i_k} a_k$, $k \geq 0$, where $a_j \in A$ and $y_{i_j} \in Y$, then its image is represented by the product term $h(a_0) y_{i_1} h(a_1) \ldots h(a_{k-1}) y_{i_k} h(a_k)$.

(ii) If a semiring-polynomial is represented by a sum of product terms $\sum_{1 \leq j \leq m} t_j$, then its image is represented by the sum of product terms $\sum_{1 \leq j \leq m} h(t_j)$.

Then, by the proof of Theorem 1.4.31 of Lausch, Nöbauer [96], this extension is well-defined and again a semiring morphism.

Theorem 7.4. *Let $h : A \to A^{Q \times Q}$ be a complete semiring morphism. Consider an A-algebraic system $y = p$ with least solution σ. Then $h(\sigma)$ is the least solution of the $A^{Q \times Q}$-algebraic system $y = h(p)$.*

Proof. Let $(\sigma^j \mid j \in \mathbb{N})$ be the approximation sequence of $y = p$. Then $(h(\sigma^j) \mid j \in \mathbb{N})$ is the approximation sequence of $y = h(p)$ and we obtain

$$\mathrm{fix}(h(p)) = \sup(h(\sigma^j) \mid j \in \mathbb{N}) = h(\sup(\sigma^j \mid j \in \mathbb{N})) = h(\sigma).$$

Here we have applied Theorem 3.2 in the second equality. Since $\mathrm{fix}(h(p))$ is the least solution of $y = h(p)$ by Theorem 3.6, we have proved our theorem. □

Corollary 7.5. *Let $h : A \to A^{Q \times Q}$ be a complete A'-algebraic semiring morphism. Then $h(a) \in \mathfrak{Alg}(A')^{Q \times Q}$ for $a \in \mathfrak{Alg}(A')$.*

Proof. By Theorem 7.4, $h(a)$, $a \in \mathfrak{Alg}(A')$, is a component of the least solution of an $\mathfrak{Alg}(A')^{Q \times Q}$-algebraic system. Hence, the entries of $h(a)$ are in $\mathfrak{Alg}(\mathfrak{Alg}(A'))$. Now apply Theorem 3.14. □

Theorem 7.6. *Let $h : A \to A^{Q \times Q}$ and $h' : A \to A^{Q' \times Q'}$ be complete A'-algebraic semiring morphisms. Then the functional composition $h \circ h' : A \to A^{(Q' \times Q) \times (Q' \times Q)}$ is again an A'-algebraic semiring morphism.*

Proof. Clearly, $h'(a)_{q_1, q_2} \in \mathfrak{Alg}(A')$ for $a \in A'$, $q_1, q_2 \in Q'$. Now, Corollary 7.5 implies that the entries of $h(h'(a))$, $a \in A'$, are again in $\mathfrak{Alg}(A')$. □

We are now ready to introduce the notions of a rational and an algebraic transducer.

An $(\hat{A}', A')$-*rational transducer*

$$\mathfrak{T} = (Q, h, S, P)$$

is given by

(i) a finite set Q of *states*,
(ii) a complete $(\hat{A}', A')$-rational semiring morphism $h : \hat{A} \to A^{Q \times Q}$,
(iii) $S \in \mathfrak{Rat}(A')^{1 \times Q}$, called the *initial state vector*,
(iv) $P \in \mathfrak{Rat}(A')^{Q \times 1}$, called the *final state vector*.

The mapping $\|\mathfrak{T}\| : \hat{A} \to A$ *realized* by an $(\hat{A}', A')$-rational transducer $\mathfrak{T} = (Q, h, S, P)$ is defined by

$$\|\mathfrak{T}\|(a) = Sh(a)P, \qquad a \in \hat{A}.$$

A mapping $\tau : \hat{A} \to A$ is called an $(\hat{A}', A')$-*rational transduction* iff there exists an $(\hat{A}', A')$-rational transducer $\mathfrak{T}$ such that $\tau(a) = \|\mathfrak{T}\|(a)$ for all $a \in \hat{A}$. In this case, we say that τ is *realized* by $\mathfrak{T}$. An (A', A')-rational transducer (in case $\hat{A} = A$ and $\hat{A}' = A'$) is called an A'-*rational transducer* and an (A', A')-rational transduction is called an A'-*rational transduction*.

An $(\hat{A}', A')$-*algebraic transducer* $\mathfrak{T} = (Q, h, S, P)$ is defined exactly as an $(\hat{A}', A')$-rational transducer except that h is now a complete $(\hat{A}', A')$-algebraic semiring morphism, and the entries of S and P are in $\mathfrak{Alg}(A')$. The definition of the notions of $(\hat{A}', A')$-*algebraic transduction*, A'-*algebraic transducer* and A'-*algebraic transduction* should be clear.

The next two theorems show that $(\hat{A}', A')$-rational (resp. $(\hat{A}', A')$-algebraic) transductions map $\mathfrak{Rat}(\hat{A}')$ (resp. $\mathfrak{Alg}(\hat{A}')$) into $\mathfrak{Rat}(A')$ (resp. $\mathfrak{Alg}(A')$).

Theorem 7.7. *Assume that $\mathfrak{T}$ is an $(\hat{A}', A')$-rational transducer and that $a \in \mathfrak{Rat}(\hat{A}')$. Then $\|\mathfrak{T}\|(a) \in \mathfrak{Rat}(A')$.*

Proof. Let a be the behavior of the finite $\hat{A}'$-automaton $\mathfrak{A} = (Q, M, S, P)$. Assume that $\mathfrak{T} = (Q', h, S', P')$. We consider now the finite $\mathfrak{Rat}(A')$-automaton $\mathfrak{A}' = (Q \times Q', h(M), S'h(S), h(P)P')$. By Corollary 4.9 we obtain $\|\mathfrak{A}'\| \in \mathfrak{Rat}(A')$. Since $\|\mathfrak{A}'\| = S'h(S)h(M)^*h(P)P' = S'h(SM^*P)P' = \|\mathfrak{T}\|(\|\mathfrak{A}\|)$, our theorem is proved. □

Theorem 7.8. *Assume that $\mathfrak{T}$ is an $(\hat{A}', A')$-algebraic transducer and that $a \in \mathfrak{Alg}(\hat{A}')$. Then $\|\mathfrak{T}\|(a) \in \mathfrak{Alg}(A')$.*

Proof. Let a be the behavior of the $\hat{A}'$-pushdown automaton $\mathfrak{P} = (Q, \Gamma, M, S, p_0, P)$. Assume that $\mathfrak{T} = (Q', h, S', P')$. We consider now the $\mathfrak{Alg}(A')$-pushdown automaton $\mathfrak{P}' = (Q \times Q', \Gamma, h(M), S'h(S), p_0, h(P)P')$. By Corollary 6.6 and Theorem 3.14 we obtain $\|\mathfrak{P}'\| \in \mathfrak{Alg}(A')$.

Since $\|\mathfrak{P}'\| = S'h(S)(h(M)^*)_{p_0, \varepsilon}h(P)P' = S'h(S)h((M^*)_{p_0, \varepsilon})h(P)P' = S'h(S(M^*)_{p_0, \varepsilon}P)P' = \|\mathfrak{T}\|(\|\mathfrak{P}\|)$, our theorem is proved. □

We now consider the functional composition of A'-rational (resp. A'-algebraic) transductions.

Theorem 7.9. *The family of A'-rational (resp. A'-algebraic) transductions is closed under functional composition.*

Proof. Let $\mathfrak{T}_j = (Q_j, h_j, S_j, P_j)$, $j = 1, 2$, be two A'-rational (resp. A'-algebraic) transducers. We want to show that the mapping $\tau : A \to A$ defined by $\tau(a) = \|\mathfrak{T}_2\|(\|\mathfrak{T}_1\|(a))$, $a \in A$, is an A'-rational (resp. A'-algebraic) transduction.

Consider $\mathfrak{T} = (Q_1 \times Q_2, h_2 \circ h_1, S_2 h_2(S_1), h_2(P_1) P_2)$. By Theorem 7.3 (resp. Theorem 7.6) the mapping $h_2 \circ h_1$ is a complete A'-rational (resp. A'-algebraic) semiring morphism. Furthermore, the entries of $S_2 h_2(S_1)$ and $h_2(P_1) P_2$ are in $\mathfrak{Rat}(A')$ (resp. $\mathfrak{Alg}(A')$). Hence, $\mathfrak{T}$ is an A'-rational (resp. A'-algebraic) transducer. Since, for $a \in A$,

$$
\begin{aligned}
\|\mathfrak{T}\|(a) &= S_2 h_2(S_1) h_2(h_1(a)) h_2(P_1) P_2 = S_2 h_2(S_1 h_1(a) P_1) P_2 = \\
&= S_2 h_2(\|\mathfrak{T}_1\|(a)) P_2 = \|\mathfrak{T}_2\|(\|\mathfrak{T}_1\|(a)),
\end{aligned}
$$

our theorem is proved. $\qquad\qquad\qquad\qquad\qquad\qquad\qquad\qquad\qquad$ □

Most of the definitions and results developed up to now in this section are due to Kuich [88]. They are generalizations of definitions and results which we will consider now. We specialize our definitions and results to semirings of formal power series (see Nivat [108], Jacob [67], Salomaa, Soittola [123], Kuich, Salomaa [92]). We make the following conventions for the remainder of this section: The set Σ_∞ is a fixed infinite alphabet, and Σ, possibly provided with indices, is a finite subalphabet of Σ_∞.

In connection with formal power series, our basic semiring will be $A\langle\langle \Sigma_\infty^* \rangle\rangle$, where A is *commutative*. The subsemiring of $A\langle\langle \Sigma_\infty^* \rangle\rangle$ containing all power series whose supports are contained in some Σ^* is denoted by $A\{\{\Sigma_\infty^*\}\}$, i. e.,

$$
A\{\{\Sigma_\infty^*\}\} = \{r \in A\langle\langle \Sigma_\infty^* \rangle\rangle \mid \text{there exists a finite alphabet } \Sigma \subset \Sigma_\infty \\
\text{such that } \operatorname{supp}(r) \subseteq \Sigma^*\}.
$$

For $\Sigma \subset \Sigma_\infty$, $A\langle\langle \Sigma^* \rangle\rangle$ is isomorphic to a subsemiring of $A\{\{\Sigma_\infty^*\}\}$. Hence, we may assume that $A\langle\langle \Sigma^* \rangle\rangle \subset A\{\{\Sigma_\infty^*\}\}$.

Furthermore, we define three subsemirings of $A\{\{\Sigma_\infty^*\}\}$, namely, the semiring of algebraic power series $A^{\text{alg}}\{\{\Sigma_\infty^*\}\}$, the semiring of rational power series $A^{\text{rat}}\{\{\Sigma_\infty^*\}\}$ and the semiring of polynomials $A\{\Sigma_\infty^*\}$ by

$$
\begin{aligned}
A^{\text{alg}}\{\{\Sigma_\infty^*\}\} &= \{r \in A\{\{\Sigma_\infty^*\}\} \mid \text{there exists a finite alphabet } \Sigma \subset \Sigma_\infty \\
&\qquad \text{such that } r \in A^{\text{alg}}\langle\langle \Sigma^* \rangle\rangle\}, \\
A^{\text{rat}}\{\{\Sigma_\infty^*\}\} &= \{r \in A\{\{\Sigma_\infty^*\}\} \mid \text{there exists a finite alphabet } \Sigma \subset \Sigma_\infty \\
&\qquad \text{such that } r \in A^{\text{rat}}\langle\langle \Sigma^* \rangle\rangle\}, \text{ and} \\
A\{\Sigma_\infty^*\} &= \{r \in A\{\{\Sigma_\infty^*\}\} \mid \operatorname{supp}(r) \text{ is finite}\}.
\end{aligned}
$$

Moreover, we define $A\{\Sigma_\infty \cup \varepsilon\} = \{r \in A\{\Sigma_\infty^*\} \mid \text{supp}(r) \subset \Sigma_\infty \cup \{\varepsilon\}\}$ and $A\{\Sigma_\infty\} = \{r \in A\{\Sigma_\infty^*\} \mid \text{supp}(r) \subset \Sigma_\infty\}$. Observe that $A^{\text{rat}}\{\{\Sigma_\infty^*\}\} = \Re\mathfrak{at}(A\{\Sigma_\infty \cup \varepsilon\})$ and $A^{\text{alg}}\{\{\Sigma_\infty^*\}\} = \mathfrak{Alg}(A\{\Sigma_\infty \cup \varepsilon\})$.

A multiplicative morphism $\mu : \Sigma_\infty^* \to (A\{\{\Sigma_\infty^*\}\})^{Q \times Q}$ is called a *representation* iff there exists a Σ such that $\mu(x) = 0$ for $x \in \Sigma_\infty - \Sigma$. Observe that if μ is a representation, there exist only finitely many entries $\mu(x)_{q_1,q_2} \neq 0$, $x \in \Sigma_\infty$, $q_1, q_2 \in Q$. Hence, there is a Σ' such that $\mu(w)_{q_1,q_2} \in A\langle\!\langle \Sigma'^* \rangle\!\rangle$ for all $w \in \Sigma_\infty^*$. A representation can be extended to a mapping $\mu : A\langle\!\langle \Sigma_\infty^* \rangle\!\rangle \to (A\langle\!\langle \Sigma_\infty^* \rangle\!\rangle)^{Q \times Q}$ by the definition

$$\mu(r) = \mu\Big(\sum_{w \in \Sigma_\infty^*} (r, w)w \Big) = \sum_{w \in \Sigma_\infty^*} \text{diag}((r, w))\mu(w), \quad r \in A\langle\!\langle \Sigma_\infty^* \rangle\!\rangle,$$

where $\text{diag}(a)$ is the diagonal matrix whose diagonal entries all are equal to a. (Observe that $\text{diag}((r, w))\mu(w) = (r, w) \otimes \mu(w)$, where $\otimes$ denotes the Kronecker product. For definitions and results in connection with the Kronecker product see Kuich, Salomaa [92].) Note that we are using the same notation "μ" for both mappings. However, this should not lead to any confusion. Observe that in fact μ is a mapping $\mu : A\langle\!\langle \Sigma_\infty^* \rangle\!\rangle \to (A\{\{\Sigma_\infty^*\}\})^{Q \times Q}$.

The following theorem states an important property of the extended mapping μ and is proved by an easy computation.

Theorem 7.10. *Let A be a commutative ω-continuous semiring. If $\mu : \Sigma_\infty^* \to (A\{\{\Sigma_\infty^*\}\})^{Q \times Q}$ is a representation then the extended mapping $\mu : A\langle\!\langle \Sigma_\infty^* \rangle\!\rangle \to (A\{\{\Sigma_\infty^*\}\})^{Q \times Q}$ is a complete semiring morphism.*

A representation μ is called *rational* or *algebraic* iff

$$\mu : \Sigma_\infty^* \to (A^{\text{rat}}\{\{\Sigma_\infty^*\}\})^{Q \times Q} \quad \text{or} \quad \mu : \Sigma_\infty^* \to (A^{\text{alg}}\{\{\Sigma_\infty^*\}\})^{Q \times Q},$$

respectively.

Theorem 7.11. *Let A be a commutative ω-continuous semiring and let $\mu_j : \Sigma_\infty^* \to (A\{\{\Sigma_\infty^*\}\})^{Q_j \times Q_j}$, $j = 1, 2$, be rational (resp. algebraic) representations. Then $\mu : \Sigma_\infty^* \to (A\{\{\Sigma_\infty^*\}\})^{(Q_1 \times Q_2) \times (Q_1 \times Q_2)}$, where $\mu(x) = \mu_2(\mu_1(x))$ for all $x \in \Sigma_\infty$, is again a rational (resp. an algebraic) representation.*

Furthermore, for all $r \in A\langle\!\langle \Sigma_\infty^* \rangle\!\rangle$, $\mu(r) = \mu_2(\mu_1(r))$.

Proof. By Theorem 7.3 (resp. Theorem 7.6), the entries of $\mu(x)$ are in $A^{\text{rat}}\{\{\Sigma_\infty^*\}\}$ (resp. $A^{\text{alg}}\{\{\Sigma_\infty^*\}\}$) for all $x \in \Sigma_\infty$. Hence, μ is a rational (resp. an algebraic) representation.

We now prove the equality of our theorem. We deduce, for all $r \in A\langle\!\langle \Sigma_\infty^* \rangle\!\rangle$,

$$\mu_2(\mu_1(r)) = \mu_2\Big(\sum_{w \in \Sigma_\infty^*} \text{diag}((r, w))\mu_1(w) \Big)$$

$$= \sum_{w \in \Sigma_\infty^*} \text{diag}((r, w))\mu_2(\mu_1(w)) = \sum_{w \in \Sigma_\infty^*} \text{diag}((r, w))\mu(w) = \mu(r). \qquad \square$$

We now specialize the notions of A'-rational (resp. A'-algebraic) trans-
ducers and consider $A\{\Sigma_\infty \cup \varepsilon\}$-rational (resp. $A\{\Sigma_\infty \cup \varepsilon\}$-algebraic) transducers $\mathfrak{T} = (Q, \mu, S, P)$, where μ is a rational (resp. an algebraic) represen-
tation. Hence, there exist finite alphabets Σ and Σ' such that $\mu(x) = 0$
for $x \in \Sigma_\infty - \Sigma$ and the entries of $\mu(w)$, $w \in \Sigma^*$ are in $A^{\mathrm{rat}}\langle\!\langle \Sigma'^* \rangle\!\rangle$
(resp. $A^{\mathrm{alg}}\langle\!\langle \Sigma'^* \rangle\!\rangle$). Furthermore, we assume that the entries of S and P are
in $A^{\mathrm{rat}}\langle\!\langle \Sigma'^* \rangle\!\rangle$ (resp. $A^{\mathrm{alg}}\langle\!\langle \Sigma'^* \rangle\!\rangle$). We call these $A\{\Sigma_\infty \cup \varepsilon\}$-rational (resp.
$A\{\Sigma_\infty \cup \varepsilon\}$-algebraic) transducers simply *rational* (resp. *algebraic*) *transduc-
ers*.

A rational or an algebraic transducer $\mathfrak{T} = (Q, \mu, S, P)$ specified as above
can be considered to be a finite automaton equipped with an output device. In
a state transition from state q_1 to state q_2, $\mathfrak{T}$ reads a letter $x \in \Sigma$ and outputs
the rational or algebraic power series $\mu(x)_{q_1,q_2}$. A sequence of state transitions
outputs the product of the power series of the single state transitions. All
sequences of length n of state transitions from state q_1 to state q_2 reading
a word $w \in \Sigma^*$, $|w| = n$, output the power series $\mu(w)_{q_1,q_2}$. This output
is multiplied with the correct components of the initial and the final state
vector, and $S_{q_1}\mu(w)_{q_1,q_2}P_{q_2}$ is said to be the *translation* of w by transitions
from q_1 to q_2. Summing up for all $q_1, q_2 \in Q$, $\sum_{q_1,q_2 \in Q} S_{q_1}\mu(w)_{q_1,q_2}P_{q_2} =
S\mu(w)P$ is said to be the *translation* of w by $\mathfrak{T}$. A power series $r \in A\langle\!\langle \Sigma_\infty^* \rangle\!\rangle$
is *translated* by $\mathfrak{T}$ to the power series

$$\|\mathfrak{T}\|(r) = S\mu(r)P = S\Big(\sum_{w \in \Sigma_\infty^*} \mathrm{diag}((r,w))\mu(w) \Big)P$$

$$= \sum_{w \in \Sigma_\infty^*} (r,w)S\mu(w)P = \sum_{w \in \Sigma_\infty^*} (r,w)\|\mathfrak{T}\|(w) \in A\langle\!\langle \Sigma'^* \rangle\!\rangle.$$

Observe that $\|\mathfrak{T}\|(r) = \|\mathfrak{T}\|(r \odot \mathrm{char}(\Sigma^*))$. Hence, in fact, $\mathfrak{T}$ translates a
power series in $A\langle\!\langle \Sigma^* \rangle\!\rangle$ to a power series in $A\langle\!\langle \Sigma'^* \rangle\!\rangle$.

Specializations of Theorems 7.7 and 7.8 yield the next result.

Corollary 7.12. *Assume that $\mathfrak{T}$ is a rational (resp. an algebraic) transducer
and that $r \in A^{\mathrm{rat}}\langle\!\langle \Sigma^* \rangle\!\rangle$ (resp. $r \in A^{\mathrm{alg}}\langle\!\langle \Sigma^* \rangle\!\rangle$). Then $\|\mathfrak{T}\|(r) \in A^{\mathrm{rat}}\langle\!\langle \Sigma'^* \rangle\!\rangle$
(resp. $A^{\mathrm{alg}}\langle\!\langle \Sigma'^* \rangle\!\rangle$) for some Σ'.*

We now introduce the notion of a substitution. Assume that $\sigma : \Sigma_\infty^* \to
A\{\{\Sigma_\infty^*\}\}$ is a representation, where $\sigma(x) = 0$ for $x \in \Sigma_\infty - \Sigma$ and the
entries of $\sigma(x)$, $x \in \Sigma$, are in $A\langle\!\langle \Sigma'^* \rangle\!\rangle$. Then the mapping $\sigma : A\langle\!\langle \Sigma_\infty^* \rangle\!\rangle \to
A\langle\!\langle \Sigma'^* \rangle\!\rangle$, where $\sigma(r) = \sum_{w \in \Sigma_\infty^*} (r,w)\sigma(w)$ for all $r \in A\langle\!\langle \Sigma_\infty^* \rangle\!\rangle$, is a complete
semiring morphism. We call this complete semiring morphism a *substitution*.
If $\sigma : \Sigma_\infty^* \to A^{\mathrm{rat}}\langle\!\langle \Sigma'^* \rangle\!\rangle$ or $\sigma : \Sigma_\infty^* \to A^{\mathrm{alg}}\langle\!\langle \Sigma'^* \rangle\!\rangle$ then we call the substitution
rational or *algebraic*, respectively. Clearly, a rational or algebraic substitution
is a particular rational or algebraic transduction, respectively.

Corollary 7.13. *If σ is a rational (resp. algebraic) substitution and r is a
rational (resp. algebraic) power series then $\sigma(r)$ is again a rational (resp.
algebraic) power series.*

We now turn to language theory (see Berstel [4]). The basic semiring is now $\mathfrak{P}(\Sigma_\infty^*)$. Let $\mathfrak{L}(\Sigma_\infty)$ be the subset of $\mathfrak{P}(\Sigma_\infty^*)$ containing all formal languages, i. e.,

$$\mathfrak{L}(\Sigma_\infty) = \{L \mid L \subseteq \Sigma^*, \ \Sigma \subset \Sigma_\infty\}.$$

A representation is now a multiplicative morphism $\mu : \Sigma_\infty^* \rightarrow \mathfrak{L}(\Sigma_\infty)^{Q \times Q}$. If the representation is rational or algebraic then the entries of $\mu(x)$, $x \in \Sigma_\infty$, are regular or context-free languages, respectively. A rational or an algebraic transducer $\mathfrak{T} = (Q, \mu, S, P)$ is specified by a rational or an algebraic representation μ as above; moreover, the entries of S and P are regular or context-free languages, respectively.

Corollary 7.14. *Assume that $\mathfrak{T}$ is a rational (resp. algebraic) transducer. Then $\|\mathfrak{T}\|(L)$ is a regular (resp. context-free) language if L is regular (resp. context-free).*

A substitution is now a complete semiring morphism $\sigma : \Sigma_\infty^* \rightarrow \mathfrak{L}(\Sigma_\infty)$ such that $\sigma(x) = \emptyset$ for $x \in \Sigma_\infty - \Sigma$. It is defined by the values of $\sigma(x) \subseteq \Sigma'^*$ for $x \in \Sigma$. Since σ is a complete semiring morphism, we obtain $\sigma(w) = \sigma(x_1) \dots \sigma(x_n) \subseteq \Sigma'^*$ for $w = x_1 \dots x_n$, $x_i \in \Sigma$, $1 \leq i \leq n$, and $\sigma(L) = \bigcup_{w \in L} \sigma(w) \subseteq \Sigma'^*$ for $L \subseteq \Sigma^*$.

A substitution is called *regular* or *context-free* iff each symbol is mapped to a regular or context-free language, respectively.

Corollary 7.15. *A regular (resp. context-free) substitution maps a regular (resp. context-free) language to a regular (resp. context-free) language.*

We now return to the general theory. Since Theorem 7.16 below is, in general, not valid for ω-continuous semirings (see Goldstern [43], Beispiel 5.11), we introduce continuous semirings.

Let A be a complete, naturally ordered semiring. Then A is called *continuous* iff, for all $a_i, c \in A$ and all index sets I, the following condition is satisfied:

$$\text{if } \sum_{i \in E} a_i \sqsubseteq c \text{ for all finite } E \subseteq I \text{ then } \sum_{i \in I} a_i \sqsubseteq c.$$

(An equivalent condition is $\sup\{\sum_{i \in E} a_i \mid E \subseteq I, E \text{ finite}\} = \sum_{i \in I} a_i$. Compare with Theorem 2.3 and see Goldstern [43], Sakarovitch [124], and Karner [69].)

Theorem 7.16. (Goldstern [43]) *Let A be a continuous semiring and Δ be an alphabet, and consider a multiplicative monoid morphism $h : \Delta^* \rightarrow A$. Then there exists a unique extension of h to a complete semiring morphism $h : \mathbb{N}^\infty \langle\langle \Delta^* \rangle\rangle \rightarrow A$. This unique extension is given by*

$$h(r) = h\Big(\sum_{v \in \Delta^*} (r, v)v \Big) = \sum_{v \in \Delta^*} \nu((r, v))h(v), \quad r \in \mathbb{N}^\infty \langle\langle \Delta^* \rangle\rangle,$$

where ν is defined by $\nu(n) = \sum_{0 \leq i \leq n-1} 1$, $n \in \mathbb{N}$, and $\nu(\infty) = \sum_{i \geq 0} 1$.

Two remarks are now in order. Firstly, if A and $\hat{A}$ are continuous semirings and $h : \mathbb{N}^\infty \langle\!\langle \Delta^* \rangle\!\rangle \to A$ and $h' : A \to \hat{A}$ are complete semiring morphisms then, for $r \in \mathbb{N}^\infty \langle\!\langle \Delta^* \rangle\!\rangle$, $h'(h(r)) = \sum_{v \in \Delta^*} \nu((r, v)) h'(h(v))$.

Secondly, if $h : \mathbb{N}^\infty \langle\!\langle \Delta^* \rangle\!\rangle \to A^{Q \times Q}$, where A is a continuous semiring, then, for $r \in \mathbb{N}^\infty \langle\!\langle \Delta^* \rangle\!\rangle$, $h(r) = \sum_{v \in \Delta^*} \operatorname{diag}(\nu((r, v))) h(v)$.

Clearly, a continuous semiring is also an ω-continuous semiring. Hence, we do not violate our convention that A always denotes an ω-continuous semiring if we make the following additional convention: In the remainder of Section 7, A always will denote a *continuous* semiring. Moreover, we make the convention that all sets Q, possibly provided with indices, are finite and nonempty, and are subsets of some fixed countably infinite set Q_∞ with the following property: if $q_1, q_2 \in Q_\infty$ then $(q_1, q_2) \in Q_\infty$.

Consider the family of all semiring morphisms $h : A \to A^{Q \times Q}$, $Q \subset Q_\infty$, Q finite. A nonempty subfamily $\mathfrak{H}$ of this family is *closed under matricial composition* iff the following conditions are satisfied for arbitrary $h : A \to A^{Q \times Q}$ and $h' : A \to A^{Q' \times Q'}$ in $\mathfrak{H}$:

(i) The functional composition $h \circ h' : A \to A^{(Q' \times Q) \times (Q' \times Q)}$ is again in $\mathfrak{H}$.
(ii) If $Q \cap Q' = \emptyset$ then the mapping $h + h' : A \to A^{(Q \cup Q') \times (Q \cup Q')}$ defined by

$$(h + h')(a) = \begin{pmatrix} h(a) & 0 \\ 0 & h'(a) \end{pmatrix}, \quad a \in A,$$

where the blocks are indexed by Q and Q', is again in $\mathfrak{H}$.

Clearly, the family of all complete semiring morphisms $h : A \to A^{Q \times Q}$, $Q \subset Q_\infty$, Q finite, is closed under matricial composition. By Theorem 7.3 (resp. Theorem 7.6), the family of all complete A'-rational (resp. A'-algebraic) semiring morphisms $h : A \to A^{Q \times Q}$, $Q \subset Q_\infty$, Q finite, is closed under matricial composition.

For the rest of this section, we assume $\mathfrak{H}$ to be a nonempty family of *complete* semiring morphisms that is *closed under matricial composition*.

An A'-rational (resp. A'-algebraic) transducer $\mathfrak{T} = (Q, h, S, P)$ where $h \in \mathfrak{H}$ is called $\mathfrak{H}$-A'-*rational* (resp. $\mathfrak{H}$-A'-*algebraic*) *transducer*. The definition of the notions of a $\mathfrak{H}$-A'-*rational* and a $\mathfrak{H}$-A'-*algebraic transduction* should be clear. By a slight generalization of Theorem 7.9, the family of $\mathfrak{H}$-A'-rational and the family of $\mathfrak{H}$-A'-algebraic transductions is closed under functional composition.

We now introduce the notion of an abstract family of elements. Given $A' \subseteq A$, we define $[A'] \subseteq A$ to be the least complete subsemiring of A that contains A'. The semiring $[A']$ is called the *complete semiring generated by A'* and is again an continuous semiring. Each element a of $[A']$ can be generated from elements of A' by multiplication and summation (including "infinite summation"):

$$a \in [A'] \quad \text{iff} \quad a = \sum_{i \in I} a_{i1} \dots a_{in_i},$$

where $a_{ij} \in A'$ and I is an index set. Any subset $\mathfrak{L}$ of $[A']$ is called A'-*family of elements*.

Let $\mathfrak{T}$ be an A'-rational transducer. Then for each $a \in [A']$, we obtain $\|\mathfrak{T}\|(a) \in [A']$. Hence, the set

$$\mathfrak{M}(\mathfrak{L}) = \{\tau(a) \mid a \in \mathfrak{L}, \tau : A \to A \text{ is an } \mathfrak{H}\text{-}A'\text{-rational transduction}\}$$

is again an A'-family of elements for each A'-family of elements $\mathfrak{L}$. An A'-family of elements $\mathfrak{L}$ is said to be *closed under $\mathfrak{H}$-A'-rational transductions* iff $\mathfrak{L} = \mathfrak{M}(\mathfrak{L})$. The notation $\mathfrak{F}(\mathfrak{L})$ is used for the smallest fully rationally closed subsemiring of $[A']$ that is closed under $\mathfrak{H}$-A'-rational transductions and contains $\mathfrak{L}$. (We have tried to use in our notation letters customary in AFL theory to aid the reader familiar with this theory. See Ginsburg [37].) An A'-family of elements $\mathfrak{L}$ is called A'-*abstract family of elements* (briefly A'-AFE) iff $\mathfrak{L} = \mathfrak{F}(\mathfrak{L})$. Observe that in the definitions of $\mathfrak{M}(\mathfrak{L})$ and $\mathfrak{F}(\mathfrak{L})$ the family of complete semiring morphisms $\mathfrak{H}$ and the subset A' of A are implicitly present. In the sequel we assume $A = [A']$. This will cause no loss of generality.

The next result is implied by Theorems 7.7 and 7.8.

Theorem 7.17. $\mathfrak{Rat}(A')$ *and* $\mathfrak{Alg}(A')$ *are* A'-*AFEs.*

We now show that $\mathfrak{Rat}(A')$ is the smallest A'-AFE.

Theorem 7.18. *Let $\mathfrak{L}$ be an A'-AFE. Then $\mathfrak{Rat}(A') \subseteq \mathfrak{L}$.*

Proof. Since $\mathfrak{L}$ is a semiring, we have $1 \in \mathfrak{L}$. Since $\mathfrak{H}$ is nonempty, there exists, for some $Q \subset Q_\infty$, a morphism $h : A \to A^{Q \times Q}$ in $\mathfrak{H}$. Let q_0 be in Q. For $a \in \mathfrak{Rat}(A')$, consider now the $\mathfrak{H}$-A'-rational transducer $\mathfrak{T}_a = (Q, h, P, S_a)$, where $P_{q_0} = 1$, $P_q = 0$, $(S_a)_{q_0} = a$, $(S_a)_q = 0$ for $q \neq q_0$. Given $b \in A$, we obtain $\|\mathfrak{T}_a\|(b) = h(b)_{q_0,q_0} a$. Hence, $\|\mathfrak{T}_a\|(1) = a$. Since $\mathfrak{L}$ is closed under $\mathfrak{H}$-A'-rational transductions, we obtain $\|\mathfrak{T}_a\|(1) = a \in \mathfrak{L}$ for all $a \in \mathfrak{Rat}(A')$. $\square$

In the sequel, $\Delta = \{\mathbf{a} \mid a \in A\} \cup Z$ is an alphabet. Here $\{\mathbf{a} \mid a \in A\}$ is a copy of A and Z is an infinite alphabet of variables. A multiplicative monoid morphism $h : \Delta^* \to A^{Q \times Q}$ is *compatible with* $\mathfrak{H}$ iff the following conditions are satisfied:

(i) The mapping $h' : A \to A^{Q \times Q}$ defined by $h'(a) = h(\mathbf{a})$, $a \in A$, is a complete A'-rational semiring morphism in $\mathfrak{H}$,

(ii) $h(\mathbf{a}), h(z) \in \mathfrak{Rat}(A')^{Q \times Q}$ for $a \in A'$, $z \in Z$, and $h(z) = 0$ for almost all variables $z \in Z$.

If $h : \Delta^* \to A^{Q \times Q}$ is compatible with $\mathfrak{H}$ and if $h_1 : A \to A^{Q_1 \times Q_1}$ is a complete A'-rational semiring morphism in $\mathfrak{H}$ then $h_1 \circ h : \Delta^* \to A^{(Q \times Q_1) \times (Q \times Q_1)}$ is again compatible with $\mathfrak{H}$.

We introduce now the notions of a *type T*, a *T-matrix*, a *T-automaton* and the *automaton representing T*. Intuitively speaking this means the following.

A T-automaton is a finite automaton with an additional working tape, whose contents are stored in the states of the T-automaton. The type T of the T-automaton indicates how information can be retrieved from the working tape. For instance, pushdown automata can be viewed as automata of a specific type.

A *type* is a quadruple

$$(\Gamma_T, \Delta_T, T, \pi_T),$$

where

(i) Γ_T is the set of *storage symbols*,
(ii) $\Delta_T \subseteq \{\mathbf{a} \mid a \in A'\} \cup Z$ is the alphabet of *instructions*,
(iii) $T \in (\mathbb{N}^\infty\{\Delta_T\})^{\Gamma_T^* \times \Gamma_T^*}$ is the *type matrix*,
(iv) $\pi_T \in \Gamma_T^*$ is the *initial contents of the working tape*.

In the sequel we often speak of the type T if Γ_T, Δ_T and π_T are understood.

A matrix $M \in (\mathfrak{Rat}(A')^{Q \times Q})^{\Gamma_T^* \times \Gamma_T^*}$ is called a T-*matrix* iff there exists a monoid morphism $h : \Delta^* \to A^{Q \times Q}$ that is compatible with $\mathfrak{H}$ such that $M = h(T)$. If $M = h(T)$ is a T-matrix and $h' : A \to A^{Q' \times Q'}$ is a complete A'-rational semiring morphism in $\mathfrak{H}$ then, by the first remark after Theorem 7.16 and by Theorem 7.3, $h' \circ h$ is compatible with $\mathfrak{H}$ and $h'(M) = h'(h(T))$ is again a T-matrix.

A T-*automaton*

$$\mathfrak{A} = (Q, \Gamma_T, M, S, \pi_T, P)$$

is defined by

(i) a finite set Q of *states*,
(ii) a T-matrix M, called the *transition matrix*,
(iii) $S \in \mathfrak{Rat}(A')^{1 \times Q}$, called the *initial state vector*,
(iv) $P \in \mathfrak{Rat}(A')^{Q \times 1}$, called the *final state vector*.

Observe that Γ_T and π_T are determined by T. The *behavior* of the T-automaton $\mathfrak{A}$ is given by

$$\|\mathfrak{A}\| = S(M^*)_{\pi_T, \varepsilon} P.$$

Clearly, for each such T-automaton $\mathfrak{A}$ there exists a $\mathfrak{Rat}(A')$-automaton $\mathfrak{A}' = (\Gamma_T^* \times Q, M', S', P')$ such that $\|\mathfrak{A}'\| = \|\mathfrak{A}\|$. This is achieved by choosing $M'_{(\pi_1, q_1),(\pi_2, q_2)} = (M_{\pi_1, \pi_2})_{q_1, q_2}$, $S'_{(\pi_T, q)} = S_q$, $S'_{(\pi, q)} = 0$, $\pi \neq \pi_T$, $P'_{(\varepsilon, q)} = P_q$, $P'_{(\pi, q)} = 0$, $\pi \neq \varepsilon$, $q_1, q_2, q \in Q$, $\pi_1, \pi_2, \pi \in \Gamma_T^*$.

The *automaton* $\mathfrak{A}_T$ *representing a type* $(\Gamma_T, \Delta_T, T, \pi_T)$ is an $\mathbb{N}^\infty\{\Delta_T\}$-automaton defined by

$$\mathfrak{A}_T = (\Gamma_T^*, T, S_T, P_T),$$

where $(S_T)_{\pi_T} = \varepsilon$, $(S_T)_\pi = 0$, $\pi \in \Gamma_T^*$, $\pi \neq \pi_T$, $(P_T)_\varepsilon = \varepsilon$, $(P_T)_\pi = 0$, $\pi \in \Gamma_T^*$, $\pi \neq \varepsilon$. The behavior of $\mathfrak{A}_T$ is $\|\mathfrak{A}_T\| = (T^*)_{\pi_T, \varepsilon}$.

In a certain sense, $\mathfrak{A}_T$ generates an A'-family of elements. Let $\hat{A} = \mathbb{N}^\infty\langle\!\langle \Delta_T^* \rangle\!\rangle$ and $\hat{A}' = \{d \mid d \in \Delta_T\} \cup \{\varepsilon, 0\}$ and consider $(\hat{A}', A')$-rational transducers $\mathfrak{T} = (Q, h, S, P)$, where $h : \Delta^* \to A^{Q\times Q}$ is a monoid morphism compatible with $\mathfrak{H}$. Given an $(\hat{A}', A')$-rational transducer $\mathfrak{T} = (Q, h, S, P)$, consider the T-automaton $\mathfrak{A} = (Q, \Gamma_T, M, S, \pi_T, P)$, where $M = h(T)$. We apply $\mathfrak{T}$ to $\|\mathfrak{A}_T\|$ and obtain

$$\|\mathfrak{T}\|(\|\mathfrak{A}_T\|) = Sh((T^*)_{\pi_T, \varepsilon})P = S(M^*)_{\pi_T, \varepsilon}P = \|\mathfrak{A}\|.$$

Conversely, for each T-automaton $\mathfrak{A}$ there exists an $(\hat{A}', A')$-rational transducer $\mathfrak{T}$ such that $\|\mathfrak{A}\| = \|\mathfrak{T}\|(\|\mathfrak{A}_T\|)$.

We define now the A'-family of elements

$$\mathfrak{Rat}_T(A') = \{\|\mathfrak{A}\| \mid \mathfrak{A} \text{ is a } T\text{-automaton}\}.$$

Hence, $\mathfrak{Rat}_T(A')$ contains exactly all elements $\|\mathfrak{T}\|(\|\mathfrak{A}_T\|)$, where $\mathfrak{T} = (Q, h, S, P)$ is an $(\hat{A}', A')$-rational transducer and $h : \Delta^* \to A^{Q\times Q}$ is compatible with $\mathfrak{H}$. Observe that in the definitions of a T-matrix, of a T-automaton and of $\mathfrak{Rat}_T(A')$, $\mathfrak{H}$ is implicitly present.

It will turn out that $\mathfrak{Rat}_T(A')$ constitutes an A'-AFE if T is a restart type. Here a type $(\Gamma_T, \Delta_T, T, \pi_T)$ is called a *restart type* iff $\pi_T = \varepsilon$ and the non-null entries of T satisfy the conditions $T_{\varepsilon, \varepsilon} = z^0 \in Z$, $T_{\varepsilon, \pi} \in \mathbb{N}^\infty\{Z - \{z^0\}\}$ and $T_{\pi, \pi'} \in \mathbb{N}^\infty\{\Delta_T - \{z^0\}\}$ for all $\pi, \pi' \in \Gamma_T^*$, $\pi \neq \varepsilon$, and for some distinguished instruction $z^0 \in \Delta_T$. Observe that the working tape is empty at the beginning of the computation.

A full proof of the next theorem is given in Kuich [88].

Theorem 7.19. *If T is a restart type then $\mathfrak{Rat}_T(A')$ is a fully rationally closed semiring containing $\mathfrak{Rat}(A')$.*

Proof. We prove only closure under star. Assume that $\mathfrak{A} = (Q, \Gamma_T, M, S, \varepsilon, P)$, where $M = h(T)$, is a T-automaton. We give the construction of a T-automaton $\mathfrak{A}' = (Q, \Gamma_T, M', S, \varepsilon, P)$ with $\|\mathfrak{A}'\| = \|\mathfrak{A}\|^+$.

Let $h' : \Delta^* \to A^{Q\times Q}$ be defined by $h'(z^0) = h(z^0) + PS$, $h'(d) = h(d)$, $d \in \Delta - \{z^0\}$. Then h' is compatible with $\mathfrak{H}$. Define $\tilde{M} \in (\mathfrak{Rat}_T(A')^{Q\times Q})^{\Gamma_T^* \times \Gamma_T^*}$ by $\tilde{M}_{\varepsilon, \varepsilon} = PS$, $\tilde{M}_{\pi_1, \pi_2} = 0$ for $(\pi_1, \pi_2) \neq (\varepsilon, \varepsilon)$. Let $M' = h'(T)$. Then we obtain $M' = M + \tilde{M}$ and $M'^* = (M^*\tilde{M})^* M^*$. We compute $(M^*\tilde{M})_{\varepsilon, \pi} = 0$ for $\pi \in \Gamma_T^+$, $(M^*\tilde{M})_{\varepsilon, \varepsilon} = (M^*)_{\varepsilon, \varepsilon}PS$, $((M^*\tilde{M})^*)_{\varepsilon, \varepsilon} = ((M^*)_{\varepsilon, \varepsilon}PS)^*$ and $((M^*\tilde{M})^*)_{\varepsilon, \pi} = 0$, $\pi \in \Gamma_T^+$. Hence,

$$(M'^*)_{\varepsilon, \varepsilon} = ((M^*\tilde{M})^*)_{\varepsilon, \varepsilon}(M^*)_{\varepsilon, \varepsilon} = ((M^*)_{\varepsilon, \varepsilon}PS)^*(M^*)_{\varepsilon, \varepsilon}$$

and

$$\|\mathfrak{A}'\| = S((M^*)_{\varepsilon, \varepsilon}PS)^*(M^*)_{\varepsilon, \varepsilon}P = \|\mathfrak{A}\|^+.$$

Since $\mathfrak{Rat}(A')$ is a subset of $\mathfrak{Rat}_T(A')$ and $\mathfrak{Rat}_T(A')$ is closed under addition, $\|\mathfrak{A}\|^+ + 1 = \|\mathfrak{A}\|^*$ is in $\mathfrak{Rat}_T(A')$. $\qquad\square$

Theorem 7.20. *If T is a restart type then $\mathfrak{Rat}_T(A')$ is closed under $\mathfrak{H}$-A'-rational transductions.*

Proof. Assume that $\mathfrak{A} = (Q, \Gamma_T, M, S, \varepsilon, P)$, where $M = h(T)$, is a T-automaton and that $\mathfrak{T} = (Q', h', S', P')$ is an $\mathfrak{H}$-A'-rational transducer. Since $h : \Delta^* \to A^{Q \times Q}$ is compatible with $\mathfrak{H}$ and $h' : A \to A^{Q' \times Q'}$ is in $\mathfrak{H}$, the monoid morphism $h' \circ h : \Delta^* \to A^{(Q \times Q') \times (Q \times Q')}$ is again compatible with $\mathfrak{H}$. We prove now that the behavior of the T-automaton $\mathfrak{A}' = (Q \times Q', \Gamma_T, (h' \circ h)(T), S'h'(S), \varepsilon, h'(P)P')$ is equal to $\|\mathfrak{T}\|(\|\mathfrak{A}\|)$:

$$
\begin{aligned}
\|\mathfrak{A}'\| &= S'h'(S)(((h' \circ h)(T))^*)_{\varepsilon,\varepsilon}h'(P)P' = S'h'(S)h'((h(T)^*)_{\varepsilon,\varepsilon})h'(P)P' \\
&= S'h'(S(h(T)^*)_{\varepsilon,\varepsilon}P)P' = S'h'(\|\mathfrak{A}\|)P' = \|\mathfrak{T}\|(\|\mathfrak{A}\|).
\end{aligned}
$$
□

Corollary 7.21. *If T is a restart type then $\mathfrak{Rat}_T(A')$ is an A'-AFE.*

In order to get a complete characterization of A'-AFEs we need a result "converse" to Corollary 7.21.

Let $\mathfrak{L}$ be an A'-AFE. Then we construct a restart type T such that $\mathfrak{L} = \mathfrak{Rat}_T(A')$. Assume $\mathfrak{L} = \mathfrak{F}(\mathfrak{R})$. For each $b \in \mathfrak{R}$ there exists an index set I_b such that $b = \sum_{i \in I_b} a_{i1} \ldots a_{in_i}$, $a_{ij} \in A'$, i. e.,

$$
\mathfrak{R} = \{ b \mid b = \sum_{i \in I_b} a_{i1} \ldots a_{in_i} \}.
$$

Such a representation of $\mathfrak{L}$ is possible since $\mathfrak{L} = \mathfrak{F}(\mathfrak{L}) \subseteq [A']$. The restart type $(\Gamma_T, \Delta_T, T, \varepsilon)$ is defined by

(i) $\Gamma_T = \bigcup_{b \in \mathfrak{R}} \Delta_b$, where $\Delta_b = \{ \mathbf{a}_b \mid a \in A' \}$ is a copy of A' for $b \in \mathfrak{R}$,
(ii) $\Delta_T = \{ \mathbf{a} \mid a \in A' \} \cup \{ z^0 \} \cup \{ z_b \mid b \in \mathfrak{R} \}$,
(iii) $T \in (\mathbb{N}^\infty \{ \Delta_T \})^{\Gamma_T^* \times \Gamma_T^*}$, where the non-null entries of T are
$T_{\varepsilon,\varepsilon} = z^0$,
$T_{\varepsilon, \mathbf{a}_b} = z_b$ for $\mathbf{a}_b \in \Delta_b$, $b \in \mathfrak{R}$,
$T_{\pi \mathbf{a}_b, \pi \mathbf{a}_b \mathbf{a}'_b} = \mathbf{a}$ for $\pi \in \Delta_b^*$, $\mathbf{a}_b, \mathbf{a}'_b \in \Delta_b$, $b \in \mathfrak{R}$,
$T_{\pi \mathbf{a}_b, \varepsilon} = \mathbf{a}$ for $\pi \in \Delta_b^*, \mathbf{a}_b \in \Delta_b$, such that $\pi \mathbf{a}_b = (\mathbf{a}_{i1})_b \ldots (\mathbf{a}_{in_i})_b$ for some $i \in I_b$, $b \in \mathfrak{R}$.

Theorem 7.22. *Assume that $\mathfrak{H}$ contains the identity mapping $e : A \to A$. Then $\mathfrak{Rat}_T(A') = \mathfrak{F}(\mathfrak{R}) = \mathfrak{L}$.*

Proof. We first compute $(T^*)_{\varepsilon,\varepsilon}$. This computation is easy if we consider the blocks of T according to the partition $\{\{\varepsilon\}\} \cup \{\Delta_b^+ \mid b \in \mathfrak{R}\} \cup \{\Gamma\}$, where $\Gamma = \Gamma_T^+ - \bigcup_{b \in \mathfrak{R}} \Delta_b^+$. The only non-null blocks according to this partition are $\{\varepsilon\} \times \{\varepsilon\}$, $\{\varepsilon\} \times \Delta_b^+$, $\Delta_b^+ \times \{\varepsilon\}$, $\Delta_b^+ \times \Delta_b^+$, $b \in \mathfrak{R}$. Hence, by Theorem 2.8, we obtain

$$(T^*)_{\varepsilon,\varepsilon} = \left(T(\{\varepsilon\},\{\varepsilon\}) + \sum_{b\in\mathfrak{R}} T(\{\varepsilon\},\Delta_b^+)T(\Delta_b^+,\Delta_b^+)^*T(\Delta_b^+,\{\varepsilon\}) \right)^*$$

$$= \left(z^0 + \sum_{b\in\mathfrak{R}}\sum_{i\in I_b} z_b \mathbf{a}_{i1}\ldots\mathbf{a}_{in_i} \right)^*.$$

We show now $\mathfrak{L} \subseteq \mathfrak{Rat}_T(A')$. Fix a $b \in \mathfrak{R}$ and let $h : \Delta^* \to A^{2\times2}$ be the monoid morphism defined by

$$h(\mathbf{a}) = \begin{pmatrix} a & 0 \\ 0 & a \end{pmatrix}, \ a \in A, \qquad h(z_b) = \begin{pmatrix} 0 & 1 \\ 0 & 0 \end{pmatrix},$$
$$h(z_{b'}) = h(z^0) = 0 \text{ for } b' \in \mathfrak{R},\ b' \neq b.$$

Since $e \in \mathfrak{H}$, h is compatible with $\mathfrak{H}$.

We obtain

$$h((T^*)_{\varepsilon,\varepsilon}) = \left(\sum_{i\in I_b} \begin{pmatrix} 0 & 1 \\ 0 & 0 \end{pmatrix} \begin{pmatrix} a_{i1}\ldots a_{in_i} & 0 \\ 0 & a_{i1}\ldots a_{in_i} \end{pmatrix} \right)^* = \begin{pmatrix} 1 & b \\ 0 & 1 \end{pmatrix}$$

and infer that $b \in \mathfrak{Rat}_T(A')$. Hence, $\mathfrak{R} \subseteq \mathfrak{Rat}_T(A')$. Since $\mathfrak{Rat}_T(A')$ is an A'-AFE, we infer $\mathfrak{F}(\mathfrak{R}) \subseteq \mathfrak{Rat}_T(A')$.

Conversely, we show now $\mathfrak{Rat}_T(A') \subseteq \mathfrak{L}$. Assume $a \in \mathfrak{Rat}_T(A')$. Then there exists a monoid morphism $h : \Delta^* \to A^{Q\times Q}$ compatible with $\mathfrak{H}$, and $S \in \mathfrak{Rat}(A')^{1\times Q}$, $P \in \mathfrak{Rat}(A')^{Q\times1}$ such that $a = Sh((T^*)_{\varepsilon,\varepsilon})P$. Consider now the entries of

$$h((T^*)_{\varepsilon,\varepsilon}) = \left(h(z^0) + \sum_{b\in\mathfrak{R}} h(z_b)h(\mathbf{b}) \right)^*.$$

The entries of $h(\mathbf{b})$ are in $\mathfrak{F}(\mathfrak{R})$, the entries of $h(z^0)$ and $h(z_b)$ are in $\mathfrak{Rat}(A') \subseteq \mathfrak{F}(\mathfrak{R})$. Since only finitely many $h(z_b)$ are unequal to zero, the entries of $h(z^0) + \sum_{b\in\mathfrak{R}} h(z_b)h(\mathbf{b})$ are in $\mathfrak{F}(\mathfrak{R})$. Since $\mathfrak{F}(\mathfrak{R})$ is fully rationally closed, the entries of $h((T^*)_{\varepsilon,\varepsilon})$ are in $\mathfrak{F}(\mathfrak{R})$. This implies $a \in \mathfrak{F}(\mathfrak{R})$. $\square$

We have now achieved our main result of Section 7, a complete characterization of A'-AFEs.

Corollary 7.23. *Assume that $\mathfrak{H}$ contains the identity mapping $e : A \to A$. Then a family of elements $\mathfrak{L}$ is an A'-AFE iff there exists a restart type T such that*

$$\mathfrak{L} = \mathfrak{Rat}_T(A').$$

We now turn again to formal power series, i. e., our basic semiring is $A\langle\langle \Sigma_\infty^* \rangle\rangle$. We choose $A' = A\{\Sigma_\infty \cup \varepsilon\}$. Then $[A'] = A\langle\langle \Sigma_\infty^* \rangle\rangle$ and each subset $\mathfrak{L}$ of $A\langle\langle \Sigma_\infty^* \rangle\rangle$ is an $A\{\Sigma_\infty \cup \varepsilon\}$-family of elements.

In the sequel, we are interested only in $A\{\Sigma_\infty \cup \varepsilon\}$-families of elements that are subsets of $A\{\{\Sigma_\infty^*\}\}$. Such a subset $\mathfrak{L}$ of $A\{\{\Sigma_\infty^*\}\}$ is now called *family of power series*.

We choose $\mathfrak{H}$ to be the family of all rational representations. By Theorem 7.11, the family of all mappings $\mu : A\langle\langle \Sigma_\infty^* \rangle\rangle \to (A\{\{\Sigma_\infty^*\}\})^{Q\times Q}$, $Q \subset Q_\infty$, Q finite, where μ is a rational (resp. an algebraic) representation, is closed under matricial composition. A family of power series $\mathfrak{L}$ is now called a *full abstract family of power series* (abbreviated *full AFP*) iff $\mathfrak{L} = \mathfrak{F}(\mathfrak{L})$.

The next theorem is implied by the equalities $\mathfrak{Rat}(A\{\Sigma_\infty \cup \varepsilon\}) = A^{\text{rat}}\{\{\Sigma_\infty^*\}\}$ and $\mathfrak{Alg}(A\{\Sigma_\infty \cup \varepsilon\}) = A^{\text{alg}}\{\{\Sigma_\infty^*\}\}$, and by Theorem 7.17.

Theorem 7.24. $A^{\text{rat}}\{\{\Sigma_\infty^*\}\}$ and $A^{\text{alg}}\{\{\Sigma_\infty^*\}\}$ are full AFPs.

Theorem 7.25. *A family of power series $\mathfrak{L}$ is a full AFP iff there exists a restart type T such that $\mathfrak{L} = \mathfrak{Rat}_T(A\{\Sigma_\infty \cup \varepsilon\})$.*

Proof. Corollary 7.21 proves that $\mathfrak{Rat}_T(A\{\Sigma_\infty \cup \varepsilon\})$ is a full AFP if T is a restart type. The converse is shown analogously to the proof of Theorem 7.22. But we have now $\mathfrak{R} = \{b \mid b = \sum_{w\in\Sigma_b^*}(b,w)w\}$ and the monoid morphism $h : \Delta^* \to A^{2\times 2}$ is defined by

$$h(\mathbf{a}) = \begin{pmatrix} a & 0 \\ 0 & a \end{pmatrix}, \ a \in A\langle\Sigma_b \cup \varepsilon\rangle, \qquad h(z_b) = \begin{pmatrix} 0 & \varepsilon \\ 0 & 0 \end{pmatrix},$$

$$h(\mathbf{a}) = h(z_{b'}) = h(z^0) = 0 \text{ for } a \in A\{\Sigma_\infty\} - A\langle\Sigma_b\rangle, \ b' \in \mathfrak{R}, \ b' \neq b. \qquad \square$$

In this survey, we do not consider regulated representations, regulated rational transducers and AFPs. A reader interested in these topics should consult Jacob [67], Salomaa, Soittola [123], Kuich, Salomaa [92] and Küster [95]. Readers interested in pushdown transducers should consult Karner [70, 71].

We now consider formal languages, i. e., our basic semiring is $\mathfrak{P}(\Sigma_\infty^*)$. Each subset of $\mathfrak{L}(\Sigma_\infty)$ is called *family of languages*. A family of languages $\mathfrak{L}$ is called a *full abstract family of languages* (abbreviated *full AFL*) iff $\mathfrak{L} = \mathfrak{F}(\mathfrak{L})$. Theorems 7.24 and 7.25 admit at once two corollaries.

Corollary 7.26. *The family of regular languages and the family of context-free languages are full AFLs.*

Corollary 7.27. *A family of languages $\mathfrak{L}$ is a full AFL iff there exists a restart type T such that $\mathfrak{L} = \mathfrak{Rat}_T(\{\{x\} \mid x \in \Sigma_\infty \cup \{\varepsilon\}\})$.*

In this survey, we do not consider AFLs. Readers interested in AFL-theory should consult Ginsburg [37]. An excellent treatment of full AFLs is given in Berstel [4]. Advanced results can be found in Berstel, Boasson [6].

8. The Theorem of Parikh

In the first part of this section we introduce generalizations of algebraic systems, where the right sides of the equations are arbitrary formal power series with commuting variables. These generalizations are called "commutative A'-systems". Their theory parallels the theory of algebraic systems as developed in Section 3. In the second part of this section we characterize $\mathfrak{Rat}(A')$ by the A'-semilinear elements of A and present the Theorem of Parikh. The proof of the Theorem of Parikh follows the ideas of Pilling [113] (see also Conway [21] and Kuich [83]).

Throughout this section, A will denote a *commutative* ω-continuous semiring. We denote by $\Sigma^{\oplus}$ the free commutative monoid generated by the finite alphabet Σ, i. e., for $\Sigma = \{x_1, \ldots, x_m\}$ we have

$$\Sigma^{\oplus} = \{x_1^{i_1} \ldots x_m^{i_m} \mid i_j \in \mathbb{N}, 1 \leq j \leq m\}.$$

Consider the power series in $A\langle\langle Y^{\oplus}\rangle\rangle$, where $Y = \{y_1, \ldots, y_n\}$ denotes as before a finite set of variables. Each $r = \sum_{\alpha \in Y^{\oplus}} (r, \alpha)\alpha \in A\langle\langle Y^{\oplus}\rangle\rangle$ defines a mapping $\bar{r} : A^n \to A$ by $\bar{r}(\sigma_1, \ldots, \sigma_n) = \sum_{\alpha \in Y^{\oplus}} (r, \alpha)\alpha(\sigma_1, \ldots, \sigma_n)$, $\sigma_i \in A$, $1 \leq i \leq n$, where $\alpha(\sigma_1, \ldots, \sigma_n) = \sigma_1^{i_1} \ldots \sigma_n^{i_n}$ for $\alpha = y_1^{i_1} \ldots y_n^{i_n}$. Let $r = (r_1, \ldots, r_n)$, where $r_i \in A\langle\langle Y^{\oplus}\rangle\rangle$, $1 \leq i \leq n$. Then r defines a mapping $\bar{r} : A^n \to A^n$ by

$$\bar{r}(\sigma_1, \ldots, \sigma_n) = (\bar{r}_1(\sigma_1, \ldots, \sigma_n), \ldots, \bar{r}_n(\sigma_1, \ldots, \sigma_n)), \quad \sigma_i \in A, 1 \leq i \leq n.$$

In the sequel, the mappings $\bar{r}, \bar{r}_1, \ldots, \bar{r}_n$ are simply denoted by $r, r_1, \ldots, r_n$, respectively.

A *commutative A'-system* (with *commuting* variables in Y) is a system of equations of the form

$$y_i = r_i(y_1, \ldots, y_n), \quad r_i \in A'\langle\langle Y^{\oplus}\rangle\rangle, 1 \leq i \leq n.$$

Let r be the vector with i-th component r_i, $1 \leq i \leq n$. Then the above commutative A'-system can also be written as the matrix equation $y = r(y)$.

A *solution* to the commutative A'-system $y = r(y)$ is given by $\sigma \in A^n$ such that $\sigma = r(\sigma)$. A solution σ of the commutative A'-system $y = r(y)$ is termed *least* iff $\sigma \sqsubseteq \tau$ for all solutions τ of $y = r(y)$. Clearly, the least solution of a commutative A'-system $y = r(y)$ is nothing else than the least fixpoint of the mapping r. An easy application of Theorems 3.2 and 3.3 proves that each mapping $r : A^n \to A^n$, $r \in (A\langle\langle Y^{\oplus}\rangle\rangle)^n$, is ω-continuous. Hence, by the Fixpoint Theorem of Section 3, the least solution of a commutative A'-system $y = r(y)$ exists and equals $\text{fix}(r) = \sup(r^n(0) \mid n \in \mathbb{N})$.

To compute this unique least fixpoint, we define the *approximation sequence*

$$\sigma^0, \sigma^1, \ldots, \sigma^j, \ldots, \quad \sigma^j \in (A\langle\langle Y^{\oplus}\rangle\rangle)^{n \times 1},$$

associated to the commutative A'-system $y = r(y)$ as follows:

$$\sigma^0 = 0, \quad \sigma^{j+1} = r(\sigma^j), \; j \geq 0.$$

Then the least solution is given by $\sup(\sigma^j \mid j \in \mathbb{N})$. This proves the first theorem of this section.

Theorem 8.1. *The least solution of the commutative A'-system $y = r(y)$ exists and is given by*

$$\sup(\sigma^j \mid j \in \mathbb{N}),$$

where $\sigma^0, \sigma^1, \ldots, \sigma^j, \ldots$ is the approximation sequence associated to it.

For the remainder of this section, A denotes an *idempotent commutative ω-continuous semiring*. That means that $a + a = a$ for all $a \in A$, which obviously is equivalent to $1 + 1 = 1$. We first prove some identities needed later on.

Theorem 8.2. *Let A be an idempotent commutative ω-continuous semiring. Then, for all $a, b_i \in A$, $1 \leq i \leq j$, the following identities are valid:*

(i) $a^* = (a^*)^k = (a^*)^*$, $k \geq 1$,
(ii) $(ab_1^* \ldots b_j^*)^* = 1 + aa^*b_1^* \ldots b_j^*$,
(iii) $(b_1 + \ldots + b_j)^* = b_1^* \ldots b_j^*$.

Proof. (i) Since A is idempotent, we obtain

$$(a^*)^2 = \left(\sum_{i \in \mathbb{N}} a^i\right)\left(\sum_{j \in \mathbb{N}} a^j\right) = \sum_{i,j \in \mathbb{N}} a^{i+j} = \sum_{i \in \mathbb{N}} a^i = a^*.$$

Hence, we obtain $(a^*)^2 = a^*$ and, by induction, $(a^*)^k = a^*$, $k \geq 2$. Moreover, we obtain $a^*a^* + 1 = a^*$. This equality implies, by Theorem 2.4 (i), $(a^*)^* \sqsubseteq a^*$. The trivial inequality $a^* \sqsubseteq (a^*)^*$ yields now $a^* = (a^*)^*$.

(ii) $(ab_1^* \ldots b_j^*)^* = 1 + \sum_{i \geq 1}(ab_1^* \ldots b_j^*)^i = 1 + \sum_{i \geq 1} a^i b_1^* \ldots b_j^*$
$\qquad = 1 + a^+ b_1^* \ldots b_j^*$ by (i).

(iii) The proof is by induction on j. The case $j = 1$ is clear. Let now $j > 1$. Then we obtain

$$\begin{aligned}
(b_1 + \ldots + b_j)^* &= ((b_1 + \ldots + b_{j-1})b_j^*)^* b_j^* = (1 + (b_1 + \ldots + b_{j-1})^+ b_j^*)b_j^* \\
&= b_j^* + (b_1 + \ldots + b_{j-1})^+ b_j^* = (b_1 + \ldots + b_{j-1})^* b_j^* \\
&= b_1^* \ldots b_{j-1}^* b_j^*.
\end{aligned}$$

Here, the first equality follows by Theorem 2.4 (iii), the second equality by (ii) and the last equality by the induction hypothesis. $\qquad\square$

We now return to commutative A'-systems.

Theorem 8.3. *Let A be an idempotent commutative ω-continuous semiring. Then the least solution of the commutative A'-system (with one variable y) $y = \sum_{j\geq 0} a_j y^j$, $a_j \in A'$, is given by $\sigma = \left(\sum_{j\geq 1} a_j a_0^{j-1}\right)^* a_0$.*

Proof. By Theorem 8.2 (i) we obtain, for all $n \geq 1$,

$$\sigma^n = \left(\left(\sum_{j\geq 1} a_j a_0^{j-1}\right)^*\right)^n a_0^n = \left(\sum_{j\geq 1} a_j a_0^{j-1}\right)^* a_0^n = \sigma a_0^{n-1}.$$

Hence, σ is a solution:

$$\sum_{j\geq 0} a_j \sigma^j = a_0 + \sum_{j\geq 1} a_j \sigma a_0^{j-1} = a_0 + \sigma \sum_{j\geq 1} a_j a_0^{j-1}$$

$$= a_0 + \left(\sum_{j\geq 1} a_j a_0^{j-1}\right)^+ a_0 = \sigma.$$

We now prove that σ is the least solution of the commutative A'-system. Let $(\tau_t \mid t \in \mathbb{N})$ be the approximation sequence associated to the commutative A'-system $y = \sum_{j\geq 0} a_j y^j$. Then we show by induction on t that, for all $t \geq 0$,

$$\sum_{0\leq n\leq t} \left(\sum_{j\geq 1} a_j a_0^{j-1}\right)^n a_0 \sqsubseteq \tau_{t+1}.$$

For $t = 0$, we obtain $a_0 \sqsubseteq \tau_1 = a_0$. Let now $t > 0$. Then we obtain

$$\sum_{0\leq n\leq t} \left(\sum_{j\geq 1} a_j a_0^{j-1}\right)^n a_0 = a_0 + \left(\sum_{j\geq 1} a_j a_0^{j-1}\right) \sum_{0\leq n\leq t-1} \left(\sum_{j\geq 1} a_j a_0^{j-1}\right)^n a_0$$

$$\sqsubseteq a_0 + \left(\sum_{j\geq 1} a_j a_0^{j-1}\right) \tau_t \sqsubseteq a_0 + \left(\sum_{j\geq 1} a_j \tau_t^{j-1}\right) \tau_t = \tau_{t+1}.$$

Here, the first inequality follows by the induction hypothesis and the second inequality by $a_0 \sqsubseteq \tau_t$ for all $t > 0$. Hence,

$$\sum_{0\leq n\leq t} \left(\sum_{j\geq 1} a_j a_0^{j-1}\right)^n a_0 \sqsubseteq \sup(\tau_t \mid t \in \mathbb{N}).$$

Since A is ω-continuous, we infer that $\sigma \sqsubseteq \sup(\tau_t \mid t \in \mathbb{N})$. Theorem 8.1 implies now that σ is the least solution of our commutative A'-system.   □

Theorem 8.4. *Let A be an idempotent commutative ω-continuous semiring. Then the least solution of a commutative A'-system (with one variable y) $y = \sum_{j\geq 0} a_j y^j$, where $\sum_{j\geq 0} a_j y^j \in \mathfrak{Rat}(A'\langle y^* \rangle)$, is in $\mathfrak{Rat}(A')$.*

Proof. It is easy to construct an $A'\langle y^* \rangle$-rational transducer (resp. an $(A'\langle y^* \rangle, A')$-rational transducer) that maps $\sum_{j\geq 0} a_j y^j$ into $\sum_{j\geq 1} a_j y^{j-1}$ (resp. a_0). Hence, by Theorem 7.7, $\sum_{j\geq 1} a_j y^{j-1}$ is in $Rat(A'\langle y^* \rangle)$ and a_0 is in $\mathfrak{Rat}(A')$. It is again easy to construct an $(A'\langle y^* \rangle, A')$-rational transducer that maps

$\sum_{j\geq 1} a_j y^{j-1}$ into $\sum_{j\geq 1} a_j a_0^{j-1}$. Theorem 7.7 implies now that $\sum_{j\geq 1} a_j a_0^{j-1}$ is in $\Re\mathfrak{at}(A')$. Hence, the least solution $\left(\sum_{j\geq 1} a_j a_0^{j-1}\right)^* a_0$ of our commutative A'-system is in $\Re\mathfrak{at}(A')$.   □

A proof by induction on the number of variables of a commutative A'-system, where the induction basis is given by Theorem 8.4, yields a generalization of this theorem.

Theorem 8.5. (Kuich [83], Theorem 4.9) *Let A be an idempotent commutative ω-continuous semiring. Then the components of the least solution of a commutative A'-system $y_i = r_i$, $1 \leq i \leq n$, where $r_i \in \Re\mathfrak{at}(A'\langle Y^{\oplus}\rangle)$, are in $\Re\mathfrak{at}(A')$.*

Corollary 8.6. *Let A be an idempotent commutative ω-continuous semiring. Then*
$$\mathfrak{Alg}(A') = \Re\mathfrak{at}(A').$$

Corollary 8.7. *Let A be an idempotent commutative ω-continuous semiring. Then*
$$A^{\mathrm{alg}}\langle\!\langle\{x\}^*\rangle\!\rangle = A^{\mathrm{rat}}\langle\!\langle\{x\}^*\rangle\!\rangle.$$

Corollary 8.8. *Every context-free language over a one-element alphabet is regular.*

We now transfer the ideas of Eilenberg, Schützenberger [27] to idempotent commutative ω-continuous semirings and characterize $\Re\mathfrak{at}(A')$ by the A'-semilinear elements of A. An element c of A of the form $c = ab^*$, $a, b \in A'$, is termed A'-*linear*. An A'-*semilinear element* c of A is of the form $c = c_1 + \ldots + c_m$, where c_i, $1 \leq i \leq m$, is linear.

Theorem 8.9. *Let A be an idempotent commutative ω-continuous semiring and A' be a subsemiring of A. Then c is an A'-semilinear element iff $c \in \Re\mathfrak{at}(A')$.*

Proof. Clearly, if c is semilinear then $c \in \Re\mathfrak{at}(A')$. We now show the converse. Each $a \in A'$ is A'-linear by $a = a \cdot 0^*$. By definition, the sum of A'-semilinear elements is again A'-semilinear. By Theorem 8.2 (iii), the product of A'-linear elements is A'-semilinear. Hence, the product of A'-semilinear elements is again A'-semilinear.

By Theorem 8.2 (ii), (iii), we obtain, for all $a_i, b_i \in A'$, $1 \leq i \leq m$,

$$\begin{aligned}
(a_1 b_1^* + \ldots + a_m b_m^*)^* &= (a_1 b_1^*)^* \ldots (a_m b_m^*)^* \\
&= (1 + a_1(a_1 + b_1)^*) \ldots (1 + a_m(a_m + b_m)^*).
\end{aligned}$$

This implies that the star of an A'-semilinear element is again A'-semilinear.   □

From now on we have only language-theoretic aims in this section. Our basic semiring will be isomorphic to the semiring $\mathfrak{P}(\Sigma^{\oplus})$, where $\Sigma = \{x_1, \ldots, x_m\}$. Clearly, the monoid $\Sigma^{\oplus}$ is isomorphic to the commutative monoid

$$\langle \mathbb{N}^m, +, (0, \ldots, 0) \rangle$$

of m-tuples over $\mathbb{N}$, where addition is performed componentwise. Hence, the semiring $\mathfrak{P}(\Sigma^{\oplus})$ is isomorphic to the idempotent commutative semiring

$$\langle \mathfrak{P}(\mathbb{N}^m), \cup, +, \emptyset, \{(0, \ldots, 0)\} \rangle.$$

Here, $S_1 + S_2 = \{s_1 + s_2 \mid s_1 \in S_1, s_2 \in S_2\}$ for $S_1, S_2 \in \mathfrak{P}(\mathbb{N}^m)$. We now choose A' to be the set $\mathfrak{P}_f(\mathbb{N}^m)$ of finite subsets of $\mathbb{N}^m$. A $\mathfrak{P}_f(\mathbb{N}^m)$-semilinear element of $\mathfrak{P}(\mathbb{N}^m)$ is now simply called a *semilinear set*.

We define, for all $n \in \mathbb{N}$ and $(i_1, \ldots, i_m) \in \mathbb{N}^m$,

$$n \otimes (i_1, \ldots, i_m) = (n \cdot i_1, \ldots, n \cdot i_m).$$

This implies $(i_1, \ldots, i_m)^n = n \otimes (i_1, \ldots, i_m)$ and $\{(i_1, \ldots, i_m)\}^* = \{n \otimes (i_1, \ldots, i_m) \mid n \geq 0\}$. Hence, a set of m-tuples is a $\mathfrak{P}_f(\mathbb{N}^m)$-linear element iff it is of the form

$$\{s_0 + \sum_{1 \leq j \leq t} n_j \otimes s_j \mid n_j \geq 0, s_j \in \mathbb{N}^m\}.$$

The semilinear sets are now finite unions of $\mathfrak{P}_f(\mathbb{N}^m)$-linear elements.

Let $\psi : \mathfrak{P}(\Sigma^*) \to \mathfrak{P}(\mathbb{N}^m)$ be the complete semiring morphism defined by $\psi(\{x_i\}) = \{(0, \ldots, 0, 1, 0, \ldots, 0)\}$, where 1 stands on the i-th position. Then $\psi(\{w\}) = \{(|w|_1, \ldots, |w|_m)\}$, $w \in \Sigma^*$, where $|w|_i$ denotes the number of occurrences of the symbol x_i in w, and $\psi(L) = \bigcup_{w \in L} \psi(\{w\})$ for $L \subseteq \Sigma^*$.

The Theorem of Parikh (Parikh [109]) is the last result of this section.

Theorem 8.10. *For any context-free language L, the set $\psi(L)$ is semilinear.*

Proof. Let $L \subseteq \Sigma^*$ be context-free. By Corollary 3.8, L is in $\mathfrak{Alg}(\{\{w\} \mid w \in \Sigma^*\})$. Since ψ is a $(\mathfrak{P}_f(\Sigma^*), \mathfrak{P}_f(\mathbb{N}^m))$-rational transduction, where $\mathfrak{P}_f(\Sigma^*)$ is the set of finite subsets of Σ^*, Theorem 7.8 implies that $\psi(L)$ is in $\mathfrak{Alg}(\mathfrak{P}_f(\mathbb{N}^m))$. By Corollary 8.6, $\psi(L)$ is in $\mathfrak{Rat}(\mathfrak{P}_f(\mathbb{N}^m))$. Hence, by Theorem 8.9, $\psi(L)$ is a semilinear set. $\square$

9. Lindenmayerian algebraic power series

Lindenmayer systems were introduced by Lindenmayer [100] in connection with a theory proposed for the development of filamentous organisms. We will consider Lindenmayerian algebraic systems (briefly, L algebraic systems) that define Lindenmayerian algebraic power series (briefly, L algebraic series).

These L algebraic series are generalizations of both algebraic series and ET0L languages. (See Rozenberg, Salomaa [118].)

In this section we give only a short introduction to the theory of L algebraic series. The interested reader should consult Honkala [57, 58], Kuich [87], Honkala, Kuich [62, 63] for more information on this theory.

Throughout this section, A will denote a *commutative ω-continuous* semiring.

An *L algebraic system* (with variables in Y) is a system of formal equations

$$y_i = p_i(y_1, \ldots, y_n, h_{11}(y_1), \ldots, h_{1s}(y_1), \ldots, h_{n1}(y_n), \ldots, h_{ns}(y_n)),$$

where $p_i(y_1, \ldots, y_n, z_{11}, \ldots, z_{1s}, \ldots, z_{n1}, \ldots, z_{ns})$, $1 \leq i \leq n$, is a polynomial in $A\langle\langle(\Sigma \cup Y \cup Z)^*\rangle\rangle$, $Z = \{z_{ij} \mid 1 \leq i \leq n, 1 \leq j \leq s\}$, and $h_{ij} : A\langle\langle\Sigma^*\rangle\rangle \to A\langle\langle\Sigma^*\rangle\rangle$ are finite substitutions, i. e., substitutions that are defined by $h_{ij} : \Sigma \to A\langle\Sigma^*\rangle$, $1 \leq i \leq n$, $1 \leq j \leq s$. We want to emphasize that when we consider the polynomial p_i we do not assume that each z_{ij} actually has an occurrence in p_i, $1 \leq i \leq n$, $1 \leq j \leq s$.

If there is no danger of confusion, we use a vectorial notation. We write y for $y_1, \ldots, y_n$, p for $p_1, \ldots, p_n$, h for $h_{11}, \ldots, h_{1s}, \ldots, h_{n1}, \ldots, h_{ns}$ and $h(y)$ for $h_{11}(y_1), \ldots, h_{1s}(y_1), \ldots, h_{n1}(y_n), \ldots, h_{ns}(y_n)$. By this vectorial notation, an L algebraic system as defined above is now written as

$$y = p(y, h(y)).$$

The development of the theory of L algebraic systems parallels that of the usual algebraic systems as given in Section 3.

A *solution* to the L algebraic system $y = p(y, h(y))$ is given by $\sigma \in (A\langle\langle\Sigma^*\rangle\rangle)^n$ such that $\sigma = p(\sigma, h(\sigma))$. It is termed the *least solution* iff $\sigma \sqsubseteq \tau$ holds for all solutions τ of $y = p(y, h(y))$. Hence, a solution (resp. the least solution) of $y = p(y, h(y))$ is nothing else than a fixpoint (resp. the least fixpoint) of the mapping $f : (A\langle\langle\Sigma^*\rangle\rangle)^n \to (A\langle\langle\Sigma^*\rangle\rangle)^n$ defined by $f(\sigma) = p(\sigma, h(\sigma))$, $\sigma \in (A\langle\langle\Sigma^*\rangle\rangle)^n$.

Since a substitution is a complete semiring morphism, we infer, by Theorem 3.2, that a substitution is an ω-continuous mapping. Since ω-continuous mappings are closed under functional composition, Corollary 3.5 implies that the mapping f is ω-continuous. Hence, by the Fixpoint Theorem of Section 3, the least solution of an L algebraic system $y = p(y, h(y))$ exists and equals $\mathrm{fix}(f) = \sup(f^n(0) \mid n \in \mathbb{N})$.

To compute this unique least fixpoint, we define the *approximation sequence*

$$\sigma^0, \sigma^1, \sigma^2, \ldots, \sigma^j, \ldots, \qquad \sigma^j \in (A\langle\langle\Sigma^*\rangle\rangle)^{n \times 1},$$

associated to the L algebraic system $y = p(y, h(y))$ as follows:

$$\sigma^0 = 0, \quad \sigma^{j+1} = p(\sigma^j, h(\sigma^j)), \ j \geq 0.$$

Then the least solution is given by $\sup(\sigma^j \mid j \in \mathbb{N})$. This proves the first theorem of this section.

Theorem 9.1. *The least solution of the L algebraic system $y = p(y, h(y))$ exists and is given by*

$$\sup(\sigma^j \mid j \in \mathbb{N}),$$

where $\sigma^0, \sigma^1, \ldots, \sigma^j, \ldots$ is the approximation sequence associated to it.

A power series is called *L algebraic* iff it is a component of the least solution of an L algebraic system. The set of all L algebraic series in $A\langle\langle \Sigma^* \rangle\rangle$ is denoted by $A^{\mathrm{Lalg}}\langle\langle \Sigma^* \rangle\rangle$. Clearly, we have $A^{\mathrm{alg}}\langle\langle \Sigma^* \rangle\rangle \subseteq A^{\mathrm{Lalg}}\langle\langle \Sigma^* \rangle\rangle$. Observe that each power series in $A^{\mathrm{Lalg}}\langle\langle \Sigma^* \rangle\rangle$ is the least solution of an L algebraic system $y = p(h(y))$, where $p \in A\langle\langle (\Sigma \cup Z)^* \rangle\rangle$. This is achieved by replacing each variable y_i, $1 \le i \le n$, by $e(y_i)$, where $e : A\langle\langle \Sigma^* \rangle\rangle \to A\langle\langle \Sigma^* \rangle\rangle$ is the identity mapping.

Theorem 9.2. *Assume that $\sigma \in (A\langle\langle \Sigma^* \rangle\rangle)^n$, where $(\sigma, \varepsilon) = 0$, is the least solution of an L algebraic system and consider an index i, $1 \le i \le n$. Then there exist an L algebraic system with one variable with the least solution $\hat{\sigma}$ and a finite alphabet Δ such that $\sigma_i = \hat{\sigma} \odot \mathrm{char}(\Delta^*)$.*

Proof. Let $\Sigma^{(i)}$, $1 \le i \le n$, be copies of Σ. We use the following notation throughout the proof: the upper index renames a symbol $x \in \Sigma$ into the corresponding symbol $x^{(i)} \in \Sigma^{(i)}$, a word $w \in \Sigma^*$ into the corresponding word $w^{(i)} \in (\Sigma^{(i)})^*$, $1 \le i \le n$, etc.

Let σ be the least solution of the L algebraic system $y = p(h(y))$. Define the morphisms

$$g_{jk}(x^{(i)}) = \delta_{j,i} h_{jk}(x), \quad 1 \le j, i \le n, \ 1 \le k \le s, \ x \in \Sigma,$$

where $\delta_{j,i}$ is the Kronecker symbol. Let z be a new variable and consider the L algebraic system with the variable z

$$z = \sum_{1 \le i \le n} p_i(g_{11}(z)^{(i)}, \ldots, g_{1s}(z)^{(i)}, \ldots, g_{n1}(z)^{(i)}, \ldots, g_{ns}(z)^{(i)})^{(i)}.$$

We claim that $\hat{\sigma} = \sigma_1^{(1)} + \ldots + \sigma_n^{(n)}$ is a solution. This is shown by the following equalities:

$$\sum_{1 \le i \le n} p_i(g_{11}(\hat{\sigma})^{(i)}, \ldots, g_{1s}(\hat{\sigma})^{(i)}, \ldots, g_{n1}(\hat{\sigma})^{(i)}, \ldots, g_{ns}(\hat{\sigma})^{(i)})^{(i)}$$

$$= \sum_{1 \le i \le n} p_i(h_{11}(\sigma_1)^{(i)}, \ldots, h_{1s}(\sigma_1)^{(i)}, \ldots, h_{n1}(\sigma_n)^{(i)}, \ldots, h_{ns}(\sigma_n)^{(i)})^{(i)}$$

$$= \sum_{1 \le i \le n} \sigma_i^{(i)} = \hat{\sigma}.$$

Consider now $\tau_1, \ldots, \tau_n \in A\langle\langle \Sigma^* \rangle\rangle$ such that $\hat{\tau} = \tau_1^{(1)} + \ldots + \tau_n^{(n)}$ is a solution of the L algebraic system, i. e.,

$$\hat{\tau} = \sum_{1 \le i \le n} p_i(g_{11}(\hat{\tau})^{(i)}, \ldots, g_{1s}(\hat{\tau})^{(i)}, \ldots, g_{n1}(\hat{\tau})^{(i)}, \ldots, g_{ns}(\hat{\tau})^{(i)})^{(i)}.$$

Assume, furthermore, that $\hat{\tau} \sqsubseteq \hat{\sigma}$. Hence, $(\hat{\tau}, \varepsilon) = 0$. Define the morphism $h_i : \left(\bigcup_{1 \le j \le n} \Sigma^{(j)}\right)^* \to \Sigma^*$, $1 \le i \le n$, by $h_i(x^{(j)}) = \delta_{i,j}x$, extend it in the usual manner to a semiring morphism and apply it to the above equality. Then we obtain

$$h_i(\hat{\tau}) = \tau_i$$
$$= \sum_{1 \le j \le n} h_i(p_j(g_{11}(\hat{\tau})^{(j)}, \ldots, g_{1s}(\hat{\tau})^{(j)}, \ldots, g_{n1}(\hat{\tau})^{(j)}, \ldots, g_{ns}(\hat{\tau})^{(j)})^{(j)})$$
$$= \sum_{1 \le j \le n} h_i(p_j(h_{11}(\tau_1)^{(j)}, \ldots, h_{1s}(\tau_1)^{(j)}, \ldots, h_{n1}(\tau_n)^{(j)}, \ldots, h_{ns}(\tau_n)^{(j)})^{(j)})$$
$$= p_i(h_{11}(\tau_1), \ldots, h_{1s}(\tau_1), \ldots, h_{n1}(\tau_n), \ldots, h_{ns}(\tau_n)).$$

These equalities prove that $\tau = (\tau_1, \ldots, \tau_n)$ is a solution of the original L algebraic system $y = p(h(y))$. Since σ is the least solution, we obtain $\sigma \sqsubseteq \tau$. Hence, $\sigma_i \sqsubseteq \tau_i$ for all $1 \le i \le n$, $\hat{\sigma} \sqsubseteq \hat{\tau}$ and we infer that $\hat{\sigma} = \hat{\tau}$.

This proves that $\hat{\sigma}$ is the least solution of the L algebraic system with the variable z constructed above. Because $\sigma_i^{(i)} = \hat{\sigma} \odot \text{char}((\Sigma^{(i)})^*)$, $1 \le i \le n$, renaming proves our theorem. $\qquad\square$

A proof analogous to that of Theorem 3.11 proves the next theorem.

Theorem 9.3. $\langle A^{\text{Lalg}}\langle\langle \Sigma^* \rangle\rangle, +, \cdot, 0, \varepsilon \rangle$ *is a fully rationally closed semiring.*

In the sequel, we denote the monoid of finite substitutions mapping $A\langle\langle \Sigma^* \rangle\rangle$ into $A\langle\langle \Sigma^* \rangle\rangle$ by $\langle \mathfrak{S}(\Sigma), \circ, e \rangle$. Here, $\circ$ denotes functional composition and e denotes the identity mapping from $A\langle\langle \Sigma^* \rangle\rangle$ onto $A\langle\langle \Sigma^* \rangle\rangle$. As defined in Section 2, $A\langle\langle \mathfrak{S}(\Sigma) \rangle\rangle$ denotes the set of power series over $\mathfrak{S}(\Sigma)$. The *application* of a matrix $M \in (A\langle\langle \mathfrak{S}(\Sigma) \rangle\rangle)^{n \times n}$ of power series of finite substitutions to a vector $\sigma \in (A\langle\langle \Sigma^* \rangle\rangle)^{n \times 1}$ of power series, denoted as $M(\sigma) \in (A\langle\langle \Sigma^* \rangle\rangle)^{n \times 1}$, is defined by its entries

$$M(\sigma)_i = \sum_{1 \le j \le n} M_{ij}(\sigma_j) = \sum_{1 \le j \le n} \sum_{h \in \mathfrak{S}(\Sigma)} (M_{ij}, h)h(\sigma_j), \quad 1 \le i \le n.$$

A *Lindenmayerian linear system* (briefly, *L linear system*) is of the form

$$y = M(y) + P,$$

where $M \in (A\langle \mathfrak{S}(\Sigma) \rangle)^{n \times n}$ is a matrix of polynomials of finite substitutions and $P \in (A\langle \Sigma^* \rangle)^{n \times 1}$ is a vector of polynomials.

A *solution* to this L linear system is given by a vector $\sigma \in (A\langle\langle \Sigma^* \rangle\rangle)^{n \times 1}$ such that $\sigma = M(\sigma) + P$. It is clear that we can consider the L linear system $y = M(y) + P$, where σ is a solution, to be the L algebraic system

$$y_i = \sum_{1 \le j \le n} \sum_{h \in \mathfrak{S}(\Sigma)} (M_{ij}, h)h(y_j) + P_i, \quad 1 \le i \le n,$$

where $\sigma_1, \ldots, \sigma_n$ is a solution. By this observation all the definitions and results for L algebraic systems transfer to L linear systems.

A power series is termed an *L rational power series* iff it is a component of the least solution of an L linear system. The set of all L rational power series in $A\langle\langle\Sigma^*\rangle\rangle$ is denoted by $A^{\mathrm{Lrat}}\langle\langle\Sigma^*\rangle\rangle$. Clearly, we have $A^{\mathrm{rat}}\langle\langle\Sigma^*\rangle\rangle \subseteq A^{\mathrm{Lrat}}\langle\langle\Sigma^*\rangle\rangle \subseteq A^{\mathrm{Lalg}}\langle\langle\Sigma^*\rangle\rangle$. Formally, the next theorem is analogous to Theorem 4.1.

Theorem 9.4. *The least solution of the L linear system $y = M(y) + P$ is given by $\sigma = M^*(P)$.*

Proof. We consider the L linear system as an L algebraic system and compute the approximation sequence $\sigma^0, \sigma^1, \ldots, \sigma^j, \ldots$ associated to this L algebraic system. We show by induction on j that $\sigma^{j+1} = \sum_{0 \le t \le j} M^t(P)$, $j \ge 0$. We obtain $M^0(P) = E(P) = P = \sigma^1$ and, for $j > 0$,

$$\begin{aligned}
\sigma^{j+1} &= M(\sigma^j) + P = M\left(\sum_{0 \le t \le j-1} M^t(P)\right) + P \\
&= \sum_{0 \le t \le j-1} M^{t+1}(P) + P = \sum_{0 \le t \le j} M^t(P).
\end{aligned}$$

Hence, $\sigma = \sup(\sigma^j \mid j \in \mathbb{N}) = \sup\left(\sum_{0 \le t \le j} M^t(P) \mid j \in \mathbb{N}\right) = M^*(P)$. $\square$

We now introduce ET0L *power series* over a finite alphabet Σ. These are the power series in

$$A^{\mathrm{ET0L}}\langle\langle\Sigma^*\rangle\rangle = \{s(\omega) \mid s \in \mathfrak{Rat}(A\langle\mathfrak{S}(\Sigma)\rangle))\},$$

where ω is a special symbol.

Theorem 9.5. $A^{\mathrm{ET0L}}\langle\langle\Sigma^*\rangle\rangle = A^{\mathrm{Lrat}}\langle\langle\Sigma^*\rangle\rangle$.

Proof. (i) Consider the ET0L power series $s(\omega)$, where $s \in \mathfrak{Rat}(A\langle\mathfrak{S}(\Sigma)\rangle))$. Then, by Theorems 4.5 and 4.6, there exist a matrix $M \in (A\langle\mathfrak{S}(\Sigma)\rangle))^{n \times n}$ and a column vector $R \in (A\langle e\rangle)^{n \times 1}$ such that $s = (M^*R)_1(\omega)$. We compute now the least solution σ of the L linear system $y = M(y) + P$, where $P_i = (R_i, e)\omega$, $1 \le i \le n$, and obtain, for $1 \le i \le n$,

$$\begin{aligned}
\sigma_i &= M^*(P)_i = \sum_{1 \le j \le n} (M^*)_{ij}(P_j) = \sum_{1 \le j \le n} (M^*)_{ij}((R_j, e)\omega) \\
&= \sum_{1 \le j \le n} (M^*)_{ij}(R_j, e)e(\omega) = \sum_{1 \le j \le n} (M^*)_{ij}R_j(\omega) = (M^*R)_i(\omega).
\end{aligned}$$

Hence, for all $1 \le i \le n$, $(M^*R)_i(\omega)$ is an L rational power series, i. e., $\sigma_1 = s(\omega) \in A^{\mathrm{Lrat}}\langle\langle\Sigma^*\rangle\rangle$.

(ii) Consider now the least solution $M^*(P)$ of the L linear system $y = M(y) + P$, where $M \in (A\langle\mathfrak{S}(\Sigma)\rangle)^{n\times n}$ and $P \in (A\langle\Sigma^*\rangle)^{n\times 1}$. Let R_i be the finite substitution defined by $R_i(\omega) = P_i$, $1 \leq i \leq n$, and let R be the column vector with components R_i. Then, for all $1 \leq i \leq n$,

$$
\begin{aligned}
M^*(P)_i &= \sum_{1\leq j\leq n} (M^*)_{ij}(P_j) = \sum_{1\leq j\leq n} (M^*)_{ij}(R_j(\omega)) \\
&= \sum_{1\leq j\leq n} (M^*)_{ij}R_j(\omega) = (M^*R)_i(\omega).
\end{aligned}
$$

Hence, $M^*(P)_i = (M^*R)_i(\omega)$ is an ET0L power series for $1 \leq i \leq n$, i. e., $M^*(P)_i \in A^{\mathrm{ET0L}}\langle\langle\Sigma^*\rangle\rangle$. □

If the construction in the proof of Theorem 9.2 is applied to an L linear system, it is transferred to an L linear system with one variable. This yields the next theorem.

Theorem 9.6. *Assume that $\sigma \in (A\langle\langle\Sigma^*\rangle\rangle)^{n\times 1}$, where $(\sigma, \varepsilon) = 0$, is the least solution of an L linear system and consider an index i, $1 \leq i \leq n$. Then there exist finite substitutions $h_1, \ldots, h_s$, a special symbol ω and a finite alphabet Δ such that*

$$
\sigma_i = (h_1 + \ldots + h_s)^*(\omega) \odot \mathrm{char}(\Delta^*).
$$

This theorem is a power series generalization of the result of Ginsburg, Rozenberg [39] stating that $\mathbb{B}^{\mathrm{ET0L}}\langle\langle\Sigma^*\rangle\rangle$ coincides (via the isomorphism between $\mathfrak{P}(\Sigma^*)$ and $\mathbb{B}\langle\langle\Sigma^*\rangle\rangle$) with the set of ET0L languages over Σ. It justifies our point of view to consider $A^{\mathrm{ET0L}}\langle\langle\Sigma^*\rangle\rangle$ as the proper generalization of the ET0L languages over Σ.

10. Selected topics and bibliographical remarks

The theory of rational formal power series in noncommuting variables was originated in Schützenberger [125, 126, 127] and a similar theory for algebraic formal power series in Chomsky, Schützenberger [20]. Kuich [83, 84, 88] has generalized in a series of ICALP-papers some results of these theories to ω-continuous semirings. These generalizations are the basis for Chapter 9.

Eilenberg [25], Salomaa, Soittola [123], Wechler [140], Berstel, Reutenauer [8] and Kuich, Salomaa [92] are books dealing with the different aspects of power series in connection with formal languages and automata. Berstel [3] contains a collection of papers dealing with power series. Hebisch, Weinert [53] and Golan [40] are books on semirings.

Due to space limitations, we have only considered a few topics on semirings and formal power series. In the remainder of this section we give a survey on some selected topics.

By Chomsky, Schützenberger [20], a formal power series r in $\mathbb{Q}^{\mathrm{alg}}\langle\langle\Sigma^{\oplus}\rangle\rangle$ is algebraic in the sense of complex analysis, i. e., there exists an irreducible polynomial $p(y) \in \mathbb{Q}\langle\langle (\Sigma \cup \{y\})^{\oplus}\rangle\rangle$ such that $p(r) = 0$. Using elimination theory and applying methods and results from algebra and algebraic geometry, Kuich, Salomaa [92] gave a procedure for the construction of the polynomial p.

Theorem 10.1. (Kuich, Salomaa [92], Corollary 16.11) *For every power series* $r \in \mathbb{Q}^{\mathrm{alg}}\langle\langle \Sigma^{\oplus}\rangle\rangle$, *an irreducible polynomial* $p(y) \in \mathbb{Q}\langle\langle (\Sigma \cup \{y\})^{\oplus}\rangle\rangle$ *can be effectively constructed such that* $p(r) = 0$.

Theorem 10.1 yields decision procedures for some language-theoretic problems that are difficult, if not impossible, to obtain by language-theoretic methods. The next three corollaries are originally due to Semenov [129] and follow easily from Theorem 10.1. (See also Salomaa, Soittola [123], Section IV.5 and Kuich, Salomaa [92], Section 16.)

Corollary 10.2. *Assume that* G *and* G' *are given unambiguous context-free grammars such that* $L(G') \supseteq L(G)$. *Then it is decidable whether or not* $L(G') = L(G)$.

Corollary 10.3. *Assume that* G *and* G' *are given context-free grammars such that* $L(G') = L(G)$ *and* G *is unambiguous. Then it is decidable whether or not* G' *is unambiguous.*

Corollary 10.4. *Given an unambiguous context-free grammar* G *and a regular language* R, *it is decidable whether or not* $L(G) = R$.

We now turn to another application of Theorem 10.1. The *structure generating function* of a language L is given by the series $f_L(z) = \sum_{n \geq 0} u_n z^n$, where u_n is the number of distinct words of length n in L. The *structure generating function* of a context-free grammar G without chain rules and ε-rules is given by the series $f_G(z) = \sum_{n \geq 0} v_n z^n$, where v_n is the number of distinct leftmost derivations of all words of length n in $L(G)$. The next theorem is due to Chomsky, Schützenberger [20] (see also Kuich [80]) and is an easy consequence of Theorems 10.1 and 3.9.

Theorem 10.5. *The structure generating function of a context-free grammar without chain rules and ε-rules is an algebraic function.*

Corollary 10.6. *If the structure generating function of a context-free language* L *is transcendental then* L *is inherently ambiguous.*

Results analogous to Theorem 10.5 and Corollary 10.6 are also valid for grammars of various kinds, e. g., tuple grammars (Kuich, Maurer [89, 90]), phrase structure grammars and state grammars with context-free control languages (Kuich, Shyamasundar [93]), and context-free grammars with regular parallel control languages (Kuich, Prodinger, Urbanek [91]).

Corollary 10.6 was used by Baron, Kuich [2], and, in a very refined manner, by Flajolet [31, 32] to prove the inherent ambiguity of certain context-free languages. See also Kemp [73, 74], Berstel, Boasson [6] and Petersen [112].

Another method for proving the inherent ambiguity of a context-free language is given in Corollary 10.8. A definition and a theorem is needed before the corollary.

Let $\Sigma = \{x_1, \ldots, x_m\}$. A language $L \subseteq \Sigma^*$ is said to be *Parikh slender* iff there exists a $k \geq 0$ such that, for all $(i_1, \ldots, i_m) \in \mathbb{N}^m$,

$$|\{w \in L \mid \psi(w) = (i_1, \ldots, i_m)\}| \leq k.$$

Theorem 10.7. (Honkala [60], Theorem 4.4) *Let $\Sigma = \{x_1, \ldots, x_m\}$ and assume that $L \subseteq \Sigma^*$ is a Parikh slender unambiguous context-free language. Then the structure generating function of L is a rational function.*

Corollary 10.8. (Honkala [60], Theorem 5.1) *Let $\Sigma = \{x_1, \ldots, x_m\}$ and consider a context-free language $L \subseteq \Sigma^*$. If there is a regular language $R \subseteq \Sigma^*$ such that $L \cap R$ (resp. $(\Sigma^* - L) \cap R$) is Parikh slender and the structure generating function of $L \cap R$ (resp. $(\Sigma^* - L) \cap R$) is nonrational then L is inherently ambiguous.*

Example 10.1. The following context-free languages L are inherently ambiguous:

(i) $L = L_1 \cup L_2$, where $L_1 = \{x_2 x_1^i x_2 x_1^{i+2} \mid i \geq 1\}^* \{x_2\}\{x_1\}^*\{x_2\}$,
$L_2 = \{x_2 x_1^2\}\{x_2 x_1^i x_2 x_1^{i+2} \mid i \geq 1\}^* \{x_2\}$.
(ii) $L = L_1 \cup L_2$, where $L_1 = \{x_1\}\{x_2^i x_1^i \mid i \geq 1\}^*$,
$L_2 = \{x_1^i x_2^{2i} \mid i \geq 1\}^* \{x_1\}^+$.
(iii) The *Goldstine language*
$L = \{x_1^{n_1} x_2 x_1^{n_2} x_2 \ldots x_1^{n_k} x_2 \mid k \geq 1, \, n_k \geq 0, \text{ and there exists a } j, 1 \leq j \leq k, \text{ such that } n_j \neq j\}$.
(iv) $L = \{w \in \{x_1, x_2, x_3\}^* \mid |w|_1 = |w|_2 \text{ or } |w|_2 = |w|_3\}$.
(v) *Crestin's language* (formed with products of palindromes)
$L = \{w_1 w_2 \mid w_1, w_2 \in \{x_1, x_2\}^*, \, w_1 = w_1^R, \, w_2 = w_2^R\}$, where w^R denotes the reverse of w.
(vi) $L = \{x_1^i x_2 w_1 x_1^i w_2 \mid i \geq 1, \, w_1, w_2 \in \{x_1, x_2\}^*\}$.   □

The next result is an application of Corollary 10.6 or Corollary 10.8 and concerns infinite words in Σ^ω. Given $w \in \Sigma^\omega$, one defines its *prefix language* and its *coprefix language* by $\mathrm{Pref}(w) = \{v \mid v \text{ is a finite prefix of } w\}$ and $\mathrm{Copref}(w) = \Sigma^* - \mathrm{Pref}(w)$, respectively.

Theorem 10.9. (Autebert, Flajolet, Gabarro [1]) *Let $w \in \Sigma^\omega$. If $\mathrm{Copref}(w)$ is context-free, but nonregular, then it is inherently ambiguous.*

For more information on this topic see Berstel [5].

A context-free grammar $G = (Y, \Sigma, P, y_1)$ is termed *nonexpansive* iff there is no derivation $y \Rightarrow^* w_1 y w_2 y w_3, \, y \in Y, \, w_1, w_2, w_3 \in \Sigma^*$, according to G.

Theorem 10.10. (Kuich [80], Theorem 4) *The structure generating function of a nonexpansive context-free grammar without chain rules and ε-rules is a rational function.*

Corollary 10.11. *Assume that L is a context-free language whose structure generating function is nonrational. Then L cannot be generated by an unambiguous nonexpansive context-free grammar.*

Example 10.2. Let L be the *restricted Dyck language* consisting of the well-formed words over a pair of parantheses, i.e., L is generated by the unambiguous context-free grammar $G = (\{y\}, \{(,)\}, \{y \to (y)y, y \to \varepsilon\}, y)$. The structure generating function of L is given by $f_L(z) = \sum_{n \geq 0} C_n z^{2n}$, where C_n is defined in Example 3.1. Since $f_L(z) = (1 - \sqrt{1 - 4z^2})/2z^2$ for $|z| < 1/2$ is nonrational, L cannot be generated by an unambiguous nonexpansive context-free grammar.   □

Theorem 10.12. (Ibarra, Ravikumar [65]) *Assume that G is a given unambiguous nonexpansive context-free grammar. If $L(G)$ is regular, a finite automaton $\mathfrak{A}$ can be effectively constructed such that $\|\mathfrak{A}\| = L(G)$.*

Ullian [135] has shown that an algorithm for the construction of $\mathfrak{A}$ does not exist, if G is an arbitrary context-free grammar. The algorithm for the construction of $\mathfrak{A}$ in Theorem 10.12 uses Corollary 10.4 and Theorem 10.10. Compare Corollary 10.4 and Theorem 10.12 with Stearns [134], Theorem 6.

The next theorem is due to Litow [101] and considers, for $i, j \geq 1$, languages $L_{i,j} = \{w \in \{x_1, x_2\}^* \mid i|w|_1 - j|w|_2 = 0\}$.

Theorem 10.13. *Assume that G is a given context-free grammar with terminal alphabet $\{x_1, x_2\}$. Then, for given $i, j \geq 1$, it is decidable whether or not*

(i) $L(G) \cap L_{i,j} = \emptyset$,
(ii) $L(G) \subseteq L_{i,j}$,
(iii) $L(G) = L_{i,j}$, *if G is unambiguous.*

The next theorem extends the Equality Theorem of Eilenberg [25] to multitape finite automata. Consider the monoid $M = \Sigma_1^* \times \ldots \times \Sigma_m^*$. For an element $w = (w_1, \ldots, w_m) \in M$, define the *length* $|w|$ of w to be $|w| = \sum_{1 \leq i \leq m} |w_i|$. Let $M' = \{w \in M \mid |w| = 1\}$. Our basic semiring is now $\mathbb{Q}\langle\langle M \rangle\rangle$ and we consider finite $\mathbb{Q}\langle M' \rangle$-automata.

Theorem 10.14. (Harju, Karhumäki [47], Theorem 3.8) *Given two finite $\mathbb{Q}\langle M' \rangle$-automata $\mathfrak{A}_1$ and $\mathfrak{A}_2$, it is decidable whether or not $\|\mathfrak{A}_1\| = \|\mathfrak{A}_2\|$.*

Corollary 10.15. (Harju, Karhumäki [47], Theorem 3.10) *Given two multitape deterministic finite automata, it is decidable whether or not they accept the same language.*

We now consider formal power series with coefficients in the tropical semiring $\langle \mathbb{N}^\infty, \min, +, \infty, 0 \rangle$. A finite $\{\sum_{x \in \Sigma \cup \{\varepsilon\}} (r, x)x \mid (r, x) \in \{\infty, 1, 0\}, x \in \Sigma \cup \{\varepsilon\}\}$-automaton $\mathfrak{A} = (Q, M, S, P)$ is called a *distance automaton* iff the entries of M (resp. of S and P) are in $\{\sum_{x \in \Sigma} (r, x)x \mid (r, x) \in \{\infty, 1, 0\}, x \in$

Σ} (resp. {$a\varepsilon \mid a \in \{\infty, 0\}$}). A distance automaton $\mathfrak{A}$ is said to be *limited in distance* iff $\sup\{(\|\mathfrak{A}\|, w) \mid (\|\mathfrak{A}\|, w) \neq \infty, w \in \Sigma^*\} < \infty$.

Theorem 10.16. (Hashiguchi [48, 50], Leung [98], Simon [133]) *Given a distance automaton $\mathfrak{A}$, it is decidable whether or not $\mathfrak{A}$ is limited in distance.*

Hashiguchi [49] used Theorem 10.16 to establish an algorithm to compute the star height of a given regular language. (See also Perrin [111].) More information on distance automata can be found in Simon [132], Krob [76], Leung [99], Weber [137, 138, 139].

Readers interested in decision problems of different kinds should consult Eilenberg [25], Salomaa, Soittola [123], Jacob [68], Salomaa [122], Culik II, Karhumäki [23] and Honkala [56, 61].

Algebraic and combinatorial properties of formal power series are studied in Shamir [131], Gross, Lentin [45], Kuich [81], Cori, Richard [22], Jacob [66], Goldman [42], Salomaa, Soittola [123], Reutenauer [114, 115, 116, 117], Diekert [24], Flajolet, Gardy, Thimonier [35], Flajolet [34], Litow, Dumas [103], Mateescu [105], Matos [106], Fernau [28], Honkala [59], Kudlek, Mateescu [79], Golan, Mateescu, Vaida [41].

Berstel, Reutenauer [9] and Honkala [55, 54] consider zeta functions of formal languages and formal power series. Berstel, Reutenauer [7], Flajolet [29, 30, 33], Bozapalidis [13, 14] and Bozapalidis, Rahonis [15] deal with formal power series on trees. Bertoni, Goldwurm, Massazza [10], Bertoni, Goldwurm, Sabadini [11] and Litow [101, 102] study the complexity of problems in connection with formal power series.

Acknowledgement. The author expresses his gratitude to Juha Honkala and Georg Karner for a very careful reading of the manuscript and for many clever technical observations. Thanks are also due to Günther Eigenthaler, Helmut Prodinger, Arto Salomaa, Friedrich Urbanek and Andreas Weber for useful discussions and comments.

References

1. Autebert, J. M., Flajolet, Ph., Gabarro, J.: Prefixes of infinite words and ambiguous context-free languages. Inf. Process. Lett. 25(1987) 211–216.
2. Baron, G., Kuich, W.: The characterization of nonexpansive grammars by rational power series. Inf. Control 48(1981) 109–118.
3. Berstel, J. (ed): Séries formelles en variables non commutatives et applications. Laboratoire d'Informatique Théorique et Programmation. Ecole Nationale Supérieure de Techniques Avancées, 1978.
4. Berstel, J.: Transductions and Context-Free Languages. Teubner, 1979.
5. Berstel, J.: Properties of infinite words: recent results. STACS89, Lect. Notes Comput. Sci. 349, Springer-Verlag 1989, 36–46.
6. Berstel, J., Boasson, L.: Context-free languages. In: J. van Leeuwen, ed., Handbook of Theoretical Computer Science, Vol. B. North-Holland, 1990, 59–102.

7. Berstel, J., Reutenauer, C.: Recognizable formal power series on trees. Theor. Comput. Sci. 18(1982) 115–148.
8. Berstel, J., Reutenauer, C.: Les séries rationelles et leurs langages. Masson, 1984. English translation: Rational Series and Their Languages. EATCS Monographs on Theoretical Computer Science, Vol. 12. Springer-Verlag 1988.
9. Berstel, J., Reutenauer, C.: Zeta functions of formal languages. ICALP88, Lect. Notes Comput. Sci. 317, Springer-Verlag 1988, 93–104.
10. Bertoni, A., Goldwurm, M., Massazza, P.: Counting problems and algebraic formal power series in noncommuting variables. Inf. Proc. Lett. 34(1990) 117–121.
11. Bertoni, A., Goldwurm, M., Sabadini, N.: The complexity of computing the number of strings of given length in context-free languages. Theor. Comp. Sci. 86(1991) 325–342.
12. Bloom, St. L., Ésik, Z.: Iteration Theories. EATCS Monographs on Theoretical Computer Science. Springer-Verlag 1993.
13. Bozapalidis, S.: Effective construction of the syntactic algebra of a recognizable series on trees. Acta Inf. 28(1991) 351–363.
14. Bozapalidis, S.: Convex algebras, convex modules and formal power series on trees. Aristotle University of Thessaloniki, 1995.
15. Bozapalidis, S., Rahonis, G.: On two families of forests. Acta Inf. 31(1994) 235–260.
16. Bucher, W., Maurer, H.: Theoretische Grundlagen der Programmiersprachen. B. I. Wissenschaftsverlag, 1984.
17. Carré, B.: Graphs and Networks. Clarendon Press, 1979.
18. Chomsky, N.: On certain formal properties of grammars. Inf. Control 2(1959) 137–167.
19. Chomsky, N.: Context-free grammars and pushdown storage. MIT Res. Lab. of Elect., Quarterly Prog. Rep. 65(1962) 187–194.
20. Chomsky, N., Schützenberger, M. P.: The algebraic theory of context-free languages. In: P. Braffort and D. Hirschberg, eds., Computer Programming and Formal Systems. North-Holland, 1963, 118–161.
21. Conway, J. H.: Regular Algebra and Finite Machines. Chapman & Hall, 1971.
22. Cori, R., Richard, J.: Enumeration des graphes planaires à l'aide des series formelles en variables non commutatives. Discrete Math. 2(1972) 115–162.
23. Culik II, K., Karhumäki, J.: A note on the equivalence problem of rational formal power series. Inf. Proc. Lett. 23(1986) 29–31.
24. Diekert, V.: Transitive orientations, Möbius functions and complete semi-Thue systems for free partially commutative monoids. ICALP88, Lect. Notes Comput. Sci. 317, Springer-Verlag 1988, 176–187.
25. Eilenberg, S.: Automata, Languages and Machines. Vol. A. Academic Press, 1974.
26. Eilenberg, S.: Automata, Languages and Machines. Vol. C. Draft of Sections I–III, 1978.
27. Eilenberg, S., Schützenberger, M. P.: Rational sets in commutative monoids. J. Algebra 13(1969) 173–191.
28. Fernau, H.: Valuations of languages, with applications to fractal geometry. Theor. Comput. Sci. 137(1995) 177–217.
29. Flajolet, Ph.: Analyse d'algorithmes de manipulation d'arbres et de fichiers. Cahiers du BURO 34/35(1981) 1–209.
30. Flajolet, Ph.: Elements of a general theory of combinatorial structures. FCT85, Lect. Notes Comput. Sci. 199, Springer-Verlag 1985, 112–127.

31. Flajolet, Ph.: Ambiguity and transcendence. ICALP85, Lect. Notes Comput. Sci. 194, Springer-Verlag 1985, 179–188.
32. Flajolet, Ph.: Analytic models and ambiguity of context-free languages. Theor. Comput. Sci. 49(1987) 283–309.
33. Flajolet, Ph.: Mathematical methods in the analysis of algorithms and data structures. In: E. Börger, ed., Trends in Theoretical Computer Science, Chapter 6. Computer Science Press, 1988, 225–304.
34. Flajolet, Ph.: Analytic analysis of algorithms. ICALP92, Lect. Notes Comput. Sci. 623, Springer-Verlag 1992, 186–210.
35. Flajolet, Ph., Gardy, D., Thimonier, L.: Birthday paradox, coupon collectors, caching algorithms and self-organizing search. Discrete Appl. Math. 39(1992) 207–229.
36. Floyd, R. W.: Syntactic analysis and operator precedence. J. Assoc. Comput. Mach. 10(1963) 313–333.
37. Ginsburg, S.: Algebraic and Automata-Theoretic Properties of Formal Languages. North-Holland, 1975.
38. Ginsburg, S., Rice, H. G.: Two families of languages related to ALGOL. J. Assoc. Comput. Mach. 9(1962) 350–371.
39. Ginsburg, S., Rozenberg, G.: T0L schemes and control sets. Inf. Control 27(1975) 109–125.
40. Golan, J. S.: The Theory of Semirings with Applications in Mathematics and Theoretical Computer Science. Pitman Monographs and Surveys in Pure and Applied Mathematics 54. Longman Sci. Tech., 1992.
41. Golan, J. S., Mateescu, A., Vaida, D.: Towards a unified theory of sequential parallel and semi-parallel processes. Technical report, 1995.
42. Goldman, J. R.: Formal languages and enumeration. J. Comb. Theory, Series A, 24(1978) 318–338.
43. Goldstern, M.: Vervollständigung von Halbringen. Diplomarbeit, Technische Universität Wien, 1985.
44. Greibach, S.: A new normal-form theorem for context-free phrase structure grammars. J. Assoc. Comput. Mach. 12(1965) 42–52.
45. Gross, M., Lentin, A.: Introduction to Formal Grammars. Springer-Verlag 1970.
46. Harrison, M. A.: Introduction to Formal Language Theory. Addison-Wesley, 1978.
47. Harju, T., Karhumäki, J.: The equivalence problem of multitape finite automata. Theor. Comput. Sci. 78(1991) 347–355.
48. Hashiguchi, K.: Limitedness theorem for finite automata with distance functions. J. Comp. Syst. Sci. 24(1982) 233–244.
49. Hashiguchi, K.: Algorithms for determining relative star height and star height. Inform. Comput. 78(1987) 124–169.
50. Hashiguchi, K.: Impoved limitedness theorems on finite automata with distance functions. Theor. Comput. Sci. 72(1990) 27–38.
51. Hebisch, U.: The Kleene theorem in countably complete semirings. Bayreuther Mathematische Schriften, 31(1990) 55–66.
52. Hebisch, U.: Eine algebraische Theorie unendlicher Summen mit Anwendungen auf Halbgruppen und Halbringe. Bayreuther Mathematische Schriften, 40(1992) 21–152.
53. Hebisch, U., Weinert, H. J.: Halbringe. Teubner, 1993.
54. Honkala, J.: A necessary condition for the rationality of the zeta function of a regular language. Theor. Comput. Sci. 66(1989) 341–347.
55. Honkala, J.: On algebraic generalized zeta functions of formal power series. Theor. Comput. Sci.79(1991) 263–273.

56. Honkala, J.: On D0L systems with immigration. Theor. Comput. Sci. 120(1993) 229–245.
57. Honkala, J.: On Lindenmayerian series in complete semirings. In: G. Rozenberg and A. Salomaa, eds., Developments in Language Theory. World Scientific, 1994, 179–192.
58. Honkala, J.: On morphically generated formal power series. Informatique théorique et Applications/Theoretical Informatics and Applications 29(1995) 105–127.
59. Honkala, J.: On Lindenmayerian algebraic sequences. University of Turku, 1995.
60. Honkala, J.: On Parikh slender languages and power series. J. Comp. Syst. Sci., to appear.
61. Honkala, J.: On images of algebraic series. University of Turku, 1995.
62. Honkala, J., Kuich, W.: On a power series generalization of ET0L languages. Fundamenta Informaticae, to appear.
63. Honkala, J., Kuich, W.: On Lindenmayerian algebraic power series. University of Turku, 1995.
64. Hopcroft, J. E., Ullman, J. D.: Introduction to Automata Theory, Languages, and Computation. Addison-Wesley, 1979.
65. Ibarra, O. H., Ravikumar, B.: On sparseness, ambiguity and other decision problems for acceptors and transducers. STACS86, Lect. Notes Comput. Sci. 210, Springer-Verlag 1986, 171–179.
66. Jacob, G.: Sur un Théorème de Shamir. Inf. Control 27(1975) 218–261.
67. Jacob, G.: Représentations et substitutions matricielles dans la théorie algébrique des transductions. Thèse de doctorat d'état, Université Paris, VII, 1975.
68. Jacob, G.: La finitude des representations lineaires des semigroupes est decidable. J. Algebra 52(1978) 437–459.
69. Karner, G.: On limits in complete semirings. Semigroup Forum 45(1992) 148–165.
70. Karner, G.: Nivat's Theorem for pushdown transducers. Theor. Comp. Sci. 97(1992) 245–262.
71. Karner, G.: On transductions of formal power series over complete semirings. Theor. Comp. Sci. 98(1992) 27–39.
72. Karner, G.: A topology for complete semirings. STACS 94, Lect. Notes Comput. Sci. 775, Springer-Verlag 1994, 389–400.
73. Kemp, R.: On the number of words in the language $\{w \in \Sigma^* \mid w = w^R\}^2$. Discrete Math. 40(1980) 225–234.
74. Kemp, R.: A note on the density of inherently ambiguous context-free languages. Acta Inf. 14(1980) 295–298.
75. Krob, D.: Monoides et semi-anneaux continus. Semigroup Forum 37(1988) 59–78.
76. Krob, D.: The equality problem for rational series with multiplicities in the tropical semiring is undecidable. Int. J. of Algebra and Comput. 4(1994) 405–425.
77. Kleene, St. C.: Representation of events in nerve nets and finite automata. In: C. E. Shannon, J. McCarthy, eds., Automata Studies. Princeton University Press, 1956, 3–41.
78. Kozen, D.: A completeness theorem for Kleene algebras and the algebra of regular events. Inf. Computation 110(1994) 366–390.
79. Kudlek, M., Mateescu, A.: On distributed catenation. Universität Hamburg, 1995.

80. Kuich, W.: Über die Entropie kontext-freier Sprachen. Habilitationsschrift, Technische Hochschule Wien, 1970. English translation: On the entropy of context-free languages. Inf. Control 16(1970) 173–200.
81. Kuich, W.: Languages and the enumeration of planted plane trees. Indagationes Mathematicae 32(1970) 268–280.
82. Kuich, W.: Hauptkegel algebraischer Potenzreihen. EIK 23(1987) 147–170.
83. Kuich, W.: The Kleene and the Parikh theorem in complete semirings. ICALP87, Lect. Notes Comput. Sci. 267, Springer-Verlag 1987, 212–225.
84. Kuich, W.: ω-continuous semirings, algebraic systems and pushdown automata. ICALP90, Lect. Notes Comput. Sci. 443, Springer-Verlag 1990, 103–110.
85. Kuich, W.: Automata and languages generalized to ω-continuous semirings. Theor. Comput. Sci. 79(1991) 137–150.
86. Kuich, W.: Automaten und Formale Sprachen. Skriptum, Technische Universität Wien, 1991.
87. Kuich, W.: Lindenmayer systems generalized to formal power series and their growth functions. In: G. Rozenberg and A. Salomaa, eds., Developments in Language Theory. World Scientific, 1994, 171–178.
88. Kuich, W.: The algebraic equivalent of AFL theory. ICALP95, Lect. Notes Comput. Sci. 944, Springer-Verlag 1995, 39–50.
89. Kuich, W., Maurer, H.: The structure generating function and entropy of tuple languages. Inf. Control 19(1971) 195–203.
90. Kuich, W., Maurer, H.: On the inherent ambiguity of simple tuple languages. Computing 7(1971) 194–203.
91. Kuich, W., Prodinger, H., Urbanek, F.: On the height of derivation trees. ICALP79, Lect. Notes Comput. Sci. 71, Springer-Verlag 1979, 370–384.
92. Kuich, W., Salomaa, A.: Semirings, Automata, Languages. EATCS Monographs on Theoretical Computer Science, Vol. 5. Springer-Verlag 1986.
93. Kuich, W., Shyamasundar, R. K.: The structure generating function of some families of languages. Inf. Control 32(1976) 85–92.
94. Kuich, W., Urbanek, F.: Infinite linear systems and one counter languages. Theor. Comput. Sci. 22(1983) 95–126.
95. Küster, G.: Das Hadamardprodukt abstrakter Familien von Potenzreihen. Dissertation, Technische Universität Wien, 1986.
96. Lausch, H., Nöbauer, W.: Algebra of Polynomials. North-Holland, 1973.
97. Lehmann, D. J.: Algebraic structures for transitive closure. Theor. Comput. Sci. 4(1977) 59–76.
98. Leung, H.: Limitedness theorem on finite automata with distance functions: an algebraic proof. Theor. Comput. Sci. 81(1991) 137–145.
99. Leung, H.: A note on finitely ambiguous distance automata. Inform. Process. Lett. 44(1992) 329–331.
100. Lindenmayer, A.: Mathematical models for cellular interactions in development, parts I, II. J. Theor. Biol. 18(1968) 280–315.
101. Litow, B.: A context-free language decision problem. Theor. Comput. Sci. 125(1994) 339–343.
102. Litow, B.: Numbering unambiguous context-free languages. Austral. Comput. Sci. Commun. 16(1994) 373–378.
103. Litow, B., Dumas, Ph.: Additive cellular automata and algebraic series. Theor. Comput. Sci. 119(1993) 345–354.
104. Mahr, B.: Iteration and summability in semirings. Ann. of Discrete Math. 19(1984) 229–256.

105. Mateescu, A.: On (left) partial shuffle. Lect. Notes Comput. Sci. 812, Springer-Verlag 1994, 264–278.
106. Matos, A. B.: Periodic sets of integers. Theor. Comput. Sci. 127(1994) 287–312.
107. Moll, R. N., Arbib, M. A., Kfoury, A. J.: An Introduction to Formal Language Theory. Springer-Verlag 1988.
108. Nivat, M.: Transductions des langages de Chomsky. Ann. Inst. Fourier 18(1968) 339–455.
109. Parikh, R. J.: On context-free languages. J. Assoc. Comput. Mach. 13(1966) 570–581.
110. Paz, A.: Introduction to Probabilistic Automata. Academic Press, 1971.
111. Perrin, D.: Finite automata. In: J. van Leeuwen, ed., Handbook of Theoretical Computer Science, Vol. B. North-Holland, 1990, 1–57.
112. Petersen, H.: The ambiguity of primitive words. STACS 94, Lect. Notes Comput. Sci. 775, Springer-Verlag 1994, 679–690.
113. Pilling, D. L.: Commutative regular equations and Parikh's theorem. J. Lond. Math. Soc., II. Ser. 6 (1973) 663–666.
114. Reutenauer, C.: On a question of S. Eilenberg. Theor. Comput. Sci. 5(1977) 219.
115. Reutenauer, C.: Sur les séries rationelles en variables non commutatives. Lect. Notes Comput. Sci. 62, Springer-Verlag 1978, 372–381.
116. Reutenauer, C.: Sur les séries associées à certains systèmes de Lindenmayer. Theor. Comput. Sci. 9(1979) 363–375.
117. Reutenauer, C.: An Ogden-like iteration lemma for rational power series. Acta Inf. 13(1980) 189–197.
118. Rozenberg, G., Salomaa, A.: The Mathematical Theory of L Systems. Academic Press, 1980.
119. Rosenkrantz, D. J.: Matrix equations and normal forms for context-free grammars. J. Assoc. Comput. Mach. 14(1967) 501–507.
120. Salomaa, A.: Theory of Automata. Pergamon Press, 1969.
121. Salomaa, A.: Formal Languages. Academic Press, 1973.
122. Salomaa, A.: Formal languages and power series. In: J. van Leeuwen, ed., Handbook of Theoretical Computer Science, Vol. B. North-Holland, 1990, 103–132.
123. Salomaa, A., Soittola, M.: Automata-Theoretic Aspects of Formal Power Series. Springer-Verlag 1978.
124. Sakarovitch, J.: Kleene's theorem revisited. Lect. Notes Comput. Sci. 281, Springer-Verlag 1987, 39–50.
125. Schützenberger, M. P.: Un problème de la théorie des automates. Seminaire Dubreil-Pisot, 13.3, Inst. H. Poincaré, 1960.
126. Schützenberger, M. P.: On the definition of a family of automata. Inf. Control 4(1961) 245–270.
127. Schützenberger, M. P.: On a theorem of R. Jungen. Proc. Am. Math. Soc. 13(1962) 885–889.
128. Schützenberger, M. P.: Certain elementary families of automata. In: Proceedings of the Symposium on Mathematical Theory of Automata, Polytechnic Institute of Brooklyn, 1962, 139–153.
129. Semenov, A. L.: Algoritmitšeskie problemy dlja stepennykh rjadov i kontekstnosvobodnykh grammatik. Dokl. Akad. Nauk SSSR 212(1973) 50–52.
130. Seneta, E.: Non-Negative Matrices and Markov Chains, Second Edition. Springer-Verlag 1981.
131. Shamir, E.: A representation theorem for algebraic and context-free power series in non-commuting variables. Inf. Control 11(1967) 239–254.

132. Simon, I.: Recognizable sets with multiplicities in the tropical semiring. MFCS88, Lect. Notes Comput. Sci. 324, Springer-Verlag 1988, 107–120.
133. Simon, I.: On semigroups of matrices over the tropical semiring. Informatique théorique et Applications/Theoretical Informatics and Applications 28(1994) 277–294.
134. Stearns, R. E.: A regularity test for pushdown machines. Inf. Control 11(1967) 323–340.
135. Ullian, J. S.: Partial algorithm problems for context-free languages. Inf. Control 11(1967) 80–101.
136. Urbanek, F.: On Greibach normal form construction. Theor. Comput. Sci. 40(1985) 315–317.
137. Weber, A.: Distance automata having large finite distance or finite ambiguity. Math. Systems Theory 26(1993) 169–185.
138. Weber, A.: Finite-valued distance automata. Theor. Comput. Sci. 134(1994) 225–251.
139. Weber, A.: Exponential upper and lower bounds for the order of a regular language. Theor. Comput. Sci. 134(1994) 253–262.
140. Wechler, W.: The Concept of Fuzziness in Automata and Language Theory. Akademie-Verlag, 1978.
141. Wechler, W.: Universal Algebra for Computer Scientists. EATCS Monographs on Computer Science, Vol. 25. Springer-Verlag 1992.
142. Weinert, H. J.: Generalized semialgebras over semirings. Lect. Notes Math. 1320, Springer-Verlag 1988, 380–416.

Syntactic Semigroups

Jean-Eric Pin

1. Introduction

This chapter gives an overview on what is often called the algebraic theory of finite automata. It deals with languages, automata and semigroups, and has connections with model theory in logic, boolean circuits, symbolic dynamics and topology.

Kleene's theorem [70] is usually considered as the foundation of this theory. It shows that the class of *recognizable* languages (i.e. recognized by finite automata), coincides with the class of *rational* languages, which are given by rational expressions. The definition of the *syntactic monoid*, a monoid canonically attached to each language, was first given in an early paper of Rabin and Scott [128], where the notion is credited to Myhill. It was shown in particular that a language is recognizable if and only if its syntactic monoid is finite. However, the first classification results were rather related to automata [89]. A break-through was realized by Schützenberger in the mid-1960s [144]. Schützenberger made a non-trivial use of the syntactic monoid to characterize an important subclass of the rational languages, the *star-free* languages. Schützenberger's theorem states that a language is star-free if and only if its syntactic monoid is finite and aperiodic.

This theorem had a considerable influence on the theory. Two other important "syntactic" characterizations were obtained in the early 1970s: Simon [152] proved that a language is piecewise testable if and only if its syntactic monoid is finite and $\mathcal{J}$-trivial and Brzozowski-Simon [41] and independently, McNaughton [86] characterized the locally testable languages. These successes settled the power of the semigroup approach, but it was Eilenberg who discovered the appropriate framework to formulate this type of results [53].

A variety of finite monoids is a class of monoids closed under taking submonoids, quotients and finite direct product. Eilenberg's theorem states that varieties of finite monoids are in one to one correspondence with certain classes of recognizable languages, the varieties of languages. For instance, the rational languages correspond to the variety of all finite monoids, the star-free languages correspond to the variety of finite aperiodic monoids, and the piecewise testable languages correspond to the variety of finite $\mathcal{J}$-trivial monoids. Numerous similar results have been established during the past fifteen years and, for this reason, the theory of finite automata is now intimately related to the theory of finite monoids.

It is time to mention a sensitive feature of this theory, the role of the empty word. Indeed, there are two possible definitions for a language. A first definition consists in defining a language on the alphabet A as a subset of the free monoid A^*. In this case a language may contain the empty word. In the second definition, a language is defined as a subset of the free semigroup A^+, which excludes the empty word. This subtle distinction has dramatic consequences on the full theory. First, one has to distinguish between *-varieties (the first case) and +-varieties of languages (the latter case). Next, with the latter definition, monoids have to be replaced by semigroups and Eilenberg's theorem gives a one to one correspondence between varieties of finite semigroups and +-varieties of languages. Although it might seem quite annoying to have two such parallel theories, this distinction proved to be necessary. For instance, the locally testable languages form a +-variety, which correspond to locally idempotent and commutative semigroups. But no characterization of the locally testable languages is known in terms of syntactic monoids.

An extension of the notion of syntactic semigroup (or monoid) was recently proposed in [112]. The key idea is to define a partial order on syntactic semigroups, leading to the notion of ordered syntactic semigroups. The resulting extension of Eilenberg's variety theory permits to treat classes of languages that are not necessarily closed under complement, a major difference with the original theory. We have adopted this new point of view throughout this chapter. For this reason, even our definition of recognizable languages may seem unfamiliar to the reader.

The theory has now developed into many directions and has generated a rapidly growing literature. The aim of this chapter is to provide the reader with an overview of the main results. As these results are nowadays intimately related with non-commutative algebra, a certain amount of semigroup theory had to be introduced, but we tried to favor the main ideas rather than the technical developments. Some important topics had to be skipped and are briefly mentioned in the last section. Due to the lack of place, no proofs are given, but numerous examples should help the reader to catch the spirit of the more abstract statements. The references listed at the end of the chapter are far from being exhaustive. However, most of the references should be reached by the standard recursive process of tracing the bibliography of the papers cited in the references.

The chapter is organized as follows. The amount of semigroup theory that is necessary to state precisely the results of this chapter is introduced in Section 2. The basic concepts of recognizable set and ordered syntactic semigroup are introduced in Section 3. The variety theorem is stated in Section 4 and examples follow in Section 5. Some algebraic tools are presented in Section 6. Sections 7 and 8 are devoted to the study of the concatenation product and its variants. Connections with the theory of codes are discussed in Section 9. Section 10 gives an overview on the operators on recognizable languages. Various extensions are briefly reviewed in Section 11.

2. Definitions

We review in this section the basic definitions about relations and semigroups needed in this chapter.

2.1 Relations

A relation $\mathcal{R}$ on a set S is *reflexive* if, for every $x \in S$, $x \mathcal{R} x$, *symmetric*, if, for every $x, y \in S$, $x \mathcal{R} y$ implies $y \mathcal{R} x$, and *transitive*, if, for every $x, y, z \in S$, $x \mathcal{R} y$ and $y \mathcal{R} z$ implies $x \mathcal{R} z$. A *quasi-order* is a reflexive and transitive relation. An *equivalence relation* is a reflexive, symmetric and transitive relation. Given a quasi-order $\mathcal{R}$, the relation $\sim$ defined by $x \sim y$ if and only if $x \mathcal{R} y$ and $y \mathcal{R} x$ is an equivalence relation, called the *equivalence relation associated with* $\mathcal{R}$. If this equivalence relation is the equality relation, that is, if, for every $x, y \in S$, $x \mathcal{R} y$ and $y \mathcal{R} x$ implies $x = y$, then the relation $\mathcal{R}$ is an *order*.

Relations are naturally ordered by inclusion. Let $\mathcal{R}_1$ and $\mathcal{R}_2$ be two relations on a set S. The relation $\mathcal{R}_2$ is *coarser* than $\mathcal{R}_1$ if and only if, for every $s, t \in S$, $s \mathcal{R}_1 t$ implies $s \mathcal{R}_2 t$. In particular, the coarsest relation is the universal relation.

2.2 Semigroups

A *semigroup* is a set equipped with an internal associative operation which is usually written in a multiplicative form. A *monoid* is a semigroup with an identity element (usually denoted by 1). If S is a semigroup, S^1 denotes the monoid equal to S if S has an identity element and to $S \cup \{1\}$ otherwise. In the latter case, the multiplication on S is extended by setting $s1 = 1s = s$ for every $s \in S^1$.

A relation on a semigroup S is *stable on the right* (resp. *left*) if, for every $x, y, z \in S$, $x \mathcal{R} y$ implies $xz \mathcal{R} yz$ (resp. $zx \mathcal{R} zy$). A relation is *stable* if it is stable on the right and on the left. A *congruence* is a stable equivalence relation. Thus, an equivalence relation $\sim$ on S is a congruence if and only if, for every $s, t \in S$ and $x, y \in S^1$, $s \sim t$ implies $xsy \sim xty$. If $\sim$ is a congruence on S, then there is a well-defined multiplication on the quotient set $S/\sim$, given by

$$[s][t] = [st]$$

where $[s]$ denotes the $\sim$-class of $s \in S$.

An *ordered semigroup* is a semigroup S equipped with a stable order relation $\leq$ on S. Ordered monoids are defined analogously. The notation $(S, \leq)$ will sometimes be used to emphasize the role of the order relation, but most of the time the order will be implicit and the notation S will be used for semigroups as well as for ordered semigroups. If $S = (S, \leq)$ is an ordered semigroup, then $(S, \geq)$ is also an ordered semigroup, called the *dual* of S and denoted $\check{S}$.

A *congruence on an ordered semigroup* S is a stable quasi-order which is coarser than $\leq$. In particular, the order relation $\leq$ is itself a congruence. If $\preceq$ is a congruence on S, then the equivalence relation $\sim$ associated with $\preceq$ is a congruence on S. Furthermore, there is a well-defined stable order on the quotient set $S/\sim$, given by

$$[s] \leq [t] \quad \text{if and only if } s \preceq t$$

Thus $(S/\sim, \leq)$ is an ordered semigroup, also denoted $S/\preceq$.

Given a family $(S_i)_{i \in I}$ of ordered semigroups, the product $\prod_{i \in I} S_i$ is the ordered semigroup defined on the set $\prod_{i \in I} S_i$ by the law

$$(s_i)_{i \in I}(s_i')_{i \in I} = (s_i s_i')_{i \in I}$$

and the order given by

$$(s_i)_{i \in I} \leq (s_i')_{i \in I} \text{ if and only if, for all } i \in I, \ s_i \leq s_i'.$$

Products of semigroups, monoids and ordered monoids are defined similarly.

If M is a monoid, the set $\mathcal{P}(M)$ of the subsets of M is a monoid under the operation

$$XY = \{xy \mid x \in X \text{ and } y \in Y\}$$

2.3 Morphisms

Generally speaking, a morphism between two algebraic structures is a map that preserves the operations and the relations of the structure. This general definition applies in particular to semigroups, monoids, ordered semigroups and ordered monoids. Given two semigroups S and T, a *semigroup morphism* $\varphi : S \to T$ is a map from S into T such that for all $x, y \in S$, $\varphi(xy) = \varphi(x)\varphi(y)$. *Monoid morphisms* are defined analogously, but of course, the condition $\varphi(1) = 1$ is also required. A *morphism of ordered semigroups* $\varphi : S \to T$ is a semigroup morphism from S into T such that, for every $x, y \in S$, $x \leq y$ implies $\varphi(x) \leq \varphi(y)$.

A morphism of semigroups (resp. monoids, ordered semigroups) $\varphi : S \to T$ is an *isomorphism* if there exists a morphism of semigroups (resp. monoids, ordered semigroups) $\psi : T \to S$ such that $\varphi \circ \psi = Id_T$ and $\psi \circ \varphi = Id_S$. It is easy to see that a morphism of semigroups is an isomorphism if and only if it is bijective. This is not true for morphisms of ordered semigroups. In particular, if $(S, \leq)$ is an ordered semigroup, the identity induces a bijective morphism from $(S, =)$ onto $(S, \leq)$ which is not in general an isomorphism. In fact, a morphism of ordered semigroups $\varphi : S \to T$ is an isomorphism if and only if φ is a bijective semigroup morphism and, for every $x, y \in S$, $x \leq y$ is equivalent with $\varphi(x) \leq \varphi(y)$.

For every semigroup morphism $\varphi : S \rightarrow T$, the equivalence relation $\sim_\varphi$ defined on S by setting $s \sim_\varphi t$ if and only if $\varphi(s) = \varphi(t)$ is a semigroup congruence. Similarly, for every morphism of ordered semigroups $\varphi : S \rightarrow T$, the quasi-order $\preceq_\varphi$ defined on S by setting $s \preceq_\varphi t$ if and only if $\varphi(s) \leq \varphi(t)$ is a congruence of ordered semigroup, called the *nuclear congruence* of φ.

A semigroup (resp. monoid, ordered semigroup) S is a *quotient* of a semigroup (resp. monoid, ordered semigroup) T if there exists a surjective morphism from T onto S. In particular, if $\sim$ is a congruence on a semigroup S, then $S/\sim$ is a quotient of S and the map $\pi : S \rightarrow S/\sim$ defined by $\pi(s) = [s]$ is a surjective morphism, called the *quotient morphism* associated with $\sim$. Similarly, let $\preceq$ be a congruence on an ordered semigroup $(S, \leq)$ and let $\sim$ be the equivalence relation associated with $\preceq$. Then $(S/\preceq)$ is a quotient of $(S, \leq)$ and the map $\pi : S \rightarrow S/\preceq$ defined by $\pi(s) = [s]$ is a surjective morphism of ordered semigroups.

Let $\sim_1$ and $\sim_2$ be two congruences on a semigroup S and let $\pi_1 : S \rightarrow S/\sim_1$ and $\pi_2 : S \rightarrow S/\sim_2$ be the quotient morphisms. Then $\sim_2$ is coarser than $\sim_1$ if and only if π_2 *factorizes through* π_1, that is, if there exists a surjective morphism $\pi : S/\sim_1 \rightarrow S/\sim_2$ such that $\pi \circ \pi_1 = \pi_2$.

A similar result holds for ordered semigroups. Let $\preceq_1$ and $\preceq_2$ be two congruences on an ordered semigroup S and let $\pi_1 : S \rightarrow S/\preceq_1$ and $\pi_2 : S \rightarrow S/\preceq_2$ be the quotient morphisms. Then $\preceq_2$ is coarser than $\preceq_1$ if and only if π_2 factorizes through π_1.

Let S be a semigroup (resp. ordered semigroup). A *subsemigroup* (resp. an ordered subsemigroup) of S is a subset T of S such that $t, t' \in T$ implies $tt' \in T$. Subsemigroups are closed under intersection. In particular, given a subset E of S, the smallest subsemigroup of S containing E is called the subsemigroup of S *generated* by E.

A semigroup S *divides* a semigroup T if S is a quotient of a subsemigroup of T. Division is a quasi-order on semigroups. Furthermore, one can show that two finite semigroups divide each other if and only if they are isomorphic.

2.4 Groups

A *group* is a monoid in which every element has an inverse. We briefly recall some standard definitions of group theory. Let p be a prime number. A *p-group* is a finite group whose order is a power of p. If G is a group, let $G_0 = G$ and $G_{n+1} = [G_n, G]$, the subgroup generated by the commutators $hgh^{-1}g^{-1}$, where $h \in G_n$ and $g \in G$. A finite group is *nilpotent* if and only if $G_n = \{1\}$ for some $n \geq 0$. A finite group is *solvable* if and only if there is a sequence

$$G = G^{(0)}, G^{(1)}, \ldots, G^{(k)} = \{1\}$$

such that each $G^{(i+1)}$ is a normal subgroup of $G^{(i)}$ and each quotient $G^{(i)}/G^{(i+1)}$ is commutative. It is a well-known fact that every p-group is nilpotent and every nilpotent group is solvable. See [143]

2.5 Free semigroups

Let A be a finite alphabet. The set of words on A is denoted A^* and the set of non-empty words, A^+. Thus $A^* = A^+ \cup \{1\}$, where 1 is the empty word. The length of a word u is denoted $|u|$. If a is a letter, $|u|_a$ denotes the number of occurrences of a in u. In particular, $|u| = \sum_{a \in A} |u|_a$. A word p is a *prefix* of a word u if $u = pu'$ for some $u' \in A^*$. Symmetrically, a word s is a *suffix* of u if $u = u's$ for some $u' \in A^*$. A word x is a *factor* of u if there exist two words u' and u'' (possibly empty) such that $u = u'xu''$. This notion should not be confused with the notion of subword. A word $a_1 \cdots a_n$ (where the a_i's are letters) is a *subword* of u if $u = u_0 a_1 u_1 \cdots a_n u_n$ for some words $u_0, \ldots, u_n \in A^*$.

The semigroup A^+ is the free semigroup on A and $(A^+, =)$ is the free ordered semigroup on A. Indeed, if $\varphi : A \to S$ is a function from A into an ordered semigroup S, there exists a unique morphism of ordered semigroups $\bar{\varphi} : (A^+, =) \to S$ such that $\varphi(a) = \bar{\varphi}(a)$ for every $a \in A$. Moreover $\bar{\varphi}$ is surjective if and only if $\varphi(A)$ is a generator set of S. It follows that if $\eta : (A^+, =) \to S$ is a morphism of ordered semigroups and $\beta : T \to S$ is a surjective morphism of ordered semigroups, there exists a morphism of ordered semigroups $\varphi : (A^+, =) \to T$ such that $\eta = \beta \circ \varphi$. This property is known as the *universal property* of the free ordered semigroup. Similarly, A^* is the free monoid on A and $(A^*, =)$ is the free ordered monoid on A.

As was explained in the introduction, there are two parallel notions of languages. If we are working with monoids, a *language* on A is a subset of the free monoid A^*. If semigroups are considered, a *language* on A is a subset of the free semigroup A^+.

2.6 Order ideals

An *order ideal* of an ordered semigroup S is a subset I of S such that, if $x \le y$ and $y \in I$, then $x \in I$. The order ideal *generated by an element* x is the set $\downarrow x$ of all $y \in S$ such that $y \le x$. The intersection (resp. union) of any family of order ideals is also an order ideal. Furthermore, if I is an order ideal and K is an arbitrary subset of S^1, then the left quotient $K^{-1}I$ and the right quotient IK^{-1} are also order ideals. Recall that, for each subset X of S and for each element s of S^1, the *left (resp. right) quotient* $s^{-1}X$ (resp. Xs^{-1}) of X by s is defined as follows:

$$s^{-1}X = \{t \in S \mid st \in X\} \quad \text{and} \quad Xs^{-1} = \{t \in S \mid ts \in X\}$$

More generally, for any subset K of S^1, the left (resp. right) quotient $K^{-1}X$ (resp. XK^{-1}) of X by K is

$$K^{-1}X = \bigcup_{s \in K} s^{-1}X = \{t \in S \mid \text{there exists } s \in K \text{ such that } st \in X\}$$
$$XK^{-1} = \bigcup_{s \in K} Xs^{-1} = \{t \in S \mid \text{there exists } s \in K \text{ such that } ts \in X\}$$

2.7 Idempotents

An element e of a semigroup S is *idempotent* if $e^2 = e$. A semigroup is *idempotent* if all its elements are idempotent. In this chapter, we will mostly use finite semigroups, in which idempotents play a key rôle. In particular, the following proposition shows that every non empty finite semigroup contains an idempotent.

Proposition 2.1. *Let s be an element of a finite semigroup. Then the subsemigroup generated by s contains a unique idempotent and a unique maximal subgroup, whose identity is the unique idempotent.*

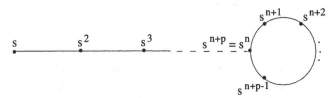

Fig. 2.1. The semigroup generated by s.

If s is an element of a finite semigroup, the unique idempotent power of s is denoted s^ω. If e is an idempotent of a finite semigroup S, the set

$$eSe = \{ese \mid s \in s\}$$

is a subsemigroup of S, called the *local subsemigroup* associated with e. This semigroup is in fact a monoid, since e is an identity in eSe.

A finite semigroup S is said to satisfy *locally* a property $\mathcal{P}$ if every local subsemigroup of S satisfies $\mathcal{P}$. For instance, S is *locally trivial* if, for every idempotent $e \in S$ and every $s \in S$, $ese = e$.

A *zero* is an element 0 such that, for every $s \in S$, $s0 = 0s = 0$. It is a routine exercise to see that there is at most one zero in a semigroup. A non-empty finite semigroup that contains a zero and no other idempotent is called *nilpotent*.

2.8 Green's relations

Green's relations on a semigroup S are defined as follows. If s and t are elements of S, we set

$s \mathcal{L} t$	if there exist $x, y \in S^1$ such that $s = xt$ and $t = ys$,
$s \mathcal{R} t$	if there exist $x, y \in S^1$ such that $s = tx$ and $t = sy$,
$s \mathcal{J} t$	if there exist $x, y, u, v \in S^1$ such that $s = xty$ and $t = usv$.
$s \mathcal{H} t$	if $s \mathcal{R} t$ and $s \mathcal{L} t$.

For finite semigroups, these four equivalence relations can be represented as follows. The elements of a given $\mathcal{R}$-class (resp. $\mathcal{L}$-class) are represented in a row (resp. column). The intersection of an $\mathcal{R}$-class and an $\mathcal{L}$-class is an $\mathcal{H}$-class. Each $\mathcal{J}$-class is a union of $\mathcal{R}$-classes (and also of $\mathcal{L}$-classes). It is not obvious to see that this representation is consistent: it relies in particular on the fact that, in finite semigroups, the relations $\mathcal{R}$ and $\mathcal{L}$ commute. The presence of an idempotent in an $\mathcal{H}$-class is indicated by a star. One can show that each $\mathcal{H}$-class containing an idempotent e is a subsemigroup of S, which is in fact a group with identity e. Furthermore, all $\mathcal{R}$-classes (resp. $\mathcal{L}$-classes) of a given $\mathcal{J}$-class have the same number of elements.

* a_1, a_2	* a_3, a_4	a_5, a_6
b_1, b_2	* b_3, b_4	* b_5, b_6

A $\mathcal{J}$-class.

In this figure, each row is an $\mathcal{R}$-class and each column is an $\mathcal{L}$-class. There are 6 $\mathcal{H}$-classes and 4 idempotents. Each idempotent is the identity of a group of order 2.

A $\mathcal{J}$-class containing an idempotent is called *regular*. One can show that in a regular $\mathcal{J}$-class, every $\mathcal{R}$-class and every $\mathcal{L}$-class contains an idempotent.

A semigroup S is $\mathcal{L}$-*trivial* (resp. $\mathcal{R}$-*trivial*, $\mathcal{J}$-*trivial*, $\mathcal{H}$-*trivial*) if two elements of S which are $\mathcal{L}$-equivalent (resp. $\mathcal{R}$-equivalent, $\mathcal{J}$-equivalent, $\mathcal{H}$-equivalent) are equal. See [75, 102] for more details.

2.9 Categories

Some algebraic developments in semigroup theory motivate the introduction of categories as a generalization of monoids. A *category* C is given by

(1) a set $Ob(C)$ of *objects*,
(2) for each pair (u, v) of objects, a set $C(u, v)$ of *arrows*,
(3) for each triple (u, v, w) of objects, a mapping from $C(u, v) \times C(v, w)$ into $C(u, w)$ which associates to each $p \in C(u, v)$ and $q \in C(v, w)$ the composition $pq \in C(u, w)$.
(4) for each object u, an arrow 1_u such that, for each pair (u, v) of objects, for each $p \in C(u, v)$ and $q \in C(v, u)$, $1_u p = p$ and $q 1_u = q$.

Composition is assumed to be associative (when defined).

For each object u, $C(u, u)$ is a monoid, called the *local monoid* of u. In particular a monoid can be considered as a category with exactly one object. A category is said to be *locally idempotent* (resp. *locally commutative*, etc.) if all its local monoids are idempotent (resp. commutative, etc.).

If C and D are categories, a *morphism of categories* $\varphi : C \to D$ is defined by the following data:

(1) an *object map* $\varphi : Ob(C) \to Ob(D)$,
(2) for each pair (u, v) of objects of C, an *arrow map* $\varphi : C(u,v) \to D(\varphi(u), \varphi(v))$ such that
    (a) for each triple (u, v, w) of objects of C, for each $p \in C(u, v)$ and $q \in C(v, w)$, $\varphi(pq) = \varphi(p)\varphi(q)$
    (b) for each object u, $\varphi(1_u) = 1_{\varphi(u)}$.

A category C is a *subcategory* of a category D if there exists a morphism $\varphi : C \to D$ which is injective on objects and on arrows (that is, for each pair of objects (u, v), the arrow map from $C(u, v)$ into $D(\varphi(u), \varphi(v))$ is injective). A category C is a *quotient* of a category D if there exists a morphism $D \to C$ which is bijective on objects and surjective on arrows. Finally C *divides* D if C is a quotient of a subcategory of D.

3. Recognizability

Recognizable languages are usually defined in terms of automata. This is the best definition from an algorithmic point of view, but it is an asymmetric notion. It turns out that to handle the fine structure of recognizable languages, it is more appropriate to use a more abstract definition, using semigroups in place of automata, due to Rabin and Scott [128]. However, we will slightly modify this standard definition by introducing ordered semigroups. As will be shown in the next sections, this order occurs quite naturally and permits to distinguish between a language and its complement. Although these definitions will be mainly used in the context of free semigroups, it is as simple to give them in a more general setting.

3.1 Recognition by ordered semigroups

Let $\varphi : S \to T$ be a surjective morphism of ordered semigroups. A subset Q of S is *recognized* by φ if there exists an order ideal P of T such that

$$Q = \varphi^{-1}(P)$$

This condition implies that Q is an order ideal of S and that $\varphi(Q) = \varphi\varphi^{-1}(P) = P$. By extension, a subset Q of S is said to be *recognized* by an ordered semigroup T if there exists a surjective morphism of ordered semigroups from S onto T that recognizes Q.

It is sometimes convenient to formulate this definition in terms of congruences. Let S be an ordered semigroup and let $\preceq$ a congruence on S. A subset Q of S is said to be *recognized* by $\preceq$ if, for every $q \in Q$, $p \preceq q$ implies $p \in Q$. It is easy to see that a surjective morphism of ordered semigroups φ recognizes Q if and only if the nuclear congruence $\preceq_\varphi$ recognizes Q.

Simple operations on subsets have a natural algebraic counterpart. We now study in this order intersection, union, complement, inverse morphisms and left and right quotients.

Proposition 3.1. *Let $(\eta_i : S \to S_i)_{i \in I}$ be a family of surjective morphisms of ordered semigroups. If each η_i recognizes a subset Q_i of S, then the subsets $\cap_{i \in I} Q_i$ and $\cup_{i \in I} Q_i$ are recognized by an ordered subsemigroup of the product $\prod_{i \in I} S_i$.*

If P is an order ideal of an ordered semigroup S, the set $S \setminus P$ is not, in general, an order ideal of S. However, it is an order ideal of the dual of S.

Proposition 3.2. *Let P be an order ideal of an ordered semigroup $(S, \leq)$. Then $S \setminus P$ is an order ideal of $(S, \geq)$. If P is recognized by a morphism of ordered semigroups $\eta : (S, \leq) \to (T, \leq)$, then $S \setminus P$ is recognized by the morphism of ordered semigroups $\eta : (S, \geq) \to (T, \geq)$.*

Proposition 3.3. *Let $\varphi : R \to S$ and $\eta : S \to T$ be two surjective morphisms of ordered semigroups. If η recognizes a subset Q of S, then $\eta \circ \varphi$ recognizes $\varphi^{-1}(Q)$.*

Proposition 3.4. *Let $\eta : S \to T$ be a surjective morphism of ordered semigroups. If η recognizes a subset Q of S, it also recognizes $K^{-1}Q$ and QK^{-1} for every subset K of S^1.*

3.2 Syntactic order

The syntactic congruence is one of the key notions of this chapter. Roughly speaking, it is the semigroup analog of the notion of minimal automaton. First note that, if S is an ordered semigroup, the congruence $\leq$ recognizes every order ideal of S. The syntactic congruence of an order ideal Q of S is the coarsest congruence among the congruences on S that recognize Q.

Let T be an ordered semigroup and let P be an order ideal of T. Define a relation $\preceq_P$ on T by setting

$$u \preceq_P v \text{ if and only if, for every } x, y \in T^1, \ xvy \in P \Rightarrow xuy \in P$$

One can show that the relation $\preceq_P$ is a congruence of ordered semigroups on T that recognizes P. This congruence is called the *syntactic congruence* of P in T. The equivalence relation associated with $\preceq_P$ is denoted $\sim_P$ and called the *syntactic equivalence* of P in T. Thus $u \sim_P v$ if and only if, for every $x, y \in T^1$,

$$xuy \in P \Longleftrightarrow xvy \in P$$

The ordered semigroup $S(P) = T/\preceq_P$ is the *ordered syntactic semigroup* of P, the order relation on $S(P)$ the *syntactic order* of P and the quotient morphism η_P from T onto $S(P)$ the *syntactic morphism* of P. The syntactic congruence is characterized by the following property.

Proposition 3.5. *The syntactic congruence of P is the coarsest congruence that recognizes P. Furthermore, a congruence $\preceq$ recognizes P if and only if $\preceq_P$ is coarser than $\preceq$.*

It is sometimes convenient to state this result in terms of morphisms:

Corollary 3.6. *Let $\varphi : R \to S$ be a surjective morphism of ordered semigroups and let P be an order ideal of R. The following properties hold:*

(1) The morphism φ recognizes P if and only if η_P factorizes through it.

(2) Let $\pi : S \to T$ be a surjective morphism of ordered semigroups. If $\pi \circ \varphi$ recognizes P, then φ recognizes P.

3.3 Recognizable sets

A subset of an ordered semigroup is *recognizable* if it is recognized by a finite ordered semigroup. Propositions 3.1 and 3.4 show that the recognizable subsets of a given ordered semigroup are closed under finite union, finite intersection and left and right quotients.

In the event where the order relation on S is the equality relation, every subset of S is an order ideal, and the definition of a recognizable set can be slightly simplified.

Proposition 3.7. *Let S be a semigroup. A subset P of S is recognizable if and only if there exists a surjective semigroup morphism φ from S onto a finite semigroup F and a subset Q of F such that $P = \varphi^{-1}(Q)$.*

The case of the free semigroup is of course the most important. In this case, the definition given above is equivalent with the standard definition using finite automata. Recall that a finite (non deterministic) automaton is a quintuple $\mathcal{A} = (Q, A, E, I, F)$, where A denotes the alphabet, Q the set of states, E is the set of transitions (a subset of $Q \times A \times Q$), and I and F are the set of *initial* and *final* states, respectively. An automaton $\mathcal{A} = (Q, A, E, I, F)$ is *deterministic* if I is a singleton and if the conditions $(p, a, q), (p, a, q') \in E$ imply $q = q'$.

Two transitions (p, a, q) and (p', a', q') are *consecutive* if $q = p'$. A *path* in $\mathcal{A}$ is a finite sequence of consecutive transitions

$$e_0 = (q_0, a_0, q_1), \ e_1 = (q_1, a_1, q_2), \ \ldots \ , \ e_{n-1} = (q_{n-1}, a_{n-1}, q_n)$$

also denoted

$$q_0 \xrightarrow{a_0} q_1 \xrightarrow{a_1} q_2 \ \cdots \ q_{n-1} \xrightarrow{a_{n-1}} q_n$$

The state q_0 is the *origin* of the path, the state q_n is its *end*, and the word $x = a_0 a_1 \cdots a_{n-1}$ is its *label*. It is convenient to have also, for each state q, an empty path of label 1 from q to q. A path in $\mathcal{A}$ is *successful* if its origin is in I and its end is in F.

The language of A^* *recognized* by $\mathcal{A}$ is the set of the labels of all successful paths of $\mathcal{A}$. In the case of the free semigroup A^+, the definitions are the same, except that we omit the empty paths of label 1.

Automata are conveniently represented by labeled graphs, as in the example below. Incoming arrows indicate initial states and outgoing arrows indicate final states.

Example 3.1. Let $\mathcal{A} = (Q, A, E, I, F)$ be the automaton represented below, with $Q = \{1, 2\}$, $A = \{a, b\}$, $E = \{(1, a, 1), (1, a, 2), (2, a, 2), (2, b, 2), (2, b, 1)\}$, $I = \{1\}$ and $F = \{2\}$. The path $(1, a, 1)(1, a, 2)(2, b, 2)$ is a successful path of

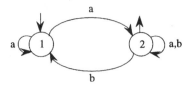

Fig. 3.1. A non deterministic automaton.

label *aab*. The path $(1, a, 1)(1, a, 2)(2, b, 1)$ has the same label but is unsuccessful since its end is 1. The set of words accepted by $\mathcal{A}$ is aA^*, the set of all words whose first letter is a.

The equivalence between automata and semigroups is based on the following observation. Let $\mathcal{A} = (Q, A, E, I, F)$ be a finite automaton. To each word $u \in A^+$, there corresponds a relation on Q, denoted by $\mu(u)$, and defined by $(p, q) \in \mu(u)$ if there exists a path from p to q with label u. It is not difficult to see that μ is a semigroup morphism from A^+ into the semigroup[1] of relations on Q. The semigroup $\mu(A^+)$ is called the *transition semigroup* of $\mathcal{A}$, denoted $S(\mathcal{A})$. For practical computation, it can be conveniently represented as a semigroup of boolean matrices of order $|Q| \times |Q|$. In this case, $\mu(u)$ can be identified with the matrix defined by

$$\mu(u)_{p,q} = \begin{cases} 1 & \text{if there exists a path from } p \text{ to } q \text{ with label } u \\ 0 & \text{otherwise} \end{cases}$$

Note that a word u is recognized by $\mathcal{A}$ if and only if $(p, q) \in \mu(u)$ for some initial state p and some final state q. This leads to the next proposition.

Proposition 3.8. *If a finite automaton recognizes a language L, then its transition semigroup recognizes L.*

Example 3.2. If $\mathcal{A} = (Q, A, E, I, F)$ is the automaton of example 3.1, one gets

$$\mu(a) = \begin{pmatrix} 1 & 1 \\ 0 & 1 \end{pmatrix} \qquad \mu(b) = \begin{pmatrix} 0 & 0 \\ 1 & 1 \end{pmatrix} \qquad \mu(aa) = \mu(a)$$

$$\mu(ab) = \begin{pmatrix} 1 & 1 \\ 1 & 1 \end{pmatrix} \qquad \mu(ba) = \mu(bb) = \mu(b)$$

Thus the transition semigroup of $\mathcal{A}$ is the semigroup of boolean matrices

$$\mu(A^+) = \left\{ \begin{pmatrix} 0 & 0 \\ 1 & 1 \end{pmatrix}, \begin{pmatrix} 1 & 1 \\ 0 & 1 \end{pmatrix}, \begin{pmatrix} 1 & 1 \\ 1 & 1 \end{pmatrix} \right\}.$$

---

[1] Given two relations $\mathcal{R}$ and $\mathcal{S}$ on Q, their product is the relation $\mathcal{R}\mathcal{S}$ defined by $(p, q) \in \mathcal{R}\mathcal{S}$ if and only if there exists r such that $(p, r) \in \mathcal{R}$ and $(r, q) \in \mathcal{S}$

The previous computation can be simplified if $\mathcal{A}$ is deterministic. Indeed, in this case, the transition semigroup of $\mathcal{A}$ is naturally embedded into the semigroup of partial functions on Q under the product $fg = g \circ f$.

Example 3.3. Let $A = \{a, b\}$ and let $\mathcal{A}$ be the deterministic automaton represented below. It is easy to see that $\mathcal{A}$ recognizes the language $A^+ \setminus (ab)^+$.

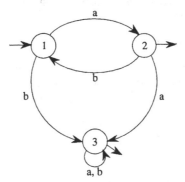

Fig. 3.2. A deterministic automaton.

The transition semigroup S of $\mathcal{A}$ contains five elements which correspond to the words a, b, ab, ba and aa. Furthermore aa is a zero of S and thus can be denoted 0. The other relations defining S are $aba = a$, $bab = b$ and $bb = 0$.

	a	b	aa	ab	ba
1	2	3	3	1	3
2	3	1	3	3	2
3	3	3	3	3	3

This semigroup is usually denoted BA_2 in semigroup theory.

Conversely, given a semigroup morphism $\varphi : A^+ \to S$ and a subset P of S, one can build a deterministic automaton recognizing $L = \varphi^{-1}(P)$ as follows. Take the right representation of A on S^1 defined by $s \cdot a = s\varphi(a)$. This defines an automaton $\mathcal{A} = (S^1, A, E, \{1\}, P)$, where $E = \{(s, a, s \cdot a) \mid s \in S^1, a \in A\}$ that recognizes L. We can now conclude.

Proposition 3.9. *A language is recognizable if and only if it is recognized by a finite automaton.*

See [102] for a detailed proof.

3.4 How to compute the syntactic semigroup?

The easiest way to compute the ordered syntactic semigroup of a recognizable language L is to first compute its minimal (deterministic) automaton $\mathcal{A} =$

$(Q, A, \cdot, \{q_0\}, F)$. Then the syntactic semigroup of L is equal to the transition semigroup S of $\mathcal{A}$ and the order on S is given by

$s \leq t$ if and only if, for every $x \in S^1$, for every $q \in Q$, $q \cdot tx \in F \Rightarrow q \cdot sx \in F$

Example 3.4. Let $\mathcal{A}$ be the deterministic automaton of example 3.3. It is the minimal automaton of $L = A^+ \setminus (ab)^+$. The transition semigroup was calculated in the previous section. The syntactic order is given by $0 \leq s$ for every $s \in S$. Indeed, $q \cdot 0 = 3 \in F$ and thus, the formal implication

$$q \cdot sx \in F \Rightarrow q \cdot 0x \in F$$

holds for any $q \in Q$, $s \in S$ and $x \in S^1$. One can verify that there is no other relations among the elements of S. For instance, a and ab are incomparable since $1 \cdot aa = 3$ but $1 \cdot aba = 2 \notin F$ and $1 \cdot abb = 3$ but $1 \cdot ab = 1 \notin F$.

4. Varieties

To each recognizable language is attached a finite ordered semigroup, its ordered syntactic semigroup. It is a natural idea to try to classify recognizable languages according to the algebraic properties of their ordered semigroups. The aim of this section is to introduce the proper framework to formalize this idea.

A *variety of semigroups* is a class of semigroups closed under taking sub-semigroups, quotients and direct products. A *variety of finite semigroups*, or *pseudovariety*, is a class of finite semigroups closed under taking subsemigroups, quotients and finite direct products. *Varieties of ordered semigroups* and *varieties of finite ordered semigroups* are defined analogously. Varieties of semigroups or ordered semigroups will be denoted by boldface capital letters, like $\mathbf{V}$. The *join* of two varieties of finite (ordered) semigroups $\mathbf{V}_1$ and $\mathbf{V}_2$ is the smallest variety of finite (ordered) semigroups containing $\mathbf{V}_1$ and $\mathbf{V}_2$.

Given a class C of finite (ordered) semigroups, the variety of finite (ordered) semigroups *generated* by C is the smallest variety of finite (ordered) semigroups containing C. In a more constructive way, the variety of finite (ordered) semigroups *generated* by C is the class of all finite (ordered) semigroups that divide a finite direct product $S_1 \times \cdots \times S_n$, where $S_1, \ldots, S_n \in C$.

4.1 Identities

Varieties are conveniently defined by identities. Let Σ be a denumerable alphabet and let (u, v) be a pair of words of Σ^+. A semigroup S *satisfies the identity* $u = v$ if and only if $\varphi(u) = \varphi(v)$ for every semigroup morphism $\varphi : \Sigma^+ \to S$. Similarly, an ordered semigroup S *satisfies the identity* $u \leq v$ if and only if $\varphi(u) \leq \varphi(v)$ for every morphism of ordered semigroups $\varphi : \Sigma^+ \to S$. If

Γ is a set of identities, the class of all semigroups (resp. ordered semigroups) that satisfy all the identities of Γ is a variety of semigroups (resp. ordered semigroups), called the *variety defined by* Γ. The following theorem, due to Birkhoff [22] and to Bloom [33] in the ordered case, shows that the converse also holds.

Theorem 4.1. *A class of semigroups (resp. ordered semigroups) is a variety if and only if it can be defined by a set of identities.*

For instance, the identity $xy = yx$ defines the variety of commutative semigroups and $x = x^2$ defines the variety of idempotent semigroups.

Since we are interested in finite semigroups, it would be interesting to have a similar result for varieties of finite semigroups. The problem was solved by several authors but the most satisfactory answer is due to Reiterman [129] (see also [125] in the ordered case). Reiterman's theorem states that pseudovarieties are also defined by identities. The difference between Birkhoff's and Reiterman's theorem lies in the definition of the identities. For Reiterman, an identity is also a formal equality of the form $u = v$, but u and v are now elements of a certain completion $\widehat{\Sigma^+}$ of the free semigroup Σ^+. Let us make this definition more precise.

A finite semigroup S *separates* two words $u, v \in \Sigma^+$ if $\varphi(u) \neq \varphi(v)$ for some semigroup morphism $\varphi : \Sigma^+ \to S$. Now set, for $u, v \in \Sigma^+$,

$$r(u, v) = \min\{\operatorname{Card}(S) | \ S \text{ is a finite semigroup separating } u \text{ and } v \ \}$$

and

$$d(u, v) = 2^{-r(u,v)}$$

with the usual conventions $\min \emptyset = +\infty$ and $2^{-\infty} = 0$. One can verify that d is a metric for which two words are close if a large semigroup is required to separate them. For this metric, multiplication in Σ^+ is uniformly continuous, so that Σ^+ is a topological semigroup. The completion of the metric space (Σ^+, d) is a topological semigroup, denoted $\widehat{\Sigma^+}$, in which every element is the limit of some Cauchy sequence of (Σ^+, d). In fact, one can show that $\widehat{\Sigma^+}$ is a compact semigroup.

Some more topology is required before stating Reiterman's theorem. We now consider finite semigroups as metric spaces, endowed with the discrete metric

$$d(x, y) = \begin{cases} 0 & \text{if } x = y \\ 1 & \text{if } x \neq y \end{cases}$$

Let S be a finite semigroup. Then a map $\varphi : \widehat{\Sigma^+} \to S$ is *continuous* if and only if, for every converging sequence $(u_n)_{n \geq 0}$ of $\widehat{\Sigma^+}$, the sequence $\varphi(u_n)_{n \geq 0}$ is ultimately constant[2]. A finite semigroup (resp. ordered semigroup) S satisfies an identity $u = v$ (resp. $u \leq v$), where $u, v \in \widehat{\Sigma^+}$, if and only if $\varphi(u) = \varphi(v)$

---

[2] that is, if there exists an integer n_0 such that, for all $n, m \geq n_0$, $\varphi(u_n) = \varphi(u_m)$.

(resp. $\varphi(u) \leq \varphi(v)$) for every continuous morphism $\varphi : \widehat{\Sigma}^+ \to S$. Note that such a continuous morphism is entirely determined by the values of $\varphi(a)$, for $a \in \Sigma$. Indeed, any map $\varphi : \Sigma \to S$ can be extended in a unique way into a semigroup morphism from Σ^+ into S. Since S is finite, such a morphism is uniformly continuous : if two elements of Σ^+ cannot be separated by S, their images under φ have to be the same. Now, a well known result of topology states that a uniformly continuous map whose domain is a metric space admits a unique continuous extension to the completion of this metric space.

Given a set E of identities of the form $u = v$ (resp. $u \leq v$), where $u, v \in \widehat{\Sigma}^+$, we denote by $[\![E]\!]$ the class of all finite semigroups (resp. ordered semigroups) which satisfy all the identities of E. Reiterman's theorem can now be stated as follows.

Theorem 4.2. *A class of finite semigroups (resp. ordered semigroups) is a variety if and only if it can be defined by a set of identities of $\widehat{\Sigma}^+$.*

Although Theorem 4.2 gives a satisfactory counterpart to Birkhoff's theorem, it is more difficult to understand, because of the rather abstract definition of $\widehat{\Sigma}^+$. Actually, no combinatorial description of $\widehat{\Sigma}^+$ is known and, besides the elements of Σ^+, which are words, the other elements are only defined as limits of sequences of words. An important such limit is the ω-power. Its definition relies on the following proposition.

Proposition 4.3. *For each $x \in \widehat{\Sigma}^+$, the sequence $(x^{n!})_{n \geq 0}$ is a Cauchy sequence. Its limit x^ω is an idempotent of $\widehat{\Sigma}^+$.*

Now, if $\varphi : \widehat{\Sigma}^+ \to S$ is a continuous morphism onto a finite semigroup S, then $\varphi(x^\omega)$ is equal to $\varphi(x)^\omega$, the unique idempotent power of $\varphi(x)$, which shows that our notation is consistent. The notation x^ω makes the conversion of algebraic ·properties into identities very easy. For instance, the variety $[\![yx^\omega = x^\omega]\!]$ is the class of finite semigroups S such that, for every idempotent $e \in S$ and for every $s \in s$, $se = e$. Similarly, a finite semigroup S is locally commutative, if, for every idempotent e, the local monoid eSe is commutative. It follows immediately that finite locally commutative semigroups form a variety, defined by the identity $x^\omega y x^\omega z x^\omega = x^\omega z x^\omega y x^\omega$. More generally, if **V** is a variety of finite monoids, **LV** denotes the variety of all finite semigroups S, such that, for every idempotent $e \in S$, the local monoid eSe is in **V**.

Another useful example is the following. The *content* of a word $u \in \Sigma^+$ is the set $c(u)$ of letters of Σ occurring in u. One can show that c is a uniformly continuous morphism from Σ^+ onto the semigroup 2^Σ of subsets of Σ under union. Thus c can be extended in a unique way into a continuous morphism from $\widehat{\Sigma}^+$ onto 2^Σ.

Reiterman's theorem suggests that most standard results on varieties might be extended in some way to pseudovarieties. For instance, it is well known that varieties have free objects. More precisely, if **V** is a variety and A

is a finite set, there exists an A-generated semigroup $F_A(\mathbf{V})$ of $\mathbf{V}$, such that every A-generated semigroup of $\mathbf{V}$ is a quotient of $F_A(\mathbf{V})$. This semigroup is unique (up to an isomorphism) and is called the *free semigroup* of the variety $\mathbf{V}$. To extend this result to a pseudovariety $\mathbf{V}$, one first relativizes to $\mathbf{V}$ the definition of r and d as follows:

$$r_{\mathbf{V}}(u,v) = \min\{\operatorname{Card}(S)\mid S \in \mathbf{V} \text{ and } S \text{ separates } u \text{ and } v\}$$

and $d_{\mathbf{V}}(u,v) = 2^{-r_{\mathbf{V}}(u,v)}$. The function $d_{\mathbf{V}}(u,v)$ still satisfies the triangular inequality and even the stronger inequality

$$d_{\mathbf{V}}(u,v) \leq \max\{d_{\mathbf{V}}(u,w), d_{\mathbf{V}}(w,v)\}$$

but it is not a metric anymore because one can have $d_{\mathbf{V}}(u,v) = 0$ with $u \neq v$: for instance, if $\mathbf{V}$ is the pseudovariety of commutative finite semigroups, $d_{\mathbf{V}}(xy, yx) = 0$ since xy and yx cannot be separated by a commutative semigroup. However, the relation $\sim_{\mathbf{V}}$ defined on A^+ by $u \sim_{\mathbf{V}} v$ if and only if $d_{\mathbf{V}}(u,v) = 0$ is a congruence and $d_{\mathbf{V}}$ induces a metric on the quotient semigroup $A^+/\sim_{\mathbf{V}}$. The completion of this metric space is a topological compact semigroup $\widehat{F}_A(\mathbf{V})$, called the *free pro-$\mathbf{V}$ semigroup*. This semigroup is generated by A as a topological semigroup (this just means that $A^+/\sim_{\mathbf{V}}$ is dense in $\widehat{F}_A(\mathbf{V})$) and every A-generated semigroup of $\mathbf{V}$ is a continuous homomorphic image of $\widehat{F}_A(\mathbf{V})$. The combinatorial description of these free objects, for various varieties of finite semigroups, is the object of a very active research [7]. A more detailed presentation of Reiterman's theorem and its consequences can be found in [5, 7, 193].

4.2 The variety theorem

The variety theorem is due to Eilenberg [53]. Eilenberg's original theorem dealt with varieties of finite semigroups. The "ordered" version presented in this section is due to the author [112].

A *class of recognizable languages* is a correspondence $\mathcal{C}$ which associates with each finite alphabet A a set $\mathcal{C}(A^+)$ of recognizable languages of A^+.

If $\mathbf{V}$ is a variety of finite ordered semigroups, we denote by $\mathcal{V}(A^+)$ the set of recognizable languages of A^+ whose ordered syntactic semigroup belongs to $\mathbf{V}$ or, equivalently, which are recognized by an ordered semigroup of $\mathbf{V}$. The correspondence $\mathbf{V} \to \mathcal{V}$ associates with each variety of finite ordered semigroups a class of recognizable languages. The next proposition shows that this correspondence preserves inclusion.

Proposition 4.4. *Let $\mathbf{V}$ and $\mathbf{W}$ be two varieties of finite ordered semigroups. Suppose that $\mathbf{V} \to \mathcal{V}$ and $\mathbf{W} \to \mathcal{W}$. Then $\mathbf{V} \subseteq \mathbf{W}$ if and only if, for every finite alphabet A, $\mathcal{V}(A^+) \subseteq \mathcal{W}(A^+)$. In particular, $\mathbf{V} = \mathbf{W}$ if and only if $\mathcal{V} = \mathcal{W}$.*

It remains to characterize the classes of languages which can be associated with a variety of ordered semigroups. For this purpose, it is convenient to introduce the following definitions. A set of languages of A^+ (resp. A^*) closed under finite intersection and finite union is called a *positive boolean algebra*. Thus a positive boolean algebra always contains the empty language and the full language A^+ (resp. A^*) since $\emptyset = \bigcup_{i \in \emptyset} L_i$ and $A^+ = \bigcap_{i \in \emptyset} L_i$. A positive boolean algebra closed under complementation is a *boolean algebra*.

A *positive variety of languages* is a class of recognizable languages $\mathcal{V}$ such that

(1) for every alphabet A, $\mathcal{V}(A^+)$ is a positive boolean algebra,
(2) if $\varphi : A^+ \to B^+$ is a semigroup morphism, $L \in \mathcal{V}(B^+)$ implies $\varphi^{-1}(L) \in \mathcal{V}(A^+)$,
(3) if $L \in \mathcal{V}(A^+)$ and if $a \in A$, then $a^{-1}L$ and La^{-1} are in $\mathcal{V}(A^+)$.

Proposition 4.5. *Let $\mathbf{V}$ be a variety of finite ordered semigroups. If $\mathbf{V} \to \mathcal{V}$, then $\mathcal{V}$ is a positive variety of languages.*

So far, we have associated a positive variety of languages with each variety of finite ordered semigroups. Conversely, let $\mathcal{V}$ be a positive variety of languages and let $\mathbf{V}(\mathcal{V})$ be the variety of ordered semigroups generated by the ordered semigroups of the form $S(L)$ where $L \in \mathcal{V}(A^+)$ for a certain alphabet A. This variety is called the variety *associated with $\mathcal{V}$*, in view of the following theorem.

Theorem 4.6. *For every positive variety of languages $\mathcal{V}$, $\mathbf{V}(\mathcal{V}) \to \mathcal{V}$.*

In conclusion, we have the following theorem.

Theorem 4.7. *The correspondence $\mathbf{V} \to \mathcal{V}$ defines a one to one correspondence between the varieties of finite ordered semigroups and the positive varieties of languages.*

A *variety of languages* is a positive variety closed under complement, that is, satisfying

(1') for every alphabet A, $\mathcal{V}(A^+)$ is a boolean algebra.

For varieties of languages, Theorem 4.7 can be modified as follows:

Corollary 4.8. *The correspondence $\mathbf{V} \to \mathcal{V}$ defines a one to one correspondence between the varieties of finite semigroups and the varieties of languages.*

There is an analogous theorem for varieties of ordered monoids. In this case, one defines languages as subsets of a free monoid and the definitions of a class of languages and of a positive variety have to be modified. To distinguish between the two definitions, it is convenient to add the prefixes $+$ or $*$ when necessary: $+$-class or $*$-class, $+$-variety or $*$-variety.

A *$*$-class of recognizable languages* is a correspondence $\mathcal{C}$ which associates with each finite alphabet A a set $\mathcal{C}(A^*)$ of recognizable languages of A^*. A *positive $*$-variety of languages* is a class of recognizable languages $\mathcal{V}$ such that

(1) for every alphabet A, $\mathcal{V}(A^*)$ is a positive boolean algebra,

(2) if $\varphi : A^* \to B^*$ is a monoid morphism, $L \in \mathcal{V}(B^*)$ implies $\varphi^{-1}(L) \in \mathcal{V}(A^*)$,

(3) if $L \in \mathcal{V}(A^*)$ and if $a \in A$, then $a^{-1}L$ and La^{-1} are in $\mathcal{V}(A^*)$.

of course, a *-*variety of languages* is a positive *-variety closed under complement, that is, satisfying

(1') for every alphabet A, $\mathcal{V}(A^*)$ is a boolean algebra.

The monoid version of Theorem 4.7 can be stated as follows.

Theorem 4.9. *The correspondence* $\mathbf{V} \to \mathcal{V}$ *defines a one to one correspondence between the varieties of finite ordered monoids and the positive *-varieties of languages.*

Corollary 4.10. *The correspondence* $\mathbf{V} \to \mathcal{V}$ *defines a one to one correspondence between the varieties of finite monoids and the *-varieties of languages.*

5. Examples of varieties

In this section, we illustrate the results of the previous section by a few examples. We present, in this order, some standard examples (Kleene's and Schützenberger's theorems), the commutative varieties and varieties defined by "local" properties. Other examples will be given in the next sections.

Let us start by some general remarks. If $\mathbf{V}$ is a variety of finite ordered semigroups or monoids, the associated positive varieties of languages are denoted by the corresponding cursive letters, like $\mathcal{V}$. There are now several equivalent formulations to state a typical result on varieties, for instance:

"The variety of finite ordered semigroups $\mathbf{V}$ is associated with the positive variety $\mathcal{V}$."

"A recognizable language belongs to $\mathcal{V}(A^+)$ if and only if it is recognized by an ordered semigroup of $\mathbf{V}$."

"A recognizable language belongs to $\mathcal{V}(A^+)$ if and only if its ordered syntactic semigroup belongs to $\mathbf{V}$."

We shall use mainly statements of the first type, but the last type will be occasionally preferred, especially when there are several equivalent descriptions of the languages. But it should be clear to the reader that all these formulations express exactly the same property.

5.1 Standard examples

The smallest variety of finite monoids is the trivial variety **I**, defined by the identity $x = 1$. The associated variety of languages is defined, for every alphabet A, by $\mathcal{I}(A^*) = \{\emptyset, A^*\}$.

The largest variety of finite monoids is the variety of all finite monoids **M**, defined by the empty set of identities. Recall that the set of *rational* languages of A^* is the smallest set of languages containing the languages $\{1\}$ and $\{a\}$ for each letter $a \in A$ and closed under finite union, product and star. Now Kleene's theorem can be reformulated as follows.

Theorem 5.1. *The variety of languages associated with* **M** *is the variety of rational languages.*

An important variety of monoids is the variety of aperiodic monoids, defined by the identity $x^\omega = x^{\omega+1}$. Thus, a finite monoid M is *aperiodic* if and only if, for each $x \in M$, there exists $n \geq 0$ such that $x^n = x^{n+1}$. This also means that the cyclic subgroup of the submonoid generated by any element x is trivial (see Proposition 2.1) or that M is $\mathcal{H}$-trivial. It follows that a monoid is aperiodic if and only if it is group-free: every subsemigroup which happens to be a group has to be trivial. Aperiodic monoids form a variety of monoids **A**.

The associated variety of languages was first described by Schützenberger [144]. Recall that the *star-free* languages of A^* form the smallest boolean algebra containing the languages $\{1\}$ and $\{a\}$ for each letter $a \in A$ and which is closed under product.

Theorem 5.2. *The variety of languages associated with* **A** *is the variety of star-free languages.*

Example 5.1. Let $A = \{a, b\}$ and $L = (ab)^*$. Its minimal (but incomplete) automaton is represented below: The syntactic monoid M of L is the monoid

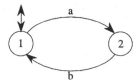

Fig. 5.1. The minimal automaton of $(ab)^*$.

with zero presented on A by the relations $a^2 = b^2 = 0$, $aba = a$ and $bab = b$. Thus $M = \{1, a, b, ab, ba, 0\}$. Since $1^2 = 1$, $a^3 = a^2$, $b^3 = b^2$, $(ab)^2 = ab$, $(ba)^2 = ba$ and $0^2 = 0$, M is aperiodic and thus $(ab)^*$ is star-free. Indeed, if R^c denotes the complement of a language R, $(ab)^*$ admits the following star-free expression

$$L = \left(b\emptyset^c \cup \emptyset^c a \cup \emptyset^c aa\emptyset^c \cup \emptyset^c bb\emptyset^c\right)^c$$

We shall come back to Schützenberger's theorem in section 8.

When a variety is generated by a single ordered monoid, there is a direct description of the associated variety of languages.

Proposition 5.3. *Let M be a finite ordered monoid, let $\mathbf{V}$ be the variety of ordered monoids generated by M and let $\mathcal{V}$ be the associated positive variety. Then for every alphabet A, $\mathcal{V}(A^*)$ is the positive boolean algebra generated by the languages of the form $\varphi^{-1}(\downarrow m)$, where φ is any monoid morphism from A^* into M and m is any element of M.*

Of course, a similar result holds for varieties of ordered semigroups.

Proposition 5.4. *Let S be a finite ordered semigroup, let $\mathbf{V}$ be the variety of ordered semigroups generated by S and let $\mathcal{V}$ be the associated positive variety. Then for every alphabet A, $\mathcal{V}(A^+)$ is the positive boolean algebra generated by the languages of the form $\varphi^{-1}(\downarrow s)$, where φ is any semigroup morphism from A^+ into S and s is any element of S.*

This result suffices to describe a number of "small" varieties of languages. See for instance propositions 5.5 and 5.6 or [119].

5.2 Commutative varieties

In this section, we will consider only varieties of finite monoids and ordered monoids. A variety of ordered monoids is *commutative* if it satisfies the identity $xy = yx$. The smallest non-trivial variety of aperiodic monoids is the variety $\mathbf{J}_1$ of idempotent and commutative monoids (also called *semilattices*)[3], defined by the identities $xy = yx$ and $x^2 = x$. One can show that $\mathbf{J}_1$ is generated by the monoid $U_1 = \{0, 1\}$, whose multiplication table is given by $0 \cdot 0 = 0 \cdot 1 = 1 \cdot 0 = 0$ and $1 \cdot 1 = 1$. Thus Proposition 5.4 can be applied to get a description of the $*$-variety associated with $\mathbf{J}_1$.

Proposition 5.5. *For every alphabet A, $\mathcal{J}_1(A^*)$ is the boolean algebra generated by the languages of the form $A^* a A^*$ where a is a letter. Equivalently, $\mathcal{J}_1(A^*)$ is the boolean algebra generated by the languages of the form B^* where B is a subset of A.*

Proposition 5.5 can be refined by considering the variety of finite ordered semigroups $\mathbf{J}_1^+$ (resp. $\mathbf{J}_1^-$), defined by the identities $xy = yx$, $x = x^2$ and $x \leq 1$ (resp. $x \geq 1$). One can show that $\mathbf{J}_1^+$ is generated by the ordered monoid $U_1^+ = (U_1, \leq)$ where the order is given by $0 \leq 1$. Then one can apply Proposition 5.3 to find a description of the $*$-variety $\mathcal{J}_1^+$ associated with $\mathbf{J}_1^+$.

---

[3] The notation $\mathbf{J}_1$ indicates that $\mathbf{J}_1$ is the first level of a hierarchy of varieties $\mathbf{J}_n$ that will be defined in section 8.4.

Let A be an alphabet and B be a subset of A. Denote by $L(B)$ the set of words containing at least one occurrence of every letter of B. Equivalently,

$$L(B) = \bigcap_{a \in B} A^* a A^*$$

Proposition 5.6. *For each alphabet A, $\mathcal{J}_1^+(A^*)$ consists of the finite unions of languages of the form $L(B)$, for some subset B of A.*

Corollary 5.7. *For each alphabet A, $\mathcal{J}_1^-(A^*)$ is the positive boolean algebra generated by the languages of the form B^*, for some subset B of A.*

Another important commutative variety of monoids is the variety of finite commutative groups **Gcom**, generated by the cyclic groups $\mathbb{Z}/n\mathbb{Z}$ $(n > 0)$ and defined by the identities $xy = yx$ and $x^\omega = 1$. For $a \in A$ and $k, n \geq 0$, let

$$F(a, k, n) = \{u \in A^* \mid |u|_a \equiv k \bmod n\}$$

Proposition 5.8. *For every alphabet A, $\mathcal{G}com(A^*)$ is the boolean algebra generated by the languages of the form $F(a, k, n)$, where $a \in A$ and $0 \leq k < n$.*

The largest commutative variety that contains no non-trivial group is the variety **Acom** of aperiodic and commutative monoids, defined by the identities $xy = yx$ and $x^\omega = x^{\omega+1}$. For $a \in A$ and $k \geq 0$, let

$$F(a, k) = \{u \in A^+ \mid |u|_a \geq k\}$$

Proposition 5.9. *For every alphabet A, $\mathcal{A}com(A^*)$ is the boolean algebra generated by the languages of the form $F(a, k)$ where $a \in A$ and $k \geq 0$.*

Again, this proposition can be refined by considering ordered monoids. Let **Acom**$^+$ be the variety of ordered monoids satisfying the identities $xy = yx$, $x^\omega = x^{\omega+1}$ and $x \leq 1$.

Proposition 5.10. *For every alphabet A, $\mathcal{A}com^+(A^*)$ is the positive boolean algebra generated by the languages of the form $F(a, k)$ where $a \in A$ and $k \geq 0$.*

Finally, the variety **Com** of all finite commutative monoids, defined by the identity $xy = yx$, is the join of the varieties **Gcom** and **Acom**.

Proposition 5.11. *For every alphabet A, $\mathcal{C}om(A^*)$ is the boolean algebra generated by the languages of the form $F(a, k)$ or $F(a, k, n)$ where $a \in A$ and $0 \leq k < n$.*

The "ordered" version is the following. Let **Com**$^+$ be the variety of ordered monoids satisfying the identities $xy = yx$ and $x \leq 1$.

Proposition 5.12. *For every alphabet A, $\mathcal{C}om^+(A^*)$ is the positive boolean algebra generated by the languages of the form $F(a, k)$ or $F(a, k, n)$ where $a \in A$ and $0 \leq k < n$.*

5.3 Varieties defined by local properties

Contrary to the previous section, all the varieties considered in this section will be varieties of finite (ordered) semigroups. These varieties are all defined by *local* properties of words. Local properties can be tested by a *scanner*, which is a machine equipped with a finite memory and a sliding window of a fixed size n to scan the input word. The window can also be moved before

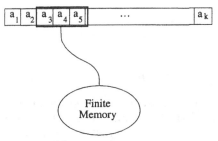

Fig. 5.2. A scanner.

the first letter and beyond the last letter of the word in order to read the prefixes and suffixes of length $< n$. For example, if $n = 3$, and if the input word is *abbaaab*, the various positions of the window are represented on the following diagram:

At the end of the scan, the scanner memorizes the prefixes and suffixes of length $< n$ and the set of factors of length n of the input word. The memory of the scanners contains a table of possible lists of prefixes (resp. suffixes, factors). A word is accepted by the scanner if the list of prefixes (resp. suffixes, factors) obtained after the scan matches one of the lists of the table. Another possibility is to take into account the number of occurrences of the factors of the word.

Local properties can be used to define several varieties of languages. A language is *prefix testable*[4] if it is a boolean combination of languages of the form xA^*, where $x \in A^+$ or, equivalently, if it is of the form $FA^* \cup G$ for some finite languages F and G. Similarly, a language is *suffix testable*[5] if it is a boolean combination of languages of the form A^*x, where $x \in A^+$.

Proposition 5.13. *Prefix (resp. suffix) testable languages form a variety of languages. The associated variety of finite semigroups is defined by the identity* $x^\omega y = x^\omega$ *(resp.* $yx^\omega = x^\omega$ *).*

---

[4] These languages are called *reverse definite* in the literature.

[5] The suffix testable languages are called *definite* in the literature.

Languages that are both prefix and suffix testable form an interesting variety. Recall that a language is *cofinite* if its complement is finite.

Proposition 5.14. *Let L be a recognizable language. The following conditions are equivalent:*

(1) L is prefix testable and suffix testable,
(2) L is finite or cofinite,
(3) $S(L)$ satisfies the identities $x^\omega y = x^\omega = yx^\omega$,
(4) $S(L)$ is nilpotent

Proposition 5.14 can be refined as follows

Proposition 5.15. *A language is empty or cofinite if and only if it is recognized by a finite ordered nilpotent semigroup S in which $0 \leq s$ for all $s \in S$.*

Note that the finite ordered nilpotent semigroups S in which 0 is the smallest element form a variety of finite ordered semigroups, defined by the identities $x^\omega y = x^\omega = yx^\omega$ and $x^\omega \leq y$. The dual version of Proposition 5.15 is also of interest.

Corollary 5.16. *A language is full or finite if and only if it is recognized by a finite ordered nilpotent semigroup S in which $s \leq 0$ for all $s \in S$.*

A language is *prefix-suffix testable*[6] if it is a boolean combination of languages of the form xA^* or A^*x, where $x \in A^+$.

Proposition 5.17. [90] *Let L be a language. The following conditions are equivalent:*

(1) L is a prefix-suffix testable language,
(2) L is of the form $FA^*G \cup H$ for some finite languages F, G and H,
(3) $S(L)$ satisfies the identity $x^\omega yx^\omega = x^\omega$,
(4) $S(L)$ is locally trivial.

Proposition 5.17 shows that prefix-suffix testable languages form a variety of languages. The corresponding variety of semigroups is **LI**.

A language is *positively locally testable* if it is a positive boolean combination of languages of the form $\{x\}$, xA^*, A^*x or A^*xA^* ($x \in A^+$) and it is *locally testable* if it is a boolean combination of the same languages. The syntactic characterization of locally testable languages is relatively simple to state, but its proof, discovered independently by Brzozowski and Simon [41] and by McNaughton [86], requires sophisticated tools that are detailed in section 6.

Theorem 5.18. *Let L be a language. The following conditions are equivalent:*

---

[6] These languages are called *generalized definite* in the literature.

(1) L is locally testable,
(2) S(L) satisfies the identities $x^\omega y x^\omega z x^\omega = x^\omega z x^\omega y x^\omega$ and $x^\omega y x^\omega y x^\omega = x^\omega y x^\omega$,
(3) S(L) is locally idempotent and commutative.

In the positive case, the identity $x^\omega y x^\omega \leq x^\omega$ must be added. Thus an ordered semigroup S satisfies this identity if and only if, for every idempotent $e \in S$, and for every element $s \in S$, $ese \leq e$. This means that, in the local monoid eSe, the identity e is the maximum element.

Theorem 5.19. *Let L be a language. The following conditions are equivalent:*

(1) L is positively locally testable,
(2) S(L) satisfies the identities $x^\omega y x^\omega z x^\omega = x^\omega z x^\omega y x^\omega$, $x^\omega y x^\omega y x^\omega = x^\omega y x^\omega$ and $x^\omega y x^\omega \leq x^\omega$.
(3) S(L) is locally idempotent and commutative and in every local monoid, the identity is the maximum element.

Example 5.2. Let $A = \{a, b\}$ and let $L = A^+ \setminus (ab)^+$ be the language of Example 3.4. Let S be the ordered syntactic semigroup of L. The idempotents of S are ab, ba and 0 and the local submonoids are $abSab = \{ab, 0\}$, $baSba = \{ba, 0\}$ and $0S0 = \{0\}$. These three monoids are idempotent and commutative and their identity is the maximum element (since $0 \leq s$ for every $s \in S$). Therefore, L is positively locally testable. Indeed, one has

$$L = bA^* \cup A^* a \cup A^* aaA^* \cup A^* bbA^*$$

One can also take into account the number of occurrences of a given word. For each word $x, u \in A^+$, let $\left[\begin{smallmatrix} u \\ x \end{smallmatrix}\right]$ denote the number of occurrences of x as a factor of u. For every integer k, set

$$F(x, k) = \left\{ u \in A^+ \mid \begin{bmatrix} u \\ x \end{bmatrix} \geq k \right\}$$

In particular, $F(x, 1) = A^* x A^*$, the set of words containing at least one occurrence of x. A language is said to be *(positively) threshold locally testable* if it is a (positive) boolean combination of languages of the form $\{x\}$, xA^*, $A^* x$ or $F(x, k)$ $(x \in A^+, k \geq 0)$.

A new concept is needed to state the syntactic characterization of these languages in a precise way. The *Cauchy category of a semigroup S* is the category C whose objects are the idempotents of S and, given two idempotents e and f, the set of arrows from e to f is the set

$$C(e, f) = \{s \in S \mid es = s = sf\}$$

Given three idempotents e, f, g, the composition of the arrows $p \in C(e, f)$ and $q \in C(f, g)$ is the arrow $pq \in C(e, g)$. The algebraic background for this definition will be given in section 6.2.

Theorem 5.20. *Let L be a recognizable subset of A^+. Then L is threshold locally testable if and only if $S(L)$ is aperiodic and its Cauchy category satisfies the following condition: if p and r are arrows from e to f and if q is an arrow from f to e, then $pqr = rqp$.*

The latter condition on the Cauchy category is called Thérien's condition.

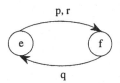

Fig. 5.3. The condition $pqr = rqp$.

In the positive case, the characterization is the following.

Theorem 5.21. *Let L be a recognizable subset of A^+. Then L is positively threshold locally testable if and only if $S(L)$ is aperiodic, its Cauchy category satisfies Thérien's condition and in every local monoid, the identity is the maximum element.*

Example 5.3. Let $A = \{a, b\}$ and let $L = a^*ba^*$. Then L is recognized by the automaton shown in the figure below.

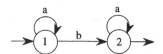

Fig. 5.4. The minimal automaton of a^*ba^*.

The transitions and the relations defining the syntactic semigroup S of L are given in the following tables

	a	b	bb
1	1	2	–
2	2	–	–

$a = 1$
$b^2 = 0$

Thus $S = \{1, b, 0\}$ and $E(S) = \{1, 0\}$. The local semigroups are $0S0 = \{0\}$ and $1S1 = S$. The latter is not idempotent, since $b^2 \neq b$. Therefore, L is not locally testable. On the other hand, the Cauchy category of $S(L)$, represented in the figure below, satisfies the condition $pqr = rqp$.

Therefore L is threshold locally testable and is not positively threshold locally testable since $b \not\leq 1$.

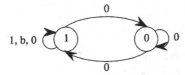

Fig. 5.5. The Cauchy category of $S(L)$.

Another way of counting factors is to count modulo n for some integer n. To this purpose, set, for every $x \in A^+$ and for every $k \geq 0$, $n > 0$,

$$F(x, k, n) = \left\{ u \in A^+ \mid \begin{bmatrix} u \\ x \end{bmatrix} \equiv k \bmod n \right\}$$

A language is said to be *modulus locally testable* if it is a boolean combination of languages of the form $\{x\}$, xA^*, A^*x or $F(x, k, n)$ ($x \in A^+$, $k, n \geq 0$).

Theorem 5.22. [171, 169] *Let L be a recognizable subset of A^+. Then L is modulus locally testable if and only if every local monoid of $S(L)$ is a commutative group and the Cauchy category of $S(L)$ satisfies Thérien's condition.*

5.4 Algorithmic problems

Let **V** be a variety of finite ordered semigroups and let $\mathcal{V}$ be the associated positive variety of languages. In order to decide whether a recognizable language L of A^+ (given for instance by a finite automaton) belongs to $\mathcal{V}(A^+)$, it suffices to compute the ordered syntactic semigroup S of L and to verify that $S \in \mathbf{V}$. This motivates the following definition: a variety of finite ordered semigroups **V** is *decidable* if there is an algorithm to decide whether a given finite ordered semigroup belongs to **V**. All the varieties that were considered up to now are decidable, but several open problems in this field amount to decide whether a certain variety is decidable or not.

Once it is known that a variety is decidable, it is usually not necessary to compute the ordered syntactic semigroup to decide whether L belongs to $\mathcal{V}(A^+)$. Most of the time, one can obtain a more efficient algorithm by analyzing the minimal automaton of the language [127].

6. Some algebraic tools

The statement of the more advanced results presented in the next sections requires some auxiliary algebraic tools: relational morphisms, Mal'cev products and semidirect products.

6.1 Relational morphisms

Relational morphisms were introduced by Tilson [188]. If S and T are semigroups, a *relational morphism* $\tau : S \to T$ is a relation from S into T, i.e. a mapping from S into $\mathcal{P}(T)$ such that:

(1) $\tau(s)\tau(t) \subseteq \tau(st)$ for all $s, t \in S$,

(2) $\tau(s)$ is non-empty for all $s \in S$,

For a relational morphism between two monoids S and T, a third condition is required

(3) $1 \in \tau(1)$

Equivalently, τ is a relation whose graph

$$\mathrm{graph}(\tau) = \{\, (s, t) \in S \times T \mid t \in \tau(s) \,\}$$

is a subsemigroup (resp. submonoid if S and T are monoids) of $S \times T$, with first-coordinate projection onto S.

It is not necessary to introduce a special notion of relational morphism for ordered semigroups. Indeed, if S and T are ordered, then the graph of τ is naturally ordered as a subsemigroup of $S \times T$ and the projections on S and T are order-preserving.

Semigroup morphisms and the inverses of surjective semigroup morphisms are examples of relational morphisms. This holds even if the semigroups are ordered, and there is no need for the morphisms to be order-preserving. In particular, if $(S, \leq)$ is an ordered semigroup equipped with a non trivial order, then the identity defines a morphism from $(S, =)$ onto $(S, \leq)$ but also a relational morphism from $(S, \leq)$ onto $(S, =)$.

The composition of two relational morphisms is again a relational morphism. In particular, given two surjective semigroup morphisms $\alpha : A^+ \to S$ and $\beta : A^+ \to T$, the relation $\tau = \beta \circ \alpha^{-1}$ is a relational morphism between S and T. We shall consider two examples of this situation in which S and T are syntactic semigroups.

Our first example illustrates a simple, but important property of the concatenation product. Let, for $0 \leq i \leq n$, L_i be recognizable languages of A^*, let $\eta_i : A^* \to M(L_i)$ be their syntactic morphisms and let $\eta : A^* \to M(L_0) \times M(L_1) \times \cdots \times M(L_n)$ be the morphism defined by $\eta(u) = (\eta_0(u), \eta_1(u), \ldots, \eta_n(u))$. Let $a_1, a_2, \ldots, a_n$ be letters of A and let $L = L_0 a_1 L_1 \cdots a_n L_n$. Let $\mu : A^* \to M(L)$ be the syntactic morphism of L. The relational morphism $\tau = \eta \circ \mu^{-1} : M(L) \to M(L_0) \times M(L_1) \times \cdots \times M(L_n)$ has a remarkable property.

Proposition 6.1. *For every idempotent e of $M(L_0) \times M(L_1) \times \cdots \times M(L_n)$, $\tau^{-1}(e)$ is an ordered semigroup that satisfies the identity $x^\omega y x^\omega \leq x^\omega$.*

Proposition 6.1 is a simplified version [126, 127] of an earlier result of Straubing [164].

There is a similar result for syntactic semigroups. In this case, we consider languages of the form $L = u_0 L_1 u_1 \cdots L_n u_n$, where $u_0, u_1, \ldots, u_n$ are words of A^* and $L_1, \ldots, L_n$ are recognizable languages[7] of A^+. Let $\eta_i : A^+ \to S(L_i)$ be the syntactic morphism of L_i and let $\eta : A^+ \to S(L_1) \times S(L_2) \times \cdots \times S(L_n)$ be the morphism defined by $\eta(u) = (\eta_1(u), \eta_2(u), \ldots, \eta_n(u))$. Finally, let $\mu : A^+ \to S(L)$ be the syntactic morphism of L and let $\tau = \eta \circ \mu^{-1}$.

Proposition 6.2. *For every idempotent e of $S(L_1) \times S(L_2) \times \cdots \times S(L_n)$, $\tau^{-1}(e)$ is an ordered semigroup that satisfies the identity $x^\omega y x^\omega \le x^\omega$.*

The product $L = L_0 a_1 L_1 \cdots a_n L_n$ is *unambiguous* if every word u of L admits a unique factorization of the form $u_0 a_1 u_1 \cdots a_n u_n$ with $u_0 \in L_0, \ldots,$ $u_n \in L_n$. It is *left deterministic* (resp. *right deterministic*) if, for $1 \le i \le n$, u has a unique prefix (resp. suffix) in $L_0 a_1 L_1 \cdots L_{i-1} a_i$ (resp. $a_i L_i \cdots a_n L_n$). If the product is unambiguous, left deterministic or right deterministic, Proposition 6.1 can be improved as follows.

Proposition 6.3. *If the product $L_0 a_1 L_1 \cdots a_n L_n$ is unambiguous (resp. left deterministic, right deterministic), then for every idempotent e of $M(L_0) \times M(L_1) \times \cdots \times M(L_n)$, $\tau^{-1}(e)$ is an ordered semigroup that satisfies the identity $x^\omega y x^\omega = x^\omega$ (resp. $x^\omega y = x^\omega$, $y x^\omega = x^\omega$).*

A similar result holds for languages of A^+ and unambiguous (resp. left deterministic, right deterministic) products of the form $L = u_0 L_1 u_1 \cdots L_n u_n$.

The product $L = L_0 a_1 L_1 \cdots a_n L_n$ is *bideterministic* if it is both left and right deterministic. Bideterministic products were introduced by Schützenberger [148] and studied in more detail in [123, 34, 35]. In particular, Branco proved an analog of Proposition 6.3 for the bideterministic product, but the condition bears on the kernel category of the relational morphism τ (see section 6.3).

Our second example, also due to Straubing [164] concerns the star operation. Recall that a language L is *pure* if, for every $u \in L$ and every $n > 0$, $u^n \in L$ implies $u \in L$. Now consider the syntactic morphism $\eta : A^* \to M(L)$ of a recognizable language L and let $\mu : A^* \to M(L^*)$ be the syntactic morphism of L^*. Then consider the relational morphism $\tau = \eta \circ \mu^{-1} : M(L^*) \to M(L)$.

Proposition 6.4. *If L^* is a pure language, then for every idempotent e of $M(L)$, $\tau^{-1}(e)$ is an aperiodic semigroup.*

---

[7] The reason of this modification is the following: a product of the form $L_0 a_1 L_1 \cdots a_n L_n$ where $L_1, \ldots, L_n$ are languages of A^* can be written as a finite union of languages of the form $u_0 L_1 u_1 \cdots L_n u_n$, where $u_0, u_1, \ldots, u_n$ are words of A^* and $L_1, \ldots, L_n$ are recognizable languages of A^+.

6.2 Mal'cev product

Let S and T be ordered semigroups and let $\tau : S \to T$ be a relational morphism. Then, for every ordered subsemigroup T' of T, the set

$$\tau^{-1}(e) = \{s \in S \mid e \in \tau(s)\}$$

is an ordered subsemigroup of S. Let $\mathbf{W}$ be a variety of ordered semigroups. A relational morphism $\tau : S \to T$ is called a $\mathbf{W}$-*relational morphism* if, for every idempotent $e \in T$, the ordered semigroup $\tau^{-1}(e)$ belongs to $\mathbf{W}$.

If $\mathbf{V}$ is a variety of semigroups (resp. monoids), the class $\mathbf{W} \,\text{\textcircled{M}}\, \mathbf{V}$ of all ordered semigroups (resp. monoids) S such that there exists a $\mathbf{W}$-relational morphism from S onto a semigroup (resp. monoid) of $\mathbf{V}$ is a variety of ordered semigroups, called the *Mal'cev product* of $\mathbf{W}$ and $\mathbf{V}$. Let $\mathbf{W}$ be a variety of ordered semigroups, defined by a set E of identities. The following theorem, proved in [124] describes a set of identities defining $\mathbf{W} \,\text{\textcircled{M}}\, \mathbf{V}$.

Theorem 6.5. *Let $\mathbf{V}$ be a variety of monoids and let $\mathbf{W} = \llbracket E \rrbracket$ be a variety of ordered semigroups. Then $\mathbf{W} \,\text{\textcircled{M}}\, \mathbf{V}$ is defined by the identities of the form $\sigma(x) \le \sigma(y)$, where $x \le y$ is an identity of E with $x, y \in \widehat{B}^*$ for some finite alphabet B and $\sigma : \widehat{B}^* \to \widehat{A}^*$ is a continuous morphism such that, for all $b, b' \in B$, $\mathbf{V}$ satisfies the identity $\sigma(b) = \sigma(b') = \sigma(b^2)$.*

Despite its rather abstract statement, Theorem 6.5 can be used to produce effectively identities of some Mal'cev products [124]. Mal'cev products play an important role in the study of the concatenation product, as will be shown in Section 7.1.

6.3 Semidirect product

Let S and T be semigroups. We write the product in S additively to provide a more transparent notation, but it is not meant to suggest that S is commutative. A *left action* of T on S is a map $(t, s) \to ts$ from $T^1 \times S$ into S such that, for all $s, s_1, s_2 \in S$ and $t, t_1, t_2 \in T$,

(1) $(t_1 t_2)s = t_1(t_2 s)$
(2) $t(s_1 + s_2) = ts_1 + ts_2$
(3) $1s = s$

If S is a monoid with identity 0, the action is *unitary* if it satisfies, for all $t \in T$,

(4) $t0 = 0$

Given such a left action[8], the *semidirect product* of S and T (with respect to this action) is the semigroup $S * T$ defined on the set $S \times T$ by the product

---

[8] We followed Almeida [5] for the definition of a left action. This definition is slightly different from Eilenberg's definition [53], where an action is defined as a map from $T \times S$ into S satisfying (1) and (2).

$$(s_1, t_1)(s_2, t_2) = (s_1 + t_1 s_2, t_1 t_2)$$

Given two varieties of finite semigroups $\mathbf{V}$ and $\mathbf{W}$, denote by $\mathbf{V} * \mathbf{W}$ the variety of finite semigroups generated by the semidirect products $S * T$ with $S \in \mathbf{V}$ and $T \in \mathbf{W}$. One can define similarly the semidirect product of two varieties of finite monoids, or of a variety of finite monoid and a variety of finite semigroups. For instance, if $\mathbf{V}$ is a variety of finite monoids and $\mathbf{W}$ is a variety of finite semigroups, $\mathbf{V} * \mathbf{W}$ is the variety of finite semigroups generated by the semidirect products $S * T$ with $S \in \mathbf{V}$ and $T \in \mathbf{W}$ such that the action of T on S is unitary.

The *wreath product* is closely related to the semidirect product. The wreath product $S \circ T$ of two semigroups S and T is the semidirect product $S^{T^1} * T$ defined by the action of T on S^{T^1} given by

$$tf(t') = f(tt')$$

for $f : T^1 \to S$ and $t, t' \in T^1$. In particular, the multiplication in $S \circ T$ is given by

$$(f_1, t_1)(f_2, t_2) = (f, t_1 t_2) \text{ where } f(t) = f_1(t) + f_2(t_1 t) \text{ for all } t \in T^1$$

In a way, the wreath product is the most general semidirect product since every semidirect product $S * T$ is a subsemigroup of $S \circ T$. It follows that $\mathbf{V} * \mathbf{W}$ is generated by all wreath products of the form $S \circ T$, where $S \in \mathbf{V}$ and $T \in \mathbf{W}$. Although the semidirect product is not an associative operation, it become associative at the variety level. That is, if $\mathbf{V}_1$, $\mathbf{V}_2$ and $\mathbf{V}_3$ are varieties of finite semigroups, then $(\mathbf{V}_1 * \mathbf{V}_2) * \mathbf{V}_3 = \mathbf{V}_1 * (\mathbf{V}_2 * \mathbf{V}_3)$.

Wreath products allow to decompose semigroups into smaller pieces. Let U_1 be the monoid $\{1, 0\}$ under usual multiplication and let $U_2 = \{1, a, b\}$ be the monoid defined by the multiplication $aa = ba = a$ and $ab = bb = b$.

Theorem 6.6. *The following decompositions hold:*

(1) Every solvable group divides a wreath product of commutative groups,
(2) Every $\mathcal{R}$-trivial monoid divides a wreath product of copies of U_1,
(3) Every aperiodic monoid divides a wreath product of copies of U_2,
(4) Every monoid divides a wreath product of groups and copies of U_2,

Statement (4) is the celebrated Krohn-Rhodes theorem [5, 53, 169]. Wreath product decompositions were first used in language theory to get a new proof of Schützenberger's theorem [44, 88]. This use turns out to be a particular case of Straubing's "wreath product principle" [159, 168], which provides a description of the languages recognized by the wreath product of two finite monoids.

Let M and N be two finite monoids and let $\eta : A^* \to M \circ N$ be a monoid morphism. We denote by $\pi : M \circ N \to N$ the monoid morphism defined by $\pi(f, n) = n$ and we put $\varphi = \pi \circ \eta$. Thus φ is a monoid morphism from A^* into N. Let $B = N \times A$ and $\sigma : A^* \to B^*$ be the map defined by

$$\sigma(a_1 a_2 \cdots a_n) = (1, a_1)(\varphi(a_1), a_2) \cdots (\varphi(a_1 a_2 \cdots a_{n-1}), a_n)$$

Observe that σ is not a morphism, but a sequential function [19]. Straubing's result can be stated as follows.

Theorem 6.7. *If a language L is recognized by $\eta : A^* \to M \circ N$, then L is a finite boolean combination of languages of the form $X \cap \sigma^{-1}(Y)$, where $Y \subset B^*$ is recognized by M and where $X \subset A^*$ is recognized by N.*

In view of the decomposition results of Theorem 6.6, this principle can be used to describe the variety of languages corresponding to solvable groups (Theorem 7.18 below), $\mathcal{R}$-trivial monoids (Corollaries 7.8 and 7.14) or aperiodic monoids (Theorem 5.2). Theorem 6.6 is also the key result in the proof of Theorems 7.12 and 7.21.

The wreath product principle can be adapted to the case of a wreath product of a monoid by a semigroup. Varieties of the form $\mathbf{V} * \mathbf{LI}$ received special attention. See the examples at the end of this section.

However, all these results yield the following question: if $\mathbf{V}$ and $\mathbf{W}$ are decidable varieties of finite monoids (or semigroups), is $\mathbf{V} * \mathbf{W}$ also decidable ? A negative answer was given in the general case [1], but several positive results are also known. A few more definitions on categories are needed to state these results precisely.

Let M and N be two monoids and let $\tau : M \to N$ be a relational morphism. Let C be the category such that $Ob(C) = N$ and, for all $u, v \in N$,

$$C(u, v) = \{(u, s, v) \in N \times M \times N \mid v \in u\tau(s)\}.$$

Composition is given by $(u, s, v)(v, t, w) = (u, st, w)$. Now the *kernel category* of τ is the quotient of C by the congruence $\sim$ defined by

$$(u, s, v) \sim (u, t, v) \text{ if and only if } ms = mt \text{ for all } m \in \tau^{-1}(u)$$

Thus the kernel category identifies elements with the same action on each fiber[9] $\tau^{-1}(u)$.

The next theorem, due to Tilson [189], relates semidirect products and relational morphisms.

Theorem 6.8. *Let $\mathbf{V}$ and $\mathbf{W}$ be two variety of finite monoids. A monoid M belongs to $\mathbf{V} * \mathbf{W}$ if and only if there exists a relational morphism $\tau : M \to T$, where $T \in \mathbf{W}$, whose kernel category divides a monoid of $\mathbf{V}$.*

There is an analogous result (with a few technical modifications) when $\mathbf{W}$ is a variety of finite semigroups. In view of Theorem 6.8, it is important to characterize, given a variety of finite monoids $\mathbf{V}$, the categories that divide a

---

[9] The action is the multiplication on the right. In some applications [120, 35], it is more appropriate to use a definition of the kernel category that takes also in account the multiplication on the left [138].

monoid of **V**. The problem is not solved in general, but one can try to apply one of the following results.

A first idea is to convert a category, which can be considered as a "partial semigroup" under composition of arrows, into a real semigroup. If C be a category, associate with each arrow $p \in C(u,v)$ the triple (u,p,v). Let $S(C)$ be the set of all such triples, along with a new element denoted 0. Next define a multiplication on $S(C)$ by setting

$$(u,p,v)(u',p',v') = \begin{cases} (u,pp',v') & \text{if } v = u' \\ 0 & \text{otherwise} \end{cases}$$

It is easy to verify that $S(C)$ is a semigroup. The interest of this construction lies in the following theorem [166, 189]. Recall that BA_2 is the five element semigroup considered in example 3.3.

Theorem 6.9. *Let* **V** *be a variety of finite monoids containing the monoid* BA_2^1. *Then a category* C *divides a monoid of* **V** *if and only if* $S(C)^1$ *belongs to* **V**.

Corollary 6.10. *Let* **V** *be a decidable variety of finite monoids containing the monoid* BA_2^1. *Then it is decidable whether a given finite category divides a monoid of* **V**.

Thus the question raised above is solved for varieties of finite monoids that contain BA_2^1, for instance the variety **A** of finite aperiodic monoids. We are now mainly interested in varieties of finite monoids that do not contain BA_2^1. These varieties are exactly the subvarieties of the variety of finite monoids in which each regular $\mathcal{J}$-class is a semigroup. This variety, denoted **DS**, is defined by the identities $\big((xy)^\omega (yx)^\omega (xy)^\omega\big)^\omega = (xy)^\omega$.

The problem is also easy to solve for another type of varieties, the *local varieties*. For these varieties, it suffices to check whether the local monoids of the category are in **V**. More precisely, a variety **V** is local if and only if every category whose local monoids are in **V** divides a monoid of **V**. Local varieties were first characterized in [183]. See also [189].

Theorem 6.11. *A non trivial variety* **V** *is local if and only if* **V** $*$ **LI** $=$ **LV**.

In spite of this theorem, it is not easy in general to know whether a variety of finite monoids is local or not, even for the subvarieties of **DS**. The next theorem summarizes results of Simon, Thérien, Weiss, Tilson, Jones and Almeida [41, 53, 183, 196, 189, 67, 6]. In this theorem, **DA** denotes the intersection of **A** and **DS**. Thus **DA** is the variety of finite monoids in which each regular $\mathcal{J}$-class is an idempotent semigroup.

Theorem 6.12. *The following varieties are local: any non trivial variety of finite groups, the varieties* $\mathbf{J}_1$, **DA**, **DS**, $[\![x^\omega = x^{\omega+1}, x^\omega y = yx^\omega]\!]$ *and the varieties* $[\![x^n = x]\!]$ *for each* $n > 1$.

Note that if **V** is a decidable local variety, one can effectively decide whether a finite category divides a monoid of **V**. Unfortunately, some varieties, like the trivial variety **I**, are not local. However, a decidability result can still be obtained in some cases. We just mention the most important of them, which concern four important subvarieties of **DS**.

Theorem 6.13. [189] *A category C divides a trivial monoid if and only if, for every $u, v \in Ob(C)$, the set $C(u, v)$ has at most one element.*

Theorem 6.14. [72] *A category C divides a finite $\mathcal{J}$-trivial monoid if and only if, for every $u, v \in Ob(C)$, for each $p, r \in C(u, v)$ and each $q, s \in C(v, u)$, $(pq)^{\omega} ps(rs)^{\omega} = (pq)^{\omega}(rs)^{\omega}$.*

Theorem 6.15. [183] *A category C divides a finite commutative monoid if and only if, for every $u, v \in Ob(C)$, for every $p, r \in C(u, v)$ and every $q \in C(v, u)$, $pqr = rqp$.*

Theorem 6.16. [183] *A category C divides a finite aperiodic commutative monoid if and only if, the local monoids of C are aperiodic and, for every $u, v \in Ob(C)$, for every $p, r \in C(u, v)$ and every $q \in C(v, u)$, $pqr = rqp$.*

Semidirect products of the form **V** ∗ **LI** deserved special attention. The key result is due to Straubing [166] and was formalized in [189]. It is a generalization of former results of [41, 86, 71].

Theorem 6.17. [166] *Let **V** be a variety of finite monoids. A semigroup belongs to **V** ∗ **LI** if and only if its Cauchy category divides a monoid of **V**.*

With all these powerful tools in hand, one can now sketch a proof of Theorems 5.18, 5.20 and 5.22. First, one makes use of the wreath product principle to show that the variety of finite semigroups associated with the locally testable (resp. threshold locally testable, modulus locally testable) languages is the variety $\mathbf{J}_1 \ast \mathbf{LI}$ (resp. **Acom** ∗ **LI**, **Gcom** ∗ **LI**). It follows by Theorem 6.17 that a recognizable language is locally testable (resp. threshold locally testable, modulus locally testable) if and only if the Cauchy category of its syntactic semigroup divides a monoid of $\mathbf{J}_1$ (resp. **Acom**, **Gcom**). It remains to apply Theorem 6.12 (or Theorem 6.16 in the case of **Acom**) to conclude.

Theorem 8.18 below is another application of Theorem 6.17.

6.4 Representable transductions

In this section, we address the following general problem. Let $L_1, \ldots, L_n$ be languages recognized by monoids $M_1, \ldots, M_n$, respectively. Given an operation φ on these languages, find a monoid which recognizes $\varphi(L_1, \ldots, L_n)$. The key idea of our construction is to consider, when it is possible, an operation $\varphi : A^* \times \cdots \times A^* \to A^*$ as the inverse of a transduction $\tau : A^* \to A^* \times \cdots \times A^*$.

Then, for a rather large class of transductions, it is possible to solve our problem explicitly. Precise definitions are given below.

Transductions were intensively studied in connection with context-free languages [19]. For our purpose, it suffices to consider transductions τ from a free monoid A^* into an arbitrary monoid M such that, if P is a recognizable subset of M, then $\tau^{-1}(P)$ is a recognizable subset of A^*. It is well known that rational transductions have this property. In this case, τ can be realized by a transducer, which is essentially a non deterministic automaton with output. Now, just as automata can be converted into semigroups, automata with outputs can be converted into matrix representations. In particular, every rational transduction admits a linear representation. It turns out that the important property is to have a matrix representation. Whether this representation is linear or not is actually irrelevant for our purpose. We now give the formal definitions.

Let M be a monoid. A *transduction* $\tau : A^* \to M$ is a relation from A^* into M, i.e. a function from A^* into $\mathcal{P}(M)$. If P is a subset of M, $\tau^{-1}(P)$ is the image of P by the relation $\tau^{-1} : M \to A^*$. Therefore

$$\tau^{-1}(P) = \{u \in A^* \mid \tau(u) \cap P \neq \emptyset\}$$

The definition of a representable transduction requires some preliminaries. The set $\mathcal{P}(M)$ is a semiring under union (as addition) and subset product (as multiplication). Therefore, for each $n > 0$, the set $\mathcal{P}(M)^{n \times n}$ of n by n matrices with entries in $\mathcal{P}(M)$ is again a semiring for addition and multiplication of matrices induced by the operations in $\mathcal{P}(M)$.

Let X be an alphabet and let $M \oplus X^*$ be the free product of M and X^*, that is, the set of the words of the form $m_0 x_1 m_1 x_2 m_2 \cdots x_k m_k$, with $m_0, m_1, \ldots, m_k \in M$ and $x_1, \ldots, x_k \in X$, equipped with the product

$$(m_0 x_1 m_1 x_2 m_2 \cdots x_k m_k)(m'_0 x'_1 m'_1 x'_2 m'_2 \cdots x'_k m'_k) =$$
$$m_0 x_1 m_1 x_2 m_2 \cdots x_k (m_k m'_0) x'_1 m'_1 x'_2 m'_2 \cdots x'_k m'_k$$

A series in the non commutative variables X with coefficients in $\mathcal{P}(M)$ is an element of $\mathcal{P}(M \oplus X^*)$, that is, a formal sum of words of the form $m_0 x_1 m_1 x_2 m_2 \cdots x_k m_k$.

A *representation of dimension* n for the transduction τ is a pair (μ, s), where μ is a monoid morphism from A^* into $\mathcal{P}(M)^{n \times n}$ and s is a series in the non commutative variables $\{x_{1,1}, \ldots, x_{n,n}\}$ with coefficients in $\mathcal{P}(M)$ such that, for all $u \in A^*$,

$$\tau(u) = s[\mu(u)]$$

where the expression $s[\mu(u)]$ denotes the subset of M obtained by substituting $(\mu(u)_{1,1}, \ldots, \mu(u)_{n,n})$ for $(x_{1,1}, \ldots, x_{n,n})$ in s. A transduction is *representable* if it admits a representation. The following example should help the reader to understand this rather abstract definition.

Example 6.1. Let $A = \{a, b\}$, $M = A^*$ and let μ be the morphism from A^* into $\mathcal{P}(A^*)^{2 \times 2}$ defined by

$$\mu(u) = \begin{pmatrix} u & \emptyset \\ \emptyset & A^{|u|} \end{pmatrix}$$

Then the following transductions from A^* into A^* admit a representation of the form (μ, s) for some series s.

(1) $\tau_1(u) = A^{|u|} u A^{|u|}$
(2) $\tau_2(u) = L_0 u L_1 u \ldots u L_k$, where $L_0, \ldots, L_k$ are arbitrary languages,

It suffices to take

$$s = x_{2,2} x_{1,1} x_{2,2}$$

in the first case and

$$s = \sum_{u_0 \in L_0, \ldots, u_k \in L_k} u_0 x_{1,1} u_1 \cdots x_{1,1} u_k$$

in the second case.

Suppose that a transduction $\tau : A^* \to M$ admits a representation (μ, s), where $\mu : A^* \to \mathcal{P}(M)^{n \times n}$. Then every monoid morphism $\varphi : M \to N$ induces a monoid morphism $\varphi : \mathcal{P}(M)^{n \times n} \to \mathcal{P}(N)^{n \times n}$. The main property of representable transductions can now be stated.

Theorem 6.18. [115, 116] *Let (μ, s) be a representation for a transduction $\tau : A^* \to M$. If P is a subset of M recognized by φ, then the language $\tau^{-1}(P)$ is recognized by the monoid $(\varphi \circ \mu)(A^*)$.*

Corollary 6.19. *Let $\tau : A^* \to M$ be a representable transduction. Then for every recognizable subset P of M, the language $\tau^{-1}(P)$ is recognizable.*

The precise description of the monoid $(\varphi \circ \mu)(A^*)$ is the key to understand several operations on languages.

Example 6.2. This is a continuation of Example 6.1. Let $\tau_1(u) = A^{|u|} u A^{|u|}$. Then, for every language L of A^*,

$$\tau_1^{-1}(L) = \{u \in A^* \mid \text{there exist } u_0, u_1 \text{ with } |u_0| = |u| = |u_1| \text{ and } u_0 u u_1 \in L\}$$

Thus $\tau_1^{-1}(L)$ is the set of "middle thirds" of words of L.
Let $\tau_2(u) = u^2$. Then, for every language L of A^*,

$$\tau_2^{-1}(L) = \{u \in A^* \mid u^2 \in L\}$$

Thus $\tau_2^{-1}(L)$ is the "square root" of L. In both cases, the transduction has a representation of the form (μ, s), where $\mu : A^* \to \mathcal{P}(A^*)^{2 \times 2}$ is defined by

$$\mu(u) = \begin{pmatrix} u & \emptyset \\ \emptyset & A^{|u|} \end{pmatrix}$$

Thus if $\varphi : A^* \to N$ is a monoid morphism and if $\varphi(A) = X$, $(\varphi \circ \mu)(A^*)$ is a monoid of matrices of $\mathcal{P}(N)^{2 \times 2}$ of the form

$$\begin{pmatrix} \{x\} & \emptyset \\ \emptyset & X^k \end{pmatrix}$$

with $x \in N$ and $k \geq 0$. Thus this monoid can be identified with a submonoid of $N \times C$, where C is the submonoid of $\mathcal{P}(N)$ generated by X. In particular, C is commutative.

We now give some examples of application of Theorem 6.18.

Inverse substitutions. Recall that a substitution from A^* into M is a monoid morphism from A^* into $\mathcal{P}(M)$. Therefore a substitution $\sigma : A^* \to M$ has a representation of dimension 1. Thus if L is a subset of M recognized by a monoid N, then $\sigma^{-1}(L)$ is recognized by a submonoid of $\mathcal{P}(N)$.

Length preserving morphisms. Let $\varphi : A^* \to B^*$ be a length preserving morphism. Then the transduction $\varphi^{-1} : B^* \to A^*$ is a substitution. Thus, if L is a subset of A^* recognized by a monoid N, then $\varphi(L)$ is recognized by a submonoid of $\mathcal{P}(N)$.

Shuffle product. Recall that the *shuffle* of n words $u_1, \ldots, u_n$ is the set $u_1 \amalg \ldots \amalg u_n$ of all words of the form

$$u_{1,1} u_{2,1} \cdots u_{n,1} u_{1,2} u_{2,2} \cdots u_{n,2} \cdots u_{1,k} u_{2,k} \cdots u_{n,k}$$

with $k \geq 0$, $u_{i,j} \in A^*$, such that $u_{i,1} u_{i,2} \cdots u_{i,k} = u_i$ for $1 \leq i \leq n$. The shuffle of k languages $L_1, \ldots, L_k$ is the language

$$L_1 \amalg \cdots \amalg L_k = \bigcup_{u_1 \in L_1, \ldots, u_k \in L_k} u_1 \amalg \cdots \amalg u_k$$

Let $\tau : A^* \to A^* \times \cdots \times A^*$ be the transduction defined by

$$\tau(u) = \{(u_1, \cdots, u_k) \in A^* \times \cdots \times A^* \mid u \in u_1 \amalg \cdots \amalg u_k\}$$

Then $\tau^{-1}(L_1 \times \cdots \times L_k) = L_1 \amalg \cdots \amalg L_k$. Furthermore τ is a substitution defined, for every $a \in A$, by

$$\tau(a) = \{(a, 1, \ldots, 1), (1, a, 1, \ldots, 1), \ldots, (1, 1, \ldots, 1, a)\}$$

Thus, if $L_1, \ldots, L_k$ are languages recognized by monoids $M_1, \ldots, M_k$, respectively, then $L_1 \amalg \cdots \amalg L_k$ is recognized by a submonoid of $\mathcal{P}(M_1 \times \cdots \times M_k)$.

Other examples include the concatenation product – which leads to a construction on monoids called the *Schützenberger product* [163] – and the inverse of a sequential function, or more generally of a rational function, which is intimately related to the wreath product. See [115, 116, 105] for the details.

7. The concatenation product

The concatenation product is certainly the most studied operation on languages. As was the case in section 6.1, we shall actually consider products of the form $L_0 a_1 L_1 \cdots a_n L_n$, where the a_i's are letters or of the form $u_0 L_1 u_1 \cdots L_n u_n$, where the u_i's are words.

7.1 Polynomial closure

Polynomial operations comprise finite union and concatenation product. This terminology, first introduced by Schützenberger, comes from the fact that languages form a semiring under union as addition and concatenation as multiplication. There are in fact two slightly different notions of polynomial closure, one for +-classes and one for *-classes.

The *polynomial closure* of a set of languages $\mathcal{L}$ of A^* is the set of languages of A^* that are finite unions of languages of the form

$$L_0 a_1 L_1 \cdots a_n L_n$$

where $n \geq 0$, the a_i's are letters and the L_i's are elements of $\mathcal{L}$.

The *polynomial closure* of a set of languages $\mathcal{L}$ of A^+ is the set of languages of A^+ that are finite unions of languages of the form

$$u_0 L_1 u_1 \cdots L_n u_n$$

where $n \geq 0$, the u_i's are words of A^* and the L_i's are elements of $\mathcal{L}$. If $n = 0$, one requires of course that u_0 is not the empty word.

By extension, if $\mathcal{V}$ is a *-variety (resp. +-variety), we denote by Pol $\mathcal{V}$ the class of languages such that, for every alphabet A, Pol $\mathcal{V}(A^*)$ (resp. Pol $\mathcal{V}(A^+)$) is the polynomial closure of $\mathcal{V}(A^*)$ (resp. $\mathcal{V}(A^+)$). Symmetrically, we denote by Co-Pol $\mathcal{V}$ the class of languages such that, for every alphabet A, Co-Pol $\mathcal{V}(A^*)$ (resp. Co-Pol $\mathcal{V}(A^+)$) is the set of languages L whose complement is in Pol $\mathcal{V}(A^*)$ (resp. Pol $\mathcal{V}(A^+)$). Finally, we denote by BPol $\mathcal{V}$ the class of languages such that, for every alphabet A, BPol $\mathcal{V}(A^*)$ (resp. BPol $\mathcal{V}(A^+)$) is the closure of Pol $\mathcal{V}(A^*)$ (resp. Pol $\mathcal{V}(A^+)$) under finite boolean operations (finite union and complement).

Proposition 6.1 above is the first step in the proof of the following algebraic characterization of the polynomial closure [126, 127], which makes use of a deep combinatorial result of semigroup theory [153, 154, 155].

Theorem 7.1. *Let* **V** *be a variety of finite monoids and let* $\mathcal{V}$ *be the associated variety of languages. Then* Pol $\mathcal{V}$ *is a positive variety and the associated variety of finite ordered monoids is the Mal'cev product* $[\![x^\omega y x^\omega \leq x^\omega]\!] \, \textcircled{M} \, \mathbf{V}$.

In the case of +-varieties, the previous result also holds with the appropriate definition of polynomial closure. The following consequence was first proved by Arfi [11, 12].

Corollary 7.2. *For each variety of languages* $\mathcal{V}$, *Pol* $\mathcal{V}$ *and Co-Pol* $\mathcal{V}$ *are positive varieties of languages. In particular, for each alphabet A, Pol* $\mathcal{V}(A^*)$ *and Co-Pol* $\mathcal{V}(A^*)$ *(resp. Pol* $\mathcal{V}(A^+)$ *and Co-Pol* $\mathcal{V}(A^+)$ *in the case of a* +*-variety) are closed under finite union and intersection.*

7.2 Unambiguous and deterministic polynomial closure

The *unambiguous polynomial closure* of a set of languages $\mathcal{L}$ of A^* is the set of languages of A^* that are finite unions of unambiguous products of the form $L_0 a_1 L_1 \cdots a_n L_n$, where $n \geq 0$, the a_i's are letters and the L_i's are elements of $\mathcal{L}$. Similarly, the unambiguous polynomial closure of a set of languages $\mathcal{L}$ of A^+ is the set of languages of A^+ that are finite unions of unambiguous products of the form

$$u_0 L_1 u_1 \cdots L_n u_n$$

where $n \geq 0$, the u_i's are words of A^* and the L_i's are elements of $\mathcal{L}$. If $n = 0$, one requires that u_0 is not the empty word.

The *left and right deterministic polynomial closure* are defined analogously, by replacing "unambiguous" by "left (resp. right) deterministic".

By extension, if $\mathcal{V}$ is a variety of languages, we denote by UPol $\mathcal{V}$ the class of languages such that, for every alphabet A, UPol $\mathcal{V}(A^*)$ (resp. UPol $\mathcal{V}(A^+)$) is the unambiguous polynomial closure of $\mathcal{V}(A^*)$ (resp. $\mathcal{V}(A^+)$). Similarly, the left (resp. right) deterministic polynomial closure of $\mathcal{V}$ is denoted D^ℓPol $\mathcal{V}$ (resp. D^rPol $\mathcal{V}$). The algebraic counterpart of the unambiguous polynomial closure is given in the following theorems [97, 120].

Theorem 7.3. *Let* **V** *be a variety of finite monoids (resp. semigroups) and let* $\mathcal{V}$ *be the associated variety of languages. Then UPol* $\mathcal{V}$ *is a variety of languages, and the associated variety of monoids (resp. semigroups) is* $[\![x^\omega y x^\omega = x^\omega]\!] \, \textcircled{M} \, \mathbf{V}$.

Theorem 6.5 leads to the following description of a set of identities defining the variety $[\![x^\omega y x^\omega \leq x^\omega]\!] \, \textcircled{M} \, \mathbf{V}$.

Proposition 7.4. *Let* **V** *be a variety of monoids. Then* $[\![x^\omega y x^\omega \leq x^\omega]\!] \, \textcircled{M} \, \mathbf{V}$ *is defined by the identities of the form* $x^\omega y x^\omega \leq x^\omega$, *where* $x, y \in \hat{A}^*$ *for some finite set A and* **V** *satisfies* $x = y = x^2$.

Recall that the variety $[\![x^\omega y x^\omega = x^\omega]\!]$ is the variety **LI** of locally trivial semigroups. Thus, the +-variety associated with **LI**, described in Proposition 5.17, is also the smallest +-variety closed under unambiguous product. This is a consequence of Theorem 7.3, applied with $\mathbf{V} = \mathbf{I}$. For $\mathbf{V} = \mathbf{J}_1$, one can show that $\mathbf{LI} \, \textcircled{M} \, \mathbf{J}_1$ is equal to **DA**, the variety of finite monoids in which each regular $\mathcal{J}$-class is an idempotent semigroup. This variety is defined by the identities $(xy)^\omega (yx)^\omega (xy)^\omega = (xy)^\omega$ and $x^\omega = x^{\omega+1}$ [148].

Corollary 7.5. *For each alphabet A, $\mathcal{D}\mathcal{A}(A^*)$ is the smallest set of languages of A^* containing the languages of the form B^*, with $B \subseteq A$, and closed under disjoint union and unambiguous product. Equivalently, $\mathcal{D}\mathcal{A}(A^*)$ is the set of languages that are disjoint unions of unambiguous products of the form $A_0^* a_1 A_1^* a_2 \cdots a_k A_k^*$, where the a_i's are letters and the A_i's are subsets of A.*

An interesting consequence of the conjunction of Theorems 7.1 and 7.3 is the following characterization of UPol $\mathcal{V}$, which holds for $*$-varieties as well as for $+$-varieties.

Theorem 7.6. *Let $\mathcal{V}$ be a variety of languages. Then Pol $\mathcal{V} \cap$ Co-Pol $\mathcal{V} =$ UPol $\mathcal{V}$.*

For the left (resp. right) deterministic product, similar results hold [97, 98]. We just state the result for $*$-varieties.

Theorem 7.7. *Let $\mathbf{V}$ be a variety of finite monoids and let $\mathcal{V}$ be the associated variety of languages. Then D^ℓPol $\mathcal{V}$ (resp. D^rPol $\mathcal{V}$) is a variety of languages, and the associated variety of monoids is $[\![x^\omega y = x^\omega]\!] \, \textcircled{M} \, \mathbf{V}$ (resp. $[\![yx^\omega = x^\omega]\!] \, \textcircled{M} \, \mathbf{V}$).*

One can show that $[\![x^\omega y = x^\omega]\!] \, \textcircled{M} \, \mathbf{J}_1$ is equal to the variety $\mathbf{R}$ of all finite $\mathcal{R}$-trivial monoids, which is also defined by the identity $(xy)^\omega x = (xy)^\omega$. This leads to the following characterization [53, 39]

Corollary 7.8. *For each alphabet A, $\mathcal{R}(A^*)$ consists of the languages which are disjoint unions of languages of the form $A_0^* a_1 A_1^* a_2 \cdots a_k A_k^*$, where $k \geq 0$, $a_1, \ldots a_n \in A$ and the A_i's are subsets of A such that $a_i \notin A_{i-1}$, for $1 \leq i \leq k$.*

A dual result holds for $\mathcal{L}$-trivial monoids.

7.3 Varieties closed under product

A set of languages $\mathcal{L}$ is *closed under product*, if, for each $L_0, \ldots, L_n \in \mathcal{L}$ and $a_1, \ldots, a_n \in A$, $L_0 a_1 L_1 \cdots a_n L_n \in \mathcal{L}$. A $*$-variety of languages $\mathcal{C}$ is *closed under product*, if, for each alphabet A, $\mathcal{V}(A^*)$ is closed under product. The next theorem, due to Straubing [160], shows, in essence, that closure under product also corresponds to a Mal'cev product.

Theorem 7.9. *Let $\mathbf{V}$ be a variety of finite monoids and let $\mathcal{V}$ be the associated variety of languages. For each alphabet A, let $\mathcal{W}(A^*)$ be the smallest boolean algebra containing $\mathcal{V}(A^*)$ and closed under product. Then $\mathcal{W}$ is a $*$-variety and the associated variety of finite monoids is $\mathbf{A} \, \textcircled{M} \, \mathbf{V}$.*

This important result contains Theorem 5.2 as a particular case, when $\mathbf{V}$ is the trivial variety of monoids. Examples of varieties of finite monoids $\mathbf{V}$ satisfying the equality $\mathbf{A} \, \textcircled{M} \, \mathbf{V} = \mathbf{V}$ include varieties of finite monoids defined by properties of their groups. A group in a monoid M is a subsemigroup of

M containing an identity e, which can be distinct from the identity of M. As was mentioned in section 2.8, the maximal groups in a monoid are exactly the $\mathcal{H}$-classes containing an idempotent. Given a variety of finite groups $\mathbf{H}$, the class of finite monoids whose groups belong to $\mathbf{H}$ form a variety of finite monoids, denoted $\overline{\mathbf{H}}$. In particular, if $\mathbf{H}$ is the trivial variety of finite groups, then $\overline{\mathbf{H}} = \mathbf{A}$, since a monoid is aperiodic if and only if it contains no non trivial group.

Theorem 7.10. *For any variety of finite groups* $\mathbf{H}$, $\mathbf{A} \textcircled{M} \overline{\mathbf{H}} = \overline{\mathbf{H}}$.

It follows, by Theorem 7.9, that the $*$-variety associated with a variety of monoids of the form $\overline{\mathbf{H}}$ is closed under product. Varieties of this type will be considered in Theorems 7.19 and 9.4 below.

7.4 The operations $L \to LaA^*$ and $L \to A^*aL$

A slightly stronger version of Theorem 7.9 can be given [98, 192, 194].

Theorem 7.11. *Let* $\mathbf{V}$ *be a variety of finite monoids and let* $\mathcal{V}$ *be the associated variety of languages. Let* $\mathcal{W}$ *be the variety of languages associated with* $\mathbf{A} \textcircled{M} \mathbf{V}$. *Then, for each alphabet* A, $\mathcal{W}(A^*)$ *is the smallest boolean algebra of languages containing* $\mathcal{V}(A^*)$ *and closed under the operations* $L \to LaA^*$ *and* $L \to A^*aL$, *where* $a \in A$.

In view of this result, it is natural to look at the operation $L \to LaA^*$ [98, 192, 194].

Theorem 7.12. *Let* $\mathbf{V}$ *be a variety of finite monoids and let* $\mathcal{V}$ *be the associated variety of languages. For each alphabet* A, *let* $\mathcal{W}(A^*)$ *be the boolean algebra generated by the languages* L *or* LaA^*, *where* $a \in A$ *and* $L \in \mathcal{V}(A^*)$. *Then* $\mathcal{W}$ *is a variety of languages and the associated variety of monoids is* $\mathbf{J}_1 * \mathbf{V}$.

Since the variety $\mathbf{R}$ of $\mathcal{R}$-trivial monoids is the smallest variety closed under semidirect product and containing the commutative and idempotent monoids, one gets immediately

Corollary 7.13. *Let* $\mathbf{V}$ *be a variety of finite monoids and let* $\mathcal{V}$ *be the associated variety of languages. For each alphabet* A, *let* $\mathcal{W}(A^*)$ *be the smallest boolean algebra of languages containing* $\mathcal{V}(A^*)$ *and closed under the operations* $L \to LaA^*$, *for all* $a \in A$. *Then* $\mathcal{W}$ *is a variety of languages and the associated variety of monoids is* $\mathbf{R} * \mathbf{V}$.

In particular, this leads to another description of the languages associated with $\mathbf{R}$ (compare with Corollary 7.8).

Corollary 7.14. *For each alphabet* A, $\mathcal{R}(A^*)$ *is the smallest boolean algebra of languages closed under the operations* $L \to LaA^*$, *for all* $a \in A$.

Operations of the form $L \to LaA^*$ and $L \to A^*aL$ were also used by Thérien [176] to describe the languages whose syntactic monoid is idempotent (see also [38, 53]). This characterization is somewhat unusual, since it is given by induction on the size of the alphabet.

Theorem 7.15. *Let* **V** *be the variety of finite idempotent monoids and let* $\mathcal{V}$ *be the associated variety of languages. Then* $\mathcal{V}(\emptyset^*) = \{\emptyset, \{1\}\}$, *and for each non empty alphabet* A, $\mathcal{V}(A^*)$ *is the smallest boolean algebra of languages containing the languages of the form* A^*aA^*, LaA^* *and* A^*aL, *where* $a \in A$ *and* $L \in \mathcal{V}(A \setminus \{a\})^*$.

The languages associated with subvarieties of the variety of finite idempotent monoids are studied in [151].

7.5 Product with counters

Let $L_0, \ldots, L_k$ be languages of A^*, let $a_1, \ldots, a_k$ be letters of A and let r and p be integers such that $0 \le r < p$. We define $(L_0a_1L_1 \cdots a_kL_k)_{r,p}$ to be the set of all words u in A^* such that the number of factorizations of u in the form

$$u = u_0 a_1 u_1 \cdots a_k u_k$$

with $u_i \in L_i$ for $0 \le i \le k$, is congruent to r modulo p. This *product with counter* is especially useful for the study of group languages. A recognizable language is called a *group language* if its syntactic monoid is a group. Since equality is the only stable order on a finite group, the ordered syntactic monoid is useless in the case of a group language.

A frequently asked question is whether there is some "nice" combinatorial description of the group languages. No such description is known for the variety of all group languages, but there are simple descriptions for some subvarieties. We have already described the variety of languages corresponding to commutative groups. We will now consider the varieties of p-groups, nilpotent groups and solvable groups.

The p-groups form a variety of finite monoids $\mathbf{G}_p$. The associated variety of languages $\mathcal{G}_p$ is given in [53], where the result is credited to Schützenberger. For $u, v \in A^*$, denote by $\binom{v}{u}$ the number of distinct factorizations of the form $v = v_0 a_1 v_1 \cdots a_n v_n$ such that $v_0, \cdots, v_n \in A^*$, $a_1, \cdots, a_n \in A$ and $a_1 \cdots a_n = u$. In other words $\binom{v}{u}$ is the number of distinct ways to write u as a subword of v. For example

$$\binom{abab}{ab} = 3 \qquad \binom{aabbaa}{aba} = 8$$

Theorem 7.16. *For each alphabet* A, $\mathcal{G}_p(A^*)$ *is the boolean algebra generated by the languages*

$$S(u, r, p) = \left\{ v \in A^* \mid \binom{v}{u} \equiv r \bmod p \right\}$$

for $0 \leq r < p$ and $u \in A^*$. It is also the boolean algebra generated by the languages $(A^* a_1 A^* \cdots a_k A^*)_{r,p}$ where $0 \leq r < p$ and $k \geq 0$.

Nilpotent groups form a variety of finite monoids **Gnil**. A standard result in group theory states that a finite group is nilpotent if and only if it is isomorphic to a direct product $G_1 \times \cdots \times G_n$, where each G_i is a p_i-group for some prime p_i. This result leads to the following description of the variety of languages $\mathcal{G}nil$ associated with **Gnil** [53, 173].

Theorem 7.17. *For each alphabet A, $\mathcal{G}nil(A^*)$ is the boolean algebra generated by the languages $S(u, r, p)$, where p is a prime number, $0 \leq r < p$ and $u \in A^*$. It is also the boolean algebra generated by the languages $(A^* a_1 A^* \cdots a_k A^*)_{r,p}$, where $a_1, \ldots, a_k \in A$, $0 \leq r < p$, p is a prime number and $k \geq 0$.*

Solvable groups also form a variety of finite monoids **Gsol**. The associated variety of languages $\mathcal{G}sol$ was described by Straubing [159].

Theorem 7.18. *For each alphabet A, $\mathcal{G}sol(A^*)$ is the smallest boolean algebra of languages closed under the operations $L \rightarrow (LaA^*)_{r,p}$, where $a \in A$, p is a prime number and $0 \leq r < p$.*

The variety of languages associated with $\overline{\mathbf{Gsol}}$ was first described by Straubing [159]. See also [178, 169]. The formulation given below is due to Weil [192, 194].

Theorem 7.19. *Let $\mathcal{V}$ be the $*$-variety associated with $\overline{\mathbf{Gsol}}$. For each alphabet A, $\mathcal{V}(A^*)$ is the smallest boolean algebra of languages closed under the operations $L \rightarrow LaA^*$ and $L \rightarrow (LaA^*)_{r,p}$, where $a \in A$, p is a prime number and $0 \leq r < p$.*

The variety of languages associated with the variety of finite monoids in which every $\mathcal{H}$-class is a solvable group is described in the next theorem, which mixes the ideas of Theorems 7.19 and 7.15.

Theorem 7.20. (Thérien [176]) *Let $\mathbf{V}$ be a variety of finite monoids in which every $\mathcal{H}$-class is a solvable group and let $\mathcal{V}$ be the associated variety of languages. Then $\mathcal{V}(\emptyset^*) = \{\emptyset, \{1\}\}$, and for each non empty alphabet A, $\mathcal{V}(A^*)$ is the smallest boolean algebra of languages closed under the operations $L \rightarrow (LaA^*)_{r,n}$, where $a \in A$ and $0 \leq r < n$, and containing the languages of the form $A^* a A^*$, LaA^* and $A^* aL$, where $a \in A$, $L \in \mathcal{V}((A \setminus \{a\})^*)$.*

7.6 Varieties closed under product with counter

Finally, let us mention the results of Weil [194]. Let n be an integer. A set of languages $\mathcal{L}$ of A^* is *closed under product with n-counters* if, for any language $L_0, \ldots, L_k \in \mathcal{L}$, for any letter $a_1, \ldots, a_k \in A$ and for any integer r such that

$0 \leq r < n$, $(L_0 a_1 L_1 \cdots a_k L_k)_{r,n} \in \mathcal{L}$. A set of languages $\mathcal{L}$ of A^* is *closed under product with counters* if it is closed under product with n-counters, for arbitrary n.

Theorem 7.21. *Let p be a prime number, let $\mathbf{V}$ be a variety of finite monoids and let $\mathcal{V}$ be the associated variety of languages. For each alphabet A, let $\mathcal{W}(A^*)$ be the smallest boolean algebra containing $\mathcal{V}(A^*)$ and closed under product with p-counters. Then $\mathcal{W}$ is a variety of languages and the associated variety of monoids is $\mathbf{LG}_p \textcircled{M} \mathbf{V}$.*

Theorem 7.22. *Let p be a prime number, let $\mathbf{V}$ be a variety of finite monoids and let $\mathcal{V}$ be the associated variety of languages. For each alphabet A, let $\mathcal{W}(A^*)$ be the smallest boolean algebra containing $\mathcal{V}(A^*)$ and closed under product and product with p-counters. Then $\mathcal{W}$ is a variety of languages and the associated variety of monoids is $\mathbf{L\overline{G}}_p \textcircled{M} \mathbf{V}$.*

Theorem 7.23. *Let $\mathbf{V}$ be a variety of finite monoids and let $\mathcal{V}$ be the associated variety of languages. For each alphabet A, let $\mathcal{W}(A^*)$ be the smallest boolean algebra containing $\mathcal{V}(A^*)$ and closed under product with counters. Then $\mathcal{W}$ is a variety of languages and the associated variety of monoids is $\mathbf{LGsol} \textcircled{M} \mathbf{V}$.*

For instance, if $\mathbf{V} = \mathbf{J}_1$ it is known that $\mathbf{LGsol} \textcircled{M} \mathbf{J}_1$ is the variety of monoids whose regular $\mathcal{D}$-classes are unions of solvable groups.

Theorem 7.24. *Let $\mathbf{V}$ be a variety of finite monoids and let $\mathcal{V}$ be the associated variety of languages. For each alphabet A, let $\mathcal{W}(A^*)$ be the smallest boolean algebra containing $\mathcal{V}(A^*)$ and closed under product and product with counters. Then $\mathcal{W}$ is a variety of languages and the associated variety of monoids is $\mathbf{L\overline{Gsol}} \textcircled{M} \mathbf{V}$.*

8. Concatenation hierarchies

By alternating the use of the polynomial closure and of the boolean closure one can obtain hierarchies of recognizable languages. Let $\mathcal{V}$ be a variety of languages. The concatenation hierarchy of basis $\mathcal{V}$ is the hierarchy of classes of languages defined as follows.

(1) level 0 is $\mathcal{V}$
(2) for every integer $n \geq 0$, level $n + 1/2$ is the polynomial closure of level n
(3) for every integer $n \geq 0$, level $n + 1$ is the boolean closure of level $n + 1/2$.

Theorem 7.1 shows that the polynomial closure of a variety of languages is a positive variety of languages. Furthermore the boolean closure of a positive variety of languages is a variety of languages. Therefore, one defines a sequence of varieties $\mathcal{V}_n$ and of positive varieties $\mathcal{V}_{n+1/2}$, where n is an integer, as follows:

(1) $\mathcal{V}_0 = \mathcal{V}$

(2) for every integer $n \geq 0$, $\mathcal{V}_{n+1/2} = \text{Pol } \mathcal{V}_n$,

(3) for every integer $n \geq 0$, $\mathcal{V}_{n+1} = \text{BPol } \mathcal{V}_n$.

The associated varieties of semigroups and ordered semigroups (resp. monoids and ordered monoids) are denoted $\mathbf{V}_n$ and $\mathbf{V}_{n+1/2}$. Theorem 7.1 gives an explicit relation between $\mathbf{V}_n$ and $\mathbf{V}_{n+1/2}$.

Proposition 8.1. *For every integer* $n \geq 0$, $\mathbf{V}_{n+1/2} = [\![x^\omega y x^\omega \leq x^\omega]\!] \, \textcircled{M} \, \mathbf{V}_n$.

Three concatenation hierarchies have been considered so far in the literature. The first one, introduced by Brzozowski [36] and called the *dot-depth hierarchy*, is the hierarchy of positive +-varieties whose basis is the trivial variety. The second one, first considered implicitly in [174] and explicitly in [163, 166] is called the *Straubing-Thérien hierarchy*: it is the hierarchy of positive *-varieties whose basis is the trivial variety. The third one, introduced in [83], is the hierarchy of positive *-varieties whose basis is the variety of group-languages. It is called the *group hierarchy*.

It can be shown that these three hierarchies are strict: if A contains at least two letters, then for every n, there exist languages of level $n + 1$ which are not of level $n + 1/2$ and languages of level $n + 1/2$ which are not of level n. This was first proved for the dot-depth hierarchy in [40].

The main question is the decidability of each level: given an integer n (resp. $n + 1/2$) and a recognizable language L, decide whether or not L is of level n (resp. $n + 1/2$). The language can be given either by a finite automaton, by a finite semigroup or by a rational expression since there are standard algorithms to pass from one representation to the other.

There is a wide literature on the concatenation product: Arfi [11, 12], Blanchet-Sadri [24, 25, 26, 27, 28, 29, 30, 31, 32], Brzozowski [36, 40, 41], Cowan [47], Eilenberg [53], Knast [71, 72, 40], Schützenberger[144, 148], Simon [41, 152, 156], Straubing [118, 120, 160, 164, 167, 170, 172], Thérien [120, 170], Thomas [185], Weil [191, 172, 195, 126, 127] and the author [97, 100, 104, 118, 120, 126, 127]. The reader is referred to the survey articles [110, 111] for more details on these results.

We now describe in more details the first levels of each of these hierarchies. We consider the Straubing-Thérien hierarchy, the dot-depth hierarchy and the group hierarchy, in this order.

8.1 Straubing-Thérien's hierarchy

Level 0 is the trivial *-variety. Therefore a language of A^* is of level 0 if and only if it is empty or equal to A^*. This condition is easily characterized.

Proposition 8.2. *A language is of level* 0 *if and only if its syntactic monoid is trivial.*

It is also well known that one can decide in polynomial time whether the language of A^* accepted by a deterministic n-state automaton is empty or equal to A^* (that is, of level 0).

By definition, the sets of level 1/2 are the finite unions of languages of the form $A^* a_1 A^* a_2 \cdots a_k A^*$, where the a_i's are letters. An alternative description can be given in terms of shuffle. A language is a *shuffle ideal* if and only if for every $u \in L$ and $v \in A^*$, $u \amalg v$ is contained in L.

Proposition 8.3. *A language is of level 1/2 if and only if it is a shuffle ideal.*

It follows from Theorem 7.6 that the only shuffle ideals whose complement is also a shuffle ideal are the full language and the empty language. It is easy to see directly that level 1/2 is decidable. One can also derive this result from our syntactic characterization.

Proposition 8.4. *A language is of level 1/2 if and only if its ordered syntactic monoid satisfies the identity $x \leq 1$.*

One can derive from this result a polynomial algorithm to decide whether the language accepted by a complete deterministic n-state automaton is of level 1/2. See [126, 127] for details.

Corollary 8.5. *One can decide in polynomial time whether the language accepted by a complete deterministic n-state automaton is of level 1/2.*

The sets of level 1 are the finite boolean combinations of languages of the form $A^* a_1 A^* a_2 \cdots a_k A^*$, where the a_i's are letters. In particular, all finite sets are of level 1. The sets of level 1 have a nice algebraic characterization, due to Simon [152]. There exist now several proofs of this deep result [3, 170, 157, 64].

Theorem 8.6. *A language of A^* is of level 1 if and only if its syntactic monoid is $\mathcal{J}$-trivial, or, equivalently, if and only if it satisfies the identities $x^\omega = x^{\omega+1}$ and $(xy)^\omega = (yx)^\omega$.*

Thus $\mathbf{V}_1 = \mathbf{J}$, the variety of finite $\mathcal{J}$-trivial monoids. Theorem 8.6 yields an algorithm to decide whether a given recognizable set is of level 1. See [157].

Corollary 8.7. *One can decide in polynomial time whether the language accepted by a deterministic n-state automaton is of level 1.*

The sets of level 3/2 also have a simple description, although this is not a direct consequence of the definition [118].

Theorem 8.8. *The sets of level 3/2 of A^* are the finite unions of sets of the form $A_0^* a_1 A_1^* a_2 \cdots a_k A_k^*$, where the a_i's are letters and the A_i's are subsets of A.*

We derive the following syntactic characterization [126, 127].

Theorem 8.9. *A language is of level* $3/2$ *if and only if its ordered syntactic monoid satisfies the identity* $x^\omega y x^\omega \leq x^\omega$ *for every* x, y *such that* $c(x) = c(y)$.

Corollary 8.10. *There is an algorithm, in time polynomial in* $2^{|A|}n$, *for testing whether the language of* A^* *accepted by a deterministic* n-*state automaton is of level* $3/2$.

We now arrive to the level 2. Theorem 8.8 gives a combinatorial description of the languages of level 2 [118].

Theorem 8.11. *The languages of level* 2 *of* A^* *are the finite boolean combinations of the languages of the form* $A_0^* a_1 A_1^* a_2 \cdots a_k A_k^*$, *where the* a_i *'s are letters and the* A_i *'s are subsets of* A.

The next theorem [118] gives a non-trivial (but unfortunately non effective) algebraic characterization of level 2. Given a variety of monoids **V**, denote by **PV** the variety generated by all monoids of the form $\mathcal{P}(M)$, where $M \in$ **V**.

Theorem 8.12. *A language is of level* 2 *in the Straubing-Thérien hierarchy if and only if its syntactic monoid belongs to* **PJ**.

Let $\mathbf{V}_2$ be the variety of finite monoids associated with the languages of level 2. Theorem 8.12 shows that $\mathbf{V}_2 = \mathbf{PJ}$. Unfortunately, no algorithm is known to decide whether a finite monoid divides the power monoid of a $\mathcal{J}$-trivial monoid. In other words, the decidability problem for level 2 is still open, although much progress has been made in recent years [25, 29, 47, 118, 167, 172, 191, 195]. This problem is a particular case of a more general question discussed in section 8.5.

In the case of languages whose syntactic monoid is an inverse monoid, a complete characterization can be given [47, 126, 127]. An *inverse automaton* is a deterministic automaton $\mathcal{A} = (Q, A \cup \bar{A}, i, F)$ over a symmetrized alphabet $A \cup \bar{A}$, which satisfies, for all $a \in A$, $q, q' \in Q$

$$q \cdot a = q' \text{ if and only if } q' \cdot \bar{a} = q$$

Note however that this automaton is not required to be complete. In other words, in an inverse automaton, each letter defines a partial injective map from Q to Q and the letters a and $\bar{a}$ define mutually reciprocal transitions.

Theorem 8.13. *The language recognized by an inverse automaton* $\mathcal{A} = (Q, A \cup \bar{A}, i, F)$ *is of level* 2 *in the Straubing-Thérien hierarchy if and only if, for all* $q, q' \in Q$, $u, v \in (A \cup \bar{A})^*$, *such that* $q \cdot u$ *and* $q' \cdot u$ *are defined,* $q \cdot v = q'$ *and* $c(v) \subseteq c(u)$ *imply* $q = q'$.

Actually, one can show [126, 127] that each language recognized by an inverse automaton $\mathcal{A}$ is the difference of two languages of level $3/2$ recognized by the completion of $\mathcal{A}$. It is proved in [191, 195] that Theorem 8.13 yields the following important corollary.

Corollary 8.14. *It is decidable whether an inverse monoid belongs to* $\mathbf{V}_2$.

Example 8.1. Let $A = \{a, b\}$ and let $L = (ab)^*$ be the language of example 5.1. Its minimal automaton satisfies the conditions of Theorem 8.13 and thus it has level 2. In fact, by observing that $\emptyset^* = \{1\}$, L can be written in the form

$$(\emptyset^* \cup \emptyset^* a A^* \cup A^* b \emptyset^*) \setminus (\emptyset^* b A^* \cup \emptyset^* A^* a \cup A^* a \emptyset^* a A^* \cup A^* b \emptyset^* b A^*)$$

Theorem 8.11 describes the languages of level 2 of A^* as finite boolean combinations of the languages of the form $A_0^* a_1 A_1^* a_2 \cdots a_k A_k^*$, where the a_i's are letters and the A_i's are subsets of A. Several subvarieties of languages can be obtained by imposing various conditions on the A_i's. This was the case, for instance for Corollaries 7.5 and 7.8. The known results are summarized in the table below.

Conditions	Variety	Algebraic description	Ref.
Unambiguous product	**DA**	Regular $\mathcal{J}$-classes are idempotent semigroups	[148]
Left det. product	**R**	$\mathcal{R}$-trivial	[97, 98]
Right det. product	**L**	$\mathcal{L}$-trivial	[97, 98]
Bidet. product	**J∩Ecom**	$\mathcal{J}$-trivial and idempotents commute	[123]
$a_i \notin A_{i-1}$	**R**	$\mathcal{R}$-trivial	[53, 39]
$a_i \notin A_i$	**L**	$\mathcal{L}$-trivial	[53, 39]
$a_i \notin A_{i-1} \cup A_i$	**J∩Ecom**	$\mathcal{J}$-trivial and idempotents commute	[17]
$a_i \notin A_{i-1}$ and $A_{i-1} \subseteq A_i$	**L$_1$**	Idempotent and $\mathcal{L}$-trivial	[119]
$a_i \notin A_{i-1}$ and $A_i \subseteq A_{i-1}$	**R$_1$**	Idempotent and $\mathcal{R}$-trivial	[119]
$A_i \cap A_j = \emptyset$ for $i \neq j$ and $a_i \notin A_{i-1} \cup A_i$	**J∩LJ$_1$∩ Ecom**	The syntactic semigroup is $\mathcal{J}$-trivial, locally idempotent and commutative and its idempotents commute	[150]

The variety of monoids associated with the condition $A_0 \subseteq \cdots \subseteq A_n$ is also described in [5], pp. 234–239, but this description, which requires an infinite number of identities, is too technical to be reproduced here.

Little is known beyond level 2: a semigroup theoretic description of each level of the hierarchy is known, but it is not effective. Each level of the hierarchy is a variety or a positive variety and the associated variety of (ordered) semigroups admits a description by identities, but these identities are not known for $n \geq 2$. Furthermore, even if these identities were known, this would not necessarily lead to a decision process for the corresponding variety. See also the conjecture discussed in section 8.5.

8.2 Dot-depth hierarchy

Level 0 is the trivial $+$-variety. Therefore a language of A^+ is of dot-depth 0 if and only if it is empty or equal to A^+ and one can decide in polynomial time whether the language of A^+ accepted by a deterministic n-state automaton is of level 0.

Proposition 8.15. *A language is of dot-depth 0 if and only if its syntactic semigroup is trivial.*

The languages of dot-depth 1/2 are by definition finite unions of languages of the form $u_0 A^+ u_1 A^+ \cdots u_{k-1} A^+ u_k$, where $k \geq 0$ and $u_0, \ldots, u_k \in A^*$. But since $A^* = A^+ \cup \{1\}$, these languages can also be expressed as finite unions of languages of the form

$$u_0 A^* u_1 A^* \cdots u_{k-1} A^* u_k$$

The syntactic characterization is a simple application of our Theorem 7.1.

Proposition 8.16. *A language of A^+ is of dot-depth 1/2 if and only if its ordered syntactic semigroup satisfies the identity $x^\omega y x^\omega \leq x^\omega$.*

Corollary 8.17. *One can decide in polynomial time whether the language accepted by a deterministic n-state automaton is of dot-depth 1/2.*

The languages of dot-depth 1 are the finite boolean combinations of languages of dot-depth 1/2. The syntactic characterization of these languages, due to Knast [71, 72], makes use of the Cauchy category.

Theorem 8.18. *A language of A^+ is of dot-depth 1 if and only if the Cauchy category of its syntactic semigroup satisfies the following condition (K): if p and r are arrows from e to f and if q and s are arrows from f to e, then $(pq)^\omega ps(rs)^\omega = (pq)^\omega (rs)^\omega$.*

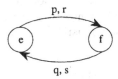

Fig. 8.1. The condition (K).

The variety of finite semigroups satisfying condition (K) is usually denoted $\mathbf{B_1}$ ($\mathbf{B}$ refers to Brzozowski and 1 to level 1). Thus $\mathbf{B_1}$ is defined by the identity

$$(x^\omega p y^\omega q x^\omega)^\omega x^\omega p y^\omega s x^\omega (x^\omega r y^\omega s x^\omega)^\omega = (x^\omega p y^\omega q x^\omega)^\omega (x^\omega r y^\omega s x^\omega)^\omega$$

Theorem 8.18 can be proved in the same way as Theorems 5.18, 5.20 and 5.22. First, one makes use of the wreath product principle to show that a language is of dot-depth 1 if and only if its syntactic semigroup belongs to $\mathbf{J} * \mathbf{LI}$. Then Theorems 6.17 and 6.14 can be applied. See also [5] for a different proof.

The algorithm corresponding to Theorem 8.18 was analyzed by Stern [158].

Corollary 8.19. *One can decide in polynomial time whether the language accepted by a deterministic n-state automaton is of dot-depth 1.*

Straubing [166] discovered an important connection between the Straubing-Thérien and the dot-depth hierarchies. Let $\mathbf{B}_n$ be the variety of finite semigroups corresponding to the languages of dot-depth n and let $\mathbf{V}_n$ be the variety of finite monoids corresponding to the languages of level n in the Straubing-Thérien hierarchy.

Theorem 8.20. *For every integer $n > 0$, $\mathbf{B}_n = \mathbf{V}_n * \mathbf{LI}$.*

Now Theorems 6.17, 6.11 and 6.14 reduce decidability questions about the dot-depth hierarchy to the corresponding questions about the Straubing-Thérien hierarchy.

Theorem 8.21. *For every integer $n \geq 0$, $\mathbf{B}_n$ is decidable if and only if $\mathbf{V}_n$ is decidable.*

It is very likely that Theorem 8.21 can be extended to take in account the half levels, but this is not yet formally proved. In particular, it is not yet known whether level 3/2 of the dot-depth hierarchy is decidable.

8.3 The group hierarchy

We consider in this section the concatenation hierarchy based on the group languages, or group hierarchy. By definition, a language of A^* is of level 0 in this hierarchy if and only if its syntactic monoid is a finite group. This can be easily checked on any deterministic automaton recognizing the language [126, 127].

Proposition 8.22. *One can decide in polynomial time whether the language accepted by a deterministic n-state automaton is a group language.*

The languages of level 1/2 are by definition finite unions of languages of the form $L_0 a_1 L_1 \cdots a_k L_k$ where the a_i's are letters and the L_i's are group languages. By Theorem 7.1, a language is of level 1/2 if and only if its ordered syntactic monoid belongs to the variety $[\![x^\omega y x^\omega \leq x^\omega]\!] \, \textcircled{M} \, \mathbf{G}$. The identities of this variety are given by Theorem 6.5.

Proposition 8.23. *A language is of level* 1/2 *in the group hierarchy if and only if its syntactic ordered monoid satisfies the identity* $x^\omega \leq 1$.

Proposition 8.23 can be converted to an algorithm on automata.

Corollary 8.24. *There is a polynomial time algorithm for testing whether the language accepted by a deterministic n-state automaton is of level* 1/2 *in the group hierarchy.*

Level 1/2 is also related with the topology of the free monoid defined by the distance $d_\mathbf{G}$. This topology is called the *pro-group* topology. An example of a converging sequence is given by the following proposition [135].

Proposition 8.25. *For every word* $u \in A^*$, $\lim\limits_{n \to \infty} u^{n!} = 1$.

As the multiplication is continuous and a closed set contains the limit of any converging sequence of its elements, it follows that if L is a closed set in the pro-group topology, and if $xu^n y \in L$ for all $n > 0$, then $xy \in L$. The converse is also true if L is recognizable.

Theorem 8.26. *A recognizable set* L *of* A^* *is closed in the pro-group topology if and only if for every* $u, x, y \in A^*$, $xu^+ y \subseteq L$ *implies* $xy \in L$.

Since an open set is the complement of a closed set, one can also state:

Theorem 8.27. *A recognizable set* L *of* A^* *is open in the pro-group topology if and only if for every* $u, x, y \in A^*$, $xy \in L$ *implies* $xu^+ y \cap L \neq \emptyset$.

These conditions can be easily converted in terms of syntactic monoids.

Theorem 8.28. *Let* L *be a recognizable language of* A^*, *let* M *be its syntactic monoid and let* P *be its syntactic image.*

(1) L *is closed in the pro-group topology if and only if for every* $s, t \in M$ *and* $e \in E(M)$, $set \in P$ *implies* $st \in P$.
(2) L *is open in the pro-group topology if and only if for every* $s, t \in M$ *and* $e \in E(M)$, $st \in P$ *implies* $set \in P$.
(3) L *is clopen[10] in the pro-group topology if and only if* M *is a group.*

Finally, condition (1) states exactly that the syntactic ordered monoid of L satisfies the identity $1 \leq x^\omega$ and condition (2) states that the syntactic ordered monoid of L satisfies the identity $x^\omega \leq 1$. In particular, we get the following result.

Corollary 8.29. *Let* L *be a recognizable language of* A^*. *The following conditions are equivalent.*

(1) L *is open in the pro-group topology,*

---

[10] clopen is a common abbreviation for "closed and open"

(2) L is of level 1/2 in the group hierarchy,
(3) the syntactic ordered monoid of L satisfies the identity $x^\omega \leq 1$.

To finish with these topological properties, let us mention a last result.

Theorem 8.30. *Let L be a language of A^* and let $\bar{L}$ be its closure in the pro-group topology. If L is recognizable, then $\bar{L}$ is recognizable. If L is open and recognizable, then $\bar{L}$ is clopen.*

A few more definitions are needed to state the algebraic characterization of the languages of level 1. A *block group* is a monoid such that every $\mathcal{R}$-class (resp. $\mathcal{L}$-class) contains at most one idempotent. Block groups form a variety of monoids, denoted **BG**, which is defined by the identity $(x^\omega y^\omega)^\omega = (y^\omega x^\omega)^\omega$. Thus **BG** is a decidable variety.

Theorem 8.31. *A language is of level 1 in the group hierarchy if and only if its syntactic monoid belongs to **BG**.*

Corollary 8.32. *There is a polynomial time algorithm for testing whether the language accepted by a deterministic n-state automaton is of level 1 in the group hierarchy.*

Several other descriptions of **BG** are known.

Theorem 8.33. *The following equalities holds:* **BG** = **PG** = **J** $*$ **G** = **J** Ⓜ **G**. *Furthermore, a finite monoid belongs to **BG** if and only if its set of idempotents generates a $\mathcal{J}$-trivial monoid.*

The results of this section are difficult to prove. They rely on the one hand on a topological conjecture of Reutenauer and the author [114], recently proved by Ribes and Zalesskii [140], and on the other hand on the solution by Ash [15, 16] of a famous open problem in semigroup theory, the Rhodes "type II" conjecture. Actually, as was shown in [106], the topological and algebraic aspects are intimately related. See the survey [62] for more references and details.

Theorem 8.30 is proved in [113]. The second part of the statement was suggested by Daniel Lascar. The study of the languages of level 1 in the group hierarchy started in 1985 [83] and was completed in [63] (see also [62]). The reader is referred to the survey article [111] for a more detailed discussion. See also [101, 106, 114, 62, 108].

8.4 Subhierarchies

Several subhierarchies were considered in the literature. One of the most studied is the subhierarchy inside level 1 of a concatenation hierarchy. Recall that if $\mathcal{V}$ is a $*$-variety of languages, then BPol$\mathcal{V}$ is the level one of the concatenation hierarchy built on $\mathcal{V}$. For each alphabet A, let $\mathcal{V}_{1,n}(A^*)$ be the boolean algebra generated by languages of the form

$$L_0 a_1 L_1 a_2 \cdots a_k L_k$$

where the a_i's are letters, $L_i \in \mathcal{V}(A^*)$ and $0 \leq k \leq n$. Then the sequence $\mathcal{V}_{1,n}$ is an increasing sequence of $*$-varieties whose union is BPol$\mathcal{V}$.

For the Straubing-Thérien's hierarchy, $\mathcal{V}_{1,n}(A^*)$ is the boolean algebra generated by the languages of the form $A^* a_1 A^* a_2 \cdots a_k A^*$, where the a_i's are letters, and $0 \leq k \leq n$. The corresponding variety of finite monoids is denoted $\mathbf{J}_n$. In particular, $\mathcal{J}_1$ is the $*$-variety already encountered in Proposition 5.5. The hierarchy $\mathbf{J}_n$ is defined in [36]: Simon proved that $\mathbf{J}_2$ is defined by the identities $(xy)^2 = (yx)^2$ and $xyxzx = xyzx$: the variety $\mathbf{J}_3$ is defined [24] by the identities $xzyxvxwy = xzxyxvxwy$, $ywxvxyzx = ywxvxyxzx$ and $(xy)^3 = (yx)^3$ but there are no finite basis of identities for $\mathbf{J}_n$ when $n > 3$ [30].

For the group hierarchy, the variety $\mathcal{V}_{1,1}$ admits several equivalent descriptions [85], which follow in part from Theorem 7.12.

Theorem 8.34. *Let $\mathbf{A}$ be an alphabet and let K be a recognizable language of A^*. The following conditions are equivalent:*

(1) K is in the boolean algebra generated by the languages of the form L or LaA^*, where L is a group language and $a \in A$,

(2) K is in the boolean algebra generated by the languages of the form L or A^*aL, where L is a group language and $a \in A$,

(3) K is in the boolean algebra generated by the languages of the form L or LaL', where L and L' are group languages and $a \in A$,

(4) idempotents commute in the syntactic monoid of K.

Thus the corresponding variety of finite monoids is the variety of monoids with commuting idempotents, defined by the identity $x^\omega y^\omega = y^\omega x^\omega$. It is equal to the variety **Inv** generated by finite inverse monoids. Furthermore, the following non trivial equalities hold [14, 82, 85]

$$\mathbf{Inv} = \mathbf{J}_1 * \mathbf{G} = \mathbf{J}_1 \, \text{Ⓜ} \, \mathbf{G}$$

More generally, one can define hierarchies indexed by trees [100]. These hierarchies were studied in connection with game theory by Blanchet-Sadri [24, 26].

8.5 Boolean-polynomial closure

Let $\mathbf{V}$ be a variety of finite monoids and let $\mathcal{V}$ be the associated $*$-variety. We have shown that the algebraic counterpart of the operation $\mathcal{V} \to \text{Pol } \mathcal{V}$ on varieties of languages is the operation $\mathbf{V} \to [\![x^\omega y x^\omega \leq x^\omega]\!] \, \text{Ⓜ} \, \mathbf{V}$. Similarly, the algebraic counterpart of the operation $\mathcal{V} \to \text{Co-Pol } \mathcal{V}$ is the operation $\mathbf{V} \to [\![x^\omega \leq x^\omega y x^\omega]\!] \, \text{Ⓜ} \, \mathbf{V}$. It is tempting to guess that the algebraic counterpart of the operation $\mathcal{V} \to \text{BPol } \mathcal{V}$ is also of the form $\mathbf{V} \to \mathbf{W} \, \text{Ⓜ} \, \mathbf{V}$ for some variety $\mathbf{W}$. A natural candidate for $\mathbf{W}$ is the join of the two varieties $[\![x^\omega y x^\omega \leq x^\omega]\!]$

and $[\![x^\omega \le x^\omega y x^\omega]\!]$, which is equal to the variety $\mathbf{B}_1$ defined in section 8.2 [126, 127]. We can thus formulate our conjecture as follows:

Conjecture 8.1. Let $\mathcal{V}$ be a variety of languages and let $\mathbf{V}$ be the associated variety of finite semigroups (resp. monoids). Then the variety of finite semigroups (resp. monoids) associated with BPol $\mathcal{V}$ is $\mathbf{B}_1 \circledM \mathbf{V}$.

One inclusion in the conjecture is certainly true.

Proposition 8.35. *The variety of finite semigroups (resp. monoids) associated with BPol $\mathcal{V}$ is contained in $\mathbf{B}_1 \circledM \mathbf{V}$.*

Now, by Theorem 6.5, the identities of $\mathbf{B}_1 \circledM \mathbf{V}$ are

$$(x^\omega p y^\omega q x^\omega)^\omega x^\omega p y^\omega s x^\omega (x^\omega r y^\omega s x^\omega)^\omega = (x^\omega p y^\omega q x^\omega)^\omega (x^\omega r y^\omega s x^\omega)^\omega \quad (8.1)$$

for all $x, y, p, q, r, s \in \widehat{A}^*$ for some finite alphabet A such that $\mathbf{V}$ satisfies $x^2 = x = y = p = q = r = s$. These identities lead to another equivalent statement for our conjecture.

Proposition 8.36. *The conjecture is true if and only if every finite semigroup (resp. monoid) satisfying the identities (8.1) is a quotient of an ordered semigroup (resp. ordered monoid) of the variety $[\![x^\omega y x^\omega \le x^\omega]\!] \circledM \mathbf{V}$.*

Conjecture 8.1 was proved to be true in a few particular cases. First, if $\mathbf{V}$ is the trivial variety of monoids, then $\mathbf{B}_1 \circledM \mathbf{V} = \mathbf{J}$. In this case, the second form of the conjecture is known to be true. This is in fact a consequence of Theorem 8.6 and it was also proved directly by Straubing and Thérien [170].

Theorem 8.37. *Every $\mathcal{J}$-trivial monoid is a quotient of an ordered monoid satisfying the identity $x \le 1$.*

Second, if $\mathbf{V}$ is the trivial variety of semigroups, then $\mathbf{B}_1 \circledM \mathbf{V} = \mathbf{B}_1$ is, by Theorem 8.18, the variety of finite semigroups associated with the languages of dot-depth 1. Therefore, the conjecture is true in this case, leading to the following corollary.

Corollary 8.38. *Every monoid of $\mathbf{B}_1$ is a quotient of an ordered monoid satisfying the identity $x^\omega y x^\omega \le x^\omega$.*

Third, if $\mathbf{V} = \mathbf{G}$, the variety of monoids consisting of all finite groups, $\mathbf{B}_1 \circledM \mathbf{G} = \mathbf{J} \circledM \mathbf{G} = \mathbf{PG} = \mathbf{BG}$ is the variety associated with the level 1 of the group hierarchy. Therefore, the conjecture is also true in this case.

Corollary 8.39. *Every monoid of $\mathbf{BG}$ is a quotient of an ordered monoid satisfying the identity $x^\omega \le 1$.*

The level 2 of the Straubing-Thérien's hierarchy corresponds to the case $\mathbf{V} = \mathbf{J}_1$. Therefore, one can formulate the following conjecture for this level:

Conjecture 8.2. A recognizable language is of level 2 in the Straubing-Thérien's hierarchy if and only if its syntactic monoid satisfies the identities

$$(x^\omega p y^\omega q x^\omega)^\omega x^\omega p y^\omega s x^\omega (x^\omega r y^\omega s x^\omega)^\omega = (x^\omega p y^\omega q x^\omega)^\omega (x^\omega r y^\omega s x^\omega)^\omega$$

for all $x, y, p, q, r, s \in \widehat{A}^*$ for some finite alphabet A such that $c(x) = c(y) = c(p) = c(q) = c(r) = c(s)$.

If this conjecture was true, it would imply the decidability of the levels 2 of the Straubing-Thérien's hierarchy and of the dot-depth. It is known that Conjecture 8.2 holds true for languages recognized by an inverse monoid [172, 191, 195] and for languages on a two letter alphabet [167].

More generally, the conjecture $\mathbf{V}_{n+1} = \mathbf{B}_1 \textcircled{M} \mathbf{V}_n$ would reduce the decidability of the Straubing-Thérien's hierarchy to a problem on the Mal'cev products of the form $\mathbf{B}_1 \textcircled{M} \mathbf{V}$. However, except for a few exceptions (including $\mathbf{G}$, $\mathbf{J}$ and the finitely generated varieties, like the trivial variety or $\mathbf{J}_1$), it is not known whether the decidability of $\mathbf{V}$ implies that of $\mathbf{B}_1 \textcircled{M} \mathbf{V}$.

9. Codes and varieties

In this section, we will try to follow the terminology of the book of Berstel and Perrin [20], which is by far the best reference on the theory of codes. Recall that a subset P of A^+ is a *prefix code* if no element of P is a proper prefix of another element of P, that is, if $u, uv \in P$ implies $v = 1$. The next statement shows that the syntactic monoids of finite codes give, in some sense, a good approximation of any finite monoid.

Theorem 9.1. [84] *For every finite monoid M, there exists a finite prefix code P such that*

(1) M divides $M(P^*)$

(2) there exists a relational morphism $\tau : M(P^*) \to M$, such that, for every idempotent e of M, $\tau^{-1}(e)$ is an aperiodic semigroup.

If $\mathcal{V}$ is a $*$-variety of languages, denote by $\mathcal{V}'$ the least $*$-variety such that, for each alphabet A, $\mathcal{V}'(A^*)$ contains the languages of $\mathcal{V}(A^*)$ of the form P^*, where P is a finite prefix code. Similarly, if $\mathcal{V}$ be a $+$-variety of languages, let $\mathcal{V}'$ be the least $+$-variety such that, for each alphabet A, $\mathcal{V}'(A^+)$ contains the languages of $\mathcal{V}(A^+)$ of the form P^+, where P is a finite prefix code. By construction, $\mathcal{V}'$ is contained in $\mathcal{V}$, but the inclusion is proper in general. A variety of languages $\mathcal{V}$ is said to be *described by its finite prefix codes* if $\mathcal{V} = \mathcal{V}'$. The next theorem summarizes the results of [84] and [79].

Theorem 9.2. *(1) Every $*$-variety closed under product is described by its finite prefix codes.*

(2) The $+$-variety of locally testable is described by its finite prefix codes.

(3) *The $*$-variety associated with the variety $[\![x^\omega y^\omega = y^\omega x^\omega]\!]$ is described by its finite prefix codes.*

(4) *Let $\mathbf{H}_1, \ldots, \mathbf{H}_n$ be varieties of finite groups. Then the $*$-variety associated with the variety $\mathbf{A} * \mathbf{H}_1 * \mathbf{A} \cdots * \mathbf{H}_n * \mathbf{A}$ is described by its finite prefix codes.*

In contrast, the varieties of languages corresponding to the varieties $\mathbf{R}$, $\mathbf{J}$, $\mathbf{DA}$ or $\mathbf{B}_1$ are not described by their finite prefix codes. Schützenberger [148] conjectured that the $+$-variety of languages corresponding to $\mathbf{LG}$ is described by its finite prefix codes. This conjecture is still open.

Prefix codes were also used to impose a restriction on the star operation. Let us say that a set $\mathcal{L}$ of languages of A^* is *closed under polynomial operations* if it is closed under finite union and product. Then Kleene's theorem can be stated as follows: "the rational languages of A^* form the smallest set of languages of A^* closed under polynomial operations and star". Schützenberger has obtained a similar statement for the star-free languages, in which the star operation is restricted to a special class of languages. A language L of A^* is *uniformly synchronous* if there exists an integer $d \geq 0$ such that, for all $x, y \in A^*$ and for all $u, v \in L^d$,

$$xuvy \in L^* \text{ implies } xu, vy \in L^*$$

The theorem of Schützenberger can now be stated. The paradoxical aspect of this theorem is that it gives a characterization of the star-free languages that makes use of the star operation!

Theorem 9.3. [146, 147] *The class of star-free languages of A^* is the smallest class of languages of A^* closed under polynomial operations and star operation restricted to uniformly synchronous prefix codes.*

Schützenberger also characterized the languages associated with the variety of finite monoids $\overline{\mathbf{Gcom}}$ [146].

Theorem 9.4. *Let $\mathcal{V}$ be the $*$-variety of languages associated with $\overline{\mathbf{Gcom}}$. For each alphabet A, $\mathcal{V}(A^*)$ is the least boolean algebra $\mathcal{C}$ of languages of A^* closed under polynomial operations and under the star operation restricted to languages of the form P^n, where $n > 0$ and P is a uniformly synchronous prefix code of $\mathcal{C}$.*

Another series of results concern the circular codes. A prefix code P is *circular* if, for every $u, v \in A^+$, $uv, vu \in C^+$ implies $u, v \in C^+$. Circular codes are intimately related to locally testable languages [49, 50, 51]

Theorem 9.5. *A finite prefix code[11] P is circular if and only if P^+ is a locally testable language.*

---

[11] This is a "light" version of the theorem, which holds for a larger class of (non necessarily prefix) codes.

Theorem 9.5 leads to the following characterization of the +-variety of locally testable languages [99].

Theorem 9.6. *Let +-variety of locally testable languages is the smallest +-variety containing the finite languages and closed under the operation $P \to P^+$ where P is a finite prefix circular code.*

Similar results hold for pure codes and star-free languages. The definition of a pure language was given in section 6.1.

Theorem 9.7. [133] *A finite prefix code P is pure if and only if P^* is a star-free language.*

Theorem 9.8. [99] *Let $*$-variety of star-free languages is the smallest $*$-variety containing the finite languages and closed under the operation $P \to P^*$ where P is a finite prefix pure code.*

However, Theorems 9.6 and 9.8 are less satisfactory than Theorems 9.3 and 9.4, because they lead to a description of the languages involving all the operations of a $*$-variety, including complement, left and right quotient and inverse morphisms.

10. Operators on languages and varieties

In view of Eilenberg's theorem, one may expect some relationship between the operators on languages (of combinatorial nature) and the operators on semigroups (of algebraic nature). The following table summarizes the results of this type related to the concatenation product.

Closure of $\mathcal{V}$ under the operations ...	**V**
Product and union	$[\![x^\omega y x^\omega \leq x^\omega]\!] \, \textcircled{M} \, \mathbf{V}$
Unambiguous product and union	$[\![x^\omega y x^\omega = x^\omega]\!] \, \textcircled{M} \, \mathbf{V}$
Left deterministic product and union	$[\![x^\omega y = x^\omega]\!] \, \textcircled{M} \, \mathbf{V}$
Right deterministic product and union	$[\![y x^\omega = x^\omega]\!] \, \textcircled{M} \, \mathbf{V}$
Product, boolean operations	$\mathbf{A} \, \textcircled{M} \, \mathbf{V}$
Product with p-counters, boolean operations	$\mathbf{LG}_p \, \textcircled{M} \, \mathbf{V}$
Product with counters, boolean operations	$\mathbf{LGsol} \, \textcircled{M} \, \mathbf{V}$
Product, product with p-counters, boolean op.	$\mathbf{L\overline{G}}_p \, \textcircled{M} \, \mathbf{V}$
Product, product with counters, boolean op.	$\mathbf{L\overline{Gsol}} \, \textcircled{M} \, \mathbf{V}$
BPol $\mathcal{V}$	$\mathbf{B}_1 \, \textcircled{M} \, \mathbf{V}$ (Conjecture)

Other standard operations on languages have been studied. A $*$-variety is *closed under star* if, for every alphabet A, $L \in \mathcal{V}(A^*)$ implies $L^* \in \mathcal{V}(A^*)$.

Theorem 10.1. [94] *The only ∗-variety of languages closed under star is the variety of rational languages.*

Similarly, a +-variety is closed under the operation $L \to L^+$ if, for every alphabet A, $L \in \mathcal{V}(A^+)$ implies $L^+ \in \mathcal{V}(A^+)$.

Theorem 10.2. *The only +-varieties of languages closed under the operation $L \to L^+$ are the variety of languages associated with the trivial variety, the variety of finite semigroups and the varieties $[\![xy = x]\!]$ and $[\![yx = x]\!]$.*[12]

A ∗-variety is *closed under shuffle* if, for every alphabet A, $L_1, L_2 \in \mathcal{V}(A^*)$ implies $L_1 \amalg L_2 \in \mathcal{V}(A^*)$. The classification of the varieties of languages closed under shuffle was initiated in [96] and was recently completed by Esik and Simon [54]. If **H** is a variety of commutative groups, denote by **Com$_H$** the variety of commutative monoids whose groups belong to **H**.

Theorem 10.3. *The only ∗-varieties of languages closed under shuffle are the variety of rational languages and the varieties of languages associated with the varieties of the form **Com$_H$**, for some variety of commutative groups **H**.*

The definition of the operator $\mathbf{V} \to \mathbf{PV}$ was given in section 8.1. The next theorem shows that length-preserving morphisms or inverse substitutions form the corresponding operator on languages [134, 161, 96]. Recall that a *substitution* $\sigma : A^* \to B^*$ is a monoid morphism from A^* into $\mathcal{P}(B^*)$.

Theorem 10.4. *Let **V** be a variety of monoids and let $\mathcal{V}$ and $\mathcal{W}$ be the varieties of languages corresponding respectively to **V** and **PV**. Then for every alphabet A,*

(1) *$\mathcal{W}(A^*)$ is the boolean algebra generated by the languages of the form $\varphi(L)$, where $\varphi : B^* \to A^*$ is a length-preserving morphism and $L \in \mathcal{W}(B^*)$*
(2) *$\mathcal{W}(A^*)$ is the boolean algebra generated by the languages of the form $\sigma^{-1}(L)$, where $\sigma : A^* \to B^*$ is a substitution and $L \in \mathcal{W}(B^*)$.*

Theorem 10.4 motivated the systematic study of the varieties of the form **PV**. At present, this classification is not yet complete, although many results are known. Almeida's book [5] gives a complete overview of this topic along with the relevant references.

11. Further topics

The notion of syntactic semigroup and the correspondence between languages and semigroups has been generalized to other algebraic systems. For instance, Rhodes and Weil [139] partially extend the algebraic theory presented in this

---

[12] A language of A^+ is recognized by a semigroup of $[\![xy = x]\!]$ (resp. $[\![yx = x]\!]$) if and only if it is of the form BA^* (resp. A^*B) where $B \subseteq A$.

chapter to torsion semigroups and torsion languages. Reutenauer [136] introduces a notion of *syntactic algebra* for formal power series in non commutative variables and proves an analog of the variety theorem.

Büchi was the first to use finite automata to define sets of infinite words. Although this involves a non trivial generalization of the semigroup structure, a theory similar to the one for finite words can be developed. In particular, a notion of syntactic ω-semigroup can be defined and a variety theorem also holds in this case. See [13, 93, 198, 199] for more details.

There are several important topics that could not be covered in this chapter. The first one is the algebraic study of varieties. In particular, there is a lot of literature on the description of the free profinite semigroups of a given variety of finite semigroups, the computation of the identities for the join, the semidirect product or the Mal'cev product of two varieties. See the book of Almeida [5] for an overview and references. See also the survey articles [7, 62] and the thematic issue **39** of *Russian Mathematics (Izvestiya VUZ Matematika)*, (1995), devoted to pseudovarieties.

The second one is the important connections between varieties and formal logic. The main results [42, 87, 185] relate monadic second order to rational languages, first order to star-free languages and the Σ_n hierarchy of first order formulas to the concatenation hierarchies. There are also some results about the expressive power of the linear temporal logic [68, 56, 46]. See the chapter of W. Thomas in this Handbook or the survey articles [107, 109].

The third one is the connection with boolean circuits initiated by Barrington and Thérien. The recent book of Straubing [169] is an excellent introduction to this field. It contains also several results about the connections with formal logic.

The fourth one is the theory of recognizable and rational sets in arbitrary monoids. In particular, the following particular cases have been studied to some extent: product of free monoids (relations and transductions), free groups, free inverse monoids, commutative monoids, partially commutative monoids (traces). The survey paper [21] is an excellent reference.

The fifth one is the star-height problem for rational languages. We refer the reader to the survey articles [37, 110] for more details. The star-height of a rational expression counts the number of nested uses of the star operation. The *star-height of a language* is the minimum of the star-heights of the rational expressions representing the language. The *star-height problem* is to find an algorithm to compute the star-height. There is a similar problem, called the *extended star-height problem*, if extended rational expressions are allowed. Extended rational expressions allow complement in addition to the usual operations union, product and star.

It was shown by Dejean and Schützenberger [48] that there exists a language of star-height n for each $n \geq 0$. It is easy to see that the languages of star-height 0 are the finite languages, but the effective characterization of the other levels was left open for several years until Hashiguchi first settled the problem for star-height 1 [59] and a few years later for the general case [60].

The languages of extended star-height 0 are the star-free languages. Therefore, there are languages of extended star-height 1, such as $(aa)^*$ on the alphabet $\{a\}$, but, as surprising as it may seem, nobody has been able so far to prove the existence of a language of extended star-height greater than 1, although the general feeling is that such languages do exist. In the opposite direction, our knowledge of the languages proven to be of extended star-height ≤ 1 is rather poor (see [120, 141, 142] for recent advances on this topic).

Acknowledgements

I would like to thank Jorge Almeida, Mario Branco and Pascal Weil for many useful suggestions on a first version of this chapter.

References

1. D. Albert, R. Baldinger and J. Rhodes, The identity problem for finite semigroups, *J. Symbolic Logic* **57** 1992, 179–192.
2. J. Almeida, Equations for pseudovarieties, *Formal properties of finite automata and applications*, J.-E. Pin (ed.) *Lect. Notes in Comp. Sci.* **386**, Springer-Verlag, Berlin 1989, 148–164.
3. J. Almeida, Implicit operations on finite $\mathcal{J}$-trivial semigroups and a conjecture of I. Simon, *J. Pure Appl. Algebra* **69** 1990, 205–218.
4. J. Almeida, On pseudovarieties, varieties of languages, filters of congruences, pseudoidentities and related topics, *Algebra Universalis* **27** 1990, 333–350.
5. J. Almeida, *Finite semigroups and universal algebra*, Series in Algebra Vol 3, World Scientific, Singapore, 1994.
6. J. Almeida, *A syntactical proof of the locality of* **DA**, *International Journal of Algebra and Computation*, 6(2) 1996, 165–178.
7. J. Almeida and P. Weil, Relatively free profinite monoids: an introduction and examples, in NATO Advanced Study Institute *Semigroups, Formal Languages and Groups*, J. Fountain (ed.), Kluwer Academic Publishers, 1995, 73-117.
8. J. Almeida and P. Weil, Free profinite semigroups over semidirect products, with J. Almeida, *Izvestiya VUZ Matematika* **39** 1995, 3–31. English version, *Russian Mathem. (Iz. VUZ)* **39** 1995, 1–27.
9. J. Almeida and P. Weil, Reduced factorization in free profinite groups and join decompositions of pseudo-varieties, *International Journal of Algebra and Computation* **4** 1994, 375–403.
10. J. Almeida and P. Weil, Profinite categories and semidirect products, to appear.
11. M. Arfi, Polynomial operations and rational languages, *4th STACS, Lect. Notes in Comp. Sci.* **247**, Springer-Verlag, Berlin 1987, 198–206.
12. M. Arfi, Opérations polynomiales et hiérarchies de concaténation, *Theor. Comput. Sci.* **91** 1991, 71–84.
13. A. Arnold, A syntactic congruence for rational ω-languages, *Theor. Comput. Sci.* **39** 1985, 333–335.

14. C. J. Ash, Finite semigroups with commuting idempotents, in *J. Austral. Math. Soc. Ser. A* **43** 1987, 81–90.

15. C. J. Ash, Inevitable sequences and a proof of the type II conjecture, in *Proceedings of the Monash Conference on Semigroup Theory*, World Scientific, Singapore 1991, 31–42.

16. C. J. Ash, Inevitable Graphs: A proof of the type II conjecture and some related decision procedures, *Int. Jour. Alg. and Comp.* **1** 1991, 127–146.

17. C. J. Ash, T. E. Hall and J. E. Pin, On the varieties of languages associated to some varieties of finite monoids with commuting idempotents, *Information and Computation* **86** 1990, 32–42.

18. Y. Bar-Hillel and E. Shamir, Finite-State Languages, *Bull. Res. Council Israel.* **8F** 1960, 155–166.

19. J. Berstel, *Transductions and context-free languages*, Teubner, Stuttgart, 1979.

20. J. Berstel and D. Perrin, *Theory of codes*, Academic Press, London, 1985.

21. J. Berstel and J. Sakarovitch, Recent results in the theory of rational sets, *Proceedings of MFCS'86, Lect. Notes in Comp. Sci.* **233**, Springer-Verlag Berlin 1986, 15–28.

22. G. Birkhoff, On the structure of abstract algebras, *Proc. Cambridge Phil. Soc.* **31** 1935, 433–454.

23. P. Blanchard, Morphismes et comptages sur les langages rationnels, *J. Inform. Process. Cybernet.* **23** 1987, 3–11.

24. F. Blanchet-Sadri, Games, equations and the dot-depth hierarchy, *Comput. Math. Appl.* **18** 1989, 809–822.

25. F. Blanchet-Sadri, On dot-depth two, *Informatique Théorique et Applications* **24** 1990, 521–530.

26. F. Blanchet-Sadri, Games, equations and dot-depth two monoids, *Discrete Appl. Math.* **39** 1992, 99–111.

27. F. Blanchet-Sadri, The dot-depth of a generating class of aperiodic monoids is computable, *International Journal of Foundations of Computer Science* **3** 1992, 419–442.

28. F. Blanchet-Sadri, Equations and dot-depth one, *Semigroup Forum* **47** 1993, 305–317.

29. F. Blanchet-Sadri, On a complete set of generators for dot-depth two, *Discrete Appl. Math.* **50** 1994, 1–25.

30. F. Blanchet-Sadri, Equations and monoids varieties of dot-depth one and two, *Theor. Comput. Sci.* **123** 1994, 239–258.

31. F. Blanchet-Sadri, Equations on the semidirect product of a finite semilattice by a $\mathcal{J}$-trivial monoid of height k, *Informatique Théorique et Applications* **29** 1995, 157–170.

32. F. Blanchet-Sadri, Some logical characterizations of the dot-depth hierarchy and applications, *J. Comput. Syst. Sci.* **51** 1995, 324–337.

33. S. L. Bloom, Varieties of ordered algebras, *J. Comput. System Sci.* **13** 1976, 200–212.

34. M. J. J. Branco, On the Pin-Thérien expansion of idempotent monoids, *Semigroup Forum* **49** 1994, 329–334.

35. M. J. J. Branco, Some variants of the concatenation product and the kernel category, 1995, to appear.

36. J. A. Brzozowski, Hierarchies of aperiodic languages, *RAIRO Inform. Théor.* **10** 1976, 33–49.

37. J. A. Brzozowski, Open problems about regular languages, *Formal language theory, perspectives and open problems*, R. V. Book (ed.), Academic Press, New York 1980, 23–47.

38. J. A. Brzozowski, K. Čulik II and A. Gabrielian, Classification of noncounting events, *J. Comp. Syst. Sci.* **5** 1971, 41–53.
39. J. A. Brzozowski and F. E. Fich, Languages of $\mathcal{R}$-trivial monoids, *J. Comput. System Sci.* **20** 1980, 32–49.
40. J. A. Brzozowski and R. Knast, The dot-depth hierarchy of star-free languages is infinite, *J. Comput. System Sci.* **16** 1978, 37–55.
41. J. A. Brzozowski and I. Simon, Characterizations of locally testable languages, *Discrete Math.* **4** 1973, 243-271.
42. J. R. Büchi, Weak second-order arithmetic and finite automata, *Z. Math. Logik und Grundl. Math.* **6** 1960, 66–92.
43. S. Cho and D. T. Huynh, Finite automaton aperiodicity is PSPACE-complete, *Theor. Comput. Sci.* **88** 1991, 99–116.
44. R. S. Cohen and J. A. Brzozowski, On star-free events, *Proc. Hawaii Int. Conf. on System Science*, Honolulu, 1968, 1–4.
45. R. S. Cohen and J. A. Brzozowski, Dot-depth of star-free events, *J. Comput. System Sci.* **5** 1971, 1–15.
46. J. Cohen, D. Perrin and J.-E. Pin, On the expressive power of temporal logic, *J. Comput. System Sci.* **46** 1993, 271–294.
47. D. Cowan, Inverse monoids of dot-depth 2, *Int. Jour. Alg. and Comp.* **3** 1993, 411–424.
48. F. Dejean and M. P. Schützenberger, On a question of Eggan, *Information and Control* **9** 1966, 23–25.
49. A. De Luca, On some properties of the syntactic semigroup of very pure subsemigroups, *RAIRO Inf. Théor.* **14** 1980, 39–56.
50. A. De Luca, D. Perrin, A. Restivo and S. Termini, Synchronization and simplification, *Discrete Math.* **27** 1979, 297–308.
51. A. De Luca and A. Restivo, A characterization of strictly locally testable languages and its application to subsemigroups of a free semigroup, *Inf. and Control* **44** 1980, 300–319.
52. S. Eilenberg, *Automata, languages and machines*, Vol. A, Academic Press, New York, 1974.
53. S. Eilenberg, *Automata, languages and machines*, Vol. B, Academic Press, New York, 1976.
54. Z. Esik and I. Simon, Modeling literal morphisms by shuffle, *Semigroup Forum*, to appear.
55. F. E. Fich and J. A. Brzozowski, A characterization of a dot-depth two analogue of generalized definite languages, *Proc. 6th ICALP, Lect. Notes in Comp. Sci.* **71**, Springer-Verlag, Berlin 1979, 230–244.
56. D. Gabbay, A. Pnueli, S. Shelah and J. Stavi, On the temporal analysis of fairness, *Proc. 7th ACM Symp. Princ. Prog. Lang.* 1980, 163–173.
57. A. Ginzburg, About some properties of definite, reverse definite and related automata, *IEEE Trans. Electron. Comput.* **15** 1966, 806–810.
58. M. Hall Jr., A topology for free groups and related groups, *Ann. of Maths,* **52** 1950, 127–139.
59. K. Hashiguchi, Regular languages of star height one, *Information and Control* **53** 1982, 199–210.
60. K. Hashiguchi, Representation theorems on regular languages, *J. Comput. System Sci.* **27** 1983, 101–115.
61. K. Hashiguchi, Algorithms for Determining Relative Star Height and Star Height, *Information and Computation* **78** 1988, 124–169.
62. K. Henckell, S. W. Margolis, J.-E. Pin and J. Rhodes, Ash's Type II Theorem, Profinite Topology and Malcev Products, *Int. Jour. Alg. and Comp.* **1** 1991, 411–436.

63. K. Henckell and J. Rhodes, The theorem of Knast, the $PG = BG$ and Type II Conjectures, in *Monoids and Semigroups with Applications*, J. Rhodes (ed.), Word Scientific, Singapore 1991, 453–463.
64. P. Higgins, A proof of Simon's Theorem on piecewise testable languages, submitted.
65. J. Howie, *An Introduction to Semigroup Theory*, Academic Press, London, 1976.
66. J. Howie, *Automata and Languages*, Oxford Science Publications, Clarendon Press, Oxford, 1991.
67. P. Jones, Monoid varieties defined by $x^{n+1} = x$ are local, *Semigroup Forum* **47** 1993, 318–326.
68. J. A. Kamp, Tense logic and the theory of linear order, Ph. D. Thesis, Univ. of California, Los Angeles, 1968.
69. M. Keenan and G. Lallement, On certain codes admitting inverse semigroups as syntactic monoids, *Semigroup Forum* **8** 1974, 312–331.
70. S. C. Kleene, Representation of events in nerve nets and finite automata, in *Automata Studies*, C. E. Shannon and J. McCarthy (eds.), Princeton University Press, Princeton, New Jersey, 1956, 3–42.
71. R. Knast, A semigroup characterization of dot-depth one languages, *RAIRO Inform. Théor.* **17** 1983, 321–330.
72. R. Knast, Some theorems on graph congruences, *RAIRO Inform. Théor.* **17** 1983, 331–342.
73. G. Lallement, Regular semigroups with $\mathcal{D} = \mathcal{R}$ as syntactic monoids of finite prefix codes, *Theor. Comput. Sci.* **3** 1977, 35–49.
74. G. Lallement, Cyclotomic polynomials and unions of groups, *Discrete Math.* **24** 1978, 19–36.
75. G. Lallement, *Semigroups and combinatorial applications*, Wiley, New York, 1979.
76. G. Lallement and E. Milito, Recognizable languages and finite semilattices of groups, *Semigroup Forum* **11** 1975, 181–185.
77. E. Le Rest and M. Le Rest, Sur le calcul du monoïde syntactique d'un sous-monoïde finiment engendré, *Semigroup Forum* **21** 1980, 173–185.
78. M. Lothaire, *Combinatorics on Words*, Cambridge University Press, 1982.
79. S. W. Margolis, On the syntactic transformation semigroup of the language generated by a finite biprefix code, *Theor. Comput. Sci.*, **21** 1982, 225–230.
80. S. W. Margolis, Kernel and expansions: an historical and technical perspective, in *Monoids and semigroups with applications*, J. Rhodes (ed.), World Scientific, Singapore, 1989, 3–30.
81. S. W. Margolis, On maximal pseudo-varieties of finite monoids and semigroups, *Izvestiya VUZ Matematika* **39** 1995, 65–70. English version, *Russian Mathem. (Iz. VUZ)* **39** 1995, 60–64.
82. S. W. Margolis and J. E. Pin, Languages and inverse semigroups, *11th ICALP*, *Lect. Notes in Comp. Sci.* **172**, Springer-Verlag, Berlin 1984, 337–346.
83. S. W. Margolis and J. E. Pin, Product of group languages, *FCT Conference*, *Lect. Notes in Comp. Sci.* **199**, Springer-Verlag, Berlin 1985, 285–299.
84. S. W. Margolis and J. E. Pin, On varieties of rational languages and variable-length code II, *J. Pure Appl. Algebra* **41** 1986, 233–253.
85. S. W. Margolis and J. E. Pin, Inverse semigroups and varieties of finite semigroups, *Journal of Algebra* **110** 1987, 306–323.
86. R. McNaughton, Algebraic decision procedures for local testability, *Math. Syst. Theor.* **8** 1974, 60–76.
87. R. McNaughton and S. Pappert, *Counter-free Automata*, MIT Press, 1971.

88. A. R. Meyer, A note on star-free events, *J. Assoc. Comput. Mach.* **16** 1969, 220–225.

89. M. Perles, M. O. Rabin and E. Shamir, The theory of definite automata, *IEEE. Trans. Electron. Comput.* **12** 1963, 233–243.

90. D. Perrin, Sur certains semigroupes syntactiques, *Séminaires de l'IRIA, Logiques et Automates,* Paris 1971, 169–177.

91. D. Perrin and J.-F. Perrot, A propos des groupes dans certains monoïdes syntactiques, *Lecture Notes in Mathematics* **855**, Springer-Verlag Berlin 1980, 82–91.

92. D. Perrin, *Automata,* Chapter 1 in Handbook of Theoretical Computer Science, J. Van Leeuwen (ed.), Vol B: Formal Models and Semantics, Elsevier Amsterdam 1990.

93. D. Perrin and J.-E. Pin, Semigroups and automata on infinite words, in NATO Advanced Study Institute *Semigroups, Formal Languages and Groups,* J. Fountain (ed.), Kluwer Academic Publishers, 1995, 49–72.

94. J. F. Perrot, Variétés de langages et opérations, *Theor. Comput. Sci.* **7** 1978, 197–210.

95. J.-E. Pin, Sur le monoïde de L^* lorsque L est un langage fini. *Theor. Comput. Sci.* **7** 1978, 211–215.

96. J.-E. Pin, Variétés de langages et monoïde des parties, *Semigroup Forum* **20** 1980, 11–47.

97. J.-E. Pin, Propriétés syntactiques du produit non ambigu. *7th ICALP, Lect. Notes in Comp. Sci.* **85**, Springer-Verlag, Berlin 1980), 483–499.

98. J.-E. Pin, Variétés de langages et variétés de semigroupes, Thèse d'état, Université Paris VI 1981.

99. J.-E. Pin, Langages reconnaissables et codage préfixe pur, Langages reconnaissables et codage préfixe pur, in *8th ICALP, Lect. Notes in Comp. Sci.* **115**, Springer-Verlag, Berlin 1981, 184–192.

100. J.-E. Pin, Hiérarchies de concaténation, *RAIRO Inform. Théor.* **18** 1984, 23–46.

101. J.-E. Pin, Finite group topology and p-adic topology for free monoids, in *12th ICALP, Lect. Notes in Comp. Sci.* **194**, Springer-Verlag, Berlin 1985, 445–455.

102. J.-E. Pin, *Variétés de langages formels,* Masson, Paris, 1984. English translation: *Varieties of formal languages,* Plenum, New-York, 1986.

103. J.-E. Pin, Power semigroups and related varieties of finite semigroups, *Semigroups and Their Applications,* S. M. Goberstein and P. M. Higgins (ed.), D. Reidel 1986, 139–152.

104. J.-E. Pin, A property of the Schützenberger product, *Semigroup Forum* **35** 1987, 53–62.

105. J.-E. Pin, Relational morphisms, transductions and operations on languages, in *Formal properties of finite automata and applications,* J.-E. Pin (ed.), *Lect. Notes in Comp. Sci.* **386**, Springer-Verlag, Berlin 1989, 120–137.

106. J.-E. Pin, Topologies for the free monoid, *Journal of Algebra* **137** 1991, 297–337.

107. J.-E. Pin, Logic, semigroups and automata on words, *Theor. Comput. Sci., Annals of Mathematics and Artificial Intelligence,* **16**, 1996, 343–384.

108. J.-E. Pin, Polynomial closure of group languages and open sets of the Hall topology, *ICALP 1994, Lect. Notes in Comp. Sci.* **820**, Springer-Verlag, Berlin 1994, 424–435.

109. J.-E. Pin, Logic on words, *Bulletin of the European Association of Theoretical Computer Science* **54** 1994, 145–165.

110. J.-E. Pin, Finite semigroups and recognizable languages: an introduction, in NATO Advanced Study Institute *Semigroups, Formal Languages and Groups*, J. Fountain (ed.), Kluwer Academic Publishers, 1995, 1–32.

111. J.-E. Pin, **BG** = **PG**, a success story, in NATO Advanced Study Institute *Semigroups, Formal Languages and Groups*, J. Fountain (ed.), Kluwer Academic Publishers, 1995, 33–47.

112. J.-E. Pin, A variety theorem without complementation, *Izvestiya VUZ Matematika* **39** 1995, 80–90. English version, *Russian Mathem. (Iz. VUZ)* **39** 1995, 74–83.

113. J.-E. Pin, Polynomial closure of group languages and open sets of the Hall topology, extended version of [108], to appear in *Theor. Comput. Sci.*

114. J.-E. Pin and C. Reutenauer, A conjecture on the Hall topology for the free group, *Notices of the London Math. Society* **23** 1991, 356–362.

115. J.-E. Pin and J. Sakarovitch, Operations and transductions that preserve rationality, *6th GI Conf., Lect. Notes in Comp. Sci.* **145**, Springer-Verlag, Berlin 1983, 277–288.

116. J.-E. Pin and J. Sakarovitch, Une application de la représentation matricielle des transductions, *Theor. Comput. Sci.* **35** 1985, 271–293.

117. J.-E. Pin and H. Straubing, Remarques sur le dénombrement des variétés de monoïdes finis, *C. R. Acad. Sci. Paris, Sér. A* **292** 1981, 111–113.

118. J.-E. Pin and H. Straubing, Monoids of upper triangular matrices, *Colloquia Mathematica Societatis Janos Bolyai , Semigroups, Szeged*, **39** 1981, 259–272.

119. J.-E. Pin, H. Straubing and D. Thérien, Small varieties of finite semigroups and extensions, *J. Austral. Math. Soc.* **37** 1984, 269–281.

120. J.-E. Pin, H. Straubing and D. Thérien, Locally trivial categories and unambiguous concatenation, *J. Pure Appl. Algebra* **52** 1988, 297–311.

121. J.-E. Pin, H. Straubing and D. Thérien, New results on the generalized star height problem, *STACS 89, Lect. Notes in Comp. Sci.* **349** Springer-Verlag, Berlin 1989, 458–467.

122. J.-E. Pin, H. Straubing and D. Thérien, Some results on the generalized star height problem, *Information and Computation*, **101** 1992, 219–250.

123. J.-E. Pin and D. Thérien, The bideterministic concatenation product, *Int. Jour. Alg. and Comp.* **3** 1993, 535–555.

124. J.-E. Pin and P. Weil, Profinite semigroups, Malcev products and identities, *Journal of Algebra*, **182**, 1996, 604–626.

125. J.-E. Pin and P. Weil, A Reiterman theorem for pseudovarieties of finite first-order structures, *Algebra Universalis*, **35**, 1996, 577–595.

126. J.-E. Pin and P. Weil, Polynomial closure and unambiguous product, *ICALP 1995, Lect. Notes in Comp. Sci.* **944** Springer-Verlag, Berlin 1995, 348–359.

127. J.-E. Pin and P. Weil, Polynomial closure and unambiguous product (full version), submitted.

128. M. O. Rabin and D. Scott, Finite automata and their decision problems, *IBM J. Res. and Develop.* **3** 1959, 114–125. Reprinted in *Sequential Machines, E. F. Moore (ed.)*, Addison-Wesley, Reading, Massachussetts, 1964, 63–91.

129. J. Reiterman, The Birkhoff theorem for finite algebras, *Algebra Universalis* **14** 1982, 1–10.

130. A. Restivo, Codes and aperiodic languages, *Lect. Notes in Comp. Sci.* **2** Springer-Verlag, Berlin 1973, 175–181.

131. A. Restivo, On a question of McNaughton and Papert, *Inf. Control* **25** 1974, 93–101.

132. A. Restivo, A combinatorial property of codes having finite synchronization delay, *Theor. Comput. Sci.* **1** 1975, 95–101.

133. A. Restivo, Codes and automata, in *Formal properties of finite automata and applications*, J.-E. Pin (ed.), *Lect. Notes in Comp. Sci.* **386**, Springer-Verlag, Berlin 1989, 186–198.

134. C. Reutenauer, Sur les variétés de langages et de monoïdes, *Lect. Notes in Comp. Sci.* **67**, Springer-Verlag, Berlin 1979, 260–265.

135. C. Reutenauer, Une topologie du monoïde libre, *Semigroup Forum* **18** 1979, 33–49.

136. C. Reutenauer, Séries formelles et algèbres syntactiques, *J. Algebra* **66** 1980, 448–483.

137. C. Reutenauer, Sur mon article "Une topologie du monoïde libre", *Semigroup Forum* **22** 1981, 93–95.

138. J. Rhodes and B. Tilson, The kernel of monoid morphisms, *J. Pure Appl. Algebra* **62** 1989, 227–268.

139. J. Rhodes and P. Weil, Algebraic and topological theory of languages, *Informatique Théorique et Applications* **29** 1995, 1–44.

140. L. Ribes and P. A. Zalesskii, On the profinite topology on a free group, *Bull. London Math. Soc.* **25** 1993, 37–43.

141. M. Robson, Some Languages of Generalised Star Height One, *LITP Technical Report* **89-62** 1989, 9 pages.

142. M. Robson, More Languages of Generalised Star Height 1, *Theor. Comput. Sci.* **106** 1992 327–335.

143. J. J. Rotman, *The theory of groups, An introduction*, Allyn and Bacon, Boston, 1973.

144. M. P. Schützenberger, On finite monoids having only trivial subgroups, *Information and Control* **8** 1965, 190–194.

145. M. P. Schützenberger, Sur certaines variétés de monoïdes finis, in *Automata Theory*, E. R. Caianiello (ed.), Academic Press, New-York, 1966, 314–319.

146. M. P. Schützenberger, Sur les monoïdes finis dont les groupes sont commutatifs, *RAIRO Inf. Theor.* **1** 1974, 55–61.

147. M. P. Schützenberger, Sur certaines opérations de fermeture dans les langages rationnels, *Ist. Naz. Alta Math. Symp. Math.* **15** 1975, 245–253.

148. M. P. Schützenberger, Sur le produit de concaténation non ambigu, *Semigroup Forum* **13** 1976, 47–75.

149. C. Selmi, Over testable languages, *Theor. Comput. Sci.*, to appear.

150. C. Selmi, Strongly locally testable semigroups with commuting idempotents and associated languages, *Informatique Théorique et Applications*, to appear.

151. H. Sezinando, The varieties of languages corresponding to the varieties of finite band monoids, *Semigroup Forum* **44** 1992, 283–305.

152. I. Simon, Piecewise testable events, *Proc. 2nd GI Conf., Lect. Notes in Comp. Sci.* **33**, Springer-Verlag, Berlin 1975, 214–222.

153. I. Simon, Properties of factorization forests, in *Formal properties of finite automata and applications*, J.-E. Pin (ed.), *Lect. Notes in Comp. Sci.* **386**, Springer-Verlag, Berlin 1989, 34–55.

154. I. Simon, Factorization forests of finite height, *Theor. Comput. Sci.* **72** 1990, 65–94.

155. I. Simon, A short proof of the factorization forest theorem, in *Tree Automata and Languages*, M. Nivat and A. Podelski eds., Elsevier Science Publ., Amsterdam, 1992, 433–438.

156. I. Simon, The product of rational languages, *Proceedings of ICALP 1993, Lect. Notes in Comp. Sci.* **700**, Springer-Verlag, Berlin 1993, 430–444.

157. J. Stern, Characterization of some classes of regular events, *Theor. Comput. Sci.* **35** 1985, 17–42.

158. J. Stern, Complexity of some problems from the theory of automata, *Inform. and Control* **66** 1985, 63–176.

159. H. Straubing, Families of recognizable sets corresponding to certain varieties of finite monoids, *J. Pure Appl. Algebra*, **15** 1979, 305–318.

160. H. Straubing, Aperiodic homomorphisms and the concatenation product of recognizable sets, *J. Pure Appl. Algebra* **15** 1979, 319–327.

161. H. Straubing, Recognizable sets and power sets of finite semigroups, *Semigroup Forum*, **18** 1979, 331–340.

162. H. Straubing, On finite $\mathcal{J}$-trivial monoids, *Semigroup Forum* **19** 1980, 107–110.

163. H. Straubing, A generalization of the Schützenberger product of finite monoids, *Theor. Comput. Sci.* **13** 1981, 137–150.

164. H. Straubing, Relational morphisms and operations on recognizable sets, *RAIRO Inf. Theor.* **15** 1981, 149–159.

165. H. Straubing, The variety generated by finite nilpotent monoids, *Semigroup Forum* **24** 1982, 25–38.

166. H. Straubing, Finite semigroups varieties of the form $\mathbf{V} * \mathbf{D}$, *J. Pure Appl. Algebra* **36** 1985, 53–94.

167. H. Straubing, Semigroups and languages of dot-depth two, *Theor. Comput. Sci.* **58** 1988, 361–378.

168. H. Straubing, The wreath product and its application, in *Formal properties of finite automata and applications*, J.-E. Pin (ed.), *Lect. Notes in Comp. Sci.* **386**, Springer-Verlag, Berlin 1989, 15–24.

169. H. Straubing, *Finite automata, formal logic and circuit complexity*, Birkhäuser, Boston, 1994.

170. H. Straubing and D. Thérien, Partially ordered finite monoids and a theorem of I. Simon, *J. of Algebra* **119** 1985, 393–399.

171. H. Straubing, D. Thérien and W. Thomas, Regular languages defined with generalized quantifiers, in *Proc. 15th ICALP, Lect. Notes in Comp. Sci.* **317** 1988, 561–575.

172. H. Straubing and P. Weil, On a conjecture concerning dot-depth two languages, *Theoret. Comp. Sci.* **104** 1992, 161–183.

173. D. Thérien, Languages of nilpotent and solvable groups, *Proceedings of ICALP 1979, Lect. Notes in Comp. Sci.* **71** Springer-Verlag, Berlin 1979, 616–632.

174. D. Thérien, Classification of finite monoids: the language approach, *Theoret. Comp. Sci.* **14** 1981, 195–208.

175. D. Thérien, Recognizable languages and congruences, *Semigroup Forum* **23** 1981, 371–373.

176. D. Thérien, Languages recognized by unions of groups, Technical Report SOCS-82.13, McGill University, 1982.

177. D. Thérien, Subword counting and nilpotent groups, in *Combinatorics on Words: Progress and Perspectives*, L. J. Cummings (ed.), Academic Press, New York 1983, 297–306.

178. D. Thérien, Sur les monoïdes dont les groupes sont résolubles, *Semigroup Forum* **26** 1983, 89–101.

179. D. Thérien, A language theoretic interpretation of Schützenberger representations, *Semigroup Forum* **28** 1984, 235–248.

180. D. Thérien, Catégories et langages de dot-depth un, *RAIRO Inf. Théor. et Applications* **22** 1988, 437–445.

181. D. Thérien, On the equation $x^t = x^{t+r}$ in categories, *Semigroup Forum* **37** 1988, 265–271.

182. D. Thérien, Two-sided wreath products of categories, *J. Pure Appl. Algebra* **74** 1991, 307–315.

183. D. Thérien and A. Weiss, Graph congruences and wreath products, *J. Pure Appl. Algebra* **36** 1985, 205–215.
184. A. Weiss and D. Thérien, Varieties of finite categories, *Informatique Théorique et Applications* **20** 1986, 357–366.
185. W. Thomas, Classifying regular events in symbolic logic, *J. Comput. Syst. Sci.* **25** 1982, 360–375.
186. W. Thomas, An application of the Ehrenfeucht-Fraïssé game in formal language theory, *Bull. Soc. Math. de France* **16** 1984, 11–21.
187. W. Thomas, A concatenation game and the dot-depth hierarchy, in *Computation Theory and Logic*, E. Böger (ed.), *Lect. Notes in Comp. Sci.* **270**, Springer-Verlag, Berlin 1987, 415–426.
188. B. Tilson, Chapters 11 and 12 of [53].
189. B. Tilson, Categories as algebras, *J. Pure Appl. Algebra* **48** 1987, 83–198.
190. P. Weil, Groups in the syntactic monoid of a composed code, *J. Pure Appl. Algebra* **42** 1986, 297–319.
191. P. Weil, Inverse monoids of dot-depth two, *Theoret. Comp. Sci.* **66** 1989, 233–245.
192. P. Weil, Products of languages with counter, *Theoret. Comp. Sci.* **76** 1990, 251–260.
193. P. Weil, Implicit operations on pseudovarieties: an introduction, in *Monoids and semigroups with applications*, J. Rhodes (ed.), World Scientific, Singapore 1991, 89-104.
194. P. Weil, Closure of varieties of languages under products with counter, *J. Comput. System Sci.* **45** 1992, 316–339.
195. P. Weil, Some results on the dot-depth hierarchy, *Semigroup Forum* **46** 1993, 352–370.
196. A. Weiss, The local and global varieties induced by nilpotent monoids, *Informatique Théorique et Applications* **20** 1986, 339–355.
197. A. Weiss and D. Thérien, Varieties of finite categories, *Informatique Théorique et Applications* **20** 1986, 357–366.
198. T. Wilke, An Eilenberg theorem for ∞-languages, in *Automata, Languages and Programming*, *Lect. Notes in Comp. Sci.* **510**, Springer-Verlag, Berlin 1991, 588–599.
199. T. Wilke, An algebraic theory for regular languages of finite and infinite words, *Int. J. Alg. Comput.* **3** 1993, 447–489.

Regularity and Finiteness Conditions

Aldo de Luca and Stefano Varricchio

The aim of this chapter is to present some recent research work in the combinatorial aspects of the theory of semigroups which are of great interest both for Algebra and Theoretical Computer Science. This research mainly concerns that part of combinatorics of finite and infinite words over a finite alphabet, which is usually called the theory of "unavoidable" regularities. The book by Lothaire [77] can be considered an excellent introduction to this theory.

The unavoidable regularities of sufficiently large words over a finite alphabet are very important in the study of finiteness conditions for semigroups. This problem consists in considering conditions which are satisfied by a finite semigroup and are such as to assure that a semigroup satisfying them is finite. The most natural requirement is that the semigroup is *finitely generated*. If one supposes that the semigroup is also *periodic* the study of finiteness conditions for these semigroups (or groups) is called the *Burnside problem for semigroups* (or groups). There exists an important relationship with the theory of *finite automata* because, as is well known, *a language L over a finite alphabet is regular (i.e. recognizable by a finite automaton) if and only if its syntactic semigroup $S(L)$ is finite*. Hence, in principle, any finiteness condition for semigroups can be translated into a regularity condition for languages. The study of finiteness conditions for periodic languages (i.e. such that $S(L)$ is a periodic semigroup) has been called the *Burnside problem for languages*.

Several finiteness conditions for finitely generated semigroups have been given in recent years based on different concepts such as: *permutation properties, iteration conditions, minimal conditions on ideals, repetitive morphisms, etc.*

These conditions are analyzed in some detail in Section 2. They are based, as we said before, on the existence of some different unavoidable regularities on very large words over a finite alphabet. As we shall see the permutation conditions are related to the Shirsov theorem and the iteration conditions to bi-ideal factorizations. A very recent result shows that these two regularities "appear" simultaneously in a suitable way in very large words. This fact gives rise to a new finiteness condition in which the elements of the semigroups can be either permutable or iterable.

We present also finiteness conditions for semigroups based on *chain conditions*. In particular, we consider some remarkable generalizations of a theorem of Hotzel and of a theorem of Coudrain and Schützenberger. From these one derives an extension of the theorem of Green and Rees relating the bounded Burnside problem for semigroups with the Burnside problem for groups, and a new simple proof of the theorem of McNaughton and Zalcstein which gives a positive answer to the Burnside problem in the case of $n \times n$ matrices with elements in a field. The proof of these results requires also some structure theorems on semigroups based on the Green relations, as the $\mathcal{J}$-depth decomposition theorem, which are given in Section 2.

Section 3 concerns the following general problem: Given a semigroup S, under which conditions can we say that the finite parts of S are recognizable sets? This is also equivalent to the following problem: Let θ be a congruence in a finitely generated free semigroup A^+; when are the congruence classes of θ in A^+ *regular languages*? A semigroup whose finite parts are recognizable is called *finitely recognizable*. Some general results on this problem relating it to the $\mathcal{J}$-depth decomposition of S are shown. In particular we refer to the case when the semigroup S is the quotient semigroup $M_n = A^*/\theta_n$, where θ_n is the congruence generated by the relation $x^n = x^{n+1}$. The problem of the regularity of the congruence classes (non-counting classes) was posed for any $n > 0$ by Brzozowsky about 25 years ago. The authors have proved that this problem has a positive answer for $n > 4$. In the proof the finiteness condition for semigroups due to Hotzel and introduced in Section 2 has been used. The proof of this result allows one to show also that the *word problem* for the semigroup M_n when $n > 4$ is recursively solvable. This result was subsequently improved by other authors for the case $n > 2$ and extended to more general cases.

Section 4 deals with the Burnside problem for languages. From the finiteness conditions for semigroups one can easily find some *uniform* conditions which assure the regularity of a periodic language. However, the Burnside problem for languages is more complicated since the regularity conditions can be presented in a *non-uniform* way, i.e. they, in general, depend on the contexts which complete the words in the language. The use of Ramsey's theorem is often a good tool to transform non-uniform conditions in uniform ones. Some important regularity conditions such as the *block-pumping* property of Ehrenfeucht, Parikh and Rozenberg and the *permutative property* of Restivo and Reutenauer are proved. Moreover, the existence of a non-uniform and *positive* block pumping property is shown.

In Section 5 we present some further combinatorial aspects of semigroups related to the notion of *well quasi-order* which gives a new insight into the combinatorics of the free monoids. Some regularity conditions based on well quasi-orders extend classical theorems of Automata Theory, such as the Myhill theorem. For instance one has that *a language is regular if and only if it is a closed part of a monotone well quasi-order*. Some applications of these notions and techniques for the regularity conditions are given.

In conclusion, this chapter, which outlines in its structure a wide monograph in preparation on the subject, aims to be a presentation of a very recent research work on those combinatorial aspects of the theory of semigroups, as unavoidable regularities, which are intimately related to the fundamental property of finite automata, namely finiteness. This relation with automata gives rise to regularity conditions for formal languages. Of course there exist many other regularity conditions, based on different techniques and concepts, which are not covered by this chapter (see, for instance, [56],[10],[44],[23]).

1. Combinatorics on words

This section deals with some properties, known as unavoidable regularities, which are always satisfied by sufficiently long words over a finite alphabet. The study of these regularities, as we shall see in the next sections, is of great interest in combinatorics on words both for the importance of the subject itself and for the applications in many areas of algebra and theoretical computer science.

In the following A will denote a finite *alphabet*, i.e. a finite nonempty set whose elements are called *letters*. By A^+ we denote the set of all finite sequences of letters, or finite *words*. A finite word, or simply word, w can be uniquely represented by a juxtaposition of its letters:

$$w = a_1 \ldots a_n,$$

with $a_i \in A$, $1 \leq i \leq n$. The *length* $|w|$ of w is n, i.e. the number of letters occurring in w. The set A^+ of all the words over A is the *free semigroup* on A, where the semigroup operation, called *product*, is defined by concatenation or juxtaposition of the words. If one adds to A^+ the *identity* element ϵ, called *empty word*, one obtains the *free monoid* A^* over A. The length of ϵ is taken to be equal to 0. For each $n \geq 0$ we denote by A^n the set of all the words of length n. Moreover, $A^{[n]}$ will be the set of all the words of length $\leq n$.

A word u is a *factor*, or *subword*, of w if there exist words $p, q \in A^*$ such that $w = puq$. If p (resp. q) is equal to ϵ, then u is called *prefix* (resp. *suffix*) of w. For any pair (i,j) of integers such that $1 \leq i \leq j \leq n$ we denote by $w[i,j]$ the factor $w[i,j] = a_i \ldots a_j$. A *language* L over the alphabet A is any subset of A^*. In the following $\mathbb{Z}$ will denote the set of the integers and $\mathbb{N}$ (resp. $\mathbb{N}_+$) the set of non-negative (resp. positive) integers. For notations and definitions not given in this chapter the reader is referred to [75] for what concerns semigroups and to [45] for automata and languages.

1.1 Infinite words and unavoidable regularities

A *two-sided infinite* (or *bi-infinite*) word w over the alphabet A is any map

$$w : \mathbb{Z} \to A.$$

For each $n \in \mathbb{Z}$, we set $w_n = w(n)$ and denote w also as:

$$w = \ldots w_{-3}w_{-2}w_{-1}w_0w_1w_2w_3 \ldots .$$

A word $u \in A^+$ is a finite *factor* of w if there exist integers $i, j \in \mathbb{Z}$, $i \leq j$, such that $u = w_i \ldots w_j$; the sequence $w[i,j] = w_i \ldots w_j$ is also called an *occurrence* of u in w. The set of all two-sided infinite words over A will be denoted by $A^{\pm\omega}$.

A one-sided (from left to right) infinite word over A is any map

$$w : \mathbb{N}_+ \to A.$$

For each $n > 0$ the factor $w[1,n] = w_1 \ldots w_n$ of length n is called the *prefix* of w of length n and will be simply denoted by $w[n]$. The set of all infinite words $w : \mathbb{N}_+ \to A$ will be denoted by A^ω. For any finite or infinite word w, $F(w)$ will denote the set of all its finite factors and $alph(w)$ the set of all letters of the alphabet A occurring in w. For any language L one denotes by $F(L)$ the set of the factors of all the words of L. A language L is *closed by factors* if $L = F(L)$.

If $w \in A^{\pm\omega}$ one can associate to it the one-sided infinite word $w_+ \in A^\omega$ defined for all $n > 0$ as $w_+(n) = w(n)$; trivially, one has that $F(w_+) \subseteq F(w)$.

The following lemma is a slight generalization of the famous König's lemma for infinite trees. It can be proved by a standard technique based essentially on the "Pigeon-Hole principle":

Lemma 1.1. *Let $L \subseteq A^*$ be an infinite language. Then there exists an infinite word $s \in A^{\pm\omega}$ such that $F(s) \subseteq F(L)$.*

Proof. For any word $w \in A^+$ we set $r_w = [|w|/2]$. Since L is an infinite language then from the "pigeon-hole" principle there will exist infinitely many words w of L having the same letter, say a_0 at the position r_w. Let us denote by L_1 this infinite subset. By using the same principle there will exist an infinite subset L_2 of L_1 whose words w are such that:

$$w[r_w, r_w + 1] = a_0a_1,$$

for a suitable letter $a_1 \in A$. By using again the same argument there will exist a letter a_{-1} and an infinite subset L_3 of L_2 such that all the words of L_3 satisfy the condition

$$w[r_w - 1, r_w + 1] = a_{-1}a_0a_1.$$

Continuing in this way one can construct the two-sided infinite word

$$s = \ldots a_{-2}a_{-1}a_0a_1a_2 \ldots,$$

which is such that any of its factors is a factor of infinitely many words of L.

Now we introduce the notion of *unavoidable regularity*. Informally a property P is unavoidable if it is not possible to construct arbitrarily long words not satisfying P.

Definition 1.1. *Let P be a property defined in the free monoid A^*. P is called an unavoidable regularity if the set $L = \{x \in A^* \mid x$ does not satisfy $P\}$ is finite. We say that P is an ideal property if $P(x)$ implies $P(lxm)$ for any $l, m \in A^*$. P is called avoidable if it is not unavoidable, i.e. there exist infinitely many words which do not satisfy P.*

Example 1.1. We consider the property P in A^* defined as follows: x satisfies P if and only if $x = luum$, with $l, m \in A^*, u \in A^+$. P is trivially an ideal property. If A has only two letters then it is easy to see that any word whose length is greater than or equal to four contains a square, so in this case P is unavoidable. If A has three letters then the property P is avoidable. Indeed, as proved by A. Thue [102], there exist infinitely many square-free words in a three letter alphabet (cf.[77]).

Proposition 1.1. *Let P be an ideal property. The following statements are equivalent:*

i. *There exists $w \in A^{\pm\omega}$ such that no factor of w satisfies P.*
ii. *P is avoidable.*

Proof. Clearly i. implies ii.. Conversely, if P is avoidable then $L = \{x \in A^* \mid x$ does not satisfy $P\}$ is infinite. By Lemma 1.1 there exists an infinite word $w \in A^{\pm\omega}$ such that $F(w) \subseteq F(L)$. Since P is an ideal property, it follows that L is closed by factors. Then one has $L = F(L)$ and $F(w) \subseteq L$.

1.2 The Ramsey theorem

One of the most important results in combinatorics is the well known Ramsey theorem. There are several formulations and proofs of this theorem and many questions and open problems are related to it (cf. [49]). In our context we are only interested in its different applications in combinatorics on words. Let E be a finite set and $P_m(E)$ the set of all subsets of E having cardinality m. In the sequel we will use the following version of the Ramsey theorem (cf. [49]):

Theorem 1.1. *Let m, n, k be positive integers, with $m \leq n$. There exists an integer $R(m, n, k)$ such that for any finite set E with $card(E) \geq R(m, n, k)$ and for any partition θ of $P_m(E)$ in k classes there exists $F \subseteq E$ such that $card(F) = n$ and $P_m(F)$ is contained a single class module θ.*

Definition 1.2. *Let A^* be a free monoid, S a set and k an integer greater than 1. A map $\phi : A^* \to S$ is called k-repetitive (resp. k-ramseyan) if there exists a positive integer L, depending on ϕ and k, such that any word w, with $|w| \geq L$, may be written as*

$$w = uw_1w_2\ldots w_kv,$$

with $w_i \in A^+$, $1 \le i \le k$, $u,v \in A^*$ and

$$\phi(w_1) = \phi(w_2) = \ldots = \phi(w_k)$$

(resp. $\phi(w_i\ldots w_j) = \phi(w_{i'}\ldots w_{j'})$, $1 \le i \le j \le k$, $1 \le i' \le j' \le k$). Moreover, if $|w_1| = |w_2| = \ldots = |w_k|$, then ϕ is called k-uniformly repetitive. One says that ϕ is repetitive (resp. uniformly repetitive, resp. ramseyan) if ϕ is k-repetitive (resp. k-uniformly repetitive, resp. k-ramseyan) for all $k > 1$.

Trivially any ramseyan map is repetitive, but the converse is not true in general. Moreover it is easy to show that any map from A^* in a finite set is repetitive; however, the following stronger result holds:

Theorem 1.2. *Let A^* be a free monoid and S a finite set. Then any map $\phi : A^* \to S$ is ramseyan.*

Proof. Let k be a positive integer. Let $n = card(S)$ and $S = \{1, 2, \ldots, n\}$. We set $L = R(2, k+1, n)$, where R is the Ramsey function. Let $w \in A^*$, $|w| = L$ and $E = \{1, \ldots, L+1\}$. The set $P_2(E)$ can be partitioned in the n classes $\theta_1, \theta_2, \ldots, \theta_n$, where for any $l \in \{1, \ldots, n\}$, θ_l is defined by

$$\{i, j\} \in \theta_l \Leftrightarrow \phi(w[i, j-1]) = l,$$

for $1 \le i < j \le L+1$. By the Ramsey theorem, there exists a subset $Y = \{i_1, i_2, \ldots, i_{k+1}\}$ of E with $i_1 < i_2 < \ldots < i_{k+1}$, and an integer $s \in \{1, 2, \ldots, n\}$ such that $P_2(Y)$ is contained in θ_s. Then for any $i, j, i', j' \in \{i_1, i_2, \ldots, i_{k+1}\}$, with $i < j$ and $i' < j'$, one has $\phi(w[i, j-1]) = \phi(w[i', j'-1]) = s$. Thus if we set

$$w_j = w[i_j, i_{j+1} - 1],$$

the word $w_1w_2\ldots w_k$ is a factor of w, such that $\phi(w_i\ldots w_j) = \phi(w_{i'}\ldots w_{j'})$, for $1 \le i \le j \le k$, $1 \le i' \le j' \le k$.

Definition 1.3. *Let $(w_1, \ldots, w_m)$ be a sequence of words. We say that $(u_1, \ldots, u_n)$ is a derived sequence of $(w_1, \ldots, w_m)$ if there exist $n+1$ integers $j_1, j_2, \ldots, j_{n+1}$ such that $1 \le j_1 < j_2 < \ldots < j_{n+1} \le m+1$, and*

$$u_1 = w_{j_1}\ldots w_{j_2-1}, \ldots, u_n = w_{j_n}\ldots w_{j_{n+1}-1}.$$

From the preceding definition one has that the word $w = w_1\ldots w_m$ can be rewritten as $w = xu_1\ldots u_ny$ with $x = w_1\ldots w_{j_1-1}$ and $y = w_{j_{n+1}}\ldots w_m$. The following lemma, which will be useful in the sequel, is another application of the Ramsey theorem.

Lemma 1.2. *Let $<$ be a total ordering on the free monoid A^*. For any integer $n > 1$ there exists a positive integer $r(n)$ with the property that if $(w_1, \ldots, w_h)$ is a sequence of h words, with $h \geq r(n)$, then there exists a derived sequence $(u_1, \ldots, u_n)$ of $(w_1, \ldots, w_h)$ such that either*

i. $\forall j \in [1, n-1]$, $u_j u_{j+1} < u_{j+1} u_j$, or
ii. $\forall j \in [1, n-1]$, $u_j u_{j+1} > u_{j+1} u_j$, or
iii. $\forall j \in [1, n-1]$, $u_j u_{j+1} = u_{j+1} u_j$.

1.3 The van der Waerden theorem

The following theorem was proved by B. L. van der Waerden in 1927.

Theorem 1.3. *For all positive integers k, n there exists an integer $W(k, n) > 0$ such that if the set $\{1, \ldots, W(k, n)\}$ is partitioned in k classes, then there exists a class containing an arithmetic progression of length n.*

In other words the theorem shows the existence of the following "unavoidable regularity": all sufficiently long sequences of elements belonging to a finite set contain arbitrarily long subsequences of identical elements in arithmetic progressions. There exist several and conceptually different proofs of van der Waerden's theorem (cf. [49],[77]). An historical account in which van der Waerden describes the circumstances of the theorem's discovery is in [104].

The van der Waerden theorem can be reformulated in terms of words as follows: Let $w = a_1 a_2 \ldots a_n \in A^*$; a *cadence* of w is a sequence $T = \{t_1, t_2, \ldots, t_r\}$ of positive integers such that $0 < t_1 < t_2 \ldots < t_r \leq n$ and $a_{t_1} = a_{t_2} \ldots = a_{t_r}$. The number r is called the *order* of the cadence. The cadence T is called *arithmetic* if there exists a positive integer q, called *rate*, such that, for any $i \in \{1, \ldots, r-1\}$, one has $t_{i+1} = t_i + q$.

Proposition 1.2. *Let A be an alphabet with k letters. For any positive integer n, there exists an integer $W(k, n) > 0$, such that any word $w \in A^*$ of length at least $W(k,n)$ contains an arithmetic cadence of order n.*

The evaluation of the van der Waerden function W, as well as the Ramsey function, has proved to be extremely difficult. The only non trivial exact values of $W(k, n)$ are: $W(2, 3) = 9$, $W(2, 4) = 35$, $W(3, 3) = 27$, $W(4, 3) = 76$, $W(2, 5) = 178$. Let us fix $k = 2$ and consider the function $W(n) = W(2, n)$. There are upper bounds to $W(n)$ which are not "reasonable", in the sense that they are expressed by functions which are not even primitive recursive [49].

Now we consider two interesting corollaries (cf.[77]) of van der Waerden's theorem. Let $\mathbb{N}_+$ be the additive semigroup of positive integers.

Theorem 1.4. *Let A be a finite alphabet. Then any morphism $\phi : A^+ \to \mathbb{N}_+$ is repetitive.*

Proof. Let us consider a new alphabet $B = \{b_1, b_2, \ldots b_m\}$, where $m = max\{|\phi(a)| \ a \in A\}$, and a morphism $\psi : A^+ \to B^+$ defined for all $a \in A$ as $\psi(a) = b_l b_{l-1} \ldots b_1$ where $\phi(a) = l$. Let $k \geq 1$ and $w \in A^+$; by van der Waerden's theorem there exists an integer n such that if $w \in A^+$ and $|\psi(w)| \geq n$, then $\psi(w)$ contains an arithmetic cadence $\{t_1, t_2, \ldots, t_{k+1}\}$ of rate q. Moreover, let b_p be the letter of B that occurs in such a cadence.

For any $i \in [1, k+1]$ let us denote by u_i the shortest prefix of w such that $|\psi(u_i)| \geq t_i$. Clearly one has that $|\psi(u_i)| = \phi(u_i) = t_i + p - 1$. We can then factorize w as $u_1 w_1 w_2 \ldots w_k$, where w_i is the word such that $u_{i+1} = u_i w_i$. Then one has

$$\phi(w_i) = \phi(u_{i+1}) - \phi(u_i) = t_{i+1} + p - 1 - (t_i + p - 1) = q,$$

which implies $\phi(w_1) = \phi(w_2) = \ldots = \phi(w_k)$.

By using the van der Waerden theorem one can prove the following remarkable result:

Theorem 1.5. *Let S be a finite semigroup. Then any morphism $\phi : A^+ \to S$ is uniformly repetitive.*

1.4 The Shirshov theorem

Recently many authors have rediscovered a combinatorial theorem of A. I. Shirshov [96],[97], which has, as we shall see in the next section, interesting applications for the Burnside problem and related questions. Originally this theorem was used for proving some properties of the algebras with polynomial identities (cf.[77], [94]).

Let A be a totally ordered alphabet. One can extend this order to A^* by the *lexicographic order* $<_{lex}$, or simply $<$, defined as: for all $u, v \in A^*$

$$u < v \iff v \in uA^* \text{ or } u = hx\xi, \ v = hy\eta, \ h, \xi, \eta \in A^*, \ x, y \in A \text{ and } x < y.$$

For any $n > 0$, S_n denotes the group (symmetric group) of all the permutations on the set $\{1, \ldots, n\}$.

Definition 1.4. *A sequence $(u_1, u_2, \ldots, u_n)$ of n words of A^+ is called an n-division (inverse n-division) of the word $u = u_1 u_2 \ldots u_n$ if for any nontrivial permutation σ of S_n one has $u_1 u_2 \ldots u_n > u_{\sigma(1)} u_{\sigma(2)} \cdots u_{\sigma(n)}$ (resp. $u_1 u_2 \ldots u_n < u_{\sigma(1)} u_{\sigma(2)} \cdots u_{\sigma(n)}$). A word u is called n-divided (inversely n-divided) if u admits a factorization $u = u_1 u_2 \ldots u_n$, with $u_i \in A^+$ $(i = 1, \ldots, n)$ where $(u_1, u_2, \ldots, u_n)$ is an n-division (resp. inverse n-division) of u.*

Example 1.2. Let $A = \{a, b\}$ be a two-letter alphabet ordered by setting $a < b$. The word $w = ababbaba$ is 3-divided by the sequence $(ababb, ab, a)$ and inversely 3-divided by $(a, ba, bbaba)$. Moreover, one can easily verify that $(ababb, ab, a)$ is the only 3-division of w and that w does not admits 4-divisions.

Definition 1.5. *A word $w \in A^*$ is an n-power, $n > 0$, if $w = u^n$ with $u \in A^*$. A word $w \in A^+$ is n-power-free $(n > 1)$ if $F(w)$ does not contain an n-power of a nonempty word.*

Theorem 1.6. *Let A be a totally ordered alphabet with k elements, and p, n be integers ≥ 1. There exists a positive integer $N(k, p, n)$ such that any word $w \in A^*$ of length at least $N(k, p, n)$, contains as a factor an n-divided word or a p-power of a non-empty word.*

Two proofs, which are different from the original one, will be given later as a consequence of much stronger results (cf. Corollary 1.1, Theorem 1.8). We premise some definitions.

Two words $f, g \in A^*$ are *conjugate* if there exist words $u, v \in A^*$ such that $f = uv$, $g = vu$. The conjugation relation is an equivalence relation in A^*. A word w is called *primitive* if the conjugation class of w contains $|w|$ elements. One can easily prove that this is also equivalent to the statement that $w \neq u^h$ for any $h > 1$ and $u \neq \epsilon$. A word $w \in A^+$ is called a *Lyndon word* if it is primitive and minimal with respect to the lexicographic ordering in the class of its conjugates (cf. [77]).

Definition 1.6. *Let $s \in A^\omega$ be an infinite word. We say that s is ultimately periodic if $s = uv^\omega$, where $u \in A^*$, $v \in A^+$, and v^ω is the infinite word $v^\omega = vvvv \ldots v \ldots$. We say that s is ω-divided if s can be factorized as $s = s_1 \ldots s_n \ldots$, with $s_i \in A^+$, $i > 0$, and for any $k > 0$, $(s_1, \ldots, s_k)$ is a k-division.*

The following theorem generalizes a result of C. Reutenauer (cf. [93]). For the proof see [106].

Theorem 1.7. *For any $t \in A^\omega$ there exists $s \in A^\omega$ such that*

i. $F(s) \subseteq F(t)$,
ii. $s = l_1 \ldots l_n \ldots$ where for any $i \geq 1$, l_i is a Lyndon word and $l_i \geq l_{i+1}$.

Corollary 1.1. *For any $t \in A^\omega$ there exists $s \in A^\omega$ such that*

i. $F(s) \subseteq F(t)$,
ii. s is ultimately periodic or s is ω-divided.

Proof. By Theorem 1.7 there exists $s \in A^\omega$ such that $F(s) \subseteq F(t)$ and $s = l_1 l_2 \ldots l_n \ldots$ where l_i is a Lyndon word and $l_i \geq l_{i+1}$ for any $i \geq 1$. Now either
    (a) there exists $k > 0$ such that $l_j = l_k$ for any $j \geq k$, in such a case s is ultimately periodic,
or
    (b) one can write $s = l_1'^{p_1} l_2'^{p_2} \ldots l_n'^{p_n} \ldots$, where l_i' is a Lyndon word and $l_i' > l_{i+1}'$ for any $i \geq 1$. As proven in [93], for any $n > 0$ the sequence $(l_1'^{p_1}, l_2'^{p_2}, \ldots, l_n'^{p_n})$ is an n-division so that, by definition, s is ω-divided.

Proof of Theorem 1.6. We suppose by contradiction that for some integers $k, n, p \geq 1$, there are arbitrarily large words in A^*, with $card(A) = k$, containing, as a factor, neither a p-power of a nonempty word nor an n-divided word. We denote by L the infinite language of these words. By Lemma 1.1 there exists an infinite word $m \in A^\omega$ such that $F(m) \subseteq F(L)$. By Corollary 1.1 there exists $s \in A^\omega$ such that $F(s) \subseteq F(m) \subseteq F(L)$ and s is ultimately periodic or ω-divided. If s is ω-divided, then $F(s)$ contains an n-divided word. If s is ultimately periodic, then $F(s)$ contains a p-power of a non-empty word. In any case we have a contradiction, since no factor of any word of L is of this kind. $\blacksquare$

Corollary 1.2. *Let A be a totally ordered alphabet with k elements, and p, n be integers ≥ 1, with $p \geq 2n$. There exists a positive integer $N(k, p, n)$ such that any word in A^* of length at least $N(k, p, n)$, contains as a factor an n-divided word or a p-power of a non-empty word v with $|v| < n$.*

Proof. We take the integer $N = N(k, p, n)$ which satisfies Theorem 1.6. Let w be a word of length at least N and suppose that w does not contain as a factor an n-divided word. By Theorem 1.6, w contains a factor of the kind $u = v^p$, with $v \in A^+$. We may assume that v is primitive. The statement is proved if we show that $|v| < n$. Suppose by contradiction that $|v| \geq n$. Since v is primitive, there are n distinct conjugates of v. Let $v_1 > v_2 > \ldots > v_n$ be n distinct conjugates of v. It is evident that for any $i \in \{1, 2, \ldots, n\}$ v_i is a factor of v^2. Since $p \geq 2n$, u contains as a factor the word $(vv)^n$ and so u contains as a factor a word of the kind

$$f = (v_1 z_1)(v_2 z_2) \ldots (v_n z_n).$$

Let $\sigma \in \mathcal{S}_n$ be a non-trivial permutation, and $i > 0$ the first integer such that $i \neq \sigma(i)$. Since $\sigma(i) > i$ one has $v_i > v_{\sigma(i)}$ so that: $(v_{\sigma(1)} z_{\sigma(1)}) \ldots (v_{\sigma(n)} z_{\sigma(n)})$
$= (v_1 z_1) \ldots (v_{i-1} z_{i-1})(v_{\sigma(i)} z_{\sigma(i)}) \ldots (v_{\sigma(n)} z_{\sigma(n)}) < (v_1 z_1) \ldots (v_n z_n)$.

Thus f is n-divided and this contradicts the fact that $F(w)$ does not contain any n-divided word. $\blacksquare$

1.5 Uniformly recurrent words

In this subsection we introduce the notion of uniformly recurrent words. These words are very important, since, as we shall see later, they present many interesting unavoidable regularities.

Definition 1.7. *An infinite word $w \in A^{\pm\omega}$ (resp. $w \in A^\omega$) is recurrent if any factor $u \in F(w)$ will occur infinitely often in w, i.e. there exist infinitely many pairs of integers (i, j) with $i \leq j$ and such that $u = w[i, j]$. An infinite word $w \in A^{\pm\omega}$ (resp. $w \in A^\omega$) is uniformly recurrent if there exists a map $k : F(w) \rightarrow \mathbb{N}$, such that: for any $u, v \in F(w)$, with $|v| \geq k(u)$, one has $u \in F(v)$.*

It is clear that if $w \in A^{\pm\omega}$ is uniformly recurrent, then $w_+ \in A^\omega$ will be so. For any uniformly recurrent word w we denote by K_w, or simply by K, the map $K : \mathbb{N} \to \mathbb{N}$, defined for all $n \in \mathbb{N}$ as:

$$K(n) = max\{k(w) \mid w \in F(w) \cap A^n\}.$$

If $v \in F(w)$ and $|v| > K(n)$, then

$$F(w) \cap A^n \subseteq F(v).$$

The function K will be called the *recurrence function* of w.

For any infinite or bi-infinite word t, we define for all $u \in F(t)$ the quantity

$$k(t, u) = Sup\{|w| \mid w \in F(t) \text{ and } u \notin F(w)\}.$$

One has, of course, that t is uniformly recurrent if and only if $k(t, u) < \infty$ for each $u \in F(t)$.

The next lemma shows that for any infinite language L there exists an infinite uniformly recurrent word whose factors are factors of the words of L. This result can be derived by arguments of symbolic-dynamics [47]. We report here a proof based on a simple combinatorial argument [66].

Lemma 1.3. *Let $L \subseteq A^*$ be an infinite language. There exists an infinite word $x \in A^{\pm\omega}$ such that:*

i. x is uniformly recurrent,
ii. $F(x) \subseteq F(L)$.

Proof. In view of Lemma 1.1 it is sufficient to prove that if $s \in A^\omega$ then there exists a uniformly recurrent word $x \in A^{\pm\omega}$ such that $F(x) \subseteq F(s)$. Let $s \in A^\omega$ be an infinite word and let $w_1, w_2, \ldots, w_n, \ldots$ be an arbitrary enumeration of the factors of s. We define an infinite sequence $\{t_n\}_{n\geq 0}$ of infinite words as follows: $t_0 = s$; for every $i > 0$ we consider the set $E_i = \{v \in F(t_{i-1}) \mid w_i \notin F(v)\}$. One has, of course that E_i is a finite set if and only if $k(t_{i-1}, w_i) < \infty$. We then pose $t_i = t_{i-1}$ if E_i is a finite set. If E_i is infinite by Lemma 1.1 there exists an infinite word, that we take as t_i, such that $F(t_i) \subseteq E_i$. Let us observe that in the latter case $w_i \notin F(t_i)$. Moreover in any case $F(t_i) \subseteq F(t_{i-1})$. Hence if $w_r \in F(t_j)$ and $j \geq r$ then one derives that $w_r \in F(t_r)$, $t_r = t_{r-1}$, and $k(t_{r-1}, w_r) < \infty$. Further, since $F(t_j) \subseteq F(t_r)$ it follows that $k(t_j, w_r) \leq k(t_{r-1}, w_r) < \infty$. Let us now choose in each t_i, $i \geq 0$, a factor u_i of length $|u_i| = i$. Let us denote by U the set $U = \{u_i\}_{i\geq 0}$. By using again Lemma 1.1 one has that there exists an infinite word $x \in A^{\pm\omega}$ such that any factor of x is a factor of infinitely many words of U. This implies, of course, $F(x) \subseteq F(s)$. Suppose now that $w \in F(x)$. One has $w = w_r$ for a suitable $r > 0$. Since there are infinitely many $j \geq r$ such that $w = w_r \in F(t_j) \subseteq F(t_r)$, it follows that $k(t_j, w) \leq k(t_{r-1}, w) < \infty$. Thus $k(x, w) < \infty$ since by construction $F(x) \subseteq F(t_j)$. Hence x is uniformly recurrent.

Many classical results on unavoidable regularities in words, are an easy consequence of the previous one.

Corollary 1.3. *Let J be a two-sided ideal of A^*. If for any uniformly recurrent $w \in A^\omega$, $F(w) \cap J \neq \emptyset$ then there exists an $n > 0$ such that $A^n A^* \subseteq J$.*

Proof. Suppose that there exist infinitely many words which belong to the set $C = A^* \backslash J$. Since C is closed by factors then by Lemma 1.3 there exists $t \in A^{\pm\omega}$ such that $F(t) \subseteq C$. By Lemma 1.3 there exists a uniformly recurrent word $w \in A^{\pm\omega}$ such that $F(w) \subseteq F(t) \subseteq C$. Hence $F(w_+) \subseteq C$ which is a contradiction since $w_+ \in A^\omega$ is uniformly recurrent.

The following result due to T. C. Brown [12] is reminiscent of van der Waerden's theorem. The proof is a trivial consequence of Lemma 1.3.

Corollary 1.4. *(Brown's Lemma) Let $t \in A^\omega$ be an infinite word on the alphabet A. Then there exists an infinite word $x \in A^\omega$ such that $F(x) \subseteq F(t)$ and for any $a \in alph(x)$ there exists $k_a > 0$ with the property that any factor of x of length at least k_a contains the letter a.*

Definition 1.8. *A finite sequence $f_1, \ldots, f_n$ of words of A^* is called a bi-ideal sequence of order n if $f_1 \in A^+$, and for any $i \in [1, n-1]$ one has: $f_{i+1} = f_i g_i f_i$, with $g_i \in A^*$. If for all $i \in [1, n-1]$, $g_i \neq \epsilon$, then the bi-ideal sequence is called a strict bi-ideal sequence. An infinite sequence $\{f_n\}_{n>0}$ of words is a bi-ideal (infinite) sequence if for all $i > 0$, $f_{i+1} \in f_i A^* f_i$.*

Bi-ideal sequences have been considered, with different names, by several authors in algebra and combinatorics (cf., for instance, [17],[61],[4],[108]). The term bi-ideal sequence is due to M. Coudrain and M. P. Schützenberger [17]. The following corollary shows that bi-ideal sequences are regularities which occur in sufficiently large words.

Corollary 1.5. *(Coudrain and Schützenberger) Let A be a finite alphabet with k elements. For any $n > 0$, there exists a positive integer $D(k, n)$, such that any word of A^* of length at least $D(k, n)$ contains as a factor the n-th term of a bi-ideal sequence of order n.*

Proof. Suppose by contradiction that for some integers k, n such an integer $D(k, n)$ does not exists. Then there exist an alphabet A with k letters and an infinite word $m \in A^\omega$ such that no factor of m is the n-th term of a bi-ideal sequence of order n (cf. Proposition 1.1). By Lemma 1.3 there exists a uniformly recurrent word $t \in A^\omega$ such that $F(t) \subseteq F(m)$. Now, by induction on n, we prove that t contains, as a factor, the n-th term of a bi-ideal sequence of order n. For $n = 1$ the statement is trivial. Let $n > 1$ and suppose that $f_{n-1} \in F(t)$ is the $(n-1)$-th term of a bi-ideal sequence. Since m is recurrent, there exists $g \in A^*$ such that $f_n = f_{n-1} g f_{n-1} \in F(t) \subseteq F(m)$. Thus m contains the factor f_n which is the n-th term of a bi-ideal sequence and this is a contradiction.

1.6 Some extensions of the Shirshov theorem

In this subsection we prove the existence of some unavoidable regularities which appear in uniformly recurrent words. These regularities include, as particular cases, the bi-ideal sequences of Coudrain and Schützenberger, and the n-divided words considered by Shirshov. To this end we need some preliminary definitions and notations. We introduce now some canonical factorizations associated with a bi-ideal sequence. Let $\{f_i\}_{i=1,\ldots,n}$ be a bi-ideal sequence of order n. We set $w_n = f_1$ and

$$w_{n-i} = f_i g_i, \ i \in [1, n-1].$$

From Definition 1.8 for all $i \in [1, n-1]$, $f_{i+1} = f_i g_i f_i = w_{n-i} f_i$, so that, by iteration, one has for all $i \in [0, n-1]$:

$$f_{i+1} = w_{n-i} \ldots w_n. \tag{1.1}$$

Moreover, since $w_i = f_{n-i} g_{n-i}$, $i \in [1, n-1]$, from Eq.(1.1) one has

$$w_i = w_{i+1} \ldots w_n g_{n-i} \in w_{i+1} \ldots w_n A^*. \tag{1.2}$$

From Eq.(1.1), $f_n = w_1 w_2 \ldots w_n$. The n-tuple $(w_1, w_2, \ldots, w_n)$ is called the *canonical factorization* of f_n. One can also introduce the inverse canonical factorization by setting $w_1' = f_1$ and

$$w_{i+1}' = g_i f_i, \ i \in [1, n-1].$$

One easily derives that for all $i \in [1, n]$

$$w_1' \ldots w_i' = f_i, \tag{1.3}$$

where for any $i \in [1, n-1]$

$$w_{i+1}' = g_i w_1' \ldots w_i' \in A^* w_1' \ldots w_i'. \tag{1.4}$$

From Eq.(1.3) $f_n = w_1' \ldots w_n'$. The n-tuple $(w_1', \ldots, w_n')$ is called the *inverse canonical factorization* of f_n.

Conversely, one easily verifies that if $(w_1, w_2, \ldots, w_n)$ is an n-tuple of words such that $w_n \neq \epsilon$ (resp. $w_1 \neq \epsilon$) and for any $i \in [1, n-1]$, $w_i \in w_{i+1} \ldots w_n A^*$ (resp. $w_{i+1} \in A^* w_1 \ldots w_i$) then the sequence of words $f_i = w_{n-i+1} \ldots w_n$ (resp. $f_i = w_1 \ldots w_i$), $1 \leq i \leq n$, is a bi-ideal sequence of order n whose canonical (resp. inverse canonical) factorization is exactly $(w_1, w_2, \ldots, w_n)$.

From now on we shall suppose that the alphabet A, even though not explicitly stated, is totally ordered. By $<$ we denote as usual the lexicographic ordering induced in A^*. The following proposition was proved in [31]:

Proposition 1.3. *Let $(w_1, \ldots, w_n)$ and $(w'_1, \ldots, w'_n)$ be the canonical factorizations of the n-th term of a bi-ideal sequence. The sequence $(w_1, \ldots, w_n)$ is an n-division (resp. inverse n-division) if and only if for all $i \in [1, n-1]$, $w_{i+1} w_i < w_i w_{i+1}$ (resp. $w_{i+1} w_i > w_i w_{i+1}$). Moreover, $(w_1, \ldots, w_n)$ is an n-division (resp. inverse n-division) if and only if $(w'_1, \ldots, w'_n)$ is an inverse n-division (resp. n-division).*

Let $(w_1, \ldots, w_m)$ be the canonical factorization of a bi-ideal sequence of order m and $(u_1, \ldots, u_n)$ be a derived sequence of $(w_1, \ldots, w_m)$. As one easily verifies, for any $i \in [1, n-1]$, $u_i \in u_{i+1} \ldots u_n A^*$ so that the derived sequence $(u_1, \ldots, u_n)$ is the canonical factorization of a bi-ideal sequence of order n. In the same way one can verify that a derived sequence $(u_1, \ldots, u_n)$ of the inverse canonical factorization $(w'_1, \ldots, w'_m)$ of a bi-ideal sequence of order m is the inverse canonical factorization of a bi-ideal sequence of order n.

Proposition 1.4. *For any integer $n > 1$ there exists an integer $r(n)$ with the property that if $(w_1, \ldots, w_h)$ is a bi-ideal sequence of order $h \geq r(n)$, then there exists a derived sequence $(u_1, \ldots, u_n)$ of $(w_1, \ldots, w_h)$ such that either*

i. $(u_1, \ldots, u_n)$ is an n-division or
ii. $(u_1, \ldots, u_n)$ is an inverse n-division or
iii. $\forall j \in [1, n-1], \quad u_j u_{j+1} = u_{j+1} u_j.$

Proof. By Lemma 1.2 there exists an integer $r(n)$ with the property that if $(w_1, \ldots, w_h)$ is the canonical factorization of a bi-ideal sequence of order h, $h \geq r(n)$, then there exists a derived sequence $(u_1, \ldots, u_n)$ of $(w_1, \ldots, w_h)$ such that either

i. $\forall j \in [1, n-1], \; u_j u_{j+1} < u_{j+1} u_j$, or
ii. $\forall j \in [1, n-1], \; u_j u_{j+1} > u_{j+1} u_j$, or
iii. $\forall j \in [1, n-1], \; u_j u_{j+1} = u_{j+1} u_j$.

Now if $u_j u_{j+1} > u_{j+1} u_j$ (resp. $u_j u_{j+1} < u_{j+1} u_j$) for all $j \in [1, n-1]$, by Proposition 1.3, the derived sequence $(u_1, \ldots, u_n)$ is an n-division (resp. inverse n-division), otherwise for any $j \in [1, n-1]$, $u_j u_{j+1} = u_{j+1} u_j$.

Definition 1.9. *An infinite word $w \in A^{\pm\omega}$ (resp. $w \in A^\omega$) is called ω-power-free if for any $u \in F(w)$, $u \neq \epsilon$, there exists an integer $p_w(u)$, or simply $p(u)$, such that $u^{p(u)} \notin F(w)$. We shall take as $p(u)$ the minimal integer for which the above condition is satisfied. Moreover for any $n > 0$ we set $P_w(n) = max\{p(u) \mid u \in F(w) \text{ and } 0 < |u| \leq n\}$. If $p(u) = p$ for any nonempty word $u \in F(w)$, then w is called p-power-free. The map $P_w : \mathbb{N}_+ \to \mathbb{N}$, that we simply denote by P, will be called the power-free function of w.*

Example 1.3. Let us recall the following important examples of p-power-free infinite words. Let $A = \{a, b\}$ be a two-symbol alphabet and f the infinite word, called the *Fibonacci word*,

$$f = abaababaabaababaababa\ldots;$$

f is defined as the limit (cf. [77]) of the infinite sequence $\{f_n\}_{n\geq 0}$ of words, where $f_0 = a, f_1 = ab$ and $f_{n+1} = f_n f_{n-1}$, for all $n > 0$; f can be introduced also as the limit sequence obtained by iterating, starting from the letter a, the morphism $\phi : A^* \to A^*$, defined as: $\phi(a) = ab$, $\phi(b) = a$. One has, in fact, $\phi(a) = ab$, $\phi^{(2)}(a) = aba$, $\phi^{(3)}(a) = abaab$, etc. and $\phi^{(n+1)}(a) = \phi^{(n)}(a)\phi^{(n-1)}(a)$, for all $n > 1$. The word f has squares and cubes. However, it is possible to show [71] that it is 4-power-free. Another very famous infinite word is the Thue-Morse word t on two symbols

$$t = abbabaabbaababba\ldots;$$

t can be introduced by iterating, starting from the letter a, the morphism $\tau : A^* \to A^*$, defined as $\tau(a) = ab$, $\tau(b) = ba$. A classical result of Thue (cf.[77]) has shown that t is overlap-free and then cube-free. Let $B = A \cup \{c\}$. The Thue-Morse word m on three symbols

$$m = abcacbabcbacabcacb\ldots$$

can be introduced by iterating, starting from the letter a, the morphism $\mu : B^* \to B^*$, defined as: $\mu(a) = abc$, $\mu(b) = ac$, $\mu(c) = b$. It is well known that m is square-free. One can also prove that the words f, t and m are uniformly recurrent.

Let $x \in A^\omega$ be a uniformly recurrent and ω-power free word and K and P be the recurrence and the power-free functions of x. Let us define the function $q : \mathbb{N} \to \mathbb{N}$ recursively as follows: $q(0) = 0$, $q(1) = 1$ and for any $m > 1$,

$$q(m) = 2\max\{q(m-1)P(q(m-1)), K(q(m-1))\} + q(m-1). \qquad (1.5)$$

The following lemma, whose proof we omit, holds [31]:

Lemma 1.4. *Let $x \in A^\omega$ be a uniformly recurrent and ω-power free word and $q : \mathbb{N} \to \mathbb{N}$ the map defined by Eq.(1.5). For every pair of integers m and i such that $i > q(m)$, $x[i]$ has as a suffix the m-th term of a bi-ideal sequence $\{f_i\}_{i=1,\ldots,m}$, where $f_1 = x_i$ and $|f_i| \leq q(i)$, for $i = 1,\ldots,m$. Moreover if $(w_1,\ldots,w_m)$ is the canonical factorization of f_m, then for any i,j,k such that $1 \leq i < j < k \leq m$, one has: $w_i \ldots w_{j-1} w_j \ldots w_{k-1} \neq w_j \ldots w_{k-1} w_i \ldots w_{j-1}$.*

Combining Proposition 1.4 and Lemma 1.4 one obtains the following:

Corollary 1.6. *Let $x \in A^\omega$ be a uniformly recurrent and ω-power free word. There exists a map $h : \mathbb{N} \to \mathbb{N}$ such that for every pair of integers n and i such that $i > h(n)$, $x[i]$ has as a suffix the m-th term of a bi-ideal sequence $\{f_i\}_{i=1,\ldots,m}$ of order m, $m \geq n$, such that its canonical factorization has a derived sequence which is either n-divided or inversely n-divided.*

Proof. Let us define $h(n) = q(r(n))$, where $r : \mathbb{N} \to \mathbb{N}$, and $q : \mathbb{N} \to \mathbb{N}$ are the functions of Proposition 1.4 and Lemma 1.4, respectively. Let $n \geq 1$ and $m = r(n)$. By Lemma 1.4, if $i \geq q(r(n))$, then $x[i]$ has a suffix f_m which is the m-term of a bi-ideal sequence of order m. Moreover if $(w_1, \ldots, w_m)$ is the canonical factorization of f_m, then for any i, j, k such that $1 \leq i < j < k \leq m$, one has:

$$w_i \ldots w_{j-1} w_j \ldots w_{k-1} \neq w_j \ldots w_{k-1} w_i \ldots w_{j-1}. \tag{1.6}$$

Since $m = r(n)$, by Proposition 1.4, there exists a derived sequence $(u_1, \ldots, u_n)$ of $(w_1, \ldots, w_m)$ which is either an n-division or an inverse n-division. In fact, by Eq. (1.6), condition *iii*) of Proposition 1.4 cannot be satisfied.

Lemma 1.5. *Let x be a uniformly recurrent and ω-power free one-sided infinite word over a finite totally ordered alphabet A. For any $n > 1$, x has a factor which is the n-th term of a strict bi-ideal sequence of order n whose canonical factorization is n-divided.*

Proof. We shall prove the statement of the lemma in the case of uniformly recurrent two-sided infinite words. Since by Lemma 1.3 for any uniformly recurrent one-sided infinite word x there exists a uniformly recurrent two-sided infinite word w such that $F(w) \subseteq F(x)$, then also x will satisfy the statement of the lemma. Let $a = min(alph(w))$, i.e. the letter occurring in w which is minimal with respect to the total order in A. We denote by L_a and M_a the sets defined as:

$$L_a = F(w) \cap (A\backslash\{a\})^*, \quad M_a = F(w) \cap \{a^h \mid 0 < h < p_w(a)\}.$$

L_a is the set of all factors of w in which the letter a does not occur; it is finite because of the uniform recurrence of w. The set M_a is also finite because w is ω-power-free.

Let $X = a^+(A\backslash\{a\})^+$ and consider the set $Y = F(w) \cap X$; Y is finite since $Y \subseteq M_a L_a$. Let B be a finite alphabet having the same cardinality of Y and consider a bijection $\delta : B \to Y$, which can be extended to a morphism of B^* in Y^*. The alphabet B can be totally ordered by $<_B$ as follows: if $b_1, b_2 \in B$ then

$$b_1 <_B b_2 \Leftrightarrow \delta(b_1) <_A \delta(b_2). \tag{1.7}$$

This order can be extended to the lexicographic order of B^*. Moreover one can easily prove that for all $u, v \in B^*$

$$u <_B v \Rightarrow \delta(u) <_A \delta(v).$$

The set Y is a code (i.e. the basis of a free submonoid of A^*) having a synchronization delay equal to 1, i. e. for any pair $(y_1, y_2) \in Y \times Y$ and $\alpha, \beta \in A^*$ if $\alpha y_1 y_2 \beta \in Y^*$ then $\alpha y_1, y_2 \beta \in Y^*$ (cf. [6]). Thus since Y is a code, δ is an isomorphism of B^* in Y^*. Moreover since $F(w) \subseteq F(Y^*)$ one has, from the synchronization property, that any factor of w can, uniquely, be "parsed" in terms of the elements of Y. One says also that w can, uniquely, be factorized by the elements of Y. Thus there exists a word $\mu \in B^{\pm\omega}$ on the alphabet B such that $w = \delta(\mu)$. The word μ is ω-power-free. Indeed, let $u \in F(\mu)$. One has that $\delta(u) \in F(w)$ so that $(\delta(u))^{p_w(d(u))} = \delta(u^{p_w(d(u))}) \notin F(w)$. This implies that $u^{p_w(d(u))} \notin F(\mu)$. The word μ is uniformly recurrent. In fact let $u \in F(\mu)$; one has that $\delta(u) \in F(w)$; moreover there exists a letter $x \in A\backslash\{a\}$ such that $x\delta(u)a \in F(w)$. Hence the factor $x\delta(u)a$ is recurrent in the unique factorization of w in terms of the elements of Y. This implies that u will be recurrent in μ. Moreover, μ is uniformly recurrent; in fact one easily derives that for any n an upper bound to $K_\mu(n)$ is given, up to an additive constant, by $K_w(nl_M)$ where $l_M = max\{|y| \mid y \in Y\}$.

The proof of the lemma is by induction on the integer n. We begin by proving the basis of the induction, i.e. the case $n = 2$. Let us consider in w any factor $u = ya$, with $y \in Y$. We can write u as $u = a^h va$, with $h > 0$ and $v \in L_a$. We can set $w_2 = a$ and $w_1 = a^h v$. One has that $w_2 w_1 = aa^h v <_A w_1 w_2 = a^h va$. Hence (w_1, w_2) is the canonical factorization of the 2-term $w_1 w_2$ of a strict bi-ideal sequence of order 2. Let us now suppose that we have proved the result up to $n - 1$ for any uniformly recurrent and ω-power-free two-sided infinite word. Thus, by induction, the word μ has a factor

$$u = u_1 \ldots u_{n-1}, \ u_i \in B^+, \ (i = 1, \ldots, n-1)$$

such that for any non-trivial permutation $\sigma \in \mathcal{S}_{n-1}$,

$$u_{\sigma(1)} u_{\sigma(2)} \cdots u_{\sigma(n-1)} <_B u_1 u_2 \ldots u_{n-1} \tag{1.8}$$

and, moreover, $u_i \in u_{i+1} \ldots u_{n-1} B^+$ for all $i \in [1, n-2]$. Let us set $\delta(u_i) = w_i$, $i \in [1, n-2]$. By Eq.s (1.7) and (1.8) for any non-trivial permutation $\sigma \in \mathcal{S}_{n-1}$, one has:

$$w_{\sigma(1)} w_{\sigma(2)} \cdots w_{\sigma(n-1)} <_A w_1 w_2 \ldots w_{n-1}$$

and, by the fact that $\delta(B^+) \subseteq aA^*$,

$$w_i \in w_{i+1} \ldots w_{n-1} a A^*, \ \text{for all} \ i \in [1, n-2]. \tag{1.9}$$

Let us then set $w_n = a$. One has that $w_1 \ldots w_{n-1} a \in F(w)$. Moreover,

$$w_n w_{n-1} = a w_{n-1} <_A w_{n-1} a = w_{n-1} w_n.$$

By Eq.(1.9) $(w_1, \ldots, w_n)$ is the canonical factorization of the n-th term of a bi-ideal sequence. Since for any $i \in [1, n-1]$, $w_{i+1} w_i <_A w_i w_{i+1}$ then by Proposition 1.3, $(w_1, \ldots, w_n)$ is an n-division. Moreover the bi-ideal sequence is strict due to the fact that it can never occur that $w_i = w_{i+1} \ldots w_{n-1} a$, for all $i \in [1, n-2]$.

A different and more constructive proof of the previous lemma is in [35]. As a consequence of Lemma 1.5 we can give the following improvement of Shirshov's theorem [31].

Theorem 1.8. *For all k, p, n positive integers there exists an integer $N(k, p, n) > 0$ such that for any totally ordered alphabet A of cardinality k any word $w \in A^*$ whose length is at least $N(k, p, n)$ is such that*

i. there exists $u \neq \epsilon$ such that $u^p \in F(w)$ or

ii. there exists $s \in F(w)$ which is the n-th term of a bi-ideal sequence of order n whose canonical factorization $(w_1, \ldots, w_n)$ is an n-division of s.

Proof. Let A be a totally ordered alphabet of cardinality k. The set of all words of A^* which satisfy either *i.* or *ii.* is a two-sided ideal $J_{k,n,p}$, or simply J, of A^*. Let x now be any uniformly recurrent word on A. Now either x has a factor which is a p-power and in this case $F(x) \cap J \neq \emptyset$, or x is p-power free. In this latter case from Lemma 1.5 it follows again that $F(x) \cap J \neq \emptyset$. Hence by Corollary 1.3, the result follows.

2. Finiteness conditions for semigroups

The study of finiteness conditions for semigroups consists in giving some conditions which are satisfied by finite semigroups and which are such as to assure the finiteness of them. In this study one of the properties which is generally required of a semigroup is that of being *finitely generated*.

These conditions are very important both in algebra and automata theory. Indeed, if one supposes that the semigroup is also *periodic* (or *torsion*), the study of these finiteness conditions for semigroups (and groups) is called the (general) *Burnside problem for semigroups* (and *groups*) (cf.[75]). Several finiteness conditions for finitely generated semigroups have been given in recent years based on different concepts such as: *permutation property, chain conditions on ideals, strong periodicity, iteration properties, strongly and uniformly repetitive morphisms* etc. (cf.[99],[22],[90],[17],[58],[29],[54],[30],[31], [84]).

The relationship with automata is based on the fact that *a language L on a finite alphabet is recognizable if and only if the syntactic semigroup $S(L)$ of L is finite*. Hence, in principle, any finiteness condition for finitely generated semigroups can be translated into a regularity condition for languages. The study of regularity conditions for periodic languages (i.e. languages whose syntactic semigroup is periodic) has been called the *Burnside problem for languages* (cf.[91]).

2.1 The Burnside problem

A semigroup (group) S is *finitely generated* if there exists a finite subset X of S such that the subsemigroup (subgroup) $< X >$ generated by X is S. S

is called *locally finite* if every finitely generated subsemigroup (subgroup) of S is finite.

A semigroup (resp. group) S is periodic (or torsion) if any element $s \in S$ generates a subsemigroup $< s >$ (resp. subgroup) of a finite order. This is also equivalent to the statement that for any $s \in S$ there exist integers i and j, $i < j$, depending on the element s, such that

$$s^i = s^j. \tag{2.1}$$

Let j be the minimal integer for which the above relation is satisfied. The integer i, which is unique, is called the *index* and $p = j - i$ the *period* of s. The order of $< s >$ is then given by $i + p - 1$. In the case of a group, due to cancellativity, condition (2.1) simply becomes $s^p = 1$, where 1 denotes the identity of the group and the period p depends on the element s.

The problem of whether a finitely generated and periodic group is finite was posed by W. Burnside in 1902 and, subsequently, extended to the case of semigroups. However, the condition of a finite generation and the periodicity are not sufficient to assure the finiteness of a semigroup or a group.

In the case of semigroups this has been shown in 1944 by M. Morse and G. Hedlund [83] as a consequence of the fact that on an alphabet with at least three distinct symbols there exist arbitrarily long words without repetitions of consecutive blocks (square-free words). Indeed these authors, starting with a problem related to "chess games" (that is whether there could exist unending games), rediscovered a result shown by Thue in 1906 (cf.[102],[103]). By means of this result they were able to give an answer to a problem of semigroup theory posed by Dilworth showing the existence of a semigroup S generated by three elements such that $s^2 = 0$ for any element s of S, without S being nilpotent (a semigroup S with a 0 element is nilpotent if there exists a positive integer k such that $S^k = 0$). This result gave, obviously, a negative answer to the Burnside problem for semigroups.

In the case of groups a negative answer to the Burnside problem was given by E. S. Golod in 1964 [48]. This author by means of a technique of proof discovered with I. R. Shafarevich based on the non-finiteness of a dimension of a suitable algebra associated with a field (cf.[70]), was able to show the existence of an infinite 3-generated p-group. Other examples of finitely generated and torsion groups which are infinite have been more recently given by S. V. Aliochine and A. I. Olchanski (cf.[70]), R. I. Grigorchuk [51] and N. Gupta and S. Sidki [53].

If a finitely generated semigroup S is such that for any s of S, $s^m = s^n$, where m and n are fixed exponents such that $0 \le m < n$, then the problem of finiteness of S is called the *bounded Burnside problem*. Also in this case there exist semigroups and groups which are infinite. A particular case which is very interesting and surprising is that of finitely generated and idempotent semigroups (i.e. such that for any s of S, $s = s^2$). In this case one can show that S is finite. This result is a consequence of a theorem of J. A. Green and D.

Rees (1952) (cf.[50]) which relates the bounded Burnside problems for groups and for semigroups. More precisely let us denote by $S(r,n)$ (resp. $G(r,n)$) the free semigroup (resp. the free group) in the variety of all semigroups (resp. groups) having r generators, satisfying the identity $x = x^{n+1}$ (resp. $x^n = 1$). One has:

Theorem 2.1. *(Green and Rees). The following two conditions are equivalent:*

i. $S(r,n)$ is finite for each positive integer r.
ii. $G(r,n)$ is finite for each positive integer r.

A more general version of this theorem will be given later (cf. Corollary 2.1) as a consequence of some finiteness conditions for semigroups based on chain conditions.

In the case $n = 1$ the group $G(r,1)$ is trivial and thus the idempotent semigroup $S(r,1)$ is finite. We recall that it has been shown that $G(r,n)$ is finite for every positive integer r, for $n = 2, 3, 4$ and 6 (cf. [70]). The problem is still open for $n = 5$. Moreover, as a consequence of some works of P. S. Novikov and S. I. Adjan (1968) and Adjan (cf.[1]) it is known that if n is an *odd* integer ≥ 665, then $G(r,n)$ is infinite for any $r > 1$. Recently I. G. Lysionok (cf.[2]) proved that $G(r,n)$ is infinite for any $r > 1$ and n *odd* integer ≥ 115. Moreover, S. V. Ivanov [59] proved that $G(r,n)$ is infinite for any $r > 1$ and n integer $\geq 2^{48}$.

The Burnside problem has a positive answer for some classes of semigroups and groups. Besides the classical results of I. Schur (1911) (cf. [95]) and I. Kaplansky (1965) (cf. [69]) relative to finitely generated torsion subgroups of matrices over the field of complex numbers and over an arbitrary field, respectively, we recall the following results [81],[101]:

Theorem 2.2. *(McNaughton and Zalcstein). A torsion semigroup of $n \times n$ matrices over a field is locally finite.*

In the next section we shall give a new proof of the preceding theorem based on a suitable finiteness condition for semigroups which generalizes a theorem of Coudrain and Schützenberger.

Theorem 2.3. *(Straubing). A torsion subsemigroup of a ring which satisfies a polynomial identity is locally finite.*

The proof is based on Shirshov's theorem utilized in the case of rings satisfying a polynomial identity. A similar result in the case of groups was proved by C. Procesi [88].

Let A be a totally ordered alphabet. We can totally order A^+ by the relation $<_a$, called *alphabetic order*, defined as: for all $u, v \in A^+$

$$u <_a v \iff |u| < |v| \text{ or if } |u| = |v| \text{ then } u <_{lex} v,$$

where $<_{lex}$ is the lexicographic order. From the definition it follows that $<_a$ is a well-order. In the following we shall identify a finitely generated semigroup S with $A^+/\phi\phi^{-1}$, where A is a finite alphabet and $\phi : A^+ \to S$ is a surjective morphism. Let $s \in S$. In the set $\phi^{-1}(s)$ there is a unique minimal element with respect to $<_a$ usually called the *canonical representative* of s. If $s, t \in S$ we say that s is a *factor* of t if $t \in S^1 s S^1$. If $t \in s S^1$ (resp. $t \in S^1 s$) then s is called *left factor* (resp. *right factor*) of t. For any $s \in S$ we denote by $F(s)$ the set of the factors of s. For any subset X of S, $F(X) = \cup_{s \in X} F(s)$. One says that X is *factorial* or *closed by factors* if $F(X) = X$. We recall some technical lemmas on canonical representatives [31].

Lemma 2.1. *Let S be a finitely generated semigroup and T any subset of S closed by factors. Then the set C_T of the canonical representatives of T is closed by factors.*

Lemma 2.2. *Let S be a finitely generated semigroup and $\phi : A^+ \to S$ the canonical epimorphism. If H is an infinite subset of S closed by factors, then there exists a bi-ideal sequence $\{f_n\}_{n>0}$ such that*

i. $\phi(f_n) \in H$, for all $n > 0$,
ii. for all i, j, $i \neq j$, $\phi(f_i) \neq \phi(f_j)$.

2.2 Permutation property

A finitely generated, torsion and commutative semigroup is obviously finite. A. Restivo and C. Reutenauer introduced in 1984 (cf.[90]) a property of semigroups, called *permutation property*, which generalizes commutativity and is such that a finitely generated and torsion semigroup is finite if and only if it is permutable. Let us give the following:

Definition 2.1. *Let S be a semigroup and n an integer > 1. A sequence $s_1, \ldots, s_n$ of n elements of S is called permutable if the product $s_1 \ldots s_n$ remains invariant under some non-trivial permutation of its factors, i.e. there exists a permutation $\sigma \in S_n$, different from the identity, such that $s_1 s_2 \ldots s_n = s_{\sigma(1)} s_{\sigma(2)} \cdots s_{\sigma(n)}$.*

We say that the semigroup S is *n-permutable*, or that it satisfies the property $\mathcal{P}_n$, if any sequence of n elements of S is permutable. The property $\mathcal{P}_2$ is obviously equivalent to commutativity. We say that S is *permutable*, or satisfies the property $\mathcal{P}$, if there exists an integer $n > 1$ such that S is n-permutable.

Theorem 2.4. *(Restivo and Reutenauer) Let S be a finitely generated and periodic semigroup. S is finite if and only if it is permutable.*

The proof is based on the Shirshov theorem. In the sequel we will prove a more general theorem (cf. Theorem 2.6) from which Theorem 2.4 is derived as a corollary. A further proof of Theorem 2.4 based on *well quasi-orders*, will be given in Sect. 5.

Permutation property has been investigated for groups. In [19] the following characterization of permutable groups is proved: *A group G satisfies the permutation property if and only if it has a normal subgroup N of finite index such that its derivate group N' is finite.* In the case of a finitely generated group G it has been shown by M. Curzio *et al.* (1983)(cf.[18]) that G satisfies the permutation property if and only if G is *abelian-by-finite*, i.e. G has an abelian (normal) subgroup N of finite index. From this one can easily prove the theorem of Restivo and Reutenauer in the case of groups.

The are many generalizations of the permutation property. Here we recall the *weak-permutation property:*

Definition 2.2. *Let S be a semigroup. For each $n > 1$ one can consider the following property $\mathcal{P}'_n$: For any sequence $s_1, s_2, \ldots, s_n$ of n elements of S there exist two permutations $\sigma, \tau \in S_n$, $\sigma \neq \tau$ such that $s_{\sigma(1)} s_{\sigma(2)} \cdots s_{\sigma(n)} = s_{\tau(1)} s_{\tau(2)} \cdots s_{\tau(n)}$.*

A semigroup S which satisfies $\mathcal{P}'_n$ is called *weakly n-permutable*. One says that S is *weakly permutable*, or satisfies $\mathcal{P}'$, if there exists an $n > 1$ such that S is n-weakly permutable. It is obvious that if a semigroup S is permutable then it is weakly permutable. One can ask the question whether the converse is true. It has been proved by R. D. Blyth (1987) that *in the case of groups the weak permutability is equivalent to permutability* [8]. The result of Blyth cannot be extended to the case of semigroups. In fact G. Pirillo [86] gave an example of a finitely generated semigroup which belongs to the class $\mathcal{P}'$ and not to the class $\mathcal{P}$. Moreover, one can ask the question whether in the Theorem 2.4 one can make the weaker hypothesis that the semigroup S belongs to the class $\mathcal{P}'$. Also in this case the answer is negative; in fact a large class of counterexamples can be obtained by considering Rees quotient monoids of finitely generated free monoids by two-sided ideals, whose growth-functions are quadratically upper bounded (cf.[24],[25]). To illustrate this result we make the following considerations.

Let S be a finitely generated semigroup and $\phi : A^* \to S$ the canonical epimorphism. The *growth-function* of S can be defined for all $n > 0$, as:

$$g_S(n) = card\{s \in S \mid \phi^{-1}(s) \cap A^{[n]} \neq \emptyset\}.$$

Let J be a two-sided ideal of A^* and set $L = A^* \backslash J$. The language L is closed by factors since any factor of a word of L is still an element of L. We can consider the Rees quotient-monoid $M = A^*/J$ of A^* by the ideal J. The monoid M has as support the set $L \cup \{0\}$, where 0 is a new element not in L and the product $\circ$ is defined as follows: for any $f_1, f_2 \in L$, $f_1 \circ f_2 = f_1 f_2$ if the word $f_1 f_2 \in L$, $f_1 \circ f_2 = 0$, otherwise.

Theorem 2.5. *(de Luca and Varricchio). Let M be a Rees-quotient monoid of A^* by a two sided ideal J. If the growth function of M is quadratically upper bounded (i.e. $g_M(n) \leq cn^2$, for all $n > 0$), then M is weakly permutable.*

A way of constructing a Rees quotient monoid of A^* is the following: Let $w \in A^\omega$ be an infinite word over the alphabet A and be $F(w)$ the set of all its factors of finite length. The *subword complexity* f_w of w is the map defined for all $n \in \mathbb{N}$ as:

$$f_w(n) = card(F(w) \cap A^n).$$

For each n, $f_w(n)$ gives the number of factors of w of length n. The set $A^* \backslash F(w)$ is an ideal J. We denote by $M(w)$ the quotient A^*/J. Let g_M be the growth function of $M = M(w)$. One easily see that for any $n \geq 0$, one has $g_M(n) = \sum_{k=0}^n f_w(k)$.

Let us recall that an infinite word w is p-power-free, $p > 1$, if it does not contain factors like u^p with $u \neq \epsilon$. We have seen in the previous section that the Fibonacci word f, and the Thue-Morse words t and m in two and three symbols are respectively 4, 3 and 2 power-free. In all these cases one can prove that the subword complexities are linearly upper bounded so that the growth functions of the monoids $M(f)$, $M(t)$ and $M(m)$ are quadratically upper bounded. Moreover, since these monoids are finitely generated, periodic and infinite, from Theorems 2.4 and 2.5 it follows that they are weakly permutable and not permutable. These results are of some interest for the Burnside problem for semigroups since they provide examples of finitely generated and torsion semigroups which are weakly permutable and infinite. Moreover, by Theorem 2.4 these monoids cannot be permutable.

We recall that Restivo proved first in [89] that $M(f)$ is 8-weakly permutable. A proof that $M(t)$ and $M(m)$ are 5-weakly permutable was given in [24]. Subsequently, these results were widely generalized by proving [28] that $M(w)$ is weakly-permutable in the case of any infinite p-power-free word w whose subword complexity is linearly upper-bounded [28]. The above results have been extended later on by F. Mignosi [82] to the case of any infinite word whose subword complexity is *linearly upper-bounded*. However, it should be remarked that in the case of an infinite word w the considered language $L = F(w)$ is right-prolongable (i.e. for any $u \in L$ there exists at least one letter $a \in A$ such that $ua \in L$), whereas in the statement of Theorem 2.5 one supposes only that the language is factorial. Moreover, Theorem 2.5 provides a further generalization since one considers a quadratic upper bound to the growth-function; this implies, in the case of prolongable languages, a linear upper bound to the subword complexity.

We shall now introduce the following concept of ω-*permutable* (resp. ω-*weakly permutable* semigroup) which is, as we shall see, more general than that of permutable (resp. weakly permutable) in the sense of Restivo and Reutenauer.

Definition 2.3. *A semigroup S is called ω-permutable (resp. ω-weakly permutable) if for any (infinite) sequence $s_1, s_2, \ldots, s_k, \ldots$ of elements of S there exist an integer $n > 1$ and a permutation $\sigma \in \mathcal{S}_n$, $s \neq id$ (resp. two permutations $\sigma, \tau \in \mathcal{S}_n$ with $\sigma \neq \tau$) such that $s_1 s_2 \ldots s_n = s_{\sigma(1)} s_{\sigma(2)} \ldots s_{\sigma(n)}$, (resp. $s_{\sigma(1)} s_{\sigma(2)} \ldots s_{\sigma(n)} = s_{\tau(1)} s_{\tau(2)} \ldots s_{\tau(n)}$).*

One easily verifies that if a semigroup is permutable (resp. weakly permutable) in the sense of Restivo and Reutenauer, then it is also ω-permutable (resp. ω-weakly permutable). We shall denote by $\mathcal{P}_\omega$ (resp $\mathcal{P}'_\omega$) the class of all ω-permutable (resp. weakly ω-permutable semigroups). The ω-permutation property, as well as the weakly ω-permutation property, has been introduced in the case of groups by Blyth and Rhemtulla [9]. The main result of these authors is the following: *A group G is ω-permutable if and only if it is permutable*. Moreover, Blyth and Rhemtulla proved also that if a group is ω-weakly permutable then in general it is not weakly permutable. In the case of semigroups the situation is quite different since an ω-permutable semigroup can be not permutable. The following holds (cf.[27]):

Theorem 2.6. *Let S be a finitely generated and periodic semigroup. S is finite if and only if S is ω-permutable.*

Proof. Let S be a finitely generated semigroup and $\phi : A^+ \to S$ be the canonical epimorphism. For any $s \in S$ let x_s be the canonical representative of s in A^+. Let H be the set of all the canonical representatives of the elements of S. We know that H is closed by factors, so that the morphism ϕ restricted to H is a bijection of H in S. By the lemma of König there exists an infinite word $t \in A^\omega$, with the property that for any $i > 0$, $t[i] \in H$; this implies that $\phi(t[i]) \neq \phi(t[j])$, for $i \neq j$. From Corollary 1.1 there exists a word $s \in A^\omega$) such that $F(s) \subseteq F(t)$ and s is ultimately periodic or s is ω-divided. Let us first suppose that s is ultimately periodic. In this case there exist words $u \in A^*$, $v \in A^+$ such that $s = uv^\omega$. Let $\phi(v) = \xi$. Since S is periodic there exist i and j, $i \neq j$ such that $\xi^i = \xi^j$. Thus uv^i and uv^j would have the same image by ϕ which is a contradiction. Let us now suppose that s is ω-divided, i.e. s can be factorized as $s = s_1 \ldots s_n \ldots$ with $s_i \in A^*$, $i > 0$, and for any $k > 0$ and $\sigma \in \mathcal{S}_k$, $\sigma \neq id$,

$$s_1 \ldots s_k > s_{\sigma(1)} \ldots s_{\sigma(k)}.$$

Let us set for any $n > 0$, $s'_n = \phi(s_n)$ and consider the infinite sequence $s'_1, \ldots, s'_n, \ldots$. Since S is ω-permutable there exist an $n > 1$ and a permutation $\sigma \in \mathcal{S}_n$, $\sigma \neq id$, such that

$$s'_1 s'_2 \ldots s'_n = s'_{\sigma(1)} s'_{\sigma(2)} \ldots s'_{\sigma(n)}.$$

Now, by construction, the sequence $s_1 \ldots s_n$ is the canonical representative of $s'_1 s'_2 \ldots s'_n$, but $s_1 \ldots s_n > s_{\sigma(1)} \ldots s_{\sigma(n)}$, which is a contradiction.

2.3 Chain conditions and $\mathcal{J}$-depth decomposition

A binary relation $\leq$ on a set S is a *quasi-order* if $\leq$ is reflexive and transitive. If for all $s, t \in S$, $s \leq t \leq s$ implies $s = t$ then $\leq$ is a *partial order*. If $s \leq t$ implies $t \leq s$, then $\leq$ is an equivalence relation. The meet $\leq \cap \leq^{-1}$ is an equivalence relation $\sim$ and the quotient of S by $\sim$ is a *poset* (partially ordered set). It is clear that any quasi-order generates a partial order over the equivalence classes, mod. $\sim$. An element $s \in X \subseteq S$ is *minimal* (resp. *maximal*) in X with respect to $\leq$ if, for every $x \in X$, $x \leq s$ (resp. $s \leq x$) implies that $x \sim s$. For $s, t \in S$ if $s \leq t$ and s is not equivalent to t mod. $\sim$, then we set $s < t$.

Now let us consider the following relations in a semigroup S defined as: for $s, t \in S$ we set

$$s \leq_{\mathcal{L}} t \Longleftrightarrow S^1 s \subseteq S^1 t,$$
$$s \leq_{\mathcal{R}} t \Longleftrightarrow s S^1 \subseteq t S^1,$$
$$s \leq_{\mathcal{J}} t \Longleftrightarrow S^1 s S^1 \subseteq S^1 t S^1.$$

One can easily see that $\leq_{\mathcal{L}}$, $\leq_{\mathcal{R}}$ and $\leq_{\mathcal{J}}$ are quasi-order relations of S. The equivalence relation $\sim_{\mathcal{L}}$ (resp. $\sim_{\mathcal{R}}$, resp. $\sim_{\mathcal{J}}$) is the Green relation $\mathcal{L}$ (resp. $\mathcal{R}$, resp. $\mathcal{J}$) (cf.[16],[75]). For $s \in S$ we denote by L_s (resp. R_s, J_s) the $\mathcal{L}$-class (resp. $\mathcal{R}$-class, $\mathcal{J}$-class) containing s. One can then partially order the $\mathcal{L}$-classes (resp. $\mathcal{R}$-classes, $\mathcal{J}$-classes) by setting for $s, t \in S$:

$$L_s \leq L_t \Longleftrightarrow s \leq_{\mathcal{L}} t, \quad R_s \leq R_t \Longleftrightarrow s \leq_{\mathcal{R}} t, \quad J_s \leq J_t \Longleftrightarrow s \leq_{\mathcal{J}} t.$$

Definition 2.4. *A semigroup satisfies the minimal condition on principal right (resp. left, two-sided) ideals if any strictly descending $\leq_{\mathcal{R}}$-chain (resp. $\leq_{\mathcal{L}}$-chain, $\leq_{\mathcal{J}}$-chain) is finite. We denote by $\min_R$ (resp. $\min_L, \min_J$) this minimal condition.*

One has that S satisfies $\min_R$ (resp. $\min_L, \min_J$) if and only if any nonempty subset of $\mathcal{L}$-classes (resp. $\mathcal{R}$-classes, $\mathcal{J}$-classes) contains a minimal element.

Definition 2.5. *An element s of a semigroup S is called right-stable (resp. left-stable) if for any $t \in J_s$, $t S^1 \subseteq s S^1$ (resp. $S^1 t \subseteq S^1 s$) implies $s \mathcal{R} t$ (resp. $s \mathcal{L} t$). An element of S is called stable if it is both right and left-stable. S is called right-stable (resp. left-stable, resp. stable) if any element of S is right-stable (resp. left-stable, resp. stable).*

One can prove that a semigroup which satisfies $\min_R$ (resp. $\min_L$) is right (resp. left) stable [16]. Moreover, a periodic semigroup is stable (cf.[33]).

Definition 2.6. *Let s be an element of a semigroup S. The $\mathcal{J}$-depth of s is the length of the longest strictly ascending chain of two-sided principal ideals starting with s. The $\mathcal{J}$-depth of s can be infinite. A semigroup S admits a $\mathcal{J}$-depth function $d_{\mathcal{J}}$ if and only if for every $s \in S$ the $\mathcal{J}$-depth $d_{\mathcal{J}}(s)$ of s is finite.*

We remark (cf.[7]) that the existence of the $\mathcal{J}$-depth function is stronger than the ascending chain condition on the $\mathcal{J}$-order and weaker than the ascending chain condition on the ideals.

Definition 2.7. *A semigroup S is weakly finite $\mathcal{J}$-above if each $\mathcal{J}$-class of S has only finitely many $\mathcal{J}$-classes above it.*

One easily verifies that a semigroup S is weakly finite $\mathcal{J}$-above if and only if for any $s \in S$ the factors of s can lie only in a finite number of $\mathcal{J}$-classes.

Definition 2.8. *A semigroup S is finite $\mathcal{J}$-above if and only if for each $s \in S$ the set $\{t \in S \mid t \geq_{\mathcal{J}} s\}$ is finite, i.e. S is weakly finite $\mathcal{J}$-above and every $\mathcal{J}$-class is finite.*

It is clear from the definitions that if S is weakly finite $\mathcal{J}$-above, then it has a $\mathcal{J}$-depth function.

Let S be a semigroup. We define inductively a sequence $\{K_n\}_{n \geq 0}$ of sets as follows: $K_0 = \emptyset$ and, for all $n > 0$,

$$K_n = \bigcup_{j=1,\ldots,n} C_j,$$

where for $j > 0, C_j$ is the set of the elements of $S \backslash K_{j-1}$ which are maximal with respect to $\leq_{\mathcal{J}}$ in $S \backslash K_{j-1}$. Moreover we set $K_S = \cup_{j>0} K_j$.

It holds the following lemma whose proof is straightforward [33].

Lemma 2.3. *Let S be a semigroup. For all $j > 0$, K_j is closed by factors and is a union of $\mathcal{J}$-classes.*

Definition 2.9. *A semigroup S has a weak $\mathcal{J}$-depth decomposition if for all $j > 0$ the sets K_j are finite. Moreover, if S is infinite then K_S has to be infinite. A semigroup S has a $\mathcal{J}$-depth decomposition if it has a weak $\mathcal{J}$-depth decomposition and $S = K_S$.*

We remark that if S is finite then $S = K_S$. Indeed, since $K_n \subseteq K_{n+1}$ for all $n \geq 0$, there must exist an integer h such that $K_n = K_h$ for all $n \geq h$. Moreover $S = K_h$, otherwise, one would have $C_{h+1} \neq \emptyset$ and $K_h \subset K_{h+1}$.

Proposition 2.1. *Let S be a semigroup. The following conditions are equivalent*

i. S has a $\mathcal{J}$-depth function and a weak $\mathcal{J}$-depth decomposition.
ii. S has a $\mathcal{J}$-depth decomposition.

Proof. i. $\Rightarrow$ ii. We have to prove that $S = K_S$. If S is finite, then the result is trivial so that we assume that S is infinite. Suppose to the contrary that $S \supset K_S$. If $s \in S \backslash K_S$, then for all $j > 0$, $s \notin K_j$. Let $n > d_{\mathcal{J}}(s)$, where $d_{\mathcal{J}}(s)$ denotes the $\mathcal{J}$-depth of s. Since $s \notin K_n$ it follows that there exists $t_{n-1} \in S \backslash K_{n-1}$ such that $s <_{\mathcal{J}} t_{n-1}$. By iteration it follows that there exists a sequence $t_0, t_1, t_2, \ldots, t_{n-1}$ of n elements of S such that $t_i \in S \backslash K_i$, $(i = 0, 1, \ldots, n-1)$, and

$$s <_{\mathcal{J}} t_{n-1} <_{\mathcal{J}} \cdots <_{\mathcal{J}} t_0.$$

Hence we have a strictly ascending $\mathcal{J}$-chain starting with s and having a length greater than $d_{\mathcal{J}}(s)$ which is a contradiction. Hence $S = K_S$.

ii. $\Rightarrow$ i. We have to prove that S has a $\mathcal{J}$-depth function. Let $s \in S$ and h be the integer such that $s \in C_h$. We prove that $d_{\mathcal{J}}(s) = h$. Since $s \in C_h$, there exists a strictly ascending chain

$$s <_{\mathcal{J}} t_{h-1} <_{\mathcal{J}} \cdots <_{\mathcal{J}} t_1,$$

with $t_i \in C_i$ $(i = 1, \ldots, h-1)$. Thus $h \leq d_{\mathcal{J}}(s)$. We want to prove that $d_{\mathcal{J}}(s) = h$. Suppose that there exists a chain $s <_{\mathcal{J}} s_{k-1} <_{\mathcal{J}} \cdots <_{\mathcal{J}} s_1$, with $k > h$. Let $s_1 \in C_p$, $p \geq 1$; one easily derives that $s \in C_q$ with $q \geq p + k - 1$. Hence in any case $q > h$, which is a contradiction.

Proposition 2.2. *If a semigroup S has a $\mathcal{J}$-depth decomposition, then S is finite $\mathcal{J}$-above.*

Proof. Let $s \in C_h \subseteq K_h$ and t be an element of S such that $s \leq_{\mathcal{J}} t$. This implies that t is a factor of s. By Lemma 2.3, K_h is closed by factors so that one derives $t \in K_h$. Since for any h the set K_h is finite the result follows.

The following basic theorem on the $\mathcal{J}$-depth decomposition holds [33]:

Theorem 2.7. *($\mathcal{J}$-depth decomposition theorem). Let S be a finitely generated semigroup, which is right stable and whose subgroups are locally finite. Then S has a weak $\mathcal{J}$-depth decomposition.*

From the preceding theorem one derives the following [29], [33]:

Theorem 2.8. *(de Luca-Varricchio). Let S be a finitely generated semigroup S whose subgroups are locally finite. If S satisfies min_R, then S is finite.*

Proof. If S satisfies min_R, then S is right-stable. Suppose that S is infinite; then by the preceding theorem so will be K_S. Since K_S is closed by factors from Lemmas 2.1 and 2.2 one derives that there exists an infinite sequence $\{f_n\}_{n>0}$ of elements of K_S such that

$$f_n = f_{n-1} g_{n-1} f_{n-1}, \quad g_{n-1} \in S^1, \ n > 1,$$

and $f_n \neq f_m$, for $n \neq m$. Since $f_n S^1 \subseteq f_{n-1} S^1$ from min_R there exists an integer k such that for all $n \geq k$, $f_n \mathcal{R} f_k$. Let j be such that $f_k \in K_j$. Since K_j is a union of $\mathcal{J}$-classes it follows that $f_n \in K_j$ for $n \geq k$, which is a contradiction since K_j is a finite set.

Theorem 2.8 is a generalization of a theorem of E. Hotzel (cf.[58],[29]). Indeed, one requires only that *finitely generated* subgroups are finite (instead of *all* subgroups as in Hotzel's theorem). This generalization is important since it allows us to derive finiteness conditions for finitely generated semigroups which can be brought back to finiteness conditions on finitely generated groups. In fact, as a consequence of this result one can find important finiteness conditions for semigroups, some of which provide significant answers to the Burnside problem(cf.[29]). We recall here the following result (cf.[105]) which gives a noteworthy generalization of the theorem of Green and Rees (cf. Theorem 2.1):

Corollary 2.1. *(Varricchio). Let S be a finitely generated semigroup such that*

i. *for any $s \in S$ there exists $k > 1$ such that $s = s^k$,*
ii. *the subgroups of S are locally finite.*

Then S is finite.

The proof is obtained by showing that if a finitely generated semigroup S satisfies condition (i), then it satisfies min_R [29].

Definition 2.10. *A semigroup S satisfies the minimal condition on principal bi-ideals if any strictly descending chain*

$$s_1 S^1 s_1 \supset s_2 S^1 s_2 \supset \ldots \supset s_n S^1 s_n \supset \ldots,$$

with $s_1, s_2, \ldots, s_n, \ldots \in S$, has a finite length.

A famous theorem of Coudrain and Schützenberger [17] states that if S is *a finitely generated semigroup satisfying the minimal condition on principal bi-ideals and all subgroups of S are finite, then S is finite.*

A remarkable generalization of this theorem has been proved in [37]. More precisely let S be a semigroup. For $s, t \in S$ we set $s \leq_B t$ if $s \in \{t\} \cup tS^1 t$; we say that S satisfies the condition min_B if and only if any strictly descending chain w.r.t. $\leq_B$ of elements of S has a finite length. One can easily prove that if S satisfies the minimal condition on principal bi-ideals then S satisfies min_B.

Let T be a semigroup and T' be a subsemigroup of T. We say that a subgroup G of T is *locally finite in T'* if any subgroup of G which is generated by a finite subset of T' is finite. The main result is the following stronger version of the Coudrain and Schützenberger theorem:

Theorem 2.9. *Let T be a semigroup satisfying min_B. Let T' be a subsemigroup of T such that all subgroups of T are locally finite in T'. Then T' is locally finite.*

The theorem of Coudrain and Schützenberger is then derived when ($i.$) $T' = T$, ($ii.$) condition min_B is replaced by the stronger minimal condition on principal bi-ideals, ($iii.$) the local finiteness of subgroups of T in T' is replaced by the finiteness of all subgroups.

We give now a new proof of the McNaughton and Zalcstein theorem (cf. Theorem 2.2) on the local finiteness of periodic semigroups of matrices of a finite dimension on a field. Essentially, we show that one can reduce the local finiteness of a subsemigroup of matrices to one of its subgroups.

Let F denote a field and $\mathcal{M}_n(F)$ the semigroup of $n \times n$ squares matrices over F. We shall identify, up to an isomorphism, $\mathcal{M}_n(F)$ with the semigroup $End_n(V, F)$ of the endomorphisms of a vectorial space V of dimension n over the field F.

Corollary 2.2. *(McNaughton and Zalcstein). Let S be a finitely generated subsemigroup of $\mathcal{M}_n(F)$. If S is periodic, then S is finite.*

Proof. Let us first remark that a simple corollary of Theorem 2.9 is that if T is a semigroup satisfying min_B, then any torsion subsemigroup T' whose subgroups are locally finite is locally finite [37]. Moreover, one can easily prove [37] that $End_n(V, F)$ satisfies condition min_B. Let S be a finitely generated and periodic subsemigroup of $\mathcal{M}_n(F)$. Since $\mathcal{M}_n(F)$ is isomorphic to $End_n(V, F)$, it follows that $\mathcal{M}_n(F)$ satisfies min_B. It is well known that any finitely generated and periodic subgroup of $\mathcal{M}_n(F)$ is finite (cf. [69]). Then all finitely generated subgroups of S are finite. From the first remark it follows that S is finite.

We recall that G. Jacob [60] gave a further proof of this theorem; moreover, he proved that it is possible to decide, under certain rather general assumptions, whether a finitely generated semigroup of matrices over a field is finite.

2.4 Iteration property

In this section we consider some finiteness conditions for semigroups based on the *iteration properties*. These properties are very important in formal language theory, since they naturally reflect the "pumping properties" of regular languages.

Definition 2.11. *Let S be a semigroup and m and n two integers such that $m > 0$ and $n \geq 0$. We say that the sequence $s_1, s_2, \ldots, s_m$ of m elements of S is n-iterable if there exist i, j such that $1 \leq i \leq j \leq m$ and*

$$s_1 \ldots s_m = s_1 \ldots s_{i-1}(s_i \ldots s_j)^n s_{j+1} \ldots s_m.$$

We say that S is (m, n)-iterable, or satisfies the property $C(n, m)$ if all sequences of m elements of S are n-iterable. We say that S is iterable, or satisfies the (central) iteration property, if there exist integers $m > 0$ and $n \geq 0$ such that condition $C(n, m)$ is satisfied.

In [22] the following stronger iteration property $D(n,m)$, called *iteration property on the right*, was considered. A semigroup S satisfies $D(n,m)$ if for any sequence $s_1, s_2, \ldots, s_m$ of m elements of S there exist i, j such that $1 \leq i \leq j \leq m$ and

$$s_1 \ldots s_j = s_1 \ldots s_{i-1}(s_i \ldots s_j)^n.$$

Let S be a finitely generated semigroup. It was proved in [22], as a consequence of the theorem of Hotzel, that $D(2,m)$ assures the finiteness of S. A more combinatorial proof of this result was obtained by K. Hashiguchi[54]. He was also able to give an upper bound for the number of elements of the semigroup. More recently it has been proved [29] also that $D(3,m)$ implies the finiteness of S:

Theorem 2.10. *Let S be a finitely generated semigroup. S is finite if and only if S satisfies the properties $D(2,m)$ or $D(3,m)$.*

A further condition that we consider is the following: we say that S satisfies the iteration condition $C(r,s;m)$ of order m if for any sequence $s_1, s_2, \ldots, s_m$ of m elements of S there exist i, j such that $1 \leq i \leq j \leq m$ and

$$s_1 \ldots s_m = s_1 \ldots s_{i-1}(s_i \ldots s_j)^r s_{j+1} \ldots s_m = s_1 \ldots s_{i-1}(s_i \ldots s_j)^s s_{j+1} \ldots s_m.$$

In [22] was posed the question whether the central iteration property implies the finiteness of a finitely generated semigroup S. A positive answer will be given Corollary 2.3 which is a remarkable generalization of Theorem 2.10.

Lemma 2.4. *Let S be a semigroup. If S satisfies the property $C(n, n+1; m)$, then all subgroups of S have finite orders. If S is finitely generated and satisfies $C(n, n+2; m)$ or $C(n+2, 2n+1; m)$, then all finitely generated subgroups of S have finite orders.*

Proof. Let S be a semigroup satisfying condition $C(n, n+1; m)$ and let G be any subgroup of S. We prove that $card(G) \leq m$. In fact let g_0 be the identity element of G and $g_0, g_1, \ldots, g_m$ be $m+1$ distinct elements of G. We consider then the following sequence of m elements of G: $h_1, \ldots, h_m$, having set $h_i = g_{i-1}^{-1} g_i$ $(i = 1, \ldots, m)$. From the condition $C(n, n+1; m)$ there exist integers i, j, $1 \leq i \leq j \leq m$, such that:

$$h_1 \ldots h_{i-1}(h_i \ldots h_j)^n h_{j+1} \ldots h_m = h_1 \ldots h_{i-1}(h_i \ldots h_j)^{n+1} h_{j+1} \ldots h_m.$$

By cancellation one derives

$$h_i \ldots h_j = g_0 = g_{i-1}^{-1} g_j;$$

hence $g_{i-1} = g_j$ which is a contradiction. One reaches the same conclusion if one supposes that condition $C(n+2, 2n+1; m)$ holds with n even integer. Suppose now that S is finitely generated and G be a finitely generated subgroup of S. In a group G the property $C(n, n + 2; m)$, as well as $C(n + 2, 2n + 1; m)$ with n odd integer, implies the following condition:

For any sequence $g_1, g_2, \ldots, g_m$ of m elements of G there exist i, j such that $1 \leq i \leq j \leq m$ and

$$(g_i \cdots g_j)^2 = g_0.$$

In [29] it is proved that a finitely generated group G satisfying the preceding condition is $2m^2$-weakly permutable and then permutable by Blyth's theorem [8]; hence, there exists an integer $h > 1$ such that any product of h elements of G can be rewritten in a non-trivial way. Since G is finitely generated and periodic then by using the theorem of Restivo and Reutenauer (cf. Theorem 2.4), it follows that G is finite.

Theorem 2.11. *Let S be a finitely generated semigroup and m, n integers such that $m > 0$ and $n \geq 0$. If S satisfies $C(n, n + 1; m)$ or $C(n, n + 2; m)$ or $C(n + 2, 2n + 1; m)$, then S is finite.*

Proof. Let S be a finitely generated semigroup and $\phi : A^+ \to S$ be the canonical epimorphism. Under the hypotheses of the theorem and by Lemma 2.4 one has that S is periodic and all finitely generated subgroups of S have finite orders. When $n = 0$ the result is trivial. Hence we suppose $n > 0$. Since S is periodic then S is stable, so that from the $\mathcal{J}$-depth decomposition theorem (cf. Theorem 2.7) one has that S is infinite if and only if $K_S = \cup_{j>0} K_j$ is an infinite subset of S. Moreover K_S is closed by factors. Thus if S is infinite then the set $C = C_{K_S}$ of the canonical representatives of the elements of K_S is infinite, and by Lemma 2.1, closed by factors. By using Lemma 1.3 there exists an infinite uniformly recurrent word $x \in A^\omega$ such that $F(x) \subseteq C$. One has then for every $i > 0$,

$$\phi(x[i])S^1 \subseteq \phi(x[i + 1])S^1.$$

Suppose now, first, that an integer h exists such that for all $i \geq h$

$$\phi(x[i])S^1 = \phi(x[i + 1])S^1.$$

This implies $\phi(x[i])\mathcal{R}\phi(x[h])$ for all $i \geq h$. Since $\phi(x[h]) \in K_S$, a positive integer j exists such that $\phi(x[h]) \in K_j$. We recall that K_j is a union of $\mathcal{R}$-classes so that

$$\phi(x[i]) \in K_j,$$

for all $i \geq h$. But K_j is a finite set; hence there must exist integers i and j, $i < j$, for which $\phi(x[i]) = \phi(x[j])$. This is a contradiction since $x[i]$ and $x[j]$ belong to C. Thus there exist infinitely many integers i for which

$$\phi(x[i])S^1 \supset \phi(x[i + 1])S^1. \tag{2.2}$$

The word x is ω-power free. Indeed, let $u \in F(x)$. Since S is periodic, there exists an integer $p(u)$ with the property that for any $h \geq p(u)$ there exists

$h' < p(u)$ such that $\phi(u)^h = \phi(u^h) = \phi(u^{h'}) = \phi(u)^{h'}$. Therefore, for any $h \geq p(u)$, $u^h \notin F(x)$ by the minimality of the factors of x as representatives, modulus the kernel of ϕ.

Suppose now that S satisfies an iteration property of order m. We can take i so large that $i > q(m + 2)$ (cf. Lemma 1.4), $x[i + 1] = x[i]a$, $a \in A$ and the preceding relation (2.2) is satisfied. Since x is uniformly recurrent by Lemma 1.4 one has that $x[i + 1]$ has a suffix f'_{m+2} which is the $(m + 2)$-th term of a bi-ideal sequence $\sigma = \{f'_1, \ldots, f'_{m+2}\}$, where $f'_1 = x_{i+1} = a$ and $f'_{i+1} = f'_i g'_i f'_i$ $(i = 1, , m + 1)$. Since f'_i terminates with a, for $(i = 1, \ldots, m + 2)$, we can set $f'_{i+1} = f_i a$, for $(i = 1, \ldots, m + 1)$. Then one has

$$f_{i+1}a = f'_{i+2} = f'_{i+1}g'_{i+1}f'_{i+1} = f_i a g'_{i+1} f_i a = f_i g_i f_i a,$$

for $(i = 1, \ldots, m)$. Hence, one derives for $(i = 1, \ldots, m)$, $f_{i+1} = f_i g_i f_i$ having set $g_i = a g'_{i+1}$. Moreover f_{m+1} is a suffix of $x[i]$, since $f'_{m+2} = f_{m+1}a$ is a suffix of $x[i]a$. In conclusion $x[i]$ has a suffix f_{m+1} which is the $(m+1)$-th term of a bi-ideal sequence $\sigma = \{f_1, \ldots, f_{m+1}\}$, where $f_{i+1} = f_i g_i f_i$ $(i = 1, \ldots, m)$, and $g_i \in aA^*$. The word f_{m+1} can be factorized as

$$f_{m+1} = w_0 \ldots w_{m-1}w_m,$$

where $(w_0, \ldots, w_m)$ is the canonical factorization of f_{m+1}. Let us then consider the sequence $\phi(w_0), \ldots, \phi(w_{m-1})$ of m elements of S. Since S satisfies an iteration property of order m then there exist integers n, r, s such that $n > 1$, $0 \leq r \leq s < m$, and

$$\phi(f_{m+1}) = \phi(h_{r,s,n}),$$

with $h_{r,s,n} = w_0 \ldots w_{r-1}(w_r \ldots w_s)^n w_{s+1} \ldots w_{m-1}w_m$. Since $n > 1$ one can write $h_{r,s,n}$ as:

$$h_{r,s,n} = w_0 \ldots w_{r-1}w_r \ldots w_s w_r h,$$

for a suitable $h \in A^*$. Moreover, for all $i \in [0, m - 1]$ one has by Eq.(1.2) that

$$w_i = w_{i+1} \ldots w_m g_{m-i},$$

so that, since $r \leq s$, one derives, by iteration of the preceding formula, that

$$w_r = w_{s+1} \ldots w_m g_{m-s}u,$$

with $u \in A^*$. Hence, one has that

$$h_{r,s,n} \in f_{m+1}g_{m-s}A^*.$$

Since $g_{m-s} \in aA^*$, it follows that

$$\phi(f_{m+1})\mathcal{R}\phi(f_{m+1}a).$$

But this implies

$$\phi(x[i])\mathcal{R}\phi(x[i + 1]),$$

which is a contradiction.

We remark that condition $C(1, n; m)$ is equivalent to condition $C(n, m)$ so that from the above result one derives [30]:

Corollary 2.3. *If a finitely generated semigroup S satisfies $C(2, m)$ or $C(3, m)$, then it is finite.*

Another important property, strictly related to the iteration property is the *strong periodicity*. Let S be a semigroup; we denote by $E(S)$ the set of its idempotent elements.

Definition 2.12. *Let m be a positive integer. A semigroup S is strongly m-periodic if for any sequence $s_1, \ldots, s_m$ of m elements of S there exist integers i and j such that $1 \le i \le j \le m$ and $s_i \ldots s_j \in E(S)$.*

A semigroup S is *strongly periodic* if there exists a positive integer m such that S is strongly m-periodic [99]. The origin of the term strongly m-periodic is due to the fact that if S is strongly m-periodic then S is certainly periodic and moreover the index and the period of any element are less than or equal to m.

Theorem 2.12. *(Simon). Let S be a finitely generated semigroup. The following conditions are equivalent:*

i. S is finite.
ii. $S \backslash E(S)$ is finite.
iii. S is strongly periodic.

The proof given by I. Simon [99] uses the theorem of Hotzel for that which concerns the implication *iii.* $\Rightarrow$ *i.*. We note that this latter implication is an obvious consequence of the fact that if a semigroup S is strongly m-periodic, then it satisfies the condition $D(2, m)$ or $C(2, m)$ (cf. Corollary 2.3). The implication *i.* $\Rightarrow$ *ii.* is trivial. Finally, the implication *ii.* $\Rightarrow$ *iii.* was proved using the theorem of Ramsey.

Using Corollary 1.6 one can prove [31] a finiteness condition that generalizes both Theorem 2.4 and Theorem 2.10:

Theorem 2.13. *Let S be a finitely generated and periodic semigroup. If there exists a pair of integers (n, k) with $k = 2$ or 3, such that any sequence of n elements of S is either permutable or k-iterable on the right, then S is finite.*

An open problem is whether in Theorem 2.13 one can replace the property of iteration on the right with central iteration.

2.5 Repetitive morphisms and semigroups

In the previous section we have shown that any finite semigroup S is *repetitive*, i.e. any morphism from a finitely generated semigroup into S is repetitive. A natural question is then whether repetitivity is a finiteness condition for semigroups. Since the semigroup N_+ of positive integers is repetitive, the answer to this question is negative. Nevertheless if one considers stronger conditions such as *strong repetitivity* and *uniform repetitivity*, then one obtains new finiteness conditions for finitely generated semigroups.

Definition 2.13. *Let S be a semigroup. We say that a morphism $\phi : A^+ \to S$ is strongly repetitive if and only if it satisfies the following condition: For any map $f : N_+ \to N_+$ there exists a positive integer M, which depends on f, such that for any $w \in A^+$ if $|w| > M$ then w can be factorized as:*

$$w = hv_1 \ldots v_{f(p)}h',$$

with $p \in N_+, h, h' \in A^*$ and $|v_i| \le p, 1 \le i \le f(p)$ and

$$\phi(v_1) = \phi(v_2) = \ldots = \phi(v_{f(p)}).$$

The following theorem of J. Justin [62] (cf. also [12]) uses the lemma of Brown (cf. Corollary 1.4):

Theorem 2.14. *(Justin). Let S be a finite semigroup. Then any morphism $\phi : A^+ \to S$ where A is a finite alphabet, is strongly repetitive.*

A consequence of the above theorem is the following important finiteness condition for semigroups due to Brown [12]:

Theorem 2.15. *Let $\phi : S \to T$ be a morphism of semigroups. If T is locally finite and if, for each idempotent $e \in T$, $\phi^{-1}(e)$ is locally finite, then S is locally finite.*

A further application to the Burnside problem of the previous concepts and results is given by the following [21]:

Proposition 2.3. *(de Luca). Let S be a periodic semigroup. S is finite if and only if there exists a finite alphabet A and a strongly repetitive epimorphism $\phi : A^+ \to S$.*

The importance of the notion of strongly repetitive morphism in the theory of locally finite semigroups is shown by the following:

Theorem 2.16. *(Justin and Pirillo). Let S be a semigroup. S is locally finite if and only if any morphism $\phi : A^+ \to S$, where A is a finite alphabet, is strongly repetitive.*

The theorem was announced in [64] and proved in [65] (an alternative proof based on a theorem of Brown [13] is in [38]). Let us now introduce the following definition:

Definition 2.14. *A semigroup S is k-repetitive, $k > 1$, (resp. uniformly k-repetitive) if for each finite alphabet A, each morphism $\phi : A^+ \to S$ is k-repetitive (resp. uniformly k-repetitive). A semigroup is repetitive (resp. uniformly repetitive) if it is k-repetitive (resp. uniformly k-repetitive) for each $k > 1$.*

Theorem 1.5 of the previous section can be restated as: *Any finite semigroup is uniformly repetitive.* The converse of this statement holds for finitely generated semigroups. Indeed, one can prove (cf. Corollary 2.5) that any finitely generated *uniformly repetitive* semigroup is finite. The following proposition is proved in [87].

Proposition 2.4. *Let k be an integer greater than 1. The following statements are equivalent:*

i. $\mathbb{N}$ is not uniformly k-repetitive.
ii. Any finitely generated and uniformly k-repetitive semigroup is finite.

Now, let us introduce the following definition:

Definition 2.15. *Two words u and v on an alphabet A are commutatively equivalent if and only if, for each letter $a \in A$, the number of occurrences of a in u is exactly that of a in v. A word w on A is called an abelian n-power if there exist n nonempty words, say $w_1, \ldots, w_n$, such that $w = w_1 \ldots w_n$ and for each $i, j \in \{1, 2, \ldots, n\}$ the word w_i is commutatively equivalent to w_j.*

The following proposition is due to F. M. Dekking [20]:

Proposition 2.5. *Let A be a two letter alphabet. Then there exists an infinite word on A no factor of which is an abelian 4-power.*

Corollary 2.4. *The semigroup $\mathbb{N}$ is not uniformly 4-repetitive.*

Proof. Let us consider the morphism $\psi : \{a, b\}^* \to \mathbb{N}$, defined by $\psi(a) = 1$ and $\psi(b) = 2$. An easy computation shows that for any two words u and v, with $|u| = |v|$, if $\psi(u) = \psi(v)$ then u and v are commutatively equivalent. By the previous proposition there exists an infinite word over $\{a, b\}$ no factor of which is an abelian 4-power. Such a word cannot contain a factor of the kind $w_1 w_2 w_3 w_4$, with $|w_1| = |w_2| = |w_3| = |w_4|$ and $\psi(w_1) = \psi(w_2) = \psi(w_3) = \psi(w_4)$, otherwise $w_1 w_2 w_3 w_4$ could be an abelian 4-power.

From Proposition 2.4 and Corollary 2.4 we have

Proposition 2.6. *Let S be a finitely generated semigroup. If S is uniformly k-repetitive for some $k \geq 4$, then it is finite.*

Corollary 2.5. *A finitely generated uniformly repetitive semigroup is finite.*

It is an open problem whether $\mathbb{N}$ is uniformly 2-repetitive or uniformly 3-repetitive. This seems to be a difficult problem in combinatorial number theory.

3. Finitely recognizable semigroups

A subset X of a semigroup S is *recognizable* if it is a union of classes of a congruence of finite index defined in it (cf. [45]). As is well known a characterization of the recognizable parts X of S can be given in terms of the *syntactic semigroup* $S(X)$ of X. More precisely there exists a maximal congruence $\equiv_X$ which saturates X (i.e. such that X is union of classes). This is called the *syntactic congruence* of X and it is defined as follows: for any $s \in S$ let $Cont_X(s) = \{(u,v) \in S^1 \times S^1 |\ usv \in X\}$; one sets for $s, t \in S$,

$$s \equiv_X t \iff Cont_X(s) = Cont_X(t).$$

The syntactic semigroup is then defined as $S(X) = S/\equiv_X$. One easily derives that *a part X of S is recognizable if and only if $S(X)$ is a finite semigroup* (Myhill's theorem). In the following the family of recognizable parts of S will be denoted by $Rec(S)$.

In a semigroup S one can introduce the so-called *rational operations* of union ($\cup$), product ($\cdot$) and ($+$) where for any subset X of S, X^+ gives the subsemigroup of S generated by X. The family $Rat(S)$ of the rational subsets of S is then defined as *the smallest family of parts of S containing the finite parts and closed under the rational operations.* When S is a finitely generated free semigroup, or free monoid, then a fundamental theorem due to S. C. Kleene states that $Rec(S) = Rat(S)$ (cf.[45]). Moreover, in this case, $Rec(S)$ coincides also with the family of parts (or languages) definable in terms of finite automata; these are usually called *regular languages*.

It is well known that in a finitely generated semigroup S the family $Rec(S)$ is contained in the family $Rat(S)$, but the converse is not generally true.

In the previous section we have considered several finiteness conditions for finitely generated semigroups. These conditions applied to the syntactic semigroup $S(L)$ of a language L give rise to regularity conditions for the language L. In this section we shall consider conditions for a finitely generated semigroup which are "weaker" in the sense that they do not assure the finiteness of the semigroup S but only that the *finite parts* of S are recognizable:

Definition 3.1. *A semigroup S is called finitely recognizable if all its finite parts are recognizable.*

Since $Rec(S)$ is closed under boolean operations one has that a semigroup is finitely recognizable if and only if all the singletons $\{s\}$, with $s \in S$, are recognizable. If $\phi : A^+ \to S$ denotes a canonical epimorphism then the condition that S is finitely recognizable is equivalent (cf. Lemma 3.1) to the statement that for any $s \in S$, $\phi^{-1}(s) \in Rec(A^+)$. When A is finite this means that all congruence classes in A^+, modulus the nuclear congruence $\phi\phi^{-1}$, are regular languages. In this way one obtains regularity conditions for a wide class of formal languages.

We recall that the problem posed by J. Brzozowski in 1979 of the regularity of aperiodic classes of order $n > 0$ consists precisely in proving that the semigroup $S = A^+/\theta_n$, where θ_n is the congruence generated by the identity $x^n = x^{n+1}$, is finitely recognizable. This problem will be considered in some detail later.

To deal with finitely recognizable semigroups one needs some algebraic concepts such as "$\mathcal{J}$-depth", "finiteness $\mathcal{J}$-above" already introduced in the previous section, "residual finiteness" and new techniques and tools as the "factor semigroup". A characterization of finitely recognizable semigroups having a finite $\mathcal{J}$-depth will be given in Theorem 3.2.

Definition 3.2. *A semigroup S is residually finite if for any pair s, t of elements of S such that $s \neq t$ there exists a congruence θ in S of finite index such that $(s, t) \notin \theta$.*

Finite semigroups are, trivially, residually finite. Also the free semigroup A^+ on the alphabet A is residually finite. To show this let $u, v \in A^+$, $u \neq v$, and let J be the ideal consisting of all the words of A^+ of length greater than the maximum of the lengths of u and v; the Rees congruence $\rho_J = id \cup (J \times J)$ is then of a finite index and separates u from v.

Theorem 3.1. *Let S be a finitely recognizable semigroup. Then S is residually finite.*

Proof. Let $s \in S$. Since $\{s\} \in Rec(S)$ the syntactic congruence $\equiv_{\{s\}}$ has a finite index. Let now t be any element of S such that $s \neq t$. One has that $\equiv_{\{s\}}$ separates s from t. In fact if one supposes that $s \equiv_{\{s\}} t$ it would follow that $s = t$.

Let us remark that the above theorem cannot in general be inverted. Indeed, there exist infinite groups which are residually finite. Moreover a group is finitely recognizable if and only if it is finite [3].

An important relation between the recognizable parts of S and the recognizable parts of A^+ is given by the following (cf.[33]):

Lemma 3.1. *Let S be a semigroup and $\phi : A^+ \to S$ be a canonical epimorphism. For a set $X \subseteq S$ the following two conditions are equivalent:*

i. $X \in Rec(S)$.
ii. $\phi^{-1}(X) \in Rec(A^+)$.

3.1 The factor semigroup

Let S be a semigroup and r be an element of S. We denote by $F(r)$, or simply F_r, the set of all factors of r:

$$F_r = \{s \in S \mid r \in S^1 s S^1\}.$$

Let $J_r = S \backslash F_r$; J_r is a two-sided ideal of S. We denote by N_r the Rees-quotient semigroup S/J_r of S by J_r. As is well known N_r is isomorphic to the semigroup having as support $F_r \cup \{0\}$ and the product $\circ$ is defined as follows: for $s, t \in S$, $s \circ t = st$ if $s, t \in F_r$ and $st \in F_r$, $s \circ t = 0$, otherwise. We call N_r also the *factor semigroup* of r.

Lemma 3.2. *Let r be an element of a semigroup S. If F_r is finite, then $r \in Rec(S)$.*

Proof. We can write $N_r = S/\rho_r$, where ρ_r is the Rees congruence $\rho_r = id \cup (J_r \times J_r)$. The index of ρ_r is equal to $card(F_r) + 1$, so that if F_r is finite then ρ_r is of finite index. Moreover r saturates ρ_r so that $r \in Rec(S)$.

Let S be a semigroup. If for all $r \in S$, F_r is finite, then from Lemma 3.2, S is finitely recognizable. The converse is not, in general, true as shown in [33].

Proposition 3.1. *Let S be a semigroup. S is finite $\mathcal{J}$-above if and only if for all $r \in S$, F_r is finite.*

Proof. Let us first observe that for any $r \in S$, F_r is a union of $\mathcal{J}$-classes. Indeed, let $u \in F_r$ and $u \mathcal{J} v$. This implies that $u = hvk$ with $h, k \in S^1$, so that $v \in F_r$. Now suppose that for all $r \in S$, F_r is finite. It follows that the number of $\mathcal{J}$-classes above the $\mathcal{J}$-class of r has to be finite. Thus S is weakly finite $\mathcal{J}$-above. Moreover, since $J_r \subseteq F_r$, one has that J_r is finite. Hence, S is finite $\mathcal{J}$-above. Conversely, suppose that S is finite $\mathcal{J}$-above and r is any element of S. By definition the number of $\mathcal{J}$-classes of the factors of r is finite. Moreover, any $\mathcal{J}$-class in F_r is finite; hence it follows that F_r is finite.

The following theorem gives some characterizations of finitely generated finite $\mathcal{J}$-above semigroups (cf. [33]).

Theorem 3.2. *Let S be a finitely generated semigroup having a finite $\mathcal{J}$-depth function. The following conditions are equivalent:*

i.   S has a $\mathcal{J}$-depth decomposition.
ii. S is finite $\mathcal{J}$-above.
iii. S is finitely recognizable.
iv. S is stable and all subgroups are finite.
v.   S is right-stable and the subgroups of S are locally finite.

Let S be a finitely generated semigroup and $\phi : A^+ \to S$ an epimorphism. We recall that the *word problem* for S consists in deciding whether two arbitrary words u, v of A^+ are congruent, mod. $\phi\phi^{-1}$. The word problem is said (recursively) *solvable* if there exists an algorithm to decide whether two arbitrary words are congruent. We recall the following theorem relative to the decidability of the word problem [46],[33]:

Theorem 3.3. *Let S be a finitely generated free semigroup in a finitely based variety (i. e. defined by a finite set of identities). If S is residually finite, then the word problem for S is solvable.*

Corollary 3.1. *Let S be a finitely generated free semigroup in a finitely based variety. If S is finitely recognizable, then the word problem for S is solvable.*

Proof. From Theorem 3.1, S is residually finite. Thus from Theorem 3.3 the result follows.

A recent survey on the main algorithmic problems concerning semigroups and other varieties of classical algebras is in [72].

3.2 On a conjecture of Brzozowski

In 1979 Brzozowski presented at the International Symposium on Formal Language theory in Santa Barbara, California, a paper entitled "Open problems about regular languages" [15] in which six open problems and conjectures are discussed.

Most of these problems such as the "star height" or the "optimality of finite and complete prefix codes" are still open, even though a considerable effort has been made to solve them. The fifth conjecture known as the "regularity of non-counting classes" can be formulated as follows: Let A^* be a finitely generated free monoid over a finite alphabet A and n a fixed positive integer. One can introduce in A^* the congruence relation θ_n generated by the relation $\{(x^n, x^{n+1})|\ x \in A^*\}$ and the quotient monoid $M_n = A^*/\theta_n$. For any word $w \in A^*$ the congruence class $[w]_{\theta_n}$ or simply $[w]_n$, of w is a *non-counting language of order* n according to R. McNaughton and S. Papert [80].

Conjecture 3.1. Let $n > 0$. For any word $w \in A^*$ the congruence class $[w]_n$ is a regular language.

Actually this Conjecture was already formulated in 1970 for the case $n = 2$ by Simon [98] in an unpublished manuscript. More recently, the following has been considered:

Conjecture 3.2. (Generalized Brzozowski's conjecture) Let $n, m > 0$. Let $\theta_{n,m}$ be the congruence generated by the relation $\{(x^n, x^{n+m})|\ x \in A^*\}$, then for any word $w \in A^*$ the congruence class $[w]_{\theta_{n,m}}$ is a regular language.

In what follows, when talking about Brzozowski's conjecture, we shall refer to Conjecture 3.2, even though the first formulation, as stated in Brzozowski's paper [15], is exactly the statement of Conjecture 3.1.

In 1990 [26, 32] the authors proved the Brzozowski conjecture for any $n \geq 5$ and $m = 1$. Actually the stronger result was proved, that for any $n \geq 5$ and for any $w \in A^*$ the set of factors of $[w]_n$ in the monoid A^*/θ_n

is finite and, by Lemma 3.2, this enforces the regularity of $[w]_n$. Also, the further result that, in the monoid A^*/θ_n, the *word problem* is solvable, for $n \geq 5$ was proved.

Slightly later and in a different way J. McCammond [79] proved the conjecture for $n \geq 6$ and for any $m > 0$. His proof, based on some automata transformations, is very difficult to understand (it is based on forty lemmas proved by simultaneous induction). The importance of the paper stands on his results on Burnside's subgroups, in particular he proved that the maximal subgroups are cyclic of order m.

Subsequently, A. Pereira do Lago [39] improved the result, basing his work on the techniques introduced in [26, 32]. On one hand he proved in a more direct way the conjecture for $n \geq 4$ and for any m, simplifying the original rewriting system, and on the other he underlined the links between such a system and Green's relations on the quotient monoid.

The very last contribution comes from V. Guba [52] who proved in 1993 the conjecture for $n \geq 3, m \geq 1$ and this is the most recent result. A more complete version of the work of do Lago containing some further results on the structure of the Burnside semigroup is in [40].

The case $n = 2$ is still unsolved and it is not clear how the techniques used up to now will be able to work it out. For $n = 1$ Conjecture 3.1 is trivially true since the monoid A^*/θ_1 is finite. Conjecture 3.2 is still open for the values of the integer m for which the monoid $A^*/\theta_{1,m}$ is infinite.

3.3 Problems and results

We shall now consider the Brzozowski conjecture in a more general context of problems that have been considered in recent contributions.

Let A be a finite alphabet and $n, m > 0$. Let $\pi \subseteq A^* \times A^*$ be the relation $\pi = \{(x^n, x^{n+m}) \mid x \in A^*\}$. Denote by $B(r, n, n+m)$ the quotient monoid $B(r, n, n+m) = A^*/\theta_{n,m}$, where $r = card(A)$. The monoid $B(r, n, n+m)$ is called the *Burnside semigroup* with r generators defined by the equation $x^n = x^{n+m}$. One can then consider the following problems:

1. Finiteness. For $r = 1$, $B(1, n, n+m)$ is always finite. For $r > 1$, $B(r, n, n+m)$ is finite for $n = 1$ and $m = 1, 2, 3, 4, 6$. This follows, since $B(r, 1, 1+m)$ coincides, up to the identity element, with $S(r, m)$, from the theorem of Green and Rees (cf. Theorem 2.1) and the fact that the Burnside group $G(r, n)$ is finite for $n = 2, 3, 4$ and 6. From this theorem and the results of Adjan, Novikov, Lysionok and Ivanov (cf. Sect. 2.) one derives that $B(r, 1, 1+m)$ is infinite for any odd integer $m \geq 115$ and, moreover, for any $m \geq 2^{48}$.

For $r > 2$, and $n > 1$, $B(r, n, n+m)$ is infinite. This is an easy consequence of the existence of infinitely many square-free words on an alphabet having at least three letters [103]. However, even if $r = 2$, and $n > 1$, it was proved by Brzozowski *et al.*[14] that $B(r, n, n+m)$ is infinite.

2. Regularity of congruence classes. By Lemma 3.1 one has that the congruence classes are regular if and only if the elements of $B(r, n, n+m)$ are recognizable, and this is equivalent to the requirement that $B(r, n, n+m)$ is finitely recognizable. As we said before it is known at the present time that for $n \geq 3$ and $m \geq 1$ the elements of $B(r, n, n+m)$ are recognizable.

3. Finiteness $\mathcal{J}$-above. By Proposition 3.1 the finiteness $\mathcal{J}$-above of $B(r, n, n+m)$ is also equivalent to the statement that for any $s \in B(r, n, n+m)$ the set F_s of the factors of s is finite. It has been proved that $B(r, n, n+1)$ is finite $\mathcal{J}$-above for any $n \geq 5$ [26, 32]. At the present time we know that $B(r, n, n+m)$ is finite $\mathcal{J}$-above for $n \geq 3$ and $m \geq 1$.

4. The word problem. This problem consists in deciding whether, given two words $u, v \in A^*$, they represent the same element of $B(r, n, n+m)$. As a result of the previous works we know that the word problem is solvable for $n \geq 3$ and $m \geq 1$.

5. Structure. The $\mathcal{H}$-classes of regular $\mathcal{D}$-classes have m elements. For $n \geq 3$ and $m \geq 1$ the group of such a $\mathcal{D}$-class is a cyclic group of order m.

6. Extensions and generalizations. One can consider the semi-Thue system $\pi = \{(x^n, x^{n+m}) \mid x \in A^*\}$ and $\Rightarrow_\pi^*$ the corresponding derivation relation. For any $w \in A^*$ let $L(w)$ be the language $L(w) = \{v \in A^* \mid w \Rightarrow_\pi^* v\}$. For given $n, m \geq 1$ one can ask whether $L(w)$ is a regular language. The only known results are in the case $n = m = 1$ (cf. Sect. 5.).

We now present some recent results which link the above problems. These results also give an idea of the approach followed in solving the Brzozowski conjecture. We first observe that any finite semigroup is trivially finite $\mathcal{J}$-above, moreover if $B(r, n, n+m)$ is finite then the equivalence classes are regular languages. Then, if we have an affirmative answer to Problem 1, we also have a positive answer to Problems 3 and 2. Unfortunately, as we said, there are only a few cases in which the semigroup $B(r, n, n+m)$ is finite. From Lemma 3.2 one has that to find an affirmative answer to Problem 2 it is enough to solve positively Problem 3.

We now observe that in the previous works the decidability of the word problem in the semigroups $B(r, n, n+m)$ has been directly proved, at the present for $n \geq 3$. The proof relies on some properties such as the Church-Rosser property, of suitable rewriting systems that have been shown to be equivalent to the Thue-system $\pi = \{(x^n, x^{n+m}) \mid x \in A^*\}$. An important fact underlined in [33] is that: if $B(r, n, n+m)$ is finite $\mathcal{J}$-above, then the word problem in $B(r, n, n+m)$ is solvable. This is a consequence of Proposition 3.1, Lemma 3.2 and Corollary 3.1.

Another question is the following: are Problems 2 and 3 equivalent? More generally we can ask whether a finitely recognizable semigroup is also finite $\mathcal{J}$-above. We have only a partial answer to this question by means of Theorem 3.2.

We observe that the subgroups of $B(r, n, n+1)$ are all trivial, since they satisfy the equation $x = 1$, obtained by cancellation from the equation $x^n = x^{n+1}$. Moreover, $B(r, n, n+1)$ is stable, since all periodic semigroups are stable. Therefore, from Theorem 3.2, in order to prove that $B(r, n, n+1)$ is finitely recognizable and therefore that the noncounting classes are regular, it is enough to show that $B(r, n, n+1)$ has a $\mathcal{J}$-depth function. In fact in [26, 32] the following stronger result is proved.

Theorem 3.4. *Let $M = B(r, n, n+1)$, $n \geq 5$. For any $m \in M$ there are only finitely many $\mathcal{J}$-classes of elements $t \in F_m$ (i. e. M is weakly finite $\mathcal{J}$-above). This implies that $B(r, n, n+1)$ has a $\mathcal{J}$-depth function.*

Proof. The proof is based on the following results:

Fact 1 *(Equivalence Theorem).* There exists a Thue-system σ equivalent to π. The elements of σ are suitable pairs of words of the kind $(x'x^{n-2}x'', x'x^{n-1}x'')$, with x' suffix and x'' prefix of x. A production of the kind $x'x^{n-2}x'' \to x'x^{n-1}x''$ (resp. $x'x^{n-1}x'' \to x'x^{n-2}x''$) is called an *expansion* (resp. *reduction*). Then one can prove:

Fact 2 *(Renormalization Theorem).* Let $[w]_n$ be a noncounting class, where w is its canonical representative. Then any word $x \in [w]_n$ can be obtained from w using only expansions of σ. Let $\Rightarrow$ denote the relation in A^* defined as: for $u, v \in A^*, u \Rightarrow v$ if and only if there exists in σ a pair $(x'x^{n-2}x'', x'x^{n-1}x'')$ and $h, h' \in A^*$ such that

$$u = hx'x^{n-1}x''h', \quad v = hx'x^{n-2}x''h'.$$

The renormalization theorem implies that the reduction (or semi-Thue) system $\Delta = (A^*, \Rightarrow)$ is *terminating* (i.e. well-founded) and satisfies the Church-Rosser property, i.e. $u \equiv v$, mod. θ_n implies that there exists $w \in A^*$ such that $u \Rightarrow^* w$ and $v \Rightarrow^* w$. From this one easily derives (cf. [32]) that the word problem for M is solvable.

Fact 3 *(Finiteness of the $\mathcal{J}$-depth).* Let $[w]_n$ be a noncounting class and w its canonical representative. Then for any $x \in F([w]_n)$ there exist $w' \in [w]_n$ and $x' \in F(w')$ such that w' can be derived from w by at most one expansion of σ and, moreover, $\phi(x)\mathcal{J}\phi(w')$.

Since there are only finitely so many words derivable from w by one expansion of σ, Fact 3 implies that for any $[w]_n$ there are only finitely many $\mathcal{J}$-classes in $F([w]_n)$. From this it easily follows that $B(r, n, n+1)$ has a $\mathcal{J}$-depth function.

Let us observe that do Lago in his proof of Conjecture 3.2 for $n \geq 4$ and $m > 0$ follows the lines of the proof in [26, 32]. He introduces a simpler rewriting system σ equivalent to $\pi = \{(x^n, x^{n+m}) | x \in A^*\}$ whose elements are suitable pairs of words of the kind $(x'x^{n-1}, x'x^{n-1+m})$ with x' suffix of

x. He proved that for $n \geq 4$ and $m \geq 1$ the reduction system associated with σ is terminating. The main difference with respect to the line of the proof in [26, 32] is a direct proof that $M = B(r, n, n + m)$ is finite $\mathcal{J}$-above without using Theorem 3.3 and properties of maximal subgroups of M, which are actually finite (cyclic of order m), as shown by the same author for $n \geq 4$ and by McCammond for $n \geq 6$.

Some results relative to Problem 6 will be considered in the Section 5.

4. Non-uniform regularity conditions

A language L over the finite alphabet A is recognizable or regular if and only if the syntactic semigroup $S(L)$ is finite. A property P of languages is said to be *syntactic* if whenever two languages L, L' have the same syntactic semigroup and L has the property P, then also L' has the property P. In this sense, commutativity is a syntactic property (L commutative if and only if its syntactic semigroup is commutative), periodicity is syntactic and also rationality is a syntactic property. We point out that when a characterization of rationality involves only properties of languages which are syntactic, then this result is rather a result on semigroups than on languages. However, when the conditions are formulated by properties which are not syntactic, then it is not possible to pass directly through the syntactic semigroup to prove the rationality of the language; the dependence on "contexts" can be overcome by very strong "forcing" arguments, like Ramsey's theorem, which will be used in the proof of some "non-uniform" regularity conditions.

The study of regularity conditions for periodic languages has been called also the *Burnside problem for languages*. As we have seen in the previous sections the "permutation" and the "iteration" conditions for semigroups can be translated in regularity conditions which are of interest for the Burnside problem for languages. In the following we shall analyze some non-uniform versions of these conditions which cannot be expressed as properties of the syntactic semigroup of the language.

The following lemma, called Nerode's criterion (cf. [45]), is an useful tool in dealing with non-uniform conditions.

Lemma 4.1. *Let P be a property defined in the class of all the languages over A and $\mathcal{L}_P$ the family of the languages satisfying P. If $\mathcal{L}_P$ is finite and for any $a \in A$, $L \in \mathcal{L}_P$ implies $a^{-1}L \in \mathcal{L}_P$, then any language $L \in \mathcal{L}_P$ is regular.*

Proof. Let $L \in \mathcal{L}_P$. By hypothesis, for any $u \in A^*$, $u^{-1}L \in \mathcal{L}_P$. Since $\mathcal{L}_P$ is finite, the sets $u^{-1}L$, $u \in A^*$ are finitely many. By the Nerode theorem (cf.[45]), L is regular.

4.1 Pumping properties

Let us now introduce some *pumping properties* inspired from the pumping lemma for regular languages (cf.[45]).

Definition 4.1. *Let L be a language and $x = uvw$. We say that the word v is a pump for x, in the context (u, w), relative to L if and only if for any $i \geq 0$*

$$uv^i w \in L \Longleftrightarrow x \in L.$$

We say that v is a positive-pump if the latter condition is satisfied for any $i > 0$.

We shall not specify the context of a pump v when there are no ambiguities.

Definition 4.2. *A language L satisfies the property D_m (resp. C_m) if for any $x, w_1, w_2, \ldots, w_m, y \in A^*$, there exist i, j, $1 \leq i < j \leq m + 1$, such that $w_i \ldots w_{j-1}$ is a pump (resp. positive-pump) for $xw_1w_2 \ldots w_m y$ relative to L. A language L satisfies the block pumping property (resp. positive block pumping property) if there exists an integer $m > 0$ such that L satisfies D_m (resp. C_m).*

Definition 4.3. *A language L satisfies the block cancellation property if there exists an integer $m > 0$ such that for any $x, w_1, w_2, \ldots, w_m, y \in A^*$, there exist i, j, $1 \leq i < j \leq m + 1$, such that:*

$$xw_1w_2 \ldots w_m y \in L \Longleftrightarrow xw_1w_2 \ldots w_{i-1}w_j \ldots w_m y \in L.$$

It is trivial from the definition that if a language satisfies the block pumping property then it satisfies also the positive block pumping property and the block cancellation property.

In the former definitions the integers i, j depend on the context (x, y) in which the block $w_1w_2 \ldots w_m$ is considered. If they do not depend on the context, then the corresponding properties will be said to be *uniform*. We observe that *the uniform block cancellation property and the uniform (positive) block pumping property are syntactic*. The first trivially assures the finiteness of the syntactic semigroup $S(L)$ of a language L and then the regularity of L. The second is equivalent to the statement that $S(L)$ satisfies the central iteration property $C(n, n+1, m)$ for a suitable $m > 0$ and for all $n \geq 0$ ($n > 0$ in the positive case). Therefore, by Theorem 2.11 or simply by Corollary 2.3, one has that any language satisfying the uniform block pumping property is regular. Conversely, from the Myhill theorem one easily derives that *any regular language satisfies the uniform block pumping property*.

The relevance of the block pumping properties for the Burnside problem for languages is due to the fact that *if a language L satisfies the property D_m or C_m, then it is periodic.* Indeed, one easily verifies that for any word w one has that $w^m \equiv_L w^{m+m!}$. Moreover, the following theorem was proved in [41] by A. Ehrenfeucht, R. Parikh and G. Rozenberg.

Theorem 4.1. *A language is regular if and only if it satisfies the block pumping property.*

Proof. Let $\mathcal{L}_{D_m}$ be the family of the languages satisfying D_m. We prove that for any $u \in A^*$, $L \in \mathcal{L}_{D_m}$ implies that $u^{-1}L \in \mathcal{L}_{D_m}$. In fact, if L satisfies D_m, then for any $x, y, w_1, \ldots, w_m \in A^*$ there exist i, j, with $1 \leq i < j \leq m + 1$, such that for any $s \geq 0$

$$uxw_1 \ldots w_m y \in L \Leftrightarrow uxw_1 \ldots w_{i-1}(w_i \ldots w_{j-1})^s w_j \ldots w_m y \in L;$$

therefore for any $s \geq 0$ one has

$$xw_1 \ldots w_m y \in u^{-1}L \Leftrightarrow xw_1 \ldots w_{i-1}(w_i \ldots w_{j-1})^s w_j \ldots w_m y \in u^{-1}L,$$

and $u^{-1}L$ satisfies D_m. Now we prove that $\mathcal{L}_{D_m}$ is finite. For this it is sufficient to prove that there exists an integer N such that for any two languages $L, L' \in \mathcal{L}_{D_m}$ one has

$$L = L' \Leftrightarrow L \cap A^{[N]} = L' \cap A^{[N]}. \tag{4.1}$$

In fact let $N = R(2, m+1, 2)$, where R is the Ramsey function as in Theorem 1.1. Suppose by contradiction that there exist two languages $L, L' \subseteq \mathcal{L}_{D_m}$ such that Eq. (4.1) is not satisfied. Then there exists a word w of minimal length in the symmetric difference $L \triangle L'$ and $|w| \geq N$. We can write $w = w_1 w_2 \ldots w_N$, with $|w_i| \geq 1$, for $i = 1, \ldots, N$. Let $X = \{1, \ldots, N\}$. Consider a partition of $P_2(X)$ in two subsets P_1, P_2 where $\{i, j\} \in P_1$ if and only if

$$w_1 w_2 \ldots w_N \in L \Leftrightarrow w_1 w_2 \ldots w_{i-1} w_j \ldots w_N \in L,$$

with $1 \leq i < j \leq N$, and $P_2 = P_2(X) \setminus P_1$. By the Ramsey theorem there exists a subset $Y = \{j_1, j_2, \ldots, j_{m+1}\}$ of X, such that $j_1 < j_2 < \ldots < j_{m+1}$ and either $P_2(Y) \subseteq P_1$ or $P_2(Y) \subseteq P_2$. Since L satisfies D_m, one has $P_2(Y) \subseteq P_1$. Let $(u_1, \ldots, u_m)$ be the derived sequence of $(w_1, \ldots, w_N)$ corresponding to the increasing sequence $j_1 < j_2 < \ldots < j_{m+1}$ (cf. Definition 1.3). By construction $w = xu_1 \ldots u_m y$ and, since $P_2(Y) \subseteq P_1$, for any $i, j \in \{1, \ldots, m\}$ with $i < j$, one has

$$xu_1 u_2 \ldots u_m y \in L \Leftrightarrow xu_1 u_2 \ldots u_{i-1} u_j \ldots u_m y \in L. \tag{4.2}$$

Since L' satisfies D_m, there exist $i, j \in \{1, \ldots, m\}$ such that

$$xu_1 u_2 \ldots u_m y \in L' \Leftrightarrow xu_1 u_2 \ldots u_{i-1} u_j \ldots u_m y \in L'. \tag{4.3}$$

Let $w' = xu_1 u_2 \ldots u_{i-1} u_j \ldots u_m y$. Since w is of minimal length in $L \triangle L'$ and $|w'| < |w|$, one has $w' \in L \Leftrightarrow w' \in L'$. Then, by Eq.(4.2) and Eq.(4.3), one derives $w \in L \Leftrightarrow w \in L'$, that is a contradiction. Thus Eq.(4.1) is proved and this implies that in $\mathcal{L}_{D_m}$ there are only finitely many sets. By Lemma 4.1, any language in $\mathcal{L}_{D_m}$ is regular.

We observe that the proof of the previous theorem uses only the block cancellation property which is weaker than the block pumping property. So we proved the following stronger result: *A language L is regular if and only if it satisfies the block cancellation property.* A problem left open in [41] is whether the positive block pumping implies regularity. A positive answer to this question is in the following [107]:

Theorem 4.2. *A language is regular if and only if it satisfies the positive block pumping property.*

Remark 4.1. If in Definition 4.2 the words $w_1, \ldots, w_m$ are *letters instead of blocks*, one defines the usual pumping property of Automata Theory. This condition is very far from implying the regularity. In fact it has been proved [41] that there exist uncountably many non-regular languages which satisfy this property.

Remark 4.2. One can consider some extensions of block pumping properties where the "pump" starts only from a given integer $s > 1$. If a language L satisfies this kind of property, then one easily derives that L is a periodic language. Moreover, in the uniform case one can prove, as a consequence of Theorem 2.11 that L is a regular language. An open problem is to see what occurs in the non-uniform case.

4.2 Permutative property

In this subsection we present some non-uniform properties that correspond to the permutation properties for semigroups.

Definition 4.4. *Let m and k be fixed integers such that $m \geq k > 1$. A language L is (m, k)-permutative if there exists a permutation $\sigma \in S_k \backslash id$ such that whenever a word w is written $w = u x_1 \ldots x_m v$ there exists a derived sequence $(y_1, \ldots, y_k)$ of $(x_1, \ldots, x_m)$ such that $x_1 \ldots x_m = u' y_1 \ldots y_k v'$, $u', v' \in A^*$, and*

$$w = u u' y_1 \ldots y_k v' v \in L \Longleftrightarrow u u' y_{\sigma(1)} \ldots y_{\sigma(k)} v' v \in L.$$

The language L is called permutative if there exist integers m and k, $m \geq k$ such that L is (m, k)-permutative.

When $k = 2$, that is, σ is the permutation (12), one obtains the *transposition property* of [92]: A language L has this property if, for some m, every time one distinguishes m successive factors in some word w, then there exist two consecutive blocks of these factors which, when transposed, give a word w' such that $w \in L \Longleftrightarrow w' \in L$. Note that the transposition property, when $m = 2$, is equivalent to the commutativity of L.

One can easily prove that a rational language is always permutative: consider the loops in a finite automaton accepting it. Furthermore, there are languages which do not have this property: for example, the set of palindromes; consider a palindrome of the form

$$aba^2b\ldots a^{m-1}ba^mba^mba^{m-1}\ldots ba^2ba$$

and set $x_1 = ab, x_2 = a^2b, \ldots, x_m = a^mb$. No permutation of the $x's$ other than the identity keeps this word a palindrome. The following characterization of rationality was proved in [91].

Theorem 4.3. *(Restivo and Reutenauer). A language is rational if and only if it is periodic and permutative.*

Proof. The proof can be articulated in the following points:

(i) Let k be a fixed integer ≥ 2 and let $\sigma \in S_k \backslash id$. For $m \geq k$ we set $n(m) = R(k + 1, m + 1, 2)$ where R is the Ramsey function. We denote by $\mathcal{L}_{m,p}$ the family of languages over the alphabet A such that:

– L is (m, k)-permutative,
– For any word x of length smaller than $n(m)$ one has

$$\forall u, v \in A^* : ux^p v \in L \Longleftrightarrow ux^{p+p!}v \in L. \tag{4.4}$$

(ii) Let L be a periodic language which is (m, k)-permutative. We show that $L \in \mathcal{L}_{m,p}$ for some p with $p + p! \geq 2n(m)$. Indeed, let W be the set of words of length smaller than $n(m)$. For each x in W, as L is periodic, there exist i_x, j_x with $j_x \geq 1$ such that

$$\forall u, v \in A^* : ux^{i_x}v \in L \Longleftrightarrow ux^{i_x+j_x}v \in L.$$

Let

$$i = sup\{i_x \mid x \in W\}, \; j = sup\{j_x \mid x \in W\}.$$

Choose p such that $p \geq i$, $p \geq j$, $p + p! \geq 2n(m)$. Then, if $|x| \leq n(m)$, i.e., $x \in W$, one has, $\forall u, v \in A^*$:

$$ux^p v \in L \Longleftrightarrow ux^{p-i_x}x^{i_x}v \in L$$
$$\Longleftrightarrow ux^{p-i_x}x^{i_x+j_x}v \in L \Longleftrightarrow ux^{p-i_x+j_x}x^{i_x}v \in L$$
$$\Longleftrightarrow ux^{p-i_x+j_x}x^{i_x+j_x}v \in L \Longleftrightarrow ux^{p-i_x+2j_x}x^{i_x}v \in L$$
$$\vdots$$
$$\Longleftrightarrow ux^{p-i_x+p!}x^{i_x}v \in L \text{ (because } j_x \text{ divides } p!) \Longleftrightarrow ux^{p+p!}v \in L.$$

Hence, L is in $\mathcal{L}_{m,p}$ with $p + p! \geq 2n(m)$.

(iii) By (ii), we know that any periodic language which is (m, k)-permutative is in some $\mathcal{L}_{m,p}$ with $p + p! \geq 2n(m)$. It will thus suffice to show that any such $\mathcal{L}_{m,p}$ is finite. Indeed, if L is in $\mathcal{L}_{m,p}$, then so is

$$a^{-1}L = \{w \mid aw \in L\}$$

(as may easily be verified), so one may apply Nerode's criterion (cf. Lemma 4.1) to conclude that any language in $\mathcal{L}_{m,p}$ is rational [45].

(iv) Let $p + p! \geq 2n(m)$, $\mathcal{L} = \mathcal{L}_{m,p}$ and $n = n(m)$. Furthermore, let $r = card(A)$ and $N = N(r, p + p!, n)$ be the integer of Shirshov's theorem. Let $L, L' \in \mathcal{L}$. We show that

$$L = L' \Leftrightarrow L \cap A^{[N-1]} = L' \cap A^{[N-1]}.$$

(i.e. if L and L' "agree" on all the words of length $< N$, then they coincide). This will ensure that $\mathcal{L}$ is finite and will conclude the proof.

(v) We order A^*, by the alphabetic order $<_a$ that we simply denote by $<$. Since this order is a well order, A^* has as a smallest element the empty word. Thus one may make induction with respect to this order.

(vi) We show by induction that, for any word w, $w \in L \Longleftrightarrow w \in L'$. We already know that $w \in L \Longleftrightarrow w \in L'$ if $|w| < N$. Let $|w| \geq N$. Suppose w contains the $(p + p!)$-th power of some nonempty word x: $w = ux^{p+p!}v$, with $|x| \leq n - 1$. As L, L' are both in $\mathcal{L} = \mathcal{L}_{m,p}$ we have, by Eq.(4.4),

$$(ux^{p+p!}v \in L \Longleftrightarrow ux^p v \in L) \text{ and } (ux^{p+p!}v \in L' \Longleftrightarrow ux^p v \in L').$$

Now, $ux^p v$ is smaller than w (with respect to the length), thus, by the induction hypothesis, we have

$$ux^p v \in L \Longleftrightarrow ux^p v \in L'.$$

Taking these equivalences together we obtain

$$w \in L \Longleftrightarrow w \in L'.$$

(vii) Suppose now that w contains no $(p + p!)$-th power of a nonempty word of length at most $n - 1$. Then, by Shirshov's theorem, w has a factor which is n-divided:

$$w = ux_1 \ldots x_n v, \tag{4.5}$$

where $(x_1, \ldots, x_n)$ is an n-division of $x_1 \ldots x_n$. Let $X = \{1, 2, \ldots, n\}$. Let $P_{k+1}(X)$ be the set of subsets of X of cardinality $k + 1$. Define a subset I of the set $P_{k+1}(X)$ as follows: $\{i_1, i_2, \ldots, i_{k+1}\}$ with $i_1 < i_2 < \ldots < i_{k+1}$, is in I if the corresponding derived sequence $(y_1, \ldots, y_k)$ of $(x_1, \ldots, x_n)$ is such that $x_1 \ldots x_n = u'y_1 \ldots y_k v'$ and

$$uu'y_{\sigma(1)} \cdots y_{\sigma(k)}v'v \in L.$$

Let us remark that, since $(x_1, \ldots, x_n)$ is an n-division of $x_1 \ldots x_n$, by Eq.(4.5), one has

$$w > uu'y_{\sigma(1)} \cdots y_{\sigma(k)}v'v;$$

hence, by induction on the order $<$, I remains unchanged if in its definition L is replaced by L'.

Let $J = P_{k+1}(X)\backslash I$. Then, by Ramsey's theorem, there exists an $Y \subset X$, with $card(Y) = m + 1$ and such that either $P_{k+1}(Y) \subset I$ or $P_{k+1}(Y) \subset J$. The word w may be rewritten as

$$w = u''z_1 \ldots z_m v'',$$

where $(z_1, \ldots, z_m)$ is the derived sequence of $(x_1, \ldots, x_n)$ corresponding to the set $Y = \{j_1, \ldots, j_{m+1}\}$ with $j_1 < \ldots < j_{m+1}$. For any $i_1, \ldots, i_{k+1}$, $i_1 < i_2 < \ldots < i_{k+1}$ the condition $P_{k+1}(Y) \subset I$ implies that the corresponding derived sequence $(t_1, \ldots, t_k)$ of $(z_1, \ldots, z_m)$ is such that $w = u''u'''t_1 \ldots t_k v'''v''$ and

$$u''u'''t_{\sigma(1)}t_{\sigma(2)} \ldots t_{\sigma(k)}v'''v'' \in L, \tag{4.6}$$

whereas the condition $P_{k+1}(Y) \subset J$ implies that

$$u''u'''t_{\sigma(1)}t_{\sigma(2)} \ldots t_{\sigma(k)}v'''v'' \notin L. \tag{4.7}$$

But L is (m, k)-permutative; hence, if $w \in L$, there are some $i_1, \ldots, i_{k+1}$ such that Eq.(4.6) holds; in this case, $P_{k+1}(Y) \subset I$. And if $w \notin L$, because of the permutative property, there are some $i_1, \ldots, i_{k+1}$ such that Eq.(4.7) holds; in this case, $P_{k+1}(Y) \subset J$. Thus, $w \in L \iff P_{k+1}(Y) \subset I$. Because of the remark made above, the same holds with L' in place of L. Hence,

$$w \in L \iff P_{k+1}(Y) \subset I \iff w \in L'.$$

This concludes the proof.

Similar to the permutative property is the *permutation property*: a language L has this property if, for some m (depending only on L), for any words $u, x_1, \ldots, x_m, v$ there exists a permutation $\alpha \in S_m \setminus id$, such that

$$ux_1 \ldots x_m v \in L \iff ux_{\alpha(1)} \ldots x_{\alpha(m)}v \in L.$$

This is just commutativity for $m = 2$. As before, the language of palindromes does not have the permutation property.

Problem. Does Theorem 4.3 still hold with the permutation property instead of the permutative property?

Remark 4.3. If, in the permutation property, the permutation α depends on $x_1, \ldots, x_m$ only (and not on the contexts (u, v)), then the property becomes syntactic and the answer to the problem is positive; indeed, the syntactic semigroup has the permutation property and, thus by Theorem 2.4, is finite.

5. Well quasi-orders

In this section we give some regularity conditions that can be expressed in terms of *well quasi-orders*. As shown by Ehrenfeucht et al. [43] regular languages can be characterized as the closed parts of well quasi-orders in the set of all the words over a finite alphabet (generalized Myhill's theorem); moreover, quasi-orders can be naturally associated with the derivation relations of suitable rewriting (or semi-Thue) systems, so that an interesting problem is to determine under which conditions a semi-Thue system, belonging to a given class, is such that its derivation relation is a well quasi-order. Ehrenfeucht *et al.*, for instance, characterized the *unitary rewriting systems* whose derivation relations are well quasi-orders [43]. A further example is given by *copying systems*. As an application of the generalized Myhill's theorem one has that languages on a binary alphabet generated by copying systems are regular. In the sequel we consider two further classes of rewriting systems. A first class (π) consists of productions of the kind: $u^m \to u^{m+k}$, $m > 0$, $k > 0$ and $u_1 u_2 \ldots u_n \leftrightarrow u_{\sigma(1)} u_{\sigma(2)} \ldots u_{\sigma(n)}$, for any sequence of words $u_1, \ldots, u_n \in A^+$, $n > 1$, and $u \in A^+$, where σ is a non-trivial permutation of $\{1, \ldots, n\}$; the permutation σ and the pair (m, k) depend on the words $u_1, \ldots, u_n$ and u, respectively. A second class (π') consists of productions of the kind $a^m \to a^{m+k}$, $m > 0$, $k > 0$ and $u_1 u_2 \ldots u_n \to u_{\sigma(1)} u_{\sigma(2)} \ldots u_{\sigma(n)}$, for any sequence of words $u_1, \ldots, u_n \in A^+$, $n > 1$, and $a \in A$, where σ is a fixed non-trivial permutation of $\{1, \ldots, n\}$. We prove that the derivation relations induced by the preceding systems are well quasi-orders. Two main corollaries are the following regularity conditions (cf. Theorems 5.9 and 5.11): i) *A permutable and quasi-periodic language is regular*, ii) *An almost-commutative language which is quasi-periodic on the letters is regular*.

5.1 The generalized Myhill-Nerode Theorem

Let S be a set and $\leq$ a quasi-order in S. A part X of S is *upper-closed*, or simply *closed*, with respect to $\leq$ if the following condition is satisfied: if $x \in X$ and $x \leq y$ then $y \in X$.

A quasi-order in S is called a *well quasi-order* (wqo) if every non-empty subset X of S has at least one minimal element in X but no more than a finite number of (non-equivalent) minimal elements. It is clear that a well ordered set is well quasi-ordered. There exist several conditions which characterize the concept of well quasi-order and that can be assumed as equivalent definitions (cf. [57]).

Theorem 5.1. *Let S be a set quasi-ordered by $\leq$. The following conditions are equivalent:*

i. $\leq$ *is a well quasi-order.*

ii. *the ascending chain condition holds for the closed subsets of S.*

iii. *every infinite sequence of elements of S has an infinite ascending subsequence.*

iv. *if $s_1, s_2, \ldots, s_n, \ldots$ is an infinite sequence of elements of S, then there exist integers i, j such that $i < j$ and $s_i \leq s_j$.*

v.   *there exists neither an infinite strictly descending sequence in S (i.e. $\leq$ is well founded), nor an infinity of mutually incomparable elements of S.*

vi. *S has the finite basis property, i.e. for each subset X of S there exists a finite subset F_X of X such that for every $x \in X$ there exists a $y \in F_X$ such that $y \leq x$.*

Let us now suppose that the set S is a semigroup.

Definition 5.1. *A quasi-order $\leq$ in a semigroup S is monotone on the right (on the left) if for all $x_1, x_2, y \in S$, $x_1 \leq x_2$ implies $x_1 y \leq x_2 y$ ($y x_1 \leq y x_2$).*

A quasi-order is *monotone* if and only if it is monotone on the right and on the left. One has, in particular, that a monotone equivalence is a congruence in S.

Definition 5.2. *A quasi-order $\leq$ in S is a divisibility order if it is monotone and, moreover, for all $s \in S$ and $x, y \in S^1$, $s \leq xsy$.*

The ordering by divisibility in abstract algebras was studied by G. H. Higman [57] who proved a very general theorem that in the case of semigroups becomes:

Theorem 5.2. *Let S be a semigroup quasi-ordered by a divisibility order $\leq$. If there exists a generating set of S well quasi-ordered by $\leq$, then S will be so.*

The Higman theorem has several applications in combinatorics on words as well as in language theory (see, for instance [100]). A remarkable consequence of it is the following. Let $S = A^+$ be the (free) semigroup generated by the alphabet A quasi-ordered by a relation $\leq$. This order can be extended to A^+ as follows. Let u and v be words; we set $u \leq v$ if some subsequence of v majorizes u letter-by-letter, i.e. $u = a_1 \ldots a_n$, $a_i \in A$, $i = 1, \ldots, n$ and $v \in A^* b_1 A^* b_2 A^* \ldots A^* b_n A^*$, with $b_i \in A$, $i = 1, \ldots, n$ and $a_i \leq b_i$, $i = 1, \ldots, n$. This is a divisibility order so that by Theorem 5.2 if A is wqo (as, for instance, when A is finite) then S is itself wqo.

A characterization of recognizable subsets of a semigroup can be obtained in terms of well quasi-orders. In fact it holds the following theorem, usually called the generalized Myhill-Nerode theorem, when S is a finitely generated free semigroup [43].

Theorem 5.3. *Let X be a part of a semigroup S. $X \in Rec(S)$ if and only if X is closed with respect to a monotone well quasi-order in S.*

One easily verifies that a congruence of finite index is a monotone well quasi-order which is an equivalence. Let $X \subseteq S$; we introduce the following relation $\leq_X$ in S defined as: For all $s, t \in S$,

$$s \leq_X t \text{ if and only if } \forall h, k \in S^1 (hsk \in X \Rightarrow htk \in X).$$

One also verifies that $\leq_X$ is a monotone quasi-order and X is closed w.r.t. $\leq_X$. Moreover, the equivalence relation $\leq_X \cap (\leq_X)^{-1}$ coincides with the syntactic congruence of X and $\leq_X$ is maximal (w.r.t. inclusion) in the set of all monotone quasi-orders with respect to which X is closed. The relation $\leq_X$ can be called the Myhill quasi-order relation relative to X.

Proposition 5.1. *Let X be a part of a semigroup S. $X \in Rec(S)$ if and only if $\leq_X$ is a well quasi-order.*

Proof. ($\Leftarrow$) It is trivial by the fact that X is closed w.r.t. the monotone well quasi-order $\leq_X$, so that by the generalized Myhill-Nerode theorem the result follows.

($\Rightarrow$) Let $X \in Rec(S)$. From Theorem 5.3 the set X is $\leq$-closed, where $\leq$ is a monotone wqo. This implies in view of Theorem 5.1 that for any infinite sequence $s_1, s_2, \ldots, s_n \ldots$ of elements of S, there exists an infinite subsequence $t_1, t_2, \ldots, t_n, \ldots$ such that: $t_1 \leq t_2 \leq \ldots \leq t_n \leq \ldots$. Since $\leq \subseteq \leq_X$ one derives $t_1 \leq_X t_2 \leq_X \ldots \leq_X t_n \leq_X \ldots$. This shows by Theorem 5.1 that $\leq_X$ is a wqo.

A partial generalization of Nerode's theorem and of Theorem 5.3 as well, is given by the following [36]:

Theorem 5.4. *Let X be a subset of S. The following conditions are equivalent*

i. X is recognizable.
ii. X is a closed part of a wqo $\leq_1$ monotone on the right and a wqo $\leq_2$ monotone on the left.

Let us observe that, differently from the congruential case, if a part X of a semigroup S is a closed part of a wqo monotone only in one direction, then, generally, X is not recognizable [36].

5.2 Quasi-orders and rewriting systems

Now we consider the case when S is the free monoid generated by a finite alphabet A. We recall that a *rewriting system*, or *semi-Thue system*, on A is a pair $\langle A, \pi \rangle$ where π is a binary relation on A^*, i.e. $\pi \subseteq A^* \times A^*$. Any pair of words $(p, q) \in \pi$ is called a *production* and denoted by $p \to q$. Let us denote by $\Rightarrow_\pi$ the regular closure of π, i.e. for $u, v \in A^*$,

$$u \Rightarrow_\pi v \text{ if and only if } \exists (p, q) \in \pi \text{ and } \exists h, k \in A^* : u = hpk \text{ and } v = hqk.$$

The derivation relation $\Rightarrow_\pi^*$ is defined as the reflexive and transitive closure of $\Rightarrow_\pi$; one can easily verify that $\Rightarrow_\pi^*$ is a monotone quasi-order. If π is a symmetric relation, then $\Rightarrow_\pi^*$ becomes a congruence relation in A^* and $\langle A, \pi \rangle$ is also called a *Thue-system*. From Theorem 5.1 and Proposition 5.1 we derive the following:

Proposition 5.2. *A language over a finite alphabet is regular if and only if it is $\Rightarrow_\pi^*$-closed with respect to a rewriting system π such that $\Rightarrow_\pi^*$ is a well quasi-order.*

Proof. ($\Leftarrow$) Obvious from Theorem 5.3 since $\Rightarrow_\pi^*$ is monotone.

($\Rightarrow$) Let L be a regular language. By Kleene's theorem L is recognizable. Thus by Proposition 5.1, $\leq_L$ is a wqo. Let us consider the rewriting system $\langle A, \pi \rangle$, with $\pi = \leq_L$. Since $\leq_L$ is monotone one has $\Rightarrow_\pi = \leq_L$. Moreover, since $\leq_L$ is reflexive and transitive, one has $\Rightarrow_\pi^* = \leq_L$. Since L is $\leq_L$-closed the result follows.

An interesting problem is to determine under which conditions a semi-Thue system π belonging to a given class is such that its derivation relation $\Rightarrow_\pi^*$ is a wqo. Let us, for instance, consider the class of *unitary* semi-Thue systems. We recall that a semi-Thue system π is called unitary when π is a finite set of productions of the kind $\epsilon \to q$, $q \in A^+$. Such a system is then determined by a finite set $I \subseteq A^+$. Let us simply denote by $\Rightarrow_I^*$ the derivation relation in these systems. If $I = A$, then $\Rightarrow_A^*$ is a wqo by the Higman theorem. In [43] Ehrenfeucht *et al.* gave the following interesting characterization of the finite sets I such that $\Rightarrow_I^*$ is a wqo.

Theorem 5.5. *The derivation relation $\Rightarrow_I^*$ of the unitary semi-Thue system associated to the finite set $I \subseteq A^+$ is a wqo if and only if there exists an integer $k \in \mathbb{N}$ such that every word in A^* whose length is greater than k has a factor in I.*

A set I which satisfies the property stated in the above theorem has been called *subword unavoidable*. Theorem 5.5 is then a non-trivial extension of Higman's theorem. A different extension was given by D. Haussler [55].

Let us denote with π the rewriting system $\pi = \{(x, xx) \mid x \in A^*\}$. The derivation relation $\Rightarrow_\pi^*$ is called *copying relation*. Copying systems and languages generated by them were introduced in [43] by Ehrenfeucht and Rozenberg. For any $w \in A^*$ the copying language $L_{w,\pi}$ is defined as

$$L_{w,\pi} = \{u \in A^* \mid w \Rightarrow_\pi^* u\}.$$

The language $L_{w,\pi}$ is upwards closed with respect to $\Rightarrow_\pi^*$. However, it has been proved in [43] that, *if w is a word containing at least three letters, then $L_{w,\pi}$ is not a regular language.* The following result has been proved in [11]:

Theorem 5.6. *Let A be a binary alphabet. The derivation relation $\Rightarrow_\pi^*$ is a well quasi-order on A^*.*

From Theorem 5.3 and Theorem 5.6 one derives:

Corollary 5.1. *Let A be a finite alphabet and $w \in A^*$. Then $L_{w,\pi}$ is regular if and only if w contains at most two letters.*

In the sequel we will give further classes of semi-Thue systems such that the derivation relation is a wqo.

Let the alphabet A be totally ordered by $\leq$ and let $\leq_{lex}$ and $\leq_a$ be the lexicographic and the alphabetic order generated by $\leq$, respectively.

We can then totally order the set Σ of all infinite sequences of words of A^+, $y : \mathbb{N} \to A^+$,

$$y = y_0, y_1, \ldots, y_n, \ldots$$

where $y_i = y(i)$ for all $i \geq 0$, in the following way. Let $y, z \in \Sigma$ and define, if $y \neq z$, $i = \min\{j \in \mathbb{N} \mid y_j \neq z_j\}$; we set $y \leq_a z$ if and only if $y = z$ or, otherwise, $y_i <_a z_i$.

Let $\leq$ be a quasi-order on A^+ and $x : \mathbb{N} \to A^+$ be an infinite sequence of words. We call x *bad* if for all i, j, $i \neq j$, one has $x_i \not\leq x_j$. Many proofs on well quasi-orders, as well as the proof of the Higman theorem, are based on the following proposition essentially due to Nash-Williams (cf. [74, 77]):

Proposition 5.3. *Let $\leq$ be a well founded but no well quasi-order. Then there exists a bad sequence which is minimal w.r.t. the order $\leq_a$.*

5.3 A regularity condition for permutable languages

We introduce a class of rewriting systems π whose productions are defined as follows. Let n be an integer > 1. For any sequence of words $u_1, \ldots, u_n \in A^+$ and $u \in A^+$

$$u_1 u_2 \ldots u_n \to u_{\sigma(1)} u_{\sigma(2)} \cdots u_{\sigma(n)}, \quad u_{\sigma(1)} u_{\sigma(2)} \cdots u_{\sigma(n)} \to u_1 u_2 \ldots u_n \quad (5.1)$$

$$u^m \to u^{m+k}, \quad m > 0, k > 0, \quad (5.2)$$

where $\sigma \in \mathcal{S}_n \setminus id$ and the pair (m, k) depend on the words $u_1, \ldots, u_n$ and u, respectively. Let us observe that each rewriting system π of the preceding class depends on the integer $n > 1$ and on the two maps $f : (A^+)^n \to \mathcal{S}_n \setminus id$, $g : A^+ \to \mathbb{N}_+ \times \mathbb{N}_+$. Hence a particular rewriting system of the class should be denoted by $\pi_{n,f,g}$. However, we drop the subscripts when no confusion arises.

Theorem 5.7. *The derivation relation $\Rightarrow_\pi^*$ is a well quasi-order.*

In order to prove the result we recall some theorems. For any $n > 0$ let us denote by L_n the set of all words on the alphabet A which do not contain n-divided factors. The following important theorem, due to Restivo and Reutenauer [92], gives a characterization of L_n in terms of bounded languages; we recall that a language L is *bounded* if for some words $w_1, \ldots, w_q$, it is contained in $w_1^* \ldots w_q^*$.

Theorem 5.8. *A language L is bounded if and only if for some integer n the set $F(L)$ of the factors of L does not contain n-divided words.*

The following lemma is a slight generalization of the famous lemma of Dickson (cf. [77]).

Lemma 5.1. *Let $q > 0$ and m_i, k_i be integers such that $m_i \geq 0$ and $k_i > 0$ $(i = 1, \ldots, q)$. Let us consider in $\mathbb{N}^q$ the relation $\Rightarrow$ defined as: let $(r_1, \ldots, r_q)$, $(s_1, \ldots, s_q) \in \mathbb{N}^q$, $(r_1, \ldots, r_q) \Rightarrow (s_1, \ldots, s_q)$ if and only if there exists $i \in [1, q]$ such that $r_i \geq m_i$, $s_i = r_i + k_i$ and $s_j = r_j$ for $j \neq i$. The reflexive and transitive closure $\Rightarrow^*$ of $\Rightarrow$ is a well quasi-order of $\mathbb{N}^q$.*

Proof of Theorem 5.7. The proof is by contradiction. Suppose that $\Rightarrow_\pi^*$, which we simply denote by $\leq$, is not a wqo. By definition of the rewriting system π, there cannot exist an infinite strictly descending chain w.r.t. the order $\leq$, so that $\leq$ is well founded. By Proposition 5.3 there exists a bad sequence $x : \mathbb{N} \to A^+$ which is minimal w.r.t. the order $\leq_a$. Let us now prove that for all $i \geq 0$, the word x_i does not contain n-divided factors. In fact suppose, by contradiction, that

$$x_i = x u_1 u_2 \ldots u_n y, \quad x, y \in A^*$$

and that for all permutations $\tau \in \mathcal{S}_n$, $\tau \neq id$, one has:

$$x u_{\tau(1)} u_{\tau(2)} \ldots u_{\tau(n)} y <_{lex} x u_1 u_2 \ldots u_n y. \tag{5.3}$$

By Eq.(5.1) $u_1 u_2 \ldots u_n \to u_{\sigma(1)} u_{\sigma(2)} \ldots u_{\sigma(n)}$ and $u_{\sigma(1)} u_{\sigma(2)} \ldots u_{\sigma(n)} \to u_1 u_2 \ldots u_n$, $\sigma \in \mathcal{S}_n$, $\sigma \neq id$.

Let us set $y_i = x u_{\sigma(1)} u_{\sigma(2)} \ldots u_{\sigma(n)} y$; one has from Eq.s (5.1) and (5.3)

$$y_i <_{lex} x_i, \quad x_i \leq y_i \text{ and } y_i \leq x_i.$$

Let us now prove that for all j, $j \neq i$, one has: $x_j \not\leq y_i$ and $y_i \not\leq x_j$. In fact if $x_j \leq y_i$ since $y_i \leq x_i$ we reach the contradiction $x_j \leq x_i$. Similarly if $y_i \leq x_j$, since $x_i \leq y_i$ we reach the contradiction $x_i \leq x_j$. It follows that the sequence z defined as $z_i = y_i$ and $z_h = x_h$, for $0 \leq h \neq i$, is a bad sequence. Since $y_i >_{lex} x_i$ one has $z <_a x$ and this contradicts the minimality of x. Let us now consider the language formed by the elements of the sequence x, i.e. $X = \cup_{i \in \mathbb{N}} \{x_i\}$. As we have seen before $F(X)$ does not contain n-divided words. From Theorem 5.8, X is a bounded language, i.e. there exist words $w_1, \ldots, w_q$, such that $X \subseteq w_1^* \ldots w_q^*$. Hence for any word $x_i \in X$ there exist non-negative integers $r_1, \ldots, r_q$ such that $x_i = w_1^{r_1} \ldots w_q^{r_q}$. Thus any element of X is uniquely determined by the q-tuple $(r_1, \ldots, r_q) \in \mathbb{N}^q$. Let (m_i, k_i) $(i = 1, \ldots, q)$ be the set of pairs of integers such that $w_i^{m_i} \to w_i^{m_i + k_i}$ according to Eq. (5.2). We can then consider in $\mathbb{N}^q$ the relation $\Rightarrow$ defined as in Lemma 5.1.

Let us observe that if $(r_1, \ldots, r_q) \Rightarrow (s_1, \ldots, s_q)$ then $w_1^{r_1} \ldots w_q^{r_q} \Rightarrow_\pi w_1^{s_1} \ldots w_q^{s_q}$. By Lemma 5.1 the reflexive and transitive closure $\Rightarrow^*$ of $\Rightarrow$ is a well quasi-order. Let the set M be defined as $M = \{(r_1, \ldots, r_q) \in \mathbb{N}^q \mid w_1^{r_1} \ldots w_q^{r_q} \in X\}$. By statement (v) of Theorem 5.1 there will exist two different q-tuples $(r_1, \ldots, r_q)$, $(s_1, \ldots, s_q) \in M$ such that $(r_1, \ldots, r_q) \Rightarrow^* (s_1, \ldots, s_q)$. Let $x_i = w_1^{r_1} \ldots w_q^{r_q}$ and $x_j = w_1^{s_1} \ldots w_q^{s_q}$; then one would obtain $x_i \leq x_j$ which is a contradiction.

From Theorem 5.7 one obtains a further proof [34] of the theorem of Restivo and Reutenauer (cf. Theorem 2.4). Indeed, let S be a finitely generated semigroup and $\phi : A^+ \to S$ the canonical epimorphism. If S is periodic and permutable, then one derives that there exists a rewriting system π, whose productions are defined by Eq.s (5.1) and (5.2), such that $\Rightarrow_\pi^* \subseteq \phi\phi^{-1}$. By Theorem 5.7, $\Rightarrow_\pi^*$ is a wqo; this trivially implies that the congruence $\phi\phi^{-1}$ is of a finite index. Hence, S is finite.

A language L is called *periodic* (*permutable*) if the syntactic semigroup $S(L)$ is periodic (permutable). Since $S(L)$ is finitely generated from Theorem 2.4 one derives that *a language L is regular if and only if L is periodic and permutable.*

Let us observe that the condition of periodicity for L can be expressed as: for all $u \in A^*$ there exist integers $n, k > 0$ such that for all $x, y \in A^*$ one has:

$$xu^n y \in L \Leftrightarrow xu^{n+k} y \in L. \tag{5.4}$$

We shall now prove that if L is a permutable language then L is regular under a condition which is weaker than that of periodicity. This condition is obtained by replacing in Eq. (5.4) the *double* implication with a *single* implication. More precisely let us give the following definition:

Definition 5.3. *A language L is called quasi-periodic if for all $u \in A^*$ there exist integers $n, k > 0$ such that for all $x, y \in A^*$ one has:*

$$xu^n y \in L \Rightarrow xu^{n+k} y \in L.$$

L is called co-quasi-periodic if the complement L^c of L is quasi-periodic.

One can easily prove that a language is periodic if and only if it is both quasi-periodic and co-quasi-periodic.

Theorem 5.9. *A language L is regular if and only if L is permutable and quasi-periodic or co-quasi-periodic.*

Proof. The "only if" part of the theorem is trivial by the fact that if L is regular then the syntactic semigroup $S(L)$ is finite and then periodic and permutable. Let us then prove the "if" part. First of all we observe that we can limit ourselves to prove the result when L is quasi-periodic. In fact suppose that we have already proved the theorem when the language is quasi-periodic and suppose that L is co-quasi-periodic. One has then that $S(L) = S(L^c)$ is permutable and L^c is quasi-periodic; hence L^c, as well as L is regular. Let us then suppose that L is permutable and quasi-periodic. This implies that there exists $n > 1$ such that for any sequence of words $u_1, \ldots, u_n$ of A^+ there exists a permutation $\sigma \in S_n$, $\sigma \neq id$, such that

$$u_1 u_2 \ldots u_n \equiv_L u_{\sigma(1)} u_{\sigma(2)} \ldots u_{\sigma(n)}; \tag{5.5}$$

moreover, for any word $u \in A^*$ there exist integers $m \geq 0$ and $k > 0$ such that for all $x, y \in A^*$

$$xu^my \in L \Rightarrow xu^{m+k}y \in L. \tag{5.6}$$

We can then introduce a rewriting system π whose productions are defined as: for any sequence of words $u_1, \ldots, u_n \in A^+$ and $u \in A^+$

$$u_1u_2 \ldots u_n \to u_{\sigma(1)}u_{\sigma(2)} \cdots u_{\sigma(n)}, \quad u_{\sigma(1)}u_{\sigma(2)} \cdots u_{\sigma(n)} \to u_1u_2 \ldots u_n \tag{5.7}$$

$$u^m \to u^{m+k}, \quad m > 0, k > 0, \tag{5.8}$$

where the permutations σ and the integers m and k are those in the Eq.s (5.5) and (5.6), respectively. From Eq.s (5.5) and (5.6) one has that L is $\Rightarrow_\pi^*$-closed, so that from Theorem 5.7 and Proposition 5.2 the result follows.

5.4 A regularity condition for almost-commutative languages

We introduce a class of rewriting systems π' as follows. Let m, k be positive integers, $n > 1$ and σ any fixed permutation $\sigma \in \mathcal{S}_n$, $\sigma \neq id$. The productions of π' are then defined as: For any letter $a \in A$ and words $w_1, \ldots, w_n \in A^+$

$$a^m \to a^{m+k}, \quad w_1w_2 \ldots w_n \to w_{\sigma(1)}w_{\sigma(2)} \cdots w_{\sigma(n)}. \tag{5.9}$$

Any particular rewriting system depends on the integers n, m, k and on the fixed permutation $\sigma \in \mathcal{S}_n \setminus id$, so that it should be denoted by $\pi'_{n,m,k;\sigma}$; however we drop the subscripts when there is no confusion. Let us denote by $\Rightarrow_{\pi'}$ the regular closure of π' and let $\Rightarrow_{\pi'}^*$ be the reflexive and transitive closure of $\Rightarrow_{\pi'}$. The following theorem was proved in [34]:

Theorem 5.10. *The derivation relation $\Rightarrow_{\pi'}^*$ is a well quasi-order.*

Definition 5.4. *A language L is quasi-periodic on the letters if for any $a \in A$ there exist integers $m > 0$ and $k \geq 1$ such that $a^m \leq_L a^{m+k}$. L is co-quasi-periodic on the letters if L^c is quasi-periodic on the letters.*

Definition 5.5. *A semigroup S is n-almost commutative, $n > 1$, if there exists a permutation $\sigma \in \mathcal{S}_n \setminus id$ such that for any sequence $s_1, \ldots, s_n$ of elements of S one has:*

$$s_1s_2 \ldots s_n = s_{\sigma(1)}s_{\sigma(2)} \cdots s_{\sigma(n)}.$$

S is said to be almost-commutative if it is n-almost commutative for a suitable $n > 1$. A language L is n-almost-commutative (almost-commutative) if its syntactic semigroup $S(L)$ is n-almost-commutative (almost-commutative).

A corollary of Theorem 5.10 is the following:

Theorem 5.11. *Let L be a language which is quasi-periodic, or co-quasi-periodic, on the letters. If L is almost-commutative, then L is regular.*

Proof. Let L (or L^c) be almost-commutative, i.e. there exist $n > 1$ and a permutation $\sigma \in S_n \setminus id$ such that for any sequence $u_1, \ldots, u_n$ of words of A^+ one has $u_1 u_2 \ldots u_n \equiv_L u_{\sigma(1)} u_{\sigma(2)} \ldots u_{\sigma(n)}$. If L is quasi-periodic on the letters one has that for any $a \in A$ there exists a pair of integers (m_a, k_a) such that $a^{m_a} \leq_L a^{m_a + k_a}$. Let us take $m = \max\{m_a \mid a \in A\}$ and $k = mcm\{k_a \mid a \in A\}$; it follows that for any $a \in A$, $a^m \leq_L a^{m+k}$. Let us then consider the rewriting system π' defined as: for any letter $a \in A$ and words $u_1, \ldots, u_n \in A^+$, $a^m \rightarrow a^{m+k}$ and $u_1 u_2 \ldots u_n \rightarrow u_{\sigma(1)} u_{\sigma(2)} \ldots u_{\sigma(n)}$. The language L is $\Rightarrow_{\pi'}^*$-closed so that by Theorem 5.10 the result follows. Let us now suppose that L is co-quasi-periodic on the letters. One has that L^c is quasi-periodic on the letters and that $S(L) = S(L^c)$ is almost-commutative. Hence L^c and then L are regular.

We want now to consider some classes of languages whose elements are quasi-periodic on the letters. Let L be a language over the alphabet A. We say that L satisfies the *iteration property* IP if the following condition is satisfied:

(IP) There exists a positive integer N such that each $w \in L$ of length at least N admits a factorization: $w = w_1 u w_3 v w_2$ satisfying the following conditions: *i)* $|uv| \geq 1$, *ii)* $|w_1 uv w_2| \leq N$ and *iii)* $w_1 u^n w_3 v^n w_2 \in L$, for all $n > 0$.

We recall that a *rational cone*, or simply *cone*, is any family of languages closed under rational transductions (cf. [5]). The following theorem is proved in [34].

Theorem 5.12. *Let $\mathcal{L}$ be a rational cone satisfying the iteration property IP. Then every almost-commutative language in $\mathcal{L}$ is quasi-periodic on the letters.*

From Theorems 5.11 and 5.12 one derives the following:

Corollary 5.2. *Every almost-commutative language in a rational cone satisfying the iteration property IP is regular.*

We remark that one-counter languages as well as the linear languages belong to rational cones (cf. [5]) which verify the property IP, so that from Corollary 5.2 we derive the following:

Corollary 5.3. *Almost-commutative one-counter languages and almost commutative linear languages are regular.*

We recall that Latteux and Rozenberg proved in [76] the following important theorem:

Theorem 5.13. *Let $\mathcal{L}$ be a family of languages satisfying the iteration property IP. Then every commutative language in $\mathcal{L}$ is regular.*

From the preceding theorem they derived:

Corollary 5.4. *Commutative one-counter languages and commutative linear languages are regular.*

We give now a direct proof of Corollary 5.2. We need the following lemma of P. Perkins[85]:

Lemma 5.2. *A semigroup S is almost-commutative if and only if there exist integers n and j such that $1 \leq j < j+1 \leq n$ and for any sequence $s_1, \ldots, s_n$ of n elements of S one has*

$$(s_1 \ldots s_{j-1})s_j s_{j+1}(s_{j+2} \ldots s_n) = (s_1 \ldots s_{j-1})s_{j+1}s_j(s_{j+2} \ldots s_n).$$

Proof of Corollary 5.2. Let L be an almost-commutative language in a rational cone $\mathcal{L}$ satisfying IP; by Lemma 5.2, there exist integers n and j such that $1 < j < j+1 < n$ and for any sequence $u_1, \ldots, u_n$ of words of A^+ one has

$$(u_1 \ldots u_{j-1})u_j u_{j+1}(u_{j+2} \ldots u_n) \equiv_L (u_1 \ldots u_{j-1})u_{j+1}u_j(u_{j+2} \ldots u_n). \quad (5.10)$$

Let $\alpha, \beta \in A^n$; we prove that the language $L_{\alpha,\beta} = (\alpha^{-1}L)\beta^{-1} \in \mathcal{L}$ is commutative, i.e. for any $x, y \in A^+$ one has that $xy \equiv_{L_{\alpha,\beta}} yx$. Indeed, let $z, h \in A^*$ and suppose that $zxyh \in L_{\alpha,\beta}$; one has then $\alpha zxyh\beta \in L$. From Eq. (5.10) since $|\alpha|, |\beta| = n$, one has that $\alpha zyxh\beta \in L$; hence $zyxh \in L_{\alpha,\beta}$. It follows then that $xy \equiv_{L_{\alpha,\beta}} yx$. In view of the fact that $\mathcal{L}$ is a cone satisfying IP by Theorem 5.13, it follows that $L_{\alpha,\beta}$ is regular. The language L is given, up to a finite set, by

$$\bigcup_{\alpha,\beta \in A^n} \alpha L_{\alpha,\beta}\beta;$$

hence L is regular since it is a finite union of regular languages.

We consider now the family Qrt of *quasi-rational* languages which is defined as the closure by substitution of the family Lin of linear languages. More precisely set $Lin_1 = Lin$ and, for any $k > 1$:

$$(Lin)_k = Lin \odot (Lin)_{k-1},$$

where $\odot$ denotes the substitution operator [5]. The family Qrt is then defined as:

$$Qrt = \bigcup_{k>0}(Lin)_k.$$

It holds the following noteworthy theorem due to J. Kortelainen [73]:

Theorem 5.14. *Any commutative quasi-rational language is regular.*

By the above theorem and the lemma of Perkins we can prove the following more general proposition in which "commutative" is replaced by "almost-commutative":

Theorem 5.15. *Any almost-commutative quasi-rational language is regular.*

Proof. We observe first that Qrt is a cone. Indeed, Lin is a cone; moreover, it is well known that if $\mathcal{L}$ and $\mathcal{M}$ are two cones then $\mathcal{L} \odot \mathcal{M}$ will be so. Thus for any $k > 0$, $(Lin)_k$ is a cone; hence it follows that Qrt is a cone. If L is an almost-commutative language in Qrt, then by Lemma 5.2 Eq.(5.10) is satisfied for a suitable $n > 1$. Using an argument similar to that in the proof of Corollary 5.2 one has that for $\alpha, \beta \in A^n$ the language $L_{\alpha,\beta} = (\alpha^{-1}L)\beta^{-1} \in Qrt$ is commutative. By Kortelainen's theorem one has that $L_{\alpha,\beta}$ is rational. Hence it follows that L is regular since it is a finite union of regular languages.

One can ask the question whether in Theorem 5.11 one can substitute "almost-commutative" languages with "permutable" languages. In general the answer to this question is negative; however, for 3-permutable languages the following holds (cf. [34]):

Theorem 5.16. *Let L be a 3-permutable language which is quasi-periodic or co-quasi-periodic on the letters. Then L is regular.*

References

1. S. I. Adjan, *The Burnside problem and identities in groups*, Springer-Verlag, Berlin 1978.
2. S. I. Adjan and I. G. Lysionok, The method of classification of periodic words and the Burnside problem, *Contemporary Mathematics* **131** 1992, 13–28.
3. A. V. Anisimov, Group Languages, *Kibernetika* **4** 1971, 18–24.
4. D. B. Bean, A. E. Ehrenfeucht and G. McNulty, Avoidable patterns in strings of symbols, *Pacific J. Math.* **85** 1979, 261–294.
5. J. Berstel, *Transductions and Context-Free Languages*, Teubner, Stuttgart 1979.
6. J. Berstel and D. Perrin, *Theory of Codes*, Academic Press, New York 1985.
7. J. C. Birget and J. Rhodes, Almost finite expansions of arbitrary semigroups, *Journal of Pure and Applied Algebra* **32** 1984, 239–287.
8. R. D. Blyth, Rewriting products of group elements, (Ph.D. Thesis, 1987, University of Illinois at Urbana-Champain), see also Rewriting products of group elements, I, *J. of Algebra* **116** 1988, 506–521.
9. R. D. Blyth and A. H. Rhemtulla, Rewritable products in FC-by-finite groups, *Canad. J. Math.* **41** 1989, 369–384.
10. L. Boasson, Un critére de rationalité des langages algébriques, in: *Automata, Languages and Programming*, pp. 359–365, (M. Nivat ed.), North Holland, Amsterdam 1973.
11. D. P. Bovet and S. Varricchio, On the regularity of languages on a binary alphabet generated by copying systems, *Information Processing Letters* **44** 1992, 119–123.
12. T. C. Brown, An interesting combinatorial method in the theory of locally finite semigroups, *Pacific J. of Math.* **36** 1971, 285–289.
13. T. C. Brown, On Van der Waerden's Theorem and the theorem of Paris and Harrington, *Journal of Combinatorial Theory, Ser. A*, **30** 1981, 108–111.
14. J. Brzozowski, K. Culik II and A. Gabriellan, Classification of noncounting events, *J. Comput. System Sci.* **5** 1971, 41–53.

15. J. Brzozowski, Open problems about regular languages, in: R. V. Book, ed., *Formal Language Theory, Perspectives and Open Problems*, Academic Press, London, New York 1980, 23–45.
16. A. H. Clifford and G. B. Preston, *The algebraic theory of semigroups*, vol.1, 1961, vol.2, 1967, Mathematical Surveys, Number 7, American Mathematical Society, Providence, Rhode Island.
17. M. Coudrain and M. P. Schützenberger, Une condition de finitude des monoides finiment engendrés, *C. R. Acad. Sc. Paris, Ser. A*, **262** 1966, 1149–1151.
18. M. Curzio, P. Longobardi and M. Maj, Su di un problema combinatorio in teoria dei gruppi, *Atti Acc. Lincei Rend.fis. VIII* **74** 1983, 136–142.
19. M. Curzio, P. Longobardi, M. Maj and D. J. S. Robinson, A permutational property of groups, *Arch.math.* **44** 1985, 385–389.
20. F. M. Dekking, Strongly non repetitive sequences and progression-free sets, *J. Comb. Theory, Ser. A*, **27** 1979, 181–185.
21. A. de Luca, A note on the Burnside problem for semigroups, *Semigroup Forum* **31** 1985, 251–254.
22. A. de Luca and A. Restivo, A finiteness condition for finitely generated semigroups, *Semigroup Forum* **28** 1984, 123–134.
23. A. de Luca, A. Restivo and S. Salemi, On the centers of a languages, *Theoretical Computer Science* **24** 1983, 21–34.
24. A. de Luca and S. Varricchio, Some combinatorial properties of the Thue-Morse sequence and a problem in semigroups, *Theoretical Computer Science* **63** 1989, 333–348.
25. A. de Luca and S. Varricchio, Factorial languages whose growth function is quadratically upper bounded, *Information Processing Letters* **30** 1989, 283–288.
26. A. de Luca and S. Varricchio, On non counting regular classes, Proc.s of the 17° ICALP, 1990, *Lecture Notes in Computer Science*, **443**, Springer-Verlag, Berlin 1990, 74–87.
27. A. de Luca and S. Varricchio, A note on ω-permutable semigroups, *Semigroup Forum* **40** 1990, 153–157.
28. A. de Luca and S. Varricchio, A combinatorial theorem on p-power-free words and an application to semigroups, *R.A.I.R.O. I. T.* **24** 1990, 205–228.
29. A. de Luca and S. Varricchio, A finiteness condition for semigroups generalizing a theorem of Hotzel, *J. of Algebra* **136** 1991, 60–72.
30. A. de Luca and S. Varricchio, Finiteness and iteration conditions for semigroups, *Theoretical Computer Science* **87** 1991, 315–327.
31. A. de Luca and S. Varricchio, Combinatorial properties of uniformly recurrent words and an application to semigroups, *Int. J. Algebra Comput.* **1** 1991, 227–245.
32. A. de Luca and S. Varricchio, On non counting regular classes, *Theoretical Computer Science* **100** 1992, 67–102; MR 93i:68136.
33. A. de Luca and S. Varricchio, On finitely recognizable semigroups, *Acta Informatica* **29** 1992, 483–498.
34. A. de Luca and S. Varricchio, Some regularity conditions based on well quasi-orders, *Lecture Notes in Computer Science*, **583** Springer-Verlag, Berlin 1992, 356–371.
35. A. de Luca and S. Varricchio, A finiteness condition for semigroups, in "Proc.s Conference on Semigroups, Algebraic Theory and Applications to Formal Languages and Codes" Luino, 22–27 June 1992, World Scientific, Singapore 1993, 42–50.
36. A. de Luca and S. Varricchio, Well quasi-orders and regular languages, *Acta Informatica* **31** 1994, 539–557.

37. A. de Luca and S. Varricchio, A finiteness condition for semigroups generalizing a theorem of Coudrain and Schützenberger, *Advances in Mathematics* **108** 1994, 91–103.

38. F. Di Cerbo, *Sul Problema di Burnside per i semigruppi*, Tesi Università di Napoli, 1985.

39. A. P. do Lago, Sobre os Semigrupos de Burnside $x^n = x^{n+m}$, Master's Thesis, Universidade de São Paulo, 1991, *see also* Proc. of Conf. Latin, 1992, *Lecture Notes in Computer Science*, **583** Springer-Verlag, Berlin 1992, 329–355.

40. A. P. do Lago, On the Burnside Semigroups $x^n = x^{n+m}$, *Int. J. Algebra Comput.*, **6** 1996, 179–227.

41. A. Ehrenfeucht, R. Parikh and G. Rozenberg, Pumping lemmas for regular sets, *SIAM J. of Computation* **10** 1981, 536–541.

42. A. Ehrenfeucht, D. Haussler and G. Rozenberg, On regularity of context-free languages, *Theoretical Computer Science* **27** 1983, 311–332.

43. A. Ehrenfeucht and G. Rozenberg, On regularity of languages generated by copying systems, *Discrete Appl. Math.* **8** 1984, 313–317.

44. A. Ehrenfeucht and G. Rozenberg, Strong iterative pairs and the regularity of context-free languages, *RAIRO, Inform. Théor. Appl.* **19** 1985, 43–56.

45. S. Eilenberg, *Automata, Languages and Machines* vol. A, Academic Press, New York, 1974.

46. T. Evans, Some connections between residual finiteness, finite embeddability and the word problem, *J. London Math. Soc.* **1** 1969, 399–403.

47. H. Furstenberg, Poincaré recurrence and number theory, *Bulletin of A.M.S.* **5** 1981, 211–234.

48. E. S. Golod, On nil-algebras and finitely approximable p-groups, *Izv. Akad. Nauk SSSR, Ser.Mat.* **28** 1964, 273–276.

49. R. Graham, B. L. Rothschild and J. H. Spencer, *Ramsey theory*, 2-nd ed., Wiley and Sons, New York, 1990.

50. J. A. Green and D. Rees, On semigroups in which $x^r = x$, *Proc. Cambridge Philos. Soc.* **48** 1952, 35–40.

51. R. I. Grigorchuk, Burnside's problem on periodic groups, *Funktsional'nyi Analiz i Ego Prilozheniya* **14** 1980, 53–54.

52. V. Guba, The word problem for the relatively free semigroups satisfying $T^m = T^{m+n}$ with $m \geq 3$, *Int. J. Algebra Comput.* **3** 1993, 335–348.

53. N. Gupta and S. Sidki, On the Burnside problem for periodic groups, *Math.Z.* **182** 1983, 385–388.

54. K. Hashiguchi, Notes on finitely generated semigroups and pumping conditions for regular languages, *Theoretical Computer Science* **46** 1986, 53–66.

55. D. Haussler, Another generalization of Higman's well quasi order result on Σ^*, *Discrete Mathematics* **57** 1985, 237–243.

56. F. C. Hennie, One-tape off line Turing machine computations, *Informations and Control* **8** 1965, 553–578.

57. G. H. Higman, Ordering by divisibility in abstract algebras, *Proc. London Math. Soc.* **3** 1952, 326–336.

58. E. Hotzel, On finiteness conditions in semigroups, *Journal of Algebra* **60** 1979, 352–370.

59. S. V. Ivanov, On the Burnside problem on periodic groups, *Bulletin of the American Mathematical Society* **27** 1992, 257–260.

60. G. Jacob, La finitude de représentations linéaires des semi-groupes est décidable, *J. Algebra* **52** 1978, 437–459.

61. N. Jacobson, *Structure of Rings*, Amer. Math. Soc. Colloq. Publ. vol. 37, 1964.

62. J. Justin, Généralisation du théoreme de van der Waerden sur les semigroupes répétitifs, *J.Comb.Theory, Ser. A*, **12** 1972, 357–367.

63. J. Justin, Characterization of the repetitive commutative semigroups, *J. of Algebra* **21** 1972, 87–90.

64. J. Justin, Propriétés combinatoires de partitions finies du demi-groupe libre, in *Séminaire d'Informatique Théorique Année 1981-82 (LITP, Univ.Paris)*, 55–66.

65. J. Justin and G. Pirillo, On a natural extension of Jacob's ranks, *J.of Combinatorial Theory*, Ser. A, **43** 1986, 205–218.

66. J. Justin and G. Pirillo, Shirshov's theorem and ω-permutability of semigroupes, *Advances in Mathematics* **87** 1991, 151–159.

67. J. Justin and G. Pirillo, A finiteness condition for semigroups generated by a finite set of elements of finite order, *PUMA*, Ser. A, **1** 1990, 45–48.

68. J. Justin, G. Pirillo and S. Varricchio, Unavoidable regularities and finiteness conditions for semigroups, in "Proc.s of third Italian Conference in Theoretical Computer Science" Mantova 2-4 November 1989, World Scientific, Singapore 1989 350–355.

69. I. Kaplansky, *Fields and Rings*, 2nd ed. University of Chicago Press, Chicago and London, 1972.

70. M. Kargapolov and Iou. Merzliakov, *Eléments de la théorie des groupes*, Editions MIR-Moscou, 1985.

71. J. Karhumaki, On cube-free ω-words generated by binary morphisms, *Discrete Applied Mathematics* **5** 1983, 279–297.

72. O. G. Kharlampovich and M. V. Sapir, Algorithmic Problems in Varieties, *Int. J. Algebra Comput.* **5** 1995, 379–602.

73. J. Kortelainen, Every commutative quasi-rational language is regular, *RAIRO, Informatique Théorique* **20** 1986, 319–337.

74. J. Kruskal, The theory of Well-Quasi-ordering: A Frequently Discovered Concept, *J. Combin. Theory*, Ser. A, **13** 1972, 297–305.

75. G. Lallement, *Semigroups and Combinatorial Applications*, John Wiley, New York 1979.

76. M. Latteux and G. Rozenberg, Commutative one-counter Languages are regular, *J. Comput. and System Sci.* **29** 1984, 54–57.

77. M. Lothaire, *Combinatorics on words*, Encyclopedia of Mathematics and its Applications, vol.17, Addison-Wesley, 1983.

78. R. C. Lyndon and P. E. Schupp, *Combinatorial Group Theory*, Springer-Verlag, Berlin 1977.

79. J. McCammond, On the solution of the word problem for the semigroups satisfying $T^a = T^{a+b}$ with $a \geq 6$, *Int. J. Algebra Comput.* **1** 1991, 1–32.

80. R. McNaughton and S. Papert, *Counter-free Automata*, MIT Press, Cambrige, MA 1971.

81. R. McNaughton and Y. Zalcstein, The Burnside problem for semigroups of matrices, *J. of Algebra* **34** 1975, 292–299.

82. F. Mignosi, Infinite words with linear subword complexity, *Theoretical Computer Science* **65** 1989, 221–242.

83. M. Morse and G. Hedlund, Unending chess, symbolic dynamics and a problem in semigroups, *Duke Math.J.* **11** 1944, 1–7.

84. J. Okniński, *Semigroup Algebras*, M. Dekker, New York 1990.

85. P. Perkins, Bases for equational theories of semigroups, *J. of Algebra* **11** 1968, 298–314.

86. G. Pirillo, On permutation properties for finitely generated semigroups, in: *Combinatorics 86*, (A. Barlotti, M. Marchi and G. Tallini, Eds.), *Ann. Discrete Math.* **37** North-Holland, Amsterdam 1988, 375–376.

87. G. Pirillo and S. Varricchio, On uniformly repetitive semigroups, *Semigroup Forum* **49** 1994, 125–129.

88. C. Procesi, The Burnside problem, *J. of Algebra* **4** 1966, 421–425.
89. A. Restivo, Permutation properties and the Fibonacci semigroup, *Semigroup Forum* **38** 1989, 337–345.
90. A. Restivo and C. Reutenauer, On the Burnside problem for semigroups, *Journal of Algebra* **89** 1984, 102–104.
91. A. Restivo and C. Reutenauer, Rational languages and the Burnside problem, *Theoretical Computer Science* **40** 1985, 13–30.
92. A. Restivo and C. Reutenauer, Some applications of a theorem of Shirshov to Language theory, *Information and Control* **57** 1983, 205–213.
93. C. Reutenauer, Mots de Lyndon et un theoreme de Shirshov, *Ann. Sc. Math. Québec* **10** 1986, 237–245.
94. L. Rowen, *Polynomial identities in Ring Theory* Academic Press 1980.
95. I. Schur, Uber Gruppen periodischer Substitutionen, *Sitzungsber. Preuss. Akad. Wiss.* 1911, 619–627.
96. A. I. Shirshov, On certain non associative nil rings and algebraic algebras, *Mat. Sb.* **41** 1957, 381–394.
97. A. I. Shirshov, On rings with identity relations, *Mat. Sb.* **43** 1957, 277–283.
98. I. Simon, Notes on noncounting languages of order 2, 1970, manuscript
99. I. Simon, Conditions de finitude pour des semi-groupes, *C.R. Acad. Sci. Paris Sér. A*, **290** 1980, 1081–1082.
100. I. Simon, Piecewise testable events, in *Lecture Notes in Computer Science* **33** Springer-Verlag, Berlin 1975, 214-222.
101. H. Straubing, The Burnside problem for semigroups of matrices, in *Combinatorics on words, Progress and Perspectives* (L. J. Cummings ed.) Academic Press, New York 1983, 279–285.
102. A. Thue, Uber unendliche Zeichenreihen, *Norske Vid. Selsk. Skr. I. Mat. Nat. Kl., Christiania* **7** 1906, 1–22.
103. A. Thue, Uber die gegenseitige Lage gleicher Teile gewisser Zeichenreihen, *Videnskabsselskabets Skrifter* 1, Mat.-Nat. K.1., Christiania 1912.
104. B. L. van der Waerden, Wie der Beweis der Vermutung von Baudet gefunden wurde, *Abhandlungen des Mathematischen Seminars der Hanseatischen Universität Hamburg* 1965, 6–15, also published as: [How the proof of Baudet's conjecture was found, *Studies in Pure Mathematics*, Academic Press, New York 1971, 251–260.]
105. S. Varricchio, A finiteness condition for finitely generated semigroups, *Semigroup Forum* **38** 1989, 331–335.
106. S. Varricchio, Factorizations of free monoids and unavoidable regularities, *Theoretical Computer Science* **73** 1990, 81–89.
107. S. Varricchio, A positive non uniform pumping condition for regular sets, *preprint LITP, University of Paris 7, 90.33, SIAM Journal of Computation*, to appear.
108. A. I. Zimin, Blocking sets of terms, *Matem. Sbornik* **119** 1982, 363–375.

Families Generated by Grammars and L Systems

Gheorghe Păun and Arto Salomaa

1. Grammar forms

1.1 Introduction: structural similarity

Grammars appear in many chapters of this Handbook, often constituting the key notion of the chapter. Usually a specific language, $L(G)$, is associated to a grammar G. The language $L(G)$ is *generated* by the grammar G, obtained from a specific starting point by rules specified in the grammar. In this chapter we take a more general point of view. A grammar G defines a collection of *structurally similar* grammars G', called *interpretations* of G. Each of the interpretations G', in turn, generates a language $L(G')$ in the usual way. In this chapter we consider the *family $\mathcal{L}(G)$ of languages $L(G')$* generated by the interpretations G' of G. The family $\mathcal{L}(G)$ is referred to as the *grammatical family* associated to G. Thus, from the point of view taken in this chapter, grammars generate families of languages rather than single languages. When grammars are considered in this way, the term *"grammar form"* rather than "grammar" will be used. As constructs grammar forms and grammars are identical. However, they are applied in different ways. We will consider also *L forms*, that is L systems applied similarly, defining language families via interpretations.

Before we continue our discussion, let us take a simple example. Consider two grammars

$$G_{REG} = (\{S\}, \{a\}, S, \{S \to aS, S \to a\}), \text{ and}$$
$$G_{CF} = (\{S\}, \{a\}, S, \{S \to SS, S \to a\}).$$

Clearly, $L(G_{REG}) = L(G_{CF}) = a^+$ and, hence, the two grammars are *equivalent* in the sense that they generate the same language. However, it is intuitively clear that the grammar G_{CF} is "stronger" with respect to the generative capacity. This essential difference becomes evident if we view the productions as *schemata* for productions, that is, any grammar G' whose productions are of types $A \to BC$ and $A \to b$, where A, B, C are nonterminals and b is a terminal, is viewed as a "pattern match" or an "interpretation" of G_{CF}. The latter term is used in the theory of grammar forms, and grammars

G' described above constitute exactly the interpretations of G_{CF}, according to the formal definition given below. Every (λ-free) context-free language can be generated by such an interpretation G' and, consequently, the grammatical family associated to G_{CF}, when G_{CF} is viewed as a grammar form, equals the family of context-free languages not containing the empty word λ. Similarly, the grammatical family associated to G_{REG} equals the family of λ-free regular languages. Thus, the generative capacity of G_{REG} is far less than that of G_{CF}, when the two constructs are viewed as grammar forms rather than grammars.

We already pointed out that the interpretations G' of a grammar form G are intended to be *structurally similar* grammars. Indeed, the basic G can be viewed as a *master grammar*, sometimes also referred to as a *form grammar* that defines a class of structurally similar grammars. As we will see, the resulting notion of structural similarity is rather strong, giving rise to natural coherent classes of similar grammars. Formal language theory is rich in other notions and ideas concerning structural similarity – we will now briefly mention some of them.

The simplest notion of structural similarity is that of *equivalence*: the grammars generate the same language. However, this notion is too primitive, since grammars which are very different in structure can be related in this way. A second notion of grammatical similarity, *structural equivalence*, was suggested in the late 60's. Essentially, two context-free grammars G_1 and G_2 are *structurally equivalent* if, for every derivation tree in G_1 with a terminal frontier there is an equally shaped derivation tree in G_2, and vice versa. Thus, not only the languages but also the derivations coincide. The same effect is obtained if we require that the *parenthesized versions* of G_1 and G_2 are equivalent in the normal sense: the right sides of all productions are included in parentheses, $A \to \alpha$ becomes $A \to (\alpha)$. A remarkable result here is that, while equivalence is undecidable, structural equivalence is decidable for context-free grammars.

The reader is referred to [50] for a discussion concerning structural equivalence and related notions, as well as for early literature in this area. The notion of *cover* and *syntax-directed translation* have had considerable importance in the area of compiler theory and practice. In both cases this involves a concept of grammatical similarity. We refer to [1] and [10] for further details. Also [50] contains a brief exposition and further references.

An idea of *topological similarity* for grammars has also been suggested, see [48]. Of all the approaches to grammatical similarity briefly mentioned here, the idea of a *grammar morphism*, first studied in [12], seems have been the most useful and, at the same time, the most natural. This approach dealt with *category theory*. A grammar defines a category, called the *syntax category*. The objects are sentential forms of a grammar G and the morphisms are derivations in G. Essentially, a grammar morphism is a functor between two syntax categories. This mathematical background confirms that two grammars, one of which is obtained from the other by a grammar morphism, are structurally similar. An interested reader can find from [50] further details and references. We only mention here that the idea of a grammar morphism comes close to the relation of a grammar being an interpretation of another grammar, defined formally in the next subsection.

The idea of structural similarity investigated in the theory of grammar forms is based on *collections* defined by a grammar: a master grammar defines a collection of structurally similar grammars. When considering L systems, we obtain in the same way a collection of structurally similar L systems. The next thought is immediate: form theory can and should be applied to the study of *dialects* of languages, both natural and programming languages, as well as of *species* of organisms. So far no significant steps have been taken in this direction.

A brief outline of the contents of this chapter follows. This first section presents the basics of *grammar form* theory, beginning with the fundamental definitions in Section 1.2. The second section discusses *L forms* theory in a similar fashion. Section 3 takes up the topic of *density*. The density of grammatical families is a surprising phenomenon, perhaps the most surprising one brought about by grammar form theory. It can be viewed as the only known instance of density obtained by purely language-theoretic means, not invoking the density of real numbers as done in the ordering of complexity classes. We will extend in Section 3 the idea of density from grammatical to so-called *linguistical* families, as well as *color families* of graphs. The form theory presented in the first three sections is restricted to the *context-free* case, which in the case of L systems means E0L and ET0L systems and forms. The final section extends the considerations to various other types of grammars.

1.2 Interpretations and grammatical families

We assume that the reader is familiar with the basic notions concerning grammars and L systems. For this purpose, Section 3 of Chapter 1 of this Handbook should be sufficient. If not, appropriate parts of Chapters 4 and 5 can be consulted. All more specific notions needed in our presentation will be explained below. We will often use the following λ-*convention* without further mention. Two languages L_1 and L_2 are considered equal if $L_1 - \{\lambda\} = L_2 - \{\lambda\}$. Two language families $\mathcal{L}_1$ and $\mathcal{L}_2$ are considered equal if, for every $L_1 - \{\lambda\} \neq \emptyset$ in $\mathcal{L}_1$, there is an L_2 in $\mathcal{L}_2$ such that $L_1 - \{\lambda\} = L_2 - \{\lambda\}$, and vice versa. Thus, the λ-convention means that we ignore the empty set in language families and the empty word in languages. One has to be cautious in applying the λ-convention. For instance, ignoring the empty word in a factor of a catenation $L_1 L_2$ may cause a drastic change in the whole catenation.

In this Section 1, all grammars considered will be *context-free* and written in the form $G = (V, \Sigma, S, P)$, where Σ is the terminal and V the total alphabet (sometimes we consider also the nonterminal alphabet $V_N = V - \Sigma$),

$P \subseteq V_N \times V^*$ is a finite set of productions and $S \in V_N$ is the start letter. We often define grammars by simply listing their productions, in which case capital letters denote nonterminals and lower case letters denote terminals. A notion needed in the following fundamental definition is that of a *disjoint-finite-letter substitution* (*dfl-substitution*). Let V_1 and V_2 be alphabets. A *dfl-substitution* is a mapping μ of V_1 into the set of nonempty subsets of V_2 such that $\mu(a) \cap \mu(b) = \emptyset$ for all distinct $a, b \in V_1$. Thus, a dfl-substitution associates one or more letters of V_2 to each letter of V_1, and no letter of V_2 is associated to two letters of V_1. Because μ is a substitution, its domain is immediately extended to concern words and languages over V_1. For a production $A \to \alpha$, we define

$$\mu(A \to \alpha) = \{A' \to \alpha' \mid A' \in \mu(A) \text{ and } \alpha' \in \mu(\alpha)\}.$$

We are now ready for the basic

Definition 1.1. *A grammar* $G' = (V', \Sigma', S', P')$ *is an* interpretation *of a grammar* $G = (V, \Sigma, S, P)$ *modulo* μ, *denoted* $G' \lhd G$ (*mod* μ), *where* μ *is a dfl-substitution on* V, *if conditions (i) and (ii) are satisfied:*

(i) $V' = \mu(V)$, $\Sigma' = \mu(\Sigma)$ *and* $S' \in \mu(S)$,
(ii) $P' \subseteq \mu(P) = \bigcup_{p \in P} \mu(p)$.

The grammar G *is referred to as the* master *or* form grammar, *while* G' *is the* image *or* interpretation grammar. *The* grammar family *and the* grammatical (language) family *of* G *are defined by*

$$\mathcal{G}(G) = \{G' \mid G' \lhd G \ (mod \ \mu), \text{ for some } \mu\},$$
$$\mathcal{L}(G) = \{L(G') \mid G' \in \mathcal{G}(G)\}.$$

A language family $\mathcal{L}$ *is termed* grammatical *if* $\mathcal{L} = \mathcal{L}(G)$, *for some* G. *Two grammars are* form equivalent *if their language families coincide. They are* strongly form equivalent *if their grammar families coincide.*   □

Operationally we obtain an interpretation grammar by mapping terminals and nonterminals of the form grammar into disjoint sets of terminals and nonterminals, respectively, then extending the mapping to concern productions and, finally, taking a *subset* of the resulting production set. The last-mentioned point is especially important: great flexibility results because we are able to *omit* productions. *Full interpretations*, where it is required that $P' = \mu(P)$, offer rather limited possibilities.

A grammar is said to be a *grammar form* if it is used within the framework of interpretations. There is no difference between a grammar and a grammar form as constructs.

Let us consider some examples. The grammatical families of the grammar forms G_{REG} and G_{CF}, discussed in Section 1.1, equal the families REG and CF of regular and context-free languages, respectively. Thus, REG and CF

are grammatical families. (Observe that we invoke here the λ-convention !) On the other hand, the grammar family $\mathcal{G}(G_{CF})$ does not contain every context-free grammar; it contains only the grammars in the Chomsky normal form. The grammatical family of the grammar form defined by the productions

$$S \to a, \; S \to aS, \; S \to aSS$$

also equals CF, and the grammar family consists of all context-free grammars in the so-called Greibach 2-standard normal form. Also the grammar form defined by the productions

$$S \to a^i, 1 \le i \le 15, \; S \to a^3 SaSa^{10}$$

gives rise to a normal form for context-free grammars. That actually every context-free language is generated by a grammar in such a normal form (and, thus, the grammatical family of the grammar form under consideration equals again CF) is a consequence of Theorem 1.18 below.

We often say briefly that G' is an interpretation of G, or $G' \lhd G$, if $G' \lhd G \; (mod \; \mu)$ holds for some μ. Let us still discuss how interpretations are constructed, trying to clarify a couple of essential issues.

Occurrences of the same letter may be mapped to different letters in interpretations, whereas different letters must always be mapped to different letters. The type of a letter (terminal or nonterminal) cannot be changed.

Let the master grammar G be defined by

$$S \to aba, \; S \to bSccSa.$$

The grammar consisting of the single rule $S \to cab$ is an interpretation of G, whereas the grammar consisting of the rule $S \to abb$ is not an interpretation of G. (Clearly no dfl-substitution can produce the latter grammar from G, since b would have to be an interpretation of both a and b in the original G.) The grammar defined by the rules

$$S \to bdc, \; S \to ada, \; A \to bdb, \; A \to cda, \; B \to cdc,$$
$$S \to dAefBa, \; S \to dBeeSb, \; S \to dSfeSa,$$
$$A \to dAffAc, \; B \to dAefSb$$

is an interpretation of G modulo the dfl-substitution μ:

$$\mu(S) = \{S, A, B\}, \; \mu(a) = \{a, b, c\}, \; \mu(b) = \{d\}, \; \mu(c) = \{e, f\}.$$

The term *uniform interpretation* is used if in each rule two occurrences of the same letter must map to the same letter. If different letters are allowed to map to the same letter, we speak of *quasi-interpretations*.

The following theorems contain some rather straightforward results.

Theorem 1.1. *The relation $\lhd$ is reflexive and transitive. Consequently, $G_1 \lhd G_2$ holds iff $\mathcal{G}(G_1) \subseteq \mathcal{G}(G_2)$. If $G_1 \lhd G_2$ holds then $\mathcal{L}(G_1) \subseteq \mathcal{L}(G_2)$.* □

The converse of the last sentence does not hold true. For instance, $\mathcal{L}(G_{REG}) \subseteq \mathcal{L}(G_{CF})$ but G_{REG} is not an interpretation of G_{CF}.

Theorem 1.2. *The relation $\lhd$ is decidable. Consequently, strong form equivalence between two grammar forms is decidable.*   □

Strong form equivalence is indeed an equivalence relation in the set of grammars. Denote the equivalence class determined by a grammar G by $[G]$. A partial order for the equivalence classes is defined by $[G_1] \leq [G_2]$ iff $G_1 \lhd G_2$. The reader is referred to [50] for additional information concerning the next theorem.

Theorem 1.3. *The equivalence classes $[G]$ constitute a distributive lattice under the partial order $\leq$.*   □

Also the following simulation results follow easily by the definitions. There is a close relationship between the derivations in an interpretation grammar and its form grammar. Indeed, each derivation in an interpretation grammar is always an image of a derivation in the form grammar. Observe that μ^{-1} is a morphism.

Theorem 1.4. *Assume that $G' \lhd G$ (mod μ) and $\alpha_0' \implies \alpha_1' \implies \ldots \implies \alpha_m'$ is a derivation according to G'. Then $\alpha_0 \implies \alpha_1 \implies \ldots \implies \alpha_m$ is a derivation in G, where $\alpha_i = \mu^{-1}(\alpha_i'), 0 \leq i \leq m$. If $L(G)$ is finite, so is $L(G')$, but not necessarily conversely.*   □

Theorem 1.5. *Assume that $G_i = (V_i, \Sigma_i, S_i, P_i), i = 1, 2$, are grammar forms such that $\Sigma_1 \subseteq \Sigma_2$ and, for each production $A \to \alpha$ in P_1, there is a derivation $A \implies^+ \alpha$ in G_2. Then $\mathcal{L}(G_1) \subseteq \mathcal{L}(G_2)$. Assume that $A \implies^+ \alpha$ is a derivation according to G_1. Then the grammar form $(V_1, \Sigma_1, S_1, P_1 \cup \{A \to \alpha\})$ is form equivalent to G_1.*   □

The interpretations discussed above were earlier referred to as *strict* interpretations or *s-interpretations*, in contrast to the so-called *g-interpretations*. For instance, this terminology was still used in [50]. The reasons were historical. The terminology of the definition above was first used in [35]. We will now discuss this matter in more detail.

Grammar forms and g-interpretations were introduced by Cremers and Ginsburg in their conference paper [3], which subsequently appeared as [4]. This approach was preceded by an attempt in [8] to study *grammar schemata*.

The definition of a g-interpretation differs from the one given above in that μ can be any finite substitution for terminal letters: terminal letters can be replaced by terminal words, and no disjointness is required. For nonterminals, μ still is a dfl-substitution. Thus, $A \to BC$ and $S \to dadSAabcd$ are both g-interpretations of the rule $S \to aSbcSb$. Essentially, only information about nonterminals is carried over from the original rule to g-interpretations. Information about terminals is lost; the only fact preserved concerns whether or not terminals may occur in a certain position.

All definitions given above carry over to g-interpretations. In particular, we may speak about *g-grammatical families* $\mathcal{L}_g(G)$. Let us still state briefly the difference between the two interpretation mechanisms. In a g-interpretation, every terminal letter is interpreted as an arbitrary terminal word, including the empty word. Moreover, different terminals can be interpreted as the same word. This implies that the interpretations μ defined above (that is, interpretations earlier referred to as strict or s-interpretations) are much more pleasing mathematically because, in them, (i) nonterminals and terminals are interpreted in the same way and (ii) μ^{-1} is a morphism.

The productions of an interpretation resemble more closely the productions of the master grammar than to those of a g-interpretation. Two very different-looking grammars can be g-interpretations of the same master grammar. Different grammatical families collapse to the same g-grammatical family. Consider, for instance, the grammar forms $G_1 - G_4$ defined by

$$G_1 \; : \; S \to aSb, \; S \to a,$$
$$G_2 \; : \; S \to aSa, \; S \to a,$$
$$G_3 \; : \; S \to aSa, \; S \to a, \; S \to a^2,$$
$$G_4 \; : \; S \to aS, \; S \to Sa, \; S \to a.$$

It is immediate that the g-grammatical family of each $G_i, 1 \leq i \leq 4$, equals the family LIN of linear languages, whereas the grammatical families constitute a strictly increasing hierarchy:

$$\mathcal{L}(G_1) \subset \mathcal{L}(G_2) \subset \mathcal{L}(G_3) \subset \mathcal{L}(G_4) = LIN.$$

Typical languages L_i in the differences $\mathcal{L}(G_i) - \mathcal{L}(G_{i-1}), 2 \leq i \leq 4$, are

$$L_2 = \{a^n ca^n \mid n \geq 1\},$$
$$L_3 = a^+,$$
$$L_4 = \{a^{2n} ca^n \mid n \geq 1\}.$$

This example illustrates the fact that the variety of grammatical families is much richer than that of g-grammatical families. Usually infinite hierarchies of grammatical families collapse into one g-grammatical family. Indeed, there is no g-grammatical family between the families of regular and linear languages, and if attention is restricted to grammar forms with only one non-terminal and only one terminal, then the only possible g-grammatical families are FIN, REG, LIN and CF. We will see below how much richer the picture is for grammatical families. Also the most interesting language-theoretic results in form theory, those concerning density, concern only grammatical, not g-grammatical families.

It was first observed in [19] that g-interpretations are unsuitable for L systems. The systematic study of interpretations as defined above (and at that time referred to as strict) was begun in [26]. Certain aspects had been investigated previously, [9]. [50] contains further references to the early developments.

The following theorem, [26], summarizes the relationships between grammatical and g-grammatical families. We use the notation

$$\mathcal{H}(\mathcal{L}) = \{h(L) \mid L \in \mathcal{L} \text{ and } h \text{ is a morphism}\}.$$

Theorem 1.6. *For any grammar form G,*

$$\mathcal{L}_g(G) = \mathcal{H}(\mathcal{L}(G)).$$

The family FIN of finite languages is the only g-grammatical family which is not grammatical. A family different from FIN and closed under morphism is grammatical iff it is g-grammatical.

Finally, let G_1 and G_2 be grammar forms such that $L(G_1)$ and $L(G_2)$ are infinite and

$$\mathcal{L}_g(G_1) \subset \mathcal{L}_g(G_2).$$

Then an infinite sequence of grammar forms $F_1, F_2, \ldots$ can be constructed such that

$$\mathcal{L}_g(G_1) \subset \mathcal{L}(F_1) \subset \mathcal{L}(F_2) \subset \ldots \subset \mathcal{L}_g(G_2). \qquad \square$$

1.3 Closure properties and normal forms

In this subsection we investigate how certain standard grammar constructions carry over to form theory. The results are from [25] and [26] and concern closure properties and reductions to normal forms.

We first introduce some terminology. We use customary terminology about grammars. Thus, a grammar is *reduced* if every nonterminal is reachable from the initial letter and derives, moreover, a word over the terminal alphabet. It is *chain-free* (resp. λ-*free*) if it has no productions of the form $A \to B$ (resp. $A \to \lambda$). It is in the *Chomsky normal form* if all productions are of the types $A \to BC$ and $A \to a$. It is in the *Greibach 2-standard normal form* if all productions are of the types $A \to aBC, A \to aB$ and $A \to a$. It is *sequential* if there is a linear ordering $A_1, \ldots, A_n$ of all nonterminals such that, whenever $A_i \to \alpha$ is a production, then α does not contain any nonterminal A_j with $j < i$. Grammar forms with just one terminal letter are termed *unary*.

We consider first the closure of grammatical families under the customary language-theoretic operations. Although some grammatical families (such as REG and CF) have very strong closure properties, very little can be said about closure properties in general. A simple example is provided by the grammar form G determined by the single production $S \to abb$. The family $\mathcal{L}(G)$ is closed under none of the following operations: (i) union, (ii) catenation, (iii) catenation closure, (iv) morphisms, (v) inverse morphisms, (vi) mirror image. For nonclosure under union it is essential that G is not unary.

This condition is not essential as regards the other operations mentioned. Let G_1 be the unary grammar form determined by the productions

$$S \to aSa|aS|a \qquad \text{(abbreviation of } S \to aSa, S \to aS, S \to a\text{)}.$$

Then $\mathcal{L}(G_1)$ is closed under none of the operations (ii)–(vi). The following theorem summarizes the closure properties obtainable for all grammatical families:

Theorem 1.7. *Every grammatical family is closed under dfl-substitution and intersection with regular languages. For a unary grammar form G, the family $\mathcal{L}(G)$ is closed under union.* $\square$

It is customary in formal language theory to construct, for a given grammar G, grammars which are in some sense simpler than G but still equivalent to G (in the sense that they generate the same language). Such constructions have been carried out also for grammar forms (and for L forms). In this case, we are looking for forms which are *form equivalent* to the given form. In many cases the proof of a normal form result consists of the verification of the fact that the standard grammatical construction works also for grammar forms. However, also some special techniques are needed. For instance, the following "lemma of nonterminal elimination" comes into use when the production $A \to BCD$ is simulated by the productions $A \to BC_1$ and $C_1 \to CD$.

Theorem 1.8. *Assume that $B \neq S$ is a nonterminal in a grammar form G_1 that does not occur on both the left and the right side of any production. Construct a new grammar form G_2 by removing from G_1 all productions involving B and adding, for all productions*

$$A \to \alpha_1 B \alpha_2 \ldots \alpha_k B \alpha_{k+1}, \ B \to \beta_1 | \ldots | \beta_m,$$

in G_1, where none of the α's contains B, all productions

$$A \to \alpha_1 \beta_{i_1} \alpha_2 \ldots \alpha_k \beta_{i_k} \alpha_{k+1}, \ 1 \leq i_j \leq m, 1 \leq j \leq k.$$

Then G_1 and G_2 are form equivalent. $\square$

Theorem 1.9. *For every grammar form G_1, a form equivalent grammar form G_2 can be constructed such that G_2 is (i) reduced, (ii) λ-free, and (iii) chain-free. For every grammar form, a form equivalent grammar form in the Chomsky normal form can be constructed.* $\square$

As regards the Greibach normal form, the situation is a little more complicated. Once *left recursions* (that is, derivations $A \Longrightarrow^* A\alpha$) have been eliminated, the standard construction for grammars becomes applicable. Left recursions can be eliminated but the proof is more involved than for grammars. The details can be found in [50]. The following theorem is an immediate consequence of the elimination of left recursions.

Theorem 1.10. *For every grammar form, a form equivalent grammar form in the Greibach 2-standard normal form can be constructed.*   □

The proof of the following theorem is again by the standard construction: a nonterminal A generating a finite language L_A is eliminated by substituting words from L_A in all possible ways for all occurrences of A.

Theorem 1.11. *For every grammar form G_1, a form equivalent grammar form G_2 can be constructed such that all nonterminals of G_2 different from the initial letter generate an infinite language.*   □

If $L(G_1)$ is finite, then Theorem 1.11 gives a form equivalent G_2 such that all productions of G_2 are of the type $S \rightarrow x$, where x is a word over the terminal alphabet. Essentially the same construction gives the following more general result. It is useful in constructions, where some exceptional "initial mess" has to be eliminated.

Theorem 1.12. *For every grammar form G_1 and every integer k, a form equivalent grammar form G_2 can be constructed such that all productions of G_2 containing the initial letter S are of the types*

$$S \rightarrow S', \ S \rightarrow x, \ |x| \leq k,$$

where x is a terminal word and, moreover, all terminal words derivable from S' (if any) are of length strictly greater than k.   □

To illustrate Theorem 1.12, consider a unary grammar form G. Let L be an arbitrary language whose length set is included in the length set of $L(G)$. Assume that, for some k, there is an interpretation G_1 of G such that $L(G_1)$ coincides with L as regards words of length strictly greater than k. Then there is also another interpretation G_2 of G such that $L(G_2) = L$. (This argument is not valid if G is not unary because we need the closure under union.)

The classical language-theoretic example, [47], can be used to prove the following result.

Theorem 1.13. *There is a grammar form possesing no form equivalent sequential grammar form.*   □

We mention, finally, a well-known result concerning grammars, for which no corresponding result can be obtained in form theory. For every triple (k, l, m) of nonnegative integers, every grammar G can be transformed into an equivalent normal form where each nonterminating production is of the type

$$A \rightarrow w_k B w_l C w_m, \ |w_k| = k, |w_l| = l, |w_m| = m,$$

and each terminating production $A \rightarrow w$ has the property that $|w|$ appears in the length set of $L(G)$. (Here the w's are words over the terminal alphabet.) This "supernormal-form theorem" was established in [38]. For the triples $(0,0,0)$ and $(1,0,0)$, this theorem gives sharpened formulations of the Chomsky and Greibach normal form results. Instead of triples, the theorem can be formulated in terms of n-tuples with $n \geq 3$. However, there are explicit examples of triples showing that the result cannot be extended to concern grammar forms.

Theorem 1.14. *There are grammar forms G for which no form equivalent grammar form in the $(1,0,1)$ normal form can be constructed.*   □

Such a G is, for instance, defined by the productions $S \rightarrow aS$ and $S \rightarrow a$. Clearly, $\mathcal{L}(G) = REG$. However, rules $A \rightarrow aBCb$ (of the required type) necessarily create pumping situations giving rise to non-regular languages.

1.4 Completeness

Consider a grammar form $G = (V, \Sigma, S, P)$ and a terminal $a \in \Sigma$. The *a-restriction G_a* is obtained from G by removing all terminals $b \neq a$ and all productions containing any of them. (It is possible that G_a contains no production or that $L(G_a)$ is empty.) a-restrictions will play an important role in the following completeness considerations.

Definition 1.2. *A grammar form G is* complete *(resp. sufficient) with respect to a family $\mathcal{L}$ of languages or, briefly, $\mathcal{L}$-complete (resp. $\mathcal{L}$-sufficient) if $\mathcal{L}(G) = \mathcal{L}$ (resp. $\mathcal{L}(G) \supseteq \mathcal{L}$). G is* complete *if it is CF-complete. A grammatical family $\mathcal{L}$ is* unary-complete *if, whenever a grammar form G satisfies $\mathcal{L}(G) = \mathcal{L}$, then there is an a-restriction G_a of G satisfying $\mathcal{L}(G_a) = \mathcal{L}$.*   □

A characterization of $\mathcal{L}$-completeness, for instance REG-completeness, amounts to describing what kinds of normal forms are possible for grammars generating languages in the family $\mathcal{L}$. We will see that such a characterization, even a relatively simple one, is possible for the basic families of regular, linear and context-free languages. However, the fundamental undecidability result in the next subsection shows that such a characterization is not possible for all grammatical families.

Unary-completeness of a family $\mathcal{L}$ means that, whenever $\mathcal{L}$ is generated by a grammar form G with more than one terminal, then some of the terminals and all productions involving them are superfluous and can be removed from G without reducing its generative capacity as a grammar form. (The language $L(G)$ may become smaller in this process.) We shall see that the families REG, LIN and CF are unary-complete. On the other hand, the family $\mathcal{L}(G_1)$, where G_1 is defined by the productions $S \rightarrow aSb$ and $S \rightarrow a$, is not unary-complete. Indeed, there is no unary grammar form that is form equivalent to G_1. Thus, two terminals are necessary to generate the family $\mathcal{L}(G_1)$. It is a rather fundamental open problem whether all examples of non-unary-completeness are of this type. In other words, are there non-unary-complete grammatical families that can be generated also by a unary grammar form ?

We discuss first *REG*-completeness of grammar forms. We also outline the proofs because this is easy for the family *REG* and the argumentation still gives an idea of the general techniques. We call a reduced grammar or grammar form *self-embedding* if it has a nonterminal A such that $A \Longrightarrow^* x_1 A x_2$, where x_1 and x_2 are nonempty terminal words. It is a consequence of Theorem 1.4 that all interpretations of a non-self-embedding grammar are themselves non-self-embedding. This fact can be used to obtain the auxiliary result: a reduced grammar form G is self-embedding iff $\mathcal{L}(G)$ contains nonregular languages. Because a^+ is a regular language and grammatical families are closed under intersection with regular languages (Theorem 1.7), the following basic result can be inferred.

Theorem 1.15. *A reduced unary grammar form G is REG-complete iff $L(G) = a^+$ and G is not self-embedding.* □

Consider now any (not necessarily unary) *REG*-complete grammar form G. Then, for some terminal letter $a, a^+ \subseteq L(G)$ because, otherwise, no language b^+ is in the family $\mathcal{L}(G)$. But this means that the a-restriction G_a satisfies the conditions of Theorem 1.15 and, hence, is *REG*-complete. Consequently, the following general result can be obtained as a corollary of Theorem 1.15.

Theorem 1.16. *A reduced grammar form G is REG-complete iff G is not self-embedding and possesses a REG-complete a-restriction.* □

Since it is decidable whether or not a grammar is self-embedding and since obviously every *FIN*-sufficient grammar form G must satisfy the condition $L(G) \supseteq a^+$, for some terminal a, we obtain also the following result.

Theorem 1.17. *The family REG is unary-complete. It is decidable whether or not a grammar form is REG-complete. The following conditions (i)–(iii) are equivalent for a grammar form G: (i) G is FIN-sufficient, (ii) G is REG-sufficient, (iii) $L(G) \supseteq a^+$, for some terminal a.* □

We now turn to the more difficult task of characterizing *LIN*-completeness. The examples $G_1 - G_3$ discussed in connection with g-interpretations should already give an idea of the difficulties. As another illustration we mention that the grammar form defined by the productions $S \to a^3 S | S a^4 | a | a^2 | a^3$ is *LIN*-complete. In the completeness considerations below, we will mostly just list results, rather than try to give any arguments. We begin with the simple case of (S, a)-*forms*. They are unary grammar forms with only one nonterminal S. The supernormal-form theorem discussed above yields the following result for the family *CF*.

Theorem 1.18. *An (S, a)-form G is complete iff G is nonlinear and $L(G) = a^+$.* □

For LIN-completeness of (S,a)-forms G it is necessary and sufficient that $L(G) = a^+$ and, moreover, G has the productions $S \to a^m S$ and $S \to Sa^n$, for some $n, m > 0$. An exhaustive characterization for grammatical families of (S,a)-forms can now be given in the following theorem, the first version of which appeared in [25].

Theorem 1.19. *Assume that G is an (S,a)-form satisfying $L(G) = a^+$. If G is nonlinear, then $\mathcal{L}(G) = CF$. If G is linear, the following three cases are possible. (i) All productions are right-linear, or else all productions are left-linear. Then $\mathcal{L}(G) = REG$. (ii) G has the productions $S \to a^m S$ and $S \to Sa^n$, for some $m, n > 0$. Then $\mathcal{L}(G) = LIN$. (iii) Neither (i) nor (ii) holds. Then $REG \subset \mathcal{L}(G) \subset LIN$.*   □

It turns out that the family LIN is unary-complete and, hence, we may restrict the attention to unary grammar forms when studying LIN-completeness. To secure LIN-completeness, you have to be able to get all word lengths and to "pump" indefinitely on both sides of the nonterminal. The latter goal is reached, for instance, if the productions $S \to a^m S$ and $S \to Sa^n$ are available. The following definition is needed to characterize the situation formally.

Definition 1.3. *A nonterminal A in a unary reduced grammar form G is left-pumping (resp. right-pumping) if, for some fixed $m, n \geq 0$, there are infinitely many values i such that*

$$A \Longrightarrow^* a^{i+m} A a^n \qquad (\text{resp. } A \Longrightarrow^* a^m A a^{n+i}).$$

A nonterminal A is pumping if it is both left-pumping and right-pumping. Let $A_1, \ldots, A_m$ be all the pumping nonterminals in G. Denote by $p(A_i), 1 \leq i \leq m$, the period of the ultimately periodic sequence consisting of the length of the terminal words generated by A_i. Let p be the least common multiple of all the numbers $p(A_i), i = 1, 2, \ldots, m$. Denote by $R_0, R_1, \ldots, R_{p-1}$ the residue classes modulo p. The residue class R_j is A_i-reachable if there are numbers r, s, t such that

$$S \Longrightarrow^* a^r A_i a^s, \quad A_i \Longrightarrow^* a^{t+np},$$

for all $n \geq 0$, and $j \equiv r + s + t \pmod{p}$. The pumping spectrum of G consists of all numbers in all A_i-reachable residue classes, where i ranges over $1, \ldots, m$.   □

The notion of a pumping spectrum, as well as the following results characterizing LIN-completeness are due to [26].

Theorem 1.20. *A unary reduced linear grammar form G is LIN-complete iff $L(G) = a^+$ and the pumping spectrum of G consists of all numbers.*   □

Nonlinearity of a reduced grammar form G does not imply that $\mathcal{L}(G)$ contains nonlinear languages. The nonlinearity of G can be "harmless" in this sense. The situation is clarified in the following theorem. Observe that the pumping spectrum was defined also for nonlinear grammar forms.

Theorem 1.21. *A unary reduced grammar form G is LIN-complete iff each of the following conditions (i)–(iii) is satisfied. (i) $\mathcal{L}(G) \subseteq LIN$. (ii) $L(G) = a^+$. (iii) The pumping spectrum of G consists of all numbers. Each of the conditions (i)–(iii) is decidable. In particular, (i) holds exactly in case G has no sentential form with two occurrences of self-embedding nonterminals. Finally, G is LIN-sufficient iff (ii) and (iii) hold.* □

The next theorem constitutes the final step in the characterization of LIN-completeness.

Theorem 1.22. *The family LIN is unary-complete. Hence, it is decidable whether or not a grammar form is LIN-complete.* □

The characterization of completeness, [35], runs along similar lines. However, the problem remained open until the supernormal-form theorem was established. Particularly important for grammar form theory is the requirement of "terminal balance": for each terminating production $A \to w$, the length $|w|$ appears in the length set of $L(G)$.

For the formal statement of the results, the notion of an *expansion spectrum* is needed. It is analogous to the notion of a pumping spectrum. A nonterminal A in a reduced grammar or grammar form is said to be *expansive* if $A \Longrightarrow^* x_1 A x_2 A x_3$, for some words x_i and, in addition, A derives some *nonempty* terminal word. The expansion spectrum is now obtained from the expansive nonterminals $A_1, \ldots, A_m$ in exactly the same way as the pumping spectrum was obtained from the pumping nonterminals.

Theorem 1.23. *A reduced unary grammar form G is complete iff (i) $L(G) = a^+$ and (ii) the expansion spectrum of G consists of all numbers. Both (i) and (ii) are decidable conditions.* □

For instance, of the two grammar forms

$$G \quad : \quad S \to aA|A|a, \ A \to AA|a^2,$$
$$G' \quad : \quad S \to A|B, \ A \to A^2|a^2, \ B \to aB|a,$$

G is complete, whereas G' is not complete because it does not satisfy condition (ii).

The following theorem is the final step in the characterization of completeness.

Theorem 1.24. *The family CF is unary-complete. The completeness of a grammar form is decidable.* □

The notion of a *generator* introduced in [25] is intimately connected with unary-completeness.

Definition 1.4. *A family $\mathcal{L}$ of languages is* unary-sufficient *if, whenever G is an $\mathcal{L}$-sufficient grammar form, then G possesses an a-restriction which is also $\mathcal{L}$-sufficient. A context-free language L is a* generator *of a family $\mathcal{L}$ of context-free languages if, for every grammar form $G, L(G) = L$ implies $\mathcal{L}(G) \supseteq \mathcal{L}$. A generator L is* proper *if $L \in \mathcal{L}$.* □

Clearly, for grammatical families, unary-sufficiency implies unary-completeness. Intuitively, a generator of $\mathcal{L}$ contains enough structural information to describe all of $\mathcal{L}$.

Theorem 1.25. *The language $L_1 = a^+$ is a generator of both FIN and REG. Moreover, every generator of one of these families contains L_1, or an alphabetic variant of it, as a subset. The family FIN has no proper generator.* □

Theorem 1.26. *A unary-sufficient family containing a nonregular language possesses no generators.* □

Theorem 1.27. *The families LIN and CF possess no generator.* □

1.5 Linguistical families and undecidability

A very basic problem in form theory is the problem of *form equivalence*: do two given grammar forms generate the same grammatical families ? Special cases of this problem were settled in the last subsection. For instance, when we decide whether or not a grammar form G is LIN-complete, we are actually deciding the form equivalence of G and the grammar form

$$G_{LIN} \; : \; S \rightarrow aS|Sa|a.$$

The general case remained open for some time, and was finally shown undecidable in [40]. An important notion needed in the proof is that of a *linguistical family* introduced in [37]. Moreover, linguistical families form an essential link to graph coloring and density problems discussed in Section 3 of this chapter.

Consider an arbitrary language $L \subseteq \Sigma^*$. A language $L' \subseteq \Sigma'^*$ is an *interpretation* of L, in symbols $L' \lhd L$, if there is a letter-to-letter morphism $h : \Sigma'^* \longrightarrow \Sigma^*$ such that $h(L') \subseteq L$. Observe that every subset of L is an interpretation of L. The family of all interpretations of L is denoted by $\mathcal{L}(L)$:

$$\mathcal{L}(L) = \{L' \mid L' \lhd L\}.$$

A family of languages is termed *linguistical* if it equals the family $\mathcal{L}(L)$, for some L. (We do not impose here any restriction on the language L, it need not be even recursively enumerable. Also languages over infinite alphabets were considered as interpretations in [37]. In our exposition the interpretations L' are languages in the ordinary sense, over a finite alphabet.) The language L is often referred to as a *language form* when the family $\mathcal{L}(L)$ is considered.

Observe that $\mathcal{L}(L)$ is, in general, a "very large" family. For instance, if $L = a^+$ then $\mathcal{L}(L)$ consists of all languages. In the sequel we consider also the subfamilies $\mathcal{L}_{REG}(L)$ and $\mathcal{L}_{FIN}(L)$ of $\mathcal{L}(L)$, consisting of all regular and finite languages in $\mathcal{L}(L)$. Thus,

$$\mathcal{L}_{REG}(L) = \mathcal{L}(L) \cap REG, \ \mathcal{L}_{FIN}(L) = \mathcal{L}(L) \cap FIN.$$

Let G_1 and G_2 be grammars. If $G_1 \lhd G_2$ then $L(G_1) \lhd L(G_2)$ but the reverse implication is not valid.

Theorem 1.28. *For any context-free grammar G, the family $\mathcal{L}(G)$ is contained in the family $\mathcal{L}(L(G))$. If $L(G)$ is finite, the two families are equal. If $L(G)$ is infinite, the containment is strict.* □

The following theorem established in [43] is an important tool in reductions.

Theorem 1.29. *Assume that G is a grammar form such that every language in the family $\mathcal{L}(G)$ is regular. Then*

$$\mathcal{L}(G) = \mathcal{L}_{REG}(L(G)).$$ □

If $\mathcal{L}(L_1) = \mathcal{L}(L_2)$ then obviously also $\mathcal{L}_{REG}(L_1) = \mathcal{L}_{REG}(L_2)$ and $\mathcal{L}_{FIN}(L_1) = \mathcal{L}_{FIN}(L_2)$. The following theorem, [37], shows that also the reverse implication holds.

Theorem 1.30. *Whenever $\mathcal{L}_{FIN}(L_1) = \mathcal{L}_{FIN}(L_2)$ holds for two languages L_1 and L_2, then also $\mathcal{L}_{REG}(L_1) = \mathcal{L}_{REG}(L_2)$ and $\mathcal{L}(L_1) = \mathcal{L}(L_2)$.* □

A notion central in the reduction arguments leading to the subsequent undecidability results is that of a *singleton-catenation*. The language

$$wL = \{ww' \mid w' \in L\}$$

is said to be a singleton-catenation of L. A language family $\mathcal{L}$ is *closed under singleton catenation* if, for every $L \in \mathcal{L}$, all singleton-catenations of L also belong to $\mathcal{L}$. (Thus, if $\mathcal{L}$ contains all singletons and is closed under catenation, it is closed under singleton-catenation as well.) The following decidability results are from [40].

Theorem 1.31. *Let the language family $\mathcal{L}$ be closed under each of the following operations: singleton-catenation, letter-to-letter morphisms and union. Then all of the following relations are decidable, or none of them is, for languages L_1 and L_2 in $\mathcal{L}$: (i) $L_1 \lhd L_2$, (ii) $\mathcal{L}(L_1) = \mathcal{L}(L_2)$, (iii) $L_1 \subseteq L_2$, (iv) $L_1 = L_2$.* □

The closure under singleton-catenation is essential in the proof of Theorem 1.31 because of the following fact. Let L_1 and L_2 be languages over the alphabet $\{a_1, \ldots, a_n\}$. Then

$$L_1 \subseteq L_2 \ \text{iff} \ a_1 \ldots a_n L_1 \lhd a_1 \ldots a_n L_2.$$

Theorem 1.32. *Let $\mathcal{L}_1$ and $\mathcal{L}_2$ be closed under singleton-catenation and letter-to-letter morphisms. Let further $\mathcal{L}_2$ be closed under intersection and union with languages in $\mathcal{L}_1$. Then the following four conditions are equivalent for languages $L_i \in \mathcal{L}_i, i = 1, 2$:*

(i) $L_1 \lhd L_2$ *and* $L_2 \lhd L_1$ *are both decidable;*
(ii) $\mathcal{L}(L_1) = \mathcal{L}(L_2)$ *is decidable;*
(iii) $L_1 \subseteq L_2$ *and* $L_2 \subseteq L_1$ *are both decidable;*
(iv) $L_1 = L_2$ *is decidable.* □

Well-known decidability results for CF and REG now yield the following corollary.

Theorem 1.33. *The following relations are undecidable for a context-free grammar G and a right-linear grammar G_1: (i) $\mathcal{L}(L(G_1)) = \mathcal{L}(L(G))$, (ii) $L(G_1) \lhd L(G)$. On the other hand, the relation $L(G) \lhd L(G_1)$ is decidable. Consequently, the equation $\mathcal{L}(L(G_1)) = \mathcal{L}(L(G_2))$ is decidable for two right-linear grammars.* □

By Theorems 1.29 and 1.30, the core of Theorem 1.33 can be presented in the following form.

Theorem 1.34. *Let G be an arbitrary context-free and G_1 an arbitrary right-linear grammar. Then the equation $\mathcal{L}_{REG}(L(G_1)) = \mathcal{L}_{REG}(L(G))$ is undecidable, whereas the inclusion $\mathcal{L}_{REG}(L(G)) \subseteq \mathcal{L}_{REG}(L(G_1))$ is decidable.* □

Theorem 1.34 is also a nice illustration of the borderline between decidability and undecidability. In this setup we reach the rather unusual situation that inclusion is decidable but equivalence is undecidable !

We now derive a connection between grammatical and linguistical families. Apart from being important in general language theory, it constitutes the main tool in proving decidability results.

The grammatical family $\mathcal{L}(G)$ is obtained from a grammar G in two steps. *First* the *interpretation* mechanism creates a family of grammars and, *secondly*, the language *generation* mechanism produces the languages of these grammars. When dealing with language forms, the same processes are used in the opposite order. *First* the *generation* mechanism is used to get the language $L(G)$, while in the *second* step the *interpretation* mechanism produces the linguistical family $\mathcal{L}(L(G))$. The two processes do not in general commute: the final results $\mathcal{L}(G)$ and $\mathcal{L}(L(G))$ are different. For any G, $\mathcal{L}(G)$ is contained in the family $\mathcal{L}(L(G))$, the containment being strict if $L(G)$ is infinite (Theorem 1.28). Whenever L is infinite, the family $\mathcal{L}(L)$ contains languages that are not even recursively enumerable. The comparison between grammatical and linguistical families becomes more reasonable if the families are intersected with some fixed family, such as REG. We already introduced the restriction $\mathcal{L}_{REG}(L)$. Similarly,

$$\mathcal{L}_{REG}(G) = \mathcal{L}(G) \cap REG.$$

Theorem 1.29 can now be expressed in the following form. For any G,

$$\mathcal{L}_{REG}(G) = \mathcal{L}_{REG}(L(G)).$$

The intersection between grammatical and linguistical families can be expressed by saying that the following diagram commutes:

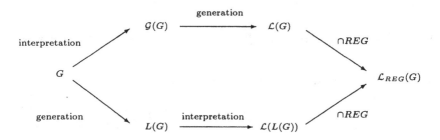

Theorem 1.34 now implies the undecidability of the inclusion problem for grammatical families.

Theorem 1.35. *It is undecidable whether or not $\mathcal{L}(G_1) \subseteq \mathcal{L}(G)$, for a right-linear grammar G_1 and context-free grammar G. Consequently, the inclusion problem is undecidable for grammatical families.* $\square$

The final step to the undecidability of form equivalence can be taken, using notions of the *superdisjoint union* of languages and the *direct sum* of grammars. The former is defined only for two languages over disjoint alphabets and is simply the union of the languages. Let $G_i = (V_i, \Sigma_i, S_i, P_i), i = 1, 2$, be grammars such that $V_1 \cap V_2 = \emptyset$ and S_i does not occur in the right side of any production. The grammar

$$G_1 \oplus G_2 = ((V_1 \cup V_2) - \{S_2\}, \Sigma_1 \cup \Sigma_2, S_1, P_1 \cup P_2'),$$

where P_2' is P_2 with S_2 replaced by S_1, is called the *direct sum* of G_1 and G_2. We have now, for any context-free grammar forms G_1 and G_2,

$$\mathcal{L}(G_1) \subseteq \mathcal{L}(G_2) \quad \text{iff} \quad \mathcal{L}(G_1 \oplus G_2) = \mathcal{L}(G_2).$$

Observe that we can always rename the alphabet of a language form (resp. grammar form) without changing its linguistical (resp. grammatical) family. Grammatical families are also closed under superdisjoint union. Theorem 1.35 now yields the following fundamental result.

Theorem 1.36. *Form equivalence is undecidable.* $\square$

This result is very natural considering the richness of grammatical families, further illustrated by the density results in Section 3 below.

2. L forms

2.1 E0L and ET0L forms

This section will be the parallel counterpart of Section 1. The sequential rewriting of grammars will be contrasted with the parallel rewriting of L systems. Many facts are still analogous to the corresponding ones discussed in Section 1 and, therefore, our exposition will be somewhat shorter in the present section.

As a construct an *E0L system* $G = (V, \Sigma, S, P)$ is almost identical with a context-free grammar, the only difference being that it is required that, for every letter a in V, terminal and nonterminal alike, P contains a production with a on the left side. This requirement reflects the *parallel* derivation mode of an E0L step: at each derivation step every letter must be rewritten. (Of course we can make E0L systems and context-free grammars identical as constructs by adding dummy rules $a \to a$ to a context-free grammar whenever necessary.)

The definition of an *interpretation*, *grammar family* and *grammatical family* is exactly the same as the one given in Section 1.2. The basic construct, G, is now an E0L system rather than a context-free grammar. Consequently, also the interpretations $G' \lhd G$ are E0L systems and must satisfy the requirement that there is at least one production for every letter. The grammatical family $\mathcal{L}(G)$ is a family of E0L languages. When the families $\mathcal{G}(G)$ and $\mathcal{L}(G)$ are discussed, then the E0L system G is usually referred to as an *E0L form*. The notions of *form equivalence* and *strong form equivalence* are defined exactly as before. The results in this subsection are mostly from [19] and [21]. [50] contains further references. Our first theorem is established in exactly the same way as the corresponding result for grammar forms, because grammar forms and E0L forms are identical as constructs.

Theorem 2.1. *The relation $\lhd$ for E0L forms is decidable and transitive. The relation $G_1 \lhd G_2$ holds iff $\mathcal{G}(G_1) \subseteq \mathcal{G}(G_2)$. The relation $G_1 \lhd G_2$ implies the inclusion $\mathcal{L}(G_1) \subseteq \mathcal{L}(G_2)$ but the converse implication is not valid in general. Strong form equivalence is decidable for E0L forms.* □

As an example, consider the E0L forms defined by

$$G_{REG} \ : \ S \to Sa, \ S \to a, \ a \to a,$$
$$G_{E0L} \ : \ S \to S, \ S \to SS, \ S \to a, \ a \to S.$$

(As before, capital letters are nonterminals and small letters terminals.) Observe first that G_{REG} and G_{E0L} are equivalent as E0L systems: they both generate the same E0L language a^+. However, they are by no means form equivalent. In fact,

$$\mathcal{L}(G_{REG}) = REG, \ \mathcal{L}(G_{E0L}) = E0L,$$

where EOL is the family of EOL languages. Both of these families possess very strong *closure properties*. (Indeed, REG is a full AFL and EOL is closed under all AFL-operations except inverse morphisms.) On the other hand, the resulting family $\mathcal{L}(G)$ is an anti-AFL (is closed under none of the AFL-operations) if G is defined by the productions

$$S \to a|cc|AAAA, \ A \to AA|b, \ a \to a, \ b \to b, \ c \to c.$$

A detailed proof of this fact, as well as of the above claims concerning G_{REG} and G_{EOL}, can be found in [19]. For instance, the languages $\{a_1, a_2a_2\}$ and $\{a_2, a_1a_1\}$ are both in $\mathcal{L}(G)$ but their union is not in $\mathcal{L}(G)$. A very useful property showing nonclosure is that no language in $\mathcal{L}(G)$ contains a word of length 3.

A result analogous to Theorem 1.7 can be obtained, [25], for *synchronized* EOL forms. By definition, an EOL system or form is *synchronized* if, for all terminals $a, a \to N$ is the only production for a, where N is a nonterminal with the only production $N \to N$. (Synchronization means that terminals must be introduced at the same step everywhere in the word, because a word containing both terminals and nonterminals can never lead to a terminal word.)

Theorem 2.2. *Let $G = (V, \Sigma, S, P)$ be an EOL form. Then $\mathcal{L}(G)$ is closed under dfl-substitution. If G is synchronized then $\mathcal{L}(G)$ is closed under intersection with regular languages. If, further, Σ consists of one letter then $\mathcal{L}(G)$ is closed under union.*   □

For every EOL system, an equivalent (that is, generating the same language) synchronized EOL system can be constructed. The same statement does not hold true for EOL forms and form equivalence. Indeed, there is no synchronized EOL form which is form equivalent to the very simple EOL form G defined by the productions

$$S \to a, \ a \to b, \ b \to b.$$

This follows because every nonempty language in the family $\mathcal{L}(G)$ contains at least two words, which phenomenon cannot be achieved by a synchronized EOL form. This is an example of a phenomenon customarily referred to as "terminal forcing" in the theory of L forms: whenever a language in $\mathcal{L}(G)$ contains a terminal word of a certain type then it necessarily contains also a terminal word of certain other type. For more sophisticated nonreducibility results, one may use in the same fashion chains consisting of more than two terminal words. The following reducibility result is a very useful technical tool.

Theorem 2.3. *For every EOL form G, one can construct a form equivalent EOL form G_1 such that every production in G_1 is of one of the following types:*

$$A \to \lambda, \ A \to a, \ A \to B, \ A \to BC, \ a \to A,$$

where the capital letters are nonterminals, not necessarily distinct. Moreover, if G is synchronized (resp. propagating), then also G_1 is synchronized (resp. propagating).   □

For every E0L system G_1, an equivalent propagating E0L system (that is, containing no λ-productions) can be constructed. The following theorem constitutes the form-theoretic counterpart of this language-theoretic result.

Theorem 2.4. *There are E0L forms G such that no propagating E0L form G_1 is form equivalent to G. For every synchronized E0L form, a form equivalent synchronized propagating E0L form can be constructed.*   □

We will give below, when discussing ET0L forms, an example of a form G mentioned in the first sentence of Theorem 2.4.

Our next topic will be *completeness*. The definitions are analogous to the ones given in Section 1. An E0L form G is *$\mathcal{L}$-complete* (resp. *$\mathcal{L}$-sufficient*), where $\mathcal{L}$ is a family of languages, if $\mathcal{L}(G) = \mathcal{L}$ (resp. $\mathcal{L}(G) \supseteq \mathcal{L}$). It is *complete* if it is *E0L*-complete.

Contrary to grammar forms, no exhaustive characterization of completeness is known for E0L forms. Several conditions are known that are either necessary or sufficient for completeness, as well as results dealing with special cases, see [50], [19], [27], [25]. It has not even been shown that completeness is a decidable property of E0L forms or synchronized E0L forms. The following characterization was given in [27] for the special case of *two-symbol synchronized E0L forms*: the only terminal is a and the only nonterminals S and N, the latter being the blocking symbol for a.

Theorem 2.5. *A two-symbol synchronized E0L form is complete iff it contains the production $S \to S$ and, for some $i \geq 2$, all of the productions $S \to S^i, S \to a, \ldots, S \to a^{i-1}$.*   □

We list below some complete E0L forms G_i, as well as some incomplete E0L forms H_i. They illustrate some frequently occurring phenomena. For more details we refer to [19] and [50].

$$
\begin{aligned}
G_1 &: \ S \to a, \ S \to S, \ S \to SS, \ a \to S, \\
G_2 &: \ S \to a, \ S \to S, \ S \to Sa, \ a \to S, \\
G_3 &: \ S \to a, \ S \to S, \ S \to aS, \ a \to S, \\
G_4 &: \ S \to a, \ S \to S, \ S \to S, \ a \to SS, \\
G_5 &: \ S \to a, \ S \to \lambda, \ S \to S, \ S \to SSS, \ a \to S, \\
G_6 &: \ S \to A, \ A \to S, \ A \to SS, \ A \to a, \ a \to A, \\
G_7 &: \ S \to a, \ S \to SSA, \ S \to S, \ a \to S, \ A \to \lambda, \\
G_8 &: \ S \to a, \ S \to S, \ S \to SS, \ a \to N, \ N \to N, \\
H_1 &: \ S \to a, \ S \to S, \ S \to aa, \ a \to S,
\end{aligned}
$$

$$H_2 \ : \ S \to a, \ S \to S, \ a \to SS,$$
$$H_3 \ : \ S \to a, \ S \to S, \ a \to S, \ S \to SSS,$$
$$H_4 \ : \ S \to a, \ a \to S, \ a \to SS, \ a \to a,$$
$$H_5 \ : \ S \to A, \ A \to S, \ S \to SS, \ A \to a, \ a \to A,$$
$$H_6 \ : \ S \to a, \ S \to SS, \ a \to S.$$

We now give a brief survey over *ET0L forms*. Again, an ET0L form is simply an ET0L system. The definitions of interpretations and the families $\mathcal{G}(G)$ and $\mathcal{L}(G)$ carry directly over. (An interpretation must have equally many tables as the original form.) So do the notions of (strong) form equivalence, $\mathcal{L}$-completeness, completeness and synchronization. The last notion plays a similar role as for E0L forms: there are ET0L forms for which no form equivalent synchronized ET0L form exists. A result corresponding to Theorem 2.1 holds for ET0L forms as well. Rather little is known about completeness of ET0L forms. The following two ET0L forms are typical examples of complete forms. They are defined by listing the productions in the tables – a table is indicated by brackets.

$$G_1 \ : \ [S \to a, S \to S, S \to SS, a \to S], \ [S \to S, a \to S],$$
$$G_2 \ : \ [S \to S, S \to SS, a \to a], \ [S \to S, a \to a], \ [S \to a, a \to a].$$

Observe that the second table of G_2 is not dummy when G_2 is viewed as an ET0L form: in interpretations it can affect various changes of the alphabet.

For the E0L form G defined by the productions

$$S \to a, \ a \to b, \ b \to b,$$

no form equivalent synchronized E0L form exists. However, the synchronized ET0L form defined by the tables

$$[S \to A, A \to B, B \to B, a \to N, b \to N, N \to N], \text{ and}$$
$$[S \to N, A \to a, B \to b, a \to N, b \to N, N \to N]$$

is form equivalent to G. Thus, synchronization is achieved by introducing a new table. However, this technique does not always work. For the E0L form defined by the productions

$$S \to a, \ a \to aa,$$

no form equivalent synchronized ET0L form exists. (This implies the nonexistence of a form equivalent synchronized E0L form; E0L forms are a special case of ET0L forms, namely, they are ET0L forms with only one table.)

The E0L form G defined by the productions

$$S \to a, \ a \to abba, \ b \to \lambda,$$

constitutes an example for the first sentence of Theorem 2.4. Moreover, no propagating ET0L form is equivalent to G. The following theorem summarizes some positive reduction results.

Theorem 2.6. *For every ET0L form, a form equivalent ET0L form can be constructed such that each of its tables has only productions of the types listed in Theorem 2.3. For every synchronized ET0L form, a form equivalent synchronized propagating ET0L form can be constructed. For every ET0L form, a form equivalent ET0L form with two tables can be constructed.*   □

The ET0L form defined by the two tables

$$[S \to S, S \to SS, a \to a], \ [S \to a, a \to a]$$

is E0L-complete. This result is of particular interest because the tables, when taken individually as E0L forms, produce very degenerate grammatical families. A similar example is provided by the complete ET0L form

$$[S \to S, S \to SS, a \to S], \ [S \to a, S \to S, a \to S].$$

Neither one of its two "E0L restrictions" is E0L-complete. (We refer to [50] for a more detailed discussion concerning examples mentioned here.) There is also a complete ET0L form, where the only rule for the terminal a is $a \to a$, defined by the three tables

$$[S \to S, S \to SS, a \to a], \ [S \to S, a \to a], \ [S \to a, a \to a].$$

The corresponding result is not valid for E0L forms: a propagating complete E0L form must have a terminal rule $a \to \alpha$, where α contains a nonterminal.

A large problem area results from the comparison of grammatical families obtainable by sequential and parallel rewriting. To avoid confusion, let us reserve in this context the term "grammatical" to the (sequential) families considered in Section 1, whereas the (parallel) families $\mathcal{L}(G)$ generated by E0L forms G are referred to as "E0L families". The first sentence of the following theorem was established in [27] and the second in [2].

Theorem 2.7. *Every grammatical family contained in LIN is an E0L family, generated by a synchronized E0L system. The family CF is not an E0L family.*   □

The converse of the first sentence is open even for subregular E0L families; it is not known whether every E0L family contained in *REG* is grammatical. It is known, [25], that the family of E0L languages possesses no generator.

We conclude this subsection with some remarks concerning decidability. Although strong form equivalence is decidable for ET0L, E0L and grammar forms, the problem is NP-complete in each case. This follows by a reduction to the clique problem of graphs, see [50]. The argument using linguistical families applied for the proof of Theorem 1.36 can be carried over to concern the form equivalence of E0L forms. The crucial interconnection is in this case

$$\mathcal{L}_{REG}(G) = \mathcal{L}_{REG}(L(G)),$$

where G is an arbitrary synchronized E0L form, and the subscript denotes the restriction of the family to regular languages. The conclusion is the following result, [40].

Theorem 2.8. *Form equivalence is undecidable for E0L forms, and even for synchronized E0L forms.*   □

2.2 Good and very complete forms

Vomplete E0L forms (short for *very complete*) were introduced already in the basic paper [19]. Later on, [22], the notion was further generalized to the notion of *goodness*. Although originally stated for E0L forms, analogous definitions can be given for any type of grammar and L forms. The results mentioned in this subsection are from [22], [23], [27] and [50].

Definition 2.1. *An E0L form G is* vomplete *if, for every E0L form G_1, an interpretation $G' \lhd G$ exists such that $\mathcal{L}(G') = \mathcal{L}(G_1)$. An E0L form G is* good *if, for every E0L form G_1 with the property $\mathcal{L}(G_1) \subseteq \mathcal{L}(G)$, an interpretation G' exists such that $\mathcal{L}(G') = \mathcal{L}(G_1)$. If G is not good, it is called* bad.   □

Thus, G is vomplete iff G is both complete and good. G being vomplete means that it is able to generate not only every *language* in the family *E0L* but also every *E0L*-grammatical *language family* via its interpretations.

The following example shows that the difference between good and bad occurs already at a very simple level and also that deciding whether a given form is good or bad might be a very challenging problem even for surprisingly "innocent-looking" forms. Consider the E0L forms

$$G_1 \quad : \quad S \to a,\ a \to N,\ N \to N,$$
$$G_2 \quad : \quad S \to a,\ a \to a.$$

Clearly, G_1 and G_2 are form equivalent, $\mathcal{L}(G_1) = \mathcal{L}(G_2)$, the family consisting of finite languages of single-letter words. The E0L form

$$G_3 \ : \ S \to a,\ a \to b,\ b \to b$$

(which is an interpretation of G_2) generates the subfamily, where each nonempty language contains at least two words (because of terminal forcing). This subfamily cannot be generated by any interpretation of G_1. This example reflects also the following fact.

Theorem 2.9. *No propagating or synchronized E0L form can be vomplete.*
□

Thus, there are complete E0L forms that are not vomplete. The following E0L form is vomplete:

$$S \to \lambda,\ S \to a,\ S \to S,\ S \to SS,\ a \to S.$$

Indeed, vompleteness follows by Theorem 2.3.

Historical remark. The rather strange term "vomplete", used quite frequently in the literature, came into existence as follows. Hermann Maurer had in Karlsruhe a very conscientious secretary, Frau Wolf, who had once produced a typo "vomplete" when typing "complete". It was later realized by the MSW-group that the typo was perhaps intentional: the E0L form in question was not only complete but also vomplete.   □

As regards synchronized forms, Theorem 2.9 can be strengthened as follows.

Theorem 2.10. *Every synchronized E0L form G such that $L(G)$ contains at least one nonempty word is bad.*   □

It is interesting that, as far as language families generated by E0L forms are concerned, synchronization seems to be a drawback rather than an advantage, as usual in customary L systems theory. However, the two-tabled synchronized ET0L form consisting of

$$[S \rightarrow S, a \rightarrow N, N \rightarrow N], \ [S \rightarrow a, a \rightarrow N, N \rightarrow N]$$

is good. No synchronized or propagating ET0L form can be vomplete. The ET0L form defined by the two tables

$$[S \rightarrow a, S \rightarrow \lambda, S \rightarrow S, S \rightarrow SS, a \rightarrow S], \ [S \rightarrow S, a \rightarrow S]$$

is vomplete. (We have left to the reader the straightforward definition of good and vomplete ET0L forms.) Goodness can also be defined relative to a particular collection of grammatical families, [23]. A special class of E0L forms, where completeness implies vompleteness, has been constructed in [27].

2.3 PD0L forms

We have already pointed out that our definition of an interpretation and related notions can be extended to concern any types of grammars and L systems. Although definitionally very simple, the theory of *D0L forms* presents a number of challenging problems. We discuss now results concerning *PD0L forms*, presented in [5]. Related material is contained also in [17] and [45]. The results concerning D0L forms are theoretically very interesting because of their connection with the celebrated D0L equivalence problem. D0L forms also open a new vista to form theory: in addition to a family of languages, a form generates a *family of sequences*. Contrary to most other classes of forms, form equivalence is decidable for PD0L forms.

Our discussion here will concern only PD0L forms. The arguments used in [5] rely heavily on the systems being propagating. Of course, all definitions can be readily extended to concern D0L forms.

We write PD0L systems as triples $G = (\Sigma, w, h)$, where $w \in \Sigma^+$ and $h : \Sigma^* \longrightarrow \Sigma^*$ is a nonerasing morphism. The *sequence* $S(G)$ of G consists of two words

$$h^0(w) = w, \; h(w), \; h^2(w), \ldots,$$

and the *language* of G is $L(G) = \{h^i(w) \mid i \geq 0\}$. A PD0L system is *strict* if $|h^i(w)| < |h^{i+1}(w)|$, for all $i \geq 0$. It is *finite* if $L(G)$ is finite. A finite system is of *constant length* k if every word in $L(G)$ is of length k. In the case of $k = 1$ we speak of *unary constant-length systems*. As a construct, a *PD0L form* is defined exactly like a PD0L system. A PD0L system $G' = (\Sigma', w', h')$ is an *interpretation* of G, $G' \lhd G$, if there is a dfl-substitution μ defined on Σ such that $w' \in \mu(w)$ and, for each $a \in \Sigma$, $\mu(a) \subseteq \Sigma'$ and finally, for each $a \in \Sigma', h'(a) \in \mu(h(\mu^{-1}(a)))$.

Observe that if PD0L systems are viewed as degenerate E0L systems, then the interpretations defined above are also interpretations in the sense of Section 2.1. However, we have now the additional requirement that the interpretations themselves must be PD0L systems. Observe also that $G' \lhd G$ iff there is a letter-to-letter morphism t such that $t(\Sigma') = \Sigma, t(w') = w$ and $t(h'(a)) = h(t(a))$, for all $a \in \Sigma'$.

Given a PD0L form G, the families of PD0L *systems*, PD0L *languages* and PD0L *sequences* generated by G (or associated to G) are defined by

$$\mathcal{G}(G) = \{G' \mid G' \lhd G\},$$
$$\mathcal{L}(G) = \{L(G') \mid G' \lhd G\},$$
$$\mathcal{S}(G) = \{S(G') \mid G' \lhd G\}.$$

PD0L forms G_1 and G_2 are *strongly form equivalent* (resp. *form equivalent, sequence equivalent*) if $\mathcal{G}(G_1) = \mathcal{G}(G_2)$ (resp. $\mathcal{L}(G_1) = \mathcal{L}(G_2)$, $\mathcal{S}(G_1) = \mathcal{S}(G_2)$).

As before, the relation $\lhd$ is decidable and transitive, which implies that strong form equivalence is decidable. (In the usage of the word "strong" we follow here [50]. The expression "strict form equivalence" has also been frequently used in the literature.) The sequence equivalence of two forms implies always their form equivalence. The families $\mathcal{L}(G)$ and $\mathcal{S}(G)$ are invariant under renaming the letters of the alphabet of G. Thus, if two forms become identical after renaming the letters, then they are sequence and form equivalent. If G is finite (resp. infinite) then every language in the family $\mathcal{L}(G)$ is finite (resp. infinite).

We say that two sequences of words u_i and v_i, $i = 1, 2, \ldots$, are *isomorphic* if there is a one-to-one letter-to-letter morphism f such that the equation $v_i = f(u_i)$ holds for all $i \geq 1$. They are *ultimately isomorphic* if there is a number i_0 such that the equation mentioned holds for all values $i \geq i_0$.

The following theorem, [17] is a special case of the general theory. It can be established directly, independently of the general theory.

Theorem 2.11. *If G_1 and G_2 are form equivalent strict PD0L forms then the sequences $S(G_1)$ and $S(G_2)$ are isomorphic. Form equivalence is decidable for strict PD0L forms. Two strict PD0L forms are form equivalent iff they are sequence equivalent.* □

The converse of the first sentence of Theorem 2.11 does not hold true. Consider the two strict PD0L forms with the axiom ab and productions defined by

$$G_1 \; : \; a \to aba, \; b \to bab,$$
$$G_2 \; : \; a \to ab, \; b \to abab.$$

Then $S(G_1) = S(G_2)$ but neither of the families $\mathcal{L}(G_1)$ and $\mathcal{L}(G_2)$ is included in the other.

The basic question is: how do form equivalent non-identical PD0L forms look like ? If the alphabet consists of at most two letters, the only examples are provided by unary constant-length forms. The forms with the axiom a and productions

$$G_1 \; : \; a \to b, \; b \to b,$$
$$G_2 \; : \; a \to b, \; b \to a$$

are indeed form equivalent. Their language family consists of all languages with cardinality at least 2 and with all words of length 1. The sequences $S(G_1)$ and $S(G_2)$ are not even ultimately isomorphic. Similar examples can be given of unary constant-length forms over any alphabet.

Consider next the three-letter alphabet $\{a, b, c\}$. Define two PD0L forms G_1 and G_2 as follows. The axiom in both forms is $w_1 ab w_2$, where w_1 and w_2 are (possibly empty) words over the alphabet $\{c\}$. The productions are defined by

$$(*) \qquad G_1 \; : \; a \to w_3, \; b \to w_4, \; c \to w_5,$$
$$G_2 \; : \; a \to w_3', \; b \to w_4', \; c \to w_5,$$

where the right sides are nonempty words over the alphabet $\{c\}$ and $w_3 w_4 = w_3' w_4'$. Then G_1 and G_2 are form equivalent. The general characterization of form equivalence given below implies that these examples exhaust the possibilities, apart from renaming of letters, of non-identical, form equivalent PD0L forms over an alphabet with cardinality at most 3. Moreover, the characterization makes it easy to list all the possibilities with respect to any alphabet of small cardinality.

The next two theorems, Theorems 2.12, 2.13, although interesting also on their own right, are the main lemmas in establishing the characterization result, Theorem 2.14.

Theorem 2.12. *If G and G' are form equivalent infinite PD0L forms, then the sequences $S(G)$ and $S(G')$ are ultimately isomorphic.*   □

Theorem 2.13. *If G and G' are form equivalent constant-length non-unary PD0L forms, then the sequences $S(G)$ and $S(G')$ are isomorphic.*   □

The following notion is crucial in the proof and reflects the intricacies of the whole argument. Consider two sequence equivalent PD0L systems

$$G = (\Sigma, w, h) \text{ and } G' = (\Sigma, w, h').$$

(Thus $S(G) = S(G')$.) We divide the letters of Σ into *good* and *bad* with respect to the pair (G, G'). Let w_i be an arbitrary word in the sequence $S(G) = S(G')$. Decompose $w_i = x_1 \ldots x_k, |x_j| \geq 1$, in such a way that, for each $x_j, h(x_j) = h'(x_j)$, whereas $h(x'_j) \neq h'(x'_j)$ for each proper prefix x'_j of x_j. (This implies that $h(x'_j) \neq h'(x'_j)$ also for each proper suffix x'_j.) A letter is *bad* if it occurs in some w_i in some x_j with $|x_j| \geq 2$. Otherwise, it is *good*.

Thus, if $h(a) \neq h'(a)$, then a is bad. The letters a and b in $(*)$ are bad provided $w_3 \neq w'_3$. However, a can be bad although $h(a) = h'(a)$. For instance, this happens if $w_i = bac$ and h and h' are defined by

$$\begin{aligned} h &: \quad a \to d, \ b \to d^2, \ c \to d, \ d \to d^2, \\ h' &: \quad a \to d, \ b \to d, \ c \to d^2, \ d \to d^2. \end{aligned}$$

Intuitively, a is good if, whenever it occurs in the sequence, it generates the same subword (including position) of the next word according to both G and G'.

Theorem 2.14. *All of the following conditions (i)–(iii) are equivalent for two PD0L forms G and G' which are not unary constant-length forms.*

(i) G and G' are form equivalent.

(ii)   The sequences $S(G)$ and $S(G')$ are isomorphic and each bad letter occurs only once.

(iii) G and G' are sequence equivalent.

Both the form equivalence and the sequence equivalence are decidable for arbitrary PD0L forms.   □

3. Density

3.1 Dense hierarchies of grammatical families

Perhaps the most interesting formal properties about grammatical families deal with density. In fact, no other collections of language families obtained by generative devices are known to possess such a density property: between any two families in the collection one can "squeeze" in a third one. Although various infinite hierarchies of language families have been presented in other contexts (for instance, in AFL-theory, see Chapter 4 of this Handbook), no nontrivial examples of dense hierarchies had been presented before the advent of grammatical hierarchies. Moreover, the interconnection between interpretations and graph coloring leads also to dense hierarchies of (finite) graphs, as will be seen below. The subsequent density results constitute undoubtedly an important aspect in the mathematical theory of context-free languages.

The following definition is stated in terms of grammatical families, in other words, grammatical families constitute the basic collection, the universe we are speaking about. However, the same definition can and will be applied to other collections as well.

Definition 3.1. *Assume that $\mathcal{L}$ and $\mathcal{L}'$ are grammatical families such that $\mathcal{L} \subset \mathcal{L}'$. The pair $(\mathcal{L}, \mathcal{L}')$ (sometimes also referred to as an interval) is dense if, whenever $\mathcal{L}_1$ and $\mathcal{L}_2$ are grammatical families satisfying $\mathcal{L} \subseteq \mathcal{L}_1 \subset \mathcal{L}_2 \subseteq \mathcal{L}'$, then there is a grammatical family $\mathcal{L}_3$ such that $\mathcal{L}_1 \subset \mathcal{L}_3 \subset \mathcal{L}_2$. The pair $(\mathcal{L}, \mathcal{L}')$ is* maximal dense *if it is dense and there is no grammatical family $\mathcal{L}''$ such that either $\mathcal{L}'' \subset \mathcal{L}$ and $(\mathcal{L}'', \mathcal{L}')$ is dense, or else $\mathcal{L}' \subset \mathcal{L}''$ and $(\mathcal{L}, \mathcal{L}'')$ is dense. A grammatical family $\mathcal{L}_1$ is* density forcing *if, whenever $\mathcal{L}_2$ is a grammatical family such that $\mathcal{L}_1 \subset \mathcal{L}_2$, then the pair $(\mathcal{L}_1, \mathcal{L}_2)$ is dense. In this case we say also that the grammar form generating $\mathcal{L}_1$ is density forcing. A grammatical family $\mathcal{L}_2$ is a* successor *of a grammatical family $\mathcal{L}_1$ (and $\mathcal{L}_1$ is a* predecessor *of $\mathcal{L}_2$) if $\mathcal{L}_1 \subset \mathcal{L}_2$ and there is no grammatical family $\mathcal{L}_3$ with the property $\mathcal{L}_1 \subset \mathcal{L}_3 \subset \mathcal{L}_2$. The terms "successor" and "predecessor" are extended to concern also the grammar forms generating the families.* $\square$

The notions defined above can be extended in a straightforward manner to concern other collections of language families. For instance, we may consider families generated by synchronized E0L forms or even arbitrary language families in place of grammatical families. Observe that if $\mathcal{L}$ is density forcing, then no superset of $\mathcal{L}$ can have a successor. It is conceivable that a superset of $\mathcal{L}$ has a predecessor. In general, the reader is warned against viewing the families between $\mathcal{L}$ and $\mathcal{L}'$ as linearly ordered – the frequently used term "dense interval" is in this sense misleading.

We already introduced in Section 1.5 the superdisjoint union of two languages, denoted from now on by $L_1 \uplus L_2$, as well as the direct sum of two grammars, $G_1 \oplus G_2$. Obviously,

$$L(G_1 \oplus G_2) = L(G_1) \uplus L(G_2).$$

A language L is called *coherent* if cannot be represented as a superdisjoint union in a nontrivial way: whenever $L = L_1 \uplus L_2$ then $L_1 = L$ or $L_1 = \emptyset$. Observe that the language family generated by the grammar form $G_1 \oplus G_2$ satisfies

$$\mathcal{L}(G_1 \oplus G_2) = \{L_1 \uplus L_2 \mid L_i \in \mathcal{L}(G_i), i = 1, 2\}$$

and that every coherent language in the family $\mathcal{L}(G_1 \oplus G_2)$ belongs to one of the families $\mathcal{L}(G_i), i = 1, 2$. If $L(G)$ (and hence every language in the family $\mathcal{L}(G)$) is finite, we say that the grammar form G is *finite* or a *finite form*.

Investigations about density were initiated in [36]. This subsection contains also material from [28], [29], [34]. We will begin here with a simple construction showing that if we add something to the length set then we reach a "successor situation" and, hence, cannot obtain density.

Theorem 3.1. *Assume that G is a grammar form and $k \geq 1$ an integer not in the length set of $L(G)$. Then $\mathcal{L}(G)$ has a successor.*   □

Proof. Let $b_1, \ldots, b_k$ be terminal letters not in the alphabet of G, and let G_1 be the grammar form determined by the production $S \to b_1 \ldots b_k$. Consider the grammar form $G' = G \oplus G_1$. Clearly, $\mathcal{L}(G) \subset \mathcal{L}(G')$. By what was said about coherent languages in the family $\mathcal{L}(G \oplus G_1)$, it is easy to argue that there can be no grammatical family strictly between the families $\mathcal{L}(G)$ and $\mathcal{L}(G')$. Thus, $\mathcal{L}(G')$ is a successor of $\mathcal{L}(G)$.   □

The next result, which can be contrasted with our subsequent density results, is an immediate corollary of Theorem 3.1.

Theorem 3.2. *There is an infinite sequence $G_1, G_2, \ldots$ of grammar forms such that $\mathcal{L}(G_{i+1})$ is a successor of $\mathcal{L}(G_i)$, for each $i \geq 1$. Moreover, we can choose each G_i to be infinite or each G_i to be finite.*   □

We now turn to the discussions sufficient for density. The following result, [36], is a central auxiliary tool.

Theorem 3.3. *Assume that $\mathcal{L}_1$ and $\mathcal{L}_2$ are grammatical families such that $\mathcal{L}_1 \subset \mathcal{L}_2$ and, furthermore, there is an infinite coherent language in the difference $\mathcal{L}_2 - \mathcal{L}_1$. Then there is a grammatical family $\mathcal{L}_3$ satisfying $\mathcal{L}_1 \subset \mathcal{L}_3 \subset \mathcal{L}_2$.*
    □

Theorem 3.4. *The pair (REG, CF) is dense. Consequently, the family REG is density forcing.*   □

Proof outline. Assuming that $\mathcal{L}_1$ and $\mathcal{L}_2$ are grammatical families satisfying

$$REG \subseteq \mathcal{L}_1 \subset \mathcal{L}_2 \subseteq CF,$$

we show that the difference $\mathcal{L}_2 - \mathcal{L}_1$ contains an infinite coherent language, after which Theorem 3.3 becomes applicable. Take any language L' from the difference; L' must be infinite because REG contains all finite languages. If L' is not coherent, it can be broken into a superdisjoint union: $L' = L_1 \uplus L_2$. Grammatical families are closed under intersection with regular languages, as well as under superdisjoint union. Consequently, both of the languages L_1 and L_2 are in $\mathcal{L}_2$, whereas one of them, say L_1, must be outside $\mathcal{L}_1$. Since $\mathcal{L}_1$ contains all finite languages, L_1 is infinite. We have now exactly the same situation as at the beginning: L_1 is an infinite language in the difference $\mathcal{L}_2 - \mathcal{L}_1$. Moreover, the alphabet of L_1 is smaller than that of L'. If L_1 is not coherent, we repeat the same procedure. We end up with a coherent language L in the difference $\mathcal{L}_2 - \mathcal{L}_1$ because we cannot reduce the alphabet size indefinitely.   □

The construction in Theorem 3.4 is effective in the following sense. Assume that $\mathcal{L}_1$ and $\mathcal{L}_2$ are given by two grammar forms G_1 and G_2. Assume, furthermore, that an oracle tells us that $\mathcal{L}_1 \subset \mathcal{L}_2$ and gives us a language L in the difference $\mathcal{L}_2 - \mathcal{L}_1$. (The oracle need not tell us whether L is coherent.) Then we can effectively construct a grammar form G_3 such that $\mathcal{L}_1 \subset \mathcal{L}(G_3) \subset \mathcal{L}_2$. The details are given in [36]. The oracular information is necessary because inclusion and equivalence problems are undecidable for grammar forms.

It will be seen below in Section 3.3 that the interval (REG, CF) is not maximal dense. On the other hand, infinite sequences of Theorem 3.2 are not possible, [36], if the alphabet size of the grammar forms G_i is bounded.

The only fact used in the proof of Theorem 3.4 about the family REG is that REG contains all finite languages. Therefore, Theorem 3.4 can be expresses in the following stronger formulation.

Theorem 3.5. *Let $\mathcal{M}$ be a collection of grammatical families and $\mathcal{L}$ a family in $\mathcal{M}$ containing all the finite languages. Then $\mathcal{L}$ is density forcing in $\mathcal{M}$.* $\square$

The setup in Theorem 3.5 is more general than, for instance, that in Theorem 3.4. Theorem 3.5 can still be considerably generalized by considering *arbitrary collections of language families* satisfying certain basic properties necessary for proving a result akin to it; such collections are termed *MSW spaces*. We will define formally an MSW space and also show how an arbitrary collection of language families can be turned into an MSW space in a particularly simple manner. This demonstrates not only that such spaces are easily obtained, but also that the abstraction is meaningful in that "most" MSW spaces are not generated by grammar forms.

We need some auxiliary notions. A language L_1 is obtained from a language $L \subseteq \Sigma^*$ by *breaking* with respect to an alphabet $\Sigma_1 \subseteq \Sigma$ if $L_1 = L \cap \Sigma_1^*$ and $L - L_1$ contains no word containing a letter of Σ_1. (Breaking is, in some sense, an operation inverse to superdisjoint union. If a language is coherent, it cannot be broken in a nontrivial fashion.) For a language L and an integer $i \geq 1$, we define the language

$$\overline{L}(i) = \{w \in L \mid |w| \neq i\},$$

that is, $\overline{L}(i)$ is obtained from L by omitting all words of length i, if any. Similarly, for a language family $\mathcal{L}$ and integer $i \geq 1$, we define

$$\mathcal{L}(i) = \{\overline{L}(i) \mid L \in \mathcal{L}\}$$

and refer to $\mathcal{L}(i)$ as an *extraction* of $\mathcal{L}$. A language family $\mathcal{L}$ is closed under *covering* if, for every infinite language L, the fact that $\overline{L}(i)$ is in $\mathcal{L}$ for infinitely many values of i implies that L itself is in $\mathcal{L}$. Finally, the *superdisjoint wedge* of two language families $\mathcal{L}_1$ and $\mathcal{L}_2$ is defined by

$$\mathcal{L}_1 \vee \mathcal{L}_2 = \{L_1 \uplus L_2 \mid L_i \in \mathcal{L}_i, i = 1, 2\}.$$

Definition 3.2. *A collection $\mathcal{M}$ of language families is an MSW space if the following conditions (i)–(iii) hold:*

(i) *Each $\mathcal{L}$ in $\mathcal{M}$ is closed under superdisjoint union and breaking.*
(ii) *$\mathcal{M}$ is closed under superdisjoint wedge.*
(iii) *For each infinite language L occurring in some language family of $\mathcal{M}$, there exist subsets L_i of L for $i = 1, 2, \ldots$ such that (a) and (b) hold:*
  (a) *L is in a language family $\mathcal{L}$ of $\mathcal{M}$ iff L_i is in $\mathcal{L}$ for all i with $L_i \neq L$.*
  (b) *If L belongs to $\mathcal{L}$ in $\mathcal{M}$, then for every i with $L_i \neq L$ there exists an $\mathcal{L}_i \in \mathcal{M}$ such that $\mathcal{L}_i \subseteq \mathcal{L}$, L_i is in $\mathcal{L}_i$, and L is not in $\mathcal{L}_i$.* □

For instance, collections consisting of

(i) all grammatical families,
(ii) families generated by finite grammar forms,
(iii) families generated by one-sided linear grammar forms,
(iv) families of arbitrary grammar forms (see Section 4.1),
(v) families generated by synchronized E0L forms,

constitute all MSW spaces. Another example is the collection

$$\mathcal{M}(\mathcal{G}) = \{\mathcal{L}(G) \mid G \in \mathcal{G}\},$$

where $\mathcal{G}$ is an arbitrary collection of CF grammars which is closed under direct sum and taking interpretations. On the other hand, the collection of all AFL's (see Chapter 4 of this Handbook) and the collection of all g-grammatical families (see Section 1) are not MSW spaces. The next theorem shows that "most" MSW spaces are obtained quite independently of form theory.

Theorem 3.6. *Assume that $\mathcal{M}$ is a collection of language families such that each family $\mathcal{L}$ of $\mathcal{M}$ is closed under superdisjoint union, intersection with regular sets and covering. Let $\overline{\mathcal{M}}$ be the closure of $\mathcal{M}$ under superdisjoint wedge and extraction. Then $\overline{\mathcal{M}}$ is an MSW space.* □

Theorem 3.6 gives the following general method of constructing MSW spaces. Start with an arbitrary (finite or infinite) collection of languages. Close each language in the collection with respect to the operations $\uplus$, intersection with regular languages and covering, yielding a collection $\mathcal{M}$ of language families. Close $\mathcal{M}$ under superdisjoint wedge and extraction to obtain $\overline{\mathcal{M}}$. Then $\overline{\mathcal{M}}$ is an MSW space.

The main result about the density properties is the following theorem.

Theorem 3.7. *Let $\mathcal{M}$ be an MSW space and let $\mathcal{F}$ be the collection of all finite languages occurring in the language families of $\mathcal{M}$. If $\mathcal{L} \supset \mathcal{F}$ is a family of $\mathcal{M}$, then $\mathcal{L}$ is density forcing.* □

Theorem 3.7 can also be applied in showing that certain collections of language families, for instance the collection of all AFL's, are not MSW spaces.

We conclude this subsection with some results, [28], concerning $\{S, a\}$-grammatical families, that is families generated by grammar forms with only one nonterminal S and one terminal a. (When we speak of density in this context, the underlying collection $\mathcal{M}$ is restricted to $\{S, a\}$-grammatical families. This is why maximality results are more easily obtainable here than in the general case.)

All of the families REG, LIN and CF are $\{S, a\}$-grammatical. However, in this context CF is a *successor* of LIN. The interval (REG, LIN) is still dense. Density means that there are "many" families of the kind considered. Therefore, the decidability of the identity of two families is rather surprising – there are not many similar examples in language theory.

Theorem 3.8. *For $\{S, a\}$-forms, the pair (REG, LIN) is maximal dense and, moreover, form equivalence is decidable for $\{S, a\}$-forms.*   □

Theorem 3.8 can be strenghtened by characterizing *all* maximal dense intervals, [28]. Essentially, every maximal dense interval is obtained by restricting (REG, LIN) to some length set.

3.2 Color families of graphs and language interpretations

The investigation of grammar forms and, in particular, language interpretations offers a link to graphs first observed in [33]. The mechanism of taking interpretations, applied to graphs, turns out to be a characterization of n-coloring and induces families of graphs (called *color families*) in a similar way as it does in the form theory. Such an interpretation mechanism can be defined in the same way in a much more general environment, for instance, for universal algebras. As regards graphs, we can apply the technique for both directed and undirected graphs. In this way a classification is obtained that generalizes the notion of coloring in a natural way. Also certain graph-theoretic problems can be identified with certain problems concerning language interpretations.

We begin with considerations concerning *finite undirected graphs*. The sets of vertices and edges of a graph G are denoted by $V(G)$ and $E(G)$. The fact that two vertices x and y are *adjacent* (that is, the edge (x, y) is in $E(G)$) is denoted by $A_G(x, y)$ or $A(x, y)$ if G is understood. We assume that no vertex is adjacent to itself, that is $A(x, x)$ never holds. For an integer $n \geq 1$ (resp. $n \geq 3$), we denote by K_n (resp. C_n) the *complete graph* (resp. the *cycle*) with n vertices. Thus, $K_3 = C_3$ and K_1 consists of one vertex and no edges. An *elementary morphism* in a graph G consists of identifying two non-adjacent vertices x and y and inserting an edge between the identified vertex $x = y$ and all vertices z adjacent to either x or y in G. A graph G' is a *morphic image* of G if it is obtained from G by finitely many elementary morphisms. G is also considered to be a morphic image of itself. A graph is *minimal* if none of its morphic images, apart from G itself, is a subgraph of G.

A graph H is *colorable according to a graph* G, in symbols $H \leq_c G$, if there is a mapping φ of $V(H)$ into $V(G)$ such that, for all $x, y \in V(H)$,

$$(*) \qquad A_H(x, y) \text{ implies } A_G(\varphi(x), \varphi(y)).$$

The mapping φ is referred to as the *coloring* of H according to G. Assume that $G = K_n, n \geq 2$. Then clearly $H \leq_c G$ iff H is n-colorable in the customary sense. Hence, the notion defined above is a natural extension of the customary notion of vertex coloring. The *color family* of a graph G is defined by

$$\mathcal{L}(G) = \{H \mid H \leq_c G\}.$$

The family $\mathcal{L}(G)$ is analoguous to the family $\mathcal{G}(G)$ associated to grammar forms. We have followed here the notation customary in the literature; for graphs there is no family corresponding to the grammatical family.

Since our graphs are finite, every color family (although infinite as a family) consists of finite graphs. If G has n vertices, the complete graph K_{n+1} is not in $\mathcal{L}(G)$. The Four-Color Theorem tells us that every planar graph is in $\mathcal{L}(K_4)$. The following inclusions are easy to verify by the definitions:

$$\ldots \subset \mathcal{L}(C_{2n+1}) \subset \mathcal{L}(C_{2n-1}) \subset \ldots \subset \mathcal{L}(C_5) \subset \mathcal{L}(C_3) =$$
$$= \mathcal{L}(K_3) \subset \mathcal{L}(K_4) \subset \ldots \subset \mathcal{L}(K_n) \subset \mathcal{L}(K_{n+1}) \subset \ldots$$

This ascending hierarchy is customarily referred to as the *basic hierarchy* of color families. Every family in the basic hierarchy contains the family $\mathcal{L}(K_2)$ consisting of all 2-colorable graphs, that is, of all graphs having no cycles of an odd length. On the other hand, let G be an arbitrary graph possessing a cycle of an odd length. Then there are m and n such that

$$\mathcal{L}(C_{2m+1}) \subseteq \mathcal{L}(G) \subseteq \mathcal{L}(K_n).$$

This is the best result we can get in the general case because, for any $m \geq 2, n \geq 3$, there is a graph G with the following properties:

(i) G is in $\mathcal{L}(K_n)$ but not in $\mathcal{L}(K_{n-1})$,
(ii) C_{2m-1} is not in $\mathcal{L}(G)$.

Consequently, each of the color families

$$\mathcal{L}(C_{2m-1}), \ldots, \mathcal{L}(C_3) = \mathcal{L}(K_3), \ldots, \mathcal{L}(K_{n-1})$$

is incomparable with the family $\mathcal{L}(G)$. (These facts are due to the graph-theoretic result: for any $m, n \geq 2$, there is a graph whose chromatic number equals n and where the length of the shortest cycle exceeds m. See [33] for details.) To summarize, we have the following result. Consider any graph G such that $\mathcal{L}(K_2) \subset \mathcal{L}(G)$. Then $\mathcal{L}(G)$ can be incomparable only with the families in some middle segment of the basic hierarchy. The middle segment can be made arbitrarily long, both as regards the C- and K-parts.

The following results are straightforward consequences of the definitions.

Theorem 3.9. *The relation $\leq_c$ is transitive and decidable. The inclusion $\mathcal{L}(H) \subseteq \mathcal{L}(G)$ holds iff $H \leq_c G$. Consequently, the relations $\mathcal{L}(H) \subseteq \mathcal{L}(G)$ and $\mathcal{L}(H) = \mathcal{L}(G)$ are decidable for graphs H and G.* □

The next two theorems, also rather straightforward, are from [33] and [49].

Theorem 3.10. *A graph H is colorable according to G iff a morphic image of H is isomorphic to a subgraph of G.* □

Theorem 3.11. *For every color family $\mathcal{L}$, there is a unique minimal graph G such that $\mathcal{L} = \mathcal{L}(G)$.* □

Uniqueness means in Theorem 3.11 that, whenever G_1 and G_2 are non-isomorphic minimal graphs, then the color families $\mathcal{L}(G_1)$ and $\mathcal{L}(G_2)$ are different. Thus, minimal graphs can be viewed as a *normal form* of graphs for the representation of color families. Let us use the notation $H <_c G$ to mean that $H \leq_c G$ but not $G \leq_c H$. Clearly, $K_1 <_c K_2$ and there is no graph G satisfying

$$K_1 <_c G <_c K_2.$$

However, according to the following *Density Theorem*, the pair (K_1, K_2) constitutes the only example where one graph is a *predecessor* of another graph in this sense of the relation $<_c$. We use the term "a nontrivial graph" to mean a graph possessing at least one edge.

Theorem 3.12. *Assume that G_1 and G_2 are nontrivial graphs satisfying $G_1 <_c G_2$. Then there is a graph G_3 such that $G_1 <_c G_3 <_c G_2$.* □

Theorem 3.12 was first established in [49]; a different proof appears in [46]. The proofs rely heavily on the notion of a *weak predecessor*. H is a *weak predecessor* of G if $H <_c G$ and, moreover, apart from H itself there is no morphic image H_1 of H satisfying $H_1 <_c G$. For instance, C_5 is a weak predecessor of C_3, and so is the following graph with eight vertices:

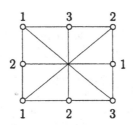

Coloring according to C_3 has been indicated by numbers. An important auxiliary tool in the proof of Theorem 3.12 is the fact that a connected graph possesses no predecessor iff it possesses infinitely many weak predecessors.

The connection between graphs and grammar forms should be now clear: *colorings* are essentially the same as *interpretations*. Here it is appropriate to consider interpretations of finite languages, as presented in Section 1.5. We consider now also *commutative languages* or *c-languages*: together with a word the language contains also all permutations of the word. (Algebraically, c-languages are subsets of the free commutative monoid.) A finite language or c-language is termed *graph-like* if it contains only words of length 2. A graph-like language or c-language is *universal* if it contains the word aa for some letter a.

The definition given above concerning coloring and the relation $\leq_c$ can be extended in a natural way to concern *digraphs*. Instead of edges (x, y) we now speak of *arrows* from x to y, and the pairs (x, y) and $(\varphi(x), \varphi(y))$ in $(*)$ are viewed as ordered pairs.

There is a natural correspondence between finite directed (resp. undirected) graphs and graph-like languages (resp. c-languages): each letter denotes a vertex, and the words in the language (resp. c-language) indicate the arrows (resp. edges). Apart from some trivial cases (we may add an isolated vertex to a graph without changing the language), this correspondence is one-to-one. Therefore, we may speak of the language (resp. c-language) *corresponding* to a given digraph (resp. graph). For instance, the c-languages

$$\{ab, bc, cd, de, ea\} \text{ and } \{ab, ac, ad, bc, bd, cd\}$$

correspond to the graphs C_5 and K_4, respectively. (We know that also ba is present with ab in a c-language; we list only one of the commutative variants.)

A language or c-language L is universal iff the corresponding digraph or graph contains a loop. In this case the family $\mathcal{L}(L)$ of interpretations consists of all graph-like languages or c-languages, that is, the language families of different universal languages coincide and consist of everything possible. Analoguous statements hold for graphs and digraphs containing loops.

The following theorem, immediate from the definitions, shows that all questions concerning hierarchies of color families of graphs or digraphs can be reduced to questions concerning corresponding c-languages or languages, and vice versa. This concerns, in particular, questions concerning dense and maximal dense intervals.

Theorem 3.13. *Let L and L' be c-languages or languages corresponding to the graphs or digraphs G and G'. Then $G' \leq_c G$ iff $L' \lhd L$.* □

Hierarchies of color families generated by digraphs are essentially richer than those generated by graphs. Contrary to the case of undirected graphs, one is able to obtain here nontrivial examples of "nexts". We will give an infinite sequence $G_0, G_1, \ldots,$ of digraphs, where each G_i is a successor of

G_{i+1}, that is, an infinite "descending" hierarchy. We define the digraphs by
the corresponding languages:

$$G_0 = \{ab, bc, cd\},$$
$$G_1 = \{ab, bc_1, b_1c_1, b_1c, cd\},$$
$$G_2 = \{ab, bc_1, b_1c_2, b_2c_2, b_2c_1, b_1c, cd\},$$
$$G_3 = \{ab, bc_1, b_1c_2, b_2c_3, b_3c_3, b_3c_2, b_2c_1, b_1c, cd\},$$

and in general G_{i+1} is obtained from G_i by replacing the word b_ic_i with the
three words $b_ic_{i+1}, b_{i+1}c_{i+1}$ and $b_{i+1}c_i$. The graphical representation of the
beginning of this hierarchy is as follows:

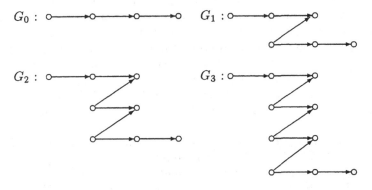

Thus, in G_i the index i indicates the number of additional levels needed
before the "exit".

Theorem 3.14. *For each $i \geq 0$, the digraph G_i is a successor of G_{i+1}.*   □

Although the membership problem is decidable for the color families
$\mathcal{L}(G)$, it is NP-complete in most important cases. Even in simple cases, such
as the cyclic graph C_5, no characterization is known for the color family.

3.3 Maximal dense intervals

Ever since the interval (REG, CF) was found to be dense, one tried to settle
the problem of its maximal density. If we restrict our attention to context-free
grammar forms, as has been done so far in this chapter, the interval cannot
be extended upwards. But can it be extended downwards ? The crucial issue
is the following. According to Theorem 3.5, a family is density forcing if it
contains all finite languages. But is this sufficient condition for being density
forcing also necessary ? The answer is "no", and so is the final answer to the
question about the maximality of the interval (REG, CF). However, maximal
dense intervals do exist but they all lie within the range of finite languages.

While the denseness of a given interval of grammatical families is undecidable, its maximal denseness is decidable. The initial tools needed here (concerning nonlooping languages) were developed in [39] and [34], and the problems were finally settled in [41] and [42].

Nonlooping languages are most significant when denseness is considered; an interval between two grammatical families is dense iff the families contain the same nonlooping languages. We now give the definitions needed.

A language L is *looping* if either L contains a word with at least two occurrences of the same letter or there exist distinct words $w_1, \ldots, w_m, m \geq 2$, in L and distinct letters $a_1, \ldots, a_m$ in the alphabet of L such that a_i and a_{i+1} occur in $w_i, 1 \leq i < m$, while a_m and a_1 occur in w_m. Otherwise, L is *nonlooping*. The intuitive idea behind a loop should be clear. The words in a language L (possible infinite) are viewed as vertices in a graph. Two vertices are connected by an edge if they have a common letter. There is also an edge from a vertex to itself if the word contains two occurrences of the same letter. A language is looping iff the graph has a loop. An important initial observation is that *all nonlooping languages are finite*.

For finite grammar forms G, the grammatical family $\mathcal{L}(G)$ coincides with the linguistical family $\mathcal{L}(L(G))$. Therefore, we consider mostly linguistical families in the sequel.

Consider a language L (possibly infinite) and its linguistical family $\mathcal{L}(L)$. We denote by $\mathcal{L}_N(L)$ the part of $\mathcal{L}(L)$ consisting of nonlooping languages in $\mathcal{L}(L)$. Two languages L_1 and L_2 are *nonlooping equivalent* if $\mathcal{L}_N(L_1) = \mathcal{L}_N(L_2)$. A language L is *minimal* if it has no proper subset L' with the property $\mathcal{L}(L) = \mathcal{L}(L')$. We are now ready to state the first result, [34]. A language L having a predecessor means that its linguistical family $\mathcal{L}(L)$ has a predecessor within the range of linguistical families.

Theorem 3.15. *A coherent minimal language has a predecessor iff it is nonlooping.*   □

Theorem 3.16. *Assume that $\mathcal{L}(L_1) \subset \mathcal{L}(L_2)$. Then the interval $(\mathcal{L}(L_1), \mathcal{L}(L_2))$ is dense iff L_1 and L_2 are nonlooping equivalent.*   □

Theorem 3.17. *The interval (REG, CF) is not maximal dense.*   □

The grammar form G defined by the productions

$$S \to a|b|ab|ba|aaA, \quad A \to a|aA$$

satisfies $\mathcal{L}(G) \subset REG$, while $(\mathcal{L}(G), CF)$ is dense.

As an example about the intricacies of finite nonlooping language forms we mention that the language $L_1 = \{ab, ba\}$ is a successor of $\{ab\}$. However, $L_2 = \{ab, bc, cd\}$ is by no means a successor of L_1. In fact, there is an infinite sequence of languages between L_1 and L_2 such that every language in the sequence is a predecessor of the language preceding it in the sequence. This was already explained in terms of digraphs at the end of Section 3.2.

We now present the main result from [41].

Theorem 3.18. *Let G_1 and G_2 be grammar forms or synchronized EOL forms. Then the interval $(\mathcal{L}(G_1), \mathcal{L}(G_2))$ is dense iff $L(G_1)$ and $L(G_2)$ are nonlooping equivalent.*   □

Theorem 3.18 should be understood in the proper sense. If G_1 and G_2 are grammar forms (resp. synchronized EOL forms), then grammatical families (resp. families of synchronized EOL forms) constitute the basic collection of language families, for which density is defined.

The following grammar forms G_1 and G_2 give a somewhat more sophisticated example of a dense interval $(\mathcal{L}(G_1), \mathcal{L}(G_2))$.

$$
\begin{aligned}
G_1 \quad &: \quad S \to a_1 S_3 b_2 | a_2 S_4 b_1 | \lambda, \\
&\quad\ S_1 \to a_1 S_3 b_2 | \lambda, \ S_2 \to a_2 S_4 b_1 | \lambda, \\
&\quad\ S_3 \to a_2 S_1 b_1, \ S_4 \to a_1 S_2 b_2, \\
G_2 \quad &: \quad S \to a_1 S_1 b_1 | a_1 S_1 b_2 | a_2 S_1 b_1 | a_2 S_1 b_2 | \lambda, \\
&\quad\ S_1 \to a_1 S b_1 | a_1 S b_2 | a_2 S b_1 | a_2 S b_2.
\end{aligned}
$$

Theorem 3.19. *It is undecidable whether or not two given context-free languages are nonlooping equivalent. Let G_1 and G_2 be two given grammar forms or synchronized EOL forms of which it is known that $\mathcal{L}(G_1) \subset \mathcal{L}(G_2)$. Then it is undecidable whether or not $(\mathcal{L}(G_1), \mathcal{L}(G_2))$ is a dense interval.*   □

Theorem 3.20. *Let G_1 and G_2 be grammar forms of which it is known that $L(G_1)$ is regular and either $\mathcal{L}(G_1) \subset \mathcal{L}(G_2)$ or $\mathcal{L}(G_2) \subseteq REG$. Then it is decidable whether or not $(\mathcal{L}(G_1), \mathcal{L}(G_2))$ is a dense interval.*   □

The following result is the basic extension theorem for dense intervals. We say that a language L is *inherently looping* if there is no nonlooping language L' such that $\mathcal{L}(L) = \mathcal{L}(L')$.

Theorem 3.21. *Given a grammar form (resp. synchronized EOL form) G with $L(G)$ inherently looping, there is a grammar form (resp. synchronized EOL form) G' such that $\mathcal{L}(G') \subset \mathcal{L}(G)$ and $(\mathcal{L}(G'), \mathcal{L}(G))$ is a dense interval.*   □

Theorem 3.22. *Assume that G_1 and G_2 are grammar forms or synchronized EOL forms such that $L(G_1)$ is inherently looping and the interval $(\mathcal{L}(G_1), \mathcal{L}(G_2))$ is dense. Then the interval mentioned is not maximal dense.*   □

Thus, the existence problem of maximal density reduces to the search of a dense interval reaching down to some nonlooping language. All nonlooping languages are finite. Consequently, we must search for an interval determined by two finite languages; in the finite case grammatical and linguistical families coincide. The following results are from [42].

Let L be a language over Σ and $k \geq 1$. We denote by $L(k)$ the set of all words of length k in L. Let $w^{(k)}$ be the kth letter of the word w. If $k > |w|$, by definition $w^{(k)} = \emptyset$.) Similarly,

$$L^{(k)} = \{w^{(k)} \mid w \in L\} \ (\subseteq \Sigma).$$

The *power language* $L^{(p)}$ of L is the language over the alphabet $\Sigma^{(p)}$, whose letters are nonempty subsets of Σ, constructed as follows: By definition

$$L^{(p)} = \bigcup_{k \geq 1} L(k)^{(p)}.$$

We still have to define the power language $L(k)^{(p)}$. We first index the words of $L(k)^{(p)}$ by nonempty subsets of $L(k)$. Thus, words in $L(k)^{(p)}$ are denoted by $\alpha(W), \emptyset \neq W \subseteq L(k)$. The ith letter of the word $\alpha(W), 1 \leq i \leq k$, is defined by

$$\alpha(W)^{(i)} = \{w^{(i)} \mid w \in W\}.$$

In other words, the ith letter in the word indexed by the words $w_1, \ldots, w_n$ is itself indexed by the set of the ith letters in the same words.

Clearly, the power language of every finite language is effectively constructable. The following theorem is valid for all languages.

Theorem 3.23. *The power language $L^{(p)}$ of an arbitrary language L satisfies the following conditions (i)–(iii): (i) An alphabetic variant of L is a subset of $L^{(p)}$. (ii) L and $L^{(p)}$ are nonlooping equivalent. (iii) Whenever $\mathcal{L}_N(L') \subseteq \mathcal{L}_N(L)$ then $\mathcal{L}(L') \subseteq \mathcal{L}(L^{(p)})$.* $\square$

This theorem has the following corollaries.

Theorem 3.24. *Let L be a nonlooping language. If $\mathcal{L}(L) \subset \mathcal{L}(L^{(p)})$ then $(\mathcal{L}(L), \mathcal{L}(L^{(p)}))$ is a maximal dense interval. If $\mathcal{L}(L) = \mathcal{L}(L^{(p)})$ then $\mathcal{L}(L)$ does not belong to any maximal dense interval.* $\square$

Theorem 3.25. *For an arbitrary context-free grammar G, it is decidable whether or not $L(G)$ is a lower limit of any maximal dense interval. For two arbitrary context-free grammar forms G_1 and G_2, it is decidable whether or not $(\mathcal{L}(G_1), \mathcal{L}(G_2))$ is a maximal dense interval.* $\square$

The second sentence of Theorem 3.25 should be contrasted with the undecidability of the question whether $(\mathcal{L}(G_1), \mathcal{L}(G_2))$ is a dense interval, or an interval. The second sentence of Theorem 3.24 shows how maximal dense intervals can possibly be constructed. The construction was carried out in [42]. In that construction the cardinality of the alphabet of L is over 200. However, the size of the alphabet of $L^{(p)}$ can be restricted to less that 1000. Thus we have the following result.

Theorem 3.26. *Maximal dense intervals of grammatical and linguistical families can be constructed.* $\square$

4. Extensions and variations

4.1 Context-dependent grammar and L forms

We have presented in the first three sections what we consider to be the core of the theory of grammar and L forms: the context-free case and the interconnections with graphs. In this environment, also the most important formal aspect of the theory, namely the results concerning density, can be best described.

By its very nature, form theory can be readily extended to concern any family of devices investigated in language theory. We can just consider any generative structure to be a "form" having "interpretations". The more complicated the structures become, the more undecidability, as well density results are obtained. [50] contains many references to the work in the non-context-free case. In the sequel we will present some glimpses over these areas. We do not try to be encyclopedic but are rather aiming at some unexpected or surprising results. Since the formal definitions are quite analogous to the ones given above, our aim is to be discursive rather than formal.

Phrase structure grammar forms (briefly, *grammar forms*) are simply phrase structure grammars – we assume that the left sides of the productions contain only nonterminals. All basic notions (interpretations, grammatical families, form equivalence, completeness, etc.) are defined as before. The following considerations are from [18], where also some other interpretation mechanisms are considered. As regards the g-interpretation discussed in Section 1 above, it is still an open problem whether or not there is a g-grammatical family strictly between CF and RE (recursively enumerable languages). Contrast this with Theorem 4.1. Recall that in a g-interpretation a terminal letter can be mapped to a terminal word, also the empty word.

Our first theorem shows that grammatical families may have "gaps": the family contains some finite languages but no "simple" infinite languages; all infinite languages in the family are non-context-free. (We now use the term "grammatical family" for the family of languages generated by a phrase structure grammar form.)

Theorem 4.1. *There exist grammar forms G such that $CF \subset \mathcal{L}(G) \subset RE$. There also exist grammar forms G such that $\mathcal{L}(G)$ contains both finite and infinite languages but all infinite languages in $\mathcal{L}(G)$ are non-context-free (resp. non-regular).*   □

In fact, the context-free grammar form defined by the productions $S \to aSb|ab$ is an example needed for the second sentence concerning non-regular languages. As regards the other example (concerning non-context-free languages), we just list the productions:

$$S \to FZ, \ Z \to MZ, \ Z \to MBBBD, \ MB \to BAM,$$
$$MA \to AM, \ MD \to D, FB \to bF, \ FA \to aF, \ FD \to \lambda.$$

The next theorem concerns a fundamental normal form.

Theorem 4.2. *For every grammar form, a form equivalent grammar form can be constructed such that all productions in the latter are of the types*

$$AB \rightarrow C, \; A \rightarrow \lambda, \; A \rightarrow a, \; A \rightarrow B, \; A \rightarrow BC. \qquad \square$$

Theorem 4.3. *Define three grammar forms $G_1 - G_3$ as follows:*

$$
\begin{aligned}
G_1 &: \; S \rightarrow \lambda|a|S|SS, \; SS \rightarrow S, \\
G_2 &: \; S \rightarrow \lambda|a|S|SS, \; SS \rightarrow SS, \\
G_3 &: \; S \rightarrow a|S|SS, \; SS \rightarrow SS.
\end{aligned}
$$

Then $\mathcal{L}(G_1) = \mathcal{L}(G_2) = RE$ and $\mathcal{L}(G_3) = CS$. $\qquad \square$

Theorem 4.4. *Completeness of a grammar form G is undecidable, and so is each of the following equations: $\mathcal{L}(G) = \{\emptyset\}$, $\mathcal{L}(G) = REG$, $\mathcal{L}(G) = CF$, $\mathcal{L}(G) = CS$.* $\qquad \square$

We mention, finally, a modified restricted interpretation, called *Q-inter-pretation* in [18]. A subset Q of the production set of a grammar form is distinguished. There are special conditions for interpreting a production $\alpha \rightarrow \beta$ in Q: (i) all occurrences of a nonterminal occurring in both α and β must be replaced by the same nonterminal (this is referred to as the weakly uniform interpretation), (ii) all (rather than some) productions $\alpha' \rightarrow \beta'$ with $\alpha' \in \mu(\alpha), \beta' \in \mu(\beta)$ are taken to be interpretations (provided that the above weak uniformity condition is satisfied). If such Q-interpretations are used, even "parallel families" such as E0L and ET0L become grammatical.

$E(m,n)L$ *systems and forms* will be discussed next. Their productions look like

$$(\alpha, a, \beta) \rightarrow w, \; |\alpha| = m, |\beta| = n.$$

Thus, the letter a can be rewritten as w in the context (α, β). Here an important variant of the interpretation is the *CC-interpretation*, an abbreviation of the words "constant context". If the rewriting $a' \rightarrow w'$ is possible in a CC-interpretation, then it must be possible in all contexts which are interpretations of the context (α, β) where the original rule $a \rightarrow w$ was given in the form.

Some results (mainly from [24], see also [50]) concerning completeness, reducibility and decidability are summarized in the following theorems. The terminology used should be self-explanatory. For instance, if an E(1,1)L form G satisfies $\mathcal{L}_{CC}(G) = E0L$, we say that G is CC-E0L-complete. Here $\mathcal{L}_{CC}(G)$ denotes the family of languages generated by CC-interpretations of G.

Theorem 4.5. *For any E0L form G, there is an E(1,1)L form G_1 such that $\mathcal{L}(G) = \mathcal{L}_{CC}(G_1)$. There are CC-E0L-complete and CC-E0L-vomplete*

E(1,1)L forms. There are complete E(1,0)L forms. There are CC-CF-complete E(1,0)L forms, as well as CC-ETOL-complete E(1,0)L forms. For every synchronized ETOL form G, there is an E(1,0)L form G_1 such that $\mathcal{L}_{CC}(G_1) = \mathcal{L}(G)$.   □

Theorem 4.6. *For every synchronized E(m,n)L form, a form equivalent synchronized E(1,0)L (or E(0,1)L) form can be constructed. There are E(1,1)L forms for which no form equivalent E(m,0)L form exists. There are EIL forms for which no form equivalent propagating or synchronized EIL form exists. There are no vomplete E(m,n)L forms.*   □

Theorem 4.7. *Form equivalence, CC-form equivalence, completeness, CC-EOL-completeness, CC-ETOL-completeness and CC-CF-completeness are all undecidable for E(1,1)L forms.*   □

4.2 Matrix forms

Recall that a (context-free) matrix grammar is a quadruple $G = (V, \Sigma, S, M)$, where V, Σ, S are as in a context-free grammar and M is a finite of *matrices* (of context-free productions), that is nonempty ordered sequences of the form

$$m = [A_1 \rightarrow w_1, \ldots, A_n \rightarrow w_n], \; n \geq 1.$$

Using such a matrix in a derivation step means to use all its rules, in the specified order, in the context-free manner. (Contrast this with the mode of applying a table in an ETOL system.) We denote by MAT^λ (resp. MAT) the family of languages generated by matrix grammars with arbitrary context-free rules (resp. with λ-free rules). One knows that

$$CF \subset MAT \subset MAT^\lambda \subset RE, \; MAT \subset CS$$

(see [6] and [7]).

All notions related to matrix grammar forms – we say, shortly, *matrix forms* – can be defined in the natural way, with the following important specification: when interpreting a matrix $m = [r_1, \ldots, r_n]$ (using a dfl-substitution μ), we obtain a set $\mu(m)$ of matrices, all of them containing n rules,

$$\mu(m) = \{[r'_1, \ldots, r'_n] \mid r'_i \in \mu(r_i), 1 \leq i \leq n\}.$$

Then, an interpretation of $G = (V, \Sigma, S, M)$ is a matrix grammar $G' = (V', \Sigma', S', M')$ with $V' = \mu(V), \Sigma' = \mu(\Sigma), S' \in \mu(S)$, and

$$M' \subseteq \mu(M) = \bigcup_{m \in M} \mu(m).$$

Thus, we are much closer to the case of grammar forms or EOL forms than to the case of ETOL forms: the number of matrices in the interpretation is not necessarily equal to the number of matrices in the form.

We briefly present here the basic results in [13], [14], [15], [16]. Some of them are also presented in [6] and [50]. The natural problems arising in this framework concern *reducibility* and *$\mathcal{L}$-completeness*, for $\mathcal{L}$ being MAT, MAT^λ or families contained in MAT, such as FIN, REG, LIN, CF.

It is known that for each matrix grammar G, an equivalent matrix grammar G' can be constructed, containing matrices of the following forms:

$$[A \to BC], \ [A \to a], \ [A \to \lambda],$$
$$[A \to BC, D \to E], \ [A \to B, C \to D],$$

where A, B, C, D, E are nonterminals and a is a terminal (when G is λ-free, then matrices of the form $[A \to \lambda]$ are not used). The reduction methods used for obtaining this result can be extended also to matrix forms.

First, it is possible to obtain

Theorem 4.8. *For each matrix form, there is an equivalent matrix form with matrices consisting of at most two productions, each production having the right side of length at most two.* □

This result can be strenghtened in several ways:

Theorem 4.9. *For each matrix form, there is an equivalent matrix form containing matrices of only one of the following forms (i)–(iii):*

(i) $[A \to BC], \ [A \to B, C \to D], \ [A \to a], \ [A \to \lambda],$
(ii) $[A \to B, C \to DE], \ [A \to a], \ [A \to \lambda],$
(iii) $[A \to BC, D \to E], \ [D \to a], \ [A \to \lambda],$

where A, B, C, D, E are nonterminals and a is a terminal. □

As a consequence, we can find MAT- and MAT^λ-complete matrix forms (which proves that MAT and MAT^λ are grammatical families – of course, with respect to matrix forms).

Theorem 4.10. *The matrix forms defined by the following matrices are MAT^λ-complete:*

$$
\begin{aligned}
G_1 \ &: \ [S \to SS], \ [S \to S, S \to S], \ [S \to a], \ [S \to \lambda],\\
G_2 \ &: \ [S \to S, S \to SS], \ [S \to a], \ [S \to \lambda],\\
G_3 \ &: \ [S \to SS, S \to S], \ [S \to a], \ [S \to \lambda].
\end{aligned}
$$

The following matrix form is MAT-complete:

$$G_4 \ : \ [S \to SS], \ [S \to S, S \to S], \ [S \to a]. \qquad \Box$$

Note that G_2, G_3, G_4 contain three matrices each. Are there MAT- or MAT^λ-complete matrix forms with two matrices ? The problem is still open. In [13] it is proved that there is no MAT- or MAT^λ-complete matrix form containing only one matrix. Using Theorem 4.10, a sufficient condition for the completeness of a matrix form $G = (V, \Sigma, S, M)$ is that $\mu(M)$ contains each of the matrices defining any one of the forms $G_1 - G_4$ in the theorem.

The characterization of $\mathcal{L}$-completeness of matrix forms, [14], for $\mathcal{L}$ one of FIN, REG, LIN, CF, is somewhat similar to the characterization of $\mathcal{L}$-completeness of grammar forms, see Section 1.4.

Theorem 4.11. *There is no matrix form which is FIN-complete.*   □

Theorem 4.12. *An (S, a)-matrix form G is REG-complete iff $a^+ \subseteq L(G)$ and G is not self-embedding.*   □

Theorem 4.13. *An (S, a)-matrix form G is LIN-complete iff the following conditions (i)–(iii) are fulfilled: (i) $a^+ \subseteq L(G)$, (ii) S is not expansive in G, (iii) for each $n, m \geq 0$, there are in G derivations of the form $S \Longrightarrow^* a^n S, S \Longrightarrow^* Sa^m$.*   □

In what concerns the CF-completeness, the results are rather surprising. Let us consider the matrix form G defined by the matrices

$$[S \to SS], \ [S \to a, S \to \lambda].$$

It is easy to see that $CF \subseteq \mathcal{L}(G)$. Unexpectedly, the inclusion is proper: the language $\{a^n b^n c^n \mid n \geq 1\}$ is generated by the following interpretation of G:

$$[S \to S_1 S_4], \ [S_4 \to S_2 S_3], \ [S_1 \to S_a S_A],$$
$$[S_A \to S_1 A], \ [S_1 \to S_a A], \ [S_2 \to S_b S_B],$$
$$[S_B \to S_2 B], \ [S_2 \to S_b B], \ [S_3 \to S_c S_C],$$
$$[S_C \to S_3 C], \ [S_3 \to S_c C], \ [S_a \to a, B \to \lambda],$$
$$[S_b \to b, C \to \lambda], \ [S_c \to c, A \to \lambda].$$

(The last three matrices check whether or not the number of occurrences of A, B, C are equal.)

In fact, in [14] it is proved that there is no CF-complete matrix form containing a matrix with at least two productions. The result is sharpened in [15] as follows:

Theorem 4.14. *A matrix form $G = (\{S, a\}, \{a\}, S, M)$ is CF-complete iff (i) every marix in M contains only one production, (ii) $L(G) = a^+$, (iii) there is a derivation $S \Longrightarrow^* a^p S a^q S a^r$ in G, for some $p, q, r \geq 0$.*   □

The result is extended to arbitrary terminal alphabets using the notion of a-restriction (see Section 1.4):

Theorem 4.15. *A matrix form $G = (\{S\} \cup T, T, S, M)$ is CF-complete iff (i) every matrix in M contains only one production, (ii) $\lambda \notin L(G)$, (iii) there is $a \in T$ such that $\mathcal{L}(G_a) = CF$.*   □

As a corollary, we obtain

Theorem 4.16. *The CF-completeness of matrix forms $G = (\{S\} \cup T, T, S, M)$ is decidable.*   □

4.3 Further variants

Uniform interpretations were first briefly considered in [19], then investigated in [20] for E0L and ET0L forms, in [31] for synchronized E0L forms and in [30] for context-free grammar forms.

As we have noted in Section 1.2, in a *uniform interpretation* the terminals are uniformly interpreted in the following sense: in each rule, two occurrences of the same terminal letter map to the same letter. (The nonterminal symbols are interpreted as in the previous sections.) The relation of uniform interpretation is denoted by $\lhd_u$ and the subscript u is also added to $\mathcal{G}(G)$ and $\mathcal{L}(G)$ when only uniform interpretations of G are considered, where G can be a grammar form or an L form.

A variant, called *weakly uniform interpretation* in [20], requires that when interpreting an E0L production $a \to w$ all occurrences of a in w have to be replaced by the same letter as a on the left side of the production. This can be extended to context-dependent productions: when interpreting $u \to v$ in the weakly uniform way, each occurrence of a terminal symbol a occurring both in u and v must map to the same letter. No restriction is imposed to interpretations of terminal letters occurring, for instance, only in v. We shall discuss here only uniform interpretations.

The difference between uniform and usual interpretations is illustrated by the following example: for the E0L form G defined by the productions

$$S \to SS, \ S \to a, \ a \to a, \qquad\qquad (*)$$

we have $\mathcal{L}_u(G) = CF$, but $CF \subset \mathcal{L}(G)$.

The following theorem is an immediate consequence of the definitions.

Theorem 4.17. *For any ET0L form G, $\mathcal{L}_u(G) \subseteq \mathcal{L}(G)$. The equality holds if no production in G contains more than one occurrence of any terminal.* □

The following theorem can be proved in the same way as the corresponding assertions in Theorems 1.1 and 1.2.

Theorem 4.18. *The relation $\lhd_u$ is decidable and transitive. For two ET0L forms G_1 and G_2, $\mathcal{G}_u(G_1) \subseteq \mathcal{G}_u(G_2)$ iff $G_1 \lhd_u G_2$. Strong uniform form equivalence between two ET0L forms is decidable.* □

In [20] one investigates the problem of reducibility, with respect to uniform interpretation of E0L and ET0L forms. The properties of a form of being reduced, separated, synchronized, short, binary, and propagating are considered. The general corollary of the results is that $\mathcal{G}_u(G)$ is an "antireducible" family of generative devices: most of the customary normal-form theorems do not hold. We refer the reader to [20] for details and we mention here only the following results.

Theorem 4.19. *(i) There are E0L (resp. ET0L) forms for which no uniform form equivalent synchronized E0L (resp. ET0L) form can be found. (ii) For any $n \geq 2$, there is an ET0L form G_n such that $\mathcal{L}_u(G_n) = \mathcal{L}_u(G)$ holds for no ET0L form G with less than n tables.*   □

In what concerns uniform completeness, from the second sentence in Theorem 4.17 we easily obtain uniformly complete (called *uni-complete*) E0L and ET0L forms. Indeed, every uni-complete form is also complete. For instance, the E0L form defined by the productions

$$S \to SS, \ S \to S, \ S \to a, \ a \to S,$$

is uni-complete, whereas for the ET0L form G determined by the tables

$$[S \to S, S \to SS, a \to S], \ [S \to a, a \to S],$$

we have $\mathcal{L}_u(G) = E0L$. On the other hand, there are complete forms which are not uni-complete. An example is the E0L form G defined by the productions

$$S \to S_1, \ S_1 \to S_1 S_1, \ S_1 \to S_1, \ S_1 \to aa,$$
$$S_1 \to aaa, \ a \to N, \ S \to b, \ b \to B, \ B \to N, \ N \to N.$$

We have $\mathcal{L}(G) = E0L$ but each string x with $|x| \leq 3$ in a language $L \in \mathcal{L}_u(G)$ contains occurrences of only one letter. In fact, we have, [31]

Theorem 4.20. *There are infinitely many complete E0L forms which are not uni-complete.*   □

Moreover, results like Theorem 4.19 prove that there is no uni-vomplete form. There are, however, E0L and ET0L forms G such that $\mathcal{L}_u(G) = CF$. We have mentioned that the E0L form G defined by productions $(*)$ above is uniformly CF-complete; the same assertion is true for the ET0L form determined by the tables

$$[S \to SS, a \to a], \ [S \to a, a \to a].$$

We close the discussion about uniform interpretation with the following theorem, [31].

Theorem 4.21. *If $\mathcal{L} \subseteq E0L$ contains all finite languages then $\mathcal{L}$ has no uniform E0L generator.*   □

As a corollary, we find that there are no uniform E0L generators for *FIN, REG, LIN, CF, E0L*. Compare Theorem 4.21 with Theorems 1.25–1.27, at the end of Section 1.4.

Derivation languages of grammar forms. Consider a (context-free) grammar $G = (V, \Sigma, S, P)$ with the rules in P labelled with symbols in a given set

Lab. The string of labels of the rules used in a derivation $D : S \Longrightarrow^* w$ in G is called the *control word* (or *Szilard word*) associated to D, and denoted by $Sz(D)$. The language of all words $Sz(D)$ associated to all terminal derivations in G is denoted by $Sz(G)$ and it is called the *Szilard* (or *derivation*) language of G. Since a grammar form is a grammar, we can associate to it the Szilard language. Moreover, we can associate to a grammar form G the family of Szilard languages corresponding to interpretations of G, denoted

$$Szilard(G) = \{Sz(G') \mid G' \lhd G\}.$$

It is known that $Sz(G)$ is not necessarily context-free for G being a context-free grammar. This can be sharpened as follows.

Theorem 4.22. *There are context-free grammar forms G such that every infinite language in $Szilard(G)$ is non-context-free.* □

An example is the grammar form with the rules

$$p : S \to ASB, \; q : S \to \lambda, \; r : A \to \lambda, \; s : B \to \lambda,$$

for which we have

$$Sz(G) \cap p^* r^* s^* q = \{p^n r^n s^n q \mid n \geq 0\}.$$

In [50] it is proved that for every $G' \in \mathcal{G}(G)$ with infinite $Sz(G')$ we have $Sz(G') \notin CF$.

A surprising result about families $Szilard(G)$ is proved in [32]:

Theorem 4.23. *The equation $Szilard(G_1) = Szilard(G_2)$ is decidable for grammar forms G_1, G_2.* □

This follows from the fact that

$$Szilard(G_1) = Szilard(G_2) \text{ iff } Sz(G_1) = Sz(G_2),$$

and the fact that the equation $Sz(G_1) = Sz(G_2)$ is decidable, see [44].

We refer to [32] for further related results (for instance, about the Szilard languages associated to leftmost derivations in grammars and grammar forms).

A notion intimately related to the derivation languages of grammar forms is that of *controlled grammar forms*. This is again an extension from regulated rewriting area, [6], to grammar form theory. The idea, investigated first in [11], is the following: take a grammar form G and a family $\mathcal{F}$ of languages; define the family $\mathcal{L}(G, \mathcal{F})$ of languages of the form $L(G', C)$ consisting of all strings generated by G' in $\mathcal{G}(G)$ using derivations D such that $Sz(D) \in C$, for given $C \in \mathcal{F}$. (This is very much similar to matrix forms, where the associated family $\mathcal{F}$ consists of all languages of the form $\{p_1 p_2 \ldots p_n \mid$ where $[p_1, \ldots, p_n]$ is a matrix$\}^*$.)

The control increases considerably the power of grammar forms:
$\mathcal{L}(G, LIN) = RE$ for grammar forms G containing only context-free rules and having rather weak additional properties – see [50] for details.

We also refer to [50] for further topics in grammar form theory, investigated so far in a smaller extent: ambiguity, descriptional complexity, pure grammar forms, programmed grammar forms, full interpretation, subset-restricted interpretation.

Acknowledgements. This work has been supported by Project 11281 of the Academy of Finland. Useful comments from Lucian Ilie and Valeria Mihalache are gratefully acknowledged. This chapter was originally assigned to another person who, due to unfortunate circumstances, could not accomplish the task. The present authors regret possible slips and omissions in the text, caused by the late alarm.

References

1. A. V. Aho, J. D. Ullman, *The Theory of Parsing, Translation, and Computing, Vol. I: Parsing*, Prentice-Hall, Inc., Englewood Cliffs, N. J. 1972.
2. J. Albert, H. Maurer, The class of CF languages is not an E0L family, *Inform. Processing Letters.*, **6** 1978, 190–195.
3. A. B. Cremers, S. Ginsburg, Characterizations of context-free grammatical families, *IEEE 15th Annual Symp. Switching and Automata Th.*, New Orleans 1974.
4. A. B. Cremers, S. Ginsburg, Context-free grammar forms, *J. Computer System Sci.*, **11** 1975, 86–116.
5. K. Culik II, H. A. Maurer, T. Ottmann, K. Ruohonen, A. Salomaa, Isomorphism, form equivalence and sequence equivalence of PD0L forms, *Theor. Computer Sci.*, **6** 1978, 143–174.
6. J. Dassow, Gh. Păun, *Regulated Rewriting in Formal Language Theory*, Springer-Verlag, Berlin 1989.
7. J. Dassow, Gh. Păun, A. Salomaa, Grammars with Controlled Derivations, in this *Handbook*.
8. A. Gabrielian, S. Ginsburg, Grammar schemata, *Journal of the ACM*, **21** 1974, 213–226.
9. S. Ginsburg, B. L. Leon, O. Mayer, D. Wotschke, On strict interpretations of grammar forms, *Math. Systems Th.*, **12** 1979, 233–252.
10. J. N. Gray, M. A. Harrison, On the covering and reduction problems for context-free grammars, *Journal of the ACM*, **19, 4** 1972, 675–698.
11. S. A. Greibach, Control sets on context-free grammar forms, *J. Computer System Sci.*, **15** 1977, 35–98.
12. G. Hotz, Eindeutigkeit und Mehrdeutigkeit Formaler Sprachen, *EIK*, **2** 1966, 235–247.
13. R. Leipälä, On context-free matrix forms, *Inform. Control*, **39** 1978, 158–176.
14. R. Leipälä, On the generative capacity of context-free matrix forms, *Intern. J. Computer Math.*, **7** 1979, 251–269.

15. R. Leipälä, Generation of context-free languages by matrix forms, *Ann. Univ. Turku*, Ser. AI, **179** 1979.

16. R. Leipälä, *Studies on Context-free Matrix Forms*, PhD Thesis, University of Turku 1979.

17. H. A. Maurer, Th. Ottmann, A. Salomaa, On the form equivalence of L forms, *Theor. Computer Sci.*, **4** 1977, 199–225.

18. H. A. Maurer, M. Penttonen, A. Salomaa, D. Wood, On non context-free grammar forms, *Math. Syst. Th.*, **12** 1979, 297–324.

19. H. A. Maurer, A. Salomaa, D. Wood, EOL forms, *Acta Informatica*, **8** 1977, 75–96.

20. H. A. Maurer, A. Salomaa, D. Wood, Uniform interpretations of L forms, *Inform. Control*, **36** 1978, 157–173.

21. H. A. Maurer, A. Salomaa, D. Wood, ETOL forms, *J. Computer System Sci.*, **16** 1978, 345–361.

22. H. A. Maurer, A. Salomaa, D. Wood, Good EOL forms, *SIAM J. Comput.*, **7** 1978, 158–166.

23. H. A. Maurer, A. Salomaa, D. Wood, Relative goodness of EOL forms, *RAIRO*, **12** 1978, 291–304.

24. H. A. Maurer, A. Salomaa, D. Wood, Context-dependent L forms, *Inform. Control*, **42** 1979, 97–118.

25. H. A. Maurer, A. Salomaa, D. Wood, On generators and generative capacity of EOL forms, *Acta Informatica*, **13** 1980, 87–107.

26. H. A. Maurer, A. Salomaa, D. Wood, Context-free grammar forms with strict interpretations, *J. Computer System Sci.*, **21** 1980, 110–135.

27. H. A. Maurer, A. Salomaa, D. Wood, Synchronized EOL forms, *Theor. Computer Sci.*, **12** 1980, 135–159.

28. H. A. Maurer, A. Salomaa, D. Wood, Decidability and density in two-symbol grammar forms, *Discrete Applied Math.*, **3** 1981, 289–299.

29. H. A. Maurer, A. Salomaa, D. Wood, MSW spaces, *Inform. Control*, **46** 1981, 200–218.

30. H. A. Maurer, A. Salomaa, D. Wood, Uniform interpretations of grammar forms, *SIAM J. Comput.*, **10** 1981, 483–502.

31. H. A. Maurer, A. Salomaa, D. Wood, Synchronized EOL forms under uniform interpretation, *RAIRO, Th. Informatics*, **15** 1981, 337–353.

32. H. A. Maurer, A. Salomaa, D. Wood, Derivation languages of grammar forms, *Intern. J. Computer Math.*, **9** 1981, 117–130.

33. H. A. Maurer, A. Salomaa, D. Wood, Colorings and interpretations – a connection between graphs and grammar forms, *Discrete Applied Math.*, **3** 1981, 119–135.

34. H. A. Maurer, A. Salomaa, D. Wood, On predecessors of finite grammar forms, *Inform. Control*, **50** 1981, 259–275.

35. H. A. Maurer, A. Salomaa, D. Wood, Completeness of context-free grammar forms, *J. Computer System Sci.*, **23** 1981, 1–10.

36. H. A. Maurer, A. Salomaa, D. Wood, Dense hierarchies of grammatical families, *Journal of the ACM*, **29** 1982, 118–126.

37. H. A. Maurer, A. Salomaa, D. Wood, Finitary and infinitary interpretations of languages, *Math. Systems Th.*, **15** 1982, 251–265.

38. H. A. Maurer, A. Salomaa, D. Wood, A supernormal form theorem for context-free grammars, *Journal of the ACM*, **30** 1983, 95–102.

39. H. A. Maurer, A. Salomaa, D. Wood, On finite grammar forms, *Intern. J. Computer Math.*, **12** 1983, 227–240.

40. V. Niemi, The undecidability of form equivalence for context-free and EOL forms, *Theor. Computer Sci.*, **32** 1984, 261–277.

41. V. Niemi, Density of grammar forms, I and II, *Intern. J. Computer Math.*, **20** 1986, 3–21 and 91–114.
42. V. Niemi, Maximal dense intervals of grammar forms, *Lecture Notes in Computer Science*, Springer-Verlag, Berlin **317** 1988, 424–438.
43. Th. Ottmann, A. Salomaa, D. Wood, Sub-regular grammar forms, *Inform. Processing Letters*, **12** 1981, 184–187.
44. M. Penttonen, On derivation languages corresponding to context-free grammars, *Acta Informatica*, **3** 1974, 285–291.
45. G. Rozenberg, A. Salomaa, *The Mathematical Theory of L Systems*, Academic Press, New York 1980.
46. A. Salomaa, On color-families of graphs, *Ann. Acad. Scient. Fennicae* **AI6** 1981, 135–148.
47. E. Shamir, On sequential languages, *Z. Phonetik Sprachwiss. Kommunikat.*, **18** 1965, 61–69.
48. H. Walter, Topologies of formal languages, *Math. Systems Th.*, **9** 1975, 142–158.
49. E. Welzl, Color-families are dense, *Theor. Computer Sci.*, **17** 1982, 29–41.
50. D. Wood, *Grammar and L Forms: An Introduction, Lecture Notes in Computer Science*, Springer-Verlag, Berlin **91** 1980.

Index

Printed in the United States
By Bookmasters